Earth Science

SECOND EDITION

Richard J. Ordway

Professor of Geology
State University College
New Paltz, New York

D. VAN NOSTRAND COMPANY
New York Cincinnati Toronto London Melbourne

To the many students who have made teaching a stimulating and rewarding experience for me

D. Van Nostrand Company Regional Offices:
New York Cincinnati Millbrae

D. Van Nostrand Company International Offices:
London Toronto Melbourne

Library of Congress Catalog Card Number: 70-181097

ISBN: 0-442-26291-4

Manufactured in the United States of America

Published by D. Van Nostrand Company
450 West 33rd Street, New York, N. Y. 10001

Published simultaneously in Canada by
Van Nostrand Reinhold Ltd.

15 14 13 12 11 10 9 8 7 6 5 4 3 2

Design by Morris Karol

Extensive, widespread progress and modification have occurred throughout the earth sciences in recent years, but the changes in some areas have been truly remarkable. Sea-floor spreading, continental drifting, and plate tectonics have revolutionized some aspects of the geological sciences. Study of the Earth from orbiting spacecraft, space exploration, the successful Apollo landings on the moon, the discovery of cosmic background radiation, quasars, and pulsars have been equally striking. Moreover, these are the years that have seen a growing awareness of the tremendous impact that man has had and is having on his environment.

These dynamic events have necessitated a particularly thorough revision of *Earth Science*. Chapter 1 is entirely new and is a brief, overall view of spaceship Earth and its environmental problems, stressing the role played by the earth sciences in, e.g., air and water pollution, waste disposal, and the depletion of mineral resources. Marine and continental geology are now so closely interwoven that they have been combined into one large unit (Chapters 2 through 15), and sea-floor geology (Chapter 11) is considered before geosynclines and mountain building (Chapter 12). Former appendixes on minerals and rocks have been deleted, and the material is now part of a new Chapter 3. The chapters on igneous activity and on geologic structures, earthquakes, and the Earth's interior have been moved ahead of the chapters on weathering, streams, subsurface water, glaciers, and the wind. Part II deals with astronomy and has been consolidated into ten chapters (16 through 25), rather than eleven chapters as in the first edition. Moreover, the units on space travel and tools of the astronomer have been moved ahead. Part III consists of four chapters (26 through 29) that deal with meteorology.

Earth science is a subject whose data lend themselves to visual treatment, and much effort has been spent in locating meaningful and striking photographs. Most of the photographs are new to this edition, with a concentration on aerial and space views. These, together with the new line drawings, total 635 illustrations.

The emphasis in the new edition is on principles and broad coverage, and the nonmathematical approach remains. However, some quantitative work has been added, especially in the astronomy chapters. The text is still intended for college students who are taking a beginning science course that includes geology, astronomy, meteorology, and oceanography.

As with the original version, the keynote of the revision is flexibility. The various sections are self-contained, permitting each instructor to take up the major topics in any sequence and to the depth appropriate for his specific course. Regardless of how the course is arranged, it would seem desirable to supplement the text with laboratory studies, geology field trips, and certain observations in both astronomy and meteorology.

Many of the changes in the second edition of *Earth Science* have resulted from the constructive criticisms of instructors who used the first edition. Their comments were helpful and are appreciated. Dr. Theodore Mehlin read each of the astronomy and meteorology chapters for the first edition and made many constructive suggestions for their improvement. His help is gratefully acknowledged. Colleagues in the Department of Geological Sciences at New Paltz have read and criticized several chapters: Dr. Frank Caruccio did this for Chapters 7 and 8, Dr. Russell Waines for Chapters 1, 11, and 12, and Dr. Glen Tague for the chapter on dry lands in the first edition. I much appreciate their assistance.

Working with the Van Nostrand Reinhold staff has been a pleasant, cooperative experience, and I want to express particular appreciation to Stephen Kraham and Aria Ruks. My wife, Mary Jane, has assisted me directly at times and indirectly by providing quiet working conditions and relatively few, flexibly scheduled household chores. Photographs and illustrative materials have come from many different sources and are acknowledged with each illustration, but I want to underline here my thanks to every friend, colleague, and institution who put resources so generously at my disposal. I wish also to thank the many students who have studied earth science with me and who have made useful suggestions about ways to present the material clearly.

Richard J. Ordway

New Paltz, New York

Contents

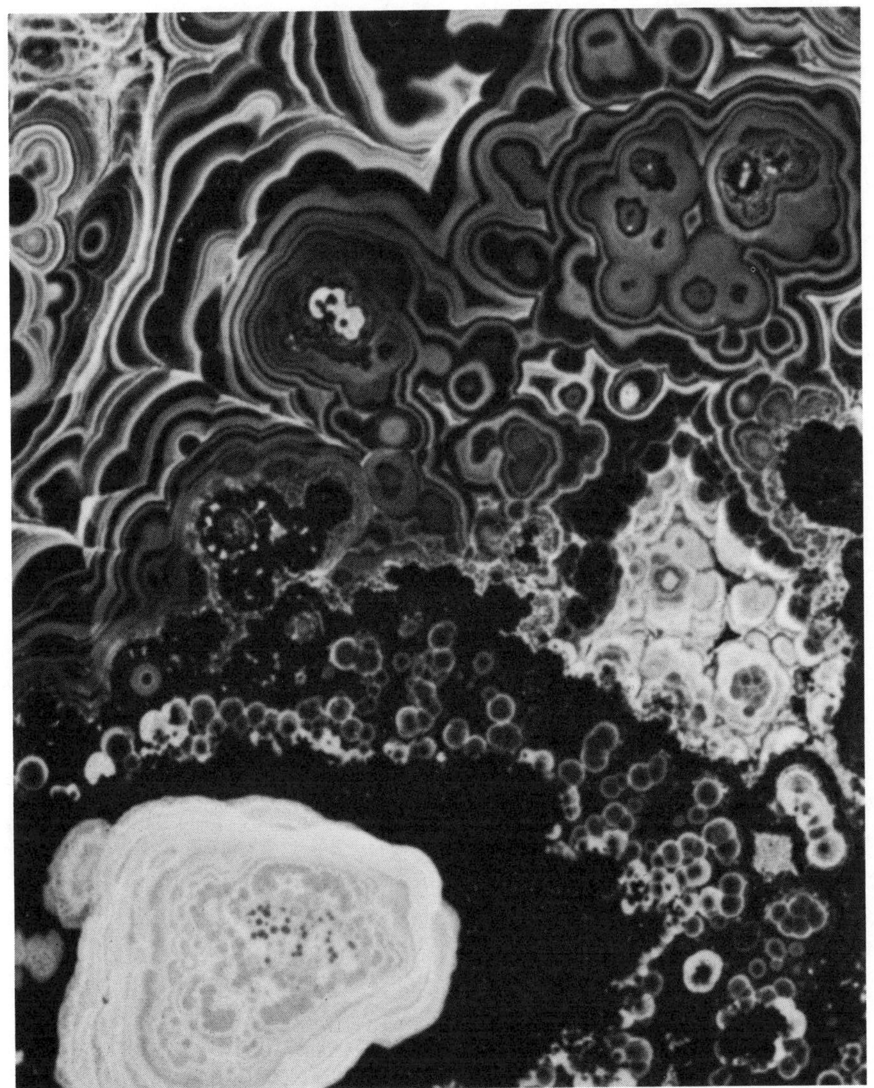

Part 1
Geology

1

Man's Environment and the Earth Sciences

By the early 1970's, it had become apparent that a most important result of the successful Apollo lunar landing program was the realization that the Earth itself was very much like a spaceship. As viewed from the vicinity of the moon (Figs. 1–1 and 1–2), the Earth was seen in a different perspective—a small, beautiful ball embedded in an immensity of exceedingly inhospitable space. Discernible beneath the white shifting swirls of its clouds was the striking blue of its waters and the subdued browns of its continents. In a way, the view from space engendered a heightened realization of the fragile nature of the Earth's life-supporting environment and underscored its possession of strictly limited quantities of the resources needed by man.

Upon closer inspection, the waters of many of the rivers and lakes of this planet called Earth were unsafe to drink, its air in places was dangerous and unpleasant to breathe (Fig. 1–3), wastes of all descriptions had accumulated untidily here and there, and its cities were nearly bursting with men, women, and children together with their possessions and the multitudes of accoutrements that form part of a twentieth-century technological society. Moreover, each day saw increases in the magnitudes of all these.

On the July 1969 weekend, when astronauts Armstrong and Aldrin walked on the moon and Collins orbited overhead, the sulfur dioxide content in the air above the East Chicago-Gary area climbed to ten times the legal limit. When it rained on Sunday, the sulfur dioxide and rainwater combined into a substance like sulfuric acid that made lawns turn brown, produced holes in leaves, and caused some birds to lose their feathers. Man had accomplished an amazing technological feat in reaching the moon, but an even greater and important task remained for him at home—that of making the Earth a safe and pleasant place for him and his descendants to dwell on.

The litter that the citizens of the United States carelessly and unthinkingly scattered across the landscape that same July weekend, added to that discarded on the other days of the year, cost perhaps half a billion

1-1. Apollo 11 photograph (July 1969) of the Mare Smythii Region of the moon with Earth in the distance. The lighter colored lunar highlands stand on the average nearly one mile above the darker colored and less cratered maria. *(NASA: AS11-44-6551.)*

dollars to pick up. If this expense had been avoided—as it would if each citizen had only tossed his wastes into a nearby receptacle or saved them until he came to a trash can—that $500 million would have been available, say, to build 20,000 homes and 100 schools.

Realization thus came to some of the inhabitants of planet Earth—in growing numbers and with increasing urgency—that the planet was now gravely threatened by overpopulation, by pollution, and by the depletion of its nonrenewable resources, just as much as by nuclear arsenals and wars of mass destruction. Thus *ecology* and *recycling* became familiar words in 1970 as action was initiated and intensified to create an "Earth National Park" and, moreover, to preserve a part of its fauna severely threatened in the coming decades by extinction—man himself.

Damage to planet Earth in the years that immediately preceded the 1970's was not generally caused by evil men, nor was it primarily the result of new ways of living or working. For example, pollutants in the rivers of Illinois may have their main source

1-2. LM and astronaut on the lunar surface in November 1969, Apollo 12 photograph. *(NASA: AS12-46-6779.)*

in the effluents from the farms that produce our corn and meat (eutrophication, see below) and only partially by industrial discharges and the disposal of town sewage. American society, as well as those of other so-called developed countries, had evolved into a system that produced, consumed, and then discarded. This resulted in an immense one-way flow of materials from factories, mines, and farms through the households and thence into the garbage dumps, air, and water. Production and consumption were emphasized, and relatively little attention was paid to the waste by-products. When numbers and amounts were relatively small, the effluents from these activities were not harmful. After all, the "secret to pollution is dilution," and the smoke from a few chimneys plus the discharge from a few sewers do not pollute the huge volumes of air and water that they spread into. Neither does the gasoline to run a few cars deplete our petroleum reserves or contaminate our atmosphere. In like manner, the birth and growth of a single boy or girl does not noticeably crowd a school or city, nor cause a food or transportation problem. However, these and similar activities, operating at their present rates, are overtaxing the resources and life-support systems of planet Earth, and man seems in danger of smothering in his own debris. In fact, man is now said to have achieved geological magnitude as a force capable of modifying the Earth's surface.

1-3. Air pollution. Many single sources are diffused by atmospheric turbulence and contribute to the total pollution produced by this industrial complex. The heated effluents cool as they rise above the chimneys, become less buoyant, and move downwind. *(Courtesy U. S. Department of Health, Education, and Welfare.)*

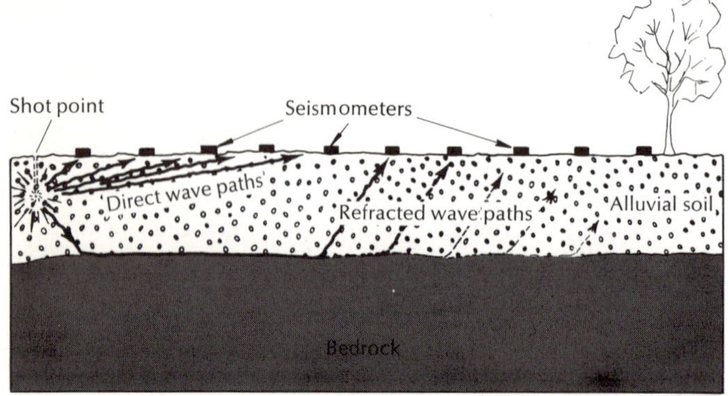

1-4. Determining depth to bedrock by seismic profiling. The shock-generated waves generally travel faster in bedrock than in the overlying soil. Thus arrival times at the seismometers depend upon depth to bedrock as well as upon the nature of the soil and rock. *(Courtesy U. S. Geological Survey.)*

Since problems can also be regarded as opportunities, the earth and biological sciences have major roles to play in the years that lie immediately ahead. For example, before the end of this century, according to one geologist, the engineers and architects of the world will have to build at least as many structures as now exist on the entire Earth: buildings, dams, roads, bridges, tunnels, airports, and the like. Some of these will replace existing structures, but many will be completely new. Since each structure will be built in contact with the ground, and since many will require excavation, geology must play a key role in the selection of safe, suitable sites (Fig. 1–4). In other words, and in a somewhat broader vein, nearly all of our man-made environment will be built anew in the next half century or so. For man to survive on planet Earth, these alterations must be made in harmony with the natural environment. Geologists also play many roles in the location and extraction of necessary minerals and fuels, and they are concerned with vital questions of water supply and pollution as well as with phenomena such as earthquakes, landslides, and volcanic eruptions (Fig. 1–5). Air pollution, of course,

1-5. Shishaldin Volcano, Aleutian Islands. *(Courtesy U. S. Geological Survey.)*

1-6. Hurricane Camille, 19 August 1969. *(Courtesy NOAA.)*

directly involves meteorology (Fig. 1–6), and a very long list is necessary to encompass all the many ways in which oceanography is of importance to man and the functioning of planet Earth. The influence of solar system astronomy is less direct but nevertheless important. Some of the ways in which the earth sciences bear most importantly on man's environment are listed or discussed briefly in this chapter, but these and others are also included in pertinent sections of other chapters.

Ecology

Ecology has been defined as the science of the complex web of interrelationships that occur among organisms and between organisms and their environment—also in 1970 as the science of survival on a polluted planet. An *ecosystem* then refers to the community of living things, together with its intimately interlinked physical environment, that exists in some one part of nature. Ecology has a number of major themes. Only one species can occupy a particular niche in a particular habitat at a particular time. Interdependence is widely prevalent, and just about every form of life is related to and dependent upon other forms of life. Moreover, these relationships are complex, and the causes and effects may be widely separated in both space and time. Ecologists also stress the limitations that govern biological systems: blueberries do not grow as large as basketballs, men do not become giants, nor can a particular species—e.g., elephants, foxes, hummingbirds, or man—grow unchecked in numbers.

In this complex web of life, plants are basic because they convert solar energy plus various inorganic materials into foods via the process of photosynthesis. Animals then consume the plants, and some animals eat other animals. Finally, various decomposing agents such as bacteria, fungi, and insects convert dead plants and animals and their waste products back into inorganic materials such as water, carbon dioxide, and mineral matter that are needed by growing plants. Thus this is an endless, recycling closed system in which none of the elements are lost, and in which each of the three groups—plants, animals, and decaying agents—plays a vital role.

Complexity lends stability to ecological systems because the improper functioning of one part of a system may be offset by adjustments in different parts. Therefore, when man destroys a forest and replaces it with a cornfield, he has traded a complex system with a built-in set of checks and balances for a relatively simple one that works nicely only as long as he controls each of the insects and diseases that might destroy the corn. Therefore, if a new disease comes along that man cannot quickly control—as it did in the United States in 1970—the corn may be damaged or even destroyed abruptly.

A few examples will illustrate this complexity and the danger of upsetting a balanced system. According to one report, a particular area in Borneo was sprayed with DDT by the World Health Organization to kill flies and roaches. Lizards then gorged on the dead insects and as a result moved too slowly to escape the cats. The cats subsequently died; therefore, rats became more numerous, which brought the threat of a plague. Apparently the death of the cats resulted from so-called biological magnification in which a persistent, toxic, highly mobile material such as DDT all goes into the fatty tissues of an organism and thus becomes ever more concentrated as it is transmitted from one group of organisms to another upward in a food chain. In other words, all the DDT in 100 pounds of insects becomes concentrated into 10 pounds of lizard, which in turn becomes part of 1 pound of cat (the numbers are used only as an illustration). It follows that if bears then ate cats, and men ate bears, the concentration of DDT in man would be higher than in cats and, unfortunately, man is located at the pinnacle of many such food pyramids.

To continue with the Borneo example, a type of parasite that fed on caterpillars was also killed by the DDT. Therefore, caterpillars multipled on the roofs of native dwellings, where they fed on the thatching. This caused some roofs to cave in.

Additional examples of man's unintentional interference with complex ecosystems were described in 1969.* As pointed out, one of the great humane ideas prevalent in the developed countries is that of sharing their "progress" by exporting their technology to less-developed countries. However, in some instances, unforeseen and undesirable events have resulted when certain ecosystems in the importing countries were subsequently upset. Moreover, some of the mistakes that have caused environmental damage at home are also being exported and repeated abroad.

To illustrate, dried milk was introduced as a food to some of the inhabitants of northeastern Brazil where a vitamin A deficiency existed—a well intentioned, seemingly innocuous and helpful type of action. However,

* "The Unforeseen International Ecologic Boomerang," *Natural History*, February 1969.

this sudden increase in protein consumption caused some individuals to grow more rapidly, which in turn led to a depletion of their already scanty supplies of vitamin A (removed from the liver as a protein complex). This then caused some individuals to become partially or totally blind.

Furthermore, it is now realized that milk intolerance may exist among a majority of the peoples living in the underdeveloped countries. The enzyme, lactase, is needed to absorb and digest milk, but lactase tends not to be produced within a person after he is weaned unless milk is consumed regularly. Therefore, if a catastrophic event such as a great storm or earthquake should devastate part of a country inhabited by lactase-deficient peoples, dried milk probably should not be included among the food supplies sent to feed the survivors. Otherwise, their inability to digest the milk might produce diarrhea and further weaken the victims.

In other areas, the use of insecticides against a certain pest has sometimes resulted in the development of resistant strains of the pest or in reducing the effectiveness of natural controls by killing some of the pest's natural enemies. In still other instances, technological assistance in drier regions may result in the construction of a large dam or in the drilling of deep wells to tap underground water supplies. Widespread irrigation may then take place and may so alter the soil that it becomes a desirable habitat for unwanted insect pests—such pests seem able to travel across many miles of dry ground to reach man-made oases in desert regions. Thus in Egypt, a close correlation has been noted between areas that are irrigated perennially and areas in which a debilitating disease, schistosomiasis, is widespread. It seems that a species of snail acts as a host for a type of worm that causes the disease, and these snails find it possible to survive and migrate within the canals and irrigation ditches. Previously, the irrigation had been seasonal and the snails would tend not to survive when they did invade the irrigation ditches.

The So-called Population Bomb

It has been said that all men are polluters and that all living Americans are big polluters. In other words, the people of a high-technology society consume many more resources and produce many more wastes per capita than do the peoples of so-called underdeveloped lands (many inhabitants are poorly fed and illiterate; agriculture is inefficient; industrialization and total gross national product are limited). To illustrate this in terms of energy, that generated in the United States in 1970 would about equal the muscular energy output of 500 men working as slaves for each American man, woman, and child. Therefore, in terms of the amount of pressure put upon the environment and the Earth's supply of nonrenewable resources, the birth of one child into an affluent American family has the same effect as the birth of several dozen or more Indian, African, or Indonesian children. In the late 1960's, Americans comprised only 6 percent of the world's population, yet on the average they consumed about 35% of all the Earth's resources used up in one year—meat, iron, copper, oil, and electric power to name a few. Therefore, one American farmer, albeit unintentionally and quite innocently, may cause fully as much damage to the natural environment as do 50 Chinese peasant families. The rather ugly term "popullution" has been suggested to denote the polluting force of too large a population.

The very rapid, relatively recent increase in the world's population (Fig. 1–7) to a 1970 total of more than 3.5 billion people was first described as a population explosion and aptly called a "population bomb" nearly two decades ago. The term was used to highlight the view that the present rate of population growth, if unchecked during another decade or so, may be just as much a threat to civilization and mankind as an arsenal of nuclear weapons. The problem of checking population growth is certainly one of the most important problems facing mankind today—if not the *most* important

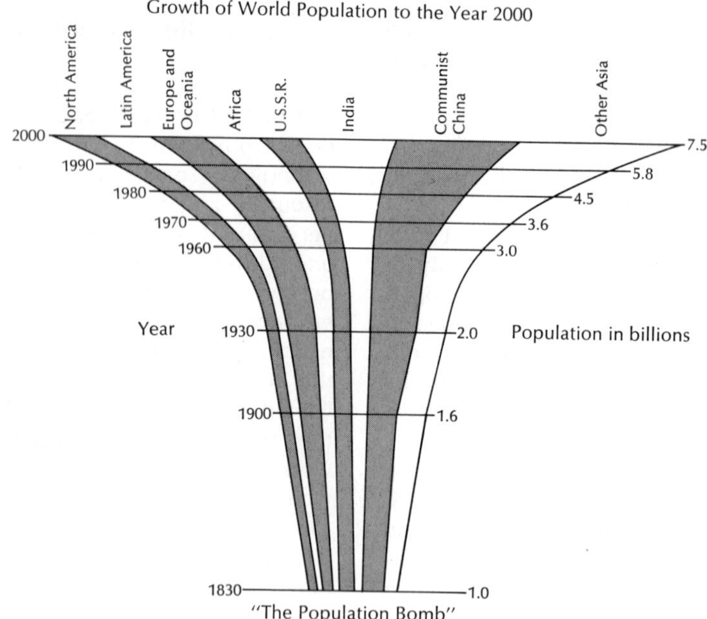

1-7. World population. *(Environmental Quality, The First Annual Report of the Council on Environmental Quality, Aug. 1970.)*

—and is an underlying factor in many of his other problems. However, in the developed countries, increases in per capita consumption far outweigh the effects of increased population in some categories. To illustrate, 90% of the increased use of electric power in the United States in the last three decades has been a per-capita increase, whereas only 10% can be attributed to a population increase. For beef consumption in the last two decades, the per capita increase has been 75%, far exceeding that produced by the increase in the number of consumers.

The Earth's population probably totalled about 250 million in 650 AD, about 500 million in 1650, about 1 billion in 1850, and about 2 billion in 1930. Thus the time needed for a doubling of world population has decreased at a geometric rate—from 1000 years, to 200 years, to 80 years, to the current figure of about 35 years. However, the doubling time tends to be shorter than this for the underdeveloped countries (e.g., an estimated 20 years for the Philippines) and longer than this for the developed countries (ranges up to about 200 years). As Ehrlich (1968) has pointed out, this means that for the Philippines even to maintain its present status as a poorly fed underdeveloped country, most of the services, materials, and foods that it now requires will have to be doubled within the next two decades—e.g., the number of doctors, nurses, teachers, schools, hospitals, houses, policemen, and buses—and all this without improving at all the lot of the average person. How much better off the average person would be if total population could stay about the same and the additional services and materials were used instead to increase per capita consumption.

World population totals are determined by the proportion of births to deaths, and general, worldwide, even startling decreases in the death rate have been a major factor in the present population explosion, especially in the underdeveloped countries. In many instances, this has been accomplished by stamping out such infectious diseases as malaria and smallpox. Another factor: populations grow at a compounded rate, which means that a 1% annual increase leads to a doubling in 70 years (not in 100 years), whereas a 2% annual rate (the present worldwide average rate) causes a doubling every 35 years. Any decrease in the rate at which the world population grows must thus come about in either of two ways: a decrease in the birth rate or an increase in the death rate. Most ominous for the immediate future, so far as births are concerned, is the fact that about 40% of the population of the underdeveloped countries is less than 15 years of age (this figure is about 30% for the United States).

It has been said that the present problems of dealing with the multitudes of individuals in cities, schools, parks, roads, and airports suggest that the optimum population for many countries is less than the present number. There are obvious pressures of overpopulation on the physical environment such as depletion of mineral resources, the necessity for more food, increased energy requirements, and increases in the quantities of discarded wastes and effluents. However, there are additional, less obvious, but just as important effects upon the quality of our lives. More crowding necessarily tends to bring more regulation and less choice, more noise and less solitude, more frustration and less happiness, and may well lead to unrest and violence.

Although attempts have been made to check population growth, the effort to date on a worldwide basis has been very small and is still very small; e.g., funds expended for research on contraceptive methods in 1970 everywhere on Earth may not have totalled more than $35 million, although time is such an important factor. To raise the standard of living of everyone on Earth a particular amount in, say, 1970 would require the expenditure of a certain quantity of materials and human energy. In 1973, however, a much larger expenditure would be needed to accomplish the same results because services and goods must then be spread among many more persons—the current increase in population is equivalent to a new United States every 3 years or so.

Certainly a minimum first step in coping with overpopulation is the development of suitable methods of birth control and the widespread dissemination of the necessary information and materials. All peoples of the Earth should be able to plan their own families (e.g., 20% of all births in the United States may be unwanted). To paraphrase: every conception should result from deliberate choice; every choice should be an informed one. However, this may not be enough, and incentives of various kinds, such as higher taxes for larger families, will probably be needed to make the average couple want a small family. Encouraging is the current rapid growth of the group advocating zero population growth (ZPG) and the slogan that will accomplish this for each married couple—stop at two. However, the difficulties in accomplishing such a program on a worldwide basis should not be underestimated. On the other hand, the alternative if it is not accomplished is so depressing and frightening that the incentive to get on with the job must become irresistible. Overpopulation in the developed countries is less of a problem now than that in the underdeveloped countries. Nevertheless, for developed countries to ignore overpopulation in this interconnected world is like saying, "Your end of the boat is sinking."

Pollution and the Citizen

Even the casual reading of a newspaper in late July of 1970 brought into sharp focus the alarming concentration of pollutants in the Earth's atmosphere (Fig. 1–8). In the vernacular of the day, our effluent, affluent society was now eyeball to eyeball with its environment. Cities (Table 1–1), of course, were worse off, but even on Whiteface Mountain in the Adirondacks—260 miles from New York City and generally the location of some of the cleanest air in the United States—pollutants were occasionally as dense as in a crowded suburban area. As an air mass stagnated over New York City and pollutants within it accumulated to a dangerous level, more than 8000 people on the other side of the Earth in Tokyo were treated in hospitals in one week for smarting eyes, burning throats, and other physical ailments caused by a persistent white smog (oxygen-vending machines were common on Tokyo's streets, and many residents wore gauze masks). Furthermore, trees and shrubs in the secluded gardens of the Imperial Palace were reported to be dying, and pollution of air, water, and soil throughout Japan as a whole was said to have reached crisis proportions (more than 100 million people crowded into an area no larger than Montana). "Progress" in the form of very large and very rapid industrial growth during the previous decade in Japan was cited as the main cause.

Meanwhile in Washington, D.C. tourists gazed at the Washington Monument through a yellowish haze and in Sydney, Australia, measures to control environmental pollution were announced following vigorous complaints caused by the discharge of a large mass of waste hydrogen sulfide gas (strong odor of rotten eggs) from an oil

1-8. *Air* pollution: principal emissions and sources in the United States, 1968. *(First Annual Report of the Council on Environmental Quality, Aug. 1970.)*

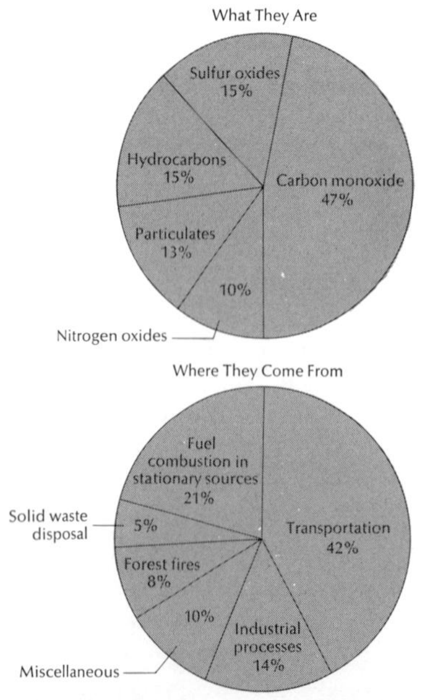

TABLE 1-1
CLIMATIC CHANGES PRODUCED BY CITIES

PARAMETER	CITY AS COMPARED WITH RURAL SURROUNDINGS
Temperature:	
Annual mean	0.9° to 1.4°F. higher.
Winter minimum	2° to 3°F. higher.
Cloudiness:	
Clouds	5 to 10% more.
Fog, winter	100% more.
Fog, summer	30% more.
Dust particles	10 times more.
Wind speed:	
Annual mean	20 to 30% lower.
Extreme gusts	10 to 20% lower.
Precipitation	5 to 10% more.

Cities tend to be "heat islands," especially during the warmer months when winds are light. Heat energy is released directly by heating, air conditioning, transportation and industrial processes. Moreover, solar energy is absorbed by pavements and buildings during the day and then reradiated as heat energy both day and night—rural areas tend to absorb a smaller percentage of the solar radiation. (First Annual Report of the Council on Environmental Quality, August 1970.)

company plant. Moreover, earlier disclosures that Russia was likewise suffering from severe pollution of its air, water, and overall natural environment indicated that industrialization itself, and not the system of government, was the cause. Russia's problems were somewhat alleviated by two factors: the immense size of the country itself and the relative scarcity of the auto, now the chief culprit in air pollution in many cities; e.g., some 60,000 cars were abandoned in New York City alone in 1969, whereas in all of Moscow there were only 80,000 privately owned automobiles, and none of these were being abandoned.

To date, degradation of the natural environment has accompanied so-called progress in the form of industrial expansion and economic growth, but severe pollution is not inevitable. Rather the natural environment must no longer be used free-of-charge for waste disposal. A so-called social balance sheet should be included in the design of each industrial plant, and the cost of clean air, clean water, and an undamaged environment must be an integral part of the final product and the price at which it sells.

If industries do not treat their own wastes but instead dump them directly into some municipal system for treatment, then each company can be taxed according to the quantity and quality of its pollutants. When seen in true perspective, such charges will not prove to be exorbitant and will be most worthwhile because many economic, personal, health, and esthetic benefits of environmental improvement are not now calculated as part of the cost structure of present-day products: e.g., the physical well-being from living in a pollution-free environment, less absenteeism from work, fewer job turn-overs, the inverse relationship that tends to exist between property values and air pollution, the opportunity to sniff a flower rather than garbage, the delight of seeing a beautiful sunset rather than a dirty smog, and the chance to swim or fish in a local river. Moreover, when waste disposal and treatment become an integral part of the costs, ways will undoubtedly be found to do the job more effectively and at less expense.

The conflict between individual profit and public welfare in a society that emphasizes freedom of choice (a similar conflict also occurs in communistic societies) has been illustrated by imagining that several men are grazing cattle in a large common pasture. When numbers become large and overgrazing results, a conflict of interests develops. The addition of an extra cow to the herd of one of the men then produces two economic results. The full value of the added animal accrues only to this man, whereas he shares the damaging effect of the increased overgrazing with each of his fellows. Thus his positive share much exceeds his negative share, and he profits individually, even though the public as a whole loses because of his actions.

When this concept is applied to pollution, we find that discarded wastes are added to the natural environment by individual persons or companies, whereas the undesirable effects of the resulting pollution are shared with all of society. Since the individual's share of this pollution is less than the cost to him of treating the wastes, he has a strong incentive to pollute. Moreover, if such costs are not included in the pricing of objects and materials such as autos, paper, and aluminum, then the demand for these end products will be greater than it would be if they had to be purchased at their true, higher unit cost. In turn, this greater demand and consumption lead to more extensive environmental degradation.

Serious problems thus arise in particular areas as soon as enough men are sufficiently active industrially to put pressure on the natural environment. Therefore, regulation becomes necessary, and on a national level, because companies that treat their wastes in one city or state will be at a disadvantage if similar companies in other cities or states are not also making similar expenditures. As President Nixon stated in February, 1970: "The tasks that need doing require money, resolve, and ingenuity—and they are too big to be done by government alone. They call for fundamentally new philosophies of land, air, and water use, for stricter regulation, for expanded government action, for greater citizen involvement, and for new programs to insure that government, industry, and individuals all are called on to do their share of the job and to pay their share of the cost."

We must not underestimate the problems that are involved (see below) nor the magnitude of the inconveniences and sacrifices that will be demanded for their solution. However, neither must we underestimate the ingenuity of man to overcome these problems—once they have been generally recognized and their solution has been urged vigorously by a well-informed, aroused society.

In an article in the *New York Times* (19 April 1970), Edwin Dale, Jr. emphasized the difficulties in controlling environmental deterioration by describing three economic laws—each quite depressing. These are: the law of economic growth, the law of compound interest, and the law of the mix between public and private spending. Dale begins with the conclusion that the growth of production—which he assumes our society is not willing to give up—is the basic cause of pollution growth. To him, economic growth seems certain to continue because the output of each worker in the United States has increased about 2 to 3% per year for several decades. Moreover, the population itself is now increasing at a rate of about 1% per year, and these two rates are compounded each year.

Thus we move from a gross national product (in constant dollars) of $100 billion in 1944, to one of $450 billion in 1957, to more than $700 billion in 1969. The gross national product consists of items such as electric power, paper, and steel which pollute, of tin cans and bottles which must be disposed of, and of cars, buses, and trucks which spew clouds of exhaust gases into the air. The gross national product has increased so rapidly because the percentage of growth is applied to a larger base each year; this is the law of compound interest, and a 2% annual growth rate causes a doubling every 35 years. Thus we have now arrived at the stage where our growth in just one year is equal to half the total output of Canada. Put in other words, the total real output of goods and services in the United States in the last two decades has equaled that produced in all the preceding centuries. For this reason, the same activities which caused no serious pollution a decade or two ago, because their effects were then smaller and thus more diluted, now cause serious environmental deterioration.

In his third law, Dale contends that the mix between public and private spending is not a major factor because each leads to economic growth, which in turn leads to more pollution. However, if we now shift from our past policy of assigning no cost to air, water, and the natural environment to one that places a price on each act of pollution, more growth may take place with less pollution. Thus the call for a fundamental new philosophy of land, air, and water use.

Air Pollution

The internal combustion engine as it functioned in autos, buses, and trucks in 1970 was a major polluter of the air in many cities, although sulfur oxides from burning fuels were perhaps even more damaging. However, progress was being made in pollution abatement—e.g., devices attached to autos to decrease emissions were now mandatory, and a shift to nonleaded gas seemed in the offing. Nevertheless these were not yet enough, and a partial or complete ban of the auto from the city—at least in certain areas at certain times—was no longer inconceivable. A persistent smog associated with a strong low-lying inversion (p. 688) now brought a city dweller face to face with a drastic choice: the auto or air fit to breathe. This in turn suggested some searching questions concerning his transportation. Why the present inadequacies of mass transportation: bus, train, and subway? Why the present glamour attached to the automobile and its rapid, built-in obsolescence? If drivers would only settle for something less than a zooming, jetlike pickup in their autos, engines could be designed that would run twice as far on a gallon of gas. This in turn might cut the pollution effects in half and stretch our oil supplies over many more decades. Furthermore, a complete substitute for city driving, such as the steam or electric auto, might then be possible. At any rate, once the citizen decides that he is willing to trade off a somewhat lower level of performance for a decreased amount of pollution, the auto pollution problem will be solved.

Specific instances of air pollution are many and varied. In one, certain buses that could not be crowded into a particular garage on Staten Island were run continuously during some winter nights to make certain that they would start in the morning. Families lived within ten feet of the exhausts, and their children at times became too sick from breathing the emissions to attend school. The question is, how long will society endure such conditions before it feels compelled to eliminate them?

Smog and polluted air are discussed in Chapter 27, but we make two points here: (1) the total damage caused by air pollution probably exceeds by far all the expenditures necessary to control it, and (2) enough is now known to correct the present situation, although newer and better methods will undoubtedly be developed, and many questions are still unanswered. It is the will to improve the environment that seems to be lacking.

A 1970 *U. S. News & World Report* study concluded that at least $71 billion would have to be spent over a 5-year period to clean up our environment ($54 billion for water, $13 billion for air, and $4 billion for solid waste disposal). However, greatly offsetting this cost was the estimated loss each year from polluted air ($13.5 billion) and water ($12 billion). Moreover, this estimate excluded effects on health and loss of farm crops as well as very important but intangible esthetic benefits. If these estimates hold true and if the all-out effort is made, then the United States could find itself 5 years from now with clean air, water, and land and a saving of more than $55 billion!

In addition to problems concerned with pollution of the air, there has been fear in some quarters that man's activities would so upset the natural environment that the Earth's atmosphere might be depleted of oxygen in a few decades or more. However, according to a 1970 report by W. S. Broecker, ". . . the molecular oxygen supply in the atmosphere and in the broad expanse of open ocean are not threatened by man's activities in the foreseeable future. Molecular oxygen is one resource that is virtually unlimited." The conclusion was based upon a comparison with measurements made some 60 years ago and certain other considerations. Furthermore, if we were to use up all known fossil fuel reserves, their oxidation would consume less than 3% of the available oxygen. In addition, although changes in the rate of primary photosynthesis may be critical to man's food supplies, they apparently have no bearing on his oxygen supply.

Water Pollution

The Rhine, economically Europe's most important river, is called the "sewer of Europe" by the Dutch who live at its noisome, downstream end. It illustrates water pollution problems in a river. According to a *New York Times* article (22 March 1970) sewage and industrial effluent now constitute one-fifth of the river at its mouth and result from recent population and industrial growth. The river's natural ability to cleanse itself through dilution, oxidation, and bacterial action has been overwhelmed by the quantity of pollutants dumped into it.

Despite large expenditures for waste and sewage treatment, the pollution has worsened over the years, a fact underscored by the death of millions of fish during June of 1969 within the German portion. Along its route, the Rhine picks up wastes from boats, farms, and cities as well as acids, dyes, pesticides, salts, and other materials from industry. Despite all this, the Dutch still use the Rhine for more than 60% of their fresh-water supply. Since the quantity of pollutants tends to increase each year, increased treatment is needed just to keep from falling further behind in the clean-up program.

West Germany has been described as a country that is particularly proficient in re-using water and in its method of taxing polluters. In the heavily industrialized Ruhr area, all groups that discharge polluted water—towns, coal mines, and factories—must join water purification associations through which they are taxed on the basis of the quantity and quality of the pollutants they give off. This insures the conservation of water supplies, and more than 60% of industrial demand is met by using recycled, reprocessed water; in fact, one cardboard mill is reported to renew its water supply only twice a year. To replenish water suitable for drinking, the West Germans filter polluted water through river banks and settling basins into infiltration basins. Seepage then transfers this cleaned water into the ground-water zone from which it can be pumped via wells.

A river can cleanse itself of most pollutants if their quantity is limited. Larger objects settle to the bottom, sunlight destroys some pollutants, and beneficial, oxygen-using bacteria consume as food the carbonaceous organic matter present in sewage. Oxygen is used up in this process (termed biochemical oxygen demand, or BOD) and is replenished by natural aeration and by oxygen given off by growing plants. However, when so much sewage and other organic wastes are present in a river that all of the available oxygen is used up, then the excess sewage can not be degraded, and the river becomes polluted. In one case of extreme pollution, an oily, waste-laden river actually began to burn in 1969 and almost destroyed two bridges (Cuyahoga River in Cleveland).

Water wastes occur as suspended solids and as dissolved organic and inorganic materials, and more than 99 percent of most sewage is water itself. The main goals of waste water treatment by communities are: to remove suspended materials, to kill harmful bacteria, and to reduce the quantity of oxygen-demanding, organic material.

However, before the late 1960's, only the major pollutants tended to be checked in routine water analyses, thus ignoring less voluminous but harmful materials such as mercury and DDT.

The basic treatment methods (Fig. 1-9) have been in use for some time and fall into two groups. *Primary treatment* is essentially a mechanical process that involves removal of large floating objects, grease, and scum, and screening out of smaller pieces,

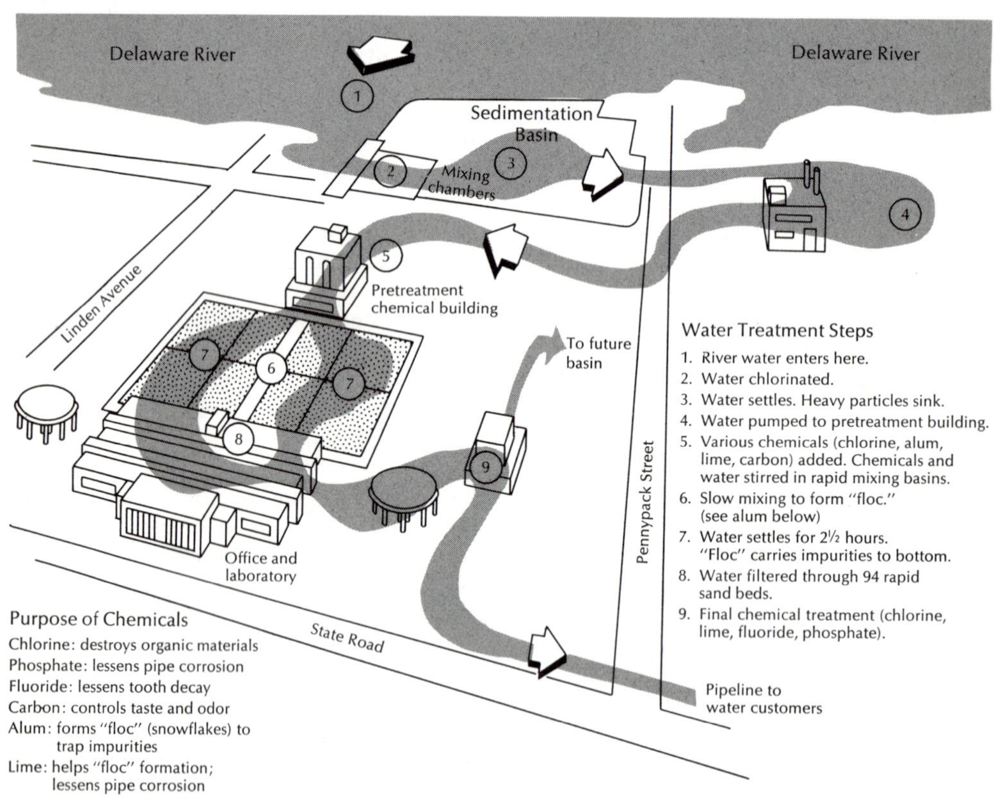

1-9. Philadelphia's Torresdale water treatment plant. *(Swenson and Baldwin, A Primer on Water Quality, U.S. Geological Survey.)*

and then detention of the waste water for a time in a primary clarifier so that tiny suspended particles settle out to form a sludge. About one-third of the polluting material is removed in this manner, and the clarified, still-polluted water is dumped into a nearby waterway.

In *secondary treatment,* the effluent from the primary treatment is acted upon by microorganisms which eat most of the remaining organic matter and some of the nutrients, thereby changing them into harmless chemical compounds. Essentially this is the same process by which bacteria in a river cleanse it naturally (biological oxidation). In one method, the effluent trickles slowly through a bed of rocks that contains the beneficial microorganisms, whereas in another method the effluent goes into bacteria-containing closed tanks which are then aerated (activated sludge process). In the late 1960's it became possible to use pure oxygen in the aeration which reduces space and pumping requirements. The water is then chlorinated or treated in some other manner to kill remaining germs and returned to waterways; it is now nearly suitable for drinking and may be used for irrigation.

Sludge disposal is a problem and may make up ¼ to ½ of the total cost of the treatment. After concentration, de-watering, bacterial action, drying, and combustion—or some combination of such processes—the sludge may be used for land fill or as a soil conditioner, incinerated, or dumped into the ocean. In general, the more the water is cleaned, the greater is the quantity of sludge left for disposal.

In the United States in 1970, about one-third of the people lived in areas without sewers, the sewage of about 5% was untreated, the sewage of nearly 25% received only primary treatment, and that of 40% underwent secondary treatment. The upgrading planned for the early 1970's would result in secondary treatment for about 90% of the total and primary treatment for the rest. However, this may not be enough because the effluent contains two elements that form key ingredients in fertilizers—nitrogen and phosphorus. These enrich the nutrient supply (eutrophication) and thus stimulate the growth and overgrowth of algae of various kinds which then compete with fish and bacteria for the available oxygen. Some fish may die and anaerobic bacteria may thrive in certain locations, which can then lead to unpleasant odors and tastes. The nitrogen and phosphorus can be controlled by adding a so-called *tertiary treatment* involving various chemical and mechanical processes, although costs are then higher However, these and other steps may also be performed earlier, and the term *advanced waste treatment* is used for such an overall system. Offsetting the increased cost is the potential availability of such reprocessed water for swimming and perhaps even for drinking. However, care must be used here because the reclaimed water may still contain substances such as pesticides, viruses, and various chemical and biological toxins. The addition of such nutrients as nitrogen and phosphorus to our water supplies also occurs naturally as a result of weathering and erosion, but the rates are much slower.

The problem of eutrophication may be lessened by a move away from high-phosphorus detergents and perhaps by using less fertilizer on farms—it is only the excess nitrogen and phosphorus not used by plants that gets into our water supplies and poses a problem. Advances have been made in recent years in waste-water treatment, and sewage plants have increased more rapidly than the population. This is encouraging, together with the rapidly growing public awareness of pollution and environmental problems. Moreover, increased research has resulted in promising new techniques, some of which are now being tried out in pilot plants. In one method, the microorganisms are dispensed with, and chemicals are added to the raw sewage to remove most of the suspended organic materials. The effluent is then filtered through granular carbon (like a filtertip cigarette) which removes much of the phosphorus together with various industrial chemicals that formerly escaped. The carbon can be cleaned and re-used.

Storm runoff is an additional problem. In many communities, sewage and runoff are moved through the same pipes. Thus when volume is greatly increased during a heavy rain, the water treatment facilities may be temporarily overwhelmed, and the excess has to be discharged untreated into nearby waterways. Temporary detention basins may alleviate this problem in some areas.

According to one view, thermal pollution of water and air are necessary by-products of the generation of electricity by both fossil fuel and nuclear energy plants, but about 50% more waste heat energy must be disposed of by nuclear plants (in producing equivalent quantities of electricity in 1971). However, this is an example of a problem that is also an opportunity, and useful ways to use the very large quantities of excess heat energy may well be found: e.g., desalting water, recycling sewage, heating cities, and growing food in greenhouses and fish ponds.

A few 1970 examples of hopeful developments in the field of water pollution are cited below. In one, pulp will be manufactured from the waste cellulose fiber in disposable diapers, from plastic-coated milk cartons, and from various paper wastes. Each ton of recycled waste paper is the equivalent of about 17 trees. In another development, a drug company described a new plant in which wastes are separated into different categories and treated by different methods. This eliminates 90% of normal wastes and produces useful by-products such as steam for power generation and a high-protein animal feed. In still another area, waste nutrient-rich water from a food-processing plant was being sprayed onto the Texas prairie. Instead of polluting local streams, this effluent fertilized grasses in the sprayed area, which were harvested as a by-product hay crop. The filtration process also cleaned the water, thus augmenting the local supply.

Waste Disposal

Waste disposal is another environmental problem exacerbated by increases in per capita consumption, by increases in total numbers, and by the concentration of people within cities. On the average in the United States, about 2.7 pounds of refuse were thrown away daily per person in 1920 and collected by private and public agencies. By 1969 this amount had about doubled, and the figure predicted for 1980 is 8 pounds per person. When industrial wastes are added to the figures for household wastes listed above, the 1969 quantity is about doubled—from about 5.3 to 10 pounds per person per day. Moreover, these figures do not include junked motor vehicles (about 7 million per year in 1968—some have been successfully dumped at sea to form fish-attracting reefs) or wastes from the mining, milling, and smelting industries (about 30 pounds per person per day in 1965).

With more people throwing away more items, waste disposal quantities mount rapidly. Other problems: "the oversized package with its undersized contents" and the tendency to sell more goods and foods in ways that involve more waste and in containers that are less readily disposed of. A large amount of this waste would certainly be unnecessary if objects and their containers were designed to make recycling and disposal more widespread and efficient and if citizens in their household, business, and recreational activities made a determined effort to reduce the volume of wastes. An incentive to do this might be a tax covering the cost of eventual disposal of each object or material that is produced. Since this tax would be paid when an object was initially sold, there would be an inducement to manufacture it in such a manner as to minimize the cost of disposal. Another suggestion involves a lower garbage-collecting fee for sorted wastes, which are more readily reclaimed, than for unsorted wastes (e.g., this could lead to greater re-use of paper wastes which make up about 60% of all residential refuse). Disposal will also be aided if experiments such as the following turn out successfully: the development of a type of paper that becomes a mulch when wet and a glass bottle container which dissolves away when a protective plastic film is broken. Thus the recent research that was aimed at creating a self-sustaining environment within a space ship by reprocessing and re-using materials needs to be applied more widely now at the Earth's surface.

By the late 1960's, waste handling and disposal in the United States were still quite primitive, but some of man's ingenuity and technological skills were beginning to be applied to these problems—thus new techniques and extensive improvements in methods of burning or burying wastes seemed sure to follow. However, to illustrate the magnitude of the problem, in 1970 nearly three-fourths of all refuse still went into open dumps which should be completely eliminated, about 15% was incinerated, and about 8% was buried in sanitary landfills. Only a small quantity was salvaged or used as compost.

Sanitary landfills have been successful and can presumably be used much more widely. Here the upgrading of marginal land can at times be a valuable by-product of the waste disposal. A thin layer of refuse is first spread across a suitable site, compacted by a bulldozer, and then covered by another layer. When the refuse layers reach a thickness of several feet or more, they are covered by a thin layer of clay or soil which is then compacted. The process is repeated until this site has been filled or piled to the desired level. It is then sealed by a layer of soil a few feet thick, and vegetation is planted over the entire area. Since the refuse is not burned and is sealed with soil or clay at the end of each day, there tends to be no difficulty with odors, flies, and rats. Microorganisms within the refuse (first the aerobic and then the anaerobic) tend to consume the organic wastes, thereby producing methane and other substances. Ground settling and slow escape of the methane tend not to be serious problems in open recreational areas. Compaction prior to disposal can produce debris that weighs 800 to 2000 pounds per cubic yard which may make long-haul rail transportation economical. It also reduces settling and extends the life of a particular site. If marginal lands such as those ruined by strip mining are chosen as landfill sites, they may be transformed into prized recreational assets such as golf courses and parks. However, care must be taken to avoid groundwater pollution.

"Mount Trashmore" near Chicago illustrates an imaginative, beneficial type of landfill project in which refuse is being heaped into a man-made hill 125 feet high—the highest land in the county. A dozen areas for tobogganing and skiing are planned for its slopes when completed. At the start, a nearby waste marshy area was excavated to form a lake. Clay from this excavation was then spread across the ground at the site of the future Mount Trashmore to form an impervious base. Additional clay is being used to cover each day's supply of refuse, and when the hill is finally completed, more clay will be spread as a seal across its flanks and summit. In the meantime, gravel is being recovered from the former marsh along with the clay and can be sold to meet some of the project's costs.

Large-volume, high-temperature incinerator technology is developing (e.g., the refuse may be sprayed with oil to make it burn more readily and completely) but care has to be taken not to substitute air pollution for solid waste disposal. Useful by-products can also be recovered such as the flyash that has been processed into large-sized, light-weight building bricks of various colors and textures. Moreover, in the water-wall type of incinerator, water is used to partially cool the incinerator and the heat energy thus absorbed can subsequently be converted into electric power or used to help run the incinerator.

Depletion of Mineral and Energy Resources

According to Park (1968) the United States has used more minerals and fuels during the last three decades than did all of the peoples of the Earth (including the United States) in all previous history. Yet two decades or so from now, the United States may be consuming twice the quantity of minerals and fuels that it is consuming today—if these are available. This is alarming from several points of view. First, these are non-renewable, one-crop resources that are present in the crust only in limited amounts. Second, many are relatively small deposits that represent extreme concentrations of metals or materials that otherwise occur only in very diffused quantities throughout the rest of the crust. Such rich, easily found, easily worked deposits represent a country's "quick assets." Since some countries have more than others, international trade is necessary to move supplies from the places where they occur to the places where they can be used. Since the industrial survival of a nation depends upon an adequate supply of such metals and fuels, vital questions of exploration, allocation, transportation, and control arise.

To illustrate, the United States has within its own borders at the present time ample supplies of only about a dozen of the 100 or so minerals that are most important in industry. Therefore, it now imports from various countries part or all of its requirements of each of the remaining 85 to 90 minerals. Included are necessary materials such as mercury, tungsten, antimony, manganese, chromium, tin, and aluminum. Furthermore, with the possible exception of the Soviet Union, the other industrial nations are likewise net importers of mineral and fuel supplies.

The impact of the recent increase in consumption—which is a very sudden and enormous change when viewed from the perspective of human history—can be illustrated by the so-called "Age of Fossil Fuels" in which we now live (Fig. 1-10). Coal, oil, and natural gas comprise the fossil fuels and form so slowly that they are essentially nonrenewable resources. They have accumulated here and there within the crust during the past 600 million years or so, which is a very long span of time. Yet we are now in the process of using up this entire accumulation within a time period of a few centuries. This is analogous to a man who saved his money carefully for 40 years or so and then spent his entire life savings in one unthinking, short-sighted spree that lasted only a few seconds. The consumption of the fossil fuels is truly a unique, one-time event in the history of the Earth and may have a significant effect upon the Earth's climate (p. 648).

To illustrate with other materials, the United States presently uses copper at the rate of about 20 pounds per person per

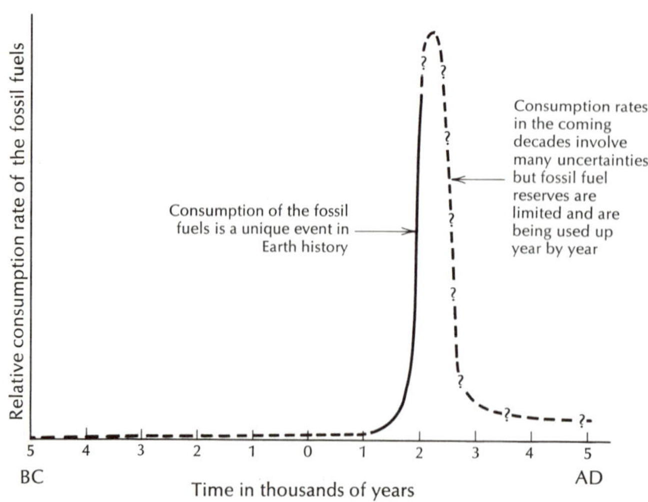

1-10. The "Age of Fossil Fuels" illustrated schematically.

year. All the other peoples of the Earth combined—more than 3 billion in number, including both developed and underdeveloped countries—average less than 3 pounds of copper per person each year. If the population of the Earth doubles by the year 2000, as is commonly predicted, and if all the peoples of the Earth then consume copper at the rate the United States does today, the required amount of copper at that time, only three decades hence, will be some 15 times the amount used today. Moreover, projections show that copper consumption per capita in the United States and in other industrialized countries is going up rapidly each year. Where is all of this copper to come from, and when will the Earth be depleted of its supply of copper (Fig. 1-11)? Similar figures and questions can be cited for lead, iron, aluminum, oil, and other materials.

Hubbert* has viewed human affairs on a time scale spanning thousands of years, and he has noted the period of rapid population and industrial growth that has taken place during the past few centuries and that seems certain to accelerate in the years that lie immediately ahead. He has concluded that this is one of the most abnormal phases in the history of mankind—such rapid growth is not the normal order of things and is not capable of extension indefinitely into the future. Rather, this period of growth is apparently a relatively brief transitional period between two very much longer periods, each essentially periods of nongrowth because rates of change were and presumably will be so slow. To follow Hubbert further, if we plot the growth of population, the per capita consumption of energy, and the use of mineral resources against a time scale of thousands of years, we find that the curves essentially coincide (Fig. 1-10). Each curve has a long, almost horizontal portion near the zero level that extends out of early human history. Each curve also began a slow upward movement a few centuries ago, and this has accelerated in recent years. The curve now extends upward along a very steep, nearly vertical path. Such rapid growth has never occurred before in the history of the Earth, apparently cannot last much longer, and seems impossible of repetition in the future.

* National Academy of Sciences, *Resources and Man*, Freeman, 1969.

If we look now at the quantity of mineral resources still left in the crust for man to discover, extract, and use, we find two sharply opposed views with, of course, various shades of opinion in between. Proponents of a very optimistic view (economists and social scientists are among the leaders of this group) contend that, given enough cheap energy, scarcities can be avoided because man will then in effect be able to apply his advanced technological skills to the grinding up of entire mountains. From these huge "open-pit mines," he will remove cubic mile after cubic mile of rock, from which he will extract the metals and other materials that occur ever so sparsely within them. Even farm land can be substituted for in this extreme view, and the magic of chemistry is expected to provide appropriate substitutes and synthetic materials when the natural deposits become depleted. Many geologists, acquainted as they are with the manner of occurrence of mineral deposits in rocks, view the problem of future mineral and fuel supplies in a quite different manner.

Geologists have calculated the average chemical composition of the Earth's crust (p. 164) and have noted that most metals are present in such small quantities that they would be unattainable if they were scattered uniformly. Aluminum and iron, comprising about 8 and 5% respectively by weight of the crust, are the most conspicuous exceptions. Fortunately, various geologic processes (p. 134) have acted to increase greatly the concentrations of these scarce materials in certain places, and such deposits are called ores if one or more useful materials (generally a metal) can be extracted from them at a profit. It follows that a whole host of factors determine whether a particular deposit is an ore or not, and that its status may change with time: e.g., cost of transportation, labor supply, price changes, value of by-products, and techniques involving the mining, milling, and smelting of a deposit. Technological advances, therefore, have changed many a lean mineral deposit into a valuable ore deposit, especially if the

1-11. This huge open-pit copper mine at Bingham Canyon, Utah, is about 1500 feet deep and two miles in diameter. Within it, strings of full-sized railroad cars appear only as thin dark lines on some of the benches. Although copper generally forms about 1% of the igneous rock being mined, more than 200,000 tons of the metal have been produced within a single year. *(Courtesy U.S. Geological Survey.)*

deposit is one having a large volume. This has led some mineral economists to assume that the total volume of the deposits containing a particular metal such as copper will increase geometrically as the percentage of the metal within the deposit decreases arithmetically. If this were true, scarcities in the future would not be serious because price increases and technological progress would make it economically feasible to extract useful metals from much larger, even though less concentrated, deposits. This assumption was developed in connection with a certain type of copper deposit where it has some validity, but it does not apply to all copper deposits nor to many other kinds of deposits. We just do not find ever-increasing tonnages of leaner and leaner mineral deposits bridging the gap between the enormous volumes of normal, relatively barren rock and the irregularly scattered concentrations that form our ore deposits.

In past decades and centuries, prospectors and geologists have discovered the richer mineral concentrations exposed at the surface. Now, greater skills must be applied to finding such deposits that lie hidden beneath tens, hundreds, and even thousands of feet of barren obscuring rock and overburden. Presumably these occur here and there beneath the lands and continental shelves (more difficult to find and exploit) but not beneath the wide expanse of deep ocean floor (underlain by younger rocks that are generally unfavorable for ore deposits). Orbiting satellites may be able to collect data that will aid in the discovery of hidden deposits. After a major deposit is discovered, several years may still go by before its product becomes available—roads or harbors may be needed as well as milling and smelting plants and extensive site preparation.

Thus a country must have a workable, farsighted minerals program involving such facets as exploration, research, conservation, and international trade. Rapid, greed-governed exploitation should be avoided in favor of long-range, carefully planned programs that aim at maximum eventual yields with minimum damage to the environment. Otherwise the richer parts of a deposit may be gouged out immediately leaving behind a ruined mine that still contains large volumes of less-concentrated minerals.

In 1969, Hubbert estimated reserves of the fossil fuels and the rates at which they are likely to be produced and used up in future years. He ignored the relatively long time periods necessary to use up the first 10%, and concentrated upon the relatively much shorter times needed to use up the middle 80%. According to his figures, 80% of world crude oil production will be used up in a period of about 60 years with peak production occurring about the year 2000. Similar figures for the United States (exclusive of Alaska) show a time period of about 65 years and a peak at about 1965. Since 70% of the world's energy production (73% for the United States itself) in 1968 came from oil and natural gas, these figures indicate the supply problems that lie immediately ahead. Moreover, world consumption of the fossil fuels has been increasing about 4% a year during recent decades. It is, of course, impossible to predict with accuracy exactly how much oil and gas occurs in the crust and how much man will eventually discover and remove. However, even if twice as much were eventually discovered as Hubbert now estimates, this would still add enough for only a few more decades. Furthermore, nearly half of the world's total reserves of recoverable crude oil—both proved and probable—seem to be located in the Middle East.

Coal, fortunately, is more abundant than petroleum and may be able to supply the entire energy needs of the world for another century or two. The bulk of the world's coal resources—both proved and probable—appear to be located chiefly in the Soviet Union and the United States (about 65% and 27% of the total respectively). Oil shales and tar sands likewise occur in huge quantities and will one day be mined for the petroleum products they contain. However, even such large deposits look small when viewed against the enormous volume of anticipated worldwide demand.

Hubbert likewise estimated the reserves of other sources of energy, such as nuclear, solar, geothermal, tidal, and water power. He sees most of these as relatively unpromising and a scarcity of high-grade uranium-235 deposits developing in another decade or two. Therefore, he concludes that our only real hope for an adequate supply of cheap energy for the centuries ahead lies in the rapid development of the so-called breeder reactor which can be fueled economically by the very low concentrations of uranium and thorium present in rocks such as shale and granite (the energy content of a high-grade granite would be nearly 100 times that of an equivalent amount of coal). Moreover, when such breeder reactors are in use, he suggests that the fossil fuels would be better utilized as chemical raw materials than as energy sources.

Hubbert also underscores the necessity of developing a comprehensive program for the safe disposal of radioactive wastes. A workable technique may consist in sealing the nonvolatile radioactive isotopes with relatively long half-life periods within ceramic materials which are then placed within beds of salt. These are impervious and do not come into contact with ground water except at their margins. If the fusion process can eventually be controlled as an energy-producing source, it will be virtually unlimited.

Like a civilization founded on fossil fuels, a high-technology civilization based upon large quantities of inexpensive energy would probably be a unique, one-time event on the Earth. If present-day society and structures were to collapse, then a new and comparable technology probably could not develop out of the ruins because the "quick-asset" rich deposits that made our present civilization possible would then be gone. The more deeply buried, leaner deposits would then need a high-technology capability for their extraction, which presumably would be unavailable.

Food

Although an adequate food supply for man does not directly concern the earth sciences, yet this pressing problem's indirect association is related through topics such as soil, water, climate, energy, and erosion. To set out the problem, at least every other person now living on the Earth either does not get enough food to eat or does not eat a nutritionally balanced diet. Today's most serious medical problem may be protein malnutrition which causes diseases of various kinds as well as mental retardation. Workers in the field disagree on the seriousness of the problem, both now and in the future, as well as on the number of people who presently starve to death each year. Data are inadequate, and the weakening effect of a prolonged lack of food makes a person susceptible to various diseases which are then listed as the primary cause.

A major difficulty, of course, is the rapid increase in population, especially in the underdeveloped countries, which apparently is going to continue nearly unchecked in the years immediately ahead. This means that in the next 15 years or so, perhaps 85 out of every 100 persons added to the world's already inflated total population will be added in such areas as India and China. It follows that a very large increase in food supplies is necessary just to keep such peoples in their present state of undernourished misery.

An increase in cereal production must be emphasized according to a World Agricultural Plan announced by the United Nations in 1969 after an intensive, 4-year, multidisciplinary period of preparation. More intensive use of lands already being farmed should also be given priority over a major effort to utilize marginal lands not now cultivated. By the late 1960's, a major breakthrough had apparently been achieved in the development of new high-yield varieties of cereals—the so-called "green revolution." More crops per year as well as more food per crop have resulted from combining these new types of seeds with fertilizers, irrigation, and pesticides. Although this welcome development does not solve the protein-deficiency problem, perhaps continued research will lead to ways of improving the protein content of certain cereals or to increasing the yields per acre from crops such as peanuts and soya that already are rich in protein.

The necessity of keeping the growing populations of the underdeveloped lands gainfully employed is nearly as important as increasing their food production. Thus programs are aimed at increasing production without a massive influx of labor-saving machinery or without importing large amounts of food from the developed countries. In fact, it is suggested that the developed countries should eventually plan to import some agricultural products from the developing countries so as to help them achieve a more balanced trade program.

What might be done to increase food production in an area such as the Ganges Plain in India is described by Young and Stout.* Paradoxically this large, potentially productive region is now the home of huge numbers of miserable starving people, who are, as Freeman described it, ". . . born to barren lives without hope . . . born to hand-to-mouth existence from hungry infancy to disease-ridden childhood to premature old age . . . old people toppling on the streets to die of starvation before the numbed gaze of those who might yet live on another day, another month." The San Joaquin Valley of California—once a desert but now green and richly productive—is compared with the Ganges Plain—still an unproductive desert except during the rainy season. Yet the soil, water, and climatic resources of the Ganges Valley exceed those of the San Joaquin. The Ganges is underlain by hundreds of feet of water-rich alluvium, and it has a year-round growing climate, although the bulk of its rainfall comes during the monsoon season. Little fertilizer and manure have been added to its soils, which are thus depleted of nutrients and are unproductive. The seeds of presently used strains are not particularly affected by fertilizer, and the new "green-revolution" seeds cannot function without fertilizer, water, and pesticides. The needed quantity of water is apparently available, although the full extent of the ground-water supply remains to be explored, but the energy to pump this water to the surface is not. If ground-water supplies do prove less than adequate, then the Ganges can be dammed to provide a year-round water supply. If the annual flow at its mouth were distributed evenly over its millions of arable acres, then the water depth would be nearly five feet. Experiments on the Ganges Plain have already shown that food output can be increased as much as ten times per acre. Four crops have been produced per year by using the new seed strains in combination with intensive irrigation and fertilization. Thus a key factor for an immense increase in food production is the availability of large amounts of relatively cheap energy, and a nuclear power station on the Ganges Plain is the suggested solution. In this plan, the entire output of electrical energy from this station should be devoted exclusively to the manufacture of fertilizer and to the pumping of water. Overall food output from the Ganges Plain with intensive year-round farming might then be increased as much as seven times.

* *Dry Land and a Hungry World,* New York Academy of Sciences, February 1969.

Young also notes that dry, barren deserts and semi-deserts form about a third of all the land areas of the Earth, whereas half of the world's population crowds together on only a fraction of this amount. He concludes that parts of such dry lands are potentially livable and productive if water and energy can be made available in sufficient quantity at a low enough cost.

Additional Considerations

Noise from city traffic, jet planes, construction work, air conditioners, and the like is widespread, has been increasing year by year, and may have doubled in the last 2½ decades. If noise continues to increase at this rate, many inhabitants of the United States in the year 2000 may well be deaf.

In one estimate, noise now costs the United States each year the equivalent of $4 billion in accidents, inefficiency, absenteeism, and the like. In addition to the annoyance and irritation caused by lesser noises, very loud noises can cause illness and even death, and the vibrations from a sonic boom may result in damage to cliff dwellings in New Mexico or to ancient cathedrals in Great Britain. Moreover, continued exposure to the musical sounds that delight teenagers today—at the volume they consider desirable—has caused damage to the hearing of some individuals.

So-called "aesthetic pollution" is also extensive and unpleasant: e.g., the insult to the eye produced by a string of unsightly billboards or a junkyard, the litter scattered carelessly and endlessly along our highways and streets, the lack of imagination shown by the design and location of many buildings and roads, the row-upon-row monotony of some housing developments, and the depressing appearance and general ugliness of many urban areas. Although in all too many instances, unfortunately, this is the way it is, yet this is not the way it has to be; e.g., one architect has described a city of tomorrow, which can be built now with today's technology. A basic part of this plan is the building of a continuous roof over city streets at about the 30-foot level thus sealing off the noise and exhausts of cars, buses, and trucks. All utility cables and pipes can be suspended from this roof, as can a monorail, and its upper surface will be carpeted with grass and trees. Terraced buildings with set-back levels will be landscaped with flowers, trees, and small ponds as havens for fish and birds, and even small man-made recycled streams may flow endlessly from upper to lower levels.

We conclude with uncertainty as to the outcome. Man's problems are many, complex, and basic, and much has yet to be learned concerning their solution. However, enough is already known to insure success in their solution, although the required effort will be long lasting and of large magnitude, and it must be accomplished by widespread and fundamental changes in philosophy and attitude. As President Nixon stated in 1970, this is "one of those rare situations in which each individual everywhere has an opportunity to make a special contribution to his country as well as his community."

Opportunities for rewarding, interesting careers in environmental work will undoubtedly grow rapidly in the years immediately ahead, and persons with broad backgrounds will be needed to fill the positions. Well-trained students of geology and the earth sciences form one of the best-equipped groups for such tasks. A little publicized "pecking order" is alleged to exist among the sciences: mathematics is placed at the top, and the series extends downward through physics and chemistry to biology and geology. In this view, a member of a particular science must be acquainted with the disciplines placed above his on the list, but he does not need to be familiar with the others. In today's specialized world, therefore, a mathematician needs to understand little if any physics and chemistry; a physicist needs to know mathematics but not necessarily chemistry; whereas a geologist, at the end of the sequence, needs to be rather fully cognizant with each of the other sciences and mathematics. Thus geologists and earth scientists would seem to stand first at the top of suitable candidates for careers in environmental work.

Exercises and Questions for Chapter 1

1. About 164 million visits occurred in 1969 to National Park Service natural recreational and historical areas, a new record high, and an increase

of nearly 9% over 1968. Describe problems created by such increased use and suggest desirable policies and programs for the National Park Service to follow.

2. The recycling of paper products in the United States during World War II reached a maximum of 35%, whereas only 19% were being recycled in 1971. If such paper recycling could be raised to 50%, it would preserve each year a forest the size of New England, New York, New Jersey, Pennsylvania, and Maryland combined (*The Conservationist,* June–July 1971). Is such a goal attainable? Is it economic? If not, what kinds of regulations would make it so? What is the significance of the following statement for the future? Although the United States is the world's greatest timber producer, 3 billion board feet were imported in 1969.
3. Discuss the energy crisis that faces the United States in the 1970s and suggest programs and policies for the United States Government to follow. As background, read the *New York Times* articles of July 6, 7, and 8, 1971, the Aug. 1, 1971, *Forbes,* Hubbert's paper in *Resources and Man,* or similar reports.
4. Why may deep, cold ocean water be the Earth's largest untapped natural resource? Reference: Gerard, R. D., and Roels, O. A., "Deep Ocean Water As a Resource," *Marine Technology Society Journal,* **4,** No. 5, (Sept.–Oct. 1970).
5. Describe some of the ways in which solar energy is now being utilized directly by man (see, e.g., *The Conservationist* for June–July 1971). Is the sun potentially a major, non polluting energy source? Is the amount of research now going on commensurate with this potential?
6. Much has been said and written about the so-called thermal pollution problem; less publicity has been given to its potential as a vast source of usable, heated water. Discuss problems and possible uses (see, e.g., *The Conservationist,* Dec.–Jan. 1970–1971).
7. Use of electric power in the United States has increased at the rate of 7% each year since 1947, and during the economic recession of 1970 it increased more than 9%. Discuss some of the problems involved and ways of attacking them: rapid use of nonrenewable fossil fuels and uranium supplies, fuel imports, choice of sites for major power plants, impact on environment, ways to reduce demand, etc.
8. In the early 1970's, autos, trucks, and buses probably produced about half of all air pollution in the United States and the proportion was considerably higher in some cities. More than half of all the petroleum consumed in the United States in 1970 was so used. What recommendations do you have concerning mass transit vs. individual auto use? The internal combustion engine vs. other sources of power?

   Perhaps you might conduct a survey in your college, neighborhood, or town concerning what trade-offs would be generally acceptable if present cars were modified to increase gas mileage and reduce emissions at the expense of a reduction in auto performance; e.g., if gas mileage were doubled, our fossil fuel reserves would immediately go up by a large fraction.
9. What recommendations do you have concerning the utilization of Alaskan north-shore fossil fuels? Should they be transported via an Alaskan pipeline? Should they be kept as a reserve for our descendants?

What recommendations do you have concerning the importation of fossil fuels into the United States?

10. What recommendations do you have concerning the widespread development of an anti-litter attitude among all the citizens of the United States and all the Earth's inhabitants?
11. What do you think should be done about "the population bomb?"

Brief Bibliography for Chapter 1

American Association of University Women, *A Resource Guide on Pollution Control,* $1.25, AAUW, 2401 Virginia Avenue, N.W., Washington, D.C. 20037.

American Chemical Society, *Cleaning our Environment: The Chemical Basis for Action,* 1969.

Common Cause, Report from Washington, published monthly except December and January.

Commoner, Barry, *Science and Survival,* The Viking Press, New York, 1967.

The Conservationist, published every second month by the New York State Department of Environmental Conservation.

De Bell, Garrett (ed.), *The Environmental Handbook,* Ballantine/Friends of the Earth, New York, 1970.

Detwyler, T. R., *Man's impact on Environment,* McGraw-Hill Book Co., New York, 1971.

Ehrlich, Paul R., and Ehrlich, Anne H., *Population, Resources, Environment,* W. H. Freeman, San Francisco, 1970.

Flawn, Peter, *Environmental Geology,* Harper and Row, New York, 1970.

Fortune, by editors of, *The Environment,* Harper and Row, New York, 1970.

National Academy of Sciences, *Resources and Man,* 1969.

Murdoch, W. W. (ed.), *Environment,* Sinsauer Associates, 1971.

Natural History, Feb. 1969.

Park, Charles F., Jr., *Affluence in Jeopardy,* Freeman, Cooper and Co., San Francisco, Calif., 1968.

Saltonstall, Richard Jr., *Your Environment and What You Can Do About It,* Walker & Co., New York.

Scientific American, Sept. 1970, "The Biosphere," and Sept. 1971, "Energy and Power."

Stewart, G. R., *Not So Rich as You Think,* Signet Books, 1967.

2

Sermons in Stones

And this our life, exempt from public haunt,
Finds tongues in trees, books in the running brooks,
Sermons in stones, and good in everything.
—*As You Like It:* Act II, Scene I

Vesuvius and Pompeii: A Record in Rock

One meaning of *"sermons in stones"* is dramatically, albeit gruesomely, illustrated by the lifelike plaster casts (Fig. 2–1) of the Pompeian men, women, and children who were suffocated in 79 A.D. when clouds of moist ash and poisonous volcanic gases enveloped them. The casts preserve facial expressions and details of musculature and clothing. Some reveal the agony of a final struggle, but others show sleeplike postures. Although the stories of Pompeii and Vesuvius are well known, newer discoveries warrant a brief retelling because they introduce some key geological concepts in a striking manner.

Vesuvius is a volcano located in Italy, and Pompeii was once a prosperous coastal city situated near its base. In 79 A.D. Vesuvius, as it is known today, did not exist. The name then belonged to a much larger, vegetation-mantled, apparently extinct volcano (later renamed Monte Somma).

Damaging earthquakes preceded the catastrophe and warned that the volcano was merely dormant—not extinct. Terrifying explosions began about noontime on August 24 and continued for several days. About half of Monte Somma was destroyed, and volcanic debris and gases blackened the sky. Fragments of shattered rock and lava were hurled down upon Pompeii and buried its buildings and its people. From 2000 to 16,000 (Maiuri, 1961) of the 20,000 inhabitants of Pompeii may have been buried.

Gravel-size fragments of volcanic debris (chiefly pumice—a frothy, cellular, glassy lava) first blanketed Pompeii to an average depth of about 12 feet. On top of this was piled about 8 feet of moist, mud-

2-1. Lifelike plaster casts of thirteen Pompeians in their final moments of flight. The man in the foreground had a bag slung over his shoulder, and two little boys lie beyond, their expressions peaceful. The final figure seems to be a man immobilizd as he struggled to rise. The bodies of the victims decayed long ago, but molds formed that preserved their shapes. *(Photograph, Lee E. Battaglia, courtesy National Geographic Magazine, copyright National Geographic Society.)*

forming volcanic ash. The ejecta were not glowing hot; the remarkably detailed, museumlike preservation of the Pompeians and their possessions shows that they were smothered, not burned. The volcanic fragments piled up inch by inch and foot by foot until first-floor doorways and windows were sealed. The ever-increasing load finally caused roofs, walls, and columns to collapse, thereby burying some Pompeians within their homes and public buildings, and poisonous volcanic gases added to the disaster. Others were immobilized in the act of fleeing by the mudlike ash which packed tightly around them to form plasterlike molds. Some perished as they attempted to shield their heads with tiles and pillows or to protect their faces from the choking, penetrating gases, dust, and ash. Observers 20 miles away reported that constant shaking was necessary to keep the hot ashes from their clothing.

About 160 acres of Pompeii have been uncovered during the past two centuries in a street-by-street, building-by-building excavation that continues today. Many details of the life and culture of Pompeii in 79 A.D. have thus been deciphered. Among the uncovered objects are: remarkably preserved paintings, statues, metal furniture, gold jewelry, and household utensils. Weathering has subsequently transformed the upper portion of the entombing volcanic debris into a topsoil, and unexcavated parts of Pompeii today are covered with vegetation.

The geological account of the Somma-Vesuvius volcanoes might begin at the time in the past when lava probably flowed out

2-2. Mount Vesuvius is partially encircled by the remains of Monte Somma. Naples is in the foreground. Hot molten material (magma, p. 128) may occur at a depth of about three miles (4.8 km) beneath Vesuvius, although its ultimate source is probably much deeper. This figure was deduced as follows. A thick pile of layered rocks underlies Vesuvius (broad synclinal structure), and a nearly vertical opening (volcanic vent) extends upward through them. This vent connects the magma below with the volcanic cone at the surface. Blocks of these layered rocks are carried upward and occur in lava flows. Blocks estimated to come from a three-mile depth are much altered, indicating proximity to the magma; blocks of overlying rocks are less altered. *(F. M. Bullard, Volcanoes: In History, in Theory, in Eruption, University of Texas Press, Austin, Texas, 1961.)*

(A) A flood has just receded, and a drowned animal's bones lie on the floodplain. The line of trees on the right marks the location of the stream channel.

(B) Sediments from successive floods cover the skeleton, which becomes fossilized. The fossil is subsequently buried far below the surface as the channel shifts laterally back and forth across the floodplain. The sediments become compacted and cemented into sedimentary rocks.

(C) Uplift of the region stops the deposition and initiates erosion. Valleys form and grow deeper.

(D) After deep erosion, a man finds the fossil projecting from a rock layer. *(Courtesy Chicago Natural History Museum. Drawings by John & Conrad Hansin.)*

2-3. A sequence of events precedes the discovery of a fossil on a valley slope.

on the floor of the Bay of Naples and piled up to form a submarine volcano. Accumulation of debris around the volcano vent made the cone wider and higher, and it became an island; continued growth then joined it to the mainland. Somma may have been partly destroyed at least once prior to 79 A.D. when its entire top portion disappeared. The new cone of Vesuvius was then built within the large depression (caldera, Figs. 2–2 and 5–14) that formed during the catastrophe. Since 79 A.D. Vesuvius has had periods of inactivity (up to five centuries) punctuated at times by explosive eruptions and lava flows. During the past three centuries, relatively gentle volcanic activity has been nearly continuous, and more violent episodes have occurred every few decades.

What lessons does Vesuvius have for the geologist? Perhaps most striking is the realization that geologic stories are recorded in rocks, particularly rocks that contain fossils. *Fossils* are the recognizable remains or traces of animals and plants that were preserved in sediments, rocks, and other materials such as ice, tar, and amber before historic times. One has little difficulty transferring our knowledge of the formation of the lifelike Pompeian molds and casts to the study of a fossil-bearing sedimentary rock (Figs. 2–3 to 2–6). A little imagination applied to the Pompeian remains reconstructs a story around the corpse of a magnificently jeweled woman discovered among a heap of skeletons in the gladiators' barracks, although the intriguing details are forever lost. One can likewise reconstruct at least a gen-

2-4. A fossil fish fourteen feet long was discovered in limestone strata in Kansas (Portheus; Late Cretaceous). Another fossil fish (Gillicus), six feet long, occurs within the remains. Portheus swallowed Gillicus head first, then died a short time later, and sank to the bottom. Burial in limy muds followed. Eventually the sediments changed to rock, the area became land, and erosion removed overlying rocks to expose the fossil. *(Collected and prepared by George F. Sternberg; photo by E. C. Almquist. Courtesy Sternberg Memorial Museum, Fort Hayes Kansas State College.)*

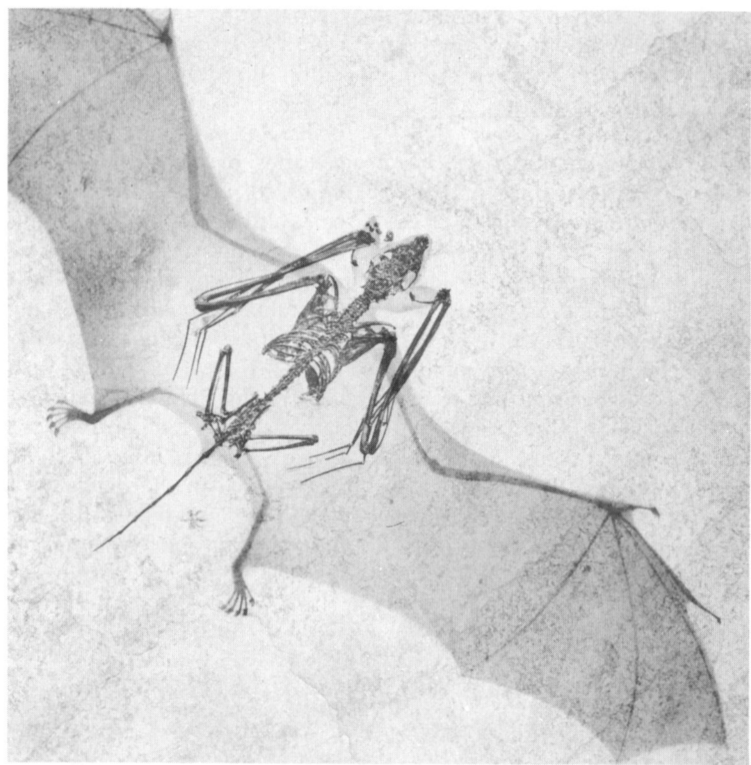

2-5. Skeleton of the oldest known fossil bat. The picture shows shadowy, theoretical restoration of the wings and feet extended as they might have been when the animal was flying in Wyoming some fifty million years ago. This vertebrate fossil from the Eocene Green River Formation is remarkably complete; e.g., small remnants of carbonized wing membranes and fragmentary residues of ingested food have been preserved. *(Courtesy Princeton University Museum of Natural History and Professor Glenn L. Jepsen.)*

eral account of the fossil palm trees and crocodiles discovered in rocks in the Dakotas, of the mastodons excavated from peat bogs in New York, and of the fossilized tropical plants of Antarctica.

Certain aspects of a central geological theme are thus illustrated. Sediments accumulate, and as they pile up under certain conditions, the remains of organisms are buried in them. The remains then become fossilized, and the sediments subsequently are transformed into sedimentary rocks. Later, the fossils may be exposed if uplift occurs and the overlying material is removed by weathering and erosion acting together as a team. *Weathering* is the nearly static part and includes all of the processes causing disintegration and chemical alteration of minerals and rocks at or near the Earth's surface. *Erosion* (Latin, *erodere,* to gnaw away), on the other hand, involves the movement of such weathered debris—i.e., the processes of loosening, removal, and transportation that wear away the Earth's surface.

2-6. Mud cracks and dinosaur footprints in Jurassic rocks in Morocco. (*Photograph, G. Termier.*)

Geology: Scope and Importance

Geology (Greek: "earth science") is concerned with the Earth: its origin, surface features, interior, and physical conditions, with the organisms that once inhabited it, and with the countless changes that have occurred in all of these during the millennia that planet Earth has orbited the sun. Geology attracts and fascinates those who are curious about the globe they live on, who come in contact with the great out-of-doors, and who want to know the what, why, and how of a geyser, an earthquake, a volcano, a waterfall, a landslide, a natural bridge, a submarine canyon, an artesian well, a towering mountain, a glacier, the stalactites in a cavern, a meandering river, and the diverse fossils found in rocks, to name some of the phenomena of geology.

Scenery and landscapes—the commonplace as well as the spectacular—provide a mature lasting satisfaction when observation is coupled with understanding. Mountains, hills, plateaus, and valleys are not permanent features of the Earth's surface but only temporary forms in an ever-changing pattern. Rocks are not inert; they do change, and they contain messages that tell us about a geologic history that stretches far, far backward into time. There really are *sermons in stones* for those who know how to decode the records.

Thomas Henry Huxley underscored strikingly the intellectual importance of geology to man when he said: "To a person uninstructed in natural history his country or seaside stroll is a walk through a gallery filled with wonderful works of art, nine-tenths of which have their faces turned to the wall."

Reading Sermons in Stones

As a model of geologic history, Hotchkiss traced the story of a millpond. The story begins and ends with unusual happenings in the history of a stream. A dam is built across the stream, and a pond forms. During ensuing years, the stream carries sediment which it drops in the quiet waters of the pond. Larger pieces are deposited first near the edge of the pond, and smaller ones are dropped farther out in deeper, quieter water. Water flowing over the dam is thus clear and relatively free of sediment. The story closes with a powerful flood that destroys the dam and pond and carves a 6-foot trench in the accumulated sediments.

If no written records or eyewitness accounts were available, we would have to turn to the sediments for clues concerning this episode in the history of the stream. These sediments consist of layers which differ in thickness, color, and particle size, and we see them edgewise along the sides of the gully. The sediments are evidence that a pond once existed because they occur in thin uniform layers unlike those deposited in the channel of a stream, and remnants of the dam indicate how the pond formed. Abundant leaves in certain layers indicates deposition during the fall, and a count of the total number of leaf-containing layers—each separated from the next by layers deposited at other times of the year—indicates approximately how long ago the dam was built.

Some layers are thicker, coarser, and lighter colored than others. Evidently they formed during spring and summer when water was abundant and the stream carried more and coarser sediment. On the other hand, when ice covered the pond during a winter, little sediment was added to the quiet waters, and the finest particles were then deposited. Previously they had been kept suspended in the more turbulent water of warmer weather, and their colors may be darker because of a higher percentage of organic matter.

If two successive leaf-containing layers are separated by exceptionally thick beds that contain sand and small pebbles, an exceptionally wet year seems indicated. On the other hand, a dry year is indicated when two such layers are separated by relatively thin, fine-grained beds. In this way, we may be able to work out the succession of wet, normal, and dry years since the dam was built. A check on our findings can be obtained by extracting a core from a nearby tree and studying its rings: wide rings tend to indicate rapid growth during favorable wet seasons, whereas narrow rings show less growth during dry seasons, although other factors also affect the growth of tree rings.

Although annual leaf-bearing layers like those described in the millpond story are quite unusual in the geologic record, somewhat similar types of annual deposits are known (Fig. 2–7). A *varve* is defined as a sedimentary deposit that has accumulated in one year. Varves deposited in lakes near glaciers tend to show alternations of thicker, lighter-colored, coarser-grained laminae (summer deposition) and thinner, darker-colored, finer-grained laminae (winter deposition). Each light-and-dark couplet thus represents the deposition of one year, and varves can sometimes be matched (correlated) from one area to another. Varves may vary in thickness from a fraction of an inch to more than a foot, but thicknesses ranging from 1 to 2 inches or less are

2-7. Varves in Pleistocene silt and clay, near the mouth of Sherman Creek, Ferry County, Washington. *(Courtesy F. O. Jones, U.S. Geological Survey.)*

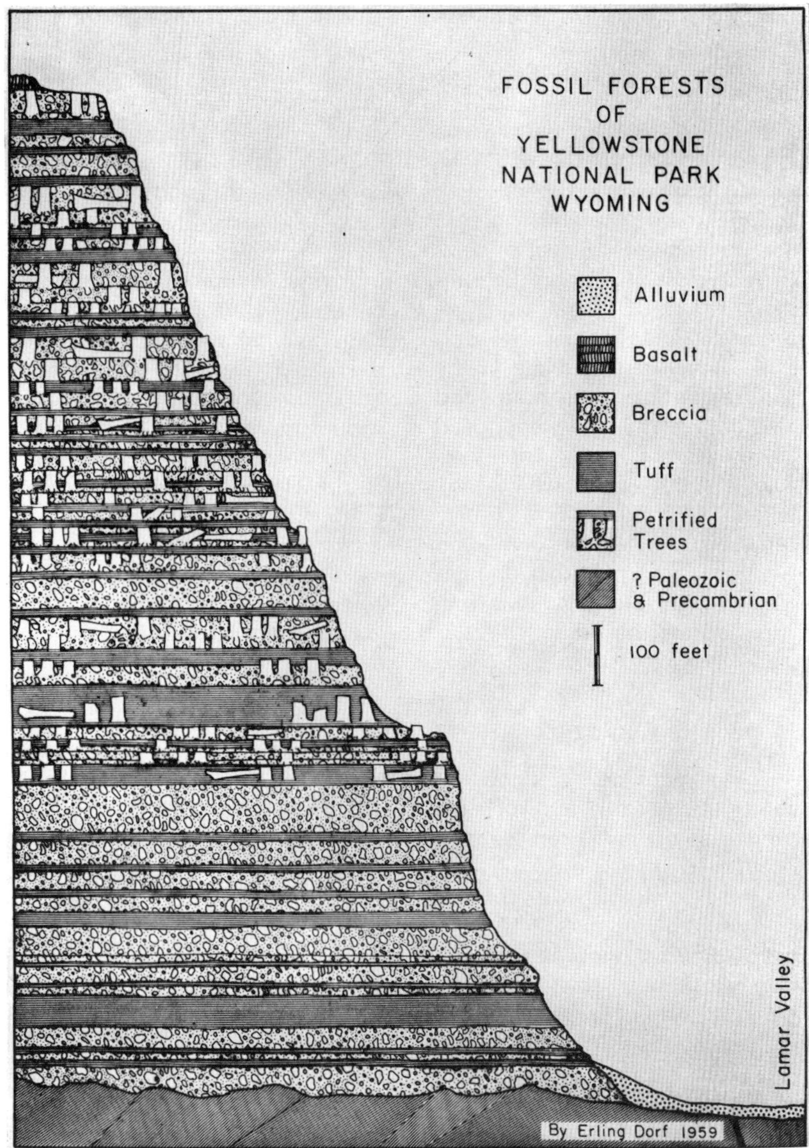

2-8. The fossil forests of Yellowstone National Park, Wyoming. The petrified trees in the twenty-seven buried forests resemble living redwoods. Tree rings show that one of the petrified trees probably lived for about 1000 years, and many trees show up to 500 annual growth rings. Some of the organic matter once present in the trees has been removed by solution and replaced by the mineral quartz, but in most cases the quartz fills cellular cavities in the wood. Each of the forests was buried by a single pyroclastic deposit (i.e., fragmental debris erupted during a volcanic explosion). Studies of volcanic areas in Mexico show that new trees can begin to grow in an area about 200 years after pyroclastic material has fallen upon its surface. However, the fossilized trees in any one of the forests were not necessarily the first generation of trees to grow after an eruption. The following sequence was repeated twenty-seven times: soil formed on the surface of the volcanic debris following an eruption; eventually a forest developed; an eruption occurred and buried part of the forest. (*Courtesy Erling Dorf.*)

most common. Pollen grains (p. 241) have been found in some varves and show that each couplet was deposited during a single year.

We now draw an analogy between the pond and a geologic epoch. The damming of the stream might correspond to the slow sinking of a large area across which the sea gradually spreads. Hundreds of feet of sediment may then be deposited upon this subsiding sea floor—mud, silt, sand, and calcareous debris which later will become sedimentary rocks. Animals live and die in the sea and on the sea bottom, their bones and shells are buried by the sediments, and some are preserved as fossils. Therefore, rocks and fossils record events which took place where and when they formed (Fig. 2–8). Next, this geologic epoch might end with a slow upward movement that changes the former sea bottom into land, following which its surface would be eroded by running water and other agents.

Utilitarian Aspects of Geology

Today's civilization relies heavily upon the natural resources of the Earth, and for such essential materials as uranium, iron, copper, lead, zinc, tungsten, manganese, oil, and coal, geology must furnish answers for important questions. Where can suitable deposits be found? How can these best be exploited? What are the reserves of these "one-crop" nonrenewable materials which can be taken out of the Earth only once? Their accumulation is too slow for new deposits to form in the centuries which lie immediately ahead.

World consumption of many mineral resources during the past two or three decades has exceeded that of all the preceding years in the history of man, and such consumption makes deposits that once seemed inexhaustible now seem quite inadequate (p. 20). Furthermore, the necessary mineral deposits are concentrated in relatively few places. No nation is self-sufficient in all the minerals that it requires, and yet industrial power is based upon ample supplies of a very great variety of mineral resources. Thus problems arise that involve international trade, politics, transportation, exploration, development, exploitation, have and have-not nations, and war or peace.

The engineering applications of geology are many and essential and seem destined for a very large and rapid increase (p. 6). Some examples include the location of satisfactory sites for dams and buildings, flood control and water supply, tunnel construction, determination of the best locations for roads, pipelines, and undersea cables, protection of coastal areas from erosion, and various military applications.

Probably more than half of all geologists in the United States are employed in the petroleum industry, but a number aid in the search for deposits of more prosaic materials: clay or shale for bricks, limestone for cement or building construction, and sand, gravel, and traprock for road-building materials. The availability of almost limitless supplies of good water is assumed as a kind of natural heritage by many—about in the same category as the air we breathe. Yet the task of finding supplies and of controlling pollution during the next few decades may involve a greater public investment than any other field of natural-resource development or conservation. Adequate knowledge of the sediments and rocks at and below the surface is an important factor here, and thus geology is directly involved.

The Earth and Its Major Subdivisions

The Earth (Fig. 2–9) is a nearly spherical planet, slightly flattened at the poles and bulging at the equator. It rotates through 360 degrees once every 23 hours and 56 minutes and revolves once about the sun in 365¼ days. The Earth has a circumference of approximately 25,000 miles (40,000 km), a polar diameter of about 7900 miles (12,714 km), and an equatorial diameter of 7927 miles (12,756 km). On a similarly shaped spheroid 100 feet across, the nearly

27-mile (43 km) difference in diameters would correspond to 4 inches, too small a difference to be noticeable to the eye.

The specific gravity of the Earth as a whole is about 5.5; i.e., an average sample contains 5½ times as much matter as does an equal volume of water. Since this is about double the specific gravity of rocks at the Earth's surface, materials in its interior are quite dense. The total mass of the Earth is estimated to be 6×10^{21} tons (6 followed by 21 zeros).

The general setting of the Earth in the universe—its relationship to other members of the solar system and to stars, nebulae, and galaxies—is discussed in Chapter 16 and in succeeding chapters dealing with astronomy.

The *atmosphere* (see Chapters 26 through 29) is important geologically, and many of the striking differences between the surface features of the Earth and moon depend upon the presence or absence of an atmosphere. Wind causes erosion directly by blowing sand grains and smaller particles against rock surfaces, and moving masses of air can transport rock fragments. As winds produce waves and currents at sea, they are the indirect cause of much erosion along its shores. Moreover, water is evaporated

2-9. This high-altitude December 1968 Apollo 8 photograph shows clouds outlining a counterclockwise circulation around a low-pressure center and a frontal system. Florida and Cuba are located within the field of view. *(NASA: AS8-16-2581.)*

from the oceans by solar energy, and some of this water vapor is transported inland, where condensation and precipitation occur. Subsequently, water returns to the oceans to complete the water cycle. The atmosphere is thus indirectly responsible for all of the eroding and transporting activities of running water. Furthermore, atmospheric gases may function directly by reacting chemically with rocks during weathering.

The *hydrosphere* includes the oceans (which submerge about 71% of the Earth's surface), lakes, rivers, water in subsurface cracks, and the water molecules that now occur in glaciers. Moreover, this water is so abundant that it would completely envelop an Earth-size sphere to a depth of more than 1½ miles.

The three main units of the Earth's interior are core, mantle, and crust (Fig. 2–10 and Chapter 6). The diameter of the core is about 4300 miles (6900 km), and iron is probably its chief ingredient. The core consists of an inner part that seems solid and an outer part that appears fluid. The *mantle* is nearly 1800 miles (2900 km) thick and makes up about 84% of the volume of the Earth. Since the volume of the core is about 16%, the crust actually makes up a very small part of the Earth as a whole. The mantle and crust are solid except for relatively small masses of magma that develop occasionally within a few tens of miles of the surface.

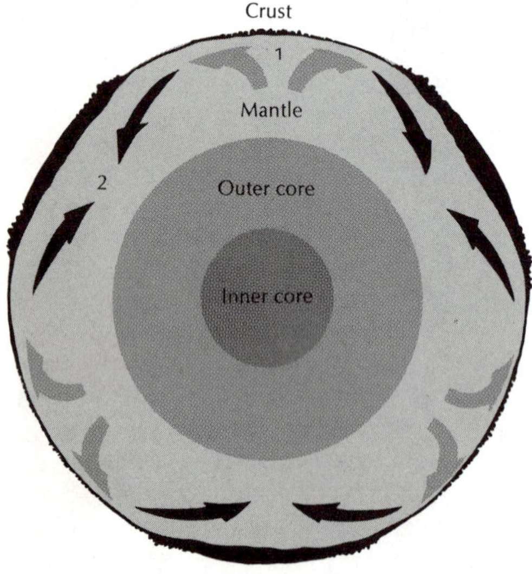

2-10. The three main units of the Earth's interior: core, mantle, and crust. Apparently the inner core is solid and has a radius of about 800 miles (1300 km) whereas the outer core is fluid and about 1350 miles (2170 km) thick. Convection currents may occur within the mantle, but this is controversial. In one view, currents rise *(1)* beneath a mid-ocean ridge and cause the ocean floor to spread apart and continents to drift. Where currents descend *(2)*, they may push against continental masses to form coastal ranges, deep-sea trenches, and island arcs. *(After ESSA.)*

On the average, the *crust* is about six times as thick beneath the continents as beneath the ocean floors—approximately 20 to 25 miles (32 to 40 km) vs. 3 to 4 miles (5 to 6.5 km). However, the continental crust is much thicker than average beneath the great mountain belts. As the term has been used recently, the *lithosphere* is the solid outer shell of the Earth, about 35 to 60 miles (50 to 100 km) or more in thickness, that occurs between the low-velocity zone and the surface. Thus it includes the crust and uppermost mantle. A major discontinuity separates crust from mantle (p. 163).

The rock most abundant in the upper part of the crust beneath the continents appears to be similar to granite in chemical composition and specific gravity (2.7; see p. 164). A heavier rock, probably similar to basalt in chemical composition and specific gravity (3.0; see p. 164), is thought to underlie the floors of the oceans and also the granitic rocks of the continents (Fig. 6–31). Thus the two-part continental crust differs fundamentally from the one-part oceanic crust; in fact, the difference accounts for the existence of continents and ocean basins.

The term crust was appropriate when it was first introduced because the whole interior of the Earth was then believed to be molten, except for a thin outer rind. Since such conditions do not occur, the term is now somewhat misleading, although it is still useful. The upper surface of the crust may be covered by water, by unconsolidated sediments, by soil and vegetation, or it may be exposed at the surface.

The Earth's surface has two fairly well defined levels: the upper level is associated with the continents which have an average altitude of about ½ mile (0.8 km) above sea level, whereas the lower level coincides with the average depth of the ocean basins, 2½ miles (4 km) below sea level. Presumably these levels occur because the thinner, denser oceanic crust is in isostatic balance with the thicker, less dense continental crust (p. 165).

The Earth's surface has three main units: *continents, ocean basins,* and the *system of mid-ocean ridges.* In addition, the continental shelves and continental slopes together form an extensive marginal zone between the continents and ocean basins (Fig. 11–3). The newly discovered system of mid-ocean ridges extends for about 40,000 miles (64,000 km) and covers nearly as much area as the continents. The Mid-Atlantic Ridge is its best-known portion.

The deepest known part of the ocean, off the Philippine coast, is about 7 miles (11 km) deep, whereas the highest mountain, Mount Everest in the Himalayas, is approximately 5½ miles in altitude. Thus the Earth's greatest relief (vertical distance between the highest and lowest points) is about 12½ miles (20 km). On a sphere with a diameter of 100 feet, this would be represented by 2 inches, and from a short distance away, the sphere would seem quite smooth.

The deepest parts of the oceans (Pacific and Indian particularly) form long, relatively narrow troughs bordering the continents. The largest of these trenches exceed 1000 miles in length and are more than 100 miles in width. As these deeps are not filled by sediments washed outward from the lands, they probably are rather youthful features, and earthquakes occurring in these areas indicate that crustal movements are still taking place. Part of the sea floor apparently is subsiding at the same time that nearby coastal regions are being uplifted.

The continental shelves are the submerged edges of the continents and comprise approximately 6% of the Earth's surface. They form the relatively smooth-floored, shallow-water zone that slopes very gradually seaward to a termination at the top of the continental slope. The average slope of the continental shelf is about 10 ft/mi (2 m/km); i.e., across a 40-mile width, the depth would change 400 feet (122 m)—a nearly flat surface to the eye. In contrast, the continental slope is much steeper. The average width of the shelf exceeds 40 miles (64 km), but it is missing in places and extends for several hundred miles in others. The average depth at its termination is about 400 to 450 feet (130 m), but it may be more or less than this.

The *continental slopes* commonly extend downward to depths exceeding 2 miles, where they merge with the floors of the ocean basins (the lower portion is now called the continental rise). The continental slopes tend to be steepest in their upper portions, but even these slopes—200 to 300 ft/mi (50 m/km)—are gentle, not nearly so steep as indicated by models or diagrams with great vertical exaggeration. Irregularities occur on the shelf and slope (see particularly submarine canyons, p. 429), and the floors of the oceans are not flat and smooth as was once thought.

The origins of the continents, ocean basins, and system of mid-ocean ridges are fundamental questions, for which revolutionary new answers seem to have been supplied during the past decade. Oceanic crust may be forming today at a mid-ocean ridge because material from the underlying mantle moves upward into its crest, along which a zone of tension and rift valleys is located. This newly formed crust then spreads laterally away from the ridge crest, which makes room for additional increments from the mantle in a process that may have been continuous for many millions of years. Eventually, this laterally spreading, relatively young ocean floor arrives at a deep-sea trench and descends to be assimilated by the upper mantle—recycling on a grand scale. In this view, the continents are drifting apart (Fig. 2–11), and the ocean basins are relatively youthful features—a remarkable change from the prevailing view of a decade or so ago that continents and ocean basins were more or less permanent features of the Earth.

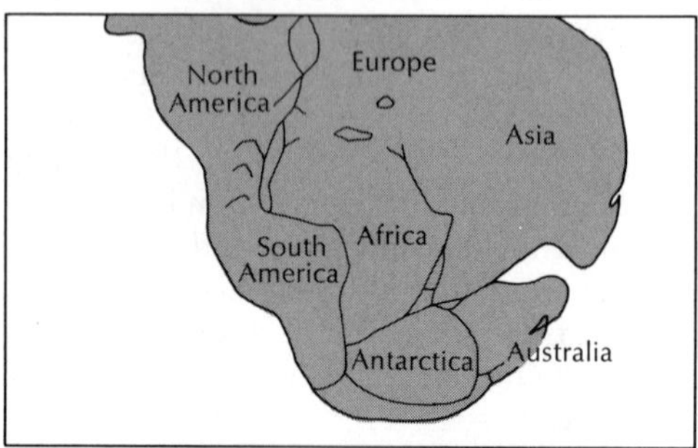

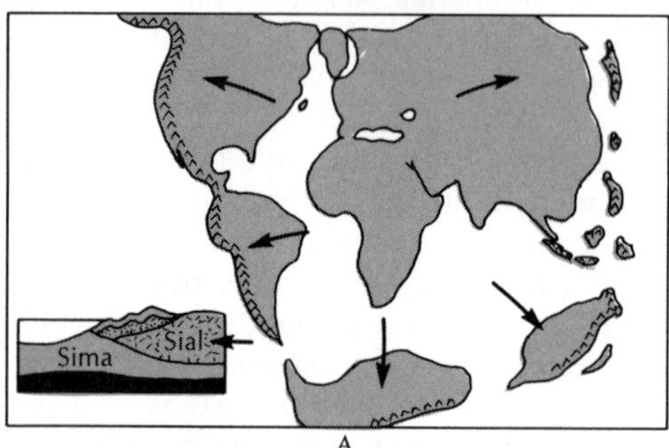

A

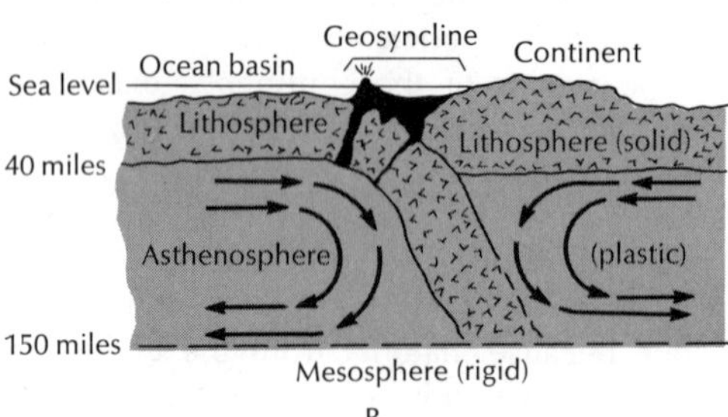

B

2-11. Part A illustrates the concept of drifting continents as proposed by Wegner in 1912 (see Chapter 12). However, Wegner's idea of the processes involved is no longer tenable (insert in lower left)—that continents (sial) move through the ocean floor (sima) like icebergs through a sea. Part B shows a tentative mechanism that is favored today. A wide section of the lithosphere—crust and upper mantle together—slide as a unit above a weaker plastic zone (asthenosphere), and convection currents may cause the movement. *(B. Mears, Jr., The Changing Earth, Van Nostrand Reinhold, 1970.)*

Important discoveries concerning the Earth's floor beneath the oceans have been made during the last two decades, and many of these fit nicely into the picture sketched briefly above—e.g., heat flow data, magnetic anomalies and reversals, sediment thicknesses, age determinations of sea floor materials, and locations of earthquakes (see Chapters 11 and 12). Therefore, although continents and ocean basins have apparently been in existence since early in the Earth's history, their numbers, shapes, and locations have probably been quite different at different times.

Preceding pages suggest the many common facets and overlapping aspects of geology, astronomy, meteorology, and oceanography—the four main fields of earth science. Moreover, the interrelationships extend far beyond these four; all of the physical and biological sciences and mathematics make contributions to earth science, and earth science in its turn to them. Discoveries in one field bring advances and changes in many of the others.

The Earth's Crust is Made of Minerals and Rocks

Bedrock is the solid rock that is exposed at the surface or immediately underlies soil and loose surface debris (Fig. 2–12). *Regolith* (Greek: "blanket of stone") is the relatively thin covering of soil and unconsolidated rock waste that hides the bedrock in most areas. Bedrock is continuous and may consist of any kind of rock, whereas regolith is discontinuous, although generally present. Regolith tends to be a few tens of feet thick or less, but may be much thicker. It may develop in place by the decay and disintegration of bedrock or consist of transported materials. *Soil* refers to the upper portion of the regolith which has been so altered by physical, chemical, and biological processes that it can support rooted vegetation. However, soil may be absent from an area.

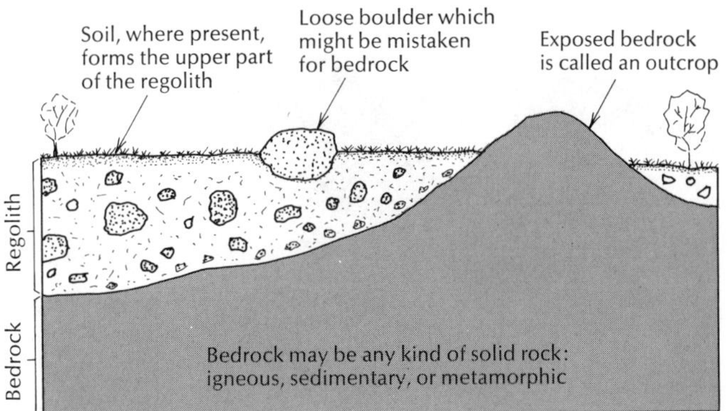

2-12. Bedrock and regolith *(transported and non-sorted in this illustration).*

By direct observation, man has access to only a very thin outer part of the crust: the deepest mines penetrate less than 2 miles (3 km) beneath the surface, and the deepest wells about 5 miles (8 km). However, certain rocks now exposed at the surface may once have been buried several miles below it.

Chemical analyses have been made for rocks of various types, and the proportions of the elements in the outer 10-mile (16 km) zone of the lithosphere have been estimated. Eight elements apparently constitute more than 98% by weight of this zone: oxygen (most abundant), silicon, aluminum, iron, calcium, sodium, potassium, and magnesium (least abundant). The following mnemonic expression arranges these eight elements in the order of their relative abundances: "Only Silly Artists In College Study Past Midnight." If the materials in the atmosphere and hydrosphere are added to those of the 10-mile zone, percentages are changed only slightly.

Of these eight elements oxygen and silicon combined as silica (SiO_2) make up about three-fourths of the total. Thus the silicate minerals are the most abundant in the crust, especially the feldspars, pyroxenes, amphiboles, micas, and quartz.

Although oxygen probably makes up about 47% of the 10-mile zone by weight, it comprises about 93% by volume, and the corresponding estimates for silicon are 28% and 1%. These differences occur because the volume of an oxygen ion is about 60 times larger than that of a silicon ion, whereas the mass of an oxygen ion is less (about half).

Atoms, atomic structure, ions, radioactivity, nuclear energy, and similar topics from the fields of chemistry and physics are discussed briefly in Appendix I. Since a firm grasp of this material is essential for an understanding of minerals, rocks, and a variety of phenomena and processes in the earth sciences, it should be studied with a degree of thoroughness appropriate to the background of the reader

Minerals

Minerals are generally solid, naturally occurring substances composed of atoms and ions that are arranged in regular, three-dimensional patterns and give each mineral characteristic physical properties and chemical compositions. Minerals are basic homogeneous units which have been combined in various ways and under different conditions to form rocks. In combination, minerals and rocks make up the lithosphere.

Most minerals consist of elements combined as chemical compounds, although a few may occur as native elements—e.g., gold, silver, and copper. One mineral may be different from another because it breaks in a special manner, develops with a distinctive shape, or has some unique property such as a peculiar taste, feel, or magnetism (Figs. 2–13 and 2–14).

2-13. Minerals may grow with distinctive shapes. *Top,* the crystals of several minerals occur in a group (about $2\frac{1}{2}$ in. across): garnet in center, black tourmaline on right, smoky quartz and rosettes of cleavelandite, a variety of feldspar. *Bottom,* a group of quartz crystals from Madagascar (the largest is $5\frac{1}{2}$ in. in length) *(Courtesy Smithsonian Institution.)*

2-14. *Top,* the large uncut specimen of rose quartz breaks along smooth curved surfaces; it exhibits one type of fracture (conchoidal). The faceted gem is 1 in. in length. *Bottom,* a polished slab of azurite and malachite shows the circular bandings that are characteristic of these brilliantly colored copper-containing minerals. Deep-blue azurite forms the darker rings, and green malachite forms the lighter areas. *(Courtesy Smithsonian Institution.)*

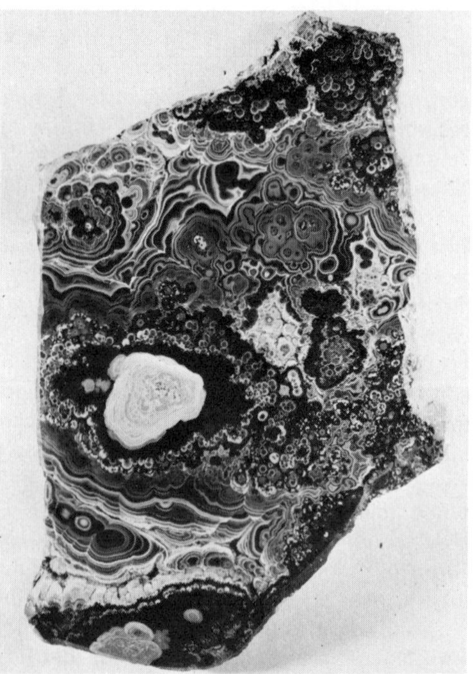

Rocks

All-inclusive, yet brief and simple, definitions are difficult to formulate, and that of a rock is no exception. *Rocks* (see Chapter 3) are commonly defined as aggregates of one or more minerals which make up essential parts of the Earth's crust. Exceptions are coal and the natural glasses, which are not made of minerals. Rocks are more variable than minerals in their physical and chemical properties, but any one type of rock, such as a granite or a sandstone, does have certain mineral associations, textures, and other properties which are characteristic of it.

An analogy of Trefethen's may clarify the distinction between a rock and a mineral. A building is constructed of wood, bricks, and glass just as a rock is composed of different minerals. Different buildings (rocks) result from varying proportions of these building materials (minerals) and from different architectural styles. Analogously, the same types of minerals (e.g., feldspar, quartz, and mica) are common in many different kinds of rocks but they occur in different proportions and arrangements. Just as a small miscellaneous collection of minerals does not constitute a rock, neither does a random collection of unused building materials make a house.

On the basis of their manner of origin, rocks are commonly subdivided into three groups: igneous, sedimentary, and metamorphic.

Igneous Rocks

Igneous rocks (Latin: "fiery") are produced by cooling and crystallization of molten rock-making material called *magma* or *lava*. Granite, basalt, and obsidian are familiar examples of this group. Dissolved gases are important constituents of magma and lava but tend to be excluded from the rock-making minerals when the hardening process occurs. Hardening may take place within or upon the Earth's crust. Thus the igneous rocks may be subdivided into the *intrusive* (from magma) and *extrusive* (from lava) groups. Intrusive rock masses are younger than the rocks they intrude, and they are exposed in places at the surface today because erosion has removed the older rocks which once were around and on top of them.

Sedimentary Rocks

The sediments in a *sedimentary* (Latin: "settling") *rock* have been produced by the weathering and erosion of igneous, sedimentary, and metamorphic rocks and of smaller, miscellaneous deposits of minerals. Sizes of the mineral and rock fragments range from huge boulders to materials carried in solution. The sediments have been transported and dropped or precipitated by such geologic agents as running water, ocean waves and currents, wind, and ice. Some of the fragments may be rounded; others are angular. Such features depend upon the distance of transportation and other factors.

Loose sediments may be changed into solid sedimentary rocks by pressure from overlying strata which are deposited later and by the precipitation of cementing material as a binder around individual grains (Fig. 2-15).

The presence of different layers, beds, or strata constitutes an outstanding feature of most sedimentary rocks and serves to distinguish them from many igneous and metamorphic rocks. Layers result from changes in conditions as deposition takes place; e.g., velocities of the transporting medium may increase or decrease so that larger or smaller particles are dropped at a given spot. Fossils are abundant in some sedimentary rocks, readily distinguishing them from most igneous and metamorphic specimens; but not all sedimentary rocks contain fossils. Fossils "label" the rocks in which they occur and yield much information about the past.

Conglomerate, sandstone, and shale—formerly gravel, sand, and mud or clay, respectively—are common types of sedimentary rocks, and the size of the average particle in the rock is the only criterion used in classification here. On the other hand, limestones are also common and are classified on their mineral content, which is chiefly calcium carbonate in the form of calcite. The calcite may have formed by precipitation from solution or by the accumulation of shell fragments (produced by or-

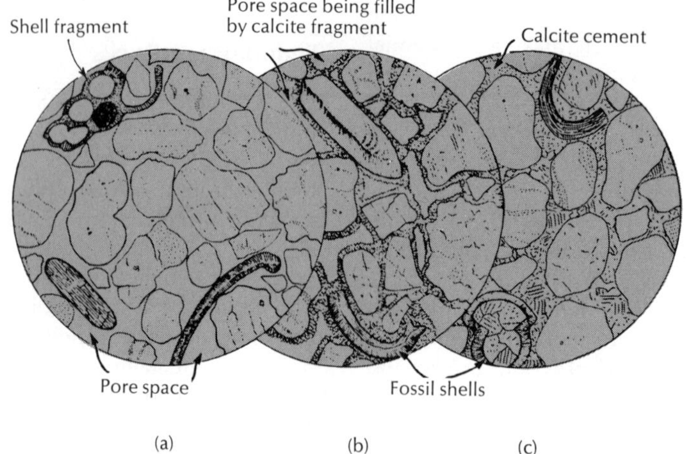

2-15. Cementation of sand as seen under the microscope: a, loose sand from an Oregon beach; b, partially cemented sandstone from a Brazilian coral reef; and c, completely cemented sandstone from Ohio. *(Gilluly, Waters, and Woodford,* Principles of Geology, *3rd ed., W. H. Freeman and Company, 1968.)*

ganisms from calcium carbonate dissolved in water).

By observing closely the manner in which sediments are produced, transported, and deposited today, geologists have gained information useful in interpreting the geologic history of ancient sedimentary rocks. The sediments come from a source area and are carried to a basin of deposition, but this may be only a temporary resting place. The sediments may again be eroded and transported to yet another basin of deposition, and the process may be repeated many times. The sea floor is the ultimate destination of most sediments, but even here the cycle does not stop because the sea floor may subsequently be uplifted to become land, and the materials again eroded, transported, and deposited.

Although the factors influencing the formation of sedimentary rocks are many and diverse, a few may be mentioned here. Perhaps the most obvious is the type of rock or rocks eroded in the source area (its lithology). Another factor is the environment of the source area and its crustal stability or instability. Different processes of weathering and erosion in different climates result in different types of sediment; e.g., warm humid climates produce decayed, chemically altered rock debris which differs greatly from that produced by the freeze-thaw effect in colder regions. The size and nature of the sediments will also be affected by crustal movements in the source area: up or down, slow or relatively rapid, large or small. The distance and type of transportation are other obvious factors; e.g., glacially eroded sediments differ from those deposited by the wind.

Finally, the nature and crustal stability or instability of the basin of deposition have important effects; e.g., deposition on a slowly subsiding, shallow sea floor will permit waves to break up the less resistant fragments into clay-size particles and to sort, round, and smooth the larger pieces. Moreover, various chemical and physical changes may occur in the sediments after they have been deposited.

Thus the geologist attempts to interpret as much information as possible concerning the nature and environment of the source area, the deposition area, and the transporting agents, i.e., a Sherlock Holmes, detective-searching-for-clues approach. It follows that a successful geologist must be a keen, imaginative observer with the ability and desire to deduce and interpret; moreover, he must be thoroughly trained in biology, chemistry, physics, and mathematics. However, all of this interpretation must be based upon evidence and subject to the strict, nonbiased limitations imposed by the methods of science.

Sedimentary rocks constitute the most readable portions of the Earth's diary that has been left for geologists in the rocks, even though some of the language has yet to be made intelligible, and many pages and even whole groups of pages have become badly tattered or are entirely missing.

Metamorphic Rocks

The *metamorphic* (Greek: "changed form") *rocks* came into being through the transformation of sedimentary, igneous, or older metamorphic rocks into new and notably different types. New kinds of minerals, or mineral particles with different shapes or orientations, can be produced in the crust by heat, pressure, and the chemical action of solutions. Some metamorphic rocks (gneiss) are characterized by the segregation of minerals of different kinds into bands, and others (schist and slate) by a tendency to break along closely spaced, nearly parallel surfaces into flat slabs. Marble is a recrystallized limestone which lacks foliation, unless impurities are present.

The study of metamorphism has been aided greatly by observations in the field (out-of-doors in contrast to the laboratory) of rocks at stages that represent transitions between nonmetamorphosed and completely metamorphosed rocks. Recognizable features in the original rocks become fainter and new minerals or structures become better developed as the metamorphic rocks are approached.

Some geological processes thus make rocks, whereas other processes break apart and destroy them. Therefore, products formed by the destruction of one kind of rock may later be combined into a new rock type, and different types of cycles may occur (Fig. 2–16).

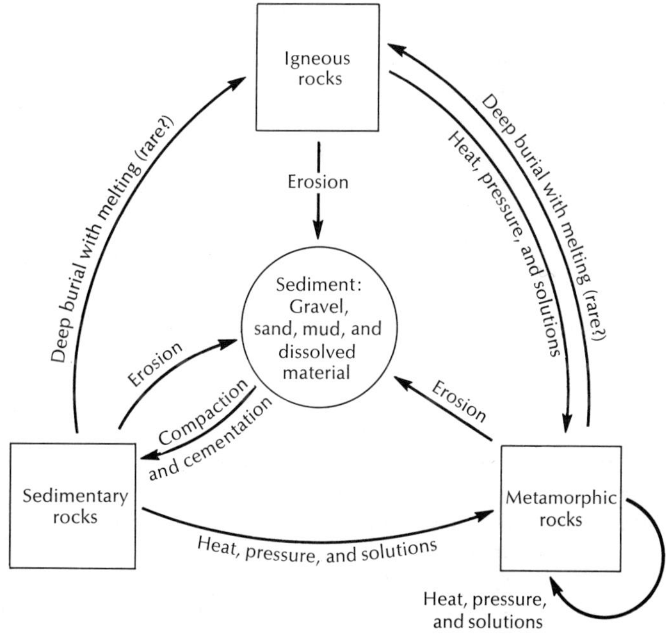

2-16. Rock cycles of different kinds may be traced by following the arrows (processes) that lead from one kind of rock to another (some become sediments along the way). If sedimentary rocks are buried so deeply that they melt, they may pass through a metamorphic stage before the melting takes place.

The Tactics and Strategy of Geology

Geology probably has more to offer the nonscientist than almost any other science, because the background necessary to appreciate and understand many of its fundamental aspects is relatively quite limited and rather easily attained. Geology, of course, is complex in many of its aspects and is rapidly becoming more so. The modern trend in geology moves strongly toward laboratory studies and experimentation, the use of complex instruments, the application of mathematics, statistics, and computers, and the quantitative studies of many geologic processes. In fact, geology has been defined as the application of other sciences—physics, chemistry, and biology—to the study of the Earth and is therefore fully as complex as the sciences applied in it. However, a number of its key concepts can be readily and rewardingly grasped. Such concepts can be active, illuminating, lifetime companions of anyone who has been even briefly exposed to geology, if he has received a proper introduction and observes the Earth around and under him.

We shall now meet some of these basic, wide-reaching concepts and conclusions—the tactics and strategy of geology. However, the list that follows is not all-inclusive, the items are not all of equal importance, and perhaps a few belong in a different grouping. Some items are discussed now, whereas a few are introduced and then discussed later. Here we collect and emphasize them.

• **1. The present is the key to the past (uniformitarianism).**

The time and manner of the Earth's origin pose questions upon which men of all centuries have expended much in the way of speculation and ingenuity. Estimates of the age of the Earth have ranged from a minimum of about 6000 years to eternity. Unfortunately, the 6000-year figure gained authority in the seventeenth century as the actual age of the Earth, and for 150 years or so, attempts were made to explain all of the Earth's features in terms of a total history of 6000 years or less. The past had to be credited with a succession of great catastrophes which had no counterparts in the present. Streams flowed in valleys formed by the sudden opening of great fissures in the Earth's crust; mountains were formed by quick cataclysmic upheavals; great floods covered large portions of the Earth, leaving behind huge boulders which the raging waters had picked up and carried.

Near the end of the eighteenth century, James Hutton, a Scottish farmer-physician, introduced a point of view which radically changed all geologic thinking. Hutton argued that the study of processes now operating on the Earth would yield information for unraveling the mysteries of the Earth's past. According to this principle, agents now at work shaping the Earth's surface have been functioning throughout the Earth's history; i.e., the fundamental laws of physics, chemistry, and biology apply now as they did in the past.

Today gravitational attraction operates, light travels at 186,000 mi/sec, two atoms of hydrogen combine with one atom of oxygen to make a water molecule, winds blow, rains fall, water flows downhill, shells are buried, rocks decay, and snow changes to ice. Of course, some of these processes have not always functioned at the same rates as they do at present; e.g., volcanic activity was probably more violent and widespread in some periods of the past than it is today, but in other periods it was less so. Yet the study of today's volcanoes is our guide in interpreting ancient volcanic activity. Given time enough, the familiar processes of the present can probably account for most of the Earth's features.

However, conditions at present may be quite different from the "normal" ones of the past. Today lands are extensive and high, and climates are extremely different at the equator from those at the poles. Milder, more equable climates and less land with fewer mountains apparently prevailed during much of the past. Moreover, continents have apparently drifted slowly from one place to another, and ocean basins may have formed and disappeared.

Hutton's *inductive* reasoning was a far cry from the preceding years of speculation. He reasoned from the particular to the general; i.e., he used the facts gained from field observations as a basis for working out general principles that then could be applied widely to new situations. Further observation was needed to check the validity of each principle. Perhaps Hutton's greatest contribution to geology was this emphasis upon painstaking field observation as opposed to the armchair deduction practiced by many of his contemporaries.

Deduction involves reasoning from the general to the particular. Some principle or generalization is assumed to be valid, and the specific consequences that should follow are predicted. Conclusions reached in this manner involved reasoning alone, without verification by observation in the field. Unfortunately, a number of basic generalizations of Hutton's predecessors were erroneous.

Concerning the Earth itself, Hutton stated that he could "find no vestige of a beginning . . . no prospect of an end." Hutton's motto—the present is the key to the past—was later championed by the great British geologist Charles Lyell and remains fundamental in geology. It is now known as the principle or assumption of *uniformitarianism,* and it is a specimen of what may be called key concepts. These are principles that guide research and give a structure to the separate facts of science.

- **2. Rocks have formed in different ways, at different times, and in different places. Fossils are the remains or traces of organisms that were preserved in sediments and other materials prior to historic times.**

After a brief introduction to the meaning of igneous, sedimentary, and metamorphic rocks and to the rock cycle, one is probably quite astonished to learn that the true nature of rocks has been generally understood only for the past two centuries or so, although many unsolved problems still exist. It seems evident that not all rocks could have formed simultaneously early in the Earth's history and have subsequently remained unchanged. Yet that all did so form was the generally accepted explanation until the latter part of the eighteenth century. Now that it has been established, today's interpretation of rocks and fossils seems quite obvious. It illustrates a helpful aspect of some of geology's key ideas—upon meeting them, we feel fully acquainted.

A milestone in the development of geology was the recognition of the true nature and significance of fossils, particularly of marine fossils exposed on the upper slopes of present-day mountains or in areas far from the sea. Much imagination is needed to reconstruct these areas as former sea floors alive with organisms of many kinds. Yet Leonardo da Vinci (1452–1519) stated the modern view; floods carry mud; the mud is deposited over animals that live in the sea near its shores; the animals die; the mud within and around their shells turns to stone; the sea withdraws; and finally, erosion exposes the fossils, which in many places show the original shells encased in stone.

- **3. There are sermons in stones.**

The history of the Earth, and of the organisms that have lived upon it, is preserved in rocks. The history is incomplete, it is imperfectly known and preserved, but it is there for the interpreting. The more ingenuity men can apply in finding and understanding the clues and data recorded in the rocks, the more completely we will know this history.

- **4. Superposition occurs as sediments are deposited.**

In many places on the Earth today sediments are being deposited layer upon layer, and presumably sediments accumulated in a similar manner in the past. Thus, the strata in a pile of sedimentary rocks are arranged in chronological order; the oldest layer is at

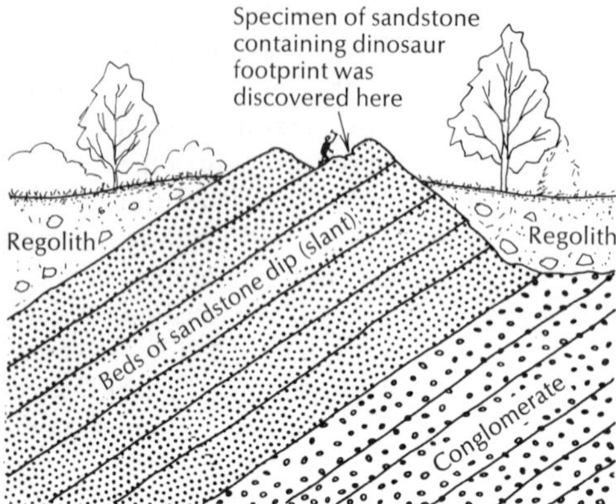

2-17. Interpretation of a sequence of geologic events. (1) Layers of gravel and then sand were transported to this area and deposited. Presumably the layers were nearly horizontal. (2) On one occasion during the deposition of the sand, a dinosaur walked across the surface. Enough moisture was present in the sand for footprints to form. The surface subsequently was consolidated (baked in sunshine), and the footprint was not destroyed when the next layer of sand was deposited. This occurred during the Mesozoic Era when dinosaurs existed. (3) The layers of sand and gravel were compacted and cemented into sandstone and conglomerate. (4) The rock layers were tilted by a crustal movement. (5) Erosion removed overlying layers and eventually exposed the dinosaur footprint. (6) Loose debris (regolith) was deposited on top of the bedrock in places.

the bottom, and the youngest layer is at the top. This principle—so evident to us now—has been generally recognized and accepted for only the past two hundred years or so. The Grand Canyon region (Fig. 4–1) furnishes a most spectacular illustration of the meaning of superposition. Exceptionally, crustal movements have overturned layered rocks and the oldest stratum is no longer on the bottom. However, intense rock deformation tends to be present and warns the geologist to expect the unusual.

• 5. Strata have a nearly horizontal orientation when they are deposited.

At present, sediments are deposited in layers which tend to parallel the Earth's surface at the places of deposition. Thus they tend to be nearly horizontal, and presumably sediments were deposited in a similar manner in the past. However, many ancient sedimentary rocks are found in a crumpled or tilted condition today (Fig. 2–17). According to the principle of *original horizontality,* these ancient strata were approximately horizontal when they formed, but subsequent crustal movements have raised, lowered, twisted, crumpled, broken, or folded them. Acceptance of this principle was postponed for some time by the sheer immensity of the crustal forces needed to produce the deformation.

• 6. Some strata were once continuous layers.

Sedimentary layers forming today commonly do not end abruptly as, for example, do the beds exposed on opposite sides of the Grand Canyon (Fig. 4–2). Instead they tend to thin or pinch out gradually. If strata are found to terminate abruptly today, we assume that erosion or offsetting by faulting has caused this truncation (Fig. 2–18). In the Grand Canyon, the series of sedimentary beds which are now exposed along the north and south rims must formerly have extended unbroken from one side to the other. Erosion has produced a great gorge where part of a plateau once existed.

• 7. Fossil succession results from superposition and biological evolution.

Fossils can be found in sedimentary beds that have been piled one on top of the other. According to the principle of superposition, these fossils represent organisms that lived at different times during the past. The oldest organisms must have been in existence when the bottom layer was deposited, but the youngest organisms could not have come into being until much later, when the top layer was formed. When fossils are collected from such rocks and arranged chronologically from oldest to youngest, a striking feature is noted. The youngest fossils resemble today's organisms more closely than do the oldest fossils, and gradations occur between them. In fact, many of the older forms long ago became

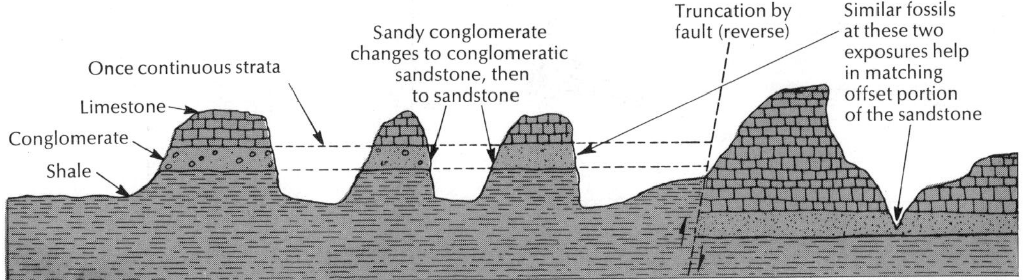

2-18. Truncation by erosion or dislocation. The sketch is highly diagrammatic. The layer of sandy conglomerate exposed in the hill at the left once extended unbroken toward the right where it changes to sandstone. Fragments dropped by a stream tend to become finer grained with increasing distance from a source area. Erosion has truncated strata along the slopes of hills and removed large volumes of rock which once existed in the spaces separating the hills. Truncation by dislocation (faulting) is shown at the right. If regolith obscured the actual trace of the fault at the Earth's surface, a geologist might still deduce that a fault existed by matching the offset fossil-bearing sandstone layers.

extinct. In general, the changes are from more simple to more complex forms, and the horse series (Fig. 2–19) is a well-known illustration.

Such fossil collections strongly support the concept of evolution (p. 343). Each type of animal or plant in existence during any one interval of geologic history apparently had reached a more or less similar stage in its development in most parts of the Earth. Each type possessed characteristic features which differed from those of its ancient relatives and from those of the generations which followed it. Some forms evidently evolved slowly through millions of years, whereas others evolved at a faster pace. Relatively rapid and widespread migration, dispersal, and mixing apparently account for this. Therefore, sediments deposited more or less simultaneously (as a geologist regards time) in various parts of the Earth should contain fossils which are somewhat similar. Sedimentary rocks are thus tagged by the fossils which they contain, and scattered outcrops may be matched or correlated on the basis of these labels.

William "Strata" Smith, an English surveyor, is generally given credit for this discovery. However, as with many other scientific discoveries, the labors and thoughts of other men had also contributed. If Smith had not discovered the principle of fossil succession at this time, the discovery might soon have been made by someone else; the season was ripe for the harvesting of this idea.

Smith studied layers of sedimentary rocks in excavations and natural exposures wherever he worked, and as a hobby, he collected the fossils which were abundant in them. Near the end of the eighteenth century, he was able to arrive at two important conclusions.

(1) In a certain area, individual layers always occur in the same succession at each outcrop; e.g., a gray cherty limestone would always underlie a cross-bedded white sandstone; or in another example, the "monuments" in Monument Valley (Fig. 2–20) all have thick resistant beds of sandstone on top and less resistant, thinner layers of shale and sandstone beneath.

(2) Each kind of sedimentary rock has a distinctive set of fossils which distinguishes it from all other kinds (oversimplified: some sedimentary rocks, in whole or in part, do not contain fossils). Smith amazed his fossil-collecting friends by his ability to examine their collections and tell them the location and stratum from which each of the fossils came.

Correlation of different sedimentary rocks

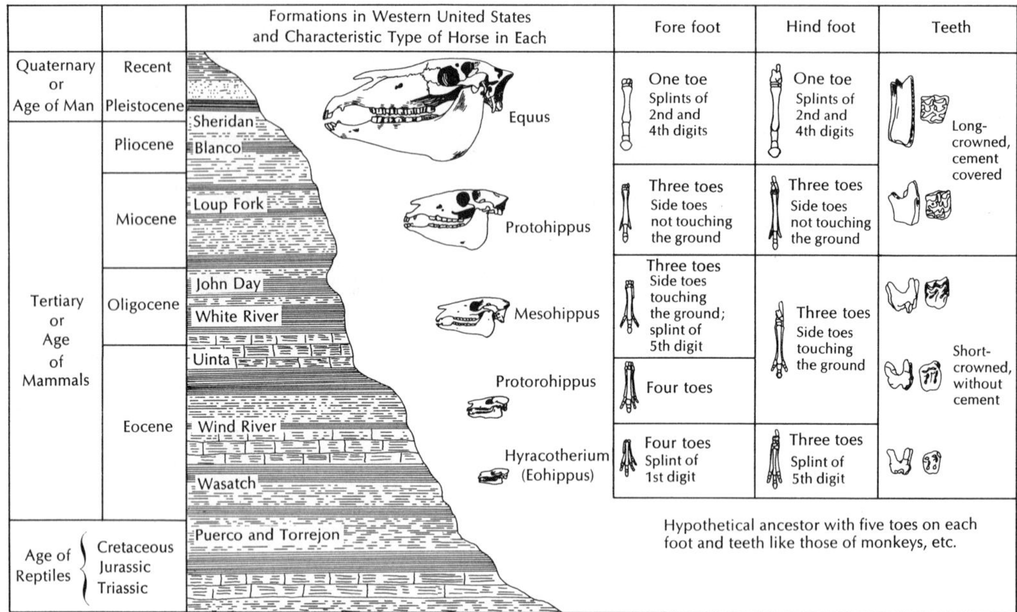

2-19. Evolution of the horse. Contrary to popular belief, the evolutionary development of the horse was not a straight-line, undeviating, progressive development which began with a small, four-toed, browsing animal that gradually became larger, lost all but one of its toes, and developed teeth suitable for grazing. Several varieties of horses apparently existed simultaneously, and of these, all but one type subsequently became extinct. For example, three-toed browsing horses lived at the same time as one-toed grazing horses; the one-toed horse evolved toward the present-day horse, whereas the three-toed horse became extinct. Some varieties of horse became smaller rather than larger. Furthermore, rates of evolutionary development apparently differed at different intervals. *(After W. D. Matthew.)*

on the basis of their fossil content is complicated by the fact that most living organisms are confined to distinctive environments; e.g., certain types of marine animals live on sandy bottoms, a different group live on muddy bottoms, and only a few species live in both environments. Identical fossil forms, therefore, cannot be found in sedimentary beds which formed simultaneously, but in different environments.

However, the age equivalence of the different fossil types can sometimes be determined by tracing the strata involved from one area to another. A series of sandstone beds may crop out in one area, and layers of shale may be exposed a few miles away. As a geologist walks from one area to the other, he may be able to observe at intervening outcrops that the sandstone changes gradually to shaly sandstone, then to sandy shale, and finally to shale. A geologist would call this a facies change, and the age equivalence of fossils found in the sandstone and the shale is thus demonstrated. A geologic chronology based upon worldwide fossil collections has enabled investigators to put the Earth's history into more or less chronological order.

Apparently life has always been in existence somewhere on Earth since its beginning, but different organisms have been dominant at different times and places. Some older forms became extinct and have not been found in younger rocks. The names of the eras and their subtitles suggest such a change: the Paleozoic Era (age of invertebrates), the Mesozoic Era (age of reptiles and dinosaurs), and the Cenozoic Era (age of mammals). However, some organisms have persisted with so little change that specimens living today are quite similar to their Paleozoic or Mesozoic ancestors; these are the so-called "living fossils."

• 8. Shallow seas were once more extensive.

Marine waters at one time or another in the past have covered most areas which are now dry land, commonly more than once, and areas now covered by the sea were formerly land. This does not mean that continents have at times become ocean basins or vice versa. The different structure of the continental and oceanic crust and the concept of isostasy (p. 165) argue against this. Rather, local sinking of the land or world-wide rise in sea level causes ocean water to advance across low-lying sections of a continent. Geologically speaking, the coastline which separates land from sea has been an extremely flexible and movable boundary.

A rise in sea level is caused by the displacement of sea water by sediments, by eruptions of lava on the sea floor, by upward movements of the sea floor, and by the melting of glaciers. If the floor of a shallow sea subsided more or less continuously, space was thereby made available for thick accumulations of sediments.

• 9. Remnants of the sedimentary record may be correlated.

Sites of deposition have shifted from one

2-20. Monument Valley in Utah. Thick resistant sandstone beds form the upper parts of the monuments, and thinner beds of sandstone and shale form the lower portions. The scattered monuments are the remnants of once-continuous layers.

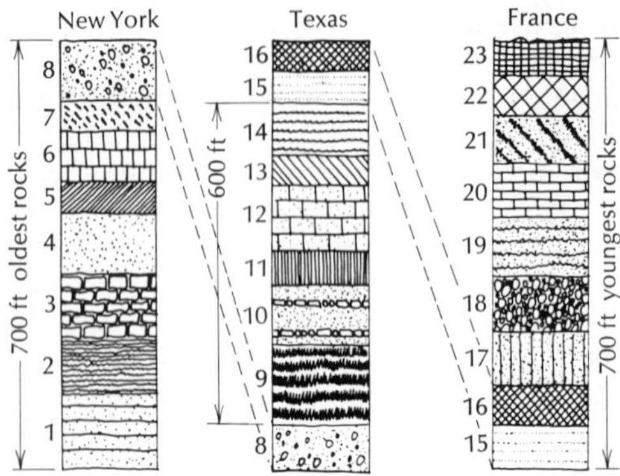

2-21. Composite nature of the sedimentary record (highly schematic). Fossils show: (a) that the New York strata are oldest; (b) that those in France are youngest; (c) that bed 8 in New York was deposited about the same time as bed 8 in Texas (layer 8 was not, of course, continuous from New York to Texas); and (d) that beds 15 and 16 in Texas match those in France. Thus the composite thickness of the three groups is 2000 feet (duplicates are not counted).

area to another throughout geologic time, and erosion commonly has begun in an area as soon as deposition ceased. Both erosion and deposition do take place in the same area, but one commonly predominates. As erosion occurs in a region, previously formed rocks are destroyed, together with their record of the Earth's history. During such a time of erosion, it is unlikely that sediments will be deposited within this region which can be preserved as permanent records of the events then taking place. Thus the determination of the Earth's geologic history must be based upon collecting and matching many scattered fragments into a more complete composite record. Arranging the numerous isolated sequences of sedimentary beds into their chronological order and deciphering the messages they contain have proved to be a monumental task made feasible only by worldwide cooperation (Fig. 2–21).

Correlation is the determination of the age equivalence or contemporaneity of fossils, strata, or events in different areas. In other words, is this particular fossil (layer, event) older, younger, or about the same age as that one? Fossils are the most useful means of correlating sedimentary rocks from one area to another, particularly from one continent to another. However, the radioactive technique for dating igneous and metamorphic rocks (p. 107), and thus indirectly the sedimentary rocks associated with them, provides another tool for intercontinental correlation. Over shorter distances, geologists may match similar types of rocks and similar sequences of rocks, or utilize other techniques (Fig. 2–22).

• **10. The age of the Earth is about 4.6 billion years.**

This conclusion is important because it provides time enough for processes such as stream erosion and deposition—which seem very slow on the scale of human lifetimes—to accomplish results of the very large magnitude shown by the geologic record. It is eye-opening to calculate how long it would take for a person to count to 4.6 billion. Let us assume a rate of 3 counts per second continuously day and night, week after week, until the total is reached. About 10 years would be required to count to 1 billion!

• **11. The Earth's crust does, did, and presumably will continue to move.**

In one type of crustal movement, very gradual upwarping or subsidence occurs and affects large regions of the Earth, either land or sea; little or no crumpling and folding of the rocks occur, although some tilting may take place. Movement of a different type, known as *orogenic* (Gr.: "origin of mountains"), affects elongated belts of the crust and results in the formation of mountains by folding, faulting, and thrusting. Both horizontal and vertical crustal movements are involved in orogenies, and the crust in the elongated belts appears to be severely compressed or squeezed at times.

• **12. The Earth's present topographic features, its scenery and landscapes, are only transient forms. They have evolved from the different shapes of the past and will pass slowly and inexorably into the still different forms of the future.**

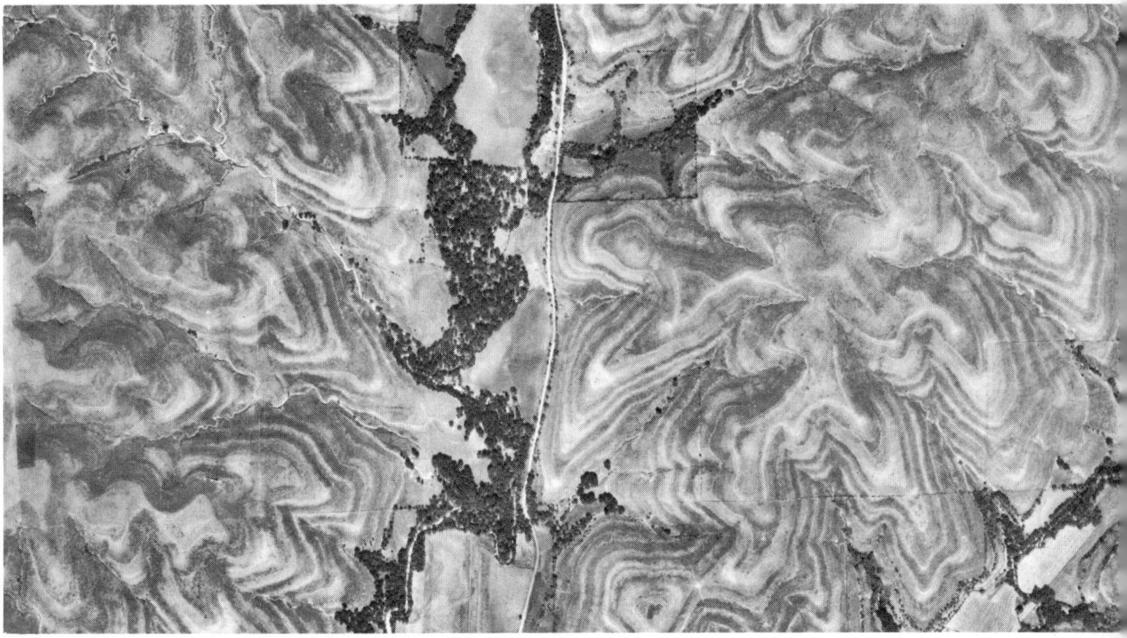

2-22. Flat-lying limestone (whitish bands) and shale beds crop out in Flint Hills in northeastern Kansas. *(Courtesy U.S. Geological Survey.)*

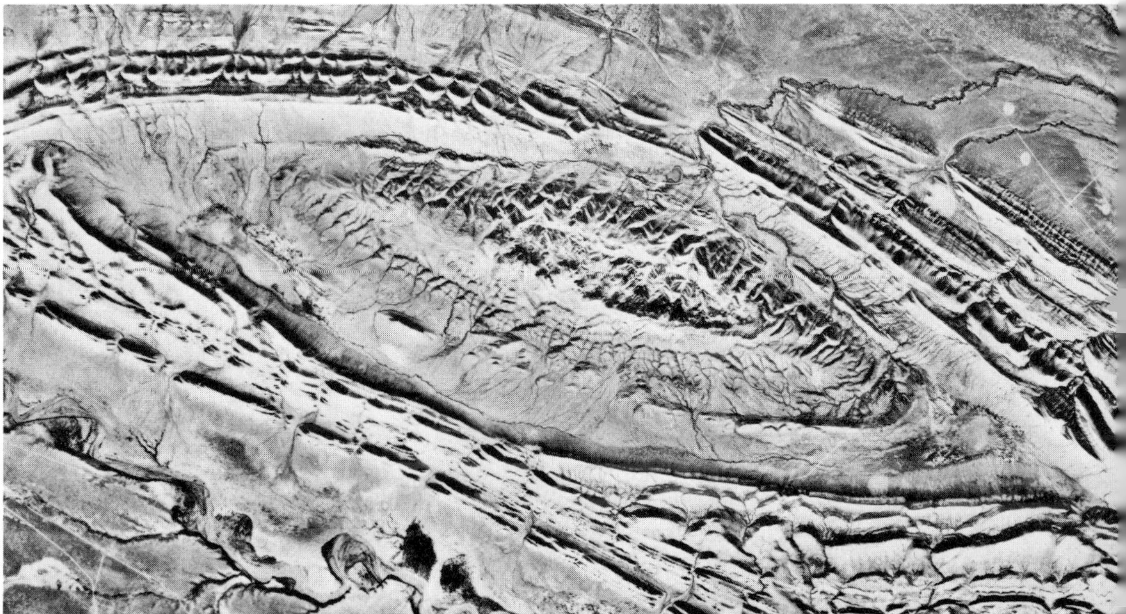

2-23. Structural, elongated dome in Mesozoic rocks in Fremont County, Wyoming. *(Courtesy U.S. Geological Survey.)*

2-24. Land forms of the United States. *(Courtesy Erwin Raisz.)*

The evolutionary development of land forms depends upon factors such as (1) the kind or kinds of rocks underlying the surface, (2) the rock structures present such as folds, faults, domes, or flat-lying beds, (3) the types of erosional processes which are occurring, (4) the time during which these various factors have been interacting, and (5) the stability or instability of the crust. The Earth's land forms, although differing widely in detail, are not haphazard. They can be classified genetically with some success, and their shapes and histories can be understood in terms of the five groups of factors just listed.

Differential erosion is evident in the Earth's surface features in all gradations of scale from the very large to the very small, because the relative resistances or weaknesses of the different types of rocks to erosion are soon made evident. On a large scale, the locations of mountains, ridges, and valleys commonly depend upon the locations of resistant and weak rocks (Fig. 2–23). Rocks which are resistant in one climate, however, may not be so under the different weathering conditions of another.

The rock structures of the crust are key factors in controlling the shape of the Earth's surface in many areas (Fig. 2–24). A flat-topped mesa, developed on flat-lying strata in a semiarid region, contrasts sharply with the elongated ridges which develop upon tilted strata and with the elliptical outline of an eroded structural dome or basin. Mountains formed by faulting differ in shape from those produced by folding, particularly the zigzag, loop-shaped mountains which are sculptured from folds whose axes plunge.

Each erosional process has its diagnostic features. An area in a climate favoring stream erosion differs in appearance from a colder region where glaciers predominate. Characteristic land forms are produced by winds in an arid region, by waves and currents along a coast, and by the action of subsurface water in a humid region underlain by soluble rocks.

The Earth's surface undergoes an orderly, sequential development as erosion proceeds in an area, although details of the sequences in different areas vary because of variations in the controlling factors listed above. Because erosion is a slow process, it is necessary to construct a sequence in theory and then to check this against the actual features present here and there on the Earth. Such features represent different stages in the development of different sequences. This has led to the concept of a *geomorphic cycle* and to the application of the terms *youth, maturity,* and *old age* to its three main stages (Fig. 2–25).

Each cycle, however, is a continuous gradual development. It can be subdivided only arbitrarily, and any one cycle may be interrupted before it has been completed. A geomorphic cycle is not defined in terms of an actual number of years (time is relative only) and its three main subdivisions are not of equal length. The term cycle is somewhat misleading because a complete sequence of events does not occur at regular intervals. Furthermore, there is so much diversity in land forms that no single geomorphic cycle can account adequately for all of the features observed. Despite these limitations, the concept of a geomorphic cycle—of the orderly sequential development of land forms—is useful and valid.

Most landscapes are not the result of a single process but of a combination of processes. Similarly, most cycles are interrupted before completion by a change in climate or by crustal movement (Fig. 2–26).

The landscapes and scenery of today tend to be relatively youthful features. Although many of the rocks, folds, and faults found in mountain belts such as the Appalachians and Rockies are quite ancient, each of these as a land form has resulted from widespread extensive uplift and subsequent differential erosion during the Cenozoic Era.

• 13. Pleistocene glaciation and climatic changes have had recent and worldwide influence.

A succession of glacial and interglacial ages occurred during the recent Ice Age and were accompanied by widespread changes that extended far beyond the regions covered di-

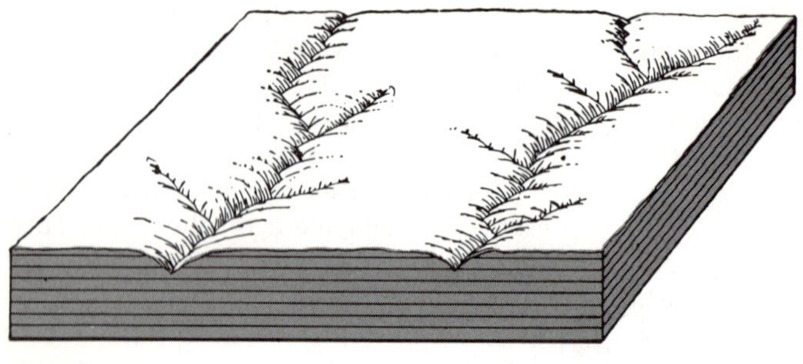

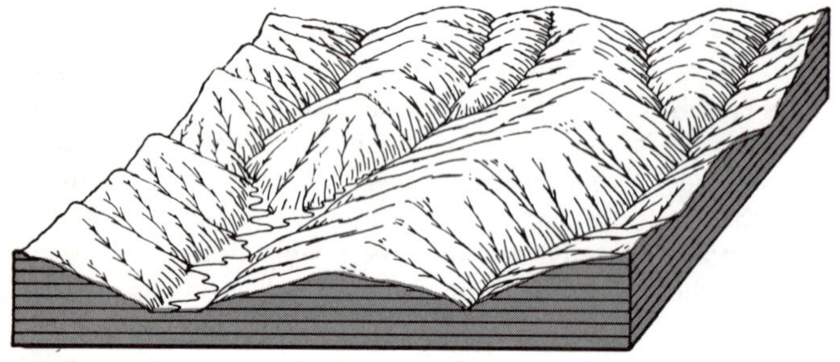

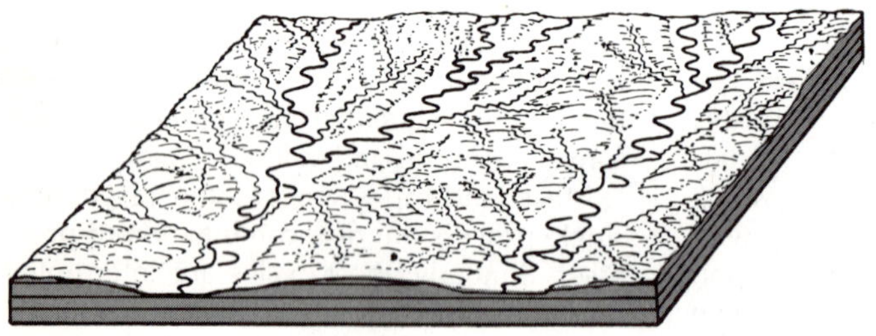

2-25. Sequential development of topography in a moderately elevated region of essentially uniform, stratified rocks devoid of important structures. *Top,* region in youth; *middle,* in maturity; *bottom,* in old age. A region in typical maturity is one of slopes with a rolling hill-and-valley topography. Drainage is good, relief is at its maximum, and erosion is most effective. Major streams have begun to develop floodplains and meanders. Uplift and other crustal movements tend to interrupt such a sequence before it reaches old age.

rectly by glacier ice. Some of these changes involved crustal movements, the volume of ocean water, and the migrations of plants and animals (p. 241). In addition, other features originated within the glaciated areas as a result of glacial erosion and deposition. Thus we may observe features today that were formed under different climatic conditions during parts of the Pleistocene. Therefore, the assumption that the present is the key to the past must be modified somewhat: e.g., to understand certain land forms located at a latitude of 40 degrees, it may be necessary to study current processes located some 10 to 20 degrees closer to the poles—a modern glacier or a permafrost region.

• **14. Minor catastrophes are allowed for in uniformitarianism.**

The cumulative effects of relatively small, more or less continuous processes such as stream and wave erosion and deposition are emphasized in uniformitarianism. However, it is now realized that in a number of processes the bulk of the erosion and deposition is accomplished during brief, infrequent storms or floods. During an exceptionally

2-26. The entrenched meanders or goosenecks of the San Juan River in southeastern Utah. The canyon exceeds 1200 feet in depth. Through it, the winding river flows six miles to travel an airline distance of one mile. The meanders developed when the river flowed on a wide floodplain. Subsequent uplift occurred, and the river entrenched its channel into the horizontally layered sedimentary strata. *(Courtesy E. C. LaRue, U.S. Geological Survey.)*

powerful flood, of the kind that occurs once, twice, or three times each decade in some valleys, much more erosion and transportation may be accomplished by a river than during the intervening months and years when low, normal, and flood stages alternate. Perhaps more wave erosion is produced along certain coasts when a tsunami (p. 417) strikes than is accomplished by all of the wave action that takes place in the intervening years.

Hurricanes, some landslides, movements along some faults, volcanic eruptions, and the occasional rain squall in a desert all indicate that one must be cognizant of the infrequent minor catastrophes in assessing the relative roles of various geologic processes. Of course, in a geologic sense, these are not catastrophes at all, but merely times of increased erosion and deposition which occur occasionally.

• 15. Isostasy and isostatic adjustment.

Isostasy (Greek: "equal standing") is a balance apparently present in the outer part of the Earth (Fig. 6–33). Isostatic balance is upset if enough matter is transferred from one area to another, and this happened during the Pleistocene when water was evaporated from the oceans and fell as snow to make glaciers on the lands. It also occurs when a large mountain range is worn down and its eroded debris is transported to an adjoining ocean area. In these circumstances isostatic adjustment occurs. At depth, rocks apparently flow slowly outward from an overloaded area, which sinks, to an underloaded area, which thus rises. This continues until balance is restored.

• 16. A sequence of geologic events may be interpreted in chronological order from geologic maps and structure sections.

This provides a framework for the inclusion of a number of important ideas and extends the concept of sermons in stones. A *geologic map* shows the distribution of rocks of different types and ages at the surface. If we assume that a vertical trench is cut along a line extending across such a surface, then a *geologic structure section* is a diagram of the rocks and structures that would occur along one of its walls (i.e., as rocks would be displayed along a straight, vertical cliff).

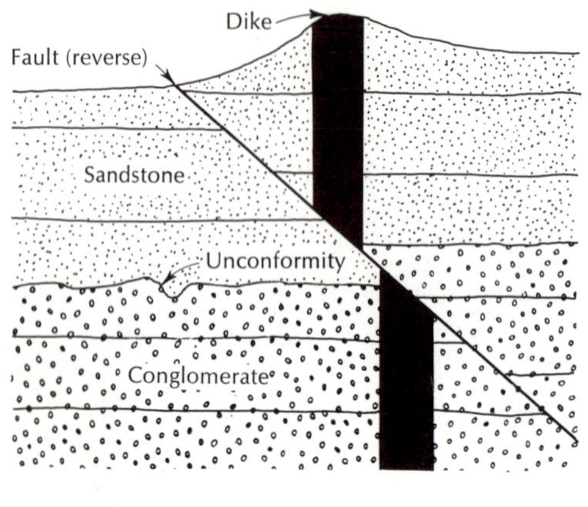

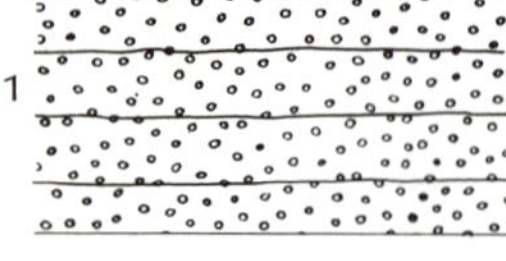

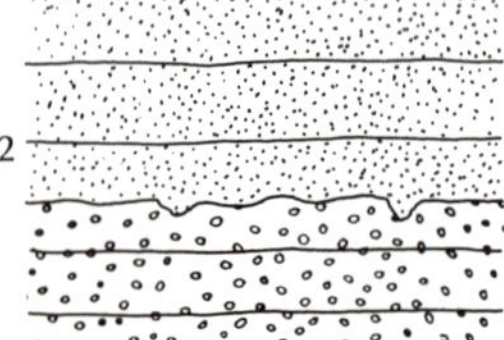

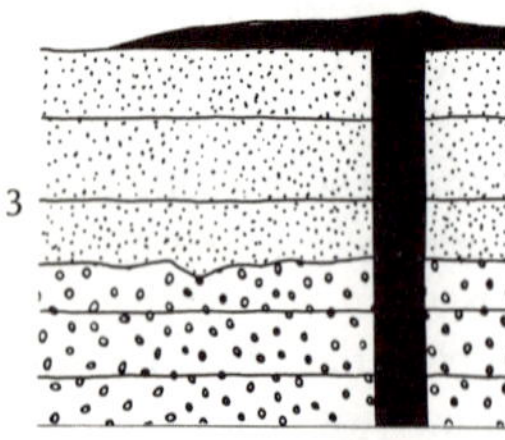

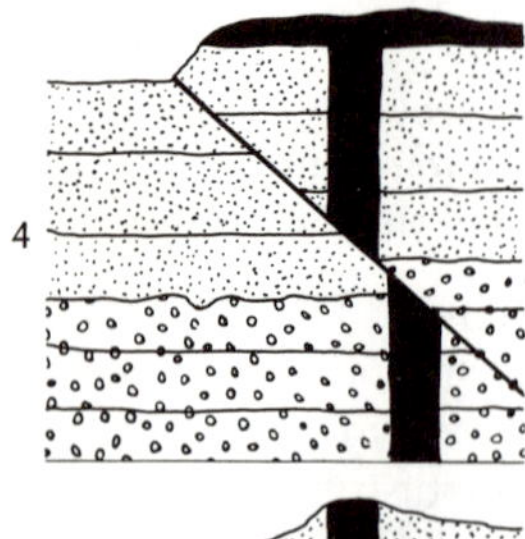

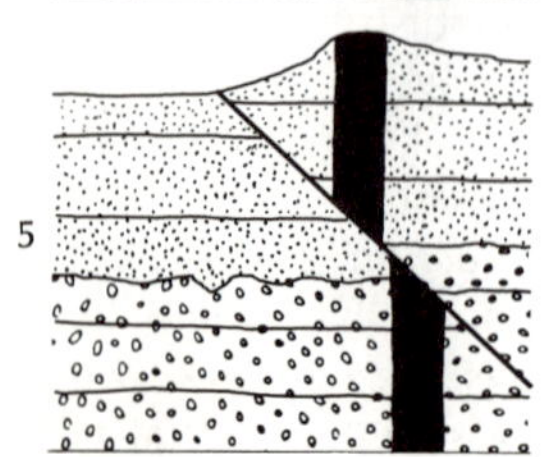

2-27. A geologist studies the large structure section at the top and interprets a sequence of events (smaller structure sections 1 to 5).

The top of a structure section is a profile of surface topography along its trend, and its base occurs at whatever depth is needed for the features we want to show. Drill holes are one method of obtaining subsurface data for use in constructing a structure section.

Among the types of geologic features that can be shown on a structure section is the *unconformity,* which is a relationship between older and younger rocks that involves a break in the geologic record (p. 149). Thus a surface of unconformity is commonly a buried surface of erosion. Unconformities may be local or widespread and may involve minor or major gaps in the geologic record. To illustrate, assume that a marine limestone —one containing fossils of organisms that lived only in the ocean—is overlain unconformably by a marine shale and that a buried subaerial erosion surface occurs between the two formations. (*Subaerial*—formed on a land surface—contrasts with *subaqueous*—formed beneath a water body.) The following sequence of events presumably occurred: (1) calcareous sediments accumulated on a sea floor and eventually were compacted and cemented to form limestone; (2) the area became land, either because the land was uplifted or because sea level fell; (3) erosion then carved out gullies and small valleys and removed some of the limestone; (4) the sea again covered the area, and mud and clay were deposited on top of this erosion surface; and (5) uplift and erosion occurred again to expose the rocks as they are observed today.

The interpretation of structure sections involves relationships such as the following. If an igneous mass is intruded into the Earth's crust, then the igneous material is younger than the rocks it has intruded. If a fault extends upward and stops abruptly at an unconformity, we know that the faulting occurred before the erosion which produced the unconformity. On the other hand, a fault is younger than any of the rocks it offsets. In general, in working out the sequence of events observed in a structure section, one begins at the bottom, following the principle of superposition, and works toward the top (Fig. 2–27). Intrusive igneous masses are exceptions here.

• 17. Both work in the field and work in the laboratory are important in geology.

Geology is primarily a field study, but laboratory experimentation has increased greatly since the 1940's, as has the use of quantitative methods. The basis of modern geology was laid by detailed field observations made mainly during the last two centuries or so by numerous diligent investigators. It has been said that one can learn geology only through his feet, and it is difficult to see how a satisfactory understanding of geology can be achieved without emphasis upon field observations. Countless pairs of shoes and boots have been worn out by geologists slogging through mud, climbing over rocks, and walking up hill and down in search of information relating to the Earth's history. Although photographs and diagrams can help bring the outdoors within four walls, field trips remain essential features of the study of this science.

On the other hand, the modern practice of geology emphasizes a quantitative approach, laboratory experimentation, and the application to geological problems of nongeological techniques taken from fields such as biology, chemistry, physics, and mathematics. Quantitative studies and the mathematical manipulation of geological data have become possible in recent decades because of the large amounts of observational data now available. These data can be correlated with the rapidly multiplying results of laboratory experiments.

The geologist is concerned with the processes that shape the Earth as well as with the Earth itself. He can study these processes as they operate in that part of nature (the geologist's main laboratory) that is accessible to him. But he can also study some aspects of a number of these processes under controlled conditions in the laboratory. In this manner, he may view the results of chemical reactions that occur at temperatures and pressures simulating those that exist at depths of several miles and more below the Earth's surface. He may then examine these end products by means of X-rays, polarized light, the electron microscope, and the mass spectrometer.

Observations and measurements have become more widespread, more varied, and more precise as new instruments have been developed: e.g., the topographic features of the sea floor are now being mapped, the waters and currents of the oceans can be studied at all depths, and slight changes in the paths of orbiting satellites (caused by the Earth's gravitational attraction) permit interpretations concerning the nature of the Earth's interior. The techniques of physics and chemistry have been emphasized greatly in geology in recent decades, and the new disciplines of geochemistry and geophysics have resulted. It is in such overlapping zones between two related sciences that discovery and progress are most rapid today.

The rapidity of geologic progress in the last several decades is underscored by the estimate that perhaps 90% of all of the geologists who ever lived are still alive and active.

Exercises and Questions for Chapter 2

1. Describe ways in which fossils aid in the interpretation of geologic history and the assumptions and concepts upon which these are based.
2. Describe some of the ways in which geology is of economic and intellec-

tual value to man and explain why it is likely to be of even greater economic importance in the future.

3. What is the difference between a mineral and a rock?
   (A) What accounts for such physical properties of minerals as cleavage, crystal shape, and hardness?
   (B) Describe a number of ways in which minerals are useful to mankind and describe the physical properties which account for these.
   (C) Why is identification of many rocks easier in the field than as small specimens in a laboratory (the absence of a microscope is assumed in each case)?
   (D) What is meant by the rock cycle?
   (E) Why are sediments deposited in layers?
   (F) How do sediments become rocks?
4. Grow artificial crystals in the laboratory or at home, and vary conditions such as the rate of cooling so that larger crystals are produced at one time and smaller ones at another time. For example, you might dissolve enough sugar in hot water to form a thick syrupy solution and allow this to cool. If a string is suspended in the solution, some crystals should become attached to it. Compare their shapes with those of crystals forming along the bottom of the container and account for any differences.
5. What strategy might a geologist follow in attempting to locate the source area of the sediments contained in a certain sedimentary formation?
6. Describe a sequence of events in the geologic history of an area that could be called a "sermon in stones." Next, describe the type of evidence one would need to find in the rocks of the area to be able to interpret these episodes in its geologic history. Illustrate by means of a structure section.
7. Give some examples of what is meant by uniformitarianism and also some examples of what is not meant by it. How did the viewpoint of the catastrophists differ from that of the uniformitarianists?
8. Why has the idea of a geomorphic cycle been criticized? Why is it of value in discussing certain aspects of geologic history?
9. How does the deductive method of solving a problem differ from that of the inductive method?

3

Minerals and Rocks

We have defined *minerals* (Fig. 3–1) as naturally occurring substances, generally solid and inorganic, whose atoms and ions are arranged in regular, three-dimensional patterns that give each mineral a characteristic set of physical properties and chemical composition. Minerals that have this orderly internal atomic arrangement are said to be *crystalline* and thus always have certain shapes or break in a certain pattern. The electrical and other forces holding these atomic groups together are very strong in some directions, but weak in others, and these directions and patterns extend throughout the mineral specimen. Thus a large crystal of a certain mineral may have the same shape as a small crystal of this mineral, and it breaks in the same way, because it consists of the same atomic groups, oriented with respect to each other in precisely the same structural pattern, throughout all its parts. Although more than 2000 mineral species are known, only two dozen or so are important as rock-making or ore minerals.

Some minerals originate by precipitation from solution in much the same way that rock candy is made or the way that salt crystals form from a water solution. The amount of a substance which may be dissolved in a solvent depends upon what it is, what the solvent is, and upon physical factors such as temperature and pressure. Ordinarily, the amount decreases at lower temperatures. Thus as high-temperature solutions rise from the depths along cracks or other passageways in rocks, they reach zones of lower temperatures nearer the Earth's surface and precipitation may occur. Chemical reactions between such solutions and the rocks with which they come in contact are also important factors in causing precipitation. If several materials are dissolved, the least soluble substance crystallizes first and the most soluble substance is precipitated last. In this manner the walls of a crack in bedrock or an opening in rocks may be lined with minerals of various kinds (Fig. 3–2). Molten rock-making material like that erupted at volcanoes is the type of solution (magma or lava) from which many

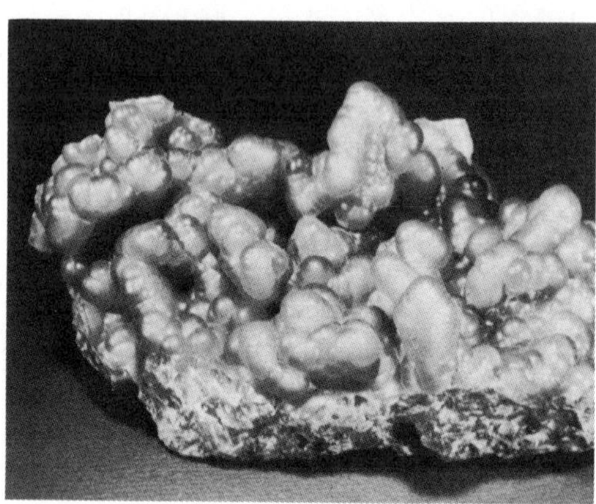

3-1. Crystals of four minerals. *Upper left:* tabular wulfenite (specimen is 6 in. across; from Glove Mine, Arizona). *Upper right:* calcite (specimen is 5½ in. across; from England). *Lower left:* halite showing simple cubic form (specimen is 3 in. across; from Searles Lake, California). *Lower right:* rounded masses of chalcedony that developed along a side of a cavity in volcanic rock (specimen is 6½ in. across; from Mexico). *(J. Sinkankas,* Mineralogy for Amateurs, *Van Nostrand Reinhold, 1964.)*

minerals have formed, and hot-water solutions are another type. Minerals have also been formed by metamorphism and sublimation (passage directly from the gaseous to the solid state).

Atoms and groups of atoms generally occur in minerals as *ions* (positively or negatively charged atoms that have lost or gained one or more electrons). We may imagine ions as small solid spheres whose sizes and electrical charges are important in determining the orderly, three-dimensional structure in which they are packed to form minerals (Figs. 3–3 and 3–4). Ions are held together or bonded in different ways, and metallic, covalent, and ionic bonds are important examples (Fig. 3–5).

In *metallic bonding* the cloud of electrons around one nucleus merges with the clouds of electrons around neighboring nuclei. In this arrangement, the nuclei have a regular spacing throughout an interwoven swarm of electrons. The mobility of the merged electrons makes metals good conductors of electricity and heat energy, and the attractions holding the nuclei in place are relatively weak. Therefore, they may be shifted rather readily if a force is supplied, but they also re-bond readily after the force has been removed. Thus a metal such as gold can be hammered into a very thin sheet.

Covalent bonding tends to be quite strong and involves a sharing of outer electrons—quartz and diamond (see below) are examples. Materials with this type of bonding do not tend to break or melt readily nor to be good electrical conductors. *Ionic bonding* involves an electrical attraction between ions of opposite electrical charge, and halite (see below) is an example.

In gases or liquids, ions are moving too rapidly for electrical charges to fit them into a rigid pattern. However, the rate of motion decreases as temperatures decrease, and sufficient cooling permits solid crystalline substances to form.

One of the simplest and most readily visualized structures is that shown by the mineral halite, which is made of ions of sodium and chlorine (Fig. 3–4). A sodium atom readily loses one electron to become a positively charged sodium ion, whereas a chlorine atom readily gains one electron to become a negatively charged chlorine ion. A sodium ion and a chlorine ion unite because of this transfer of electrons and the electrical attraction of their opposite charges. The ions pack into a three-dimensional structure called a *space lattice* that is cubical in form for halite. In this type of packing, any one chlorine ion has six equidistant sodium ions symmetrically arranged around it; likewise, any one sodium ion has six chlorine ions symmetrically arranged around it, and all of the angles in the lattice are 90 degrees. Ions of sodium and chlorine thus alternate in three mutually perpendicular planes as if they were in contact. As this crystal lattice structure extends continuously throughout the mineral, we cannot speak of an individual sodium chloride molecule as we can speak of a molecule as a separate particle in a liquid or a gas. The cubical lattice of halite is the cause of its cubical crystals (Fig. 3–1) and excellent cubical

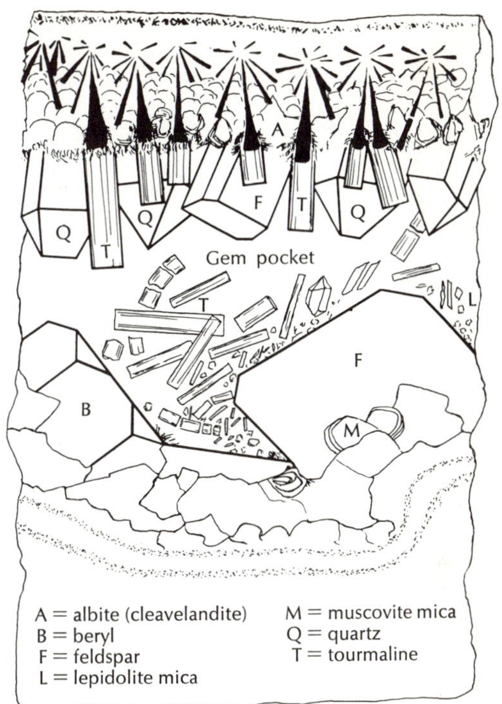

3-2. A cross section through a gem pocket. The cavity was filled with water when the different crystals formed. *(J. Sinkankas,* Gemstones of North America, *Van Nostrand Reinhold, 1959.)*

cleavage; when shattered, halite separates along three planes oriented at 90-degree angles to one another. X-ray photographs are a basic tool in the study of crystalline structures.

The basic building unit in the Earth's crust is a tetrahedron of silicon and oxygen. Each minute silicon ion is at a center among four equidistant and equispaced oxygen ions; in other words, the oxygen ions are located at the corners of a tetrahedron and the silicon ion is at its center.

With respect to electric charges, a silicon atom can lose four electrons and become a silicon ion with four plus charges. Each oxygen atom readily gains two electrons to form an oxygen ion with a negative electric charge of 2. As the combination of one silicon ion and four oxygen ions results in a total of four plus charges as opposed to

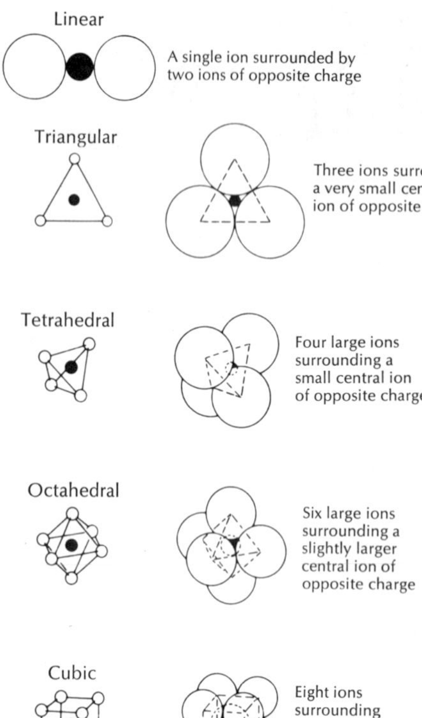

3-3. Above, ions are packed together in various patterns to produce minerals. Basic three-dimensional patterns tend to be repeated without change throughout all parts of a crystal. Patterns differ because ions have different sizes and are attracted (bonded) in different ways. Below, model of halite (salt). The model shows the relative size and arrangement of the ions of sodium (small spheres) and chlorine (large spheres) in a crystal of salt. (J. Sinkankas, *Mineralogy for Amateurs,* Van Nostrand Reinhold, 1964. *(Courtesy American Museum of Natural History.)*

eight negative charges, it is necessary for some of the oxygen atoms to be shared or for additional positive ions to be packed into the lattice.

In quartz, each oxygen atom is shared by two neighboring silicon atoms. This results in the chemical formula of SiO_2. The tetrahedra may be joined in various geometric patterns (Fig. 3–6) such as chains, rings, and sheets in which from one to four of the oxygen ions of each tetrahedron are shared.

An ion may replace any other ion that has a similar electric charge and size (the ionic radius is the more important factor); e.g., sodium and calcium have nearly identical ionic radii, as do iron and magnesium. Thus sodium may take the place of calcium or vice versa, and iron and magnesium ions may be interchanged. Because of such alternative components, the chemical formulas of many minerals vary within certain limits.

Mineral Properties

Each mineral has a distinctive set of physical properties (see Tables 3–2 and 3–3), and usually a combination of only a few properties suffices to distinguish one mineral from the rest. With practice, an observer learns which properties are diagnostic of a mineral and which properties show considerable variation. Memorization of a long list of properties for each mineral is unnecessary and unwise, although one or two features must be remembered for each. You can identify a friend by the way she walks or talks, and you do this without consciously

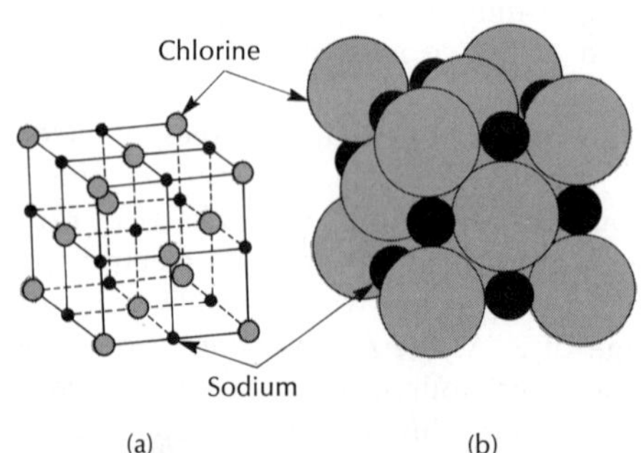

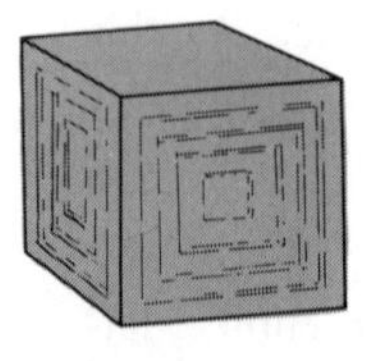

3-4. Halite: (a) space lattice; (b) packing diagram; (c) external form. (B. Mears, Jr., *The Changing Earth,* Van Nostrand Reinhold, 1970.)

Metallic bond

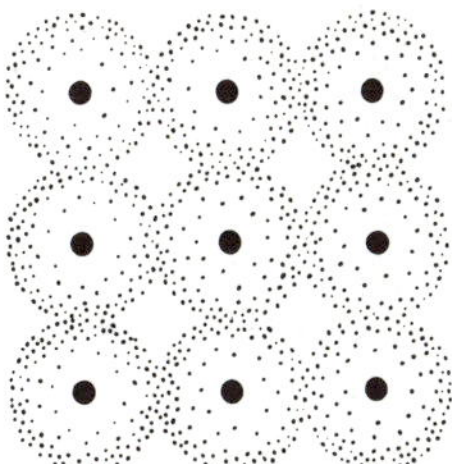

Atomic nuclei surrounded by clouds of mobile electrons.

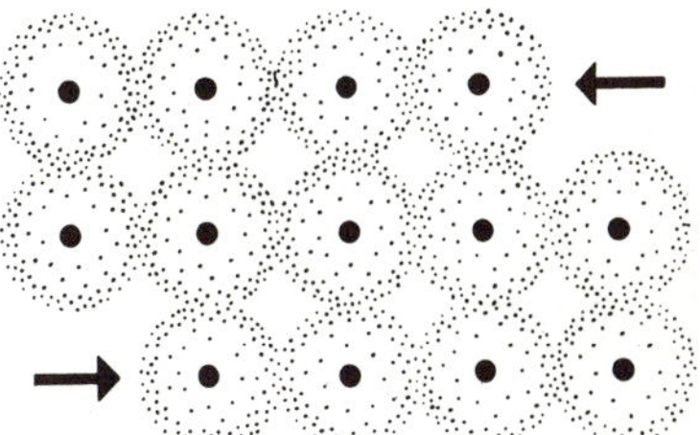

Shear forces easily slip atoms because bonds are not very strong, but atoms quickly re-bond in the new position. This results in malleability.

Covalent bond in carbon (diamond)

Schematic of electron arrangement around bonded carbon atoms showing how each atom is surrounded by eight electrons.

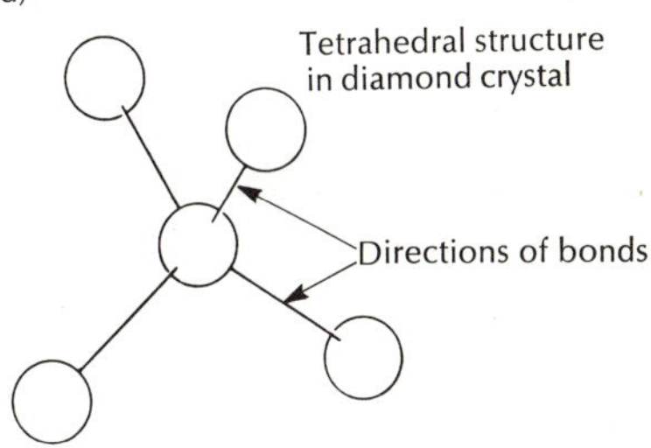

Actual arrangement of atoms in three dimensions. The outer atoms are bonded to others, repeating the tetrahedral pattern to the limits of the crystal.

Ionic bond in calcium fluoride (fluorite)

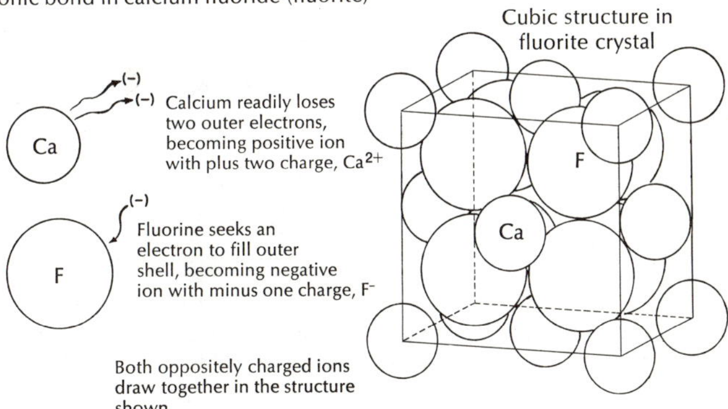

3-5. Three important types of bonding between ions in minerals are shown schematically. (J. Sinkankas, *Mineralogy for Amateurs*, Van Nostrand Reinhold, 1964.)

listing as identifying features her age, height, and weight. Your identification is successful despite the fact that she may wear rumpled jeans and a pleasant smile one day or peer at you sophisticatedly from a blue sheath dress the next. Furthermore, the absence of several pounds after a successful diet does not prevent your identification.

Thus it is with minerals; specimens may be large or small, and the colors of some minerals range from white to black, but there is something diagnostic about each. The principal useful properties for field identification are color, streak, hardness, luster, crystal form, cleavage or fracture, specific gravity, and miscellaneous characteristics

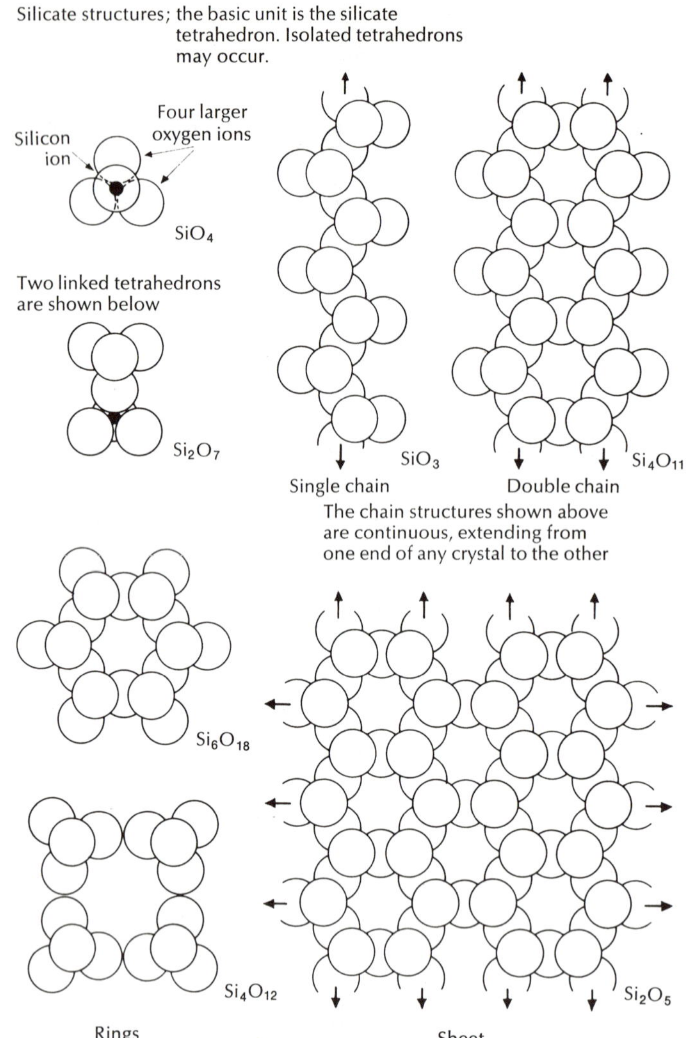

3-6. Silica tetrahedra are joined together in different patterns to form the silicate group of minerals. (J. Sinkankas, *Mineralogy for Amateurs*, Van Nostrand Reinhold, 1964.)

involving reaction to acid, taste, feel, odor, and magnetism.

Diamond and graphite (Fig. 3–7) illustrate distinctive physical properties of the kind that mark most minerals. Each is composed of carbon, yet one is the hardest of known minerals and transparent, whereas the other (graphite) is one of the softest, is opaque, and marks paper. Such differences in properties show that the carbon atoms themselves are not the cause of differences such as hardness—the carbon atoms in graphite are exactly like the atoms in diamond. They are arranged in diamond in a different pattern from that in graphite. In graphite, the atoms occur in parallel planes that pull apart easily to produce the tiny smooth flakes that make graphite a good lubricant. In diamond the same atoms are combined firmly in a tetrahedral network in which neighboring

Carbon polymorphs

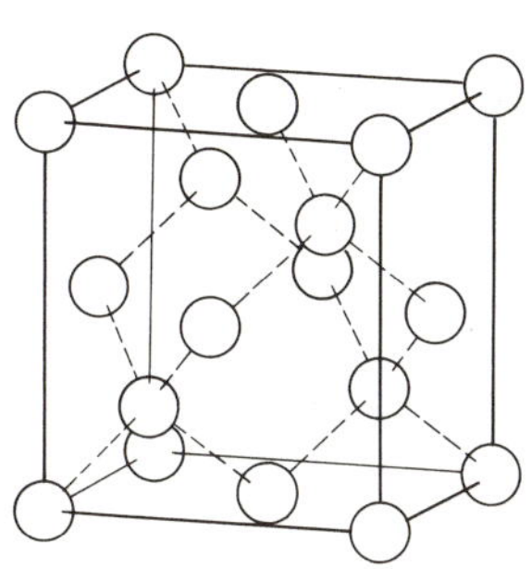

Diamond

Each carbon atom is at the center of a tetrahedron of four other carbon atoms. This arrangement continues throughout the crystal

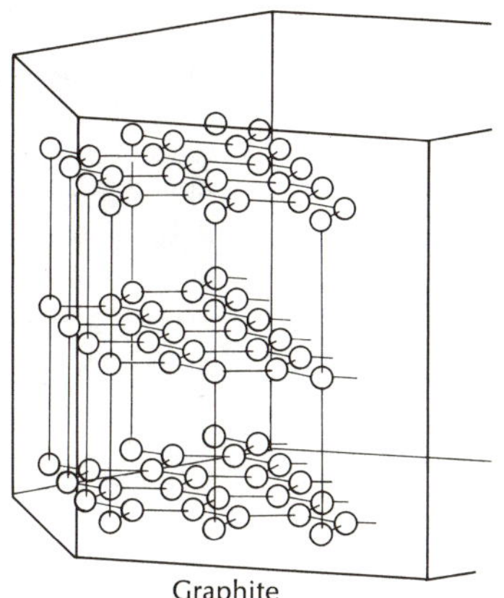

Graphite

Carbon atoms are closely packed in sheets, but sheets are separated by a considerable distance. Bonds are strong within sheets but very weak between them

3-7. Schematic drawing of the crystalline structures of diamond and graphite. The regular arrangements extend throughout each specimen.. The sketch is exploded; i.e., the ions have been separated widely so that their three-dimensional arrangements can be observed. Ions are imagined to be in contact. (J. Sinkankas, *Mineralogy for Amateurs,* Van Nostrand Reinhold, 1964.)

carbon atoms share electrons. When diamond is rubbed against some other mineral, the diamond lattice structure always proves more durable than the packing arrangement of the other mineral.

Color and Streak

The color of a mineral is probably its most obvious physical property and may be useful in mineral identification, or it may be deceptive. An inherent color depends upon the kinds and arrangement of the atoms in the mineral and is diagnostic for some minerals, chiefly those with a metallic appearance (e.g., the brass yellow of pyrite, the golden yellow of chalcopyrite, and the azure blue of azurite). Colors of other minerals vary widely because they depend upon the chance presence of impurities or of fractures (e.g., quartz and calcite). Moreover, the color of a surface alteration should be distinguished carefully from the true color of the mineral.

The color of a mineral in powdered form is called its *streak.* To find the streak, rub the mineral specimen on a hard surface as you would rub chalk on a blackboard, and the equivalent of a blackboard is commonly a piece of white unglazed porcelain called a streak plate. The color of the streak is likely to be more constant than the color of the mineral; e.g., the streak of hematite is always reddish brown, whereas the streak of limonite is yellowish brown. Streak is not important in the identification of a number of minerals, particularly those which are quite hard. Short streaks on a white streak plate in the geology laboratory are as diagnostic as long ones and mean less frequent washings for the streak plates. Begin at one edge of a plate and use the space systematically.

Luster

This is the appearance of the unweathered surface of a mineral in ordinary reflected light. Some minerals, such as pyrite and galena, have a metallic luster (like that of a shiny polished metal surface); others have nonmetallic lusters—glassy, dull, greasy, or pearly.

Hardness

The hardness of a mineral is its resistance to abrasion or scratching (its "scratchability") and is a very useful property. Most minerals have a characteristic hardness that can be checked very readily. The Mohs' hardness scale is commonly used and has mineral

TABLE 3-1
THE MOHS' SCALE OF HARDNESS

NUMBER AND MINERAL ON MOHS' SCALE	COMPARABLE HARDNESSES
1. Talc	(Softest)
2. Gypsum	A finger nail is 2-2½
3. Calcite	A copper coin is 3
4. Fluorite	
5. Apatite	
	Common window glass is 5½
6. Feldspar	A knife blade is 5½-6
	A steel file is 6½-7
7. Quartz	
8. Topaz	
9. Corundum	
10. Diamond	(Hardest)

representatives from 1 to 10 (Table 3–1), although hardness does not increase by precisely uniform steps from 1 to 10; e.g., the jump from 1 to 9 may be proportionally as large as the jump from 9 to 10. A soft mineral is scratched by the fingernail, whereas hard minerals scratch glass and cannot be scratched by a knife. Minerals of medium hardness do not scratch glass and can be scratched by a steel nail or knife blade, but not by the fingernail. Skill in determining the ease with which glass is scratched can be acquired and will help in distinguishing among several hard minerals. Thus a steel nail, a small piece of window glass (not those in the geology laboratory), a quartz crystal, and one's fingernail are all that is necessary to test the approximate hardnesses of minerals. Small hardness sets containing nine specimens from 1 to 9 on the Mohs' scale are also commonly available for testing in the laboratory.

A soft mineral may leave a mark on a hard mineral, but this mark can be rubbed off. A true scratch can generally be detected with the fingernail as a tiny groove, although observation through a hand lens is sometimes necessary. Short scratches on the glass are as useful as long ones, make the glass plate last longer, and suggest the work of a systematic confident individual.

Crystal Form

The delight of the mineral collector is the crystal (Figs. 2–13 and 3–8). A *crystal* is a solid bounded by smooth plane surfaces that always meet at precisely the same angles for all specimens of any one mineral (similar angles on the different specimens must be compared). However, the plane surfaces or faces may have different areas in different specimens and thus may not have a uniform appearance. The shapes depend upon the arrangement of the atoms in the space lattice of the mineral; in fact, discovery that the angles between the faces of a crystal are constant for any one species of mineral led to the idea that minerals have definite internal structures. If space is available during crystallization, minerals are precipitated from solution with characteristic shapes. Thus different minerals have different crystal shapes, and this is a useful property in distinguishing one mineral from another. However, if many tiny crystals form here and there in a solution and grow larger, they eventually interfere with one another's growth, and a granular aggregate results. This is a more common occurrence than the development of perfectly shaped crystals.

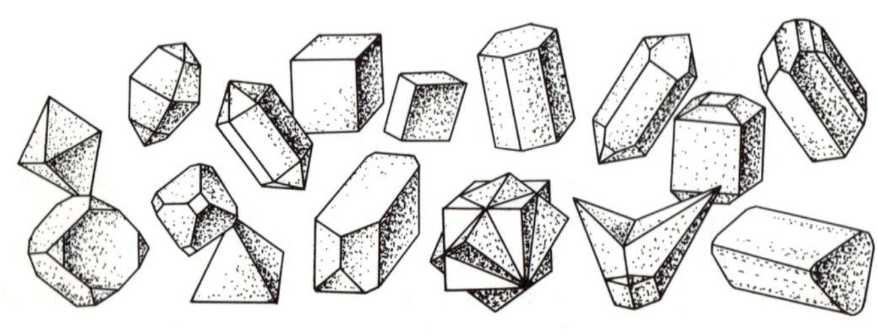

3-8. Shapes of some crystals. (B. Mears, *The Changing Earth*, Van Nostrand Reinhold, 1970.)

3-9. Mineral cleavage. *Top left,* mica illustrates nearly perfect cleavage in one direction. *Top right,* a crystal of feldspar (orthoclase) shows cleavage surfaces in two directions that intersect at about 90 degrees; it also fractures in other directions. *Bottom left,* calcite cleaves in three directions, but the intersections do not form 90-degree angles. *Bottom right,* fluorite cleaves in four different directions. The specimen on the left has been cleaved into a nearly symmetrical form. The large faceted gm is 1$\frac{5}{16}$ in. in length. *(Courtesy Neil Croom; Ward's Natural Science Establishment, Inc.; J. Sinkankas,* Gemstones of North America, *Van Nostrand Reinhold, 1959.)*

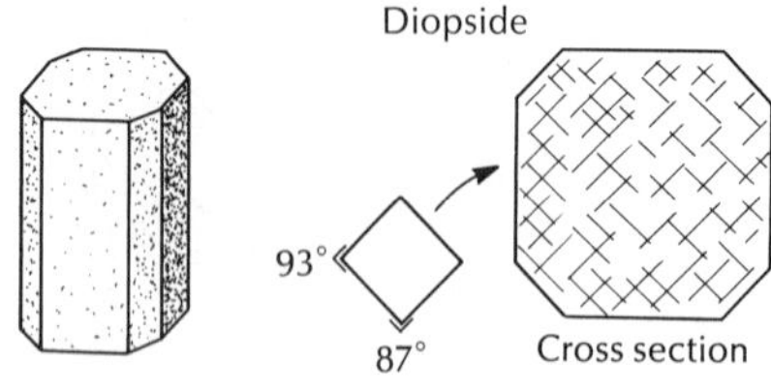

Prismatic Cleavage of Pyroxenes

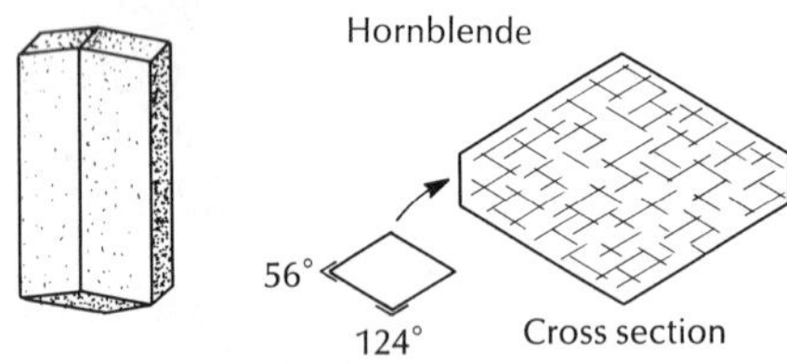

Prismatic Cleavage of Amphiboles

3-10. Cleavage. (J. Sinkankas, *Mineralogy for Amateurs*, Van Nostrand Reinhold, 1964.)

Manner of Breaking: Cleavage and/or Fracture

The manner in which a mineral breaks is determined by its space lattice structure and may be very useful in identification. If the break produces one or more smooth plane surfaces, the specimen is said to have *cleavage* (Figs. 3–9 and 3–10). On the other hand, *fracture* results in irregular or curved surfaces. Different types of cleavage and fracture occur, and variations in cleavage fall into three categories: (1) minerals may cleave in different directions (from 1 to 6), (2) the angles at which different cleavage planes intersect may be different for different minerals, and (3) the cleavage surfaces range from very smooth planes to slightly irregular surfaces.

Mica illustrates practically perfect cleavage in one direction, and countless thin films may be peeled from a specimen of mica, but all of the cleavage planes are parallel and thus constitute only one direction of cleavage. Calcite has three cleavage planes which do not meet at 90-degree angles, whereas galena and halite break along three smooth planes mutually at right angles (cubical cleavage). A cleavage fragment of halite may thus resemble a box. Its top and bottom are smooth flat cleavage surfaces (one direction). Its sides mark another direction of cleavage, and its ends the third. The three directions are mutually perpendicular at any corner of the box. Quartz does not exhibit cleavage but commonly breaks with a distinctive fracture that is curved and undulating like the inside of a shell (conchoidal fracture, Fig. 2–14).

The presence or absence of cleavage can be recognized most readily by observing the manner in which light is reflected as a specimen is rotated. If cleavage surfaces are present, light will flash as if a number of tiny mirrors were oriented at definite angles to each other. The distinction between a cleavage surface and a crystal face is sometimes difficult (it should be; each results from the same orderly space lattice arrangement). Crystal faces always occur on the exterior of a mineral, whereas cleavage surfaces originate when a mineral is broken and its interior is exposed. Such a broken surface may show a number of stairlike treads and risers, produced when a break shifts from one plane of cleavage to another, and galena is an excellent example. On the other hand, crystal faces tend to be smooth and flat.

Miscellaneous Properties

Halite (salt) may be readily identified by its taste and calcite by its effervescence in cold dilute hydrochloric acid. This distinguishes calcite from dolomite (which bubbles only when powdered) and from other minerals such as feldspar and quartz that do not fizz at all. *For obvious reasons, specimens tested with acid should be washed immediately and thoroughly in water.* Magnetite is attracted by a magnet. Test for magnetism by balancing a magnet so that it teeters on the edge of a table. Next, bring the specimen beneath and close to the unsupported end of the magnet. For smaller specimens or scattered magnetite grains in a large specimen, break off a few specks

TABLE 3-2
SIMPLIFIED MINERAL IDENTIFICATION KEY

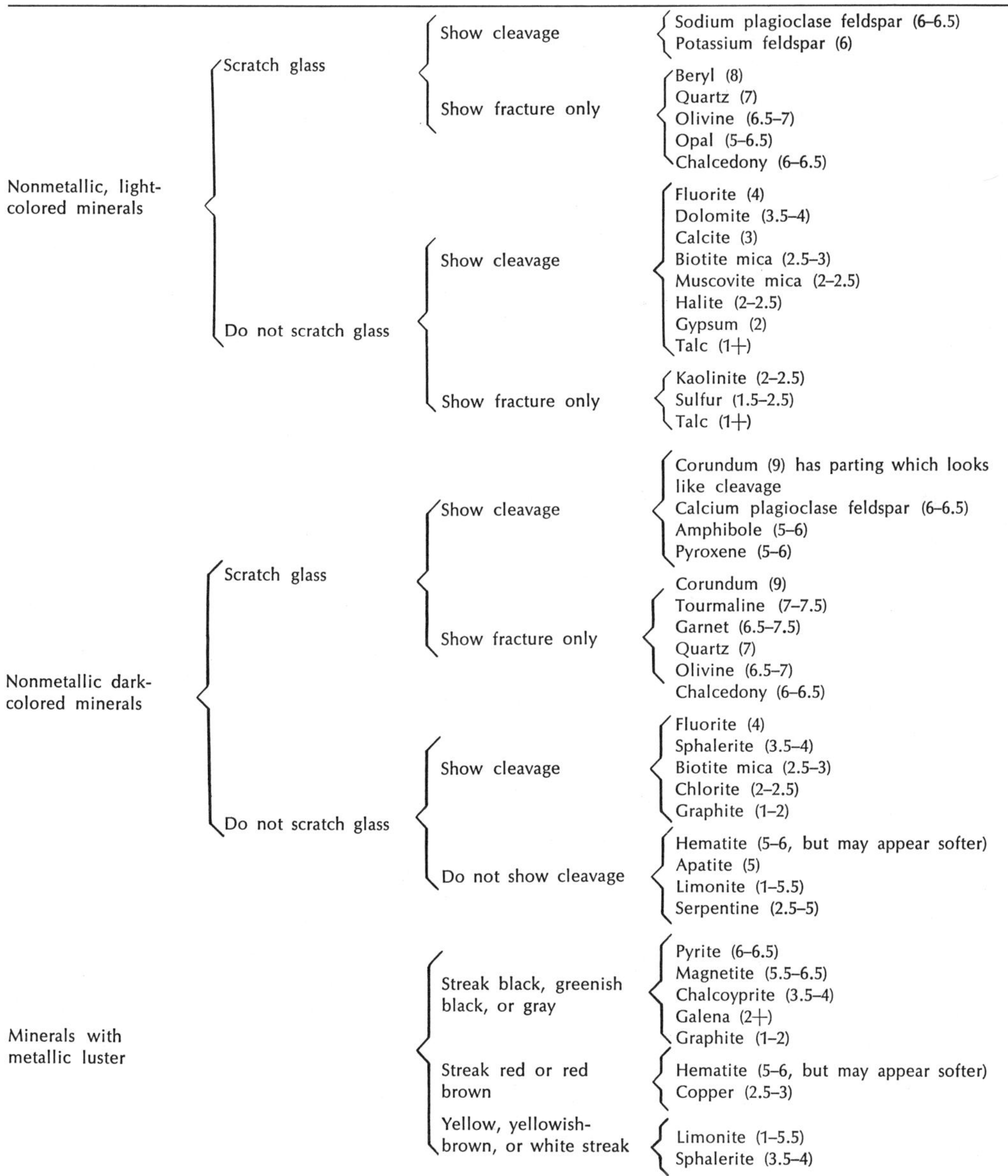

Group	Hardness	Cleavage / streak	Minerals
Nonmetallic, light-colored minerals	Scratch glass	Show cleavage	Sodium plagioclase feldspar (6–6.5) Potassium feldspar (6)
		Show fracture only	Beryl (8) Quartz (7) Olivine (6.5–7) Opal (5–6.5) Chalcedony (6–6.5)
	Do not scratch glass	Show cleavage	Fluorite (4) Dolomite (3.5–4) Calcite (3) Biotite mica (2.5–3) Muscovite mica (2–2.5) Halite (2–2.5) Gypsum (2) Talc (1+)
		Show fracture only	Kaolinite (2–2.5) Sulfur (1.5–2.5) Talc (1+)
Nonmetallic dark-colored minerals	Scratch glass	Show cleavage	Corundum (9) has parting which looks like cleavage Calcium plagioclase feldspar (6–6.5) Amphibole (5–6) Pyroxene (5–6)
		Show fracture only	Corundum (9) Tourmaline (7–7.5) Garnet (6.5–7.5) Quartz (7) Olivine (6.5–7) Chalcedony (6–6.5)
	Do not scratch glass	Show cleavage	Fluorite (4) Sphalerite (3.5–4) Biotite mica (2.5–3) Chlorite (2–2.5) Graphite (1–2)
		Do not show cleavage	Hematite (5–6, but may appear softer) Apatite (5) Limonite (1–5.5) Serpentine (2.5–5)
Minerals with metallic luster		Streak black, greenish black, or gray	Pyrite (6–6.5) Magnetite (5.5–6.5) Chalcoyprite (3.5–4) Galena (2+) Graphite (1–2)
		Streak red or red brown	Hematite (5–6, but may appear softer) Copper (2.5–3)
		Yellow, yellowish-brown, or white streak	Limonite (1–5.5) Sphalerite (3.5–4)

The minerals are subdivided into three main groups on the basis of luster (metallic or nonmetallic) and color (light or dark). Further subdivision is based upon the presence or absence of cleavage. In the last column on the right, the minerals are arranged in order of decreasing hardness. See Table 3-3 for a list of some of the physical properties of a number of the more common minerals. A few minerals are listed in more than one place: e.g., some specimens of a certain mineral may be light, whereas others are dark colored; some specimens of another mineral may have a metallic luster, whereas others do not. You should not expect to detect relatively small differences in hardness: e.g., the difference between 3.5 and 4; but you should easily distinguish between 5 and 7 or 2 and 4. (After Grout, *Kemp's Handbook of Rocks,* Van Nostrand Reinhold, 1940)

TABLE 3-3
PROPERTIES OF SOME COMMON MINERALS

The minerals are arranged alphabetically, and the most useful properties in identification are printed in italic type. Most minerals can be identified by means of two or three of the properties listed below. In some minerals, color is important; in others, cleavage is characteristic; and in others, the crystal shape identifies the mineral.

NAME AND CHEMICAL COMPOSITION	HARDNESS	COLOR	STREAK	TYPE OF CLEAVAGE	REMARKS
Amphibole (complex ferromagnesian silicate)	5–6	*Dark green to black*	Greenish black	Two directions at angles of 56° and 124°	Viterous luster. Hornblende is the common variety. Long, slender, six-sided crystals. *Black with shiny cleavage surfaces at 56° and 124°.*
Apatite (calcium fluophosphate)	5	Green, brown, red, variegated	White	Indistinct	Crystals are common as are granular masses; vitreous luster.
Biotite mica (complex silicate)	2.5–3	Black, brown, dark green	Colorless	*Excellent in one direction*	*Thin elastic films peel off easily.* Nonmetallic luster.
Calcite ($CaCO_3$)	3	Varies	Colorless	*Excellent, three directions, not at 90° angles*	*Fizzes in dilute hydrochloric acid. Hardness.* Nonmetallic luster.
Chalcedony (SiO_2)	6–6.5	Varies from white to black	None	Conchoidal fracture	So-called crypto-crystalline quartz. Dull luster as compared with opal.
Chalcopyrite ($CuFeS_2$)	3.5–4	*Golden yellow*	Greenish black	None	*Hardness and color distinguish from pyrite.* Metallic luster.
Chlorite (complex silicate)	1–2.5	*Greenish*	Colorless	Excellent, one direction	*Nonelastic flakes, scaly, micaceous.*
Copper (Cu)	2.5–3	*Copper red*	Red	None	*Metallic luster on fresh surface. Ductile and malleable. Sp. gr. 8.5 to 9.*
Corundum (Al_2O_3)	8	Dark grays or browns common	Colorless	Parting resembles cleavage	*Barrel-shaped, six-sided crystals with flat ends.*
Diamond (C)	10	Colorless to black	Colorless	Excellent, four directions	Hardest of all minerals.
Dolomite ($CaMg(CO_3)_2$)	3.5–4	Varies	Colorless	*Good, three directions, not at 90°*	*Scratched surface fizzes in dilute hydrochloric acid. Cleavage surfaces curved.*
Feldspar (potassium variety) (silicate)	6	*Flesh, pink, and red are diagnostic;* may be white and light gray	Colorless	*Good, two directions, 90° intersection*	*Hardness, color, and cleavage.*

Feldspar (sodium plagioclase variety) (silicate)	6	*White to light gray*	Colorless	*Good, two directions, about 90°*	*If striations are visible, they are diagnostic.* Nonmetallic luster.
Feldspar (calcium plagioclase variety) (silicate)	6	*Gray to dark gray*	Colorless	*Good, two directions, about 90°*	*Striations commonly visible;* may show iridescence. Associated with augite, whereas other feldspars are associated with hornblende. Nonmetallic luster.
Fluorite (CaF_2)	4	Varies	Colorless	*Excellent, four directions*	Nonmetallic luster. In cubes or octahedrons as crystals and in cleavable masses.
Galena (PbS)	2+	*Bluish lead gray*	Lead gray	*Excellent, three directions, intersect 90°*	*Metallic luster.* Occurs as crystals and cleavable masses. *Very heavy.*
Garnet (silicate)	6.5–7.5	Varies from red, brown, yellow, to black	None	Fractures: conchoidal and uneven	Crystals with 12 and 24 faces are common, also in granular masses. Resinous or glassy luster.
Gold (Au)	2.5–3	*Gold*	Gold	None	Malleable, ductile, *heavy*. Metallic luster.
Graphite (C)	1–2	*Silver gray to black*	Grayish black	Good, one direction	Metallic or earthy luster. *Foliated, scaly masses common. Greasy feel, marks paper.* This is the "lead" in a lead pencil (mixed with clay).
Gypsum (hydrous calcium sulfate)	2	White, yellowish, reddish	Colorless	*Very good in one direction*	Vitreous luster. *Can be scratched easily by fingernail.*
Halite (NaCl)	2–2.5	Colorless and various colors	Colorless	*Excellent, three directions intersect at 90°*	*Taste, cleavage, hardness.*
Hematite (Fe_2O_3)	5–6	*Reddish*	*Reddish*	None	Sp. gr. 5.3. Metallic luster (also earthy). May appear softer.
Kaolinite (hydrous aluminum silicate)	2–2.5	White	Colorless	None (without a microscope)	Dull, earthy luster. Claylike masses.
Limonite (group of hydrous iron oxides)	1–5.5	*Yellowish brown*	*Yellowish brown*	None	Earthy, granular. Rust stains.
Magnetite (Fe_3O_4)	5.5–6.5	*Black*	Black	None	Metallic luster. Occurs in eight-sided crystals and granular masses. *Magnetic. Sp. gr. 5.2.*

TABLE 3-3 (Continued)

NAME AND CHEMICAL COMPOSITION	HARDNESS	COLOR	STREAK	TYPE OF CLEAVAGE	REMARKS
Muscovite mica (complex silicate)	2–2.5	Colorless in thin films; yellow, red, green, and brown in thicker pieces	Colorless	*Excellent, one direction*	*Thin elastic films peel off readily.* Nonmetallic luster.
Olivine (iron-magnesium silicate)	6.5–7	*Yellowish and greenish*	White to light green	None	*Green, glassy, granular*
Opal (hydrous silica)	5–6.5	Varies	Colorless	None	*Glassy and pearly lusters, conchoidal fracture.*
Pyrite (FeS_2)	6–6.5	*Brass yellow*	Greenish black	None	*Cubic crystals* and granular masses. Metallic *luster.* Crystals may be striated. *Hardness important.*
Pyroxene (complex silicate)	5–6	Greenish black	Greenish gray	*Two, nearly at 90°*	*Stubby eight-sided crystals and cleavable masses. Augite* is common variety. Nonmetallic.
Quartz (SiO_2)	7	Varies from white to black	Colorless	None	Vitreous luster. *Conchoidal fracture. Six-sided crystals common.* Many varieties. Very common mineral. *Hardness.*
Serpentine (hydrous magnesium silicate)	2.5–4	*Greenish (variegated)*	Colorless	Indistinct	*Luster resinous to greasy. Conchoidal fracture.* The most common kind of asbestos is a variety of serpentine.
Sphalerite (ZnS)	3.5–4	Yellowish brown to black	White to yellow	*Good, six directions*	*Color, hardness, cleavage, and resinous luster.*
Sulfur (S)	1.5–2.5	*Yellow*	White to yellow	Indistinct	Granular, earthy.
Talc (hydrous magnesium silicate)	1+	White, green, gray	Colorless	Good, one direction	Nonelastic flakes with *greasy feel. Soft.* Nonmetallic luster.
Topaz (complex silicate)	8	Varies	Colorless	*One distinct (basal)*	Vitreous. *Crystals commonly striated lengthwise.*

and test with a magnet. Graphite has a greasy feel, and a few minerals have characteristic odors.

The specific gravity of a mineral specimen may be distinctive; this compares the mass (weight) of a substance with that of an equal volume of water. Most metallic minerals are heavy, whereas most nonmetallic minerals have a low specific gravity; e.g., halite, 2.1; quartz, 2.6; pyroxene, about 3.0 to 3.5; sphalerite, about 4; magnetite, hematite, or pyrite, about 5; and galena, about 7.5.

Rocks

The three main groups of rocks (Fig 3–11) were discussed briefly in Chapter 2, and in studying a rock specimen, one should attempt first to identify it as sedimentary, igneous, or metamorphic. Identification of a rock in the field in its natural setting among other rocks is commonly much easier than the determination of a small chip in the laboratory; e.g., stratification in a thick-bedded sandstone may be readily apparent in field exposures, and fossils may be present in overlying or underlying beds. However, evidence of bedding and fossils may be entirely missing in a small specimen taken from one of the layers.

Igneous Rocks

Igneous rocks differ in texture and mineral composition and are classified by these criteria, although different levels of sophistication are possible. We shall use a simplified version applicable in the field or laboratory without the use of chemical analyses or a microscope, but a small hand lens will be useful at times. Classification is difficult because continuous gradations occur in both texture and mineral content, and subdivision must be somewhat arbitrary.

Texture

Common minerals in igneous rocks are: feldspar, pyroxene, amphibole, quartz, and mica. The shapes, sizes, and arrangement of individual particles (grains) of such minerals in a rock determine its *texture* (Latin: "weave"). The grains are so shaped and oriented that they mutually interlock and leave few openings, if any (Fig. 3–12). A growing crystal entirely surrounded by fluid can extend its boundaries outward unhindered and develop with its characteristic crystal shape. However, when many different crystals develop at about the same time throughout a fluid medium, their outward growth is mutually hindered and few, if any, crystals can form.

On the other hand, some substances dissolved in magma are more soluble than others. Therefore, as cooling occurs, the least soluble substances are precipitated first as small crystals that grow larger, and the most soluble substances crystallize last. Accordingly, the first mineral grains to crystallize out of a liquid medium develop with their characteristic shapes. The next mineral grains to form have somewhat modified crystal shapes because their growth is hindered by the first-formed grains. Mineral grains precipitated last are formless masses filling all remaining spaces (Fig. 3–13).

Order of Crystallization

In a very simplified account of the order of crystallization from a magma, dark-colored minerals that are relatively low in silica tend to crystallize first (e.g., some pyroxenes, some feldspars, and dark mica). Light-colored feldspars and mica (relatively high in silica) form next, and quartz (100% silica) forms last. Thus low-silica magmas (and lavas) tend to produce dark-colored rocks upon cooling, whereas high-silica magmas tend to produce light-colored rocks.

In the precipitation of minerals from solution, as in the baking of cookies, all of the ingredients must be present in the proper proportions. When one of the necessary materials runs out—it might be iron, calcium, chocolate, or sugar—a particular mineral or cookie can no longer be made, although nature (by substituting ions of

3-11. Igneous, sedimentary, and metamorphic rocks. *Top left,* conglomerate. *(Courtesy Ward's Natural Science Establishment, Inc.) Top right,* micaceous schist containing garnet crystals. *(Courtesy The Smithsonian Institution.) Bottom left,* a polished slab of graphic granite, a type of igneous rock. The light-colored mineral in the rock is feldspar and the darker-colored one is quartz. *(Courtesy Ward's National Science Establishment, Inc.) Bottom right,* an igneous rock (trachyte porphyry) with a porphyritic texture. The large light-colored minerals (phenocrysts) are feldspar. The darker part of the rock consists of much smaller grains of feldspar and other minerals. *(Courtesy Ward's National Science Establishment, Inc.)*

about the same size and electric charge) and housewives can improvise somewhat. In other words, certain kinds of pyroxene tend to precipitate from a magma first and extract from it all of the silica necessary to combine with the other elements. Abundant grains of this type of pyroxene can form only from a low-silica magma rich in the necessary elements. On the other hand, quartz cannot be present in rocks formed from low-silica magmas because all of the silica is used up in making the minerals that precede quartz in the order of crystallization.

Grain Sizes

Slow cooling under a thick cover of rocks produces large mineral grains, whereas a sudden chilling (e.g., when lava reaches the surface) results in the formation of many small particles. For large mineral grains to form, ions must migrate through a magma and encounter and become attached to an enlarging grain. Such ions in a magma immediately surrounding any one grain are rather rapidly used up and must be replaced by others if the grain is to continue to grow. This means movement of the ions or grains or both so that unwanted ions are replaced by the necessary ones. Thus any factor that facilitates the migration of ions also promotes the growth of large mineral grains, and maintaining the magma in a fluid state is most important. The fluid state is prolonged if a magma cools slowly, if it is inherently fluid (low-silica magmas), or if dissolved gases are abundant. In general, intrusive rocks are coarse-grained, and extrusive rocks are fine-grained.

A *pegmatite* is a type of igneous rock with a very coarse-grained, uneven texture. In fact, some pegmatite minerals have been known to grow into truly gigantic crystals that measure 10 to 20 feet and more in length, although these are exceptional—imagine a book of mica that is 15 feet across, a log of beryl nearly 20 feet in length and 3 to 4 feet in diameter, or a feldspar crystal that weighs hundreds of tons; all of these and more have been found in pegmatites. Apparently pegmatites form from magmas that contain relatively large amounts of water and other volatile substances such as fluorine, lithium, and boron (p. 134). These keep the magma in a highly fluid condition so that ions can move readily through it, and large mineral grains can grow relatively rapidly. Most pegmatites tend to

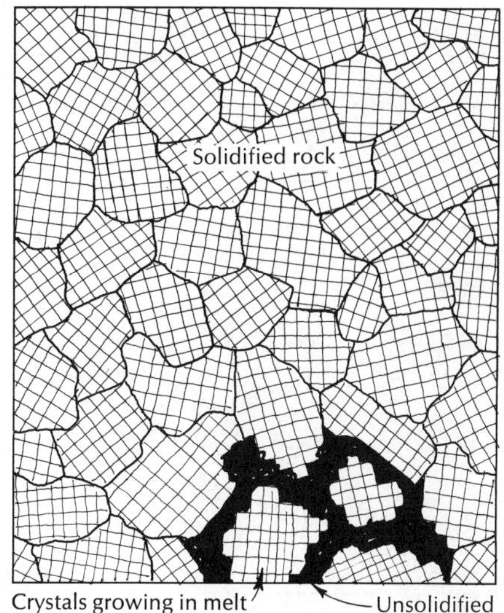

3-12. An interlocking texture is commonly produced by crystallization of a magma or lava (schematic). (J. Sinkankas, *Mineralogy for Amateurs,* Van Nostrand Reinhold, 1964.)

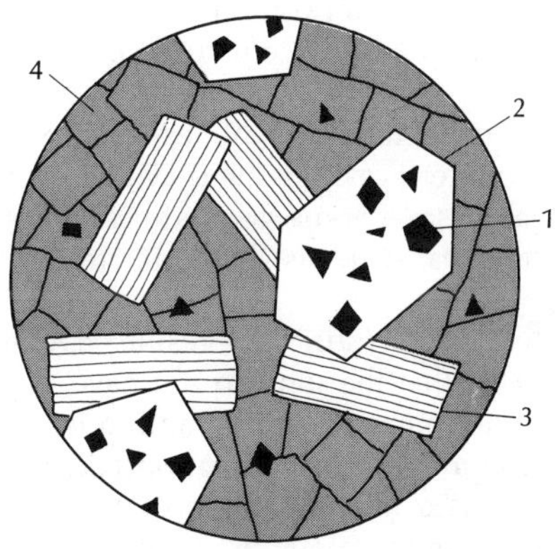

3-13. Interlocking texture and order of crystallization in an igneous rock. *(After Knopf.)*

3-14. Pegmatite dike in Precambrian gneiss, Bear Creek Canyon, Front Range, Colorado. *(Courtesy U.S. Geological Survey.)*

be granitic in composition and to occur as dikes (Fig. 3–14) or lenticular bodies, but some have a complex mineralogy that is a mineral collector's delight, especially when he encounters cavities lined with beautiful, unusual crystals. Some pegmatites are mined for minerals such as feldspar, mica, and beryl.

Natural glass *(obsidian)* results if chilling is so rapid that the ions do not have time to unite into mineral grains, and this occurs most commonly when extrusion takes place. Obsidian is a frozen sample of the original molten material minus its gases (unless differentiation or contamination has occurred). Pumice is obsidian froth which is so full of gas bubble holes (vesicles) that it is light enough to float in water.

Porphyritic Texture

Some igneous rocks are composed of mineral grains of two sharply different sizes, one much larger than the other, and such a texture is called *porphyritic* (Fig. 3–11). One manner in which porphyritic texture may develop involves two stages of cooling. A rising magma may stop far below the surface and cool slowly. The first minerals to crystallize can thus develop into large, well shaped grains which may remain suspended in the fluid magma. Later the magma may resume its rise toward the surface and be erupted as lava, whereupon it cools rapidly. Many small particles then crystallize and surround the larger ones *(phenocrysts)* which had formed previously below the surface.

Classification of Igneous Rocks

Tables 3–4 and 3–5 have the same basic organization. Rocks with the same texture are aligned horizontally in the same row. (1) *Granular* texture: mineral grains are large enough to be visible and identified; these rocks are intrusive with perhaps a few exceptions. (2) *Aphanitic* texture: mineral grains range from specks too small to be identified without a microscope to still smaller invisible grains, but porphyritic aphanitic rocks have phenocrysts that are large enough to be identified. (3) *Glassy* texture: no mineral grains occur unless phenocrysts are present; the atoms were immobilized (relatively) in a haphazard uncombined state as a natural glass. (4) *Pyro-*

TABLE 3-4

The gradational nature of the classification of igneous rocks is illustrated schematically. The graph does not show all of the possible combinations of minerals that have been found in igneous rocks. A vertical line through the graph at any one place indicates the approximate mineral content of a certain type of igneous rock. Only a few types are shown. Granite, diorite, and gabbro in the top row are all coarse grained, and felsite and basalt are fine grained. Felsite thus differs from granite only in grain size. From the graph one can determine that granite is a coarse-grained igneous rock composed chiefly of potassium feldspar with some quartz and minor amounts of ferromagnesian minerals and plagioclase feldspar (potassium feldspar and quartz are the diagnostic minerals). Gabbro is made of ferromagnesian minerals with the plagioclase type of feldspar. *[After Pirsson.]*

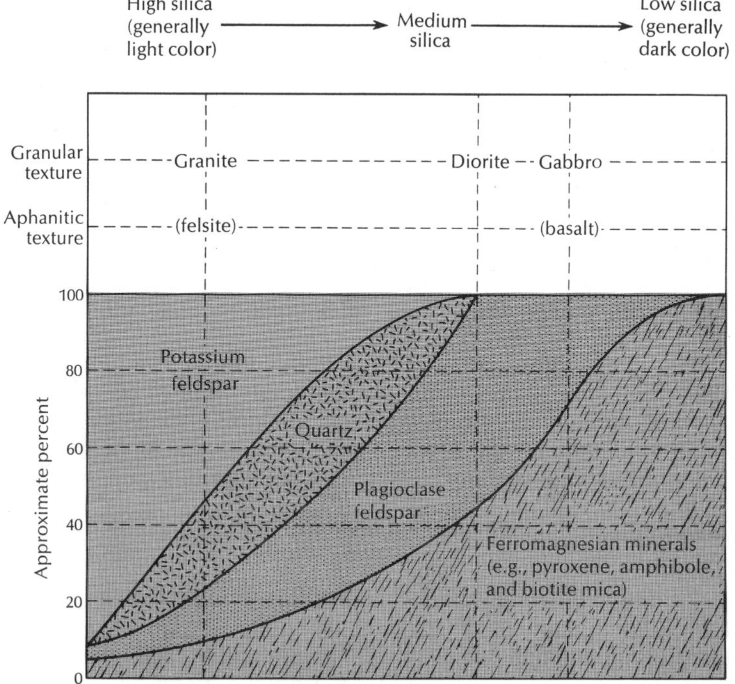

TABLE 3-5
CLASSIFICATION OF IGNEOUS ROCKS
(AFTER FORBES ROBERTSON IN PART)

Silica content	High (70–80%)		Low (40–50%)
Texture (each rock in a horizontal row has same texture)	Potassium feldspar is dominant. Quartz is abundant. Some sodium plagioclase, hornblende, and mica may be present.	Sodium plagioclase is dominant. Hornblende is abundant.	Calcium plagioclase and augite are dominant.
Granular (all rocks in this row may be porphyritic)	Granite	Diorite	Gabbro
Aphanitic porphyritic	Rhyolite (phenocrysts of potassium feldspar and/or quartz)	Andesite (phenocrysts of sodium plagioclase and/or hornblende)	Basalt (dark colored)
Glassy	Obsidian (although dark colored commonly, most obsidians form from high-silica lavas; they may be porphyritic) Pumice (obsidian froth)		
Pyroclastic	These rocks may be divided into two groups on the basis of particle size—silica and mineral content are not factors: Volcanic tuff: average fragment is less than 4 mm (1/6 inch) Volcanic breccia: average fragment is more than 4 mm—1/6 inch)		

1. The feldspars must be identified to use this classification, and the following criteria will be useful.
   (1) POTASSIUM FELDSPARS: diagnostic colors are flesh, pink, and red, may also be white or light grey, but without striations; hornblende and quartz tend to be associated minerals.
   (2) SODIUM PLAGIOCLASE FELDSPAR: colors are commonly white to light gray; striations, if visible, are diagnostic, hornblende is a common associate.
   (3) CALCIUM PLAGIOCLASE FELDSPAR: diagnostic colors are gray to dark gray, some with bluish iridescence; augite is a common associate; striations may be visible.

2. Several additional igneous rocks are described briefly below. They illustrate the gradational nature and diversity of the igneous rocks. All are granular in texture. The first four are included in the term granitic.
   (1) SYENITE: potassium feldspar is dominant; some sodium plagioclase may be present, also hornblende; quartz is missing; however, quartz syenite is a transition to granite; a relatively rare rock.
   (2) MONZONITE: potassium and sodium plagioclase about equal; if quartz is present, the rock is called a quartz monzonite.
   (3) GRANODIORITE: sodium plagioclase is more abundant than potassium feldspar; some quartz is also present.
   (4) QUARTZ DIORITE: feldspar is all plagioclase, and some quartz is present.
   (5) Rock consists almost entirely of ferromagnesian minerals:
      a. PERIDOTITE: made of olivine and pyroxene.
      b. DUNITE: chiefly olivine.
      c. PYROXENITE: chiefly pyroxene.

clastic texture: fragments were erupted from a volcano as solids or as liquids that cooled in flight. Pieces of sedimentary and metamorphic rocks torn from the walls of volcanic vents may be included. *Volcanic tuff* contains average-sized particles 4 mm or less in diameter (about 1/6 of an inch); *volcanic breccia* is made of larger pieces. These rocks accumulated more or less as sediments but are classified here with the igneous rocks.

Rocks of the same mineral and chemical content (except the glasses and pyroclastic rocks) are aligned in vertical rows. High-silica rocks at the left are commonly light-colored and grade gradually into low-silica rocks at the right that commonly are dark-colored.

Sedimentary Rocks

Sedimentary rocks (p. 45) are estimated to underlie about 75% of the Earth's land surface and constitute about 10 to 15% of the outer 10-mile zone of the lithosphere by volume; i.e., they constitute a relatively thin discontinuous skin on the outside of the Earth. On the basis of their manner of formation, sedimentary rocks may be subdivided into groups: (1) the clastic (Greek: "broken" or "fragmental") group, which show rounding or other evidence of transportation before deposition, and (2) the chemical and organic group. Some rocks are a combination of the two.

The Clastic Sedimentary Rocks

This group is subdivided into rock types chiefly on the basis of particle size (Table 3–6). The major units are: *conglomerate, sandstone,* and *shale* which are compacted and cemented gravel, sand, and mud-clay respectively. Since mud (silt and clay) is the major sediment added by large rivers to the oceans, marine shales are by far the most common type of sedimentary rock. In fact, shale, sandstone, and limestone together may comprise about 99% of all sedimentary rocks. Thus they are the principal rocks that cap the lithosphere and that we see exposed at the surface. Mineral matter precipitated as a cement around the clastic particles is commonly calcium carbonate, silica, and iron oxide, and all gradations exist between loose sediments and firmly consolidated sedimentary rocks. Compaction tends to reduce the original thickness of a sedimentary layer, especially the smaller sizes (e.g., a section of shale may be only half as thick as the clay from which it formed).

TABLE 3-6
THE CLASTIC SEDIMENTARY ROCKS

SEDIMENTS	CONSOLIDATED ROCK	AVERAGE DIAMETERS IN MILLIMETERS (25.4 mm = 1 inch)
Gravel	Conglomerate	Greater than 2 mm
Sand	Sandstone	2 mm to 1/16 mm
Silt	Siltstone	1/16 mm to 1/156 mm
Clay	Shale	Less than 1/256 mm

The diameter of an irregularly shaped particle can be considered as equal to the diameter of a sphere of equal volume. In many shales, about half or more of the particles are of silt size.

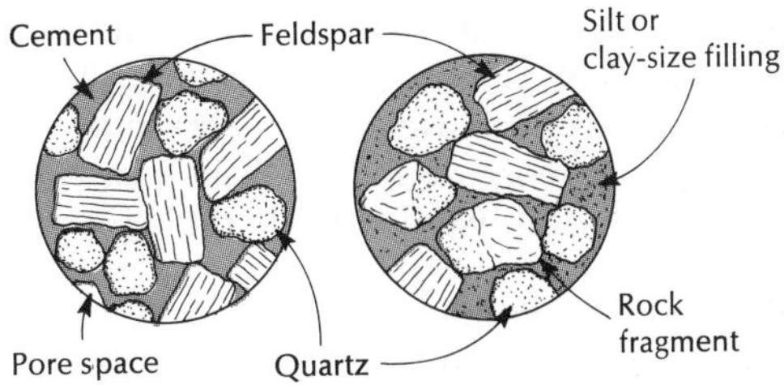

3-15. Magnified views of two sandstones. *(B. Mears, Jr.,* The Changing Earth, *Van Nostrand Reinhold, 1970.)*

If the gravel-size fragments are angular rather than rounded—probably because they were transported a relatively short distance—the rock is called a *sedimentary breccia* (brĕchĭ-á). Other kinds also form: e.g., volcanic breccia and fault breccia. Most sandstones consist chiefly of quartz, although some contain considerable feldspar and are called *arkose* (Fig. 3–15). Another type of sandstone is *graywacke* which commonly contains angular rock fragments and clay as well as quartz and feldspar and tends to be rather dark in color. We might describe it as a "dirty" sandstone.

Mixtures of different sizes are comon and may be described by terms such as sandy shale, conglomeratic sandstone, and shaly limestone. A sandy shale is chiefly shale, whereas a shaly sandstone is primarily sandstone.

The particle sizes of sediments tend to grade from coarse to fine away from a source area (Fig. 3–16). Evidently the source ar•a for the sediments in this illustration was some distance away because the particles

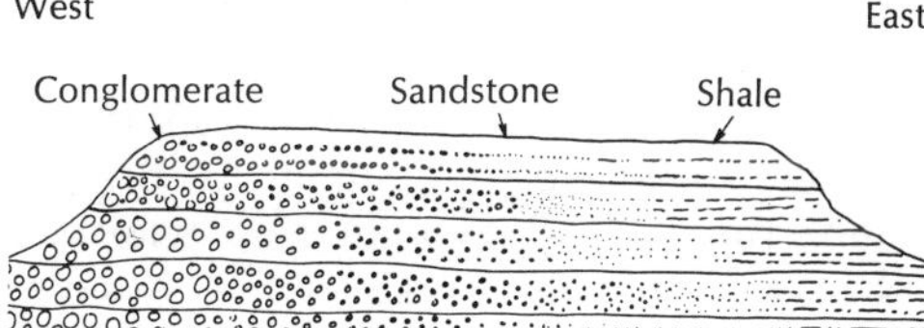

3-16. Size gradation is from coarse to fine away from a source area (thicknesses of individual beds is greatly exaggerated). One type of rock grades gradually into the next across the area.

in the conglomerate and sandstone are well rounded and almost entirely quartz. Thus larger pieces of feldspar and other minerals —originally more abundant than quartz in the source area—were reduced during transportation to finer particles, and these were then carried farther eastward to produce the shale. The size of the durable quartz pebbles in the conglomerate implies a moderately high land mass such as could give streams flowing down its slopes sufficient speed to transport the pebbles. We assume that the source area has subsequently been destroyed by erosion, and that the sedimentary rocks in the sketch are now the only evidence of its former existence.

Chemical and Organic Sedimentary Rocks

Most of these rocks have formed from sediments that were precipitated from saturated solutions or that accumulated as a result of the actions of living organisms; in contrast to the clastic group, they have undergone little transportation after deposition (however, prior to this, material may have been carried a long distance in solution). Some, such as coal, represent accumulations of organic matter.

Dissolved substances are brought to the oceans each year by rivers—chiefly calcium carbonate, silica, and sodium chloride. If water containing such dissolved salts is evaporated, the least soluble salt is precipitated first and the most soluble salt last. Such precipitated grains are commonly small and have an interlocking texture similar to that shown by most igneous rocks. Many valuable deposits have been made in this manner. When the volume of sea water is reduced about one-half by evaporation, calcium carbonate (limestone) is precipitated. The mineral gypsum is precipitated when the volume is reduced still more; thus it may accumulate in layers to form the sedimentary rock also called gypsum. Sodium chloride is finally precipitated when the volume is about one-tenth of the original amount.

Limestones

Limestones are made almost entirely of calcite (calcium carbonate—$CaCO_3$) and may be clastic, nonclastic, or a combination of the two. The source of the calcite may be shell fragments or material precipitated from solution by chemical and/or organic processes. Calcium carbonate remains in solution only if dissolved carbon dioxide is present. Therefore, deposition is caused by the life functions of plants and animals which withdraw carbon dioxide from water.

Many limestones are so fine grained that individual particles cannot be seen with the unaided eye, and such rocks in small specimens look somewhat like basalt, felsite, and certain types of quartz. However, other limestones are composed of larger calcite grains and look somewhat like marble. Limestones are scratched easily by a knife, do not scratch a glass plate if pure, and effervesce readily in cold dilute hydrochloric acid (calcite cement is common in some clastic sedimentary rocks, and these also effervesce somewhat in acid).

Limestones may be almost any color, but grays are common. Clastic material such as clay may be washed into a basin where calcite is accumulating and results in the formation of an impure limestone. Thus all gradations are possible from pure limestones, through impure limestones, into calcareous shales, to shale. Limestones do not break into flat slabs as readily as shale.

Many limestones contain irregularly shaped nodules and lenses of a very fine-grained type of quartz (chalcedony) or of tiny siliceous shells. These are commonly aligned in certain beds or occur as continuous or discontinuous layers. Chert also occurs in more extensive beds not directly associated with limestone.

Dolomite is a sedimentary rock similar to limestone but is composed of the mineral dolomite (calcium magnesium carbonate) instead of calcite. It does not effervesce in cold dilute hydrochloric acid, but if scratched by a knife, its powder does fizz. Some geologists refer to this rock as dolostone.

3-17. These light-colored, flat-lying sedimentary rocks (Beacon Sandstone) in South Victoria Land, Antarctica have been intruded by a dark-colored igneous rock similar to basalt and gabbro. The magma squeezed in parallel to the sandstone layers to form a sill (bottom) but cut across them elsewhere (slopes from top downward to the right). *(Courtesy Warren Hamilton, U.S. Geological Survey.)*

3-18. Bryce Canyon in southern Utah. Horizontal sedimentary rocks are cut by systems of vertical joints which separate the strata into columns. Erosion has worn away the edges of the columns more rapidly than the sides and produced the rounded shapes we see today.

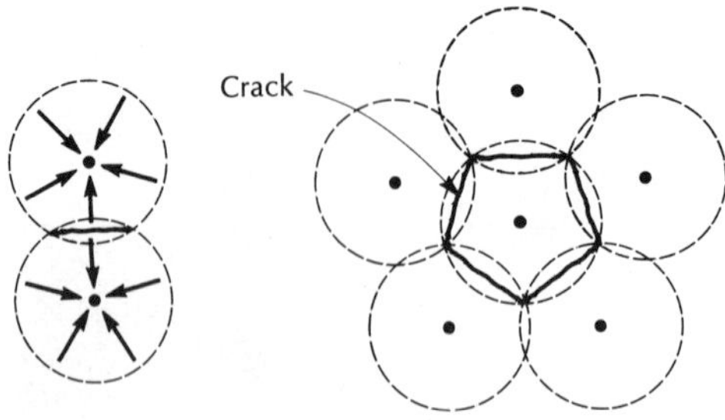

3-19. Origin of shrinkage cracks. In this view, shrinking is toward centers of cooling that are more or less uniformly spaced. (B. Mears, Jr., *The Changing Earth,* Van Nostrand Reinhold, 1970.)

Properties of Sedimentary Rocks

Sedimentary rocks are composed of layers (Figs. 3–17 and 3–18) which differ from each other in color, texture, or composition, or in all of these. Layered rocks break most readily along bedding surfaces and thus tend to form slablike pieces. Change is the keynote of stratification, one of the two outstanding properties of sedimentary rocks (presence of fossils is the other).

The colors of sedimentary rocks vary widely, and the color of an outer weathered surface should be distinguished carefully from the color of a fresh surface—the true color of the rock. Carbonaceous matter gives rocks dark colors, and iron oxides tend to color rocks yellowish brown or red, but only a small percentage of the coloring agent is necessary; e.g., a red sandstone may contain as little as 1% of finely divided hematite.

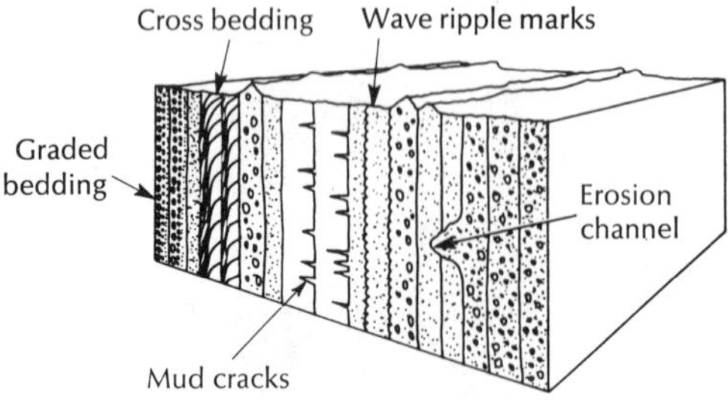

3-20. Some methods of determining the top from the bottom of a steeply dipping layer: fossils, graded bedding, cross-bedding, mud cracks, wave ripple marks, and erosion channels. Another method involves the matching of a sequence of beds at one site with undisturbed layers some distance away. Graded bedding can be illustrated by tossing a number of handfuls of mixed sand, silt, and gravel into a deep pool of water. The coarser pieces commonly reach the bottom first.

Almost everyone has seen mud cracks formed by the drying and shrinking of sediment as a puddle of water evaporates (Fig. 3–19). Similar cracks have formed many times in the past, and raindrop prints are sometimes associated with them. Very large mud cracks may reach a width of several inches across the top and attain a depth ten times greater. Sand and silt blown by the wind or transported by water may fill such cracks before the next layer of sediment is deposited, thus preserving them. Because mud cracks are wedge-shaped in cross section, the point of the wedge toward the bottom, they are useful in telling tops from bottoms in steeply dipping sedimentary rocks (Fig. 3–20). They also indicate that the enclosing sedimentary rocks formed in shallow water. Ripple marks and animal trails may form in a similar manner.

Some sedimentary rocks have a conspicuous minor layering oriented at an angle to the main trend of the beds, and this feature is called *cross-bedding* (Fig. 10–10). It may form in the bed of a stream where a depression is being filled rapidly by deposition from upstream. Although the main volume of sand dropped by a stream in its channel has a horizontal trend, the surfaces along which the sand is deposited are commonly tilted. Cross-beds tend to be concave upward, and these minor layers generally thin out gradually toward the bottom and stop abruptly at the top. Erosion by a stream (or by wind if in a sand dune) commonly removes the top part of the cross-beds before overlying sediments accumulate. Cross-bedding is very well developed in some dune sands and in deltaic deposits made by rivers, but it may form wherever sediments are deposited rapidly over fairly steep slopes.

Cross-bedding may be measured in fractions of an inch or in tens of feet (Fig. 7–32).

Metamorphic Rocks

Through time, living organisms become adapted to their environments. Analogously, rocks also become adapted to their environments, although this is less well known. During this adaptation, which is a slow and sometimes incomplete process, older rocks are changed into new, notably different kinds of rocks; i.e., they are metamorphosed until they attain a more or less stable form capable of resisting further change under existing conditions.

Changes Produced by Metamorphism

Metamorphic transformations involve the development of new textures, new structures, new minerals, or any combination of these. The original characteristics of a rock may be destroyed completely by metamorphism while the bulk of the rock materials remains solid at all times.

Great heat and pressure may cause the atoms present in a rock to be regrouped into new minerals, and commonly these are dense, space-saving types which are stable in the new environment. Certain minerals are found only in intensely metamorphosed rocks, whereas others are limited to mildly metamorphosed rocks. Such minerals constitute valuable "geologic thermometers" which measure the degree of the metamorphic changes. However, some materials may be transformed much more readily than others.

Agents Causing Transformations

Heat has a major role in metamorphism because a rise in temperature makes most chemical changes proceed at a faster rate. Furthermore, certain chemical reactions can occur at high temperatures which are impossible at lower ones.

Pressure is another factor in metamorphism. Powerful forces may crush, pulverize, and break rocks by squeezing and twisting them. In addition, great pressures applied slowly may deform rocks *plastically;* i.e., a slow permanent change in shape is produced without rupture.

Solutions aid in the formation of new minerals by adding and removing various substances or by facilitating the recrystallization and rearrangement of mineral materials already present. Many chemical reactions are slow or impossible between dry substances but take place rapidly when a little water is added.

Foliation

Foliation (Latin: "leaf formation") or rock cleavage is a new structure produced during metamorphism and refers to the capacity of some metamorphic rocks to split into thin slabs along closely spaced, nearly parallel, relatively smooth surfaces (Figs. 3–21 and 3–22). Metamorphic rocks may be subdivided into two groups by the presence or absence of foliation. Slate, schist, and gneiss are common examples of foliated metamorphic rocks; hornfels, marble, and quartzite represent nonfoliated types. Some of the factors involved in the development of a slaty cleavage (foliation) are discussed below.

Sedimentary rocks such as shale are not in fact foliated, though they break along bedding surfaces to produce thin slabs. A bedding surface should be distinguished from a foliated surface, although this is sometimes difficult. When shale is squeezed by great compressive forces during one kind of metamorphism, the layers are bent and folded. During this folding, existing mineral grains are granulated, recrystallized, and reoriented, and new platy or elongated minerals may develop. Solution tends to be most intense at points or surfaces of greatest pressure, whereas precipitation occurs where pressure is least. Most of the resulting flaky mineral particles are aligned with their flat surfaces parallel to each other and perpendicular to the axis of greatest compression.

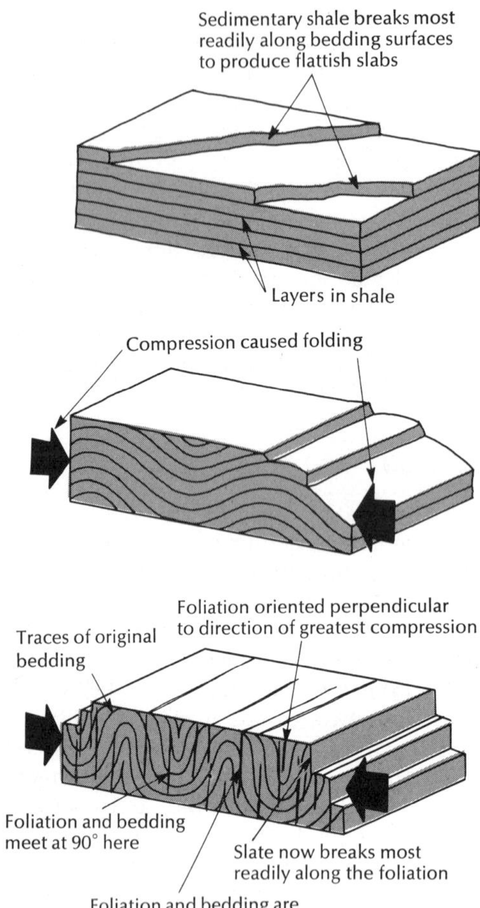

3-21. A shale is metamorphosed into slate. *Middle,* after folding and erosion, the rock is still shale. *Bottom,* after more intense folding, metamorphism, and erosion, the rock is now slate.

Imagine the microscopic flaky minerals that cause this foliation to be playing cards. The flat sides of the cards are aligned parallel with the foliation and would be seen edgewise when viewed at a 90-degree angle to the foliation. The resulting metamorphic rock is a slate, and its excellent foliation results from the combined cleavages of its tiny flaky constituents.

All traces of the original bedding have been destroyed in some slates, but bedding may be preserved in others. Bedding and foliation may be parallel or intersect at any angle because the original bedding tends to cut across the crest or trough of a fold at a 90-degree angle, whereas it nearly parallels the bedding along the sides of a fold. Gradations between shale and slate are known in nature, and such partially metamorphosed rocks provide valuable clues concerning the manner in which slates and other metamorphic rocks originate.

When shale is metamorphosed by heat and/or solutions in the absence of compressive forces (greater pressure in one direction than in others), parallel orientation of platy minerals is lacking and the resulting rock has no cleavage. This rock is called a hornfels, and it may form from other rocks than shale.

Foliation cannot develop unless platy or elongated minerals are present or can form. Thus sandstones (chiefly quartz, no cleavage) and limestones (calcite, cleavage in three directions) change to quartzites and marbles which are nonfoliated unless they contain impurities such as clay.

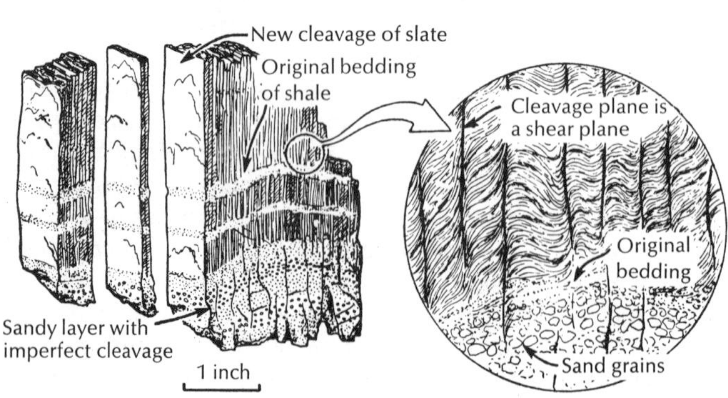

3-22A. Slate vs. shale. Relics of original bedding are shown in the slate. Before metamorphism the rock would have broken into flat pieces parallel to the sedimentary layers. After metamorphism, the rock breaks parallel to the foliation (rock cleavage). The sandy layer has imperfect cleavage because it lacks flaky minerals. (Gilluly, Waters, and Woodford, *Principles of Geology,* W. H. Freeman and Co., 1968.)

3-22B. Bedding (nearly vertical) and slaty cleavage (slants gently to right) in Newport County, Rhode Island. *(Courtesy J. B. Woodworth, U.S. Geological Survey.)*

Some Common Metamorphic Rocks

Slate is a fine-grained, well foliated rock which can be split into thin smooth slabs for blackboards, roofing, and a number of other purposes. The distinction between shale and slate is sometimes difficult. Traces of the original sedimentary layers oriented at an angle to the foliation may be found in the slate and make identification certain. However, bedding and foliation may be parallel, or no traces of the original stratification may remain. In these cases the following three criteria are helpful, although exceptions are known for each (especially if foliation is imperfect): (1) slates tend to break with flat surfaces, whereas many shales split along gently curved surfaces; (2) slates may have a higher luster than shales; (3) in addition, a slate may ring under the blow of a metallic object, whereas a shale may sound dull. Shale and slate may be black, gray, red, and other colors.

Schist is a coarse-grained, foliated metamorphic rock which splits easily into thin slabs—its grains are visible to the unaided eye. Schists may appear to be composed almost entirely of the micaceous or elongated minerals exposed along their wrinkled, foliated surfaces. However, observation of a specimen in cross section shows that other minerals such as feldspar and quartz are also present. A shale may be metamorphosed first to slate (low rank), and later the slate, by more intense metamorphism, may be changed to schist (high rank). Schists may be described by the minerals that are abundant in them and cause the foliation: e.g., mica schist, hornblende schist, and chlorite schist.

Gneiss (pronounced "nice") is a coarse-grained metamorphic rock which has poorly developed foliation and a banded, streaked appearance (Fig. 3–14). Gneisses and schists are high-rank metamorphic rocks which grade into each other, and they result from the metamorphism of a variety of other kinds of rocks—e.g., many gneisses were once granites. The streaked appearance of a gneiss depends upon the segregation of its materials into bands which commonly are crumpled and contorted.

Hornfels is a dense nonfoliated metamorphic rock which has resulted from the thorough recrystallization of older rocks, commonly around the margins of igneous intrusives. The compression necessary to produce parallel orientation of platy or elongated minerals was absent. Hornfelses may be fine grained or have a sugary texture visible to the unaided eye. They are hard and tough and vary in color. Some hornfelses break with a smooth, curved fracture and closely resemble rocks of other kinds (e.g., limestone, felsite, and basalt) and some varieties of quartz. Identification of certain specimens is not possible without a microscope. However, field relationships may furnish important clues.

Quartzite is a nonfoliated rock which is made of quartz grains that have been cemented firmly together; the rock breaks through, not around, the individual grains. Some quartzites have formed by metamorphism of a quartz-rich sandstone, but others are sedimentary rocks (see next paragraph). Recrystallization may take place and produce an interlocking texture. Quartzites commonly have smooth, glassy surfaces. In contrast, sandstones tend to break around the grains and have rough surfaces which may feel like sandpaper when rubbed. Both sandstone and quartzite readily scratch a glass plate.

Some quartzites are sedimentary rocks that formed by the deposition of a silica cement around the quartz grains in a sandstone and may occur as part of a group of superposed sedimentary rocks. Metamorphic quartzites (metaquartzites) cannot be distinguished from sedimentary quartzites (orthoquartzites) in hand specimens with the unaided eye, but associated rocks in the field should provide necessary clues.

Marble is commonly a nonfoliated metamorphic rock composed almost entirely of calcite (also of dolomite) and is produced by the metamorphism of limestone and similar rocks. Marble can be scratched easily with a knife and effervesces briskly in cold, dilute hydrochloric acid (dolomitic marble does not). Marbles differ from most limestones in being visibly crystalline. Metamorphism of a limestone that contains a considerable amount of clay may result in the formation of a foliated marble; atoms in the clay minerals may recombine into flaky minerals such as mica (resembles part of a schist). Foliated quartzites may form in a similar manner. Marbles tend to sparkle when a specimen is rotated because light is reflected from cleavage surfaces on the calcite grains.

Distinctions Among Rocks

Sedimentary rocks characteristically are layered, and some specimens are fossilferous. Rounded particles are part of some clastic sedimentary rocks, and features such as mud cracks, ripple marks, and cross-bedding may be present. If a high percentage of quartz occurs in a specimen, or if calcite is present, it can be distinguished from an igneous rock (Fig. 3–23).

Igneous rocks commonly have distinctive sets of minerals with characteristic interlocking textures. Marbles and quartzites may likewise have interlocking textures but have distinctive compositions. Visibly crystalline marble can generally be distinguished from dense limestone on the basis of grain size. However, some limestones are coarse grained and may be indistinguishable in hand specimens from marble unless undeformed fossils are present.

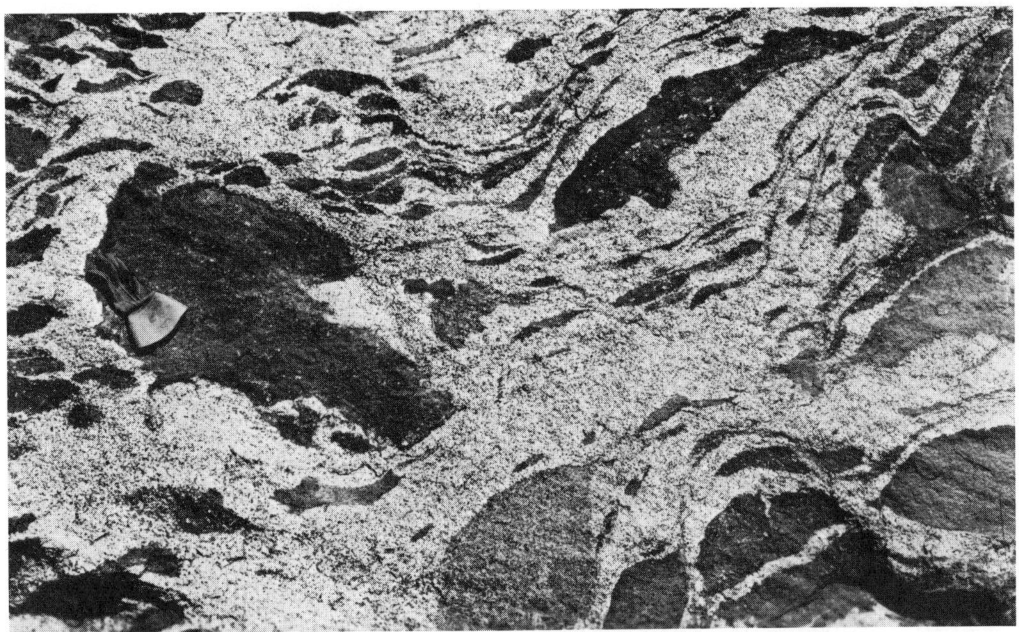

3-23. Inclusions of diorite (dark) in granite in Yosemite National Park, California. *(Courtesy F. C. Calkins, U.S. Geological Survey.)*

Foliated metamorphic rocks may generally be distinguished from igneous rocks on the basis of foliation and from layered sedimentary rocks because this foliation depends upon the parallel orientation of platy or elongated minerals and not upon an alternation of various layers or beds. Hornfelses take more skill in recognizing. Symbols generally used for the more common types of rocks are shown in Table 3–7.

Exercises and Questions for Chapter 3

1. Describe how the members of each of the following pairs differ one from another:
   (A) metallic bonding vs. covalent bonding
   (B) crystalline vs. noncrystalline substances
   (C) plagioclase vs. potassium feldspar
   (D) calcite vs. halite
   (E) shale vs. slate
   (F) foliated vs. nonfoliated metamorphic rocks
   (G) pegmatite vs. obsidian
   (H) porphyritic vs. equigranular igneous rocks
2. How can the texture of an igneous rock help in determining:
   (A) whether the rock is intrusive or extrusive?
   (B) the order in which the mineral grains crystallized?

TABLE 3-7
SYMBOLS USED FOR SOME COMMON ROCKS AND REGOLITH. (U. S. GEOLOGICAL SURVEY.)

REGOLITH

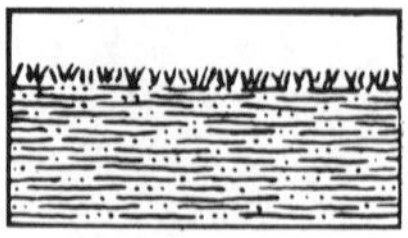

Soil, etc.

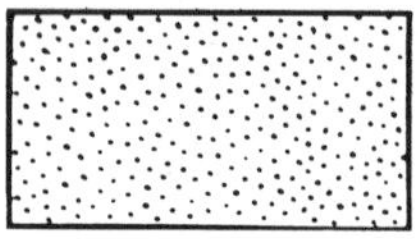

Sand

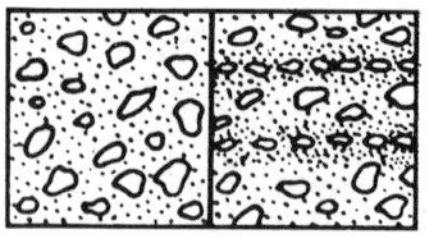

Gravel, etc.

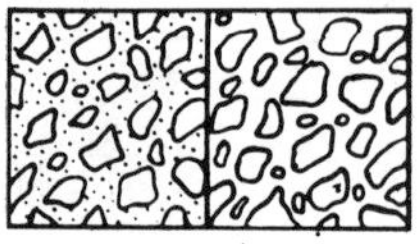

Glacial till

SEDIMENTARY

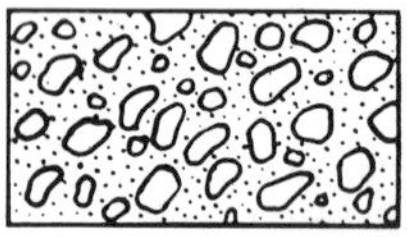

Conglomerate

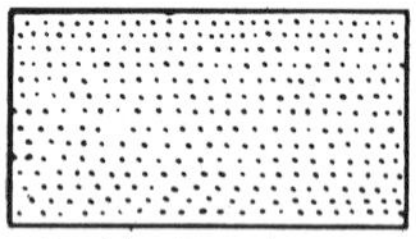

Bedded sandstone

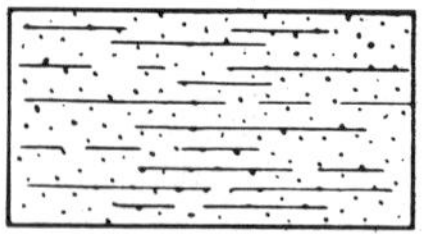

Thin-bedded or shaly sandstone

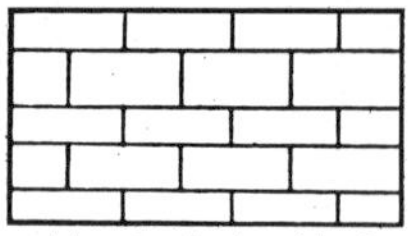

Massively bedded limestone

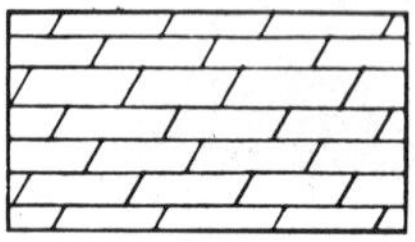

Dolomite

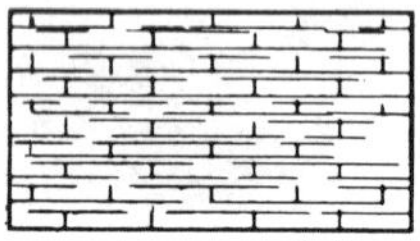

Calcareous shale or shaly limestone

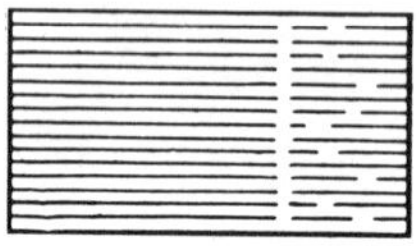

Shale

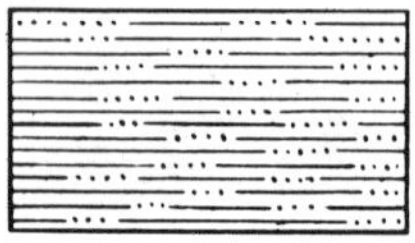

Sandy shale

METAMORPHIC

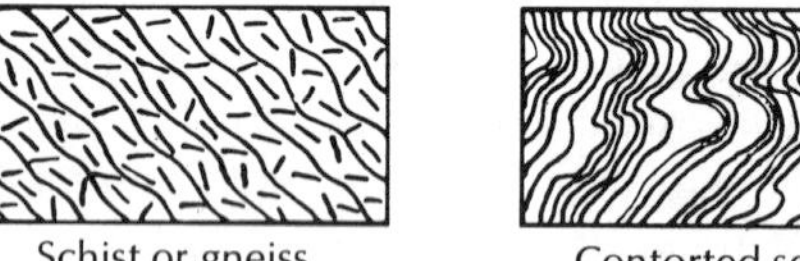

Schist or gneiss

Contorted schist

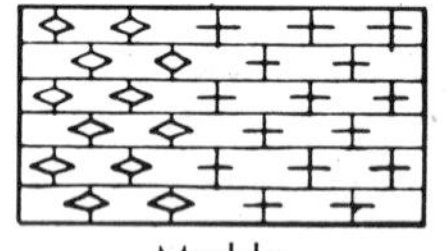

Marble

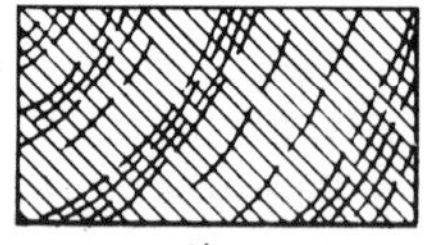

Slate

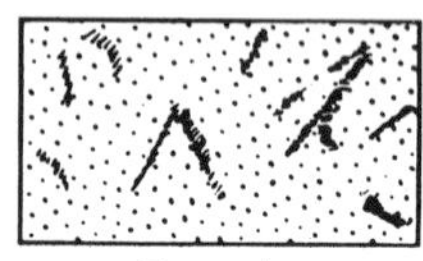

Quartzite

IGNEOUS

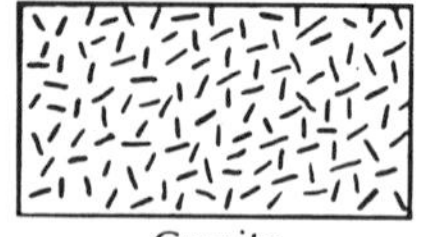

Granite

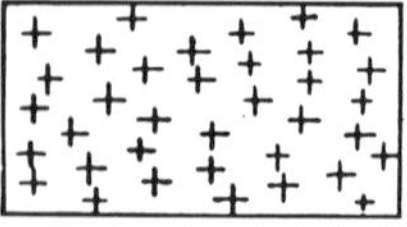

Massive igneous rock

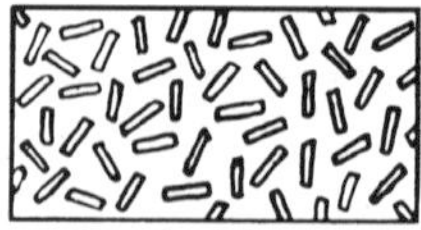

Porphyritic rock

Basaltic flow

3. Describe criteria that can be used to determine the original order of superposition in sedimentary beds that now stand vertically.
4. What minerals and textures would identify each of the following: granite, rhyolite, syenite, gabbro, diorite, and porphyritic granite?
5. Assume that the rocks in an area consist chiefly of two types: granite occurs in the eastern part and limestone occurs in the western part. In the region where the limestone and granite come into contact, different age and structural relationships are possible. Describe three possible kinds of relationships and the criteria that might be used to distinguish one from another.

4

Crustal Movements and Geologic Time

Geologic time is so long, and crustal movements have been so extensive, that we find difficulty in grasping their immensity. Yet such great magnitudes are inherent in a science that has as a subject the whole Earth throughout all its existence. A developing awareness of the breadth and depth of geology is the aim of this chapter, and such awareness can come best from a field trip. In fact, one can not truly understand geology without field trips. Moreover, one's travels across or above the Earth's surface will have a new significance once a geological perspective has been achieved, even in rudimentary form. Of all the field trips on which a geologist would delight to take his students, none perhaps ranks higher than one to the Grand Canyon (Figs. 4–1 and 4–2). So let us begin this journey now, in imagination for the moment, but with a firm resolve that one day we too will see this majestic spectacle—the opened book of the geological ages.

The Grand Canyon

In describing the geological story of the Grand Canyon, some local details are necessary. However, it is suggested that these be de-emphasized, at least at this time. Rather, let us concentrate on the general principles and ideas that grow out of this story and have wide application. For example, let us search for evidence that: (a) mountains once existed in an area where none now occur, (b) the Earth's surface in a region was at one time lifted 1½ miles (e.g., marine fossils may occur at 1½ miles above sea level today), whereas at another time the surface was depressed this much and more, (c) an unbroken plateau existed where a great valley now occurs, and (d) conditions in a region change with time—a sea floor, by uplift and climatic change gradually becomes a desert; a mountain is slowly worn to a plain; organisms of one type slowly develop into those of another type.

A story that extends for some time requires a cal-

4-1. The Grand Canyon and the Colorado River. The unconformity between horizontal Paleozoic sedimentary rocks and Precambrian igneous and meamorphic rocks is located near the top of the inner gorge (about 1000 feet deep) and is identified by dashes along one portion. Before deposition of the Paleozoic sedimentary rocks, erosion had removed a thickness of about two miles of Precambrian sedimentary rocks as well as an unknown thickness of Precambrian igneous and metamorphic rocks. The Redwall limestone is a conspicuous, cliff-making part of the Paleozoic rocks. *(Courtesy George Grant, U.S. Geological Survey.)*

endar, and geology has such a calendar (p. 106). Here we refer only to its four main divisions or eras: Precambrian, Paleozoic, Mesozoic, and Cenozoic. Probably these began about 3500, 600, 230, and 65 or so million years ago respectively.

We can start with the dimensions of the Grand Canyon—of the order of 200 miles in length, nearly 1 mile in depth, and 4 to 18 miles in width from rim to rim—but the spectacle from the brink of the north or south rim is one of the wonders of the world, and the feelings that overwhelm a visitor have never found adequate expression in words. Many miss completely the significance of this enormous valley and the layered rocks which make cliffs and slopes of red, white, and gray along its walls. Others understand it partially—e.g., the farmer from a soil-eroded farm in the semiarid West who exclaimed: "Golly, what a gully!" Yet the carving of the canyon by running water and other erosive agents is only the most recent of a sequence of geologic events that the trained eye can see clearly in the canyon walls.

The story begins more than a billion years ago with the origin of the Precambrian igneous and metamorphic rocks now exposed in the inner gorge (Fig. 4–3). Since some of these ancient metamorphic rocks seem once to have been sedimentary, the history of the Grand Canyon begins after geologic processes had been in operation for some time, because such sedimentary rocks are made of fragments of still older rocks.

Similar igneous and metamorphic rocks are found today only in mountainous regions or in areas that geologists think once were mountainous. Thus their presence implies a range of mountains in this area long before the origin of the Colorado Plateau and Colorado River. Although erosion has subsequently removed all but the stumps of these mountains, their former existence—but not, of course, their precise shapes or altitudes—seems as certain as does that of the forest which the imagination can so easily reconstruct for a freshly lumbered hillside.

The upper parts of these Precambrian rocks are everywhere beveled to a remarkably flat surface (Fig. 4–1), apparently the end product of a very long period of erosion during which the mountains were worn down to an approximately level surface near sea level (a peneplain).

The tilted layers of ancient sedimentary rocks (younger Precambrian) are next in the sequence because they rest unconformably upon the still older igneous and metamorphic rocks. They accumulated when the area gradually subsided sometime after peneplanation. More than 12,000 feet of sedi-

In the diagram above the vertical scale is exaggerated ten times.

4-2. The Grand Canyon region.

ments were eventually deposited upon this slowly sinking, nearly horizontal surface (Fig. 4–3c) and later were compacted and cemented into sedimentary rocks. The ultimate cause of the subsidence, of subsequent major crustal movements in this area, and of similar movements in other parts of the Earth is unknown. We assume that the strata were deposited originally in a nearly horizontal position because sediments pile up in that manner today.

Later, this 2-mile thickness of sedimentary rocks was broken into separate blocks along great fractures (faults) in the Earth's crust. One side of each block was pushed upward a great distance to form a mountain, whereas the other side was tilted downward (Fig. 4–3d). This second generation of mountains in the Grand Canyon area was then destroyed by erosion during another, almost inconceivably long stretch of time (Fig. 4–3e). The uptilted, cut-off, younger Precambrian sedimentary rocks are evidence for these steps.

Next in the sequence of events comes the formation of the horizontal Paleozoic strata, about 4000 feet thick, which today rest unconformably upon the older rocks below. Subsequently many additional layers of sediments accumulated during the Mesozoic and Cenozoic Eras, and they are exposed today in the spectacular scenery of Zion and Bryce Canyons. In a number of places, the horizontal Paleozoic rocks rest directly upon eroded Precambrian igneous and metamorphic rocks (Fig. 4–1). In such places, all of the events shown in steps C, D, and E of Fig. 4–3 have been wiped completely from the geologic record: the deposition of a 2-mile pile of sediments, the faulting that

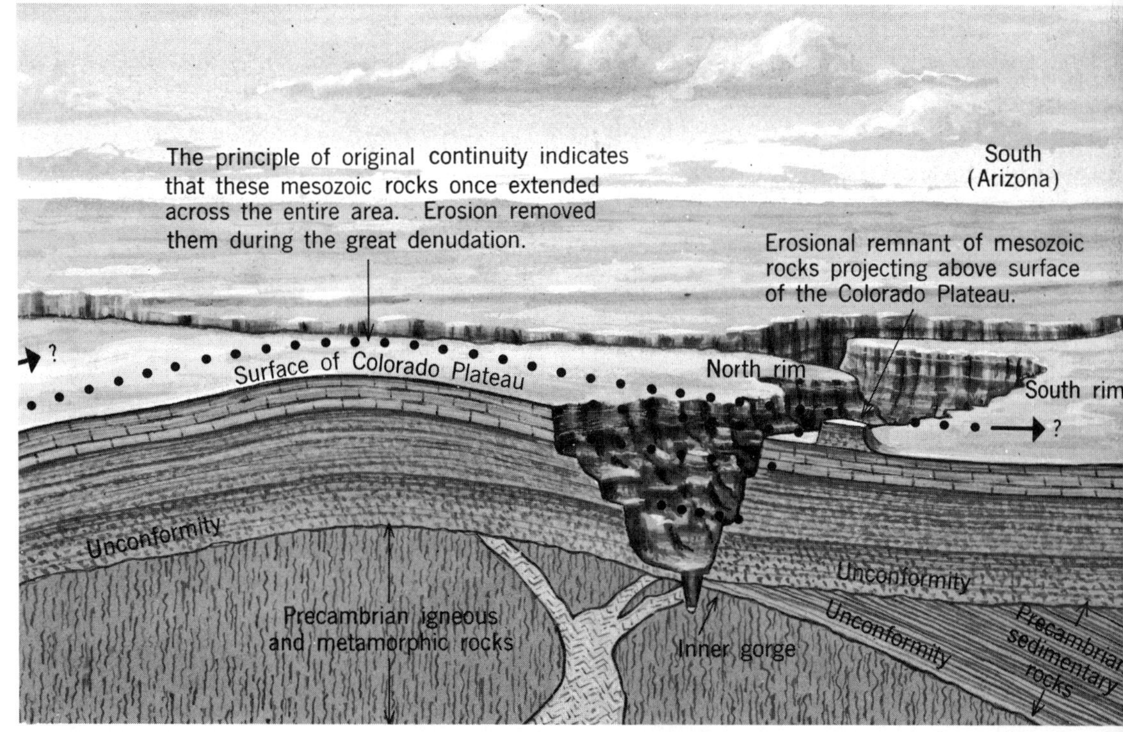

The silhouette below is correctly proportioned.

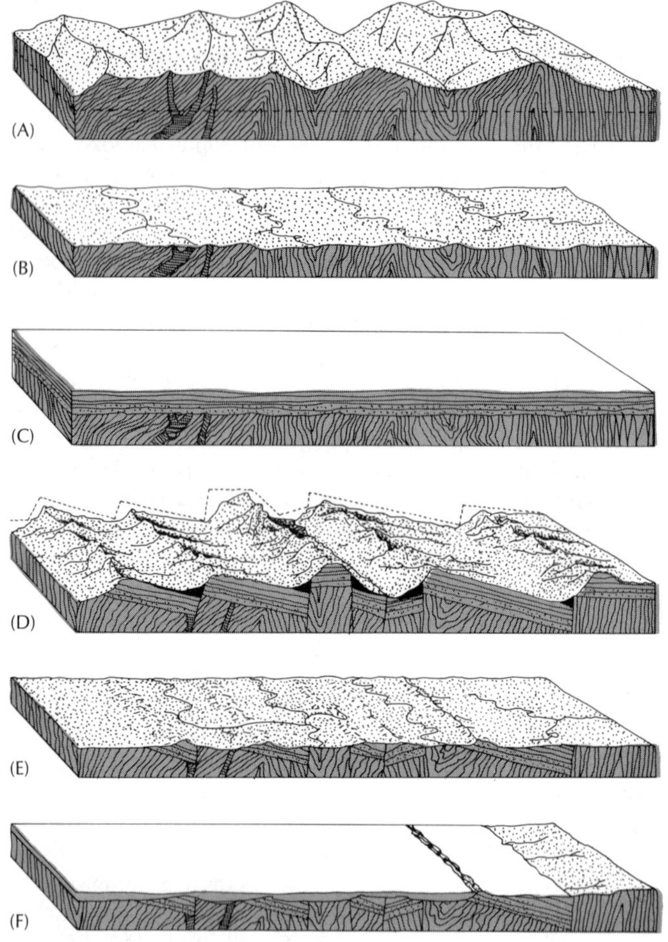

4-3. Five stages in the Precambrian history of the Grand Canyon region. The view is northward, and the sections represent an east-west distance of about fifteen miles. The solid black shading in block D represents alluvium. (A) Precambrian rocks are formed by sedimentation, folding, metamorphism, and igneous activity. Mountains result. (B) Peneplanation leaves only stumps of the mountains. (C) Precambrian sedimentary rocks are formed. (D) Mountains are produced by faulting. (E) Peneplanation occurs and again only the stumps of the mountains remain. In fact, in some parts of the Grand Canyon region, all of the Precambrian sedimentary rocks have been eroded away. (F) Sedimentation is resumed at the start of the Paleozoic. (C. O. Dunbar, *Historical Geology*, John Wiley & Sons, Inc., 1949.)

produced mountains, and the subsequent erosion that wore away these mountains. Impressed geologists have called this a profound unconformity.

A number of the Paleozoic beds contain abundant fossils of marine origin; other layers include fossils of land organisms; still others—the white sandstones exposed in the 300-foot cliff near the top of the canyon—resemble the wind-blown sands which accumulate in dunes in desert areas today. Apparently, therefore, the Grand Canyon area was covered by the sea during several long intervals within the period of time represented by the formation of these sedimentary rocks. Dry land occurred at other times when the sea receded from the area, although its altitude seems always to have been rather low, and once the region may have been a desert.

The present-day topographic features of the Grand Canyon region were produced by erosion and uplift during the Cenozoic Era. Since some Mesozoic rocks now rest as erosional remnants on the Colorado Plateau, the entire region must have been formerly blanketed by them, and perhaps by some of the Cenozoic rocks also (this follows the principle of original continuity). During a time picturesquely named the Great Denudation these strata, hundreds of feet in thickness, were stripped from above the Paleozoic rocks that now form the surface of the Colorado Plateau. As erosion wore away the rocks toward the north, the edges of resistant beds formed cliffs and the steplike topography shown in the Pink, White, and Vermilion Cliffs today (Fig. 4–2). The downward tilt of the sedimentary strata toward the north also explains why one encounters successively younger rocks in each cliff without much overall increase in altitude.

Some details of the subsequent canyon cutting are yet to be agreed upon. According to an older account, the river now called the Colorado began to flow across this region when the surface was still at a low altitude. As the region was gradually uplifted a mile or more to form the present dome-shaped plateau, the river flowed more swiftly. Like a saw cutting through a board, the river eroded downward in its channel through the central portions of the rising dome. As evidence, the river now flows from lower ground in the north, through the elevated plateau region. into lower ground to the southwest. If the plateau had been in existence first, presumably the river would have flowed around and not across it. Down-cutting by the river and its tributaries enabled weathering and various types of mass-wasting (Fig. 7–23) to widen the canyon by moving disintegrated, decayed rock debris down the canyon walls to the channel where the river functions as a conveyor belt to transport it toward the sea.

However, stream capture (p. 186) may have played an important role. In a newer view, two independent stream systems once existed, and one of these slowly cut its way headward (upstream) across the plateau dome, much as the location of a waterfall may retreat upstream. This stream eventually intersected and captured the water of the other stream system and together they formed the present Colorado River which continued to carve the canyon deeper.

In contemplating the scene, the Colorado River may look very small and the Grand Canyon itself very large—as they do. This emphasizes anew the very long time intervals needed for completion of such a sequence of events. To indicate how this might occur, we note that the Colorado River now transports nearly a million tons of sediment through the canyon every 24 hours, and that in 3½ centuries, this would approximate about 1 cubic mile of material. Of course, this debris would come from the entire drainage area of the Colorado River, not from just the canyon itself. In a million years, this rate would amount to nearly 3000 cubic miles, and on such a scale the origin of the canyon by the slow-acting processes of today seems within reach.

Many of the Paleozoic, Mesozoic, and Cenozoic strata bear fossils; e.g., trilobites (extinct members of the phylum Arthropoda, p. 348) and certain kinds of primitive fish and plants are found only in the Paleozoic rocks, whereas dinosaur fossils are confined to Mesozoic beds, and certain mammals occur only in the rocks of the Cenozoic. Thus the rocks of the Grand Canyon region also furnish a thick picture book of the life of the past.

Instability of the Earth's Crust

Let us now marshal evidence showing that crustal movements of different kinds and magnitudes have occurred—large, small, quick, slow, up, down, and sideways (see Chapter 6). Mean sea level is a satisfactory reference surface for measuring vertical movements because the oceans are interconnected, and mean tide is nearly level throughout the world. However, sea level is not a fixed measuring surface. Continents seem to drift, ocean basins may widen and then narrow, and the total volume of water in the oceans has perhaps increased with time. Moreover, vertical movements of the ocean floors and the addition of sediments and lavas to the ocean basins likewise cause sea level to change, and tides, winds, and fluctuations in atmospheric pressure also produce distortions in the ocean surface. However, such changes are relatively small, very slow, or nearly uniform throughout the ocean-covered areas of the Earth. With reservations, therefore, we can refer vertical movements of the land to sea level as a standard.

Earthquakes can produce notable, but relatively small scale, topographic changes (Fig. 4–4). In 1899 in the Yakutat Bay region of southern Alaska, several great earthquakes resulted from the sudden lifting of part of the coast a vertical distance of nearly 50 feet. Simultaneously another part of the same coast moved downward. After an earthquake in Chile in 1882, coastal towns were located 3 to 4 feet higher above sea level than before the shock.

In the greatest quakes, the ground is reported to have moved up and down in waves, causing fissures to open and close in the soil and trees to tilt at various angles. Accounts of such fissures in these and other earthquakes are very commonly exaggerated because the terror, fright, and emotional excitement experienced during an earthquake are conducive to distortions. Fissures and cracks may be produced in surficial debris during the passage of earthquake waves, and others may develop in debris that is jarred loose from a nearby mountainside and slides violently down a slope. But the Earth's crust is not pulled apart to leave gaping chasms which later are closed.

A powerful earthquake shook the Hebgen Lake region of southwestern Montana at about 11:40 p.m. on August 17, 1959. Small cliffs, up to 20 feet high, were produced by the faulting, which represented renewed movement along old faults. The Earth's surface was tilted in the vicinity of Hebgen Lake; one shoreline was submerged about 10 feet, whereas the opposite shoreline emerged about 10 feet. However, the most disastrous effect of this earthquake was a gigantic landslide in the canyon of the Madison River (Fig. 7–13).

Slower, less conspicuous movements have also been observed by man. Abundant marine shells on land surfaces bordering the Baltic coast indicated that uplift was occurring, and marks were placed along the

4-4. Wreckage of Government Hill School, Anchorage, Alaska caused by the "Good Friday" earthquake (27 March 1964). Note undamaged water tower. Graben in foreground is about twelve feet deep. *(Courtesy W. R. Hansen, U.S. Geological Survey.)*

coast to measure this. Some stakes, once at mean tide level, are now several feet above sea level and more than a mile inland from the present shoreline. A maximum uplift of about 7 feet has occurred in northern Sweden in the last two centuries, but the movement is so slow and gradual that it is imperceptible to inhabitants, except for such measurements.

A classic example of movement of land with respect to sea level involves a building constructed about 2000 years ago by Romans along the seashore west of Naples. Three of its columns still stand, and each contains holes (within 18 feet of the base) bored by small marine clams; shells remain in some of the holes. How can these facts be interpreted?

The floor of the building was originally constructed above sea level and is above sea level now. Clam shells found in the holes belong to animals whose living relatives exist only in the ocean, and presumably such clams have been ocean-dwellers during the past 2000 years. Thus the land must once have been submerged until about 18 feet of sea water covered the bases of the columns. Afterward the land emerged again above sea level.

Sediments containing the shells and bones of marine animals were deposited in this area when the building was partly submerged and are exposed today. In fact, the three columns were discovered in the latter part of the eighteenth century and had to be excavated from sediments which had buried their bases. The sedimentary deposits, up to 20 feet in thickness, corroborate nicely the evidence from the columns.

Was this movement an actual sinking of the land followed by a later uplift, or did sea level rise and then fall? If sea level rose and fell through a vertical distance of nearly 20 feet within the last 2000 years, the evidence should be worldwide, because the oceans are linked together. A check readily reveals the local nature of the movement and indicates that the land itself moved.

Ancient docks built on part of the Isle of Crete in the Mediterranean about 2000 years ago are today as much as 27 feet above sea level, but structures elsewhere on Crete are below sea level today. Assume that the Cretan docks continue their upward movement at the same rate for a million years—not a long interval from the geological point of view. Their altitude would then be 13,500 feet above sea level. Thus small movements or forces, acting through a very long period of time, produce major results—an important concept in geology.

Rocks Show Evidence of Crustal Movements

Our strategy here involves the assumption that the present is the key to the past. To illustrate, along the seacoast today, we find certain topographic features associated closely with sea level: beaches, cliffs cut by the surf, and smooth, beveled rock surfaces (surf-eroded benches) sloping gently out to sea. Uneroded remnants (stacks) may project above the surface of a surf-eroded bench. After extensive traveling and observation, we conclude that certain topographic features form only along the coast line near sea level. Their marks on the lands are somewhat analogous to the dirt line on a tub showing the former depth of water (Junior's after a Saturday morning scrimmage on a muddy football field). However, such features may also be found today above sea level (Fig 4–5), which implies changes in the relative positions of land and sea.

Extensive travels over land and sea and continued field study—this time of shells and animal remains—leads to the conclusion that recognizable differences exist today between marine and nonmarine forms; i.e., certain types of animal life live only in the sea, whereas other types live only on land.

Assume now that we find unconsolidated beach sands 100 feet or more above sea level and several miles inland. These sands are littered with shells that seem identical to those of present-day marine organisms. We do not hesitate to conclude that the

4-5. Surf-eroded benches, recently uplifted, on San Clemente Island, California. A surf-eroded bench formed when the crust was relatively stable in this area, and an uplift then followed. This sequence has been repeated several times. *(R. S. Dietz,* Geol. Soc. Am. Bull., **74,** *Aug. 1963.)*

land was elevated after deposition of the sediments. We next make similar discoveries elsewhere at higher altitudes and farther inland, and we reach similar conclusions concerning their origin. Eventually we encounter firmly consolidated sedimentary rocks, containing abundant marine fossils, exposed on mountain slopes high above sea level (several miles in the Himalayas), and we must conclude that such areas were also once beneath the sea. Thus fossils in sedimentary rocks provide clues concerning the past distribution of land and sea and of crustal movements that may have raised or lowered such strata through great distances after their formation.

Where sediments such as sand or mud are deposited in extensive layers today, they tend to form beds which parallel the underlying surface and are nearly horizontal. Yet

4-6. Steeply dipping strata exposed in the wall of a cirque in Gallatin County, Montana (part of the Skyline anticline). *(Courtesy I. J. Witkind, U.S. Geological Survey.)*

some ancient sedimentary rocks, like sandstone or shale, now occur with their layers tilted at a steep angle to the horizontal, or perhaps even standing vertically (Fig 4–6). We conclude that such rocks have been tilted after deposition by powerful crustal movements.

The discovery of once-horizontal sedimentary strata which have been bent and folded is striking and convincing evidence of crustal movements. Furthermore, rock formations have been broken and offset along great fractures (faults) in the crust (Figs. 4–7 and 4–8). In addition to vertical movements, some folds and faults result from great compressive forces in the crust.

Other, less direct evidence also indicates that crustal movements have been widespread, long continued, and very extensive. If erosion has been taking place unendingly since the first rains fell, why have the continents not been worn away? Forces must exist which oppose those of erosion and act to increase and elevate the land areas. To illustrate, on the average the land surface in the United States is perhaps being lowered at present at a rate of about 2 to 3 inches per 1000 years (p. 170). In 20 million years, this would amount to about 4000 feet, which exceeds the average level of the land. However, the rate of erosion should decrease as slopes decrease, although just how much is difficult to estimate. Thus the overall picture seems clear. Some millions of years hence the United States would be a nearly featureless, low-lying plain—no Rockies, no Appalachians, no Colorado Plateau—if uplift did not offset erosion. Such rejuvenation, however, has occurred throughout known geologic history, and we confidently expect it to do so in the future.

4-7. A fault scarp along the south side of Macdonald Lake (near Great Slave Lake). Ancient Precambrian granite occurs on the left, and less ancient, more easily eroded, Precambrian sedimentary rocks occur on the right. The present-day cliff results from differential erosion along the line of the fault—not directly from crustal movement. It is 900 feet high in places, and the fault has been traced for a few hundred miles. *(Photograph, Royal Canadian Air Force; courtesy Department of Mines and Technical Surveys, Canada.)*

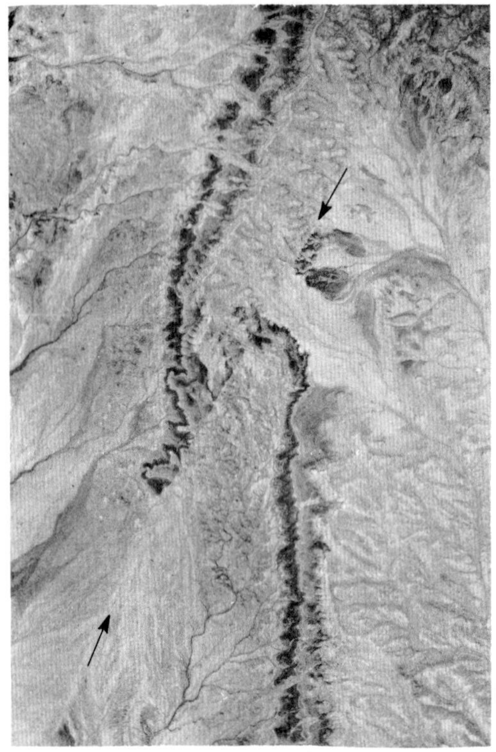

4-8. Gently dipping sedimentary rocks are offset by a near-vertical fault (arrows mark its trend) in Utah. The dark-toned layers are shale. The fault can be traced across the alluvium in the lower left. *(Fig. 83,* U.S. Geol. Surv. Profess. Paper *373.)*

Measurement of Geologic Time

Methods for calculating geologic time have certain features in common: (1) attention is fixed upon some change or process now occurring upon the Earth; (2) this change is assumed to have been going on uniformly since very early in the history of the Earth, or an average rate is calculated; (3) the total change is measured; and (4) the annual quantity of the change is determined. Dividing the total change by the annual change then gives the age of the Earth (Fig. 4–9).

Sodium played an important role in one, now-discarded attempt to measure the age of the Earth. Sodium is dissolved from rocks and minerals during weathering and is then transported to the sea by running water. Water is recirculated (water cycle, Fig. 8–2), but the dissolved sodium remains in solution because the oceans are not saturated with it. Thus the oceans are becoming saltier every hour because of the inflow of fresh water—fresh water, however, that contains dissolved mineral matter.

The sodium now in solution in the oceans may total about 16×10^{15} tons (the total volume of ocean water can be approximated, and its sodium content seems everywhere constant). About 16×10^{7} tons of sodium are thought to be added each year (estimated by determining the sodium content and annual volume of certain large rivers and extrapolating the results worldwide). After the data had been collected (about 1900) the total amount was divided by the annual increment and gave an age for the oceans of about 100 million years —much older than previously estimated (the Earth and oceans were assumed to have nearly the same ages).

But how reliable is this sodium-in-the-sea method? Certain assumptions have been necessary, and we name several: (1) the original ocean consisted of fresh water; (2) all of the sodium brought to the oceans has remained in solution; (3) the annual increment of sodium at present is a fair average for all of geologic time; and (4) the ocean basins and ocean water have had their present volumes since early in the Earth's history. These assumptions either cannot be verified or now appear incorrect; e.g., some sodium has not remained in the oceans, and the present annual rate of erosion is probably far above the average for the geologic past. Therefore, the present annual addition of sodium is probably much too high for an all-time average, but how much can only be guessed. It might be 2, 6, 10, 15, or more times too much. Thus the method has little quantitative value at present.

Sedimentation played a key role in another attempt to measure the age of the Earth. Sediments have been eroded, transported, and deposited since early in the

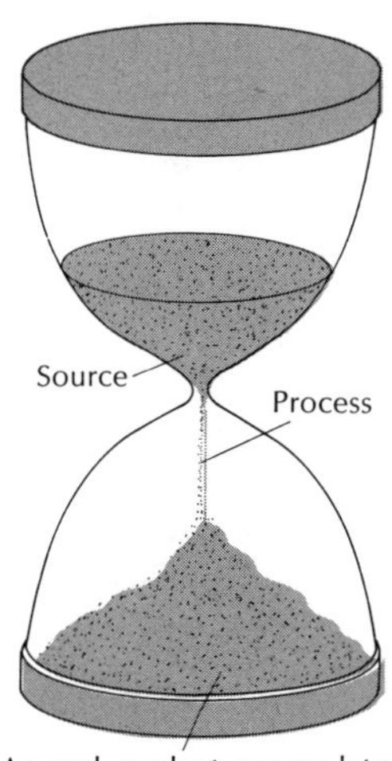

Source

Method 1: Sodium is a common element in many rocks

Method 2: Rocks of the Earth's crust are eroded to produce sediments

Method 3: Radioactive elements are unstable and disintegrate spontaneously into new elements

Process (annual rate is measured)

Method 1: Sodium is carried in solution by rivers and added to the oceans

Method 2: Sediments are transported from source areas to areas of deposition

Method 3: The end member of a radioactive series is stable and accumulates with time

An end product accumulates as time passes (total end product is measured)

Method 1: Total quantity of sodium now in the oceans is measured

Method 2: A total thickness of about 100 miles of sedimentary rocks has been correlated

} Rates of accumulation have varied and today are probably much higher than average

Method 3: Total quantity of nonradioactive end member of a radioactive decay series is measured

} Rate of disintegration appears to be uniform

4-9. Three methods of estimating the age of the Earth are illustrated schematically.

Earth's history, and these processes have been going on at different places, at different times, and at different rates. Older sedimentary rocks were subsequently eroded, but some remnants remained here and there, and geologists have attempted to correlate these scattered fragments into a composite geologic record (p. 54). To date, they have measured a total thickness of about 100 miles (160 km). Of course, this is a composite record; only a small fraction has accumulated at any one place, and many gaps such as unconformities still occur within it. Since it consists of shales, sandstones, limestones, and similar rocks, an average annual rate of deposition is difficult to determine, because a foot of sand may accumulate in one place in the same interval that 20 feet of gravel or 0.1 inch of calcareous material pile up elsewhere. Thus no accurate estimate of an all-time average rate seems possible. In addition, many of the oldest known rocks are metamorphic types which originally were sedimentary strata, and no accurate estimate is possible for their thickness. Therefore, the bulk of the 100-mile column is made up of sediments deposited since the Precambrian, a relatively small fraction of total geologic time. Again we lack a reliable quantitative result, but the technique has been of great importance in determining relative ages and in working out a geologic calendar (p. 106).

Geologic Chronology

Both *relative ages* (i.e., whether a certain rock, fossil, or event is younger or older than another) and *absolute ages* (the actual number of years since the origin of a particular rock, fossil, or event) are important in geology. In general, we can determine relative ages more readily, because they are based upon the principle of superposition and other relationships among rocks (p. 49). The 100-mile composite geologic column was originally worked out and subdivided on the basis of relative ages; however, this was done while the pile was much smaller.

The whole range of geologic history has been divided into four major units called *eras:* Precambrian, Paleozoic, Mesozoic, and Cenozoic. In turn, each era is composed of a number of smaller units known as *periods.* We can think of the eras as four volumes of a multivolume set, whereas the periods are like the several chapters which make up a single volume. Volumes and chapters vary greatly in length, and the number of chapters in a volume is far from uniform (Figs. 4–10 and 13–2).

The major units were delineated chiefly by nineteenth century workers in geology on the basis of physical and biological changes that they observed in the rocks in the local areas they investigated (largely in western Europe and Great Britain). The most profound changes occurred where major unconformities separated thick units of folded, faulted, deformed rocks from thick sequences of younger and much less deformed rocks. Fossils found in the rocks above the unconformities were much different from those in the older rocks below. Therefore, the rocks occurring below a major unconformity were said to have formed in one era, and those above in another. This constituted a natural basis for subdividing the rock record. Unconformities of lesser magnitude and extent, and associated with smaller fossil differences, were then used to subdivide the rocks of each era into smaller units, and each such unit was said to have accumulated during a geologic period.

The names of the eras reflect the fossil differences: *Paleozoic* (ancient life), *Mesozoic* (middle life), and *Cenozoic* (recent life).

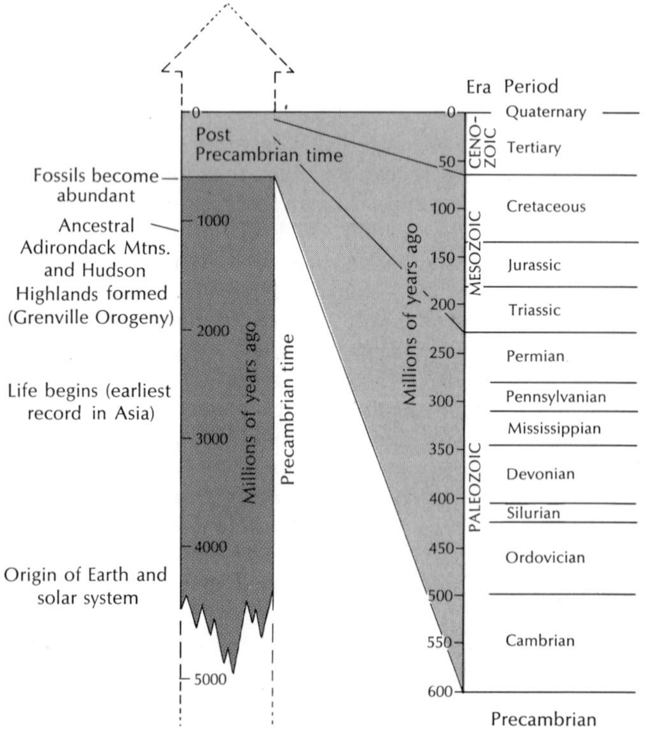

4-10. Geologic chronology *(Broughton, Fisher, Isachsen, and Rickard, The Geology of New York State, 1962.)*

The *Precambrian* Era (eras?) begins with the beginning of geologic history (i.e., with the records contained in the very oldest rocks) and includes everything that preceded the Cambrian Period—the oldest period in the Paleozoic Era. The Precambrian probably spans some five-sixths of the rock record and should eventually be subdivided into a number of eras. Names for the periods generally have local geographic significance (e.g., "Cambrian" is from Cambria, the ancient name for Wales).

Disagreement and uncertainty subsequently developed when these units—locally natural and logical where first applied—were looked for on other lands. According to an older view, world-wide natural physical breaks occur in the rock record and might be recognizable. Proponents of this view reasoned that mountain building may have been so extensive at times that it affected large parts of the Earth almost simultaneously. At such times, crustal movements deform rocks, elevate lands, and restrict seas, perhaps in part by deepening the ocean basins. Erosion thus tends to predominate everywhere on the lands, few permanent records are formed which are later accessible to man, and older records are destroyed.

Climatic changes caused by such mountain building may also speed up evolutionary development, because organisms must adapt themselves successfully to changed environments or perish. Restricted seas crowd shallow-water marine animals into smaller areas where many perish. When erosion finally succeeds in wearing down the continents, or if subsidence takes place, seas again encroach upon the lands, and sediments deposited in such shallow seas may be preserved as permanent records. These sediments rest unconformably upon the underlying rocks and contain distinctly different fossils from those in the older rocks beneath. Thus, these geologists have pointed out, great biological changes tend to correspond to great physical changes on the Earth.

However, geologists today tend to disagree with these conclusions. They think that orogenies and unconformities (even major ones) were not simultaneous, worldwide phenomena. In their view, natural breaks occur in the rocks on all the continents, but these developed at different times in different places. Thus the gap produced by an unconformity in one country may be represented by a thick pile of fossiliferous sedimentary rocks in another country—their geologic histories are not duplicates.

Although disagreement still exists, most geologists use the chronology as originally developed and try to subdivide the rocks in their areas into similar units primarily by matching fossils (p. 50).

Determining Geologic Time by Radioactivity

The chronology discussed in the preceding section was primarily a relative one; i.e., the Paleozoic was known to be older than the Mesozoic, but just when it began or ended in millions of years was not known, because establishment of absolute dates for the geologic eras and periods had to await the discovery of radioactivity.

The nuclei of a number of elements are radioactive naturally; i.e., they spontaneously emit alpha, beta, and gamma radiations (p. 773). *Alpha* particles are the nuclei of helium atoms and each consists of two protons and two neutrons. Therefore, emission of an alpha particle from a radioactive nucleus reduces its mass number by four and its atomic number by two—the nucleus is thus transformed into a new element. *Beta* particles are electrons that are ejected from nuclei. For each beta particle, a neutron within the nucleus was changed into a proton (remains within the nucleus) and an electron. The mass number of the nucleus is unchanged, but its atomic number has been increased by one—it too is a new element. *Gamma* radiation consists of very high-energy, very short wave, electromagnetic radiation; its emission releases energy but does not change the atomic or mass numbers. A radioactive nucleus may emit just one of the three types of radiation or any combination of the three and at very different rates.

Nothing known to man can alter the rates of these spontaneous disintegrations. The radioactive decay of a particular isotope proceeds at a uniform rate; i.e., its half-life period remains constant. Here is a geologic process which apparently has not varied in rate of activity during the long past, and apparently it provides an accurate measurement of the ages of rocks. The Earth's age is greater than that of the oldest rock measured by the radioactive method (about 3.5 billion years), and meteorites provide an additional clue. Their age determinations average about 4.6 billion years. If meteorites are representative of the material from which the planets of the solar system were built, the age of the Earth should be about the same as the ages of the meteorites. Moreover, the oldest samples of lunar rocks likewise formed 4.6 billion years ago.

Radioactive isotopes of uranium, rubidium, and potassium are important in determining the ages of rocks, but we shall concentrate upon the uranium-238 isotope as an example. In nature, uranium is found as a constituent of certain kinds of minerals which have formed in different ways, including precipitation from solution as crystals (Fig. 3–2). Since precipitation, the number of uranium atoms in such crystals has steadily decreased, whereas the number of lead atoms (the end product of uranium disintegration) has steadily increased. Thus the age of the crystals can be determined (i.e., the time when it was precipitated from the solution) because older crystals contain proportionally more lead.

The uranium-238 isotope spontaneously undergoes a series of changes during which it emits eight alpha particles and six beta particles. This results in the successive formation of fourteen new isotopes, the fourteenth of which is stable lead-206. Thus a uranium-238 isotope emits an alpha particle and becomes an isotope of thorium; this in its turn emits a beta particle and becomes an isotope of protactinium, which in its turn . . . until the fourteenth isotope, stable lead-206, is reached. The number of uranium-238 isotopes decrease at a known uniform rate, and the stable end-product lead-206 isotopes accumulate at a known uniform rate. Therefore, the proportion of uranium-238 to lead-206 in a rock measures the age of that rock, because the oldest rocks have the highest proportion of lead. Thus when a uranium-bearing mineral forms within rocks, it is as if a clock had been placed within the rock—a clock that has ticked away the time ever since.

For the radioactive clock to produce reliable results, several conditions are necessary. (1) The disintegration must be uniform (this seems to hold true), and its rate must be known. (2) We must determine with great precision the amounts of the radioactive isotope and the nonradioactive end product present in a specimen. (3) The system must have remained a closed one chemically; i.e., weathering should not have removed either the radioactive "parent" atoms or their end product "daughter" atoms. (4) The specimen should not contain atoms of so-called common origin (i.e., that did not form by radioactive disintegration) that are identical to the daughter atoms.

The rate of disintegration can be determined in several ways, and a statistical approach and the half-life period of an isotope are involved. In life insurance it is impossible to predict exactly which individual members of a group of 60-year-old men will die in the next year. However, based upon past experience, one can predict the number of deaths quite reliably, if the total group is large enough. In a similar manner, one cannot determine which of the isotopes of a radioactive element will change to another isotope, or when this will occur; but one can predict the length of time necessary for half of any large number to change (any other major fraction could be used—half is convenient). Thus in approximately 4½ billion years, one-half of the uranium-238 isotopes in any given quantity will have disintegrated into the isotopes of other elements; a large portion will be lead, but not half, because in-between members occur in the uranium-238 decay series. One-half of the remainder will be changed in another 4½ billion years and so on. This relationship is known as the *half-life period,* and each radioactive isotope has one of a different length.

Let us illustrate the meaning of half-life period with a fictitious example. Imagine that a certain radioactive isotope has a half-life period of 10 years and that 100 lb of this was placed in a safe in 1900. How much would be left in 1950? In 1910, 50 lb would remain; in 1920, 25 lb; in 1930, 12½ lb; in 1940, 6¼ lb; and in 1950; 3⅛ lb.

The original quantity of uranium in a mineral specimen can be determined by adding the total number of uranium atoms now in the specimen to the total number of atoms in the specimen which have resulted from its disintegration. Each of these atoms was once a uranium atom. However, the specimen must have remained a closed chemical system. Our results would be inaccurate if we included in our count any common-origin lead that may subsequently have been added to the lead of radioactive origin, or that may have been present in the specimen originally. Fortunately, such contamination, if it occurs, can usually be detected (see below). On the other hand, some of the radioactive lead isotopes that formed in the specimen may have left it, but this too can usually be detected and allowed for.

To avoid the effects of contamination and solution, the best specimens to use for age determinations are single unweathered crystals of uranium-bearing minerals. The lead isotopes which form by disintegration of the uranium are trapped in these crystals. However, newer techniques have been developed, and whole rock specimens can now be analyzed and dated radioactively without separating the minerals that contain the radioactive isotopes from the rest of the specimen.

Tests of various uranium-bearing minerals from different locations have given results which range from a few million years to a few billion years. This spread indicates that uranium-bearing minerals have formed at many different times in the past, and the oldest rocks yet measured by this method are about 3.5 billion years old. Perhaps somewhat older rocks will be found in the future, but present theory suggests that the 4.6 billion year age for meteorites is a limit.

There are checks on the accuracy of the radioactive methods of measuring the ages of rocks, e.g., different radioactive isotopes disintegrate at different rates into different end products and several may be found in a single specimen (Table 4–1). This is analogous to finding several clocks which all give the same time. One assumes the time is correct. Thus the methods, half-life periods, and results seem to be relatively reliable but only, it should be emphasized, in terms of millions of years.

The ratio of lead-207 to lead-206 in a single uranium-bearing specimen also provides reliable age determinations. Their relative proportions in the specimen change with time because their parent uranium isotopes disintegrate at different rates. Furthermore, even if part of this lead is subsequently removed during weathering, their proportions in the remaining lead would tend to be unchanged.

Lead isotopes also provide another means of age determination (the so-called common lead method). We assume that the four lead isotopes (204, 206, 207, and 208) were present originally in the gaseous nebula from which the Earth was formed (p. 309), and that their proportions at that time were about the same as those in iron meteorites today (because these contain

TABLE 4-1
ISOTOPES MOST COMMONLY USED IN THE RADIOACTIVE DATING OF ROCKS

PARENT ISOTOPE	DAUGHTER ISOTOPE	HALF-LIFE PERIOD IN BILLIONS OF YEARS
Uranium–238	Lead–206	4.5
Uranium–235	Lead–207	0.7
Rubidium–87	Strontium–87	47.0
Potassium–40	Argon–40	1.3

little if any uranium and thorium—a source of lead-208). So far as is known, lead-204 is not formed by radioactive disintegration. Therefore, the total quantity of lead-204 on the Earth must have remained constant since the Earth formed, whereas the total quantities of the other three lead isotopes on the Earth must have steadily increased via radioactive disintegration but at different rates for each isotope. Correspondingly, the supply of the radioactive isotopes must have steadily dwindled; e.g., half of the original quantity of uranium-238 should have disintegrated by now and about 99% of the

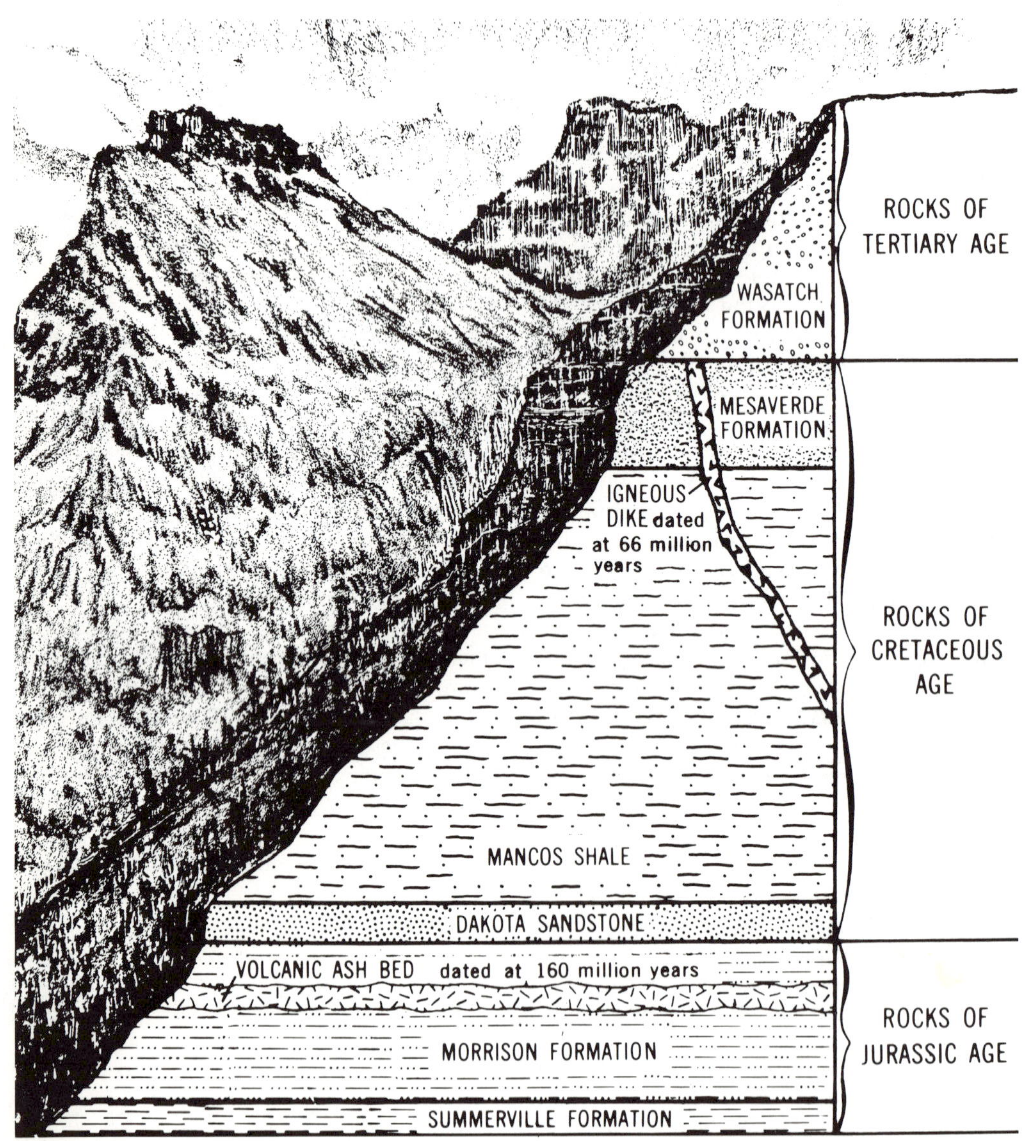

4-11. The approximate absolute ages of some sedimentary rocks may be determined by their relationships with igneous rocks. Thus the Mesaverde Formation's age is between 66 and 160 million years. *(Courtesy U.S. Geological Survey.)*

uranium-235 (shorter half-life). Assume that lead isotopes were produced by radioactive disintegration of uranium and thorium in a particular part of the crust long ago. Assume further that these were then dissolved and subsequently precipitated in a new location as crystals of the mineral galena (lead sulfide). The proportions of the isotopes in the galena should depend upon the time in the past when this occurred, and thus provide a means for dating the galena.

Radioactive rubidium and potassium occur in common minerals such as mica and feldspar and are thus found in many rocks. Although the argon from potassium-40 is a gas (inert), it tends to be trapped in surrounding lattice structures unless subsequent heating drives it away. The potassium-argon method may be used to date rocks that formed as recently as a few hundred thousand years ago or less.

Although the ages of many igneous and metamorphic rocks can now be determined, sedimentary rocks are more difficult to date directly by the radioactive method. A pebble in a sedimentary rock may contain suitable radioactive elements, but its age determination would give the time that the source rock formed and not the more recent time when the sediments were deposited. Geologists obtain the ages of most sedimentary rocks by their relationship to igneous rocks which may have intruded into them or which may be correlated with them in some other way (Fig. 4–11). However, if the cementing material of a sedimentary rock contains a radioactive isotope and accumulated at the same time as the sediments, then its dating will give an age for the sedimentary rock. Glauconite is such a mineral (a hydrous, potassium-iron silicate) because it contains potassium, forms on the sea floor, and accumulates with the sediments and organisms that are piling up at the same time and place. However, glauconite tends to lose some of its argon daughter atoms when buried under a thick pile of sediments. It also may be added to the sediments after their deposition, so such age determinations tend to be minimum ones.

Picturing Geologic Time

The immensity of geologic time is quite unimaginable, but the relative length of its subdivisions can be grasped with the help of analogies. Two are described here.

Imagine that you are walking into the past at the rate of 100 years for each 3-foot step. Walking a mile (1760 steps) on this scale would be the equivalent of going back 176,000 years in time. The first step takes you back to about the Civil War, and the second step approximately to the Revolutionary War. Eighteen more steps would total 2000 years and carry you back to the time of Christ. If you were walking in the northern part of the United States after about 110 steps (11,000 years), you would be stepping on a continental ice sheet a few thousand feet thick, and during the first several miles you would be passing through the Ice Age.

However, to walk back in time far enough to see a dinosaur would require a 360-mile hike (about 65 million years ago at the end of the Mesozoic Era). For the next 950 miles or so you would be passing through the Mesozoic Age of Reptiles, and a total of about 3000 to 3500 miles—more than the distance from New York to San Francisco—would have to be walked to arrive at the beginning of the Paleozoic Era, some 600 or so million years ago. On the scale we have chosen, rocks that are 3½ billion years old would be the equivalent of going from ocean to ocean across the 48 contiguous United States about seven times—each 3-foot step equivalent to 100 years!

Again, suppose that all of the Earth's 4 billion or so years of conjectured history is compressed into a single year. On this scale, the Paleozoic Era, which contains the oldest rocks with abundant fossils, would not even begin until 10½ months had passed. Mammals would appear on the scene about the second week in December. Apparently man would arrive shortly before midnight on December 31, and all of recorded history would be represented by the last 90 seconds!

The Carbon-14 Method of Measuring the Age of Dead Organic Matter

Akin to the methods of measuring the ages of rocks is a technique to measure the age of dead organic matter (i.e., it indicates the approximate number of years which have elapsed since the source organism stopped living). The technique was discovered somewhat incidentally rather than as the result of a planned search—such unexpected fruitful results occur fairly commonly in scientific research. It provides an excellent illustration of *science in action:* the careful unbiased observation of natural phenomena and accumulation of data, the coming of ideas not necessarily in view when the work started, the formulation of hypotheses to be verified, modified, and expanded or discarded, and the development of new instrumentation. All of these are merged into a close-knit interwoven pattern of activity that is not readily organized into a set one-two-three series of steps which can be called *the* scientific method. Its goal is the development of principles, concepts, laws, and understandings of natural phenomena which make a meaningful whole out of otherwise seemingly unrelated, uninteresting factual knowledge.

The carbon-14 method began as an outgrowth of research by Libby and others on cosmic radiation. During their bombardment of the upper atmosphere, cosmic rays produce secondary neutrons which react with some of the nitrogen atoms in the atmosphere to make a carbon-14 isotope which is radioactive. Common carbon has an atomic weight of 12 and is not radioactive (a carbon-13 isotope also exists but can be ignored here). A single carbon-14 atom does not exist long because it disintegrates back into nitrogen by ejecting a beta particle. However, additional carbon-14 atoms are constantly forming in the air.

Libby theorized that all living organisms should contain some carbon-14 and that a constant ratio of carbon-14 atoms to carbon-12 atoms should be present in the atmosphere and in all living organisms including man. He reasoned somewhat as follows. Carbon-14 is constantly forming in the Earth's atmosphere; some of it unites with oxygen to form carbon dioxide; plants assimilate carbon dioxide during their life functions; and animals eat plants.

The two hypotheses were at once tested. Samples of such diverse organic matter as sewage, trees, seal oil, and sea shells were obtained from all parts of the Earth and checked. Carbon-14 was present and the proportions of carbon-14 to carbon-12 were found to be approximately the same in all instances. The worldwide distribution of specimens was made to determine whether or not the strong variation of cosmic radiation at different latitudes would have a noticeable effect, and it did not. Although more carbon-14 atoms are produced at certain times, latitudes, and altitudes than at others, sufficient mixing seems to occur in the atmosphere and hydrosphere to make the carbon-14 content about uniform (but see below).

A carefully measured quantity of the sample is processed chemically (which destroys the sample) before the amount of carbon-14 in it can be measured with a Geiger counter. The radioactivity is directly proportional to the quantity of carbon-14 atoms present in the specimen, and the level of activity in living organisms is about 14 disintegrations per minute per gram of carbon. This is exceedingly small when compared with the normal background radioactivity and necessitates a very sensitive measurement and ingenious shielding devices; e.g., the researcher is more radioactive than the sample of dead tree that he may be checking.

Now a third hypothesis was formulated. In living organisms, a constant proportion of carbon-14 to carbon-12 is maintained because new supplies of carbon-14 are continuously being taken in to replace those lost by disintegration. But after death, replenishment ceases, whereas disintegration continues. The proportion of carbon-14 left in a dead organism, therefore, should indicate the length of time that has gone by since the organism died. A small proportion of carbon-14 to carbon-12 would show that the organism died long ago.

Recent measurements indicate a half-life period for carbon-14 of about 5700 years (however, 5568 years has been used to date in almost all age determinations—multiply published ages by 1.03 to convert). Therefore, if only half of the original quantity of carbon-14 remains, the specimen is about 5700 years old; if one-quarter remains, the age is 11,400 years, etc.

This hypothesis was tested against known historic records—e.g., on wood taken from coffins that contained Egyptian mummies, the dates of whose deaths were known. Presumably the wood for a coffin was cut at about the time the death occurred. The carbon-14 age determinations corresponded closely to the dates based on recorded history.

To be sure, the method requires some precautions, particularly with the oldest specimens. Cosmic radiation may have varied in intensity in the past. Young carbon may contaminate old samples; e.g., the addition of only 1 percent of modern carbon to a 57,000 year-old sample would reduce its age to about 35,000 years, and such contamination is difficult to detect. Allowances must be made for two relatively recent changes: (1) much so-called dead carbon (it lacks carbon-14) has been put into the atmosphere during the past century or so by the increased burning of the fossil fuels (p. 648), and (2) considerable carbon-14 has been added since 1954 from nuclear explosions in the atmosphere.

The carbon-14 method has a time limitation of about 40,000 years, though further refinements may lengthen this somewhat. Laboratory techniques are precise enough to extend this still further, but such relatively ancient samples are very likely to be contaminated by additions of younger carbon. Further extensions would probably result in inaccurate dates attained by very precise measurements.

Imagine the eagerness with which an archaeologist might wait for the results. Perhaps he submitted a sample of charcoal from an ancient campfire around which he had found the bones of a few extinct animal species and the arrowheads and artifacts of prehistoric human culture. Before the discovery of the carbon-14 method, an exact age determination would have been impossible. Now the date when the campfire was built (actually when the firewood died) is determinable within a small percentage of error. Or the archeologist may have cut a trench into the floor of a cave inhabited by man for many centuries and he may have collected samples of charcoal and artifacts at different levels. Now these levels can be dated; proportionally less carbon-14 occurs in specimens from lower and lower levels.

The specimens tested are so numerous, varied, and interesting that everyone should find some favorites among them, and a few samples are listed here. The Dead Sea scrolls (including a nearly complete text of the Book of Isaiah) were written about 1900 years ago. Charred bread from a house in Pompeii has been dated by the carbon-14 technique as has the heartwood of a very large redwood tree cut down in 1874. The tree rings were counted and totaled about 2900, approximately the same age as determined by its carbon-14 content. This shows that the carbon in the heartwood accumulated very early in the life of the tree and that its dwindling supply of carbon-14 was not replenished.

Some 300 pairs of sandals were found in Fort Rock Cave in Oregon. The sandals were woven out of grass rope and are attractively designed and shaped. Of course, one is curious about the time they were made, and carbon-14 tests give an age of about 9000 years. As they were piled neatly in the cave, they give evidence of an American cobbler of 7000 B.C. or so.

Exercises and Questions for Chapter 4

1. Questions relating to the geologic history of the Grand Canyon region:
   (A) Describe a number of basic geologic concepts illustrated by this history.
   (B) If you studied the rocks on the north and south rims, what evidence could you find that an uplift of about 1½ miles had occurred?
   (C) What evidence indicates that mountains were formed at two different times and subsequently were worn away?
   (D) What is meant by the "great denudation" and what is the evidence for it?
2. Assume that you have a skeptical roommate who refuses to believe that crustal movements involving hundreds and thousands of feet of displacement have occurred in the past. What evidence can you cite to prove that such crustal movements have occurred, and how can you organize this evidence in a convincing manner? Is it also necessary for you to furnish evidence that the Earth is very old?
3. How can the relative ages of a number of sedimentary rocks occurring in different areas be determined? How can their absolute ages be determined? What assumptions or concepts are involved? How accurately are such ages known?
4. List a number of events that might occur in the geologic history of an area. Next, draw and label a structure section that illustrates this sequence of events and give it to a classmate to interpret.
5. Assume that a tree was buried by glacial debris about 22,400 years ago. How should its carbon-14 content compare with that of a tree living today? Assume the tree has remained a closed chemical system and use 5600 years as the half-life period of carbon-14.
   (A) Why is there a time limit of about 40,000 years on age determinations made using the carbon-14 technique?
   (B) Why has it been said that age determinations exceeding this limit would probably represent inaccurate results achieved with great precision?
   (C) The development of the carbon-14 method has been described as an instructive case history illustrating scientific discovery and the methods of science. Discuss.
6. In what way is the sodium-in-the-sea method of estimating the age of the Earth similar to that of the radioactive method? In what significant way do they differ?

5

Igneous Activity

A most exciting and spectacular geological event occurred again on 20 February 1943—the boisterous birth of a brand-new volcano (Fig. 5–1). On this occasion the place was a cornfield near the village of Paricutin about 200 miles west of Mexico City. Indians working in the cornfield noticed a fissure that suddenly belched smoke, and the ground around thundered and hissed. During the preceding two weeks, earthquakes had occurred in increasing intensity and numbers, and a volcano had been predicted by an engineer. In panic the workers hurried from the cornfield which spouted steam, ashes, and rocks. Nearby trees caught fire. That night the whole area was lighted up as molten lava and red-hot cinders erupted every few seconds from the ground. The fragmental debris fell to the surface and built a cone-shaped pile around the opening. It was 30 feet high by the next morning.

In one week the volcanic cone grew to an altitude of nearly 500 feet, and within a year it towered about 1100 feet above the surrounding lands. During its growth the volcano erupted gaseous, liquid, and solid material, and lava flowed out from its base and flanks. At first the explosions were like cannon fire and occurred every few seconds, while thuds of falling fragments punctuated this background of explosive sounds. As volcanic debris fell back to the earth, it occasionally smashed against upthrown pieces to add to the dust in the air. Clouds of condensed steam and other gases boiled upward to heights of 3 miles or more and varied from white to black as their ash content changed.

At night the spectacle attained its dramatic peak and became a display of fireworks as glowing clots of erupted lava flew through the air and later bounced and rolled down the flanks of the volcano to streak the slopes with red and orange colors. The cinder cone grew wider as it piled higher, although the slope of its flanks remained at about the same angle—the angle of repose for solid chunks of volcanic debris.

As the months went by, Paricutin's surroundings were buried beneath dust and lava. Flow after flow

issued from the base of the volcano and moved outward for distances up to six miles and overwhelmed nearby villages. As the flows cooled into dark rock, they were covered by new flows, until a superposed pile several hundred feet thick had accumulated around the base of the cone. The dust made a much larger area uninhabitable and was thickest near the volcano (up to 10 feet or more). Occasionally it drifted down on areas more than 100 miles away.

In 1952 the volcano became inactive, and probably it will not erupt again. Its present height above the former cornfield is about 1400 feet. It was measured, sampled, and observed frequently during its active years and provided geologists with a fine laboratory for studying the birth, development, and death of a small, short-lived volcano. According to one estimate, about 2½ billion tons of dust and larger fragments and 1½ billion tons of lava were erupted during the nine years of Paricutin's active existence. Paricutin is located in an elongated area that contains thousands of similar volcanoes, and several hundred are situated within a distance of 75 miles. The history of Paricutin has paralleled that of its many neighbors.

New volcanoes also form on the sea floor, and some build upward to form islands (Fig. 5–2). Volcanic eruptions were observed off the southern coast of Iceland in water about 400 feet deep on 14 November 1963, and in 10 days enough volcanic debris had piled up to form an island about ½ mile long, its summit nearly 350 feet above sea level. It was given the name Surtsey. Thunderstorms were apparently initiated by violent eruptions, and vivid flashes of lightning illuminated the volcanic clouds almost continuously at times. The volcanic eruptions and lightning displays were photographed and observed from land, sea, and air by teams of scientists during the following months. Violent eruptions and lightning stopped in April 1964, when an accumulation of scoria and lava filled the crater. Previously the crater had frequently been

5-1. Paricutin in eruption in 1944. Incandescent volcanic bombs and clots of molten lava make the light streaks. *(Courtesy Tad Nichols.)*

flooded with sea water, and the transformation of some of the water into steam had enhanced the vigor of the eruptions. Some activity continued into 1965.

Volcanoes consist of two parts: an external hill or mountain which commonly is heaped on the surface as a conical mass (Figs. 5–3 and 5–4), and a more important internal part consisting of a cylindrically shaped vent or a fissure (more than one opening may also occur) which leads from the surface to a source of magma far below.

Distribution of Volcanoes

Several hundred volcanoes are active now or have erupted during historic times, and thousands of others have been active in the recent geologic past. Erosion has had time to modify them only slightly. Volcanoes are widespread on the lands and oceans, where some rise above the surface as islands, but many more do not. However, most are grouped in relatively narrow, elongated belts, the most pronounced of which rings the Pacific in a "girdle of fire" on the east, north, and west. The Andes, Aleutians, and Japanese Islands form part of this. Another great belt extends from the Alpine-Mediterranean area eastward to the Indonesian states, and a third major volcanic zone, chiefly submerged, occurs along the system of mid-ocean ridges (p. 294). Many volcanoes also have a linear arrangement on a small scale, apparently because they are aligned along crustal fractures.

Earthquakes occur most frequently in these same volcanic areas, and two-thirds of all earthquakes probably originate in the circum-Pacific belt. The major volcanic and earthquake belts also tend to coincide with the mountainous zones that have formed relatively recently on the Earth. Thus a close relationship is apparent between volcanoes, earthquakes, and mountain belts, and today we think that such active zones may mark the margins of very large, slowly moving,

5-2. Lightning in a volcanic cloud over Surtsey Island, Iceland, on 1 December 1963 (90-sec. exposure; latitude N 63° 81′, longitude W 20° 37′). The altitude of the top of the picture is about 5 miles (8 km). *(Courtesy Sigurgeir Jonasson.)*

5-3. Iliamna volcano in Alaska. Note gases escaping from two vents (fumaroles) near the summit and the effects of glaciation. *(Courtesy U.S. Air Force.)*

lithospheric slabs (see plate tectonics, p. 314).

Volcanoes originating in historic times have been confined to known volcanic areas, as should volcanoes forming in the decades and centuries immediately ahead. However, almost all parts of the Earth's surface have experienced igneous activity at some time in the past. In short, although no area is immune to volcanic action, some places are much more susceptible than others.

Volcanic Eruptions

Volcanoes may explode violently, or lava may flow out at the surface rather quietly. Moreover, a single volcano may erupt explosively at one time and quietly at another; but in many volcanoes, one type of activity or the other predominates. The explosive activity of a volcano seems to depend upon the amount of water vapor and other gases in the magma beneath it and the manner in which that gas escapes (Figs. 5–5 and 5–6). An explosion cannot occur unless gas is abundant and has been trapped to build pressures to the bursting point. Much of this gas was formerly dissolved in the magma below the volcano.

As magma cools, a sequence such as the following seems likely. Different minerals crystallize (are precipitated) from the molten solution, but water vapor and other dissolved gases tend to be excluded from such mineral grains. Hence, the volume of liquid material in the magma body dwindles steadily as crystallization proceeds, and the relative proportion of gas in the remaining liquid increases. When the gas content becomes too great, gas bubbles separate from the liquid solution and stream upward in a magma chamber to zones of lower temperature and pressure near its roof. In a fluid magma—one that is very hot or that has a relatively low silica content—such gas may be able to escape readily and continuously without creating explosive pressures. As the gas expands, it forces the lava quietly upward to the surface through cracks. If the magma is viscous (nonfluid) or if the volcanic vent has been plugged by solidification of once molten material, the gas cannot escape and pressures increase until an explosion occurs.

Some gases in a magma may come from the vaporization of ground water encountered by the magma in its upward journey or from rocks that have been melted and assimilated; e.g., limestone may add much carbon dioxide gas in this manner. Thus the eruptive action of a volcano may change during its lifetime because the composition, temperature, and amount of dissolved gas and magma are all variable factors.

Major volcanoes may be underlain at a relatively shallow depth (3 to 4 miles, Fig. 5–7) by a chamber in which magma accumulates prior to an eruption. Apparently the magma originates at greater depths and flows upward to the chamber just prior to an eruption. A magma chamber is probably lenticular in shape because swelling and shrinking are observed before and during an eruption. Precise measurements have shown that the entire top of a volcano such as Kilauea commonly bulges upward for several feet and tilts outward during the months before an eruption. Presumably this occurs as the underlying chamber is filled and inflated by accumulating magma. Shrinkage follows swelling, and the two may alternate somewhat prior to an eruption. However, the rate of swelling may increase rapidly immediately before an eruption, and

5-4. Active, dangerous Asama Volcano, Honshu Island, Japan. Ash covers most of the surface near the crater which is partially encircled by the rim of a caldera. A lava flow (1738) is conspicuous. *(Courtesy U.S. Geological Survey.)*

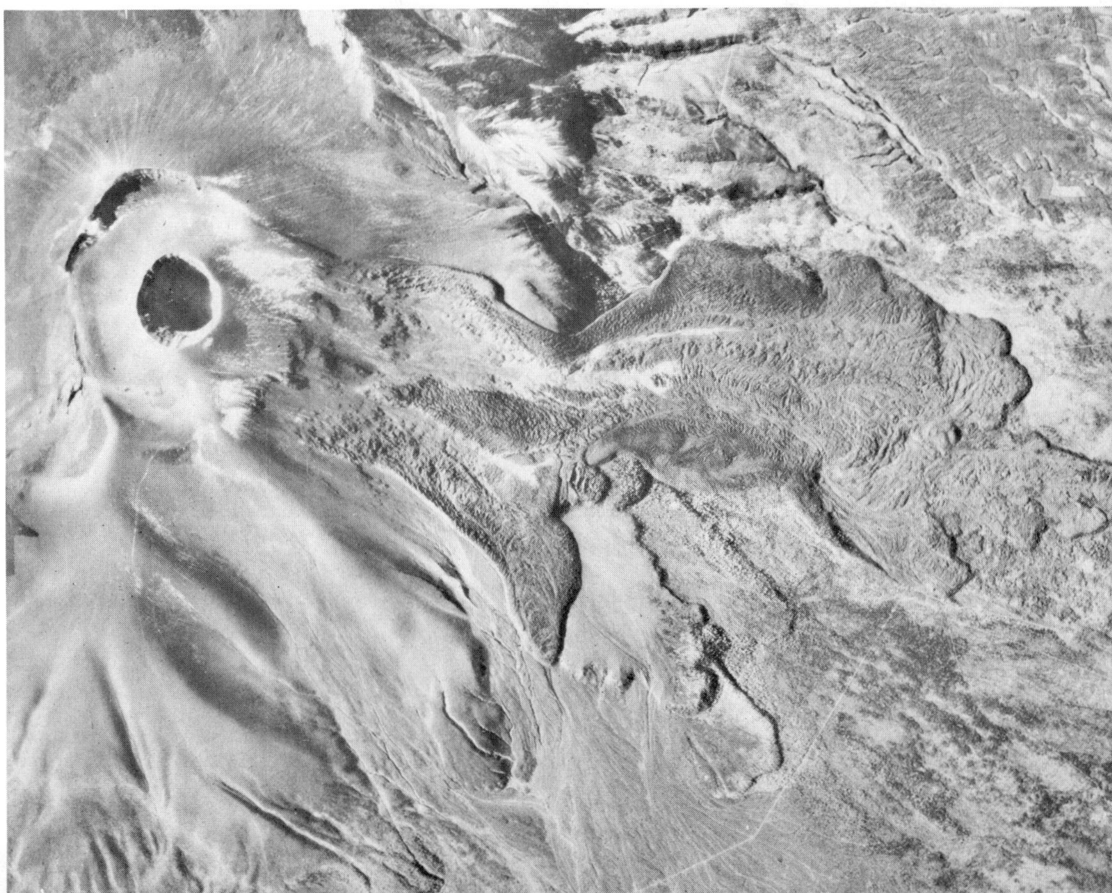

5-5. Lava fountain, Kilauea, Hawaii, 1960. *(Courtesy U.S. Geological Survey.)*

small earthquakes may then become very numerous. Perhaps the times of future eruptions can be predicted fairly accurately for volcanoes of this type.

Uncontaminated samples of volcanic gases are difficult to obtain for chemical analysis, but steam is by far the most abundant material given off. Other gases include carbon dioxide, carbon monoxide, hydrogen sulfide, sulfur dioxide, sulfur, hydrochloric acid, and ammonium chloride.

Heavy rains are commonly associated with volcanic eruptions, and the precipitation has two immediate sources: huge quantities of steam may be erupted and then condense; and air above a volcano may be blown violently upward, which causes expansion, cooling, and condensation of much of its moisture. Torrents of water sometimes pour down on loose fragments resting on steep volcanic slopes and result in disastrous mudflows.

Materials erupted from volcanoes include solids and liquids as well as gases. The solids, and the blobs of liquid lava that solidify during flight, have a wide range in size, shape, and composition and are lumped together as *pyroclastic debris* ("fire broken"). This may also include pieces of sedimentary and metamorphic rocks torn from the walls of a volcanic vent as well as fragments of igneous rocks that had previously solidified in and around a vent. When the ash, cinders, and other fragments become a consolidated rock, the material is

5-6. Eruption of Myojin Reef Volcano, 170 miles south of Tokyo. *(Official U.S. Navy Photograph.)*

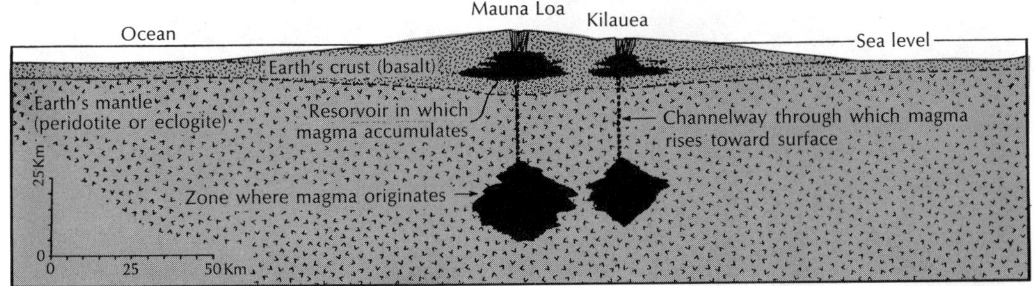

5-7. Hypothetical structure section through Mauna Loa and Kilauea volcanoes, Hawaii. [*G. A. Macdonald,* Science. **133,** *677 (10 March 1961).*]

called *tuff* if the average size is less than 4 mm and *volcanic breccia* if the average size is greater than 4 mm.

Tiny pieces of glass and pumice are commonly abundant in ash, and if these are still hot when deposited, the pliable particles tend to join together and interlock into a firmly consolidated rock that may look like obsidian or rhyolite. It is called *welded tuff* and apparently is much more widespread than once believed—in fact, many rocks formerly identified as rhyolite lava flows have now been relabelled as welded tuffs. Deposits from fiery clouds (see below) may cover a wide area, and the fine-grained portions may form welded tuffs.

Obsidian is natural volcanic glass and tends to have an overall chemical composition (high in silica) like rhyolite and granite. *Pumice* is glass that was so distended by expanding gases that it frothed before it hardened; it is very porous because of its many gas bubble holes and can float on water.

A mass of liquid lava may be heaved violently into the air during a volcanic eruption. It may spin and twist during flight and cool into a rounded shape with pointed, twisted ends—a *volcanic bomb* (Fig. 5–8). Some bombs weigh many tons, whereas others are as small as pebbles. If a bomb is still plastic when it strikes the surface, its shape is distorted; in fact, some bombs have been molded around tree limbs. Cracks or holes in a bomb result from expansion of the gas that was once dissolved in it.

5-8. Volcanic bombs and impact trails photographed at Paricutin volcano in Mexico. (Fred M. Bullard, *Volcanoes: In History, In Theory, In Eruption,* University of Texas Press, 1962.)

Fiery Clouds

A particularly destructive and explosive type of volcanic eruption is that known as the fiery cloud (*nuée ardente*) in which huge quantities of gas and incandescent volcanic debris are blasted upward or laterally through the side of a cone if its vent is plugged. This material forms a dense, seething, turbulent mass that hurtles like an avalanche down a volcanic slope at speeds that may exceed 100 mi/hr. The extreme mobility of such a turbid cloud is caused by the pres-

5-9. Fluid lavas congeal with a smooth to ropy surface (pahoehoe lava in Hawaii). *(Courtesy U.S. Geological Survey.)*

ence of vast quantities of gas that bubbles, froths, and expands continuously and explosively at the reduced pressures at the Earth's surface. One obtains a somewhat similar reaction by vigorously shaking and then uncapping a warm bottle of ginger ale. Rock fragments are readily transported within the lower part of a fiery cloud, because it is much denser than air. Thus it moves rapidly down a volcanic slope as a pyroclastic flow, and such a cloud from Mount Pelée overwhelmed the town of St. Pierre on the island of Martinique in the West Indies in 1902. It is reported that all of the inhabitants (about 28,000) were killed except two, a prisoner in a dungeon and a young shoemaker, both severely injured.

Lava Flows

A tunnel may form in a lava flow because the central region remains warm and fluid after its top and sides have cooled and hardened. Therefore, when the supply of lava is eventually cut off, the last of it drains out of the central portion of the flow and leaves a tunnel. Large tunnels extend for a few miles and are 50 or more feet high. Some lunar rilles may have originated in this manner (p. 553).

The formation of a hardened crust over still flowing lava also accounts for the blocky, jagged, highly irregular surfaces, characteristic of some viscous flows. Such crusts may also be shattered by the explosive escape of gases from underlying hot lava. On the other hand, the surfaces of fluid flows tend to be relatively smooth and become wrinkled into ropy, rounded, corded shapes (Fig. 5–9). Lava flow surfaces commonly contain many spheroidal holes (vesicles) produced by cooling and hardening of the lava around bubbles of gas that had formed and expanded in it.

Some lava flows show *columnar jointing* (Fig. 5–10) produced by shrinkage during cooling (somewhat similar to the contraction and cracking of mud as it dries out, see Fig. 3–19). Such features may also form in tabular intrusive rock bodies. When cooling occurs under uniform conditions against a flat surface like the floors of some lava flows, the shrinkage tends to be about equal in all directions—as if uniformly spaced centers of cooling had developed. Cracks tend to form midway between the centers of cooling and then extend into the cooling body from the cooling surface. The joints may thus delineate columns that are perpendicular to the cooling surface. Six-sided columns tend to develop ideally, but columns with a greater or smaller number of sides are common.

Lava pillows (Fig. 5–11) have been observed to form where lava flows into water. Rapid cooling ensues, and part of a flow may separate into discrete rounded units a few feet or less in size. Volcanic glass, resulting from the rapid cooling, forms a rind on many pillows, and openings among pillows may be filled with sediments or lava fragments.

5-10. Columnar jointing in a basalt flow, Devil's Post Pile National Monument, California. *(Courtesy U.S. Geological Survey.)*

Lava may also well upward through long fissures and pour out quietly without building cones at the surface. The lava then flows to lower elevations and fills in depressions. Commonly it is of the highly fluid basaltic type that can readily flow for many miles. At times in the past, enormous volumes of such material have flooded entire regions and built plateaus (Fig. 5–12) on the Earth and produced the maria on the moon.

Volcanic Topography

Characteristically, the external parts of volcanoes have conical shapes, consist of lay-

5-11. Pillow lavas, Ingot Island, Prince William Sound, Alaska. *(Courtesy F. H. Moffit, U.S. Geological Survey.)*

5-12. Layered basalt flows of the Columbia River Plateau, Hell's Canyon, Oregon-Idaho border. Approximately 35,000 cubic miles of fluid lava poured out of fissures in the latter part of the Cenozoic Era and spread widely. Flow piled on top of flow until thicknesses totalled hundreds of feet. Single flows have been traced for 100 miles. Lake- and stream-deposited layers of clay, sand, and gravel occur between some of the lava layers. *(Courtesy U.S. Geological Survey.)*

ers of volcanic materials that dip outward from the center, and are marked by craters at their tops. Eruptions throw masses of molten lava and rock fragments high into the air because volcanic vents commonly are vertical, and the bulk of this material then falls around the opening to form a cone. In a single eruption, large pieces tend to fall first and smaller particles later, thus producing a crude stratification. Successive eruptions vary in magnitude, as do the sizes of the fragments thrown out. Between explosions, lava may flow down the surface of a volcano from its top or, more frequently as it grows larger, from secondary fissures on its flanks or near its base. The continuous upward movement of volcanic debris from the central opening during an eruption prevents most of the material from falling back directly into this opening, and thus a crater is formed at the top; its diameter tends to approximate that of the volcanic vent. When the explosions cease, some material falls into the crater, and other debris slumps in from the sides.

Volcanic cones differ widely in shape and size, but three types are most common—the cinder cone, shield, and composite types.

Cinder cones are relatively small and steepsided, have uniform slopes of 30 to 40 degrees, and were built chiefly by fragments erupted explosively from a central vent. Paricutin is an example (also Wizard Island, Fig. 5–14).

Shield volcanoes have been formed mainly by the piling up of flow after flow of fluid lava to produce rounded, gently sloping domelike masses. The slopes are about 10 degrees or less near the summit and about 2 degrees near the base. The Hawaiian Islands furnish excellent examples (Fig. 5–13). Instead of rising up a central pipelike vent, as occurs in cinder cones, the lava apparently rises through numerous narrow cracks that radiate outward from the center of a volcano as spokes do in a wheel. The top of Mauna Loa is nearly 14,000 feet above sea level, but its base is some 16,000 feet below sea level. Thus it towers nearly six miles above the surrounding ocean floor, one of the tallest and largest piles of volcanic debris known on the Earth.

High-silica viscous lava may also flow out of a vent and build a lava cone, but its shape is quite different from that of a shield volcano. The lava cannot flow far and builds a steep-sided bulbous dome over the vent. Such cones may grow chiefly by addition of lava inside as pressure from the rising magma expands the outer hardened shell, which is cracked and pushed aside.

As indicated by its name, a *composite cone* is formed partly by explosive eruptions, which are chiefly responsible for increasing its height and steepening its summit slopes, and partly by lava that flows forth quietly and widens its base. Slopes vary from about 30 degrees near the summit to 5 degrees near the base. The symmetry and beauty of cones of this type are admired throughout the world and many examples are well known: Fujiyama in Japan, Mayon in the Philippines, Mount Rainier in Washington, Vesuvius in Italy.

The circular depressions that cap some volcanoes are much larger than the average crater; in fact, the entire top sections of such volcanoes are missing. These depressions are called *calderas,* and Crater Lake (Figs. 5–14 and 5–15) occurs in a well-known and strikingly beautiful example, and Vesuvius has grown within another. The diameter of a caldera is many times wider than the diameter of its volcanic vent. Probably a caldera forms upon collapse of the top of a volcano along steep-sided circular faults. As evidence, rocks on the floors of some calderas may be matched with rocks that are exposed high on the caldera walls, and circular fault surfaces have been observed. Presumably, space was made available for the collapsed mountain top by the eruption of magma and also by its lateral outward flow from beneath the volcano.

Many volcanoes contain a system of vertical fractures, now filled by hardened magma (dikes), which lead radially outward from their central vents. Such radial fractures seem to be the main channelways in the shield volcanoes, whereas the fractures apparently are subordinate to the central vent as a passageway for magma in other volcanoes. Deep erosion has exposed the central volcanic neck and radiating dikes in some areas (Fig. 5–16).

Intrusives: Dikes, Sills, and Batholiths

Intrusive masses of igneous rocks are classified chiefly on the basis of their shapes and relationships to surrounding rocks (Fig. 5–17), and dikes, sills, and batholiths are three common examples. Dikes and sills are tabular-shaped masses which formed when magma was squeezed under pressure into cracks or weak zones in rocks. They tend to be shaped somewhat like sheets of plywood and thus have two large dimensions and one small. They may be quite large and extend for miles or may be very small.

Magma may squeeze between two layers of stratified rocks to form a *sill*—a tabular-shaped intrusive igneous rock body that is parallel to the layers of rock that enclose it (Figs. 5–18 and 3–17). A sill may have any orientation in the crust; if the enclosing layers are vertical, tilted, or horizontal, then the sill is also vertical, tilted, or horizontal. A sill tends to resemble a buried lava flow that occurs in layered rocks. Since each may consist of the same type of relatively fine-grained rock (e.g., basalt), distinction must be based upon differences due to their different manner of formation. (1) A sill can alter (bake or metamorphose) the strata along each margin (contact), whereas a flow can alter only the rocks along its bottom contact—air formed the upper contact at the time of origin. (2) Tongues (small dikes) may extend into the rocks on each side of a sill, whereas such tongues can develop only at the base of a flow. (3) Chunks of the overlying rock may break off and become a part of a sill, whereas pieces of the top part of a lava flow may be broken off and incorporated into the sediments that are eventually deposited on top of it. (4) The upper surface of a flow may have many gas bubble holes.

5-13. The snow-covered summit of Mauna Loa in Hawaii. The oval-shaped caldera at its summit is about 3 miles long by $1\frac{1}{2}$ miles wide. Part of the summit collapsed to produce the caldera. Three craters occur in the foreground. *(Courtesy U.S. Air Force.)*

If magma is intruded in the form of a tabular body that cuts across enclosing layers or is squeezed into a crack in nonstratified rocks, it forms a *dike* (Fig. 3–14). Such intrusive rocks may be more resistant to erosion than the materials they intrude, or less resistant, and thus they develop into various topographic forms as differential erosion proceeds in a region (e.g., a dike may form a ridge).

A dike has some resemblance to a vein (p. 136) in that each fills a fissure in rocks. However, a dike is a body of intrusive igneous rock, whereas a common type of vein consists of a suite of minerals that were precipitated out of water solutions that once filled a fissure.

A batholith (Greek: "deep rock") has been defined as a huge intrusive mass of granitic rock that exceeds 40 square miles (100 sq km) in surface area—an arbitrary figure customarily used to distinguish it from a similar but smaller intrusive mass called a stock. Batholiths tend to cut across the rocks they have invaded (Fig. 5–17), but in places their margins parallel the stratification or other structures in neighboring rocks. Batholiths may be very large; some are exposed across tens of thousands of square miles of the Earth's surface and through a vertical range of more than a mile (from mountain top to valley bottom). Data from surface exposures and drill holes show that many batholiths become larger at increasing depths, but this applies only to the top few miles or less. At still greater depths, geophysical evidence suggests that a number of batholiths may be sheetlike or taper downward like a gigantic tooth.

Batholiths consist of coarsely granular rocks with a high or intermediate silica content, and large batholiths are made up of several kinds of these granitic rocks. Batholiths formed far below the surface and are located in the cores of mountain belts or in eroded zones that were probably once mountainous. They tend to be found only on the continents. Thus batholiths are large deep-seated masses that raise many major questions concerning their origin. These are

5-14. Crater Lake in Oregon is one of the world's most scenic spots—fully as beautiful as Oregonians and others say it is. The lake occurs within the caldera of a topless volcano that was once high enough to have a radial system of valley glaciers. Two of the rounded glaciated valleys show in the photo (see also Fig. 5-15). *(Courtesy U.S. Air Force.)*

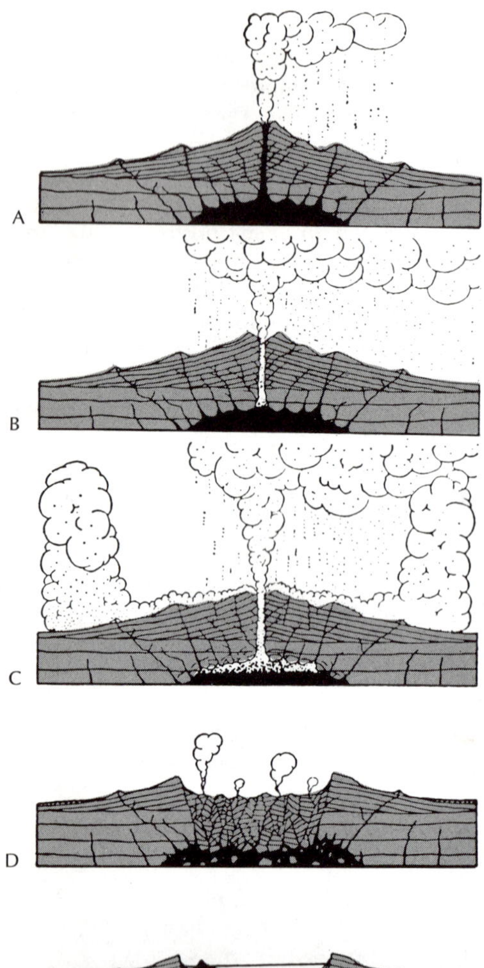

5-15. The evolution of Crater Lake (A) Beginning of culminating eruptions after growth of a large composite volcanic cone. Magma is high in the conduit and there is a mild eruption of pumice. (B) Activity increases in violence. Showers of pumice are more voluminous and the ejecta are larger. The magma level lowers to the top of the feeding chamber. (C) Activity approaches the climax. There is a combination of vertically directed explosions and glowing avalanches (nuées ardentes). The chamber is being emptied rapidly, and the roof is beginning to fracture and founder. Magma is also being drained from the chamber through fissures at depth. The carbon-14 content of trees buried by the climatic eruptions indicates that this event occurred about 6600 years ago. (D) The cone collapses into a jumble of enormous blocks, some of which are shown sinking through the magma. Fumaroles appear on the caldera floor. (E) Crater Lake today. Post-collapse eruptions have formed the cone of Wizard Island, and another cinder cone that is completely submerged; lava covers part of the lake bottom. Magma in the chamber has subsequently hardened into rock. *(Courtesy Howell Williams.)*

tied in with the origin of magma and of granite (see below).

Identification of a rock outcrop as part of an intrusive mass leads to the following interpretation: (1) the rock was once molten magma, (2) it is younger than the rocks into which it was intruded, and (3) erosion has removed the rocks which once covered this outcrop.

Origin of Magma

Lava (Italian: "flood") is the molten rock-making material, together with its associated gases, that flows out of volcanic vents and fissures, but the term is also applied to the rock that solidifies from this material. *Magma* (Greek: "dough") is the name used for this material when it occurs below the surface where the gas content tends to be greater. Both lava and magma may contain solid mineral grains which have crystallized from them. A magma, therefore, is hot; it is mobile; and it may consist entirely of a natural silicate melt or be a mixture of liquid and solid in which the solid material predominates.

Magma and lava are solutions in which the silica (SiO_2) content may vary from about 40 to 80%; aluminum, iron, calcium, sodium, potassium, and magnesium are the other abundant elements. Although all variations have been found, one may speak of a high-silica magma (about 70% or more of silica), a low-silica magma (some 50% or less of silica), and an intermediate-silica magma. The aluminum content is fairly constant, whereas the iron, calcium, and magnesium vary inversely with the silica. Low-silica magmas are fluid and harden into

5-16. Shiprock, New Mexico. As a stump records the former presence of a tall tree, so this volcanic neck and radiating dikes are remnants of a once-majestic volcano. The neck extends more than 1000 feet above the surrounding countryside. *(Courtesy W. T. Lee, U.S. Geological Survey.)*

dark-colored rocks such as basalt, whereas high-silica magmas tend to be viscous and harden into light-colored rocks such as granite.

Granitic rocks are far more voluminous than all other kinds of intrusive igneous rocks, and basalts have a similar position among the extrusive group (basalt may also be intrusive). This is anomalous in the eyes of some geologists who assume that the extrusive igneous rocks should correspond in composition to the intrusive group. Of course, basaltic magmas are more fluid and thus more likely to be extruded. Moreover, basaltic and granitic magmas may form in different ways.

Lavas range in temperature from about 1100° to 2200°F (600° to 1200°C) and low-silica basaltic lavas generally have the highest temperatures. Specimens of basalt and granite can be melted in a laboratory at somewhat higher temperatures, but the presence or absence of water is important here; in general, the addition of water lowers the melting point, and basalt melts at a higher temperature than granite. Magma temperatures far below the surface must be higher than those for surface lavas because of the greater pressures, but how much higher is uncertain as is the quantity of dissolved water.

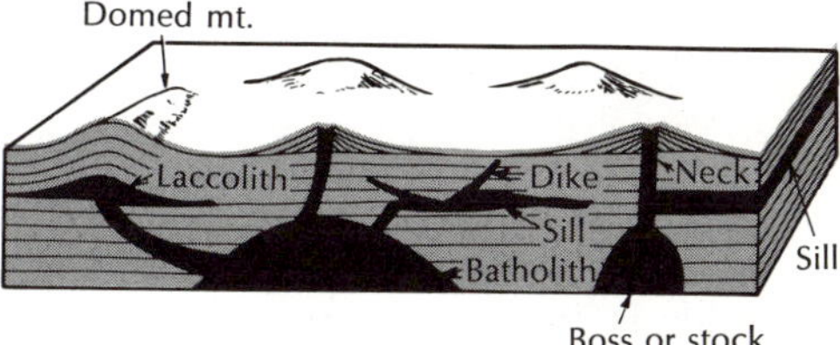

5-17. Types of igneous intrusives. A laccolith resembles a sill because the magma is squeezed between rock layers. However, the laccolith is lens-shaped and domes the rocks above it. A stock is a small batholith. (Namowitz and Stone, *Earth Science,* Van Nostrand Reinhold, 1965.)

Plateau Basalts

At times basaltic lava has been extruded widely on the lands and on the ocean floor and in enormous volumes; e.g., such flows in western India once covered an area of nearly 500,000 square miles to a maximum thickness of almost 2 miles. Lava flow after lava flow, averaging about 15 feet in thickness, emerged in the distant past from fissures in the Earth's crust, piled one on top of the other, and spread widely. Similar flows have occurred from time to time in many parts of the Earth—the Columbia Plateau in the northwestern United States is such a feature (Fig. 5–12)—and a world-

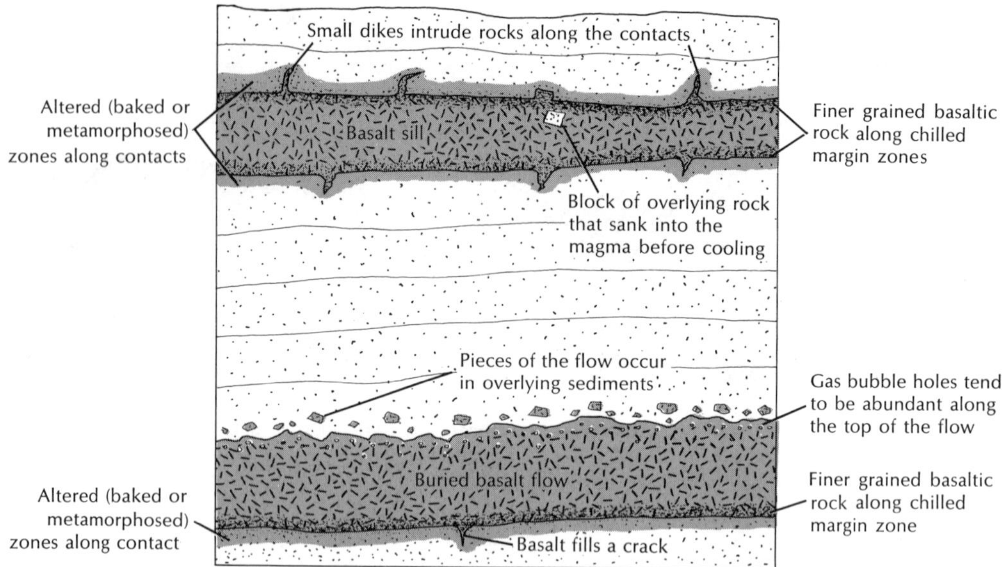

5-18. Some distinctions between a sill and a buried lava flow in horizontally layered sedimentary rocks. Note that the upper contacts are more crucial than the lower ones.

wide source is assumed to exist at depth (probably in the upper mantle).

Localized Fusion (Melting) Produces Basaltic Magma

Let us now consider the origin of basaltic magma and pressures and temperatures in the crust and upper mantle. Measurements in wells and mines in the outer 5 miles of the Earth's crust show that temperatures increase downward at an average rate of approximately 1°F every 50 to 150 feet (1°C per 30 to 100 meters; the geothermal gradient). Hot springs, geysers, and volcanic phenomena also show that temperatures rise with increasing depth. If this average increase in temperature continues downward, the source rock assumed to exist in the upper mantle may be hotter than the temperature at which it would melt at the surface. However, for most solids, melting necessitates expansion, and pressures at such depths may be so great that melting cannot occur.

Accordingly, a worldwide layer of solid low-silica rock may exist in the upper mantle at temperatures that are above its melting point at the surface. If this is correct, then the origin of basaltic magma probably involves the localized fusion of this source material either by localized heating or by localized reduction in pressure.

Fusion might occur within the source rock without an increase in temperature if the pressure could be reduced enough locally to permit the expansion that would allow melting to occur, and several possibilities have been suggested: great cracks may develop in the outer part of the Earth in earthquake areas, erosion may lighten a once-mountainous region, or upwarping of the crust may occur. Reduction in pressure might also allow a quantity of water vapor and other gases to separate out of some mineral grains, and their presence should then lower the melting point still further. However, local reduction in pressure may be unlikely at great depths where hydrostatic (lithostatic) equilibrium presumably exists; i.e., the confining pressures on any one point are the same in all directions. Isostatic adjustment is one indication of such an equilibrium.

If local reduction of pressure does not solve the problem, then localized increase in temperature presumably occurs, but we can only speculate about causes. Local concentrations of radioactive elements could produce the necessary heat energy, yet tests with Geiger counters on flowing lava show no such concentrations. Friction along fault surfaces at depth has been proposed, but this seems inadequate for the generation of very large volumes of magma.

Another area of uncertainty involves the fusion process—it may be complete or selective; i.e., basaltic magma might form by the complete fusion of rock with the same chemical composition as basalt, or by the selective melting of the more readily fusible minerals in a source rock of somewhat different chemical composition. Laboratory experiments have shown that a rock does not melt as a unit at any one temperature; rather, there is differential melting throughout a range of a few hundred degrees before complete fusion has taken place. Moreover, if fusion of a homogeneous source rock occurs at different depths, different magmas will probably form because the melting points of some minerals do not change uniformly with increases in pressure.

Thus basaltic magma probably forms because localized regions of the upper mantle are heated and melt. However, we do not know how they are heated or whether the melting is selective (seems more likely) or complete. Perhaps the cause is some deep-seated movement such as a convection current that occurs in the mantle.

Depth at Which Fusion May Occur

We have conflicting, inconclusive evidence concerning the depth at which fusion occurs. In Hawaii, volcanic tremor, a rhythmic trembling of the ground, originates at a depth of a few miles in association with moving magma (Fig. 5–7). Similar tremor has been detected recently that originated at depths of about 30 miles (50 km), which is also the depth at which many earthquakes occur in these areas. However, in other parts of the Earth the source region may be about twice as deep (see low-velocity zone, p. 165). Thus basaltic magma may originate at different depths in different regions, which may account for certain differences in the chemical compositions of basaltic magmas.

Once magma has formed, it is less dense than the solid material surrounding it, and the weight of the overlying rocks may then force the magma upward through a crack or weak zone. Analogously, if cracks develop in the ice covering a lake, the weight of the ice may push downward on each side of the crack and force water onto the surface. In addition, as magma moves upward, its dissolved gases expand and aid the lifting effect. The upward-moving magma may cool and harden before it reaches the surface to form intrusive rocks, or it may be extruded at the surface.

Origin of Granite

Granitic magma apparently has formed in two ways, and granite itself probably has three different modes of origin. Granitic magma has probably formed by (1) the partial or complete fusion of a portion of the crust made up of rocks such as sandstone, shale, and gneiss (huge volumes may have formed in this manner) and (2) by the modification (differentiation) of basaltic magma, but the capacity of this process seems limited.

Granitic rocks may have formed by crystallization from either of these magmas and also by the metamorphism *(granitization)* of older rocks without the presence of a magmatic phase. Thus granite may be an igneous rock or a metamorphic rock, and definite criteria are lacking to tell one from the other in hand specimens and even in the field.

Magmatic Differentiation May Produce Granitic Magma

Continuous gradations in chemical composition exist between low-silica magma that hardens to basalt or gabbro and high-silica material that produces granite when it cools slowly far below the surface. Accordingly, it has been proposed that basaltic magmas may change into other kinds of magmas through a process called *magmatic differentiation.* Imagine that a large volume of low-silica basaltic magma moves upward through a weak zone in the crust, and that its upward movement stops several thousand feet below the surface. Here the top and outer margins of the basaltic mass cool (Fig. 5–19), and crystallization begins within these more rapidly cooling marginal zones. The first minerals to crystallize tend to be types that take large proportions of iron, calcium, and magnesium, but relatively little silica; these solid crystals then sink. As only a relatively small amount of silica is removed, the proportion of silica in the remaining magma increases within the regions where the crystallization is occurring. Thus the percentage of silica in the magma near the top of the igneous reservoir may gradually increase until eventually it becomes granitic in composition.

At any stage in this process, and from any depth, part of the magma may resume its upward movement. If the resurgent magma comes from the upper portions of the igneous chamber after crystallization has been going on for some time, its silica content will be higher than that of the original basaltic material. Therefore, any gradation between a high- and a low-silica magma is theoretically possible. In addition to differentiation, the chemical composition of a magma may be modified if some of the rocks it encounters are broken off, melted, and assimilated.

Presumably the dense, first-formed low-silica minerals sink in the molten matter until they are melted or come to rest on the floor of the igneous chamber. But other crystals that are less dense may also form and float to the top. Therefore, an intrusive igneous body, such as a thick sill, may show a crudely layered structure in which certain minerals are more concentrated at certain levels than at others. The Palisades along the Hudson River in New York is such a sill (in places it cuts across the enclosing beds and becomes a dike), and the concept of magmatic differentiation grew out of studies of similar structures.

If a series of intrusive or extrusive igneous masses have formed in an area, the lower-silica members tend to be older than the higher-silica ones. This general order of intrusion or extrusion likewise supports the concept of magmatic differentiation be-

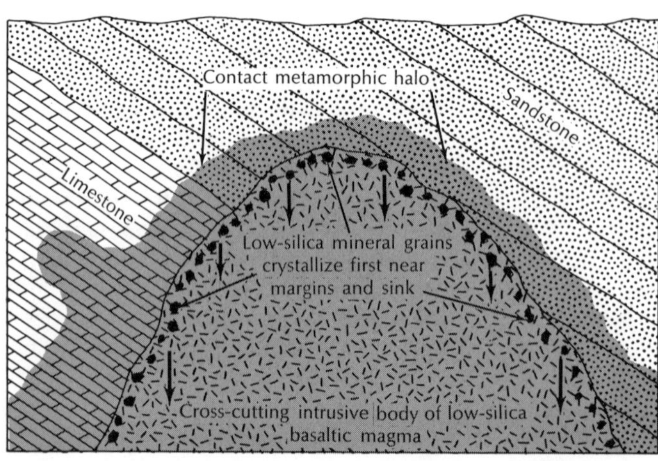

5-19. Magmatic differentiation illustrated schematically. Tilted sedimentary rocks have been intruded by a mass of basaltic magma and metamorphosed along its margins. The metamorphic halo enclosing the intrusive is thicker in the more easily metamorphosed limestone than in the sandstone.

cause it indicates that the silica content of the parent magma has increased during crystallization.

Thus, convincing evidence shows that differentiation has occurred, but the volume of granitic magma formed in this manner is uncertain. Differentiation seems quite inadequate to produce the thousands of cubic miles of granitic rocks that occur in some batholiths. One would expect that differentiation in a basaltic magma would produce a relatively small volume of granitic magma, a greater volume of medium-silica magma, and that very large volumes of low-silica magma would remain. Thus the medium-silica intrusive rocks should be more abundant than granite, and the low-silica intrusives should be the most voluminous of all. This is not the case.

Granitic Magma Produced by Fusion in Mountain Belts

Most granitic magma may form independently of basaltic magma and may involve the melting of the lower part of the Earth's continental crust within a great mountain belt. Sediments have piled up in such elongated zones to thicknesses of several miles, and subsequent crustal movements have shifted them still deeper (see geosynclines, Chapter 12). Fusion may have occurred locally along the base of such a mass and may involve sedimentary, metamorphic, and older igneous rocks that have approximately the same proportions of silica and other elements as granite. The fusion may be complete or it may involve only the selective melting of the more fusible components to produce liquids here and there in the interstices in the once-solid rock.

When enough liquid has accumulated, the mass becomes mobile and may move upward as a plastic mixture of liquid and solid materials, i.e., it has become a magma. However, much of it may also crystallize subsequently in or near the zone in which it formed, which might solve the problem of batholithic emplacement (see below). Such a sequence of events may account for the origin of the huge granitic batholiths that are restricted geographically to mountain belts on the continents.

A major problem involves the emplacement of large granitic batholiths. If granite is an intrusive igneous rock, then the space now occupied by a granitic batholith must once have been occupied by other rocks. How did the granitic bodies find or make room for themselves? No known process or combination of processes seems really adequate because thousands of cubic miles of rocks are involved in some batholiths. The problem is unsolved unless much of the granitic rock has been generated about in the place it is found today, and such may be the case.

Origin of Granite by Metamorphism (Granitization)

The discovery of granitic rocks that contain traces of sedimentary stratification and certain other features has led to still a third explanation for the origin of granitic rocks—their formation by metamorphism or granitization. One group of geologists, facetiously called the Soaks, maintain that some granites are metamorphic rocks that have formed by replacement, much as wood is petrified, when emanations from below transformed rocks in the crust into granite. Perhaps this is accomplished in rocks far below the surface by the addition of certain ions, such as those of sodium and aluminum, via gaseous or liquid solutions. Other ions, such as those of iron and calcium would have to be removed by the solutions. The diffusion might also occur in the solid state, although this process may be effective only across small distances.

In this controversy, the field relationships between a granite batholith and its neighboring rocks are interpreted differently by different observers. One observer (a Soak) sees the gradation from a massive granite, through a granite with faint sedimentary relic structures, into an unchanged sedimentary rock as that from a completely granitized rock, through a partially granitized zone, to the unchanged sedimentary rock. A magmatist, on the other hand, explains the same rocks as having been formed by the emplacement of an intrusive granite body and its metamorphism of the invaded sedimentary rocks, including some granitization.

As supporting evidence, the magmatist can point to high-silica lava flows—of unquestioned magmatic origin—that have minerals, textures, and other features very similar to those of many granites, except that the grain sizes are smaller. In addition, many granite bodies have sharp margins that are traceable for miles—these seem unlikely to be produced by granitization—and thin tongues of the granite have intruded outward from the main body as dikes and sills. Furthermore, the homogeneity observable across large portions of a batholith is also a difficult feature to explain by the granitization of different types of rocks.

Most geologists are agreed that granite has formed in the three ways described above and that the chief problems lie in determining which manner of origin is the more common and how each process operates.

Relationship of Some Ore Deposits to Magmas

An *ore body* is a rock or mineral deposit from which one or more metals can be extracted profitably. Most ore bodies represent concentrations of relatively rare metals that at one time were widely dispersed throughout a much larger volume of rock. Thus the geology of ore deposits is fundamentally involved with the processes that have caused such concentrations.

According to one view, magmas are the source material from which most primary ore deposits have formed, and a number of processes are probably involved in the concentration of ore minerals: magmatic segregation, contact metamorphism, hydrothermal solutions, fissure filling, and replacement are some examples.

Let us look briefly at some aspects of the theory that links the origin of primary ore deposits with magmas. Our story can begin with the intrusion of a fairly large mass of high- to medium-silica magma into sedimentary rocks such as limestone and shale. The magma consists primarily of mutually dissolved silicates of aluminum, iron, calcium, sodium, potassium, and magnesium, but scattered widely in it are dissolved gases and ions of such metals as gold and copper. However, the kinds of metals present, as well as their relative abundances, vary from magma to magma.

As cooling and crystallization proceed, minerals such as pyroxene, feldspar, and mica form. However, these rock-forming minerals do not generally add the scattered metals and gases to their developing crystals (mica uses water). Thus the volume of magma shrinks during crystallization, and the gases and metals are concentrated in the dwindling portion that remains. During this crystallization, *magmatic segregation* may occur; i.e., certain valuable minerals, such as those containing chromium, copper, and iron, may crystallize early, settle out, and accumulate into a valuable deposit. Or the valuable metals may be segregated later, perhaps as a residual liquid, and collect within the intrusive or be expelled into neighboring rocks.

Pegmatites (p. 79) seem to form during a late magmatic stage, especially from high-silica magmas, and appear to be transitional between an earlier, magmatic, rock-making phase and a later phase involving hot water solutions. Some contain valuable minerals.

Heat from the magma metamorphoses the invaded rocks to produce an enveloping *contact metamorphic halo* (Fig. 5–19). But more important for the formation of ore deposits, gases and liquids escape under pressure from the intrusive mass and transport outward some of the valuable metals formerly dispersed in the magma. As the fluids move through the contact metamorphic zone, their dissolved metals may be deposited here and there as other materials are removed to make space available (Fig. 5–20). This *replacement* of part of a contact

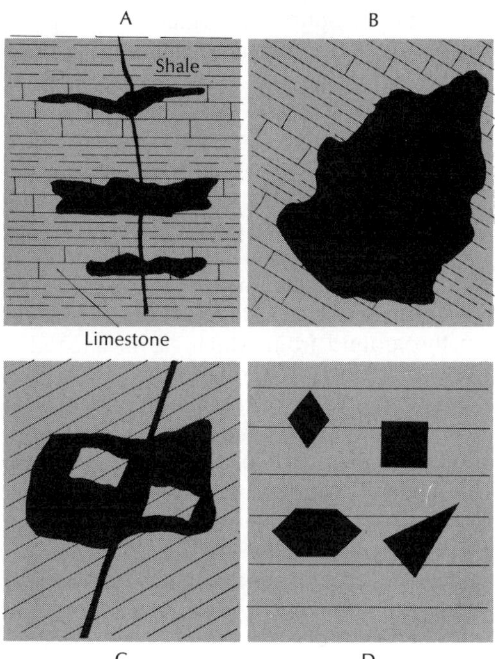

5-20. Evidence that some ore deposits form by replacement (i.e., the mineral grains that previously occupied part of a rock have been removed and replaced by the new minerals that make up the ore deposit). In A, limestone is replaced more readily than shale; therefore, the ore deposits occur only within the limestone. In B, traces of the original stratification are preserved as evidence of replacement. In C, the process of replacement was incomplete. If the ore deposit had formed as a cavity filling, these unreplaced portions would not be aligned with the surrounding rock. The crystals shown in D are younger than the rock that encloses them. *(After Bateman.)*

metamorphic zone by an ore deposit occurs because certain minerals are more stable than others under the existing conditions. Thus a mineral such as calcite may be dissolved and one such as garnet or galena may replace it in a volume-for-volume substitution. As some rocks are more susceptible of replacement than others, the metallic deposits are scattered here and there within the favorable rocks of the contact metamorphic halo like the nuts in a pecan ice cream.

Near the end of the process of crystallization, the once widely dispersed gases and metals may have collected near the top of the intrusive body. In moving up, the gases ooze and stream through the magma and pick up some of the metals. At this stage, pressure may force the gases and their dissolved rare elements to leave the magma chamber and to move along zones of weakness toward the surface (Fig. 5–21). Such fluids may begin their journey upward as liquids, or as gases which later become liquids, and water is their chief constituent. These hot water (hydrothermal) solutions rise into zones of lower pressures and temperatures where they can no longer hold in solution all of the dissolved materials. Chemical reactions with wall rocks also cause precipitation. The substances in solution are precipitated, perhaps in fissures to form veins (Fig. 5–22), and other deposits form by replacement of the wall rocks along the fissures. Such deposits represent very

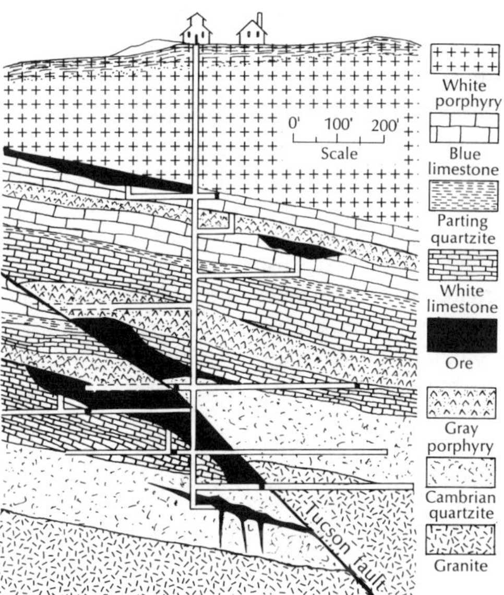

5-21. Structure section showing occurrence of lead and zinc ore bodies at the Tucson Shaft, Leadville, Colorado. The ore-bearing fluids apparently moved upward along the fault surface and tend to occur within the limestones because these are more readily replaced. *(After Argall.)*

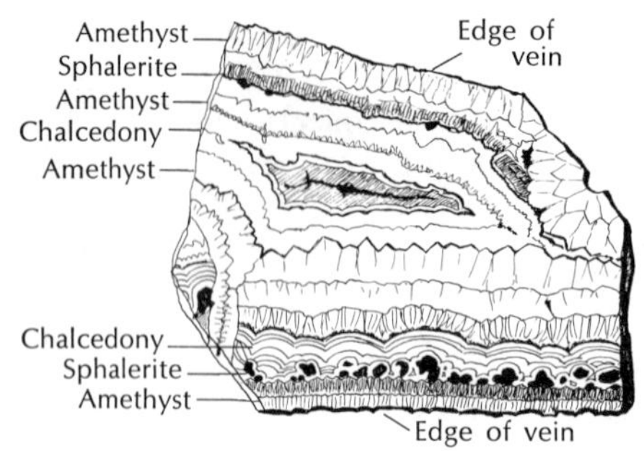

5-22. Ore vein occurs in former cavity in rocks. The cavity was once filled with water and the different minerals were precipitated from solution in succession as conditions changed (e.g., temperature, pressure, and chemical content of the water). (J. Sinkankas, *Mineralogy for Amateurs,* Van Nostrand Reinhold, 1964.)

great concentrations of materials once widely dispersed, and according to this theory, magma is the source material.

This explanation of the origin of ore deposits developed partly from studies made at fumaroles (see below) and hot springs. During observations, cracks were filled by minerals such as galena and hematite, and these were deposited from gases that carried lead, iron, sulfur, and other elements in solution. Most deposits are also associated spatially with intrusives and certain gradual changes can be observed as deposits are traced outward from the parent igneous material. However, some ore bodies have no direct known relationship to a parent magma, and circulating ground water seems to have played a key role in their origin.

Geysers, Fumaroles, and Hot Springs

Volcanoes and intrusive igneous material may give off hot gases which produce a number of interesting phenomena as they emerge from openings in the ground. Geysers, fumaroles, and some hot springs originate in this manner, and steam is the chief gas involved. Some of the steam has a magmatic source, but much more probably originates by vaporization of ground water upon contact with upward-moving hot magmatic gases. However, not all hot springs are related directly to igneous activity. Ground water may percolate downward considerable distances to zones of higher temperature, and later the heated waters may rise along a fault or other channelway and emerge at the Earth's surface.

Fumaroles (Fig. 5–3) are holes or fissures in the ground from which steam and other gases issue. Hot springs (Fig. 5–23) and fumaroles are of practical importance to man: in laundries, baths, and the heating of buildings (e.g., half of Iceland's population has central heating from hot spring sources), and as sources of steam for the generation of electricity (Fig. 5–24).

Geysers are hot springs which occasionally erupt columns of hot water and steam, and most geysers occur in three areas: Yellowstone National Park, New Zealand, and Iceland. Geysers interest many people, and on a small scale they illustrate some of the spectacular effects of volcanic activity. Old Faithful in Yellowstone National Park (Fig. 5–25) is justly famous as the tourist's friend. Approximately 66 minutes elapses between eruptions that throw thousands of gallons of hot water higher than 100 feet into the air, and Old Faithful has been performing at these intervals winter and summer since its discovery by white men in 1871. Other geysers have since become extinct and new ones have formed. Altogether there are several thousand geysers, hot springs, and fumaroles in Yellowstone National Park, but some geysers are transformed into hot springs during the wet season, just as certain hot springs become fumaroles during dry seasons. A number of changes resulted from an earthquake in 1959 because some underground channelways were plugged and new ones were formed at this time.

The action of a geyser probably depends upon the relationship between the boiling point of water and pressure. An increase in pressure raises the boiling point, and a decrease lowers it. Familiar examples are the ease with which water in a radiator boils at high altitudes (low pressure allows a low boiling point) and the short time interval in which foods can be made edible in a pressure cooker (high pressure produces a high boiling point).

Below a geyser vent, a series of interconnected fissures and openings are filled with hot water, chiefly ordinary ground water (Fig. 5–26). Heat is supplied by steam rising from below. At first the steam is assimilated, and the temperature of the water rises until it boils. However, because pressure increases downward from the surface, the boiling point is reached only at higher temperatures at greater depths (e.g., at a depth of 500 feet, water will not boil until the temperature is about 428°F). If the passageways are tortuous enough to prevent

5-23. The Paint Pots in Yellowstone National Park, Wyoming. Hot water rises through fine-grained materials, such as clay produced by the weathering of volcanic rocks, to form a muddy mixture. Hot gases escape as bubbles that pop here and there at the surface and remind one of hot cereal cooking on a stove *(Courtesy U.S. Geological Survey.)*

5-24. Natural steam development at The Geysers, Sonoma County, California, 1964. This is a region of fumaroles, hot springs, and volcanic rocks along a fault zone. Deep wells encounter an inclined fault at depths where pressures and temperatures are greater (rocks are at temperatures of about 600°F at 600 feet, but this varies). Steam is piped from geothermal wells to the electrogenerating plant. *(Courtesy Pacific Gas and Electric Co.)*

5-25. Old Faithful in Yellowstone National Park. *(Courtesy Northern Pacific Railroad Co.)*

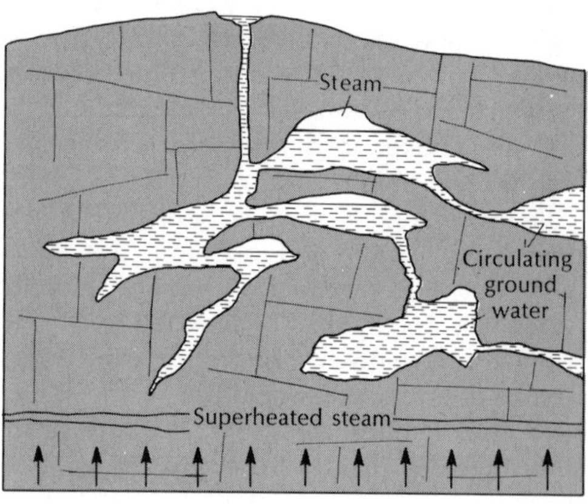

5-26. The origin of geysers. *(After Day and Allen; courtesy The Carnegie Institution of Washington.)*

convection, water in a geyser tube will everywhere be heated to its boiling point. When all of the water in the geyser tube is boiling (at different temperatures at different depths), steam must accumulate. When sufficient steam has accumulated, its expansive force causes overflow at the surface and the consequent upward movement of water throughout the system. This upward movement transports water into zones of lower pressures where it can boil at lower temperatures. It flashes instantaneously into steam and erupts.

Exercises and Questions for Chapter 5

1. Volcanic cones and eruptions:
   (A) What causes volcanic eruptions? Why do eruptions differ in violence from one volcano to another? from one time to another at the same volcano?
   (B) Why do volcanic cones differ in shape?
   (C) Why do craters tend to occur at volcano tops? How does a crater differ from a caldera?
   (D) How might one account for a cinder cone that has an elliptically shaped ground plan?
   (E) What is the relationship of some dikes to volcanic cones?

(F) Can volcanic eruptions be predicted?
(G) Why are volcanoes likely to occur along a line in any one region?
(H) Why are the surfaces of some lava flows conspicuously different from those of other flows? Where does lava usually flow out of a volcanic cone?

2. Make a topographic map of an island that is formed by the top portion of a volcanic cone. The top of the island is 910 feet high and has a crater 70 feet deep at the top. Use a 20-foot contour interval and choose a convenient horizontal scale.
3. How does a dike differ from a sill? from a vein?
4. What problems are involved in the emplacement of a granitic batholith?
5. Discuss some of the factors involved in the origin of basaltic magma and describe the evidence that supports your statements.
6. Discuss the origin of granite and describe some of the evidence that supports each of the three main hypotheses involved.
7. Why have magmas been called the parent materials of some ore bodies?
8. Do fossils occur in igneous rocks?
9. What problem occurs in the classification of rocks that form from debris erupted from volcanoes?
10. Describe one way in which the eruption of a geyser may be accounted for.

6

Rocks Bend, Break, and Flow

Powerful forces have acted upon the Earth's crust many times in the geologic past and are acting upon it today. Some parts of the crust have been squeezed together, but other parts have been stretched apart; some areas have been uplifted, whereas others have been depressed; and such movements have been slow or fast as well as large or small. The crust itself is apparently much more mobile than was once thought likely, and processes like continental drifting and ocean floor spreading now have many enthusiastic advocates and strong supporting evidence (Chapter 12).

Depending upon conditions of pressure, temperature, composition, solutions, and time, rocks have reacted to these forces by cracking, faulting, folding (Fig. 6–1), and becoming metamorphosed. A sudden rupture of part of the crust or upper mantle produces an earthquake, and some of the waves from a powerful earthquake pass through the Earth. Their records at different stations provide the most illuminating evidence available concerning the nature of the interior of the Earth. In effect, they X-ray the Earth and show its core, mantle, and other features.

In addition, high-temperature, high-pressure laboratory experiments have been made on rocks and minerals and may duplicate conditions occurring tens of miles below the surface. Furthermore, the Earth's behavior as a planet provides additional data on the nature of its interior.

Folded and Tilted Strata

Most of us think of rocks as hard, brittle substances incapable of any change of shape without fracturing. Yet even at the Earth's surface, rocks have been known to bend a little without breaking; e.g., in some monuments, limestone slabs supported only at their ends have sagged slightly under their own weight in a few hundred years. Under powerful confining pressures in laboratory experiments, rocks of certain kinds have been made to flow *plastically* (i.e., to change shape by slow continuous movement without breaking). When crustal forces act slowly, layered rocks under great pressure tend to be deformed by warping or folding. In contrast to the warping that produces relatively slight, irregular bends in rocks, folds commonly show a pairing of crests and troughs, more uniformity of structure, and a greater amount of bending.

Folding in layered rocks may take place by plastic flow, by microscopic fracturing and slipping, and by the sliding of one layer over another. In the latter process, which is an important one in stratified rocks, any one stratum arches upward over the bed lying beneath it. What happens can be illustrated if a package of file cards is bent by pushing inward on the ends. The top layer slides the greatest distance. However, folds also occur in igneous and metamorphic rocks. Major folds result from the compression of an area by powerful forces which act more or less parallel to the Earth's surface. Crustal shortening occurs in the folded area, but this may affect only a relatively thin portion of the upper crust.

Upfolds forming arches are called anticlines and downfolds forming troughs are termed synclines. As the names indicate, beds incline away from the crest of an anticline and toward the trough of a syncline. Anticlines and synclines are common kinds of folds and tend to occur together like waves on a lake surface. Such folds vary in scale from small ones that are measured in inches to huge ones several miles and more across. Folds may be upright and symmetrical, or one side may be steeper than the other (Fig. 6–2).

The *axis* of a fold is a line along the crest of an anticline or along the trough of a syncline, and this axis may be horizontal or it may slant or plunge (Fig. 6–3). After erosion has partly dissected a series of folds,

6-1. Striking folds in the Humber Arm Formation at Port au Port Bay, Newfoundland. (Geological Survey of Canada photo no. 120457 by L. M. Cumming.)

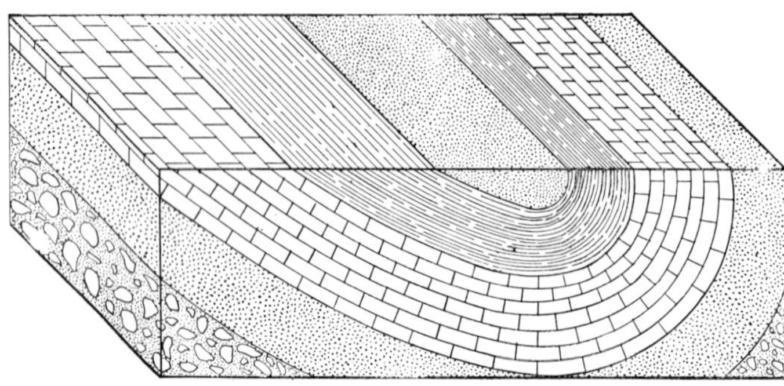

6-2. Block diagram of an asymmetrical syncline with a horizontal axis. Note that the youngest layers are exposed along the axis. *(J. M. Trefethen,* Geology for Engineers, *Van Nostrand Reinhold,* 1959.)

we note that the youngest layers in a syncline are exposed along its trough, whereas the oldest layers in an anticline occur along its eroded crest. The squeezing of rocks during folding may cause plastic flow that thickens beds along crests and troughs and thins them along the sides (limbs) of folds.

Folds seem generally to form at some distance below the surface where rocks are less brittle. Thus they were not necessarily associated closely with the surface topography that existed above them at the time of their formation; i.e., an anticline was not necessarily a ridge at the time of its formation. Furthermore, following prolonged erosion, a syncline may form the crest of a mountain and an anticline underlie a valley (Fig. 6–4).

A series of ridges and valleys are produced by the erosion of a sequence of tilted strata that differ in their resistances to erosion. The strata may be tilted on the flanks of folds, or they may have no association with folds. Ridges are located where resistant rocks project out at the surface, and valleys develop along the zones underlain by weaker rocks.

The shape of a ridge may be determined primarily by the dip or slant of the strata that underlie it, and this can be at any angle from zero degrees (horizontal layers) to 90 degrees (vertical layers). Erosion of horizontal beds in a dry region tends to produce flat-topped steep-sided hills or plateaus. Erosion of gently dipping strata (perhaps 15

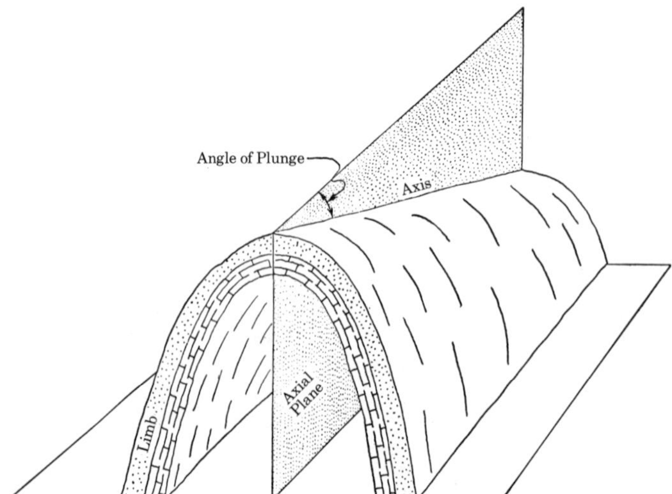

6-3. Terms used in describing a fold. The axial plane divides a fold into two halves that are as nearly symmetrical as possible. The line of intersection between this plane and any one bed thus marks the axis of a fold. An anticline with a plunging axis is shown. *(J. M. Trefethen,* Geology for Engineers, *Van Nostrand Reinhold,* 1959.)

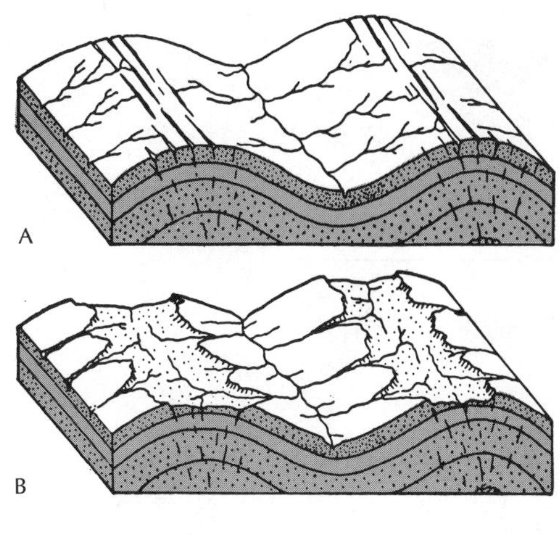

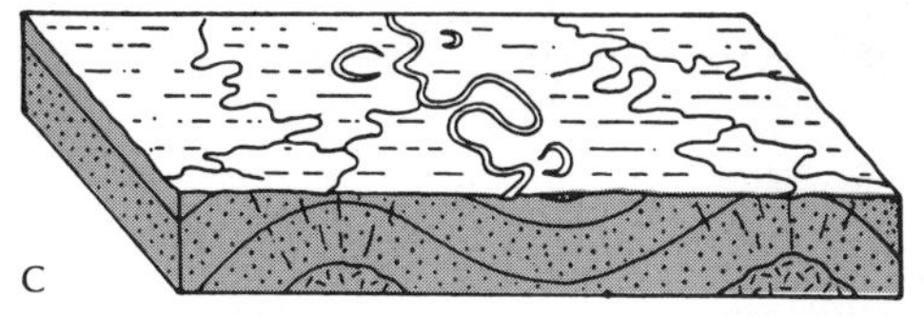

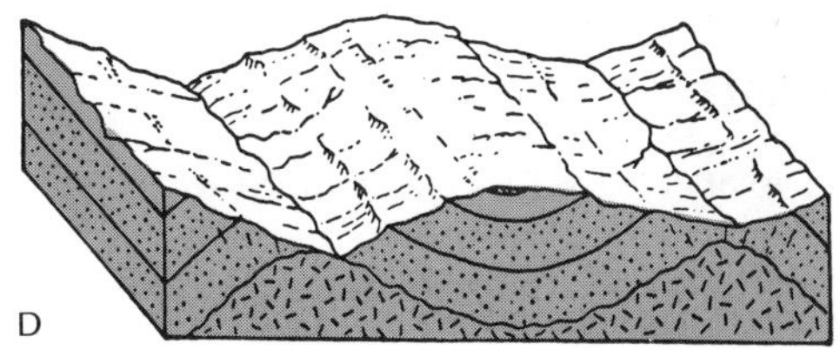

6-4. Origin of a synclinal mountain. (A) An anticline occurs on either side of a syncline; the fold axes are horizontal. (B) Erosion has broken through and worn away the layer of resistant rock that previously formed the crest of each anticline. (C) Continued erosion has worn down the area to a nearly featureless plain; a strip of the resistant rock still occurs along the trough of the syncline. (D) Uplift and more erosion has produced a synclinal mountain; its crest is formed by the strip of resistant rock. *(After Miller.)*

to 20 degrees or less) commonly results in the formation of asymmetrical ridges known as *cuestas.* The gentler slope of a cuesta parallels the dip of the underlying rocks and is an example of a *dip slope:* a land surface whose slope conforms approximately to that of the underlying rocks. On the other hand, erosion of steeply dipping strata tends to produce more symmetrically shaped ridges, and all gradations occur between steep and gentle dips and in the shapes of the ridges that result. Ridges likewise occur in humid regions along belts of resistant rocks, but the differences in resistance to erosion may be less conspicuous because of the type of weathering and mass-wasting that predominates.

The orientation of the axis of a fold has an important influence upon the landscape where a pile of sedimentary rocks of different resistances to erosion has been folded and eroded. If the axis is horizontal, the ridges and valleys tend to be parallel (Fig. 6–4), but if the axis plunges, loop-shaped or zigzag patterns of ridges and valleys result (Fig. 6–5).

Smaller folds can be seen in their entirety in a single outcrop, and a large fold, structural dome, or structural basin may be a striking sight on an aerial photograph (Fig. 6–6). However, others can sometimes be determined only by mapping and matching outcrops scattered over many square miles (Figs. 6–7 and 6–8). Topographic features have different shapes where flat-lying stratified rocks are being dissected by erosion.

In measuring the orientation of a certain bed, a fault surface, or any other planar structure in the crust, geologists refer to its strike and dip (Fig. 6–9). *Strike* is the direction of a horizontal line in the plane of a particular bed (fault, joint, foliation surface, etc.), and *dip* is the angle between this plane and the horizontal when measured at 90 degrees to the strike. In other words, strike is the direction or trend of an inclined layer across a horizontal surface (the trend of a layer across a sloping surface generally does not parallel its strike). If a sheet of plywood (representing a sedimentary bed) is tilted and then partly inserted into a pool of water, its line of intersection with the water is its strike. Dip is the angle between the water surface and the plywood sheet, and the direction of dip is that followed by a ball rolling down the sheet of plywood.

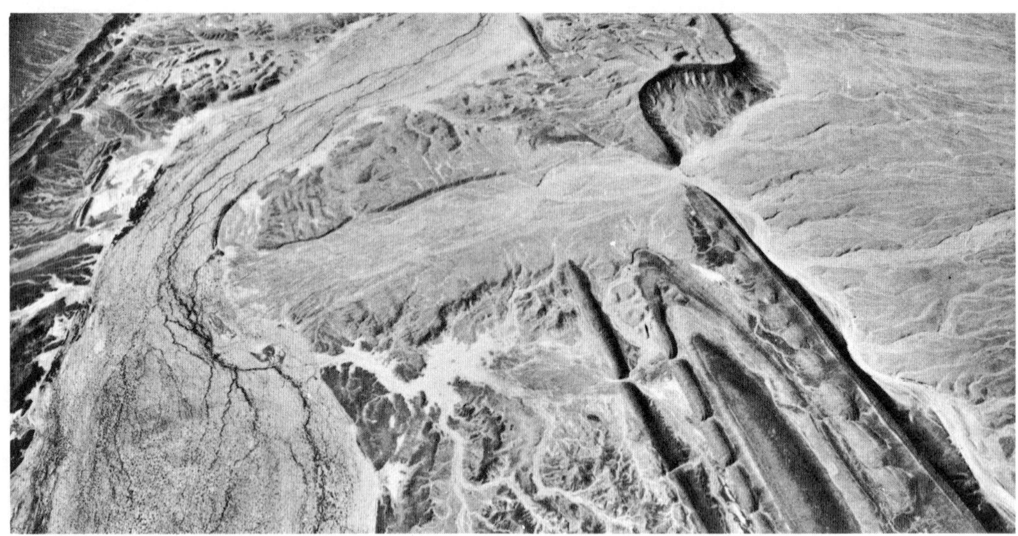

6-5. Land forms produced by erosion of folded rocks in Africa. The sedimentary rocks differ in their resistances to erosion, and ridges are formed where upturned resistant beds crop out. The ridges and valleys in the lower right outline a plunging anticline cut by water gaps, and a prominent water gap appears near the center of the photograph. (Photograph, U.S. Air Force: N. 28° 28′, W. 09° 35′.)

Joints

A *joint* (Fig. 6–10) is a fracture or break in a rock along which no appreciable displacement has occurred parallel to the joint surface, and in this absence of offsetting movement, it differs from a fault. Many joints probably result from crustal movements that have raised, lowered, squeezed, extended, and twisted the jointed rocks at

6-6.—An eroded structural basin in Africa. The youngest exposed rocks are located at the center of the basin, and successively older rocks occur outward from the center. *(Courtesy U.S. Air Force: N 32° 30′; W 03° 35′.)*

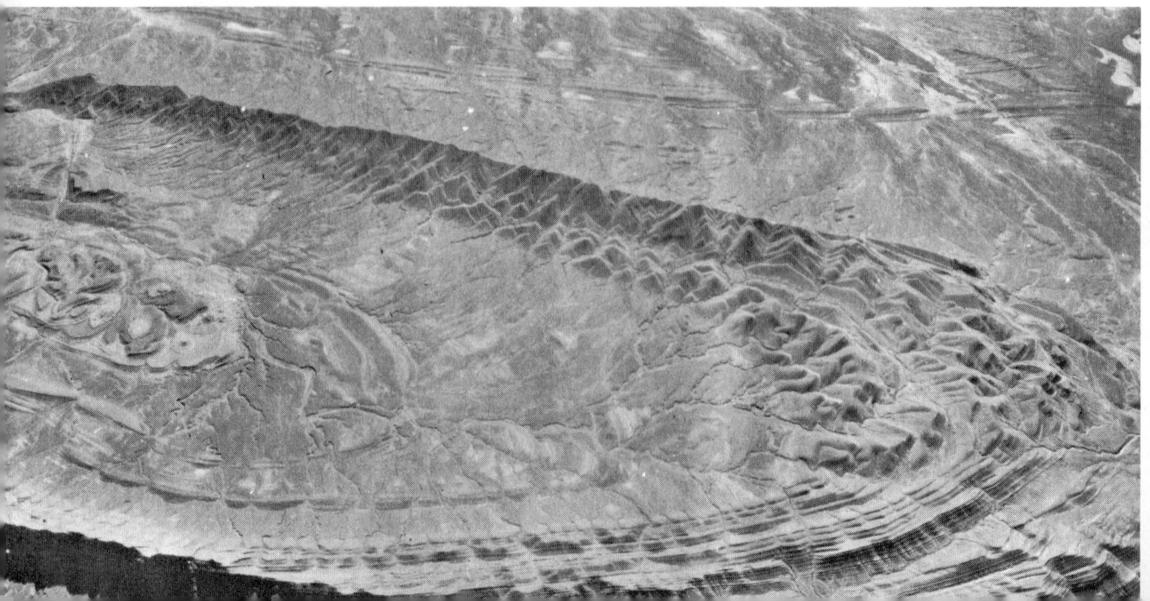

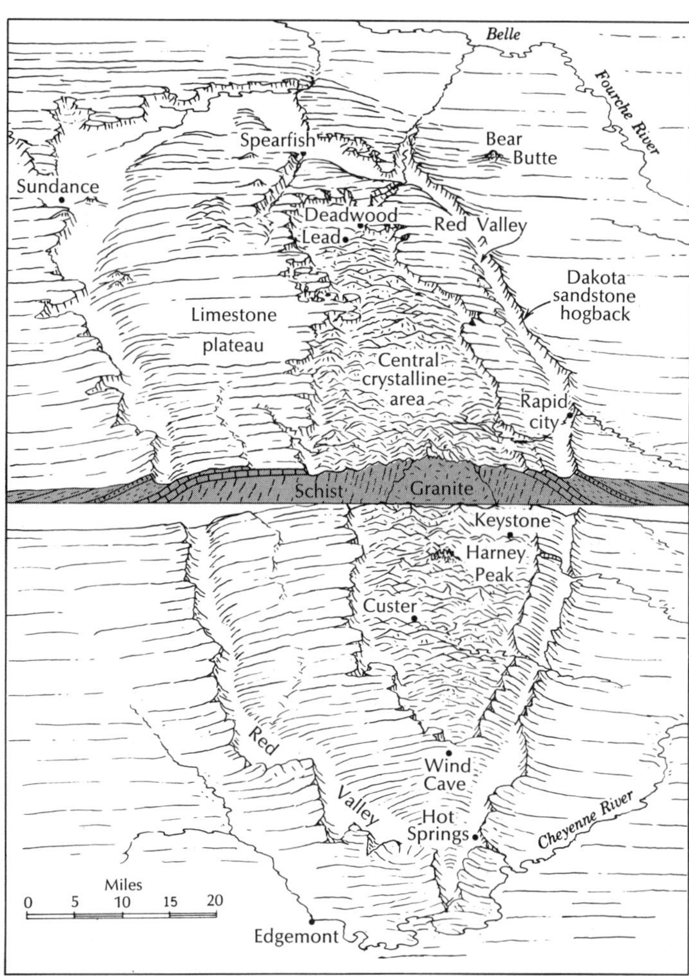

6-7. The Dakota Black Hills consist of a broad, flat-topped dome deeply eroded to expose a core of crystalline rock. Ridges occur where resistant rocks crop out and their steeper slopes face inward. Intervening low lands occur where less resistant rocks crop out. (A. N. Strahler, *Physical Geography,* John Wiley & Sons, Inc., 1951.)

one time or another. In igneous rocks, many joints are caused by the decrease in volume that occurs during cooling from a liquid to a solid (see columnar joints, p. 122). Joints may be scarce or numerous. They tend to occur in sets that are separated by distances commonly ranging from a few inches to several feet, and the joints in any one set are approximately parallel. Many rocks exhibit more than one set of joints, and two mutually perpendicular sets are common. Joints aid the processes of weathering by furnishing passageways for air, water, and other fluids to enter.

The arches in Arches National Monument (Fig. 6--11) illustrate the importance of joints in weathering and erosion. Here gently dipping layers of sandstone are cut by numerous, nearly parallel vertical joints. Apparently water enters the joints and dissolves some of the cementing material. This frees some of the sand grains, which are then removed by wind and water. The vertical cracks slowly become wider and develop into narrow steep-walled canyons separated by narrow steep-sided ridges (fins). More rapid weathering of one section of a fin eventually produces a hole that then develops into an arch. Delicate Arch is thus an isolated remnant of a fin; the rest of this

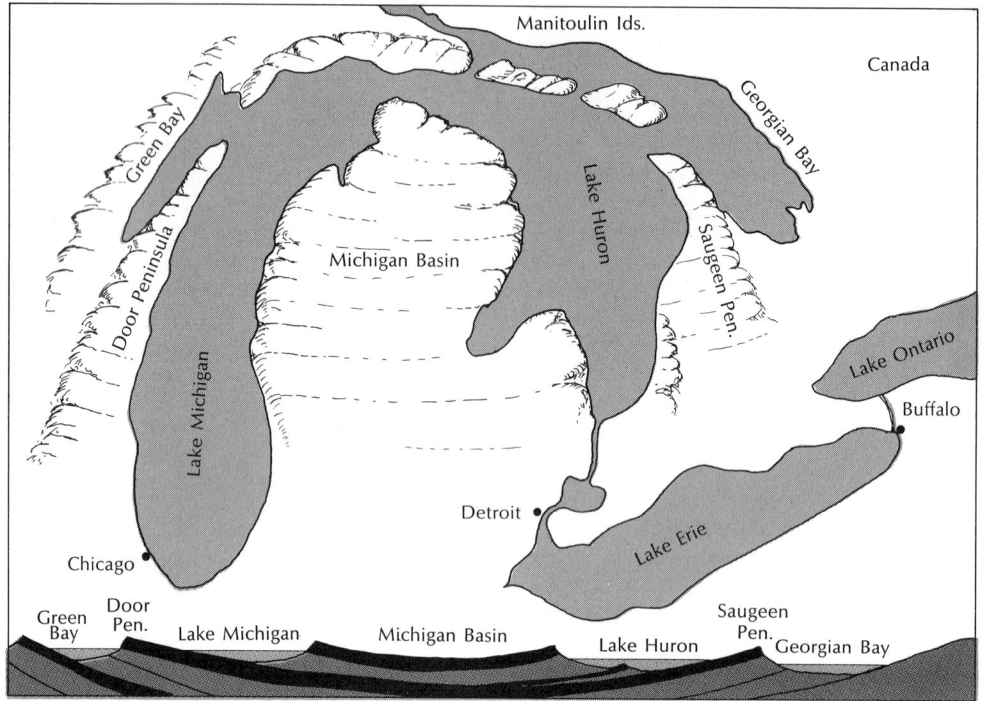

6-8. The relationship between rock structures and topography in the Great Lakes region. Note that most features in the eastern half have equivalents in the western half: e.g., Saugeen and Door Peninsulas, Green and Georgian Bays. Stratified rocks form a broad shallow structural basin and are arranged somewhat like a stack of saucers. *(A. K. Lobeck,* Things Maps Don't Tell Us, *The Macmillan Company, 1956.)*

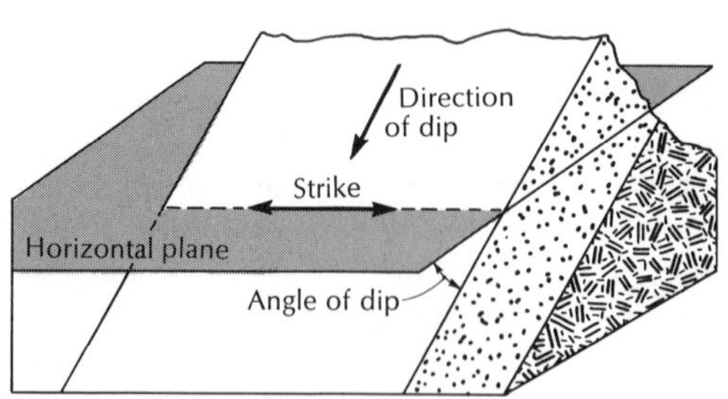

6-9. Strike and dip. Strike is the direction of a horizontal line in the plane of the bedding. Dip is the angle between the horizontal and the plane of the bedding. Strike and dip are always measured at right angles to each other. If north is toward the top of the sketch, then the sedimentary rocks strike east-west and dip to the south about 30 degrees.

fin, as well as the fins and arches that once were its neighbors, have been entirely removed by erosion. In the monument, one can see fins and arches in all stages of development (natural bridges form in a different manner).

Faults

Rocks tend to break or rupture rather than fold when crustal forces are applied relatively rapidly, when confining pressure is sufficient for plastic flow, or when rocks are of a type resistant to flow (e.g., crystalline, nonstratified rocks). A *fault* is a fracture in rocks along which one side has moved relative to the other (Figs. 6–12 through 6–14); formerly contiguous points along the

6-10. Prominent, widely spaced joints in sandstone. (Fig. 42, *U.S. Geol. Surv. Profess. Paper* 373.)

6-11. Delicate Arch is 60 feet high and perches on the edge of a high cliff. It is located in Arches National Monument in southeastern Utah.

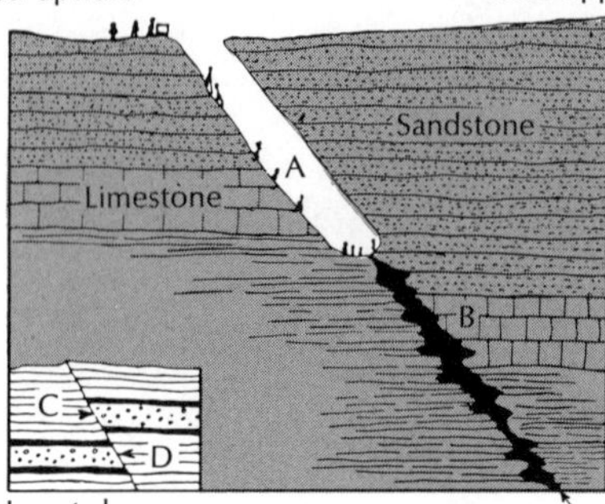

6-12. Normal fault (large) and reverse fault (insert in lower left). We assume that mining operations are proceeding downward along a mineralized zone (vein) along the fault surface. The hanging-wall side (right) apparently moved downward making this a normal fault. Formerly contiguous rocks such as the edges of the limestone have been separated so that a gap now exists; i.e., they seem to have been pulled apart (points A and B). In reverse faults, formerly contiguous points now overlap (points C and D in insert). In each fault, erosion has worn down the uplifted side.

crack have been offset by the movement. Like folds, faults may be measured in inches, yards, or miles, and they may be visible in a single outcrop or require careful mapping over large areas for detection.

Two common kinds of faults, normal and reverse, were originally defined by miners. As shown in Fig. 6–12, miners would walk along one side (the foot-wall) of an inclined fault surface, whereas the other side (the hanging-wall) would hang over their heads. In the majority of faults, the hanging-wall side apparently had moved downward, so the miners called this type a *normal fault*. However, some faults apparently had the opposite direction of movement—the hanging-wall side apparently moved upward, and these came to be known as *reverse faults*. Normal faults tend to be caused by tensional (pulling apart) forces, whereas reverse faults tend to result from compression. However, forces may be applied in other ways to produce the effects of tension and compression.

The word "apparent" in the definition of a normal or reverse fault is necessary; e.g., although the hanging-wall side (above the fault surface) of a normal fault appears to have dropped, the actual movement may have been quite different: (1) both sides may have moved downward, and the hanging-wall block may have moved a greater distance; (2) each block may have moved upward, but the foot-wall block (below the fault surface) may have covered the greater distance; (3) the hanging-wall block may have remained motionless while the foot-wall block moved upward.

The development of a large fault undoubtedly requires much time. Movements of thousands of feet along fault surfaces have been observed, and presumably these are the end results of a number of smaller movements, each involving a few tens of feet at the most; each quick movement perhaps caused an earthquake. However, crustal movements are bending pipes and casing in drill holes in a California oil field and show that faulting can also occur gradually without causing earthquakes. Fault surfaces commonly are curved.

Movement along fault surfaces may be sideways as well as up or down, and any combination of these is possible. If the block of rocks on one side of a fault moved horizontally and parallel with the strike of the fault (or if both blocks moved in this manner, then we have a *strike-slip fault*. To describe the direction of movement, a geologist stands on one block and looks across the fault surface at the other block. If it has moved to the right, he calls it a *right-lateral fault*; if it has moved in the opposite direction, it is a *left-lateral fault*.

Large-scale, low-angle reverse faults are

6-13. A normal fault in sandy shale showing a displacement of several feet. The hanging-wall side is at the left and appears to have moved downward. *(Courtesy A. Keith, U.S. Geological Survey.)*

6-14. A reverse fault (Alabama). The hanging-wall side (right) appears to have moved upward. *(Courtesy W. H. Monroe, U.S. Geological Survey.)*

known as *thrust faults* (Fig. 6–15), and on some thrusts, the hanging-wall side may have been shoved upward and forward for many miles. Such movements are common during some kinds of mountain building.

Fault surfaces may be *slickensided;* i.e., they have been polished, scratched, and grooved by abrasion between the moving blocks, and the grooves are parallel to the direction of movement. If a finger is rubbed along some fault surfaces, it feels smoother in one direction (the last movement along the fault) than in the opposite direction. This may be useful in determining whether a certain fault is normal or reverse, but it also may not be reliable. Thus faults may be recognized by the presence of slickensided surfaces or by the discovery of once-contiguous rocks that are now offset.

Unconformities

An unconformity records a significant story about Earth history (p. 60). Sediments are conformable when they accumulate layer upon layer without important delay or change occurring during deposition. As strata pile up, fossils may be buried with them, and thus they record information concerning events of the time of burial. However, conditions may change in an area; e.g., an uplift may occur which causes erosion to become predominant over deposition. Tiny valleys are carved out of the recent products of sedimentation, and these grow larger and become more numerous. If in a later age the Earth's crust subsides, deposition will again become predominant in this location. The newly deposited strata are then said to be unconformable upon the older strata be-

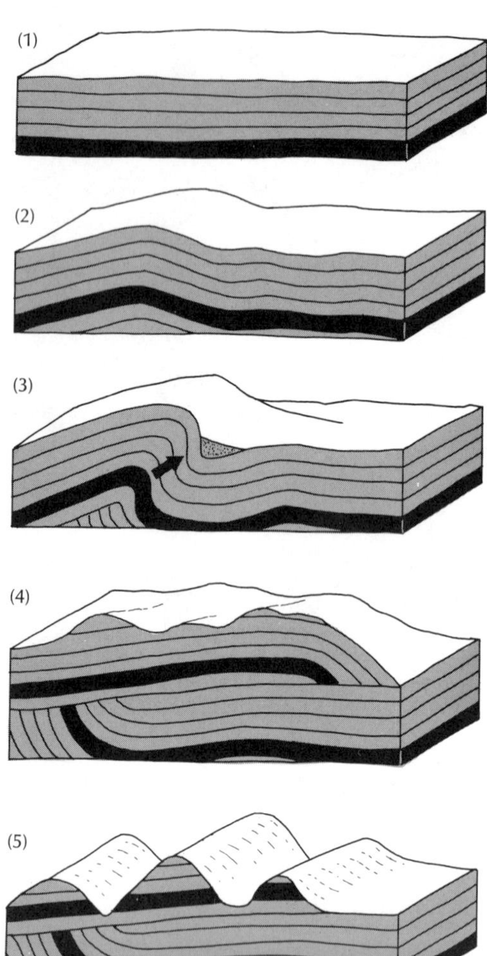

6-15. Development and erosion of a large thrust fault. (1) Horizontal sedimentary rocks are shown; the oldest rocks are on the bottom. (2) Compression has produced an anticline. (3) The anticline has become asymmetrical. (4) Continued compression has caused the anticline to rupture and older, once deeper rocks from the left have been shoved upward toward the right where they now rest upon younger rocks—an exception to the principle of superposition. (5) Erosion has produced ridges and valleys. *(After Hussey.)*

neath them, because a lack of continuity occurs between the two groups of rocks.

Accordingly, an *unconformity* is a relationship between older and younger rocks that involves a break in the geologic record, and

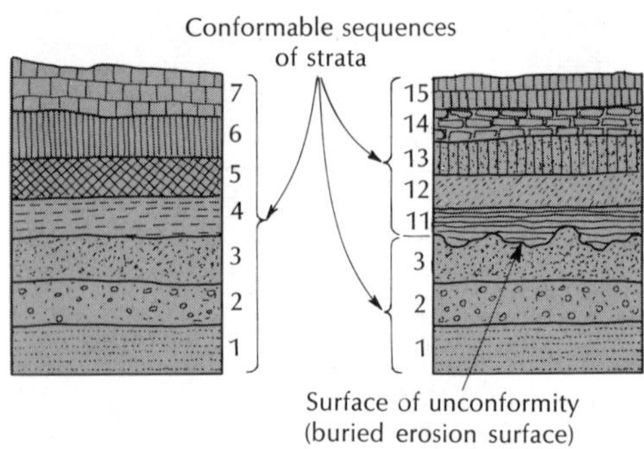

6-16. Unconformity. Strata are numbered in chronological order, and number 1 is the oldest rock exposed here. Layers 1 through 7 were deposited (left), and erosion then proceeded to remove layers 7, 6, 5, and 4 from this area. During this interval of erosion, beds 8, 9, and 10 were deposited in an adjoining area. Bed 11 was deposited when sedimentation began again in this area. Thus, the unconformity below shows a local gap in the geologic record represented elsewhere by beds 4 through 10. There are two causes for the break: (1) erosion removed previously formed rocks, and (2) deposition generally did not occur in this area while erosion was going on.

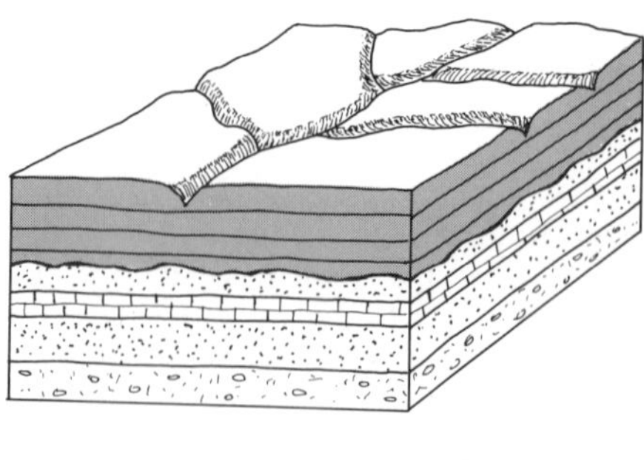

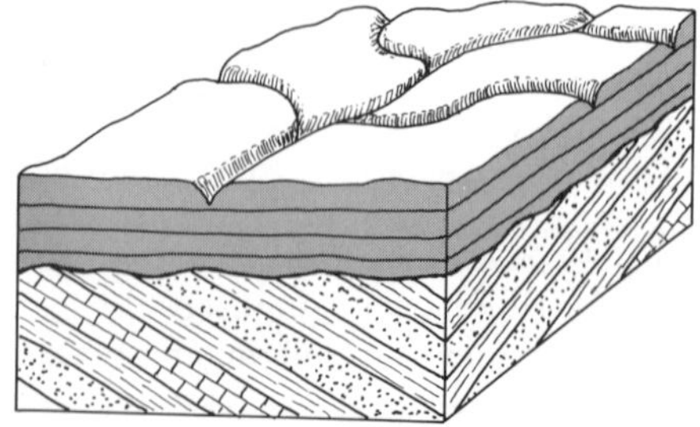

6-17. Unconformities. Older rocks beneath an unconformity may or may not be folded, faulted, or otherwise deformed and then eroded before younger rocks form on top of them.

a *surface of unconformity* is commonly a buried surface of erosion (Fig. 6–16). Crustal movements, such as folding or faulting, may occur before the younger, overlying layers are deposited. If such movements do occur, then older rocks beneath the surface of an unconformity are deformed, whereas younger rocks above this surface are not (Figs. 6–17 and 6–18). Lost time intervals represented by unconformities in the geologic record are both numerous and long.

Interpretation of a Sequence of Events from Rocks

Part of the fascination of geology lies in the attempt to interpret from rock outcrops the sequence of events which occurred long ago in an area (Figs. 2–27 and 6–19). A few of the relationships between geologic structures and mining are described below to illustrate some of the practical applications of geology.

Geologic Sleuthing Locates Hidden Deposits

Fig. 6–20 illustrates a simplified version of a practical problem frequently encountered in mining. A horizontal series of sedimentary rocks contains a valuable iron ore deposit which is being mined far below the surface. Work has proceeded from a shaft located to the left of the diagram. The mine has proved to be a rich one, but one day the miners find that the ore body ends. Because the deposit stops abruptly, it may have been truncated by a fault. If so, is the offset part upward along the fault surface or downward, and how far has it moved? Exploratory drilling, especially in wrong directions, is expensive. A hurried call goes out to the company geologist who has previously studied the rocks in the area. Following a careful inspection, the geologist is able to state that the missing bed will be found some 350 feet down the fault surface. Work proceeds and the geologist is proved correct. How was

6-18. Striking unconformity. Horizontal sands and gravels rest upon tilted and eroded shales (Cenozoic). *(Courtesy G. W. Stose, U.S. Geological Survey.)*

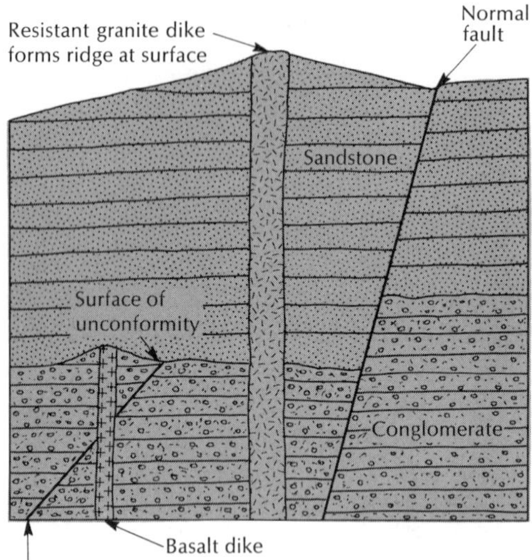

6-19. Geologic history determined from a structure section. Events occurred in the following chronological order: (1) The conglomerate formed first. (2) A fault cut across the conglomerate (type unknown because layers cannot be matched on opposite sides of fault). (3) A basalt dike was intruded into the conglomerate and across the fault. If the fault had come after the dike, the dike would be offset by it. (4) Erosion occurred. (5) Sand was deposited on the erosion surface and later became sandstone; an unconformity now occurs between the conglomerate and sandstone. (6 and 7) A granite dike was intruded, and a normal fault developed (order cannot be determined because we lack cross-cutting relationships). (8) Erosion occurred, and a small valley has developed along the fault surface because erosion has proceeded more rapidly in the crushed-rock zone along the fault.

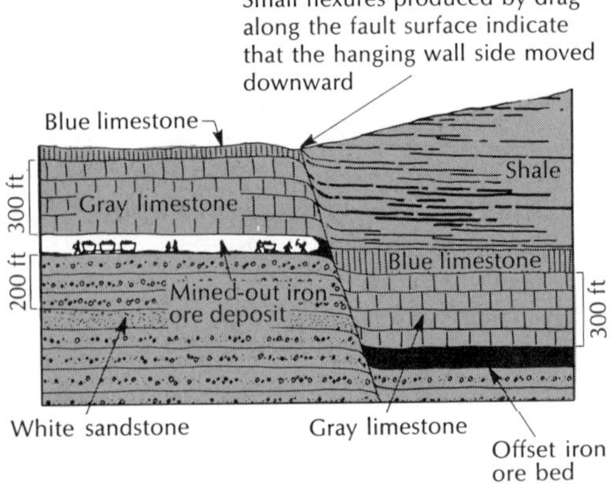

6-20. Ore deposit cut off by a fault. Two signs show this to be a normal fault: matching strata and small flexures along the fault (in this example, the flexures bend upward along the hanging-wall side of the fault showing that it moved downward).

this done? Note that the ore deposit occurs among a number of different kinds of rock layers, many of which can be recognized and matched. For example, the iron-ore bed occurs 200 feet stratigraphically above a white sandstone but 300 feet below a blue limestone. Distinctive fossils in these strata aid in their correlations. Thus identification of the bed exposed in the mine on the opposite side of the fault surface indicates whether the offset ore deposit is above or below, and about how far.

Small flexures may be formed as the edges of layers are dragged up or down a fault surface and may aid in determining whether the hanging-wall side went up or down (Fig. 6–20). Such flexures bend toward each other on opposite sides of a fault surface in both normal and reverse faults. A simple experiment illustrates why the flexures bend as they do. Slant the palm of your hand as a fault surface and move the edge of a piece of stiff paper (held horizontally) in contact with your palm up and then down. Note the direction in which the edge of the paper bends because of friction between it and your hand.

Faults are common in mining areas, and if several faults intersect mutually, the ingenuity of the mining geologist is particularly tested. Other types of geologic structures may also need to be discovered and interpreted correctly before valuable ore bodies can be found (Figs. 6–21 and 6–22).

Geology and the Search for Petroleum

Geology plays an important role in the search for oil and natural gas. Four pre-

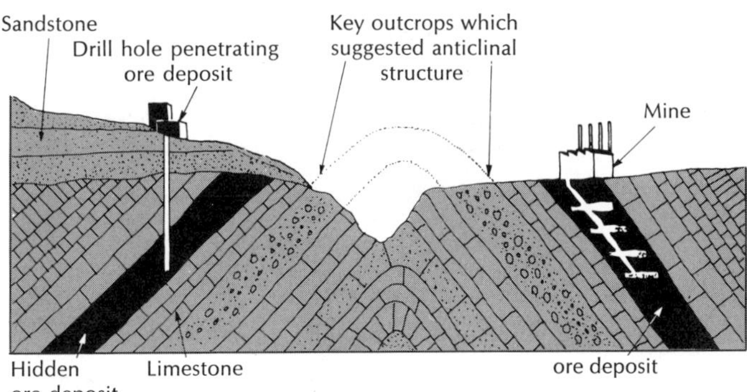

6-21. Discovery of a hidden ore deposit. The following geological events are recorded by the structure section: (1) Origin of the older sedimentary rocks and the ore body. (2) Folding. (3) Erosion. (4) Deposition of horizontal beds of sandstone which probably once covered the entire area. (5) Erosion to the present surface.

requisites are necessary for oil (and gas) to accumulate in commercial quantites in an area: (1) The oil originates in a source bed, and a marine shale, once a black mud rich in organic compounds, is thought to be a common source rock. (2) The oil then migrates to a permeable reservoir rock, and to do this it may travel for long distances both vertically and horizontally. Oil cannot move through the tiny openings of the shale source beds rapidly enough to be extracted profitably. (3) A nonpermeable layer must occur above a reservoir bed. Since oil is lighter than water, it tends to move upward through openings and cracks until it encounters impervious beds that it cannot penetrate. The oil may then accumulate beneath the impervious layers. Some gas occurs in solution within the oil, and if enough is present it separates out to occupy the uppermost region of such a trap. (4) A favorable structure must exist to concentrate the oil (Fig. 6–23) and anticlines, salt plugs, and faults are common examples. A fault zone may itself be impervious, or faulting may have shifted an impervious bed so that it now blocks a reservoir bed. Stratigraphic traps tend to be more difficult to locate and may form where tilted reservoir beds are overlain unconformably by impervious layers or where the reservoir beds become thinner up-dip and wedge out within enclosing impervious beds. Thus oil that was once distributed in sparse amounts throughout a very large volume of rock may now be richly concentrated within the

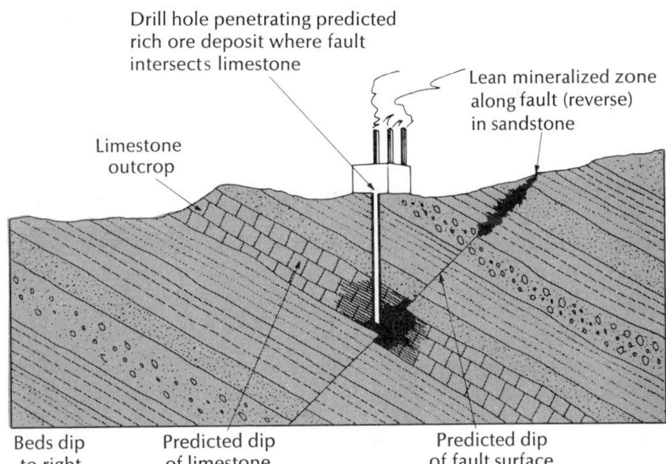

6-22. An ore deposit predicted on the basis of surface outcrops. Most of the sedimentary rocks shown are insoluble in water, and the mineralized zone is lean along the fault within them. However, limestone is readily soluble in water. The geologist who found the outcrop of limestone hoped that it had been replaced by the ore minerals along the fault to form a rich deposit. He calculated where the fault should intersect the limestone, and the drill hole proved his hypothesis correct. Another ore deposit occurs on the foot-wall side of the fault in the offset part of the limestone.

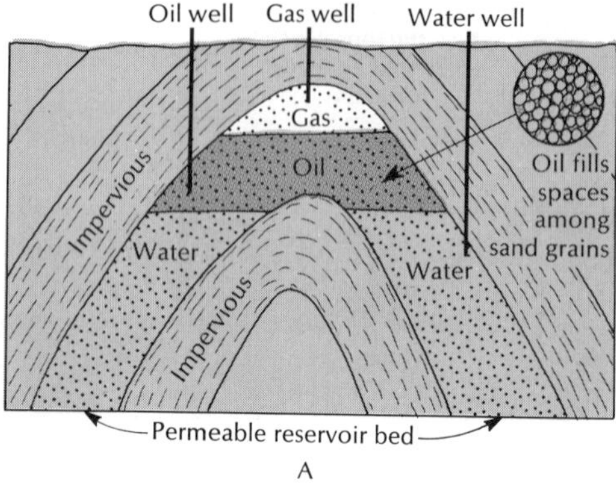

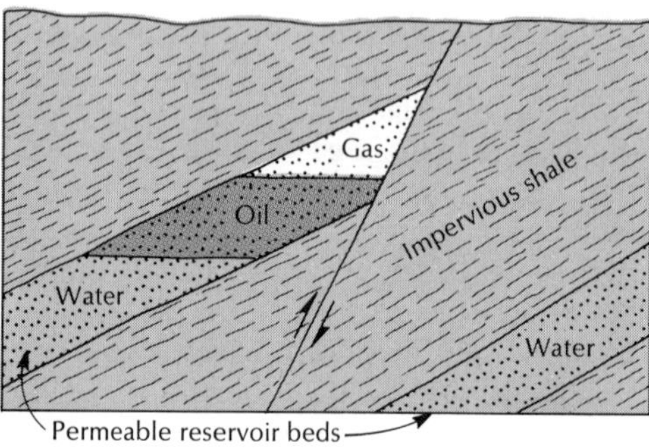

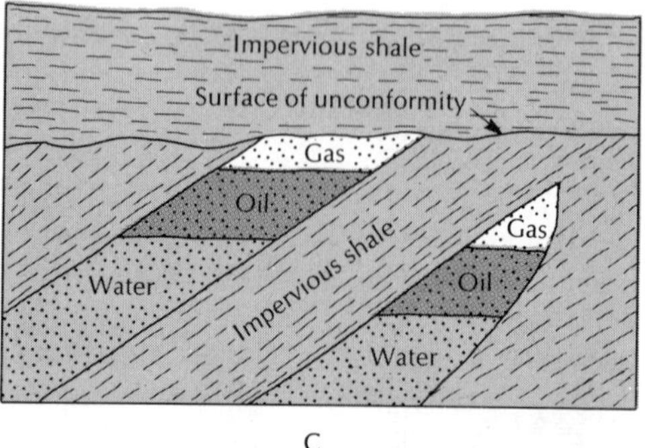

6-23. Schematic sections through three common kinds of oil-trapping structures. In each case, a permeable sandstone reservoir bed occurs between layers of impervious shale, and pore spaces among the sand grains are filled with gas (top), oil, and water (bottom), respectively. In C, note lack of surface clues to oil hidden below.

uppermost portions of favorable reservoir rocks.

The task of the geologist is the location of promising structures in regions where rocks are favorable for the occurrence of the other prerequisites. Drilling a hole is then the only known method of determining whether or not oil is present in the structure.

Earthquakes

The term *earthquake* refers to groups of elastic shock waves which radiate outward in all directions from a suddenly disturbed zone in the Earth's crust or mantle, and which cause rocks and regolith to shake and vibrate. The Earth reacts as if it had been violently jarred by a blow from Paul Bunyan's hammer. Some of these waves go through the Earth, whereas others are confined to the crust and go around the Earth. As the waves pass any one place, they cause the Earth to shake and vibrate; they produce an earthquake.

More than a million natural earthquakes occur each year, and almost 700 of these are strong enough to cause considerable damage in the regions where they take place. Their study is called *seismology* (Gr., *seismos* means earthquake), and sensitive instruments called *seismographs* (Fig. 6–24) record the arrival of these waves. Even a passing railroad train or heavy truck will' cause the Earth to tremble locally a little.

Although seismographs have now become quite sophisticated, a key part of each instrument is a heavy weight whose inertia keeps it nearly motionless, together with a recording device such as a rotating drum which shakes with the Earth and on which the vibrations of the different waves are traced. Records today are commonly made photographically or electronically (computers assist in analysis).

Seismology has made spectacular progress in recent years in developing instruments and techniques that can be used to distinguish the seismic waves produced by underground nuclear explosions from those produced by natural earthquakes (in most instances, this can be done). Advances here in what might be called applied research have had direct transfer value to the study of seismology in general, and much new and detailed information has been obtained concerning the nature of the Earth's crust and upper mantle.

Seismic waves produced by small man-made explosions have also been used successfully during the past few decades by oil-exploration geologists to determine the depths to certain distinctive sedimentary rocks, and reflection shooting is such a method (Figs. 1–4 and 14–8. In this, an explosion is set off at a particular location, and this disturbance produces waves which move outward in all directions. Some of the wave energy is then reflected upward to the surface by certain underlying layers and recorded at wave-detecting instruments located at different distances from the shot point. The signals reach the more distant recorders later, and thus the times and distances can be interpreted to show the depths to strongly reflecting layers. In this way, a buried anticline or other favorable oil trap may be discovered.

Earthquakes have terrified mankind for thousands of years, and for many the term evokes a picture of death and devastation: huge waves race across an ocean and overwhelm coastal regions, villages are buried by landslides, fissures open and spout water, rivers and springs go dry, and all of this is imagined to occur amid the rumble of earth sounds. Although earthquakes have relatively little effect in shaping the Earth's topography, some of them have brought disaster upon man, both directly and indirectly, by causing destructive waves, landslides, fires, broken water mains, and diseases produced by contaminated water supplies.

The opening and closing of fissures during earthquakes is not so common as is popularly believed and occurs chiefly in loose debris, but it does grip the imagination. A woman working in a rice paddy in Japan in 1948 reportedly sank to her neck

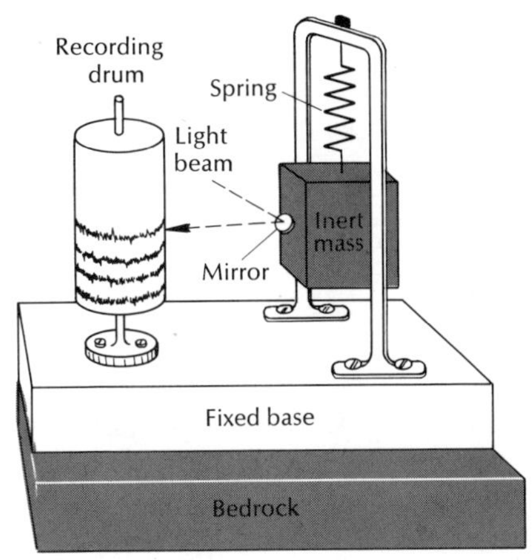

6-24. A simple seismograph. A seismogram is the wavy line traced on the slowly rotating, vibrating drum during an earthquake. Many seismographs today are kept in darkened rooms so that the records may be made on photographic paper wound around a drum. *(B. Mears, Jr., The Changing Earth, Van Nostrand Reinhold, 1970.)*

in a fissure that opened beneath her and was crushed when it closed. During the 1950 earthquake in Assam, four men and a mule are said to have perished in a similar manner. In the center of the earthquake area, people, rocks, and small buildings were apparently "tossed into the air like peas on an enormous kettledrum."

Great loss of life has accompanied some earthquakes; in the 1 September 1923 Japanese catastrophe, approximately 140,000 persons were killed, and property damage was estimated at 3 billion dollars. The earthquake occurred at noon when meals were being cooked, and hundreds of fires broke out at once as houses and stoves were knocked down. Although many buildings were damaged (a reported 90% in some villages) and some loss of life occurred, the total immediate devastation was small compared to that which followed. Fires in Tokyo and in other cities and towns quickly spread out of control because the city water systems had been disrupted by the earthquake. The intensely heated air over Tokyo rose in a violent updraft, formed clouds, and

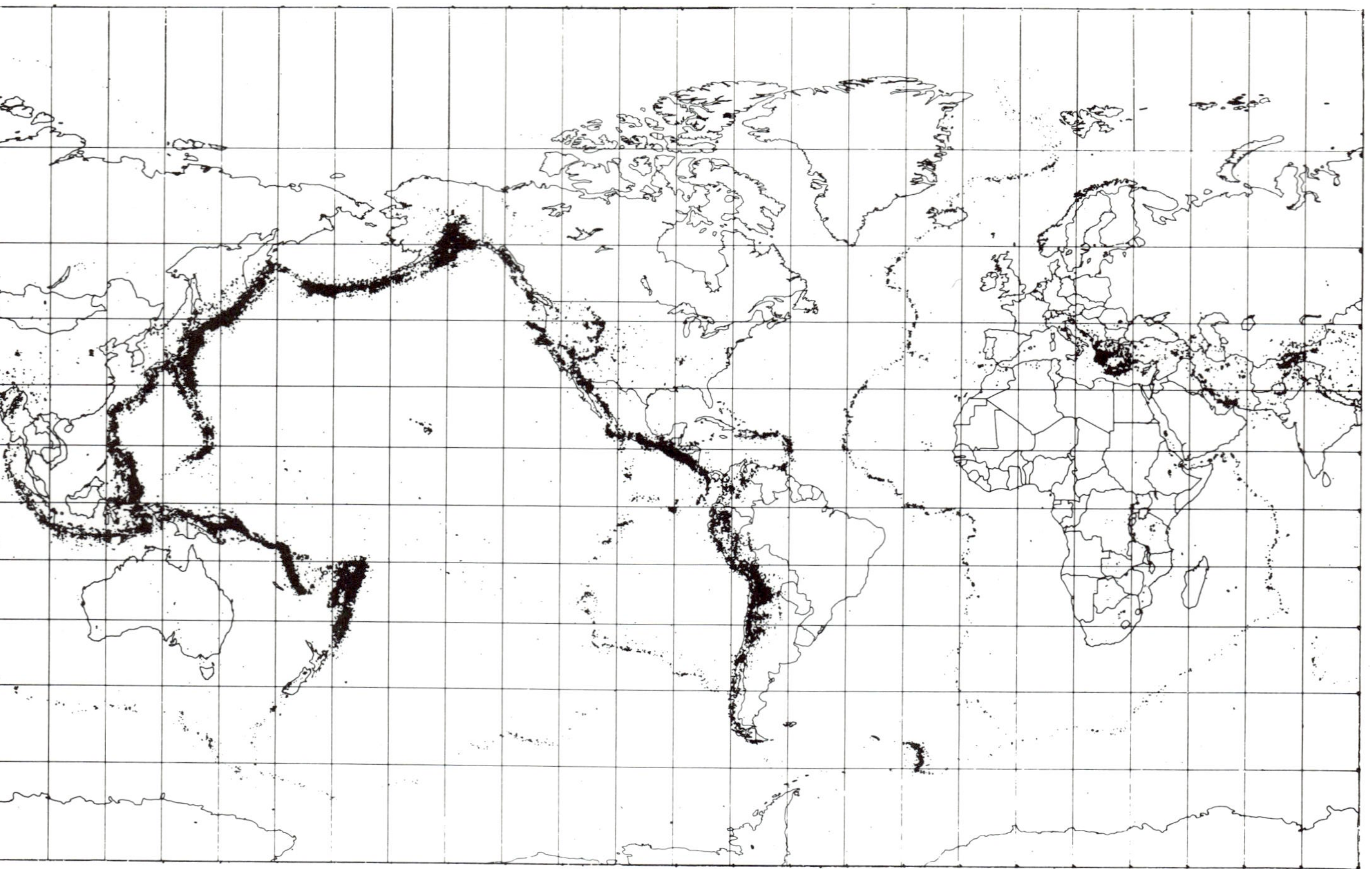

6-25. The illustration shows about 42,000 earthquake epicenters for the period 1 January 1961 to 30 September 1969, for all depths. Epicenter clusters indicate major seismic zones. From west: the circum-Pacific seismic belt, the Earth's most active seismic feature; the mid-Atlantic Ridge; and the Alpide Belt, which links up with the circum-Pacific Belt at New Guinea.

spawned powerful whirlwinds. Thus wind directions and velocities changed frequently, and the rapidly advancing flames encircled thousands of people, who were burned to death as they huddled on bridges, in parks, and in other open areas. Destruction by fire far exceeded that caused directly by the earthquake. Seismic sea waves (tsunamis, p. 417) and landslides increased the magnitude of the catastrophe.

Earthquakes tend to be most frequent and violent in elongated belts that coincide approximately with the regions in which volcanoes are or have recently been most active and also with young mountain systems. Of these, the circum-Pacific belt is by far the most important (about 80% of all earthquakes, Fig. 6–25). Another belt extends from the Mediterranean area eastward to the Indonesian States (about 15% of all earthquakes), and a third belt follows the crest of the system of mid-ocean ridges. According to plate tectonics (p. 314), such belts occur along the interacting margins of great, slowly moving lithospheric plates. Destructive seismic sea waves are produced by some submarine earthquakes.

Causes of Earthquakes

According to the *elastic rebound theory*, most earthquakes, including all major ones, are caused by the passage of a group of elastic waves produced by quick movements of parts of the Earth's crust and mantle along fault surfaces (Fig. 6–26). Volcanic eruptions and the underground shifting of magma probably cause most other earthquakes, all relatively minor. During the years preceding an earthquake, slow movements in the crust—in opposite directions on opposite sides of a stressed zone—have bent the rocks gradually until they are strained to and finally beyond their breaking point. An earthquake is produced by their sudden rupture and abrupt return approximately to their prestressed positions. Most earthquake-producing faults do not displace the Earth's surface, and most occur beneath the oceans.

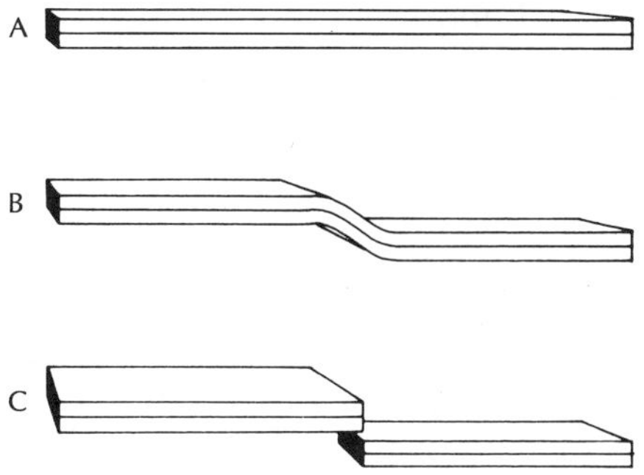

6-26. Elastic rebound theory. A, B, and C are blocks of the Earth's crust. A is unaffected by stresses. B shows bending in response to vertical stresses which have been applied slowly. In C the strain has been relieved by a sudden movement along the fault surface. A group of elastic shock waves was thus produced, and they moved outward in all directions from this place of origin.

Although most earthquakes originate within the outer 35 miles (55 km) or so of the Earth's spherical body, some are created by disturbances that are located at greater depths, and the deepest occur about 435 miles (700 km) below the surface. According to one theory, rocks at such depths should flow plastically under stress; they should not be brittle. However, the behavior of a block of wax under stress may illustrate how an earthquake can be produced under such conditions. If pressure is exerted on the wax slowly, it flows; if pressure is exerted abruptly, the wax fractures. Another theory attributes such deep-focus earthquakes to a sudden, pressure-induced collapse of a relatively small volume of rocks at depth (i.e., minerals undergo abrupt phase changes, p. 164, which involve space-saving, denser rearrangements of their three-dimensional lattice structures).

Quantitative data have been obtained concerning the slow shifting of rocks on opposite sides of a fault surface during the years prior to the abrupt rupture which causes an earthquake. Such measurements have been made along the San Andreas fault in California which is approximately vertical, passes near San Francisco, and can be traced for many tens of miles both northwest and southeast of the city. In April 1906 abrupt horizontal movement occurred along more than 250 miles of this fault. Roads and fences along the line of the fault were offset as much as 21 feet.

Careful measurements which span several decades show that land west of the San Andreas fault is moving northward, whereas land on the east side of the fault is moving southward (i.e., it is a right lateral fault). The rate of movement is about 2 inches a year and is greatest nearest the fault zone, which may mark the boundary between two large, laterally moving, rigid plates of the lithosphere. In one place, a building rests directly on the fault zone, and part of it moves northward, while the other part moves southward. Needless to say, repairs must be frequent. After such movement has continued for a number of decades, the rocks have been shifted several feet and can bend no farther. Therefore, they break, the previously bent rocks straighten out with a snap, and the energy that had been accumulating for years is suddenly released to cause an earthquake.

The amount of bending necessary to cause the next break may be more or less than that which caused the preceding one. Thus the exact time when a major earthquake will occur in the future is difficult or impossible to predict, but that such earthquakes will occur seems inevitable. Moreover, as people and buildings continue to multiply rapidly in the earthquake areas, the most devastating earthquakes seem likely to be those of the future; e.g., the San Fernando Valley earthquake of 9 February 1971 in California was of moderate size only, and occurred about dawn when most inhabitants were still in their homes, yet 64 people died, and damage may equal 1 billion dollars. However, precautions can be taken to reduce damage, and we list a few: a building should be constructed so that it can vibrate as a unit, and buildings should be fastened firmly to bedrock where possible—buildings resting upon loose, water-saturated fill shake as if built upon a quivering bowl of jelly. One should eliminate architectural ornaments, overhanging balconies, and other objects which might be shaken loose from second stories and taller buildings. Since damage tends to decrease with distance from an epicenter, one should try to avoid locations close to active fault zones, although this is difficult if one wants to live in such an earthquake-active region as California (like playing "earthquake roulette").

Repeated movements tend to occur along the same fault zone because this is the weakest portion of the area being deformed. However, the slow bending eventually stops in any one area because many inactive faults are known. Total horizontal movement along the San Andreas fault may amount to many miles, and this undoubtedly has resulted from numerous, relatively small movements spread across many many centuries.

The ultimate forces causing such movements of rocks are unknown, although we shall consider some tentative hypotheses in Chapter 12. However, man's activities have triggered some earthquakes in recent years; e.g., contaminated water pumped downward into deep wells recently in the Denver, Colorado area apparently lubricated old fault zones sufficiently to permit slippage which caused earthquakes. In another type of activity, huge bodies of water have piled up behind some large dams and have overloaded the crust enough locally to cause renewed movement along old faults, although the weight of water behind other large dams has not affected local seismograph records. Lake Mead on the Colorado River has shown such a correlation between depth of water and number of earth tremors. Moreover, a large earthquake apparently was produced in this way at Koyna Dam south of Bombay, India in 1967 and resulted in 200 deaths in a nearby village and extensive, widespread damage.

Seismologists have discussed the possibility of attempting to prevent some major earthquakes by the occasional artificial triggering of small earthquakes along large, active faults—this might prevent accumulation of sufficient energy to cause a really disastrous earthquake by releasing it in more frequent, but smaller amounts. However, this might not happen, and a major earthquake might also be triggered, and so we have a dilemma. Do we suffer through artificially induced, more frequent, small earthquakes so that our children or grandchildren may avoid a really disastrous one? Would such triggering activities be so expensive that we would be better off expending our resources in preparation for the large earthquakes that are sure to come in the decades ahead? In fact, do we yet know enough about the causes of earthquakes to pose such questions?

The Interior of the Earth

Inaccessible it is, yet information concerning the nature of the Earth's interior can be gleaned in a variety of ways including the behavior of seismic waves, meteorites, high-temperature, high-pressure laboratory experiments on minerals, astronomical observations on the shape and motions of the Earth, and data interpreted from the nature of the Earth's gravitational and magnetic fields and thermal properties.

Earthquake waves probably provide the most important data about the Earth's interior and are of three main types: primary, secondary, and long waves. (1) The *primary* or *P* wave travels fastest; it passes through the Earth and arrives first at a distant station (Fig. 6–27). It is similar to the familiar sound wave (primary waves are audible at times), and the medium through which it passes is alternately compressed and expanded: an apt term is a push-pull wave. Such waves can travel through solids, liquids, and gases. (2) The *secondary* or *S* wave is recorded next at a distant station and likewise passes through the Earth. It is a transverse wave that moves because particles in the transporting medium vibrate back and forth in planes that are perpendicular to the direction in which the wave itself moves. Such waves can travel only through solids. (3) The *long* (main) waves are produced where *P* waves and *S* waves reach the surface or some other interface, are complex, and travel around the outside of the Earth confined to the crust. One type involves both vertical and horizontal motions, whereas another involves only horizontal motions.

P waves and *S* waves generally increase in speed with depth (Fig. 6–32), and the speed of the *P* waves at any one depth is about double that of the *S* waves. This distorts their wave fronts from circular to elliptical shapes (refraction, Fig. 19–5), and thus the waves follow curved paths through the Earth. When they reach a different medium, they are reflected (Fig. 6–27) and also produce other waves that then travel along the boundary zone (discontinuity) between the two media.

The distance from a seismograph station to the place of origin of an earthquake can be determined by the time interval between the arrivals of the first *P* and *S* waves—the more distant the earthquake, the longer is the interval (Fig. 6–28). This is somewhat similar to a type of problem in which two cars start together and move down the same road at constant rates of 20 mi/hr and 10 mi/hr respectively. The faster car arrives at a station first, and if it arrives 5 hours ahead of the slower car, one can calculate that it has traveled 100 miles.

To illustrate, assume that the first *P* wave arrives 10 minutes before the first *S* wave. Reference to the travel-time curves for these waves in Fig. 6–28 (ignore the twice-reflected *PP* and *SS* waves) indicates that the earthquake must have occurred at a distance of about 9000 km (5580 miles). If this difference had been 5 minutes instead of 10 minutes, the distance would have been about 3500 km (2170 miles).

The *focus* is the place within the Earth where an earthquake originated, and the *epicenter* is the place on the Earth's surface directly above this focus. An epicenter may be located by the three-circle method if its distance from each of three stations is

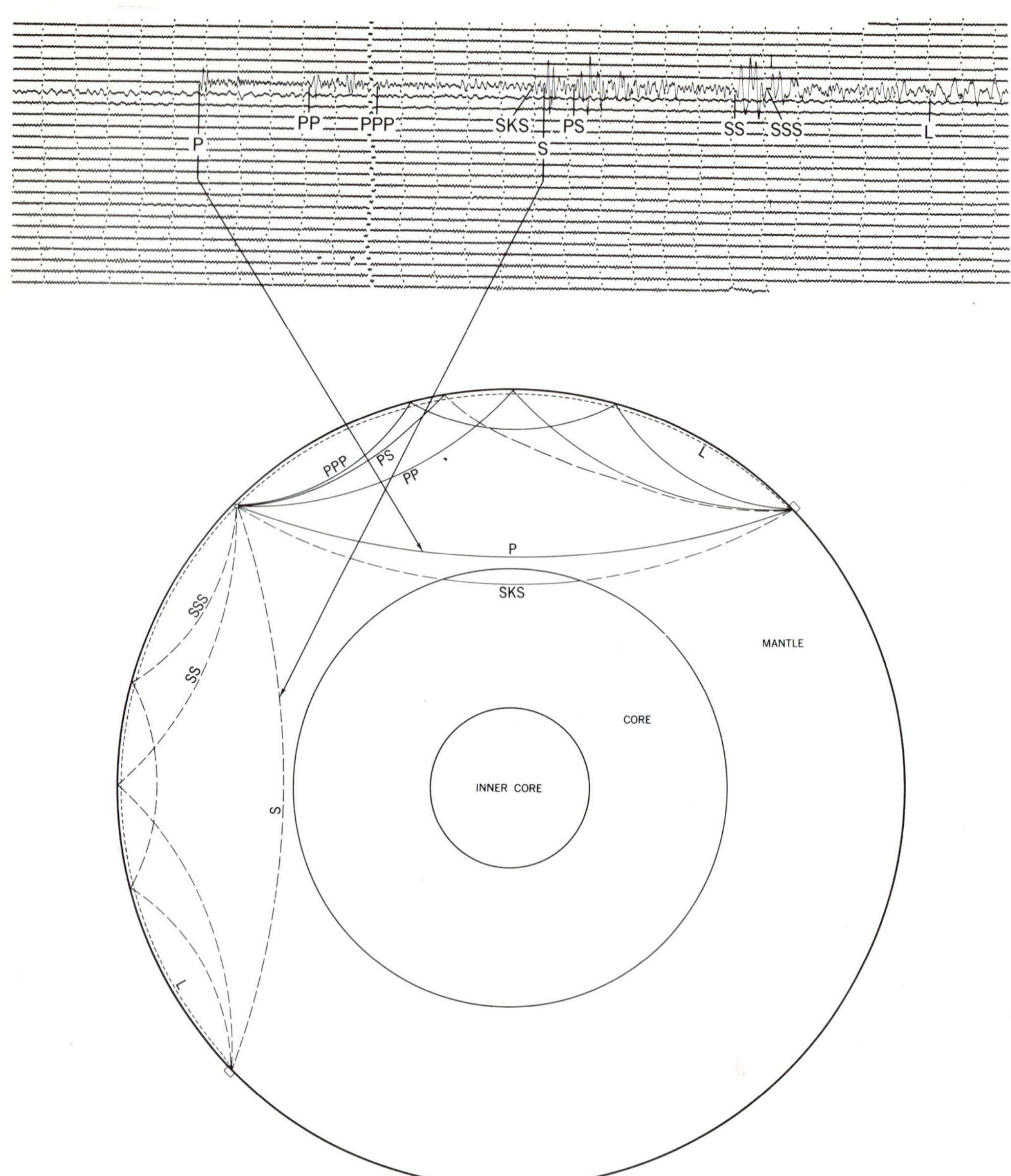

EARTHQUAKE WAVES RECORDED ON TYPICAL SEISMOGRAM
AND PATHS OF PROPAGATION THROUGH THE EARTH

6-27. Paths of principal earthquake waves. An earthquake is assumed to occur at the tiny circle in the upper left, and two seismograph stations are shown at equal distances (small rectangles at surface in lower left and upper right). *P* waves and *S* waves travel similar paths, but to avoid duplication, *P* waves are shown only by lines to the right and *S* waves by dashes downward and to the left (*PP* waves are reflected twice, and *SSS* waves are reflected three times, but we ignore these here). Only a few waves are shown as samples; other waves travel similar paths but reach the surface at different locations, either closer to the epicenter or farther from it. *(Courtesy U.S. Geological Survey.)*

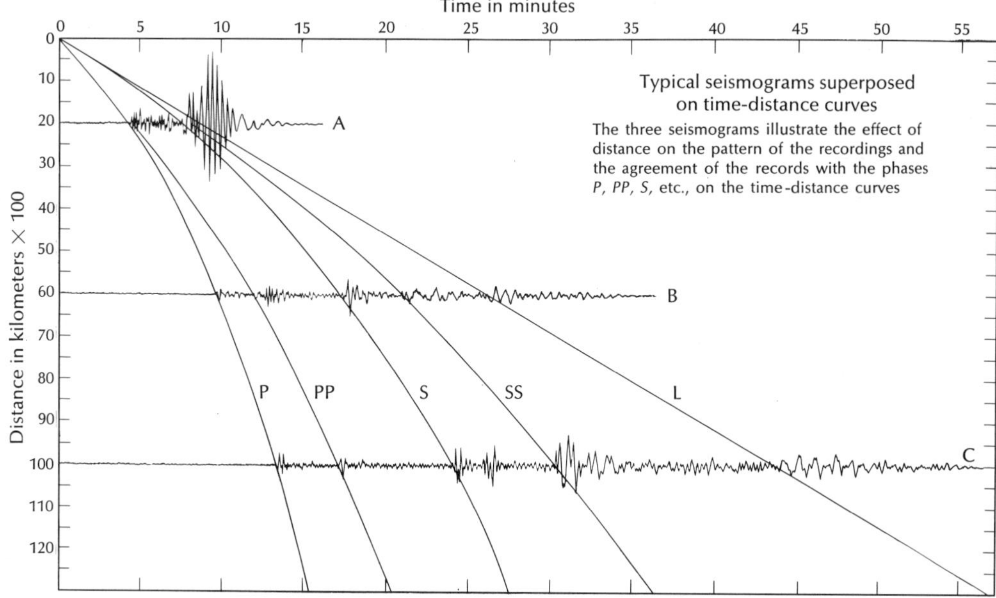

6-28. Travel-time curves are shown with idealized seismograms. The time interval between the first arrival of each of the three principal waves at any one station increases with distance. *PP* waves and *SS* waves have each been reflected twice. *(Courtesy U.S. Coast and Geodetic Survey.)*

known. To illustrate, three circles are drawn to scale on a map. Each station is the center of a circle, and the radii are equal to their respective distances to the earthquake. The three circles then intersect at the epicenter. This method is especially useful in locating the epicenters of earthquakes that occur beneath the oceans, but it is not precise for deep-focus earthquakes.

Earthquake waves show that the Earth's interior is separated into three main parts by two major, nearly spherical, relatively sharp zones of discontinuity: the core, mantle, and crust. The upper discontinuity marks the base of the crust and is located within a few tens of miles of the surface. The lower discontinuity marks the top of the core and occurs at a depth of about 1800 miles (2900 km). At these discontinuities, earthquake waves undergo abrupt changes in speed and/or direction below all parts of the Earth's surface.

By volume, the core constitutes slightly more than 16% of the Earth, the mantle somewhat more than 83%, and the crust about 0.5%. However, the much greater density of the core changes these proportions when masses are compared—the core is then nearly one-third of the Earth's total mass, and the mantle is slightly more than two-thirds.

The Core

The Earth has a core about 4300 miles (6900 km) in diameter whose existence is revealed by a shadow zone that occurs in association with each major earthquake. Seismograph stations located in any direction from a major earthquake receive *P, S,* and *L* waves if they are located within a distance of about 7000 miles (11,270 km) of the place of origin (we shall now ignore the long waves which tell us little about the Earth's interior). However, no direct *P* or *S* waves are received by any stations located within a belt nearly 3000 miles (4830 km) wide that is situated approximately 7000 to 10,000 miles (16,100

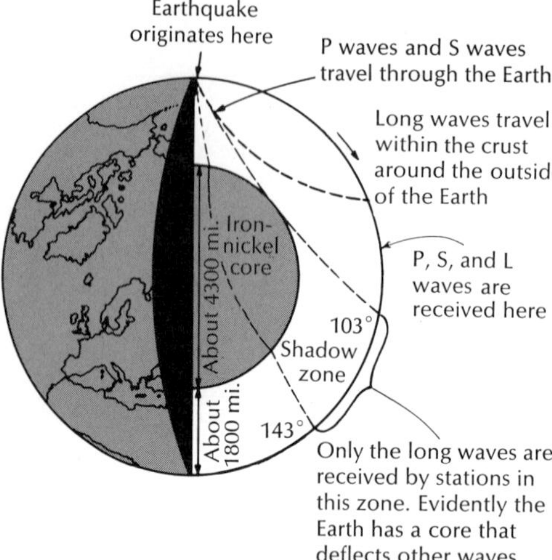

6-29. The Earth's interior. The *P*-wave shadow zone is a belt nearly 3000 miles wide (about 40 degrees) which extends around the Earth. It begins at about 73 degrees from the place of origin of an earthquake and extends to about 143 degrees from that place. Its location shifts with the location of the earthquake, which indicates the presence of a spherical core within the Earth.

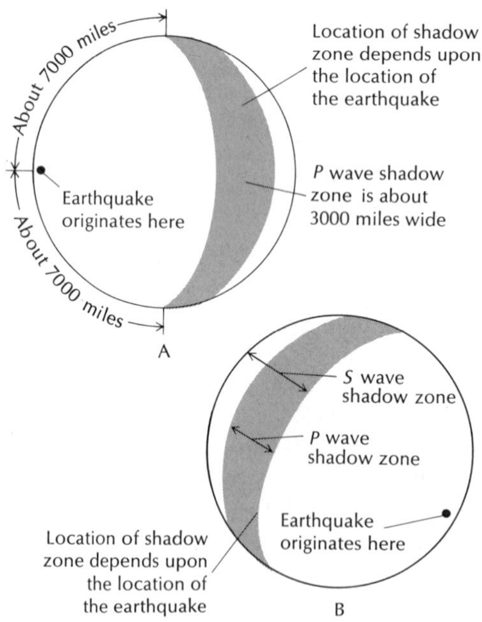

6-30. The locations of the *P*-wave and *S*-wave shadow zones depend upon the location of an earthquake.

km) from the place of origin of an earthquake (*P* waves are recorded on some very sensitive seismographs). This belt, 3000 miles in width, extends all of the way around the Earth and is aptly called the *P wave shadow zone* (Figs. 6–29 and 6–30). Its location is unique for each earthquake; it is always the same width and occurs at the same distance. The *S wave shadow zone* is more widespread because it includes all of the Earth's surface beyond the 7000-mile distance.

What causes shadow zones? *P* and *S* waves pass through the Earth along curved paths, and more distant stations receive *P* and *S* waves that have traveled more deeply through the Earth. Waves received at a station located about 7000 miles from an epicenter were about 1800 miles below the surface at the deepest part of their curved paths. Thus a change must occur at a depth of about 1800 miles: *P* waves are deflected at this depth, their speeds are decreased abruptly, and the *S* waves are completely eliminated. This means that the Earth has a core which is spherical in shape because the same data are obtained at the same distance from earthquakes located anywhere on the Earth. Stations beyond 10,000 miles from the place of origin of an earthquake receive both *P* waves and *L* waves, but no *S* waves.

The change between mantle and core at the 1800-mile depth is thought to be both chemical and physical. The outer portion of the core appears to be nonrigid; i.e., it behaves as if it were a liquid. As evidence, the *S* waves do not pass through the core, and they cannot penetrate nonrigid material. The inner part of the core is probably solid because the *P* waves appear to increase abruptly in speed part way through it, and the greater rigidity of solid material would cause such an increase in speed. The radius of the inner core is about 800 miles (1260 km).

Both parts of the core may consist chiefly of iron (about 85%) alloyed with some nickel and other elements. This hypothesis is based partly upon an analogy with meteorites, which are of two main kinds (p. 587): metallic meteorites are chiefly iron and are much less abundant than stony meteorites, which are similar to certain low-silica igneous rocks (somewhat like gabbro). Perhaps meteorites represent the type of debris out of which the Earth was made. If so, the core is the logical place for the large volumes of iron that should be present within the Earth, and the mantle may be similar in composition to the stony meteorites.

The magnitude of the Earth's equatorial bulge implies the existence of a heavy core, as does the average specific gravity of the Earth as a whole (5.5), which is about double that of crustal rocks (2.7 to 3.0). Specific gravities presumably increase from the Earth's surface to its center, but not at a uniform rate. Some calculations suggest that the mantle's specific gravity may average nearly 5 (range: about 3.3 at the top to about 6 at the bottom). On the other hand, the specific gravity of the core may average approximately 11 (range: about 10 at the top to about 12 near the center).

How temperatures change with depth within the Earth is quite uncertain. Radioactive elements—especially uranium, thorium, and potassium—release heat energy during disintegration but are far more abundant in granitic rocks than in the materials that presumably occur at greater depths. Thus the relatively high geothermal gradient (1°F per 50 to 150 feet) observed near the surface presumably decreases with depth. Temperatures at the Earth's center may be of the order of 4000°C (7000°F)—a lower figure than thought likely some years ago.

The Crust

The Earth's crust beneath the oceans differs in important aspects from that beneath the continents, particularly in thickness and chemical composition (Fig. 6–31). The base of the crust is marked by a sharp discontinuity that occurs at an average depth of 20 to 25 miles (32 to 40 km) beneath the continents but apparently at nearly twice this depth beneath some great mountain belts (at this discontinuity the speed of the *P* waves changes abruptly from less than 7 km/sec to slightly more than 8 km/sec). This same discontinuity occurs beneath the ocean floors but at an average depth of about 3 to 4 miles. Thus an average portion of the continental crust is some six times as thick as an average section of the oceanic crust. The sharp discontinuity that marks the base of the crust is called the Moho after Mohorovičić, the Yugoslav seismologist who discovered it.

Beneath the continents the crust seems to

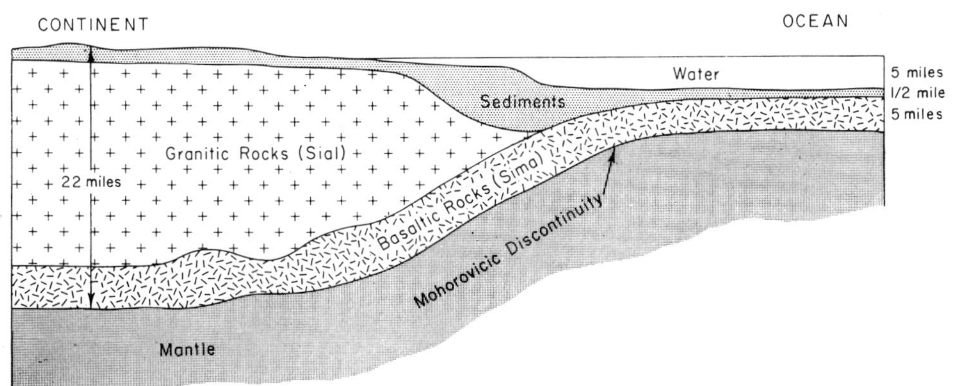

6-31. Idealized structure section of the Earth's crust near a continental margin. *(After Stokes. Courtesy Fred M. Bullard,* Volcanoes . . . *University of Texas Press.)*

consist of two layers (three where thick piles of sedimentary rocks occur) which may merge according to some seismologists but are separated by another discontinuity according to other seismologists. The upper layer of the continental crust is known as the granitic or sialic portion. In this sense, the term *granitic* encompasses deep-seated igneous or metamorphic rocks that have chemical and mineral compositions similar to granite—thus it includes many schists and gneisses as well as rocks such as granodiorite. Moreover, these other types of rocks may be far more voluminous than granite itself in some sections of the granitic crust. Because oxides of silicon and aluminum are abundant in these granitic rocks, the name *sial* was coined for this upper layer. Its average specific gravity is about 2.7. The sial appears to be missing beneath the deep ocean floor.

The lower layer of the continental crust is apparently similar to that which forms the bulk of the oceanic crust and is commonly called the basaltic-gabbroic crust or *sima*. Oxides of silicon and magnesium are most abundant in this part of the crust, which has a specific gravity of about 3.0. Basalt and gabbro have this chemical composition and are the common rocks found on oceanic islands and dredged from parts of the deep-sea floor. However, some geologists consider the oceanic crust to be chiefly serpentine, which can form by the alteration of a very low-silica igneous rock called peridotite—and peridotite may be the chief material forming the upper mantle.

Heat-flow measurements, the oceanic crust, and differences in the upper mantle beneath continents and oceans are considered elsewhere (p. 296).

The Mantle

The mantle is about 1800 miles (2900 km) thick and occurs between the crust and core. It may consist very largely of four elements: oxygen, silicon, iron, and magnesium (similar in composition to the stony meteorites). The upper mantle, immediately below the Moho, may be made chiefly of the rock peridotite. Although the mantle is a distinctive unit, some seismologists subdivide it into four secondary units based largely upon changes in the speeds of *P* waves and *S* waves with depth: the upper mantle, a low-velocity zone (see below) within the upper mantle, a transitional zone, and the lower mantle. In rounded-off numbers, the transitional zone is about 600 km (370 miles) thick and occurs between depths of 400 km (250 miles) and 1000 km (620 miles). The speeds of *P* and *S* waves and densities all seem to increase more rapidly with depth than expected within the transitional zone—thus changes of some kind presumably occur there. However, the transitional zone may not be conspicuously different from the rest of the mantle.

Seismologists use a technique somewhat as follows in interpreting conditions within the Earth. After the velocities of seismic waves have been estimated for different depths (Fig. 6–32), informed guesses are made concerning the minerals and rocks which might occur within the Earth and account for these speeds—at the temperatures and pressures that seem most likely at each depth. Differences in the speeds of *P* and *S* waves in some instances can be explained either by changes in chemical composition or by changes in physical structure. The latter are called *phase changes* and involve rearrangements in the three-dimensional patterns in which atoms and ions are packed in minerals without any change in the overall chemical composition of the mineral.

High-temperature, high-pressure laboratory experimentation is playing an important role here. It is now possible to simulate the temperatures and pressures that may exist throughout much of the upper mantle (however, the amount of material subjected to such conditions is quite small). Even higher pressures, perhaps comparable to those within the core, have been achieved for a tiny fraction of a second by using explosives or by arranging a collision between two rapidly moving projectiles. During such experiments, the lattice structures of certain familiar minerals have been changed into entirely new and denser materials. Possibly these new materials resemble those existing at comparable depths within the mantle.

Phase changes probably account for the differences observed in the transitional zone within the mantle, whereas chemical changes seem the most likely causes for the sharp discontinuities at the mantle's boundaries with the crust and core (a change from solid to liquid also occurs at the core). However, a phase change has also been suggested to account for the Moho (e.g., basalt changes into a denser rock called eclogite at high temperatures and pressures).

Low-Velocity Zone

The speeds of the *P* and *S* waves do not increase all of the way to the core. They increase for a certain depth beneath the crust, but below this for a distance the *P* wave speed stays about the same, whereas the *S* wave speed decreases to a minimum at a depth of about 85 to 90 miles (140 km). Why this happens is not known, but it may be that rocks are "softer" in this zone, which is world-wide in extent in the upper mantle. This *low-velocity zone* or plastic layer in the upper mantle exceeds 100 miles (160 km) in thickness and its upper boundary is located at a depth of about 35 to 40 miles (60 km)—thus its lower boundary is at about 155 miles (250 km). Perhaps the rocks are very close to their melting points at the pressures at these depths. If so, this could be the zone where at least some basaltic magma originates, where isostatic adjustment takes place (see below), and along which great slabs of the lithosphere slide slowly (p. 314). The low-velocity zone seems to be somewhat closer to the surface beneath the continents than it is beneath the oceans.

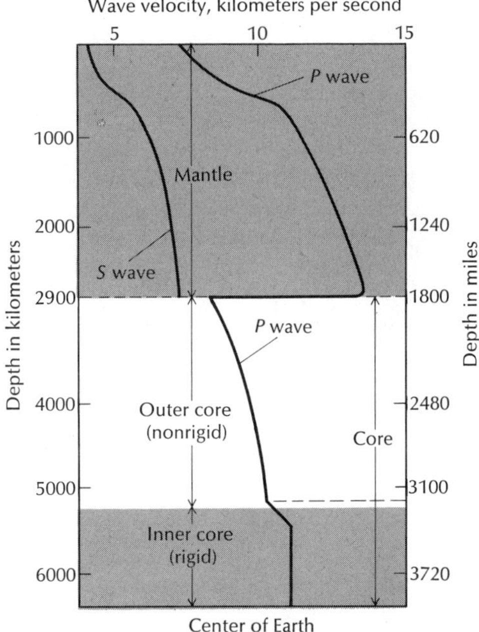

6-32. Estimated travel-time curves for *P* and *S* waves at different depths within the Earth. At the boundary zone between the mantle and core, *S* waves stop and the speed of *P* waves drops abruptly. *(After Longwell and Flint.)*

Isostasy

Isostasy (p. 59 and 239) means "equal standing" and refers to the vertical balance that occurs in the outer part of the Earth between its larger units such as mountain belts vs. extensive low-lying plains and continents vs. ocean basins (Fig. 6–33). This state of balance between major units of the crust implies that vertical movements will occur if certain areas become overloaded or underloaded—somewhat as a canoe rides lower in the water when loaded and higher when empty. This occurred during the Pleistocene Ice Age when great ice sheets formed and overloaded the lands beneath them—these regions then sagged (p. 239). However, the ice sheets have since disappeared, and today these same regions are rising. More recently—on an unexpectedly small and rapid scale—Boulder Dam was built, Lake Mead formed, and the crust was sufficiently overloaded with water to subside locally (about ½ inch per year over a decade and a half; earthquakes also resulted from the overload, p. 239).

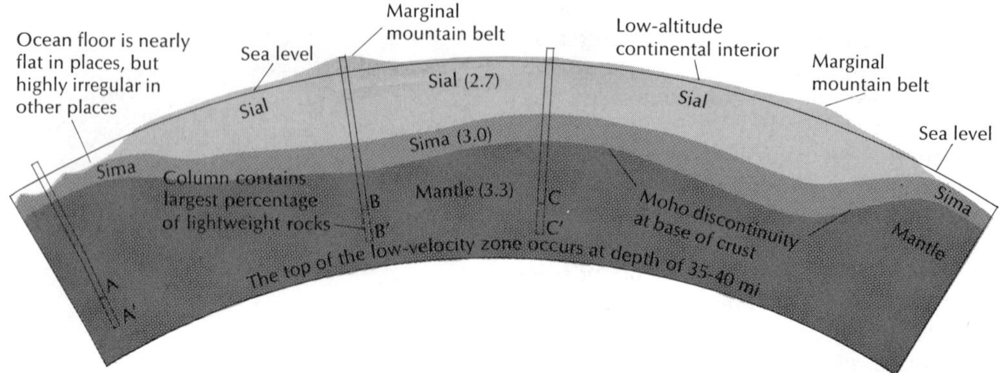

6-33. Schematic structure section across a continent illustrating isostatic balance and the increase in thickness of the crust beneath a great mountain belt. Three hypothetical columns (A, B, and C) are imagined to be equal in length, cross section, and mass. However, each contains a different proportion of sial, sima, and upper mantle (specific gravities are 2.7, 3.0, and 3.3 respectively). Although column A is formed partly of air and water, it has the largest proportion of the densest type of rock and thus has the same weight (mass) as columns B and C. If the columns were extended equal distances to A′, B′, and C′ their masses would still be equal because the same quantity of mantle would be added to the base of each column.

Such vertical shifting implies the existence of compensating lateral movements at depth because no large openings exist that might collapse to make space available for subsidence, nor can rocks be compressed this much. It now seems likely that such lateral movement probably takes place by slow plastic flow within the low-velocity zone a few tens of miles below the surface. Such vertical and lateral movements are part of the process of *isostatic adjustment.*

The concept of an isostatic equilibrium within the outer Earth grew out of a land survey in India more than a century ago. Astronomical positions were made by reference to a plumb bob (essentially a weight hanging on a string—the string should be vertical) and checked against positions mapped by triangulation along the surface (p. 528). The two sets of positions did not check because the plumb bob was pulled laterally by the gravitational attraction of the massive Himalayas (not discovered until later). Subsequent calculations then showed that this lateral pull was less than it should have been (assuming uniform conditions beneath the surface); i.e., the mountains were less massive than expected. This in turn led to the idea that great mountains are buoyed up by large root zones of lightweight rocks; in other words, the sial is thickened beneath them. In a similar manner, continents project above sea level because they consist of plates of lightweight rocks that are in isostatic balance with the heavier rocks located beneath the floors of the oceans.

In a somewhat analogous manner, icebergs project much deeper below the surface than they do above it, and those with tops towering highest above the water also extend farthest below the surface. Presumably, therefore, the simatic floors of the deep oceans cannot be uplifted to form continents, nor can large sections of the continents be depressed to form deep ocean basins.

As shown in Fig. 6–33, we may imagine that vertical columns, of equal cross sectional area, extend downward through the atmosphere, hydrosphere, and lithosphere at different geographic locations. The proportions of sial, sima, and mantle in each of these columns is such that all push downward on the top of the low-velocity zone with equal force—this is then a level of uniform pressure, and such levels should occur all of the way to the center. However, just how high into the lithosphere such levels of uniform pressure extend is uncertain.

Exercises and Questions for Chapter 6

1. Draw a structure section which shows that the following events have occurred in the geologic history of a certain area:
   (A) Sandstone formed.
   (B) Basalt dike formed.
   (C) A normal fault occurred.
   (D) Erosion continued for a long time.
   (E) Conglomerate formed.
   (F) A granite dike formed.
   (G) Erosion occurred.
2. Make structure sections to show several different ways in which an unconformity may develop.
3. What kinds of evidence could you look for to determine whether a certain crack in rocks represents a joint or a fault?
4. Folded rocks:
   (A) What is the difference between a topographic basin and a structural basin? Make sketches to illustrate.
   (B) What is meant by a synclinal mountain?
   (C) Assume that you fly over an area and note nearly parallel ridges and valleys. These trend across the area to form a zigzag or loop-shaped pattern. What is the probable cause of such features?
   (D) What sort of drainage pattern would you expect to develop in an area where sedimentary rocks have been deformed into a structural dome? Assume that sedimentary beds resistant to erosion alternate with others that are less resistant.
5. What conditions favor rock deformation by folding? by faulting?
6. In a certain area, a series of sedimentary rocks strike north-south and dip 45 degrees toward the east. The rocks consist of conglomerate (oldest), sandstone, shale, and limestone (youngest).
   (A) Which type of rock is located farthest toward the east? the west?
   (B) If a vertical hole is to be drilled in such a manner that it will penetrate all four sedimentary rocks, where must it be located at the surface?
   (C) Sketch an east-west structure section through these rocks.
7. In a certain area you find three types of rocks that trend in a north-south direction and dip about 30 degrees toward the west. Marble occurs in the west; quartzite occurs in the east; and schist occurs between the marble and quartzite.
   (A) Sketch an east-west structure section through these rocks.
   (B) Is the association of marble, schist, and quartzite in this manner one that you might expect to find or is it one that you would regard as quite unusual? Explain.

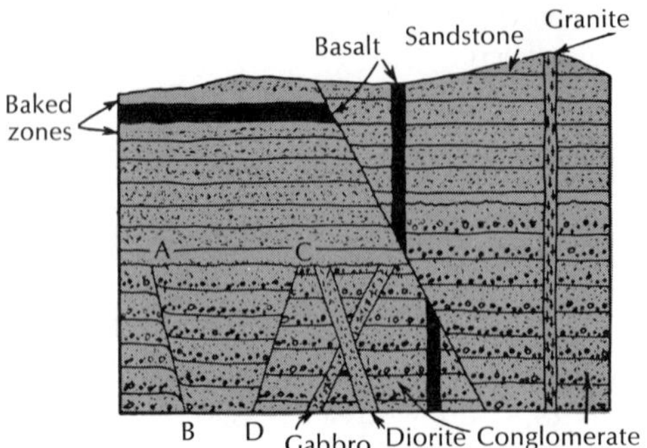

8. Refer to the structure section on this page and list in chronological order (from earliest to latest) the events that have occurred in the geologic history of this area. Describe the evidence showing that a given geologic event must have occurred before, about the same time as, or after another geologic event. However, not enough information is given in two instances for you to decide which of two or more events came first. Describe these.
9. What is the immediate cause of an earthquake (not the ultimate sources of the forces that deform rocks)?
   (A) What precautions can be taken to minimize damage when buildings are constructed in earthquake areas?
   (B) How can the distance to an earthquake epicenter be calculated from records received at a single seismograph station?
   (C) How can the location of the epicenter of an earthquake be calculated from records received at three stations?
10. What evidence indicates that the Earth has a core? that this core is spherical? that it may consist chiefly of iron? that its outer part consists of nonrigid material (i.e., that it behaves as a liquid)?

7

Weathering, Mass-Wasting, and Stream Erosion

There rolls the deep where grew the tree.
O earth, what changes hast thou seen!
There where the long street roars, hath been
The stillness of the central sea.
The hills are shadows, and they flow
From form to form, and nothing stands;
They melt like mist, the solid lands,
Like clouds they shape themselves and go.
—Tennyson: *In Memoriam*

The "everlasting hills" are really not everlasting at all. As a result of weathering, rocks break apart, decay, and crumble; the resulting fragments and dissolved material then move from higher to lower ground (mass-wasting); and eventually the sediments are transported from the area by some agent such as running water or wind. Burial in the sea is the ultimate fate of most sediments. How such vast quantities of rock can be removed from an area and how the Earth's surface can be sculptured into mountains and valleys is the theme of this chapter.

Moving from scene to scene (Figs. 7–1 and 7–2), the traveler often feels frustration as he lets his eyes fall on diverse landscapes—cliffs and valleys and badlands—and gropes vaguely for explanation of how they came to be. The pages that follow tell the story of the formation of such features by the "team" of weathering, mass-wasting, and stream erosion. Land forms sculptured by other erosional processes, such as subsurface water, wind, and ice are discussed in succeeding chapters. However, stream erosion, aided by weathering, sheetflow, and mass-wasting, is by far the dominant erosive force on the Earth today. If isostatic adjustment (p. 166), other movements of the crust, and igneous activity did not take place, the continents would eventually be worn down to almost featureless surfaces located near sea level (in about 20 million years by one estimate, but several times longer than this according to others).

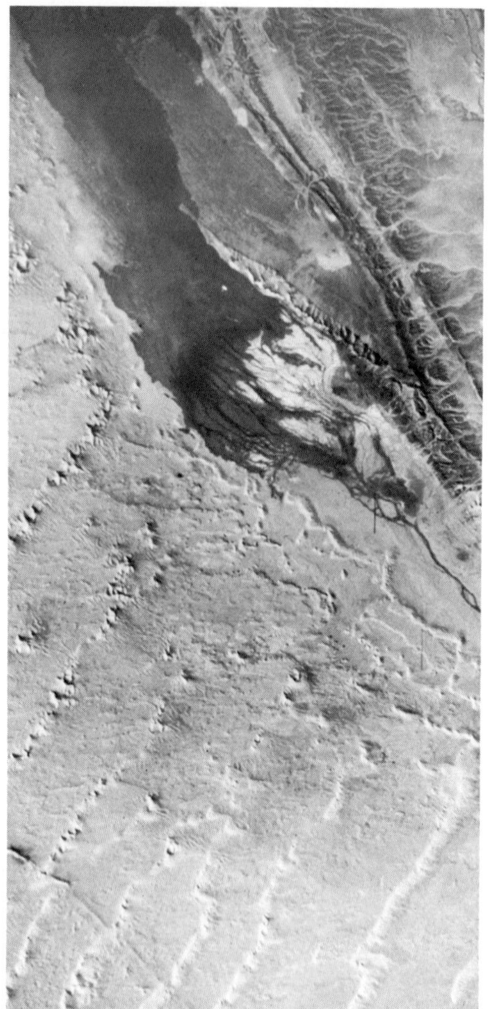

7-1. Sand dunes in Oued Saoura, western Algeria, viewed from an altitude of 100 to 200 miles. The ephemeral lake at the top was produced by runoff from the Atlas Mountains to the northwest, although the area is usually dry. Note rock structures produced by folding. *(Gemini VII, 5 Dec. 1965, S65-63830, NASA.)*

Judson* has discussed rates of erosion and man's newly discovered role as an important geologic agent in speeding up the erosion of the lands—by some two to three times on a worldwide scale. We may cite studies involving bristlecone pines (by Eardley in Utah and California) as an example of measurements made in a single small area. The original ground level, and the amount that it has been lowered by erosion (denuded) during the life of a tree, are shown by the depth of the root system that is now exposed. The ages of the trees can be determined by counting tree rings in special small-diameter cores that can be removed without damaging a tree—some of the oldest trees have an age of 4000 years. Average rates of denudation were calculated as 2 cm per 1000 years on 5-degree slopes and 10 cm (4 inches) per 1000 years on 30-degree slopes—thus a correlation exists between slope and rate.

For larger areas, denudation rates may be estimated by measuring the amount of sediment carried from a particular drainage area by its main stream (where it leaves the area), or by measuring the amount of sediment deposited by this stream in a reservoir or lake. Studies of this type, comparing drainage areas of approximately equal sizes, have shown a correlation between precipitation and erosion—maximum rates take place in areas having about 10 inches (25 cm) of precipitation per year, although other factors are also involved.

Such river-based data for the United States as a whole indicate a current denudation rate of about 6 cm (2.3 inches) per 1000 years. On the other hand, no net lowering is occurring in areas of interior drainage (areas that lack outlets to the oceans) which reduces this figure somewhat. However, this rate is not at all representative of the immediate geologic past because man's agricultural and other activities have speeded up the rates of denudation. For the United States, the rate may have doubled (i.e., it was 3 cm per 1000 years some centuries ago), and worldwide the current rate may be two to three times greater than it was before man's intervention. Thus prior to man's development as an important factor in erosion, the continents were probably being denuded at about 2.4 cm (nearly 1 inch) per 1000 years.

* Sheldon Judson, "Erosion of the Land," *American Scientist,* Winter 1968; the next few paragraphs are based upon this article.

Denudation rates of wind and glacier ice are difficult to estimate but are relatively insignificant overall when compared to the erosion accomplished by streams. However, the rate of lowering by wind and ice may each be of the order of ½ to 1 mm (1/100 to 1/25 of an inch) per thousand years.

Weathering

From the time someone first bounced a stone off a tender shin, most people have been thoroughly impressed by the hardness of rocks. Rock outcrops in one's local neighborhood show no readily discernible changes during a lifetime, and stone monuments and buildings are accepted as permanent features. Yet the action of the weather in changing shiny iron nails to rust is familiar to all, and careful observation will show that rocks do crumble and decay in buildings and in nature.

The color of the weathered surface of a rock may be strikingly different from that of a fresh surface. The outer parts of some rock exposures are clayey and can be crumbled by the fingers; but like the rusty nails, they too were once solid. This can be demonstrated by cutting downward through the weathered zone where the rock gradually becomes firmer and takes on a fresh appearance. Minerals such as feldspar and biotite mica become recognizable. Evidently they were altered to clay during the process of chemical decay.

Weathering is the nearly static part of the general process of erosion; it is a name for all processes which combine to cause the disintegration and chemical alteration of rocks at or near the Earth's surface. In this sense, it contrasts with metamorphism which involves deep-seated changes at high temperatures and pressures. Weathering takes place everywhere on the lands where rocks are exposed to air, water, and organisms, and it produces the debris that is then transported away by other processes.

Two general kinds of weathering are recognized. In *chemical weathering,* chemical reactions bring about the decay and rotting of rocks and their transformation into new substances (such as clay) which are stable under the chemical-physical conditions existing near the Earth's surface. Such rocks may have formed far below the surface at temperatures and pressures quite different from

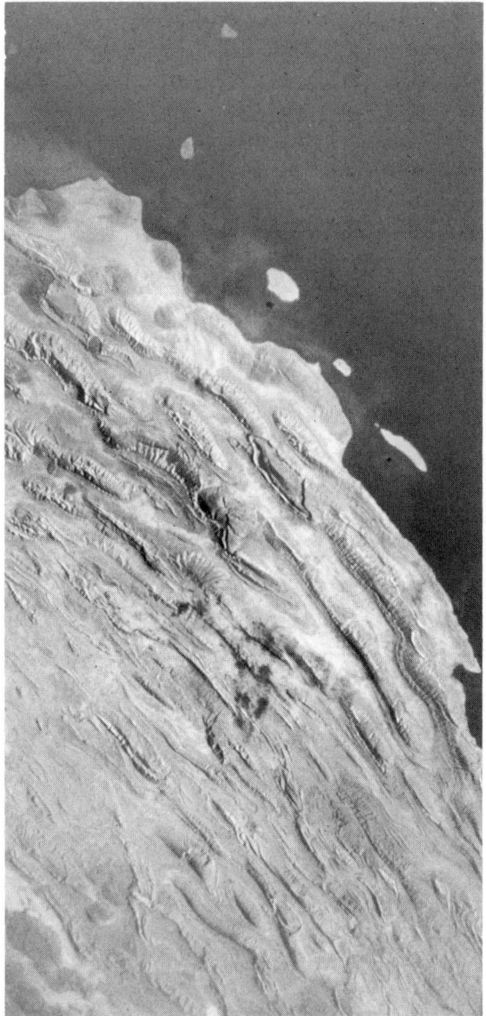

7-2. Anticlines in the Zagros Mountains in Iran and the Persian Gulf viewed from an altitude of 100 to 170 miles. Near the coast in the upper left, the dark circular or elliptical masses are salt plugs exposed at the surface (possible only in a dry climate). *(Gemini XII, 13 Nov. 1966, S66-63483, NASA.)*

those at the surface. In a sense, then, rocks tend to equilibrate with their environment, although the adaptations may be very slow and incomplete. The advisability of using copper pipes instead of iron under certain conditions provides a familiar example of some types of materials that change chemically more rapidly than others.

In *mechanical weathering,* rocks are reduced to smaller pieces without chemical decomposition. Although one type of weathering may predominate under certain conditions, the two are intimately related, and one generally aids the other. For example, the mechanical disintegration of rocks furnishes additional surfaces for attack by chemical action; yet some chemical reactions result in an increase in volume which causes disintegration. Weathering thus produces three types of materials: broken rock and mineral fragments, residual decomposition products, and dissolved substances.

Regolith is the name that we use for the loose debris that forms a widespread, relatively thin, discontinuous cloak over the bedrock and underlies the soil in many areas. We speak of transported regolith if the loose debris was formed in another area and was then carried and deposited in its present location (Fig. 2–12). A sharp break occurs between it and the underlying bedrock (common in a recently glaciated area). On the other hand, residual regolith is produced essentially in its present location by the disintegration and decay of bedrock. Thus no sharp line separates this loose debris and bedrock, each merging gradually into the other.

Rocks differ greatly in their rates of weathering, and identical rocks may weather quite differently under different climatic conditions. Differences become apparent when comparisons are made between rock types, between different sections of a single type of rock, and even between neighboring mineral grains that differ in composition. Weathering is thus responsible for etching out much of the finer detail of the Earth's scenery. *Differential erosion* (Fig. 7–3) by wind, running water, ice, and waves combines with differential weathering in shaping the larger units of topography; e.g., certain layers in a sedimentary rock may be cemented more firmly, be more soluble or permeable, or be more susceptible to chemical or mechanical change, than layers above

7-3. Differential erosion in angular folds in northern Chile. Beds are silty limestone of Early Cretaceous age. *(Courtesy Kenneth Segerstrom, U.S. Geological Survey.)*

and below them. Differences in resistance become particularly prominent if the weathered debris is removed relatively rapidly as occurs on steep slopes with little vegetation in a dry region.

In some instances, permeability may be important; e.g., if one type of rock is very permeable, most of the water that falls on its surface will sink into the ground, little runs off, erosion is slow, and the area remains as high ground because adjoining areas are eroded more rapidly. Thus weathering is influenced by many factors.

Chemical Weathering

Warm moist climate, gentle slopes, and abundant vegetation are most effective for chemical weathering. Water is important because many substances can react together chemically when they are wet, whereas they do so only very slowly or not at all when they are dry. Many of the reactions are complex. Water combines with carbon dioxide in the air and with the humus of decayed vegetation to form various acids which can dissolve and remove many rock materials, a process known as *leaching*. An increase in temperature speeds up most chemical reactions and causes some which cannot take place at lower temperatures. The commonplace process of rusting involves the chemical union of oxygen and water with iron to form a group of hydrous iron oxide minerals, collectively called limonite.

On steeper slopes, the loosened fragments tend to be pulled by the Earth's gravitational atraction and they shift, slide, or bounce to lower altitudes. This tends to expose fresh bedrock to continued weathering, and the regolith tends to be thin. In contrast, on gentle slopes, the loosened debris tends to remain more nearly in place, and long-continued chemical weathering may result in a thick regolith.

Abundant joints and rock openings of all sizes enhance weathering because they allow penetration by air and water. Thus mechanical weathering greatly furthers the chemical processes, and the disintegration of a rock into many small fragments during mechanical weathering increases the amount of surface area exposed to chemical alteration. To illustrate, a 1-inch cube has 6 square inches of surface area. If it is subdivided into smaller and smaller cubes, the total volume of the cubes remains constant at 1 cubic inch no matter how small the cubes finally become. However, the amount of surface area increases sharply; e.g., if the 1-inch cube is subdivided into eight smaller cubes, each ½ inch on a side, the total surface area will be 12 square inches, twice the original amount (each of the eight cubes has ¼ square inch of surface area on each of its six sides). Continued subdivision into cubes ¼ inch on a side produces 24 square inches of surface area, and still further subdivision into ⅛ inch and 1/16 inch cubes exposes 48 and 96 square inches of surface area respectively. Each time the linear dimension of the cubes is reduced by half, the surface area doubles.

Because feldspars, pyroxenes, amphiboles, quartz, and mica are estimated to constitute over 90 percent of the minerals of the crust, the manner in which they tend to weather is important. Feldspars decompose chiefly to clay minerals (hydrous aluminum silicates); pyroxenes, amphiboles, and biotite mica may yield clay and hydrous iron oxide; quartz and muscovite mica are highly resistant to chemical weathering.

The susceptibility of minerals to weathering more or less parallels the order in which they tend to crystallize from a magma to form an igneous rock (p. 77). The first-formed minerals tend to develop under the most deep-seated conditions involving very high pressures and temperatures. Exposure at the surface represents a change in their environment that is very much different from the one in which they formed, and thus they are quite susceptible to weathering. Therefore, olivine and calcium plagioclase feldspar tend to weather readily, hornblende and potassium feldspar tend to change more slowly, and muscovite mica and quartz (crystallize last under deep-seated condi-

7-4. The Old Man of the Mountain in Franconia Notch, New Hampshire. The freezing and thawing of water in joints has been important in shaping this rock outcrop. *(Courtesy New Hampshire Dept. of Resources and Economic Development.)*

tions) tend to be most resistant to weathering.

Quantitatively, weathering tends to have two chief end products: the formation of clay by chemical alteration and the loosening and freeing of quartz grains (in rocks containing quartz). Since quartz is very resistant to both physical and chemical changes, the quartz grains of one rock (say a granite) may become the quartz grains of another rock (say a sandstone) after erosion and sedimentation. This sandstone in turn may eventually disintegrate, and some of its quartz grains (made smaller by abrasion) may become part of yet another sandstone. Thus shales (composed of clay) and sandstones (made chiefly of quartz grains) apparently make up nearly 80% of all sedimentary rocks.

7-5. Spheroidal weathering in gabbro. *(Courtesy W. T. Schaller, U.S. Geological Survey.)*

Limestones are abundant, and these three kinds together—shales, sandstones, and limestones—may comprise about 99% of all sedimentary rocks. When limestones weather in a climate where at least a moderate amount of moisture is available, they tend to dissolve, and insoluble impurities such as clay—once only a minor constituent and scattered throughout the limestone beds—tend to accumulate at the surface. Eventually enough impurities may collect to form a thick clayey regolith.

Mechanical Weathering

The freezing and thawing of water confined in rock openings is the most important purely mechanical process involved in the disintegration of rocks—the process of "making little ones out of big ones." On freezing, water expands about 9% by volume and exerts a pressure which may exceed many hundreds of pounds per square inch. Confinement can occur after the water in a crack freezes from the top down, thereby sealing off the water beneath it. Steep slopes with little vegetation, in regions where temperatures fluctuate frequently back and forth across the freezing mark, promote this freeze-thaw type of weathering (Fig. 7–4).

In many instances, rock decay involves an increase in volume (e.g., by the chemical addition of water) and this swelling then causes disintegration. This process seems to account for one type of *exfoliation*, i.e., the spalling or flaking off of concentric shells from massive types of rock such as granite.

The rocks are weathered and decomposed into shells a few inches or less in thickness. Spheroidal boulders (Fig. 7–5) are the result of this process, which may take place at the surface or several feet below it. The rate of weathering of a rectangular block is least along its sides, more rapid along the edges, and fastest at the corners where the rock is attacked from three sides simultaneously. If evenly spaced joints occur in a massive rock, the initial blocks will have approximately cubical shapes, and nearly spherical exfoliated shells tend to be produced as the cubes are weathered.

Much thicker shells (Fig. 7–6) and sheet structures in massive, relatively unweathered rocks tend to be parallel to the present surface, and may be caused by release of pressure due to unloading—the removal of the weight of overlying rocks by erosion. In this view, such rocks formed far below the surface, and as erosion removed the tremendous mass of rocks that once rested upon them, they expanded, and the concentric shells formed.

7-6. Thick exfoliated shells exposed on Half Dome in Yosemite National Park, California. For the scale, try to find the man standing on the rounded mass of rock in the lower right. *(Courtesy U.S. Geological Survey.)*

The wedging action of rooted vegetation growing in cracks likewise reduces rocks to smaller pieces. Perhaps you know of a tree-lined sidewalk where you are most likely to stub your toe near each tree because there the growing roots have heaved up the concrete blocks.

Frost heaving damages roads and is most effective in fine-grained materials in moist regions with frequent frosts. Lenses and layers of ice form at different levels below the surface by drawing up water from below like a blotter. Frost heaves up to 18 inches have been measured. Sometimes the amount of heave has equaled the combined thicknesses of the ice lenses. The same process evidently accounts for the presence of some of the rock fragments in your garden each spring. Frost polygons or stone rings (Fig. 7–7) contribute an interesting aspect to some of the colder regions of the Earth.

Rocks such as granite disintegrate into their constituent mineral grains in many desert areas. This phenomenon was formerly explained somewhat as follows. Extreme temperature changes occur between day and night in desert areas and might be expected to cause individual mineral grains in a thin surface zone to expand and contract under the alternate heating and cooling. As different minerals swell and shrink at different rates, strains might be set up in the rocks which could cause eventual disintegration.

This logical hypothesis has been tested in the laboratory. Different rock specimens were alternately heated and cooled thousands of times at temperature variations greater than those occurring in any desert, and the microscope revealed no disrupting effects. Doubt was thus cast upon the importance of such temperature changes in causing rock disintegration directly. Volume changes caused by certain chemical reactions are apparently more important in the breaking of rocks in such areas. However, the experiments described above were performed in the absence of water. When the experiments were repeated, water was sprayed on the heated surfaces to cool them, and disintegration was observed in the specimens.

Flaking and disintegration are produced, however, by the extreme temperature changes which occur when a forest fire burns an area and is extinguished by a heavy rain. In fact, building a fire on an exposed rock surface was an ancient quarrying technique; the outer few inches of the rock expanded and could then be broken loose by hammering.

7-7. Polygonally patterned ground in permafrost area along a major stream in northern Alaska. The polygons tend to develop along old channel scars representative of old slip-off slopes. Ice-wedge formation is an important factor in the origin of the polygons. Note the oxbow lake. *(Fig. 109,* U.S. Geol. Surv. Profess. Paper *373.)*

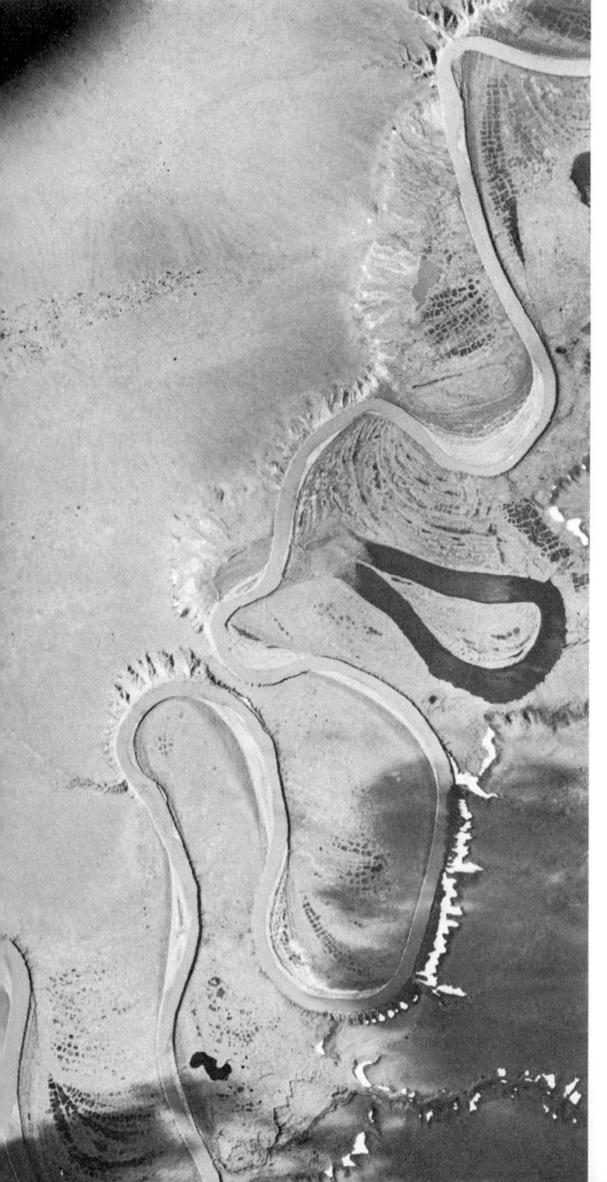

Soils

Soils can support rooted vegetation and tend, where present, to form the upper part of the regolith. Soils thus consist of decomposed rock debris and decayed organic matter *(humus)* which have been produced by weathering, and mature soils generally have a layered structure (Figs. 7–8 and 7–9). Humus is abundant in the upper portion, the *topsoil,* which contains new minerals and substances formed by weathering. Certain soluble materials and tiny particles have been extracted from the topsoil by leaching and slow downward sifting. Some of the small clay and silt particles have also been united during weathering to form larger aggregates or crumbs. The granular structure thus produced permits air, water, and tiny rootlets to penetrate the soil readily and makes it most useful to man. Such a soil can hold water between rains but permits the excess to drain through. If maintained properly during farming, it will have a balanced supply of nutrients, and it will not wash away during heavy rains nor blow away during strong winds.

Three major horizons or zones are recognizable in most mature soils: the A (top), B, and C horizons. These have formed by weathering of the underlying parent material which may be either bedrock or regolith (e.g., a granite, limestone, basalt, gravel, or tuff) and the addition of organic matter to it (most abundant in the topsoil). The original minerals in the topsoil, or *A horizon,* have been entirely weathered to new secondary minerals, except for the most resistant ones such as quartz. Some of the material from horizon A has been leached out and deposited in the underlying subsoil, or *B horizon.* The finest particles also tend to be moved downward out of the A horizon by percolating water. The B horizon is thus a zone of accumulation for some of the material removed from the A horizon, which we can regard as a zone of leaching (in more humid areas). In this light, we can now redefine soil as the A and B horizons of the regolith. Materials in the lowest, or *C horizon*, are only partially weathered, are not soil, and form a transitional zone between the soil above and the unweathered parent material below. With the passage of time, each zone becomes thicker at the expense of the underlying one.

Thousands of years may be required for the formation of a thick fertile topsoil from a naked rock outcrop, although soils have been known to develop on volcanic debris in a few hundred years or less—to a thickness of about 1 foot in 45 years in one instance. Certain rock outcrops (Fig. 7–10) show that rates of soil formation and weathering can be very slow when certain types of rocks and climates are involved; e.g., some glacially polished and scratched bedrock surfaces are fresh-looking today and nearly unchanged after thousands of years of weathering. Partly because they have had more time to mature, soils in the nonglaciated southern part of the United States tend to be thicker and better developed than soils in the glaciated northern portions and in Canada where development has been limited to the last 10,000 years or so—glacial erosion removed older soils that had formed prior to glaciation. However, less mature soils 2 feet and more in thickness have developed in these areas since deglaciation.

Abundant humus colors soils dark, whereas lighter grayish colors may be common where humus is not abundant. Iron oxides impart yellowish, brownish, or reddish colors.

Some important factors in the development of soil are the kind or kinds of parent rock material, the climate, the types of animals and plants present, the shape of the surface, and time. Formerly the view prevailed that each type of parent rock would produce a distinctive type of soil, but it is now known that climate and time are even more significant. Weathering for a long enough time under one type of climate can produce similar soils from widely different

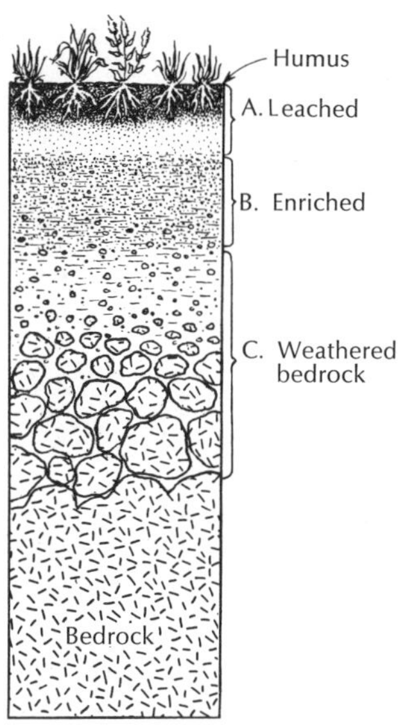

7-8. Soil horizons in cool humid region (idealized). *(B. Mears, Jr.,* The Changing Earth, *Van Nostrand Reinhold, 1970.)*

7-9. Profile of a very fine loam in Walsh County, North Dakota. The dark surface soil is fairly deep. *(Courtesy U.S. Dept. of Agriculture.)*

7-10. Perched boulders (erratics) rest upon polished, grooved, glaciated surface ("Glacier Garden," Lucerne, Switzerland). Note pothole (p. 188) in left center and fresh appearance of bedrock surface after many centuries of exposure to the weather. *(R. W. Fairbridge, ed.,* The Encyclopedia of Geomorphology, *Van Nostrand Reinhold, 1968.)*

parent rocks. Therefore, by comparing ancient soils, now preserved as part of a sedimentary rock, with modern soils that are known to have formed under particular climatic conditions, we may be able to interpret the type of climate that existed when an ancient soil formed.

Soil classification is quite complex because so many varieties are known (e.g., 10 soil orders and 40 suborders are recognized in the U. S. Dept. of Agriculture's "Seventh Approximation"), and only a few types are described below. In more humid areas such as the eastern United States, the A horizon may have a dark organic-rich surface portion that is underlain by material of a lighter color, and the B horizon may be brownish or reddish because it has been enriched in iron and clay. More soluble materials may have been entirely removed from all three horizons, although the greatest amount of leaching occurs near the surface.

Some soils in the drier western half of the United States, especially their B horizons, tend to be enriched in calcium carbonate and in other carbonates because less solution and more evaporation of water take place. Less clay also tends to be present because less clay can form in the drier climate. Some carbonate material can also be carried upward in solution from the lower horizons by capillary action (p. 218). Evaporation then tends to occur in the upper horizons or at the surface, and a hard whitish, impervious layer may form in this manner, or the carbonate material may be present in irregularly shaped, smaller, discontinuous masses.

Laterite ("brick") is a reddish-brown soil or residual regolith that has undergone extreme leaching and that develops most readily in tropical humid areas with pronounced wet and dry seasons. Leaching tends to remove soluble materials such as sodium, potassium, and calcium as well as silica which is generally insoluble under other climatic conditions. During the wet season, solution and precipitation occur; during the dry season, air can penetrate the soil, and oxidation occurs. Nearly insoluble oxides and hydrous oxides of iron and aluminum tend to accumulate, and plant nutrients may be scarce. If the parent rocks contained abundant iron or aluminum, then long-continued laterization may produce ores of these metals.

Burrowing organisms aid in the development of soil—they may mix, break, and alter the regolith, or transport some of its particles to the surface where they can be more readily weathered. Among these, the earthworms play an important role, because they literally eat their way through the soil and cause chemical and mechanical modifications in so doing. In humid temperate regions, earthworms may completely work over a soil layer from 6 to 12 inches (15 to 30 cm) thick every 50 years. Multitudes of smaller insects and microscopic bacteria also live in the soil and play their important roles in its formation.

Practical Aspects of Weathering

Without question, the formation of soil is the most important single result of weathering to mankind, but weathering has also produced many valuable mineral deposits. Either an originally useful mineral is concentrated by removal of waste products, or valuable minerals are produced by weathering and then accumulate; e.g., ore deposits of iron, copper, and aluminum have formed in this way.

Knowing how various rocks weather under different climatic conditions should enable a contractor to make wise choices of building stones. In a warm moist climate, a sandstone naturally cemented by calcium carbonate or iron oxide would be unsatisfactory, because the cement would either dissolve or stain the rock with rust spots. As another example, a badly fractured silica-rich sandstone would not be suitable in a cool moist climate, because water would enter the cracks, freeze, and disintegrate the rock. On the other hand, a soluble limestone makes an excellent building stone in dry areas.

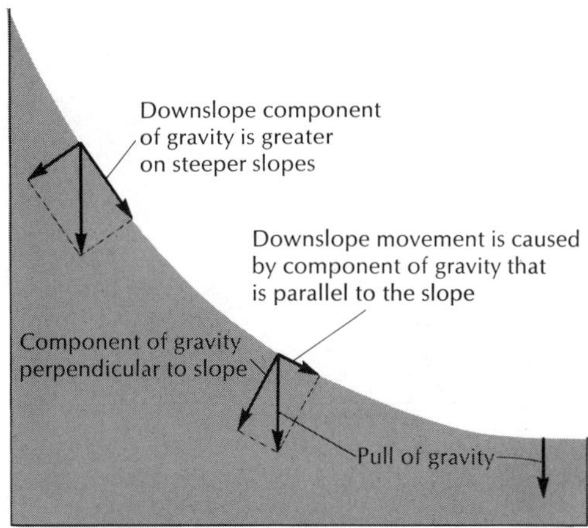

7-11. Downslope component of gravity vs.. slope. Arrows are vectors in this application of the parallelogram of forces.

Mass-Wasting

In *mass-wasting,* gravity (Fig. 7–11) causes large masses of regolith or bedrock or both to move downslope either slowly or rapidly. It encompasses a group of processes that play a leading role in the general wasting away and leveling of the lands. The processes of weathering loosen, decompose, and dissolve rock debris everywhere on the lands, and the processes of mass-wasting then move this debris to lower altitudes where streams, ice, and wind can transport it away. Weathering and mass-wasting are nearly ubiquitous on the lands, whereas streams are concentrated in gullies and valleys and cover a much smaller area. The great effect of mass-wasting in wearing down the lands has been only recently appreciated due to the almost imperceptible rate at which some types of mass-wasting occur. Gravity is the direct controlling agent of movements which may be slow or fast. Rockfalls, landslides, soil slump, creep, mudflows, and slope wash are examples of mass-wasting.

Rather than attempt to classify mass-wasting into specific processes, we shall emphasize its overall nature and importance and describe only a few of the more familiar types. At one end of the mass-wasting spectrum, we have processes such as rockfalls and landslides that involve steep slopes and debris loads that are very large relative to the amount of water present. At another extreme, we have gentle slopes in combination with debris loads that are small relative to the total amount of water present—these grade through sheetwash (p. 204) into stream systems as the amount of water increases. However, other combinations of the variables also occur to effect movement (e.g., slope, rate, load, and amount of water present).

Two general types of movement are involved in mass-wasting, either of which may be very fast—fast enough to bring death and destruction—or almost imperceptibly slow. One type is represented by the familiar *landslide,* in which a mass of bedrock and regolith slides as a unit upon some sort of steeply inclined, weakened, or lubricated surface (Fig. 7–12). However, some falling, bouncing, and flowage may also occur. Landslides tend to occur on steep slopes underlain by strata or structures (e.g., joints and foliation) that dip parallel to the surface. Water adds weight, reduces the cohe-

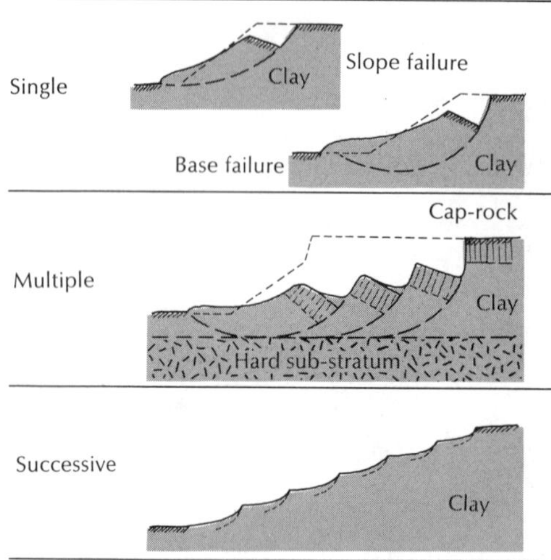

7-12. Main types of rotational landslip. *(R. W. Fairbridge, ed.,* The Encyclopedia of Geomorphology, *Van Nostrand Reinhold, 1968.)*

sion among the particles in surficial materials, and may reduce internal friction. Thus landslides tend to be most common in some areas in the spring when frequent rains combine with melted snow to add water to the regolith and bedrock. Hummocky topography is characteristic of some landslide areas.

Landslides are frequently set off by earthquakes, and a gigantic one (Fig. 7–13) occurred in August 1959 in southwestern Montana in the canyon of the Madison River 6 miles below Hebgen Dam. In a minute or so, 35 million cubic yards of broken rock slid into the canyon and covered a 1-mile span of the river and highway to a depth of 100 to 300 feet. Three weeks later a lake 175 feet deep and nearly 6 miles long had formed upstream from the landslide area. The sliding occurred along cracks which

7-13. This gigantic landslide in the canyon of the Madison River below Hebgen Dam was set off by the earthquake of 17 August 1959. The lake formed upstream from the slide. *(Photo by J. R. Stacy, U.S. Geological Survey.)*

7-14. A rock avalanche resulted from the 1964 Alaskan earthquake. Sherman Glacier is 1.5 miles (2.4 km) wide here. Shattered Peak is the new name for the mountain peak that was shaken apart. *(Photograph by Austin Post, U.S. Geological Survey.)*

slanted downward toward the river and were nearly parallel to one steep wall of the canyon. The slide started about 1300 feet above the river, swept downward and across the valley, and extended as much as 400 feet above the river on the opposite side. From this unstable position, some debris subsequently slid back into the valley. A number of people lost their lives in this disaster.

Vibrations produced by the 27 March 1964 Alaskan earthquake (perhaps double the intensity of the 1906 San Francisco earthquake) caused part of a high mountain peak to break away abruptly. The freed rock debris then moved quickly—at a rate that may have exceeded 100 mi/hr—by falling, sliding, and flowing down a tributary valley and out across the Sherman glacier (Fig. 7-14). This was a veritable avalanche of rock debris, and it now forms a widespread blanket over the glacier some 3 to 10 feet thick on the average.

Remarkable was the manner in which the debris traveled. It moved up and over a 450-foot-high ridge in its path without destroying vegetation on the lee side. Loose snow and ice along the surface of the glacier now rest relatively undisturbed beneath this blanket of rock debris. Such rock avalanches may acquire so much momentum that they can leave the ground at a break in slope and override and trap a cushion of compressed air upon which they can then slide at high speed with little friction. We picture a gigantic "hovercraft" of shattered rock debris spread out like a flying carpet and skimming not along the surface of the ground, but some feet above it—along some of its path then, the debris literally flew through the air.

The other general type of movement in mass-wasting is exemplified by the less familiar *creep* (see below) in which irregular movement occurs throughout a slowly moving mass of regolith—no conspicuous surface of slippage occurs at its base. Gradations occur between these two types of movement.

Rock glaciers form another facet of mass-wasting (Fig. 7–15)—a relatively minor and specialized one. They consist chiefly of rock debris shaped into elongated masses like small glaciers and are located in steep-sided, high-altitude mountain valleys. Rock debris falls and slides down steep valley slopes to accumulate in piles (talus cones, Fig. 7–16) along its course. Such piles may extend and merge on a valley floor and form a tongue-like lobe that creeps gradually down the valley. In some rock glaciers, considerable ice may be present, and this aids the movement. Some rock glaciers may be all that is left of a mountain glacier that was heavily loaded with debris and has melted away and dropped its load, while others may have been produced by large rockslides.

Mudflows form yet another facet of mass-wasting. Favorable factors for their origin are steep slopes, an occasional heavy rain, and little or no vegetation to bind the regolith debris together. Thus mudflows are most abundant in the mountainous parts of semiarid regions where muddy streams tend to form during storms. Such a stream will subsequently be transformed into a mudflow if it picks up enough debris as it flows along. Huge boulders can be carried by such a dense muddy fluid. Beyond the mouth of a canyon on gentler slopes, a mudflow spreads out into a wide thin tongue. With less water and slower movement, a mudflow takes on the characteristics of an earthflow.

Creep

The cloak of loose material above bedrock moves slowly downward along even the gentlest slopes. A covering mat of vegetation decreases the rate but does not halt the movement. Evidence of several kinds indicates that regolith moves imperceptibly but relentlessly downward (Fig. 7–17): roads, tunnels, and railroad tracks may be shoved out of line, and fence posts, monuments, and buildings may be tilted or disrupted by the irregular movements. The surface part of the regolith tends to move downslope more rapidly than its deeper part. Thus a downslope tilt develops in fence posts and power-line poles which have been sunk for some depth into the loose material. The trunks of some trees are distinctly curved (convex downslope) as the result of

7-15. A rock glacier in the Alaska Range. *(Wahrhaftig and Cox,* Bull. Geol. Soc. Am., *April 1959.)*

7-16. Two large talus cones, Park County, Wyoming. Rock fragments tend to fall from a cliff or steep slope and pile up at its base to form a sloping heap of rock debris called a talus. If the debris is funneled through narrow openings, cone-shaped heaps result. *(Courtesy T. A. Jaggar, Jr., U.S. Geological Survey.)*

a compromise between this downward tilting and their tendency to grow vertically.

The causes of creep are numerous, but each is small, and only their combined activity operating for a long time produces noticeable results; e.g., frost heaving tends to push loose pebbles at the surface a short distance outward at right angles to the slope. When the ice beneath a pebble melts, the pebble falls vertically downward, and thus it has moved a tiny distance downslope. Volume increases caused by some chemical weathering may have a similar effect and can occur under different climatic conditions. Weathering produces loose fragments which may fall or roll down a slope. Cavities formed by the dissolving of soluble materials, by the decay of tree roots, or by the burrowing activities of animals are all eventually filled by the downward movement of upslope material. Water adds weight and reduces the cohesion among particles. Constantly repeated, these forces—each of them exerting a tiny effect—combine to move material downslope.

Movement at the rate of 1 foot every 5 to 10 years has been measured and may be representative, but the range is very wide. Creep is certainly humdrum and unspectacular when contrasted with a rock avalanche or a landslide, but creep may be far more important quantitatively—it may cause the

7-17. Creep. The upper part of the regolith moves slowly downslope, but more rapidly than the lower part. Objects which project downward into the regolith for some distance thus tend to be tilted. The photo shows hillside creep in the Yukon Region, Alaska. *(Courtesy W. W. Atwood, U.S. Geological Survey.)*

downslope movement of much more material than all of the slides and falls put together.

Solifluction

Solifluction means soil flowage and refers to soil and regolith which become saturated with water and flow slowly down slopes that may be quite gentle. This form of mass-wasting is most common in the colder, high-latitude regions of the Earth that have permafrost, perhaps about one-fifth of the land area of the Earth. The term *permafrost* refers to the permanently frozen ground underlying the surface in such areas—ice, not water, occurs in pores and cracks in regolith and bedrock, and its thickness exceeds 1000 feet in some sections. Only the upper few feet of the permafrost thaws during the warmer months, and the resulting water cannot penetrate the underlying frozen zone. Thus water, in combination with melted snow and rain water, so saturates the upper few feet that the entire mass flows, but at an imperceptibly slow rate. Movement tends to occur on all slopes in such areas; it is not confined to channels as is a mudflow, and consequently no pronounced topographic shapes result.

The effects of solifluction may be observed in regions which do not have permafrost today, but which presumably did during the colder parts of the Pleistocene. Frost polygons (Fig. 7–7) are features associated with solifluction.

Consider some of the problems encountered by man in permafrost regions: obtaining a supply of water (pipes freeze both above and below the surface), sewage disposal, the effect of a heated building upon the frozen ground beneath it, and the installation of a large pipeline carrying heated oil.

Rounded vs. Angular Topography

Weathering and mass-wasting tend to produce rounded topographic forms in humid areas and angular shapes in arid and semiarid regions. In humid temperate zones a thick protective mat of vegetation covers the regolith and checks gullying. The entire regolith creeps slowly downslope; it fills in irregularities and hides differences between resistant and weak rock formations. Furthermore, chemical weathering so weakens the rocks that steep slopes cannot commonly be supported.

On the other hand, mechanical weathering is relatively more important in drier areas; rockfalls, landslides, and mudflows tend to leave slopes steep. The creeping of the regolith is too slight to mask differences in bedrock resistance by filling in low areas, and resistant rocks stand out as ledges or ridges. Thus the differences in shape are caused by somewhat different processes of weathering and mass-wasting functioning in different climates; the underlying rocks and rock structures may be identical.

The Work of Running Water

Little drops of water,
  Little grains of sand,
Run away together
  And destroy the land.
  —Robert E. Horton

From the atmosphere, water molecules fall upon the Earth's surface as rain, snow, hail, and sleet. Some water evaporates or is taken up by plants, some runs off immediately into streams (flow is concentrated within channels), and the remainder sinks into the ground (see water cycle, p. 213). Much ground water (Chapter 8) later emerges at the surface at a lower altitude and becomes runoff. Streams carry excess water from the land to the sea. In doing so, they erode valleys and help shape the Earth's surface. They transport rock debris and dissolved materials, and eventually they deposit most of their sediment in the oceans. Stream activity, in combination with weathering and mass-wasting predominates by far over other types of erosion such as wind, ice, or marine.

Streams are important to man whether he uses them as sources of drinking water, irrigation, or electric power, as scenic inspiration, or as places in which to fish, swim, or dump sewage. Valleys furnish the most convenient courses for many roads and railroads. The location of a number of important cities depended upon the navigability of large rivers. Civilization flourished first on fertile floodplains. Bridges, dams, and reservoirs have to be built. Frequent floods that cause loss of life and destruction of property emphasize the importance of streams to man.

Valleys Are Eroded by Streams

With few exceptions, streams have excavated the valleys in which they now flow (i.e., valleys are not great gashes in the Earth's surface that formed by faulting or other crustal movements). Different kinds of evidence indicate that this is so, and among these are the distinctive patterns generally formed by systems of streams. Little streams flow into big streams, which in turn are tributary to still bigger streams, and these different segments tend to be joined together in regular patterns. In regions of flat-lying rocks, or of uniform massive rocks, a *dendritic* drainage pattern commonly forms—a branching, treelike arrangement in which each tributary intersects its master stream at an angle that is acute (less than 90 degrees with the upstream portion, Figs. 7–18 and 7–21).

Under different conditions, a rectangular stream pattern develops, and still other orderly patterns are known. Each pattern is determined by the rocks and rock structures underlying an area, by its climate, and by other factors which are all so common that relatively few basic drainage patterns are observed. Aerial views or maps show these to best advantage.

In general, the size of a valley is proportional to the size of the stream flowing in it. Each stream tends to flow on a steeper slope near its head (upstream where it begins) and on a gentler slope near its mouth (downstream where it ends); thus it forms a profile which is concave upward. Commonly, tributaries meet larger streams at just the proper level (i.e., without a lake or waterfall at the intersection), and this accordance at intersections is maintained throughout the system. We readily observe that sediments are transported by rivers, which indicates that excavation has gone on somewhere upstream, and we can see the evidence of stream erosion in their channels when the water is low. Moreover, extensive gully systems have been seen to develop in some farming regions within just a few generations. Taken together, these observations indicate that valleys slowly change their shapes with time (Fig. 7–19).

Stream Erosion and Transportation

Sediments carried by a stream make up its *load*: some is in solution, some is in suspension (the finer particles), and some (its bed

7-18. Dendritic drainage pattern and stream capture on nearly flat-lying strata of the Arabian Peninsula's Hadramawt Plateau. The light-colored areas at the top are sand dunes and the dark-colored areas near the Gulf of Aden are igneous and metamorphic rocks. *(Gemini VII, 13 Dec. 1965, S65-64010, NASA.)*

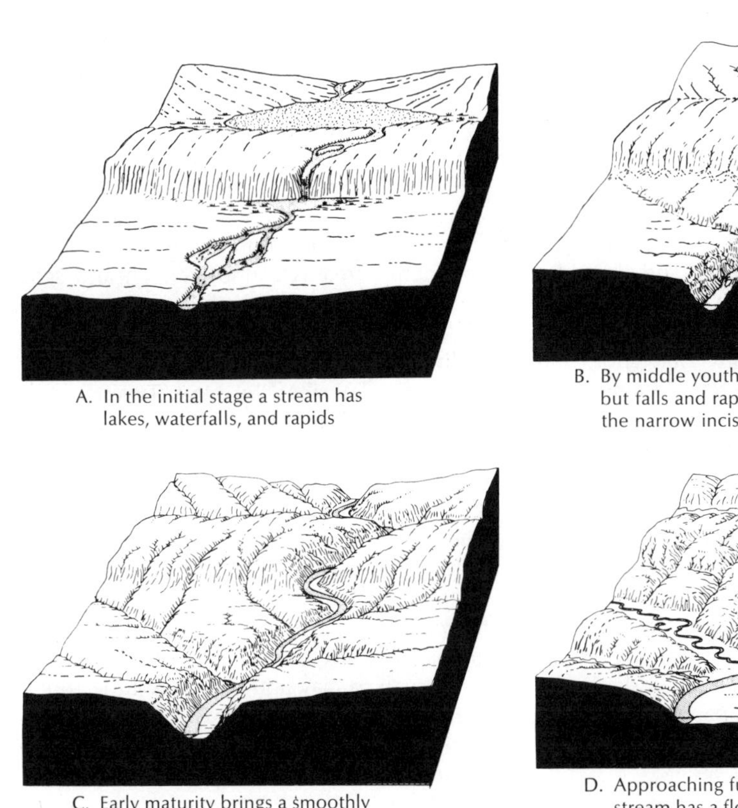

A. In the initial stage a stream has lakes, waterfalls, and rapids

B. By middle youth the lakes are gone, but falls and rapids persist along the narrow incised gorge

C. Early maturity brings a smoothly graded profile without rapids or falls, but with the beginnings of a floodplain

D. Approaching full maturity, the stream has a floodplain almost wide enough to accommodate its meanders

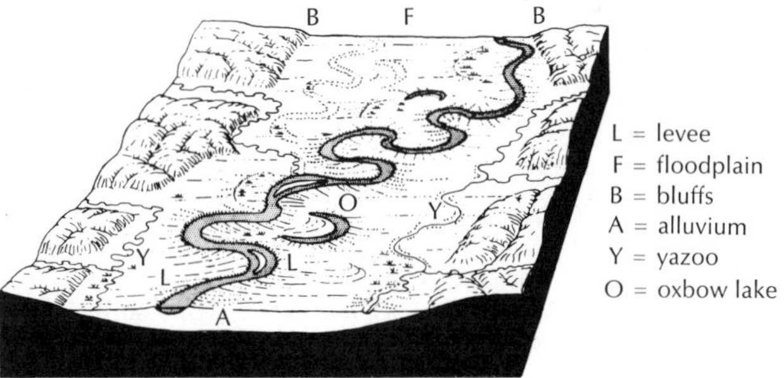

E. Full maturity is marked by a broad floodplain and freely developed meanders.

7-19. Stages in the sequential development of a stream valley. Fig. 2-25 shows the sequential development of an area. *(After E. Raisz, Courtesy A. N. Strahler,* Physical Geography, *John Wiley & Sons, Inc., 1951.)*

load) moves intermittently along the bottom by rolling, sliding, and occasional jumps. Particles that may be too small to be suspended at low-water stages may later be carried in suspension as velocities increase during high-water stages. A study of about 70 rivers in the United States showed that approximately 20% of the total measured load was carried in solution (range 1 to 64%).

Only a very small fraction of a stream's potential energy (determined by volume of water and altitude above sea level) is available for the work of erosion and transportation. Perhaps 95% and more of this energy is wasted by friction within the flowing water and between the water and the sides and floor of a channel. In turbulent flow, the speed and direction of water movement change continuously at any one spot, and eddies are common; such turbulence increases with velocity, channel roughness, and other factors. At slow speeds through smooth channels, water may flow along nearly straight paths that parallel the channel boundaries (called laminar flow). A stream's velocity depends upon volume of water, shape and roughness of channel, load, and gradient—i.e., the slope down which the stream flows, commonly measured in feet per mile. A stream's discharge is the quantity of water passing a certain point in a unit of time (commonly measured in cubic feet per second).

As a stream's velocity increases, proportionally much greater increases occur in the size of the largest particle that it can move and in the total load of sediments of all sizes that it can carry. Thus huge boulders can be moved during floods when velocities may exceed those of low-water stages by 10 or 20 times. Such boulders are immobile during low-water stages. Accordingly, streams appear to do the bulk of their erosion and transportation during the few days of flood conditions which occur occasionally. A distinction should also be made between the exceptional floods that may occur decades apart and the more normal floods that tend to occur each year.

A relatively few large rivers on any one continent account for most of the total discharge of river water into the oceans; e.g., an estimated 133 cubic miles of water flows into the ocean each year from the Mississippi, which is about one-third of the total discharge of all rivers in the 48 conterminous United States. However, the Amazon, the Earth's largest river, probably discharges nearly one-fifth as much as the combined total of all the rivers on Earth.

Stream erosion is achieved by abrasive impacts of transported fragments on the beds and sides of channels, by the solvent action of water (relatively small), and by the lifting effect of running water. Without sediment, streams cannot scratch and scour their channels, but a turbulent river readily picks up the smaller sediment sizes. A sediment-laden river, the muddy Missouri, was aptly described by Mark Twain as "too thick to navigate, but too thin to cultivate."

With increased length of transportation downstream, particles generally become smaller, probably owing in the main to selective transportation—the smallest particles are picked up most readily and carried the longest distances. Gravel shows progressive rounding downstream, but sand grains may not do so, and very small particles such as clay and silt show little rounding even after being carried many miles. Their small size and the water around them prevent effective impacts.

Potholes (Fig. 7–20) may form in the channel of a stream where eddies whirl gravel and sand around in a small circular area and gradually bore a cylindrical hole in the bedrock. New supplies of gravel and sand replace older pieces as they are worn out. Potholes may form rapidly or slowly: one 10 feet deep was observed to form in limestone in 18 months; one 5 feet deep formed in 75 years in granite, but rock resistance is only one of the factors involved.

Apparently most sand grains do not originate by the abrasion of pebbles into sand-sized particles; there are not enough pebbles for this, and most sand grains are too angular. Rather, sand apparently originates in two ways: by the fragmentation of larger pieces during transportation by streams and during the weathering of rocks such as granite and gneiss when their constituent grains of feldspar, quartz, and other minerals are released.

To determine the effect of abrasion on sand-sized particles, cube-shaped pieces of quartz and feldspar have been rolled around a circular moat of concrete in a current propelled by a sort of churn. Relatively little abrasion occurred, even on the less resistant feldspar—not enough to account for the rounding observed on sand grains. However, considerable abrasion did occur when similar grains were placed in a wind tunnel. Evidently wind causes quartz grains to lose about 100 to 1000 times more mass than does water transport over the same distance, which suggests that wind may be a more effective agent in the rounding of sand grains than had previously been considered. However, sand grains are probably transported much longer distances by running water than by wind even in arid lands.

Sand grains may participate in more than one phase of a rock cycle; e.g., sand grains may be compacted and cemented into a sandstone, which is later weathered and eroded. Sand grains thus released commonly retain the rounding that formed during the previous cycle, and the degree of rounding tends to increase with each cycle. During these cycles, the proportion of quartz among the sand grains increases and at present it may be about double that in the source rocks.

A stream normally originates as a tiny gully in a depression at the Earth's surface where runoff is concentrated. With continued flow, this miniature valley then grows deeper and wider; it also becomes longer as erosion slowly extends its upper end farther and farther upslope *(headward erosion,* Fig. 7–21). More water generally enters at the head of a valley than at any one place along its sides to cause this. However, large rivers such as the Mississippi were not formed by headward erosion. Here small segments formed at different times and places, grew larger and longer, and later joined together.

At first, water flows down a valley only after a rain, but eventually it erodes downward far enough to reach the zone in which all open spaces are filled by ground water (p. 216). The stream then becomes permanent, shrinking in size during dry spells and enlarging greatly after heavy rains.

7-20. Potholes in the granite bed of the James River, Virginia. *(Courtesy U.S. Geological Survey.)*

Steep slopes, high velocities, great volumes of water, and weak bedrock all increase the erosive capacity of a stream.

The *long profile* of a stream (the slope of its surface from head to mouth) is commonly concave upward (Fig. 7–22). In a stream's mountainous headward portion, coarser sediments must commonly be transported, and a steep gradient makes the shallow water sufficiently turbulent to move them. However, sediment sizes are smaller near the mouth of a major stream, and its volume has increased (from its many tributaries—generally the volume of water increases more rapidly than the load). Friction tends to be reduced because channels tend to be deeper and smoother. Thus greater volume, reduced friction, and finer grained load commonly combine to cause a major stream (river) to flow more rapidly on a gentler gradient near its mouth than it does on a steeper gradient near its head.

Base Level

A stream cannot cut downward indefinitely. If a river enters the sea, it may scour its channel a little below sea level in the vicinity of the coast, but no deeper. Upstream, the channel rises above sea level to furnish a slope down which the water can flow. The lowest level to which a land surface

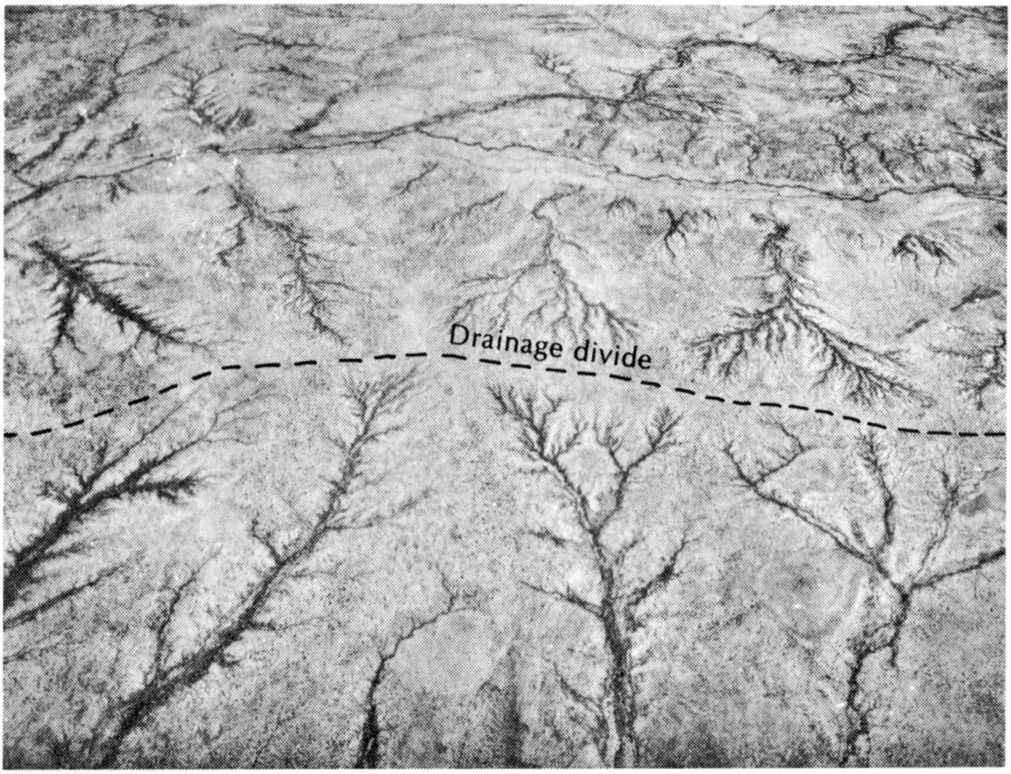

7-21. Dendritic drainage and headward erosion in Africa. A divide separates two drainage systems; we may think of it as a line that follows the crest of the higher land between any two neighboring valleys. [*Courtesy U.S. Air Force: N 14° 10′, W 23° 30′.*]

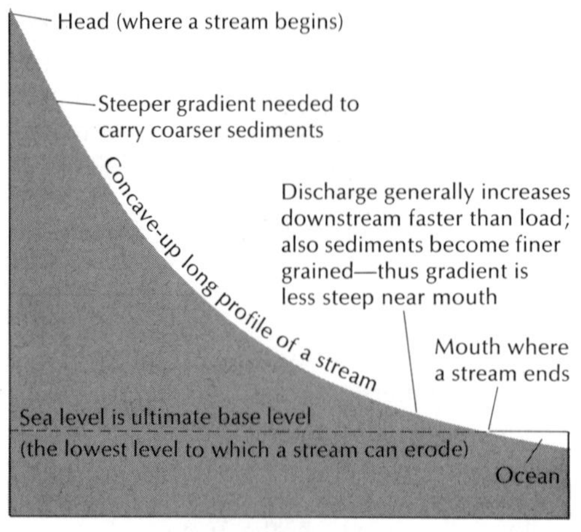

7-22. Idealized long profile of a graded stream.

can be reduced by running water is called *base level.* Sea level projected inland as an imaginary surface generally provides a lower limit for stream erosion (ultimate base level), but streams located far from the sea in the interior of a continent probably do not lower their drainage basins very close to sea level. Lakes, dams, and resistant rock formations may form local, temporary base levels. In a region of interior drainage where extensive faulting has lowered part of the surface, base level may locally extend below sea level (e.g., in Death Valley, California).

Graded Streams

At any one time and place, a stream has a certain quantity of energy available for erosion and transportation. As a stream cuts downward (especially near its head), its gradient decreases while its load tends to increase. Thus the rate of downcutting eventually diminishes because more of the stream's energy is absorbed in the transportation of its bedload. Eventually an approximate balance is reached between the load a stream carries at any one point and the volume, channel shape, and gradient at that point. Such a stream is said to be *graded*. Put in another way, the same amount of water and load enters a graded stream as leaves it.

To reach this graded condition, a stream has cut downward in many sections along its course and has deposited sediments to build up its channel floor in other sections. If the state of balance is disturbed by a change in any one of the many factors involved, the other factors change in an offsetting combination that tends to restore the balance. Thus a graded stream's slope is delicately adjusted to provide just the necessary velocity to move its load. The graded condition may be more closely related to so-called normal flood conditions than to those of low-water stages. When a stream has attained grade, minor changes in the controlling factors continue to occur, and the balance may be upset temporarily, first in one direction and then in another. It has been said that in speaking of a graded stream, one refers to its "climate" and not to its "weather." A stream may be graded even though its long profile is irregular.

Downcutting continues after a graded condition has been attained, but it then becomes exceedingly slow. The graded condition may be reached fairly early in the development of a stream while coarse material is still being brought to it by weathering and mass-wasting along the valley sides. However, as the valley walls retreat from the channel and become less steep, the size and amount of the debris that gravitates toward the stream decrease. Because this finer debris can be carried by a stream on a gentler gradient, a graded stream can continue to cut downward.

A graded condition may be upset by major changes that exceed those occurring during the daily, monthly, and yearly fluctuations; e.g., a stream's load may be increased greatly, or its volume may be decreased. In each case the stream would deposit more sediment all along the floor of its channel, thereby steepening its gradient (we assume the altitude of its mouth remains unchanged). At the greater velocity thus produced, the stream would once again carry a full load and attain a graded condition.

On the other hand, if the load of a stream is decreased substantially, the stream acquires excess energy and erodes its channel. By decreasing the gradient and increasing the load, a graded condition is again attained.

Valleys with Wide Flat Floors

Wide, flat-floored valleys are common in some areas at low altitudes and contrast sharply in shape with narrow-bottomed, steep-sided, high-altitude mountain valleys, although all gradations between these two types are observed on the lands. In fact, the gradations led to the idea of a genetic connection between the two, i.e., the wide valleys with flat floors were probably once V-shaped in cross section, with steeper gradients, and with falls, rapids, and lakes occurring along their paths. The term V-shaped actually includes a variety of cross sectional shapes, but steep valley sides and narrow bottoms are characteristic features.

To indicate how a V-shaped valley may form and be slowly transformed into a wide flat-floored valley, let us begin with a stream that is cutting downward (other valleys under other conditions may follow a different sequence). Downcutting by itself would leave vertical banks, but these are attacked as they form by weathering, mass-wasting, and sheetwash, which push, drop, and carry disintegrated and decayed rock materials downward to the stream channel (Fig. 7–23). The upper part of a valley, where

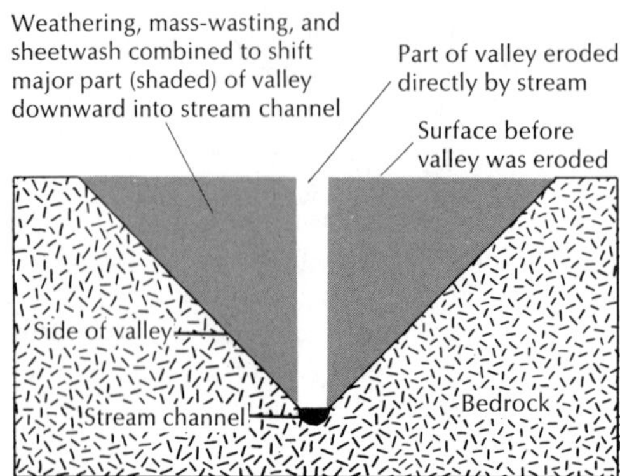

7-23. Schematic illustration of roles played by weathering, mass-wasting, and a stream in the formation of a valley. Among other factors, the shape of such a valley depends upon: rate of downcutting, rock resistance, and climate.

the widening process has occurred longest, is thus broader than the lower part. In unconsolidated or less resistant materials, the process operates relatively rapidly; in resistant rocks, the process is slow, and the valley walls remain steep. Continued deepening by the stream is thus a controlling factor because it functions as a conveyor belt in removing the sediments brought down to it. The V-shaped valley has steep sides, and the channel occupies most of its floor.

The long profile of the stream at this time is probably irregular and should remain so for some time. The irregularities are produced by initial unevenness in the land being eroded, by lakes and waterfalls that develop along its course, and by rocks of differing resistances to erosion that are encountered (or uncovered in a particular place as the stream cuts deeper).

As erosion proceeds in an area, streams might develop as follows. Numerous small streams form first on slopes and flow downward into depressions, creating lakes. The lakes subsequently rise high enough to spill over at some places along their shores, and water then flows downward along the lowest ground into valleys that have formed at lower altitudes. Initially a number of small gullies may develop along any one slope, but certain channels are favored and grow larger by capturing (Fig. 7–18) the drainage of neighboring gullies. These grow longer by headward erosion, and the capture process may be repeated again and again on a larger and larger scale. This causes additional irregularities in the long profiles of the major rivers which form eventually by union of numerous segments that developed locally; such major rivers tend to flow from continental interiors to the sea.

Where streams flow from more resistant to less resistant rocks, rapids and waterfalls develop (e.g., the Fall Line near the western margin of the Atlantic coastal plain). As streams cut downward toward base level, lakes are drained, and rapids and waterfalls slowly disappear. By erosion here and deposition there, master streams eventually become graded. Commonly a graded profile is attained first near the mouth and is slowly extended upstream. A fairly smooth, concave-up long profile is thus produced in a valley that may still be relatively narrow at the bottom.

Wide flat valley bottoms, along which streams flow in large sweeping curves, may develop in quite different ways. One method seems to involve lateral erosion by a stream that is downcutting very slowly, or perhaps not at all. Irregularities are common along the path of a stream and produce bends here and there. At any one bend, the deepest, swiftest, and most turbulent part of the channel tends to occur along the outer margin, whereas slack, less turbulent water occurs along the inside (maximum turbulence occurs a short distance downstream from the center of a bend, not at its exact center). The decrease in velocity may cause deposition along the inside of a bend, which tends to reduce the cross-sectional area of the channel. This in turn may cause the stream to increase its flow and undercut the outer and downstream side of a bend, where its velocity is greatest. Flat crescent-shaped areas may thus develop on the inside of each bend, and sediments are deposited upon these during floods. They coalesce eventually to form a continuous flat floor, because each bend migrates slowly downstream as well as laterally. This wide, flat, laterally planed valley floor has a relatively thin covering of river-deposited sediment whose upper surface is called a *floodplain*. Streams tend to flow along floodplains in winding, sinuous curves that are called *meanders* (Fig. 7–24).

The sediment blanketing the beveled bedrock floor of such a valley has accumulated in two different ways: deposition occurs at the inside of each bend, and finer sediments are spread over most of the floodplain during overbank floods. As meanders have migrated both laterally and downstream, older deposits have been reworked. The channel has thus shifted so widely that at some time or other it has been present on all portions of the floodplain. The location of a river channel on a floodplain is an ever-changing one.

However, for a beveled bedrock surface to form along the bottom of a valley by lateral planation, the crust in the area must remain stable for a very long time, so long in fact that this seems to have occurred infrequently. Rather, the wide flat floors of most of the major valleys existing today have been formed primarily by aggradation rather than by lateral planation. (Fig. 7–25—aggradation refers to the building up of a surface by the deposition of sediments along it). Sediments have been deposited in huge quantities to fill the entire bottom portions of such valleys, and Pleistocene glaciers may have contributed much of the sediment. Furthermore, sea level rose a few hundred feet as glacier ice melted following the end of the fourth main glacial age. This would raise base level, which would reduce gradients and cause deposition. The term *alluvium* is used for such sediments that have been deposited relatively recently by

7-24. A meandering river and its floodplain. The meander of Crooked Creek, California, in the left foreground shows an undercut outside bank and a gently sloping inside bank. *(Courtesy U.S. Geological Survey.)*

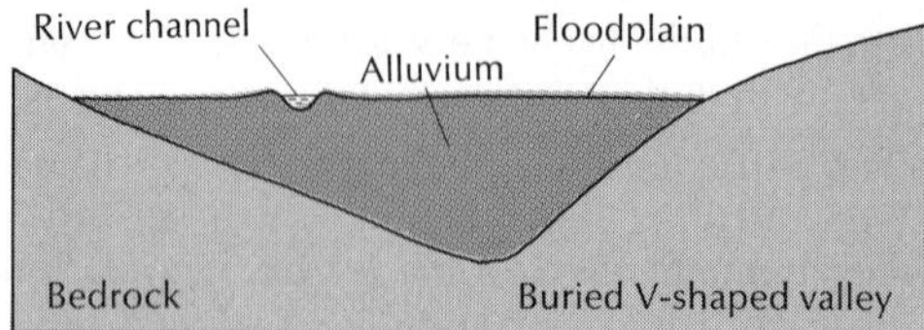

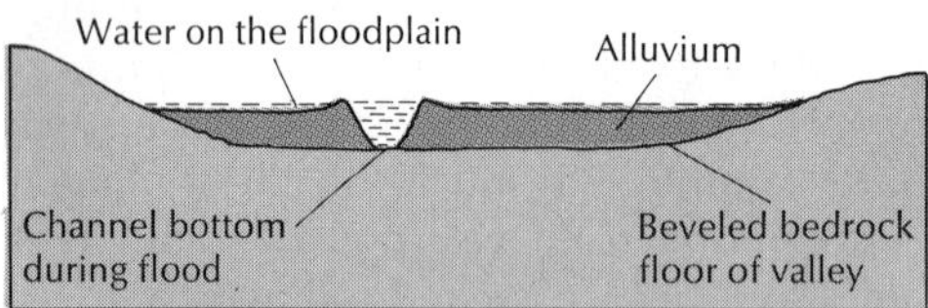

7-25. Floodplains of two types. In *B* the buried, beveled bedrock floor of the valley has been formed by side-cutting (lateral erosion) by the river. During a flood (shown in *B*), a river may deepen its channel and erode its bedrock floor. During normal-water stages, the floor of the channel is formed by alluvium.

streams, excluding those that form in lakes and seas.

Meanders are not accidental features of a stream, but much has yet to be learned about their origin. Experiments with large-scale models (Fig. 7–26) show that meanders tend to form in streams that flow on gentle gradients across fine-grained sediments that are easily eroded and transported. Bank caving seems to produce the irregularities in the channel which initiate the formation of meanders; i.e., a pile of caved-in sediments along one side of a channel deflects the current against the other bank, which is then undercut. The size of a meander is proportional to the discharge; large streams have large meanders; small streams have only small meanders. However, some streams are "born meandering" and develop meanders early in their history while they are still downcutting.

Natural Levees

When a river overflows its banks during a flood, its speed is checked abruptly beyond the margins of its channel where the water becomes shallow. This sudden decrease in velocity causes deposition of sediments in the shape of a very thin wedge that tapers away from the channel. Thicker accumulations of coarser materials are deposited at once along the banks of the channel, and thinner amounts of finer sediment are spread over the rest of the floodplain. Numerous repetitions of this process may produce low ridges that slope very gently away from a channel along each side. Such ridges are called *natural levees* (Fig. 7–19). Commonly they rise only a few feet above a floodplain, yet during extensive floods they may form long, low islands which parallel the channel and constitute the only dry land along the valley floor. Natural levees can build up only during floods high enough to flow over them.

Oxbow Lakes

As meanders migrate down a valley and shift laterally, the neck portion of a particular meander may narrow and eventually be cut through. Perhaps this occurs because the downvalley migration of the leading side of a meander is slowed by more resistant material in its path, and its upstream side then gradually catches up and impinges upon it. Eventually the neck is cut through, and the river temporarily straightens and shortens its course. However, new loops soon form to replace those that have been abandoned. Sediment may subsequently be deposited at the entrances to an abandoned channel loop, and a crescent-shaped *oxbow lake* may result (without the lake, it is called an *oxbow).*

Another kind of cutoff may also produce an oxbow or oxbow lake. This involves the sediments that are deposited within a channel as crescent-shaped sandbars on the insides of bends (Fig. 7–27). These constitute the main bulk of the sediments laid down on erosional-type floodplains. At a particular bend, they form a series of low concentric ridges with intervening troughs because the channel shifted its position time after time. During a flood, a river may cut across the inside of a meander by following a depression (chute) located between two channel-deposited ridges. This has a steeper gradient than the old channel because of its shorter distance across the bend. As a result, the older channel may be abandoned.

Meander shifting is sometimes inconvenient for man. A bridge in Oklahoma across a meander was left high and dry when the meander shifted, and the river flowed around the bridge, not under it. A town was once separated from its airfield by a meander shift, and another town discovered that its sewage was being dumped into a dry channel after a section of a meander was abandoned.

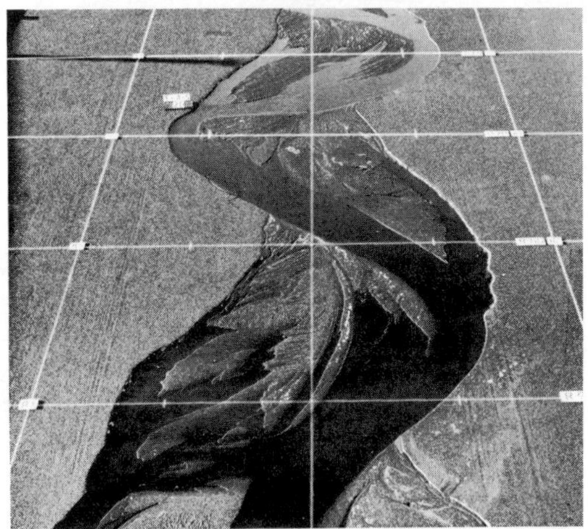

Braided Streams

If a stream flows in a wide shallow channel and has a heavy bed load, a braided condition may develop (Fig. 7–28). Deposition here and there subdivides the channel into a complex, haphazard, interlaced network of channels separated by bars. Conditions which cause braiding occur in glaciated areas where much sediment is added by the melting ice, in semiarid areas where evaporation and seepage reduce the size of a stream, on floodplains or the surfaces of deltas where the banks consist of very weak, easily caved sediment, and during periods of sufficiently reduced flow along other streams.

Alluvial Fans

An *alluvial fan* is a type of landform that is built chiefly of stream-deposited sediments and that tends to be located at the junction of a mountainous area with a low-

7-26. Meanders produced by large-scale models. Mississippi River sands and similar artificial materials were used in tests. A fixed condition of slope and discharge was maintained during this sequence, but material was added during the test. *(Courtesy U.S. Army.)*

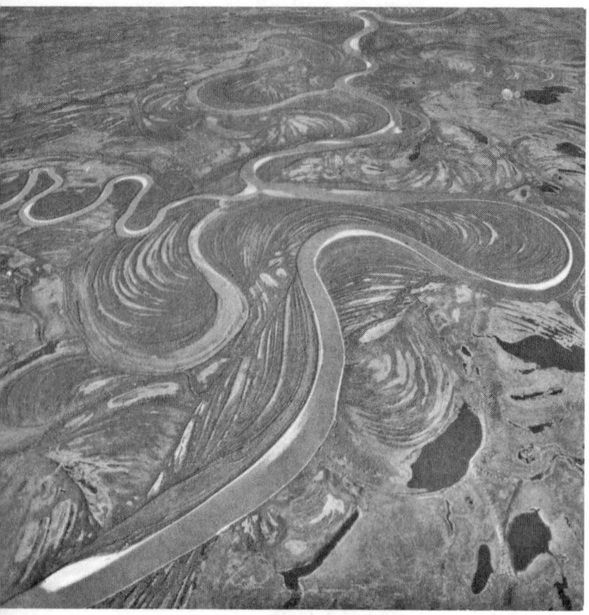

7-27. Erosion and deposition on a wide flat-floored valley in Alaska. Note the meander scars, oxbows, oxbow lakes, and sediments (white areas) deposited along channels, particularly along the inside banks of meanders. *(Courtesy U.S. Air Force: N 65° 55′, W 156° 3′0.)*

land (Fig. 7–29). Mudflows and other types of mass-wasting also contribute some sediment. An abrupt loss in transporting power occurs at the junction, apparently because the channel changes shape suddenly—the confined channel within a narrow-bottomed, steep-sided, mountain valley gives way to a wide, shallow channel on the adjacent, gently sloping lowland or plain. Also, a loss in volume will occur if infiltration takes place. Previously, deposition had been attributed to an abrupt change in a stream's gradient at the base of a mountain, but measurements have shown this not to be the case—at least generally.

The abrupt decrease in transporting power causes a stream to drop most of its load of sediments near the base of a mountain, and when enough sediment has piled up at one location, the stream shifts laterally to lower ground in a process that is repeated many times. Eventually a fan-shaped mass of debris is built up, with the apex of the fan located near the mouth of the valley. Such fans may be small or large, and they are more common in drier regions than in

7-28. Braided river channel (indicates relatively heavy load and fluctuating discharge) and fill terraces on the Rakaia River, South Island, New Zealand. Alluvial fans occur along the base of the Mt. Hutt Range in the distance. *(Courtesy V. C. Browne.)*

humid regions. Because deposition is rapid, sorting and stratification are generally not well developed. Coarser sediments such as gravel tend to predominate, especially in the upstream portion, and a braided pattern is common on the surface of a fan. A number of fans along the base of a mountain may coalesce to form a continuous alluvial slope (Fig. 7–30). If deposition is particularly abrupt, the mass of sediments becomes cone-shaped rather than fan-shaped.

Deltas

A *delta* is a deposit of sediments that forms near the junction of some rivers with a standing body of water, such as a lake or the ocean, in a manner analogous to the formation of an alluvial fan on land (Fig. 7–31). The velocity of a river decreases at the junction, though less abruptly than occurs in the formation of a fan, and sorted, stratified sediments accumulate up to and even slightly above the water level. The edge of the land thus encroaches on the water body, and large deltas such as that of the Mississippi have extended their coastal areas for hundreds of square miles into the sea. Favorable factors for the origin of a delta are abundant sediment, absence of powerful waves or shore currents, and a stable body of water. In addition, dissolved salts in sea water cause coagulation of the very fine sediments and this produces particles large enough to be deposited.

Small deltas may exhibit a characteristic type of stratification (Fig. 7–32) not present in many large deltas built into the ocean. Thicker layers of coarser-grained sediment *(foreset beds)* pile up on the sloping bottom close to shore, whereas finer sediment is deposited in thinner layers farther out *(bottomset beds)*. The bottomset beds are actually continuations of the foreset beds. Gradu-

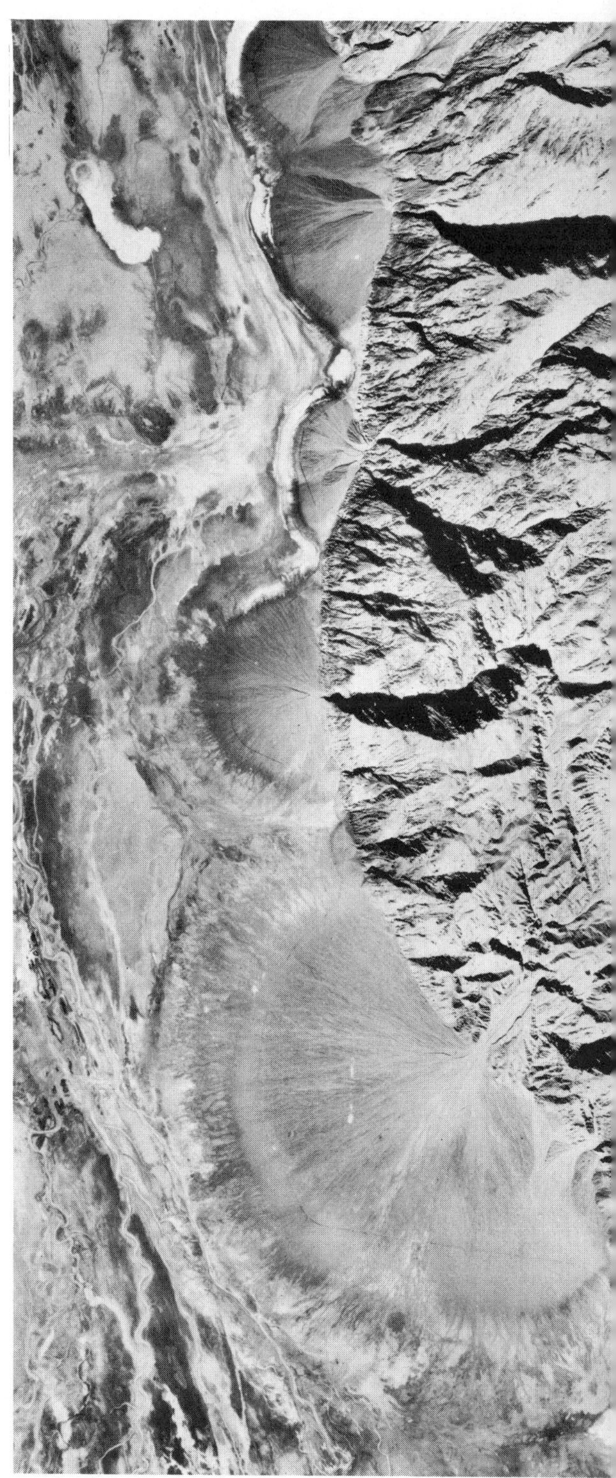

7-29. Alluvial fans are scattered along the edge of a steep, gullied fault scarp in the Death Valley-Black Mountain area, California. *(Courtesy U.S. Geological Survey.)*

ally the coarser, sloping foreset beds extend out on top of the more nearly horizontal, thinner bottomset beds. As a delta becomes larger and thicker, the stream shifts and erodes back and forth across the tops of the foreset beds; this is analogous to the shifting stream of the alluvial fan. Sediments are deposited horizontally above the eroded tops of the inclined foreset beds and make up the *topset beds* of the delta. These topset, foreset, and bottomset beds are types of cross-bedding (p. 86).

Stream Terraces

Nearly level benches or *terraces* (Fig. 7–33) may be observed along the margins of some wide-floored valleys; these may be at the same altitude on each side (paired terraces) or at different altitudes (nonpaired terraces). Each terrace is a remnant of a former valley floor, i.e., part of a former floodplain that once extended all the way across the valley, but above the level of the present floodplain. Such terraces may be produced in different ways.

Paired terraces may result from an uplift that rejuvenated a stream and enabled it to cut downward and laterally. Thus its former floodplain is eroded away, except for remnants here and there along the margins of the valley. After uplift ceases, a younger and narrower floodplain can develop at a lower altitude. This process may be repeated to form a number of paired terraces. Paired terraces can also be formed by any change that causes a relative increase in the transporting ability of a stream—e.g., by a sharp reduction in load or by an in-

7-30. A steep fault scarp trends north-northeast in Antofagasta Province, Chile. Granite occurs west of the scarp, whereas sedimentary rocks and volcanics occur to the east. A stream flows south-southwest within a narrow fault valley along the scarp and detours around two alluvial fans. Elsewhere a narrow, steeply sloping bahada occurs along the scarp. Streams are also building fans into the fault valley from the east (right), *(Courtesy U.S. Geological Survey, 14 April 1955.)*

7-31. Vegetation darkens the surface of the Nile Delta, about 500,000 square miles in area, as viewed in its entirety from the Gemini IV spacecraft. Adjoining light-colored areas are deserts. *(Courtesy NASA.)*

crease in discharge produced by a change in climate.

Nonpaired terraces may be formed in similar ways, but the slow changes are more nearly continuous. Therefore, a stream cuts downward during the time required to shift laterally from one side of a valley to the other. Floodplain remnants are thus at different levels on opposite sides, and the highest terrace is the oldest one. Terraces may form in association with both types of floodplains (aggradation and lateral erosion).

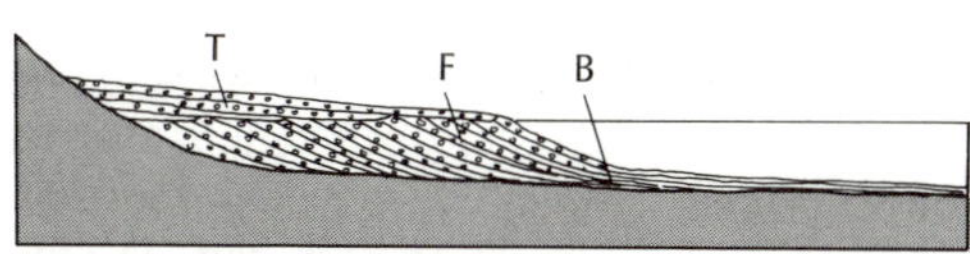

7-32. Idealized structure section through a small delta. Top-set beds (T), fore-set beds (F), and bottom-set beds (B) are shown. *(Modified after G. K. Gilbert.)*

Adjustment of Streams to Rocks and Rock Structures

As streams develop, grow larger, obtain more tributaries, and join with other streams, their channels tend to shift. The streams tend to flow on weak rocks and to cross belts of resistant rocks at their narrowest or weakest places; e.g., a so-called *subsequent stream*

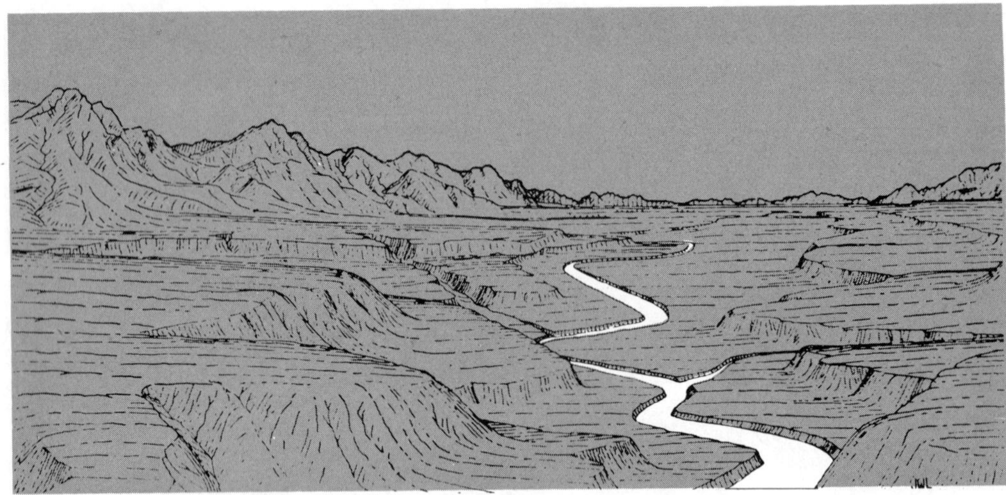

7-33. Stream terraces.

develops by relatively rapid erosion along belts of weak rocks or structures, and it becomes longer by headward erosion. In contrast, a neighboring stream flowing on resistant rocks develops more slowly and some of its drainage may be *captured* by the headward erosion of the subsequent stream.

If streams are not adjusted to the rocks beneath them, the discordance may result from a relatively recent event involving such processes as superposition, stream capture, and glaciation. The *superposition* of drainage may be the most widespread cause of discordance. It results wherever streams have developed channels on younger rocks (perhaps flat-lying) that rest unconformably upon older rocks that have different structures and resistances to erosion. The initial location of the stream channel in the overlying rocks may have little or no relationship to the underlying structures and resistances. Thus as downcutting continues, the stream penetrates into the older rocks and may flow indiscriminately across folds and along belts of resistant rocks.

If a river is superposed across a ridge of resistant rock, it eventually erodes a wide valley upstream and downstream from the ridge, but the valley through the resistant rocks in the ridge remains narrow and steep sided, it is a *water gap* (Fig. 6–5). If at some later date the river ceases to flow through this gap—perhaps it is captured by a subsequent stream eroding headward parallel to the ridge—it then is called a *wind gap*. After prolonged erosion has occurred, the areas underlain by weaker rocks on each side of such a ridge may have been worn down considerably, whereas the ridge top itself may have been lowered but little. Thus the ridge will stand higher above its surroundings and its wind gaps will appear as notches in the crest of the ridge.

To prove that superposition has occurred, it is necessary to locate remnants of the flat-lying rocks that once buried the structures now exposed. Major rivers such as the Delaware, Susquehanna, and Potomac flow across the Appalachian Mountains and across the underlying folds and faults; according to one explanation, they are superposed streams (Figs. 7–34 and 2–24). However, if this is the case, erosion has removed all traces of the coastal plain sediments that once buried the structures. Superposition apparently has been widespread in the Rocky Mountain region where a number of rivers flow across mountain ranges in deep narrow canyons.

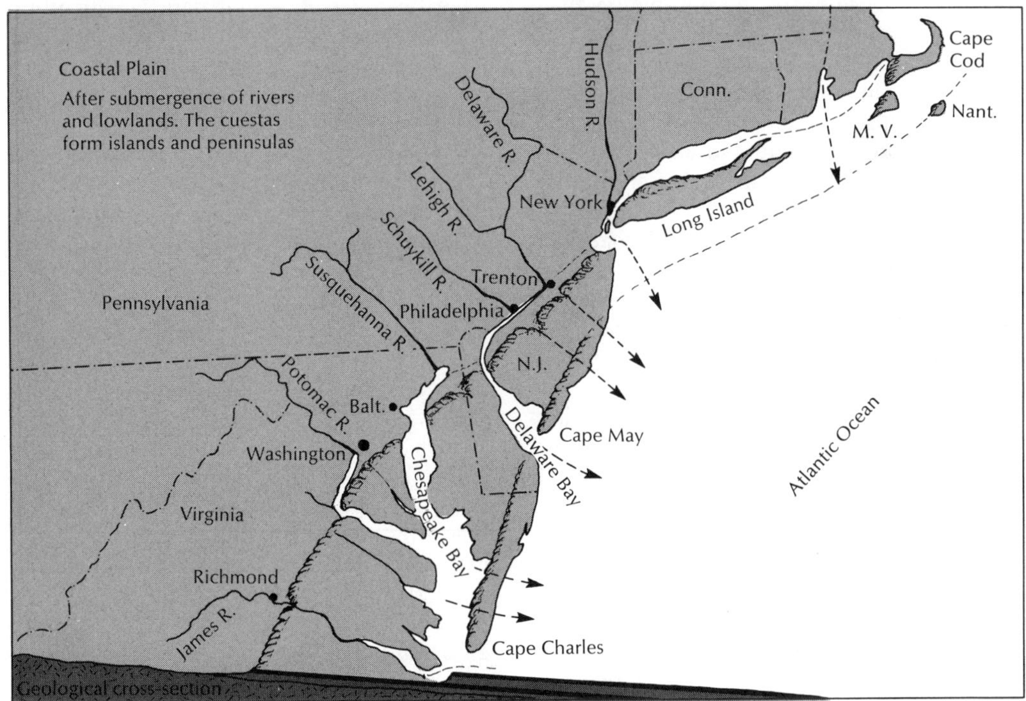

7-34. Atlantic Coastal Plain structure and drainage. At one time, rivers such as the Potomac and Delaware apparently flowed along subparallel courses toward the ocean without the right-angle bends they have today. The simplified geologic structure section at the bottom of the diagram shows sedimentary strata that slant gently downward toward the ocean. Some strata are more resistant to erosion than others. With the passage of time, northeast-southwest lowlands were produced along belts of weaker rocks. Subsequent streams grew larger and longer (by headward erosion) within these belts. Eventually they intersected and captured some of the upper portions of the major rivers. This accounts for the 90-degree bends in many of the rivers. Two main lines of cuestas have formed along the upturned edges of two resistant formations. The outer cuesta forms Cape May, Cape Charles, and Nantucket. The inner cuesta forms Long Island, Martha's Vineyard, and part of Cape Cod. (A. K. Lobeck, *Things Maps Don't Tell Us,* The Macmillan Company, 1956.)

Sequential Sculpture of Valleys and Regions

If erosion is not interrupted by some geological change such as an uplift, single valleys and whole regions pass slowly through a series of orderly, predictable, continuous changes that have been subdivided arbitrarily into three stages: *youth, maturity,* and *old age*. These stages are not marked by absolute numbers of years, because the time needed to reach a given point in a sequence in a particular area depends upon the kinds and structures of the rocks in this area, their altitudes above sea level, the distance from the sea, the quantity and distribution of rain, and many other factors.

The three stages are determined by the amount of erosion that a valley or region has already experienced relative to the amount of erosion that remains to reduce the valley or region close to its base level. This is called the *erosion cycle,* a term which has been sharply criticized. Sequential development tends not to occur in regular cycles, and only infrequently has the crust remained stable long enough for the old-age stage to develop in a region. Thus incomplete "cycles" are more common than completed ones, and the effects of two or more cycles are commonly superposed.

Diversity in land forms from region to region is caused by differences in rocks and rock structures, the rate and amount of uplift, the type of climate, time, and other factors. Therefore, a sequential erosional development cannot be described that will apply under all conditions and to all regions. However, an orderly sequential evolution of landscapes does occur, and thus the concept in generalized form remains valid and useful.

We shall now sketch with broad sweeping strokes some of the changes which occur as weathering, mass-wasting, sheetwash, and streams sculpture a single major valley and an entire region. However, first we must assume a set of geological conditions under which these processes will function: a large area is underlain by nearly horizontal marine sedimentary rocks; the climate is humid; the surface is newly exposed to erosion; uplift of this former sea floor is relatively rapid and extensive, but crustal stability then prevails for the time necessary to complete a cycle. If different conditions are assumed, then different landscapes will result; these would be variations in the erosion cycle.

Erosion Cycle in a Valley

Until a stream has attained grade, its valley remains *youthful*. Characteristic features are a narrow-bottomed, steep-sided cross profile (commonly V-shaped), little or no floodplain, few if any meanders, prominent downcutting, lack of an integrated drainage system, and an irregular long profile interrupted here and there by lakes, rapids, and waterfalls (Fig. 7–19).

A *mature* stream meanders on a floodplain on which natural levees and oxbow lakes may be present, and the floodplain is wide enough that the meanders can migrate downstream. The drainage is integrated, and the streams have relatively smooth, concave-up, long profiles because the rapids, waterfalls, lakes, and other irregularities of youth have been largely eliminated. Downcutting still occurs but is no longer prominent. Lateral planation is now dominant. Additional tributaries have developed, and the major ones may be graded or nearly so.

The *old-age* stage develops slowly and cannot be distinguished sharply from that of maturity. One might say that it has the same features as a mature valley, but in exaggerated form: meanders, oxbow lakes, and natural levees are prominent on a very wide floodplain that contains swamps, lakes, and smaller streams and is not well drained. The distant valley walls are lower and slope gently.

Such changes are continuous and only arbitrarily subdivided; therefore, it may be useful in studying different valleys to modify the three stages by terms such as early, middle, or late. The time a valley spends in each stage is far from equal. The figures for one estimate are: youth 5%, maturity 25%, and old age 70%, but this can only suggest the actual magnitudes.

At any one time, a particular valley may be in different stages in different parts of its course; e.g., a valley may be youthful in its upper reaches, mature in its middle portions, and old near its mouth where erosion has continued for the longest period of time and less downcutting had to occur. Variations can also occur if a river passes through rocks that differ conspicuously in their resistances to erosion. For example, a wide valley can develop in weak rocks both upstream and downstream from a V-shaped stretch underlain by resistant rocks.

Erosion Cycle In a Region

An entire region may also develop slowly through stages which can be characterized as youth, maturity, and old age (Fig. 2–25). Under the conditions assumed for this region, the *youthful* stage in its erosion cycle is featured by broad flat-topped divides that separate a few V-shaped valleys. The area is chiefly upland.

As streams enlarge their valleys and become more numerous, the amount of upland decreases, slopes (sides of valleys and mountains) predominate, and the area becomes *mature*. At this stage, most of the major streams flow on belts of weaker rocks or structures (i.e., segments developing on weaker rocks have grown relatively rapidly and diverted drainage from more resistant rocks).

As the region is very gradually eroded closer to base level, the amount of so-called bottom land becomes greater, and the *old-age* stage is reached. Just how some of this bottom land forms, however, is uncertain. Lateral planation by streams produces some. Retreat of slopes by mass-wasting and sheetwash produces more, but the relative proportions of the two processes are uncertain, as is the precise manner by which slopes retreat (see pediments, p. 274).

In a very generalized fashion then, upland predominates in youth, slopes predominate during maturity, and bottom land is extensive in old age. Furthermore, we must distinguish the stage reached in the sequential development of a particular major valley from that reached by an entire region (may contain a number of major valleys). The two stages may be different; e.g., young stream valleys may exist in a region which has reached maturity.

Peneplain ("almost a plain") is the term used for the nearly smooth surface eroded by streams and mass-wasting that develops in late old age, that covers a large area, and that has relatively low relief and altitude. Its surface may be dotted here and there by a few small, rounded hills or mountains (monadnocks) that are underlain by resistant rocks or are located far from the main streams. The bedrock surface of a peneplain shows beveled rock formations and geologic structures—as if a gigantic carpenter's plane had been pushed back and forth across the region.

Peneplains seem to develop infrequently because crustal movements tend to interrupt a cycle before its completion, and none are known to exist intact today. However, peneplaned surfaces may subsequently be uplifted and dissected by erosion, and remnants of such surfaces probably exist in various places on the Earth (p. 96).

Prevention and Control of Soil Erosion

The natural processes of soil formation from bedrock and regolith are very slow when measured by a man's lifetime. This makes essential the preservation of the soil that now exists—soil which is one of man's most important natural resources, as necessary as the air we breathe or the water we drink. Yet approximately one-half of the total quantity of agricultural land in the United States has lost much of its topsoil in the last two centuries through erosion, is in danger of doing so, or has already been destroyed. Fertile lands are also lost each year when roads, airports, and buildings are located on them. Stated in another way, topsoil in the United States may have averaged about 9 inches in depth before human settlement. Today this thickness is estimated at 6 inches.

In addition to the direct loss of crop-producing capacity, soil erosion increases the destructiveness of floods and decreases the storage capacity of water in reservoirs. Perhaps one-half of the water supply reservoirs in the United States will be useless within 100 years because they will be filled by sediment.

A protective mat of vegetation prevents or greatly retards soil erosion by wind and water (creep still occurs, however). Of first importance, therefore, is the restoration or maintenance of such a cover on slopes which have been deforested, overcropped, overgrazed, or burned over. Submarginal soils, especially those on steep slopes, should not be plowed. A double loss occurs when good soil is washed down a slope, mixed with infertile material in the process, and then dropped on top of fertile soils at lower altitudes.

The very important erosional effect of raindrops that strike soil on sloping land with little vegetation has been recognized

7-35. The fall, impact, and explosive burst of a raindrop on moist soil. Bits of soil are splashed upward and outward. *(Courtesy U.S. Department of Agriculture.)*

only recently (Fig. 7–35). Tiny soil particles may be moved a few inches or a few feet upon each impact and the violent disruption and pounding of a heavy rain may result in the shifting about of more than 100 tons of soil per acre. On level ground struck by vertically falling drops, the soil is splashed around, but the effects are usually canceled by the random motions. However, on sloping ground, raindrops combine with sheetwash to cause much soil erosion. Such erosion is particularly damaging if it is ignored because large gullies are not forming. Detachment of tiny particles, in which raindrop impact is most important, and transportation where sheetwash predominates are the two main steps involved.

The particles loosened and shifted by the raindrops are carried downslope by a very thin sheet of water which moves along the surface. The impacts of the raindrops increase the turbulence and transporting capacity of this unchannelized *sheetwash,* which may cover more than 95% of the surface of a slope. Soon the sheetwash becomes concentrated downslope by irregularities into tiny rivulets which can then erode deeper. The rivulets in turn join together to form slightly larger channels until gullies are formed. In this process, more and more water erodes and transports on less and less surface. This results in the downward cutting of the gullies. On slopes without gullies, superficially not undergoing serious erosion, the valuable topsoil may actually be wasting away more rapidly than on gullied slopes.

As raindrops batter and strike a surface like tiny bullets, finer particles in the soil may be shifted between larger ones to form an impervious crust, and the soil may be compacted. An impervious surface crust may thus form, and its presence may increase surface runoff and the development of gullies. As erosion breaks through this crust, the underlying unconsolidated particles may be splashed away, leaving a tiny soil pedestal (Fig. 7–36). Larger pedestals form in a somewhat similar manner, although more than raindrop splash is involved.

Various techniques can prevent or decrease the downhill movement of soil along a slope. Drainage may be controlled by a system of ditches or small dams. Small gullies may be healed by the planting of quick-growing vines and shrubs. Row crops may be planted parallel to the contours of a slope, not up and down the incline. Crops planted in any given field may be varied from year to year. Strip cropping on a slope intersperses areas of row crops such as corn with protective strips of thick-growing grains and grasses. Terracing (Fig. 7–37) of a slope permits the tilling of gently sloping or flat areas which are separated by more steeply sloping sections that may be kept permanently covered by a mat of vegetation or held in place by retaining walls. Materials such as hay or straw can be scattered about between rows of corn or among other crops and prevent much soil erosion. Such mulches have the added advantage of slowing the growth of weeds.

According to the U. S. Conservation Service, good practice involves using each acre according to its capabilities and treating each acre according to its needs.

Floods and Flood Control

Streams perform most of their work of erosion, transportation, and deposition during floods (Fig. 7–25). In fact, in the few days of a powerful flood, a river may accomplish more in these respects than it does during the rest of the year or perhaps even for several years. Some large rivers have been known to deepen their channels more than 100 feet when waters are exceptionally high. Such large rivers flow upon unconsolidated sediments that bury the bedrock floors of their valleys, and they can erode the bedrock only during floods. As a flood subsides, other sediment is deposited, refilling the bottom part of a channel.

Since floodplains are very fertile areas for farming and furnish flat land for roads and building purposes, many farms, towns, and cities are located on them, even though they are subject to periodic floods (Fig. 7–38). In one sense, these works of man are located in rivers, not just near them, because floodplains were built by rivers and are part of them during exceptional floods.

To prevent water from spreading over some sections of a floodplain, artificial levees were built to confine floodwaters to river channels. Unfortunately, this construction inaugurated a vicious cycle. During a flood, a river carries an extra load of sediment, part of which is normally spread over its floodplain. Because a river is confined to its channel, this extra sediment is deposited within the channel during the waning stages of each flood, and the level of the water is thus raised. This in turn means that the levees must be made higher, and in some valleys the tops of the levees are above the roofs of houses built on the adjacent floodplains.

7-36. Soil pedestals caused by raindrop splash. Raindrop splash may remove as much as two inches of topsoil in one heavy rain of high intensity. Plant cover controls splash erosion by intercepting the falling raindrops and absorbing their energy. The pebble-capped pedestals show that the force of the raindrop was applied from above. The soil not protected by pebbles was detached and washed away. *(Courtesy U.S. Department of Agriculture.)*

Floods and flood control affect the lives of all citizens today, either directly or indirectly, because most of the cost of major flood control structures is paid by federal funds. According to the U.S. Department of Agriculture, the total average annual damage from floods and sediment in the United States is about 1 billion dollars. This damage may be separated into *upstream areas,* which lie above existing or proposed major flood control structures, and the *downstream areas* which lie below them.

More than half of the total flood damage probably takes place in the upstream areas from frequent, small, unpublicized, but destructive floods. Few, if any, of the existing or proposed major flood-control structures aid in preventing these floods. Perhaps more than $60 has been spent on downstream structures for each dollar spent for upstream flood control—not a proper balance. More than two-thirds of the upstream damage is agricultural: damage to crops, roads, and buildings, erosion of good farmland, deposition of infertile sediment on productive soil, and indirect losses such as delays in planting or marketing.

Flood control is a many-faceted complex problem. A large river system which extends throughout several states is a natural unit and must be treated as a unit—omissions or commissions in one part of the system can affect remote areas vitally. The multiple uses of a stream system must be considered: floods, power, water supply, navigation, irrigation, recreation, sewage disposal, fish, and wildlife. Sometimes these interests conflict; e.g., reservoirs need to be kept nearly empty for maximum efficiency in preventing floods, but they need to be full to serve as adequate water supply.

The more extensive damage caused by recent floods results from more extensive use of floodplains by man and from his destruction of the cover of vegetation which formerly protected slopes throughout the drainage areas of rivers. When water falls on slopes which are covered by thick grass and numerous trees, much of it sinks into the ground, and very little runs off immediately—almost none from a forested area. Thus reforestation and soil conservation throughout the entire area drained by a river system appear to constitute a very important portion of a satisfactory long-range solution of the flood problem. Rapid runoff must be halted. Straightening and dredging channels and strengthening levees are temporary expedients.

Although reforestation is important in preventing or limiting numerous small upstream

7-37. A remarkable example of terracing of a steep valley, in Lebanon. *(Courtesy U.S. Department of Agriculture.)*

7-38. A city built "in a river" (i.e., on a floodplain). This is an aerial view of Paducah, Kentucky, at the junction of the Tennessee and Ohio rivers in flood. *(Courtesy U.S. Air Force.)*

floods, it apparently has much less effect in preventing major disastrous downstream floods. Reforestation throughout a drainage basin would not "soak up" these major floods before they started. Such floods occurred before man entered the scene and cut down the forests. They take place when heavy rains fall on already saturated ground.

A system of dams and reservoirs, therefore, is likewise essential in the control or prepention of floods. If the many tributaries that feed a large river can be prevented from flooding simultaneously, the danger will be averted, and the water can be released from the reservoirs later at times of drought. It is the simultaneous flooding of many tributaries that causes a disastrous flood downstream. The effect of a flood on any one tributary may be barely noticed far down the valley where channel capacity is much greater.

Yet agitation by the population of a large city for the building of dams and reservoirs above it may arouse the hostility of the people who live and farm in the areas which would have to be condemned and flooded to make room for the reservoirs. The annual income from an area which is to be flooded by a proposed reservoir may exceed the total benefits to be derived from its construction.

It seems that rather small dams in the headwater region of a stream serve a purpose different from that of a much larger dam farther downstream. The small dams aid in the prevention of floods and soil erosion in their upstream locations. Thus they are needed. Even very large numbers of small upstream dams, however, would apparently not be effective in preventing the really disastrous floods that must be anticipated on occasions in the downstream areas. They would not control a large enough portion of the total drainage area. Thus large dams on major tributaries are needed also.

Settlement control should play a vital role in any long-range plan for controlling floods. To keep flood waters entirely confined to a river channel is extremely difficult and expensive. However, if valuable properties could be zoned to higher altitudes or to areas somewhat removed from a river channel, much property damage would be prevented. Dikes could be located far enough from a channel to provide space for overflow in time of floods. Only temporary or less valuable buildings would be permitted in the zone between the dikes and river channel. Somewhat similar zoning regulations could be applied to farm buildings constructed on floodplains. Such zoning would tend to eliminate the two most important reasons for the construction of many huge dams and reservoirs.

Exercises and Questions for Chapter 7

1. Assume that a cube 1 meter on a side is subdivided into smaller cubes each 1 centimeter on a side. How much greater is the total surface area of all of the 1 cm cubes than that of the original cube? What application does this have to weathering?
2. Describe briefly as many processes as you can that act to reduce the altitude of the Earth's surface. Describe other processes that act to build up the surface.
3. Assume that you are in an area containing a number of large, glacially transported boulders. Some of these have been partly exposed by erosion and project from the ground in much the same manner as does a small outcrop of bedrock. How could you distinguish these boulders from outcrops of bedrock?
4. Describe the important processes involved in chemical and mechanical weathering.
   (A) What conditions are most favorable for chemical weathering? for mechanical weathering?
   (B) In what type of climate would you expect each of the following to weather relatively rapidly? to weather relatively slowly?
      a. granite
      b. limestone
      c. sandstone cemented chiefly with calcium carbonate
      d. sandstone cemented chiefly with iron oxide
      e. badly fractured quartzite
5. How does soil originate?
   (A) Why is a mature soil zoned? What are the distinguishing features of each of the three main zones?
   (B) What kinds of data might be collected in a volcanic region that would lead to estimates concerning the length of time necessary

for soils to form? Would these rates necessarily apply to other types of rocks?
(C) Describe some methods of soil conservation.

6. Describe the roles played by weathering, mass-wasting, and streams in the sculpturing of the lands. Is the analogy of a stream with a conveyor belt a useful one?
7. What is meant by differential weathering and differential erosion? On what scales do these processes operate? Look for examples in your neighborhood and in your travels.
8. List some of the names applied to different types of mass-wasting.
   (A) What do these tend to have in common?
   (B) In what ways do these tend to differ?
9. Stream flow:
   (A) How does a stream transport its load of sediments?
   (B) Describe some of the factors that affect the velocity of a stream.
   (C) How does stream flow tend to differ when flow around a bend is is contrasted with flow in a straight stretch?
10. Describe the changes that tend to occur in the development of a stream valley as it gradually evolves through the stages of youth, maturity, and old age.
11. Assume that erosion has reached the mature stage in three different regions (A, B, and C below). Assume that conditions are similar in the three areas except the rock structures (i.e., the climates are warm and humid, the areas have similar altitudes and distances from the ocean, and some layers of rock are much more resistant to erosion than others). Describe in general terms some of the more conspicuous differences in topographic features that you would expect to find in these three areas.
   (A) Area A is underlain by horizontal sedimentary rocks.
   (B) Area B is underlain by folded sedimentary rocks; the axes of the anticlines and synclines are horizontal.
   (C) Area C is underlain by folded sedimentary rocks; the axes of the anticlines and synclines are not horizontal—they plunge (dip or slant) downward.
12. Discuss some of the factors involved in floods and flood control.

8

Subsurface Water

Water enclosed in sediments and rocks beneath the Earth's surface (Fig. 8–1) is one of man's most vital natural resources. but its importance is generally unappreciated by inhabitants of humid areas with ample supplies of surface water. Worldwide, very large numbers of men, women, and children and their animals and crops depend completely upon supplies of subsurface water, and about 20% of all the water used in the United States comes from subsurface sources. Moreover, vast supplies of subsurface water go unused today in certain crowded, underdeveloped countries—the full utilization of these supplies could aid greatly in eliminating poverty and hunger in such areas.* Some understanding of underground water by the citizen is thus both relevant and important. Currently, more than 70 nations are participating in the International Hydrological Decade which began in 1965 and is sponsored by the United Nations.

According to U. S. Geological Survey estimates, overall surface and subsurface fresh water resources of the 48 conterminous United States are very large and average about 1200 billion gals/day (Table 8–1). This is twice the projected needs for 1980 and about four times the usage of the middle 1960's. Of the more than 300 billion gallons of fresh water used daily in the United States in 1965, estimated totals were about: 46% for irrigation, 46% for self-supplied industries, 1% for rural needs, and 7% for public water supplies, a major portion of which was also industrial. This excludes water used in the generation of electric power which is non-consumptive and is merely diverted through turbines.

Thus a very small amount of the total water used in the United States actually goes for the multitude of familiar personal and household purposes such as bathing, drinking, and washing dishes and cars. The average cost of this water at the faucet for municipal supplies is about 30 cents per 1000 gallons (around

* Gale Young, "Dry Land and a Hungry World," *Transactions of The New York Academy of Sciences*, February 1969.

7 cents per ton since there are 240 gallons in 1 ton of water), which makes water cheaper than almost anything else except the air we breathe. The average cost of irrigation and industrial water is much less than this.

Subsurface water has for thousands of years been a vital need in many areas of the world, and the search for it was shrouded in magic and mystery. Yet the superstition of the water diviner's forked stick is still with us, as is the ignorance that condones the juxtaposition of the outhouse and well in many farming areas, and supports belief in springs occurring on mountain tops or in great underground rivers (some do occur, especially in karst areas, p. 225). Even educated people are likely to know little more about their water supply than that a turn of the faucet brings forth clear cold water. Yet questions of ground-water supply have gained rather than lost importance in modern civilization, and its conservation has become a vital matter

8-1. Stalactites, stalagmites, and columns in Carlsbad Caverns, New Mexico. *(Courtesy National Park Service.)*

TABLE 8-1 United States Water Supply and Projected Water Use

Water Supply
Mean annual natural runoff
(1794)

Water Use: 1965, 1980, 2000
Total withdrawals in billion gallons per day

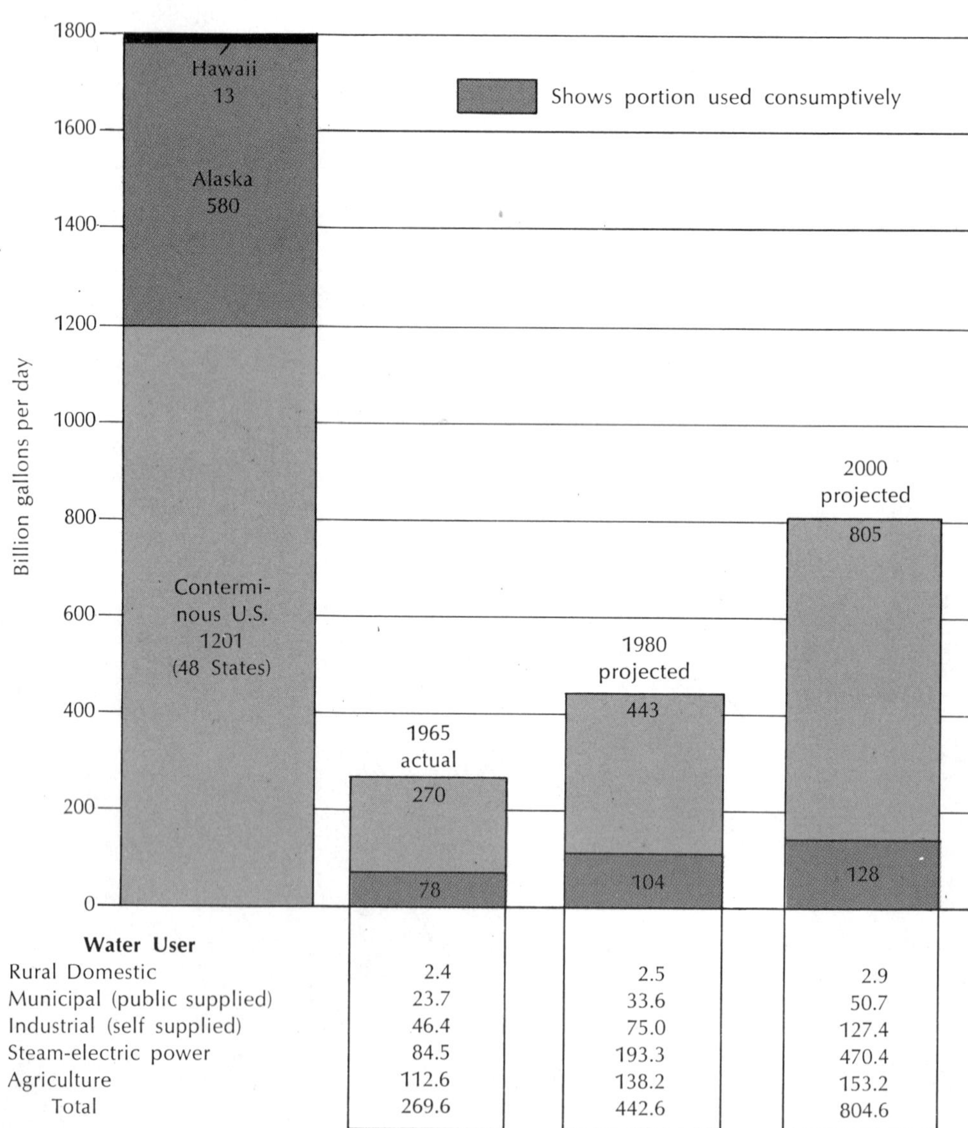

Water User	1965 actual	1980 projected	2000 projected
Rural Domestic	2.4	2.5	2.9
Municipal (public supplied)	23.7	33.6	50.7
Industrial (self supplied)	46.4	75.0	127.4
Steam-electric power	84.5	193.3	470.4
Agriculture	112.6	138.2	153.2
Total	269.6	442.6	804.6

Source: Water Resources Council, 1968 National Water Assessment.

Water use is rising; e.g., in 1969 there was a 1.5 percent increase in domestic and agricultural usage and a 2.5 percent increase in industrial usage. However, water costs are very low, and thus water is used very freely. Even a relatively small increase in unit cost may lead to a reduction in some uses and make it profitable to recycle water used chiefly for cooling. (First Annual Report of the Council on Environmental Quality, August 1970.)

that needs our understanding, participation, and support.

Of interest to many of us are caverns, sink holes, stalactites, and disappearing streams which are associated with subsurface water in humid regions underlain by soluble rocks (Fig. 8–1). Although ground water and subsurface water are used synonymously here, technically *ground water* refers only to water beneath the *water table,* the surface below which sediments and rocks are saturated with water (p. 216).

The Hydrologic Cycle

The oceans are the great reservoirs of the world's water supply. Solar energy causes evaporation at their surfaces, and some of the resulting water vapor is carried by the atmosphere over the lands where it may be precipitated or condensed as rain, snow, hail, frost, or dew (Fig. 8–2). Part of this water may be evaporated again directly or via the transpiration of plants (e.g., tree roots withdraw water from the soil; this water moves up the trunk as part of the sap, and it is later transpired by the foliage). Another part of the precipitated water runs off immediately to join streams and lakes, and a third part sinks downward into openings in sediments and rocks to become ground water. Much of this emerges at the surface again at lower altitudes.

Thus water moves endlessly between sea, air, and land in this circulation known as the *hydrologic cycle,* and a drink of water in a city such as Boston or Chicago may consume water molecules that were evaporated from a maple tree, a tobacco patch, or the Atlantic or Pacific Oceans. This may have occurred recently or centuries ago, because the water may previously have been stored for a time in the ground or have been part of a glacier. Before this, the far-traveling water molecules may have participated in

8-2. The hydrologic cycle. *(Courtesy U.S. Geological Survey.)*

countless phases of the water cycle, on land and sea, and in the northern and southern hemispheres.

The average annual precipitation in the United States is equivalent to about 30 inches of rainfall.* Of this, about 21 inches (70%) return directly to the atmosphere by evaporation and transpiration. A field of vigorously growing plants on a summer day probably furnishes nearly as much moisture to the air as is evaporated from the surface of a water body of the same size (e.g., an acre of corn may transpire 3000 to 4000 gallons of water in 24 hours), nearly the quantity that would evaporate from a nearby acre of water in the same time.

The remaining 9 inches or so are the source of all surface and ground water and would be an ample supply if distributed uniformly in time and space. Viewed from one point, this 9-inch annual supply averages about 7500 gallons per person per day, nearly enough to fill a room measuring 10 feet by 12 feet with an 8-foot ceiling. Today we use about one-fourth of this potential supply—more than 90% of it for irrigation and industry. Problems arise because too much water or too little water occurs at a certain place at a certain time, and these problems are increasing in number and magnitude each year (more people, greater consumption per person, and more building on floodplains).

Knowledge of the hydrologic cycle and of the nature and movement of ground water has developed chiefly within the last 250 years. According to some ancient sages, rainfall was insufficient to account for all subsurface water, and rain water could not penetrate deeply into the ground. Such opinions, untested by experiments, were widely held for many hundreds of years. Toward the end of the seventeenth century, experiments were performed to study the origin and movement of ground water. A French scientist (Perrault) measured rainfall and discharge in the Seine drainage basin for a period of three years. During this time, about six times as much rain and snow fell in the drainage area as left it via the Seine River (the transpiration of plants and evaporation consume a great deal of water). Halley, another scientist, was among the first to explain the water cycle correctly; he showed experimentally that evaporation from the oceans was sufficient to account for all surface and subsurface water.

* Estimated quantities for Earth's water cycle (in gals/day x 10^{12}):
Total rainfall on lands = 72
Total rainfall on oceans = 268
340
Total evaporation from lands = 48
Total evaporation from oceans = 292
340

Distribution of Subsurface Water

When water is precipitated from the atmosphere onto the lands, it may run off immediately, evaporate, be transpired by vegetation, or sink below the surface. Heavy rains on steep slopes underlain by relatively impervious (impenetrable) materials result in a high percentage of runoff. Furthermore, interception by thick foliage and subsequent evaporation may prevent much rain from even reaching the surface. On the other hand, the slow steady addition of water to pervious materials on gentle slopes causes a large fraction of the water to infiltrate into the ground.

Porosity and Permeability

The porosity and permeability of the regolith and bedrock are important factors in determining the amount of infiltration. *Porosity* refers to the percentage of interstitial space in a given volume of rocks and regolith, and porosities from 5 to 15% are considered average in rocks, although higher porosities tend to occur in unconsolidated sediments. *Permeability,* on the other hand, refers to the relative ease with which water may move through these interstices. Thus a rock must be porous to be permeable, but a porous rock is not necessarily permeable.

Interstices (Fig. 8–3) may consist of the

spaces or pores between the particles of a sediment or sedimentary rock (reduced by the amount of interstitial cement, Fig. 2–15), spaces within the particles themselves, the gas-bubble holes (vesicles) in a lava flow, spaces formed by solution in soluble rocks, and cracks and fissures of widely different sizes in all kinds of rocks. If sorting is poor and a wide range of sizes occurs in a sediment, smaller particles will fill spaces among larger ones and reduce the porosity.

On the other hand, in a well-sorted sediment, all particles are approximately the same size, and the porosity is greater. Spheres of uniform size, when arranged in the open packing shown in part A of Fig. 8–4, have the maximum possible porosity. Quicksand is sand in such open packing and is generally caused by upward-moving water that lifts the sand grains into this unstable arrangement in which each sphere touches six other spheres and in which lines connecting the centers of the spheres meet at 90-degree angles. This holds true for spheres of any size; they may all be 2 inches in diameter, or 1 centimeter, or 1 millimeter. However, the larger openings associated with the larger spheres make them more permeable. On the other hand, when spheres of uniform size are packed as closely together as possible, the porosity is about 26%. In this arrangement, called rhombic packing, each sphere touches twelve others.

In rocks composed of interlocking grains (most igneous and metamorphic types, and some sedimentary rocks) little or no space is left among the grains. However, such rocks may crack and develop porosity. If the cracks are interconnected, the rocks are also permeable.

Porosity and permeability are thus different though related properties, and a high porosity does not assure a high permeability. The size of the interstices is a controlling factor (see capillarity, p. 218), and a high permeability is associated with larger grains which can permit larger interstices to exist among them. Thus sand is more permeable than clay, although clay-sized particles have a higher porosity than sand (Fig. 8–5). Un-

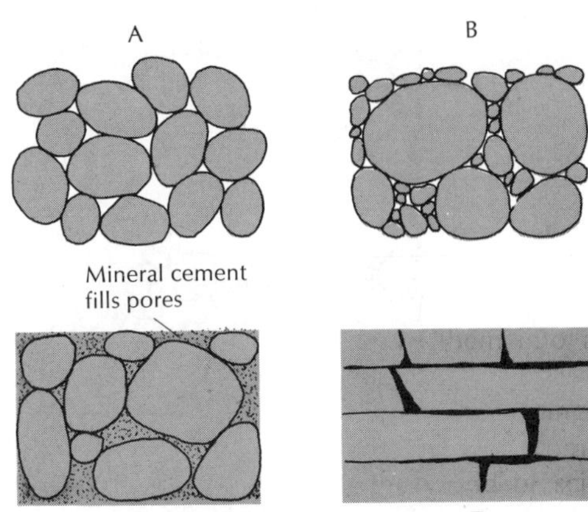

8-3. Some factors determining the porosity of a rock or sediment. (A) Spaces of different sizes and shapes may occur among particles. (B) If sorting is poor, then smaller particles fill in spaces among larger ones and reduce the porosity. (C) Mineral cement reduces the porosity. (D) Water can occur in joints and along bedding surfaces.

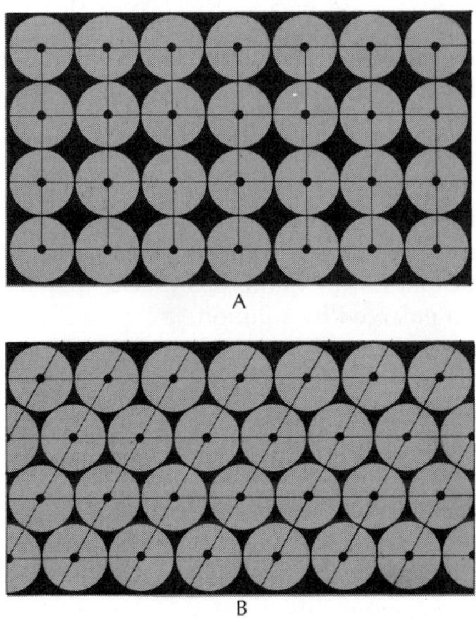

8-4. For spheres of uniform size, interstitial space is at a maximum in open packing (A; 48%) and at a minimum in rhombic packing (B; 26%).

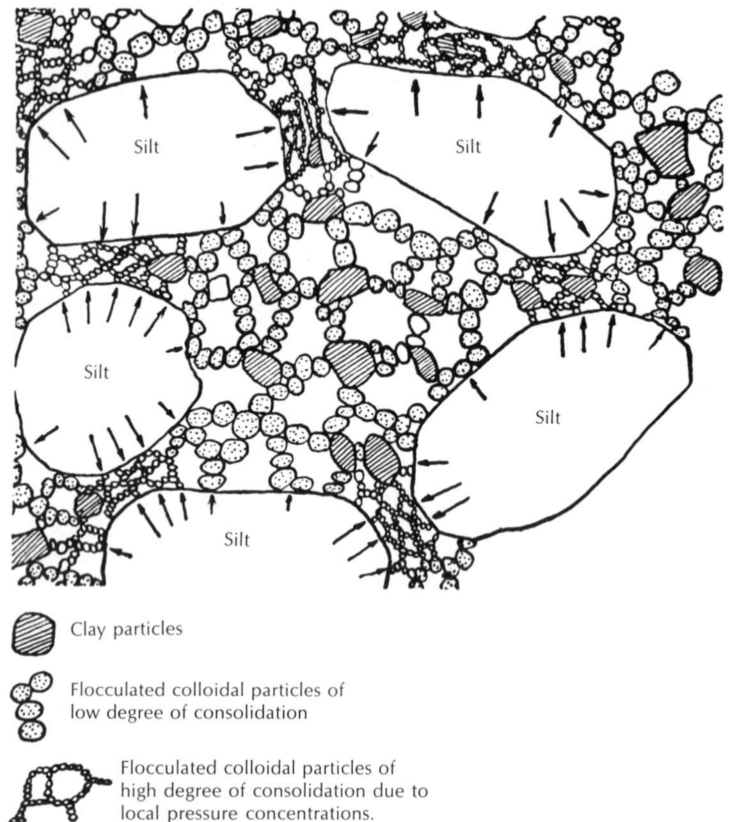

8-5. Packing in undisturbed marine clay. Silt-sized particles (range 1/16 to 1/256 mm) make up a larger proportion of clay than do clay-sized particles. A clay may contain a considerable amount of water which may subsequently be squeezed out and lead to compaction. If this occurs beneath a building, the building will settle. *(Courtesy Boston Society of Civil Engineers.)*

consolidated, well-sorted gravels are highly permeable, as are some vesicular lavas and limestones with numerous cracks that have been enlarged by solution.

The Water Table

Below a certain variable depth, interstices are filled with water (or filled locally with some other fluid such as oil or gas). The upper surface of this water-saturated zone, where the pressure is equal to 1 atmosphere, is called the *water table* (Figs. 8–6 and 8–9). The water table tends to be at a shallower depth in humid regions than in dry regions, and it rises in wet seasons and falls in dry seasons, although the pores and cracks below this surface are always filled with water. Above the water table, the interstices may contain air or water. Wherever the water table intersects the Earth's surface, the result is seepage of some kind: a swamp, a lake, a stream, or a spring (Fig. 8–7).

Most of us have pleasant memories of seashore or lakeshore outings during which we dug holes in the sand and watched their lower portions fill with water; we had penetrated the water table. Similarly, water fills the bottom portion of an ordinary well that has been dug or drilled in permeable materials below the water table. If the well is not deep enough, it may go dry during a drought if the water table falls below the well's base. In a similar manner, a shallow lake or pond may disappear during a dry season, or a swamp may dry up, if the water table falls below the bottom of the water body. The water table is not a fixed or level surface. In homogeneous rocks in humid regions, it forms a subdued replica of the topography, being farthest from the surface under hills and nearest the surface in valleys (Fig. 8–10). Under local geologic conditions, a *perched water table* may form (Fig. 8–11).

The water table is not a plane surface as is the surface of a pail of water, or of a lake, for a number of reasons: the Earth's surface is irregular, and thus water enters the ground at different altitudes in different areas; variability in rainfall and permeability cause uneven infiltration; in most aquifers, water moves slowly through pores and cracks and a complete leveling of the water table would take a very long time (karst areas, p. 225, may be exceptions). New supplies of precipitated water tend to be added frequently enough to prevent this from occurring.

A much greater volume of water is present in the pores and cracks found underground than exists in all of the lakes and streams above the surface. It is estimated that within 100 feet of the surface of the United States there is sufficient water to cover the entire area to a depth of approximately 17 feet. Unfortunately, a considerable volume of subsurface water occurs in impervious materials such as clay and shale or the quality of the water is unsatisfactory. Subsurface water, therefore, does not constitute a vast interconnected underground reservoir that is available for domestic use. The bulk of the ground water withdrawn each year in the United States comes from unconsolidated sands and gravels located near the surface.

Water is a serious problem in many mining operations, and a number of mines have been closed before exhaustion because pumping costs became too great. However, deep mines tend to be dry and hot (the deepest mine extends about 1¾ miles below the surface) because pressures increase with

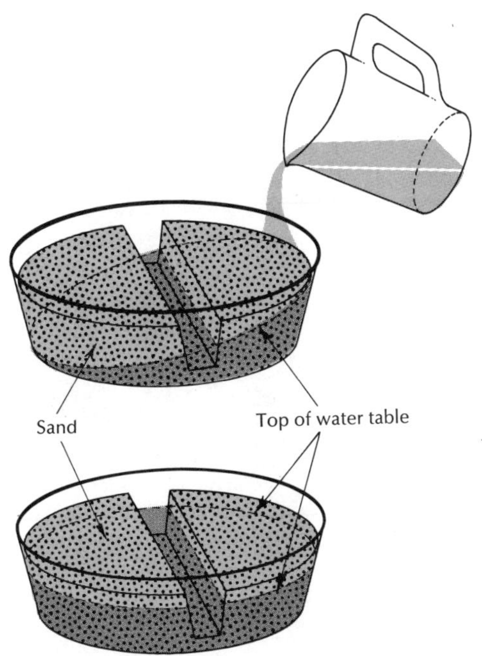

8-6. Experiment illustrating a water table in sand in a transparent container. The water table is not level in the upper sketch while the water is being poured. However, it has leveled out in the lower sketch because of subsurface movement, and the lower part of the "stream" channel is filled with water. *(After Leopold and Langbein.)*

depth, whereas porosities tend to decrease. About half of all ground water probably occurs within ½ mile (0.8 km) of the surface, and can be used by man, whereas the other half occurs farther down and is too dispersed for general use. Rock flowage takes place at still greater depths—beginning at about 10 miles (16 km)—and the porosity presumably is zero.

Movement of Subsurface Water

As a result of the Earth's gravitational attraction, rain falls onto the lands (and oceans), and water moves down slopes into streams. Similarly, gravity is the force that causes movement of ground water, but *capillarity* is also involved. This is the tendency of a

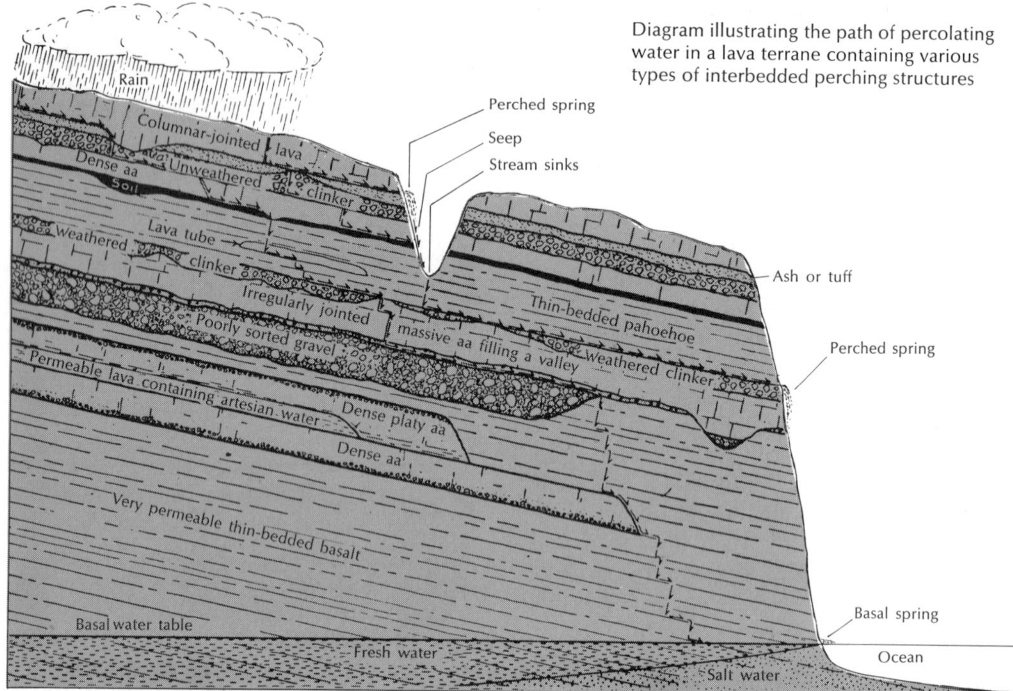

8-7. Typical occurrence of subsurface water in the Hawaiian Islands. Thus some springs, swamps, and intermittent streams may occur above the main water table, especially if impervious barriers occur below them. Aa and pahoehoe are Hawaiian names for different types of basaltic lava flows: viscous aa flows have rough, blocky, jagged surfaces (formed partly by the breaking up of a hardened surface by still-flowing lava beneath); in contrast, the more fluid pahoehoe flows have smooth ropy surfaces (Fig. 5-9). *(From Stearns and McDonald, Division of Hydrography, Hawaii.)*

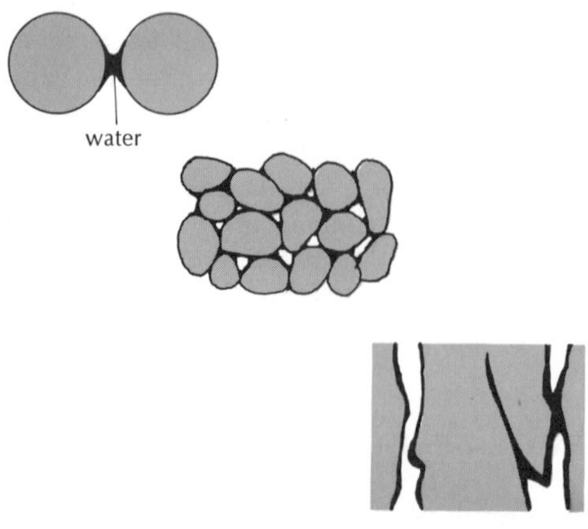

8-8. Capillarity. Molecular attraction causes thin films of water to envelope grains and occur along the walls of cracks. If openings are small enough, the films overlap, and water movement is greatly hindered. Gravity cannot remove such films, but evaporation and roots can.

liquid to cling to a solid surface and is caused by molecular attraction. In very tiny tubes or interstitial openings, capillarity can cause a liquid to rise against the pull of gravity. Capillarity (Fig. 8–8) is a very important phenomenon and can be illustrated in many ways. If one end of a towel is immersed in water, the water will rise within the towel. In a like manner, excess ink rises into a blotter, and the liquid fuel in a cigarette lighter moves up a wick. Capillary movement may be up, down, or sideways.

A zone of capillary water overlies the water table (Fig. 8–9) and may extend upward from this surface for several feet or more (but commonly less than this). The lower part of this capillary zone is commonly saturated, but water does not flow from it into a well—in this it differs from the saturated ground–water zone beneath the water table. A zone of soil moisture tends to occur immediately below ground level.

In the movement of water below the ground, capillarity causes a thin film of water to adhere to and coat the surface of each grain and crack. Such films have a top-priority claim on the supply of water available and cause moisture to be pulled from the wet grains to the dry ones in a downward, lateral, or upward direction. When all the films have been formed and the moisture requirements have been met, the soil is said to be at field capacity. It is only under these conditions that water can collect in the larger pore openings and flow downward in response to gravity. Gravitational attraction cannot pull capillary water away from the grains. Thus clays and shales tend to be impermeable because their interstitial openings are very small, and many of the capillary films around the particles may be in contact, which tends to hold the water in place.

Evaporation and the roots of plants can remove some of this capillary water. Shortly after a rainstorm, soil may be moist near

8-9. Modes of occurrence of water in soil, after F. Zunker. (*A. R. Jumikis,* Soil Mechanics, *Van Nostrand Reinhold,* 1967.)

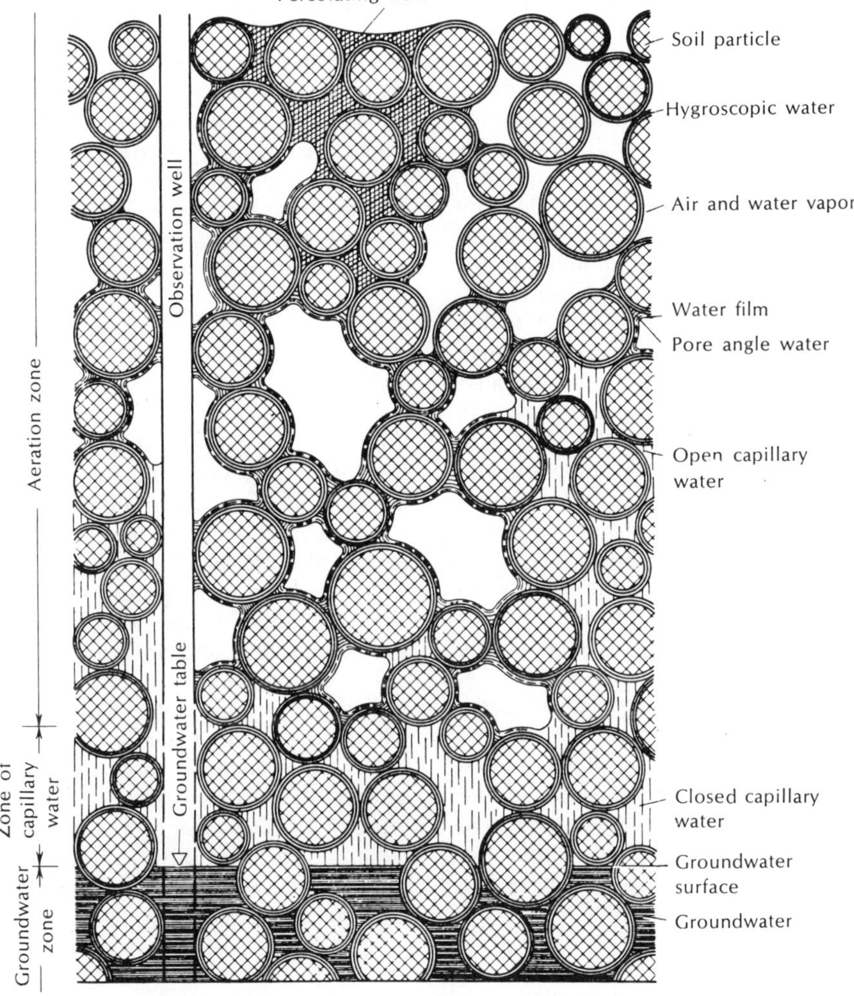

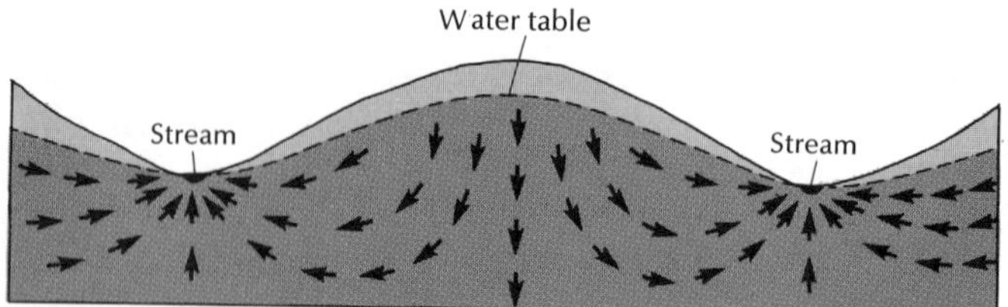

8-10. Movement of ground water in a humid area underlain by homogeneous material. If new supplies of water were not added by occasional rains, the water table would gradually flatten out and approach a level surface.

the surface but relatively dry at depth because not enough rain has fallen to produce capillary films to a greater depth. As plant roots take up capillary water near the surface, this subsurface region dries out through time, and in so doing water is brought up from below by capillary action. The next rain must replace all of this expended capillary water before any water can sink deeper. Consequently, if there are considerable intervals between rains, little or no water may sink to the water table.

In humid regions with seasonal climates, the ground-water supply may be replenished chiefly in late winter and early spring due to a combination of rains and melting snow, relatively little evaporation, and dormant vegetation. Some water may also be added during the fall and colder months as the growth of vegetation slows, and less evaporation occurs. Little water is added to the water table in the summer, even though more precipitation may occur, because evaporation and transpiration are too great.

Movement of water from the surface to the water table is dominantly straight down. When water reaches the saturated zone beneath the water table, it oozes slowly downward and laterally, generally in much the same direction as runoff at the surface. In so doing, the ground-water movement tends to be along curved paths that converge toward an outlet such as a stream (Fig. 8–10). One may think of the curved paths as the resultant of two forces: gravitational attraction which acts downward, and a force causing movement down the slope of the water table. Put in another way, at any one level within the ground-water zone, the hydrostatic pressure is greatest where the water table is highest and least where the water table is lowest. Therefore, at a particular level, such pressures tend to be greatest beneath a hill, where the water table is highest, and smallest beneath a valley. Such differences in pressure cause water to move downward beneath a hill, upward beneath a water body, and along curved paths between.

The rate of ground-water movement varies greatly: 420 feet per day is one of the fastest by field test; 1 foot in 10 years is one of the slowest by laboratory test. Less extreme rates range from a few feet per day to a few feet per year (1 foot per week might be average). A drop of water may thus move through the ground and return to the surface at a lower altitude, but this journey may take a few days or a few centuries, and it may cover a few yards or a few hundred miles.

Wells

There is a familiar problem in areas whose inhabitants rely upon individual wells. Why is one well successful at a depth of 50 feet, a neighboring one at 30 feet, and two others nearby at more than 100 feet each? Why do some of these wells go dry and others do not? Why does water flow to the surface from some wells and not from others?

The ordinary well is simply a hole dug or drilled below the water table in an area where the ground water is unconfined. Water collects in the hole to a level that defines the water table, which fluctuates through the year, rising in wet seasons and falling in dry seasons. In such wells the water must be pumped to the surface. Therefore, successful wells extend below

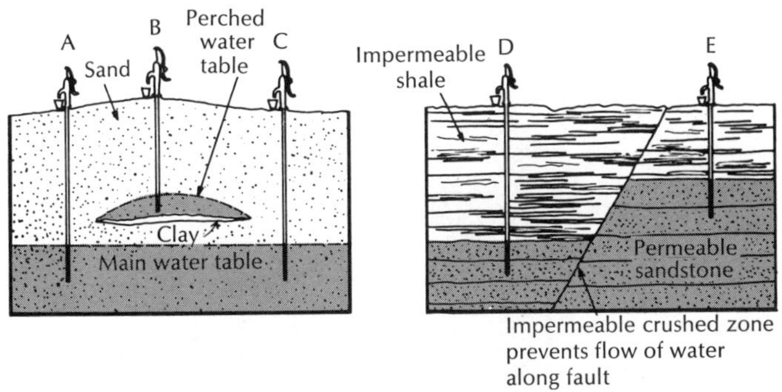

8-11. Successful wells vary in depth. At the left is a perched water body which may occur above the main water table in an area and cause variation in the depths of successful wells. A lens of impermeable clay is located in permeable sand in the sketch at the left. The main water table is situated a number of feet below the clay. However, the clay traps any water that enters the ground above it and percolates downward. Thus a local saturated zone forms above the clay. This perched water table would flatten out if frequent replenishment did not occur. At the right, successful wells on either side of a fault. Water enters the sandstone in adjoining areas where it crops out at the surface at a higher altitude.

the water table and penetrate permeable materials containing water of satisfactory quality. An *aquifer* may be defined as any sediment or rock from which supplies of ground water may be obtained economically.

Some of the causes of unsuccessful wells, as well as reasons for a variation in the depth of successful wells, are shown in Figs. 8–11 and 8–12; these include perched water bodies, faulting, and variations in regolith and bedrock, both laterally and vertically. In crystalline rocks, circulation of ground water is confined mainly to cracks, and a successful well depends upon the more or less chance intersection of an adequate network of cracks and fissures. Since openings decrease in size and number with increasing pressures at greater depths, it is sometimes more practical to start a second well at a new location than to extend a first unsuccessful one ever deeper.

A drilled well, especially a deep one, is likely to be called an artesian well, but this usage is incorrect. An *artesian* well is defined correctly as one in which confined water under pressure rises above the aquifer which contains it, and this rise is commonly above the local water table. An artesian

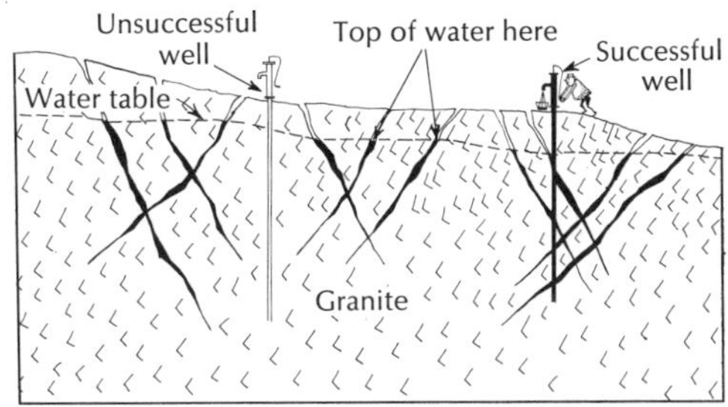

8-12. Well-drilling problems in areas of rocks with interlocking mineral grains. (Width of cracks is greatly exaggerated.)

spring is produced if an aquifer is exposed locally at the surface. Necessary conditions for one kind of artesian system are shown in Figs. 8–13 and 8–14: (1) a permeable bed must be present, and (2) impermeable beds must occur above and below this aquifer; (3) the beds must slant steeply enough to establish a hydraulic gradient, and (4) the aquifer must be exposed at the surface so that it receives supplies of ground water. In Fig. 8–14, rain falls on the permeable sandstone which crops out at a high altitude (at the left in the sketch). Some of this rain then sinks into the sandstone and oozes

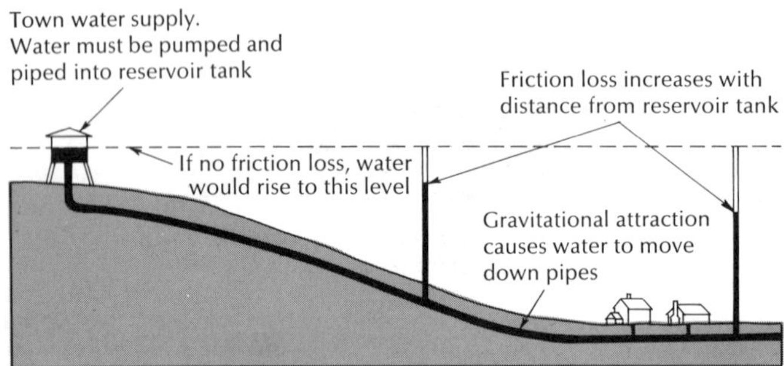

8-13. A town water supply illustrates some of the conditions necessary for an artesian circulation. *(After Leopold and Langbein.)*

slowly down the bed under the pull of gravity. When the sandstone aquifer is tapped by the well, pressure forces water up the hole. In this instance, pressure is sufficient to cause the well to flow. At the right, water seeps upward along a fractured fault surface to form an oasis. Artesian systems are less common in igneous and metamorphic rocks.

The Dakota artesian system is one of the most important in the United States. A permeable sandstone formation is exposed to rainfall in the Black Hills, Big Horn Mountains, and Rocky Mountains and extends eastward beneath the surface. Parts of Kansas, Nebraska, the Dakotas, Wyoming, and Montana are underlain at considerable depth by these sandstone beds, which are capped by impermeable layers. More than 15,000 wells have been drilled into this formation, and artesian flow was powerful enough from some of the first wells to drive small flour mills and generate electricity. Unfortunately, so much water has been withdrawn from the wells that fresh supplies from the intake areas cannot sustain the flow, pressures have dropped, and some wells that formerly had free-flowing water must now be pumped. Originally it was thought that recharge occurred where the sandstone aquifer was exposed. However, according to a more recent view, the aquifer is recharged from cavernous limestones that occur a short distance below the sandstone. These limestones are exposed in the Black Hills where many well known caverns are located. Large supplies of water come from streams that flow across these limestones and supply the ground-water recharge. Thus the ground-water supply may be much larger than previously thought because of the extensive nature of the limestones.

When an artesian system is first tapped, prodigious quantities of water may be made available. However, in dry areas the annual addition of water to the system from recharge areas may be relatively small and the water in an aquifer may be virtually irreplaceable—almost a one-crop resource. Regulation is thus necessary to maintain a balance between the natural supply and the total yields of the wells.

The important Illinois-Wisconsin artesian system formerly supplied the city of Chicago

Mountainous area with
abundant rainfall
Flowing
artesian
well
Spring and
oasis
Desert
Aquifer
Impermeable
beds of shale
lie above and below
the sandstone
Permeable
sandstone

8-14. One type of artesian system.

with a large portion of its water requirements. Permeable sandstones dip southward toward Chicago from their intake area in Wisconsin, and a drop of water may require about 200 years to travel the several hundred miles along the aquifer from the intake area to the Chicago wells (however, leakage of ground water from overlying rocks also recharges the aquifer). Under such conditions, Chicagoans were drinking "fossil" water. Perhaps the water drawn from a faucet in Chicago fell during a violent thunderstorm in Wisconsin two centuries ago and bounced off the back of a buffalo before it entered the ground!

To find out the rate of movement of ground water in such a system, we can put a colored dye down one well. When the water in a second well downdip first shows the color, we register the time and measure the distance between wells to get the rate of travel. Several such measurements made at strategic locations give an average for the entire artesian system (other methods are also used).

The age of a water sample, i.e., the length of time since its last direct contact with the atmosphere, can be measured by its radioactivity. Cosmic radiation produces in the atmosphere a rare isotope of hydrogen, called tritium. This isotope is radioactive, with a half-life period of about 12 years, and has an atomic mass number of 3 (one proton and two neutrons in each nucleus). Tritium is added to the ground-water supply in the raindrops that fall on the surface and sink into the ground. Therefore, the longer ago the water entered the ground, the less tritium is present due to its natural radioactive disintegration. However, the half-life period is so short that measurements are limited to water that is several decades or less in age.

The circulation of ocean water may also be studied and dated by this process. Surface water sinks in certain areas because it is denser than the surrounding water (low temperatures and greater salinities are two causes); a certain amount of tritium is thus carried downward and the tritium content is gradually reduced by radioactive decay as the water moves slowly far below the surface.

The tritium-dating method was checked in an unusual manner. To obtain samples of old, dated water, wines were used that had been bottled in different years in different countries, and which were made from surface waters. The method seems to be accurate, because the older wines were found to contain less tritium. For example, a 1948 wine would contain half as much tritium as one bottled in 1960, but twice as much as a 1936 vintage wine. Unfortunately, an entire case of fine quality wine had to be consumed—by the instruments—for each measurement.

Subsurface Water in Soluble Rocks

Limestones, marbles, and related calcareous rocks are soluble in water containing carbon dioxide which comes from the atmosphere or decaying organic material (Fig. 8–15). Although such rocks are dense with relatively few interstices among the mineral grains, they are commonly cut by cracks or fractures along which subsurface water can move. As water moves along such cracks, the more soluble parts of the rock slowly dissolve and the cracks widen. Solution is greatest at the surface, where carbon dioxide in the water is most abundant, and thus the cracks tend to widen near the surface.

Where two cracks intersect, a funnel-shaped opening may develop and grow larger to form a *sinkhole*. Other sinkholes develop where solution-formed openings near the surface have become large enough for the overlying rock and regolith to collapse. Sinkholes vary from tiny pits to large holes more than a mile in diameter, but they commonly measure a few tens of feet deep and a few acres or less in area. If floored with impervious material or cut below the water table, sinkholes may contain lakes or ponds. They may be circular in outline or quite irregular in shape. In some areas they

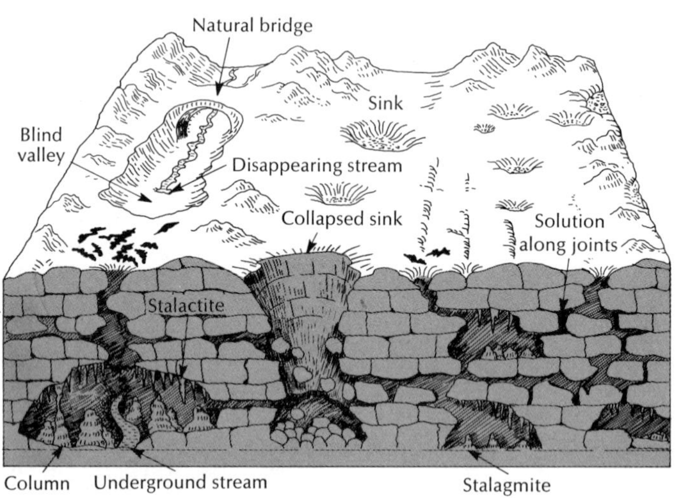

8-15. Karst topography and cave features. (*B. Mears, Jr.,* The Changing Earth, *Van Nostrand Reinhold, 1970.*)

are very numerous, and more than a thousand have been counted within a single square mile. Several smaller sinkholes may coalesce to form a very large one or even a valley, if they develop in a line along a fracture zone. The unweathered rock left between two sinkholes may form a natural bridge.

Most of the ground water that sinks below the surface in limestone areas tends to move downward and laterally, and eventually it joins streams in the deepest valleys of the area. The soluble rocks may be cut by two sets of joints. In combination with bedding surfaces along which water can move slowly, a three-dimensional network of cracks and fissures develops to carry ground water through the rocks in the area. Solution along these cracks and bedding planes gradually enlarges them into a series of interconnected passageways, and continued solution and roof collapse enlarge these into caves and caverns (Fig. 8–1). Surface streams may enter such a system and aid in the work of cave formation. Undermined parts of the system may collapse and enlarge the caverns or even create holes at the surface, and remnants of such collapsed roofs may form some natural bridges.

The dissolving action of water may take place both above and below the water table, and underground streams may further the erosion of caverns. The relative importance of these is uncertain and may well have varied from one area to another. Probably most large caverns were formed by solution below the water table and were filled with air later when the water table was lowered sufficiently by downcutting by streams. Potholes, grooves, and sediments encountered in some caverns show that stream erosion has taken place, but such erosion seems generally to have been less important than solution in the formation of the caves.

Erosion and solution in a limestone area gradually shift the drainage underground and produce a number of distinctive topographic features collectively described as a *karst* landscape (Karst is the name of the type region in Yugoslavia): abundant sinkholes, disappearing streams, valleys without streams, springs, and caverns. One kind of sequential development of a karst area is shown in Fig. 8–16.

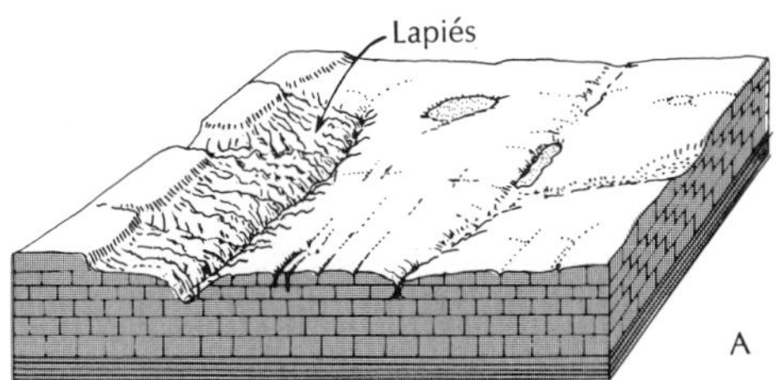

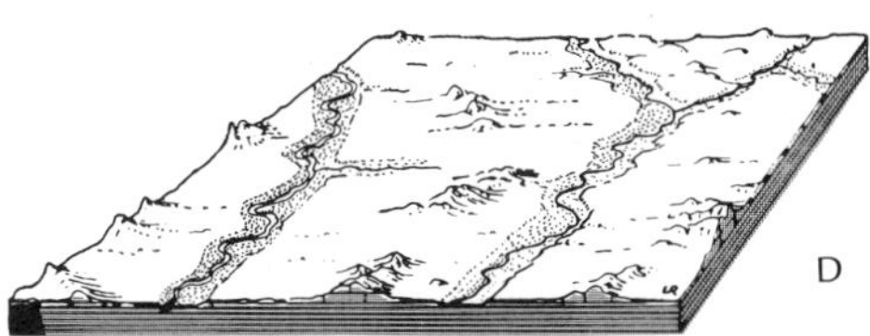

8-16. Possible stages of evolution of a karst landscape. Increasing relief and cavern development are followed by decreasing relief and the removal of the limestone mass. Note the sinkholes and natural bridge. Exposed limestone surfaces in *A* are deeply grooved and fluted into lapies. *A* doline in *B* is a deep, steep-walled funnel-like sinkhole. In *C*, a polje is an open, flat-floored valley formed by the coalescence of several sinkholes. *(A. N. Strahler,* Physical Geography, *John Wiley & Sons, Inc.)*

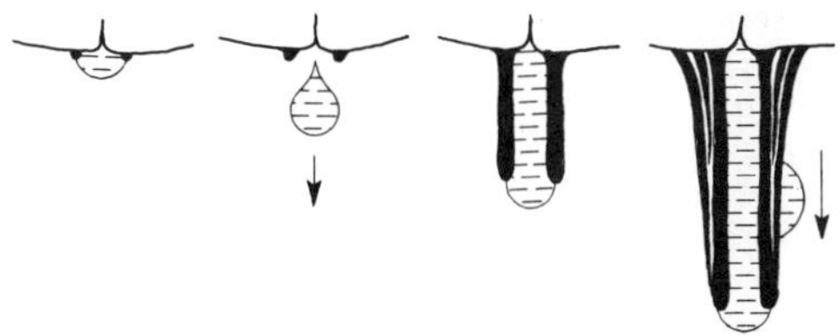

8-17. Origin of a stalactite. A small ring of calcite forms by precipitation from a drop of water that is temporarily suspended at the roof of a cave. By repetition, this ring may slowly develop into a tube which may taper downward because calcite has been precipitated from water that flowed down its outer part. *(R. W. Fairbridge, ed.,* The Encyclopedia of Geomorphology, *Van Nostrand Reinhold, 1968.)*

Water containing calcium carbonate in solution may descend along a crack to the roof of a cave where some of the calcium carbonate is precipitated, generally as calcite (Fig. 8–17). Carbonate precipitation apparently occurs because carbon dioxide is lost to air within a cave rather than by evaporation of the water—humidity tends to be about 100% within most caves. Repetition of this process eventually results in iciclelike projections from the room of the cave called *stalactites*. Part of the water may fall to the floor, where additional precipitation may occur. Eventually another iciclelike body is formed which grows upward from the floor and is called a stalagmite.* Stalactite and stalagmite may meet and form a column or pillar. Precipitation may also occur in layers to form terraces and many weird shapes. Calcium carbonate precipitation rates vary and may stop in some seasons or years; in fact, previously deposited calcium carbonate may even be removed by solution at times. Among the factors involved are: the amount of water, its rate of flow, and the carbon dioxide content of the water and cave air.

According to one estimate, solution in limestones in northern Kentucky may be occurring rapidly enough to remove an

* It has been said that stalactites and stalagmites have a certain resemblance to "ants in the pants." The "mites" go up and the "tites" come down.

average of about 1 foot of rock every 2000 years, but solution is generally much slower in other areas and in other rocks.

Water Conservation

Ground water forms a great reservoir that commonly extends downward from the water table to depths of 2000 to 3000 feet. If all of this ground water could be extracted, it would cover the land areas of the Earth to a depth of about 100 feet.

However, much ground water is in impervious beds, the quality is not everywhere desirable, and the supply is not inexhaustible. Conservation of both surface and subsurface water supplies is necessary. Not all areas have the desired amount of water at the desired time—neither too much nor too little.

Mining vs. Sustained Yield

In some areas, water is being withdrawn from the ground much more rapidly than it is being replenished or recharged, water tables have dropped alarmingly and continue to fall affecting both artesian and nonartesian sources. In these areas, water is virtually an irreplaceable natural resource; e.g., water is being "mined" today in parts of Texas, Arizona, and California (withdrawals are some twenty times greater than replenishment). Some of the aquifers now being drained may have been filled during the Pleistocene glacial ages when these regions had relatively cool moist climates.

Withdrawal of this fossil water certainly constitutes the mining of a one-crop resource. There are plainly social and political aspects to any decision to mine such stored water—entirely, in part, or not at all. However, long-term figures are needed; e.g., a certain part of Kansas experienced falling water tables for a number of years, enough to suggest that the water was being mined. Yet subsequent heavy rains raised the water tables higher than they had been twenty years earlier. There is some evidence that most recharge in arid and semiarid areas occurs during an occasional wet year.

If ground water is removed more rapidly than recharge takes place, the Earth's surface may sink as compaction occurs in an aquifer, and its permeability may be seriously and permanently damaged.* An extreme example is Mexico City where some of the sinking and compaction was caused by the consolidation of thick underlying masses of clay. About 9000 water wells were developed between 1910 and 1952 in unconsolidated materials underlying the city, and in places the surface has subsided as much as 16 feet. Some buildings are also sinking into the ground—a former ground-level entrance is now a basement which has been reinforced to prevent further collapsing. The maximum rate of building sinkage in 1953 was 20 inches per year. New buildings are now located on firmer ground, on piles, or have a floating type of foundation; but the problem remains very serious. To prevent further sinkage the pumping of ground water in the central part of Mexico City is no longer permitted. Every time a person has a drink of water in Mexico City, they say, the city sinks a little.

Withdrawals of oil may likewise cause local subsidence (as much as 26 feet at Long Beach, California), but water may be pumped into an aquifer to replace the oil taken out and thus prevent the subsidence. As a bonus, this water flooding may also increase greatly the total amount of oil that can be withdrawn from an aquifer (p. 153).

Even under natural conditions, where ground water is not being withdrawn by man in an area, some water still drains out of underground reservoirs each year and enters streams as part of the water cycle. Thus a certain fraction of the ground-water supply cannot be conserved simply by nonuse. What is needed is a long-term balance between withdrawals and recharge.

* S. S. Marsden, Jr. and S. N. Davis, "Geological Subsidence," *Scientific American*, June 1967.

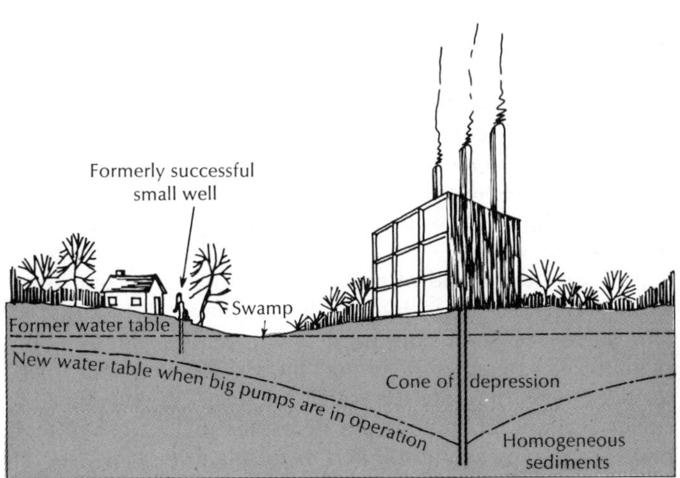

8-18. Withdrawal and recharge of ground water. In the area illustrated, the water table originally was near the surface. A swamp was located in one place, and a shallow domestic well yielded water when pumped. A large industrial concern moved to the area and drilled a deep well for a large water supply. Pumping was almost continuous and resulted in the development of a large cone of depression. The swamp and domestic well dried up. The problem was solved when the company was persuaded to return its uncontaminated water to the swamp, where it seeped into the ground and raised the level of the water table.

Withdrawal and Recharge

When water is pumped from a well in homogeneous materials, a *cone of depression* develops as the water table is lowered around the well (Fig. 8–18). The size and shape of the cone depend upon several factors. Water enters a well from the sides and bottom of the hole. As water leaves the interstices near the well, it is replaced by water from adjoining and overlying interstices. In their turn, these are filled and this transfer eventually reaches the surface of the water table, which is depressed in the area encircling the well. Water moves down the inward sloping sides of this cone of depression toward the well, and it moves more rapidly as the slope steepens (Darcy's law). This steepening continues until equilibrium is attained. If pumping then slows down and stops, the cone of depression gradually flattens out and eventually disappears.

A large withdrawal, or withdrawal from a less permeable aquifer, causes a steep-sided cone of depression. Slow withdrawal from permeable materials produces a wide shallow cone. Commonly a cone does not extend outward for more than a few hundred yards, but cones of depression from neighboring wells may overlap as determined by the porosity and permeability of the aquifer and the quantity and rate of withdrawal.

Quarrying in certain inhabited areas has created ground-water problems. The floor of a large deep quarry may extend tens of feet below the depth of the original water table. Thus water enters the quarry and is pumped and piped out of the area (i.e., the quarry acts as a giant well). This lowers the water table, and nearby shallow domestic wells go dry.

Natural recharge occurs in areas of abundant rain and snow where conditions are favorable for considerable infiltration. However, replenishment of ground water by natural recharge is not uniform throughout an area, and just as there are cones of depression, there are the reverse, water table mounds. The mounds tend to flatten out if new supplies are not frequently added.

Lakes and ponds do not cause recharge in humid areas because they themselves are fed by ground water. However, in a dry area the water table is commonly highest beneath a stream channel and recharge occurs when water occasionally flows down the channel and filters into the unsaturated materials containing the channel.

Artificial recharge utilizing both surface and ground water has been practiced successfully in many areas and is becoming more important in regions having increasing needs for water. Water may be injected under pressure down a well from which it oozes into adjacent rocks to become part of the local supply of ground water. But precautions are demanded to avoid contaminating the local water supply by injecting polluted water.

In a second method of artificial recharge, waste water is directed into shallow basins, into ditches of permeable rock, or into the regolith. Slow percolation through the soil interstices tends to purify the water, but occasional cleaning operations are necessary to prevent clogging of the soil and to maintain an efficient filtering action. Stream flow may also be regulated by dams to prevent the alternation of times of low water with those of great flow during floods when water rushes through an area and does not have an opportunity to sink into the ground to form ground-water storage supplies for future use.

Water Quality

A town's water supply may come from surface and/or subsurface sources, and the natural and human factors that influence the quality of both are considered briefly here. Water quality refers to physical properties such as taste, odor, "hardness," and appearance. These vary depending upon the amount of dissolved and suspended mineral, organic, and waste materials and gases, rate of flow, temperature, and other factors.

Subsurface water tends to have a higher dissolved mineral load and a more consistent chemical quality than surface water. Water has been called an inert solvent because it can dissolve many substances and still remain water; thus it is available for use again. In fact, water is so effective a solvent that even raindrops contain dissolved salts. Larger than average amounts of dissolved calcium and magnesium make water hard (i.e., soap does not readily lather) and are the main cause of the deposits that tend to coat boilers, hot-water heaters, and plumbing systems. According to a U. S. Geological Survey report, hard water roughens clothes as well as hands and costs the average family in the United States about $170 each year by shortening fabric life. Subsurface water in limestone-dolomite regions tends to be hard.

Limited amounts of wastes can be added to both surface and subsurface waters without causing pollution, which has been defined simply as "any impairment in water quality that makes the water unsuitable for beneficial use." Water dilutes the wastes, organic matter is oxidized, and most pollutants are ameliorized; e.g., in the middle 1960's a tributary carried into the Ohio River each year the equivalent of 200,000 tons of sulfuric acid, the effects of which were neutralized after 170 miles of downstream flow. Waste organic matter such as sewage can be decomposed and made unobjectionable by aerobic bacteria, but an adequate oxygen supply is needed.

Water that has been used and dumped into a river may be re-used farther downstream, but only if the wastes contained in the water are treated properly. Since "not all of us can live upstream," water must and can be used again. A dirty shirt or towel is laundered, not thrown away—so it can be with water, if the wastes are treated properly (Fig. 1–9).

Aeration, sedimentation, filtration, chlorination, and oxidation of harmful organic matter are some of the steps involved in the treatment of water. Experiments have shown that polluted water can be purified by slow percolation through rocks and sediments, if the openings are small enough to afford adequate filtration. Sewage-polluted river water has been purified by slow lateral filtration after moving through several hundred feet of sand. The openings among sand grains are large enough to be permeable, yet small enough to purify the water: the harmful bacteria are filtered out during percolation, are destroyed by oxidation, or are assimilated by other organisms. However, dissolved chemical wastes are a problem because percolation alone cannot remove them. This is particularly serious during dry spells when the volume of water in a river is too small to provide sufficient dilution of the wastes.

Increasing Usable Water Supplies

Usable supplies of water, both surface and subsurface, may be conserved in many ways, and water may be used and re-used with very great savings if circulation is arranged according to purity requirements—the water is used first where it must be cleanest and last where purity is less important. Thus, the Fontana steel plant of California consumes about 1400 gallons of water per ton of steel instead of the 65,000 gallons more common elsewhere.

The presence or absence of vegetation has great and controversial effects upon water supplies. Unusable vegetation in seventeen western states may absorb annually an amount of water equivalent to 1½ times the yearly flow of the Colorado River. How much of this could be saved by the destruction of the waste vegetation without causing harmful erosion is uncertain, but the disagreement that has occurred concerning the uses of Colorado River water emphasizes the importance of saving even a fraction of it.

Evaporation and transpiration together use up nearly three-quarters of the total precipitation in the United States—21 inches of the 30-inch total. As yet man has done little to prevent this loss, and perhaps there is little that he can do. As an indication of the quantities involved, remember that on a hot summer afternoon in the United States, the amount of water being added to the atmosphere by evaporation and transpiration from all available sources probably exceeds by ten times the maximum flow of the Mississippi. Experiments in which a thin film of a plastic material was sprayed on the surface of a reservoir to reduce evaporation suggest one possible solution.

Contrary to many opinions, forestation does not always increase usable supplies of water. Experiments performed in certain areas in Colorado and California have shown that deforestation has actually increased the total quantity of surface and subsurface water available in the test area, because the very large amounts used by the growth of vegetation each year are saved. Snow also falls to the ground that otherwise might collect on trees and sublimate. In test basins, as natural reforestation occurred, the total runoff decreased, suggesting an increased recharge to the ground-water supply as well as increased transpiration.

Weather modification (see p. 708) has met with some success and may also provide a method of increasing our supply of water.

Fresh Water from the Oceans

A really outstanding opportunity to increase fresh-water supplies involves the conversion of ocean or brackish water into fresh water by decreasing the dissolved salt content. For coastal areas, this would mean a practically limitless supply, and would be important even inland where much brackish water exists. Conversion of salt water into fresh water has been possible for some time; the feasibility is a matter of cost. Can huge volumes of ocean water be desalted at costs comparable to present costs for fresh water?

Here the figures cited have commonly been distorted. Thirty cents per 1000 gallons is estimated as an average water-supply cost in the United States, but this is for city water at the faucet, which constitutes only a small fraction of the total water used. Costs for water used in irrigation and industry are commonly less than 10 cents per 1000 gallons—sometimes as low as 1 cent per 1000 gallons. Conversion of salt water for household and municipal use for water-short coastal areas seems economically feasible, but substitution of it as a source for the vast quantities of cheap water used in irrigation and industry is much less promising—the very great cost of distributing the water must be added to the desalting cost. However, possibilities do exist for very large-scale, nuclear-generator plants to produce and sell electricity and desalted water in dry land regions adjacent to an ocean. Other sources and projects seem potentially more productive: manipulation of upland vegetation, decrease and control of pollution, sewage treatment, reservoir storage and evaporation control, increased ground-water recharge, and the re-use of water. Large supplies of ground water and surface water remain untapped in the United States today and represent potential supplies.

The Problem of Seabrook Farms

Some years ago Seabrook Farms in southern New Jersey faced the daily problem of getting rid of as much as 10 million gallons of water polluted with vegetable scraps. Except for some dissolved sugar and suspended starch, the Seabrook farmers removed all foreign matter by filtering and chlorinating the water. However, objections were raised to the addition of this material to the local water supply. The farmers then tried to return the water to the earth in unfarmed acres. They started a spraying operation on an unused field covered with a sparse red clover crop. It soon changed the surface to a sandy soup. Next the sprayers were moved to the edge of a scrubby white-oak forest. For two days the ground soaked up water as fast as it was sprayed on Then 50 inches of water were poured on in the next 10 hours. Still the earth absorbed all of the water. The forest floor had the capacity to soak up almost limitless amounts of water, which filtered down through alternating layers of sand and gravel to some black muck at a depth of about 200 feet.

Seabrook Farms then solved its water disposal problem by scattering rotary sprayers strategically through the woods. Their precipitation averaged more than 50 inches per week and probably made this the wettest forest in the world. Yet the forest seemed no wetter than before except for droplets of water glittering on the leaves. It did not become a swamp or a lake. Readings at observational wells have shown an upward movement of the water table, but enough springs have developed along hillsides to establish a steady state condition. Drinkable water is flowing out of the woods as fast as the waste water is being pumped in. Experiments have indicated that this general technique can also be applied in suitable areas to reclaim water from sewage-disposal plants.

Long Island's Water Supply

Long Island furnishes a good example of intensive use of ground water by many people in a relatively small area. Long Island is separated from the rest of the Atlantic Coastal Plain and forms an island about 115 miles long by 20 miles wide with a maximum altitude of 420 feet. The island is underlain by clays, sands, and gravels which dip gently seaward and contain several

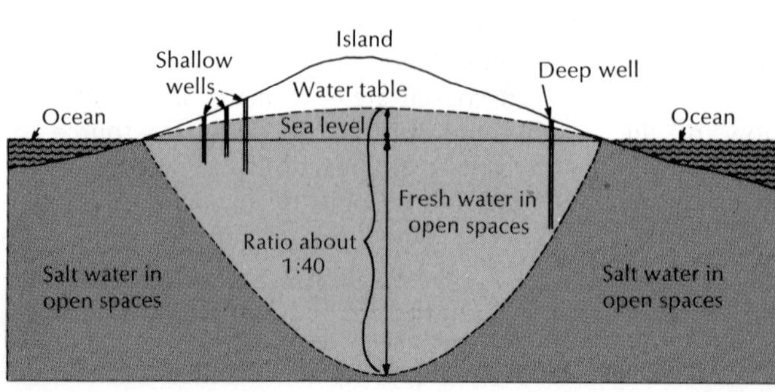

8-19. Fresh vs. salt water beneath the surface of a small island underlain by permeable homogeneous materials. As the area became settled, many shallow wells were drilled, and great volumes of water were used. This lowered the water table enough to let salt water enter the deep well. Although a cone of depression commonly does not extend more than a few hundred yards from a well, it may extend for several miles during heavy pumping in highly permeable sediments.

water-bearing permeable beds. Ground water recharge comes from precipitation, which averages 40 inches of rainfall per year. An estimated one-third to one-half of this rainfall becomes ground water. Approximately one-half of the water used on Long Island is supplied by ground water; the other half comes from the mainland via pipes and pumps.

Fresh ground water extends for considerable depths below sea level in interstices in sediments and rocks and floats upon the heavier salt water beneath it, although density differences are slight. An equilibrium between the fresh and salt ground water is commonly reached under the following conditions. If fresh ground water in a certain area extends about 1 foot above sea level, then fresh ground water probably extends about 40 feet beneath sea level in this same area (Fig. 8–19). This 1-to-40 ratio indicates that if the water table is 20 feet above sea level at one point, fresh water may extend to a depth of 800 feet below sea level at that same point. However, if the water table is lowered 1 foot by pumping, the bottom of the fresh-water zone may rise 40 feet. Consequently, excessive withdrawal results in contamination by salt water of the deepest wells, and this has occurred on Long Island. Conservation practices and induced infiltration on the island crest have partially corrected this situation (Fig. 8–20).

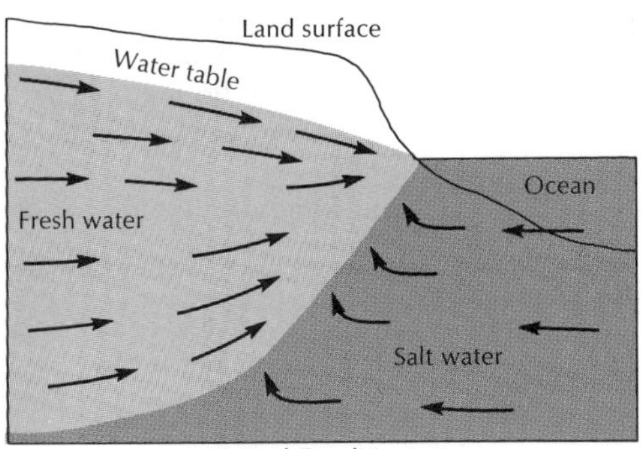

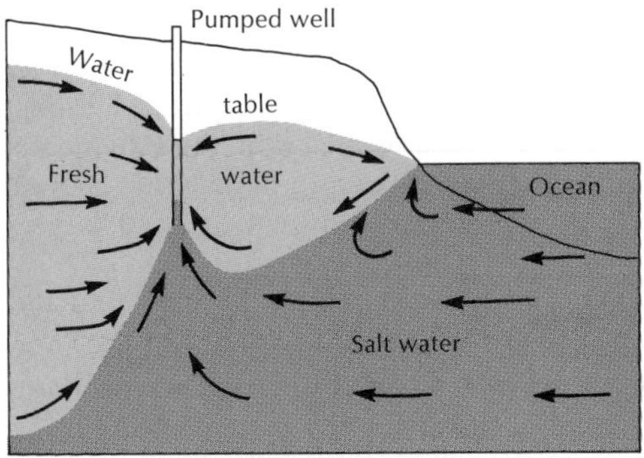

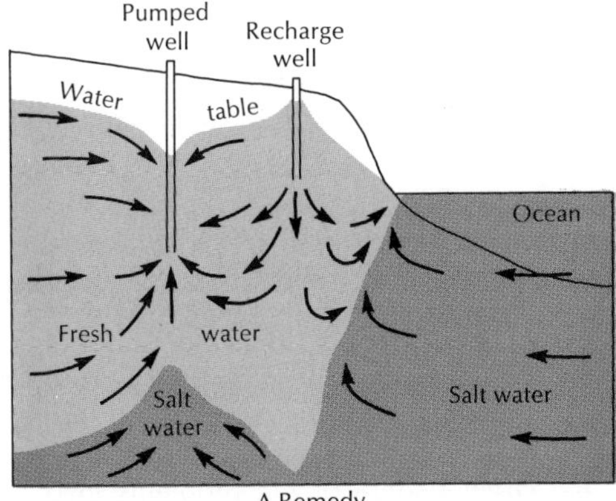

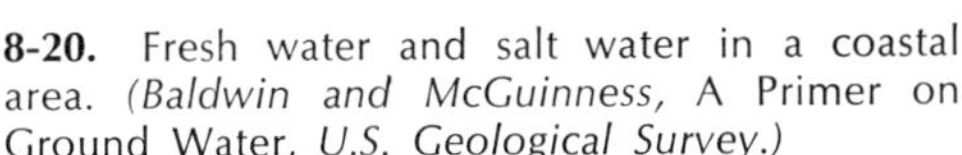
8-20. Fresh water and salt water in a coastal area. *(Baldwin and McGuinness,* A Primer on Ground Water, *U.S. Geological Survey.)*

Exercises and Questions for Chapter 8

1. What is meant by the water table? Why does its depth beneath the surface vary both areally and seasonally?

(A) In a particular location in an area with a climate similar to that of New York, when would you expect the water table to be farthest from the surface during a year? Why?
(B) Discuss capillarity as a factor in the movement of water beneath the surface.
(C) Describe some of the types of openings that may occur in rocks.
(D) Some materials have low permeabilities and yet are quite porous. Why?

2. Describe the shapes of the cones of depression that will develop under each of the following conditions.
   (A) A well located in permeable materials is pumped at a moderate rate.
   (B) A well located in materials of a rather low permeability is pumped at a rapid rate.
3. Artesian systems:
   (A) Describe the necessary conditions.
   (B) How does an artesian well differ from an ordinary well?
   (C) What is meant by the statement that one drinks "fossil water" from a certain artesian well?
   (D) How can tritium be used to measure the rate of movement in an artesian system?
4. Describe two different types of conditions that could cause neighboring wells to vary considerably in depth in order to obtain adequate supplies of ground water.
5. Water conservation:
   (A) When the water supply of a town or city becomes low during a drought, a common practice to conserve water is that of not serving a glass of water with a meal in a restaurant. Discuss the usefulness of this practice.
   (B) What is meant by the "mining" of ground water? What effect does this tend to have upon an aquifer?
   (C) Describe ways in which usable supplies of ground water may be increased.
   (D) Polluted river water moves laterally through sand in one area, through fractured limestone in another area, and through well sorted pebbles in a third area. In which area is the water likely to become drinkable? Why?
6. Why are dowsers commonly able to locate supplies of ground water?
7. Why are deposits around hot springs commonly thicker than those around cold springs?
8. Discuss some of the problems involved in obtaining a supply of fresh ground water on a small island in the ocean or in a coastal area.
9. Describe some of the topographic differences that can develop in two regions in warm humid climates if one area is underlain directly by soluble rocks and the other is not.
10. What is a geyser and what is a probable explanation for the eruption of a geyser?

9

Glaciation

The story of the recent Ice Age (Pleistocene) captures the imagination and was worked out by the many geologists who have traced the courses of the glaciers (Fig. 9–1). At maximum extent, glacier ice buried almost a third of the Earth's land surface and covered about half of the states in the United States—some completely, others partially. Evidence of the former presence of the ice is abundant, widespread, and clear. It is little affected by the weathering, erosion, burial, and metamorphism which have obscured so much of the older parts of the geologic record.

Approximately 10% of the Earth's land-surface is still covered by glacier ice (Fig. 9–2). The Antarctic Ice Sheet covers an area of about 5 million square miles and includes more than 90% of the total volume of existing glacial ice (equals 1½ times the area of the 48 conterminous United States). The Greenland Ice Sheet covers about 670,000 square miles and includes 8% of the total volume. Nearly all of Antarctica is covered by ice, which is about 14,000 feet thick in one place, and shelf ice is present in the ocean along its edge. The mean altitude of Antarctica's bedrock surface beneath the ice sheet may exceed 1 mile, but parts are depressed well below sea level.

Neither of these ice sheets was probably much more extensive when glaciers were most widespread during the recent Ice Age. Therefore, glaciers have shrunk the most or disappeared entirely in the upper middle latitudes, especially in the Northern Hemisphere.

A *glacier* has been defined as a mass of ice that formed from compacted, recrystallized snow and refrozen meltwater, which is moving or has moved, and which lies entirely or partly on land. Valley glaciers and ice sheets are the two most important kinds, and piedmont glaciers are transitional between them. A *valley glacier* moves down a channel previously eroded by a stream. It may be wide, thick, and tens of miles long, with numerous tributary glaciers, or it may be quite narrow and short. Commonly the ice fills the entire lower part of a valley from wall to wall.

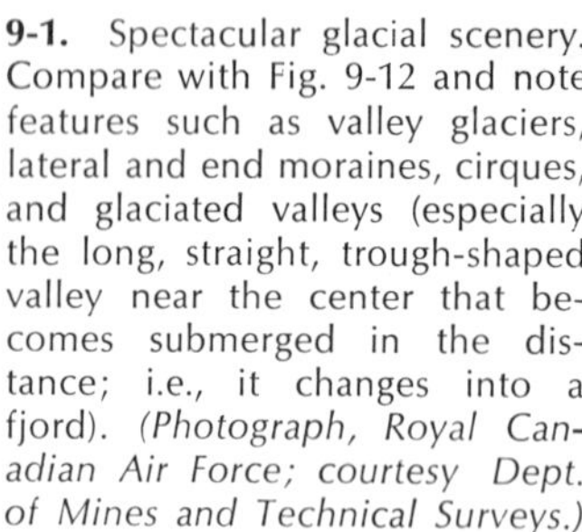

9-1. Spectacular glacial scenery. Compare with Fig. 9-12 and note features such as valley glaciers, lateral and end moraines, cirques, and glaciated valleys (especially the long, straight, trough-shaped valley near the center that becomes submerged in the distance; i.e., it changes into a fjord). *(Photograph, Royal Canadian Air Force; courtesy Dept. of Mines and Technical Surveys.)*

9-2. An expanded-foot glacier, Victoria Land, Antarctica. Its vertical terminal walls are about ten meters high, and the relative scarcity of morainal debris along its margins suggests advance rather than recession or stagnation. *(Courtesy U.S. Geological Survey, 1 Jan. 1958.)*

In contrast, an *ice sheet* is a very extensive mass of ice that spreads radially outward from a central area and rests like a blanket upon the surface; it is not confined to a single channel. Continental ice sheets may be a mile or more thick and completely bury millions of square miles of the Earth's surface, except for an occasional projecting mountain peak. The term *ice cap* is applied to a small ice sheet.

A *piedmont glacier* (Fig. 9–3) is gradational between a valley glacier and an ice sheet and forms on lower ground along the base of a mountain by the piling up of ice from one or more valley glaciers. If growth continues, both areally and vertically, and if a number of piedmont glaciers coalesce, an ice sheet may eventually be formed.

The Ice Age may have begun some 2 to 3 million years ago, and it may not yet have ended. During the Ice Age, great changes took place above, at, and below the Earth's surface both in the glaciated areas and outside them. The changes involved climates, animals and plants, crustal movements, sea water, erosion, and deposition. The Pleistocene (Greek: "most recent") epoch is the name given by geologists to the time in which these changes occurred. From now on, we shall speak of Pleistocene glaciation rather than of the Ice Age, the popular term.

9-3. A piedmont glacier in Alaska (Malaspina, covers about 800 square miles). Stagnant ice occurs along its outer margin, and ice flowing into it from tributary valleys is squeezed into complex folds made conspicuous by rock debris (dark areas). *(Courtesy U.S. Air Force: N 59° 57′, W 140° 33′.)*

Evidence of Glaciation

Louis Agassiz, a Swiss naturalist (1807–1873), is often credited with the development of the glacial concept about a hundred years ago; but a number of Agassiz's contemporaries, scientists of still earlier days, and many observant individuals all deserve to share the credit—as is often true of key discoveries in science.

The basic idea came from people who lived in areas of existing valley glaciers. Erosion and deposition by the ice created features which they could observe and compare with similar features located beyond the present ice margins. From the plain hint that glaciers once extended farther down their valleys, there slowly developed the concept of continental glaciation. But this idea was accepted only after much argument and discussion.

At first, Agassiz was skeptical whether glacier ice had once been so widespread, and he studied existing glaciers and supposedly glaciated areas with the intention of disproving the whole idea. However, the field evidence was so convincing that he became the leading figure in spreading and developing the glacial concept.

There were rival theories. British geologists had noticed large foreign boulders that were scattered widely, some at higher altitudes than their distant sources. They had observed scratched and polished bedrock surfaces (Fig. 9–4) and had studied wide-

9-4. A Pleistocene glacier left this grooved, striated, and polished rock wall at the side of its valley in Glacier Creek, Flathead National Forest, Montana. *(Photo by Asahel Curtis; courtesy U.S. Forest Service.)*

spread nonstratified deposits of boulders, gravel, sand, and mud (Fig. 9–5). These were quite unlike the sorted layers of sand and silt they could see being deposited by streams. Furthermore, these deposits occurred on hilltops as well as in low areas.

9-5. Till is material that has been transported by glaciers and dropped directly from the ice into unsorted piles; it is unstratified glacial drift. *(Photograph G. Termier.)*

Such unusual phenomena called for an unusual answer. Being familiar with the sea and not with glaciers, the British geologists theorized that a great flood had carried the huge boulders from their source area and had scratched the bedrock surfaces. The unsorted sedimentary deposits, which they termed "drift," had thus been picked up and dumped by a great rush of waters—perhaps the Noachian deluge. Similar ideas had also developed in other countries.

Worldwide Effects of the Recent Ice Age*

Glaciation has occurred several times during the Earth's geological history, but that of the Pleistocene is the most recent and provides the clearest record. Still extant glaciers were then larger, and regions which today are without ice were then covered by it. Three huge ice sheets formed more or less simultaneously in the Northern Hemisphere. (1) At its maximum size, the North American Ice Sheet covered about 5 million square miles and extended southward to Long Island and the channels of the Ohio and Missouri Rivers (Fig. 9–6). It joined the glacier complex that covered parts of the Rockies and the West Coast. (2) The smaller Scandinavian Ice Sheet once extended from northwestern Europe, through Denmark, to the northern parts of Germany and European Russia. (3) A less well-known Siberian Ice Sheet also existed. With the exception of Antarctica, ice sheets of comparable size were not present in the Southern Hemisphere.

Glacial and Interglacial Ages

Ice sheets and other glaciers probably made four major advances during the Pleistocene (Fig. 9–7). Each advance was followed by a time of major retreat or shrinkage of glacier ice, but advances and retreats of lesser mag-

* R. F. Flint, *Glacial and Pleistocene Geology,* Wiley, New York, 1957, Ch. 1.

9-6. Map of North America showing major existing glaciers and the area covered by glacier ice at one time or another during the Pleistocene. *(R. W. Fairbridge, ed.,* The Encyclopedia of Geomorphology, *Van Nostrand Reinhold, 1968.)*

nitude were superposed on the four main ones and have led to a difference of opinion concerning the total number. A glacial age occurred near the beginning of the Pleistocene epoch and was followed by an interglacial age. Then in succession came a second glacial age, a second interglacial age, a third glacial age, a third interglacial age, a fourth glacial age, and fnally the present, which may or may not be a fourth interglacial age.

The interglacial ages probably lasted much longer than the glacial ages, and conditions during parts of the interglacials were probably somewhat similar to present conditions. In fact, during at least one interglacial age, temperatures seem to have been higher than they are today. The latest glacial maximum apparently occurred about 16 to 19 thousand years ago, but this was followed by a number of lesser fluctuations. Temperatures apparently increased rather abruptly about 15,000 years ago, and glaciers began to shrink relatively rapidly (however, a date of 11,000 years has also been used here).

Volume Changes in Sea Water

Glacier ice is metamorphosed snow which has fallen as part of the water cycle. Because the total amount of water at the Earth's surface in the Pleistocene probably remained about constant, the volume of ocean water was reduced by the amount that became frozen on the lands as ice. Thus sea level fluctuated as the volume of land ice fluctuated, and it may have been 300 to 500 feet below its present level during the maximum extent of the glaciers. On the other hand, complete melting of all existing glacier ice might add a layer of water more than 200 feet thick to the oceans. However, sea level might not rise by this amount because isostatic adjustments (p. 165) should occur. The Earth's surface under the oceans should sink beneath the extra load (perhaps by nearly

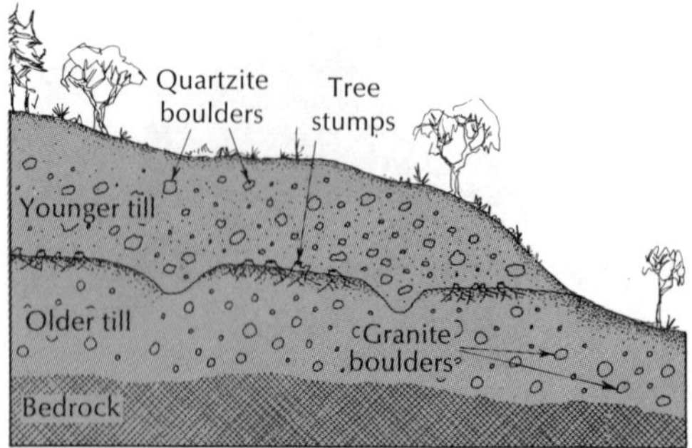

9-7. Evidence indicating multiple glaciation: (1) Granite boulders predominate in the older till, whereas boulders of quartzite predominate in the younger till. Evidently the glacier came from two somewhat different directions, and thus picked up pieces of a different kind of bedrock, on each of its two trips to this area. Near its outer margin a glacier may ride over regolith without much erosion. (2) Stumps of trees at the top of the older deposit indicate that sufficient time elapsed between glaciations for soil to develop and a forest to grow. (3) Two V-shaped valleys are shown eroded in the older glacial deposit. These are filled by the younger material. (4) The soil is thickest on the right where the younger material does not rest upon the older. In this location weathering has been producing soil for a longer time: burial under the younger glacial deposit stopped the process elsewhere. *(Modified from Leet and Judson.)*

one-third the thickness of the added water) and the lands should rise (maximum amounts in Antarctica and Greenland).

Various kinds of geologic evidence indicate that sea level fluctuated during the Pleistocene, although exact amounts are uncertain; e.g., the channel of the Hudson River can be traced seaward on the Atlantic floor for some 90 miles to a depth of about 240 feet. Presumably such a channel was cut when sea level was lower. Similar channel continuations occur in association with other rivers. Furthermore, near their mouths, rivers such as the Hudson flow on thick piles of sediments; their bedrock floors are now as much as 200 feet or more below sea level.

An erosional bench associated with shallow-water fossils has recently been discovered off the eastern United States in water nearly 500 feet deep; presumably it was formed by waves along a coast at sea level. This suggests, but does not prove, that a 500-foot change of sea level may have occurred (however, the sea floor might have moved downward locally).

In another approach to this problem, an estimate is made of the total volume of ice that existed on the lands when the glaciers were largest. One then calculates how much sea level would have to fall to supply this volume of ice.

During one or more of the interglacial ages in the Pleistocene, sea level was also higher than it is now, but correlation from one continent to another or even along the coast of one continent is complicated by local movements of the crust. As an illustration, two nearly parallel strandlines, now above sea level. extend in a north-south direction along the east coast of the United States. Like the dirt line in a tub, a *strandline* is the line along which a standing body of water meets the land, and features such as wave-cut cliffs, deltas, and beaches may be present to identify it. One strandline is about 20 to 30 feet above sea level; the other, and apparently older, is 90 to 100 feet above sea level.

According to Fairbridge* a connection may exist between the calamitous floods and deluges that are part of the legends and folklore of many ancient peoples and Pleistocene climatic and glacial changes. Such legends might have been inspired by the experiences of primitive peoples who lived near the edge of the sea. A major and relatively abrupt increase in temperature following the fourth main glacial age may have resulted in two types of flood: one type involves a relatively rapid rise in sea level, whereas the other is associated with times of very high water in river valleys produced by increased rainfall and rapidly melting snows. The rise in sea level may have culminated some 6000 years ago with a rate that at times approached 30 feet per century. If floods of the two types occurred

* Rhodes W. Fairbridge, "The Changing Level of the Sea," *Scientific American,* May 1960.

simultaneously in a seacoast area, there would be so drastic a change in man's environment as to be long remembered. It follows, also, that the best sites for archaeological exploration may well occur beneath the surface of the sea along the inner sections of the continental shelf, because the seacoast was a favored location for ancient peoples.

Crustal Movements Caused by Growth and Shrinkage of an Ice Sheet

The Earth's crust sags beneath the tremendous weight of an ice sheet, and this downwarping probably occurs simultaneously with a slow plastic flow of rock materials outward from this overloaded area (see isostatic adjustment, p. 166). At the depths where this flow occurs (probably within 100 miles of the surface), temperatures are high enough for rocks to be deformed slowly without breaking. Apparently the weight of a growing ice sheet squeezes rock material from beneath it, and the space thus made available permits the Earth's surface to bend downward. As an ice sheet melts, the crust warps upward, and rock material at depth returns slowly under the glaciated area. Similar warping of the crust was associated with each advance and retreat of each ice sheet. Because the ice retreated only recently in terms of geologic time, the Earth's surface is still rising slowly in these areas. Similar crustal movements are to be expected beneath the oceans.

Evidence of crustal movement as a direct result of glacial loading and unloading is especially well exposed in Scandinavia. First, the elliptically shaped area known to have been glaciated coincides closely with the area in which uplift has occurred and continues today. Second, the maximum rate of uplift occurs in the central portion of this area, where the ice apparently was thickest and the crust was depressed the most. The rate is about 3 feet per century at present, but seems to have been faster in the past. This is also the location where the greatest amount of uplift has already taken place—perhaps 900 feet or more—and where the largest upward movement is still to come—possibly another 900 feet or so. Thus the center of a dome-shaped upwarp coincides closely with the center of the former ice sheet.

Estimates concerning the amount of uplift still to come depend upon assumptions concerning the strength of the crust, the density of the displaced rock far below the surface, and the thickness of the ice sheet. Although these are not known with any precision, the relative specific gravities of the ice (less than 1) and the displaced rock (probably about 3.4) indicate that several thousand feet of ice are needed to depress the surface a thousand feet. How much ice is needed to begin the process is uncertain, but the Earth's crust can support the extra load of smaller glaciers without noticeable sagging.

However, the surface beneath Lake Mead is reported to have sagged as much as 5 inches during the first six years that followed the filling of this large reservoir with water. The affected area is about 30 miles in diameter and is now the site of more frequent earthquakes than occurred prior to the development of the reservoir. An estimated 12 billion tons of water were added to this small area in a short time.

The picture is less complete for crustal movements caused by the North American Ice Sheet. The Earth's surface evidently sagged hundreds of feet, deepest in the centrally located Hudson Bay region. As the front of the ice sheet retreated toward the north, a number of large lakes formed in this depressed area in the southern part of Canada and the northern part of the United States. Upward movement of the surface apparently did not take place immediately upon release of the load. Around the shores of these lakes there formed surf-eroded cliffs, beaches, and other lake-level features (strandlines). These once-level shore lines have since been bent upward progressively, and the greatest uplift has occurred in the north where the center of the ice sheet was formerly located. Uplift continues today.

After the ice front had retreated north of the St. Lawrence Valley, sea water entered this depressed area and spread into the Champlain Valley. Because of this and subsequent uplift, sediments containing marine shells and the bones of whales have been found a few hundred feet above sea level near the Vermont-Quebec border.

Pluvial Lakes in Nonglacial Areas

Climatic changes also occurred during the Pleistocene in nonglaciated areas because climatic belts were shifted equatorward and narrowed. Today areas such as Utah and Nevada are semiarid and arid; but during the glacial ages of the Pleistocene, they were

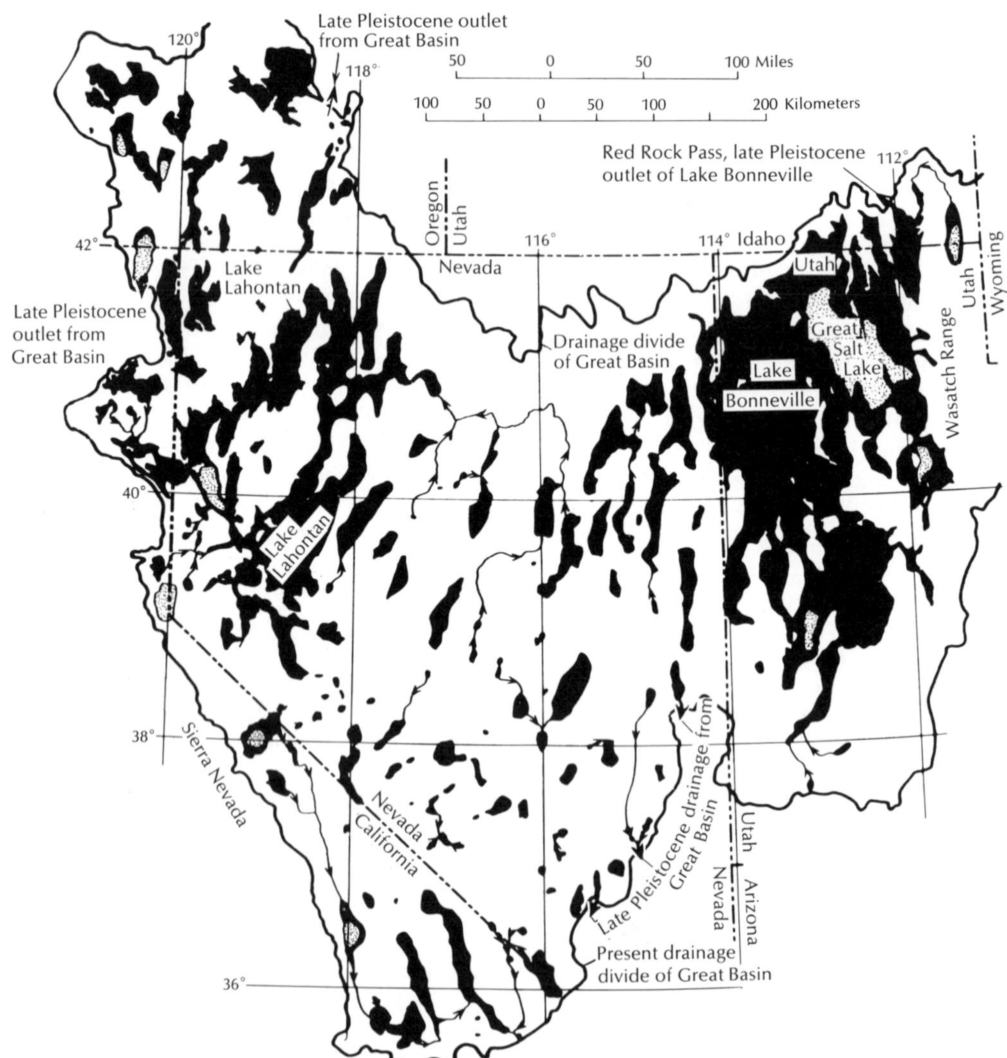

9-8. Pluvial lakes at their maximum extent within the Great Basin during the late Pleistocene. Lake Bonneville covered more than 20,000 square miles and exceeded 1000 feet in depth. Great Salt Lake now occupies a small portion of its former floor. The weight of the water was so great that the crust beneath it sagged some tens of feet. *(R. W. Fairbridge, ed.,* The Encyclopedia of Geomorphology, *Van Nostrand Reinhold,* 1968.)

cooler and moister and had less evaporation. Accordingly, present-day lakes were much larger during the glacial ages, and lakes existed in basins that are now dry. We can observe the sediments and strandlines of these now-vanished lakes, which have been called *pluvial lakes* (Latin: "of rain") because they formed at times of increased rainfall and reduced evaporation (Fig. 9–8). Apparently these pluvial lakes grew and shrank several times, corresponding presumably to the climatic changes that governed the advances and retreats of the ice sheets (Fig. 9–9).

Migrations of Animals and Plants

As the Pleistocene ice sheets gradually advanced, animals and plants apparently migrated to warmer climates; but this must have been a slow process, each generation inhabiting an area a little nearer to the equator than its predecessor. As evidence, fossil reindeer and the woolly mammoth have been found in southern New England, walrus along the Georgia coast, white spruce in Louisiana, and musk oxen in Arkansas and Texas. However, some of these organisms may have lived under climatic conditions different from those of their present-day descendants, and plants are more reliable indications of climate than large mammals. Musk oxen in Texas do not imply arctic conditions there.

As the ice fronts retreated, descendants of these animals and plants could return to former habitats, again a process stretching across the centuries. Therefore, in sediments deposited at any given location, assemblages of fossil animals and plants indicating a cool climate may be found above or below fossil assemblages indicating a warm climate.

Conclusions concerning the migrations of flowering plants and climatic changes can be drawn from a study of fossil pollen grains and other spores (palynology*). Pollen grains are the microspores produced in flowering plants as part of the reproductive process. Countless pollen grains are distributed widely during the spring and summer by air and water currents, and they form a veritable pollen rain upon the Earth's surface that is perhaps most familiar as a yellowish scum on ponds and lakes.

* Davis, M. B., "Palynology and Environmental History during the Quaternary Period," *American Scientist, Autumn 1969.*

9-9. Former levels of Lake Bonneville. Bars, beaches, deltas, terraces, and cliffs formed around the margins of the lake when it remained for a time at one level. After these features had formed at one altitude, the water level of the lake gradually changed. Eventually it became stable at a new altitude, and a new set of shoreline features then formed at this new level. *(Courtesy U.S. Geological Survey.)*

These tiny, nearly indestructible organic particles eventually settle to the bottom of a pond along with other sediment and organic matter in a process that is repeated year by year. Since the pollen grains remain intact and can be identified, they provide a method of determining which flowering plants were abundant in a certain area at a certain time in the past. A core is taken of the sedimentary layers that piled up year by year in the bottom of a bog or elsewhere. The pollen is separated from the mineral matter at different horizons, and the individual grains from different plants are identified and counted.

The ages of the pollen grains can be determined by using the carbon-14 method on the enclosing organic matter in each horizon. Correlations and comparisons can then be made among cores taken at different places. The advance or retreat of a particular type of forest across a wide region as the climate of this region gradually changed can be mapped and dated.

At any one level in any one area, certain types of pollen grains predominate, and it is generally assumed that the parent plants of these grains were most abundant in that area at that time in the past. However, relationships are complex because some plants are more prolific producers of pollen than others, and the pollen grains of some plants are better airborne travelers than others.

Study of fossil pollen and spores present in clays deposited in the deep ocean (i.e., in cores extracted from the sea floor) provides a promising new technique for advancing our knowledge of the stratigraphy and chronology of deep-sea sediments. Reversals of the Earth's magnetic field also provide a new method for dating deep-sea sediments (p. 299).

Instruments can now extract cores as long as 70 feet from the ocean floors at depths of thousands of feet (Fig. 14–12; also, see deep drilling program, p. 374), and these cores have shown alternations of the remains of warm- and cool-water animals. The animals lived near the surface and their discarded shells sank to the bottom. The alternations reflect temperature changes related to the four major glacial and interglacial ages as well as to lesser fluctuations.

Two isotopes of oxygen with mass numbers of 16 and 18 occur in water, and the proportion of one to the other varies with the temperature. A larger percentage of the oxygen-18 isotope occurs in warmer waters, because the lighter oxygen-16 isotope escapes into the atmosphere more readily at higher temperatures. Therefore, when an organism extracts calcium carbonate from sea water to construct a shell, the proportion of oxygen-16 to oxygen-18 in the shell records the approximate temperature in that part of the ocean at that time.

This method was applied to a 150-million-year-old belemnite fossil and indicated that the animal was born in the summer and died in the spring four years later. Seasonal temperature changes caused slight variations in the proportions of the oxygen isotopes, and these showed up in the skeletal layers secreted during the different seasons.

In the tropics, the surface temperature of the ocean apparently varied about 6 Celsius degrees (11 Fahrenheit degrees) between the Pleistocene glacial and interglacial ages—from about 5°C colder than it is today to about 1°C warmer. Averaged over the entire Earth, mean annual air temperatures may have fluctuated by similar amounts, but temperature changes were apparently somewhat greater over land at higher latitudes.

Summary of Major Worldwide Effects of Pleistocene Glaciation

Pleistocene glaciation originated when continents were more extensive and mountainous than they had been for much of the geologic past. As temperatures dropped, more precipitation occurred as snow, and less snow melted each summer. Three huge ice sheets developed in the Northern Hemisphere, and many smaller glaciers formed or grew larger in other areas. The major times of glacier growth and retreat appear to have been synchronous in all parts of the Earth that were affected by glaciation.

As the ice sheets grew, the following changes occurred gradually: the volume of sea water became smaller and sea level fell, the crust beneath the ice sheets warped downward, rainfall increased in areas beyond the margins of the ice sheets to form large pluvial lakes, and animals and plants migrated equatorward to warmer climates.

Temperatures then rose, the ice sheets dwindled in size, and a corresponding reversal occurred in the changes listed above. At least four major glacial ages occurred, and fluctuations of a lesser magnitude also took place. In the glaciated areas, characteristic features of erosion and deposition were formed; however, some of the features that formed during one advance were presumably destroyed during a subsequent advance. These effects continue today. Because a large ice sheet requires many years to disappear, a glacial age can be said to end at different times in different places.

After the last glacial maximum some 16,000 to 19,000 years ago, temperatures apparently became warmer until about 6000 years ago when it was probably warmer than today. A number of middle latitude valley glaciers may even have disappeared at this time. Since this warmer interval, it has again become cooler. Accumulations of snow and ice have again formed valley glaciers in these middle latitude areas or have caused those that had not disappeared to grow larger.

Lesser fluctuations have followed and continue today; e.g., some areas which were inhabited during the Middle Ages have since been covered by ice. Today the ice is retreating from these areas and exposing glacier-buried buildings. Evidently the climate was milder during the Middle Ages but then cooled. Around the beginning of the seventeenth century, temperatures seem to have dropped, and glaciers began to expand. This change was reversed during the last century, and a new cooler trend may even now be taking place.

What of the future? The fluctuations of the last several millennia certainly emphasize the difficulty or impossibility of long-range forecasting.

Glaciers Grow and Shrink

The snow-covered parts of a region are, of course, more extensive in winter than in summer, and in some areas on the Earth's surface, not all of a winter's snowfall melts or sublimates during the succeeding summer (Fig. 9–10). In the polar regions, such places occur at every altitude, but in lower latitudes they occur only near the tops of high mountains. The *snowline* refers to the lowest altitude in a region that remains covered with snow throughout a year. Thus each year in certain parts of these areas, the snow piles higher and higher. During glacial ages in the middle latitudes, the snowline was probably some 3000 to 4000 feet lower than it is today.

The total weight of many snowfalls, combined with moisture from melting snow and rain, gradually compacts the snow at the bottom of a pile, squeezes out the air, and produces a granular texture similar to that of many snowdrifts during spring. Each snowflake tends to become a tiny ball of ice, in part because water molecules are transformed by sublimation or melting from the outer margins of ice crystals, where pressure is greatest, toward their centers where pressures are reduced, and the molecules again become ice. Continued compaction by more snowfalls changes the granular material to solid ice.

Commonly this ice is layered, evidence of its formation from many separate falls of snow. However, after movement begins, the stratification may be obscured by the recrystallization that accompanies flow. Snowflakes and tiny ice granules are thus analogous to sediments; stratified snow-ice that has undergone little downslope movement is akin to a sedimentary rock; and glacier ice, recrystallized into an interlocking texture, can be considered a metamorphic rock.

Dyson* has described two glaciers that have formed on Mount Katmai, a volcano on the Alaska Peninsula, since its catastrophic eruption in June of 1912. They may well be the Earth's newest glaciers. The summit of the volcano was destroyed at this time, and valley glaciers on its slopes were beheaded, their source of supply gone. However, portions of the newly formed crater wall were still above snowline. On these, probably within the next decade or two, two small glaciers have originated and grown larger. They have become several hundred feet thick, and one is nearly a mile in length.

A glacier moves as a unit by sliding along its bedrock surface and also internally by movements involving its interlocking grains. Such movements may be more concentrated toward the base, especially in thinner, colder glaciers. When sufficient ice has piled up on a slope, gravity moves it gradually to a lower altitude (Fig. 9–11). A thickness of 100 feet or more usually accumulates before movement occurs; how much depends upon the steepness of the slope, the temperature of the ice, and other factors. The weight and motion of glacier ice set up stresses which cause individual ice grains to grow in size, to twist and rotate, and to slide along planes of weakness in the crystals, much as a pack of cards slides when pushed by the hand. Perhaps skating illustrates how part of the flow takes place. A skater actually glides on a thin film of water which forms momentarily beneath the blades of the skates because of their pressure. When ice changes to water, the volume is reduced; therefore, pressure on ice promotes liquefaction. The water freezes again immediately after the pressure is removed, as when a snowball is made. Stresses in glacier ice cause momentary liquefaction, the transfer of water downslope, and immediate refreezing.

Another type of movement is especially prominent near the front of a glacier and is

* James L. Dyson, *The World of Ice*, Knopf, New York, 1962.

9-10. Snowline on Mount McKinley in Alaska. Note the braided stream and the alluvial fan in the lower right. *(Courtesy U.S. Air Force.)*

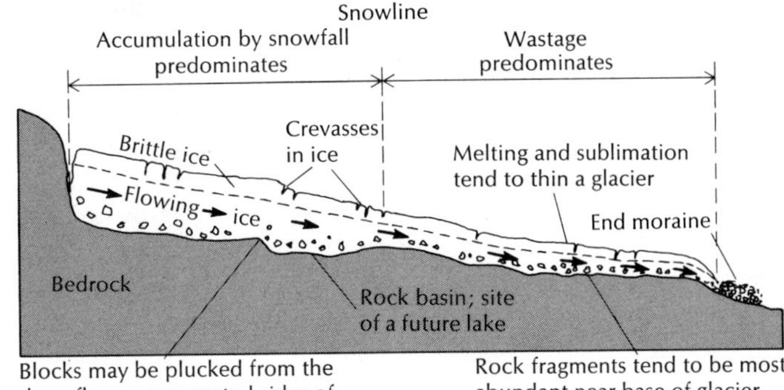

9-11. The snowline is the approximate boundary between the zone of nourishment and the zone of wastage in a valley glacier.

similar to faulting. Thin plates of ice are shoved forward and upward, one above the other, along surfaces that are concave upward. Part of this "faulting" may be caused by slower movement of the debris-loaded basal portion of the ice; the faster moving, cleaner ice just above slides along it.

Ice under great pressure at the bottom of a glacier moves plastically, but the upper part, some 100 feet or more in thickness, is too brittle and rigid to flow. It is carried by the flowing ice below. Ice cubes, we know, are brittle, but if ice cubes were piled high enough, the lower part would flow and the entire mass would move downslope. Even if the ice cubes were piled on a level surface, flow would still take place; the weight of the ice in the center would cause radial outward flow in the plastic ice near the bottom. Ice sheets and ice caps seem to move in this manner.

The rate of flow of glacier ice varies from a maximum measured rate of over 100 feet a day (even more during a glacial surge, Fig. 9–16) to a general average of a few feet to a few inches per day. The surface of a valley glacier moves more rapidly in the center than along its sides because less friction occurs there. This fact was first determined by placing a straight line of stakes across a valley glacier and measuring their subsequent movements. Agassiz made such measurements and learned that a glacier commonly shrinks less by upslope retreat of the front than by thinning, i.e., by melting and sublimation over the entire surface, especially at lower altitudes. He placed a line of stakes across a glacier in 5-foot holes in one year and found that they had toppled over by the next year. At least 5 feet of thinning had occurred. In his next experiment, Agassiz placed 18-foot stakes in holes 18 feet deep. In the following year, the ice thinned 10 feet; he found each stake projecting about 10 feet above the ice and the once straight line had changed into a crescent. If the ice thins sufficiently, the brittle zone extends to the bottom, and flow stops.

Glacial Erosion and Deposition

Plow, File, and Sled

Three short words succinctly describe the erosional and depositional work of glaciers: plow, file, and sled. Like a plow, a glacier pushes and shoves loosened debris ahead of it. The rock-studded bottom of a glacier is an effective file or rasp and polishes, scratches, and abrades the surfaces over which it moves. A glacier gathers debris from the regolith or bedrock in its path and from rock materials that slide upon it from valley walls, a process that is less important for ice sheets. Glaciers also pluck out blocks of bedrock loosened by the freezing and thawing of water in fractures beneath the ice; thus a glacier acts as a sled.

In Arctic areas, where some glaciers end at the water's edge in cliffs, the transported materials are exposed. Large blocks of ice may break off the end of a glacier (calving) to form icebergs which later melt and deposit their rock loads far from land in places ordinarily reached by only the finest of materials. The origin of such deposits—fine-grained strata containing an occasional large boulder—puzzled geologists for some time.

Blocks of rock transported by glaciers were once frozen firmly in the ice, and many were dragged under great pressure over bedrock. This abrasion may produce polished, scratched, and faceted surfaces on the blocks. Later if the ice melted temporarily around a block, it could rotate or overturn to a new position and then be given an additional smoothed surface. The scratches (*striations*; Fig. 7–10) on bedrock

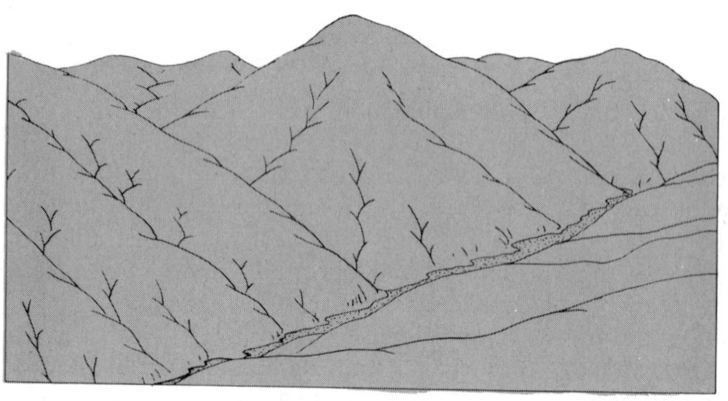

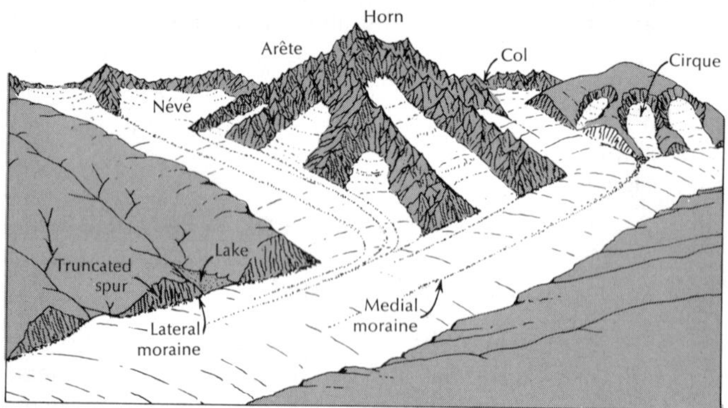

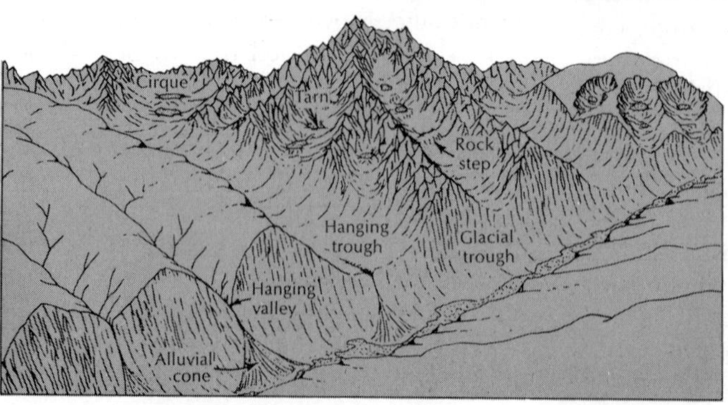

9-12. Land forms produced by valley glaciers. *Top, before* glaciation sets in, the region has smoothly rounded divides and narrow, V-shaped stream valleys. *Middle,* after glaciation has been in progress for thousands of years, new erosional forms are developed. *Bottom,* with the disappearance of the ice a system of glacial troughs is exposed. *(After W. M. Davis and A. K. Lobeck; reprinted from A. N. Strahler,* Physical Geography, *John Wiley & Sons, Inc.)*

9-13. Yosemite Valley, California. A former V-shaped valley may have been deepened by some 1500 feet and has been transformed by glacial erosion into a steep-sided U-shaped trough. *(Courtesy Santa Fe Railway.)*

are nearly parallel across very large areas (see a copy of the Glacial Map of North America). The finest scratches are probably formed by silt and sand grains held like an abrasive between a boulder and the bedrock.

Some very large blocks of bedrock have been picked up and carried by ice: e.g., a village is built on one in England, and quarries have been located in others. However, these are exceptionally large.

Features of a Glaciated Valley

Valley glaciers reshape their valleys by widening, deepening, and straightening them. In so doing the typical V-shaped cross section of a youthful mountain valley is altered to a steep-sided, flat-floored U shape (Fig. 9–12), and Yosemite Valley in California is a striking example (Fig. 9–13). If some sections of the bedrock floor of a valley are more jointed than other parts, or if they are underlain by weaker rocks, they will be eroded more, and these depressions may become lakes after the ice disappears. (Fig. 9–14).

Furthermore, constricted parts of a valley may be deepened more than wider parts because the ice flows more rapidly through them. This may explain the pronounced deepening of the bedrock floor of the Hudson River Valley between Newburgh and Peekskill, New York, where the river passes through the rugged Highlands of the Hudson. In the lowlands, both upstream and downstream from these cities, the bedrock floor is buried by about 300 feet of sediment. Within this constricted section, the sediment is two to three times thicker, and the bedrock floor actually forms an enclosed basin.

Tributary glaciers generally contain much less ice than the main valley glacier into which they flow, and as a result, they are unable to cut downward as rapidly. Their floors thus tend to be located at a higher level than the floor of the main valley (aided also by the retreat and steepening

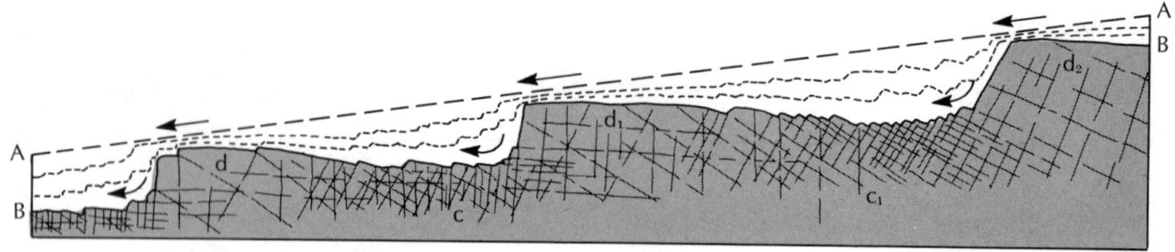

9-14. Development of a "glacial stairway." AA is a longitudinal profile along the valley floor prior to glaciation; BB is the profile after glaciation. Broken lines (short dashes) show successive in-between stages. Arrows show direction of ice movement. Bedrock was quarried more readily at c and c_1, where joints are closely spaced, than at d, d_1, and d_2, where joints are scarce *(Matthes, 1930).*

of its walls). When the ice eventually wastes away, such tributaries are commonly left hanging above the main valley, and beautiful waterfalls may occur at their mouths.

A large depression, shaped like half a bowl, is commonly located at the head of a glaciated valley and is called a *cirque* (pronounced "surk"). However, cirques may also develop along the side of a valley. A cirque has a steep headwall that is relatively free of talus at its base and may be half a mile in height. Its basin floor may contain a lake.

The development of a cirque is not entirely understood, but perhaps it forms and is enlarged somewhat as follows. A snowbank on a slope near the head of a mountain valley may initiate the development. The rock surface beneath and around this snowbank is gradually broken up by the freeze-thaw process and then deepened as the fragments of bedrock are removed (plucked or quarried). As this depression grows larger, year by year, snow and ice accumulate to greater thicknesses until some remains throughout the year. The snow at the bottom of this pile gradually changes to ice, and when enough ice has accumulated, it flows downslope as a glacier. The rigid upper part of the ice is pulled away from the head of the depression, thus forming a curved crevasse or a series of closely spaced crevasses (bergschrund). When the crevasses are open (as in the summer), meltwater and rainwater pour into them from above. Frequent freeze-thaw occurs along the valley wall, and many large blocks of rock are loosened. These fall to the bottom of the crevasse, are incorporated into the mass of ice, and move with it slowly down the valley. A steep cliff is thus produced at the head of the valley, and this cliff is eroded headward. As the cliff retreats, its base meets the abraded floor of the valley at a sharp angle. The cliffs gradually become higher and are slowly worn back to form a steep-sided semicircular basin (a cirque). But this is only a partial explanation, because problems remain: e.g., such crevasses are not present at the heads of all glaciers and do not extend to the great depths necessary to account for a headwall 3000 feet high.

A ridge between two valley glaciers may be reduced to a knife-edge crest (an arête) as the glaciers erode their valley walls (Fig. 9–12B). A depression (col) may form subsequently along such a ridge at the place of intersection of two cirques which have eroded headward from opposite sides of the ridge. Three or more cirques working headward toward a mountain peak may shape it into a jagged summit (a horn) with triangular-shaped cliffed faces like the Matterhorn in the Alps and the Grand Teton in Wyoming.

In summary, a typical glaciated valley is fairly straight, has a U-shaped cross section, heads in a large bowl-shaped depression, has several rock basin lakes along its course, and possesses hanging tributary valleys.

Forms Produced Chiefly by Glacial Deposition

Moraines are topographic features ranging from rather thin, irregular sheets called ground moraines to low ridges. The sheets and ridges are composed of rock debris that has been transported and deposited by the ice, and commonly the material has not been reworked very much by streams. Moraines are classified as ground, lateral, medial, and end (terminal and recessional). Additional features formed chiefly by deposition from glaciers or by streams associated with the ice are drumlins, eskers, kames, kettle holes, and outwash plains (Fig. 9–15).

Glacially transported debris is called drift and may be stratified or unstratified. If dropped directly by the ice, it is unsorted (*till*, Fig. 9–5); but if subsequently reworked by streams, it becomes stratified. Although till lacks obvious structures, its boulders are sometimes aligned with their longest diameters parallel to the direction of ice movement. However, relatively few large boulders occur in tills eroded from weak rocks such as shale. Much finely ground debris of silt and sand sizes (rock flour) is produced by

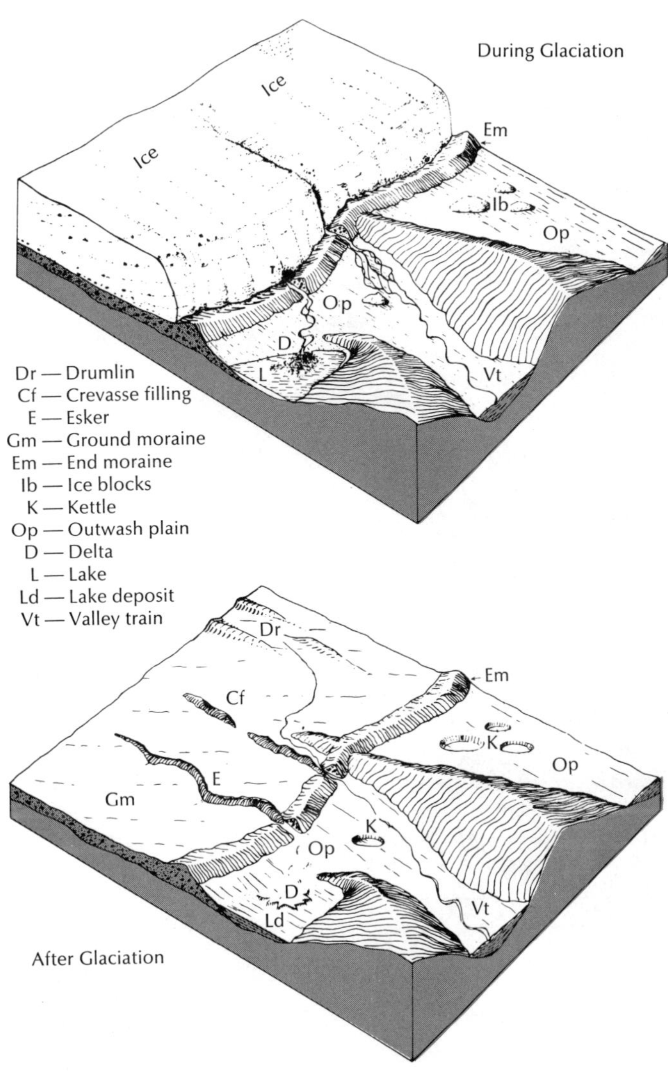

9-15. Topographic features and deposits commonly produced by an ice sheet. *(Courtesy R. W. Fairbridge, ed.,* The Encyclopedia of Geomorphology, *Van Nostrand Reinhold, 1968.)*

glacial abrasion and consists chiefly of unweathered materials. This indicates that freshly eroded bedrock is the chief source. Streams flowing from a glacier commonly have a milky appearance because of the abundance of this material.

Lateral and Medial Moraines

Lateral and medial moraines are associated with valley glaciers, but may also form near the margin of an ice sheet where the ice flows outward through valleys. Glacial ero-

9-16. Tikke Glacier in northern British Columbia. Between 1963 and 1966, ice within the glacier flowed rapidly (surged) a distance of ten miles, producing the highly crevassed surface, contorted medial moraine, and marginal lakes. *(Courtesy U.S. Geological Survey.)*

sion steepens valley walls and thus increases the amount of rock debris that tumbles, slides, or is avalanched onto the margins of the ice to form low ridges or *lateral moraines*. At the junction of two valley glaciers, their inner lateral moraines combine to form a *medial moraine*. Thus the lateral moraines of tributary glaciers subsequently become the medial moraines of a main glacier (Fig. 9–16). These stripes of rock debris may slow down the wastage of the underlying ice so that rapid thinning of the intervening cleaner ice by sublimation and melting leaves these moraines as ridges, which may exceed 100 feet in height. But the rock debris itself may not be 100 feet thick. Like the frosting on a cake, it cloaks a ridge of ice beneath.

However, in some glaciers the rock debris of a lateral moraine may be more or less continuous with the glacier's debris-loaded basal portion. Since a small tributary glacier may join a main glacier by flowing beside it or on top of it, the smaller glacier's rock debris may be distributed at different levels within the larger one. Such lateral and medial moraines do extend below the ice surface. In some instances the type of rock that predominates in a medial moraine is distinctive enough to identify it as the continuation of a certain lateral moraine farther up the valley.

Terminal and Recessional End Moraines

These may be deposited either by valley glaciers or by ice sheets, each of which may advance at one time and retreat at another. However, a retreating glacier is not actually flowing backward toward its source. The forces of nourishment which make a glacier larger oppose the forces of wastage which make it smaller (Fig. 9–11). If nourishment gains the upper hand, the terminus of a glacier advances. When wastage exceeds nourishment, the front of a glacier recedes. If forward flow and backward wastage are in equilibrium, the front of a glacier remains in the same location, even though the ice continues to flow forward from the source region.

New supplies of rock debris are thus continuously brought by the moving ice to this location and are dumped in a process that reminds one of debris piling up at the end of a conveyor belt. This material is augmented by that previously shoved ahead of the ice. A ridge of material, an *end moraine*, thus forms along the margin of an ice sheet or along the down-valley terminus of a valley glacier (Fig. 9–17). At any one point the ridge is oriented at an angle of about 90 degrees to the direction of flow.

The end moraine of a valley glacier is in part a continuation of its lateral moraines and forms a crescent because ice flows fastest in the central part of the valley (Fig. 9–18). End moraines are also curved around the margins of an ice sheet, because large ice lobes develop there like the ends of giant stubby fingers. End moraines tend to be discontinuous ridges: streams flowing outward from a glacier may subsequently remove parts of a ridge; in addition, an end moraine may have been discontinuous originally because equilibrium was never attained in some sections. An end moraine may be pitted by depressions, and a widespread outwash plain of stratified drift may extend from it beyond the ice margin (Fig. 9–19).

The outer margin of a glacier may subsequently retreat some distance from its newly formed terminal end moraine before an equilibrium between nourishment and wastage is again achieved. At this new location, a second end moraine thus forms, a recessional end moraine (Fig. 9–18). Thus several end moraines may form, and the terminal moraine is the oldest and most distant (in the direction of flow).

Drumlins

These are features associated chiefly with ice sheets. The rock debris in an ice sheet may be deposited in different ways, each with characteristic properties. If clay-rich debris is abundant near the base of an ice sheet, it may be plastered above and around any knob-like obstruction that is located in

9-17. *The* terminus of a valley glacier in Alaska. The following features can be observed: a stream at the lower left becomes braided in the lower right where glacial outwash enters it; an older outwash fan in the lower left and a younger one in the lower center; a partially dissected crescent-shaped recessional end moraine; kettle holes; and lateral and medial moraines. *(Courtesy U.S. Air Force: N 61° 25′, W 145° 10′.)*

the path of the ice. The flowing ice then molds this debris into a low, streamlined hill called a *drumlin* (Fig. 9–20). Ideally, a drumlin is shaped like an upside-down canoe or an inverted spoon. Its steepest slope tends to face the direction from which the ice came.

A typical drumlin might be a mile or less in length, half a mile or less in width, and some 50 to 100 feet high. However, drumlins show a wide range of shapes and sizes. Drumlins may be produced in swarms under a vigorously flowing ice sheet, and their long axes are aligned in the direction the ice flowed. New York State alone is estimated to have more than 10,000 (Fig. 9–21). Some drumlin-like features consist chiefly of bedrock hills that were eroded by the ice into oval shapes and then coated by a thin sheet of drift. Some drumlins consist entirely of drift, and all gradations occur between these extremes; furthermore, one cannot be distinguished from the other by casual surface observations.

Eskers, Kames, and Kettle Holes

These topographic features are made chiefly of stratified drift deposited by streams in contact with glacier ice. The streams may have been under, in, or upon the ice, which was thin and stagnant. Slumping occurred as the ice melted away.

An *esker* (Fig. 9–15) is a sinuous ridge that is more or less symmetrical in cross section and composed chiefly of stratified sand and gravel. It may wind across the Earth's surface for many miles and pass across higher ground from one low area to another. Eskers generally parallel the direction of ice movement. Many eskers apparently were formed by the deposition of sediments from streams that flowed in tunnels along the bottom of a thin stagnant sheet of ice, and perhaps also in crevasses. Ice thus formed the "valley walls" of these streams, and contained the sediment as it piled up; when the ice subsequently melted, slumping and sliding occurred. Other eskers seem to have formed from sediments deposited by streams flowing on the surface of thin stagnant ice near the margin of a glacier. When the ice melted, this material piled up on the surface as an elongated ridge.

Kames and kettle holes commonly occur together. *Kames* are small rounded hills consisting chiefly of stratified sand and gravel. *Kettle holes* are basins or depressions

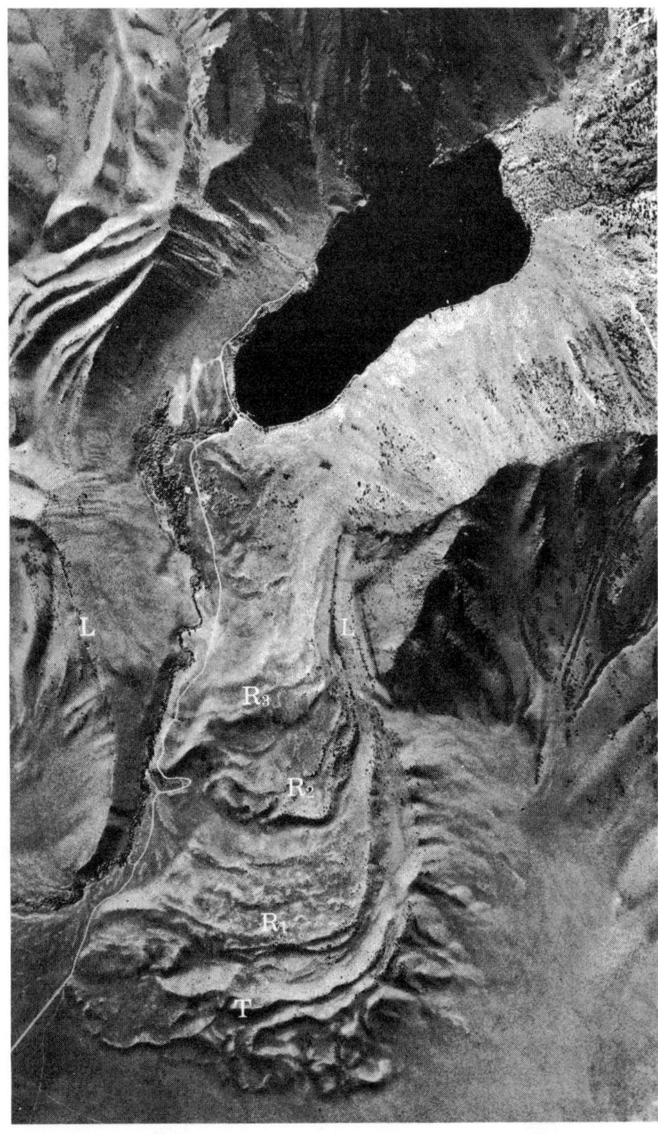

9-18. Convict Lake (nearly 7600 feet above sea level; eastern front of the Sierra Nevada) occurs in back of an end moraine. Its basin was formerly occupied by a valley glacier which flowed out upon the valley floor for about $1\frac{1}{2}$ miles as shown by the moraines: terminal end moraine T, recessional end moraines R_1, R_2, and R_3, and multiple lateral moraines L and L. The outlet creek has cut through the moraines on the left. *(Courtesy Fairchild Aerial Surveys.)*

Stages in the development of Long Island

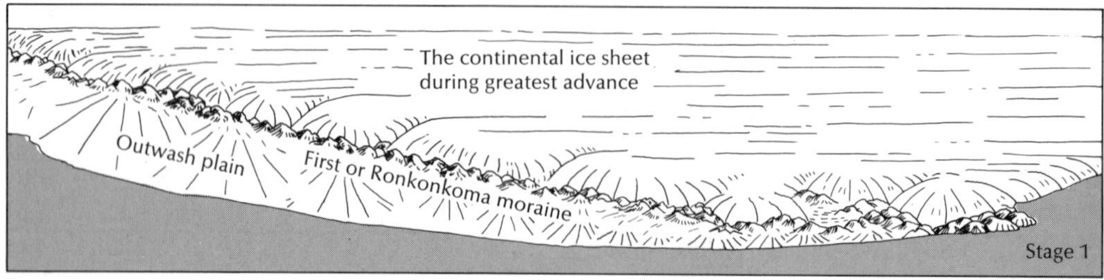

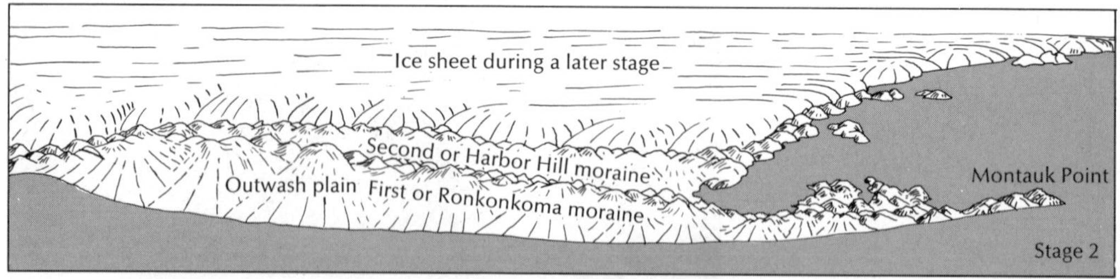

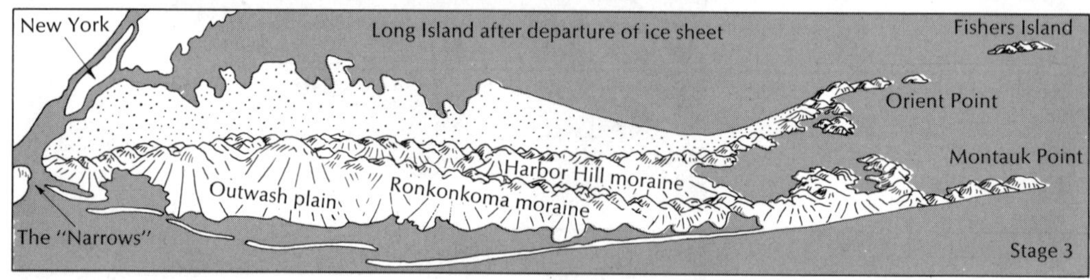

9-19. Stages in the development of Long Island. The Ronkonkoma end moraine formed first (stage 1). Lake Ronkonkoma is today located in a kettle hole in the middle of this moraine. An outwash plain is shown. The Harbor Hill end moraine is shown in stage 2. Its formation caused the destruction of the western end of the Ronkonkoma moraine (stage 3). These two moraines form the two "tails" of Long Island (Montauk and Orient Points) and some of the islands toward the east and northeast. *(A. K. Lobeck, Example 27,* Things Maps Don't Tell Us, *The Macmillan Company, 1956.)*

that have formed in stratified drift, or less commonly in till, by the melting of a block of glacier ice that had been separated from its parent body by wastage and then was partially or completely buried by sediments. Kames may form wherever streams on glacier ice are made to drop their sediments abruptly—e.g., by flowing into a hole or off the edge of the ice. Very thin stagnant ice may be separated into blocks by an interlocking network of cracks and crevasses, and these may become partially filled by stratified drift. Eventual melting of the ice under these conditions produces many irregular and rounded hills, ridges, and depressions. Irregularities also result from deposition on an uneven surface and from deposition that varied in quantity from one place to another.

Outwash

Glacier ice transports very large quantities of rock debris that is subsequently picked up by streams and carried far beyond the edge of the ice. Such streams commonly

9-20. Drumlins show direction of glacier movement (diagonally from upper right to lower left) in northern Saskatchewan. Small, short eskers occur here and there among the drumlins. *(Courtesy National Air Photo Library, Canada.)*

begin at a considerable distance back from the margin of the ice. They may flow in tunnels beneath the ice along relatively steep gradients, and their channels are confined by ice walls or are located between a valley side and the ice. Beyond the edge of the ice, gradients are commonly less steep, and channels are wider; thus deposition takes place. If an end moraine is forming, streams may flow through gaps in the end moraine

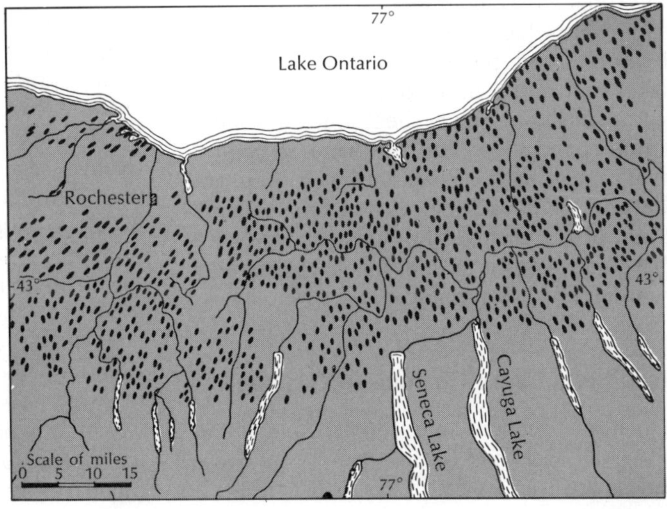

9-21. Drumlins are very numerous south of Lake Ontario. The Finger Lakes are situated in valleys that were eroded by northward-flowing streams. These valleys were then widened, deepened, and straightened by southward-flowing ice. Bedrock consists chiefly of thin-bedded shales that dip gently toward the south. Thus ice flowed across the upward slanting edges of these beds, and blocks were plucked out readily. Some valleys were deepened greatly (e.g., the bedrock floor of Seneca Lake is 1000 feet or more below the lake surface). The mouths of tributary streams "hang" above the bedrock floors of some of these lakes. *(Monnett and Brown,* Principles of Physical Geology, *Ginn and Company.)*

and deposit sediments as alluvial fans; the point of each fan is situated at a gap. A series of alluvial fans may form in front of such an end moraine and coalesce to produce an *outwash plain.* Streams shift back and forth in a braided pattern across its gently sloping, nearly smooth surface. Long Island shows such features (Fig. 9–19). Retreat of the ice results in the deposition of stratified drift around and behind the end moraine, which may eventually even be buried. Kettle holes may pit the surface of the outwash plain in such an area.

Downstream the sizes of the transported fragments become smaller and more rounded, and tributaries enter with sediment not transported by the ice. The outwash material may extend for hundreds of miles down major valleys and gradually merges with nonglacial alluvium. Many of the major rivers in the United States flow on floodplains underlain by such material. Outwash is an excellent source of sand and gravel for construction work and may form valuable aquifers.

Modification of the Earth's Surface by Ice Sheets

The capacity of an ice sheet to gouge away the surface is commonly overrated. The surfaces over which glacier ice has moved have been polished, scratched, and grooved, and lakes and ridges have formed from erosion and deposition. An ice sheet tends to round, smooth, and streamline the shapes of mountains and valleys in its path, but it does not destroy them, and it does not create them. Thin, nearly stagnant portions of ice sheets have been known to ride over loose material without great disturbance.

Ice Sheets Cause Changes in Drainage

Drainage changes are numerous within glaciated areas and outside them. The Great Lakes and Niagara Falls (Figs. 6–8 and 9–22) formed during the Pleistocene and owe their existence partly to glaciation. Lakes and swamps are very abundant in glaciated areas (swamps cover a greater area than lakes). Water has filled a multitude of depressions that were formed as the ice gouged deeply in one place and deposited debris in another. Pluvial lakes developed in non-glaciated areas. Many rivers or parts of rivers owe their locations to the former presence of an ice sheet which blocked and ponded their drainage (they flowed toward it). Thus they were made to seek new paths along the edge of the glacier. The Ohio and Missouri Rivers illustrate well the development of ice-marginal streams.

The familiar Missouri River of today (Fig. 2–24) did not exist before the Pleistocene glaciation. In the Montana-Dakota region, several rivers flowed eastward and northward toward Hudson Bay, e.g., the headwaters of the present Missouri River and the Yellowstone. South of this, other rivers flowed generally toward the east. The leading edge of the North American Ice Sheet moved slowly southward, with an occasional halt or retreat, until wastage became too great for further advance. In so doing, it blocked the drainage of these rivers. Ponds developed at these intersections, and the water rose high enough to overflow along the margin of the ice. Drainage from different rivers followed the ice margin and united; the combined waters subsequently eroded the channel that now forms the middle portion of the Missouri River as we know it today. The former channels were filled with glacial drift (note on a map the relative absence of tributaries from the west). At Kansas City, the Missouri River makes a sharp turn toward the east; here it probably follows the preglacial channel of the Kansas River that joins the Missouri River at Kansas City.

A similar story accounts for part of the valley of the Ohio River. The preglacial drainage was generally toward the northwest, and when ponded, it shifted along the ice margin toward the southwest. The combined waters of these rivers joined a preglacial section of the Ohio River somewhere in southeastern Indiana. Information from thousands of wells has enabled geologists to locate the abandoned river channels, which are now filled by glacial drift and in many cases cannot be detected from the surface. Such buried channels make excellent aquifers.

Determining Direction of Movement of an Ice Sheet

Glacial striations (Fig. 9–4) tend to parallel the direction of ice movement, but they can vary greatly in orientation because of irregular movements near the ice margins. Better

9-22. Niagara Falls and part of Niagara Gorge. After the North American Ice Sheet retreated northward from this area, the Great Lakes formed, and water flowed from Lake Erie to Lake Ontario across the Niagara escarpment to produce Niagara Falls. The falls when first formed were about seven miles downstream from their present position. During the past century the falls have retreated about four feet per year. The seven-mile gorge that has been formed as the result of this retreat indicates that this process has been going on for thousands of years. The bedrock in the area consists of sedimentary rocks that dip gently toward the south. A resistant formation forms the lip of the falls at present. It is undermined by the turbulent waters at the foot of the falls and breaks off occasionally in large blocks. As the falls retreat upstream, the dip of this resistant formation takes it to a lower altitude. Thus the falls should eventually change into a rapids before Lake Erie is reached (in an estimated 25,000 to 30,000 years). *(Courtesy Niagara Falls Chamber of Commerce.)*

indicators are striking linear features on a much larger scale—grooves, ridges, and some lakes—that develop in drift and bedrock and give a distinct grain to the surface.

Ice may move across knobs of bedrock and abrade them asymmetrically; i.e., smooth gentle slopes face the direction from which the ice came, whereas steep slopes face the opposite direction (*roche moutonnée*). A glacier more readily plucks blocks of rock from the unsupported lee sides of such outcrops.

Drumlins and end moraines provide additional evidence. The ice moved parallel to the long axes of drumlins, but it moved perpendicular to the trend of an end moraine.

Indicator stones are signposts along the glacial highway. Unique types of rock may occur as ice-transported boulders. If the boulders and their distant sources have both been located, the direction of movement of the ice in that area has been determined. When their place of origin is known, such transported boulders are called indicator stones.

A farmer found an unusual stone in his fields in Wisconsin one day, which subsequently proved to be a 16-carat diamond. About a dozen additional diamonds, most smaller than the first, have also been found in Wisconsin, Michigan, Indiana, and Ohio in ice-transported drift. Their locations have been plotted on a map, but the stones are few, and the area is large. The source is located somewhere north of the Great Lakes and has yet to be found.

Origin of Glaciation

In the past, extensive glaciation seems to have occurred only when lands were high, and land areas today apparently are more extensive and higher than they were throughout much of geologic history. Yet extensive glaciation has not occurred every time that lands have been high and widespread. Moreover, glaciers formed and advanced, and then wasted away and retreated several times during the Pleistocene, while the lands remained high. In addition, the growth and shrinkage of glaciers occurred at about the same time in both the Northern and Southern Hemispheres.

Temperature fluctuations during the Pleistocene apparently caused these changes, but why temperatures became cold enough to bring on a glacial age at one time, or warm enough to produce an interglacial age at another time is uncertain or unknown (see p. 523 for Earth-sun changes). Moreover, such temperature fluctuations followed a general lowering of temperatures that began about 50 million years or so ago when temperatures were probably some 8° to 10°C (14°F to 18°F) warmer than they are today. However, part of this temperature drop was probably caused by an increase in the average height of the continents during this interval.

Many hypotheses have been developed to account for the phenomena of Pleistocene glaciation, and none seems to be generally acceptable today. We now examine briefly, as a sample, one of these hypotheses. The North American Ice Sheet may have originated when a drop in mean annual temperatures occurred (perhaps caused by a variation in the absolute amount of solar radiation or in the amount reaching the Earth's surface). The cooling trend caused the development of valley glaciers, or increased their sizes, in the high-altitude regions of northeastern North America (Fig. 9–6). Such highlands are present in northern Quebec and Labrador and in the Canadian arctic islands. Some glaciers exist in these areas today, and mountains range up to nearly 2 miles in altitude. The drop in temperatures may have caused a small overall decrease in precipitation, but the proportion of snowfall was probably higher, and less wastage occurred.

Most of the moisture-bearing air probably came from a general southwesterly direction as it does today. Therefore, as the air rose over the mountains, more precipitation occurred on the windward sides than on the lee sides (Fig. 27–7). Over the centuries, additional valley glaciers formed, and all glaciers grew larger. The glaciers flowed out from their valleys onto lower ground near the mountains and united into piedmont glaciers. These in turn coalesced into a small ice sheet, or perhaps into several small ice sheets that subsequently combined into one.

As the glaciers grew and advanced toward the southwest, the moisture-laden, prevailing winds were probably chilled by the ice long before they reached the mountains. Therefore, the maximum amount of snowfall was shifted from the mountains to the south and southwest margins of the growing ice sheet. As the snow piled up here and changed to ice, its weight caused further flowage.

Although the western margin of the ice sheet may not have had as much snowfall as the margins on the southwest and south, wastage was probably less. Thus the glacier ice advanced toward the Rockies across an area that today has relatively little precipitation. Eventually it merged with a complex mass of valley and piedmont glaciers that formed in the Rockies and Pacific Coastal Ranges.

Snow and ice reflect more solar radiation than do rock and soil. As a result, the air above a glacier is chilled more than air over rock and soil, and the snowfall over the ice is greater. It seems quite probable that the ice became thicker toward the southwest, and perhaps it piled up higher than some of the mountains where the glaciation began; in fact, some of these mountains may well have been buried. Smaller glaciers presumably formed ahead of the ice sheet in various areas such as the White Mountains and Adirondacks and became a part of the ice sheet when it reached them. Advance was finally checked when the ice sheet had moved far enough southward for wastage to predominate.

Late Paleozoic Glaciation

Continental drifting apparently plays an important role in accounting for the extensive glaciation that took place in the latter part of the Paleozoic era. Deposits made by ice sheets have been found in Brazil, South Africa, Madagascar, India, and Australia—all areas that today are relatively near the equator and in which ice sheets would not form with a drop in temperature (no ice sheets formed in these areas during the Pleistocene). The deposits are widespread, and some are interbedded with marine deposits indicating their formation at low altitudes. Furthermore, in South Africa, flow seems to have been away from the equator. According to the recently strengthened hypothesis involving continental drift, these lands were covered by one or more ice sheets when they were located much nearer the south pole late in the Paleozoic Era (p. 317). Subsequently, these lands slowly moved to their present positions.

Exercises and Questions for Chapter 9

1. What kinds of evidence indicate that an ice sheet once covered an area?
2. What kinds of evidence support the concept of multiple glaciation during the Pleistocene (i.e., the alternation of glacial and interglacial stages)?
3. Discuss some of the major effects of glaciation during the Pleistocene:
   (A) Changes in sea level and in the volume of ocean water.
   (B) Crustal movements (isostasy and isostatic adjustment).
   (C) Migrations of animals and plants.

(D) Growth and shrinkage of pluvial lakes.
(E) Erosional and depositional changes produced directly by glaciers—such effects are chiefly modifications of topographic features already in existence.

4. Has Pleistocene glaciation ended?
5. How can sediments transported and deposited by streams commonly be distinguished from those carried and dropped by a glacier?
6. Discuss some of the factors involved in the movement of glacier ice.
7. How do glaciated mountain valleys tend to differ from nonglaciated valleys?
8. What is an end moraine?
   (A) Why are many end moraines curved like crescents?
   (B) Distinguish between a terminal end moraine and a recessional end moraine.
9. What is the evidence that parts of the Missouri and Ohio Rivers have resulted from glaciation?
10. Why are lakes common in glaciated areas?
11. How can pollen grains, in combination with carbon-14 age determinations, be used to study plant migrations?
12. How can cores of sediments extracted from the ocean floor aid in a study of Pleistocene glaciation?
13. How can the direction of flow of an ice sheet be determined?

10

Winds and Dry Lands

In May of 1934, dust in prodigious quantities was swirled from the parched surfaces of Colorado, Kansas, Oklahoma, Texas, and other parts of the southwestern United States and carried eastward by prevailing westerly winds. Hours later, some of this dust darkened the sky and obscured the sun in New York City and fell on ships several hundred miles out in the Atlantic. Measurements subsequently showed that 300 million tons or more of clay- and silt-sized particles were carried out of the source areas and dropped on the lands and ocean to the east during this single storm, which was only one of many that occurred in the 1930's. Thirty million 10-ton trucks would be needed to transport such a load. If we imagine these to be spaced at 50-foot intervals—six lined up down the length of a football field, or about 100 trucks per mile—they would span approximately 300,000 miles. The Colorado River now takes more than a year to transport an equivalent amount of sand and silt.

Yet the 300 million tons represent a small part of the total quantity of sediment moved by this storm. For every ton of dust carried out of the area, probably some three tons or so of similar or coarser particles were moved short distances and piled up in depressions and dunes (Fig. 10–1). This is similar to rainwash and rain splash on a slope—for every ton of soil washed into a nearby stream, many tons are splashed around and shifted farther down the slope.

Transportation of Sediments by Wind

Wind velocity commonly increases from the surface of the ground upward, and the movement is turbulent. Upward, downward, and sideward movements are superposed on the general forward motion as can be noted by watching the irregular path of a tiny piece of paper blown along in a brisk wind.

Laboratory experiments and field observations show

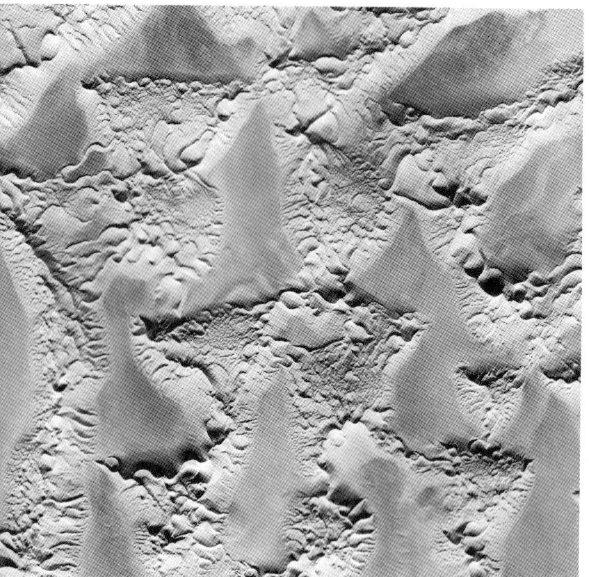

10-1. Large complex sand dunes in Saudi Arabia. Dune peaks project some 300 to 600 feet above their surroundings (photo taken from altitude of six miles). Complexity results from various combinations of direction and magnitude of both source and wind. *(Courtesy Aero Service Corp.)*

that wind transports sediment by two methods that are analogous to the suspended load and the bedload of a stream. Tiny particles of the size of clay and silt can be carried long distances in suspension. They fall more slowly than larger pieces, and as they move along, they are likely to encounter enough upward currents here and there to keep them from sinking all the way to the ground. The tabular shapes of many dust specks also increase their buoyancy. Housewives know that air carries dust inside houses even in humid regions.

On the other hand, sand-sized and larger pieces cannot be lifted or supported by the average strong wind and are rolled or pushed, or move by jumps along the ground (saltation). Of course, this limitation does not apply to the furious force of the hurricane or tornado which may blow the roofs off buildings or pick up a cow or car and set it down some distance away.

If a grain of sand cannot be lifted by average winds, how does it get into the air? Several points are involved. First, sand grains do not travel very far off the ground. One can stand on a sand dune in a desert and safely face a strong wind with eyes unprotected. Most of the sand moves along the ground in a zone that is a foot or less deep, and few grains strike one above the knees, although one's ankles are peppered continuously. Thus powerline poles in a desert may be adequately protected from sand abrasion by stones piled a foot or two high around the bases of the poles. However, if dust is also being transported, the eyes must be protected, and if the winds are very strong, sand grains may reach several feet above the surface.

Winds may roll sand grains along the surface until they hit some object, such as other sand grains or a pebble, and carom into the air or knock the obstructing sand grains into the air (Fig. 10–2). These grains fall back to the surface downwind along a path determined by the Earth's gravitational attraction, the force of the wind, the size of the particle, the angle of impact, and other factors. A well-stroked forehand drive in tennis slants downward near the baseline in a similar manner. The impacts of these falling grains cause them or other grains to bounce into the air. Thus some sand grains

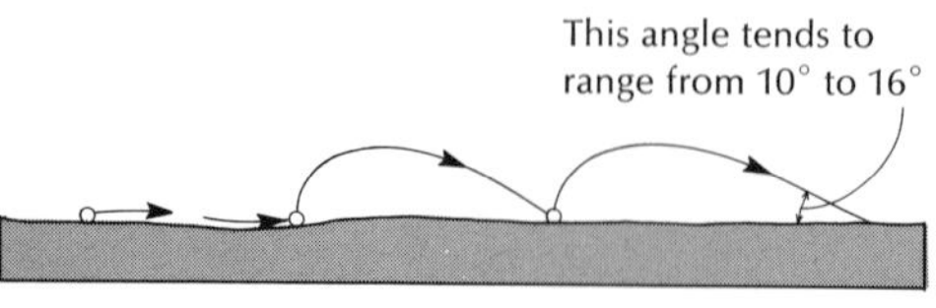

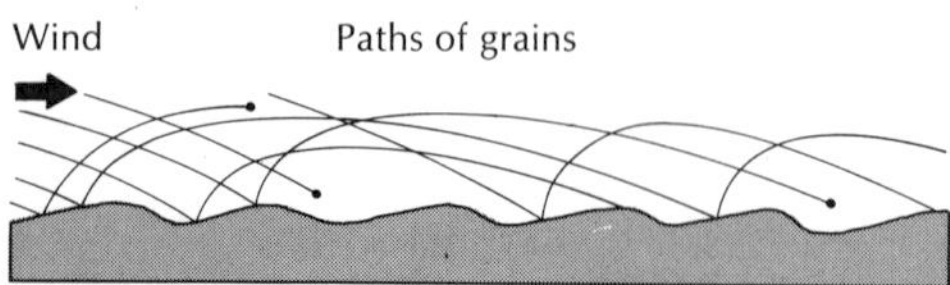

10-2. Sand moves by jumps along the surface. Winds may roll grains along the surface; their impacts kick other grains into the air; and these in turn "splash" additional ones when they fall.

are rolled or pushed along the ground, but most jump or bounce from one spot to another. Pebbles may also be rolled or pushed along the surface, and small boulders may be shifted somewhat by the removal of smaller supporting particles around their bases.

Wind is one of the most effective geologic agents in sorting sediments into sizes: silt and clay are carried in suspension, sand rolls or hops along the ground, and pebbles and larger pieces are left behind. Winnowing of the clay, silt, and sand from a material such as gravel eventually results in the complete coverage of the surface by rock fragments that are too large to be moved by the wind. These gradually settle into stable positions and form a protective armor (desert pavement) "one pebble deep" that prevents the removal of finer underlying sediments (raindrop splash combined with sheetwash may produce a similar type of surface).

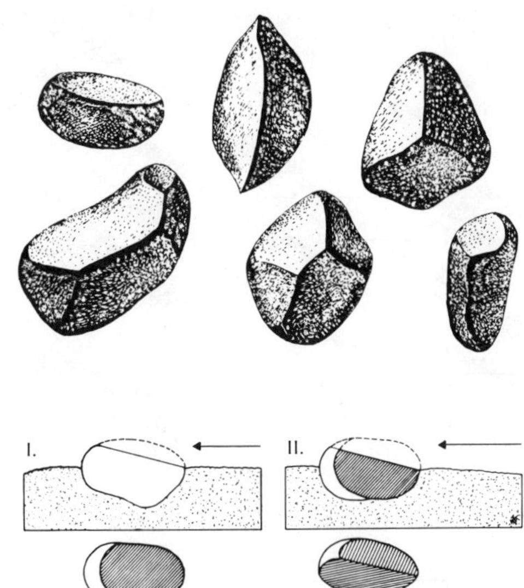

10-3. Wind-faceted pebbles. *(After Kettner, 1960; courtesy R. W. Fairbridge,* The Encyclopedia of Geomorphology, *Van Nostrand Reinhold, 1968.)*

Wind Erosion

Wind erosion occurs in two ways: (1) by the less important effect of *abrasion* (Latin: "scrape-off") and (2) by the much more important activity of *deflation* (Latin: "blow-away"). There is a sandblast effect from the wind-driven, rolling, hopping sand grains that move along the ground in a sand storm. Unprotected fence posts or powerline poles may be worn through near the ground, and notches may be cut along the bases of some rock outcrops.

Wind-blown sand may abrade, polish, and distinctively shape the surfaces of pebbles and small boulders. These are called *ventifacts* (Latin: "wind-made"). If a pebble does not move during this sandblasting, it may eventually be planed off parallel to the ground. However, if a pebble's position is shifted one or more times, or if wind directions change, a varied combination of faceted surfaces can develop (Fig. 10–3). These tend to meet at sharp angles, and commonly they are curved.

Wind abrasion, however, is probably not the main agent responsible for the many weirdly shaped rocks and holes that may be seen in some arid and semiarid regions. Rather, differential weathering has loosened and disintegrated some sections of a rock formation more than others, and the wind has aided in the removal of these fragments.

Deflation occurs with greatest effect in areas where fine dry sediments have been loosened and are unprotected by vegetation (Fig. 10–4). These were the conditions in the southwestern United States in the 1930's. Areas with marginal rainfall had been farmed, and several dry years were climaxed by severe droughts in 1933 and 1934. After the vegetation had died and exposed the soil, deflation could occur, and in the 1930's a few feet of soil actually were blown off the surface over wide areas. As another example, an average of about 8 feet of sediment may have been removed by wind from the surface of the Nile delta in 2600 years.

Thousands of rather small, shallow, in-

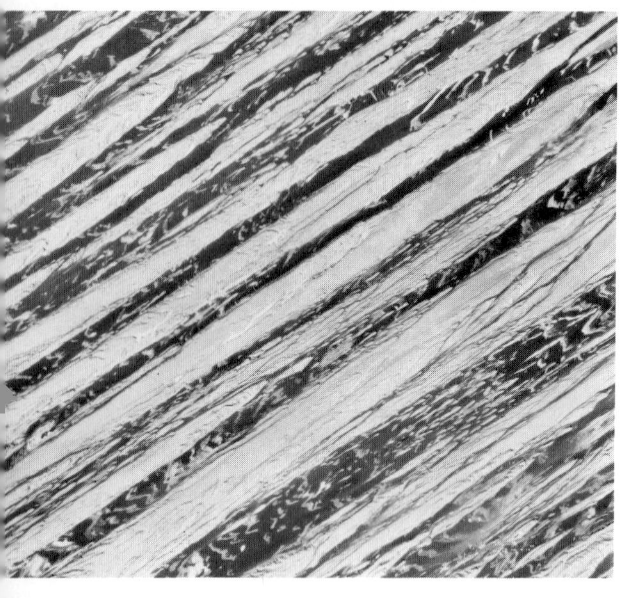

conspicuous depressions in the drier regions of the western United States probably have been produced by deflation (Fig. 10–5). Some of these may contain water or vegetation in wet years but become parched and bare and a source of wind-blown dust during the intervening dry years.

Contrary to what one might expect, a smooth surface made of tiny, loose particles of clay or silt is actually stable in a rather strong wind. This occurs partly because smooth surfaces reduce turbulence, and

10-4. These troughs of the Shahr-e Lut in Iran apparently have formed by wind erosion and deflation. The intervening, narrow, steep-sided residual ridges (yardangs) range up to 100 feet in height, 1500 feet in width, and a few miles in length. *(Courtesy U.S. Geological Survey, 22 August 1956.)*

10-5. Shale and thin sandstone interbeds are capped by poorly consolidated gravel and sand, and locally cemented with calcareous material (Texas). Fine-textured drainage (i.e., numerous, closely spaced valleys) has developed on the left where relatively impermeable shale occurs. Lack of surface drainage to the right suggests that the capping material is highly permeable. The numerous shallow depressions have resulted chiefly from wind action, although some solution of local calcareous cementing material may also have taken place. The light-toned areas in the depressions consist partly of washed-in fine materials that were dry and thus light colored at the time of the photograph; the dark-toned areas are caused by vegetation. Man-made patterns are also present: roads border squares 1-mile on a side (sections); some crops have been planted along contours, whereas others are in straight rows. *(Fig. 49,* U.S. Geol. Surv. Profess. Paper *373.)*

wind speeds are much slower near the ground then several feet above the surface. However, if such a surface is disturbed, as when animals or jeeps cross a dusty area, some of the dust particles are kicked upward into the turbulent zone and then swirl away. Downward-moving air currents that strike the surface may also do this.

Wind does not tend to pick up loose silt and clay particles directly because obstacles produce a zone in which little air movement occurs. This so-called, dead-air layer rests directly upon the ground and has a thickness that is about 1/30 of the average height of the obstacles. Furthermore, the thickness of this dead-air space is the same for winds of different speeds. If the scattered obstacles are sand grains 2 mm in diameter, a dead-air zone only 1/15 mm in thickness would occur. But silt and clay particles are less than 1/15 mm in diameter and thus would be entirely immersed within the dead-air space. Therefore, the loose silt-clay particles are not picked up, even by a strong wind, unless they are first knocked upward into turbulently flowing air by impacts from jumping or rolling sand grains.

The total amount of erosion by the wind is controversial, but it is far less than that accomplished by running water, although it may approximate that performed by glaciers. The story of the May 1934 dust storm probably suggests far more activity than commonly occurs, and land forms in deserts have been shaped primarily by weathering, mass-wasting, and stream erosion. However, wind may have been more important at certain times in the past (Fig. 10–10), and wind may be an important agent in the rounding of sand grains (p. 189).

Wind Deposition

Sand is dropped by the wind in localized piles and ridges to form sand dunes (Fig. 10–1), whereas the suspended silt and clay is spread as a relatively thin sheet over wide areas. Volcanic ash is another type of wind-deposited material and may also be spread very widely—in fact, across the entire Earth from explosions of the magnitude of Krakatoa in August 1883. A layer of volcanic ash from this explosion may be located in the Greenland and Antarctic Ice Sheets (expected at depths of 150 feet and 60 feet respectively because more snow falls and changes to ice annually in Greenland than in Antarctica). Thus discovery and dating of layers of volcanic ash in these ice sheets will provide a means of studying ice accumulation rates.

The problem of radioactive fallout became acute for most peoples on the Earth beginning in the late 1950's and illustrates the extensive nature of wind transportation and deposition. Radioactive dust is blown into the air by the explosions of nuclear bombs (Fig. 27–9), is carried far and wide by moving air masses, and gradually settles out. The heaviest concentrations of radioactive dust are dropped downwind from the explosion sites. However, the finest debris is blown upward into the stratosphere, disperses in a few years to nearly all parts of the Earth, and eventually sifts downward to the surface. Thus radioactive dust from a nuclear explosion in Siberia may eventually contaminate a monkey in an equatorial jungle or a penguin in Antarctica.

As with other transporting agents except ice, deposition of air-borne particles occurs when wind velocity decreases. Rain and snow may also remove dust from the atmosphere, and the presence of red silt in the air during some snowstorms has been responsible for "red snow."

Loess

Loess (Fig. 10–6; pronounced lûs) is a sediment that was probably transported and deposited by wind and commonly has the following properties: it is buff-colored, cliff-forming (when eroded), unconsolidated, nonstratified, fresh or slightly weathered, and is made of angular grains chiefly of silt size. The wind-blown origin of loess is indicated by evidence such as the following: (1) land snails are abundant in some of the

10-6. Loess deposits at Vicksburg, Mississippi. *(Courtesy U.S. Geological Survey.)*

deposits as are vertical tubules left by the decay of roots; (2) the loess is deposited as sheets that blanket hills and depressions alike; and (3) in some regions the deposits thin out downwind from source areas such as deserts, glacial deposits, and the flood-plains of certain rivers. Loess deposits are widespread in the Mississippi Valley and seem to be associated with fine, glacially ground-up rock and mineral debris. Previously this sediment had been spread out on valley floors by heavily loaded streams, became dried out, and was then carried eastward by prevailing westerly winds. The deposits become thinner to the east and cover all parts of the topography, high as well as low. The silt particles were probably knocked upward into turbulent air by the impacts of sand grains in the source areas.

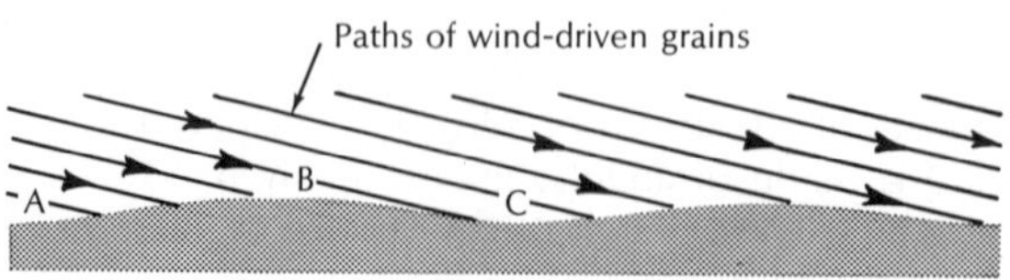

10-7. Formation of ripples. *(After Bagnold.)*

However, after they had been transported some distance and dropped—sifting downward through vegetation in many areas—the fine particles tended to cohere, were no longer bombarded by sand grains, and remained in place. Fortunately, although dust is readily transported by air, it is not readily picked up by air; otherwise, dirt and dust would pose a greater problem for us in keeping ourselves and our possessions clean. Commonly the loess is no thicker than a few tens of feet, but some deposits in China are a few hundred feet thick. Loess deposits are inconspicuous because they blanket preexisting topography without producing new distinctive land forms. Loess typically is exposed in cliffs, perhaps because its silt particles are angular, which provides an interlocking texture that can support steep faces; the presence of vertical tubules left by decayed roots may also help.

Sand Dunes

Ripples (Fig. 10–7) are readily formed by moving sand in gentle to moderate winds. As sand moves along a surface, individual grains hop downwind for different distances and strike a surface at different angles. However, they tend to have an average travel distance, velocity, and angle of impact at any one time and place. This means that more impacts per unit area will occur on the windward side of any slight rise than on the lee side or on a level surface. These impacts cause more grains to move up the windward side than down the lee side, and thus a ripple increases in height. Most grains land an average distance of one jump length downwind from such irregularities and produce new ones spaced at regular intervals. Thus ripples tend to form as tiny ridges oriented perpendicular to the wind direction, become uniformly spaced, migrate slowly downwind, and grow upward into stronger winds that prevent further piling up. Still stronger winds destroy the ripples. Ripples show that a flat, smooth, bare surface of sand is unstable in a wind. Coarser sand grains collect near the crests of wind-

10-8. Sand dunes in the New Mexico Desert. Prevailing winds are from the right. *(Courtesy Soil Conservation Service.)*

formed ripples (smaller grains are removed more readily), whereas finer grains collect near the crests of water-formed ripples.

Dunes are larger than ripples and of many shapes and sizes (Figs. 10–8 and 10–9). The largest ridges may be a mile in width, tens of miles in length, and several hundred feet high. Important factors in the formation of sand dunes are velocity, direction, and variability of the wind, the amount of loose sand, and the presence or absence of vegetation. Although dunes are abundant in desert areas, sand covers only a part of their surfaces (up to one-third of the total area for the sandiest deserts but less than 1% for desert areas in the United States).

Dunes are not confined to deserts. The necessary conditions of fairly strong winds and ample supplies of dry, loose, unprotected sand are found along the present and former shorelines of oceans and large lakes, on the floodplains of rivers in dry areas, in dry regions where weakly cemented sandstones crop out, and in areas of abundant glacial outwash (e.g., dunes occur along the eastern shore of Lake Michigan and along parts of the Atlantic Coast of the United States). The most abundant material in dunes is quartz sand, but dunes are also made of sand-sized particles of gypsum, calcite, volcanic debris, and even ice.

Some sort of obstruction is necessary for

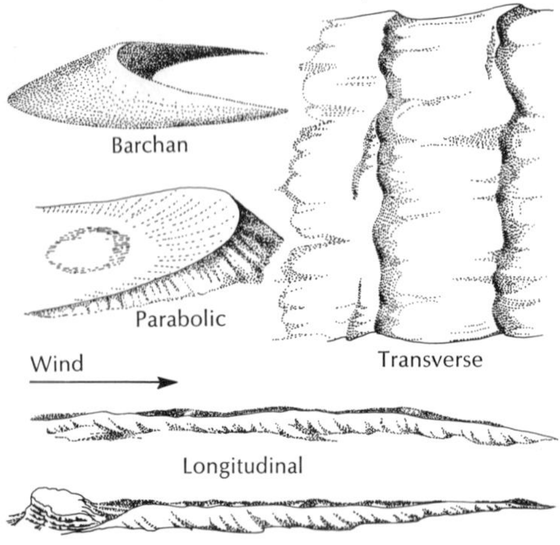

10-9. Some kinds of sand dunes (not proportional in scale). (*B. Mears, Jr.,* The Changing Earth, *Van Nostrand Reinhold, 1970.)*

a dune to form: e.g., a clump of vegetation, a boulder on a flat surface, or some sort of topographic irregularity. On a perfectly uniform surface, the sand would spread more or less as a sheet until an obstacle was reached (such a sheet, of table-top flatness and apparently about 1 foot thick, covers several thousands of square miles along the border between Sudan and Egypt).

10-10. Cross-bedding in Zion National Park, Utah.

Airspeed is reduced immediately in front of an obstacle and also beyond it where eddies develop. Sand piles up in each place and may eventually bury the obstruction. Once formed, the pile of sand is a barrier to the wind, accumulates more sand, and grows as it migrates slowly downwind. Sand grains bounce much more readily from bare rock surfaces or from desert pavement than they do from impacts on the loose grains in a dune. Thus they tend to form into discrete piles instead of being spread out as a continuous blanket.

Sand is blown up the gentle windward slope of a dune and across the top. It accumulates on the lee slope because of reduced velocity there and piles up until this slope becomes steep enough for slipping to occur. The lee slope then becomes a *slip face* and makes an angle of about 30 to 35 degrees with the horizontal. Dunes thus are cross-bedded distinctively, and this can be seen in some ancient, well-sorted sandstones (Fig. 10–10). Since a slip face slants downwind, we have information about ancient wind directions (provided original orientations have not been disturbed).

Military operations and the exploration for mineral resources, particularly for petroleum, have resulted in the last few decades in much travel in and over desert regions. The sizes and shapes of sand dunes have been found to be highly varied, and each desert seems to have some forms peculiar to it. Perhaps the most familiar form is the crescent-shaped *barchan* (Fig. 10–9) with its horns pointing downwind. Necessary conditions seem to include a somewhat limited quantity of sand, the absence of vegetation, and wind that blows primarily from one direction across a fairly flat, firm surface. Barchans have been observed to migrate a few tens of feet or more per year. Small barchans can migrate faster than large ones because their slip faces are shorter, and less sand is needed to shift them ahead a certain distance. If sand is abundant, more or less continuous ridges (transverse dunes) form at right angles to the wind.

Longitudinal dunes are ridges that are elongated with the prevailing wind, but cross winds may also be important in their growth. Where some vegetation is present, blowouts in the less protected parts of transverse dunes may produce crescentic dunes in which the points face upwind, the reverse of the barchan. Where winds are variable and sand is abundant, dune shapes become complex as different types form and overlap (Figs. 10–11 through 10–13).

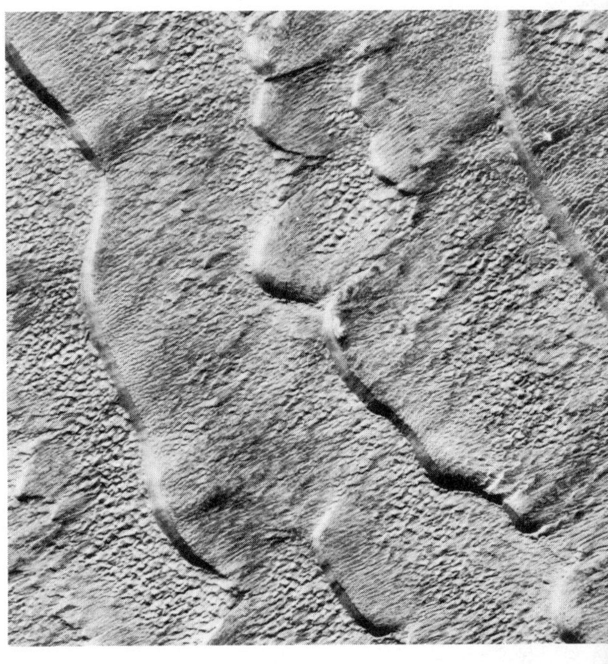

10-11. Complex transverse sand dunes in West Pakistan. The steep, slip-off slopes are spaced about one-half to one mile apart and face downwind. Smaller secondary dunes occur on the gentle windward slopes of the major dunes. *(Courtesy U.S. Geological Survey.)*

10-12. Transverse and linear dune complexes in Saudi Arabia. An exploration party camp of the Arabian American Oil Company is shown in the foreground. *(Courtesy Arabian American Oil Company.)*

10-13. Long, high, parallel dune ridges in Saudi Arabia elongated in direction of prevailing winds. The ridges attain heights of 300 feet or more, extend as far as 125 miles, and are $\frac{1}{2}$ to 1 mile or so in width. *(Courtesy Arabian American Oil Company.)*

Some Features of Dry Lands

Two points concerning the geologic activity of wind and the development of landscapes in dry lands should be made here. First, erosion and deposition by the wind are not confined to dry land regions, although the effects are more prominent there than elsewhere. Second, wind action is far from the most important process involved in the evolution of landscapes in dry regions; the combined processes of weathering, mass-wasting, and running water have sculptured the deserts (Fig. 10–14) as well as the humid regions. We must also bear in mind that some now arid regions had more precipitation during parts of the Pleistocene. Thus certain of the topographic features that we see in deserts today may have formed or been modified under somewhat different climatic conditions.

The terms desert, arid, and semiarid are difficult to define precisely, but together they apply to more than a third of the land areas of the Earth, or to more square miles than all of the humid regions combined. Thus the erosional and depositional processes that occur in them are of great importance in geology. *Desert* can be equated with arid whether the area is uninhabited or not (e.g., Las Vegas is far from deserted), and the term *arid* may be applied to a climate in which evaporation exceeds precipitation, although arid lands have also been defined as regions having less than 10 inches of precipitation per year, not enough to support much vegetation. In arid lands, the potential loss of moisture by evaporation and transpiration (i.e., if unlimited quantities were available) exceeds the actual amount of precipitation by many times. If rainfall is taken as the chief criterion, the *semiarid* areas have 10 to 20 inches of precipitation per year and support a discontinuous cover of short grasses and other vegetation.

Drought, high temperatures, and scarcity of vegetation are characteristic of arid lands, although "cold deserts" occur at high latitudes where both precipitation and evaporation are very low (widespread permafrost keeps moisture near the surface; thus vegetation is abundant, although trees are not). The even or uneven distribution of precipitation throughout a year is also an important factor in the growth of vegetation, as is altitude since in deserts, as in humid regions, more precipitation commonly occurs at higher altitudes.

Arid and semiarid lands have other common features: most of the precipitation occurs in infrequent, unevenly spaced, localized storms; the percentage of runoff is high because relatively little vegetation is present to aid infiltration; the potential evaporation exceeds the actual precipitation; relative humidity is low; daily temperature changes are large; and interior drainage is common except for a few major rivers such as the Nile and Colorado which head in more humid areas and flow through the arid lands. Contrary to some popularly held but erroneous ideas: deserts are not entirely blanketed by sand; they are not always hot; they are not completely devoid of vegetation or animal life; their land forms are not produced chiefly by wind erosion; and mechanical weathering may not be more important than chemical weathering.

Origin and Location of Dry Lands

The distribution of arid and semiarid lands is partially determined by the circulation of the atmosphere and by two meteorological

10-14. Running water plays a dominant role in the erosion of deserts. The figure shows the east face of Panamint Range, Death Valley, California. Note steep eastward dip (to the left) of the sedimentary rocks and that some beds can be traced along their strike from one ridge to the next. Dark areas on the fans are desert pavement. *(Courtesy U.S. Geological Survey.)*

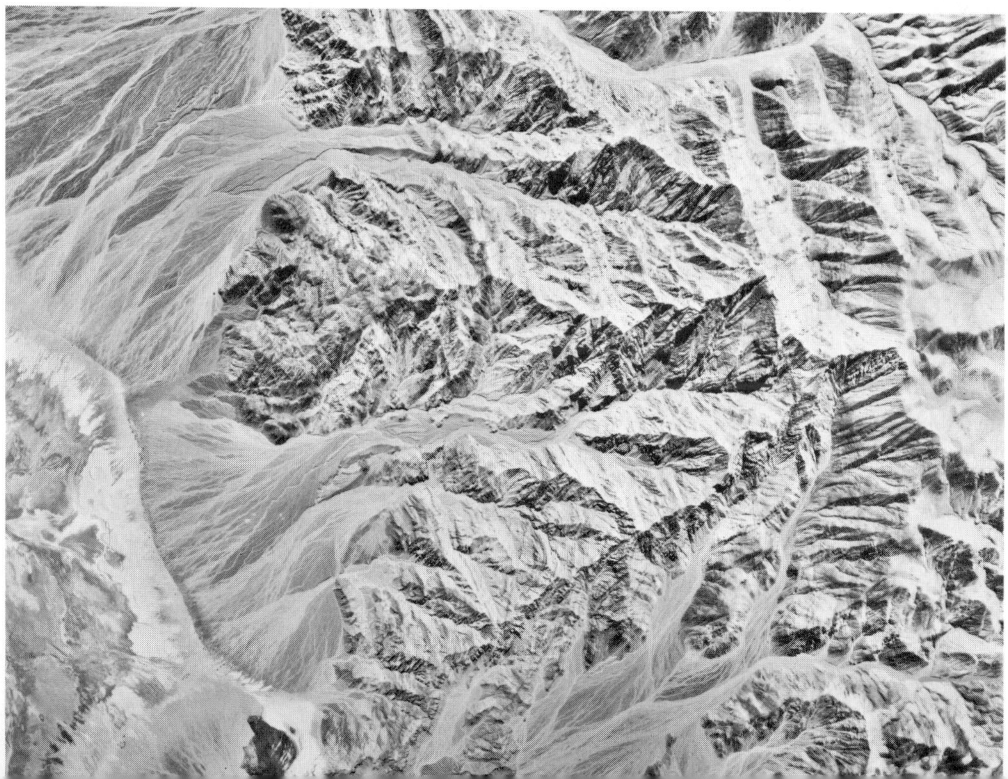

phenomena: as air sinks, it becomes compressed and warmed (p. 683); and warm air can hold more moisture than cold air (p. 677). As a part of the global circulation of the atmosphere, relatively dry air sinks and is warmed at latitudes of about 30 to 35 degrees in both the Northern and Southern Hemispheres. Some of this air moves along the surface toward the equator and is warmed further. Thus precipitation tends to be low, and many of the great deserts and near-deserts of the Earth are situated chiefly between latitudes of 15 and 35 degrees. Other deserts such as the Great Basin and Gobi occur in middle latitudes in continental interiors. These have cold winters.

A more local cause of aridity occurs in continental interiors where prevailing winds cross mountain barriers. Areas of slight precipitation ("rain shadows," Fig. 27–7) are located downwind from the mountains in contrast to the areas of abundant precipitation that occur on the windward slopes. In the western United States, dry lands occur for this reason in the area immediately east of the Sierra Nevada and coastal ranges.

Arid lands may also occur in subtropical or middle latitudes where prevailing winds blow across a cool coastal ocean current onto a continent (e.g., parts of Chile and Peru). The current chills the air and condensation occurs over the ocean. Then, as this air passes over the land it is warmed, and very little precipitation takes place.

Erosion and Deposition in Dry Lands

The traveler from a humid to a dry region may contrast the generally rounded, smoothed topography typical of many humid areas with the angularity and ruggedness that prevail in arid and semiarid sections. As he travels on within the dry lands he will observe the prevalence of three aspects of desert scenery. (1) Picturesque, steep-sided mountains of bare rock dominate the landscape, are cut by numerous stream-eroded valleys, and are nearly ideal geological laboratories because of their almost continuous bedrock exposures. (2) Flat-floored basins occur among the mountains and are particularly widespread in areas with interior drainage. (3) Relatively smooth, gravel-covered, gently sloping plains join the mountains and basins. Thus nearly flat or gently sloping plains tend to be the most widespread land form in arid and semiarid regions and may constitute about three-fourths of the total area if structural plains or dip slopes are also included (i.e., the surfaces of horizontal or gently dipping sedimentary formations coincide with the Earth's surface, as along parts of the Colorado Plateau).

As in the humid lands, the features of dry lands result chiefly from the combined action of running water, weathering, and mass-wasting. However, these processes function differently in the absence of abundant precipitation and vegetation, partly because the binding effect of roots and the chemical action of organic and other acids are reduced. In humid areas, bedrock is nearly everywhere covered by regolith, soil, and vegetation that creep slowly from higher to lower ground. Chemical weathering predominates and weakens most rock surfaces so that steep cliffs and sharp angles are relatively uncommon. The nearly ubiquitous downward-creeping regolith fills depressions and reduces and softens the effects of differential weathering and erosion.

In dry lands, the overall rate of weathering is slower, and mechanical weathering becomes relatively more important. Larger fragments predominate in the resulting rock debris, and slopes are steep.

The steepness of a slope tends to be related directly to the average size of the debris that moves down it—the coarser the debris, the steeper the slope needed to move the fragments. Thus slopes tend to change at the contacts between rocks of different types (which yield debris of different sizes). Rocks are cut by intersecting networks of joints or cracks, some along bedding surfaces, and as blocks break off along these cracks, they leave steep angular surfaces.

Precipitation in arid lands tends to be highly irregular, local, and unpredictable. An inch or more of rain may fall locally during a thunderstorm, but months or years may go by before the next major rainstorm occurs. However, these infrequent rains fall on relatively unprotected, often crusted surfaces littered with loose debris, and the proportion of runoff to infiltration is high. Therefore, raindrop splash, sheetwash, and rill wash are important in moving material down slopes. Runoff water becomes heavily loaded at once, and extensive downcutting is prevented except in the mountain valleys. Instead, water tends to flow down slopes as a thin, debris-laden sheet and in numerous small interconnected rills or channels. Mudflows are also common in some semiarid regions.

In some arid and semiarid regions, crustal movements have raised certain blocks along faults to form mountains and lowered other sections to create intervening basins. The Basin and Range country of Nevada and neighboring states illustrates this combination, and its appearance on a map has been described as that of a group of caterpillars crawling toward Mexico (Fig. 2–24). In such areas, heavily loaded streams flow down mountain valleys and dump their debris at the mouths of canyons to form alluvial fans. Such streams are relatively short and flow only after rains. The fans along the edge of a range eventually coalesce to form a continuous plain or *bajada* (pronounced "ba-hah'-da") that slopes gently downward toward the center of a basin and gradually encroaches upon it. As time passes, sediments pile up slowly on the basin floor. Deep channels do not form on the surfaces of the fans because the sediment load, evaporation, and infiltration are all too great. Tiny channels form and are filled with sediment, and new channels develop. Their locations shift back and forth across the surface, and a braided network of tiny rills and larger gullies results.

Because rainfall is slight, evaporation is rapid, the water table is far below the surface, and permanent lakes do not form in the different basins. Otherwise, such lakes would eventually overflow, connecting channels would be eroded, and an integrated drainage system would develop. Instead, drainage is downward into the centers of numerous isolated basins, and the area is said to have interior drainage. Large rivers (exotic streams) may head in humid regions and flow through such arid lands on their way to the oceans, but they are exceptional and receive little additional water in their passage through the dry lands (e.g., the Nile does not have a major tributary flowing into it for a stretch of more than 1000 miles of desert country).

The finest sediment is washed into the lowest part of a basin from all sides and collects there to form a flat-floored *playa* (pronounced "plah'-ya") which may contain a shallow body of water, a *playa lake,* immediately after a heavy rain. However, the water commonly evaporates quickly; silt, clay, and dissolved salts are deposited; and the floor is exposed to the sun and becomes mud-cracked.

Evolutionary Sequence in Arid and Semiarid Lands

If faulting has resulted in a number of basins and ranges and in disconnected interior drainage in arid or semiarid regions, erosion and deposition lead first to the formation of playas, bajadas, and steep mountain fronts. An abrupt break in slope occurs where a bajada intersects the front of a mountain, but this is more pronounced in resistant rocks than in weaker ones. As erosion continues, the steep mountain slope retreats, and the slope seems to be maintained at approximately the same angle as it retreats. This leaves behind a rather gently sloping, beveled bedrock surface, thinly covered by gravel, that is called a *pediment**

* R. F. Hadley, "Pediments and Pediment-Forming Processes," *Jo. of Geological Ed.,* April 1967.

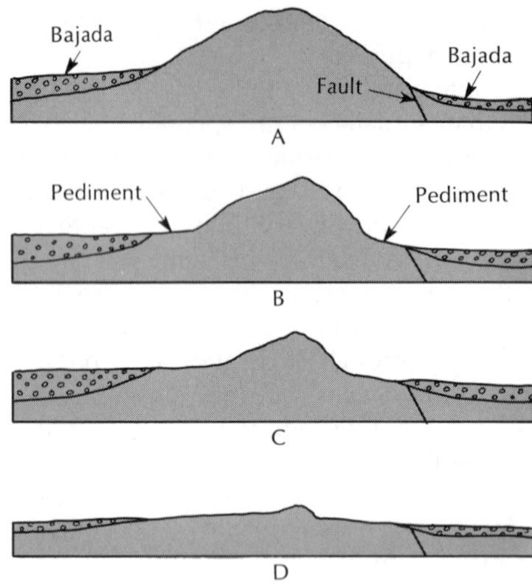

10-15. Development of a pediment and bajada. A thin veneer of alluvium covers the beveled surface of a pediment (an erosional surface) which merges imperceptibly into the surface of a bajada (a depositional surface).

(Figs. 10–15 and 10–16). The mountain slope and pediment intersect at a sharp angle.

A pediment continues the slope of a bajada, and the two cannot be readily distinguished at the surface because one merges with the other. However, deep wells drilled on a bajada show that the underlying bedrock is covered by a thick alluvial fill and that the bajada itself is a depositional feature. On the other hand, the bedrock floor of the pediment has been formed by erosion, and close inspection shows that it is only thinly covered by alluvium.

Precisely how a pediment forms and widens is uncertain, but several processes seem to be involved. Some geologists have suggested that the bedrock has been beveled chiefly by lateral planation by streams flowing on the surfaces of alluvial fans; i.e., at times a stream would flow along the margin where a fan and mountain intersect and could then undercut the adjacent mountain slope. Others have stressed the action of backweathering, sheetwash, and rill wash. Each process may well play a role in the origin of a pediment, but their relative importance still seems undetermined.

According to one view, weathering of a mountain front produces rock debris that falls or slides to the base of the mountain and piles up. The slope formed by this

10-16. Mingus Mountain, near Prescott, Yavapai County, Arizona, June 1950. The foreground shows a pediment surface. *(Courtesy L. C. Huff, U.S. Geological Survey.)*

debris is steeper for coarser debris. Along any one mountain range, the size of the rock fragments may be more or less the same, and thus its slope is about uniform. Further weathering at the base of the mountain slope reduces the sizes of the pieces, and they can then be moved downward across the fan by sheetwash and rill wash. However, new supplies of the coarser debris continue to be added from above, and the mountain slope maintains a steep angle as it retreats. Thus a pediment widens (Fig. 10–15C and D). This differs from the manner in which slopes are thought to retreat in humid regions: as these slopes retreat, the overall surface is lowered, becomes gentler, and meets lowland areas in a smooth curve.

If pediments are forming and widening on either flank of a mountain range, they may eventually cut through the crest here and there, although they are commonly at different altitudes on opposite sides. At this stage, a few steep-sided knobs are all that is left of the original mountains; the planed-off surfaces of the pediments represent their "roots." This is analogous to the peneplain stage of the humid region, and the term *pediplain* has been suggested for the land form produced by the coalesced surfaces of a number of pediments. Except for deflation by the wind, any further erosion must be very slow.

If interior drainage occurs in dry regions, sea level as an ultimate base level has little meaning. The lowest part of a basin is the local base level, and only deflation can lower this (crustal stability is assumed). However, instead of being lowered, the base level actually rises as sediments from a mountain are washed across a pediment and bajada into a playa. After pediments coalesce, a higher pediment may be rebeveled to merge with the surface of a lower pediment.

Pediments apparently have been much more important in the formation of landscapes than was appreciated until rather recently, and a number of erosion surfaces that were formerly called peneplains are now being interpreted by some geologists as pediplains. Such erosion surfaces formed long ago and have subsequently been uplifted and dissected by erosion.

Some differences between the pediplains of dry regions (perhaps these occur in humid regions also, but are more difficult to recognize there) and the peneplains of humid regions follow. (1) Pediments form early in the erosional sequence and widen slowly as steep mountain slopes retreat and remain steep. The top of a mountain or plateau undergoes relatively little erosion as a pediment grows larger. In contrast, as valley slopes retreat in humid regions, they become gentler and rounded, and the overall surface is gradually lowered. Wearing down and wearing back proceed simultaneously to produce a subdued eroded landscape. (2) A pediplain has been beveled to a rather smooth surface, and its altitude is not determined directly by sea level; in fact, local base level has risen as the landscape evolved. On the other hand, a peneplain probably is flatter overall than a pediplain, its surface has more irregularities, and it must be near sea level along a coast, although it rises gradually inland. (3) Deeply weathered regolith forms the surface of a peneplain, whereas relatively fresh bedrock with a thin covering of gravel forms the surface of a pediment. (4) Pediplains can be seen at different places on the Earth today; but few, if any, unquestioned peneplains exist at present.

Although pediplains and peneplains thus differ in many important respects, perhaps their similarities should be stressed. Each has resulted primarily from the combined action of weathering, mass-wasting, and running water, and each is the product of a long evolutionary sequence of landscape development.

Exercises and Questions for Chapter 10

1. Make a series of sketches with appropriate labels that illustrate the development of a desert pavement on a deposit of gravel.
2. Consider the transportation and deposition of sediments by wind, stream, and glacier ice. Describe criteria which might be used in determining which of these agents was responsible for the origin of an ancient sedimentary rock.
3. How does a sand dune form? Why do sand dunes have different shapes? Why are some dunes elongated more or less parallel to the prevailing wind direction whereas others trend at a steep angle to this direction?
4. Discuss the origin of wind ripples.
5. What characteristics identify a certain region as an arid land?
   (A) Where do the arid lands of the Earth tend to be located? Why?
   (B) What are the relative roles of the various erosional agents in sculpturing the arid lands? Do you have to consider the effects of the Pleistocene here?
6. What is the evidence that loess is primarily a result of wind action?
7. Describe the manner in which topographic features tend to develop and evolve in arid lands. How does the evolutionary cycle of an arid region tend to differ from that of a humid region?

11

Sea-Floor Geology

During the past two decades or so, the sediments and rocks beneath the oceans (Fig. 11–1), and the topographic features and geologic structures that compose and underlie their floors, have provided key data that have revolutionized the science of the Earth—in combination with geologic findings already won from the continents. For example, perhaps half of the deep-sea floor (one-third of the entire surface of the Earth) has originated only within the last 150–200 million years—not early in the Earth's long history as once thought. Unexpected features and processes have been discovered, and a grand, comprehensive synthesis of continental and oceanic geologic history may be almost at hand. We now focus on some of the features and processes that provide a background for understanding current ideas about the nature and functioning of the sea floor.

A number of older views concerning this or that feature or process of the oceans have become untenable in the light of rather recent discoveries, many of which have resulted from the application of geophysical techniques to the sea floor—e.g., those that involve gravitational attraction, heat flow, magnetism, and seismology (also the development of the seismic reflection profiler and the success of the deep-sea drilling program). To illustrate, the ocean floor has conspicuous relief features and seems exceptionally mobile—it is not quiet with a subdued topography as once thought—nor do sediments decrease in grain size outward across the submerged margins of the continents. Moreover, sediments on the deep-sea floor are much thinner and younger than expected, and turbidity currents (p. 288) are an important, previously unknown, transporting agent. The oceanic crust itself is both thinner and younger than had been predicted. An older view that continents drift across the face of the Earth now has strong new evidence and is widely and enthusiastically accepted. It seems destined to exert a strong influence on much of geologic thinking about the Earth's past.

Despite the extent and fruitfulness of the many

11-1. USS Missouri hits a wave during hurricane, 14 August 1953. Note blurring of air-water interface. *(Official U.S. Navy photograph.)*

new discoveries, oceanographic studies have scarcely begun, and data obtained from instruments operated by remote control and across great depths of water are subject to different interpretations. Changed viewpoints and still more discoveries undoubtedly lie ahead.

By topography and structure, the continents subdivide naturally into major units such as shield areas and marginal mountain belts, and these in turn subdivide into lesser units: coastal plains, plateaus, mountains, valleys, interior lowlands, and the like. In a somewhat similar fashion, the major units of the oceans are three in number (Fig. 11–2): the marginal or transitional zone between the continents and ocean basins, the deep ocean basins, and the system of mid-ocean ridges. Somewhat lesser units include the continental shelves, continental slopes, continental rises, submarine canyons, deep-sea trenches, abyssal plains, mid-ocean canyons, fracture zones, rift valleys, sea mounts and guyots (Fig. 11–3).

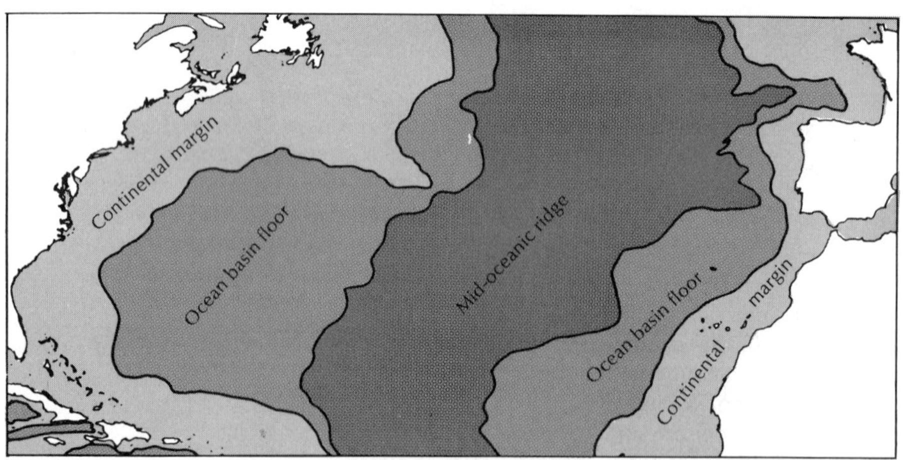

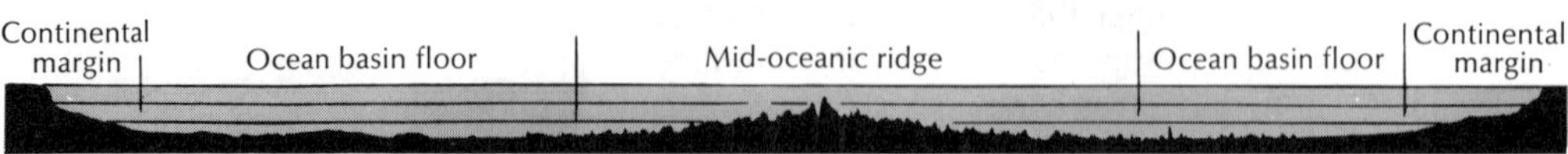

11-2. Major units of the North Atlantic Ocean. Profile is typical for a region between New England and the Sahara Coast. Vertical scale is exaggerated forty times. (Heezen, Tharp, and Ewing, "The Floors of the Oceans," *Geol. Soc. Am. Special Paper* 65, 1959.)

In many profile sections and physiographic diagrams of the ocean floor, the vertical scale must be exaggerated by twenty, forty, or more times to make certain topographic features conspicuous. However, the horizontal scale is unchanged by such a vertical exaggeration because each point on a profile is shifted vertically upward, and its distance from either end of the profile remains the same. Thus slopes on the sea floor are not nearly so steep as suggested by vertically exaggerated profiles, and vertical dimensions are actually quite limited when considered in true horizontal perspective. For example, if all of the relief features on the profile in Fig. 11–2 were reduced to natural scale—i.e., by shrinking all vertical dimensions to one-fortieth the distances shown—then the width of a pencil line drawn across the base of the section would include all of the features shown with room to spare.

Two Dominant Levels

The continental crust and oceanic crust (Fig. 11–4) differ from each other conspicuously in thickness, position relative to sea level, age, and chemical composition, although each is relatively uniform within its boundaries (more true for the oceanic crust). Because of these basic differences, more square miles of the Earth's surface are located near sea level and at about 3 miles below sea level than at any other altitudes—the two dominant levels of the crust (Fig. 11–5)—the areas consisting of very high mountains or very deep ocean are actually quite limited in extent. Apparently these two levels dominate because the continents and ocean basins are approximately in isostatic balance (p. 165). Thus the 6-times thicker, light-weight continental crust projects on the average about 3 miles above the thinner, but heavier, oceanic crust.

When their mean depths are compared to their widths, the oceans are seen to be relatively quite shallow, and ratios of the order of 1000 to 1 and 3000 or 4000 to 1 occur for the Atlantic and Pacific respectively. In perspective, therefore, if the Pacific Ocean were scaled down to a pond 1 mile wide, it would average about 1 to 2 feet in depth.

The Continental Shelf

The *continental shelf* (Figs. 11–6 and 11–7) is part of a transitional zone between continents and ocean basins—a gently sloping, relatively smooth, submerged surface that extends from a coastline towards an ocean basin. A shelf* is present off nearly all the lands, but its width varies from a few miles or less to several hundred miles or more, and in some areas, it forms the submerged extension of a coastal plain. The shelves are underlain by continental-type crust that is somewhat thinner than average. Topographically and structurally, therefore, they are part of the continents, as are the continental slopes (next section). Thus lightweight, sialic, continental-type crust underlies about 35% of the Earth's surface, although only 29% is land.

According to the 1958 Geneva Convention, "the consent of the coastal State shall be obtained in respect of any research concerning the continental shelf and undertaken there." Unfortunately, oceanographic research on the shelves off some countries has since been forbidden. The shelf was defined as extending "to a depth of 200 meters or, beyond that limit, to where the depth of the superjacent waters admits of the exploitation of the natural resources of the said area." Legal, economic, and other problems are raised by this definition.

Oceanographers place the boundary between the continental shelf and continental slope at the location where the very gentle gradient of the shelf changes rather abruptly to the relatively much steeper gradient of

* K. O. Emery, "The Continental Shelves," *Scientific American*, September 1969. This large, single-topic issue was devoted entirely to oceanography and has since been published as a book. It includes a series of handsomely illustrated, well-written articles by leading oceanographers.

11-3. Physiographic diagram of the bottom of the North Atlantic by Marie Tharp and Bruce C Heezen. The vertical scale is exaggerated about twenty times. (B. C. Heezen, "The Rift in the Ocean Floor," *Scientific American,* Oct. 1960.)

the continental slope. In some areas more than one change of gradient occurs, and this makes defining the edge of the shelf more difficult. Commonly the water is less than 600 feet deep where this change in slope occurs, and it averages some 400 to 450 feet (133 meters). Since the shelves average about 45 to 50 miles in width (78 km), their

average slope is about 10 ft/mi, a gradient of about 1:500. In other words, the vertical distance changes about 1 foot in a horizontal distance of 500 feet—to the eye a flat surface. Moreover, portions of the shelf have even lesser gradients averaging about 1:1000. In contrast, the relatively much steeper gradient of the continental slope exceeds 1:40

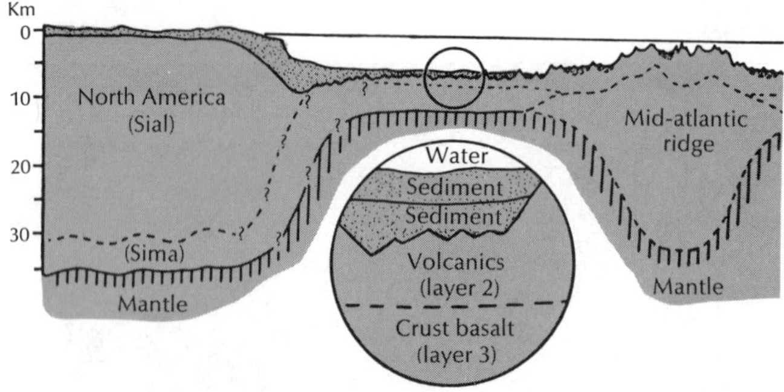

11-4. Schematic diagram of continental and oceanic crustal structure. The two sediment layers (lower is more consolidated) make up layer 1 of the oceanic crust. (W. M. Ewing, in Higginbotham, *Man, Science, Learning, and Education,* Rice University, 1963.)

or about 130/ft mi, and yet this is barely noticeable to the eye. Although hills, ridges, and depressions occur here and there along the shelf, these are subdued topographic features, and the overall impression of the shelf is one of uniformity and smoothness.

The shelves are important to man as sources of food, petroleum, sand and gravel, and other materials (p. 365). Petroleum occurs in enormous quantities in the sediments and sedimentary rocks beneath the shelf and has been extracted through holes

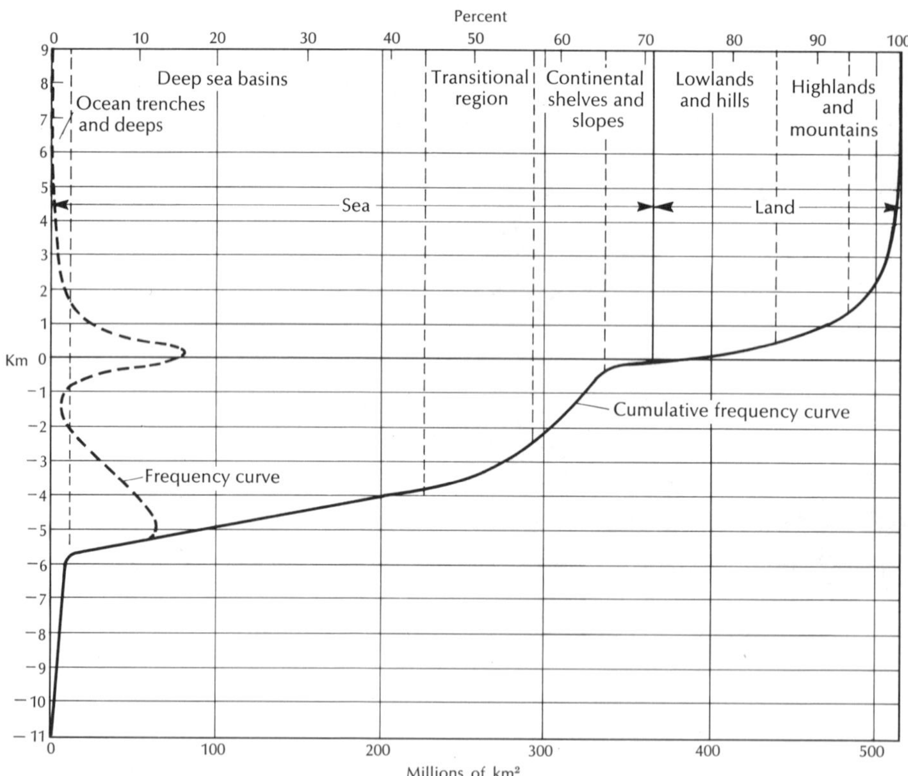

11-5. Estimated percentages of the Earth's surface occurring within certain altitude zones either above or below sea level. Kilometers can be converted into miles approximately by multiplying by 0.62. *(Courtesy William Maurice Ewing.)*

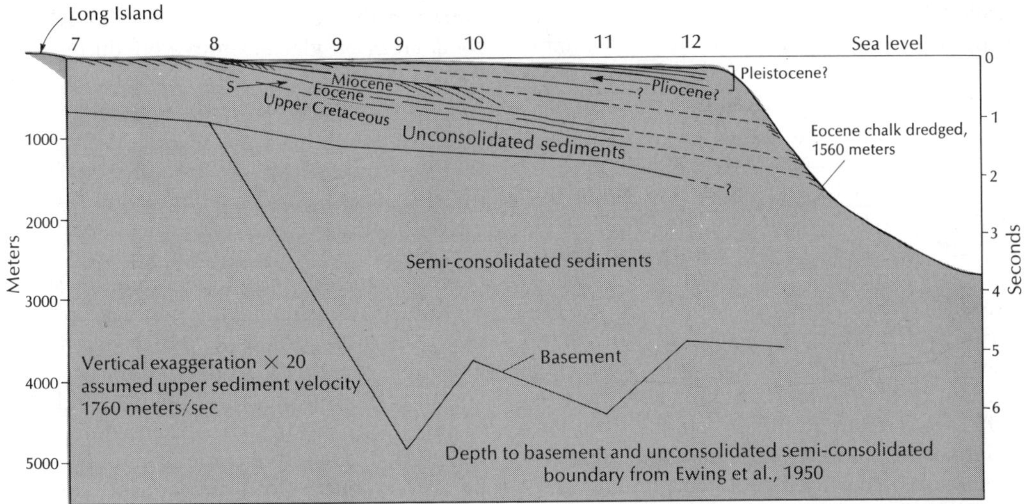

11-6. Structure section southeast of Long Island. Data from seismic refraction and seismic reflection profile studies (after Chelminsky and Fray, 1966). *(Courtesy Maurice Ewing and J. Lamar Worzel, Lamont-Doherty Geological Observatory.)*

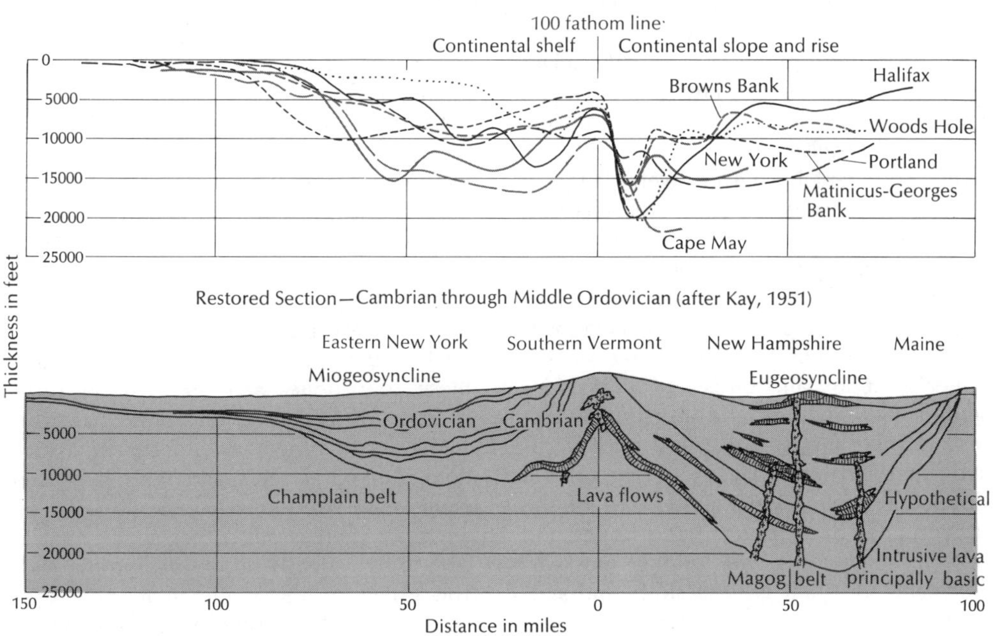

11-7. *Top:* total thicknesses of sediments and sedimentary rocks are plotted along seven different profiles. Maximum thickness occurs beneath the upper part of the rise (exceeds 20,000 feet at Cape May). *Bottom:* restored section of Cambrian-Ordovician rocks. Note apparent correlation of continental shelf with miogeosyncline and continental rise with eugeosyncline (after Drake et al. 1959). *(Courtesy J. Lamar Worzel, Lamont-Doherty Geological Observatory.)*

drilled from large platforms that are either supported on tall legs or float. In the late 1960's, about one-fifth of the world's production of oil and gas came from the continental shelves, and a much larger proportion seems to lie immediately ahead.

Food for plants and animals is abundant along shallower portions of the shelf where currents are stronger and supplies of nutrients are constantly renewed. Thus these waters contain immense food supplies for man, which today are only partially harvested. The dollar value of the worldwide harvest of marine food resources in the late 1960's was about $8 billion (chiefly from the continental shelves), approximately double the value of the shelf-produced oil and gas.

The military services and shipping industry need accurate information concerning these shallow waters and the configuration of the underlying sea floor. In another vein, many of the ancient sedimentary rocks investigated by geologists apparently accumulated in shallow seas that once covered parts of the continents, and today sediments are being deposited on the continental shelves in apparently similar environments—we study them in order to apply the principle of uniformitarianism.

The continental shelves were affected greatly by Pleistocene sea level changes. In fact, during the glacial ages, when sea level was some 300 to 400 feet lower than at present, much of the present shelf area was exposed to the air. Forests, meadows, ponds, and swamps then developed on the former sea floor, and birds and animals inhabited the latitudes and longitudes that fish and other marine organisms once occupied—and inhabit today. Dredging on the continental shelf off the East Coast of the United States has produced the bones and teeth of mammoths, mastodons, musk-ox, horse, and the giant ground sloth, and similar finds have been made on the shelves off other countries. Samples of pollen-containing fresh-water peat have also been dredged and cored from the present shelf. Superposition of the pollen grains, plus carbon-14 dating, show that changes occurred in the vegetation that dominated the shelf areas as glaciers retreated during the past 15,000 years or so. Tundra was replaced by spruce-pine forests, which in their turn were supplanted by deciduous trees. Man presumably lived and hunted on water-free parts of the shelf (like living on the present coastal plain).

A number of shelf or shelflike types have been recognized, and these commonly show a close relationship to the geology and topography of the adjacent coastal regions. To illustrate, off the glaciated coast of New England, the sea floor is irregular and has features such as troughs, basins, hills, and ridges that suggest erosion and deposition by an ice sheet. On a similar land surface, such troughs and basins would form lakes, and some of the hills and ridges would resemble drumlins and end moraines. Thus Georges Bank off Cape Cod, and similar banks off the coast to the northeast, form elongated ridges that rise above the sea floor on either side. They may well represent sedimentary debris that collected along the margin of the North American Ice Sheet in this region.

Farther south along the eastern coast of the United States, the shelf floor is marked by numerous low ridges and troughs that parallel the present coastline. They resemble the barrier islands and associated troughs that occur near shore today. Perhaps they formed when the edge of the sea shifted back and forth across this area as sea level fluctuated during the Pleistocene. Some of these low ridges shift in position during violent storms and form shipping hazards.

Shelves of other types occur in association with coral reefs and with large sediment-laden rivers that have built deltas into the sea, whereas still other shelves are underlain by igneous and metamorphic rocks. The origin of the continental shelves and slopes, and the distribution of sediments on them, are discussed later. However, sand and gravel sizes predominate on the shelf wherever bottom currents tend to be strong enough to move the smaller silt and clay sizes. When we collect samples at a particular location, we tend to find sediments that accumulated under conditions different from those of today, and Pleistocene sea level fluctuations are the chief cause.

Continental Slope and Rise

From the edge of the continental shelf, the sea floor commonly descends to the ocean basins along surfaces known as the *continental slope* (upper, steeper portion) and *continental rise* (Fig. 11–8). In other areas, deep-sea trenches, island arcs, and fault surfaces intervene between the continental slope and the ocean basins. Relative to the continental shelf and the continental rise, the continental slope is quite steep. Its gradient ranges from 1:40 to 1:6 (about 130 ft/mi to 900 ft/mi), although locally the slope is steeper than this. The continental slope occurs off all the lands and is one of the most marked relief features on the Earth. It commonly has a straight or gently curving trend, and its average width of 10 to 20 miles is narrower than either the continental shelf or rise. It begins in water some 300 to 600 feet deep and ends at depths of 1 to 2 miles.

The continental slope is most impressive off mountainous coasts bordered by deep-sea trenches. Here the continental rise and continental shelf may be absent or quite narrow—the side of a trench merges into a continental slope, which in turn grades into an adjacent mountain side towering above sea level. Such a slope begins along the high peaks of the Andes and extends almost unbroken through a vertical distance exceeding 40,000 feet (about 8 miles or 13 km) to the floor of the trench that makes a deep gash in the adjacent sea floor. Apparently a large plate of oceanic lithosphere is descending beneath continental lithosphere along the trench (p. 314).

The continental slopes are not smooth but are scarred here and there by submarine canyons (p. 429), some of which rival the Grand Canyon in size, and by smaller valley-like depressions.

The continental rise consists of a thick wedge of gently sloping sediments (Fig. 11–9) that tends to be present at the bottom of a continental slope wherever deep-sea trenches do not occur. Commonly a fairly well defined change in slope occurs between rise and slope, but in some places one merges imperceptibly into the other. The continental rise has a relatively smooth surface and varies in width from about 60 to 600 miles (100 to 1000 km). Its average gradient approximates 1:300 (about 20 ft/mi) which lies midway between those of the slope and shelf.

The continental rise apparently exists because sediments have piled up at the base

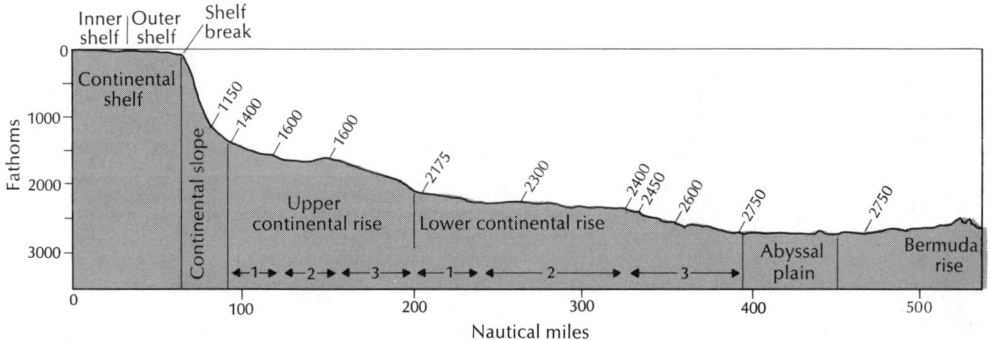

11-8. Continental margin provinces as shown on a type profile off the northeastern United States. (Heezen, Tharp, and Ewing, "The Floors of the Oceans," *Am. Geol. Soc. Spec. Paper* 65, 1959.)

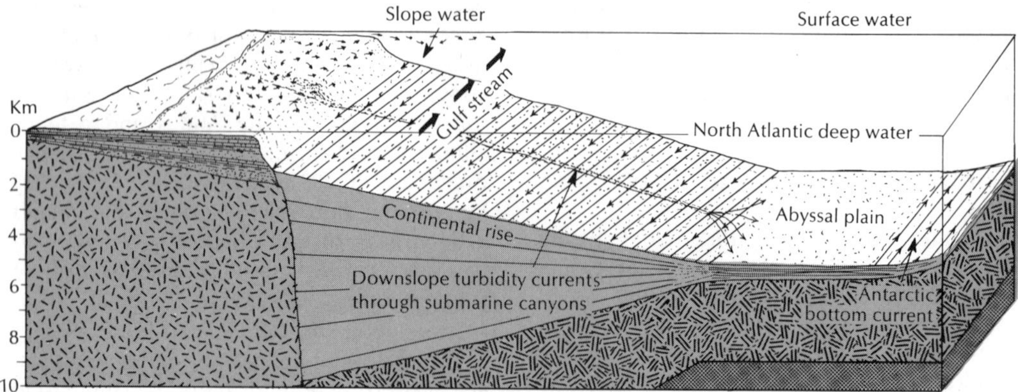

11-9. Shaping of continental rise by bottom-flowing currents (schematic). Arrows show prevailing bottom currents. Turbidites occur beneath the abyssal plain which is underlain by oceanic crust and mantle (black). The continental shelf and slope are underlain by continental crust. (Courtesy Heezen, Hollister, and Ruddiman, *Science,* 22 April 1966, p. 507.)

of the continental slope in a manner somewhat analogous to the formation of alluvial fans along the margin of a mountain range. The sediments in these coalesced deep-sea fans came from the land and may consist largely of clay-sized particles that were transported seaward of the shelf and slope; turbidity current deposits seem to be relatively uncommon. The pile of sediments is thickest—as much as 6 miles (10 km) in places—beneath the upper part of the rise and thins seaward.

According to a rather recent discovery, such land-derived sediments have been re-

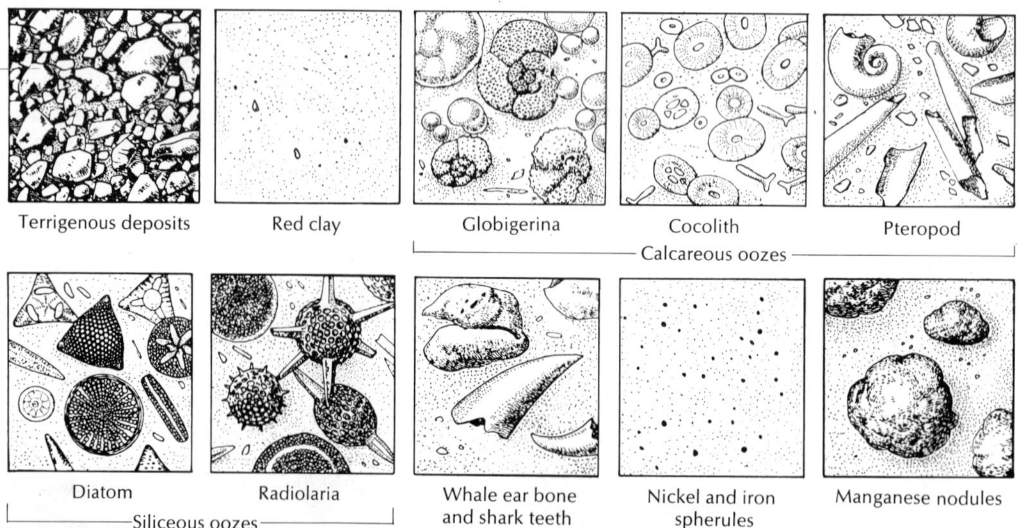

11-10. Some types of ocean-floor sediments; most are magnified, but not all are on the same scale; e.g., the manganese nodules, whale bone, and shark teeth are reduced. With one exception (extreme left) all are pelagic deposits. (John L. Mero, "Minerals on the Ocean Floor," *Scientific American,* Dec. 1960.)

distributed after reaching the continental rise by bottom-flowing currents that move parallel to contour lines along its surface Clay-size sediments, transported and deposited by such currents, have thus smoothed and shaped the surface of the rise. The Coriolis deflection (p. 658) shifts such currents toward the land, and thus the strongest currents and greatest sediment thicknesses occur near the top of a rise. Continental rises in different parts of the Earth have apparently been formed in a similar manner to those off the East Coast of the United States and have similar characteristics.

Ocean-Floor Sediments

Sediments from different sources (Fig. 11–10) have reached the ocean floor by diverse routes and have piled up slowly for many millions of years. Total thicknesses have been measured (Fig. 11–11), and current rates of sedimentation have been estimated. When these quantities are related to the 4.6-billion-year age estimated for the Earth, the total volume of sediments carpeting the ocean bottom has turned out to be surprisingly small, and the sediments are also surprisingly young—moreover, the ocean floor beneath the sediments is also youthful.

For some decades now, samples have been dredged from the ocean floors, and cores have been extracted that range up to 30 feet and more in length (p. 374). However, very great strides in sampling ocean-floor sediments were taken in the late 1960's as part of a United States deep-sea drilling project (JOIDES). A number of holes were drilled at various locations in water as much as a few miles deep, and some of these holes penetrated for hundreds of feet—all of the way through the sediments into the underlying igneous rocks. The oldest sediments cored to date are about 150 million years old (Jurassic), and cores forming a nearly complete record, in aggregate, have been obtained for much of this time span. Results to date strongly support the view that the ocean basins are relatively young features, that the sea floor is spreading away from the mid-ocean ridges, and that continents are drifting.

Ocean sediments are commonly classified as terrigenous (from the land) and pelagic (belonging to the deep sea), but the distinction is sometimes blurred. The great bulk of the sediment has come from the land either directly or indirectly. Rivers have transported very large quantities to the sea, surf action on coastal areas has produced other quantities, and waves and currents have distributed this material widely. However, the coarsel sediments and the greater thicknesses have generally accumulated along the continental margins. Sand- and gravel-size particles typify the continental shelves, whereas clay-

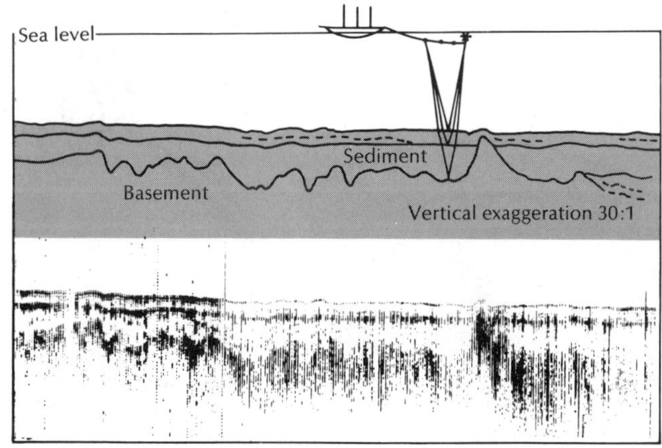

11-11. Seismic reflection profiling at sea: sketch above and actual record below. Horizontal distance is about 110 miles, and average sediment thickness is about 500 meters. Sound signals are emitted at frequent intervals while a ship is under way; thus a continuous profile results. Some of the signal is reflected by the sea floor (like an echo sounder) and some at other interfaces below the sea floor. (Courtesy R. W. Fairbridge, *The Encyclopedia of Oceanography,* Van Nostrand Reinhold, 1966.)

size particles are most common on the continental slopes. Wind has transported other sediment that was lifted into the air from the drier regions of the Earth or added by volcanic eruptions. Submarine volcanoes have contributed additional material, and icebergs peppered with rock debris have melted and dropped their loads far at sea. Meteorite fragments have been added steadily, but probably in minor quantities.

As an indirect land-derived source, substances such as calcium carbonate and silica are transported from the lands in solution and are extracted from sea water by a host of organisms to make shells. These shells—many of them very small—descend one by one to the sea floor and form a large portion of pelagic sediment. Their accumulation has been described picturesquely as the "long snowfall" and the "rain of death," but the latter term is somewhat misleading, since many shells are discarded by living animals who then grow new shells.

Ocean-floor sediments have accumulated as a result of processes of two strikingly different types. Over all parts of the oceans, particle after tiny particle has settled slowly to the bottom—clay-sized pieces that have drifted far from land as well as shells of minute, floating organisms. These accumulate steadily and can pile up on hills and ridges as well as in depressions and on smooth submarine plains. However, deep-water currents may be strong enough in places to keep these from accumulating, and some slopes are too steep.

Turbidity current deposits form the other type (Fig. 11–12) which tend to be limited to the lower areas of the sea floor and occur sporadically. They are one of the surprising discoveries of the last two decades concerning ocean-floor sediments (limited thickness and relatively youthful age are others).

Turbidity Currents and Deposits

Turbidity current deposits, or *turbidites*, are relatively coarse-grained and consist of silts, sands, and even an occasional gravel. Characteristically, the sediments in any one layer show graded bedding, i.e., coarser sediments were dropped first as a current slowed down and occur on the bottom (Fig. 11–13). Sediment sizes then decrease upward to the top of a layer. The bottom of a graded bed may be sharply defined from the sediment beneath it, whereas its finer-grained top portion may merge with the overlying layer. Turbidity current sediments may contain fragments of animals and plants that live

11-12. A small density current produced by squirting ink from a pen.

only in the shallower waters of the continental shelves and tend to have a higher lime content than deep-sea clays.

Many ancient coarse-grained sedimentary rocks contain fragments of shallow water fossils and had long been interpreted as having formed near shore. Now these are being re-examined and reinterpreted as the records of ancient turbidity currents. Evidence of strong current activity and ripple marks also used to be taken as evidence of shallow-water origin; but they have been found on the sea floor in deep water.

Turbidity current deposits have been described as displaced sediments because they contrast so conspicuously with their deep-water environment in their biological, chemical, and physical properties. Such sediments accumulate temporarily on a continental shelf, on the upper part of a continental slope, and in the heads of submarine canyons. On steeper gradients, they may pile upward until they reach an unstable angle and slumping occurs; or the sediments may be jarred loose from regions of gentler gradients by earthquakes that occur beneath the sea floor. A mass of sliding turbulent sediment tends to disperse throughout the water immediately adjacent to it, and this suspended material increases the density of the water. Thus this mass of muddy, discolored, sediment-laden water slides downslope as a gravity-pulled density current (Fig. 11–14).

On its descent, such a current is deflected by submarine hills or ridges in its path and flows along the sea floor until some gap is reached that leads to a deeper depression. Eventually the current slows down, and its suspended sediments are gradually spread out on the sea floor. They cover abruptly and contrast conspicuously with the clay-sized sediments that had accumulated slowly during the preceding years as particle after particle drifted downward. In turn, other clay particles and shell fragments are dropped on top of a turbidity current deposit. Sediments transported by such cur-

11-13. Part of a deep-sea core 460 cm long, showing a turbidite layer with graded bedding; taken at a depth of 715 meters southwest of Jan Mayen Island (70°33′ N, 11°14′ W). *(Courtesy Lamont-Doherty Geological Observatory.)*

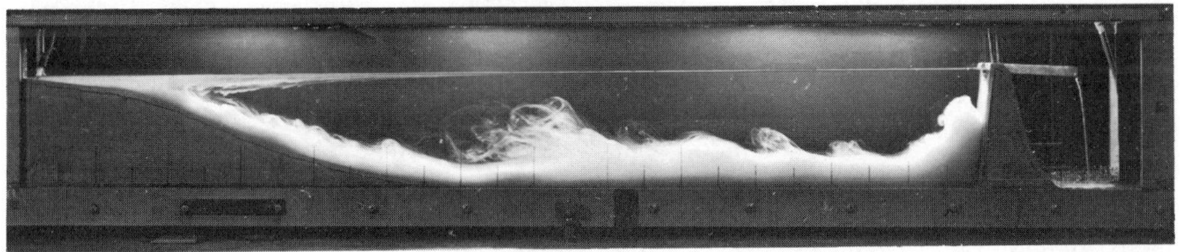

11-14. A turbidity current. *(Courtesy V. A. Vanoni.)*

rents form part of the continental rise (probably minor) and abyssal plains. In addition, turbidity currents have kept some submarine canyons from filling with sediments and may have eroded such canyons.

Burrowing organisms in places stir and mix the upper several inches of sediment on the ocean floor and destroy its stratification. Since particle by particle deposition is very slow, the resulting sediment can be reworked more or less continuously and thus lack stratification throughout a considerable thickness. Turbidity currents, on the other hand, deposit their sediments rapidly, and the lower portion of a thick deposit (below a depth of 8 inches or so) tends to show stratification. The upper portion may show little because it has been subsequently reworked by burrowing animals (by a new group, because such rapid influxes of sediment probably overwhelm the bottom dwellers in an area).

Turbidity current deposits are in turn slowly covered by tiny particles that constantly drift downward from above. Sediment samples collected from the deep-ocean floors during the earlier days of oceanographic investigation did not show turbidity current deposits (at least they were not recognized) because such samples came only from the uppermost layers. Cores tens of feet in length can now be extracted and many of these show a succession of turbidity current deposits interlayered with pelagic clays and shell fragments. Evidently turbidity current activity was at its height during the glacial stages of the Pleistocene, since such deposits appear to be relatively thick and numerous and occur far from land in areas not reached by recent turbidity currents. Perhaps the lowered sea levels of the glacial ages were largely responsible, because more than half of the present continental shelves were then above sea level. Rivers would presumably transport their newly exposed sediments into the sea, and sediments that had previously been dropped in deeper water near the outer edge of a continental shelf would be stirred by waves and currents.

Clays and Oozes

Exclusive of turbidity current deposits, material dropped by melting ice, and manganese nodules, sediments mantling the deep-sea floor far from land are very fine-grained and consist of two general kinds: calcareous and siliceous oozes and red clay (Fig. 11–10). Soft, cold, slimy oozes are formed primarily from the skeletal remains of tiny planktonic organisms, and most of these shells originated near the surface in the mixed, aerated zone that is warmed and illuminated by the sun. Such shells have settled through 2 to 3 miles of water to reach the deep ocean floor. Since the organisms vary somewhat in size, oozes are less well sorted than clays. Many fragments are of silt size, and some are of sand size. The tiny shells of these minute, floating animals and plants are made chiefly of calcium carbonate (more abundant) and silica.

If the sediment contains more than 30% of skeletal material, it is arbitrarily called an *ooze* and generally named after the most abundant type of animal or plant represented: thus we have globigerina ooze (commonly whitish and may cover nearly half of the deep ocean floor), pteropod ooze, coccolith ooze, radiolarian ooze, and diatom ooze. If the sediment contains less than 30% of skeletal material, it is called *red clay,* although many clays are not red (see below). Thus oozes and clays grade one into the other.

The distribution of shells on the ocean floor (Fig. 11–15) does not necessarily coincide with their distribution in the upper waters because calcium carbonate is soluble in water containing carbon dioxide. Calcareous shells may thus be dissolved during their slow descent and also at the bottom until they are buried. Such shells tend to be absent in waters deeper than 14,000 to 15,000 feet and at high latitudes where the water is colder and contains a higher proportion of carbon dioxide. Other tiny, abundant, planktonic organisms make shells of silica, which dissolves very slowly. Thus radiolaria (tiny animals) and diatoms (tiny

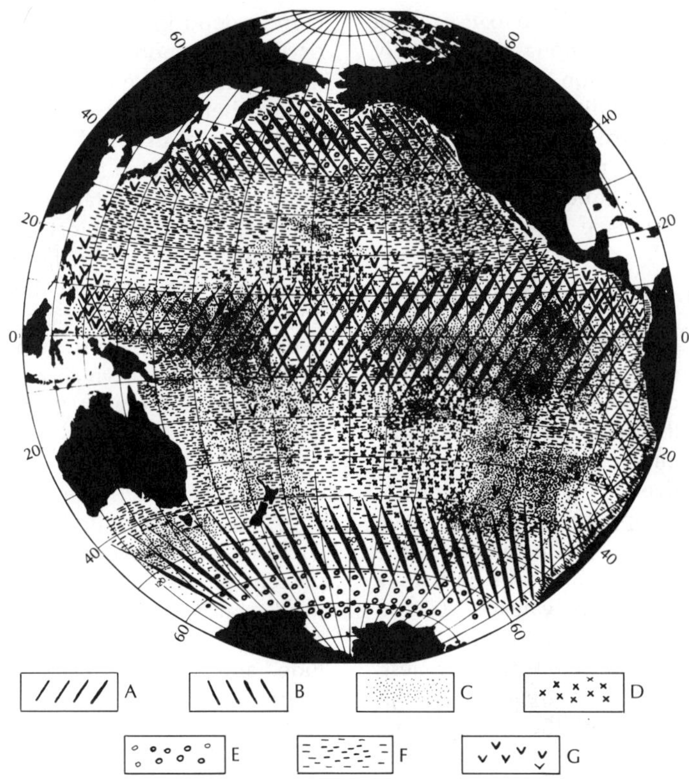

11-15. Distribution of major sedimentary components in Pacific sediments (exclusive of ferromanganese nodules). A, radiolaria; B, diatoms; C, calcium carbonate, mainly foraminifera and nannoplankton; D, zeolite; E, ice-rafted debris; F, fine-grained silicates, mainly clay minerals, quartz, and feldspar; G, fresh volcanic material. (Courtesy R. W. Fairbridge, *Encyclopedia of Oceanography,* Van Nostrand Reinhold, 1966.)

plants—algae) have accumulated in quantity on parts of the sea floor to form siliceous sediments. Generally these organisms are outnumbered in the plankton by other organisms that produce calcareous shells, and they tend to predominate only in deeper, colder waters where the calcareous forms are dissolved.

Red clay is made of very tiny particles of quartz, mica, and various clay minerals. Although brown and other colors are as common as red or more common, the first samples happened to be reddish, hence the name. The bulk of the red clay probably consists of land-derived sediments that settle very slowly because of their shapes and tiny sizes. Thus they are distributed widely by currents during the long time necessary for them to reach bottom. The largest clay particles may settle at a rate approximating 1 ft/day, but smaller sizes descend much more slowly. The reddish color, where present, comes from oxidized iron. In waters nearer shore, organic matter is likely to be present, and this uses up the oxygen supply so that bluish muds accumulate. Red clays include meteorite fragments and relatively insoluble organic remains such as the teeth of sharks and the inner ear bones of whales; some are coated with manganese. Extinct species are represented and occur in great abundance in certain areas (as many as 1500 sharks' teeth have been pulled up in a single dredge haul). Therefore, such organic remains must have accumulated over a long period of time, which emphasizes the very slow rate of sedimentation on the deep-sea floor (exclusive of turbidity current deposits). The organic content of ocean sediments is higher in areas where the oxygen content is low and where sedimentation is rapid enough to bury it before destruction.

Thicknesses and Rates of Deposition

The mid-ocean ridges appear to be nearly devoid of sediment, especially along their crests, and the distribution and thicknesses of sediments in the deep basins are variable. The Atlantic is marked by many miles of coastline, large rivers, and few deep marginal trenches to trap sediment near land. Thus sediments appear to be thicker in the Atlantic than in the Pacific, and perhaps a typical thickness in deep water is 2000 to 3000 feet or less, whereas that of the Pacific approximates 1000 to 1500 feet or less. Thicknesses several times greater than this had been expected, and the thinness and youthfulness of sea-floor sediments had been a puzzle until the concept of sea-floor spreading provided an explanation.

Thickness estimates are somewhat uncertain; e.g., lava flows have occurred on the sea floor at different places and times in the past and may constitute "false bottoms" for seismic signals. Moreover, the nature of layer 2 of the oceanic crust is somewhat controversial. It may represent firmly consolidated sediments, as some have suggested, but it more probably consists of volcanic rocks or of a mixture of different rock types. Compaction of more deeply buried sediments also decreases thicknesses. Thus 3000 feet of original clay deposition might consolidate under pressure into about 1500 feet, but calcareous ooze shrinks much less upon consolidation.

Present rates of accumulation are known only tentatively. Perhaps 1 inch of calcareous ooze will accumulate in about 2000 to 6000 years, whereas 1 inch of red clay may take 5000 to 50,000 years, but these estimates have little value in areas where turbidity current deposits occur. There, several inches or a foot or more of sediment may pile up in a matter of hours, yet many years may elapse before the next turbidity current reaches such an area.

Analysis of deep-sea cores indicates that sediments show certain changes related to latitude, and this is most evident along the equator in the mid-Pacific where the influence of sediments from the lands is small, and deposition of pelagic sediments is relatively rapid. Reasons: upwelling occurs along the equator which produces a belt of abundant nutrients; thus plankton are abundant.

Certain relatively rare elements dissolved in ocean water are extracted by organisms and thus concentrated in their remains. Such elements occur in abnormally high percentage in sediments deposited along the equator, but seem to occur in lesser quantities at 5 degrees north and south latitude, and in still smaller quantities at 10 degrees. Furthermore, at any one latitude, certain substances appear to have been concentrated by organisms more abundantly at one time than at another. If such relationships also occur at greater depths, and if cores several hundred feet long can be extracted at different latitudes, then the sediments may show whether or not the location of the geographic equator has remained unchanged in the past—relative to polar wandering and continental drifting.

Sediments of the deep-ocean floors also contain information about climatic and other changes on the Earth; e.g., temperature fluctuations are recorded by alternations of warm-water organisms with cool-water organisms in succeeding layers in cores, and these seem related to Pleistocene glacial and interglacial ages (p. 242).

According to David Ericson, if the texture of a core is uniform from top to bottom, its sediment has accumulated at a reasonably constant rate—determined by carbon-14 age measurements in the upper parts of the cores (limited to the last 40,000 years or so). Furthermore, when several cores were compared, those with similar textures had similar accumulation rates. Thus rates of accumlation determined for the upper parts of cores with uniform textures can be extrapolated to the lower portions that are too old to be dated by the carbon-14 method.

Temperatures of surface waters at different times in the past may be estimated by determining the proportion of oxygen-18 to oxygen-16 in the shells of the deep-sea oozes (p. 242). Such shells were built by tiny planktonic creatures floating near the surface.

The Ocean Basins

The ocean basins are located between the continental margins and the mid-ocean ridges and contain a number of distinctive topographic units: abyssal plains, seamounts and guyots, mid-ocean canyons, and hills and rises that project somewhat above the general level of the ocean basins (rises have larger areal extent than hills).

Abyssal Plains

The abyssal plains (Fig. 11–3) constitute the flattest areas on the Earth's surface and apparently are constructional features that have been built up chiefly by sediments dropped by turbidity currents. Cores tens of feet in length—hundreds of feet in the deep-drilling program—have been extracted from the abyssal plains and include many relatively thick layers of coarse sand that show graded bedding and contain shallow-water fossils. These are interbedded with smaller quantities of clay-size particles that settled slowly and continuously, particle by particle, onto the sea floor. The sediments evidently were deposited on a sea floor that was as rough topographically as the irregular surfaces that now surround them, and hills only partially buried by the sediments project here and there above the plains.

Presumably the irregularities of the ocean floor have resulted from igneous activity and crustal movements; e.g., some of the hills may result from the intrusion of laccoliths into layer 2 of the oceanic crust. Similar topographic features have formed on land but have been rather quickly subdued because erosion predominates over deposition. On the deep-sea floor, however, where erosion is relatively slow and deposition perdominates, such irregular topography tends to be preserved relatively unchanged except for the gradual blanketing effect of ocean-floor sedimentation.

Abyssal plains are absent off continental margins that contain deep-sea trenches not yet filled by sediments. These deep troughs in the sea floor parallel the continental margins and trap the sediments transported by turbidity currents down the adjacent continental slope. Thus abyssal plains are more widespread in the Atlantic, which has fewer of the deep troughs, than in the Pacific, but they have been found in all of the oceans (Fig. 11–16). Surfaces of abyssal plains tend

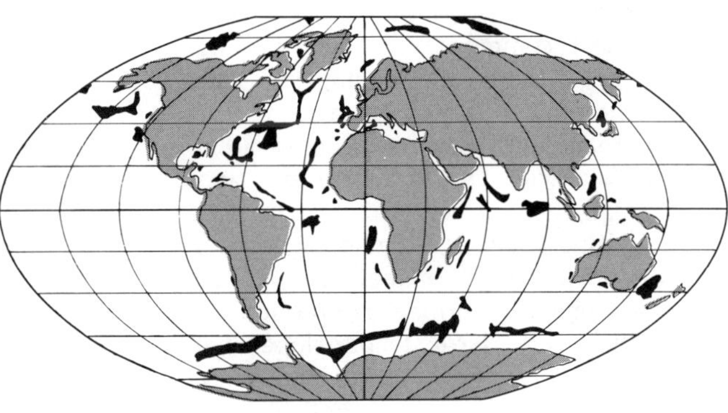

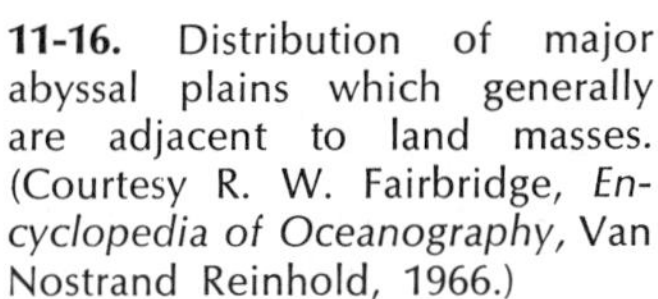

11-16. Distribution of major abyssal plains which generally are adjacent to land masses. (Courtesy R. W. Fairbridge, *Encyclopedia of Oceanography,* Van Nostrand Reinhold, 1966.)

to slope very gradually away from main supply areas such as the mouths of large submarine canyons, and two neighboring abyssal plains may be joined at a low area in the sea floor. Evidently the turbidity currents were funneled through this gap as they moved from the higher plain to the lower one.

Mid-Ocean Canyons

These mark the ocean basin floor in places but have a misleading name because they are rather shallow, steep-sided, flat-floored troughs (Fig. 11–3). A *mid-ocean canyon* may have a width of 3 to 5 miles, a depth of several hundred feet or less, and extend for hundreds of miles along a very gentle gradient. Wide, low, leveelike ridges tend to margin the channels on each side, and some Northern Hemisphere levees are larger on the right side—perhaps caused by the Coriolis deflection (p. 658). A number of mid-ocean canyons have tributary channels and extend down a continental rise and across an abyssal plain. Presumably their origin is associated with the flow of turbidity currents.

Seamounts and Guyots

A *seamount* has been defined as any isolated elevation which rises some 3000 feet or more above the surrounding sea floor, and a *guyot* (pronounced to rhyme with "Leo") is a seamount with a flat top. Seamounts occur in the lower portion of the continental margin zone, on the ocean basin floors (most common), and on the mid-ocean ridges. Moreover, the tops of some seamounts project above sea level to form islands. Seamounts seem to be most abundant in the Pacific Ocean, where thousands have already been discovered. Many have a conical shape similar to that of a volcano, and presumably most of them are actually volcanic cones that formed by eruptions on the sea floor. They have steeper slopes than comparable volcanoes on land because lava cools quickly under water, and underwater eruptions do not propel debris as far.

The flat tops of the guyots have probably been produced by erosion by surf (predominates) and streams near sea level at some time in the past. Perhaps a typical guyot is a volcanic cone that was built on the sea floor. Eventually the volcanic debris piled high enough to emerge as an island in the sea. Subsequently its top was planed off by erosion near sea level, and following this, the sea floor subsided beneath it. Thus its flat summit may now be covered by ½ to 1 mile or more of water.

Probably guyots have formed at different times in the past, and the older ones may have subsided more. However, the rate of sinking may also vary somewhat in different parts of the oceans. Furthermore, one has to distinguish between widespread regional subsidence of the ocean floor and the sinking of a single cone. Such localized downward movements may occur because the crust beneath a cone was left somewhat unsupported by the extraction of the magma to build it. Some guyots are distributed in a linear pattern that suggests regional subsidence.

Cores from seamounts and guyots have contained volcanic rocks, Mesozoic and Cenozoic fossils (including some corals which probably once lived in shallow water), and sediments showing ripple marks at depths of ½ to 1 mile or so. These suggest that loose debris is kept from accumulating by current action at such depths.

Mid-Ocean Ridges

The system of mid-ocean ridges, the continents, and ocean basins form the three main units exposed at the Earth's surface—based upon topography, structure, areal extent, and geologic importance—and all three units are approximately in isostatic balance. The Mid-Atlantic Ridge (Fig. 11–17) illustrates well the features of mid-ocean ridges and is the best-known portion of this 40,000-mile long, world-encircling system

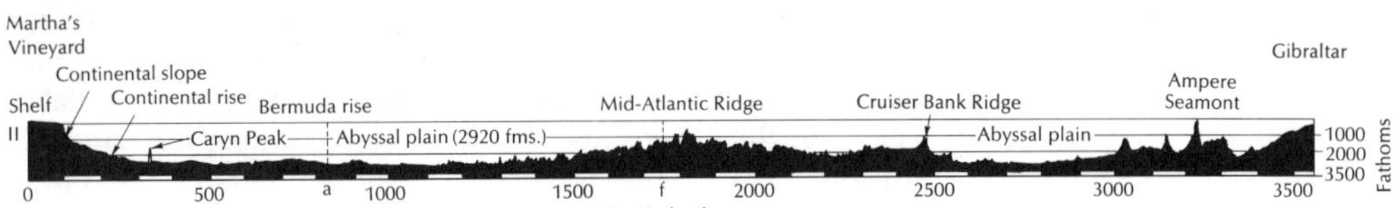

11-17. Trans-Atlantic topographic profile from Martha's Vineyard to Gibraltar. Vertical scale is exaggerated forty times. (Heezen, Tharp, and Ewing, *Am. Geol. Soc. Spec. Paper* 65, 1959.)

that approximately equals the continents in areal extent (however, the system may not be continuous, nor everywhere of the same age).

The Mid-Atlantic Ridge is centrally located between the eastern and western margins of the Atlantic Ocean and is about 1000 miles wide. It is elongated in a general north-south direction and follows a sinuous course roughly parallel to the present coastline. In areal extent, it thus forms about one-third of the entire Atlantic Ocean. The Mid-Atlantic Ridge rises some 2 miles above the ocean basin floor on either side, and a few of its highest peaks project above the surface as islands (e.g., the Azores, Ascension, St. Helena, Tristan da Cunha, Gough Island, and St. Paul's Rocks). An east-west profile across the Mid-Atlantic Ridge shows an enormous broad arch in the Earth's surface. But the surfaces that slope downward from the high central portion are very rough, jagged, and irregular.

An unexpected and significant part of the Mid-Atlantic Ridge is a large, deep, relatively narrow valley, or a zone of parallel, steep-sided valleys with intervening ridges, that extends in a north-south direction along the mid-portion of the Ridge (Fig. 11–3). Along part of the Ridge, this central zone consists of a single huge trench (rift) located on the ridge crest and averaging about 10 to 30 miles in width and 1 to 1½ miles in depth. Shallow-focus earthquakes (centers located within about 20 miles of the surface) occur frequently along this central rift zone. In fact an apparently continuous 40,000-mile long belt of earthquake epicenters follows the median line of the system of mid-ocean ridges (Fig. 6–25)—the epicenter belts apparently mark the boundaries of great lithospheric plates.

The East African rift valleys are similar features: large, elongated, lake-filled, steep-sided troughs that apparently have formed by the downward movement of part of the Earth's crust along steeply dipping marginal faults (normal) because of tensional forces (pulling apart) affecting these areas. Lake Tanganyika, Lake Nyasa, and the Red Sea are major examples, and earthquakes occur in association with them. Analogously, the central rift valley of the Mid-Atlantic Ridge is also the result of tensional forces that are pulling apart the Mid-Atlantic Ridge along its median line. Iceland, noted geologically for its igneous and seismic activity, is located on the crest of the Ridge, and the rift valley structures extend through it.

Oceanic Crust

A typical section of deep-ocean crust (sima, p. 164), such as occurs beneath abyssal plains, consists of three layers (Fig. 11–4). Beneath water that is 2½ to 3 miles deep occurs layer 1, which consists of sediment and sedimentary rock that is commonly less than ½ mile in thickness.

The underlying, little-known layer 2 probably averages about 1 mile in thickness, and *P* waves (p. 159) move within it at an average rate of about 5 km/sec. Perhaps it consists of a variety of rocks (e.g., volcanics and consolidated sedimentary rocks). Seismic profiler records indicate that the top of layer 2 tends to be a rough surface, which suggests an igneous composition for it.

Layer 3 averages about three miles in

thickness and is commonly assumed to be of basaltic-gabbroic composition; it has a *P* wave velocity averaging about 6.7 km/sec. The Mohorovičič discontinuity occurs at the base of layer 3 (i.e., at the base of the crust) where *P* wave velocities increase abruptly to approximately 8.1 km/sec in the upper mantle. According to another view, layer 3 may consist chiefly of serpentine (p. 304).

Typical *P* wave velocities do not occur beneath the Mid-Atlantic Ridge, and sediments tend to be very thin or missing. A layer of rocks about 1½ to 2 miles thick tends to form the top of the ridge, has a *P* wave velocity of about 5.1 km/sec, and is assumed to consist of volcanic rocks. Beneath these volcanic rocks occurs a thick zone of controversial rocks which have a *P* wave velocity of about 7.3 km/sec (midway between speeds for layer 3 and the mantle). Some oceanographers interpret the 7.3 km/ sec material as a mixture of mantle and crustal rocks, whereas others interpret it as part of the upper mantle. In this view, the slower *P* wave velocity is caused by higher-than-average temperatures and by fracturing in the rocks (evidence: high heat flow measurements obtained from the crest of the mid-ocean ridges).

Heat Flow

On the average, heat energy is flowing from the continents, the ocean basins, and the system of mid-ocean ridges at about the same rate, although measurements are somewhat limited in number (about 3000 by 1969) and in geographic coverage. On the other hand, heat flow is much greater than average in a rather narrow zone located along the crest of the mid-ocean ridges, whereas it tends to be less than average on the flanks of the ridges. However, one about offsets the other. Low heat flow is also associated with the deep-sea trenches.

Some 90% of all heat-flow measurements to date have been made on the deep ocean floors, partly because water temperatures below about 1000 fathoms tend to be approximately uniform, equilibrium conditions apparently pertain, and more reliable results can be obtained than on land. In one method, special temperature-measuring devices are attached at regular intervals to a core barrel which is lowered from a ship and penetrates a number of feet into the sea-floor sediments. A sediment sample is thus obtained simultaneously with measurements of the increase in temperature with depth. This measured temperature gradient, together with the thermal conductivity of the sediment sample, provide data for the heat-flow calculation.

Although heat flow is about the same from continents as from oceans, the sources are somewhat different. Heat energy is liberated by the disintegration of radioactive elements, and these are much more abundant in sial (upper part of continental crust) than in sima (oceanic crust). Therefore, less than half of the total heat flow from the continents comes from the underlying mantle, whereas ocean heat flow comes chiefly from the underlying mantle. Thus the upper mantle beneath continents probably differs somewhat from that beneath oceans. Another apparent difference: the low-velocity zone seems to be more prominent beneath the oceans than it is beneath the continents and also to be nearer the surface.

Fracture Zones

Fracture zones (Fig. 11–18) are intriguing, unexpected recent discoveries that were unexplained for a time but are now interpreted as major strike-slip fault zones that offset the system of mid-ocean ridges in a number of places. Since the fracture zones are nearly straight, the faults presumably are nearly vertical surfaces, and the displacements are chiefly horizontal. Fracture zones are long narrow regions of the ocean floor characterized by irregular topography, fault surfaces, elongated troughs and ridges, submarine volcanoes, and as much as 2 miles of vertical

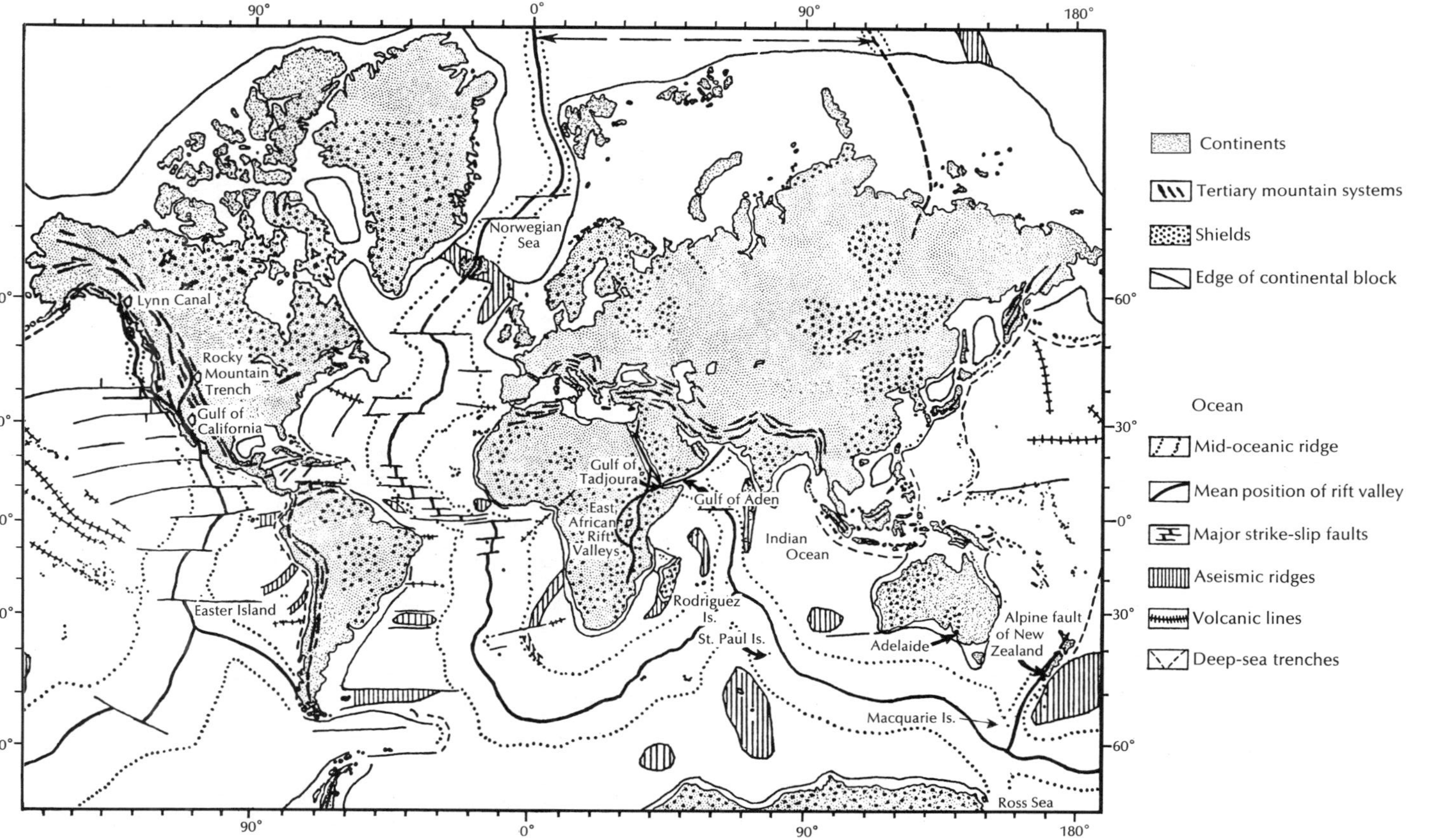

11-18. Major tectonic features of the Earth (after Heezen, 1962). (Courtesy Manik Talwani, *Lamont Geological Observatory Contribution* No. 1171, 1968.)

relief. They are somewhat mountainous belts that may extend in an east-west direction for 1000 to 2000 miles or more and are about 100 miles wide or somewhat less. The sea floor tends to be deeper on one side of a fracture zone than on the other. Such large linear fracture zones are obviously important factors in a study of the Earth's structure and history. Since magnetic measurements of the sea floor reveal striking patterns that are offset along the fracture zones, we turn now to the Earth's magnetic field and to recent discoveries concerning paleomagnetism and magnetic reversals.

Earth Magnetism

The Earth has a magnetic field which resembles that produced by a bar magnet. However, this bar magnet would be inside the Earth, several thousand miles in length, and tilted at present about 11.5 degrees to the Earth's axis of rotation (Fig. 11–19). Thus the geographic poles are offset from the magnetic poles (about 1200 miles) as is the geographic equator from the magnetic equator. Magnetic lines of force are vertical to the surface at the magnetic poles and parallel to it above the magnetic equator. Thus the north-seeking end of a freely suspended compass needle points vertically downward at the magnetic north pole but vertically upward at the magnetic south pole. Such a needle shows no dip at locations midway between the two magnetic poles because it is aligned parallel with the surface. The field decreases in intensity outward from the Earth (about as the cube of the distance) but is very weak even at the surface.

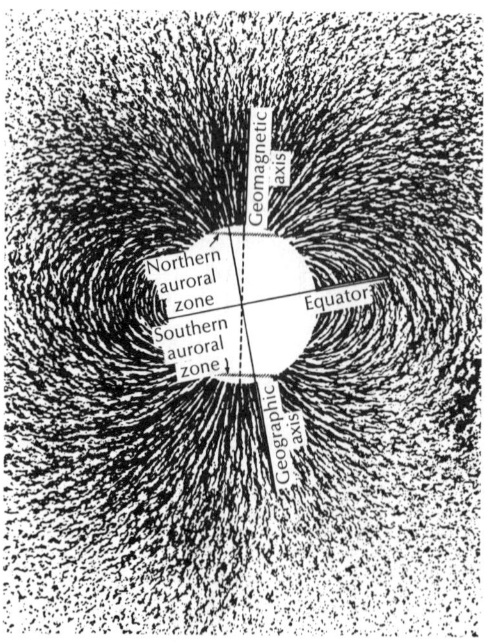

11-19. Earth's magnetic field as illustrated by iron filings in the field of a small bar magnet. *(Courtesy Carnegie Institution.)*

Measurements spanning more than three centuries show that the Earth's magnetic field changes slowly, but the cause of this, and of the field itself, is somewhat uncertain. The source is certainly within the Earth and seems to be associated with slow-moving currents in the outer, fluid part of the Earth's core. Comparison with the moon, Venus, and Mars (bodies which lack comparable magnetic fields) suggests that the presence of a dense core (chiefly iron?) and rapid rotation are two necessary prerequisites. Charged particles from the sun and space tend to follow the lines of force and produce the magnetosphere (p. 655) and the auroras (p. 601).

It is generally assumed that the Earth has had such a dipolar magnetic field throughout most of its geologic history and that its magnetic axis has always corresponded closely to its axis of rotation. This is expected on theoretical grounds and seems to have been demonstrated for the latter part of the Pleistocene (see fossil magnetism below). However, the effects of short-term variations have to be averaged out, which seems to happen over a period of a few thousand years. Thus the present displaced locations of the Earth's magnetic poles apparently are not typical of the average locations in the past—the two sets of poles should more nearly coincide.

When lava cools below a critical temperature (Curie: range is about 700° to 200°C for different minerals) the magnetic fields of iron-bearing minerals within the lava align themselves parallel with the Earth's field at that time and place. However, this magnetic orientation does not coincide with the external orientation of the mineral particles themselves, because these had crystallized previously at a temperature above the Curie point. Likewise, when tiny iron-containing particles settle slowly toward the sea floor, their minute magnetic fields tend to become aligned with that of the Earth. Enough of them retain this alignment during deposition for the magnetic field of a section of sea-floor sediments to coincide with that of the Earth at this location. The inclination of the magnetic field thus frozen within any one body of rock or sediment is called its remanent (fossil) magnetism and is a measure of the latitude at which it formed—providing the body still has its original orientation.

A surprising discovery occurred when studies were made of the fossil magnetism at different levels within superposed flows of lava in Hawaii and elsewhere and in long cores extracted from the sea floor. In some layers, the inclination was about 180 degrees from what it was in other layers—evidently the Earth's magnetic field completely reverses itself on occasions (Fig. 11–20). In other words, at a certain time in a certain geographic location, the north-seeking end of a compass points toward the north pole. At this same location, but at a different time

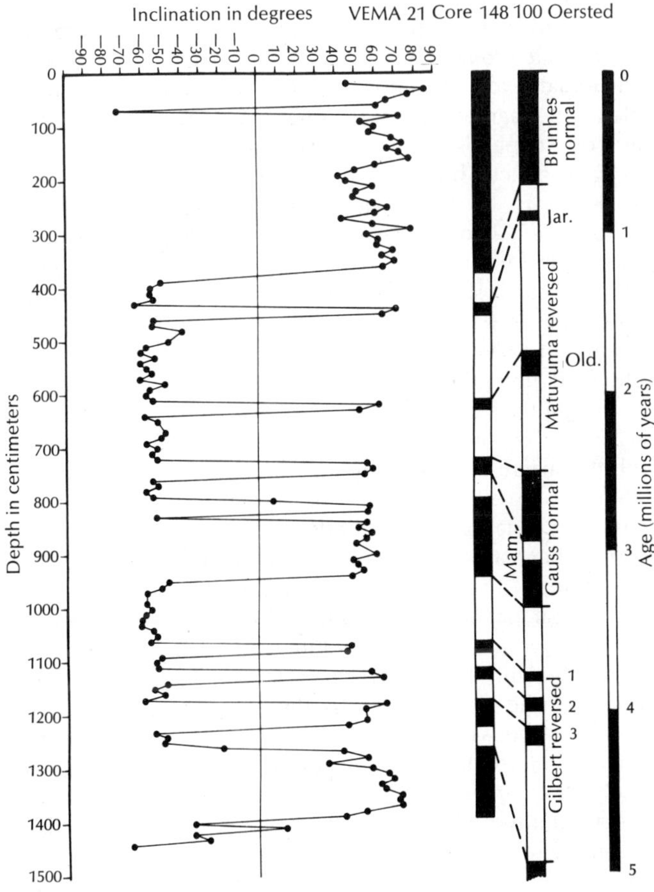

11-20. The magnetic reversals shown in this core seem to correlate with magnetic reversals in lava flows whose absolute ages are known (after Opdyke, in preparation). *(Courtesy Carnegie Institution.)*

in the past, the needle would have pointed toward the south! Moreover, these directions have alternated a number of times. Since the specimens can be dated by various radioactive clocks within them, the times of reversal of the polarity of the Earth's magnetic field have been worked out. Evidently the reversal itself occurs during a relatively brief geologic interval (say 5000 years). Since the changes are worldwide in scope, a new chronology has become available for dating and correlating the past; e.g., nine reversals may have occurred during the last 4 million years or so—each at the same time everywhere on the Earth.

Reversals of the Earth's magnetic field apparently account for the large, linear magnetic anomalies that make stripes on the sea floor, and the anomalies were detected in the following manner. The strength of the Earth's magnetic field changes at a uniform rate from a minimum at the magnetic equator to a maximum at the magnetic poles. Its magnitude at any one location, therefore, depends upon latitude and upon any local effects, but precise measurements are necessary to detect variations because the total field is quite small. Measurements were made by towing a magnetometer behind a research ship along a series of rather closely spaced east-west tracks, thus making a continuous record of the Earth's total magnetic field along these tracks. Next, theoretical magnitudes were calculated for the Earth's field at such latitudes and correlated with the magnitudes actually measured. Any differences were regarded as anomalies caused by variations in the underlying rocks or by other causes, and anomalies indeed were found. They have a striking north-south linear pattern that is offset along the great east-west, strike-slip fracture zones. If the north-south lineations were not interrupted at the fracture zones, they would extend for thousands of miles. If these patterns are interpreted correctly as horizontal offsets, the largest displacements measure several hundred miles, and the smaller ones extend for tens of miles and more.

Positive magnetic anomalies apparently occur above parts of the sea floor in which the fossil magnetism is aligned parallel with the present direction of the Earth's magnetic field—the effect of the fossil magnetism is added to that of the Earth's present field. On the other hand, belts of negative magnetic anomalies apparently occur where the inclination of the fossil magnetism is reversed from that of the present field—here the fossil magnetism subtracts from that of the Earth's field. Such anomalies had in fact been predicted by advocates of a new hypothesis—now apparently confirmed—that involves the spreading apart of the sea floor along the crest of the mid-ocean ridges.

Sea-Floor Spreading

According to a widely held view, convection currents within the mantle (Fig. 11–21) provide the most likely explanation for the discoveries made to date (see p. 317 for a different view). The currents are assumed to rise beneath the crest of a mid-ocean ridge, then to diverge sideways, and finally to sink beneath a trench (also at other sites). The existence of a mid-ocean ridge as a great elongated bulge in the crust is, in this view, associated with the upward movement of the currents beneath it and with the intrusion and extrusion along the crest of material that comes from the underlying mantle. Divergence of the currents outward from a crest creates tension that produces the central rift-valley zone and the earthquakes associated with it. High heat flow occurs within a rather narrow zone along the crest above a hot upward-moving current, whereas lower heat flow occurs along its flanks where some subsidence is taking place.

In this view, new oceanic crust forms along the axis of a mid-ocean ridge as material from the mantle rises into the axial region, cools, and hardens. Perhaps the process is somewhat similar to the intrusion of a huge vertical dike and/or the piling up of a succession of lava flows within a long, narrow, vertically oriented zone. At any rate, new oceanic crust forms at and beneath a ridge crest, subsequently splits apart, and then each half moves laterally away from the crest. However, no gap develops because new oceanic crust forms more or less continuously in the central region where a gap would otherwise develop.

The direction of the Earth's magnetic field at this time (assume it produces a positive anomaly) and place is recorded by the fossil magnetism frozen into iron-containing minerals within the newly formed crust (after sufficient cooling). Since slow lateral creep away from a ridge crest takes place at more or less equal rates on either side, each half of this zone of positive anomaly will move farther away from the ridge crest with time. However, the direction of the Earth's magnetic field becomes reversed at intervals, and thus a zone of negative magnetic anomaly will eventually form along a crest and in its turn move laterally away. Therefore, linear zones with different directions of fossil magnetism alternate to produce the observed negative and positive magnetic anomalies that parallel a mid-ocean ridge on either side (Fig. 11–22). In other words, a belt of negative anomaly on one side is matched by a corresponding belt of negative anomaly at an equal distance from the ridge crest on the other side, and each is flanked on either side by belts of positive anomaly—one older (farther from the ridge) and one younger (closer to the ridge).

Recent observations of different kinds apparently confirm the existence of sea-floor spreading; e.g., sediments increase in both age and thickness with increasing distance from a ridge crest, where little or no sediment occurs (Fig. 11–23). In other words, a hole drilled through sea-floor sediments at, say, a distance of 300 miles from a ridge crest, should penerate sediment that increases in age with depth, and the age of the sediment in the bottom of the hole should be about the same as that of the oceanic crust immediately beneath it. For a hole drilled at a greater distance, the bottom-most sediments (and crust) would be older; for a hole drilled at a lesser distance, the ages would be less, and matching ages should occur at comparable distances on the opposite side of the ridge crest—such seems to be the case (Figs. 11–24 and 11–25). Spreading rates probably vary along the mid-ocean ridge from about 1 to 9 cm/yr (about 0.4 to 3.5 in/yr). However, since this rate of movement occurs on each side, the central zone itself is being widened at double this amount.

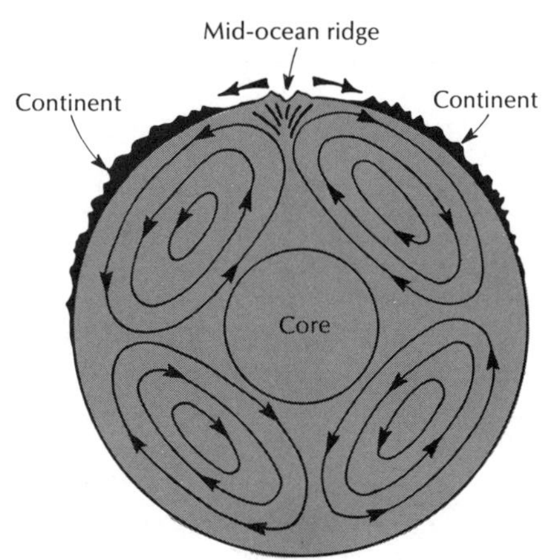

11-21. Convection cells may occur in the Earth's mantle, but perhaps they are limited to just the upper mantle rather than extending throughout the mantle as shown here. *(Courtesy W. M. Ewing.)*

The oceanic crust seems to sink slowly, and apparently at about a uniform rate over large areas, as it spreads outward from the crest of a mid-ocean ridge (thus the lower heat flow along the flanks of the ridges). Therefore, the rate of spreading may determine the steepness of the slopes on the flanks of a mid-ocean ridge: wide gentle slopes correlate with rapid spreading (e.g., the East Pacific Rise), whereas steeper slopes are produced by slower spreading (Mid-Atlantic Ridge).

Recently, more precise locations of sea-floor earthquake epicenters have become available (Fig. 6–25). The epicenters tend to occur in a narrow zone along the crest of a mid-ocean ridge (expected because of crustal movement associated with the formation

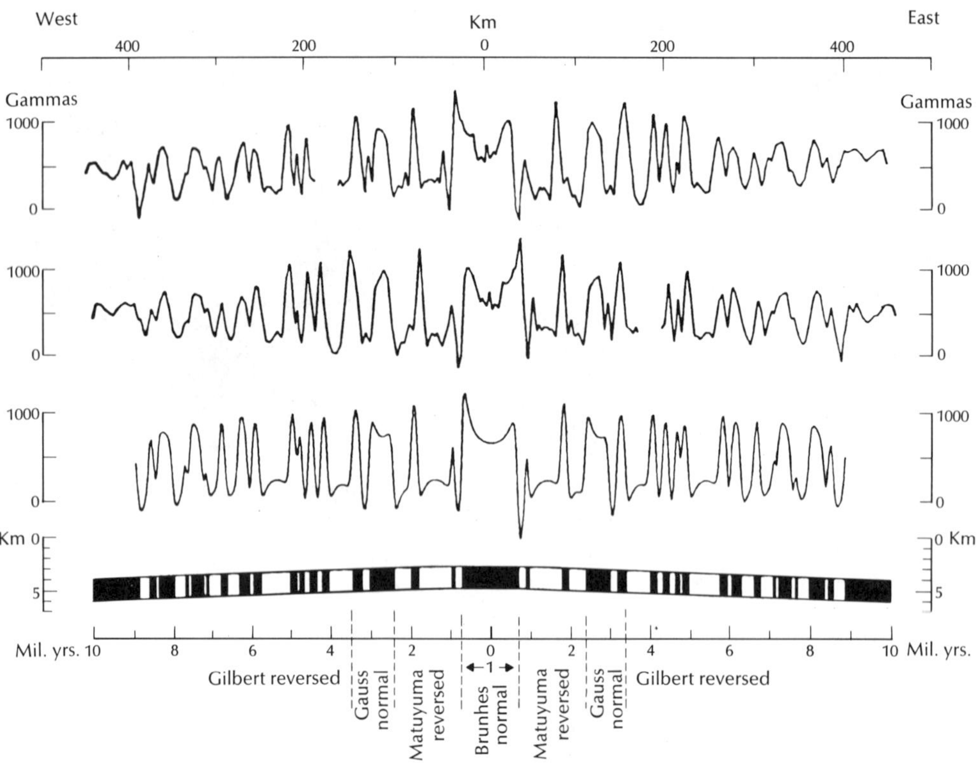

11-22. Profile section of magnetic anomalies across a mid-ocean ridge. Top and middle profiles are reversed; otherwise they are the same; i.e., east is to the left in one and to the right in the other. The lower profile is a calculated one. At the bottom, shaded areas represent present polarity, and unshaded areas represent reversed polarity. The spreading rate here is 4.5 cm/yr (after Pitman and Heirtzler, 1966). *(Courtesy Carnegie Institution.)*

of new oceanic crust) and along the section of a fracture zone that separates two ridge crests. It is only within this part of a fracture zone between two offset ridge crests that the spreading movement is in opposite directions on opposite sides of the fault zone (Fig. 11–26; the term *transform fault* applies here). Outside of this zone, the spreading movement is in the same direction on the two sides of the fracture zone, and thus few earthquakes are produced.

Wilson* has pointed out the strong supporting evidence for sea-floor spreading provided by three different features of the Earth that everywhere change in the same ratio. The first of these involves the magnetic-reversal chronology worked out from superposed lava flows at different locations. The second involves the strips of magnetic anomalies that occur on either side of a mid-ocean ridge. These vary in width because the rate of spreading varies from one part of an ocean to another, but the ratio of the widths is everywhere the same. The third involves the depths at which magnetic reversals occur in deep-sea cores. These take place at different depths at different locations because rates of deposition vary; however, the ratio of the depths is the same for all cores.

According to sea-floor spreading, the

* J. Tuzo Wilson, "A Revolution in Earth Science," *Geotimes*, December 1968.

continents bordering the Atlantic Ocean on the east and west were once joined together. They split along the line of the present Mid-Atlantic Ridge, and each side has moved outward at approximately equal rates; thus the Mid-Atlantic Ridge is centrally located. Continental drifting (next chapter), therefore, forms an essential part of this hypothesis.

Presumably, upward-moving convection currents in the mantle slow down and one day stop beneath a mid-ocean ridge. The ridge should then subside and become inactive, and perhaps this old-age stage is shown by the Mid-Pacific Rise. This extends from the Marianas eastward toward Chile and consists of long, narrow, seismically inactive ridges capped by guyots which have flat tops located ½ to 1 mile below sea level. Since this area is now approximately

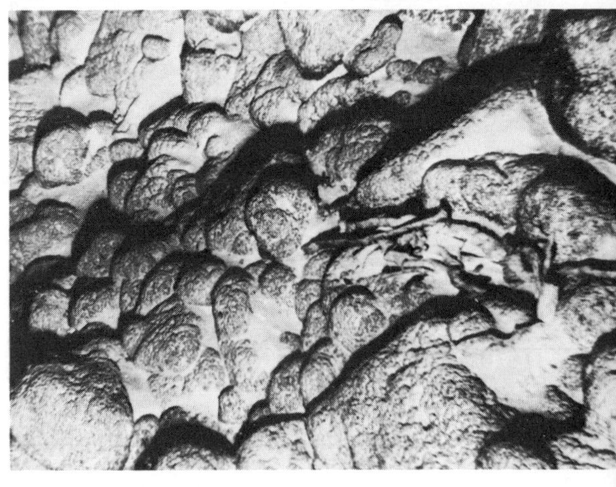

11-23. Rocks exposed near the crest of the Mid-Atlantic Ridge in the South Atlantic. Note pillow lava structure (p. 123). (After Ewing, Ewing, and Talwani, 1964; *courtesy Lamont-Doherty Geological Observatory.)*

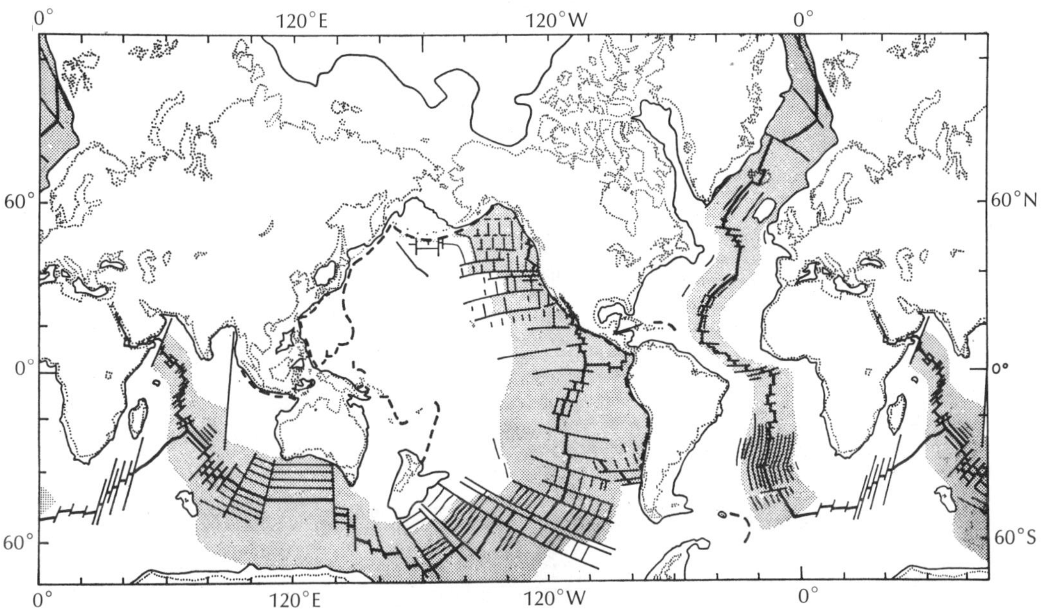

11-24. Tentative delineation of oceanic and continental crust. Symbols: thick dashed lines for deep-sea trenches; thick solid lines for ridge crests; thin solid lines transverse to ridge crests for fracture zones; and thin solid lines parallel to ridge crests for linear magnetic anomalies that can be correlated (Heirtzler et al., 1968). Shaded zone shows oceanic crust thought to have formed during the last 65 million years (Cenozoic). (F. J. Vine, *Nature,* 5 Sept. 1970.)

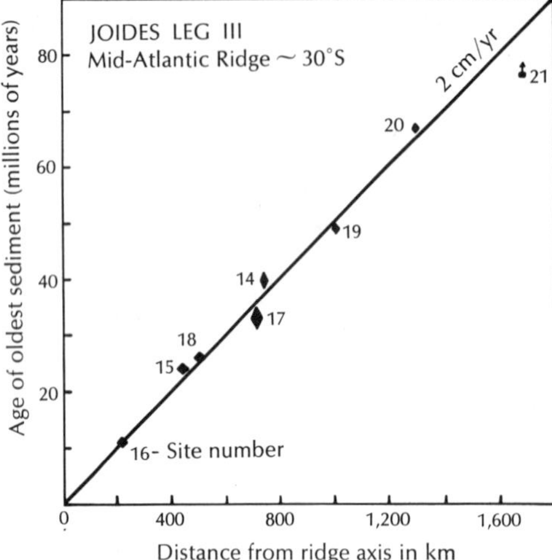

11-25. Age of oldest sediment recovered at eight sites plotted against distance from ridge crest. Holes were drilled in South Atlantic across Mid-Atlantic Ridge at about 30 degrees latitude. Sites 17 and 18 are west of axis; the others are to the east. Basalt basement was not reached at site 21. (F. J. Vine, *Nature*, 5 Sept. 1970.)

in isostatic equilibrium, it perhaps once was uplifted and then subsided.

The ocean basins, continents, and ocean water are all quite ancient as viewed by this hypothesis. However, the ocean basin floors are relatively young, and the continents and ocean basins have not always had their present shapes or locations.

Hess had different views about the nature of the oceanic crust, which he assumed to be serpentinized peridotite that had formed by alteration of the upper mantle. He attached great significance to the uniform thickness of layer 3 of the oceanic crust (about three miles thick). Hess pointed out that an oceanic crust resulting chiefly from basaltic lava flows should be thicker above the fissures that fed the flows and thinner away from them. Therefore, he attributed the uniform thickness of layer 3 to an alteration that depends upon temperatures and pressures; these should have approximately the same values at the same depth everywhere beneath the ocean floors.

In this assumed alteration, the mineral olivine combines with water to produce serpentine. Olivine is abundant in peridotite, which presumably forms the material of the upper mantle, and the mantle is thought to be the source of the water. This olivine-serpentine alteration is reversible, and laboratory experiments show the critical temperature to be about 900°F at the pressures calculated to exist at the base of the

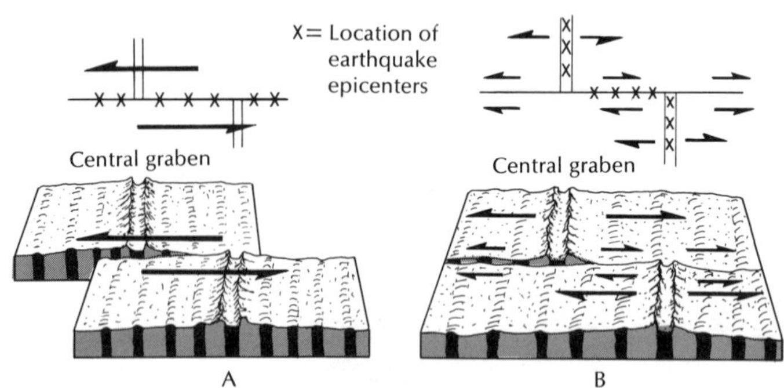

11-26. A fault offsets a midocean ridge. (A) If an ordinary strike-slip fault were to offset a ridge, crustal blocks on either side would move in opposite directions all along the fault zone, and earthquakes would occur all along this zone. (B) Spreading occurs away from a ridge axis that is offset by a so-called transform fault. Note that opposite movement along the fault is confined to the section between the offset ridge crests and that earthquakes tend to be restricted to this section as well as to a narrow zone along the ridge crests. Thus the observed distribution of earthquake epicenters supports interpretation (B). (B. Mears, Jr., *The Changing Earth,* Van Nostrand Reinhold, 1970.)

oceanic crust. At temperatures less than 900°, the denser olivine changes into less-dense serpentine (specific gravities: about 3.3 vs. 2.6). At higher temperatures, the serpentine gives off water and reverts to olivine.

In this view, new simatic oceanic crust is produced by serpentinization of the upper mantle along the crest of the mid-ocean ridges where upward-moving convection currents occur. This newly formed, horizontally spreading crust then collects a blanket of sediments as it moves toward the continental margins. This sialic sediment is subsequently compressed, thickened, and welded onto the edges of the continental crust. The convection currents are assumed to subside beneath the edge of the ocean basins, and this drags down the ocean crust to form deep-sea trenches such as those that margin the Pacific (next section). At higher temperatures and greater depths beneath the trenches, the serpentine changes back into olivine. Although basaltic material is more common than serpentine along the mid-ocean ridges, some serpentinized peridotite has been dredged from the sea floor, and inclusions of olivine and peridotite occur in some of the basaltic rocks found there.

Deep-Sea Trenches and Island Arcs

Greatest depths occur in long narrow trenches located along the ocean margins, particularly around the Pacific (Figs. 11–27 and 11–28). These striking linear depressions are associated with island arcs, volcanoes, earthquakes, young mountain belts, abnormally low gravitational attraction, and lower-than-average heat flow. They constitute highly active and significant portions of the Earth's crust.

The island arcs commonly consist of elongated, crescent-shaped chains of volcanic mountains which project upward as islands from long narrow submarine ridges (examples: Aleutians, Kuril Islands, Japan, the Philippines, the Indonesian states, and the West Indies). Island arcs are located on the landward side of the line of intersection of continental crust with oceanic crust, and they curve convexly toward the oceans. Their intrusive and extrusive igneous rocks tend to have a medium-silica content, as do the lavas that occasionally issue from their active volcanoes and harden to form a rock known as andesite (composing much of the Andes).

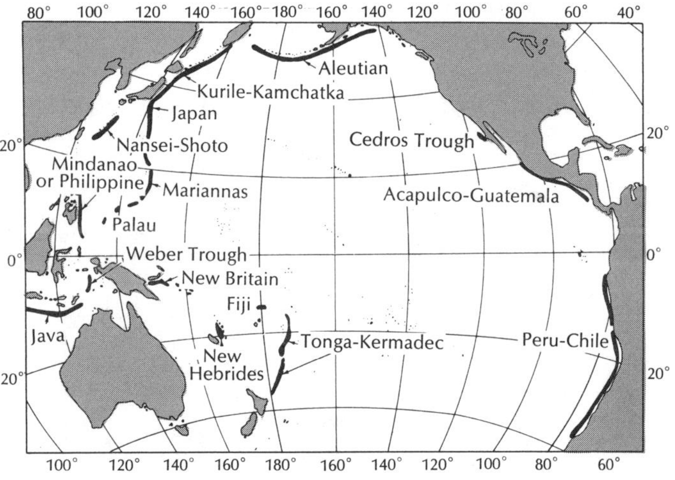

11-27. The trenches of the Pacific. (H. B. Stewart, Jr., *The Global Sea*, Van Nostrand Reinhold Co.)

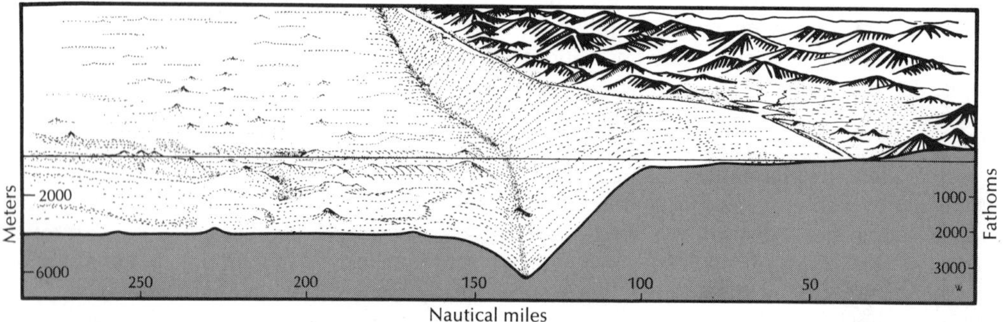

11-28. The Middle America Trench as viewed toward the northwest off the Mexican Coast. Vertical exaggeration is ten times. The trench is deeper than 2400 fathoms for a distance of 1200 miles and deeper than 3000 fathoms over one stretch of nearly 400 miles. In places the trench is V-shaped, but in other places it has a flat bottom, apparently a sedimentary fill. (R. L. Fisher, "Middle America Trench: Topography and Structure," *Geol. Soc. Am. Bull.* **72,** May 1961.)

Deep-sea trenches are located on the ocean side of island arcs or parallel coastal mountain belts such as the Andes (which have some of the fundamental features of island arcs but are not island archipelagos). The trenches may extend for distances that exceed 2000 miles and have maximum depths below sea level of about 36,000 feet (nearly 7 miles; in comparison, the Grand Canyon is about 1 mile deep). However, considerable variation in depth occurs along the trenches, which in their deepest portions actually form a series of discontinuous, sliver-shaped depressions. In some places the trenches have a V-shaped cross section, whereas in others they have flat floors produced by sedimentation. Undeformed sedimentary layers occur in the central and outer parts of some of the trenches, but narrow zones of much deformed sediments have been located recently on their inner sides (toward the island arcs).

Earthquakes in great numbers are associated with the trenches, and an important relationship exists between the depth of origin of an earthquake (its focus) and its

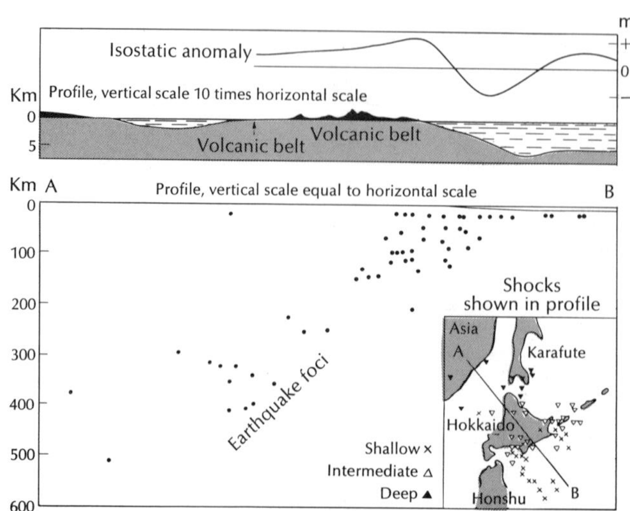

11-29. Crustal section for typical island arc (Gutenberg and Richter). In the map *(right)* deep foci are marked by black triangles. (M. Ewing and F. Press, "Structure of the Earth's Crust," *Lamont Geological Observatory Contribution* No. 143.)

location relative to the trenches and island arcs (Figs. 11–29). Most earthquakes originate from faulting that occurs within 20 miles or so of the surface (shallow-focus type), and these are the only kind that occur beneath the trenches. Earthquakes originating at intermediate depths are located beneath the island arcs, whereas the deepest known earthquakes occur inland from the island arcs at a maximum depth of about 400 miles (only a small percentage of the total). Thus the earthquake foci tend to be located along a plane that extends at a steep angle from the base of a trench downward below a continent.

According to the concepts of plate tectonics (next chapter), two great plates of the lithosphere may converge and one plate may descend beneath the other (the plane of earthquake foci shows its trend)—in such a mobile belt we find arcs, trenches, volcanism, the greatest earthquakes, and maximum topographic relief. Moreover, a correlation exists between andesite-erupting volcanoes and such mobile belts—andesites are apparently produced only at such locations where sediments and crust have been recently thrust downward into the mantle.

The gravitational attraction of the Earth can be calculated theoretically for any given spot on the basis of its latitude and position relative to sea level. Next, the actual gravitational pull of the Earth at this place is measured, and any difference between calculated and measured quantities forms an anomaly that we then try to account for. Such gravitational data show that deep-sea trenches have abnormally low gravitational attraction, whereas island arcs tend to have above-average gravitational attraction. Evidently the mass of the rocks beneath a trench is less than its volume would indicate, or the crust is being pulled downward to cause the below-average gravitational pull.

Exercises and Questions for Chapter 11

1. To illustrate the vertical exaggeration common in oceanographic profiles, construct an imaginary but approximately scaled topographic profile. Include familiar objects such as a hill, building, and small cinder cone. Use a natural scale (vertical scale is same as horizontal). On this profile, superimpose three additional profiles with vertical exaggerations of 5, 10, and 40 times.
2. What are the two dominant levels of the Earth's surface and how may they be accounted for?
3. How much of the Earth's surface is truly continental? What criteria may be used to distinguish continents from ocean basins? Which criterion seems to be the most basic or significant?
4. Discuss the continental shelves: general characteristics, economic importance, changes caused by Pleistocene sea level fluctuations, methods of exploring, and origin.
5. Discuss the sediments that occur on the sea floor:
   (A) Their sources and the processes involved in their transportation and deposition.
   (B) Methods of sampling.
   (C) Thicknesses and methods of estimating. What problem is posed by current estimates of sediment thicknesses?
   (D) What relationship exists between latitude, the depth of sea water,

and the presence or absence of certain types of oozes on the sea floor?

- (E) In what ways may sediments on the sea floor be of economic importance?
- (F) Describe some of the characteristic features of turbidity current deposits. How did the discovery of such deposits upset certain geologic ideas concerning the manner of origin of some types of ancient sedimentary rocks?

6. Describe some of the characteristic features of each of the following and discuss their probable manner of formation.
   - (A) continental rise
   - (B) abyssal plains
   - (C) seamounts and guyots
   - (D) mid-ocean ridges
   - (E) oceanic crust
   - (F) magnetism of the sea floor
7. How does sea-floor spreading seem to be related to:
   - (A) the system of mid-ocean ridges?
   - (B) reversals of the Earth's magnetic field?
   - (C) heat flow?
   - (D) continental drifting?
   - (E) locations of earthquake epicenters?
8. Describe the general features of the deep-sea trenches and island arcs. How do they seem to be related to the location of earthquake foci? To gravitational anomalies? To sea-floor spreading?

12

Origin and Development of the Earth and Its Crust

We live at an exciting, stimulating time in the development of the geological sciences—one filled with many rapid, far-reaching, even revolutionary happenings and discoveries. Continental drifting (Fig. 12–1), sea-floor spreading (p. 300), and the new global plate tectonics stir the imagination. As background, we look first at the origin of the Earth, its air and water, and then we discuss some of the processes, discoveries, and interpretations that promise to make the 1960's and 1970's high-water marks in the history of the geological sciences.

Origin of the Earth and Solar System

The regularity and uniformity shown by planetary motions indicate that the Earth and other planets originated at more or less the same time and in the same general manner, although Pluto may once have been a satellite of Neptune. All hypotheses involve the rearrangement of matter and energy previously in existence; ultimate origins are not attempted here. Two general sources for this matter have been postulated: in one view, the materials making up the present planets were once part of the sun or of a companion star of the sun (the sun would thus be much older than the planets); in the second, the materials were part of an interstellar nebula from which all of the members of the solar system subsequently formed (the sun and planets would thus have about the same age).

The nebular hypothesis of Kant and Laplace illustrates this second view and was generally accepted during the nineteenth century. A great cloud (nebula) of hot, gaseous, slowly rotating material was thought to exist. As the cloud cooled, it shrank and rotated more rapidly. However, its equatorial regions contracted more slowly than its polar regions, which changed the cloud into a flattened, lens-shaped disk. Continued contraction increased the rate of spin (see below), and a rotating ring of material was left behind.

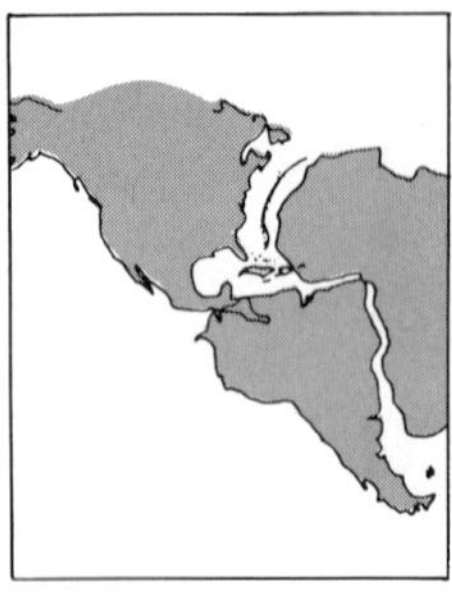

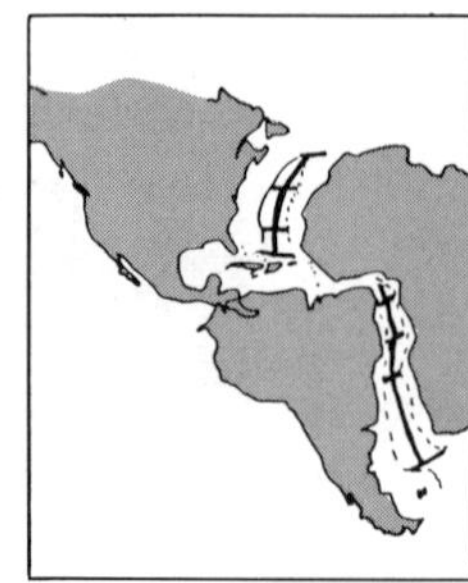

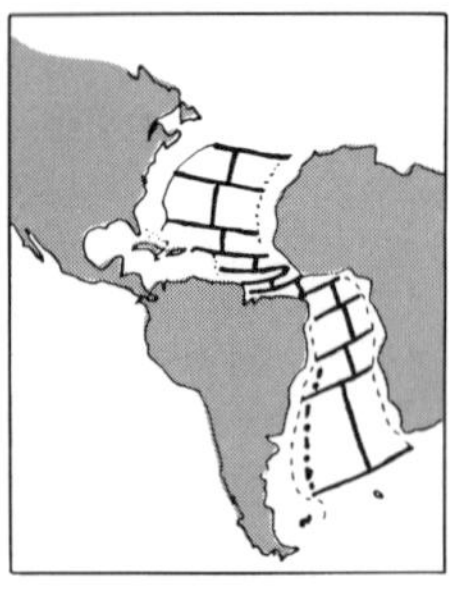

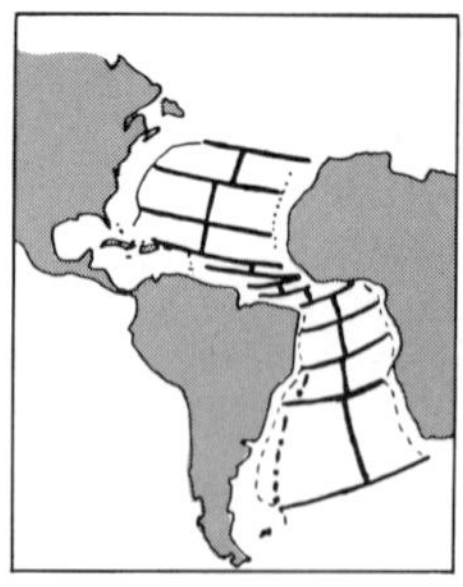

12-1. Sea-floor spreading in the Atlantic and continental drifting. Assumed conditions: upper left, 120 million years ago; lower right, 90 million years ago. Active ridge crests and fracture zones are shown by a thick continuous line; an abandoned ridge crest is shown by a thick dash-dot line. (Courtesy Xavier LePichon, *J. Geophys. Res.* 15 June 1968.)

as the rest of the cloud shrank inside the ring. The process was repeated many times as the cloud continued to contract, and a new ring was formed for each planet. These rings then changed into rotating globular gaseous masses which cooled, shrank, and threw off more rings to form the satellites. The planets were first gaseous, became liquid as they cooled, and then solidified (at least in part).

A number of serious objections have since been raised to this hypothesis, and we mention a few: the sun should be spinning rapidly and preparing to cast off another ring; all satellites should be revolving in the same west-to-east direction, and the gases in the rings should disperse into space instead of collecting into planets.

During the present century, attempts have been made to explain the origin of the planets by the ejection of material from the sun or from a companion star of the sun. The tide-producing effect of a passing star is important in some hypotheses, but the probability of such a close encounter is very slight and would imply that our planetary system is almost unique in the galaxy. Moreover, hot gases should disperse explosively into space if they are pulled out of the sun's interior.

Another difficulty: the sun should have most of the angular momentum of the solar system, whereas it has only about two percent. The *angular momentum* of a planet or other object revolving in space about a central body equals the mass of the planet times its orbital speed times its distance from the center. According to the principle of *conservation of angular momentum,* in an isolated system in space, the total angular momentum remains constant, although it can be shifted from one part of the system to another. A skater whirls faster by bringing her arms closer to her body and illustrates this principle in operation.

The more recent *protoplanet* hypothesis explains many of the phenomena of the solar system, and we shall discuss it briefly. The starting point is a wide, relatively thin, gaseous disk perhaps one-tenth as massive as the present sun. This was thought to be located in the equatorial plane of a large globule of gas and dust that subsequently became the sun. Composition of the disk and globule was similar to that of the sun today: chiefly hydrogen, with some helium, and one to two percent of heavier elements. As the disk contracted and flattened, it became unstable, and divided into a number of separate clouds or protoplanets. Cold particles joined together to form a solid globular nucleus within each protoplanet and became surrounded by a very large gaseous envelope (Fig. 12–2). At this stage, the protoplanet from which the Earth formed may have been some 1000 times more massive than the present Earth, and it had not yet become differentiated into core, mantle, and crust. The protoplanets were of somewhat different sizes, but all were probably larger

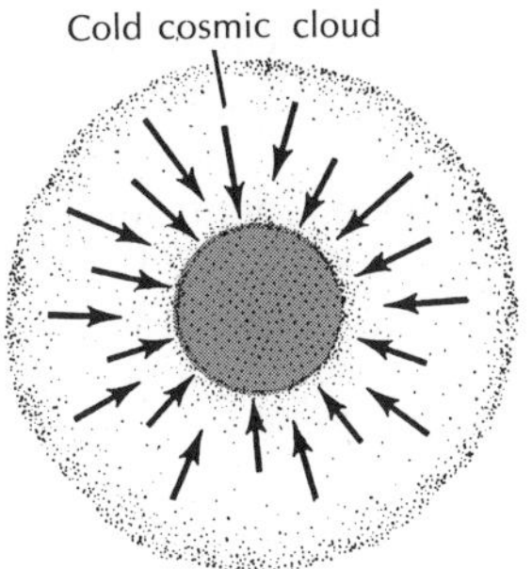

A. Initial accretion.
Evenly distributed elements

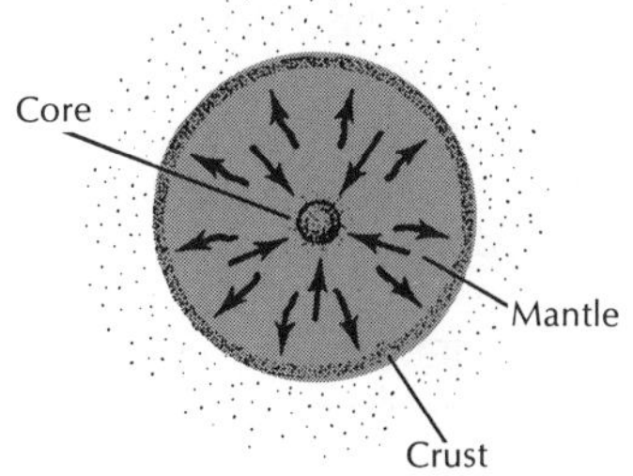

B. Zonation. Heated Earth developing zones of core, mantle, and crust

12-2. Likely events in the early history of the Earth. (B. Mears, Jr., *The Changing Earth,* Van Nostrand Reinhold, 1970.)

and more massive than the present planets.

Temperatures apparently were quite low as the particles of gas and dust accreted, otherwise the volatile materials that now occur in the oceans and atmosphere would presumably have escaped (a volatile substance is one that is readily vaporized at a relatively low temperature). These volatiles may have been trapped in interstices among the cold particles or have been present in chemical union with other elements.

The satellites probably formed in a similar manner as the protoplanets contracted, but they may have been relatively much closer to their parent bodies. Therefore, their rotation was slowed by tidal friction until they rotated and revolved at the same rate and in the same direction (most continue to do so today). Thus they remained spherical and did not subdivide further.

Since the original disk rotated in a counterclockwise direction, the planets follow counterclockwise orbits today. Moreover, the sun's tidal attraction stretched the protoplanets into elongated shapes and kept their long axes always pointed toward the sun. This made the direction of rotation the same as the direction of revolution (Venus and Uranus are exceptions). In other words, their periods of rotation and revolution were once equal.

By this time the central mass had contracted enough to become a star. As the sun's temperature rose, its radiations and ejected particles ionized the gases around it. These gases then interacted with the sun's magnetic lines of force to reduce the sun's rotation and transfer most of its angular momentum to the particles in the disk; they moved faster as a result of the transfer. This "solar wind" of radiations and ejected particles gradually swept off into space the remaining portion of the disk and most of the lighter gases of the protoplanets. A comet's tail is directed away from the sun for the same reason.

Only a small fraction of the original disk remains as the masses of the present planets. The planets nearest the sun lost very high proportions of their protoplanets, and thus they consist primarily of the heavier elements that make up only one to two percent of the sun—their masses are small and densities high. On the other hand, Jupiter and Saturn were massive and distant enough to retain more of their protoplanets (perhaps ahout one-twentieth for Jupiter), thus they contain a greater proportion of hydrogen and helium and have much lower densities. As the protoplanets shrank in size and mass, they rotated more rapidly to produce the spins we observe today (remember the swirling skater and the conservation of angular momentum).

As the masses of the protoplanets decreased, their gravitational attractions became less, and their satellites were able to revolve at increasing distances. Some eventually escaped, and a number may later have been recaptured, thus accounting for the orbital irregularities shown by some of the outer satellites.

A major protoplanet may not have developed in the belt now occupied by the asteroids. Instead, a number of smaller protoplanets developed, and these eventually produced the larger asteroids that are a few hundred miles in diameter. Collisions among some of these, and additional collisions among the resulting pieces, produced the smaller asteroids and the fragments that form most of our meteorites.

The comets probably formed beyond Pluto and have since been scattered widely. The bulk of them may now exist far out in space as frozen masses (comet nuclei) a mile or more in diameter, with stony and metallic chunks probably embedded in the ices. If disturbed by the gravitational influence of a passing star, or perturbed by the planets, a comet may swing in toward the sun.

Origin of the Atmosphere and Hydrosphere

According to an older, now discarded view, the atmosphere and hydrosphere of the Earth are residual, and the Earth has passed in turn from an original, very hot, gaseous stage through a hot molten stage into its present solid state (pockets of magma exist here and there in the crust and upper mantle, and the outer part of the core behaves like a liquid). This explained the Earth's density stratification: the heaviest materials collected at the center during the molten stage (as in a blast furnace) and successively lighter materials surrounded them in shells to form the core, mantle, and crust (Fig. 12–2B). Less conspicuous zones also formed within the main units.

Water could not have existed as a liquid at the high temperatures of the gaseous and molten stages. However, after the Earth had crusted over, raindrops could eventually form, although they were probably vaporized upon contact with the hot crust. Thus for a time, precipitation and vaporization took place. Eventually, temperatures became low enough for liquid water to accumulate on the surface, and it flowed from higher to lower ground and collected in the major depressions to form oceans (continents and ocean basins are assumed to have existed at this stage). In this view, the oceans had their present volume at a very early stage in the history of the Earth, and the Earth has had an atmosphere since its origin, although some additions and losses presumably have occurred.

This hypothesis now seems quite unlikely. Various lines of evidence indicate that the Earth's first atmosphere subsequently was lost and that its present air and water have probably originated from materials emanating from the mantle, chiefly as a result of igneous activity (this is the *degassing* hypothesis), either gradually through time or perhaps more rapidly at a certain time, or times, in the Earth's development. Gases from volcanoes, fumaroles, and some hot springs are apparently being given off at present in suitable proportions and quantities to account for all of the air and water on the Earth, provided that they have continued in this manner since early in the Earth's history.

As evidence that an original atmosphere escaped from the Earth, we note that the heavier inert gases (neon, krypton, and xenon) are apparently much less abundant on the Earth than they are in the sun and elsewhere in the universe. Silicon is used as a basis for comparison, and helium and argon are omitted since each is produced by radioactive disintegrations, and some helium escapes at present. Moreover, the deficiency is greatest for neon, which is lightest and thus should escape most readily, and least for xenon, the heaviest. Presumably the missing quantities long ago escaped from the Earth, perhaps expelled by the so-called solar wind during the proto-Earth stage or subsequently when the Earth may have become at least partially molten. The rest of the Earth's original atmosphere probably escaped at about the same time. In it, light-weight gases such as hydrogen and helium should have been far more abundant than the inert gases and should also have escaped far more readily. Ammonia and methane may also have been abundant in this original atmosphere.

Additional evidence that most or all of the Earth's original atmosphere escaped into space long ago (a secondary one has since accumulated) involves the so-called excess volatiles; e.g., water, nitrogen, chlorine, sulfur, and carbon dioxide are examples (volatiles are substances that are gaseous at low temperatures, or become so in the presence of oxygen or hydrogen). Other elements dissolved in the oceans or present in the atmosphere can be readily explained as products of the weathering and erosion of igneous rocks throughout geologic time, but not these volatiles (degassing from the mantle is their apparent source). Moreover, if all the present constituents of the atmosphere had been present originally, then certain difficulties arise; e.g., the amount of carbon dioxide would have been enormous. Enough should have been present originally to account for all the carbon dioxide that has subsequently become tied up in limestones and dolomites during all of geologic time. Furthermore, such huge quantities of carbon dioxide would presumably have made calcareous rocks very comon in the Precambrian, which is not the case. In addition, the quantity of chlorine thus called for would have made the first oceans highly acidic (the chlorine at that time would not have been neutralized by sodium, because the sodium had yet to weather from igneous rocks). Furthermore, although lighter atoms and molecules tend to escape more readily from an atmosphere than heavier ones, the volatiles show no correlation between their atomic weights and their abundances.

Modern speculation about the Earth's early history might continue somewhat as follows. The density stratification of the Earth's interior indicates a once-molten condition (at least for the deep interior). Moreover, a one-billion-year gap seems to occur between the time of Earth formation (about 4.6 billion years ago based upon the ages of meteorites and lunar rocks) and the ages of the oldest known rocks (about 3.5 billion years). Presumably the molten stage occurred during this one-billion-year interval, and two sources of heat energy seem available. Heat energy is released during radioactive disintegration, and radioactive elements must have been considerably more abundant when the Earth formed than they are at present; e.g., twice as much U-238 must have existed, and other elements with shorter half-life periods must have been still more abundant. Heat energy may also have been generated by high-speed impacts between accreting particles and the Earth's surface, i.e., so-called "gravitational heating" since the Earth's gravitational attraction was an important cause of the high-speed collisions. Some of this heat energy would have been radiated away, but some should gradually have moved toward the deep interior. For a planet the size of the Earth, such gravitational heating might possibly equal that from the radioactive source.

However, if the Earth had once been completely molten, a great loss of volatiles would presumably have occurred. Therefore, since such volatiles are still with us, the outer part of the Earth may not have melted. Perhaps the melting occurred only in the deeper interior—enough to account for the Earth's present zonation.

A second high-temperature event in the Earth's history may have occurred long ago if the moon once approached much closer to the Earth (p. 566). At such a time, heat energy produced by enormously increased tidal friction might have caused some melting and speeded up the degassing process.

With time, a secondary atmosphere accumulated around the Earth, but it presumably lacked free oxygen at first. Photosynthesis by plants has apparently been a major source of free oxygen, but it had to wait until primitive plants had become established (perhaps nearly 3½ billion years ago). Other free oxygen has come from the dissociation of water vapor by solar radiation. The oxygen thus released would probably have combined with any available ammonia (NH_3) and methane (CH_4) to produce free nitrogen, carbon dioxide, and water. Oxidation of iron and other elements would also require oxygen. However, eventually some free oxygen could accumulate and thus become available for the first animals. When free oxygen first became available in quantity is uncertain, but some have tied it to the initial abundance of fossils at the beginning of the Paleozoic Era. However, others point to the enormous, distinctive deposits of Precambrian banded iron ores and conclude that free oxygen began to accumulate in the atmosphere about 1.8 billion years ago. Ocean water first had to be saturated with free oxygen, which may have taken a very long time (by algal photosynthesis), perhaps from about 3.5 to 1.8 billion years ago; i.e., during the period in which these unique iron ores formed.

We do know that some of the oldest rocks on Earth appear to have formed by the metamorphism of sedimentary rocks. Therefore, running water for transportation must have been present at least 3.5 billion years ago, which implies the presence of an atmosphere (the water cycle), weathering, erosion, and bodies of water in depressions.

Firm conclusions concerning the origin of the atmosphere and hydrosphere are unwarranted at the present time, but it does seem that degassing through time of the Earth's mantle has provided the bulk of the present atmosphere (except for the free oxygen) and ocean water.

Global Plate Tectonics*

Some concepts of global plate tectonics are included in Chapter 11 under topics such as sea-floor spreading, origin of oceanic crust, transform faults, and magnetic reversals. This remarkable new synthesis of Earth history and structures also involves the origin of continents, continental drifting, and the formation of the Earth's great mountain belts—topics which we will take up later in this chapter. The ideas of plate tectonics are far reaching and very different from previously held views.

In this new synthesis, three units in the Earth's interior are emphasized (Fig. 12–3). The *asthenosphere* corresponds approximately to the low-velocity zone (p. 165), occurs in the upper mantle, has a thickness of a few hundred kilometers, and is considered to be a zone of weakness in which slow plastic flow can occur. The *lithosphere*, some 50 to 100 km (35 to 60 miles) or more in thickness, lies on the asthenosphere, is made of crust and uppermost mantle, and is a relatively strong unit. The *mesosphere* comprises the rest of the mantle below the asthenosphere and may not be actively involved in the tectonic processes that occur nearer the surface.

Recent, more precise data on the locations of earthquake epicenters (Fig. 6–25) show them aligned in relatively narrow, nearly continuous belts, whereas earthquakes tend to be infrequent within the wide expanses between the belts. This observation, together with other clues, has led to the widely held view that the lithosphere is subdivided into a number of large movable plates that can slide gradually across the top of the underlying weak asthenosphere; the lines of earthquake epicenters mark the boundaries of these plates. The lithospheric plates differ in both size and shape, and some uncertainty exists concerning precise shapes and exact numbers—as few as six in one view, but as many as twenty in another (Figs. 12–4, 11–18, and 11–24). Since the boundaries of the plates do not necessarily coincide with present-day boundaries between continents and oceans, the top of a lithospheric plate may be formed by a continent, by part of the sea floor, or by a mixture of continent and ocean. Moreover, since plates may form in one place (at a mid-ocean ridge) and be destroyed in another (at a deep-sea trench), the plates

* Isacks, Bryan and Oliver, Jack, "Seismology and the New Global Tectonics," JOUR. OF GEOPHYSICAL RESEARCH, 15 September 1968.
2. Vine, F. J., two articles on sea-floor spreading, JOUR. OF GEOLOGICAL EDUCATION, February 1969 and March 1970.
3. Hammond, Allen L., two articles on plate tectonics, SCIENCE, 2 July 1971 and 9 July 1971.

of today are presumably not the plates of long ago.

Since about all of the deep-sea floor seems to have formed during the Cenozoic and Mesozoic Eras, it follows that an equivalent amount of older sea floor has been destroyed during this same interval—recycling on a really grand scale.

Apparently these huge lithospheric plates tend to interact in three general ways. Two or more plates may move laterally away from a spreading ocean ridge where new oceanic crust is being formed continuously. Earthquake activity is confined to a rather shallow surface zone along such belts of divergence. Elsewhere, two plates may shift toward each other and converge along a deep-sea trench and island arc. Here one of the plates tends to move downward at a slant beneath the arc and trench, which may explain why earthquake foci tend to occur along a plane that slants at about 45 degrees from a trench downward beneath an adjacent continent (Fig. 11–29). In this re-

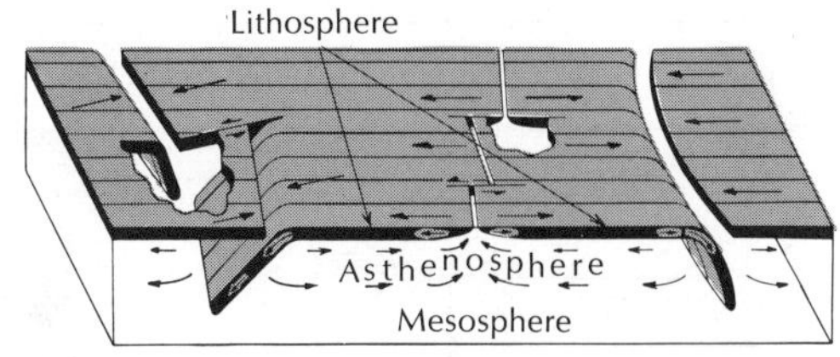

12-3. Schematic view of the new global tectonics. (Isacks, Oliver, and Sykes, *J. Geophys. Res.*, 15 Sept. 1968.)

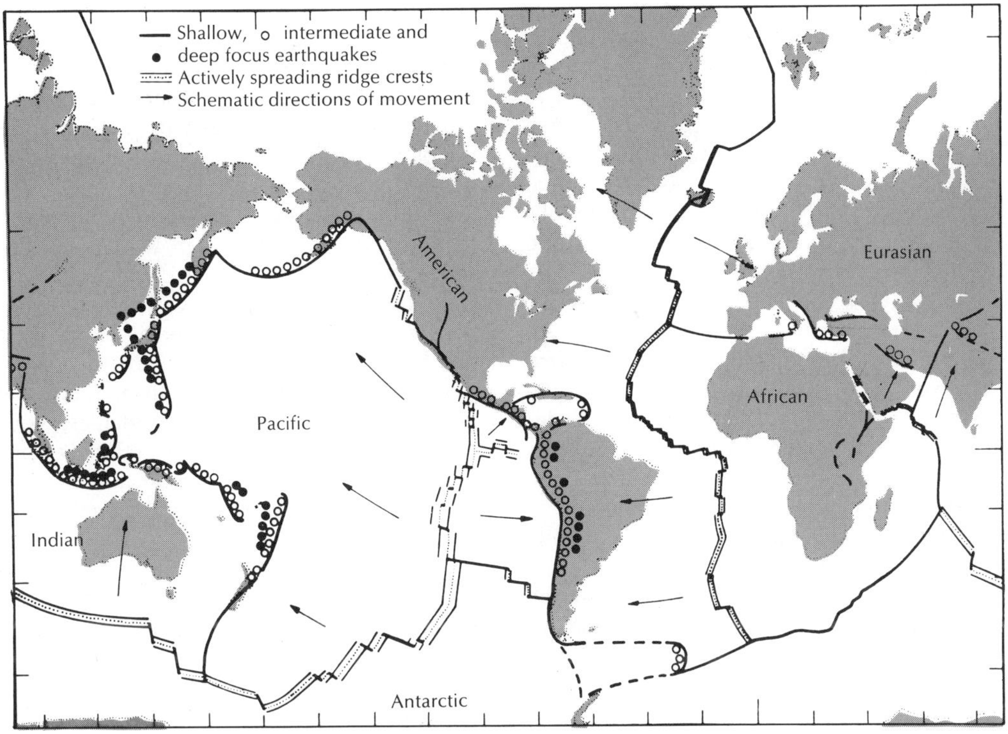

12-4. Summary of seismicity of Earth (from M. Barazangi and J. Dorman, 1969) and thus of extent of lithospheric plates bounded by active ridge crests, transform faults, trench systems, and zones of compression. Six major plates are named and six minor ones are numbered. Spreading rates at ridge crests are indicated schematically and vary from 1 cm/yr/ridge flank in the vicinity of Iceland to 9 cm/yr in the equatorial Pacific Ocean. (F. J. Vine, *Nature*, 5 Sept. 1970.)

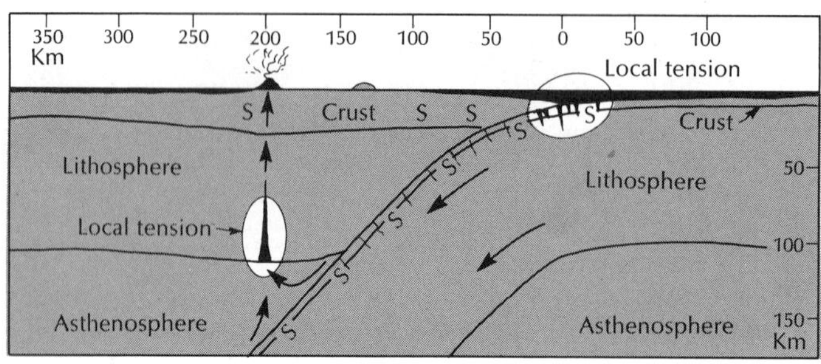

12-5. Schematic section through an island arc. S indicates seismic activity. No vertical exaggeration. (Isacks, Oliver, and Sykes, *J. Geophys. Res.*, 15 Sept. 1968.)

gion of convergence we find the deepest earthquakes, great topographic relief, and intensive volcanic activity. In the third general type of interaction, lithospheric plates slide past one another along major strike-slip faults (see fracture zones and transform faults). The San Andreas fault zone seems to be one example.

Study of earthquake-producing motions along faults supports the view that underthrusting and compression occur at a trench and island arc. However, the sharp bend in the lithosphere on the ocean side of a trench produces tension near the surface along its convex side, and such tensional effects (e.g., normal faults) have been observed (Fig. 12–5). This seems to solve one of the problems involving trenches, i.e., compressional effects had been expected, whereas tensional effects were observed. Another problem involved the flat-lying, more or less undisturbed nature of the sediments within the trenches. Much deformation had been expected. However, such highly contorted sediments have been discovered recently, but they are difficult to detect by seismic profiling and occur only along the inner wall of a trench.

Research involving seismic waves in the Fiji-Tonga area has shown an anomalous zone about 100 km (60 miles) thick that extends at a 45 degree slant down under the Tonga arc to a depth of about 700 km (430 miles). The earthquake foci are aligned in a plane that coincides closely with the upper surface of this anomalous zone. Interpretation: the leading edge of a plate of lithosphere is here being thrust beneath an arc. The deep earthquakes at 700 km are apparently caused by sudden shearing movements and are not due to sudden volume changes (explosion or implosion) as previously suggested.

Although the existence and movement of large lithospheric plates seems well established, the cause of the movement is uncertain. It is widely held that convection currents produce sea-floor spreading, continental drifting, and compression at deep-sea trenches; however, a number of problems remain. Slow plastic flow within the mantle is possible according to some geophysicists but not possible according to others. Lesser discontinuities seem to occur within the mantle, i.e., the mantle may consist of a series of concentric shells, which in one view appears inconsistent with mantle-wide convection currents. On the other hand, it has been proposed that the phase change that seems to occur at a depth of about 400 km (250 miles) might actually increase convectional movement. If convection is confined largely to the asthenosphere, then very widespread, very shallow, chiefly horizontal convection currents have to occur. Such currents might be more likely to subdivide into smaller cells than to extend unbroken from a mid-ocean ridge to a deep-sea trench.

In another model, a number of so-called hot spots are assumed to occur at fixed locations in the lower mantle. Material rises slowly above any one hot spot and moves to the upper mantle where it then spreads out and drags the lithospheric plates laterally. Later this material subsides in the regions around the hot spots. Since the hot spots are assumed to occur beneath active volcanic regions, a chain of volcanic islands can be produced as a plate moves slowly across a hot spot.

Orowan* has suggested an origin for mid-ocean ridges and for continental drifting that does not depend on convection currents. He assumes that a rift develops at some location along the ocean floor, that water and other volatile substances then migrate toward this rift, and that a transformation subsequently occurs which changes olivine (which probably predominates in the mantle) into serpentine with a resulting 25% volume increase (p. 304). In this view, the olivine-to-serpentine volume increase along the rift produces the bulge that is a mid-ocean ridge. Next, gravity causes lateral flow away from the crest of the bulge, and this leads to sea-floor spreading and continental drifting. Layer 3 of the oceanic crust should consist chiefly of serpentine according to this interpretation, whereas it is generally assumed to be basaltic. Serpentinites have recently been found in fracture zones along parts of the Mid-Atlantic Ridge, but the data may be interpreted in two ways. In one, the Mid-Atlantic Ridge is made chiefly of serpentinite overlain by a thinner basaltic zone, and the serpentinite is exposed in the bottom of the fracture zones. In the other view, the serpentinite is part of a narrow vertical zone that passes upward through a basaltic layer 3.

Continental Drifting

The drifting of continents (Fig. 12–1) is an older concept that was long shunned by many geologists; now the concept has large numbers of enthusiastic supporters, and recent discoveries involving sea-floor spreading, deep-sea drilling, and paleomagnetism have supplied convincing new evidence. Older evidence had involved factors such as: the close fit in the shapes of continents on opposite sides of an ocean; similar matches between the kinds of rocks and rock structures on different continents in areas presumed to have once been contiguous; ancient climates; and the distribution of certain fossil animals and plants. For these factors too, strong new supporting discoveries have been made. Let us consider some of the reasons why "drifters" are now so numerous.

We begin with the striking match between the coastlines of Africa and South America on opposite sides of the Atlantic. An even better fit occurs at a depth of about a mile where the true margin of a continent is located (i.e., along the continental slope). The kinds, ages, and structures of many rocks found in South America have counterparts in Africa. The matching sequencies seem too numerous and widespread to be coincidence; they suggest an origin when South America and Africa were part of one continent.

Deposits (tillites) made by the ice sheets that existed during the latter part of the Paleozoic Era (p. 259) have been found at low latitudes where ice sheets cannot possibly exist today; i.e., in Brazil, South Africa, Madagascar, India, and Australia. Striated surfaces under some of the deposits show unlikely directions of movement based upon the present locations of these lands, e.g., away from the equator or away from an ocean. No really adequate explanation has been proposed for the occurrence of these glacial deposits other than that they formed at a time in the past when these lands were all joined together (Gondwanaland, Fig. 12–6) in a location near the South Pole.

The existence of distinctive types of rock sequences containing distinctive types of fossil animals and plants in these glaciated lands had also been noted. For example, a sequence of rocks occurs in the Gondwana region of India. These are about horizontal, chiefly nonmarine, have a maximum thickness of about 4 miles, and range in age from Late Paleozoic well into the Mesozoic. A fernlike plant called *Glossopteris* is one of a large number of characteristic fossils. Similar sequences occur on the other lands that

* Egon Orowan, "The Origin of the Oceanic Ridges," *Scientific American*, Nov. 1969.

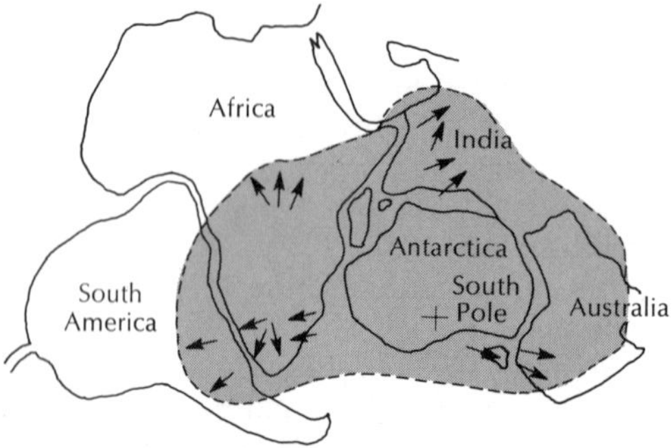

12-6. Schematic view showing DuToit's arrangement of the continents during the glaciation that occurred in the latter part of the Paleozoic Era. Dashed lines suggest the boundaries of the ice sheet, and arrows indicate probable directions of flow. (B. Mears, Jr., *The Changing Earth,* Van Nostrand Reinhold, 1970.)

were glaciated during the Late Paleozoic, and the fossil and rock similarities indicate one large supercontinent. In other words, if these lands had always been single units separated by wide expanses of ocean as they are today, then conspicuous differences should have developed in their animals, plants, and rocks. Such differences do not occur prior to the Late Mesozoic but do occur in younger rocks.

Another type of evidence favors the existence of a Gondwanaland during most of the Mesozoic and its subsequent fragmentation. The Mesozoic (Age of Reptiles) lasted more than twice as long as the Cenozoic (Age of Mammals). Yet the number of orders of Mesozoic reptiles is about one-half the number of orders of Cenozoic mammals; a more nearly equal number might be expected. One interpretation of this apparent discrepancy is that the continents were joined together during much of the Mesozoic and environmental conditions were somewhat uniform; thus fewer reptilian orders evolved. Subsequently, the continents became separated and then drifted apart during the Cenozoic. This seems to have resulted in more rapid and extensive environmental changes that in turn led to a speeding up of evolutionary development and diversification. Therefore, mammals could evolve into twice as many orders as the reptiles in less than half the time.

What has been called one of the "truly great fossil discoveries of all time" was made in Antarctica on a Thursday in December of 1969. Identification was made on the spot, and the news was dispersed so quickly that one read all about it in his Saturday morning paper! The fossil was described in the newspapers as the reptilian counterpart of the hippopotamus (*Lystrosaurus,* Triassic Period of the Mesozoic Era). Its remains are abundant in some South African layers where it is a key index fossil (it also occurs in Asia). These fossils from Africa and Antarctica are apparently much too similar to have resulted from parallel evolutionary development on separate continents, and the animals could not cross a wide ocean. Therefore, these two lands presumably were united in the Triassic as part of Gondwanaland.

If continents have moved, then at least two quite different types of movement seem possible. In one—*polar wandering*—it is assumed that the Earth's rotation has remained unchanged and that the location of its magnetic poles has tended always to coincide closely with the geographic poles. In the polar-wandering view, the outer part of the Earth shifts its position relative to the interior; perhaps the entire lithosphere as a single unit slides on the upper boundary of the asthenosphere. In this type of movement, the continents are imagined to maintain fixed relative positions, but as sliding goes on, a particular continent may be near the equator one time and near a polar region at a subsequent time. The continents are thus like logs frozen into a large sheet of drifting ice; the entire mass of ice moves as a unit, but the scattered logs have no individual movements of their own within it.

In the second type of movement, the continents are imagined to move, one relative to another (e.g., two continents drift away from a mid-ocean ridge or toward a deep-sea trench). Combinations of these two motions seem possible as well as the rotation or twisting of a particular continent as it drifts in a certain direction.

Paleomagnetism (p. 299) contributes important new evidence and seems to show that both polar wandering and an actual drifting apart of the different continents have occurred. Magnetic latitudes, as determined by fossil magnetism, seem to have remained about as they are now during the latter part of the Cenozoic but show divergence in older rocks. Measurements of the fossil magnetism of successively older rocks from North America indicate different positions for the north magnetic pole at different times. When the successive positions are plotted on a map, they fall on a curved path that was located in what is now the Central Pacific during the first part of the Paleozoic Era. The path then leads to Japan; following this, it can be traced across Siberia to its present position. Similar measurements in Europe show similar results prior to the middle part of the Mesozoic Era, but they show diverging positions for the north magnetic pole after this date. The magnetic data, therefore, seem to indicate that North America and Europe were part of one supercontinent until the Mesozoic, and that they subsequently separated and drifted slowly apart.

Fewer data are available from the other continents, but they tend generally to support the concepts of polar wandering and continental drifting. For example, fossil magnetism measurements made on rocks of the same age on different continents may indicate a different position for the north magnetic pole for each continent. Interpretation: since only one north magnetic pole is presumed to have existed at this time, the continents themselves probably had different positions at this date in the past. The proportion of polar wandering vs. continental drifting in such movements is still uncertain.

According to one interpretation, the distribution of similar and dissimilar fossil animals and plants is explained best by the former existence of two supercontinents. Laurasia comprised what is now North America, Greenland, Europe, and Asia north of India. Gondwanaland, the other supercontinent, included Antarctica, South America, Africa, India, and Australia. However, only one supercontinent is imagined by others. Of course, one supercontinent might have existed first; it then subdivided (perhaps about 200 million years ago), and the two parts subsequently fragmented into the continents we observe today.

The mechanisms and ultimate forces causing polar wandering and continental drifting are unknown, although considerable speculation has occurred. We no longer imagine that the continents drift like huge rafts through the oceanic crust nor that whole continents have subsided to become ocean bottom. Present knowledge of the nature of oceanic and continental crust suggests that this is most unlikely. In fact, suggested mechanisms of this sort apparently kept continental drifting from being more generally accepted by geologists of the Northern Hemisphere for so many decades. The actuality of continental drifting (which now seems highly probable) was not separated from the causes of continental drifting, which are still uncertain.

Origin of Continents: North America As An Example

The origin of continents is basically the problem of the origin of the isolated sialic portions of the crust, and most hypotheses fall into two groups. (1) A relatively thin granitic crust may have formed long ago at about the time the material of the Earth separated into core, mantle, and crust. Perhaps originally this granitic material formed a worldwide, scumlike layer that subsequently was broken into segments. Later these were piled together to form the continents. (2) The original crust may have been chiefly simatic, with perhaps scattered sections (the nuclei of the present continents) that were somewhat thicker and more sialic. In this view, the continents have developed by processes that tended to thicken these nuclei, enlarge them laterally, and change their composition gradually from that of sima to sial. In one version, *continental accretion* has occurred (Fig. 12–7), i.e., mar-

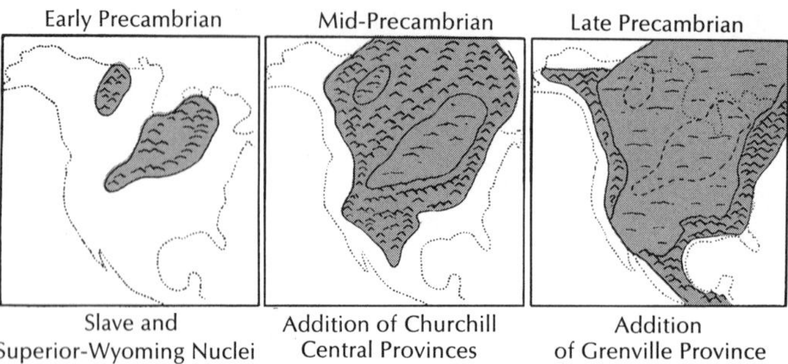

Idea of presentation from R. S. Dietz in 1966

12-7. Continental accretion as it might have occurred in North America during the Precambrian. (Idea of presentation from R. S. Dietz in 1966; Courtesy B. Mears, Jr., *The Changing Earth,* Van Nostrand Reinhold, 1970.)

ginal mountain belts have been added in succession to the edges of the continents.

Since plate tectonics and continental drifting have apparently played important roles in the last 200 million years of Earth history, they very likely also operated farther back in geologic time, perhaps even more effectively because initially the crust may have been thinner than now and less rigid. Thus two or more smaller continents (or nuclei) may have united as collisions occurred between plates, whereas on other occasions continents were rifted apart. Therefore, plate collisions, plate rifting, and position changes may have played important roles in the origin of the continents.

12-8. Physiographic and tectonic provinces of part of eastern North America. (Broughton, Fisher, Isachsen, and Rickard, *The Geology of New York State,* 1962.)

Geological Framework

The central part of North America forms a huge tract of comparatively low, topographically subdued country which stretches from the Gulf of Mexico northward into the Arctic. It has been called the *Central Stable Region* because it forms the nucleus of North America and has been relatively undeformed during the Paleozoic, Mesozoic, and Cenozoic Eras. Throughout much of the region, Precambrian igneous and metamorphic rocks are exposed at the Earth's surface or occur beneath a cover of younger sedimentary rocks.

The Central Stable Region can be subdivided into two parts: the Canadian Shield and the Interior Lowlands (Fig. 12–8). The *Canadian Shield* forms the northeastern part of the Central Stable Region in which the Precambrian igneous and metamorphic rocks are commonly exposed at the surface. It includes most of the eastern and central portions of Canada and extends southward into the United States in the Adirondack and Lake Superior areas. Each of the continents has a nucleus (or nuclei; Fig. 11–18) of this sort, and the name *"shield"* was derived from a certain topographic resemblance (a rounded convex surface) to the shields used for protection in ancient wars. Shields seem to have a somewhat lower heat flow than other sections of the continents, which may show that their overall chemical composition is more basaltic and less granitic than previously thought.

Around the margins of the Canadian Shield on the south and west are the *Interior Lowlands* where the Precambrian igneous and metamorphic rocks are buried by a relatively thin cover of nearly flat-lying sedimentary rocks of Paleozoic (chiefly marine), Mesozoic, and Cenozoic (chiefly nonmarine) ages. Broadly viewed, the marginal zone where the Canadian Shield meets the Interior Lowlands is marked by the location of a number of large lakes that extend from the Great Lakes northwestward and northward to Great Bear Lake. In general, these lakes are located in broad shallow depressions (Fig. 6–8) produced by the wearing away of gently dipping, easily eroded, sedimentary rocks of the Interior Lowlands. Glacial erosion during the Pleistocene was important in enlarging these depressions, but they had been etched out long before by mass-wasting and stream action.

The Earth's crust since the Precambrian has apparently been somewhat less stable in the Interior Lowlands, where it has been regionally depressed or uplifted more frequently and with greater magnitude than in the Canadian Shield. Isolated outcrops of Precambrian rocks may be observed in some of the unwarped areas.

Precambrian Geologic History

The North American continent may have originated as a relatively small land mass (a continental nucleus or a group of such nuclei) that projected above its surroundings near the beginning of the Precambrian; its chemical composition was probably more similar to sima than to sial. Presumably much less water existed then at the Earth's surface than occurs in the present oceans, but some must have been present because certain once-layered ancient rocks required water for their formation.

According to the hypothesis of continental accretion, elongated subsiding troughs (geosynclines, p. 324) then developed at different times around the margins of the nucleus. Large volumes of sediments were washed into the geosynclines from the nucleus and from uplifted portions of the geosynclines. In addition, lava flows and volcanic debris piled up with the sediments in huge quantities. Subsequently these sediments and lava flows were squeezed, crumpled, metamorphosed, uplifted, and eroded to form mountains. Furthermore, some of the deeply buried rocks along the bottoms of the geosynclines probably became hot enough to melt and form magma which later cooled as granitic batholiths.

A marginal mountain belt of igneous and metamorphic rocks thus forms a thickened section of crust which has been accreted and welded onto the continental nucleus to make it larger. Numerous repetitions of this process here and there along the continental margins—each new marginal belt was situated seaward from the preceding one—have resulted in the enlargement of a relatively small continental nucleus into the large Central Stable Region that exists today (Fig. 12–9).

As these processes operated, the chemical composition of the rocks in the geosynclinal belts apparently became more sialic. Perhaps materials of a sialic nature migrated upward from the mantle as magmas and/or gases and were added to the marginal mountain belts. On the other hand, materials of a simatic nature (e.g., iron and magnesium) may have been removed in solution from the source areas for the geosynclinal sediments, thus increasing the proportion of sialic materials.

Evidence for continental accretion is more clear-cut for the younger mountain belts that have formed since the Precambrian around the margins of the Central Stable Region: the Appalachians (Paleozoic) occur to the east and south; the Rocky Mountain

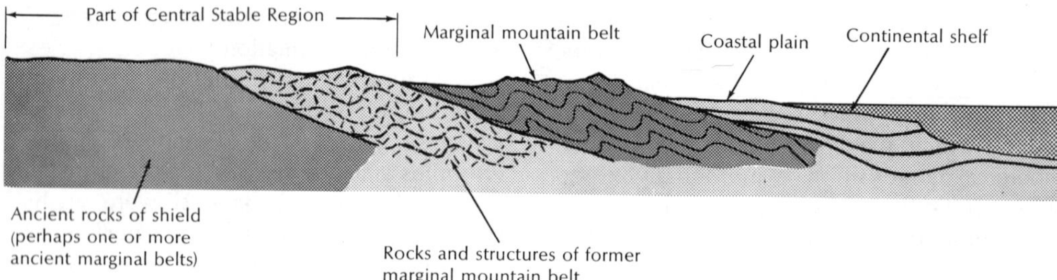

12-9. Idealized structure section from the Central Stable Region to the continental margin illustrating the hypothesis of continental accretion. *(After Philip B. King.)*

system (Mesozoic-Cenozoic) occurs to the southwest and west; and lesser known mountain systems occur to the north and northeast. In the continental accretion view, each such belt was once a geosyncline that was subsequently folded, faulted, metamorphosed, and intruded by granitic batholiths; it was accreted to the Central Stable Region.

The ancient Precambrian mountain belts may have been similar features, but subsequent erosion and burial by younger rocks has tended to obscure the evidence. However, radioactive age determinations seem to show that the oldest rocks occur centrally within the Canadian Shield and that successively younger rocks tend to be encountered as one goes radially outward (Fig. 12–10). Successive belts are arranged in a more or less concentric pattern, with the oldest on the inside and the youngest on the outside. Each belt of igneous and metamorphic rocks of about the same age apparently represents a former marginal mountain belt. Furthermore, geologic mapping and chemical analysis appear to show that each outer and younger belt has an overall chemical composition that is more sialic than its adjacent neighbor toward the continental nucleus.

12-10. Gross patterns and ages of geologic provinces in North America, as defined by major granite-forming, mountain-building events. (B. Mears, Jr., *The Changing Earth,* Van Nostrand Reinhold, 1970.)

Other continents have similar shield areas and perhaps have had more or less similar histories, although the accreted belts tend to be less conspicuous in some or perhaps not evident at all. North America may not be a typical continent, and the concentric pattern and age relationships of its belts may have been oversimplified. Moreover, all the present continents were apparently once part of one or two so-called supercontinents, which subdivided about 200 million years ago into the present continents; these continents then shifted to their present positions. However, individual continents may have grown by accretion before they merged to form the supercontinents.

The Appalachian Mountains

We now choose the Appalachians as a sample of a complex mountain belt—the concept of geosynclines originated from its study—and describe some of its rocks and structures as well as some of the ways in which geologists have thought about its formation. Then we question the amount of modification necessitated by the revolutionary new ideas of plate tectonics, sea-floor spreading, and continental drifting. The Appalachian Mountains form an elongated belt along the outer margin of the Central Stable Region that trends northeastward from Alabama for hundreds of miles into the State of New York and beyond. Northeastward from New York across New England into Nova Scotia and Newfoundland, the structures (stumps) of the mountains can be traced—the folds, faults, and batholiths—but the great mountains that once existed have been largely worn away. Thus, geologically, the Appalachians extend from Alabama to Newfoundland, and once extended even farther, because they are now covered by water or sediments at either end; topographically, however, the Appalachians do not trace readily beyond Pennsylvania and New York.

The central part of the Appalachians consists of conspicuous parallel ridges interspersed with valleys which trend northeastward in large sweeping curves and exhibit numerous sharp hairpin turns. A thick pile of Paleozoic sedimentary rocks underlies this section, has been folded into anticlines and synclines, and the axes of some of these folds plunge. Erosion of plunging folds made of alternately resistant and weak rocks produces valleys and ridges that trend in a zigzag, loop-shaped pattern. West of this portion of the Appalachians, the Paleozoic sedimentary rocks are less thick and are nearly flat-lying (Fig. 12–11). East of this folded belt, mountains and lower areas occur, and they are underlain by Precambrian and Paleozoic igneous and metamorphic rocks. Thus a section from northwest to southeast almost anywhere across the trend of the Appalachians shows that rock deformation—folds, faults, and metamorphism—increases toward the southeast.

Throughout its extent the mountainous belt parallels the margin of the Atlantic Ocean and can be subdivided into four units on the basis of topography, rocks, and rock structures (Fig. 12–8): the *Appalachian Plateau* forms the westernmost unit, next is the *Valley and Ridge* subdivision, then the *Blue Ridge*, and finally the *Piedmont* unit along the eastern margin. Between these four subdivisions and the Atlantic Ocean lies the *Coastal Plain*, a nearly featureless low-

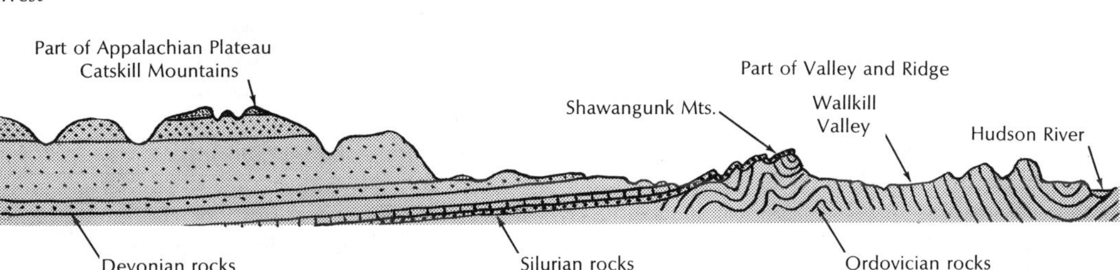

12-11. Structure section through part of the Appalachian Plateau and Valley and Ridge in southeastern New York. Ordovician rocks were folded during the Taconian orogeny. Subsequently, erosion occurred; Silurian and Devonian strata were then deposited. Some of the rocks were subsequently deformed during the Appalachian orogeny. Note the unconformity in the Shawangunk Mountains between Ordovician and Silurian rocks (this has been exaggerated somewhat). Effects of the deformation decrease toward the west (left). *(After Darton.)*

land underlain by geologically youthful, unconsolidated sedimentary layers that dip gently to the east.

Observations by American geologists in the middle 1800's showed that total accumulations of Paleozoic sedimentary rocks correlated with the location of the Appalachian Mountains: greatest thicknesses were found within the Valley and Ridge; lesser thicknesses occurred under the Plateau Region; and still smaller accumulations were observed farther westward in the Mississippi Valley area. These findings eventually led to a hypothesis about the origin of great mountain belts—based upon the Appalachian Mountains as a type example.

According to this view, no Appalachian Mountains existed at the beginning of the Paleozoic Era. Instead, during much of this era, a large elongated trough in the Earth's crust occupied the very zone out of which the Appalachian Mountains would eventually grow. This trough was called the Appalachian geosyncline, and it was assumed to have subsided slowly during most of the Paleozoic Era; occasional upward movements and times of quiet interrupted the sinking. Shallow seas commonly covered the trough or advanced and retreated slowly within it. Sediments were carried by rivers and spread out along the floor of the shallow sea. As the floor of the sea subsided, room was made available for the deposition of additional sediments. Deposition and subsidence continued during most of the Paleozoic Era, and a very thick pile of sediments accumulated. Perhaps the maximum thickness approached 40,000 feet, although the average may have been half this amount. Such an elongated, subsiding, sediment-collecting belt—commonly the site of a future great mountain chain—is called a *geosyncline*.

Features associated with shallow-water deposition (e.g., certain types of fossils, ripple marks, and mud cracks) are found here and there in the thick pile of Paleozoic strata and indicate that the floor of the sea was covered much of the time by relatively shallow water. However, recent discoveries in oceanography indicate that this shallow-water aspect may have been overemphasized.

Similar belts of thick sedimentary accumulations have since been found in other great mountain chains like the Rockies, Andes, Alps, and Himalayas. Why such great mountain chains developed out of areas which were once subsiding sea floors was one of the great mysteries of geology.

Most of the clastic sediments deposited in the Appalachian geosyncline show a gradation from thick and coarse in the east to thinner and finer grained in the west. Thus the sediments came chiefly from the east, and the source area was named Appalachia. It was imagined as a continuous highland mass of Precambrian crystalline rocks which paralleled the Appalachian geosyncline on the east. Thus it occupied an area of unknown extent that is now the Piedmont, Coastal Plain, and continental shelf. The volume of sediments deposited in the Appalachian geosyncline during the Paleozoic Era is enormous. Therefore, the source area was presumed to have been very large, or if smaller, its volume above sea level was renewed frequently by uplift. Since no high land mass occupies this region today, it was postulated that Appalachia subsided below sea level following the end of the Paleozoic Era.

Similar geosynclines and source-area borderlands were thought to have existed along the southern and western borders of the United States, and they were also described on other continents. Geological thinking has sometimes been governed by fashions, and the borderland fashion was applied widely in its day.

Subsequent field studies have made the assumption of continuous borderlands untenable and have modified greatly certain views about geosynclines; e.g., the fundamental differences between continental and oceanic crust were discovered. In combination with the concept of isostasy, this indicates that a large part of the lightweight continental crust apparently cannot subside to form a deep sea floor; the continents and ocean basins are approximately in balance. Furthermore, many of the igneous and metamorphic rocks exposed in the Piedmont and underlying the sediments of the Coastal Plain are not Precambrian in age but Paleozoic. Thus a continuous borderland of ancient rocks (Appalachia) apparently did not exist near the beginning of the Paleozoic Era. Rather, throughout the Paleozoic there must have been in this region basins of deposition, sedimentation, and igneous activity involving lava flows, volcanic eruptions and intrusions.

An island arc might be a present-day counterpart of the source area that once existed to the east (see below), but additional modification seems called for by recent discoveries concerning the ages of ocean basins, plate tectonics, and continental drifting. For example, the Atlantic Ocean apparently did not exist some 200 million years ago, nor for that matter did the Indian Ocean. The two formed subsequently as one or more supercontinents rifted apart. Therefore, the Pacific Ocean was probably once larger than it is today, and it became smaller as the Atlantic and Indian Oceans formed. In other words, the total area of the oceans some 200 million years ago may have been similar to the total area occupied by oceans today. However, an ancestral Atlantic Ocean probably did exist during part of the Paleozoic Era, and it might have closed and opened more than once. Thus the source area to the east may be tied in with the drifting and colliding of continents (Fig. 12–19).

THE APPALACHIAN GEOSYNCLINE The Appalachian geosyncline can be subdivided into an inner, or western, non-volcanic part and an outer, or eastern, volcanic part (Figs. 12–12 and 11–7). The western part (called a miogeosyncline by some) encompassed the area now occupied by the Valley and Ridge province and the eastern portion of the Appalachian Plateau province. The eastern part (called the eugeosyncline by some) occupied the space in which we now find the Piedmont and Coastal Plain provinces and perhaps part of the present continental shelf; i.e., it substitutes geographically for the borderland Appalachia. The boundary zone between the two parts of the Appalachian geosyncline thus coincided more or less with the Blue Ridge and Green Mountains. One part may have merged with the other, or the sea floor may have been higher between them to form two separate troughs, and variations may have occurred with time.

The two parts have characteristic features which have also been recognized in other geosynclines. The inner part of the Appa-

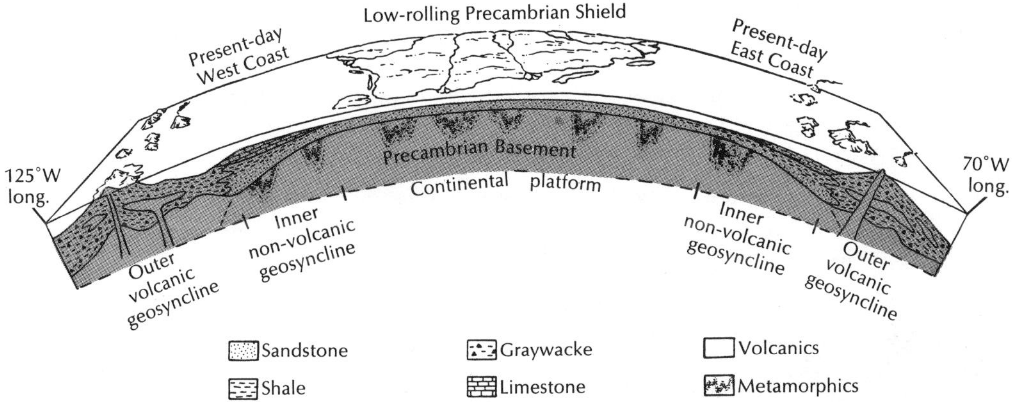

12-12. Geosynclines and part of North America as they may once have appeared. (B. Mears, Jr., *The Changing Earth,* Van Nostrand Reinhold, 1970.)

lachian geosyncline commonly lacks volcanic rocks and granitic batholiths. It contains a lesser thickness (up to 30,000 feet or so) of generally well-sorted, clastic sedimentary rocks such as sandstone and shale, as well as numerous layers of limestone. The sand grains tend to be well rounded and contain much quartz (chemically and mechanically resistant). This part apparently subsided less and more slowly than the outer part and was affected less by crustal movement and metamorphism. As exemplified by the Appalachians and Rockies, the rocks of the inner part tend to be folded and thrust-faulted only near the end of the life of a geosyncline. In contrast, such movements, accompanied by the intrusion of granitic batholiths and metamorphism, tend to affect the outer part a number of times. Thus the inner part of the Appalachian geosyncline experienced only one main time of mountain building (the Appalachian orogeny) whereas its companion part to the east was affected by three major times of mountain building: the Taconian (Fig. 12–13), Acadian, and Appalachian orogenies.

The eastern part of the Appalachian geosyncline apparently subsided more extensively and continuously than the western part, and thus more space was available for sediments to pile up to greater thicknesses. Deposition was also more rapid, pronounced intrusive and extrusive igneous activity occurred, and crustal movements were more extensive, intensive, and deep-seated. Perhaps the eastern part contained a chain of active volcanoes resembling those in the island arcs we can observe today. Evidence consists of the great quantities of volcanic rocks, many produced by explosive eruptions, that have been found interbedded with Paleozoic sedimentary rocks throughout this zone. Lava flows and explosive eruptions, apparently built the volcanoes up above sea level. In addition, folding and faulting probably raised some parts of the sea floor to form long narrow ridges, whereas other areas were moved downward. Sediments were thus transported more or less continuously from the higher source areas to the depressions. Although such activity may have taken place almost continuously, it seems to have affected different sections at different times and to have varied greatly in intensity. At maximum, such movements became sufficiently widespread and intense to be called orogenies (see below).

In addition to greater total thickness and the inclusion of huge quantities of volcanic rocks, sediments of the outer part of the Appalachian geosyncline typically are coarser grained, less well sorted, more angular, and generally less mature (i.e., they contain less weathered material) than are inner-part sediments. Such features imply relatively rapid erosion, short transportation, and deposition on a rapidly subsiding floor. Otherwise the attrition produced by transportation and wave action would sort and round the sediments and reduce the less resistant material to very tiny particles.

OROGENY *Orogeny* involves the formation of mountains, particularly the folding and faulting that affect relatively long narrow belts. In contrast, *epeirogenic* crustal move-

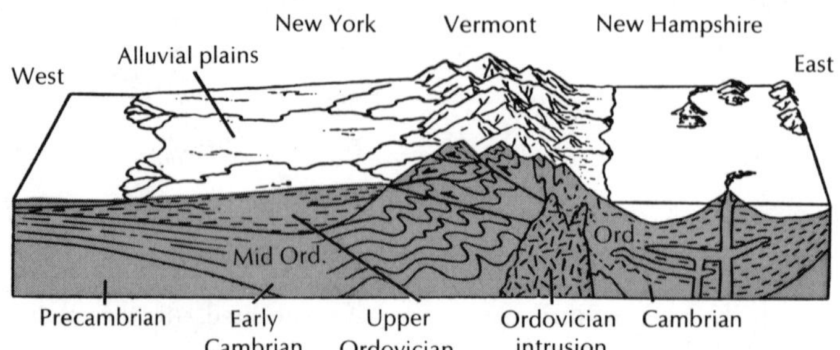

12-13. Mountains and other features at time of Late Ordovician Taconian orogeny (schematic). (B. Mears, Jr., *The Changing Earth,* Van Nostrand Reinhold, 1970.)

ments affect large portions of a continent and tend to be chiefly vertical, either up or down; such broad gradual movements cause relatively little rock deformation. Evidence that an orogeny has occurred is furnished by the deformed rocks themselves—their anticlines, synclines, thrust faults, etc. —and by the metamorphic effects and intrusive bodies that are commonly associated. An orogeny may be dated if an unconformity is discovered that shows older deformed rocks overlain by younger undeformed ones.

Additional evidence for an orogeny involves so-called orogenic sediments and their interpretation. When lands are high, streams flow swiftly down their slopes and carry large loads of coarse sediments which are deposited in great thicknesses in a belt adjoining the newly formed mountain (however, epeirogenic uplift in a source area may result in the origin of similar sediments). In contrast, when lands are low, little sediment is transported, and materials accumulating on the floor of a nearby sea may consist chiefly of calcium carbonate in the form of shell fragments or precipitated matter. Later this consolidates to form limestone. The age of such orogenic sediments—if they are interpreted correctly—dates the crustal movement because deposition follows elevation of the source area.

An orogeny tends to extend over a period of time and involve a number of pulses of activity. Evidence may consist of a number of orogenic sediments of different ages piled one on top of the other in a thick sequence. Evidence is also furnished by orogenic sediments or surfaces of unconformity that have themselves been folded or faulted.

Near the end of the Paleozoic Era, a major orogeny occurred in the Appalachian geosyncline for the third time and affected both the western and eastern parts, but not all areas simultaneously nor equally in magnitude. Most of the folds and faults now observed in the rocks of the Valley and Ridge province were produced at this time. Some of the folds were overturned to the northwest, and movement along the thrust faults was also toward the northwest. The structures suggest that the mass of the outer part of the Appalachian geosyncline was shoved against the sedmientary rocks of the inner part, with attendant crumpling and squeezing. Effects fade out toward the northwest where the rocks of the Appalachian Plateaus are nearly horizontal. In other words, the intensity of the metamorphism increases toward the southeast (across the trend of the Appalachians) into the Blue Ridge and Piedmont provinces.

Near the end of the Paleozoic Era, therefore, a chain of lofty mountains (but not the present Appalachians) apparently stretched about 2000 miles from Alabama to Newfoundland. Their eastern portion occupied a region that had been mountainous twice before as a result of the Taconian and Acadian orogenies. They had been accreted to the North American continent (Fig. 12–14).

MESOZOIC AND CENOZOIC EVENTS The chain of mountains produced during the Appalachian orogeny was subsequently lowered by erosion, and before the first part of the Mesozoic Era had ended, the entire area had apparently been worn down until only stumps of the mountains remained. Following this, in the latter part of the Triassic Period, normal faulting on a large scale affected parts of the Piedmont extending from the Carolinas to Nova Scotia. The down-faulted blocks formed basins, and the up-faulted sides furnished reddish clastic sediments in large quantities to pile up on the subsiding blocks (Fig. 12–15). In contrast to the compressive forces involved in the Paleozoic orogenies, these movements were chiefly vertical. However, some were quite extensive because the Late Triassic sedimentary rocks probably have a maximum thickness of three to four miles near the major faults.

The faults or other fractures evidently acted as passageways for the rise of basaltic magma because numerous dikes, sills, and lava flows formed. These igneous intrusive and extrusive rocks tend to be more resistant to erosion than the sandstones and shales that enclose them, and they have been

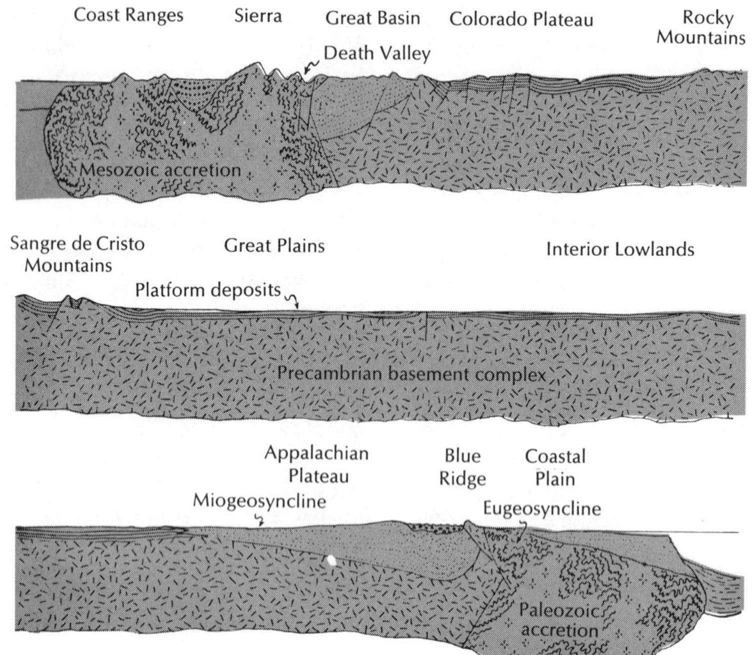

12-14. Diagrammatic structure section from the Pacific Ocean across the United States to the Atlantic Ocean. (Modified from a drawing by L. H. Nobles. Robert S. Dietz, Collapsing continental rises: an actualistic concept of geosynclines and mountain building, *J. Geol.* **71,** May 1963.)

etched out into conspicuous topographic features.

Disagreement exists concerning the interplay between uplift and erosion that ensued from the Triassic to the present. The entire area was again worn down by erosion at least once, and the extensive, low-lying, relatively smooth erosion surface that resulted was subsequently warped upward epeirogenically for hundreds of feet and more. The total uplift apparently was the end result of several times of lesser uplift, each followed by a quiet period. This upwarped surface was then dissected by erosion, which etched out the ridges (along belts of resistant rocks) and valleys (along belts of less-resistant rocks).

The Geosynclinal Cycle Summarized

We summarize now the major features of the geosynclinal cycle as they were gen-

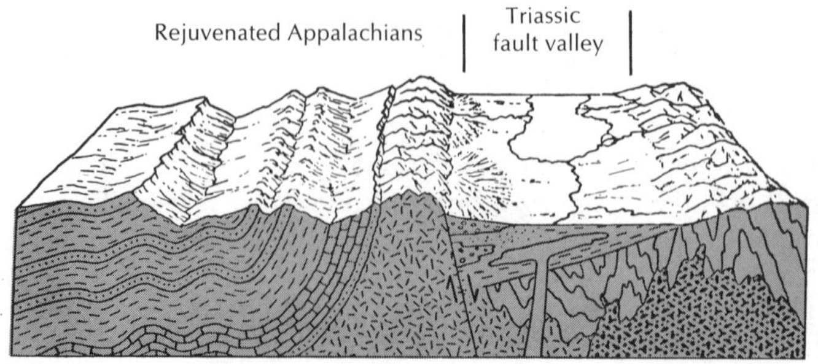

12-15. Appalachians during Late Triassic (Palisades orogeny). (B. Mears, Jr., *The Changing Earth,* Van Nostrand Reinhold, 1970.)

erally pictured by geologists a few years ago. Complex mountain systems seem to share a number of sequential steps in their development, although details vary considerably from one system to the next. (*1*) A necessary initial phase appears to be the formation of a geosyncline, its subsidence over a long period of time, and the accumulation within it of a thick belt of sediments. In addition, lava flows and volcanic debris are voluminous in some sections (the outer part). (*2*) Eventually the sedimentary and igneous rocks in the geosyncline are compressed by powerful crustal forces which act chiefly in a horizontal direction. However, such compression may sometimes have resulted from gravitational sliding down the sides of large areas that had been uplifted greatly. More than one major orogeny may occur during the life of a geosyncline, and the orogenies tend to affect different areas at different times. (*3*) During or shortly after the folding and faulting, huge batholiths of granitic rocks originate in some sections (the outer part). (*4*) Next, a long period of crustal quiet occurs during which erosion wears away the mountains. (*5*) Block faulting next affects parts of the area. Down-faulted blocks become basins in which clastic sediments pile up to great thicknesses and into which basaltic magma rises to form dikes, sills, and lava flows (e.g., the events of the Late Triassic in New Jersey). (*6*) Subsequently the entire region is arched upward and is then dissected by erosion. Weaker rocks are removed at a relatively rapid rate, and areas underlain by resistant rocks are isolated to form mountains. Thus the rocks and rock structures that had formed long ago now become etched into conspicuous topographic features by differential erosion.

Mountains

"How old is that mountain?" is a common question, but it is one that sometimes cannot be answered simply; the history of complex mountain belts like the Appalachians tells us this. We can determine the ages of the rocks making up a mountain, and this yields one answer. However, the present-day mountain did not exist when its rocks were formed. If age refers to a time in the past when orogenic movements folded, faulted, and uplifted a region, then a second and different answer is required. However, the mountains resulting from such an orogeny may have been worn down by erosion long ago. The present-day mountain may not have formed until this eroded surface was subsequently uplifted and then dissected by erosion, which gives a third age. An informed reply manifestly is an explanation involving all these factors.

Other mountains, of course, are much less complex, and perhaps simplest of all is the volcano—the mountain of accumulation (Fig. 12–16). Products of volcanism heap up quietly or explosively about an opening that leads downward to a magma reservoir. As a volcano grows larger, it is attacked ever more vigorously by erosion, and its shape at any one time is a compromise between the additions of new material from below and erosion at the surface.

Thus mountains are large isolated land masses that project conspicuously above their surroundings, and crustal movements, igneous activity, and differential erosion are important factors in their origin. For example, parts of the crust may be elevated along great fractures to form fault-block mountains, and compression may fold and dome the crust. However, some geologists do not consider the isolated remnants of a dissected plateau (made of flat-lying rocks) as mountains (e.g., the Catskills). To them, the rocks in an isolated land mass must contain structures such as folds and faults to qualify as a mountain.

Convection Currents and Geosynclines

The forces that have crumpled geosynclines, and the causes of the other events common in the history of most great mountain systems, are mysterious and have been the subjects of much speculation. The convection current hypothesis is one attempt to explain the geosynclinal cycle and has met with

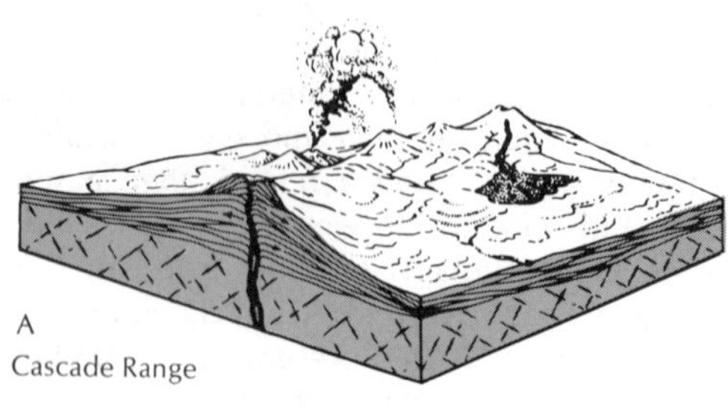

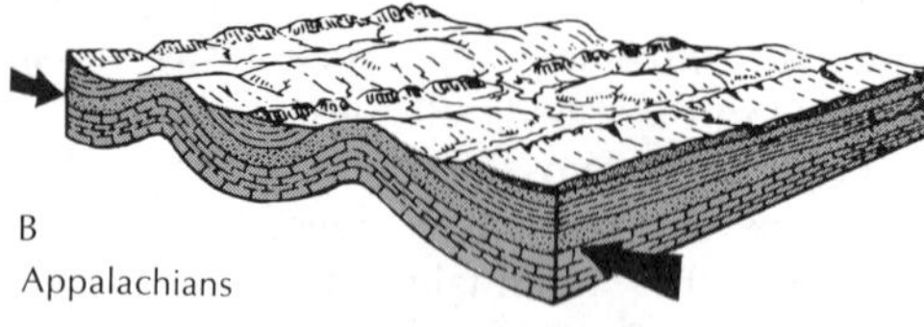

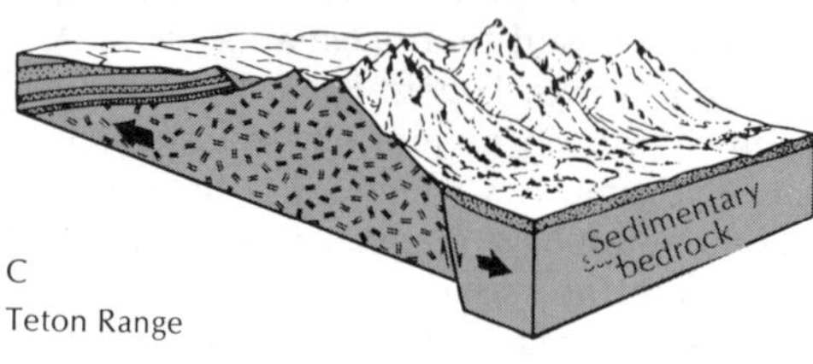

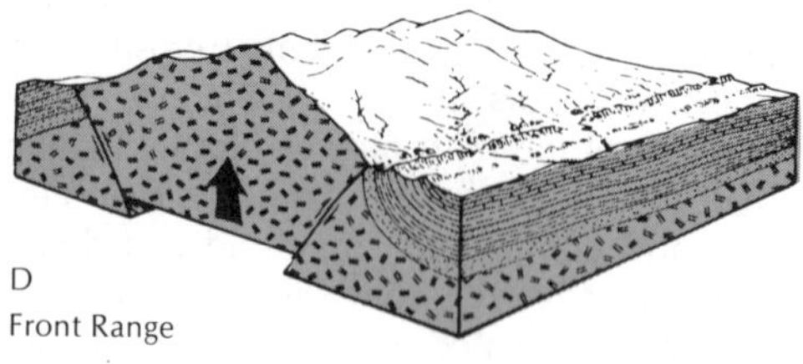

12-16. Different processes predominate in the origin of mountains: A, volcanic activity; B, folding; C and D, faulting and uplift. *(Courtesy U.S. Geological Survey.)*

some success. Recently it has been coupled with continental drifting and plate tectonics in a promising new synthesis. Additional hypotheses have involved contraction of the Earth's surface on an interior that is cooling and shrinking (one objection: the Earth's interior may not be cooling off), an overall expansion of the Earth, and gravitational sliding of relatively thin masses of sedimentary rocks down the slopes of an uplifted region (adequate only for some aspects of mountain building).

In an older version of the convection hypothesis, it was assumed that no major discontinuities existed in the mantle and that very slow plastic flow could occur within it. Some parts of the core were assumed to be hotter than adjoining parts (cause unknown), which meant that the base of the mantle overlying a hot part of the core should become heated, expand, and rise. Surrounding parts of the mantle would then move laterally toward this location, thereby setting up a gigantic, exceedingly slow convection current.

Such an upward-moving current would eventually reach the Earth's crust (or lithosphere), and move laterally. If two such currents move horizontally beneath the crust and meet, they would probably sink and drag the crust down in that area (Fig. 12–17). Clever and carefully scaled models were made in which convection currents were produced by rotating drums in a tank of glycerin. Floating at the surface of the liquid was a mixture of sawdust and oil that simulated the strength and thickness of the crust. These models showed that a trough develops where the downward moving currents meet, and perhaps the development of this trough is comparable to the formation of a geosyncline.

Convection currents, if they exist, may act cyclically. Perhaps a long initial phase of slow movement is followed by a shorter period in which currents move more rapidly. At this time, very hot material would be rising and relatively cool material sinking. The longer slower movement might produce the long-continued, general subsidence of a geosyncline, whereas a "faster" period of movement may account for a major orogeny.

A great downward bulge (root zone) of lightweight crustal rocks may thus form below a geosyncline. Crustal balance is thereby upset, and the lightweight rocks would rise if they were not prevented from rising by the downward drag of the convection currents. At this stage, the large granitic batholiths that occur only in the cores of complex mountain systems may originate. The sial (upper part of the continental crust) approximates granite in chemical composition, and in laboratory experiments, granitic rocks melt at lower temperatures than do basaltic rocks. Thus the sial near the bottom of a downward-bulging root zone may melt, because temperatures increase with depth. On the other hand, temperatures may not be high enough to melt the sima which underlies the sial and borders a root zone on either side. The molten sialic material would form granitic magma which could then move upward and squeeze into deformed rocks in the mountainous area.

Peneplanation of the original mountains may occur subsequently as the currents gradually slow down. When they finally stop, the thickened zone of lightweight sialic rocks should move slowly upward, and perhaps this corresponds to the uparching which commonly has occurred after peneplanation. A crude analogy is the reaction of a rubber ball that is held beneath the surface of a body of water and then released.

Geosynclines vs. Plate Tectonics

Long thick linear belts of sediments, of geosynclinal dimensions, occur today in association with the continental rises and with what seem to be filled deep-sea trenches (Fig. 11–7). Are these the modern counterparts of the ancient geosynclines? Will they one day be deformed into a mountain belt that is accreted to a continent? Some speculate that this may be so (Fig. 12–18).

How much of Appalachian geologic history as sketched on preceding pages, and how much of the whole concept of a geosynclinal cycle, can we retain today in the light of plate tectonics, drifting continents, and ocean basins that open and close? It may be too soon to draw firm conclusions, but it does seem as if extensive modification

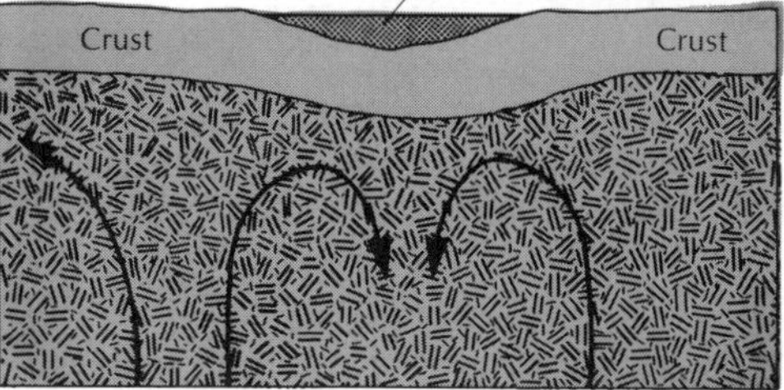

Solid material is assumed to flow slowly in the mantle

A

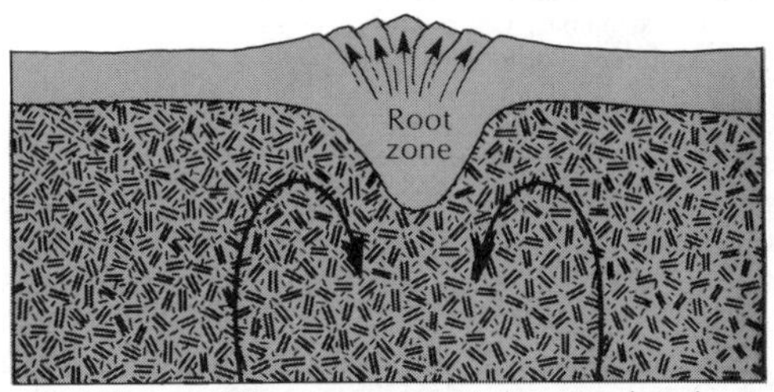

More rapidly moving currents have produced a root zone and faults and folds

B

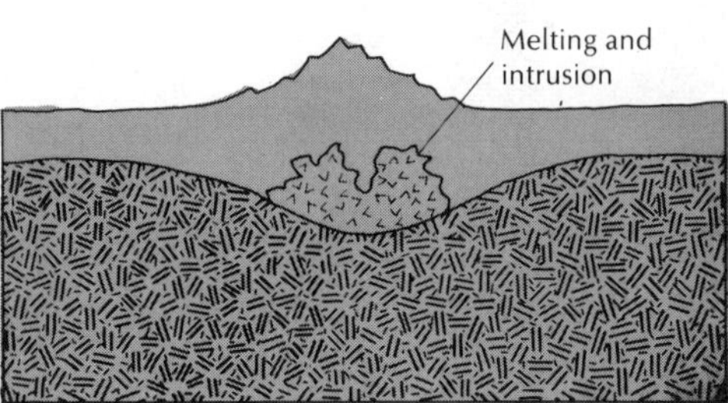

Uplift has occurred because convectional circulation has stopped. Granite batholiths have formed

C

12-17. Mountain building and convection currents (highly diagrammatic). (A) Slowly rising warm currents spread beneath the crust, cool, and descend. (B) The currents move more rapidly, the crust is greatly compressed, and its lightweight rocks are dragged deep into the mantle. (C) As convection slows and then ceases, the down-pulled part of the crust tends to move upward. A mountain belt is formed when this uplifted region is subsequently carved into isolated topographic units by erosion. *(After Griggs.)*

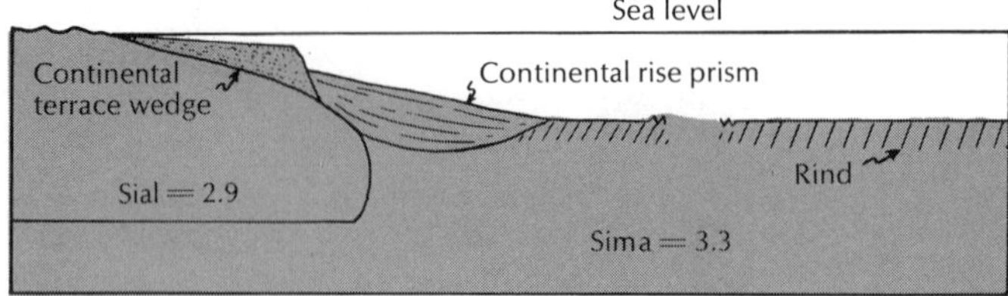

A. Geosynclinal accumulation phase — convection inactive

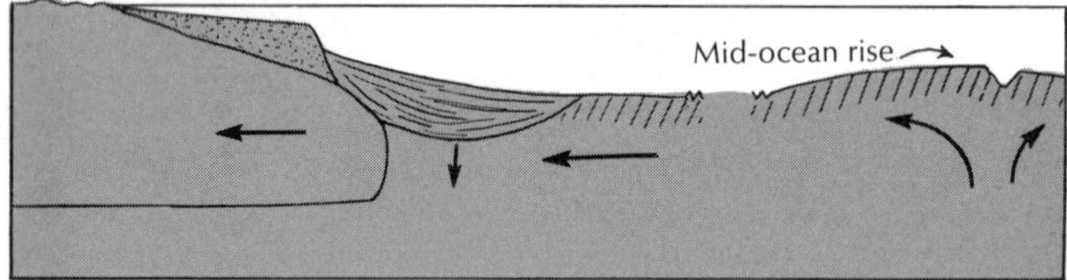

A'. Geosynclinal accumulation phase — sima active but coupled with sial — continental drift

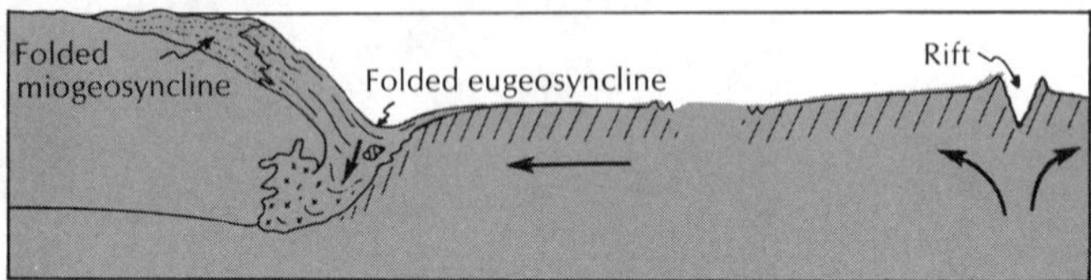

B. Orogenic phase — sima shearing beneath continental raft

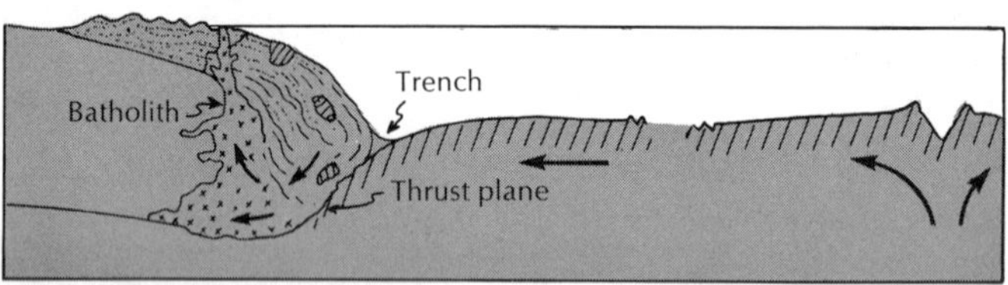

C. Late orogenic phase — plutonism

12-18. A schematic series of diagrams showing the possible development and deformation of a geosyncline and postulated relationship of its inner and outer parts to present-day continental shelves and continental rises. Thermal convection in the mantle is assumed to cause the compression, and the convection currents are assumed to extend all of the way to the sea floor. Sea-floor sediments and other materials are dragged under a continent, become part of the sial, and are accreted to the continent. (R. S. Dietz, "Collapsing Continental Rises: An Actualistic Concept of Geosynclines and Mountain Building," *J. Geol.* **71,** May 1963.)

may be needed. However, the model of a mountain belt forming along the margin of a continent fits readily into the older view as does the compression and deformation that would be produced by colliding continents. Since the same rocks are still present in the same places, we must still associate great mountain belts with elongated regions in which sediments piled up to great thickness, although now we may interpret their overall environment differently (Fig. 12–19).

Apparently, mountain belts result from the collision of two lithospheric plates, e.g., the Himalayas apparently have originated because the Asian and Indian plates collided. Since each consisted of light-weight continental material, neither plate could be dragged downward for any great distance into the mantle, and thus a massive pile of continental rock resulted. In a similar manner, the Alps seem to have resulted from collisions that occurred more than once between the African and European plates. In like manner, the Andes have apparently resulted from the underthrusting of one plate beneath another along a deep-sea trench. Such relatively youthful mountains occur now along the boundaries of plates that exist today. Presumably more ancient mountain belts formed along the boundaries of the different plates that existed when they formed long ago. Thus geologists can continue to apply uniformitarianism in their study of the Earth.

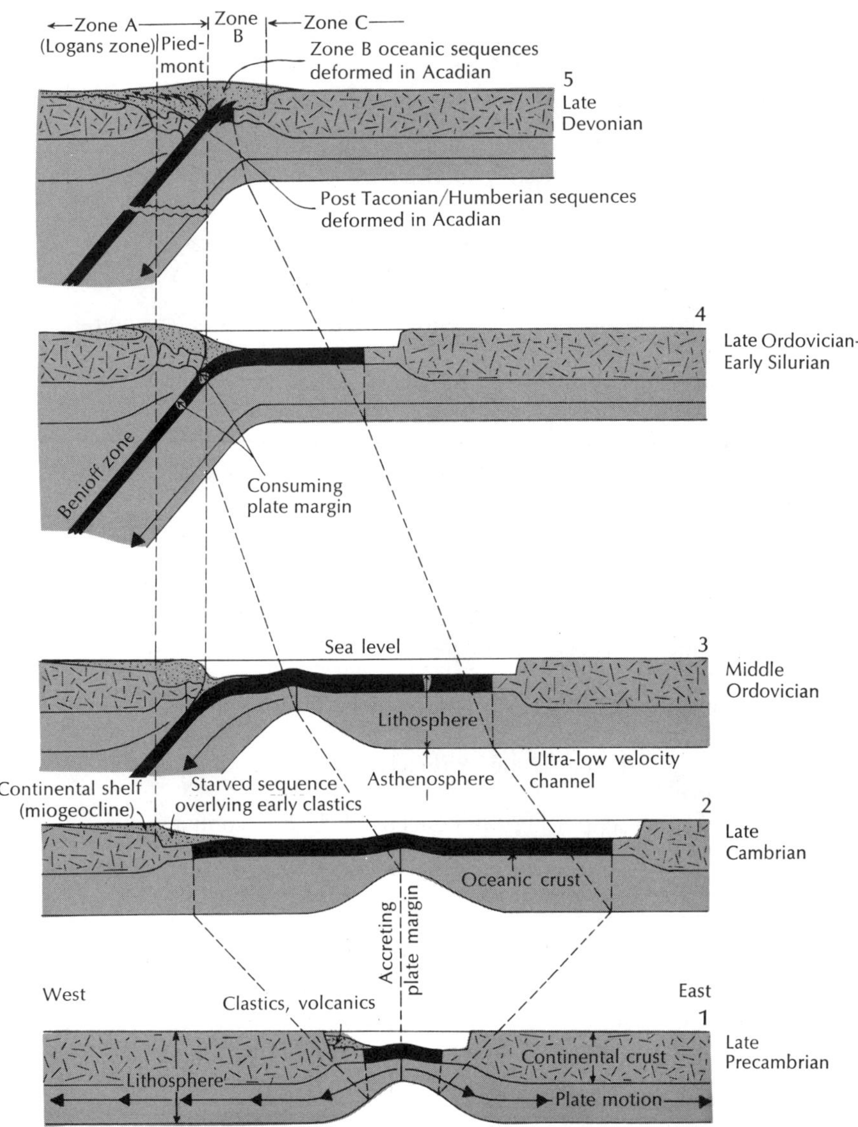

12-19. Events that may have occurred within and along the margins of an "Ancestral Atlantic Ocean" from Late Precambrian to Late Devonian. The North American continent is on the left (west) and Europe-Africa is to the right. The Ancestral Atlantic expands (1), reaches maximum size (2), and then contracts (3) and (4) with oceanic lithosphere sliding down beneath North America—the Taconian orogeny results. Later, collision between the continents (5) causes the Devonian orogeny. *(Courtesy John M. Bird and John F. Dewey.)*

Exercises and Questions for Chapter 12

1. How does the protoplanet hypothesis account for the fact that the sun, Earth, and Jupiter now differ greatly in chemical composition, although each may have formed at about the same time from the same kind of interstellar material?
   (A) Why do most planets rotate in the same direction that they revolve?
   (B) Why does Jupiter rotate more rapidly than the Earth?
   (C) Some satellites revolve in the retrograde direction around their planets. Why?
   (D) Why do the planets possess most of the angular momentum in the solar system?
2. What evidence indicates that the Earth's first atmosphere probably escaped? How did the present atmosphere probably originate? the hydrosphere?
3. What evidence suggests that the North American continent may have formed by accretion?
4. Discuss some of the major processes involved in the origin of mountains.
5. What are the major events in the geosynclinal cycle?
   (A) How does the convectional circulation hypothesis attempt to explain each of these?
   (B) What modifications are necessitated by newer ideas involving global plate tectonics?
6. What is an orogeny and what is the evidence that many orogenies consist of a series of pulses?
7. Discuss some evidence supporting the concept of drifting continents.

13

The Fossil Record

So far we have dealt largely with the inanimate side of the Earth's long, impressive and eventful past: with the advances and retreats of ancient seas which have made the world's coastlines ever-shifting and transient; with drifting continents and paleomagnetic reversals; with the igneous activity that has frequently punctuated Earth history, here in noisy, spectacular, destructive outbursts, there in quiet, deep-seated, and voluminous change; with the growth of complex mountain chains out of elongated, subsiding sections of ancient sea floors; and with the constant wasting away of higher land masses by the persistent forces of erosion and the counterbalancing renewal of land areas by upward movements of the Earth's crust.

Life (Fig. 13–1) now moves to the center of our stage, and the Earth's history assumes a new focus of interest. Our subject is the countless multitude of animals and plants—huge and tiny, spectacular and commonplace, marine and nonmarine—that lived and died during the immensity of geologic time stretching back from the present to that shadowy reach when life itself first arose. The former existence of these organisms is recorded by fossils found in rocks, and some of the more significant aspects of the fossil record constitute the subject matter of this chapter.

Whether the history is physical or biological, its keynote is unending, leisurely change—change imperceptible to untrained eyes during lifetimes measured in terms of a few scores of years, but whose cumulative effect has been very great. These changes continue today: valleys grow deeper, glaciers shrink, the Earth's crust moves, organisms migrate, and new types of animals and plants slowly develop. Although the changes of today are not precisely those of the past—in kind, location, magnitude, and rate—their observation and measurement by geologists have resulted in the development of techniques and concepts that have been applied with considerable success in elucidating the events of the past (this is the assumption of uniformitarianism which is basic in geology).

For convenience, geologists have subdivided the

13-1. A Late Mesozoic landscape in Wyoming. We see rhinoceros-like Triceratops, the terrible Tyrannosaurus, a flying reptile, and spike-armored and duck-billed dinosaurs. *(Part of a mural by Rudolph F. Zallinger in Peabody Museum, Yale University.)*

long, detailed account of past events into several major units *(eras),* each of which includes a number of smaller units *(periods)* in much the same manner as books consist of chapters (Fig. 13–2 and p. 106). Most of the pages of the first long volume are missing, tattered, or scattered widely around the world. They are a little like the bedraggled remains of an ancient book found in a leaky, wind-swept barn frequently visited by vigorous, inquisitive children. However, the less ancient portions of the geologic record become more legible and detailed, and in this respect the Earth's history parallels the story of man.

The Earth itself probably originated some 4.6 billion years ago, about 1 billion years before the oldest known rocks. Thus the *Precambrian* Era, which begins with the beginning of the rock record, started some 3.5 billion years ago, and a 1-billion year gap presently occurs in Earth history. The *Paleozoic, Mesozoic,* and *Cenozoic* Eras probably began about 600, 225, and 65 million years ago respectively, whereas the Pleistocene Ice Age began some 2 to 3 million years ago.

Fossils

Fossils (Latin: "something dug up") are the recognizable remains or traces of animals and plants that were preserved in sediments, rocks, and other materials such as ice, tar, and amber prior to historic times. Fossils provide a record of past life which is at best fragmentary. Consider the multitude of organisms in existence at any one time. A very small fraction of these will subsequently become fossilized, and only a tiny proportion of the fossilized forms will eventually be exposed by erosion at the Earth's surface and be discovered by man (Fig. 13–3). However, field trips and museum collections testify that fossils may occur in remarkable numbers and in astonishingly good condition.

Two factors are favorable for the preservation of organisms as fossils: one is the

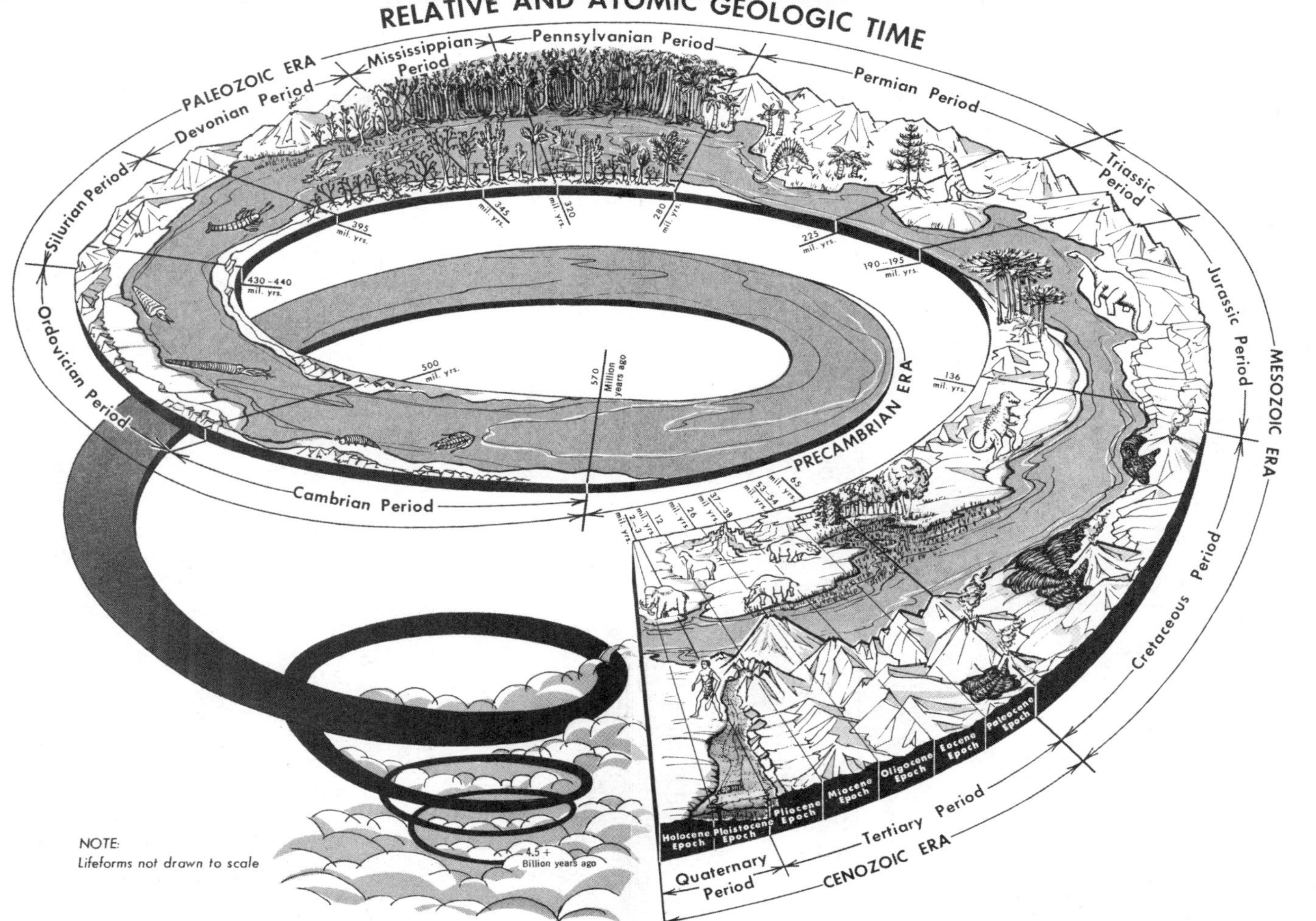

13-2. Graphic illustration of geologic time scale. (*Courtesy U.S. Geological Survey.*)

13-3. A large fossil cephalopod, about $4\frac{1}{2}$ ft. across, discovered near Fernie, British Columbia (Jurassic; *Titanites occidentalis*). *(Courtesy Geological Survey of Canada.)*

possession of hard parts such as shells and bone; the other is quick burial of the remains to prevent destruction by scavengers and decay. Although animals like worms and jellyfish have much less chance for preservation than animals like clams and oysters, some soft-bodied organisms have been fossilized. On the other hand, the possession of hard parts by no means guarantees fossilization in the future. Until late in the 1800's, huge herds of buffalo roamed the prairies of the western United States. Today the buffalo have all but vanished and hardly a trace remains of their former presence. Their carcasses were not covered rapidly by sedimentation following death, scavengers and bacteria quickly destroyed the flesh, and the hard parts soon weathered to dust—all this in a few decades.

Burial to preserve organic remains may occur in a number of ways, but fossilization in sediments on the sea floor has been by far the most common manner of origin. The seas today are relatively restricted, and climatic conditions are more diverse and extreme than they were on the average in the past, at least since fossil evidence became abundant following the Precambrian. Large portions of the present continents, perhaps as much as one-quarter to one-half or more, have at times been covered by shallow seas, and these portions varied from one age to the next as the waters gradually advanced or retreated. Countless animals with hard parts lived on the limy, muddy, or sandy bottoms of these seas. After death, their shells were covered rapidly by sediment which accumulated in layers on the sea floors. Other organisms were buried alive by sudden influxes or shifts of sediment caused by submarine landslides, by currents, or by violent storms that produced large, bottom-stirring waves. As a result, some marine rocks are extremely fossiliferous.

Fossil-bearing beds should be seen in the field for a true appreciation of the significance of fossils to geology. For example, on a field trip in the Nevada desert, you may climb an isolated, barren, flat-topped hill whose surface is formed by a flat-lying layer of limestone that is littered with crinoid stems and other fossils. You are standing on an ancient sea floor—one look convinces you of this—but the present climatic and topographic setting is strikingly different from that of the past.

At another time and place, you may drive along a straight, heavily traveled road not far from a busy city and encounter layers of sedimentary rock that trend parallel with the road. Rock outcrops form almost continuous low cliffs and in them are exposed millions of fossilized shells. Here again is an ancient sea floor, but this one may have existed some 200 million years prior to that in Nevada.

In still another area, you may observe large numbers of well-preserved fossil corals projecting from a steeply dipping limestone layer that constitutes the roof of an abandoned mine. This is all that remains of another ancient sea floor, now hardened rock and tilted at a steep angle, and you view it with fascination and deep emotion if you understand clearly its sermon in stones. On the other hand, many sedimentary rocks, marine as well as nonmarine, contain few or no fossils.

Organisms may be buried, partially preserved, and fossilized in a number of nonmarine environments: e.g., in sediments deposited on a river floodplain or within its channel; in the muds and clays that accumulate along the floor of a fresh water lake; in a limestone cavern where calcium carbonate may be precipitated as stalactites, stalagmites, and incrustations. Falls of volcanic ash may bury organic remains quickly, and much less frequently, a lava flow may overwhelm and harden around an organism before it is destroyed by the heat. A mold in the lava indicates the former presence of the animal or plant and provides an exception to the general rule that igneous rocks do not contain fossils.

For example, the mold of a rhinoceros was discovered a number of years ago in Washington. The bloated shape preserved by the mold (a plaster of Paris cast was made) indicated that the animal was dead and probably floating in water when it was buried by a flow of lava. The water chilled the lava quickly enough to preserve a mold, and a few bones were found within the mold.

Animals have likewise been buried rapidly when they were trapped in peat bogs and asphalt pits. At the famous La Brea asphalt pit in Los Angeles, oil seeped from the ground, collected in a pool at the surface, and was altered to sticky tar by evaporation of its more volatile constituents. Animals ventured on the smooth, solid-looking surface of the pool and were quickly trapped. Scavangers were probably attracted to the pool by cries of the victims and were also mired. Thus the asphalt has proved to be a happy hunting ground for fossil hunters, and excavations have uncovered the bones of horses, camels, elephants, giant bison, antelope, great ground sloths, vultures, saber-toothed tigers, bears, lions, and giant wolves as well as less conspicuous creatures such as mice, insects, and small birds. Such large numbers of carnivorous animals from one location are unusual.

Fossilization

Fossilization may occur in several ways (Fig. 13–4): preservation of an entire organism (partially decayed), preservation of the hard parts alone, petrifaction, and the formation of impressions, molds, casts, carbon residues, tracks, and trails.

Preservation Entire

An entire animal, or a portion of it, may be preserved unchanged except for some decay of the soft parts, but this is rare and occurs only for organisms that lived in relatively recent times (as measured on the geologic scale). A familiar example involves the woolly mammoths that have been preserved in the permanently frozen ground of Siberia; e.g., one woolly mammoth was discovered in the bank of a stream by a Russian hunter. Sudden violent death is indicated by the presence of unswallowed food in its mouth,

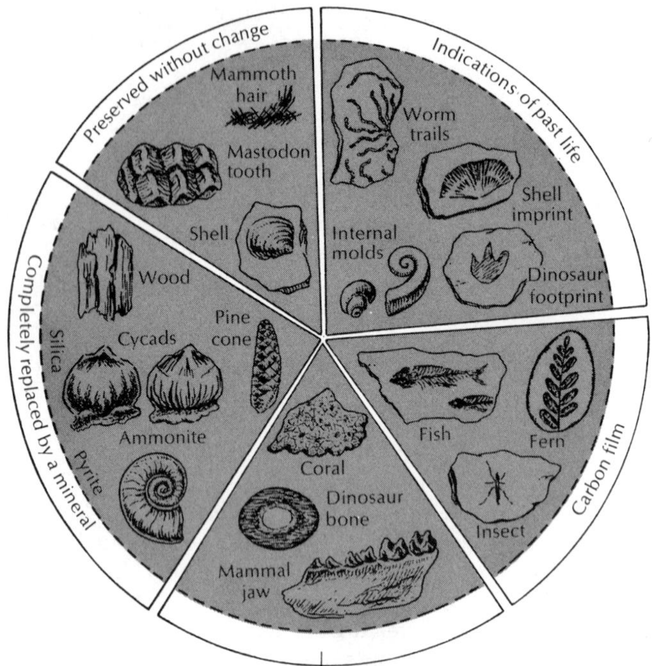

13-4. Origin of fossils. (B. Mears, Jr., *The Changing Earth,* Van Nostrand Reinhold, 1970.)

a broken hip, and clotted blood in its chest.

Perhaps the scene can be reconstructed. The animal browsed along a high river bank, moved too near the edge, caused a landslide, and was killed and buried during the resulting fall of dirt and debris. Later the material froze solid, and the mammoth was held in cold storage because the soil remains permanently frozen in this cold climate (i.e., ice and not water fills all cracks and spaces) and only the upper few feet thaw during the summer. Eventually the river undercut its bank and shifted its channel laterally far enough to expose the mammoth. When the animal was discovered, some of its flesh was still red and was eaten by wild dogs, but its stench was overwhelming. The flesh of the mammoth had partially rotted before it froze (the frozen ground around the specimen also smelled).

A frozen baby mammoth was likewise uncovered in Alaska during placer mining in gravels, and other mammoths have been buried by sediments when they broke through thin ice covering a river channel, became mired in a peat bog, or fell into a steep-sided gully. Organisms may also be preserved nearly unchanged in antiseptic oil seeps; e.g., skeletons of rhinoceroses have been found in oil seeps in Poland with some flesh, skin, and hair still attached to the bones.

Preservation of Hard Parts

More commonly, the flesh and soft tissues disappear soon after death, and only the hard parts of animals are preserved unchanged; e.g., although thousands of pairs of tusks of the woolly mammoth have been sold from Siberia as ivory, the flesh and soft parts of these huge beasts have been found preserved in only a few cases. The shells of many marine animals have been kept intact in the sediments in which they were buried on the sea floor, and hard parts have been preserved in this manner essentially unchanged for millions of years.

Petrifaction

Still more commonly, the bones and shells of animals are changed into a stony substance and thus preserved, although their original colors tend to be destroyed. Porous bone structures may be filled by mineral matter, or organic materials may actually be replaced little by little by such substances as silica, calcium carbonate, and iron oxide. In such cases, the organisms have literally been turned to stone i.e., they have been petrified, and replacement is often complete to the tiniest detail of structure.

Some of the most exciting fossil hunting occurs in limestones that contain silicified fossil shells. In a humid climate, limestone weathers relatively rapidly, whereas the silica composing the shells undergoes little change—a fine example of differential weathering. As time passes, the shells project conspicuously from the rock at an outcrop, and eventually some shells become free and fall to the ground below, where they can easily be picked up. Specimens from such a rock outcrop show diverse reactions to acid: the limestone can be completely dissolved, whereas the silicified shells are not harmed. In this manner, silicified shells with attached delicate structures may be freed undamaged.

Impressions, Molds, Casts, Footprints

These terms are nearly self-explanatory. Impressions may be formed and preserved in the sand and mud in which animals are buried; e.g., a shelled animal, say one of the snails (gastropods), may die, be buried in sediments, and lose its soft internal parts by decay. An internal mold (of the inside of the shell) of this animal forms subsequently if mineral matter or sediments fill the interior of the shell and harden before the shell itself is dissolved. An external mold may form in the entombing sediments if the entire organism—shell as well as soft parts—disappears. Mineral matter of some kind may later accumulate in the mold and harden to form a cast (Fig. 13–5).

Remarkable lifelike molds of insects turn up in amber. The insects were trapped and buried in the sticky resin of conifer trees or on the ground where the resin fell and accumulated. Later this hardened and fossilized as amber. The insects decayed and disappeared, but the hollow molds remain in the amber and faithfully reproduce the delicate hairlike appendages and other parts of their bodies. Ants, flies, mosquitoes, feathers, leaves, and pollen have all been found in amber.

Footprints and tail marks may form if an animal crosses an area of soft mud or sand that is subsequently baked and hardened by the sun's rays and then buried by additional sediments.

Carbon Residues

Organic materials are composed largely of carbon, hydrogen, and oxygen. Although the hydrogen and oxygen volatilize and disappear soon after the death of an organism, a thin film of carbon may be left to record the former presence of the animal or plant. If an impression has been formed by the organism, the carbon covers that impression, and even animals without hard parts have been preserved in this manner as "carbon copies."

Fossils and Ancient Geography

Fossils are essential in determining the locations of ancient lands and seas because they commonly distinguish marine from nonmarine sedimentary rocks (Fig. 13–6). Careful mapping of rock formations, and the matching or correlation of various strata, enable the geologist to work out past distributions of land and sea.

Fossils show that a land bridge once connected Siberia to Alaska in the Bering Strait area, and a drop in sea level of about 150 feet would produce one today. Part of the evidence involves the sudden appearance in North America of elephants and certain other animals after they had developed for

13-5. Casts of rodent burrows (Daemonelix from Nebraska). (Photo by J. B. Hatcher; *courtesy Carnegie Museum,* Pittsburgh, Pa.)

a long time in Eurasia. In other words, older rocks in Siberia contain a sequence of fossils of primitive and more advanced types of elephants, whereas these are absent in rocks of a similar age found in Alaska. Therefore, when these rocks formed, the two areas apparently were separated by water as they are today. This conclusion is based upon the general principle that two interconnected areas with similar environments should contain more or less similar types of animals (but not necessarily the same species) at any one time if they can wander freely back and forth. Since younger rocks in Alaska and Siberia both contain fossils of the more advanced types of elephants, a land connec-

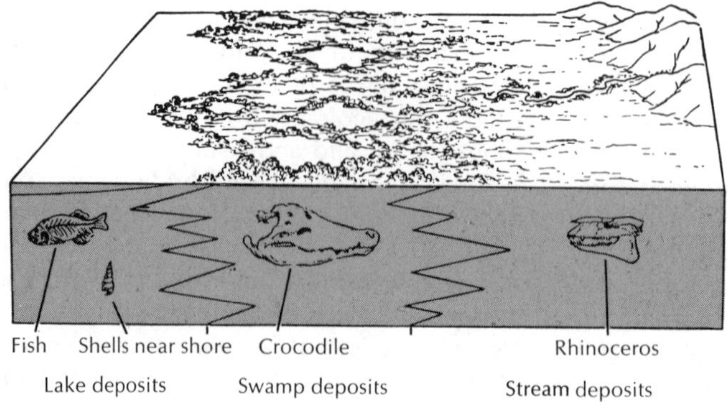

13-6. As shown by some Wyoming fossils, different organisms occupy different environments. Thus they tend not to occur together within the same kind of rock even though they have the same age. (After a *University of Wyoming Geology Museum* exhibit by P. O. McGrew.)

tion must have formed between the two which permitted migration from one continent to the other.

The direction from which the sea flooded an area can sometimes be determined; e.g., modern marine shells differ considerably along the Atlantic and Pacific coasts because a land barrier exists. In one study in the Panama area, several hundred species of shell-bearing invertebrates were collected on the Atlantic side, and a similar number of species were found on the Pacific side. However, only two dozen or so were common to both sides. Such evidence indicates that identical species are restricted to certain faunal realms even though fauna throughout the world have similar characteristics at any one interval in geologic time. Therefore, a comparison of fossil types aids geologists in determining the direction from which the interior of a continent was once flooded.

Interpretation of Former Climates

Most animals and plants of today are restricted to certain climates and types of environment, and probably their ancient relatives existed under similar climatic restrictions. It is true that animals and plants adapt themselves to widely different environments, yet whole assemblages of animals and plants seem to have strong climatic implications. Thus large reptiles and amphibia live today only in areas where temperatures remain above freezing, and ancient large reptiles and amphibia probably existed under similar climatic conditions. Therefore, fossil palms and crocodiles found in the Dakotas imply that at one time the climate in that region was warmer than it is now.

Evolution

How did the many and diverse creatures of the present come into being? This is a question that is fundamental in its importance and exceedingly broad in its sweep, and one that has stimulated searching thoughts and deep emotions. The answers have varied greatly and have come from such divergent fields as the physical and biological sciences, philosophy, and religion. Some answers have drawn material from one field alone and excluded the others, whereas other answers to this profound question have taken account of elements from each of the fields and have attempted to weave them into a meaningful, satisfying whole.*

Basically only two kinds of beginnings have been imagined by man. One type, completely rejected today by the world of science, involves the special creation at a not-too-remote time in the past of all of the ancestors of the animals and plants of today. These ancestors are presumed to have resembled closely the organisms of today, and each type of organism has descended through time, generation following upon generation, with essentially no change.

The second type of beginning involves the concept of *organic evolution* (Latin: "an unrolling"): existing animals and plants have developed from previously existing forms by slow orderly changes which, in general, have produced more complex organisms from less complex ones. Descent with change is the keynote here (e.g., the evolution of the horse, (Fig. 2–19).

According to this concept, if one could go backward into time, he would encounter animals and plants progressively more different from the familiar forms of today. Eventually on this imaginary trip, one would observe remote ancestors so different from their present-day descendants that one would fail to see any connection between them at all, had he not previously viewed the many transitional steps. The in-between forms are needed to show that descent with modification has occurred, and the fossil record plays its major role in evolution by furnishing such evidence. As this imaginary journey continued, one would eventually

* Anyone troubled by the relationship of evolution to religion is advised to read the final chapter in *Introduction to Evolution*, by Paul A. Moody, 2d ed., 1962, published by Harper and Row.

encounter types of organisms that have no living descendants—they became extinct long ago. On the other hand, certain organisms would show remarkably little change even though the process of birth, growth, reproduction, and death has been repeated a prodigious number of times.

At the conclusion of the journey, a keen observer would realize that certain important trends had occurred.

(1) Throughout known geologic time, life has tended to increase in total numbers, in the types of organisms in existence at any one time, and in the variety of environments inhabited—there has been a tendency for any one group to spread out, diversify, and occupy different types of environmental niches.

(2) Older groups of organisms have been replaced by newer and generally more complex groups, and many extinctions have occurred.

(3) Generalized forms have tended to persist longest on the Earth, whereas other forms, which became more and more specialized for life under certain conditions, could not survive when those conditions subsequently changed.

(4) Rates of modification have varied from time to time, place to place, and group to group.

(5) The vast majority of the members of any one phylum and class have remained part of that phylum and class even though new families, genera, and species have been produced within the major group. However, at certain key times in evolutionary development, a few members of a certain group may have begun a series of changes that, after much time, deviation, and random selection have resulted in the development of an entirely different group. For example, apparently only one type of invertebrate (unknown), at apparently one time in the past (early Paleozoic), developed into the first primitive vertebrates (a variety of fish). As another example, millions of years later after amphibians had come into existence, it was apparently only one type of amphibian that evolved into a primitive kind of reptile. Most amphibians remained amphibians, although modifications occurred that produced different species, genera, families, and orders of amphibians. Seemingly at only one time in history did one type of amphibian have the capability and opportunity of changing into a reptile.

(6) One is led, according to evolutionary theory, backward through time to some obscure, undated time and place where the first one-celled forms of life originated. Prior to this no organisms existed, only lifeless chemical substances.

That change occurs and produces new types of organisms is familiar to us all; e.g., consider the varieties of dogs existing today. They have all originated by selective breeding from a common ancestral stock, with most change occurring during the last few centuries. Great variations in other domesticated animals such as the horse, cow, and chicken and the immense diversity of cultivated plants—all manipulated into existence under the guidance of man—underscore the capacity of organisms to change form. A theory of evolution is accepted almost unanimously by all who have examined the evidence bearing on it, even though the exact evolutionary mechanisms are not fully understood.

Central themes in the current explanation of evolution include the following (modified considerably from Darwin's original proposal about a century ago):

(1) All organisms tend to produce more offspring than can survive within limitations set by amounts of food and space available.

(2) Because of this overpopulation, a competition for the means of existence occurs, and some organisms perish without reproducing themselves, or else reproduce in inferior numbers. This competition occurs among members of a group and also between different groups.

(3) Offspring tend to inherit the characteristics of their parents, but they vary in many minor ways; no two individuals are exactly alike. However, from time to time, notable changes show up in offspring, produced apparently in the reproductive cells

by a process of gene mutation (inheritable variability also occurs in other, but less important ways).

(4) Changes caused by gene mutations are transmitted to the next generation of offspring. On the other hand, relatively minor changes may occur in the parents during their lifetimes which are caused by their environment. Such changes are not generally passed on to their offspring.

(5) Some of the changes produced by gene mutations are advantageous in a given environment, but most handicap the offspring. A gene mutation can be regarded as a mistake in the copying portion of the reproductive process. If the variation results in organisms which are better adapted to their environment, they will tend to live longer and produce more offspring than their less fortunate relatives. The interaction of these factors eventually produces changed individuals that are remarkably fitted for survival under the conditions in which they exist. However, survival of the fittest has been misinterpreted. The fittest are those that survive, and these are not necessarily the strongest members of a group; e.g., some birds might mate more frequently because of attractive plumage, or certain insects may survive longer because of protective coloration.

(6) If the environment changes, or if the organisms move to a different environment, they may not survive under the new conditions unless their mutations produce a new and better-adapted type from the original stock. Environments have changed many times in geologic history, and thus animals and plants have evolved into many diverse forms, as one or another mutated form was favored.

Note that according to current ideas of evolution, the individual generally does not pass on the qualities he has acquired to his offspring, but only those he has inherited. The modern giraffe has a long neck, but not because its ancestors stretched for foliage high in the air in times of food shortage. What is thought to have happened is more nearly this: sudden changes in germinal material (gene mutations) created a few antelope that varied from their parents by having unusually long necks. These antelope-giraffes were better able to survive periods of food shortage than their relatives with short necks. The long-necked antelope-giraffes produced offspring who inherited the long necks of the parents because gene mutations are transmittable to future generations; and among these, those with the very longest necks thrived best. Thus antelope-giraffes with short necks became scarce, whereas long-necked antelope-giraffes increased in number.

However, the food shortage did not cause the gene mutations which produced long necks, although it did act as a selective factor that eliminated giraffes without long necks (natural selection). Perhaps gene mutations producing long necks had occurred in the antelope-giraffe species a number of times in the past when food was abundant. As there would be no advantage for long necks at such times, a breed of long-necked giraffes could not become established.

An impressive body of data supports the concept of evolution. Most convincing are the evolutionary changes attested by fossil sequences found in rock layers. Such sequences are based upon the principle of superposition, and the horse series furnishes a classic example. The rhinoceros, elephant, and camel have undergone somewhat similar development.

Fossil Record Prior to Vertebrates

Precambrian Rocks

Fossils are very uncommon in Precambrian rocks, which contrasts strikingly with their common occurrence in many Cambrian and younger rock layers. In most areas, Cambrian and younger strata are separated by a major unconformity from older, underlying Precambrian rocks which tend to be deformed and crumpled members of the igneous and metamorphic groups. One does not expect to find many fossils in such rocks. However, thick accumulations of Precambrian sedimentary rocks of different ages do occur, and these also tend to lack fossils. Moreover, in some places such Precambrian sedimentary rocks appear to merge comformably upward into fossiliferous Cambrian strata. Thus two major unsolved problems concern Precambrian life: how did it originate and why are its fossilized remains so scarce? For a number of reasons, we assume that such life did exist.

Very simple plants and microorganisms form the most common kinds of Precambrian fossils. In these generally barren, ancient rocks there occur rounded, concentrically layered masses of different sizes and shapes (stromatolites), and the oldest of these approximate 2½ billion years in age. These laminated deposits (Fig. 13–7) are widespread and relatively abundant˙ and have apparently been produced by simple plants known as algae. Similar calcareous deposits are also found in many younger rocks and are forming today. Colonies of the algae may trap sediment or cause calcium carbonate to precipitate from the water and cement it with organic material. This forms a layer which seems to duplicate the shape of the colony and repetition produces the observed laminations. However, the algae themselves are not fossilized, and the organic material decomposes and tends to be missing in ancient deposits.

The oldest known of the fossil microorganisms resemble bacteria and algae and are 3 to 3½ billion years old. Until recently, other direct evidence of Precambrian life had consisted of a few sponge spicules (now discredited ?), some wormlike trails, and a few markings or deposits of a dubious, controversial nature. However, primitive, thin-shelled brachiopods were found in the middle 1960's in Precambrian shales on Victoria Island, Canadian Archipelago, and radioactive age determinations on associated rocks suggest an age of about 700 million years.

Indirect evidence of the existence of life in the Precambrian is furnished by huge volumes of calcareous materials in ancient limestones and marbles, of carbon in graphite deposits, and of enormous Precambrian sedimentary iron ore deposits. Such deposits may owe their origin to the accumulation of organic materials or to the functions of living organisms which caused their precipitation.

Man has long searched for well-preserved Precambrian fossils and he may have found some in a most significant discovery made recently in South Australia (Ediacara Hills). The fossils occur in a ripple-marked sandstone which occurs stratigraphically below limestone layers that contain early Cambrian fossils. A 500-foot thickness of nonfossiliferous sedimentary strata separates the sandstone and limestone, and the entire sequence appears conformable. However, the difference in age between the limestone and sandstone is uncertain. The newly discovered fossils are unlike any recognized Cambrian fossils and may well be "pre"-Cambrian in age.

All of the first-found South Australian fossils represent soft-bodied creatures that lacked hard parts (some contained a few spicules): jellyfish, corals, segmented worms, wormlike creatures, and worm burrows have been described, as well as two animals completely different from any known creatures. These are somewhat circular in outline, about an inch or less in diameter, and have markings that suggest tentacles, legs, or gills. Altogether, these constitute an unusual group of fossils whose relationship to younger fossils has yet to be determined.

Evolutionary Significance of the Oldest Paleozoic Fossils

Many of the oldest Paleozoic rocks are abundantly fossiliferous (Fig. 13–7), and most of the phyla except the vertebrate chordates are represented among these fossils. These same phyla also have representatives in all of the ages from the Cambrian to the present, although membership in the phyla—e.g., the species, genera, and families—has varied greatly through time because of evolutionary changes. Some ancient Cambrian fossils, such as members of the phylum

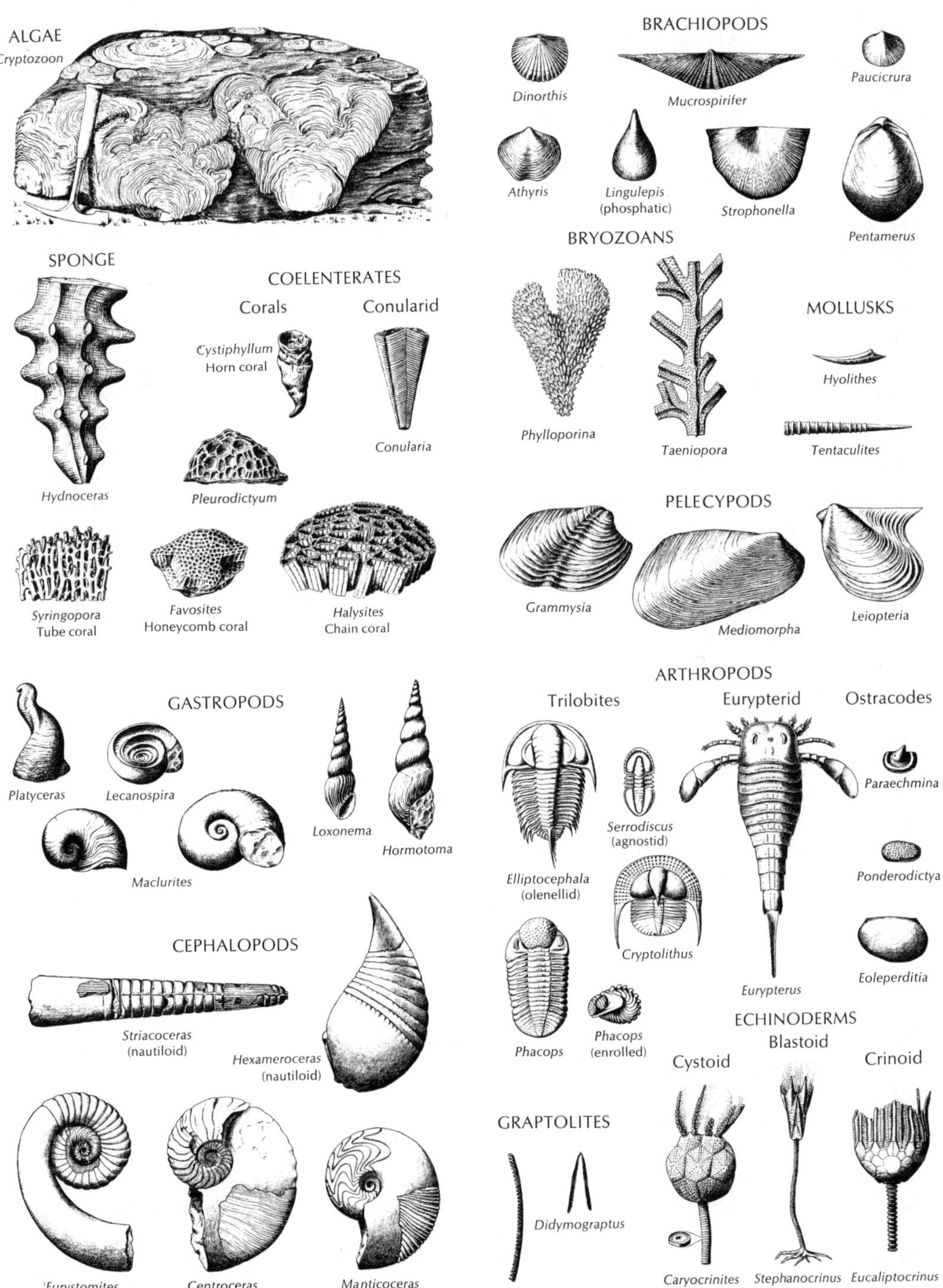

13-7. Typical fossils from Paleozoic rocks in New York. (Broughton, Fisher, Isachsen, and Rickard, *Geology of New York State*, 1962.)

Arthropoda (e.g., crabs, insects, and the extinct trilobites), were relatively complex creatures with well-developed nervous reproductive, and digestive systems. In fact, if life has evolved from a hypothetical, initial, one-celled stage to the varied creatures of today (many one-celled forms still exist), then well over half of the evolutionary development apparently occurred during the Precambrian, although we have almost no record of this. Thus our fossil record begins in uncertainty.

Whence came the first, hypothetical, one-celled organisms that presumably began the long evolutionary development that has led through geologic time to the creatures of today? According to one explanation, they were formed from nonliving chemical substances at a time in the Precambrian when conditions on the Earth were quite different from those of today (e.g., perhaps methane and ammonia were abundant in the atmosphere, and the Earth's surface was not shielded effectively from ultraviolet radiation). Radiations, plus electrical discharges, may have acted upon simple, carbon-containing compounds such as methane that were dissolved in the ocean waters, and this produced inorganic compounds of increasing complexity, until eventually the ocean waters became a sort of "chemical soup." Presumably some of these complex chemical compounds then developed the capacity to duplicate themselves and changed into the simplest of living creatures.

Laboratory experiments have attempted to duplicate this process. Gases such as methane and amonia have been dissolved in water and subjected to electric discharges and ultraviolet radiation, and traces of amino acids were produced. Such acids constitute one of a group of complex substances that make up living matter. Therefore, some scientists expect that living material will some day be produced from nonliving chemicals in laboratory experiments—an exciting, perhaps disturbing idea.

The complex structures of Cambrian invertebrates implies a long preceding evolutionary development during the Precambrian, but almost no record of this has been found. Perhaps these organisms lacked hard parts and thus were unlikely to be fossilized. Furthermore, the development of hard parts may have taken place during a geologically brief time interval that occurred near the end of the Precambrian (represented by a major unconformity in many areas). Following this, fossils would be abundant, and thus the fossil record would start in a relatively abrupt manner. However, the Cambrian is a long period (100 million years) and not all groups were present at its beginning; rather they became part of the record at different times during the Cambrian. The fossil record does not begin as abruptly as it is sometimes pictured. Along these lines, it has been suggested that the first organisms lived near the surface of the ocean, and that hard parts would be a handicap in their efforts to stay afloat. However, when this zone became too crowded, some of the organisms shifted to an ocean-floor environment, and certain ones began to prey on others. Under such conditions, protective hard parts would be an advantage, and they would evolve rapidly.

At any rate, animals with hard parts dominate among the oldest Paleozoic fossils. Trilobites are the most abundant, and brachiopods too are numerous (Fig. 13–7). However, an exciting fossil discovery in certain Cambrian shales in British Columbia indicates that soft-bodied forms probably were also abundant even though they left almost no record. About 130 species of animals have been described from these rocks, and sponges, jellyfish, worms, and shrimplike animals are preserved as thin films of carbon along bedding surfaces. Many of these organisms have not been found as fossils elsewhere, and they underscore the extreme bias of the fossil record—against soft-bodied creatures and in favor of those with hard parts.

According to another view, evolutionary changes in the Earth's atmosphere (p. 312) may have controlled evolutionary changes in the Earth's organisms. Free oxygen may have been absent in the early Precambrian atmosphere and may have been produced gradually through time by photosynthesis. Eventually, when free oxygen became sufficiently concentrated, it could provide a protective shield against deadly solar ultraviolet radiation (p. 651) as well as a supply for the development of oxygen-breathing organisms. Subsequent evolutionary development might be relatively rapid and so be reflected in the fossil record.

Geologists can only guess at the appearance of early Paleozoic land surfaces because fossil land plants have not been found in rocks older than the Devonian. However, it seems likely that vascular plants (have specialized conductive tissues such as roots, stems, and leaves) were present during the first part of the Paleozoic Era. Devonian plants were complex forms, which implies a long previous period of evolution. Perhaps early Paleozoic vascular plants were restricted to a water environment. At any rate, vegetation flourished during the latter part of the Paleozoic Era as is evidenced by the vast quantities of coal that formed at this time.

Order of First Appearance of Vertebrate Classes

A few scattered bony plates and scales have been found in fresh-water Ordovician sedimentary rocks, and these constitute the oldest-known vertebrate fossil remains. Presumably some one group of invertebrates evolved into these first primitive fish, but the identity of this group is uncertain—a real "missing link" occurs here in the fossil record.

However, many succeeding steps in the evolutionary history of the vertebrates seem to be shown clearly by fossils in Paleozoic and younger rocks. The order of increasing complexity among the five classes of vertebrates follows: fish (simplest), amphibians, reptiles, and birds and mammals (most complex). Significantly, this sequence parallels their first appearances in the geologic record. The oldest fossil fish are found in Ordovician rocks, but none of the higher vertebrates existed at this time. Furthermore, fish have been found as fossils from the Ordovician to the present, although their numbers and types have changed greatly from time to time. On the other hand, amphibians first became part of the fossil record in the latest Devonian, and they have persisted—with many changes—to the present. Reptiles have not been encountered in rocks that formed before the Pennsylvanian Period, whereas mammals and birds have not been found in any Paleozoic rocks. They first appeared in the Mesozoic Era, although not prominent in the life of that time. Future fossil discoveries may, of course, produce somewhat older representatives of each class, but this appears unlikely in some instances. These parallel sequences—the order of first appearance and of increasing complexity—suggest convincingly that evolutionary development has taken place.

The first appearance of a member of one of the vertebrate classes is sometimes followed by a fossil record showing that a few unspecialized ancestral forms spread out widely, were adapted to different environments, and developed numerous new types. Many groups of animals have tended to spread, diverge, and specialize until a maximum development has been attained. Such culminations have then been followed by the decline and eventual extinction of many forms. Study of the record suggests that conservative, unspecialized forms may persist longest on the Earth in each of the phyla, and that highly specialized forms may become extinct rather rapidly.

So far as is known, evolutionary development is not reversible, and evolution has been succinctly termed a one-way road. But the road winds and detours. and it has many dead-end branches. "One-way" road is also a somewhat misleading expression. Although more complex forms do not evolve retrogressively into simpler forms, certain trends, such as a gradual increase in size, may be reversed, or certain features, such as limbs, may be reduced or lost.

Fish

Fossil fish are uncommon in rocks formed before the Devonian Period (sometimes called the Age of Fish because of an initial abundance). Among the diverse kinds of fish existing in the Devonian were two groups which have special evolutionary significance: the lobe-finned fish, or Crossopterygii, and the lungfish, or Dipnoi (each is a member of the subclass Choanichthyes).

Although lungfish are rare today, a few members still live in areas that annually undergo long dry spells, but once they were more numerous and varied. Apparently the type has experienced relatively little change in the 350 million years that have elapsed since the Devonian. These lungfish have been found in Africa, Australia, and South America and have a swimming bladder that opens into the throat. It can be used as a crude lung, and such lungfish have been known to live out of water for more than a year.

Amphibians

The lobe-finned crossopterygians apparently evolved into a short-legged, sprawling, primitive type of amphibian late in the Devonian Period, and the evidence is quite convincing. These lobe-finned fish apparently had three structures that made this evolutionary step possible: an air bladder was connected to the throat and might have functioned as a primitive lung; their lobe fins contained bones that could support primitive limbs; and a nostril-like opening existed from the outside into the mouth. A comparison of the skeletal structures and teeth of the first amphibians with corresponding features possessed by the lobe-finned fish shows too much similarity for mere coincidence.

The evolutionary development of a swimming bladder into a lung and of fins into limbs appears to be an adaptation to recurrent conditions of dryness. As the amount of water in a fresh-water lake or stream diminished during an annual dry spell, continuous water bodies may have been reduced to isolated pools of stagnant water. Most fish would die under these circumstances. According to the modern theory, however, if gene mutations had previously produced a few fish with stronger fins and more efficient air-breathing apparatus, such fish might be able to wriggle and crawl from a waterless depression to a nearby pool of water. The mating of such survivors would produce offspring with the characteristics of their parents, and perhaps the amphibians evolved from the lobe-finned fish in such a manner, with occasional detours into unproductive forms.

Fossils found in younger Paleozoic rocks show that the amphibians subsequently became more diverse as they gradually dispersed into other areas and became adapted to different environments. As a class, the amphibians attained greatest development late in the Paleozoic Era and then declined.

Descent from fishlike ancestors is likewise strongly suggested by the life cycle of living amphibians. Members of this class of vertebrates still return to the water to lay their eggs. These then hatch into youthful, appendageless, fishlike forms which breathe by means of gills, but later in life, limbs grow and lungs commonly develop.

The question has been asked: Why do simple forms of life exist today if the evolutionary trend has been in general from the simple to the complex? For example, why have not all fish evolved into amphibians? This question seems to ask, "Why are there any fish left?" and the answer is not far to seek. So long as the waters of the Earth furnish abundant means of fish life, there will be abundant forms of fish life there. Fish no doubt continue to develop forms even better adapted to life in this and that environment in the water. But why does not a new form of amphibian crawl onto the land and start a new race of land-living forms? There is nothing in the theory of gene mutations to deny this possibility, and yet the opportunity may never again be what it was for the first lobe-finned fish that flopped out upon a land where no other vertebrate had previously existed. Today so ill-adapted a land creature would hardly survive. The first creature to produce the needed mutations for an existence on the land gave its descendants a head start that is not easily overcome. Most fish today lack both the opportunity and the necessary organic structures, such as primitive lungs and lobe fins, to initiate such an evolutionary change.

Until rather recently, the lobe-finned fish were believed to have become extinct during the Mesozoic Era. However, several living members of the coelacanths, a branch of the crossopterygians, have been found over the past few decades. The fins, a distinguishing feature of this group, have fleshy middle sections with supporting bone structures that resemble very primitive limbs. Results of dissections, however, indicate that the fins of the coelacanths are not directly allied to the limbs of amphibians and that the coelacanths were not the direct ancestors of the amphibians. Coelacanths apparently should be regarded as "the fish that time forgot"; they have undergone remarkably little change since the Devonian.

Reptiles

The oldest fossil reptiles have been found in sedimentary rocks which formed during the Pennsylvanian Period, and evidence is clear and convincing that the stem reptiles (cotylosaurs) evolved from a type of amphibian (labyrinthodonts). Many similarities in skeletal structures have been observed in the fossil representatives found in Pennsylvanian rocks. Transitional forms have been discovered which fit so nicely between reptiles and amphibians that some paleontologists classify them as reptiles, whereas others classify them as amphibians. From the stem reptiles came the turtles, certain huge marine reptiles, the flying reptiles, mammallike reptiles (therapsids), the pelycosaurs, the thecodonts (ancestral stock of the dinosaurs and birds), and other forms. Some pelycosaurs were unique in their possession of spines that projected upward from the vertebral column to a maximum length of about 3 feet. Membranous material probably connected the spines.

Fossils taken from Mesozoic rocks, the Age of Reptiles, (Figs. 13–8 and 13–9), make spectacular exhibits in museums and exciting specimens to discover in the field. Reptiles were present during the latter part of the Paleozoic Era and still exist today. But they culminated in numbers, varieties, and evolutionary development during the Mesozoic Era, when they dominated land, sea, and air. The dinosaurs (Greek: "terrible lizards") were confined to the lands and varied in bulk from the size of a rooster to ponderous giants that approached a length of 85 feet and a weight of 50 tons.

The greatest meat-eater of all times undoubtedly was *Tyrannosaurus rex,* who lived near the end of the era (Fig. 13–1). This creature had two powerful hind legs armed with claws 6 to 8 inches long, a powerful tail, short neck, two tiny forelegs, and a skull which measured 4 feet in length and 3 feet in width. It was armed with jaws that contained teeth as long as 6 inches. This carnivorous monster may have reached a length of 50 feet, walked with his head about 20 feet above the ground, and weighed as much as 10 tons. One set of footprints shows a stride more than 13 feet in length. However, *Tyrannosaurus* lived in the latter part of the Mesozoic Era and was quite different from the first relatively unspecialized carnivorous dinosaurs that appeared before the end of the Triassic Period.

Certain plant-eating dinosaurs such as *Brontosaurus* carried a huge body on four elephantlike legs; they had long tails, very long necks, and small heads. If these animals had walked head to tail as in a circus parade, seventy of them would have spanned a mile. Probably such heavy creatures lived in swampy areas to take advantage of the buoyant effect of water in moving about. The combination of such a hulk with the tiny head seems grotesque, indeed ridiculous. Keeping the huge stomach filled was probably difficult even if the animal ate almost continuously and was characterized by a slow body metabolism. Fortunately, the tiny head with its 2-ounce brain would hardly have the means to worry.

Brachiosaurus was the real giant, and perhaps he attained a length of 80 feet, much of it neck. This animal probably could have peered over the top of a three-story building, and its nostrils were located on a sort of bump on the top of its head. Evidently *Brachiosaurus*, except the top of this bump, could remain submerged in water and yet breathe.

13-8. A Late Paleozoic landscape at the right merges into a Mesozoic (Jurassic) landscape at the left. Characteristic animals and plants are shown. On the opposite page, fin-backed reptiles and a giant dragonfly move among scale trees. At the top of this page the neck of *Brontosaurus* arches over *Stegosaurus* with his ridge of plates, while *Allosaurus* feeds nearby, as shown at the bottom. *(From a mural by Rudolph Zallinger in Peabody Museum, Yale University.)*

Other types of dinosaurs also lived during the Mesozoic Era: the duckbilled, plated, armored, and horned varieties (Fig. 13–1). Furthermore, the seas were dominated by several huge reptiles and the air by others. One of the batlike flying reptiles had wings that measured 26 feet from tip to tip, but its lightweight body was about the size of a goose.

At the end of the Mesozoic Era, the dinosaurs, flying reptiles, some of the large marine reptiles, the toothed birds, and other forms of life became extinct, and this interval in geologic history has been picturesquely called "the time of the great dying." However, many of the forms which became extinct, like *Tyrannosaurus rex,* appeared to be at their zenith development—they were not decadent creatures ready for extinction. Several hypotheses attempt to explain these mysterious disappearances, but none is very satisfactory.

Birds

The oldest bird fossils have been found in Jurassic rocks and constitute a convincing link between later birds and their reptilian ancestors. Fossils show that this first bird (*Archaeopteryx,* Fig. 13–10) was about the size of a crow, had feathers, and could fly. Three features, however, were retained from its reptilian predecessors: its jaws were set with true teeth; three claws projected from each wing; and feathers were arranged pinnately on a long tail—not in the shape of a fan as in modern birds.

Evidently the toothed birds and the flying reptiles each evolved from the ancestral stock of the dinosaurs. However, the toothed birds apparently did not evolve directly from the flying reptiles.

More modern ancestors of present-day birds had evolved from the toothed birds before their extinction late in the Mesozoic Era. Although, fossils of birds are relatively rare, some unusual specimens have been found—e.g., the giant moas, a large flightless bird that lived during the Cenozoic and became extinct several thousand years ago. Some specimens weighed several hundred pounds and exceeded 7 feet in height.

Mammals

Although the first mammals apparently had evolved from reptiles (therapsids) before the middle of the Mesozoic, they remained

13-9. Mesozoic (Cretaceous) sea floor showing algae, clams, snails, straight and coiled ammonites, and belemnites (swimming). *(Courtesy Smithsonian Institution.)*

inconspicuous throughout the remainder of the era. Mammals assumed their present dominant role in the biological world during the Cenozoic Era (Age of Mammals) when they occupied many of the environmental niches previously held by reptiles in the Mesozoic. Four major trends occurred in the mammalian evolution of the Cenozoic: an increase in size and brain power, and a specialization of feet and teeth. The horse series (Fig. 2–19) is a classic illustration of these specializations.

Some mammals became efficient carnivores like the lion, tiger, and wolf; others took to the oceans like the whale and the porpoise; and still others developed the ability to run rapidly like the horse and the antelope. Some forms became extremely specialized and then extinct.

During the Cenozoic Era, North America was the home of many animals (Fig. 13–11) which have since become extinct or have moved to another continent. The woolly mammoth, the mastodon, and the rhinoceros have all vanished as have the hippopotamus, camel, giant pig, giant ground sloth, giant bison, and saber-toothed tiger. Man may have migrated to North America about the time of some of these extinctions and he may have caused some of them.

Man

Man belongs with the primates (Latin: "first"), a group of mammals that includes the chimpanzee, gorilla, and monkey. This classification is based upon studies of man's skeletal structures, muscles, organs, embryological development, vestigal organs, blood tests, and the fossil record. All men today belong to the same genus and species (*Homo sapiens:* man the wise).

The origin of man and his role upon the Earth are, of course, topics of profound sig-

13-10. A Mesozoic landscape showing the oldest known bird *(Archaeopteryx)*, a flying reptile (top), and two small dinosaurs (lower left). *(Painting by Charles R. Knight, Chicago Natural History Museum.)*

nificance and have been considered at great length, with deep emotion, and from widely divergent points of view (philosophy, religion, science). The fossil record, although fragmentary in nature, indicates that man has evolved in a manner that follows the general pattern of other groups—descent with modification. If one traces man's lineage backward into time (i.e., downward through the thick piles of superposed, fossil-bearing rocks) he passes in turn from mammal, to reptile, to amphibian, to fish, to invertebrate.

13-11. A Pleistocene landscape showing mammoths, mastodons, giant bison and sloths, large beaver and wolves, saber-toothed tigers, the modern horse (Equus), as well as the "mammalian tanks," the Glypotodonts. *(Courtesy Peabody Museum, Yale University.)*

For some, unfortunately, the evolution of man has meant conflict with religion and the erroneous idea that man has descended from present-day apes. However, the conflict with religion has subsided somewhat, and it has become more generally known that one cannot evolve from one's contemporaries. Rather, man and other present-day primates at some remote time in the past had a common ancestor, and each has descended with modification from this ancestor.

The classification of certain fossil discoveries as man or ape depends in some instances upon one's definition of man, and the capacity to use tools, the size of the brain, the nature of the jaws and teeth, and erect posture are important criteria. Among the primates, man seems most closely akin to the gorilla and chimpanzee. Fossil evidence of the primates is confined to the Cenozoic Era, and that of man has been restricted largely to the Pleistocene.

A large, well-developed brain is an outstanding primate characteristic. The average cranial capacity of man is 1350 cubic centimeters (range: 900 to 2300 cc) which is about two and one-half times that of a much bulkier, adult gorilla (average: 550 cc, normal variation 300 to 650 cc). A comparison of the skull of a man with that of a gorilla shows similarities and the following differences: the thicker skull of the gorilla exhibits a protruding mouth with projecting canine teeth, no chin, a prominent eyebrow ridge, low sloping forehead, a saggital crest, and other differences in the size and arrangement of the teeth. Certain of these features appear in some of the fossils of early man and in fossils of others probably best described as near-men.

As determined from artifacts excavated from superposed layers of sediment, man's culture has been subdivided into three main phases: the Stone Age, the Bronze Age, and the Iron Age, although a number of more recent ages can be added. At first in the

Stone Age, stones were used as tools and weapons by man as he found them (eoliths); next they were shaped only by chipping (paleoliths); and finally late in the Stone Age, they were ground and polished as well as chipped (neoliths). Next, man used implements made of metal. However, these cultures were not attained simultaneously throughout the world, and thus they have no absolute time significance (e.g., certain primitive tribes still have a stone-age culture today).

Early man evidently did not consider sanitation a problem, and refuse collected in great amounts around certain favored camp sites such as caves which were used in succession by different tribes; e.g., about 70 feet of crudely layered refuse accumulated at one cave in Palestine, and it represents a number of Stone and Bronze Age cultures.

Four types of early man (or near-men) are discussed below to illustrate the fossil record, but dozens of additional discoveries are omitted. The peoples represented by such finds have commonly been named for the localities in which they were found, and many of the discoveries are very fragmentary —part of a jaw and a limb bone, a few teeth, or bits of a shattered skull furnish the only evidence now available for the former existence of many of these types. Reconstructions and interpretations made on such limited data should be flexible enough to allow for future discoveries. Many of the fossil finds have been classified as different genera and species and have been given different names, which may have complicated the evolutionary picture unnecessarily and obscured relationships. Perhaps a much wider range of variation and geographic extent should be expected within a particular genus or species.

Australopithecus ("Southern Ape")

The first specimen of this group was discovered in South Africa in 1924, and a rather large number of similar fossils have since been found in this and other parts of Africa. Both the dating and classification of *Australopithecus* have been controversial. Part of an arm bone found recently in Kenya has been dated at 2½ million years, as have some stone flakes and pebble tools also found in Kenya. Other specimens found in East Africa (Olduvai Gorge, see below) are dated at 1¾ million years.

A typical australopithecine representative probably was not much more than 4 feet in height, had a small brain (cranial capacity apparently ranged from 450 to 700 cc), large teeth and jaws, eyebrow ridges, protruding mouth, and no chin. On the other hand, he walked in an erect manner and had features of dentition, skull, and pelvis that resemble those of man. Thus this creature had characteristics of both man and apes (it has been called both ape-man and man-ape). One group of australopithecines had exceptionally large jaws and huge grinding teeth and may have lived chiefly on vegetation, whereas another group had smaller jaws and teeth and presumably had an omnivorous diet.

Fossils of man and near-man have been found in East Africa by Dr. and Mrs. L. S. B. Leakey in lake-deposited sediments in Olduvai Gorge, a fossil-hunting site for the Leakeys since 1931. Within these lake sediments are layers of volcanic ash which can be dated by radioactive isotopes contained within them. Eventually the lake disappeared, and a river-eroded valley (Olduvai Gorge) was cut through the sediments. The Leakeys have made their fossil discoveries in this pile of superposed strata.

A few remains of a younger fossil man that lived in this area about 400,000 years ago have been found in association with a number of giant fossil mammals: pig, antlered giraffe, sheep (the horns stretched more than 6 feet from tip to tip), and a type of extinct elephant. Fossils at different levels also provide evidence of climatic changes.

The most significant and exciting primate finds (because they are so old) were made in underlying sediment. The first (*Zinjanthropus*) probably should be classified as an australopithecine. It has been described as a prehuman, brutish, vegetarian man-ape, but others consider it more man than ape.

The fossils may be 1,750,00 years old according to potassium-argon age determinations on associated layers of volcanic ash.

Homo habilis (Latin: "handy man") was found next and appears to have been a short, small-brained, upright tool-maker. First reports indicate that *Homo habilis* was 3 to 4 feet in height, had a cranial capacity of about 650 to 700 cc, made tools, built shelters, and perhaps was a contemporary of *Zinjanthropus*. The relationship of these two discoveries to the evolutionary development of man seems somewhat uncertain at the present time.

The time when the australopithecines existed is of importance in assessing their role in human evolution, but their range in time is uncertain. They are generally dated as Early and Middle Pleistocene, but recent finds suggest that they may also have existed in the latter part of the Pliocene. They may or may not be direct ancestors of man. If the australopithecines are not, then they probably resemble closely a similar type that existed earlier and was ancestral to both man and apes.

Pithecanthropus Erectus ("Erect Ape-Man")

A widely publicized discovery of part of a skull and several bones occurred in Java in 1891 and several additional finds have since been made. These are generally considered to be the remains of an early type of man and were named *Pithecanthropus erectus*. *Pithecanthropus* was about 5 feet tall and had certain apelike characteristics as well as others that were manlike. Although the brain case (about 800 to 900 cc) is smaller than that of present-day man, it is much larger than that of an existing great ape. The receding chin and forehead, the protruding mouth, and heavy brow ridges are apelike, whereas the relatively large brain, teeth, and erect posture are manlike. Associated mammalian fossils indicate a Middle Pleistocene age.

Fossil fragments have been found near Peking, China of a few dozen individuals of a type similar to *Pithecanthropus* but with somewhat larger cranial capacity (850 to 1300 cc). The remains are chiefly skulls and jaws, and each skull has been fractured in a manner which suggests cannibalism. Apparently fire was used by this type of man, and numerous flint tools were found associated with the fossil remains. Later discoveries in other areas suggest that several pithecanthropoid types lived on other continents at various times during the middle part of the Pleistocene, and the present tendency is to classify all of these fossils as *Homo erectus*.

Neanderthal Man

This prehistoric type of man lived during the latter part of the Pleistocene in Europe, western Asia, and along the northern coast of Africa. According to fairly abundant fossil remains, Neanderthal man had a short stocky body that averaged about 5 feet in height, a receding chin and forehead, heavy brow ridges, and its big toe was offset against the others in an apelike manner. Although the brain was as large as modern man's, it apparently was smaller in the parts devoted to thinking (no correlation appears to exist between brain size and intelligence in modern man). Neanderthal man used fire, buried his dead, was a good hunter, and made stone implements with considerable skill. According to one view, Neanderthal man was not the direct ancestor of modern man and became extinct, but it has also been suggested that modern man in Europe resulted from the intermixing of Neanderthal man with Cro-Magnon man.

Cro-Magnon Man

At some time during the last glacial age, a modern type of man came to southern Europe. Many skeletons show that these individuals were tall and straight and had high foreheads, prominent chins, and modern-size brain cases. The Cro-Magnons appear to be the direct ancestors of the present southern Europeans and apparently did not become extinct as did a number of earlier types. The Cro-Magnons used fire, were good hunters, buried their dead, made excellent stone implements, and possessed clothing and ornaments. In combination with the numerous multicolored paintings found on the walls of caves, these show a well-developed art and culture. As evidence that the Cro-Magnons were good hunters, the bones of about 100,000 horses were found in piles around one camp site in France.

Early Man in North America

The history of early man in the Americas is still quite uncertain but seems to involve only the last part of the Pleistocene epoch. Apparently he came from the Old World (via Siberia?) before the extinction of certain types of elephant, camel, horse, and bison, because his stone implements and some human bones are associated with them. Oldest so far is a 30,000-year date for some charcoal found in association with fossil mammoths on Santa Rosa Island off the California coast, but some do not accept the charcoal as evidence of a man-built campfire. As another illustration, some broken skull caps and other skeletal materials (Marmes Man), together with artifacts and the bones of an elk, were discovered recently near a rock shelter in the State of Washington. They have been dated tentatively as 11 to 13 thousand years of age, and the site has been occupied at different times since this date.

Tentative Conclusions on the Origin of Man

The fossil record and other evidence indicate that man evolved from the primate group of mammals, although many details are lacking. Apparently man and the great apes had a common ancestor during the latter part of the Cenozoic Era from which each subsequently diverged. This much is generally accepted today.

Some years ago it was rather generally assumed that man had evolved from this common ancestor in an undeviating progressive development from primitive to modern types. Likewise, the most primitive types (i.e., unlike modern types) of fossil men were thought to have been the oldest and most brutal ones. These ideas led to a search for a series of "missing links" which, when found, would determine the places of all fossil men in this progression. However, recent evidence throws doubt on the validity of these assumptions; e.g., certain modern types of fossil men are now known to have lived long before other types of a so-called more primitive nature.

Fossils of man illustrate an important principle in evolutionary development. Evolutionary changes apparently do not take place little by little with each part of an organism changing at the same rate. In other words, a transitional stage midway between an ancestral species and a descendant species would not be midway in the development of all of its features. More probably, there would be a number of midway organisms with various combinations of more advanced and more primitive features, and many of these midway forms become extinct; e.g., in man the limbs apparently reached their present stage of evolutionary development before the brain, skull, and teeth. In fact, the development of an erect posture apparently was a necessary prerequisite for the other developments. This freed the hands to fashion and use tools, and this in turn resulted in a relatively rapid enlargement of the brain.

Unfortunately, many details concerning man's evolutionary history are still shrouded in the mists of antiquity and the fogs of too few fossil remains. His (and her) present position of dominance in the biologic world on the Earth comes only at the end of a long succession of changes. Apparently man has succeeded to a position formerly held in turn by other mammals, and by reptiles, amphibians, fish, and invertebrates during an immensity of time that is nearly beyond comprehension.

Exercises and Questions for Chapter 13

1. Describe some of the ways in which fossils form.
   (A) Does the appearance of a fossil in a rock depend upon whether you observe the surface or edge of the enclosing layer?
   (B) How may undamaged silicified fossils be obtained from a limestone? Is an understanding of differential weathering of value in searching for fossils in such areas?
   (C) Make a collection of fossils that illustrates the concept of fossil succession.
2. Describe several different types of fossils which would give useful information about the environment in which they lived.
3. Describe some of the key ideas involved in evolution.
4. What implication does the fossil record of the oldest Paleozoic rocks have for the Precambrian? What unsolved problems occur here?
5. Describe briefly some of the important evolutionary steps in the origin of the vertebrates. Which of these steps are well supported by fossil evidence? Which are not?
6. Which came first the chicken or the egg? Does the following statement have some validity: the first bird hatched from an egg laid by a reptile?
7. If evolutionary development has generally proceeded from the simple to the complex, why are any simple forms left?
8. Why are certain organisms today called "living fossils"?
9. Obtain a strip of paper or cardboard that exceeds 46 cm in length and mark the approximate location of each of the following on a chronological scale in which 1 cm equals 100 million years:
   (A) Origin of the Earth.
   (B) First direct evidence of life.
   (C) Fossils first become abundant.
   (D) Oldest known amphibian.
   (E) Beginning and end of "Age of Reptiles."
   (F) First man.

14

The Water Planet

The next rain or snow provides you with an opportunity to reflect broadly and adventuresomely upon the varied routes traveled by water molecules that fall from the sky onto land and sea (Fig. 14–1). Too frequently we look upon precipitation as something that spoils a picnic, cancels a ball game, or calls forth the snow shovel. Instead we should regard it—especially a steady quiet rainfall—as one of nature's truly great gifts to man: because of it, plants grow, substances dissolve, animals live, rivers flow, and our water reservoirs are replenished. We drink this unique material and use it in countless ways in our daily lives: to wash our faces, dishes, and clothes; to make our gasoline, beer, and paper; to dispose of our garbage and wastes; to fight our fires; and for swimming, boating, and fishing.

The nucleus around which millions of water molecules have clustered to form a single raindrop may be a fragment of a meteor, a salt crystal from the sea, a tiny part of the Earth's crust, or a smoke particle formed by the burning of a fossil fuel—e.g., from a tree that grew in a swamp and became part of a bed of coal long before dinosaurs walked across the lands.

The water molecules themselves may have participated many times in many cycles; e.g., some may have been evaporated from the surface of an ocean, only later to be dropped upon this or some other ocean, a process which may be repeated many times. Other water molecules may form part of a mass of ocean water that becomes dense enough to sink. The water may then move slowly and at great depths; centuries later and hundreds of miles away, it may rise to the surface where some of its molecules may again be lifted into the air by energy radiated from the sun.

Some water molecules fall upon land and flow back to the sea in a river or tarry awhile in an underground reservoir, later perhaps to be pumped to the surface and become part of an animal or plant. At lower temperatures and in other regions, certain water molecules

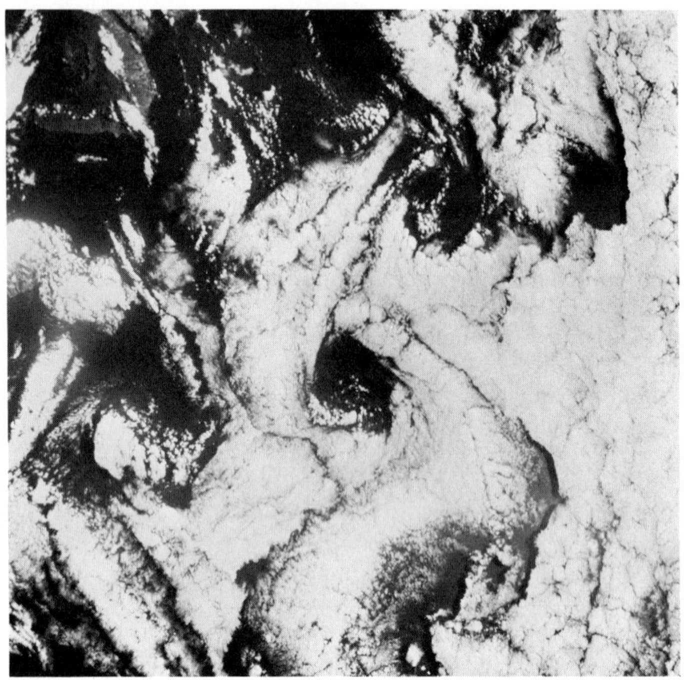

14-1. Striking pattern of stratocumulus clouds reveals a large eddy (vortex) in the atmosphere over the Canary Islands. Clear central area is about thirteen miles across. Volcanic island in upper left is Tenerife. *(Gemini VI, 16 Dec. 1965, S65-63149.)*

may form snowflakes that fall upon and become part of a glacier. They are thus removed from the various cycles of water until that time in the future when the ice has flowed to a lower altitude and melts or sublimates, and this may occur on land or on sea. Still other water molecules may be trapped among sand grains and become part of a sandstone that is subsequently folded during a period of orogeny. Later it is eroded, the water molecules are released again for more evaporation, transportation, and precipitation. These examples suggest some of the diverse journeys made by water molecules of the Earth. Furthermore, if the sun is really a "third-generation" star (p. 614), then all of the atoms that now form molecules on the Earth were once parts of stars other than the sun. Certainly one has an opportunity to let his mind range widely and imaginatively as he contemplates the falling raindrops.

Oceans and Seas

The Earth has been called "the water planet" and this term is most appropriate (Fig. 14–2) for an abundant supply of water makes the Earth unique among the planets of the solar system. Approximately 71% of its surface is covered by a single, worldwide, interconnected body of sea water—there are nearly 3 square miles of water for each square mile of land. Of the Earth's total surface area of 197 million square miles, about 139 are covered by the oceans and their marginal seas. In proportion, then, the continents are like gigantic islands that project here and there above the surface of a single widespread ocean. However, oceans that were once formidable barriers between continents—as they were with the transportation available in the days before Columbus—now serve as connecting links between the continents. On the other hand, oceanic crust is basically different from that under the continents, and the lands and seas are not distributed uniformly around the Earth.

The waters of the Earth are joined together in one continuous mass and form essentially one ocean—an immense area that has been subdivided into oceans and seas in different ways by different people. Some speak of three main oceans: the Atlantic, Pacific, and Indian. Others add the Arctic and Antarctic for a total of five oceans. Still others subdivide the Atlantic and Pacific at the equator into northern and southern parts to make seven oceans, but these are not the seven seas of antiquity. That term was current before all of the oceans were known and seems to have had no precise meaning.

The nearly circular Pacific Ocean is by far the largest of the oceans. It covers about one-third of the Earth's surface and is about double the size of either the Atlantic or Indian Ocean. The Atlantic Ocean is somewhat S-shaped and elongated in a north-south direction; it has more coastline than the Indian and Pacific Oceans combined. If the Antarctic is considered as a separate ocean, it forms a ring-shaped, cirmumpolar body that surrounds the continent of Antarctica. It must then be delineated very arbitrarily at its northern boundary from the three main oceans (at 60 degrees south latitude by some, but by others farther north, near the southern tips of the continents). The Arctic Ocean is considered a separate ocean by some, as part of the Atlantic Ocean by others, and also as a sea.

According to one suggestion, the seas are smaller than oceans and may be subdivided as follows. Deep seas such as the Mediterranean and Gulf of Mexico occur between

14-2. Clouds outline a huge counterclockwise swirl in this Apollo 11 photograph. *(NASA: AS11-5298.)*

continents and are underlain by true oceanic crust. Marginal seas such as the Bering Sea may be deep or shallow and are separated from the main oceans by island arcs. Shallow seas such as Hudson Bay and the North Sea occur within depressed parts of a continent.

The equator conveniently separates the Earth into two parts, and nearly 40 percent of the Northern Hemisphere is land, about double that of the Southern Hemisphere. However, the contrast in the proportions of land to sea in the zones most suitable for habitation by man is even greater. In the middle latitudes between 23½ and 66½ degrees, land and sea are about equal in extent in the Northern Hemisphere; whereas in the Southern Hemisphere the proportion is about 1 to 8.

Importance of Oceanography

The oceans constitute a great challenge and opportunity for man, and in many ways they are still unknown and mysterious. The challenge is similar to that of outer space, but in the 1960's it was less publicized and was appreciated and supported by few men, small funds, and scanty equipment, relatively speaking. The values of the oceans to man (see p. 284) can be grouped under many and overlapping categories such as scientific, military, economic, political, and recreational—from vital matters such as war and peace, desalination, waste disposal, tidal energy, mineral resources, vital source of animal protein, communication, transportation, and natural resources to the less tangible pleasure of observing the waves beat against a coast at sunset.

Oceanography is not a single scientific discipline but rather the coordinated application of the other sciences and mathematics to the oceans in all of their many aspects: astronomy, biology, chemistry, geology, meteorology, and physics. One cannot fully understand the Earth and its history without an adequate knowledge of the 71% that is covered by the waters of the oceans.

Over the decades and centuries, basic research performed to gain an understanding of natural phenomena has commonly led to discoveries that have been useful to man. Oceanography has already produced important examples and seems destined to provide many more because the third dimension of the oceans—that along the vertical—is truly a new frontier for man. In the early development of other areas of science, basic knowledge was pursued for its own sake—because these areas were largely unknown and their problems chiefly unsolved or perhaps not even formulated. Eventually such studies led to an understanding of principles and data and to the development of instruments and techniques which subsequently proved to have great practical value. Thus efforts expended in basic research were repaid many times. The study of the oceans is at such an early stage that the probabilities of fundamental new discoveries, of so-called "scientific breakthroughs," seem very high; in fact, a number have already been made (see Chapter 11).

In another vein, man would like to have a better understanding of storms and weather changes and perhaps even to influence them in some ways. However, attempts to change natural phenomena require the ability to predict future changes in them, which is possible only after a thorough understanding of the phenomena themselves has been achieved. This leads us to the oceans which have a major effect upon the weather and climate of the lands. They are the main source of the water that is precipitated upon the lands, and the evaporation, transportation, and condensation of water vapor is most important in the distribution of heat energy throughout the lower part of the atmosphere; e.g., about one-third of all the solar radiation that reaches the Earth's surface is expended in the evaporation of sea water. Thus certain aspects of oceanographic research may result in a better understanding of weather changes.

The development of oceanography is of

great significance militarily and politically as indicated by the nickname "the wet war." Surface ships and submarines have been important for decades and longer, but the development of the nuclear-powered, missile-firing submarine has created a problem of nightmarish proportions for the peoples of the Earth. The offensive and defensive deployment of such submarines—goal: to observe without being observed—involves precise navigation, and accurate topographic maps of the ocean floor aid in this navigation. Communication, navigation, and detection in the upper waters depend in part upon the movement of sound and other waves through these waters, and the passage of such waves in turn is influenced by the physical properties of the water. The instruments cannot be used with precision without a thorough knowledge of the waters through which they function. Thus many oceanographic observations have been (and are being) made to provide the needed data. The capabilities of the nuclear-powered submarine were underscored in August 1958, when the *Nautilus* made the first under-the-ice crossing of the Arctic Ocean via the North Pole, and by the circumnavigation of the Earth by the *Triton* in 1960. Without surfacing completely and without being detected, this submarine more or less retraced the path left by Magellan and his seamen in their epoch-making three-year voyage around the Earth (1519 to 1522). The *Triton,* however, covered more than 30,000 miles in about 60 days at an average rate of approximately 21 mi/hr.

The oceans likewise constitute a vast, but not limitless, storehouse of foods (p. 388), chemicals (p. 398), and minerals (see below) needed by man. In the late 1960's, world-wide production of useful resources from the oceans amounted to about $8 billion for food supplies for man, animals and fertilizers, $4 billion for offshore oil and gas, $200 million for sand and gravel, $175 million for common salt, and lesser amounts for magnesium and bromine. These and other materials can be obtained from the oceans in much larger quantities than are now being extracted, but they must be able to compete economically with similar materials obtained from the lands. Therefore, as the resources of the lands dwindle both in total supply and in the number of highly concentrated deposits, the resources of the oceans—as diluted as some of them are—must eventually increase in value. However, legal ownership of ocean resources is only one of several problems involved, and perhaps the most valuable material that will eventually come from the oceans is water itself (after desalting).

The study of oceanography also provides an opportunity for increased international cooperation and the chance thereby to improve international relationships. A need exists for coordinated specimen collections and synoptic measurements—measurements taken more or less simultaneously from widely scattered stations—and these can be made on an ocean-wide scale only by the coordinated efforts of all of the available oceanographers. In a somewhat different vein, advances in science and technology have an increasingly great significance in international affairs because of the prestige currently being accorded to excellence in these fields. Thus oceanographic research has many tangible and intangible ramifications.

As an example of the mineral resources of the deep-sea floor, we turn to manganese nodules which are a minor but economically important type of deep-sea sediment. They occur in great abundance over parts of the sea floor, have been photographed, publicized, and sampled (Fig. 14–3), and appear to constitute a renewable, recoverable natural resource. Many of the nodules are potato-shaped and average 1 to 2 inches in diameter, but a 1700-pound specimen was once pulled to the surface after it had become entangled in cables laid along the sea floor. Onionlike layers are common inside many nodules, and some object such as a shark's tooth generally occurs at the center. Evidently the nodules form very slowly by precipitation of manganese dioxide around isolated hard objects in areas

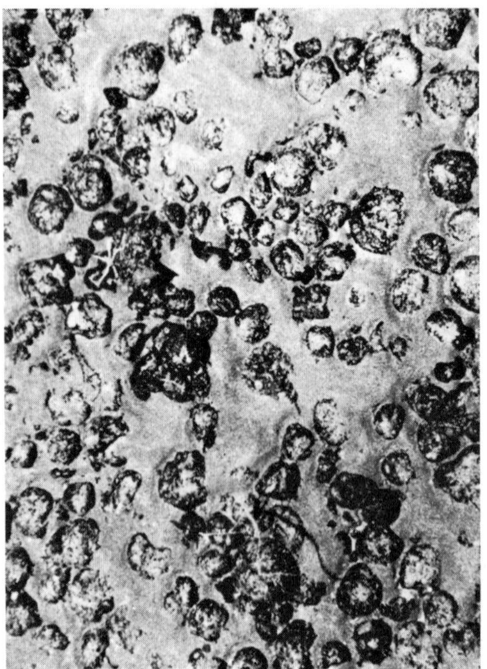

14-3. Manganese nodules at 445 fathoms on the Blake Plateau off the southeast coast of the United States. Although the nodules are commonly round, occasionally they form large slabs. They occur here in a nondepositional environment caused by ocean currents. (R. M. Pratt, "The Ocean Bottom," *Science* **138,** 492, 26 Oct. 1962.)

where deposition of clay particles and other sediment is scarce. Faint scour marks occur around some nodules, which suggests that slow currents keep sediments from accumulating. So far as is known, the nodules occur only as a single layer.

Although deposition is slow, overall quantities are so large that perhaps three times more manganese may accumulate in one year on the sea floor than is used by man worldwide in that same year. Analysis of a number of samples from the Pacific showed average percentage concentrations by weight of: manganese (24), iron (14), cobalt (0.35), nickel (0.99), and copper (0.53). On the other hand, maximum percentages were considerably higher: manganese (41), iron (26.6), cobalt (2.3), nickel (2.0), and copper (1.6). Photographs show concentrations as much as 5 to 7 pounds of nodules per square foot of sea floor in some areas (average is about 11 kg/sq meter for large sections of the Pacific floor); e.g., a 4 million square mile section of the Pacific floor west of Hawaii may contain more than 50 billion tons of nodules in concentrations greater than 1 lb/sq ft, a large part in water a mile or less in depth. Thus the deposits are enormous, and their locations and concentrations can be mapped and estimated via photography. Various methods of mining the nodules have been suggested (Fig. 14–4). There is also the metallurgical problem of extracting the different metals from the nodules once they have been brought to the surface.

Other useful materials also occur in large quantities on the sea floor; e.g., red clay contains considerable aluminum, and some oozes are chiefly calcium carbonate. Phosphorite can be used to obtain phosphate for fertilizers and in certain areas occurs in layers in relatively shallow water near shore. Heavy, durable materials such as gold, tin, and diamonds have been found concentrated in so-called placer deposits on the lands and also occur in beach and sea-floor sediments.

Some Historical Developments

According to one view, oceanography has had many beginnings; e.g., one measures the depth of water when he anchors a boat, and he may collect a sediment sample when he retrieves the anchor. If a keen observer walks along a seashore, swims or fishes in the oceans, or sails or flies across them, he can make observations and records that are of some value to oceanography. However, these are informal and unorganized.

According to another view, a history of oceanography begins appropriately with accounts of some of the early explorers and their theories concerning the Earth and its oceans. Here one can describe trips by the Phoenicians before the birth of Christ, including a voyage into the Sargasso Sea in

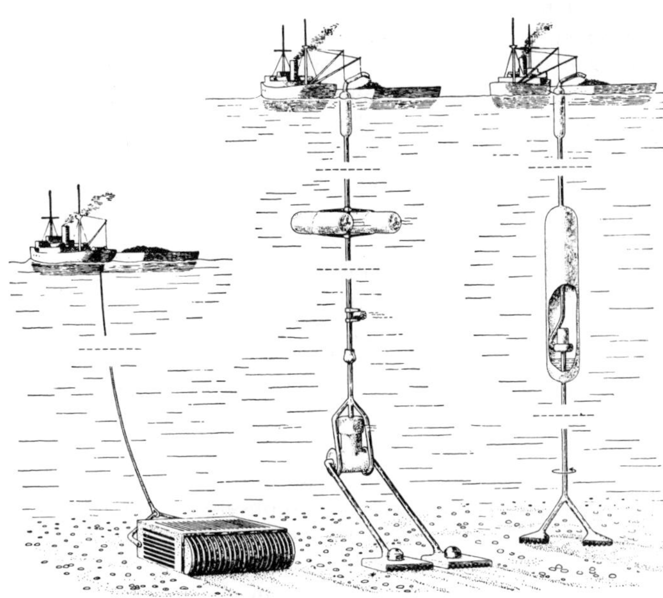

14-4. Three types of deep-sea dredges that might be used in the future to mine manganese nodules on the sea floor. Left, a simple drag device for use in shallow water. Center, hydraulic dredge that might function as a huge vacuum cleaner. Tanks on the pipelines are floats to buoy it. Pumps and motors are located near the bottom above the suction heads. Right, motor and pump are located within a single, vertically oriented float. Ocean currents might aid in moving the dredges along the sea floor. (J. L. Mero, "Minerals on the Ocean Floor," *Scientific American,* Dec. 1960.)

the Atlantic and another around the African continent. Voyages by the Vikings to Iceland, Greenland, and North America are also noteworthy, and in this group must be included the courageous, adventuresome trips of Columbus and Magellan. But the purpose of these trips was to move from one place to another. They used the sea as a means of transportation and not as an object of study.

Any account of the history of oceanography must include the accomplishments of Matthew Maury, an American naval lieutenant (1806–1873). Maury's active sailing was stopped by an injury, and he became the keeper of records (logs) of currents and winds encountered at sea during the voyages of many ships. His organization of these data revealed previously unknown patterns, and his charts were used to reduce considerably the time required to sail from one place to another; e.g., the voyage from England to California was shortened by 30 days. The success of the charts in navigation resulted in a great cooperative effort by officers on numerous ships to record data which would make the charts still more useful and which would fill in previously blank areas. Maury collected data on the Gulf Stream and prepared the first bathymetrical (Gr.: "deep measurement") map of the North Atlantic, which showed its depth via contour lines at 1000-fathom intervals (1 fathom equals 6 feet and is the approximate distance spanned by a man with outstretched arms, his "wing-spread"; thus a man could measure the length of a rope being lowered during a sounding as so many wingspreads).

Other historians date the study of oceanography from the 3½-year, 69,000 mile British *Challenger* expedition in the 1870's, which was one of the first to emphasize a three-dimensional approach to oceanic studies. Eventually a 50-volume report was published on the results of observations made in the Atlantic, Antarctic, and Pacific Oceans. The oceans were measured, observed, and sampled systematically from top to bottom from a small ship that rolled like a barrel (it tilted readily 46 degrees on one side and 52 degrees on the other). The ship was equipped with laboratories and considerable equipment (e.g., nearly 150 miles of sounding rope). Expedition members measured the water depth in many places.

The German *Meteor* expedition in the 1920's used depth-sounding equipment (p.

370) and has been called the first modern oceanographic trip. It exemplifies the older type of oceanographic survey involving systematic observation and sampling at many different locations. The data were then studied carefully in laboratories on shore during the years that followed.

In the modern approach to oceanographic research, systematic observations and samples are still made, but more emphasis is given to theory and imagination. Deductions and hypotheses are formulated concerning certain aspects of oceanography, and these are based upon the principles of physics, chemistry, biology, and other disciplines. Measurements are then made specifically to gather data which will prove or disprove the ideas involved. Imagination and observation are thus closely interwoven to the enhancement of each.

In contemplating technological advances during the 1960's, Bascom listed five important developments each of which increased at least ten times in magnitude. (1) Superships or supertankers have been constructed that can carry more than 300,000 deadweight tons, and still larger ones seem certain of being built. The 10,000-mile voyage of the 115,000-deadweight-ton *Manhattan,* a tanker converted into an icebreaker, illustrates such advances. This ship successfully crunched its way through the frozen seas of the Arctic searching for a Northwest Passage along which oil could be transported from Alaska's North Slope to the East Coast of the United States. (2) Deep-diving submarines and a host of submersible vehicles have been developed (Fig. 14–15). (3) Extensive advances have been made in deep drilling from floating platforms (p. 374). Holes as much as ½ mile in length have been drilled through sea floor sediments from ships located in water more than 3 miles deep and cores of the sediments have been extracted. (4) The positions of ships far from land can now be located with precision. (5) At present, the sea bottom can be studied by television cameras and side-looking sonar much more effectively than before the 1960's.

Other developments involve advances in the means and techniques of oceanographic research; e.g., a number of instruments are attached to a cable so that diverse measurements accrue from a single lowering. Other data are collected while a ship is under way—important because ship time is expensive. Data are also collected in a form suitable for computer processing, and considerable data processing is now done at sea. Unmanned oceanographic stations have been set up on the sea floor, and measurements of various kinds have been made and recorded automatically.

Some Oceanographic Equipment and Techniques

A great deal of imagination and ingenuity has been applied in developing instruments and techniques to sample and probe the oceans and their inhabitants, from wave-tossed surfaces to sediment-cloaked bottoms. It is an immense task to measure and observe the vast water-covered areas of the Earth in three dimensions and over periods of time. During the last decade or two, geophysical techniques, utilizing sound and electromagnetic waves as well as gravitational, magnetic, and electrical attractions, have made the greatest contributions to oceanographic knowledge (see heat-flow, p. 296; magnetic anomalies, p. 299; seismic profiling, p. 371; seismology, p. 154; and gravity, p. 307).

Prior to the emphasis upon geophysical techniques, the great bulk of measurements for oceanographic research was made with instruments lowered over the side of a ship at the end of a long rope or wire, and many measurements are still made in this manner. The wire- or rope-attached instruments are lowered through the water, suspended for a time, and then pulled back to the surface. They may sample the sediments on the sea floor or the water and organisms at various depths. They may also measure the temperature of water at different depths and the amount of heat energy released through the sea floor, photograph the bottom, and make many other types of measurements. The oceanographer must try to interpret from the reactions of his equipment on the ship end of the cable what is happening at the other end. In addition to the difficulties involved in operating equipment blindly at the end of a long wire which may curve or twist in an unknown manner, the oceanographer has to know his precise location at the surface at the time the measurements are made.

Ascertaining Position

One problem for the researcher, as for any navigator, is to ascertain the ship's position. In coastal areas, various devices are available and relatively accurate positioning is possible; e.g., a triangulation technique can be used if one is in sight of the coast and has a map on which certain landmarks are located accurately. The directions from the ship to these landmarks can be plotted on the map, and the ship is located at the spot where the direction lines intersect. Radar signals can be used in a somewhat similar manner and function when visibility is limited. In yet another technique, the ship may receive radio signals beamed from shore-based transmitters, and these can be interpreted to give the directions to the transmitters.

The position of a ship at sea can be located by use of its echo sounder (see below) and navigation based upon an accurate map of the topographic features of the sea floor —if discrete, recognizable features exist on the ocean bottom in the area. One can also navigate via dead reckoning from a known starting position: the velocity of the ship and the direction and rate of current-caused drifting are recorded and plotted. Astronomical observations have long served to determine the position of a ship at sea; e.g., at a certain time a star or the sun is observed to have a certain altitude and direction from the ship. This means that the ship is in a certain position on the Earth's surface; otherwise the direction and altitude of the celestial object at that precise time would be different. However, an error of 2 miles or so is common.

The problem of determining the precise location of a ship at sea has long troubled oceanographic research and navigation, but it has now been solved for ships that carry the necessary equipment to obtain signals and fixes from navigational satellites in orbit about the Earth (Fig. 14–5). The location of a ship 1000 miles from land can now be pinpointed within about 0.1 mile, and worldwide fixes are possible at intervals of about 1½ hours.

Radio acoustic ranging has also been used to locate a ship's position. A small explosive is detonated near a ship at a precisely recorded time. Sound waves then travel away from the ship through certain layers in the water and eventually are detected at receiving stations at known locations. The time required for the signal to be received at a

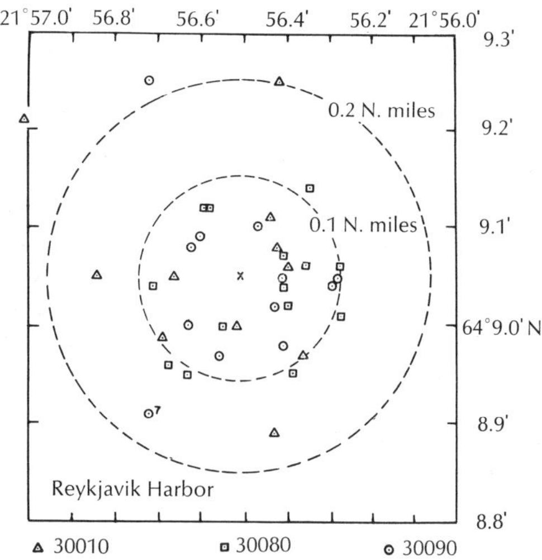

14-5. Fixes obtained from orbiting satellites by R/V Vema while docked in Reykjavik Harbor in 1966. The star shows ship position according to a British Admiralty Chart; the cross represents the mean position of the fixes (open circles) and the dots show less accurate fixes when the satellites were within 5 degrees of the horizon. Half of the fixes fall within 0.07 nautical miles of the mean position. *(Courtesy Lamont-Doherty Geological Observatory.)*

station is a measure of the distance of the ship from this station (communication between the ship and station occurs via radio waves that travel at about 186,000 mi/sec through air, whereas the velocity of sound in water is about 4900 ft/sec). On a map, the ship's navigator then draws a circle which has a receiving station at its center and a radius equal to the distance between the ship and the station. When circles have been drawn for a number of receiving stations, the ship's location is the one place on the map where all of the circles meet. Sounds have been detected under favorable circumstances across distances up to 9000 miles or more.

If a ship is to stay in one general area, one or more "pingers" (transmitters which emit sound waves at periodic intervals) may be anchored to the bottom. The ship's position may subsequently be located with respect to the pingers, whose locations are first carefully determined.

The Echo Sounder

Until the development of the echo sounder in the 1920's, measuring water depths had been a laborious task, particularly in the deeper parts of the ocean. At first glance this seems to be a simple measurement, providing one has available enough rope and a weight to attach at one end. But in deep water, 1½ hours was sometimes required for a rope to descend to the bottom, and a longer time was needed to reel it in. Furthermore, the weight of a few miles of rope was so great that the rope continued to sink after its sinker had hit bottom—fortunately, however, at a slower rate. Thus by timing the rate of descent, one could tell approximately when the bottom was reached. The sinker automatically fell away upon contact with the bottom so that the rope with its attached thermometers and water bottles could be retrieved more readily.

An experience of a group of Russian oceanographers illustrates the time that can be involved. An attempt was made to collect samples of sediments and organisms from the ocean floor at the bottom of a trench in water approximately 36,000 feet deep. About 12 hours were required to lower and raise the dredge this distance, and the dredge was towed along the bottom for about 1 hour. Unfortunately, as the dredge reached the surface after its 13-hour trip, a wave washed away its contents. The feelings of the oceanographers can be readily imagined!

The continuous echo sounder functions because sound waves spread outward from a source throughout a cone-shaped region. Some waves are then reflected back toward the source as the familiar echo if they strike an appropriate obstruction such as a cliff or the wall of a building (sound signals may also be sent out as a narrow pencil-shaped beam that is aimed vertically downward). In oceanographic work, the reflecting surface is the sea floor, and the sound waves travel through the water at about 4900 ft/sec. A ship must have suitable instruments attached to its hull to produce the sound waves and to receive the reflected signals, and an accurate clock is needed to measure the time interval. Ultrasonic waves (too short for a human ear to detect) are commonly used because they can be distinguished from the audible sounds produced during the operation of the ship. In its earlier "knock and listen" stage, a "ping" was emitted and the time for the echo to return was noted. Now, continuous automatic records are made from ships and submarines, and an imposing quantity of depth records have been accumulated. However, many of the older records were made from ships whose positions were known only approximately.

A precision depth recorder may measure a relative change in depth of about 3 feet in water that is 3 miles deep. However, the actual depth cannot be determined as accurately as this because the velocity of sound is different at different temperatures and salinities. Furthermore, erroneous results occur where slopes are steep (Fig. 14–6), and some echoes are produced by objects above the bottom (see deep-scattering layer, Fig.

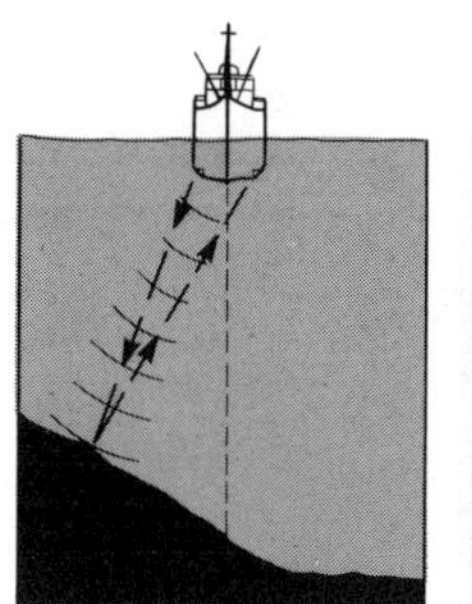
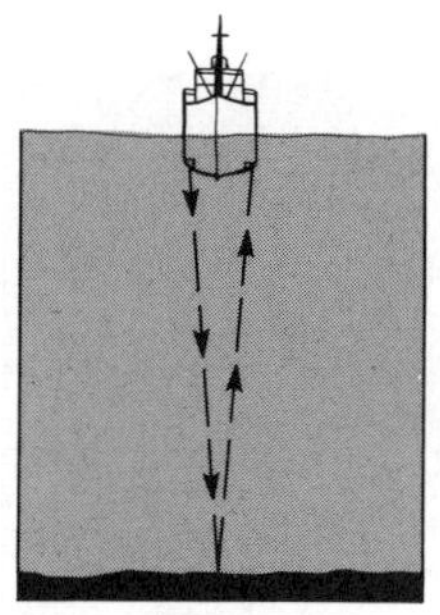
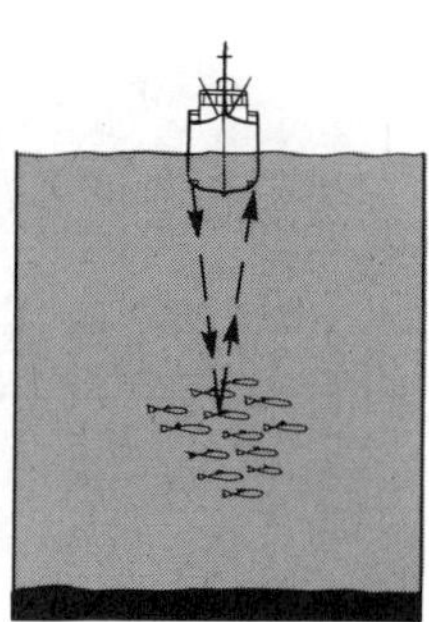

14-6. An echo sounder measures the depth of water beneath a ship.

14–17). Thus one can move across the level surface of the sea and yet know from the record made by a continuous echo sounder that miles beneath him the ocean floor has topographic features that rival those of the lands in diversity and magnitude: in turn he may pass above a submerged volcano, a submarine canyon, a deep trench on the ocean floor, a great chain of mountains, a nearly level plain, or the wreck of a famous ship (Fig. 14–7).

Reflection and Refraction

Thicknesses and structures of sediments on the floors of the deep ocean basins can be measured by studying the reflection and refraction of sound waves produced near the surface by the detonation of explosives and other means. The reflection method is similar to echo sounding, but some of the sound-wave signal may penetrate as much as 3 miles below the ocean floor before it is reflected back (Fig. 14–8). A receiving hydrophone is towed behind a moving ship (the line is slackened just before the signal is emitted). The signal may be produced by a mechanical device such as an electrical sparker or a compressed air gun (these can emit a series of signals in rapid-fire succession) or by an explosive which fires automatically after it has been released (older method that can produce a much more powerful signal for greater penetration). Longer wavelengths penetrate deeper, but shorter wavelengths give better near-surface detail. A continuous record of the sea floor and of the layers and structures of the sediments and rocks below the sea floor can thus be made from a research ship moving at normal cruising speeds (in early work, the ship had to stop for each reflection shot). The *continuous seismic reflection profiler* thus marks a major advance in oceanographic technique.

Figs. 14–9 and 14–10 illustrate the capability of seismic-reflection profiling. A number of small hills project above an abyssal plain in the Gulf of Mexico, and the profile shows that submerged domes also occur below the flat surface of the plain. These apparently represent masses of salt that were intruded in an igneouslike manner into the layers of sea-floor sediment. Similar salt-dome structures are rather common around the margins of the Gulf of Mexico (both on land and in shallow water) and have

14-7. Echo sounder record of the sea floor shows the wreck of the Lusitania. *(After Barnes.)*

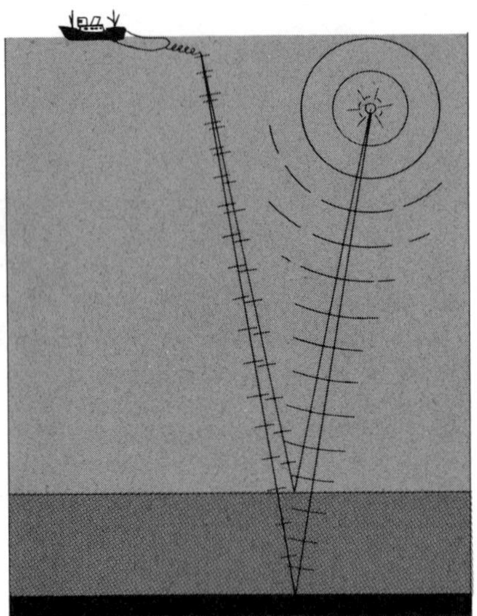

14-8. Reflection shooting at sea. Part of a sound wave signal is reflected from the ocean floor, but part penetrates deeper and is reflected from underlying layers. *(After Ewing and Engel.)*

been the source of large quantities of petroleum and sulfur. Cores extracted from the Challenger Knoll as part of a deep-drilling program contained typical salt-dome caprock minerals, sulfur, and petroleum. Thus petroleum and sulfur can evidently form and be trapped in a deep-water environment. A major problem is the origin of the salt which must come from an underlying layer—such deposits have generally been thought to form by evaporation in shallow water.

Refraction shooting (Fig. 14–11) involves a somewhat different technique and deeper penetration. It provides information about the crust and uppermost mantle but less detail concerning the top-most zone of sediments. Explosives are dropped into the water and detonated by a "shooting ship," and signals from the sound waves thus produced are picked up by hydrophones at a "listening ship." Waves go out in all directions from the point of the explosion, and some are reflected from near-surface zones, but many petetrate downward. Some of these waves are bent as they pass downward into a different zone, and then they travel along the interface between the two layers of different velocity—part of this signal is continuously refracted upward to the surface. The oceanic crust has a layered structure, and the sound waves travel faster in the deeper layers. Thus the first signal from an explosion received at a listening ship depends upon its distance. For a moderate distance, the first signal might be one that traveled laterally for a distance in the uppermost layer before being refracted back

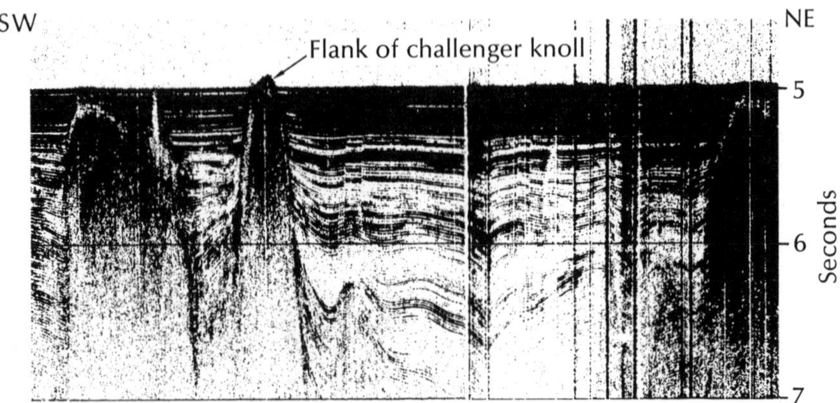

14-9. Seismic-reflection profile made in water about 12,000 feet (3600 meters) deep in the Gulf of Mexico. Profile is about 10 km (6 miles) in length. (*Initial Reports of the Deep Sea Drilling Project,* Vol. 1, National Science Foundation, 1969.)

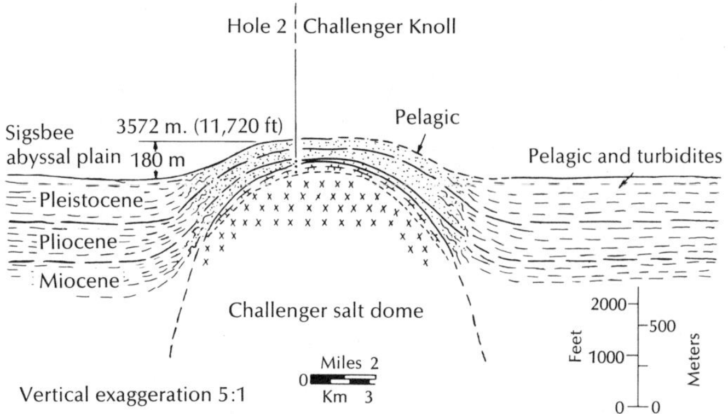

14-10. Composite profile and structure section through a small sea-floor hill and adjoining abyssal plain near the center of the Gulf of Mexico. The absence of turbidites above the knoll shows that it has projected above the sea floor since the late Miocene. (*Initial Reports of the Deep Sea Drilling Project,* Vol. 1, National Science Foundation, 1969.)

to the surface (such refraction occurs continuously as a signal moves along an interface, but it weakens with distance). For a greater distance, the first signal will be one that moved at a faster rate along a deeper interface before it was refracted back to the surface. Therefore, when records made at different distances are compared, they may be interpreted to give the depths to the different interfaces.

Sediment Samples

Samples of the sediment or rocks that compose the sea floor can be collected by using a grab of some type to sample one spot or by pulling a dredge along the bottom. If the dredge catches on the bottom, its chain is designed to break before the towing cable parts. Coring devices furnish a much better record because some of them can obtain cores of bottom sediments that are essentially undisturbed and tens of feet in thickness. Most coring tubes now have plastic inner liners that can be stoppered upon reaching the surface to keep the sediments moist. To extract a core, the hollow metal coring tube is lowered at the end of a cable (Fig. 14–12). It is open so that water can pass through it during the descent, but closed during the ascent to hold the sediment load. A tube may be shoved downward through sediments on the ocean floor by heavy weights attached to its upper end. The weights drive the corer through the sediments because the tube is allowed to fall freely during the last part of the descent (fins keep it vertical). In an older, dangerous, and now-abandoned method, a small explosive attached to the top of a coring tube was detonated as the tube hit bottom. In each method, the sediments were compressed as the metal pipe was jammed through them.

In more recent corers, arrangements have been made to create a "suction" inside a tube as it passes downward through the sediment. In one such arrangement, a piston is located inside the lower end of a tube and is attached separately to the cable.

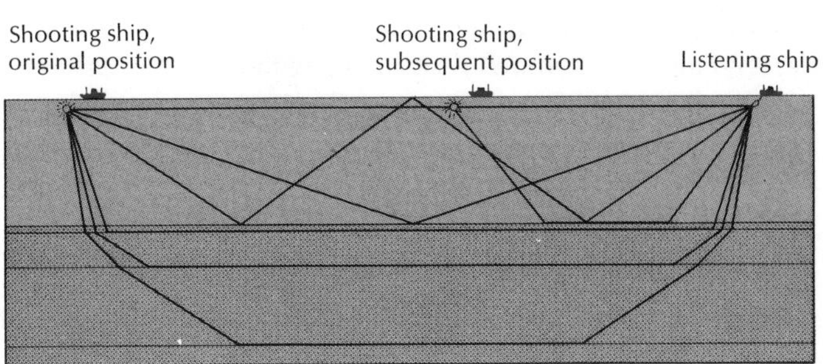

14-11. Refraction shooting at sea. (*After Ewing and Engel.*)

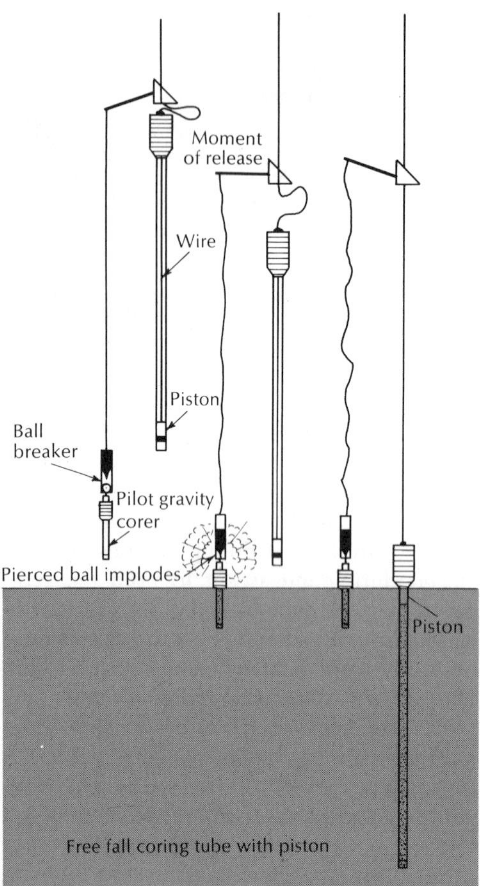

14-12. How the free-fall method of coring operates. The piston makes it possible to secure a long core by reducing friction on the inside of the core tube. A ball breaker is part of the apparatus. (Courtesy F. P. Shepard, *Submarine Geology*, 2d ed., Harper & Row.)

When the bottom of the coring tube starts to penetrate the sediments, the piston is stopped by its cable attachment. Thus, to prevent a vacuum from occurring inside the descending tube, sediment must rise within it. A similar effect would occur if the tube remained motionless and the piston were pulled upward.

Russian oceanographers have obtained the same effect by means of a large tank attached to the top of a coring tube. This is sealed at atmospheric pressure at the surface and is opened automatically when the bottom of the tube begins to penetrate the sediments on the sea floor. The pressure inside the tank is relatively so low that a core of sediments is forced upward through the tube.

Very great technological advances have been made recently in deep-sea drilling from floating platforms. Here a string of drill pipe, as in an oil well, substitutes for a cable. It is now possible to maintain a platform over a drill site in water a few miles deep, and holes have been drilled through sediments ½ mile in thickness. Cores of the sediments have been extracted, and the original igneous-rock floor of the ocean has been penetrated and sampled at a number of sites. By 1970, an almost complete, composite, geologic record had been obtained of sea floor sediments spanning the Cenozoic and extending into uppermost Mesozoic.

Furthermore, by 1970 new technical developments had made it possible for the first time to remove a drill string from a hole on the sea floor and then to guide the string back into the same hole. The job was done from a ship floating in water 2 miles deep, and the feat was likened to the first docking of two orbiting spacecraft. Before drilling began, a funnel was attached to the drill string and pushed into the sea floor. The drill was then pushed downward through the funnel, and a hole was bored into the underlying sediments and rocks. When the bit became dull, it was hauled up, leaving the funnel on the sea floor.

To change the bit, the string of drill pipe was lifted a short distance, and the topmost length of drill pipe was unscrewed and stacked on the deck. This procedure was then repeated time after time until the entire drill string had been disconnected and the last length of pipe with its attached bit was on the ship. A new bit containing a sonar device was then attached to this length of pipe. To get back to the funnel, the hauling up procedure was repeated in reverse. The rotating sonar beam within the new bit produced especially strong signals from three so-called corner reflectors on the rim of the funnel, and thus it could be guided to the 10-inch hole for re-entry. In the meantime, the ship was kept in position over the hole by using signals from sonar beacons previously placed on the sea floor near the hole and special propellers (computer operated) located at strategic places on the ship's hull.

Water Samples

Water samples at specific depths may be taken in different ways, but for samples at greater depths the preferred device over the decades has been similar to the Nansen bottle (Fig. 14–13). A metal bottle and the thermometers fastened to it are attached to a wire and lowered into the water. During the descent the bottle is open and water passes freely through it. Other bottles are attached at regular intervals to the wire, and when all of the bottles have been lowered, a small weight or "messenger" is attached to the wire and released. The messenger slides downward along the wire until it strikes the bottle nearest the surface. This impact releases a catch, the bottle turns upside down, and lids at each end of the bottle are closed by springs triggered by the messenger. The thermometers are reversed with the bottle, and the mercury columns break at a constriction in each tube. The quantity of mercury on each side of a constriction is a measure of the temperature at that depth.

One of the thermometers is not protected against water pressure, which thus squeezes more mercury pass the constriction than occurs in the other thermometer, which is protected from the pressure within a heavy glass tube. Mercury also fills the space between the thermometer and the glass tube so that the temperature of the mercury within the bulb can come into equilibrium with the temperature of the sea water. The difference in reading between the two thermometers depends upon the pressure, which is related directly to the depth. Thus the depth and temperature of a water sample are both obtained.

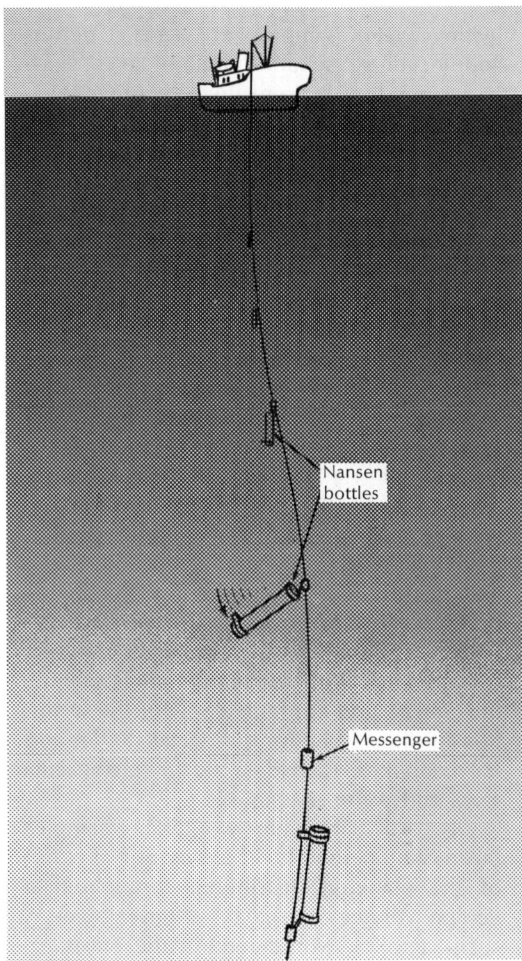

14-13. Water samples are obtained at different depths by using a string of Nansen bottles.

Another messenger that had been attached to the wire below the top bottle is released by the impact of the first messenger, and this slides downward to strike the second bottle. The sampling procedure is repeated, a third messenger is released, and eventually all of the water bottles have been reversed. The wire is then raised, the bottles removed, and the water samples analyzed.

Samples of Organisms

The very smallest of marine creatures pass through nets and must be obtained by collecting water samples at specific depths and locations; separation then takes place in a laboratory by centrifuge or other technique. Specimens of larger organisms may be obtained at various depths from the sea floor to the surface by lowering and raising netlike devices that can be opened, towed, and closed at specific depths and by other means. A meter may be attached to a net to record the volume of water that passes through the net as it is towed.

A continuous plankton sampler is described briefly as an example. Plankton is made up of small animals and plants that have very limited ability to move and are thus transported via currents. Oceanographers need to know the distribution in depth as well as in areal extent of different types of plankton in ocean water at any one time and also how these change with time—hourly, daily, seasonally, and from one year to the next. Perhaps such changes can then be correlated with the temperature, salinity, movement, and nutrient supply of the water. Besides their value for biological studies and food supplies, such data aid in tracing currents and water masses.

Of particular value would be samples taken simultaneously (the synoptic picture) of the plankton distributed throughout a certain volume of ocean water. However, such sampling would require the efforts of many men and ships, more than are available. Fortunately, a satisfactory substitute is afforded by the *continuous plankton recorder*. This instrument is lowered on a cable at the start of a voyage and raised at the end. Thus it can be towed by any ship; a specially equipped oceanographic vessel and trained personnel are unnecessary. During a voyage the instrument is towed at a constant depth which is controlled by the length of the cable. Sets of stabilizing fins tend to offset variations in the ship's velocity, and a rudder keeps the instrument out of the ship's wake. Water enters the streamlined instrument through a narrow opening at the front and passes out at the rear. Inside the instrument, the water goes through a gauze on which some of the plankton are filtered off and stored; the size of the mesh in the gauze determines the nature of the catch, and clogging occasionally is a problem. The gauze is automatically shifted from one spool to another. A second strip of gauze is added as a protective cover over that containing the plankton, and the two are wound around the collecting reel as a "plankton sandwich."

The gauze is later unwound and cut into strips. Therefore, each strip represents the catch of a certain number of miles. A sampling or reconnaissance technique can be used to speed laboratory studies—specimens may be identified only at certain intervals along the gauze strip, or some specimens are identified and others are omitted. The gauze strips can then be stored and studied in more detail at a later time.

Measurement of Ocean Currents

Ocean currents may be measured in numerous ways, some of which are much more sophisticated than others. In many methods, either an object moves within a current, and its path is determined, or an object is fixed in a particular location and records the speed and direction of a current that moves around it. The Swallow float is one of the newer devices used to measure deep ocean currents and is described here as a sample (see p. 404 for other methods). It consists of a long metal cylindrical buoy which is weighted to float at a specific depth. A transmitter within the buoy regularly emits sound waves, and these signals can be picked up by an instrument on a ship. The location of the buoy can thus be tracked, and a number of buoys floating at different depths provide data concerning the deep-water circulation of the oceans. By such means it has been learned that beneath the Gulf Stream there is a second and deeper current that flows in the opposite direction.

Experiments with "sonic tags" were begun in the summer of 1962 and utilize a similar technique. The tags weigh about 1 pound and consist of a tiny sound transmitter enclosed in an aluminum capsule which is attached to a large fish such as a tuna that is captured alive and then released. A 300-pound sand shark was followed for several hours in one experiment, and the technique seems promising. It is hoped that useful information will be obtained concerning the speed and migratory habits of large food fish and predators such as sharks. Perhaps the effectiveness of shark repellents can be studied in this manner. The original transmitter had a life of less than two weeks, but this probably can be extended.

Man Under Water

Man has been diving and swimming beneath the ocean surface for many years, but only rather recently have techniques and equipment been developed that permit him to make extended stays at continental-shelf depths when he is not within a submarine or similar protective vehicle. To do this, a man needs a continuous supply of life-sustaining oxygen, and he must be able to function under great pressures. Although sea level pressure is nearly 15 lbs/sq in (called *one atmosphere*), a man is not crushed because he breathes air that is equally compressed. Thus internal and external pressures are the same. A column of water 10 meters in length likewise exerts a pressure of about 15 lbs/sq in. Therefore, pressure at a depth of 10 meters is about twice that at the surface, and the pressure increases about one atmosphere for each additional 10 meters of depth (e.g., at a depth of 90 meters, the pressure is 10 atmospheres, or about 150 lbs/sq in). A man is not crushed at such depths if he breathes a suitable mixture of gases compressed to the same pressure because inside and outside pressures are equal.

Finding a suitable mixture of gases has been a problem. Pure oxygen becomes toxic at a depth of about 10 meters, whereas nitrogen becomes toxic at a somewhat greater depth. However, a mixture of helium and oxygen had proved satisfactory by the late 1960's, and divers were using it to work on oil well equipment and to perform other sea-floor tasks at depths exceeding 600 feet. Breathing such compressed gases poses a problem, however, since some of the gas gets into a diver's blood and fatty tissues. The amount of absorbed gas increases with time until a diver becomes saturated. If a diver ascends slowly enough (or is decompressed slowly within a pressure chamber), the dissolved gases bubble out harmlessly. However, if a diver ascends too rapidly from a great depth, the gases escape vigorously and painfully, and a diver is said to have "the bends." Decompression must thus be done with care.

The compressed gases may be pumped from a support ship through a hose to a diver who is enclosed within a helmeted diving suit with weighted shoes and attached by a lifeline to the ship. However, scuba diving (*s*elf-*c*ontained *u*nderwater *b*reathing *a*pparatus) is a newer and now widely used method in which a diver backpacks his own supply of compressed air and is thus free to swim and move without encumbering lifelines and air hoses. However, he does encounter difficulties: it becomes colder at greater depths (an electrically heated suit may help here), visibility is generally poor, currents may interfere, and dangerous marine creatures can cause problems.

Actual observation over a period of time in the natural environment is the important advantage of this method, and geologists and biologists have done a good deal of underwater field work. Geologists have studied underwater outcrops, measured strikes, dips, and thicknesses, taken samples, and made correlations; biologists have filmed the activities of fish and studied their reactions to the nets used in commercial fishing.

A major, rather recent discovery involves so-called saturation diving. The amount of gas absorbed into a diver's blood and tissues reaches saturation after about 24 hours, and thus the time for decompression is the same after a 24-hour stay at a certain depth as it is for a stay of a week, month, or longer at this same depth. This has led to various underwater, man-in-the-sea projects in which living quarters have been established on the sea floor at depths exceeding 400 feet, and teams of divers have lived in these for as long as 60 days. Compressed air is piped downward from a surface ship and keeps sea water from entering the sea floor dwellings through open hatches at the bottom. Thus pressures inside the living quarters are the same as those outside.

Men can live, sleep, eat, cook, and relax inside the dwellings, some of which have been air-conditioned (teams of women have also done this). Each day, with the aid of familiar scuba diving equipment, they can swim and work outside on the sea floor for a few hours. At the end of their stay on the sea floor, the divers undergo lengthy decompression, but this lasts no longer after a one-month sea-floor visit than for one of a few hours. Thus divers can spend much more actual working time on the sea floor, and at greater depths, than they could by descending and ascending each day. With this technique, man apparently has the potentiality of actually exploring and working on most of the continental shelves—a region comparable in area to the entire African continent.

A submersible decompression sphere may be used in diving. With divers inside and air hose attached, the sphere is lowered by cable to any desired depth. The sphere is then pressurized for this depth, and the divers leave by a hatch that they open at the bottom of the sphere (the compressed gases inside the sphere prevent flooding). They may use scuba gear or breathe from hoses attached to the sphere, and the divers may return to the sphere for food or rest. At any desired time, the pressurized sphere (with hatch now closed) can be pulled back to the surface and locked onto a larger, similarly pressurized decompression chamber on the ship. Within this, the divers may undergo a lengthy decompression. However, if their tasks are not finished, the divers stay at this same pressure and eat, sleep, and rest in preparation for a return trip to this same depth. Thus a diver may make a number of trips up and down without decompression until his job is done.

A trained porpoise (dolphin) named Tuffy was very useful in one man-in-the-sea project; e.g., he carried messages and lightweight objects from the surface to the divers or vice versa. Since getting lost in dimly lit, murky waters was an ever-present danger, Tuffy was trained to come to a signal emitted by a special buzzer. Thus one end of a line could be attached to Tuffy's harness at the underwater base, and Tuffy would then go to a buzzer signal emitted by a lost diver, who then followed the line back to the base. Other porpoises have been trained to go to an acoustic beacon and thus may be useful in discovering sunken submarines, planes, and missiles that have such sound-emitting devices attached to them. In one technique, a porpoise uses his snout to carry a special hoop to the signal, and then he tosses the hoop toward the beacon. The hoop then breaks in half, and one part sinks (it may be magnetized to attach itself to a metallic object). The other part floats upward to become a buoy at the surface. Since the two parts are attached by a wire, the sea-floor location of the beacon is marked by the buoy.

Underwater Photography and Television

Photographs have been made of the sea floor by diver-held cameras (some are waterproof and require no housing) as well as by automatically operated cameras in protective housing in water thousands of feet deep. Visible light is transmitted selectively through the water (p. 394) and blue predominates at greater depths. Except for photographs made near the surface, artificial lighting is necessary, and light scattering is a problem. For this reason, the light source is placed at some distance from the camera and as close as possible to the object being photographed (Fig. 14–14). In a common type of camera, the apparatus is lowered at the end of a cable, and the photograph is snapped when a triggering device touches the bottom. Lighting is furnished by self-contained electronic flash equipment, the film is wound automatically after each exposure, and two dozen or more photographs may be taken on one trip. The camera can be lifted off the bottom, moved to another place by the ship, and then lowered until the triggering device again activates the flash and shutter.

Bascom (p. 368) listed advances in underwater television as one of the five major developments in ocean technology that occurred during the decade of the 1960's. One advance involves new tubes that can amplify the available light by some 30,000 times. In shallower locations, this eliminates the need for artificial lighting and the obscuring backscattering thereby produced. Fouling (p. 387) of an underwater television lens can occur rapidly, however, and transparent, long-lasting, protective coatings must be developed before television monitoring stations can be used effectively.

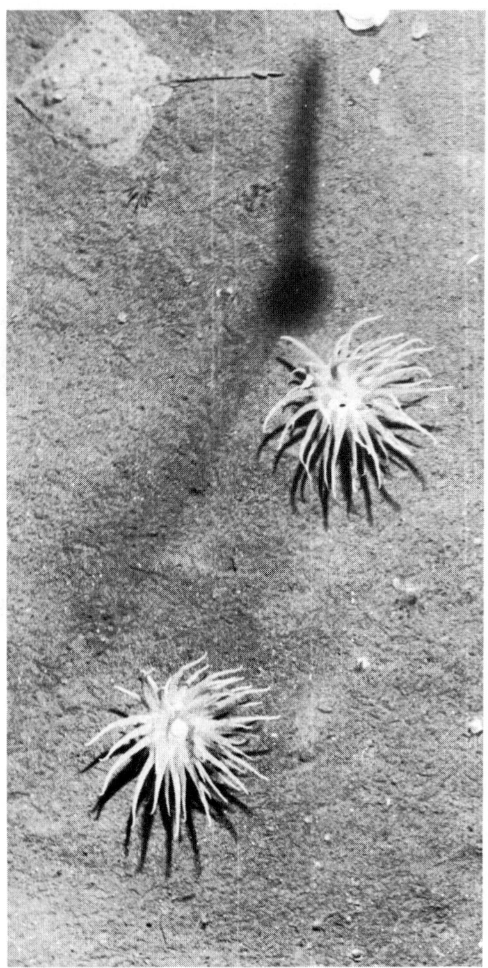

14-14. Sea bottom photograph obtained from U.S. Coast and Geodetic Survey ship *Explorer* in July 1961 about 80 miles south-southwest of Nantucket Island, Massachusetts. Water depth was about 260 fathoms (1560 feet). A skate is located in the upper left (note his protective coloration); the nearby light spot is part of the camera equipment. Some sea anemones are shown. *(Courtesy U.S. Coast and Geodetic Survey.)*

Underwater Vehicles

As a limited sampling of the newer types of oceanographic vehicles that have increased greatly in recent years in number, variety, and capability, we describe briefly the bathyscaph (first dive in 1948), the diving saucer, the manned vertical buoy, and a research submarine.

The *Trieste* (Fig. 14–15) is a bathyscaph (Gr. "deep boat") and it descended 35,800 feet to the floor of the Mariana Trench off Guam in January of 1960, at that time the deepest dive made by man. The trip took about nine hours and began and ended at a wave-roughened surface that contrasted strikingly with the quiet waters beneath. The bathyscaph is an untethered, down-and-up vessel with limited horizontal mobility. Its vertical movements are controlled by balancing the buoyant effect of its gasoline float against a load of ballast which can be dropped. The crew of two is housed within a steel-alloyed spherical cabin with an inside diameter of about 6½ feet. The cabin is suspended beneath a large compartmentalized float that is filled with gasoline (about 30,000 gallons are used; the gasoline is less dense than water and more compressible). The bottoms of the compartments are open, and the gasoline floats on water which rises into the compartments as the gasoline is compressed during a dive. In contrast, the

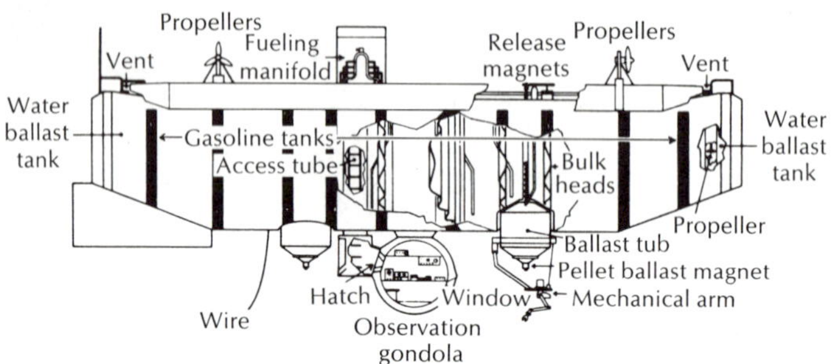

14-15. Bathyscaph Trieste. *(Courtesy U.S. Navy.)*

cabin is pressurized and its atmosphere is controlled; e.g., oxygen is added, carbon dioxide is removed, and the moisture content is kept at a certain level. The *Trieste* is a clumsy craft at the surface and can be towed only at slow speeds.

To begin a dive, an air compartment at each end of the float is flooded with water. Thereafter the crew controls the direction and rate of vertical movements by jettisoning either gasoline or ballast. The ballast consists of 9 tons of iron balls (16 tons in a more recent model) that are held in place in two narrow-bottomed bins by magnetic fields produced by electromagnets. When the electrical current activating the magnets is cut, the iron balls fall out of the bins. Thus if the power supply fails, it must "fail safe"; furthermore, the heavy bins themselves can be dropped. Because the density of the gasoline increases as a bathyscaph descends, it becomes less buoyant at greater depths, and the rate of sinking would increase if ballast were not dropped. Nearly 1 ton of ballast may be dropped for each 3000 feet of descent. A heavy chain suspended at the base of the cabin strikes the bottom first and slows the descent as the chain piles up on the ocean floor. The cabin has plexiglass portholes and searchlights and can hold some 2 tons of scientific equipment. The craft has a horizontal range of about 4 miles and is powered by electrical batteries and propellers, but it moves slowly.

Darkness became total at a depth of about 1500 feet during the 1960 deep dive of the *Trieste*, and several thermoclines were encountered near the surface (a *thermocline* is a relatively narrow transition zone between an overlying, higher-temperature, lighter-weight water layer and the colder layer that occurs beneath it). The descent illustrated relative movement in an interesting manner; the men in the *Trieste* peered through the portholes as they descended and watched organisms of various types stream upward past their windows—but it seemed to them that the *Trieste* was moving upward!

The sea floor was finally reached after a descent of nearly 7 miles at about 3 ft/sec—the speed of an average elevator. Among the interesting sights was that of a fish about 12 inches long and 6 inches wide swimming along the bottom, not at all bothered by the tremendous pressure of the water—about 16,000 lbs/sq in.

The diving saucer made its debut in 1960—a small, highly maneuverable, baby-submarine type of craft with a maximum operational depth of about 1000 feet (newer and larger models can descend to still greater depths). All of the continental shelf, or about 6% of the sea floor, as well as the uppermost part of the continental slopes are within its range. The craft is propelled and steered by battery-powered water jets and consists of a plexiglass and steel, saucer-shaped cabin in which a crew of two can lie prone. Mercury can be pumped from one interior reservoir to another to control the trim of the saucer, and there is a mechanical arm that can pick up objects. Striking photographs and movies of underwater objects have been made from the saucer. Marine animals appear to be attracted by the unusual craft.

The Flip Ship (*fl*oating *i*nstrument *p*latform, Fig. 14–16) is a manned buoy from which scientists can observe the upper 300 feet or so of ocean water. The first Flip Ship was completed in 1962 and is about 350 feet long. The ship floats horizontally as it is towed to an observation site, where the cigar-shaped, weighted stern portion is flooded with water and sinks. This tilts the ship, and the angle of tilting increases until the ship is oriented at a 90-degree angle to the surface of the water; it floats like a bottle in the water and is anchored for continued observation in one location. The bow portion projects above the water like a building four or five stories in height. Observations and measurements can be made in various places throughout the submerged portion as well as above sea level. Boats can be used to supply and transfer personnel and materials. Other manned buoys may be developed to drift with the currents, and unmanned buoys may be located at the

14-16. The Flip Ship is a new type of oceanographic research vessel. It can literally flip from a horizontal to vertical position while at sea if its long aft section is flooded with sea water. Air at high pressure blows out this water to return *Flip* to a horizontal position. *Flip* is being used for studies of wave motion, marine biology, internal waves, sound waves, and other phenomena. Two watertight cylindrical tubes permit the crew to descend to 150 feet below the ocean surface.

surface or at different depths. These can be designed to make various measurements automatically and to transmit the data to ships, planes, or other receiving stations.

On the land, periodic measurements from one location are a standard technique in meteorological and other studies. On the sea, such measurements have been uncommon, but are now possible from the Flip Ship, from anchored unmanned buoys, and from the Texas towers, or artificial islands, which have been stationed in various locations on the continental shelf as part of a military warning system or to use as drill-rig platforms in the search for oil. Ships can also be anchored in very deep water because "weightless" plastic anchor rope has been developed which has the same density as sea water; previously, long anchor lines parted under their own weight.

In the summer of 1969, the 50-foot long, 130-ton Grumman-Piccard research submarine made a 30-day, 1500-mile drift within the Gulf Stream (p. 407) off the east coast of the United States. The submarine was named the *Ben Franklin* in honor of one of the first men to chart the Gulf Stream. The trip had several goals: make studies of the biology, chemistry, and physical properties of the waters of the Gulf Stream; study the deep-scattering layers (p. 386; unfortunately not observed on this trip); and study the behavior of the 6-man crew (three "Peeping Tom" cameras photographed the crew every two minutes during the long, confined voyage).

A surface support ship kept in contact with the submarine by radiophone, gave it navigational fixes, and participated in some experiments. The submarine drifted passively in complete silence within the Gulf Stream at depths between 500 and 2000 feet; her engines were shut off. Floodlights and cameras were mounted outside, and 29 portholes made viewing possible in most directions (surface waves were once observed from a depth of 330 feet). Slow evaporation of liquid oxygen provided oxygen to breathe during the voyage, and carbon dioxide was absorbed by lithium hydroxide. Most of the food was freeze-dried and could be reconstituted by adding hot water (heated before the trip and stored in insulated tanks). Compartments were flooded to descend, and the water was blown out by compressed air to rise. Six tons of iron shot ballast could be dropped if the electricity failed. At one time a swordfish attacked the *Ben Franklin,* and at other times it moved up and down the crests and troughs of internal waves 60 to 100 feet and more in amplitude.

Marine Life

The subject of marine life is diverse, complex, and very extensive, but it can be treated only cursorily here because physical oceanography is our main topic. However, marine biology is too important and rich in interest to be omitted entirely, and the following topics are discussed briefly: the distribution of plants and animals in the oceans, some unique aspects of the marine environment, the food pyramid, some denizens of the deep sea, bioluminescence, the deep scattering layers, a life cycle, marine fouling and deterioration, and marine organisms as a food supply for man.

Marine organisms range in size from some of the smallest to the very largest of living creatures. In fact, whales apparently are the largest animals that ever lived on the Earth at any time. Some organisms are sessile creatures that remain in one location, although they commonly pass through an unattached, floating egg and larval stage. Others are passive floaters and drifters or can move feebly, and these are carried by currents from one place to another. Still others are strong swimmers. Vertical migrations are much less extensive than horizontal ones because the physical properties of water tend to change more rapidly in a vertical direction, and pressure is a conspicuous example. Living plants are confined to the upper sunlit waters, but animals have been seen and collected from the ocean surface to the bottom of the deepest trench. However, distribution is not uniform. Certain depths and areas have life in great abundance, especially at certain seasons, but in other parts of the oceans, life is less abundant. Although certain parts of the ocean are called biological deserts, like deserts on land, they are inhabited.

On the land, organisms are restricted to certain environments by obvious barriers of topography and climate. Somewhat similar environmental restrictions occur in the oceans, but these are less obvious and are governed by factors such as salinity, temperature, depth, and food supply. Day-to-day and season-to-season changes in the environment of an organism in the sea tend to be much less than the weather changes experienced by an organism on the land. However, relatively enormous changes in pressure are sustained by fragile organisms that migrate daily through a vertical distance of a few hundred feet.

The gravitational pull of the Earth bulks very large in the lives of most marine creatures, and they need some adaptation that functions constantly to keep them from sinking. For many marine organisms, there are no hiding places from predators, and these must survive, as long as they do, more or less by chance. "Eat, spawn, and be eaten" suggests the continuous cycle involving birth, growth, reproduction, and death. On the other hand, marine organisms need no specialized adaptations to keep their bodies from drying out, and the plants live completely immersed in their food supply (very dilute, however).

Food Chains, Plankton, and Recycling

Plants form the food supply of all animals either directly or indirectly, because only the green plants can build carbohydrates (various sugars and starches made of carbon, hydrogen, and oxygen) from carbon dioxide and water in the presence of sunlight. This is the familiar, vital, but not completely understood process known as photosynthesis ("putting together with light") which is an energy-absorbing one and stores some of the energy of sunlight in the plant. The simpler carbohydrates are transformed into more complex foods such as fats and proteins by additions of nitrates, sulfates, and phosphates. The energy stored in these foods is subsequently utilized by all organisms as they grow, function, and reproduce. Thus plants release oxygen and take in carbon dioxide, whereas animals release carbon dioxide and use oxygen in an endless, vital, complementary cycle.

Phosphorus, nitrogen, silicon, copper, and iron are abundant elements in the nutrients absorbed by plants, but a number of other elements are also essential, even though they exist in exceedingly dilute amounts. For example, certain tiny plants cannot live in a "sea water" synthesized artifically in the laboratory, even though the liquid has all of the known ingredients of natural sea water in precisely their natural proportions. Yet such plants can live if a small portion of natural sea water is added.

Thus an endless, vital cycle of food supply occurs in ocean water. It involves foods synthesized from inorganic nutrients by plants, the consumption of plants by animals, the eating of some animals by other animals, the death of animals and plants, the oxidation of their remains, and the transformation of these organic wastes by bacteria into nitrates, phosphates, and other nutrients that are needed by the plants.

As one illustration of a relatively short food chain, we can cite the whale, a marine mammal. Some whales subsist chiefly on small crustaceans, whereas others eat fish and squids. According to one estimate, a single whale may require several thousand herring-size fish for one meal. Each herring in turn needs several thousand small crustaceans to supply its energy for a day, and each tiny crustacean may have consumed tens of thousands of minute plants in the same time interval. Thus billions of tiny plants are required to support one whale for one meal.

Plankton ("wanderers") includes unattached animals (*zooplankton*) and plants (*phytoplankton*) as well as the eggs and larvae of other animals, that float passively or swim very feebly in the waters of the Earth. Planktonic organisms constitute the bulk of organic matter in the oceans. Most are very tiny in size and invisible to the unaided eye as discrete individuals. Some plankton can be collected in fine-mesh nets, but many are too small for this method. For these a sample of water must be obtained and evaporated or centrifuged. However, some floaters, such as jellyfish and the sargassum weed, are relatively large.

The plant members of the plankton can live only in the upper 100 meters or so of water that is penetrated with sufficient intensity by sunlight to support photosynthesis (the so-called euphotic zone). Commonly the depth is much less than this, because suspended material, especially near shore, reduces the depth of light penetration. The herbivorous zooplankton are likewise most abundant in this sunlighted zone because they develop best where their food supply is most concentrated.

Planktonic organisms tend to be small and simple for a number of reasons. A small object has a relatively large surface area in comparison to its volume; e.g., reducing the radius of a sphere by half, decreases its surface area four times, and its volume eight times (surface area and volume vary as the square and cube of the radius respectively). A relatively large surface area helps an organism to stay afloat and to absorb nutrients from the enclosing waters. Some organisms can change their sizes by increasing or decreasing the lengths of fingerlike protuberances, and thus they compensate for water that becomes more dense or less dense (higher temperatures or lower salinities). Thus in warmer, less viscous water, a certain species will tend to be smaller than other members of the same species who live in colder water because natural selection has occurred as generation followed generation. In this constant struggle to stay afloat, a certain percentage of the small planktonic plants do sink below the reach of sunlight, and thus they die. Some eggs have an attached droplet which makes them buoyant.

Phytoplankton is sometimes called the "grass of the sea" because it fills a role similar to that of the grasses and cereals on the lands. These generally minute organisms float and drift passively in the uppermost waters and furnish directly or indirectly a food supply for all of the animals in the sea. Thus they form the base of all marine food chains and control the maximum food supply that can be harvested from the oceans. Moreover, it is estimated that at least 70% of the total amount of oxygen produced each year by photosynthesis occurs in the oceans and chiefly by phytoplankton.

Phytoplankton can proliferate at a prodigious rate at certain seasons and places where strong sunlight and a rich supply of nutrients combine because they have very short life cycles; e.g., certain forms have been known to increase by 300% within 24 hours. In some areas and with certain organisms, this proliferation is greatest in the spring, with another and lesser increase occurring in the fall. Reduced concentrations occur in winter because growth and reproduction are retarded by low temperatures and reduced intensity of sunlight. In the spring, sunlight and temperatures have increased, winds are strong, waters are mixed to bring nutrients to the surface, and an abrupt increase in numbers occurs—a seasonal "blooming." The seas may be colored red, brown, or some other color by the bodies of myriads of organisms.

Still higher growth rates might be expected in the summer because of higher temperatures and intense sunlight. However, as the surface waters become warmer and less dense, mixing is reduced, and the phytoplankton gradually deplete the supply of nitrates, phosphates, and other nutrients in the surface waters. Thus their numbers are limited. Lower temperatures occur again in the fall, surface waters become chilled in some places, and these sink. This results in mixing which replenishes the supply of nutrients in the surface waters, and sunlight is still intense enough to cause a lesser blooming.

Cold surface waters at high latitudes, even those beneath ice, contain an abundant supply of plankton because larger quantities of oxygen and carbon dioxide can occur in solution at low temperatures than at high. On the other hand, phytoplankton is not especially abundant in warm tropical waters if mixing and upwelling are absent.

Many of the animals that eat phytoplankton are themselves very tiny, and perhaps the most abundant of these are the crustaceans known as copepods ("oar-foot": many members have two elongated appendages). Copepods occur in fresh water and at all depths in the oceans. The life cycles of many of the planktonic copepods span less than two weeks.

Animals may exist at all depths in the ocean, but the plant eaters who live below the top sunlit waters must wait for plant remains that descend to them; dead animals also sink. These organic remains are gradually destroyed by oxidation and by bacteria as they drift slowly downward. Therefore, the largest pieces generally have the best chance to reach the ocean floor in deep water. Many deep-water fish are organized to take full advantage of a large occasional meal—huge mouths and long teeth are combined with small bodies (commonly less than a foot in length), and some have a stomach that can be extended outward to wrap around an object too large to be swallowed. In an interesting development, a ray of the dorsal fin has been modified into a fishing rod in some of these creatures. Its location has shifted to the head, it has become much longer, and it commonly has a bioluminescent organ at the tip. Evidently the light-tipped appendage is waved to and fro to attract passing prey.

Bacteria occur in all depths of ocean water and are particularly abundant in bottom sediments. They are called scavengers of the sea and make it "the largest and most efficient septic tank on Earth." Bacteria transform the solid, liquid, and gaseous remains of plants and animals into nitrates, phosphates, carbon dioxide, and other nutrients needed by plants. Thus bacteria form an essential part of an endless cycle. However, for these bacteria-freed nutrients to reach the sunlit surface zone where plants can assimilate them, upwelling must occur. In such zones of upwelling, life in the sea attains its greatest abundance—also in the shallower waters over the continental shelves where sufficient mixing occurs to keep a rich supply of nutrients in the sunlit zone.

Fish and other organisms dredged from great depths appear related to relatively young shallow-water forms, and this suggests colonization of the depths by modification and adaptation. Deep-water organisms are not little-changed descendants of very ancient forms of life as was once predicted. Perhaps large-scale deep-water changes have occurred and resulted in many extinctions. Vacated environmental niches would subsequently then be filled by migrations of younger surviving forms from shallower waters.

Bioluminescence

Bioluminescence refers to the light produced by some organisms through chemical transformations that occur in some parts of their bodies. The firefly is probably the most familiar example, but organisms as diverse as bacteria, fish, crustaceans, corals, sponges, and fungi produce such light. However, the most highly organized animals and plants are not bioluminescent. The light-producing process is an efficient one that generates little heat energy (so-called cold light). Although not completely understood, the main steps and substances involved have been worked out, the light-producing compound in the firefly has been synthesized, and its light-stimulating enzyme has been isolated. The compound and enzyme are brought together to produce the light.

Many organisms in the ocean—from the surface to great depths—have bioluminescent organs or furnish homes for organisms such as bacteria that do. To illustrate, luminous bacteria live in small areas located under the eyes of one kind of fish and emit light continuously. However, part of the fish's skin can be folded across these areas to cover the bacteria completely. Bioluminescence is particularly common among the organisms that live in deeper water below the depth of light penetration, and perhaps two-thirds are bioluminescent. The ability to emit such light presumably is advantageous, but it is produced by some organisms without eyes. In some instances, the light attracts prey, in other organisms such as the firefly it attracts a mate, and in still others such as some squids it is a defensive device. Some regard bioluminescence as a vestigial system in organic evolution.

Bioluminescence has many interesting illustrations.* Tiny organisms near the sea surface may give off sufficient light to illuminate fish swimming among them, and when disturbed by the passage of a ship, they may make the wake glow as if the sea were burning. In fact, enough bioluminescent light is sometimes produced so that a person on the ship's deck can read at night. For a military ship trying to remain undetected on a nighttime cruise, this can be a most important phenomenon. Occasionally one observes parallel luminescent strips crossing the ocean surface. Evidently the strips are located along the lines of convergence of wind-produced convectional cells—planktonic organisms are most concentrated along the lines of convergence which parallel one another.

Bioluminescent effects on land are likewise intriguing; e.g., bacteria may sometimes provide a glow for a piece of decaying meat, and fireflies have collected in groups in trees in Thailand and flashed their lights off and on in unison. Glowworm larvae have been observed in great abundance on cave ceilings in New Zealand, and these will shut off their lights simultaneously when disturbed sufficiently; subsequently they flash their lights again and light up the cave roof. The larva of a Brazilian beetle is known as a "railroad worm" because it has green lights along its side and red lights on its head.

* McElroy, W. D., and Seliger, H. H., "Biological Luminescence," *Scientific American,* December 1962.

Vertical Movements of Organisms

Certain planktonic organisms—the "commuters"—daily move upward and downward in the ocean waters, and these inhabit various zones within half a mile or so of the surface. Why such movements occur is not entirely clear, but they correlate best with changes of light intensity; the movements are upward at night and downward during the day. Some measurements have shown a lesser upward movement on clear nights with a full moon than on a moonless or cloudy night, but other factors also appear to be involved. Perhaps this mechanism makes it possible for descendants of passive planktonic drifters and feeble swimmers to remain in one part of the ocean. Why are these passive drifters not eventually carried by surface currents entirely out of an area? Superposed shell fragments on the sea floor seem to show that certain planktonic creatures have inhabited certain regions of the sea surface over a long period of time. A daily vertical migration of a few hundred feet may partially explain this persistence in geographic location. Horizontal ocean currents move at different rates at different depths, and this tends to disperse organisms that pass upward and downward through them. Furthermore, such currents are actually quite complex in detail. Eddies and meanders occur as well as narrow currents that flow opposite to the general direction. Such variations can cause much dispersion.

The Deep Scattering Layers of the Sea (DSL's)

Echo sounders have obtained reflections of sound waves from certain organisms such as schools of fish, but they have also recorded echoes from several mysterious layers that are located during the day at depths of approximately 700 to 2400 feet (225 to 800 meters).* These are called the *deep scattering layers* and commonly they number from three to five (Fig. 14–17). The layers do not

* Dietz, Robert S., "The Sea's Deep Scattering Layers," *Scientific American,* August 1962.

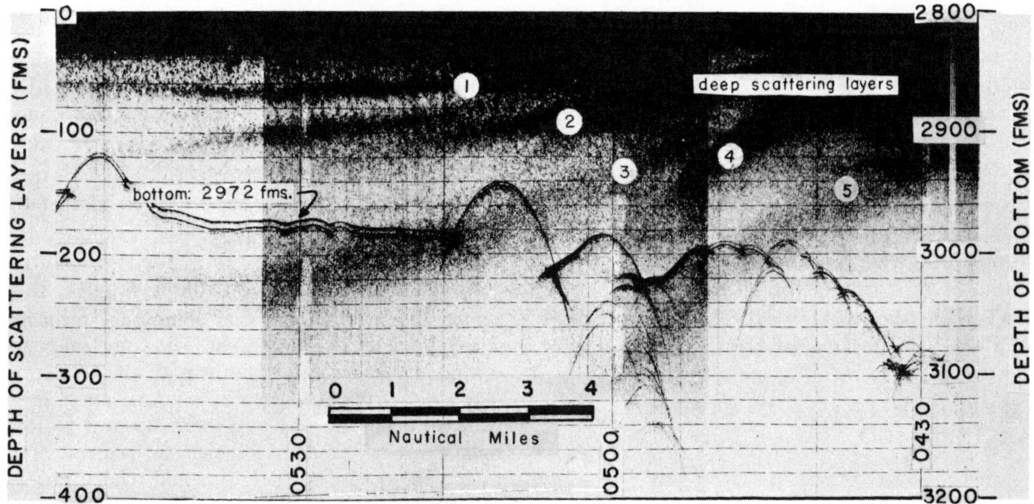

14-17. Five deep scattering layers (DSL's) at sunrise. Echogram was recorded on the R/V Conrad in the southwestern Pacific on 2 April 1965. The DSL's are shown on one scale and the bottom—some 3 miles farther down—on another. *(Courtesy M. Ewing, B. C. Heezen, and A. Lonardi.)*

reflect all of the sound waves, and the ocean floor can be detected through them. The layers tend to migrate upward at night and downward during the day, and thus they apparently are caused by organisms.

Reflections from the deep scattering layers have been attributed to the swim bladders of fishes such as the small mystophids or lantern fish, to colonies of gas-filled jellyfish, to the hard shells of shrimplike crustaceans (euphausiids), and to other organisms. Perhaps different organisms form DSL's in different locations. Possibly the animals migrate in this manner to feed within the food-rich surface zone at night and to hide from predators during the day. A camera has been lowered through a deep scattering layer and observed on an echo sounder during its descent. The DSL dispersed from the vicinity of the camera and then formed again after the camera had passed through.

According to Dietz, the top DSL rises to the surface on a dark, moonless night but does not rise this high when moonlight illuminates the sea surface. During the day the lowest DSL may sink to a depth of 2400 feet, but it tends to rise if clouds obscure the sun. The layers do not cross or mix as they rise or fall. Thus the depth of light penetration as well as the time of sunrise and sunset strongly influence the vertical migrations.

DSL's are worldwide in distribution but are most distinct where organisms are most abundant. They have been observed below the Arctic pack ice where they migrate vertically on an annual cycle; the DSL's are near the surface during Arctic winter and some 50 to 200 meters below the surface during Arctic summer. Weak nonmigratory DSL's have also been observed both during the day and at night.

Fouling and Deterioration

Marine organisms of various kinds cause extensive damage to ships, underwater equipment, and coastal installations; e.g., the annual cost to the U. S. Navy alone in the middle 1960's was estimated at $100,-000,000. Some boring organisms penetrate and weaken materials such as wood, plastic, stone, and concrete, whereas others like the barnacles attach themselves to underwater surfaces such as the hull of a ship or sonar equipment. Marine bacteria also play a harmful role; e.g., they may spread across an anti-fouling chemical that had been painted on a ship and thus provide a foundation for the attachment of organisms; or they may destroy fishing nets, wooden structures, and rubber hoses, or indirectly cause corrosion.

To decrease damage and keep ships, installations, and underwater equipment in good operating condition, it is necessary to search for long-lasting, anti-fouling agents of various kinds. However, such discoveries are necessarily based upon a knowledge of the life cycles of the organisms that cause the damage.

A Life Cycle

As an illustration of the life cycle of a marine organism, consider the eel. Most eels live in salt water, but the life history of the fresh-water eel involves both fresh and salt water. Its life cycle is interesting but not typical of that of most marine fish: only the eel, salmon, and shad are known to live in fresh water and salt water during different periods of their development. Adult eels, usually about 3 feet in length, are found in the streams of Europe and eastern North America, and each of these types of eel has a breeding ground near the Bermuda Islands. In the fall when the females are ready to lay eggs, they swim downstream and are joined by the males, who live in the tidal waters. The parents spawn in the Sargasso Sea near Bermuda and then die. During the following spring, thin and transparent young appear, and when the young are about 3 inches long, they are found in the streams of America from which their parents came. How such trips are navigated is a mystery. The young of the European eels do not appear in European rivers until the third year, and during this interval they have participated in the slow clockwise swirl that enclosed the Sargasso Sea in the North Atlantic. Thus these eels spend the major part of their lives in fresh water, but the vital functions of birth and death occur in the oceans far from their homes in fresh-water rivers.

Food from the Sea

More than half of the men, women, and children on the Earth today do not have enough to eat, and they particularly lack protein foods. On the other hand, the oceans can yield more food than is harvested today and can be especially valuable in supplying proteins. In the late 1960's, the annual worldwide food harvest from the oceans was estimated at about 55 million metric tons (live weight; more than 90% finfish) which produced an income of about $8 billion for the world's fishermen, about 1/50 of the food energy requirement for 3.5 billion people, and about 1/10 of the world's supply of animal protein. The amount harvested has tended to increase recently at a rate of about 4 to 5% or more a year—since the rate is compounded, the 1968 total is about double that of 1958. Among recent trends, the percentage of the total harvest that is used for livestock feed has increased from less than 10% in the 1930's to about 50% in 1967, and the percentage of herring and similar fishes has increased to about half of the total. These utilize phytoplankton directly for a substantial part of their energy requirements and thus form a short, more efficient, food chain.

The total potential sustainable catch that may some day be harvested from the oceans probably lies somewhere between two and four times that being harvested today. In other words, the total food supplies that we can expect from the oceans seem definitely limited when matched against the needs of present and future populations. However, the potential of the oceans as a source of animal protein is greater than this.

To increase the amount of food taken from the oceans requires knowledge about the life cycles of edible organisms: geographic distribution, abundance, growth rates, food supplies, migration habits, diseases, spawning, and reactions to changes in environment. Furthermore, one must know, which organisms are edible and which are not. New methods of fishing must be developed as well as methods of preserving catches until they can be used. Perhaps man can learn to farm the sea in a manner somewhat akin to the raising of wheat or cattle on land. Perhaps certain undesirable predators or organisms ("weeds") can be removed from a part of the sea because they compete with desirable organisms for food supplies. Certain desirable physical characteristics may be established by the application of breeding techniques. Certain elements, needed only in trace quantities, may be added locally to sea water.

Fishing techniques today are still quite primitive, involve much risk and chance, are governed by customs and traditions, and are little changed from the centuries-old methods of the hunter. Therefore, the quickest increase in food supplies from the sea probably depends upon improvements in fishing techniques, also in locating new areas for fishing; e.g., the Indian Ocean accounted for less than 5% of the total catch in 1968. Perhaps "bubble fences" can be substituted for nets in some instances, or light may be used to attract fish to the funnel-shaped opening of a pipe up which they can then be pumped. Only a small percentage of oceanic organisms are today hunted and harvested. Some forms are known to be poisonous and others are inedible, but presumably many unknown or untried oceanic sources of food remain to be utilized by man.

Improved methods of preserving foods could greatly increase usable supplies, but these methods must be inexpensive to be practical for the peoples who need these foods the most. Refrigeration is too costly for some. Spoilage could be prevented by freezing fish and other organisms at sea, but the local freezers and other equipment needed to distribute the frozen foods are unavailable in many underdeveloped areas. Perhaps irradiation will eventually solve this problem.

Fish protein concentrate is a most promising new development and involves the production of a high-protein flour (fish meal) from trash fish and the leftovers from the processing of commercial fish. These materials are dried, pulverized, and changed into a white, odorless, high-protein meal that is inexpensive and does not need special preservation. By 1967, fish protein concentrate made from specific varieties of fish had been approved in the United States as a household additive to supplement the amount of protein in food, and research was underway in different countries to discover better ways of processing the fish and to increase the varieties of fish that can be utilized.

Regulations govern the taking of certain fish—sizes, numbers, and seasons—but many other factors intervene between the egg and the adult, and some of these may be more important in controlling the number of adult fish available for man to harvest. Enormous numbers of eggs are laid when fish spawn, but some of these are eaten before they hatch, and immature fish are constantly preyed upon by many organisms. Thus relatively few fish become adults. The species achieve "survival by fertility." If man could lessen the effects of some of these hazards to the eggs and immature fish, very large changes might result in the numbers that would eventually reach adult size. In another vein, certain desirable fish have been transplanted successfully from one area to another. Perhaps overcrowding exists in one area and this results in such pressure on the food supply that growth is slowed and reduced. In new homes with adequate food supplies, the transplanted fish grow at faster rates.

Eventually man may farm the sea and eliminate some of the risks and chance inherent in present methods of locating and capturing oysters, lobsters, shrimps, and other sea foods. In overcrowded areas of the sea floor, certain undesirable, food-consuming animals perhaps can be thinned, as oyster gatherers on the Atlantic coast undertake to destroy the starfish that destroy oysters. Mariculture, or sea farming, is practiced along some coastal areas today, but detailed studies and experimentation have been limited. In the future, man may be able to breed more efficient animals and plants in the sea, and perhaps he can add nutrients in certain areas or cause upwelling to occur (e.g., the deep-sea siphon, p. 409). The field seems open and interesting and the potential rewards appear large.

Another hope of some oceanographers is to harvest food closer to the plant portion of the food pyramid because matter and energy are lost (so far as a food supply is concerned) each time one organism eats another organism. A certain fish may eat its own weight in smaller animals each day and "waste" this food supply as it swims. According to one estimate, half a ton of tiny diatoms (plants) may be consumed by various organisms during the chain of events leading to the production of 1 pound of commercial fish, and 10 pounds of fish may be needed to make 1 pound of man. It might be more efficient to utilize the lower parts of the food chain directly. Yet these organisms are very tiny, and perhaps the larger fish concentrate them more economically than man can do.

A Promising Idea

Two Lamont oceanographers (Gerard, R. D. and Worzel, J. L.) have recently proposed a plan to produce both fresh water and air conditioning as well as increase the local marine fish harvest. Necessary geographic conditions are: an arid or semiarid island or coastal area, a steady flow of moist air, and deep ocean water immediately offshore (a pilot plant is planned for St. Crois, Virgin Islands, Fig. 14–18). To begin the process, large diameter pipes extending far below the surface will be used to pump deep cold ocean water through condensers located on shore in the path of prevailing winds. Here the moisture-laden air is cooled enough to cause condensation, and the resulting fresh water is collected and stored for future use. The air, now cooler and drier, may then be used to air condition nearby buildings. Meanwhile the cold ocean water is discharged into nearby, shallow, closed-off lagoons. Since this once-deep ocean water tends to contain at least 10 to 15 times more inorganic nutrient material than occurs in the sunlit surface zone (maximum might be 200 times more), the phytoplankton should bloom mightily within the lagoons. In turn, this should lead to a blooming of the zooplankton, and this increase in abundance should continue up the food chain to fish and other organisms useful as a food supply for man. Perhaps the lagoons can be used as breeding areas for various kinds of fish and shell fish—edible varieties that would furnish maximum harvests under these conditions.

According to an additional proposal (by Gerard, R. D. and Roels, O. A.), deep, cold (5° to 10°C at a depth of 1000 meters), nu-

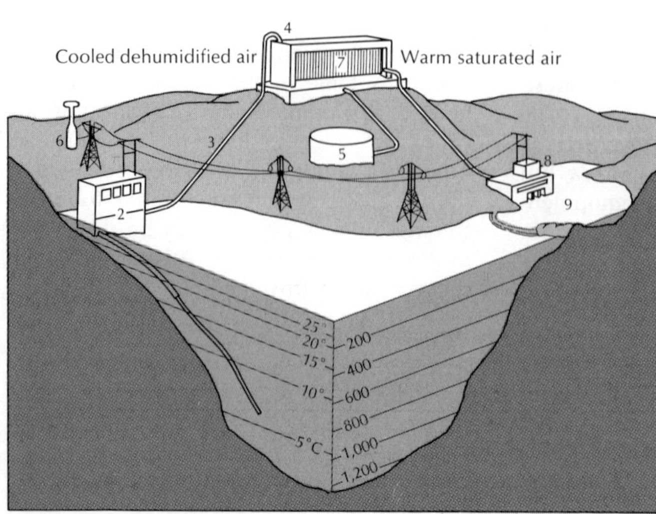

14-18. Artificial upwelling system (schematic). Numbers refer to the following: 2, pumping station; 4 and 7, fresh-water condenser; 5, fresh-water storage; 6, wind-driven generator; 8, sea water into lagoon; 9, lagoon for mariculture. (O. A. Roels, "Marine Proteins," *Chem. Eng. Progr.*, Sept. 1969.)

trient-rich, ocean water may be the world's most abundant renewable resource. For coastal areas or islands with such deep water offshore, it may be piped to shore installations for use in cooling nuclear electrical power plants, for desalination, and for mariculture programs (see paragraph above). Very great amounts of heat energy must be dissipated during the operation of a large nuclear power plant (the so-called thermal pollution problem, p. 18). However, by using for this purpose water that is initially quite cold, the water that is discharged after cooling can be at the same temperature as the nearby sea surface—thermal pollution would thus be completely eliminated. However, the desirability of doing this in areas near coral reefs has been questioned—these need clear water, and the water would probably not remain clear if its nutrient content were increased many times.

Exercises and Questions for Chapter 14

1. Use your imagination to reconstruct some of the possible movements and changes of state of a well traveled water molecule.
2. How many oceans are there? Where are the boundaries drawn?
3. Discuss some of the reasons why oceanography is of importance to man.
4. What is the current status of the various programs and methods designed to transform ocean water into "fresh" water?
5. Locate appropriate source material in a library and read about particular oceanographic explorations that especially interest you: e.g., Magellan's circumnavigation, the drift of the *Fram,* the voyage of the *Beagle,* the *Kon-Tiki* expedition, and the cruises of nuclear-powered submarines such as the *Nautilus* or *Triton.*
6. Describe some of the methods devised by oceanographers to measure or observe each of the following:
   (A) Oceanic depths and sea-floor topography.
   (B) Kinds and thicknesses of sediments on the sea floor.
   (C) Location of an observing ship at sea.
   (D) Ocean water at various depths.
   (E) Samples of organisms at various depths.
7. Describe some of the vehicles and equipment used in underwater exploration and the use of photography and television in oceanographic research.
8. Describe advantages of the saturation-diving technique and recent progress in various man-in-the-sea programs.
9. Marine life:
   (A) Why are certain regions of the sea so-called deserts whereas others support abundant life?
   (B) Describe the food cycle involving dissolved nutrients, plants, and animals.
   (C) What is plankton? Why may plankton increase abruptly in abundance at times?
   (D) Why are certain planktonic organisms called commuters? How may they be related to the deep scattering layers?
10. Discuss the sea as a source of food for man.

15

The Restless Waters

We now focus upon ocean water (Fig. 15–1) and some of its varied motions—motions that may be deep and unnoticeably slow, or wildly turbulent at a storm-tossed surface; motions that may be large or small as well as up, down, or sideways; motions that may be periodic or sporadic and understood or mysterious. It is primarily solar energy that moves these waters although its application is at times indirect; e.g., some currents are initiated by winds, whereas others form because excessive evaporation makes surface water so salty and dense that it sinks. Such movements may begin at one time and place and cause motions in other directions, at other depths, at a later time, and at a distant location. An understanding of the restless nature of the ocean waters depends upon physical properties such as temperature, pressure, and salinity which govern their migrations and separate them into large masses with diagnostic properties. Dissolved and suspended materials in the sea have both organic and inorganic sources. Currents, tides, and waves (surf) interact with the edges of the lands and produce distinctive topographic features. Ice forms at higher latitudes and icebergs break away from land-based glaciers and drift slowly through the upper waters. All of these constitute the subject matter of this chapter.

Temperature and Color

The atmosphere and hydrosphere are heated by radiations emanating from the sun, but the air is warmed chiefly from the bottom, whereas ocean water is warmed primarily at the top. However, in other regions or times, ocean water is also chilled at the top. The oceans lose heat energy in three ways: by evaporation (slightly more than half of the total), by conduction to the overlying air (relatively small) and by radiation. The rate of evaporation tends to be greatest when the difference in temperature between air and water is greatest, and the air is colder than the water. This tends to occur in winter. The actual temperature is less important than the difference in temperatures.

Most ocean water properties are produced by processes that operate near the surface. Such properties may then be retained, with gradual modification, as a mass of water subsides far beneath the surface and moves to some other latitude or location. Since air and water interact in many and important ways along their sea level boundary, we review some of the pertinent phenomena (see p. 674 for material on properties and phase changes of water and p. 646 for the nature of solar radiation). About half of the solar radiation reaching the Earth's surface consists of invisible infrared and longer radiations, whereas the other half is chiefly visible light. The specific heat of water is very high, and large quantities of energy are either absorbed or emitted when phase changes take place (heat of fusion and heat of vaporization). Thus an enormous amount of energy is transferred by currents from one location to another and by water molecules that shift from the atmosphere to the hydrosphere and vice versa. Energy may be absorbed at one place because heating, melting, evaporation, or sublimation occur and then released at another time and place via cooling, freezing, and condensation. The mobility of water as a gas and liquid is most important here, and even as solid ice, it floats and drifts.

Solar radiation is concentrated most on the sea surface in the equatorial regions and least in the polar regions, although daily and seasonal variations occur at all latitudes. Measurements show that more heat energy is absorbed at low latitudes than is radiated back to space, whereas less heat energy is absorbed at high latitudes than is radiated into space. This means that a net transfer of heat energy occurs from low latitudes to high, and nearly one-third of this heat energy is transported by ocean currents. The ocean also plays a role in the other two-thirds that is carried by the atmosphere since about one-third of all the solar radiation that reaches the Earth's surface is expended in the evaporation of sea water.

Some solar radiation is reflected from the sea surface, and the amount depends upon its angle of approach (reflection increases greatly at low angles) and upon the condition of the surface (less radiation is reflected from a calm surface). When the sun

15-1. Sea smoke 23 miles off Cape Kennedy, Florida, December 22, 1960. This type of evaporation fog tends to develop if cold air moves across a warmer water body and rapid evaporation from the water saturates the air. Temperature differences between air and water should be approximately 20° F or more. *(Official U.S. Navy photograph.)*

angle is 40, 30, 20, 10, and 5 degrees, the amount reflected is about 4, 6, 12, 27, and 42 percent respectively.

Radiation that penetrates beneath the ocean surface is eventually either absorbed or scattered, and in each case the process is a selective one. Absorption involves conversion to longer wavelengths and heats the water where it takes place—longer wavelengths (red) are absorbed more readily than shorter wavelengths (blue). On the other hand, scattering produces only a change in direction, and in water, the shorter wavelengths are scattered much more readily than the longer wavelengths. The angle of approach is much less important than wavelength in determining the depth of penetration because refraction is very great; e.g., even when the sun is on the horizon, its rays are bent downward upon entering the water at an angle of almost 50 degrees. Shorter wavelengths penetrate much deeper than longer ones, and almost all of the red and infrared wavelengths are absorbed in the upper few feet, as is much visible light. Only the blue-green colors can penetrate to a depth of a few hundred feet and only a very small percentage of the original radiation reaches this depth; however, the wavelengths most important in photosynthesis penetrate deepest. Therefore, two changes occur as solar radiation passes downward into ocean water: the quantity of radiation decreases with depth, and its character changes. Since the proportion of blue-green wavelengths increases with depth, underwater natural-light color photographs tend to be blue-green.

Ocean water may be bluish or greenish, and occasionally it is tinged by colors such as red and yellow. Blue sky and bluish ocean water have a similar cause—bluish wavelengths penetrate deepest into the ocean, are scattered and diffused most readily, and thus tend to enter our eyes as we look downward into a deep body of clear water. Deep blue ocean water, therefore, is an "ocean desert"—nutrients and living organisms are relatively scarce in it.

On the whole, suspended organic and inorganic material tends to scatter all wavelengths about equally, whereas it tends to absorb shorter wavelengths more readily than longer wavelengths. Thus this selective absorption of the shorter wavelengths by suspended particles tends to reduce the quantity of blue colors available for scattering. In other words, yellows and greens are proportionally more abundant in the scattered light, and such waters appear greenish or yellowish. On the other hand, ocean water may have a brownish or reddish tinge at times because tiny colored organisms abound in it, and under certain conditions such organisms can proliferate very rapidly (in a few days).

Sea water is completely devoid of sunlight below a certain depth which varies with conditions. Sunlight is sufficiently intense for photosynthesis in clear water to a depth of about 100 meters, but its intensity is much reduced at this depth (maximum depth for photosynthesis depends also upon factors such as type of plant, clarity of water, and incident angle of radiation). Different techniques have been used to measure the depth of light penetration. In one, an observer lowers a white disk on a wire and measures the greatest depth at which it remains visible (e.g., a 1-foot diameter Secchi disk was once visible at a depth of 217 feet in the Sargasso Sea). In another method, photographic plates may be lowered and exposed at different depths; e.g., an 80-minute exposure of a plate at a depth of 1000 meters produced a slight darkening, whereas a longer exposure at a 1700-meter depth produced no effect at all. Direct observations may also be made from submersibles, and under very favorable conditions light penetration is greater than previously thought.

A photoelectric cell equipped with suitable filters and connected to a galvanometer at the surface can also be used to measure the intensity of certain wavelengths at certain depths. The deep waters of the oceans are not utterly black, however, because many deep-water organisms emit feeble light (bioluminescence, p. 385). In fact, one measurement from a photoelectric cell showed several faint flashes of light per second—a subdued replica of a Fourth-of-July fireworks display.

Although surface waters are heated primarily by the solar radiation they absorb, they are also heated by contact with the air (i.e., by conduction) and by condensation, but these changes are relatively minor. The vast bulk of ocean water (more than 95% of the total) occurs beneath the shallow, top-heated zone and is warmed by slow downward conduction and by mixing brought about by winds and by density differences related to temperature and salinity.

The temperature of the surface water of the oceans is generally highest in the equatorial regions and lowest near the poles, although currents and mixing produce some irregularities. Lines of equal temperature (isotherms) may be drawn on a map, and these generally cross the oceans in an east-west direction, but they curve northward or southward near coastal areas where warm or cold currents flow (Fig. 15–2). The belt of warmest water tends to occur in the low latitudes of the Northern Hemisphere, and its location shifts somewhat during a year.

Highest surface temperatures may exceed 85°F (30°C; limited somewhat by evaporation which absorbs energy and increases at higher temperatures), whereas 28°F is the lowest temperature (−2°C; sea water freezes between 28 to 31°F, and this releases heat energy). Daily fluctuations in surface water

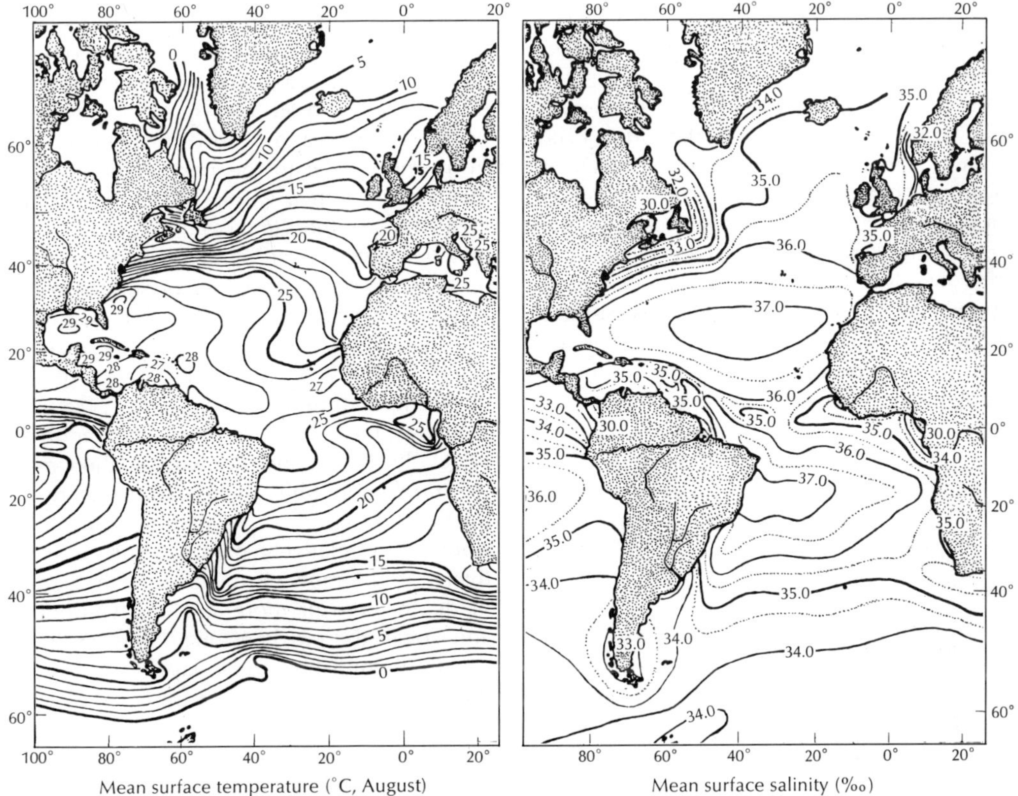

15-2. Surface temperatures and salinities in the Atlantic, which is the warmest of the oceans and has the highest salinity. Left: mean surface water temperature, in °C in August (after Sverdrup). Right: mean surface salinity, in ‰ (after Sverdrup). (R. W. Fairbridge, ed., *Encyclopedia of Oceanography*, Van Nostrand Reinhold.)

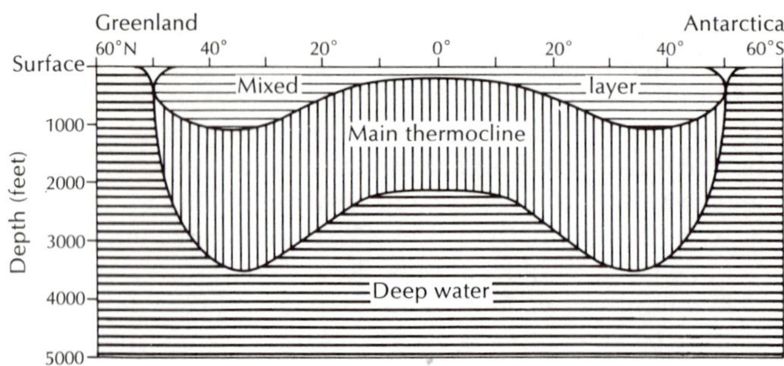

15-3. A schematic representation of the three-layered ocean. *(J. Williams,* Oceanography, *Little, Brown and Company, 1962.)*

temperatures in the open ocean at any one latitude are extremely small (a fraction of 1 degree), and even the annual change is quite limited. It tends not to exceed 15°F (7°C) in the lower middle latitudes where the annual range is greatest, and it varies much less than this in the equatorial and polar regions.

Daily and seasonal temperature changes in air immediately above the surface in the open ocean are much less than those over land areas at comparable latitudes. The thermal capacity of the upper ocean is so much greater than that of the lower atmosphere that air tends to take on the temperature of the underlying water. In other words, if ocean water radiates away enough heat energy to lower its surface temperature slightly, it has actually released a vast quantity of heat energy into the overlying air.

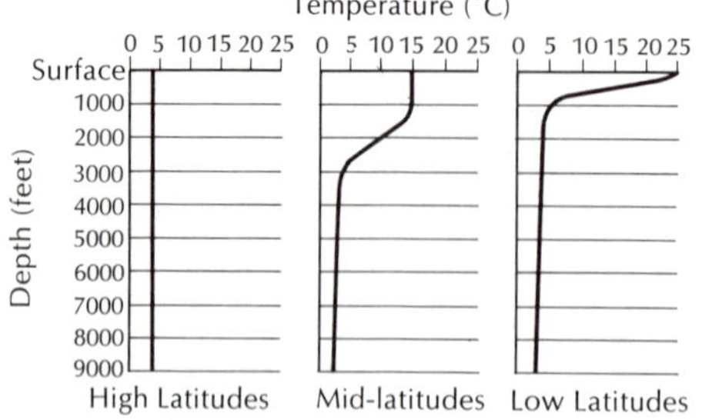

15-4. Temperature vs. depth curves for high, middle, and low latitudes during the winter months. *(J. Williams,* Oceanography, *Little, Brown and Company, 1962.)*

Differences in air-water temperatures produce fogs at sea and in coastal areas, particularly where warm moist air moves across cold water (Fig. 15–1 and p. 694).

From the ocean surface downward, the water tends to become colder, but the rate of temperature change varies greatly. Broadly viewed, three units can be recognized (Figs. 15–3 and 15–4). Mixing occurs within a surface zone that may have a depth of several hundred feet or more. This is warmest and varies most in temperature from one season to the next. Temperatures may be isothermal within this surface zone (i.e., they are the same from top to bottom) if it has been mixed sufficiently by vertical movements begun by winds or surface cooling. Such isothermal zones tend to be thickest in winter and spring and in regions where strong winds and relatively large annual temperature fluctuations occur.

Below the mixed zone occurs another relatively thin zone known as the *main thermocline* in which temperatures decrease relatively rapidly. A thermocline (Gr.: heat slope) is a relatively narrow zone of relatively rapid temperature change with change of depth. One or more smaller, temporary, and local thermoclines may develop seasonally within the surface layer. The top of the main thermocline is nearest the surface at low latitudes; i.e., the zone of mixed surface water is thinnest here. However, a surface zone and main thermocline do not occur at high latitudes where the water is cold from top to bottom.

Below the main thermocline occurs the

great bulk of ocean water, which is very cold at all latitudes and throughout thousands of feet of depth (see p. 391 for its potential as a major natural resource). Latitudinal temperature differences decrease with depth and have essentially disappeared at a depth of 1 mile or so. Temperatures increase very slightly at the bottom of deep trenches (probably from heat energy escaping upward through the ocean floor).

The temperature of surface water is readily obtained by collecting a sample and measuring its temperature, but temperatures can also be determined from a moving plane or orbiting spacecraft by an instrument that measures the infrared radiation emitted from the water. More radiation is emitted at higher temperatures, and the wavelength of maximum intensity also varies with temperature (Fig. 19–4).

Water temperatures within 200 to 300 meters or so of the surface can be measured rapidly and automatically by an instrument called a bathythermograph (Gr.: deep heat-writer) that is lowered and then raised from a moving ship (can be done in 5 min). The instrument gives a temperature vs. depth record because it has a pressure-sensitive element that changes with depth as well as a temperature-sensing one.

At greater depths more accurate measurements are needed because changes are small, and thermometers are used; commonly these are attached to water bottles suspended on a wire (Fig. 14–13).

The speed of sound in sea water is about 4.5 times that in sea level air and tends to increase with increases in temperature, pressure and salinity. Temperature tends to have the greatest effect near the surface, and pressure is the main factor at depth. Typically a minimum sound velocity zone occurs within a mile or so of the surface (Figs. 15–5 and 15–6). Sound waves produced at various levels tend to be refracted toward this minimum sound velocity zone and then tend to move along it. Wherever waves begin to leave this level, refraction occurs to bend them back toward it.

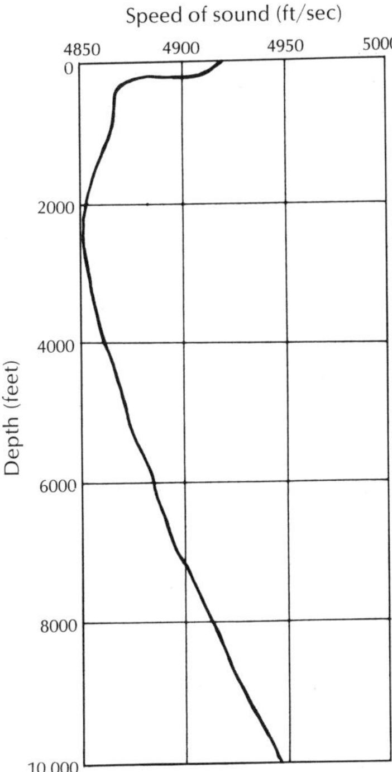

15-5. Typical variation of speed of sound with depth in the ocean. (American Practical Navigator, *U.S. Navy Hydrographic Office, 1962.)*

Materials Dissolved in Sea Water

An estimated 360,000 cu km of water evaporates each year from the oceans; this quantity would average about 1 m in thickness if taken evenly from the entire expanse of ocean surface. However, the process does not operate uniformly, and some regions

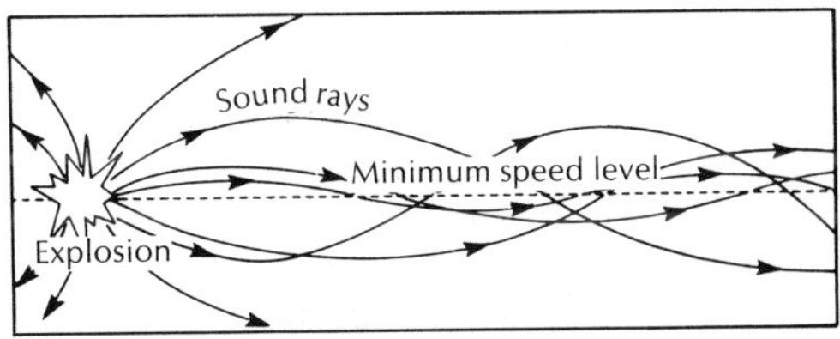

15-6. Transmission of sound waves along a minimum sound velocity level. (American Practical Navigator, *U.S. Navy Hydrographic Office, 1962.)*

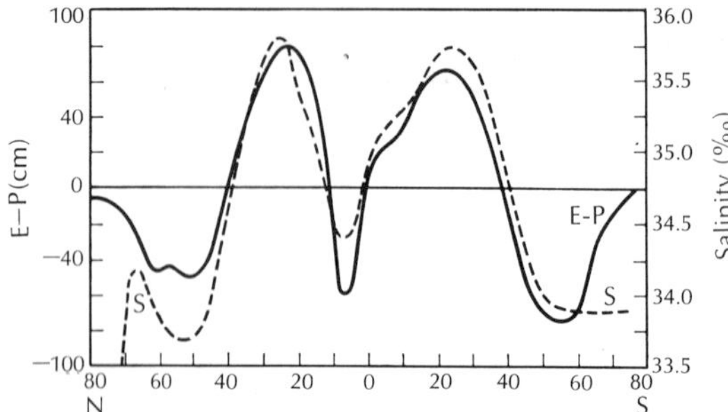

15-7. Average values for salinity (dashed line) and evaporation minus precipitation (solid line) for each parallel of latitude in the open ocean. *(After Wüst, courtesy P. Groen,* The Waters of The Sea, *Van Nostrand Reinhold, 1967.)*

have an excess of evaporation over precipitation, whereas others have more precipitation than evaporation (Fig. 15–7). These variations influence salinity. Precipitation over the oceans returns about 329,000 cu km of water per year. The difference of 31,000 cu km between evaporation and precipitation is made up by additions from the lands where precipitation exceeds evaporation and transpiration by 31,000 cu km.

As fresh river waters flow into the sea, they carry dissolved mineral matter weathered from the lands (e.g., about one-fourth of the sediment load of the Mississippi River is carried in solution). This water subsequently evaporates during one phase of the water cycle, but the dissolved material remains to make the seas saltier. Although the total tonnage of dissolved solids brought to the oceans each year by rivers is large, it is a very small fraction of the total amount now in solution in ocean water. Perhaps enough material is added in 6 million years to change the salinity of the oceans by 1 percent.

Volcanic eruptions on the sea floor and on land add material, as does biological activity in the oceans. Thus all elements and substances that occur in some abundance on the lands eventually become dissolved in the oceans. If all of this dissolved material were extracted from sea water and piled evenly on the land areas of the Earth, it might form a layer nearly 600 feet thick. Although many of these substances are present in extreme dilution (Table 15–1), total quantities are large because there are about 330 million cubic miles of sea water. For example, gold may be present in a weight ratio of only about 4 parts of gold in 100 billion parts of sea water, yet this totals about 16 biilions pounds dissolved in the oceans, or about 5 pounds for every man, woman, and child on the Earth today.

If one were to evaporate 1000 pounds of average sea water, he would have a residue of about 35 pounds of a multitude of materials that had previously been in solution in the water. This is the average *salinity* of sea water and is commonly expressed in parts per thousand by weight (symbol: ‰; compare with the parts per hundred symbol used in percentages—%). The common range in salinity is 33 to 37‰, and the overall average is just under 35‰ (equivalent to 3.5 parts per hundred or 3.5%).

Four ions make up about 97% by weight of all the different materials that are dissolved in sea water. These are: Cl^-, Na^+, $SO_4{}^{2-}$, and Mg^{2+} with percentages of 55, 30.6, 7.7, and 3.7 respectively. Since chlorine is nearly twice as abundant as sodium, one can not regard the material simply as dissolved sodium chloride (NaCl).

Salinity increases in regions of greater-than-average evaporation (Fig. 15–7), particularly in areas where mixing with adjoining regions tends to be restricted—the Red Sea is an example and has salinities that may exceed 40‰. Salinity, therefore, is high where evaporation is great (e.g., the horse latitudes) and where ice forms in large quantities (the dissolved material is excluded from the ice). On the other hand, salinity tends to be low where dilution by fresh water occurs (melting ice, influx of river water, and regions of heavy precipitation—particularly along the equator). Poleward from the middle latitudes, salinity tends to be low because reduced evaporation offsets reduced precipitation.

Although the salt content of surface waters varies somewhat, the salinity of the

TABLE 15-1
COMPOSITION OF SEAWATER*

ELEMENT	SEAWATER CONCENTRATION (μg/liter)**	PRINCIPAL DISSOLVED SPECIES
H	1.1×10^8	H_2O
He	7×10^{-3}	He (gas)
Li	1.7×10^2	Li^+
Be	6×10^{-4}	—
B	4.5×10^3	$B(OH)_3$, $B(OH)_4^-$
C	2.8×10^4	HCO_3^-, CO_3^{2-}
C (organic)	2×10^2	—
N	1.5×10^4	N_2(gas)
N	6.7×10^2	NO_3
O	8.8×10^8	H_2O
O	6×10^3	O_2
O	1.8×10^6	SO_4^{2-}
F	1.3×10^3	F^-
Ne	0.12	Ne (gas)
Na	1.1×10^7	Na^+
Mg	1.3×10^6	Mg^{2+}
Al	1	—
Si	3×10^3	$Si(OH)_4$, $SiO(OH)_3^-$
P	90	HPO_4^{2-}, $H_2PO_4^-$, PO_4^{3-}
S	9.0×10^5	SO_4^{2-}
Cl	1.9×10^7	Cl^-
Ar	4.5×10^2	Ar (gas)
K	3.9×10^5	K^+
Ca	4.1×10^5	Ca^{2+}
Sc	$<4 \times 10^{-3}$	$Sc(OH)_3^\circ$
Ti	1	$Ti(OH)_4^\circ$
V	2	$VO_2(OH)_3^{2-}$
Cr	0.5	CrO_4, Cr^{3+}
Mn	2	Mn^{2+}
Fe	3	—
Co	0.4	CO^{2+}
Ni	7	Ni^{2+}
Cu	3	Cu^{2+}
Zn	10	Zn^{2+}
Ga	3×10^{-2}	—
Ge	7×10^{-2}	$Ge(OH)_4$
As	2.6	$HAsO_4^{2-}$, $H_2AsO_4^-$
Se	9×10^{-2}	SeO_4^{2-}
Br	6.7×10^4	Br^-
Kr	0.2	Kr (gas)
Rb	1.2×10^2	Rb^+
Sr	8×10^3	Sr^{2+}
Y	1×10^{-3}	$Y(OH)_3^\circ$
Zr	3×10^{-2}	—
Nb	0.01	—
Mo	10	MoO_4^{2-}
Ru	—	—
Rh	7 to 11×10^{-3}	—
Pd	—	—
Ag	0.3	$AgCl_2^-$

TABLE 15-1 (Cont.)

ELEMENT	SEAWATER CONCENTRATION (μg/liter)	PRINCIPAL DISSOLVED SPECIES
Cd	0.1	Cd^{2+}
In	4×10^{-3}	—
Sn	0.8	—
Sb	0.3	—
Te	—	—
I	60	IO_3^-, I
Xe	5×10^{-2}	Xe (gas)
Cs	0.3	Cs^+
Ba	20	Ba^{2+}
La	3×10^{-3}	$La(OH)_3^\circ$
Ce	1×10^{-3}	$Ce(OH)_3^\circ$
Pr	0.6×10^{-3}	$Pr(OH)_3^\circ$
Nd	3×10^{-3}	$Nd(OH)_3^\circ$
Sm	0.5×10^{-3}	$Sm(OH)_3^\circ$
Eu	0.1×10^{-3}	$Eu(OH)_3^\circ$
Gd	0.7×10^{-3}	$Gd(OH)_3^\circ$
Tb	1.4×10^{-3}	$Tb(OH)_3^\circ$
Dy	0.9×10^{-3}	$Dy(OH)_3^\circ$
Ho	0.2×10^{-3}	$Ho(OH)_3^\circ$
Er	0.9×10^{-3}	$Er(OH)_3^\circ$
Tm	0.2×10^{-3}	$Tm(OH)_3^\circ$
Yb	0.8×10^{-3}	$Yb(OH)_3^\circ$
Lu	0.1×10^{-3}	$Lu(OH)_3^\circ$
Hf	$<8 \times 10^{-3}$	—
Ta	$<3 \times 10^{-3}$	—
W	0.1	WO_4^{2-}
Re	0.008	—
Os	—	—
Ir	—	—
Pt	—	—
Au	1×10^{-2}	$AuCl_2^-$
Hg	0.2	$HgCl_4^{2-}$, $HgCl_2^\circ$
Tl	0.01	Tl^+
Pb	0.03	$PbCl_3^-$, $PbCl^+$, Pb^{2+}
Bi	0.02	—
Po	—	—
At	—	—
Rn	6×10^{-13}	Rn (gas)
Ra	1×10^{-7}	Ra^{2+}
Ac	—	—
Th	$<5 \times 10^{-4}$	$Th(OH)_4^\circ$
Pa	2.0×10^{-6}	—
U	3	$UO_2(CO_3)_3^{4-}$

*Courtesy *Marine Chemistry,* National Academy of Sciences.
**μg refers to microgram (1/1,000,000 of a gram).

deeper waters of the oceans is very nearly uniform (Fig. 15–8)—evidence that ocean waters have been thoroughly mixed. Notice that both temperature and salinity change relatively rapidly near the surface within zones known as the thermocline and *halocline* (Gr.: salt slope), respectively. Changes below these zones are very small.

Actual water samples are needed for salinity determinations (obtained in water bottles, Fig. 14–13). Measuring the tiny amount of each of the dozens of elements dissolved in a water sample would be a laborious task. Fortunately, this is not necessary because ratios among the more abundant materials remain almost constant, even though the total salinity varies within certain limits. This relationship has been called the *law of relative proportions,* and it is most significant in the chemistry of ocean water; e.g., magnesium makes up about 3.75% by weight of the total dissolved load at all depths whether the salinity is 33‰, 38‰, or some other quantity.

The amount of chlorine present (the *chlorinity* of the sample) is used as a standard and is measured carefully for each sample. The quantity of each of the other dissolved elements can then be calculated because its ratio relative to chlorine is the same at all salinities. Salinity is approximately equal to 1.8 times the chlorinity, and the chlorine content can be determined by several quick, inexpensive methods (e.g., by electrical conductivity or refractive index).

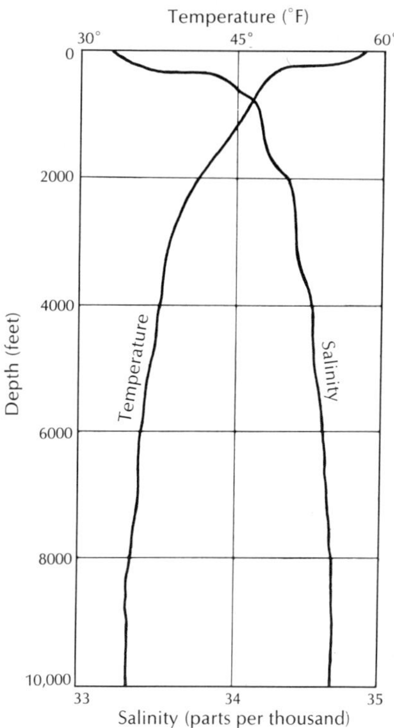

15-8. Typical variations of temperature and salinity with depth at one locality. *(American Practical Navigator, U.S. Navy Hydrographic Office, 1962.)*

The most abundant gases dissolved in sea water are oxygen, carbon dioxide, and nitrogen. Since the quantity of dissolved gases tends to increase with decreases in temperature and salinity (less important), the gases tend to be more abundant in the surface waters of high latitudes because larger quantities can be dissolved in the colder waters. Oxygen is essential for almost all organisms, and carbon dioxide is needed in the process of photosynthesis by green plants. Thus life can exist in abundance only where these two gases are likewise abundant. The oxygen content of the oceans is much less than the oxygen content of the atmosphere, but the carbon dioxide content is much greater. Carbon dioxide is produced at all depths —by animals during respiration and during the oxidation of organic compounds.

Oxygen is added to ocean water in large quantities only near the surface and only by two processes—by contact with the air and by plant photosynthesis. The oxygen content is thus highest in surface waters and tends to decrease downward, but not regularly, and the least amount occurs more commonly at mid-depth than along the bottom. Some oxygen is used by animals during respiration, whereas other quantities are used during the oxidation of organic matter, and each process can occur at all depths. The oxygen content of a mass of ocean water, therefore, furnishes a clue concerning its migrations—it is greatest at the surface and diminishes steadily after the water has left this surface zone. Since dissolved oxygen occurs in some of the deepest ocean waters—even within some deep-sea trenches—circulation on an ocean-wide scale must be occurring. In some parts of the ocean, however, the oxygen is scarce or depleted, and life is scarce or cannot exist.

Density and Pressure

The density of sea water is slightly greater than that of fresh water, and it increases at lower temperatures, at higher salinities, and with depth (see below). Therefore, density is decreased by additions of fresh water and by heating, whereas it is increased by cooling, evaporation, and the formation of ice. Fresh water is densest at about 39.2°F (4°C, see p. 676), but salt water increases steadily in density until it freezes between 28° and 31°F. Temperature differences are more important than salinity differences in causing density changes in the upper portion of the ocean. Therefore, the densest surface waters form where temperatures are very cold at high latitudes. These dense surface waters subsequently sink and spread laterally at some intermediate depth or along the bottom to form ocean currents.

Densities tend to increase most rapidly with depth within the main thermocline where temperatures are decreasing most rapidly (this zone is called the *pycnocline* or "density slope"). On the other hand, tremendous pressures at great depths increase densities only slightly because water is nearly incompressible, and temperatures and salinities remain about the same. However, if water in the ocean depths were not compressed at all by the mass of overlying water, then sea level would probably be about 90 feet higher than it is now.

The pressure on ocean water increases steadily downward, and at a depth of 1 mile it exceeds 2000 lb/sq in. For each additional mile of depth, the pressure increases by more than 1 ton/sq in. Expressed in other units, the pressure increases downward at the rate of about 1 atm (nearly 15 lb/sq in) for each 10 m of depth. Therefore, pressure at a depth of 10 m is equal to 2 atm, whereas at a depth of 990 m, it equals 100 atm, or 1500 lb/sq in.

Organisms can exist at all depths in the oceans because their bodies are in hydrostatic equilibrium with the pressures at the depth they inhabit—pressures inside their bodies equal the pressures outside of them. However, such organisms cannot survive large abrupt changes in depth, although some whales have been known to dive half a mile down and then return to the surface —apparently they experienced no ill effects from an abrupt pressure change of 1000 lb/sq in, but such feats are exceptional. If a deep-water fish is pulled rapidly to the surface, its internal organs swell greatly and then burst under the steadily diminishing pressure.

Ice

Two kinds of ice float on ocean waters in the colder regions of the Earth. *Icebergs* are broken-off portions of glaciers that formed on land and flowed downward into the sea. Icebergs contain no dissolved salts but do include rock debris picked up from the land. Such debris increases their density and is dropped on the sea floor when they melt (icebergs probably account for about three-fourths of the total wastage of glaciers per year). On the other hand, melting and sublimation along fractures may honeycomb sections of an iceberg and reduce its density. Since an iceberg is slightly less dense than water, it floats with one part above the surface and another and much greater part below the surface. The ratio of these two parts (i.e., their vertical extent above and below sea level) depends greatly upon the shape of an iceberg as well as upon the densities of the ice and water. The proportions approximate 1 to 7 for a tabular berg and 1 to 2 for a pinnacled berg.

Most icebergs come either from Greenland or Antarctica, and their shapes are distinctly different (Fig. 15–9). Some Antarctic icebergs form where the edge of the continental ice sheet has advanced into the sea, but the largest were once part of a floating ice shelf. Shelf ice has several sources: freezing of ocean water, accumulation of snow on its surface, and the addition of glacier ice that merges gradually with it. Such icebergs tend to have flat tops and bottoms. They may be 500 to 1000 ft thick, and their lengths may exceed 200 miles. A huge one observed in 1956 was twice the size of Connecticut and measured 208 miles by 60 miles. Icebergs in the Northern Hemisphere are likely to be much smaller and more irregular, but one was 7 miles by 3 miles, and another towered nearly 450 ft above sea level. Most icebergs are much smaller than the examples described above, and their average life span is about four years.

Icebergs that form off the west coast of Greenland tend to drift northward, then westward, and finally southward in a two-year journey through Baffin Bay and Davis Strait before they begin to endanger the main shipping lanes in the North Atlantic. Their numbers vary greatly and average about 400 per year at a latitude of 48 degrees (chiefly from April through June), but the range is from more than 1000 to less than 10.

Sea ice forms at the ocean surface between 28° and 31°F (the lower temperatures are associated with higher salinities). Dissolved salts are excluded from the growing ice crystals; thus they become concentrated in the remaining liquid and may be trapped in openings within the ice. At lower temperatures, freezing is more rapid, and more brine is trapped. After a layer of ice has formed at the surface, it commonly thickens by growth at the top (from additions of snow) as well as at the bottom. However, average thicknesses in the Arctic range between 2 to 3.5 m (10 ft or so) because the ice insulates the water beneath it from further cooling. Also, some melting and sublimation occur at the surface during the warmer months. Currents and winds may jam this layer of ice into irregular ridges and hummocks. Repeated melting and freezing of raised portions of sea ice permit the trapped brine to escape. Thus ice that is a few years old commonly melts into fresh water; in fact, pools of fresh water may form in depressions on the surface.

Ice islands up to 10 miles or so in length have been discovered in the pack ice of the Arctic Ocean. They project some 20 to 30 feet above sea level and float with the currents in great sweeping curves through the Arctic Ocean. They have been occupied by teams of scientists who have measured and probed the ocean waters and sea floor beneath them. Ice islands may consist of detached pieces of shelf ice that formed along

15-9. A spectacular iceberg, probably once part of the Greenland Ice Sheet. *(Official U.S. Navy photograph.)*

a coast (not as part of a glacier) and became greatly thickened before they were detached. Perhaps they formed during a colder part of the Pleistocene and have been drifting around the Arctic Ocean ever since.

Ocean Currents

The waters of the oceans have been thoroughly mixed by currents that move vertically, horizontally, and at all depths. As evidence, the salinity of the deep ocean water is nearly everywhere the same, and its variation at the surface is relatively small. When viewed broadly and in the light of measurements averaged over a number of years, some of the currents form part of an orderly global pattern of movement. However, when synoptic measurements are studied (measurements made simultaneously at a number of locations) the pattern of simple, regular, large-scale movement is lost and in its place we find a number of smaller, changeable, complex, and sometimes more rapid movements. Thus ocean currents have both a climate (long-term pattern) and a weather (short-term changes).

Ocean Current Measurement

Ocean currents may be measured or calculated in a number of ways and with varying degrees of precision. Two techniques are widely used: if an object moves with a current, its path can be plotted; on the other hand, a current's speed and direction may be measured from a stationary instrument located within it. Such current meters may be used at different depths and involve the turning of a propeller; they have some similarity to a weather vane and anemometer.

Surface currents are the most readily studied and best known. Drifting objects indicate surface currents, if their movements are not also influenced by winds. Drift bottles have long been used and may be weighted so that they offer little wind resistance and can be released at different times and places. Each contains a card that a cooperative finder fills out and mails (a 10% response may be average). Data are thus collected inexpensively but do not show exactly when or by what route a bottle arrives on some shore.

The movement of a ship can be used to measure currents, particularly if it is near shore; e.g., a north-bound ship may be moving at a rate of 10 knots relative to objects on land, whereas its velocity relative to the water around it is 15 knots (1 knot is 1 nautical mile or about 6080 ft/hr). A southward-flowing current of 5 knots is thus indicated. Farther from land, a ship may proceed from a known location at a certain rate in a certain direction, and its path is mapped. If a subsequent navigational fix on a celestial object shows the map position to be incorrect, then currents have caused a deviation, and their average direction and velocity can be estimated after allowance for changes caused by winds. A dye or radioactive material may be released at a certain point, and samples subsequently collected in the surrounding area, but the material must be capable of detection when much diluted. Buoys have been designed that emit sound waves periodically and float at a specific depth (p. 376).

A method to measure deep bottom currents was discovered accidentally when a series of photographs was once made of part of the ocean floor (automatically at evenly spaced intervals). The camera rigging hit bottom and stirred up a cloud of sediment. This mass of suspended sediment then drifted out of the field of view, but its direction and rate of movement could be traced for a time from one photograph to the next.

Another method of determining ocean currents involves measuring the faint electric current produced by a mass of water moving at a certain rate through the Earth's magnetic field. The difference in potential between two electrodes towed behind a ship can be used to obtain a continuous record of currents the ship is passing through (geomagnetic electrokinetograph).

Very large water masses with diagnostic physical properties flow slowly through the oceans, and these are detected by measurements of their physical properties—temperature, salinity, oxygen content, and certain contained nutrients. Movement of such a mass can be traced by noting the slight changes that occur with increased distance from its source area (e.g., by a gradual decrease in oxygen content).

Wind and Density Currents

Winds are the main cause of ocean currents, particularly of surface currents, and the distribution of the major surface currents of the oceans (Fig. 15–10) coincides approximately with the global pattern of atmospheric circulation (Fig. 26–21). Winds may produce a surface current by exerting a stress directly on the water beneath them or less directly by piling water against a coast. Since the sea-level surface slopes gently downward away from this piled-up region, outward flow results. Wind-produced surface currents may converge in some areas and diverge in other areas, and this leads to sinking and upwelling respectively.

Other ocean currents are gravity-controlled movements initiated by density differences between certain regions (the so-called, deep-water, thermohaline circulation). Low temperatures and high salinities increase water densities, and a dense mass of water tends to sink because it exerts a greater downward pressure at any one depth than the water adjoining it. If water sinks in one area, other water beneath it must be displaced laterally, and water in some other area must rise. A convectional circulation develops.

Deflection

The Earth's rotation causes a deflection in the direction of motion that is to the right in the Northern Hemisphere and to the left in the Southern Hemisphere (Coriolis effect, p. 658). Thus surface currents tend to move at an angle of about 10 to 45 degrees to the winds that produce them (greater deflections may occur at higher velocities). The wind-moved surface layer in turn exerts a frictional force on the water immediately beneath it, and this water also flows laterally but is deflected at a still greater angle (the Ekman spiral, Figure 15–11). As this process extends downward through successive layers, the force is reduced, speeds decrease, but the deflection continues to grow. At a depth of 100 meters or so, water flows very slowly at an angle of 180 degrees (in the opposite direction) to the surface current. The average deflection for the bulk of this near-surface lateral flow (the so-called Ekman layer) is about 90 degrees. Such deflection of wind-driven surface currents can cause upwelling along some coasts (e.g., off California, Fig. 15–12).

Convergence of surface waters results from the wind pattern of a large, well-developed, high-pressure system in the atmosphere, and divergence is caused by a low. Since clockwise circulation occurs around an atmospheric high in the Northern Hemisphere, surface waters on all sides of the high are deflected at a 90-degree angle to the wind, and thus they converge toward its center. Therefore, subsidence tends to occur beneath a large high-pressure center, whereas up-welling often takes place beneath a low.

For a steady flow in a particular direction to develop, something must counteract the effect of the Coriolis deflection. This can be achieved if less dense water occurs on the right flank of a current (facing downcurrent in Northern Hemisphere). The sea surface is then higher above the less-dense water, and it slopes gently downward across the current. Therefore, it acts laterally to oppose the Coriolis deflection (a slope of 1 to 100,000 is sufficient for a 2-knot current).

Surface Currents

The trade winds and belts of westerlies combine over the Atlantic and Pacific Oceans to produce four gigantic, slow-moving, circular whirls or eddies (gyres). This is the generalized and simplified picture, the climatic

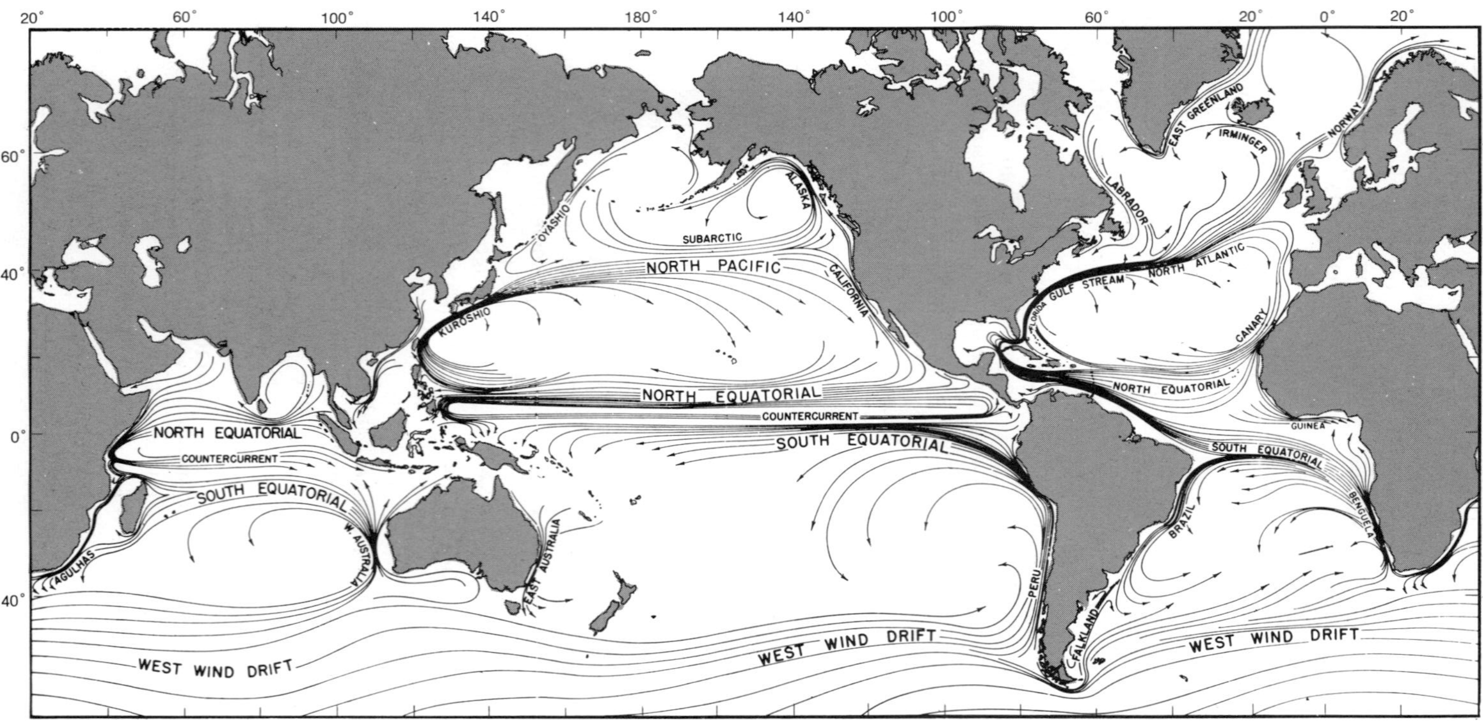

15-10. Major surface currents of the world. (Oceanography for the Navy Meteorologist, *U.S. Navy Weather Research Facility.*)

view of these ocean currents. The trade winds blow quite steadily within their zones and drag water westward to form the *north* and *south equatorial currents* (average rate less than 1 mi/hr). These shove water against the coastal areas on either side of the equator, and thus sea level rises slightly from east to west along the equatorial currents (about 3 inches per 1000 miles—the trade winds have been described as blowing uphill). Along the western margins of the oceans where water has been raised against the lands, it is deflected, and some flows eastward in a narrow belt between the two equatorial currents. This is the *equatorial countercurrent* which is located beneath the doldrums and tends to occur in the low latitudes of the Northern Hemisphere along the so-called meteorological equator.

However, most of the water moves poleward away from the equator, and we shall follow its path in the North Atlantic. Northward-moving water off Venezuela is separated by the West Indies, and part moves through the Caribbean into the Gulf of Mexico. Here it impinges on Florida and raises sea level by several inches on Florida's west coast. Water moves rapidly eastward and northeastward through the Florida Straits to form the *Florida Current.* Since the current extends far below the surface, its total flow exceeds the combined flow of all the rivers on Earth by some twenty-five times. The Florida Current is now joined by the remaining portion of northward-flowing equatorial water that had been diverted along the eastern side of the West Indies. Together they form the *Gulf Stream* which moves northeastward off the eastern margin of the United States. The Gulf Stream also has a third source of water—part of the equatorial current in the Southern Hemisphere which was deflected by the bulge of Brazil into northward and southward moving components.

The Gulf Stream is deflected eastward off Cape Hatteras where it encounters the belt of westerlies, and it spreads across the northern part of the Atlantic to form the somewhat diffused and branching *North Atlantic Current.* Part of this warm water continues northeastward and warms the coast of Norway and extends into the Arctic Ocean. Another part swings eastward and

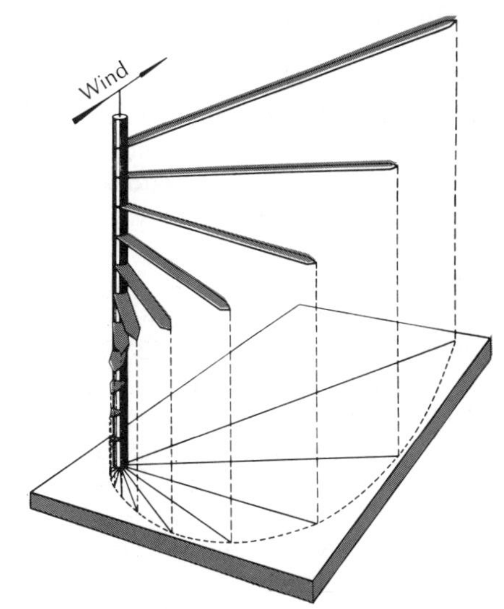

15-11. Ekman spiral. Markers project from a vertical post to illustrate the directions and velocities (proportional to the lengths of the arrows) of surface currents at different depths. (Oceanography for the Navy Meteorologist, *U.S. Navy Weather Research Facility.)*

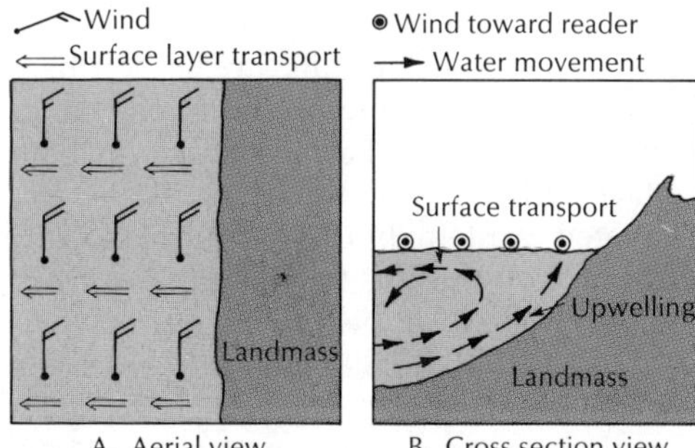

15-12. Upwelling. In the aerial view (left), wind is from the north, and surface water on the average is deflected toward the west. Colder water at greater depths thus wells up along the coast. Fogs may form if moisture-laden winds blow eastward onto the lands. (Oceanography for the Navy Meteorologist, *U.S. Navy Weather Research Facility.)*

southeastward (Canary Current) to complete a great clockwise circulation in the North Atlantic.

The Labrador Current flows southward in the western Atlantic and is one of several cold currents that move equatorward from the polar regions. Its boundary with the Gulf Stream (deep blue) may be abrupt and conspicuous (greenish and less transparent), and a ship astride the boundary once measured a temperature difference of 12°C between thermometers at its bow and stern. However, the Gulf Stream is not a current of warm water that flows through colder water on all sides. Water on its right flank (the Sargasso Sea) is warmer at all levels below the surface, and the two are nearly equal at the surface.

The Florida Current, Gulf Stream, and North Atlantic Drift do not form a single, fast-flowing, uniform, warm-water current. Rather they consist of numerous elongated tongues or filaments that interfinger and meander in a complex fashion and shift position from time to time. Neighboring fingers may be separated by currents that flow in the opposite direction. Thus a ship may navigate by a "climatic" map of the Gulf Stream and head eastward to take advantage of the following current. At certain times, however, it may be located in a current flowing in the opposite direction.

The Gulf Stream exemplifies the general intensification that occurs in the Northern Hemisphere along the western boundaries of the Atlantic and Pacific Oceans in both surface and deep-water currents. These concentrated western boundary currents seem somewhat analogous to the jet streams of the atmosphere. A deep-water current occurs some 2000 to 3000 meters beneath the Gulf Stream but flows in the opposite direction along the continental margin (Fig. 11–9). Nearly stationary water occurs at a depth of about 1500 to 2000 meters between these opposite-flowing currents. The currents that complete the surface swirls tend to be much less concentrated and more widespread.

The Sargasso Sea occurs within the clockwise gyral of the North Atlantic and is a large oval area of warm, slowly swirling water characterized by a relative scarcity of organisms and the presence of the sargassum weed (but scattered in patches so that it is not a shipping hazard). The Ekman-layer flow at a 90-degree angle to the wind direction causes convergence toward the center of the Sargasso Sea (the belt of westerlies lie to the north and the tradewinds to the south). Convergence in the Sargasso Sea leads in turn to a general subsidence beneath it, and upwelling is limited or absent in this vast area (about 2000 miles by 1000 miles) and mixing is slow. Thus the nutrients necessary for plant growth in the upper waters are not replaced rapidly enough to support an abundant supply of organisms, and the water tends to be a deep blue, which is one sign of an oceanic desert. Its salinity is high because evaporation is great, precipitation is low, and little fresh water is added by rivers or melting ice.

Water in the gyre center may be 3 to 4 feet higher than mean sea level. Such estimates in the open ocean are made indirectly by assuming that the pressure at a certain depth within the ocean is everywhere the same (e.g., 1 mile below mean sea level). It may be helpful to imagine vertical columns of ocean water pressing downward upon this level. The masses of the columns are equal (assume equal cross sections), and a column made of less dense water stands higher than a column made of more dense water. Since temperature and salinity tend to determine density, their measurement at different depths provide the basic data for calculating the vertical position of the ocean surface.

Somewhat similar sets of currents occur also in the South Atlantic (a counterclockwise swirl) and in the North and South Pacific. The Indian Ocean is located largely in the Southern Hemisphere, is affected by the Monsoons, and has a somewhat different pattern of surface currents. The Antarctic Ocean is unbroken by land barriers and has a steady, eastward-flowing current associated with winds that blow steadily from the west at latitudes of 40 to 60 degrees. This is the *West Wind Drift;* drifting objects moving at about 8 mi/day have completed one circuit in 3 to 4 years.

A strong undercurrent has been discovered rather recently in both the Atlantic and Pacific Oceans. It occurs within the westward-flowing south equatorial current and moves eastward below the surface along the equator. Its greatest speeds (up to 3 knots) occur at depths of 50 to 150 meters. This *equatorial undercurrent* may be some 300 meters thick and as much as 400 kilometers in width. At times it may extend upward to the surface.

Deep-Water Currents

Two major sources of deep-water currents exist, one in the polar region of each hemisphere. The salinity of chilled polar water is increased when ice forms, and thus it becomes dense and tends to sink. However, some Arctic water is partly diluted by additions of fresh water and tends to flow southward along the surface of the Atlantic until it meets the North Atlantic Drift. In this zone of convergence, the colder denser water from the north sinks and at depth it spreads out as a layer. Still denser water forms in the Antarctic Ocean. This sinks slowly to the bottom and then moves equatorward. Eventually some of it creeps along the ocean floor into the Northern Hemisphere. The Coriolis deflection tends to make such deep currents strongest along the western margins of continents. Such a current flows southward along the slope and rise beneath the Gulf Stream system and a counterpart flows northward in the South Atlantic.

Equatorward from the zone where Antarctic water subsides is a zone of heavy precipitation associated with the belt of westerlies. This decreases the density of the surface water, which tends to move equatorward because it is deflected to the left by the Southern Hemisphere belt of westerlies. Deeper, cooler, nutrient-rich waters well up in this zone of divergence, and organisms abound (e.g., whales). Such movements have so mixed and chilled the deep waters of the oceans that their salinities are nearly uniform and their temperatures are very low.

Another layer of subsurface water develops in the Mediterranean. In this warm dry climate, evaporation is great and salinity increases, especially in the isolated eastern portion. These saline waters are chilled during the winter, become even denser, and sink. They move westward along the bottom and pass into the Atlantic across the relatively shallow floor of the Mediterranean at Gibralter (the water is less than 1000 feet deep). Less dense surface water enters the Mediterranean at Gibralter to replace it. After flowing out of the Mediterranean, the cold saline water sinks into the Atlantic until it reaches an equilibrium level, and then it spreads slowly as a layer of water with diagnostic physical properties that has been traced southward around Africa and into the Indian Ocean.

Man-Made Currents

It has been proposed that upwelling on a small scale might be initiated artificially by means of a large, very long, vertical pipe with open ends. In some sections of the ocean, surface waters are warmer and have somewhat higher salinities than occur at depth. If one imagines that water is pumped steadily from the top of such a pipe under such conditions, then water flows upward within the pipe and becomes warmed as it ascends. When once-deep water finally reaches the top, it may be nearly as warm as the surrounding surface water. However, its lower salinity has remaind unchanged, and it is relatively less dense. The water within the pipe thus stands higher and overflows. Once such a deep-water siphon has been started, it might continue indefinitely without further pumping. Deep, nutrient-rich waters might thus be brought to the surface, which could result in a great increase in plant growth. In turn, this might lead through various food chains to an increase in the edible protein foods for man to harvest in the vicinity of the pipe. However, at present the deep-water siphon is an unproved, imaginative proposal.

Emery has described small but important man-made currents off the California coast that illustrate some of the conditions involved in the origin of major ocean currents: differences in density, lateral flow from high-to low-pressure areas, and Coriolis deflections. Warm waste water from sewage-disposal and power plants is discharged into the Pacific at Los Angeles and tends to form a very shallow lens of lightweight surface water. The central portion of the lens is thickest and thus is very slightly higher than mean sea level. This causes an outward flow that is deflected to the right to form a clockwise circulation around the center of the lens. Somewhat polluted water thus moves shoreward on the north side of the lens.

Water Masses

Although the deep waters of the oceans are nearly uniform, they are not identical in their physical properties. Many measurements at many depths have made it possible to recognize large sheetlike masses of water that differ slightly from other masses located above and below them and in other oceans. Since most of the physical properties of such water masses are produced by processes that operate near the surface, correlation between surface properties and deep-water properties may permit an oceanographer to trace the slow movement of a water mass. Furthermore, carbon-14 age determinations may be used to calculate its rate of movement. The oxygen content is of particular importance since it is added at or near the surface and steadily diminishes during movement away from the surface. Such water masses are somewhat analogous to the air masses that occur in the atmosphere. From top to bottom, oceanographers distinguish surface, upper, intermediate, deep, and bottom waters, and each of these is subdivided further into geographic units. Thus we may refer to Antarctic bottom water, North Atlantic intermediate water and Arctic surface water.

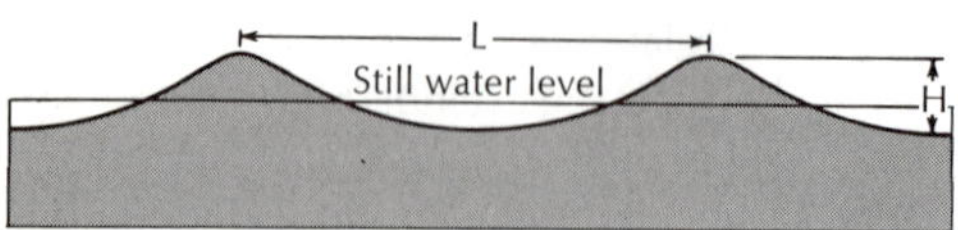

15-13. A typical sea wave. Note that crests are narrower and steeper than troughs and that the still water level is closer to the bottom of a trough than to the top of a crest. The wave crests are assumed to be parallel and oriented at a 90-degree angle to the wind direction. (American Practical Navigator, *U.S. Navy Hydrographic Office, 1962.)*

Waves

Waves and swell undulating gracefully across the sea surface or thundering and splashing against a coastline fascinate most people, whether they are regarded as things of beauty, agents of erosion, barriers to an amphibious landing, or transport for an exciting surfboard ride. The same waves at sea may produce a queasy feeling in the stomach and constitute a real hazard to shipping when they approach a maximum development. Most waves are born of the wind, and their shapes and dimensions are described by certain terms: crest, trough, length, height, and period (Fig. 15-13).

Long low waves (swell) in the open sea approach closest to this idealized picture. However, most waves exhibit various irregularities because they consist of a superimposed combination of a number of such regular waves, each of which has a different length, height, velocity, and shape. *Wavelength* is the horizontal distance between two adjacent crests or troughs, and *wave height* is the vertical distance from the bottom of a trough to the top of a crest. *Wave period* is the time necessary for a wave to move one wavelength.

The turbulence present in most winds soon changes a smooth water surface into a complex pattern of tiny hills and valleys, and the frictional surface drag of a wind blowing steadily in one direction soon organizes these into a regular series of waves, although the precise mechanism is uncertain. Smaller waves tend to disappear as they break or are overtaken by larger, faster-moving waves, and some of their energy is captured by the larger ones, which thus increase in size. New small waves are also produced. Air moving generally parallel with the sea surface and impinging upon the windward side of a ripple has one vector component that tends to push upward along the slope and thus to increase its height.

In *deep water* (defined as greater than one-half a wavelength) the form and energy of a progressive oscillatory type of wave move forward with the wind, but the water particles near the surface move in circular orbits and have only a slight net forward motion (Fig. 15-14). This was learned by observing the bobbing motion of small drifting objects and by experiments in glass-walled tanks in which waves were produced artificially. Tiny colored particles were added and filmed and revealed the oscillations at different depths. At the crest of a wave, water particles move forward in circular oscillatory paths aligned in vertical planes that parallel the direction of wave travel. Then they descend, move backward in a trough, and rise to form the succeeding crest. Near the surface the diameter of such a circular orbit is equal to the height of a wave, but it decreases rapidly at depth. At a depth equal to about one-half a wavelength, the circular orbital motion is so small that it can be disregarded. Unless waves are very long relative to their height, they do not exhibit a regular sinusoidal shape. Rather, they are so shaped that troughs are wider than crests and also closer to the still-water line (they form a trochoidal curve).

In shallow water (Fig. 15-15), bottom interference changes the oscillatory pattern into an ellipsoidal one, the orbits flatten out, and particles near the bottom move back and forth about in a horizontal plane. When a wave breaks, both water and wave energy move forward against the shore (called a wave of translation, see below).

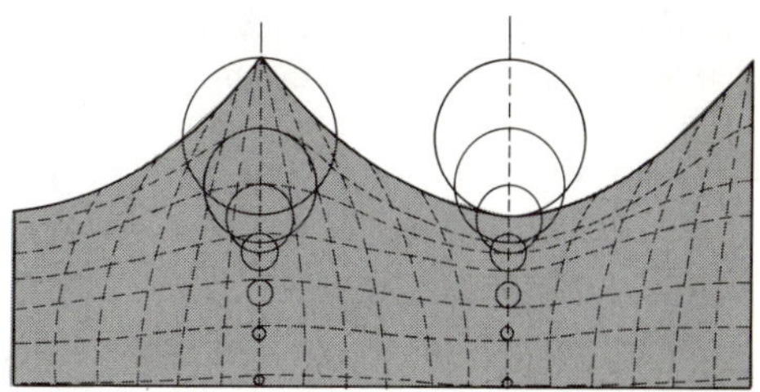

15-14. Oscillatory motion of water particles in a wave in deep water. *(After Gaillard.)*

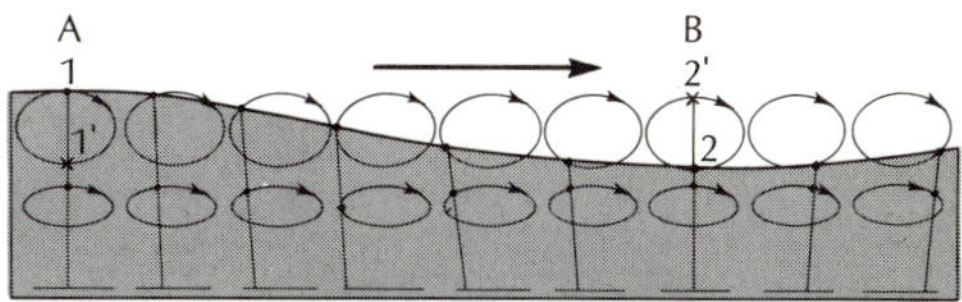

15-15. Paths of water particles in waves in shallow water (i.e., depth is less than half the wavelength). Arrows point in direction of motion. Half a wave period after present positions the water particles at locations 1 and 2 will have shifted to 1′ and 2′, respectively, and the wave crest will have moved from A to B (a trough wil then be at A). (P. Groen, *The Waters of the Sea,* Van Nostrand Reinhold, 1967.)

Wave Interference

In a windy area where waves are generated, waves of different length and velocity are produced. These travel in different directions and create a confused, chaotic, wave-disturbed surface because one wave interferes with another (Fig. 15-16). *Con-*

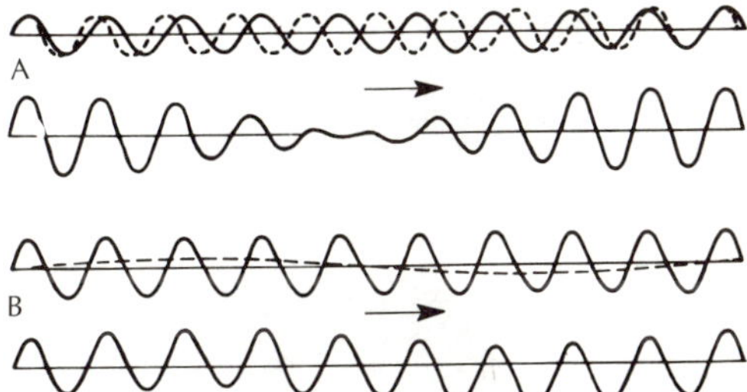

15-16. Wave interference. In (A) two waves of equal height and nearly equal length travel in the same direction and produce the wave pattern shown. In (B) short waves and long swell (dashed line) interfere. (American Practical Navigator, *U.S. Navy Hydrographic Office, 1962.)*

structive interference occurs when the crest of one wave arrives at a certain place at the same moment as the crest of another wave. The two combine their energies to produce a higher wave, and perhaps the wave breaks to produce a white cap. However, if the crest of one wave arrives at a spot simultaneously with the trough of another wave *(destructive interference),* the wave height is decreased, or perhaps the wave disappears entirely at that spot. Waves may approach a shore from different directions because they formed in different areas, and thus the sizes of successive waves may vary as interference occurs. To study such composite waves, oceanographers separate them into their component parts: a number of simple, regular waves each of a certain size and traveling in a certain direction at a certain velocity (Fig. 15-17). The effect of each on the actual composite wave must then be estimated statistically.

15-17. An irregular wave pattern (bottom) generally consists of a large number of superimposed simple waves (a wave spectrum) that differ in length, height, speed, and direction. *(Courtesy U.S. Naval Oceanographic Office.)*

Wave Velocity and Size

Waves of longer wavelength move faster than waves of shorter wavelength, and thus they become separated as distance increases from a source area. Waves may move out of a windy area where they originated and are then called *swell* (*"sea"* refers to the waves within the source area). Wave energy may thus move as swell for hundreds of miles along the ocean's surface (more than 5000 miles in some instances, and even across the equator from one hemisphere to the other). Origin in some distant storm, therefore, accounts for the large waves (actually swell) that may thunder against a beach on a day with light winds or with wind direction not aligned with wave direction. Swell is less common on the east coast of the United States than on the west coast where prevailing winds blow off the ocean. The distance to a source area possibly can be calculated by noting the time that elapses between the arrival of swell waves of different sizes: the longer the time interval, the more distant the source area (also by matching swells with the known locations of storms at particular times as shown on photographs made from orbiting spacecraft).

When wavelength is measured in feet, velocity in knots, and the period in seconds, the following approximate relationships exist (Fig. 15-18 and below; $\approx$ means proportional to):*

* *Sea and Swell Observations,* U. S. Navy H. O. Pub. No. 606e, p. 2.

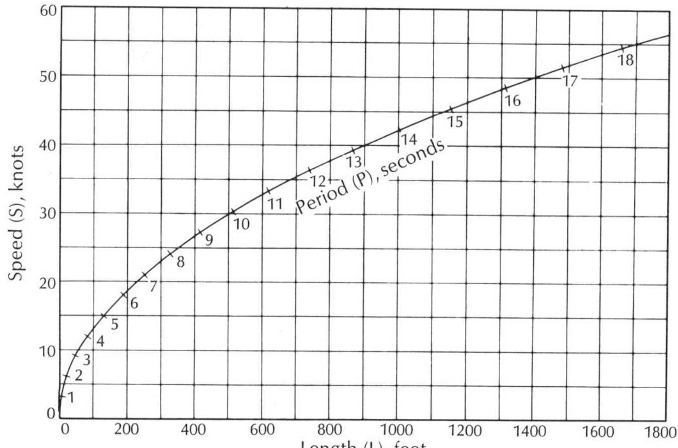

15-18. This relationship between the speeds, lengths, and periods of waves in deep water is based upon the theoretical relationship between period and length. *(American Practical Navigator, U.S. Navy Hydrographic Office.)*

$$\text{Velocity} \approx 1.3 \cdot \sqrt{\text{length}} \approx 3 \cdot \text{period}$$
$$\text{Length} \approx 0.6 \cdot (\text{velocity})^2 \approx 5 \cdot (\text{period})^2$$
$$\text{Period} \approx 0.4 \cdot \sqrt{\text{length}} \approx 0.3 \cdot \text{velocity}$$

As waves move outward from a windblown area where they formed, their periods, wavelengths, and velocities all increase. However, a group of waves travels at about one-half of the velocity of the individual waves making up the group (Fig. 15-19). Groen has likened this to the propagation of a swell which advances in ever-widening circles from a stormy source area across a relatively calm sea. However, the front of this swell advances at group velocity. Therefore, individual wave crests move through the swell in succession to become the leading wave; but beyond the front, each in turn loses height and soon disappears. Thus the waves (swell) that beat upon a shore are only the distant descendants of those formed in a remote generating area.

The energy of a wave is about half potential (involves the elevation and depression of the sea surface) and half kinetic (involves the circular motions of water particles beneath the waves). In deep water, the bulk of this kinetic energy remains with the circling water particles and is not transferred forward as is the potential energy. However, the leading wave at a front must set in motion the quiet water ahead of it, which uses up some of its potential energy. Therefore, the amount of potential energy available to produce wave motion becomes progressively smaller with increasing distance beyond the wave front, and the advancing waves decrease in size and eventually disappear.

Reports as to the sizes of large waves tend to exaggerate, but wave size can be very great; commonly, size refers to wave height. According to a navy veteran, during one violent storm at sea, a flight of stairs on his ship did not maintain its normal inclination but shifted from the nearly vertical to the nearly horizontal! Perhaps wave heights exceeding 100 feet have been observed (Fig. 15-20). The height of a wave depends upon the velocity of the wind, the duration of the wind, and the length of

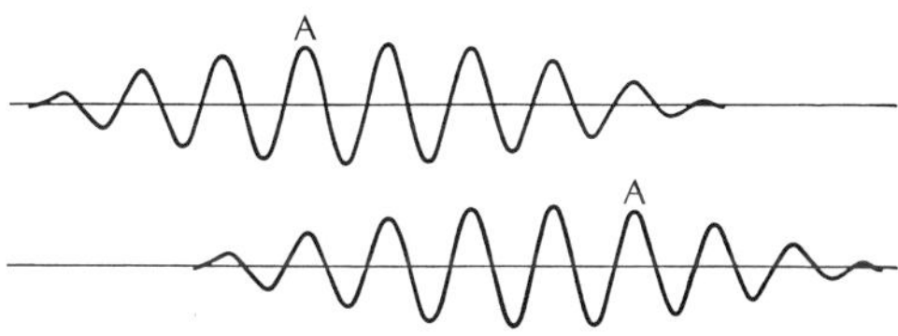

15-19. A group of waves moves at half the speed of a single crest. In this example, the entire group moves two wavelengths in the same time that a single crest (A) moves four wavelengths. *(P. Groen,* The Waters of the Sea, *Van Nostrand Reinhold, 1967.)*

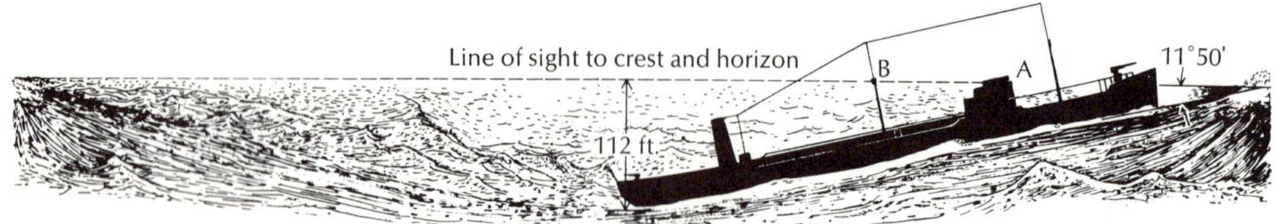

15-20. A wave 112 feet high, perhaps the largest ever measured in the open sea, was observed from a navy tanker in the Pacific in 1933. An observer at A looked toward the stern and saw the crow's-nest at B in his line of sight to the crest of a wave which had just come in line with the horizon. *(W. Bascom, "Ocean Waves,"* Scientific American, *Aug. 1959.)*

open sea across which the wave can move. Up to a point, waves increase in size with increases in each of these three factors. However, they tend to increase in size rather rapidly at first and then only gradually. Generally, the duration and fetch are too limited for waves to reach the maximum sizes that very strong winds can produce. In addition, such winds tend to blow off the wave tops.

If waves strike a ship in phase with its roll (side to side) or pitch (bow to stern), they may increase this to-and-fro movement to dangerous proportions. Thus they may act on a ship in the way that one pushes a child on a swing. Small, well-timed, periodic pushes make the swing go higher and higher. Another danger to a ship occurs if the bow is lifted by one wave at the same time that the stern is lifted by another. The middle of the ship may sag over the intervening trough.

A scale (Beaufort, Table 15-2) has long been used for estimating wind speeds at sea and on land. However, wave size is determined by factors such as duration, fetch, tides, and precipitation as well as by wind speed.

Breakers and Wave Refraction

Waves or swell encounter shallow water as they near a shore, and certain important changes occur that alter the direction of approach (*refraction*) and cause the waves to become steeper. Each wave in its turn becomes oversteepened, breaks, and is transformed into a turbulent mass of actively advancing water (*surf*) that is capable of eroding the sea floor, producing shore currents of several kinds, and transporting sediment (Figs. 15-21 and 15-22). Generally, just before breaking, the wave height increases, the wavelength and velocity decrease, whereas the period remains the same. During these changes, the velocity of the upper part of a wave becomes greater than that of the wave as a whole, and this unsupported upper portion then topples forward. A wave tends to break before the wave height equals one-seventh of the wavelength. Long low waves may thus double or triple in wave height, and breakers more than 40 feet in height have been measured.

In some breakers (*plunging*) the top of an oversteepened wave curls over and collapses abruptly around a trapped cylindrically shaped mass of air which is compressed and then expands violently to disrupt the wave. This disturbance progresses steadily along the crest of the wave. It tends to occur where the water changes depth rapidly. In other breakers (*spilling*) the wave may break here and there along a crest and do so gradually over a considerable distance.

Energy derived from the atmosphere over a wide expanse of sea, and transferred for many miles through its upper waters, is thus finally concentrated upon the edge of the land as a thin, knife-edged surf zone. Furthermore, refraction concentrates wave energy more on certain sections of a coast than on others (Fig. 15-23). Waves approach-

TABLE 15.2
BEAUFORT SCALE WITH CORRESPONDING SEA STATE CODES[a,b]

BEAUFORT NO.	WIND SPEED (MPH)	WIND SPEED (M/SEC)	WIND SPEED (KM/HR)	SEAMAN'S TERM	ESTIMATING WIND SPEED: EFFECTS OBSERVED AT SEA	ESTIMATING WIND SPEED: EFFECTS OBSERVED ON LAND	DOUGLAS SEA SCALE: TERM AND HEIGHT OF WAVES (ft)
0	Under 1	0.0–0.2	Under 1	Calm	Sea like mirror	Calm; smoke rises vertically	Calm, 0
1	1–3	0.3–1.5	1–5	Light air	Ripples with appearance of scales; no foam crests	Smoke drift indicates wind direction; vanes do not move	Smooth, less than 1
2	4–7	1.6–3.3	6–11	Light breeze	Small wavelets; crests of glassy appearance, not breaking	Wind felt on face; leaves rustle; vanes begin to move	Slight, 1–3
3	8–12	3.4–5.4	12–19	Gentle breeze	Large wavelets; crests begin to break; scattered whitecaps	Leaves, small twigs in constant motion; light flags extended	Moderate, 3–5
4	13–18	5.5–7.9	20–28	Moderate breeze	Small waves, becoming longer; numerous whitecaps	Dust, leaves, and loose paper raised up; small branches move	Rough, 5–8
5	19–24	8.0–10.7	29–38	Fresh breeze	Moderate waves, taking longer form; many whitecaps; some spray	Small trees in leaf begin to sway	
6	25–31	10.8–13.8	39–49	Strong breeze	Larger waves forming; whitecaps everywhere; more spray	Larger branches of trees in motion; whistling heard in wires	
7	32–38	13.9–17.1	50–61	Moderate gale	Sea heaps up; white foam from breaking waves begins to be blown in streaks	Whole trees in motion; resistance felt in walking against wind	Very rough, 8–12
8	39–46	17.2–20.7	62–74	Fresh gale	Moderately high waves of greater length; edges of crests begin to break into spindrift; foam is blown in well-marked streaks	Twigs and small branches broken off trees; progress generally impeded	
9	47–54	20.8–24.4	75–88	Strong gale	High waves; sea begins to roll; dense streaks of foam; spray may reduce visibility	Slight structural damage occurs; slate blown from roofs	High, 12–20
10	55–63	24.5–28.4	89–102	Whole gale	Very high waves with overhanging crests; sea takes white appearance as foam is blown in very dense streaks; rolling is heavy and visibility reduced	Seldom experienced on land; trees broken or uprooted; considerable structural damage occurs	Very high, 20–40
11	64–72	28.5–32.6	103–117	Storm	Exceptionally high waves; sea covered with white foam patches; visibility still more reduced	Very rarely experienced on land; usually accompanied by widespread damage	Mountainous, 40 and higher
12–17	73–136	32.7–61.2	118–220	Hurricane	Air filled with foam: sea completely white with driving spray; visibility greatly reduced		Confused

[a] Since January 1, 1955, weather map symbols have been based upon wind speed in knots, at 5-knot intervals, rather than upon Beaufort number.
[b] From U.S. Navy Hydrographic Office (1958).

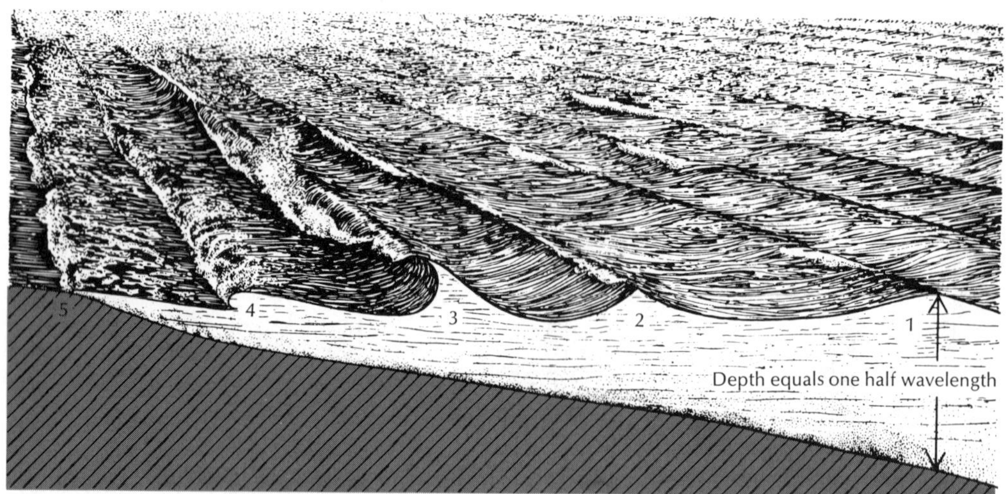

15-21. A wave breaks against a shoreline when swell moves into water shallower than half the wavelength (1). At (2) the shallow bottom has raised the wave height and decreased the wavelength. The wave form breaks at (3) where particles of water no longer have room to complete their oscillations. A foam line forms at (4) and water particles move forward. The low remaining wave moves up the face of the beach as a gentle wash (5). *(W. Bascom, "Ocean Waves,"* Scientific American, *Aug. 1959.)*

ing a shore at an oblique angle tend to bend and become more nearly parallel to it. Bending occurs because the part closest to shore is slowed first (assumption: water depth increases uniformly with increased distance from shore), whereas the part farther out in deeper water continues to move with undiminished velocity. Wave refraction concentrates wave energy against the point of a peninsula (promontory, headland) and on the shore adjacent to a submerged ridge that is oriented at a steep angle to the shoreline. In some instances, breakers may be ten times higher over such a ridge than above an adjacent valley. On the other hand, wave energy becomes less concentrated near the head of a bay or above a submarine valley. Knowledge of these relationships among wave energy, refraction, coastal irregularities, and underwater topography is necessary in selecting wisely a site to anchor a ship, build a pier, or construct some near-shore building.

15-22. Wavelength decreases as shore is approached. (Sea and Swell Observations, *U.S. Navy Hydrographic Office Pub. No. 606-e.)*

Internal Waves

Internal waves may develop and move along the boundary between two layers of water with different physical properties. Such waves may be very large (wave heights of a few hundred feet) and yet occur in deep water and have little effect upon the

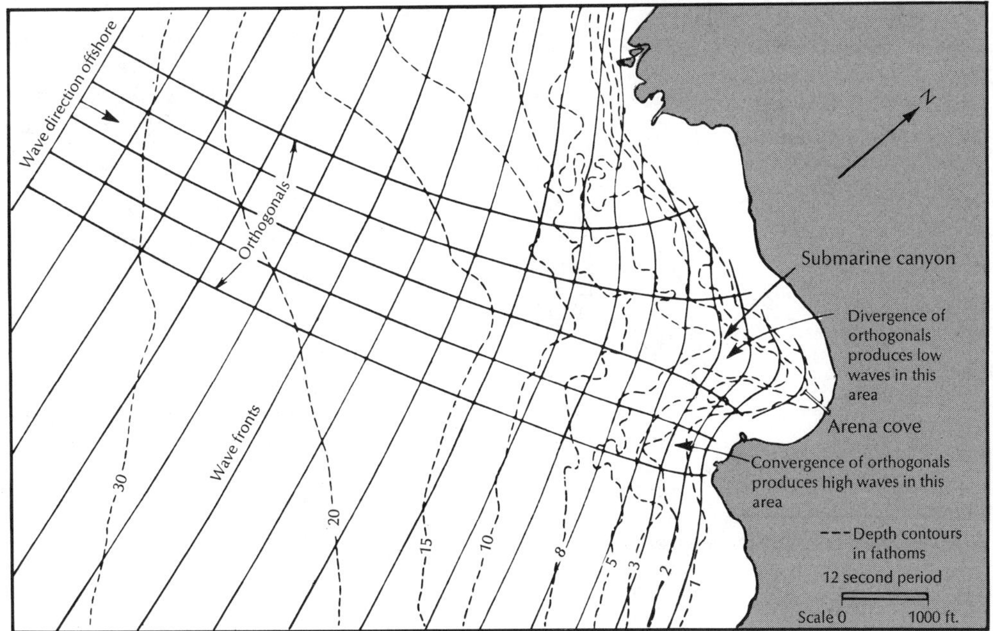

15-23. Variations in bottom topography cause wave convergence and divergence. The orthogonal lines are drawn perpendicular to wave fronts and are evenly spaced in deep water. Thus equal amounts of wave energy occur between them. Note how energy is concentrated on a headland and diffused across a cove. (American Practical Navigator, *U.S. Navy Hydrographic Office.)*

surface, although some slicks may be produced by them. On a small scale, such waves may be produced experimentally in a laboratory tank. Internal waves can be detected by observations made from an otherwise stationary submersible and by making a series of temperature measurements at a certain spot (the two layers commonly have different temperatures). Therefore, as a wave passes any one point, it causes large oscillatory orbital motions in the water near the boundary, and thus temperatures rise and fall periodically. Perhaps such waves account for ripple marks that have been photographed on the tops of guyots in water a few thousand feet deep.

Tsunamis

The most destructive sea waves are those produced by major earthquakes originating beneath or near the ocean at depths less than 30 miles (50 km), and less frequently by submarine volcanic eruptions, and perhaps by huge slides on the sea floor. These are called *tsunamis* or *seismic sea waves* (also "tidal waves," a misleading term, since tides do not cause them). Although most submarine earthquakes do not produce tsunamis, some do. Perhaps some result from the sudden dropping or upthrusting of part of the ocean floor along a fault; the overlying water is thereby disturbed violently all of the way to the surface. The waves thus produced may measure 100 miles or so in length and yet be only a few feet high in the open ocean. Their velocities may equal 450 mi/hr. At any rate, the path of a tsunami across an ocean tends to resemble that of the ripple formed by tossing a stone into a wide shallow body of water (Fig. 15-24).

Some tsunamis apparently consist of a single crest preceded by a broad trough,

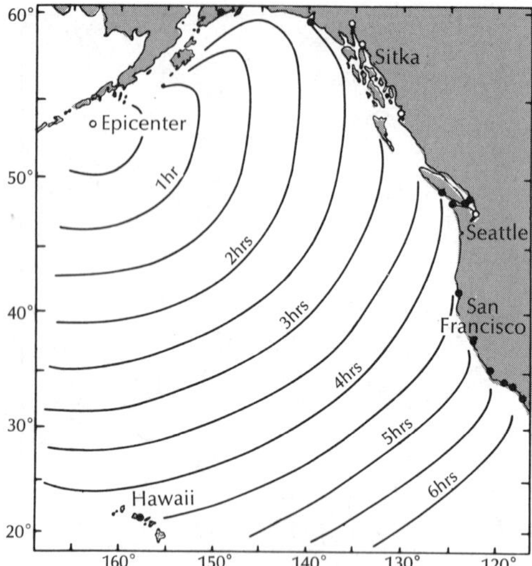

15-24. Wave-front propagation for tsunami of 1 April 1946 (Fox Islands). Large waves were recorded at some tide stations (solid circles) and not at others (open circles). *(Courtesy R. W. Fairbridge, ed.,* The Encyclopedia of Oceanography, *Van Nostrand Reinhold, 1966.)*

whereas others include a series of waves. Furthermore, the first wave is not always the largest. Such a wave can pass unnoticed beneath a ship at sea or even beneath one in a harbor. In one tsunami, people on such a ship were startled when they saw the adjacent shore attacked by a monstrous wave tens of feet high. It had been lower and undetected when it passed beneath them. Such a wave forms when a long, low, fast-moving tsunami reaches shallow water, slows down, becomes shorter, and has its height increased many times as water is piled against the shore. The configuration of the sea floor and of the coastline results in greater damage by a tsunami at some places (e.g., over an undersea ridge or against an exposed headland) than at others. Coral reefs make good barriers and a submarine valley in the path of a tsunami diverges its energy.

Some tsunamis have caused more damage at distant locations than at places closer to the source, probably owing to local conditions such as those just listed. On Hawaii in 1946, one wave struck the far side of the island (the side opposite the source area near the Aleutians) some 18 hours after the same tsunami had arrived at the near side. Presumably it was reflected by one or more topographic barriers after it passed by Hawaii.

Tsunamis are most abundant in the Pacific, which has an unstable margin that includes deep-sea trenches and island arcs, but they occur also in other oceans. The arrival of a monstrous tsunami is commonly preceded by a broad trough that causes water to withdraw far from the beaches. People have walked onto the newly exposed sea floor only to be overwhelmed some minutes later by rapidly advancing water. A network of warning stations has been established which records earthquake waves that might be associated with the origin of a tsunami. Since these travel at several miles per second, they arrive before any tsunamis that might have originated simultaneously. It has been suggested that the most disastrous tsunamis have not yet occurred. Although tsunamis in the future may not actually be larger than those of the past, people by the thousands are crowding into coastal areas located only slightly above sea level, and thus the potential for a major catastrophe grows. Recent evidence indicates that tsunamis may cause more erosion during their sporadic occurrences than do ordinary waves which function continuously.

In July 1958 a 100-foot wave was produced by an earthquake near the head of Lituya Bay on the south coast of Alaska. Three fishing boats were located about 6 miles from the place of origin and two were lifted by the advancing wave over 80-foot trees that grew on a spit behind them. Two of the boats sank, but the crew of one escaped. Along the sides of the bay, trees less than 100 feet above sea level were knocked down. At one place a large landslide occurred and caused a swash that climbed 1700 feet up a nearby ridge. Water still ran down the ridge when observed from a plane a few hours later.

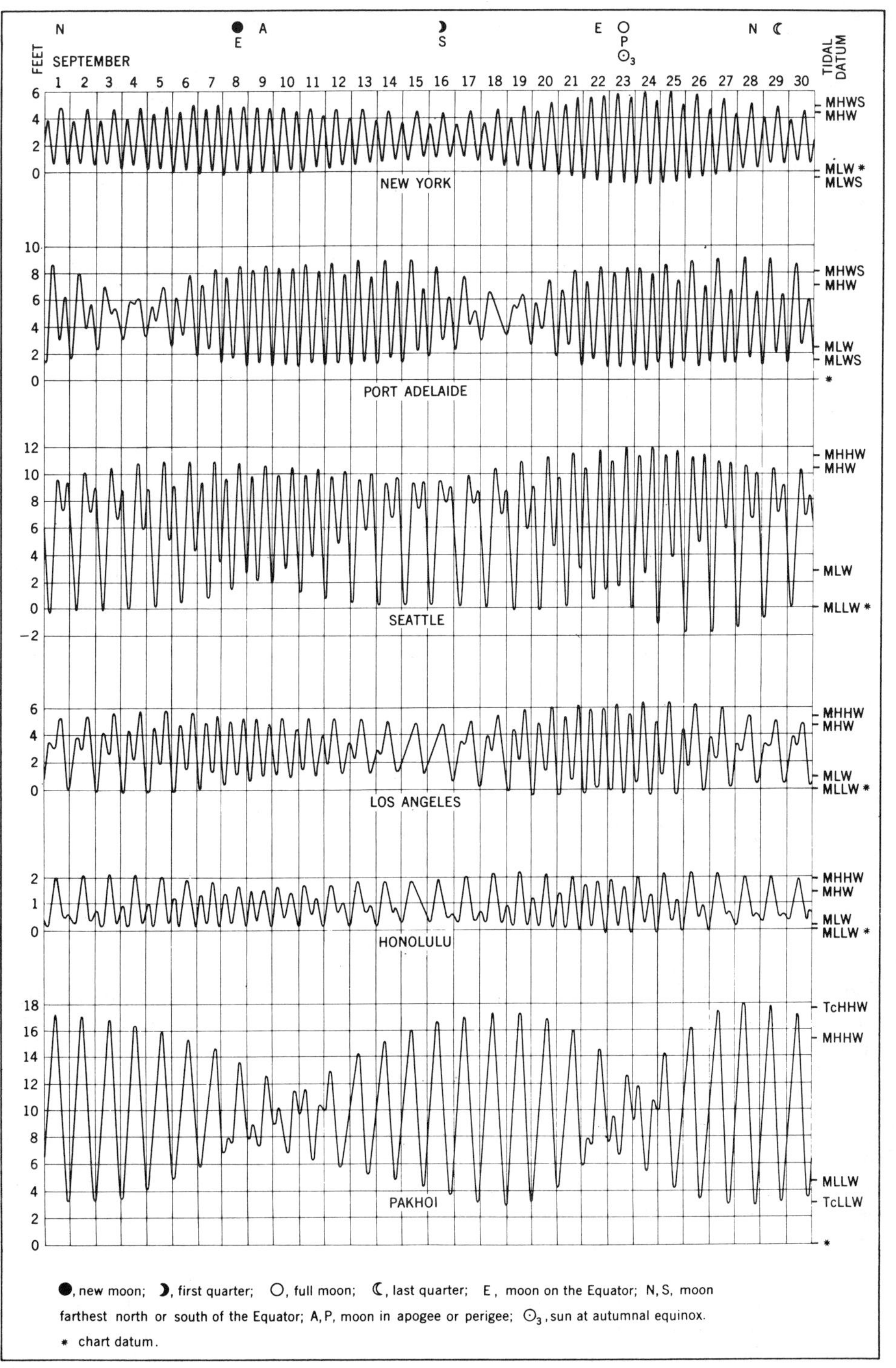

15-25. Tidal variations at different locations during a month. Note effects of moon's phases and distance. MHWS stands for mean high-water surface. (American Practical Navigator, *U.S. Navy Hydrographic Office.*)

Tides

Although two high tides and two low tides are common every 24 hours and 50 minutes, the intervals between two successive high tides may be quite different from 12 hours and 25 minutes, as may the heights of successive low or high tides (Fig. 15-25). High tides are produced simultaneously on the near and far sides of the Earth by the gravitational attraction of the moon (see Chapter 22). However, tides tend to be most extreme when the effect of the sun reinforces that of the moon at times of new and full moon, especially when the moon is closest to the Earth. Although the surface in mid-ocean may rise and fall only a few feet, the funneling effect of bays and other factors cause a much greater change in sea level along certain coasts, and changes as much as 70 feet have been reported.

Erosion and Deposition Along the Edge of the Sea

The wind-derived energy of waves and swell is finally expended against the edge of the sea by the turbulent waters of the surf which cut laterally into the lands. Surf originates at the line of breakers where the oscillatory wave form collapses and the water in a relatively narrow and thin zone advances actively upon the lands. It then retreats as part of a cyclic to-and-fro movement that is repeated many times in a single hour. Erosion is accomplished by rock fragments that are slammed against a shore and against other sediments, by the lifting effect of the turbulent waters, and by compression of air in cracks in coastal rocks; such greatly compressed air widens and extends the cracks. The energy that is concentrated against a shore during a major storm can be enormous (varies as the square of the wave height); e.g., a 135-pound boulder was once tossed high enough to fall downward through a lighthouse roof 100 feet above sea level. Perhaps this energy totals several thousand pounds per square foot at maximum. Therefore, more erosion may occur along a coast during a single major storm than takes place at other times when less-energetic waves function continuously.

Lateral planation by the surf causes cliffs to form and retreat inland. In places, and only temporarily, isolated remnants of former lands exist as stacks and arches (Fig. 15-26). A beveled, surf-eroded bench or terrace forms essentially at sea level, is partly visible at low tide, and widens as the surf-undercut cliffs retreat inland. Subsequent uplift may occur and elevate such sea level surfaces. Moreover, if times of uplift have been interspersed with times of crustal stability, then a series of surf-eroded terraces form (Fig. 4-5). Cliffs retreat at different rates depending upon their resistances to erosion, the stability of the crust, the energy of the surf, and the manner in which this energy is concentrated at certain points. Probably the great bulk of lateral planation occurs in water that is some 30 feet or less in depth. Thus it occurs essentially at sea level. The inland extension of a surf-eroded bench is limited unless sea level rises very slowly, because more of the wave energy is expended in crossing the ever-widening bench.

However, some coral terraces have formed in a different manner (Fig. 15-27).

Beaches

A beach (Fig. 15-28) is probably the most familiar geologic coastal feature. It is formed by a mass of sediments that has been piled up against a coast and includes a submerged portion that extends across the surf zone out to depths of about 30 feet. The sediments in a beach tend to shift along a shore as well as back and forth as the surf advances and retreats. Thus any given beach tends to change its dimensions from one season to the next, and the individual particles in a certain beach this year tend not to be the same particles at this same site a year or more ago.

Beaches commonly consist of sand-sized

15-26. Sea stacks sculptured from flat-lying sedimentary rocks at Bay of Islands, Australia. The exposed portion of a sandy beach is partially hidden in strong shadows along the base of the surf-eroded cliff. As time passes, the stacks become smaller and the line of cliffs retreats. *(Courtesy U.S. Air Force.)*

particles among which quartz predominates (both abundant and resistant), but some beaches consist of pebbles and cobbles. Other beaches are formed chiefly of coral fragments, and one (located near a coastal dump) consists chiefly of battered cans. Although most beaches have colors like those of New York, Florida, and California, certain beaches are made of black sands derived from volcanic debris. Some beach sediment comes from erosion of the adjacent coast, but much is brought to a coast by rivers. Surf and currents combine to spread most of the sediment parallel to a coast, but some, especially the finer particles, does move seaward. The slope of a beach depends partly upon the sizes of the particles composing it: steeper slopes are generally associated with coarser particles. A *berm* is a flat or nearly flat section of the exposed portion of a beach. More than one berm may occur, and the highest ones are formed by the exceptionally large waves that occasionally break against a shore.

A low, submerged ridge or bar, or a series of such ridges with intervening troughs, may occur within the surf zone of coastal areas with rather gently sloping sea floors, particularly during the winter storm season. Such a ridge is elongated parallel to the shoreline and may extend for many miles. How bars form is uncertain, but they are

15-27. A series of uplifted coral terraces in the Marianas. These did not form by surf erosion. Rather, corals and other organisms built a terrace upward to sea level, and uplift followed. *(Courtesy U.S. Air Force.)*

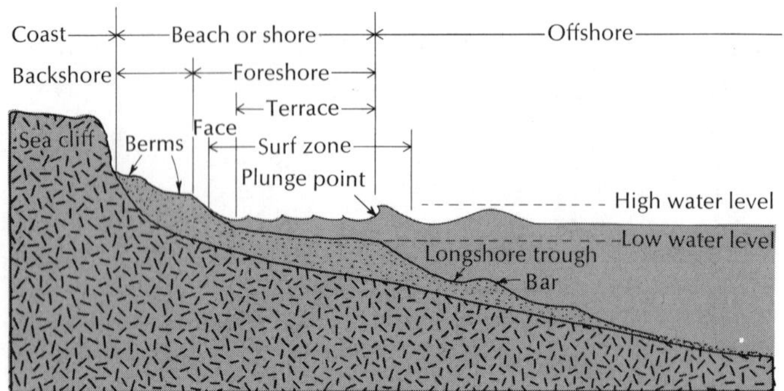

15-28. Profile of the beach and nearshore region modified after Inman (1962). According to Dietz, the principal effect of waves is to push sand against a sea cliff rather than to construct a wave-built terrace. An active wedge of beach sediment is shown pinching out offshore into a relict shelf. *(R. S. Dietz,* Geol. Soc. Am. Bull. **74,** *983, Aug. 1963.)*

involved in exchanges of material between the exposed part of a beach and its submerged portion. When particularly large waves strike a beach (more common during winter), the berm is eroded, and its sediment is transported outward by strong seaward flow. This sediment accumulates to form a bar near the outermost line of breakers and may also pile up to form other bars closer to shore. The outermost bar separates passing waves by size. The largest waves break above it, whereas the smaller ones do not. Then these smaller waves in their turn break along other bars in still shallower water closer to shore. Furthermore, large waves that break in deeper water may reform as new and smaller waves (less energy is available now) that break closer to shore.

During the season of small waves (commonly summer), the sediment in the bars is shifted shoreward, troughs are filled in, and the berm is widened by addition of material on its seaward slope (Fig. 15-29). The exposed shoreward portion of a beach may thus have a very different shape in winter (much reduced in size) than in summer. Sand washed out to sea during one storm may be gradually returned before the next one, although very powerful waves (e.g., tsunamis) may shift some beach sediment out into deep water, where it remains.

The permeability and degree of saturation of the sediment in a berm is also a factor in its growth or shrinkage. Sediment-laden surf water surges up the slope of a berm, and some water commonly sinks into the sand. The reduced amount of water in the backwash (although it now moves down-

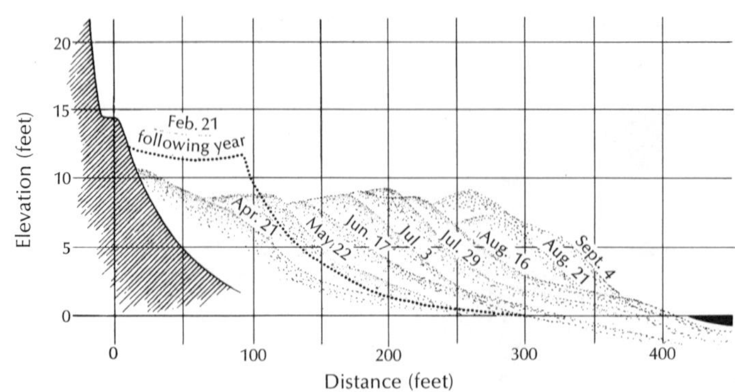

15-29. Growth of a berm at Carmel, California is shown during spring and summer. The profiles are based upon actual measurements. The vertical scale is exaggerated 10 times. The dotted line shows how the berm was cut back during the storms of the following winter. *(W. Bascom, "Beaches,"* Scientific American, *Aug. 1960.)*

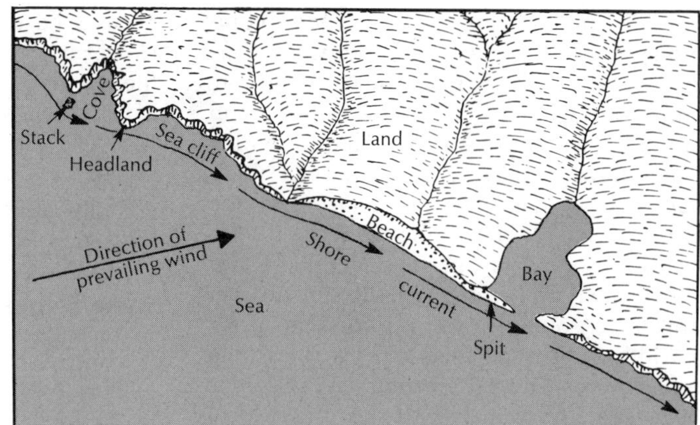

15-30. Shoreline features produced by erosion and deposition.

slope) may mean that some of the sediment just carried up the slope cannot be transported back down it. If considerable water penetrates the surface, less sediment can be carried by the backwash, and the berm grows wider. On the other hand, if the water cannot penetrate below the berm surface, then more sediment will be transported by the backwash than was carried during the uprush, and the berm is gradually reduced in size.

Despite refraction, most waves approach a shore diagonally, and particles tend to be washed obliquely up the slope of a berm. However, the backwash moves the particles directly down a slope. Thus sand grains make repeated, zigzag trips to and fro across a beach and are gradually shifted along it. Ripple marks are common on beaches.

Longshore Currents

Longshore (littoral) currents may occur in the surf zone and tend to flow parallel to a shoreline. They may form where waves approach a shore obliquely (Fig. 15-30) or where refraction has piled excess water against some section of a coast (water flows along the shore in opposite directions away from this zone). Although these currents are rather slow, they can transport sediments kept in suspension by the turbulent surf.

Rip currents (Fig. 15-31) may occur here and there along a beach where the backwash has been concentrated and eroded

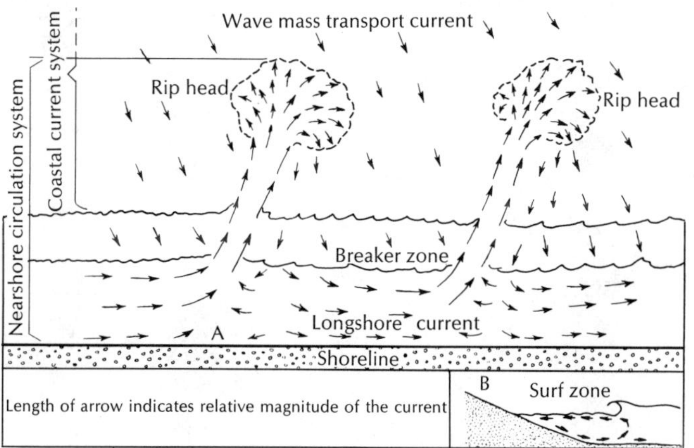

15-31. Rip currents. *(Oceanography for the Navy Meteorologist, U.S. Navy Weather Research Facility, 1960.)*

gaps in the longshore bar (or bars) commonly present in the surf zone. Coastal irregularities may also deflect a longshore current away from shore to produce a rip current. Agitated, sediment-laden, muddy water flows rapidly seaward through these gaps, faster than one can swim, and the current extends from the surface to the bottom. Beyond this narrow neck the water slows down and fans out along the surface. This larger, outer, slower portion may attain a width of one-half mile and extend a mile from shore. The rip current is dangerous because it can carry a swimmer out to sea, even as he swims vigorously toward shore. However, no "undertow" exists here, and the swimmer can save himself by moving across the current, which is quite narrow. A single rip current does not last long (several minutes to a few hours) but new ones may form in succession, and the locations shift somewhat.

Some Depositional Features

A *spit* is a narrow ridge of beach-type sediment that extends from land out into a body of water (Figs. 15-30 and 15-32). Spits are

15-32. Part of the Australian Coast. Note the following depositional features: beaches and a spit along the present shoreline; old beaches to the left of the present one; and several former spits near the center of the photograph (note hooked ends). *(Courtesy U.S. Air Force.)*

15-33. Shoreline features in Alaska. Note the wide flat-floored valley that extends from the center toward the top right and the muddy river water flowing into the ocean. Though sediment-laden, the river water is less dense than the salt water and tends to flow on top. Bars, beaches, and spits can be seen. *(Courtesy U.S. Air Force.)*

submerged at their open ends and commonly form where a beach has been extended into deeper water by longshore currents. Commonly this takes place at a bend in a shoreline, e.g., at the entrance to a bay. Spits tend to curve or hook rather sharply at their outer ends. Sand ridges or bars (bay-mouth bars) may extend across or nearly across the mouth of a bay. Perhaps such a bar originated as a spit and was gradually lengthened. Tidal or river currents may be powerful enough to maintain an opening at one side of a bay (Fig. 15-33). Much larger ridges known as *barrier islands* are located at a considerable distance offshore, extend parallel to a shore, and enclose a lagoon. Examples occur intermittently from New York to Mexico. (Fig. 2-24) and their manner of origin is uncertain. Some may represent greatly lengthened spits subsequently separated from a coast by erosion. On the other hand, perhaps they form along the zone where large waves break against a gently sloping sea floor and cause sediments to pile up. Submerged barrier islands appear to be common on some continental shelves.

Beach Conservation

The constant shifting of sediment along a beach creates problems for beach-loving man, and various methods have been used in attempts to stabilize or build up certain beaches. The steady erosion of certain coastal areas also endangers near-shore buildings. A *seawall* may be built along a shore to protect the area behind it (care must be taken that it not be undermined),

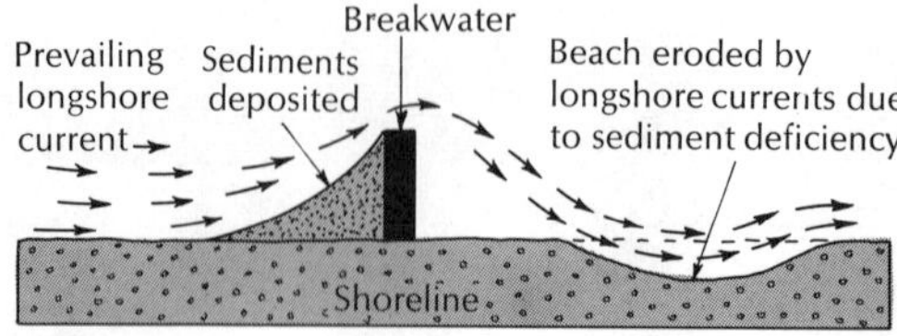

15-34. Beach modification by a breakwater. This kind of breakwater may also be called a groin. (Oceanography for the Navy Meteorologist, *U.S. Navy Weather Research Facility, 1960.)*

but this offers no protection for the beach in front of it. *Breakwaters* may be constructed offshore and oriented at different angles to blunt the attack of waves. However, occasionally a breakwater built to create a protected harbor may in fact so interfere with currents and waves that sediments in large quantities are deposited in the protected zone, and the harbor thus becomes too small and shallow. Dredging is a temporary solution for such unwanted fill.

Groins are long narrow barriers built from a shore outward at more or less a 90-degree angle into the water (Fig. 15-34). They trap sediments drifting parallel to the shore, and permeable groins seem to be most successful. However, if a certain quantity of sediment is deposited upcurrent from a groin, then a similar quantity tends to be removed from the shore beyond it. This creates problems where different parts of a shore are owned by different people. At present it seems that beach nourishment programs are generally more effective and less expensive than groin construction. In beach nourishment, sand of a suitable size is transported from some inland source and dumped on a beach or at the upcurrent end of a longshore current which then distributes the sand along the shore.

Do Wave-Built Terraces Exist?

According to many geologists, a wave-built terrace occurs on the seaward side of a surf-eroded terrace, and the sloping surface of one merges imperceptibly into that of the other (Fig. 15-35) One surface is built upward and outward by deposition, whereas the other is formed by erosion downward and inland. Such a composite surface was presumed to be one of equilibrium and the result of erosion and deposition over a long period of time. The surf-eroded terrace formed first and gradually became wider (asuming crustal stability and no change in the volume of sea water). More energy was expended in friction by water moving across the widened, surf-eroded terrace and less energy was available to attack the shore. Thus further growth of a surf-eroded terrace would be very slow. Sediments would be sorted and rounded as water moved alternately shoreward and seaward, and finer sediments would gradually be shifted outward and dumped in the deeper water beyond the edge of a surf-eroded terrace. Sediments would here pile upward until they reached this surface of equilibrium. Large waves could no longer stir and move them.

Dietz, however, has stated recently that such wave-built terraces do not exist, although an equilibrium condition can be attained in the nearshore zone (Fig. 15–28). He maintains that wave action results chiefly in moving sand onto a beach rather than sweeping sand outward across a shelf. According to Dietz, an active sand wedge pinches out toward the sea, and does not extend into a wave-built terrace. He suggests that some delta terraces (upper surface formed at sea level) may have subsequently been drowned and taken for wave-built terraces.

Erosional Sequences in Coastal Areas

Geologists have so far been unsuccessful in formulating a generally acceptable classification of coastlines. However, all agree upon a basic point: coastal areas do pass through an evolutionary development, and the resulting topographic shapes and rates of change depend upon factors such as initial coastal topography, types of rocks, nature of the sea floor, crustal stability or instability, wave energy, and time. Each coast has a somewhat different set of conditions from all other coasts. Although some details of the evolutionary development are uncertain, erosion and deposition should combine to reduce coastal irregularities and transform a coastline into a straight or broadly curved feature.

We can examine briefly possible developments along two coastal areas that initially have dissimilar conditions. If a rugged land area (made so by streams and other erosive agents) is submerged relatively rapidly, a highly irregular shoreline will result, and deep bays and islands will form (assume no further vertical shift in land vs. sea). Wave refraction concentrates wave energy upon the headlands, cliffs develop, and these become longer and higher as they retreat inland. A surf-eroded terrace forms and widens. Sea stacks and arches may project temporarily above this beveled surface, probably because they are more resistant or more sheltered than adjoining sections of a coast. Beaches develop here and there, and longshore drifting produces spits and bars of various types. After these processes have continued for a long enough time, indentations of a coastline will have been filled by deposition and headlands will have been worn back; thus a straight or broadly curving shoreline results (crustal stability has been assumed).

A quite different set of initial conditions involves a smooth, gently sloping sea floor that extends outward from a coastal plain. Here a barrier island tends to form and is separated by a lagoon from the low coast. Gaps in the bar here and there permit tidal waters to enter and leave the lagoon. As time passes, the bar migrates inland, and the shrinking lagoon is slowly filled by additions of sediment from lands and growth of vegetation. Eventually the bar is pushed against the shore which now is relatively straight. Water offshore is now deeper, and waves do not break until they are close to shore.

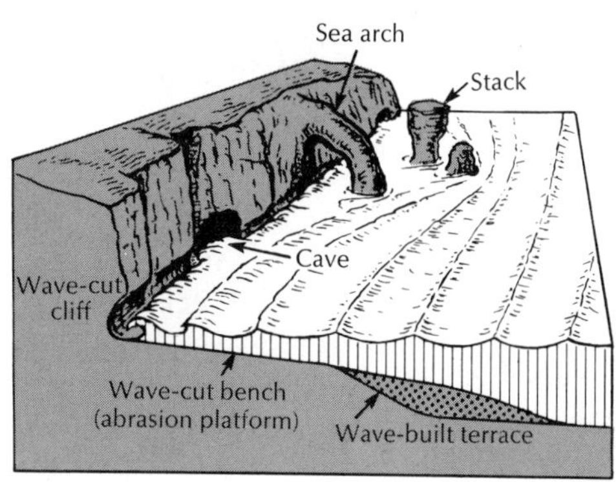

15-35. Erosional and depositional features on a rugged cliffed coast. *(B. Mears, Jr.,* The Changing Earth, *Van Nostrand Reinhold, 1970.)*

Coral Reefs and Atolls

Coral reefs form important portions of the sea floor in the warmer parts of the oceans, particularly in the tropical waters of the Pacific. They are of significance to man for a number of reasons: e.g., they constitute a dangerous underwater hazard to ships at sea and to military landing craft, but they also provide safe harbors; their manner of origin has challenged man's curiosity, and their beauty has stirred his aesthetic sense; furthermore, their porous materials sometimes serve as reservoirs for storing great quantities of petroleum. In searching for buried coral reefs that may contain petroleum, the geologist must know the types of rocks commonly associated with coral reefs and how these may be recognized at great depths by their reactions to seismic, gravitational, or other measurements. Once such a structure has been discovered at depth, its extent and trend must be determined. Such information depends upon an understanding of present-day reefs.

A *coral reef* is a limestone rock structure that has been built by organisms and fashioned into diverse topographic shapes such as ridges, platforms, and mounds. It is constructed mainly from the skeletal remains of sedentary organisms such as corals and algae which form an ever-growing pile as living organisms build upon the remains of the dead. The upper surface of a coral reef is near sea level at the time of origin, and it is firmly enough consolidated to act as a living breakwater to wave attack. The name coral reef is common, but misleading because corals form only a portion of a reef. However, they do form the main reef framework and are its most conspicuous part. The coral skeletons, shells of different organisms, and sediment added by the waves are cemented together chiefly by plants known as calcareous algae. The corals and algae tend to grow as porous, branching masses or to spread as a thin creeping blanket. Their remarkable resistance to wave erosion probably results in part from their branching porous structures which divert, blunt, and soak up the impacts of the waves.

Corals are known to live at different depths, temperatures, and latitudes in the oceans. However, reef-building corals are restricted in their environment: they thrive only in warm, shallow, relatively clear, normally saline water that is less than 150 to 200 feet deep and at a temperature of about 65° to 70°F or higher. Very heavy rains may kill sections of a coral reef by diluting the sea water or by washing sediment over it. The sedentary nature of the reef-building organisms causes them to grow best where wave action and circulating water contain oxygen and food supplies, and this may occur on the outer margin, particularly on the windward side.

The three main types of coral reefs are fringing, barrier, and atoll (at'all). As suggested by their names, *fringing reefs* develop along a land area, whereas a *barrier reef* forms an offshore ridge that tends to parallel a coast and is separated from it by a lagoon. Each is colonized by reef-building organisms whose distant ancestors traveled to their present locations during their free-swimming or floating stages. An *atoll* consists of a discontinuous chain of low islands arranged in a circular, elliptical, or irregular pattern that encloses a central lagoon (Fig. 15–36). Com-

15-36. Air view of an atoll in the South Pacific. *(Courtesy U.S. Army Air Force.)*

monly it has at least one deep passageway to permit the escape of sea water that flows or splashes inward over the reef.

Debate concerning the origin of coral reefs has centered on the questions of a possible genetic connection among them and upon the relative roles of subsidence, Pleistocene changes in sea level, and water temperatures. Darwin suggested in the 1830's that a genetic connection did exist among the three (Fig. 15–37). He proposed that a fringing reef developed first around the margin of a volcanic island that had been built upward from the sea floor in a suitable environment. Next, as the island slowly subsided, corals and other organisms built the reef upward and outward to form a barrier, because they grew best on its outer edges (questioned by some). Continued subsidence lowered the central island below sea level, and the barrier island developed into an atoll. The lagoon was partially filled by debris washed into it from the island and from the encircling reef. Darwin referred to atolls as tombstones over subsiding land and apparently formulated his hypothesis after he had seen only one barrier reef and one atoll. This is still the most satisfactory general explanation for the origin of many barrier reefs and atolls, although some atolls apparently have not developed from fringing and barrier reefs. Rather they began as atolls around the margins of submerged platforms and grew upward as subsidence occurred.

Deep drilling and seismic studies on a number of atolls (Fig. 15–38) have shown that subsidence has occurred. Holes have penetrated 4000 feet and more of coral reef limestone that overlies the basaltic rock of an extinct volcano.

In summary, Darwin's hypothesis still stands, but it has been modified considerably. Extensive subsidence has occurred, but this does not necessarily mean that present-day atolls once passed through fringing and barrier reef stages. Pleistocene climatic and sea level changes had a modifying effect upon reef development but probably were not dominant causes. If reef growth cannot keep up with subsidence, a guyot (p. 294) may develop instead of an atoll.

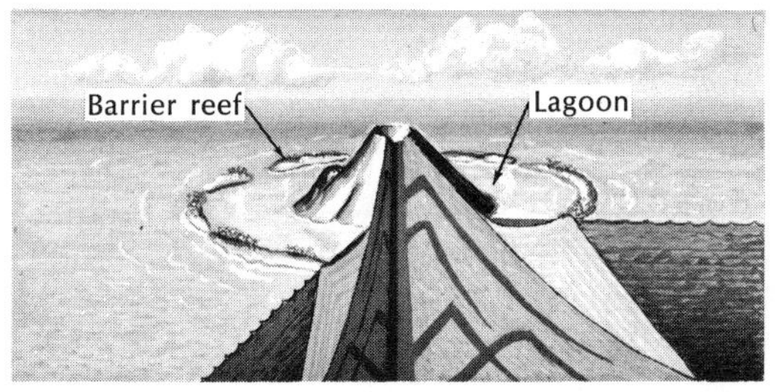

15-37. Darwin's views on the origin of barrier reefs and atolls.

Submarine Canyons

The great canyons (Fig. 11–3) that cut into the continental slopes here and there around the margins of the oceans have generated much interest and speculation, and they can be studied profitably as a case history of scientific investigation and hypothesis. Their exploration illustrates well the interplay that

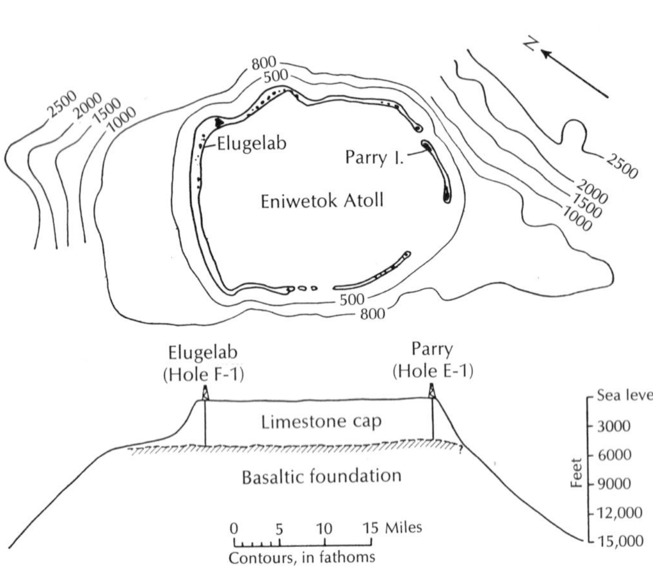

15-38. Generalized chart and section of Eniwetok Atoll. The contours were taken from a chart prepared by K. O. Emery, 1954; the section is after Ladd and Schlanger. Some samples from the drill holes contained land shells (a type living on higher islands rather than on atolls) as well as spores and pollen from a deciduous tropical forest that presumably once cloaked emerged parts of the atoll. Although the major crustal movement was downward for some 50 million years, there apparently were several periods when the island tops were hundreds of feet above sea level. Zones of leaching and recrystallization which presumably can form only during emergence have also been found and corroborate the fossil evidence. *(H. S. Ladd, "Reef Building,"* Science, **134,** *713, 16 Sept. 1961.)*

occurs when scientists tackle a major problem: a discovery takes place, measurements and observations are made, hypotheses are formulated, more data are accumulated to test the hypotheses, certain hypotheses are discarded or modified, new instruments are developed, and discoveries in related areas of science have their impact.

The term *submarine canyon* should probably be restricted to certain large, steep-sided valleys that resemble land canyons and extend for some tens of miles or less down the continental slopes along winding, sinuous courses. In general, these have V-shaped cross sections, concave-up longitudinal profiles (generally with steeper gradients than land valleys), and a number of entering tributaries, some with a dendritic pattern. A number of the canyons rival the Grand Canyon in size and shape. Thus we eliminate features such as the mid-ocean canyons (p. 294), certain trough-shaped depressions that trend diagonally down slopes, others that resemble fault valleys on land, and numerous small gullies.

Rocks of diverse ages and lithologies have been dredged from the canyon walls: Mesozoic and younger, igneous and sedimentary, hard and soft. Some of the canyons occur offshore from the mouths of large rivers, whereas others are quite unrelated to any present-day rivers on adjacent coastal areas. The canyons commonly terminate at a huge fan-shaped pile of sediment that has been transported, chiefly by turbidity currents, down the canyons toward the deep ocean floor. Off the California coast, the volume of sediment in the fans tends to exceed the volume of the canyons themselves by many times.

Some submarine canyons consist of three different parts each of which may have formed in a different way at a different time, and the Hudson Canyon (Fig. 15–39) furnishes a striking example. It has a shallower inner portion on the continental shelf, a large deep gorge on the continental slope, and an outer, relatively shallow, steep-sided, flat-floored trough (several miles wide and several hundred feet deep) that extends across the continental rise onto the adjacent abyssal plain. Low, levee-like ridges sometimes border such troughs.

A number of hypotheses developed to explain the origin of submarine canyons have subsequently become untenable: e.g., faulting is incompatible with their winding courses, tributary systems, and V-shaped cross sections. Submarine erosion by artesian springs has also proved unsatisfactory, as has the proposal of canyon erosion by currents set up by tsunamis and upwelling. Subaerial erosion (i.e., on land rather than under water) when sea level was lower during the Pleistocene glacial ages was also discarded as a general explanation for all submarine canyons when the canyons were subsequently traced to much greater depths. However, some canyons, such as those off Corsica (north shore of Mediterranean) may well have formed subaerially and then became submerged by downward movement along faults.

Two hypotheses remain to account for the origin of submarine canyons: one involves erosion beneath sea level by turbidity currents and other agents (generally favored), and the other is a modified subaerial version. It is known that some erosion occurs within the canyons at present and that large quantities of sediments are being transported through them. In the second view, the main central portions of the three-part canyons may have been eroded subaerially by rivers at different times in different places. Thus some canyons show no relationship to present-day rivers because they formed long ago. Following excavation, such regions were downwarped to become parts of the sea floor (considerable subsidence is known to have occurred off many coasts such as the eastern United States). Younger sediments deposited on the sea floor on either side of a subsiding valley would make a canyon deeper and give it a deceptively youthful age. The lower trough-shaped outer portion of a three-part submarine canyon may have been eroded by turbidity currents, whereas its inner shallower continental-shelf portion may have formed by headward erosion during times of lower sea levels in the Pleistocene.

Submarine landslides may have been an important factor in this headward erosion and, in combination with turbidity currents, in keeping the canyons flushed of sediment since their origin in the remote past (some may have been filled). Some of the erosion occurring today in the upper part of a canyon apparently is caused by the downward creep of sediment and organic debris within the channel. This polishes and gouges its floor somewhat like a valley glacier. Other erosion is performed by flowing sand and by currents of different kinds.

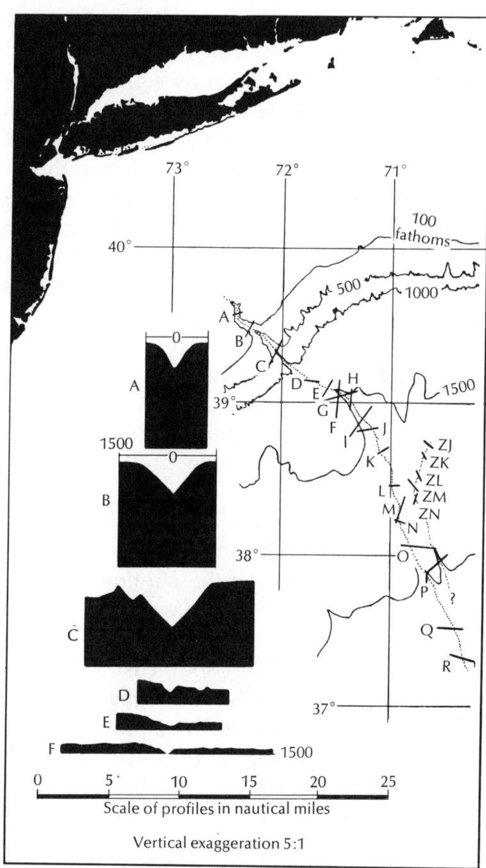

15-39. Cross sections of the Hudson Submarine Canyon which begins on the continental shelf and extends for almost 200 miles down the continental slope and across the continental rise. Finally it disappears at a depth of nearly three miles. Turbidity current deposits up to 20 feet thick have been cored from its floor. *(Heezen, Tharp, and Ewing, "The Floors of the Oceans,"* Geol. Soc. Am. Special Paper *65.)*

According to the other and more widely accepted view, these canyons have formed primarily through erosion by submarine turbidity currents (p. 288 and Fig. 11–14). Such density currents exist and are capable of transporting vast quantities of sediments (e.g., the abyssal plains appear to have been constructed chiefly by them). Furthermore, in the view of one oceanographer, if every 10,000 particles moved by turbidity currents through a canyon can erode just one particle from it, then this process can explain the origin of the canyons. His measurements show that the sediments piled up beyond the mouth of one submarine canyon exceed the volume of the canyon itself by 10,000 times.

Evidence that turbidity currents can erode has been furnished by studies of earthquake-caused cable breaks on the continental margins; e.g., a number of cables linking North America with Europe cross the shelf, slope, and deep-ocean floor southeast of Newfoundland. On 18 November 1929 a powerful earthquake originated near the edge of the Grand Banks, and some thirteen of the transatlantic cables broke, but not all simultaneously. The cables close to the place of origin broke at once, but those located farther downslope broke at successively later times until the last one, nearly 300 miles away, parted more than 13 hours after the earthquake occurred. The locations and times of the breaks were determined accurately and involved portions of the cables 100 miles or so in length. The earthquake apparently caused landslides that moved large masses of loose sediments abruptly to lower levels and produced turbidity currents that snapped each cable in their paths in succession. Parts of the cables were moved downslope and buried by sediments. The turbidity currents may have flowed at nearly 60 mi/hr initially but slowed to a 15 mi/hr rate at the last cable. Subsequent cores taken from the sea floor apparently show that a layer of sediment, averaging 3 feet in thickness, was spread by the turbidity currents over some 75,000 square miles of the sea floor.

Such currents presumably can erode the sea floor as well as transport sediment, although these conclusions have been questioned. Can turbidity currents erode resistant rocks such as granite and firmly consolidated sandstone which form the walls of some canyons? In some cores, turbidity current deposits rest upon deep-sea clays that show little disturbance. On the other hand, certain experiments in tanks, together with theory, suggest that turbidity currents are effective agents of erosion if their velocities exceed a certain critical quantity. Above such a threshold velocity, turbidity currents may pick up additional sediment, become denser, and thus flow faster if slopes remain uniform.

Turbidity current activity may have been greater during the Pleistocene than it is at present, although this, too, has been questioned. Large portions of the present continental shelves were then exposed when sea levels were lowest. Sediments on the shelves were moved seaward as major rivers extended their channels to the new coastlines (near the edges of the present continental slopes). New stream channels also formed and wave action could more readily place sediment into suspension. Thus the quantity of muddy sediments piling up on the outer edges of the shelves and upper continental slopes probably exceeded present amounts. Landslides, some triggered by earthquakes, occurred in these unstable masses and probably produced more frequent and powerful turbidity currents than occur today. Recent turbidity current deposits do not extend as far as those of the Pleistocene.

Origin of Continental Shelves and Slopes

The continental shelves and slopes are located along the boundary zone between the two dominant levels in the Earth's crust, one nearly 3 miles above the other, and this is a fundamental factor in a consideration of their origin. These two levels reflect the great differences in density and thickness that exist between the oceanic and continental crusts. Since the two types of crust are in isostatic balance, great slopes must occur at their intersection and extend from the continents downward to the ocean basins. The continental slopes tend to have straight or gently curved trends and narrow widths (about 10 to 20 miles). Many other factors, such as Pleistocene changes in sea level, have influenced the transitional zone between the continents and ocean basins. However, crustal movements and long-continued sedimentation appear to be most important.

According to Emery (p. 279), continental shelves subdivide into two main groups: one is underlain by igneous and metamorphic rocks, whereas the other and far larger group is underlain by thick accumulations of sediments and sedimentary rocks. In many instances, geologic dams or barriers of various kinds have kept these sediments piled against the lands; i.e., faulting, folding, intrusions of salt and magma, and/or the activities of organisms such as algae and corals produced barriers that were located off a shore and extended parallel with it (Fig. 15–40). These then trapped the sediments being transported seaward from the lands.

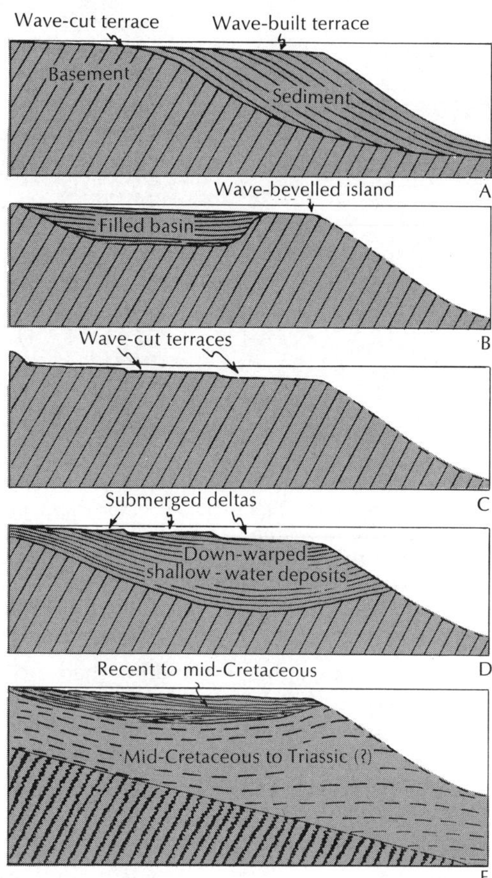

15-40. Views on the origin of the continental shelf as presented by Shepard in 1948. *(R. S. Dietz,* Geol. Soc. Am. Bull. **74,** *Aug. 1963.)*

Off the East Coast of the United States, sediments have filled and then overflowed one such dam to form a continental slope (Figs. 11–6 and 15–41). Seismic reflection profiles show nearly horizontal interfaces between sedimentary layers beneath the shelf and slope in some places, interfaces that parallel the continental slope in other places, and both kinds in still other places. Evidently this continental shelf was built upward by sedimentation as the surface beneath it subsided. A series of elongated flat zones, called benches or terraces, have been found at different levels on this continental slope and form a giant staircase that extends down the continental margin to the deep ocean floor. The benches (the treads) probably represent the upper surfaces of nearly horizontal, resistant layers of sedimentary rock. In this view, the steeper slopes between the benches occur where less resistant rocks intersect the slope. The benches are extensive and conspicuous and can be correlated from one place to another along the continental margin. Such rock outcrops indicate that this continental slope is too steep today for sediments to accumulate to any appreciable thickness; landslides, slumping, and turbidity current activity keep the sediments from piling up on it. Sediments have accumulated in very large quantities, however, at the base of this continental slope to form a continental rise.

In an older view developed before wide-

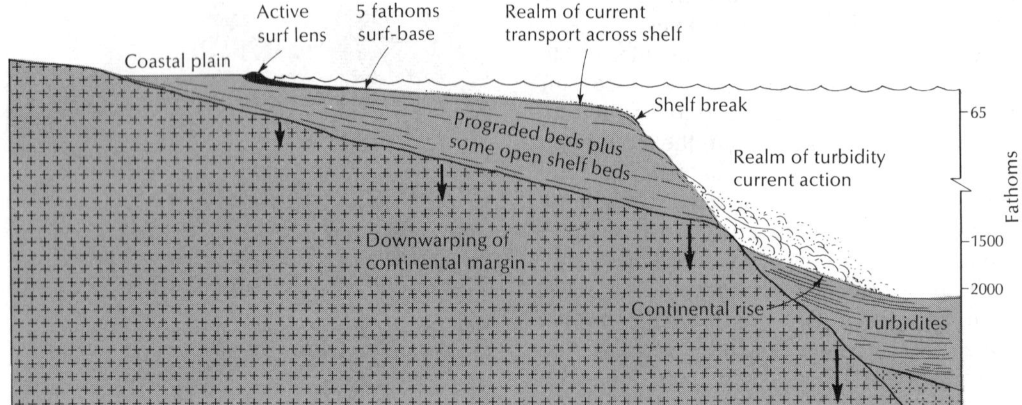

15-41. Simplified and schematic cross section of a continental margin such as that off the eastern coast of the United States. Dietz emphasizes prograding by deposition chiefly within the surf zone. Subsequent downwarping and burial by younger sediments have preserved the sediments. Prograding is the seaward advance of a shoreline resulting from nearshore deposition of sediments transported to the sea by rivers (Cotton). *(R. S. Dietz,* Geol. Soc. Am. Bull. **74.**)

spread oceanographic exploration had occurred (Fig. 15–40a), the inner part of a continental shelf was thought to have formed as an erosional feature, whereas the outer part, as well as the continental slope, originated chiefly as depositional features. Effective wave base was then thought to occur at a depth of 600 feet, and the edges of most shelves were believed to be located at this depth. Subsequent discoveries have shown that most shelves are depositional features and that the lowest limit of vigorous surf abrasion is probably about 30 to 60 feet. However, surf erosion may be the most important process in the origin of some continental shelves off coasts that are being uplifted (Figs. 15–40 and 15–42). The shelf off southern California apparently formed in this manner, because oil company exploration has shown that such beveled surfaces occur beneath the shelf sediments. These surf-eroded surfaces have been submerged and covered with sediments during the recent rise of sea level. The drowned surfaces of deltas (which form essentially at sea level) seem to form some continental shelves (Fig. 15–40d). On the other hand, shelves may be narrow or absent in some areas because strong ocean currents have prevented sediments from piling up or because crustal movements have affected them so recently that shelves have not had time to form.

Continental slopes apparently cannot generally be explained as the outer part of a seaward-growing mass of sediments washed outward from the lands. Depositional slopes such as those off the Mississippi delta are much gentler than the gradient of the present continental slopes, and some slopes steepen at depth instead of becoming gentler (expected at the forward margin of a fanlike mass of sediment).

Dietz has proposed that continental slopes have formed primarily by the compressional collapse and folding of continental rise sediments (his eugeosyncline) against a continental block. Such a mass of sediments had previously accumulated at the base of an older continental slope, chiefly by turbidity current activity. After a eugeosyncline had been deformed into a marginal mountain belt and welded to a continent, its seaward flank would form a continental slope (Fig. 12–18). Sedimentation and erosion subsequently would modify these primary continental slopes and account for their wide diversity.

15-42. Surf-eroded terraces, Middleton Island, Alaska. Note the steeply dipping sedimentary rocks that have been truncated by erosion. Uplift apparently occurred after each terrace formed. *(Courtesy S. R. Capps, U.S. Geological Survey.)*

However, some oceanographers think that faulting has been important in the origin of the continental slopes (Fig. 15–43). Unfilled, deep-sea trenches occur along the mountainous margins of some continents, and faulting may have occurred on a very large scale in such areas, although downdrag by convection currents is another explanation. In certain other areas, the continental slope cuts across the trends of elongated structures projecting oceanward from the lands, and this suggests faulting. However, the majority of oceanographers are probably not yet ready to attribute the continental slopes primarily to faulting. Thus uncertainty shadows the question of origin of the continental shelves and slopes, and answers may lie in the newer concepts of plate tectonics and continental drifting.

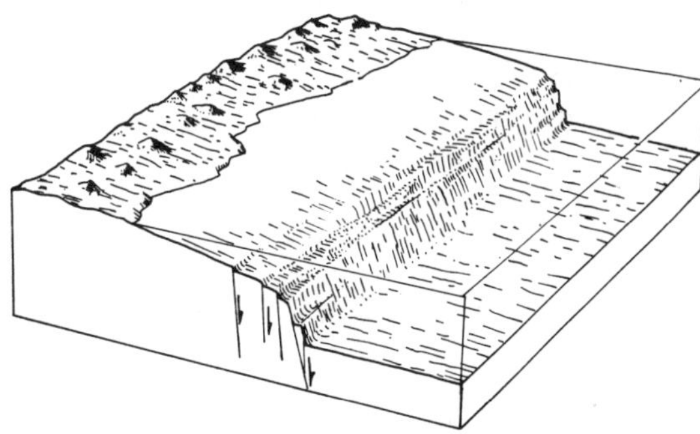

15-43. The continental slope as a fault scarp. According to Dietz, this is a commonly preferred view but cannot be the primary cause of slope relief because of isostatic considerations. However, some faulting probably occurs along the continental slope. (R. S. Dietz, *"Origin of Continental Slopes,"* American Scientist, *March 1964.)*

Exercises and Questions for Chapter 15

1. Describe water as a factor in the transfer of heat energy from one latitude to another.
2. Account for the different colors of ocean water.
3. Oxygen and carbon dioxide occur as gases dissolved in sea water.
   (A) What are the sources of these gases?
   (B) At what latitudes and depths do they tend to be more abundant? Less abundant? Why?
   (C) What roles do they play affecting organisms and organic matter?
4. Assume that several icebergs have the same thickness. Must each necessarily project the same distance above sea level? What factors are involved?
   (A) Experiment with blocks of ice of different shapes and measure the vertical distance that each block extends both above and below the water surface. What ratios do you observe?
5. How is the salinity of ocean water related to its chlorinity? What factors tend to produce a high salinity? How does the salinity of ocean water furnish proof of its thorough mixing?
6. Calculate the time needed by a signal from an echo sounder to go from a ship to the bottom and return over the deepest part of the ocean.
7. What is a thermocline and how does one develop?
8. Ocean currents:
   (A) Describe several ocean currents and some of the ways in which they affect man.
   (B) Describe some methods of measuring ocean currents.
   (C) Describe some of the factors causing surface and deep-water currents.
   (D) Describe some of the factors that cause upwelling or sinking in certain areas.
9. Ocean waves:
   (A) What relationships are involved in deep water waves concerning wave height, wavelength, and speed ?
   (B) Describe some of the phenomena that occur as a wave approaches a shore.
   (C) Discuss tsunamis.
10. Describe each of the following and discuss its origin:
    (A) Beaches.
    (B) Spits.
    (C) Longshore currents.
    (D) Barrier islands.
    (E) Beach conservation.
    (F) Rip currents.
    (G) Continental shelves.
    (H) Continental slopes.
11. What was Darwin's explanation of the relationship between a coral reef, a barrier reef, and an atoll? How much modification of this hypothesis has been necessary because of subsequent discoveries?
12. Describe some of the characteristic features of submarine canyons and the two main hypotheses concerning their origin.

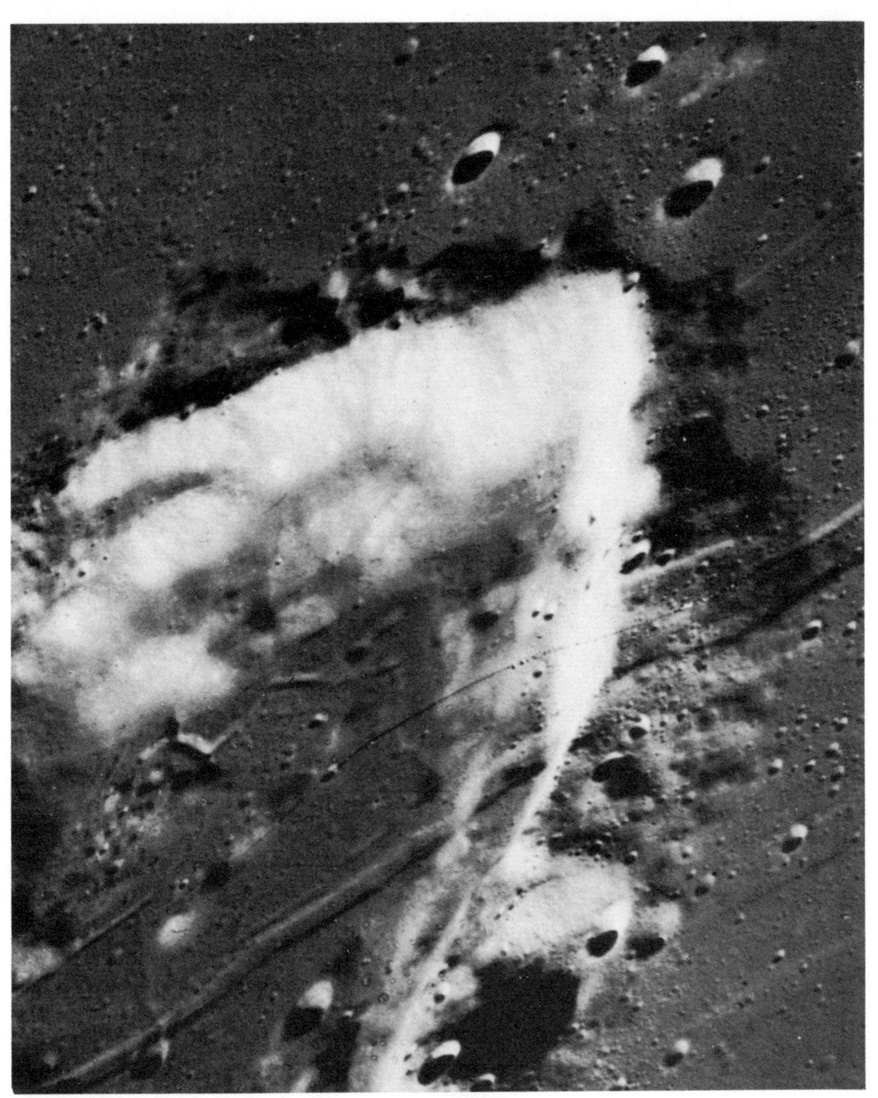

Part 11
Astronomy

16

A Journey Through Time and Space

The Solar Family

For an overall view of the general setting of the Earth in the universe, we embark on a quick journey through space to the galaxies. The nearest celestial objects (Fig. 16–1) are in the solar system, and nearest of them all is the planet we live on, the Earth—a nearly spherical body about 7900 miles (12,742 km) in diameter, slightly flattened at the poles and bulging at the equator. With respect to the stars, the Earth *rotates* once in 23 hours and 56 minutes (i.e., it spins on an axis through 360 degrees). As viewed from space looking down on the north pole, the direction is counterclockwise (imagine a giant clock face-up at the north pole). The Earth rotates from west to east, and this causes an apparent, daily *(diurnal),* east-to-west movement of all celestial objects, including the sun—thus we have day and night.

The Earth also *revolves* around the sun in a nearly circular orbit. It completes one orbit of 360 degrees every 365¼ days, thereby dividing time into years (Fig. 16–2). The Earth's mean distance from the sun, nearly 93 million miles (150 million km), is called 1 *astronomical unit* (equals the semimajor axis of the Earth's elliptical orbit, p. 467).

The Earth's axis makes an angle of 66½ degrees with the plane of its orbit about the sun (the plane of the ecliptic); it is tilted 23½ degrees from a perpendicular to this plane. At the present time, one end of this axis points almost at the North Star (Polaris), and continues to do so all year long because Polaris is very far away (about 300 light-years). Thus observers who point at Polaris from anywhere in the Northern Hemisphere, or from any orbital location, are all pointing along parallel lines in essentially the same direction (similarly, parallel railroad tracks seem to meet in the distance).

The sun's diameter (864,000 miles or 1,391,000 km) is 109 times longer than that of the Earth but is less than a hundredth of the distance to the Earth. Some 108 suns, strung like beads, would be required to span

1 astronomical unit. On a scale that shrinks the Earth into a golf ball, the sun becomes a 15-foot globe located about 1500 feet away.

There are eight other major planets in the solar family (Fig. 16–3). Some are larger than the Earth (Jupiter, Saturn, Uranus, and Neptune); some are smaller (Mercury, Mars, and Pluto), and Venus is nearly the same size. Some planets rotate more rapidly than the Earth and others more slowly. Thousands of small planets also occur in the solar system, chiefly between the orbits of Mars and Jupiter. These are the *asteroids* (planetoids), and the largest are several hundred miles in diameter. To recall the relative distances of the planets from the sun, remember: "Mary's Violet Eyes Make Anguished (as-

16-1. LM and astronaut on the lunar surface, Apollo 11 photograph, July 1969. *(NASA: AS-40-5931.)*

teroids) John Stay Up Nights Permanently." Satellites (moons) revolve about some of the planets (Jupiter is known to have twelve) but not about others (Mercury, Venus, and Pluto). All the planets revolve around the sun in the same west-to-east direction (counterclockwise), and their orbital planes all pass through the sun and are only slightly inclined to each other.

The sun—an enormous, gaseous, intensely hot body—contains some 333,000 times the mass (weight) of the Earth and comprises about 699/700 of the matter in the entire solar system. It is the ultimate source of most of the energy in the solar system. With its retinue of planets, asteroids, comets, and meteors, it speeds through space at about 12 mi/sec toward the bright star Vega. This motion is relative to the stars nearest the sun.

However, despite its great size, the sun is not outstanding. Many stars are larger, hotter, and brighter than the sun, but an even greater number are smaller, cooler, and fainter. Some stars are much denser than the sun, but others are less dense. On the one hand, the sun is so huge that if the Earth were placed at its center, the

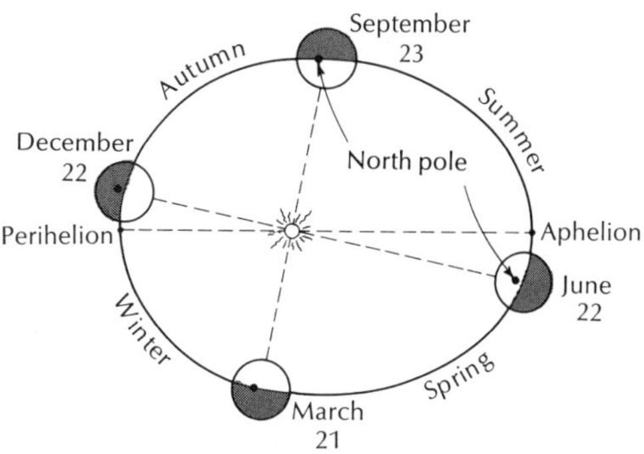

16-2. Seasons in the Northern Hemisphere and Earth's elliptical orbit about the sun (at a focal point). The Earth is closest to the sun and revolving fastest when it reaches perihelion early in January. The Earth is farthest away and orbiting most slowly at the aphelion position early in July. Note that the seasons are of unequal lengths because of variations in the distance to be covered and in orbital speed. *(R. H. Baker and L. W. Fredrick,* An Introduction to Astronomy, *Van Nostrand Reinhold, 1968.)*

16-3. Mariner VII photo of Mars, 4 August 1969, at a distance of 293,200 miles. The circle of Nix Olympica is conspicuous in the upper left. The very bright region to the right is Tharsis, perhaps one of Mars' highest regions. The south polar cap is at the bottom. *(Frame 73, NASA.)*

moon (about 240,000 miles from the Earth) would lie about halfway to the sun's surface. On the other hand, the sun is so tiny that if it were placed at the center of the giant star Antares, the orbits of Mercury, Venus, the Earth, and Mars would all fit inside Antares, with room to spare!

On a scale in which 1 yard represents 1 million miles, the sun could be a balloon about 30 inches in diameter. Mercury, Venus, the Earth, and Mars would then be about 36, 67, 93, and 141 yards away respectively. On this greatly reduced scale, Mercury and Mars could be represented by corsage pinheads, and Venus and the Earth by peas. Jupiter, Saturn, Uranus, Neptune, and Pluto would be located at distances of approximately ¼, ½, 1½, and 2 miles respectively. Oranges about 3 inches in diameter could represent Jupiter and Saturn; plums could represent Uranus and Neptune; Pluto is probably smaller than the Earth. The moon would be a large sand grain 9 inches from the pea representing the Earth. Thus the Earth forms only a tiny part of the solar system. Its distance from the sun is such that it is neither too hot nor too cold for life.

The Light-Year

Seeking a unit of distance adequate for the vast reaches of interstellar space, astronomers have had recourse to the *light-year*. This is the distance that light travels in 1 year while moving, as it does, at the astonishing rate of about 186,000 mi/sec (300,000 km/sec). The product 186,000 × 60 × 60 × 24 × 365¼ comes to about 6 trillion miles for the light-year. If 1 astronomical unit (93,000,000 miles) were reduced in scale to an inch, then a light-year would equal about 1 mile in length. At the unimaginable speed at which light travels, a beam of light sent from the Atlantic to the Pacific across the United States (about 3000 miles) and back again would make about 31 round trips in just 1 second.

The star nearest to the Earth (other than the sun) is approximately 4.3 light-years away. This star, Alpha Centauri, is the third brightest star in the Earth's sky and is located about 30 degrees from the south celestial pole. It is not visible from the latitude of New York. (Alpha Centauri is the most massive member of a triple star system that appears as one star to the unaided eye. Proxima Centauri, another member of the system, is closer to the Earth at times because the three stars revolve about a mutual center of mass.)

Despite the huge sizes of some stars, all are located at such tremendous distances from the Earth that they are mere pinpoints of light even in the largest of telescopes. Thus astronomers have no direct means of determining whether any of the stars, even the nearest ones, have planetary systems of their own. Indirectly, a few nearby stars have very slight sinuous paths, apparently caused by the gravitational attraction of planet-sized objects orbiting about them (p. 604). However, current ideas about the origin of stars and planets suggest that planetary systems may be rather common.

Seeing the Past

A fascinating aspect of celestial distances is the realization that light from celestial objects reaches the Earth after journeys that have taken a very long time. Even the light from the sun requires 8 minutes to travel to the Earth. Thus we always see the sun as it actually appeared 8 minutes earlier.

The light we see when we look at the nearest star left that star about 4.3 years ago and has been traveling at the tremendous speed of 186,000 mi/sec ever since. If a terrific explosion occurred on this star two years ago, astronomers will not find out about it for another two years! Thus in 1973, we see this star as it was in 1969; in 1974, we see it as of 1970, etc. Other stars are measured as hundreds and thousands of light-years away. Whenever we look at the stars, we are in a sense seeing the past —we have available a special kind of astronomical "time machine." We may be seeing one star as it appeared a generation ago, another as it appeared twenty centuries ago, and still another perhaps as it appeared in the remote geologic past.*

Thus our knowledge of the distances separating objects involves both space and time. If the distance to a certain galaxy is measured as 1 billion light-years, we know that the location of this galaxy 1 billion years ago was 1 billion light-years from our present location. Where we were in space when the light left this galaxy is uncertain, just as is the present location of the galaxy. In effect, we see remote celestial objects by "fossil starlight."

Stars and Clouds of Gas and Dust

Stars are not scattered at uniform distances from each other in space. They occur in pairs (binaries, p. 614) and multiple-star systems in which two or more stars revolve about the center of mass of the system. Larger numbers of stars occur together in clusters like the Pleiades (p. 623) in which the different members have similar velocities. Globular star clusters (p. 610) are larger yet. But the main unit in the universe is the *galaxy*—a gigantic assemblage of stars (numbered in the billions) and interstellar matter held together by mutual gravitational attraction.

Vast clouds of gas and dust, called *nebulae,* occur within some galaxies (Figs. 16–4 and 16–5), and though these clouds may be many light-years across, they are extremely thin and diffuse; an average sample is more nearly a vacuum than can be produced on Earth. *Bright nebulae* are illuminated by the light of stars located within them, whereas *dark nebulae* obscure the light from the stars beyond them. Relatively few stars occur between us and the dark nebulae or within them. In making time-exposure, telescopic photographs of distant nebulae and other faint celestial objects, nearby bright stars become overexposed and appear as much-enlarged circles instead of their actual pinpoint size. In some photographs, such stars may also show four spikes caused by the diffraction of light around supports for a mirror located inside a telescope.

Hydrogen and helium together may constitute approximately 96 to 99% by weight of all the matter in the nebulae and in the entire universe (hydrogen may be about three times more abundant than helium). All of the other elements combined may make up only 1 to 4% of the total.

Stars apparently have formed, and are forming now, wherever gravitational attraction can cause part of a cloud of gas and dust to contract into a sphere. Temperatures and pressures increase in the central portion of the gaseous, contracting sphere until thermonuclear reactions can occur—as if a number of so-called hydrogen bombs were being exploded each second. At this stage the sphere of gas can be called a star and contraction ceases. The gravitational attraction that tends to cause contraction is in equilibrium with the thermonuclear reactions, high temperatures, and radiation pressure that tend to cause expansion.

The amount of matter in the clouds of gas and dust is fundamental in speculations concerning the future of the universe. New stars cannot form in regions where this interstellar matter is absent or too diffused. How much exists? Some estimates suggest that only 1 to 2% of all the matter in the universe is interstellar matter, but other estimates are considerably higher. In some regions of space—e.g., within the arms of a spiral galaxy—interstellar matter may make up about one-quarter of the total, and in such regions stars can continue to form and evolve. Is matter also present within galaxies in some other form—perhaps invisible to our instruments—such as black "white dwarfs" (p. 613) or so-called "black holes" (p. 618)? Moreover, what of the spaces between the galaxies? Does intergalactic

* Approximate distances in light years to some of the brightest stars follow (arranged in order of decreasing brightness): Sirius, 9; Vega, 26; Capella, 46; Arcturus, 36; Rigel, 650; Procyon, 11; Altair, 17; Betelgeuse, 650; Aldebaran, 68; Spica, 160; Antares, 170 and Deneb, 558.

16-4. Clouds of gas and dust in space form dark and bright nebulae in the Horsehead Nebula in Orion. *(Courtesy Mount Wilson and Palomar Observatories.)*

matter occur? If so, is it in the form of stars, nebulae, or scattered atoms? The answers are not known, but some observations point toward the existence of intergalactic matter—perhaps in very large quantities.

Galaxies

Although galaxies differ in shape, many of them—the spiral varieties—have a thin, disklike form and spiral arms; in edge view, they look like two saucers placed rim to rim with the bottoms outward (Figs. 16–5 and 16–6). Other galaxies are globular, elliptical, and irregular in shape, and gradations occur among the different types. The gradations suggest that galaxies may evolve from one type to another, but other evidence suggests that they do not, and many uncertainties presently attend the question of galactic evolution.

The Milky Way—the familiar faint band of light in our sky that is visible to the unaided eye—marks the equatorial plane of the galaxy which contains "our" solar system and which we commonly call the Milky Way Galaxy or "our" galaxy (Fig. 24–1). Its diameter may span 100,000 light-years, and it appears to be a giant spiral resembling its neighbor, the Andromeda Galaxy (Fig.

16-5. An edgewise view of a spiral galaxy (NGC 4565 in Coma Berenices; NGC stands for New General Catalogue). Our galaxy would probably show a similar shape if seen edge on. Individual stars seen in the photograph belong to our galaxy, and we look past them to see this galaxy far beyond. The dark strip represents interstellar clouds of gas and dust concentrated in the equatorial plane of the galaxy. *(Courtesy Mount Wilson and Palomar Observatories.)*

25–4). The solar system seems to be located near the equatorial plane in one of its spiral arms, perhaps 33,000 light-years outward from the center. The sun and other stars are apparently revolving in the equatorial plane in somewhat circular orbits about the center; the sun may need some 200 million years to complete one revolution.

Thus the solar system forms a very tiny part of a single galaxy. All the stars we ever see at night with the unaided eye are part of this galaxy and are relatively near to us. Our sun is merely one of the billions of stars in our galaxy, not conspicuously different from many of its distant companions in size, composition, temperature, or other known properties—except, of course, in its possession of a system of planets, on one of which we live. Planetary systems may be commonplace, and life may exist on other planets, but we lack direct evidence of this.

Stars are the main units of the galaxies and occur within them in countless numbers. Yet the ratio of stellar diameters to interstellar distances (perhaps several light-years on the average) is so small that collisions among stars rarely occur. Stars within

16-6. A spiral galaxy viewed at a 90-degree angle to its equatorial plane (M101 in Ursa Major; M stands for Messier, an astronomer who catalogued the locations of various celestial objects). Our galaxy may have a similar pinwheel shape. *(Courtesy Mount Wilson and Palomar Observatories.)*

a galaxy might be likened to golf balls several hundred miles apart.

On the other hand, galaxies are the main units comprising the universe. Galaxies may range in diameter from dwarfs that are several thousand light-years across to giants that span 150,000 light-years and more. Average distances between adjacent galaxies may be of the order of 1 to 2 million light-years. The ratio of galactic diameters to intergalactic distances, although relatively much larger than that for stars, is still very small. Average intergalactic distances may be fifty times longer than average galactic diameters. Like stars, galaxies are not strewn uniformly across space. Rather, they tend to occur in pairs and in clusters, and the number of galaxies in any one cluster ranges from a few to a thousand and more. The dimensions of the universe are thus on a scale that far outdistances the range of the imagination.

One of the really fascinating aspects of the universe is the concept that it may be expanding. According to this view, the distances between clusters of galaxies are increasing, and at rates that are proportional to their distances from us: relatively nearer galaxies are moving away from us at slower rates than remote galaxies. The evidence for this involves a phenomenon known as the red shift (p. 508). Competing hypotheses —the Big Bang, the Oscillating Universe, and the Steady State—attempt to explain this relationship. We shall discuss these later (p. 630).

As astronomy developed, it upset prevailing ideas concerning man's place and role in the universe. Once the Earth was thought to be the center of a rather small universe. Then the sun became the center and the Earth was reduced in status to that of a rotating, revolving planet. At this stage, the universe was still considered relatively small and the sun was near its center. In the next revolutionary discovery, the sun was shifted to a position as a peripheral star in a galaxy containing billions of other stars. Still later it was learned that other galaxies existed and in vast numbers and at great distances. Then came the concept of an expanding universe. Now we are in the midst of yet another great change in our ideas about the universe. Are we alone and unique in space, or do intelligent beings inhabit other planets orbiting around other stars?

The Constellations

Groups of brighter stars within our own galaxy form recognizable patterns for the skywatcher—dippers, crosses, squares, circles—and are called *constellations*. Most of the familiar constellations were named long ago by people with vivid imaginations who traced in the sky the lines of earthly symbols—bears, chariots, dragons, heroes (Fig. 16–7). It need hardly be said that the resemblances are not as striking in all instances as they are in some.

Yet constellations and bright stars are useful in designating areas in the sky, much as states and cities serve to locate places on the Earth's surface. In modern usage the constellations represent definite areas bounded by straight lines, like some state boundaries. The stars in a constellation are commonly designated by small letters of

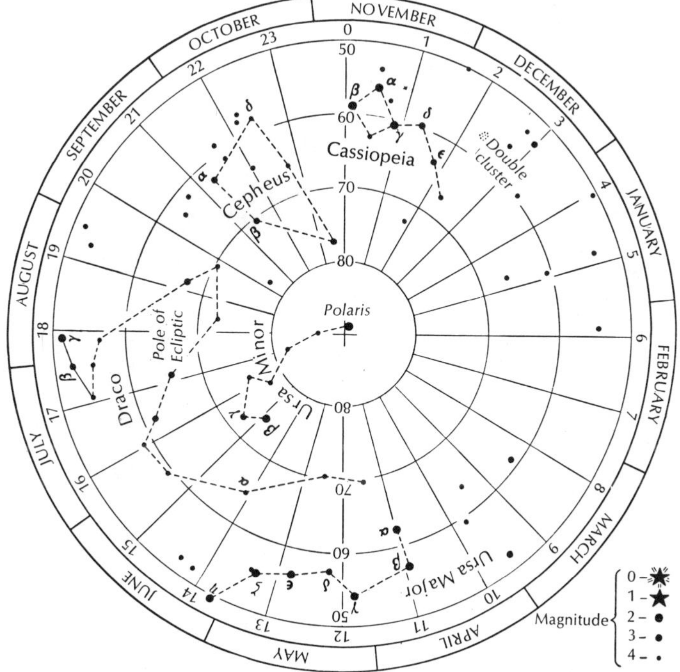

16-7. The northern constellations as viewed from 40 degrees north latitude. Hold the map so that the current month is at the top to see the constellations as they appear at 9:00 PM standard time. For any time other than 9:00 PM, rotate the map through the proper number of hours, counterclockwise for a later time and clockwise for an earlier time. For example, the Big Dipper will be high in the northern sky at 9:00 PM in May or at 9:00 AM in November. *(R. H. Baker, An Introduction to Astronomy, Van Nostrand Reinhold, 1964.)*

the Greek alphabet in order of brightness: the brightest star is alpha, the next brightest is beta, etc. Some of the brighter stars have also been given special names such as Sirius, Arcturus, and Vega.

To become familiar with the outstanding constellations is a pleasant and rewarding task. The joy of a clear, star-lit evening will forever be enhanced by this familiarity, once gained. Inexpensive star charts are available which show where the constellations appear in the sky at any hour during the year.

The stars in a given constellation need not be as closely associated in space as they are for the eye. Two stars shining side by side may look like intimate neighbors and yet be hundreds of light-years apart; one may be much closer to us than the other, even though they are located in the same direction from us in space. When a given celestial object is said to lie in a particular constellation, this means that it is visible in the general direction of the constellation. The object itself may be situated between the Earth and the greater number of the stars in the constellation, or it may be far out in space beyond them.

Diurnal Motion and the Celestial Sphere

We speak of "sunrise" and "sunset," and each day we apparently see the sun move westward across the sky. When visible, the moon, planets, stars, and other celestial objects have similar diurnal motions. However, we learn early in life that this motion is actually caused by the Earth's eastward rotation. Let us use a top-view sketch (Fig. 16–8) to illustrate diurnal motion and the direction of a celestial object from the Earth. In a top-view two-dimensional sketch, we use a circle for the Earth and imagine an observer standing on the surface of the Earth at a low latitude. For this observer, north is inward (and upward) toward the center of the circle representing the Earth. Therefore, due south is outward (and downward) from the center. When the observer

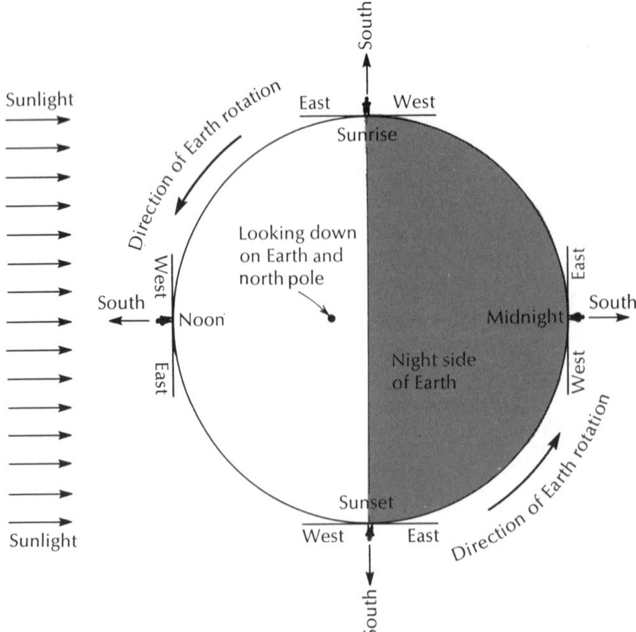

16-8. Westward diurnal motion is caused by Earth's eastward (counterclockwise) spin. Schematic view looking down on the Northern Hemisphere and a low-latitude observer at four different times: sunrise, noon, sunset, and midnight. The observer's horizon and compass directions are also shown.

faces north, west is always to his left and east to his right. However, another observer in space looking down on the Earth would see that east and west are not fixed directions but vary with the Earthly observer's location. For the observer on Earth, these directions are fixed relative to his horizon.

To account for the sun's diurnal westward motion, we note in Fig. 16–8 that an observer at sunrise sees the sun on his eastern horizon. By noontime, however, the Earth has rotated this observer—together with his horizon and compass directions—around so that the sun now appears above the horizon and due south. Continued rotation then brings this observer to the sunset position where the sun appears on the western horizon. It follows that all other celestial objects also take part in this diurnal merry-go-round (unless they are located directly above the Earth's axis).

Thus the Earth's eastward rotation accounts for the apparent diurnal westward movement, and the eastward or counterclockwise direction is by far the most common within the solar system: planets all revolve in this direction; the sun, moon, and most planets rotate in this direction; similarly, most satellites revolve and rotate eastward.

Although some stars appear brighter than others, our eyes have no accurate depth perception at great distances, and so one star does not seem closer to the Earth than another. All stars appear to be at the same distance from us, and together they seem to form a gigantic inverted bowl. Since we can make this same observation from any position on the Earth, the Earth seems to be at the center of a huge hollow sphere, and the name *celestial sphere* has been given to this nonexistent shell of infinite radius—a useful figment of the astronomer's imagination. Stars appear to line the inside of the sphere which seems to turn on an axis daily from east to west (Fig. 16–9). In

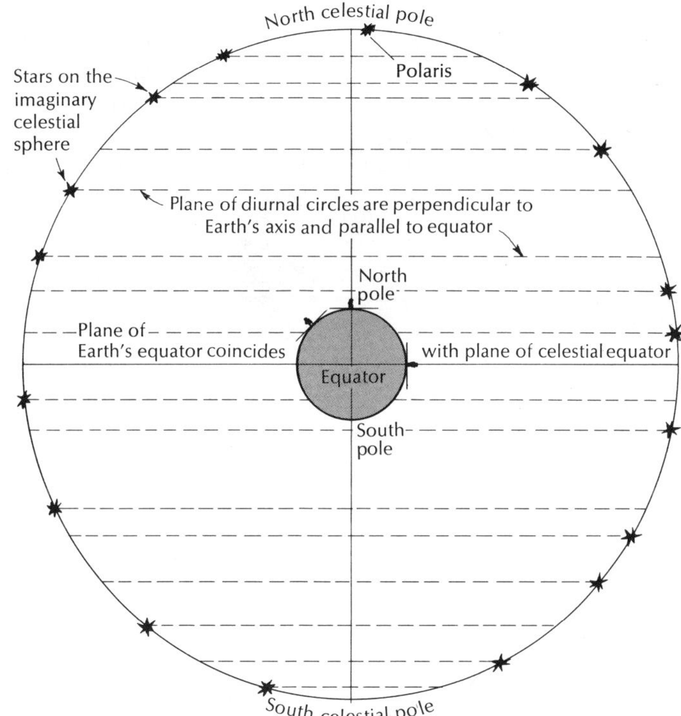

16-9. The celestial sphere and the diurnal motions of stars are shown schematically. The celestial sphere—imagined to have an infinite radius and the Earth at its center—does not exist. Three observers and their horizons are shown on the Earth. The diurnal motions of stars are parallel to the horizon at the poles, perpendicular to it at the equator, and at a slant to the horizon between the two.

the Twentieth Century, Polaris, the North Star, is about 1 degree from the northern end of this axis (the *north celestial pole*) and follows a small circular orbit in the sky. However, no similar bright star now occurs near the *south celestial pole*. The *celestial equator* is the great circle on the celestial sphere that is located midway between the two celestial poles (if a plane passes through the center of a sphere, its intersection with the sphere forms a *great circle*). Thus the plane of the Earth's equator coincides with the plane of the celestial equator.

In discussing diurnal motion, it is useful to refer to the *celestial meridian* of an observer, which is the great circle on the celestial sphere that passes through his zenith, the north and south celestial poles, and the north and south points on his horizon. From the latitude of New York the sun, moon, and planets are always highest in the sky and due south of the observer when they cross the meridian; in fact, the A.M. and P.M. of our clocks (ante meridiem and post meridiem) refer to the sun's passage across the meridian.

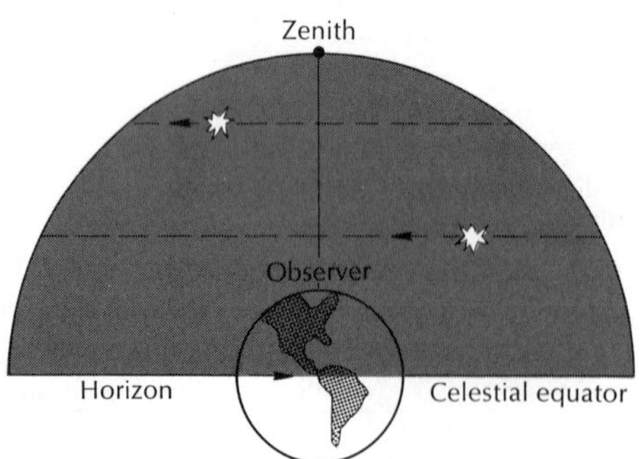

16-10. Motions of stars as viewed from the North Pole. The semicircle represents half of the imaginary celestial sphere. The observer's horizon at the North Pole is parallel to the equator and actually coincides with it—one plane instead of two as suggested by the distorted scale of the sketch—because the Earth is so small relative to the distances to the stars. As seen from the North Pole, stars follow circular paths that parallel the horizon. Stars north of the celestial equator never set, and stars south of the celestial equator never rise. *(R. H. Baker,* An Introduction to Astronomy, *Van Nostrand Reinhold, 1964.)*

Distances separating celestial objects are measured by the angles they subtend on the celestial sphere. Perhaps it is helpful in estimating the angle between two stars to imagine that a string extends in a straight line from each star to your eye. Obviously, it is meaningless to state that a certain star is "100 feet above the horizon," but the altitude of the star can be given in angular degrees, and we can describe the locations of stars by the angles that separate them on the celestial sphere. The angular diameters of the sun and moon are each about ½ degree; the pointers in the Big Dipper are about 5 degrees apart; and the square of Pegasus is about 15 degrees on a side. If a number of celestial objects have the same angular diameters, then their linear diameters are directly proportional to their distances.

The size of the diurnal circle of a star depends upon its distance from the celestial poles—the circle is small for a star such as Polaris that is near a pole, and the circle is largest of all for a star that is on the celestial equator. The orientation of the diurnal circles of stars (Figs. 17–1 and 24–8) also depends upon an observer's latitude and direction in which he looks: north, south, east, or west. Since Earthly observations are made from the center of the imaginary celestial sphere, when we observe a star near a celestial pole, we are looking at a 90-degree angle to the plane of its diurnal circle. On the other hand, when we observe a star near the celestial equator, we look edge-on at its diurnal circle (i.e., from the center of the circle outward toward its circumference), and such stars seem to follow parallel, nearly straight lines in the night sky. It follows that the planes of the diurnal star circles must always be perpendicular to the axis of the celestial sphere and parallel to the celestial equator (i.e., perpendicular to the Earth's axis and parallel to its equator).

As seen from the north or south geographic poles, therefore, the celestial poles would be located directly overhead (at the zenith), and stars would follow circular paths that parallel the horizon; they would not rise or set (the stars are all circumpolar, Fig. 16–10).

However, when viewed toward the east or west from the equator (Fig. 16–11), stars are seen to rise and set along parallel paths that are perpendicular to the horizon. Each day each star is above the horizon for 12 hours and below it for 12 hours, and all of the stars rise and set (disregard the effects of atmospheric refraction, p. 525). From the equator one can eventually see all of the visible stars in the sky. From a ship at sea at the equator, Polaris would always be within 1 degree of due north and the horizon, and stars would follow concentric circular paths about it. Different stars would follow similar paths above the southern horizon.

When we observe the sun, we note that the plane of its diurnal path through the sky always forms an angle with the horizon that is equal to 90 degrees minus the latitude of the observer. Thus at the equator, the sun always rises along a path that is perpendicular to the horizon. Therefore, it is above the horizon for 12 hours and below it for 12 hours each day (refraction increases the time the sun is above the horizon by 6 to 7 minutes in the middle latitudes). However, the sun's noon altitude at the equator varies from 66½ degrees (at the solstices) to 90 degrees (at the equinoxes). Because its path is always perpendicular to the horizon, twilight is always shorter at the equator than elsewhere on the Earth (p. 524).

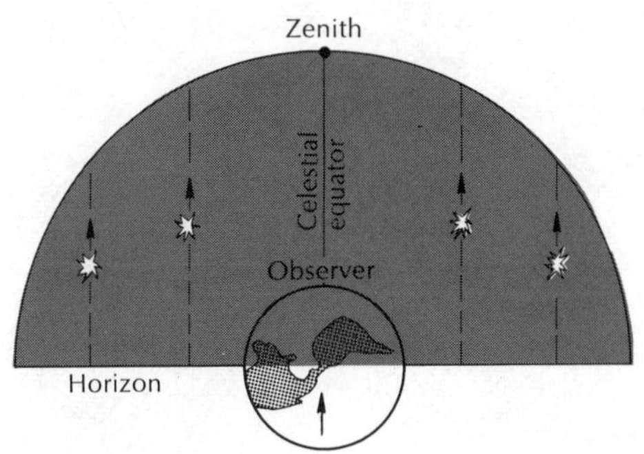

16-11. Motions of stars as viewed from the equator (see legend for Fig. 16-10). *(R. H. Baker, An Introduction to Astronomy, Van Nostrand Reinhold, 1964.)*

From latitudes between the poles and the equator, one observes stars that slant up in the east and down in the west. From

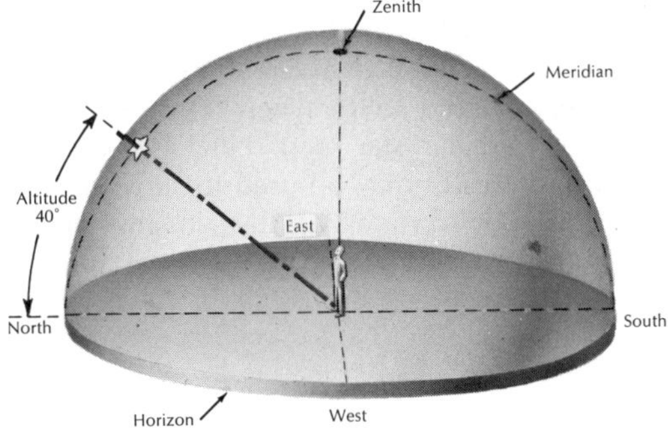

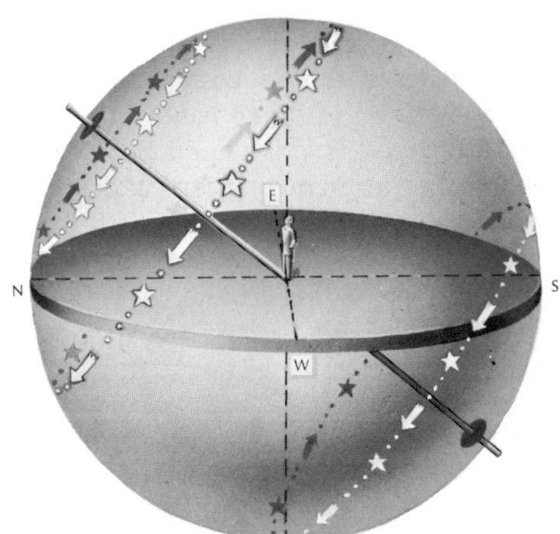

16-12. The Earth, sky, and stellar motions as they appear to an observer at 40 degrees north latitude. The Earth seems flat and the sky looks like the inside of a sphere (only half of which is visible from a single location). The North Star remains about 40 degrees above the horizon at all seasons of the year, and other stars appear to move from east to west in circular paths. Stars near Polaris never set, whereas stars farther away (i.e., making a larger angle with it) rise and set. *(After Rey.)*

16-13. Circumpolar star trails. Polaris is the bright star near the center (the north celestial pole). The number of degrees in one's latitude equals the number of degrees the center of this circumpolar motion is located above the horizon: e.g., at the north pole, the center is 90 degrees above the horizon; at 55 degrees latitude, it is 55 degrees above the northern horizon; and at the equator it is on the horizon. Approximately what time exposure was used? *(Courtesy Lick Observatory.)*

40 degrees north latitude (Fig. 16–12), stars slant up toward the south at a 50 degree angle to the horizon; they subsequently cross the celestial meridian and then slant down at a 50 degree angle toward the western horizon. At 70 degrees north latitude, Polaris would be 70 degrees above the horizon, and star trails would make 20 degree angles with the eastern and western horizons. On the other hand, at a latitude of 15 degrees, Polaris would be 15 degrees above the northern horizon, and star trails would make 75 degree angles with the eastern and western horizons. Thus the altitude of Polaris varies directly with the latitude (Figs. 16–13 and 20–5), but the slant angles of star trails with the eastern and western horizons vary as the complement of the latitude.

From south latitudes, star trails are similar except that they slant up toward the north instead of the south when they rise.

Semicircular star map sketches (Figs. 16–14 and 16–15) may be used to show the approximate locations of stars in any half of the sky (this is ¼ of the celestial sphere).

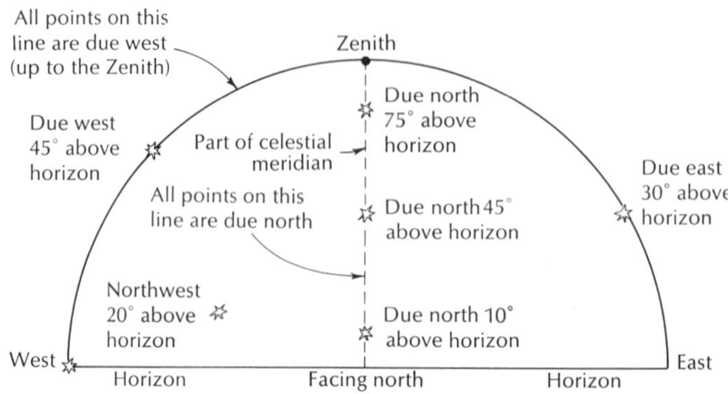

16-14. Star map sketch.

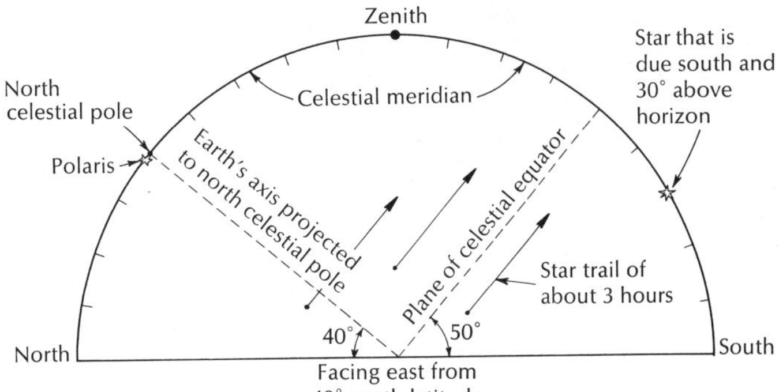

16-15. Star map sketch. Arrows showing 3-hour star trails would have somewhat different lengths in different parts of the sketch (they become shorter toward Polaris).

Assume that you want to plot the locations of a few constellations in the northern or eastern halves of the sky so that you can compare them with positions on other dates and times. In two dimensions, the straight line across the bottom of a semicircle represents the curved horizon. The semicircle itself represents a line that extends across the sky from east to west (if facing north) and passes through the zenith. A star's approximate direction and height above the horizon may be plotted on such a sketch as well as the position of the celestial meridian, star trails of different durations, etc.

Exercises and Questions for Chapter 16

1. Get a Star and Planet Finder (e.g., from Edmund Scientific Company, Barrington, N. J. 08007; 50¢ each in 1969) and learn how to "dial" on the finder the approximate location of a constellation or star for any date and hour.
   (A) Try to observe from areas where lights do not interfere. If some lights are nearby, try to block them out with a notebook or your arm.
   (B) Your eyes will adjust to the darkness over a period of several minutes or more. To maintain this adjustment, use a small flashlight with a red bulb to check the Star Finder (or cover the glass of the flashlight with red nail polish or cellophane). On some Star Finders, stars are marked by luminous paint and glow in the dark.
   (C) However, a larger flashlight, say one with five cells, makes a good "pointer" for a guided tour of the night sky.
2. Observe at least one constellation in each of the four main compass directions—north, east, south, and west. Make a star-map sketch of each of these and label the date and hour of observation.
   (A) Observe these four constellations at different hours during a single night. Plot their positions (perhaps on the maps you made in the exercise above) and make comparisons. Determine direction of movement and calculate the angular rate.
   (B) Make similar observations during subsequent months at the same

hour and plot positions on star maps. Again determine direction of movement and calculate angular rate.

(C) Perhaps you can contact students who live at different latitudes elsewhere in the United States or in other countries. Persuade them to make star-map sketches of the same constellations and then exchange copies. Explain any differences.

3. Calculate the approximate speed of light in mi/hr.
4. At a uniform rate of 5 mi/sec how long would it take to move from the sun to Pluto? From the sun to Alpha Centauri?
5. Assume that the Earth's orbit about the sun is a circle with a radius of 93,000,000 miles.
   (A) In terms of astronomical units, how far does the Earth travel in completing one orbit?
   (B) Calculate the Earth's average orbital speed in mi/sec.
6. Calculate the approximate dimensions of each of the following on a scale of 1 inch = 1000 miles. On this scale, the Earth is 7.9 inches, somewhat smaller than a basketball. This is an exercise to develop an awareness of relative sizes and distances of various celestial objects within our galaxy. Therefore, do not perform tedious additions or multiplications. Round off quantities wherever convenient (e.g., change a number such as 383,450 to 400,000). Use 5000 ft/mi instead of 5280 feet and use 60,000 in/mi instead of 63,360 inches:
   (A) Diameter of the moon (2160 miles).
   (B) Distance from the Earth to the moon (239,000 miles).
   (C) Diameter of the sun (864,000 miles).
   (D) Distance from the Earth to the sun (93,000,000 miles).
   (E) Distance from the sun to Pluto (3,670,000,000 miles).
   (F) Distance from the sun to Alpha Centauri (about 4.3 light-years; 1 light-year is about 6,000,000,000,000 miles).
   (G) Approximate diameter of our galaxy (i.e., the distance across the equatorial plane from one edge, through the nucleus, to the opposite edge); use 100,000 light-years as the diameter.
7. What is the approximate ratio between the diameters of the sun and moon? Between their distances from the Earth? Do these ratios explain why the sun and moon appear to be about the same size?
   (A) A ball with a 2-inch diameter is 10 feet away. At what distance would a 20-inch ball have the same angular diameter?
8. Show each of the following on semicircular star map sketches.
   (A) Assume an observer is at 40° north latitude and is facing south.
      (1) zenith
      (2) visible part of the celestial meridian
      (3) star that is due south and 30° above the horizon
      (4) star that is due east and 45° above the horizon
      (5) a few star trails of about 6 hours duration
   (B) Assume an observer is at 40° north latitude and is facing west.
      (1) Polaris
      (2) northern half of the axis of the celestial sphere (Earth's axis projected toward Polaris)
      (3) equatorial plane of the celestial sphere
      (4) a star that is due west and 60° above the horizon

    (5) a star that is due south and 10° above the horizon
    (6) a few star trails of about 4 hours duration
(C) Assume that an observer is at the north pole.
    (1) Polaris
    (2) some star trails of a few hours duration
(D) Assume that an observer is at the equator and is facing north.
    (1) Polaris
    (2) a star that is due east and 30° above the horizon
    (3) some star trails of 9 hours duration

17

Celestial Objects Viewed from a Moving Earth

Solar and Sidereal Days

Assume that a telescope is mounted so that it points due south continuously and that the sun is on the celestial meridian and centered in its cross hairs one noontime. About 24 hours later because of the Earth's rotation, the sun is again centered in the cross hairs, but continued measurements show that the time interval varies from a little less than 24 hours to a little more than 24 hours. This time interval between successive passages of the sun across the celestial meridian is known as the *solar day* and averages exactly 24 hours over a year.

If we repeat the observations but do so at night (Fig. 17–1) and use a remote object such as a star or a galaxy in place of the sun, we find that the time interval remains nearly uniform throughout the years at about 23 hours and 56 minutes—the so-called *sidereal day*.

Why does a 4-minute difference occur between the solar and sidereal days and why does the solar day vary in length? The orbital motion of the Earth about the sun causes these changes, whereas the Earth's rotation is not a factor (taken as uniform for this discussion, although the rate of spin is very gradually slowing, p. 566, and is very slightly irregular at times). Since the Earth moves about 1 degree each day in its orbit, the sun appears to move eastward about 1 degree a day with respect to the fixed stars (Fig. 17–11). The *ecliptic* is the name used for this apparent annual eastward path of the sun among the stars, but it may also be defined as the great circle on the celestial sphere formed by a projection of the Earth's orbital plane. The sun, of course, is always on the ecliptic, the planets are generally near it, and eclipses occur only when the moon is on or very near it—thus the name.

The Earth spins through 360 degrees in 23 hours and 56 minutes at the rate of 1 degree in 4 minutes or 15 degrees in 1 hour. Thus the daily 1-degree eastward shift of the sun means that the Earth must rotate

17-1. Star trails in the northern sky. The photograph was made in two parts: (1) A time exposure of about $1\frac{2}{3}$ hr was made through a motionless camera pointed at the northern sky (note Polaris, the Big Dipper, and part of the Little Dipper—it hangs down from Polaris). A plane moved across the field of view during the time exposure. (2) The camera was made to rotate with the stars to keep light from any one star focused on the same spot on the film. *(Photographed by John Stofan, Teaneck, N.J.)*

through 361 degrees—not just 360 degrees—to complete a solar day (Fig. 17–2); this takes 24 hours.

The length of a solar day (does not refer to amount of daylight) varies for two reasons. The Earth's orbital speed varies (p. 469), and when it revolves fastest (i.e., when the distance between A and B in Fig. 17–2

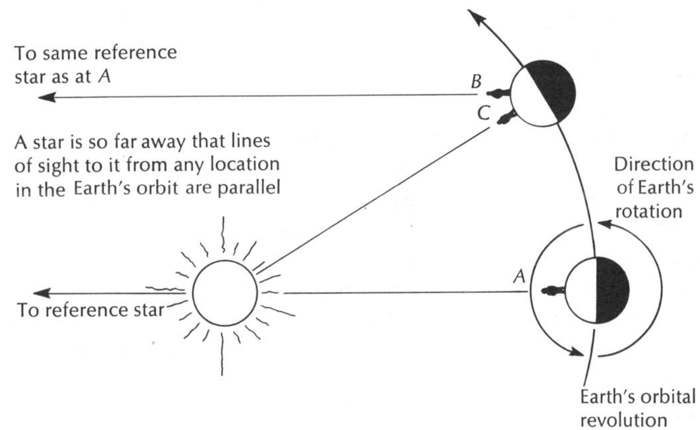

17-2. Solar vs. sidereal day. Orientation: in space looking down on the north pole. Schematic. An observer on the Earth at A at noon sees the sun and a reference star lined up due south on the celestial meridian. The observer is at B 23 hr and 56 min later after the Earth has rotated once through 360 degrees. The same reference star is again due south (lines of sight to the star are parallel from all parts of the Earth's orbit). However, the sun has not yet reached a due-south position. The Earth must spin about 4 min longer, or 1 degree farther, to move the observer from B to C.

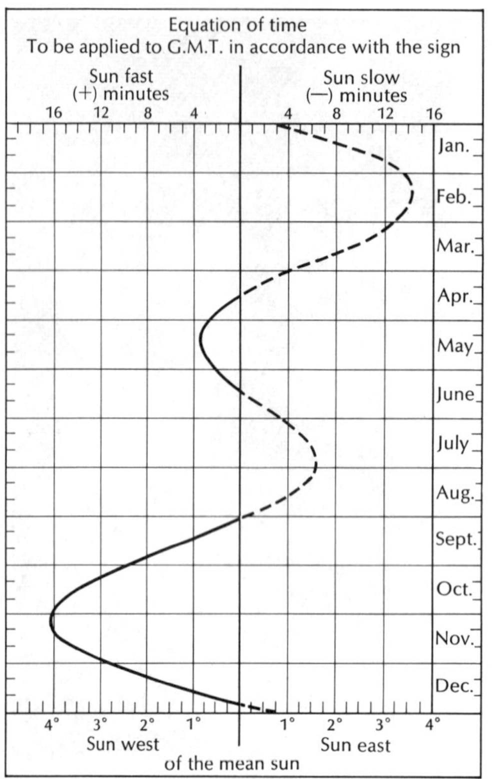

17-3. Equation of time. As shown, the sun is due south on the local celestial meridian at about 11:44 AM early in November and at about 12:14 PM early in February. *(G. W. Mixter, Primer of Navigation, Van Nostrand Reinhold, 1967.)*

is longer than average) the Earth must spin through slightly more than 361 degrees to complete a solar day. In other words, angle CSA varies. When it is largest, the Earth needs more than 4 minutes to rotate through it to complete a solar day; conversely, when angle CSA is smallest, the Earth needs less than 4 minutes to rotate through it. Since the Earth revolves fastest when it is at perihelion in early January, this tends to make solar days longer in winter than in summer.

In addition, the sun's eastward shift along the ecliptic must be projected to the celestial equator, because the solar day measures the Earth's period of rotation using the sun as a marker. Since the ecliptic and celestial equator are nearly parallel at the solstices (Fig. 20–10), the sun's eastward shift projects nearly undiminished to the celestial equator. However, at the equinoxes where the two planes intersect at an angle of 23.5 degrees, foreshortening occurs when the sun's eastward shift is projected to the celestial equator. It follows, therefore, that this projection effect tends to make solar days longest at the solstices (larger eastward shift) and shortest at the equinoxes (shorter eastward shift). The two effects tend to offset each other at times but are in step at other times.

Solar days, when averaged over a year, are exactly 24 hours in length. Therefore, for convenience in time keeping, we use a fictitious mean sun and pretend that it is due south on an observer's celestial meridian at exactly 12:00 noon on every day of the year. However, the true or apparent sun is not due south on the local celestial meridian at mean noon, except on four days each year, and it may arrive there early or late by as much as about 15 minutes (Fig. 17–3).

Diurnal and Annual Motions of Stars

The 4-minute difference between the sidereal and solar days causes stars to shift westward relative to the sun. As our clocks keep time with the mean solar day, each evening any given star is observed to rise 4 minutes earlier than it rose on the preceding evening and also to set 4 minutes earlier. If a certain star appears above the eastern horizon at 9:00 P.M. on, say, September 1, it will appear at 8:56 P.M. on the evening of September 2, and at 7:00 P.M. on the evening of October 1 (30 days at 4 minutes per day). At 9:00 P.M. on October 1, the star will be located 1/12 of a complete circle, or 30 degrees west of the position it occupied a month earlier at 9:00 P.M.

Because this daily westward shift of the stars is caused by the Earth's revolution around the sun, the stars will have shifted

360 degrees in a year, and this same star will again appear above the eastern horizon at 9:00 P.M. the following September 1. Constellations which are prominent in the early evening hours during the winter months are known as winter constellations; those prominent in the early evening hours of the summer are known as summer constellations.

If we observe the positions of stars on a certain hour and date, we can predict their positions at any other hour and date during the year (Figs. 17–4 and 17–5). Since stars complete their daily paths through the sky in about 24 hours, they shift 1/12 of the distance or 30 degrees in 2 hours, ¼ in 6 hours, and ½ in 12 hours. This is in a counterclockwise direction when facing north and clockwise when facing south (because direction clock faces has been reversed).

If observed at the same hour night after night, stars are also seen to shift their positions through 360 degrees once each year; they shift 1/12 of the distance in 1 month, ¼ in 3 months, and ½ in 6 months. The annual shift is in the same direction as the daily shift. To illustrate, we see Cassiopeia in a certain position at 8:00 P.M. tonight. Three months later and at 2:00 A.M. Cassiopeia will be shifted halfway around or 180 degrees from its present position.

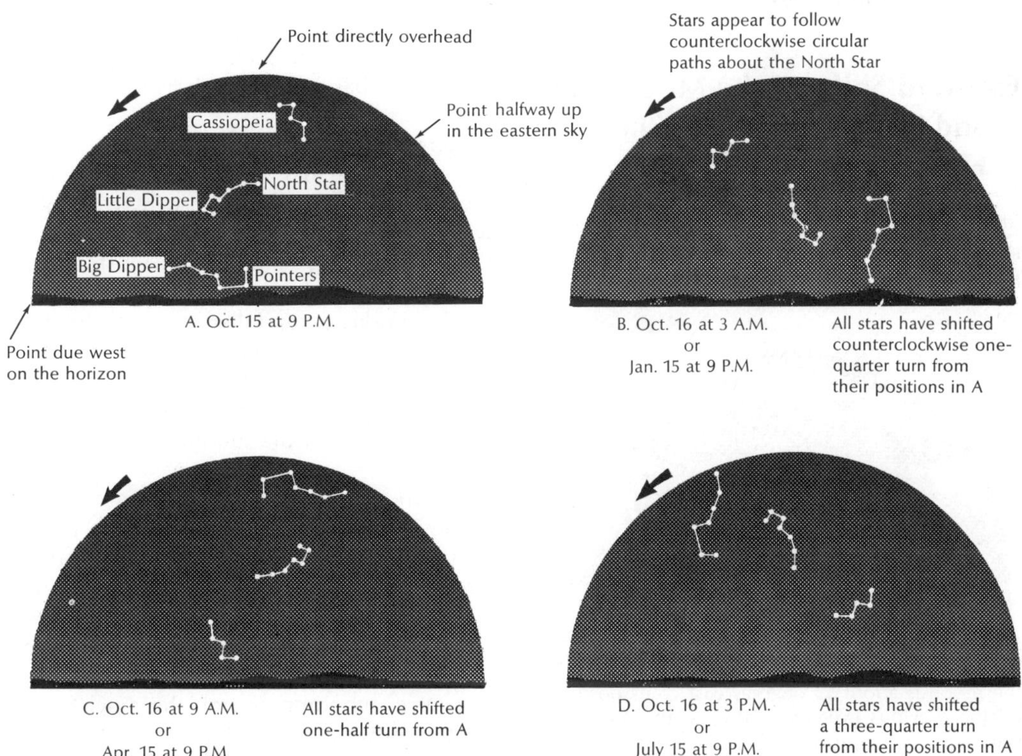

17-4. Star maps of the northern sky (see Fig. 16-14 for an explanation). Three familiar constellations are shown as observed from New York (40° latitude) approximately at the times listed, but the sketches are not to scale. Other stars are visible in this region of the sky, but they are not shown. The star maps show two motions: (1) the daily rotation of the celestial sphere, and (2) the annual shifting of the celestial sphere. These stars are near Polaris and stay above the horizon at all times, they do not rise or set and are called circumpolar.

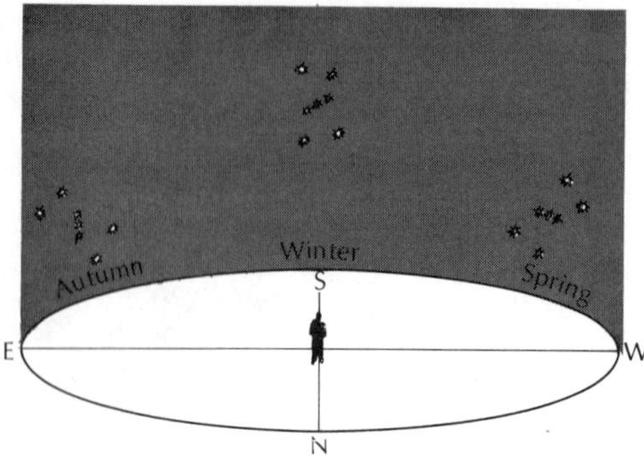

17-5. Orion as observed at the same hour (about 9:00 PM) in different seasons. The diagram can also be used to show positions on different hours in a single season: e.g., in the autumn, Orion is high in the south shortly after midnight and is low in the west at about sunrise. *(R. H. Baker,* Introduction to Astronomy, *Van Nostrand Reinhold, 1964.)*

Eastward Shifts of the Moon, Sun, and Planets Relative to the Fixed Stars

Stars actually are moving through space, some at rates of tens of miles each second, but their enormous distances prevent these movements from being discernible, except over the centuries. Within a lifetime, the shapes of the constellations do not change noticeably, and the stars are termed "fixed" since they maintain the same relative positions. This distinguishes stars from planets, which look like stars in the sky to the unaided eye but are not "fixed." *Planetes* is the Greek word for wanderer.

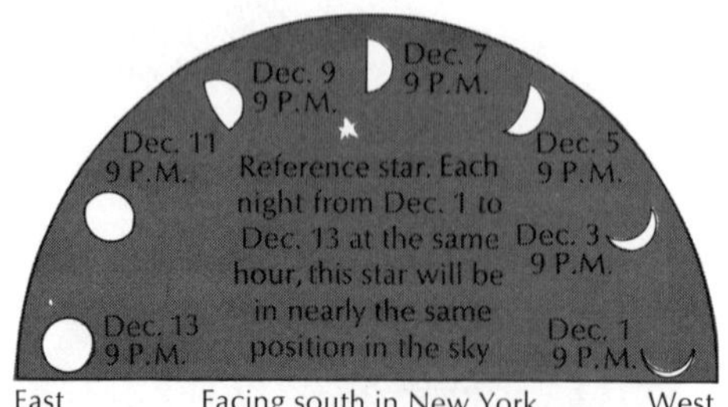

17-6. The moon's eastward shift relative to the background of stars. Each evening at 9:00 PM the moon is about 13 degrees east of its position on the previous evening relative to the reference star. The rate approximately equals the width of the full moon each hour. In about 27 days the moon completes an eastward swing of 360 degrees among the stars and returns approximately to its starting point. This eastward shift is superimposed upon the daily westward rotation of the heavens. The phases occur on different dates each year.

Because we observe the sun from a moving Earth, the sun appears to move always to the east relative to the stars. If stars were visible when the sun is observed, this motion would be quite apparent. However, we can detect it by noting the stars which appear in the eastern sky shortly before sunrise or in the western sky shortly after sunset (also by noting the antisun position at midnight). Such stars advance with the seasons, and precisely one year is needed for the stars to return to their original positions. This is the basis for the subdivision of time into years.

Stars are, of course, still present in the sky during the day. A constellation that is high in the sky during a winter evening will be high in the sky during the morning in the following summer. But just as the the stars appear dimmer on successive nights as the moon waxes fuller, so also do they become completely invisible in the much brighter glare of the sun's light. Stars can be seen during the day with a telescope, and they are visible during a total solar eclipse.

The moon circles the Earth in a counterclockwise direction as seen from the north. Like the sun, it appears to move rather steadily, and always to the east, about the width of a full moon per hour (Fig. 17–6). The eastward motions of the moon and sun, as well as the wandering of the planets, are, of course, superimposed on the much more rapid westward diurnal motion which results from the daily rotation of the Earth.

Over a long period of time, the average motion of each planet is to the east (*direct*), but over shorter periods, planets sometimes move with *retrograde* (westward) motion (Figs. 17–7 and 17–15). The length of time a planet spends within the boundaries of a particular constellation varies with its distance from the Earth: Mars averages 2 months; Jupiter, 1 year; Saturn, 2 years; and Neptune, 13 years.

Planets look like stars in the sky, although some are occasionally brighter than any of the stars. But Uranus is barely visible to the unaided eye, and Neptune and Pluto can be seen only through a telescope. The brightness of a planet varies, depending in part upon its constantly changing distance from the Earth. An observer may distinguish a planet from a star. (1) Planets tend to shine with a steady light, especially when they are high above the horizon, and stars tend to twinkle (p. 525). However, small planets also twinkle vigorously when seen near the horizon. (2) In a telescope, planets look like disks or parts of disks, whereas stars are mere pinpoints of light. (3) Planets wander slowly against the background of fixed stars, but this cannot be observed in a single night.

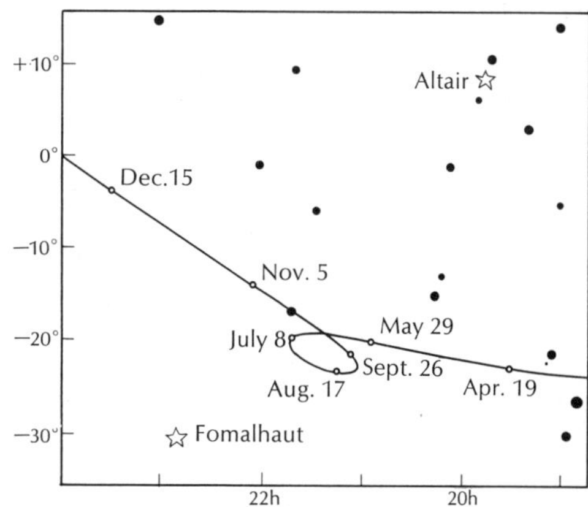

17-7. The path of Mars among the stars before and after opposition in 1971. Eastward is toward the left. See Fig. 17-15 for explanation. The westward part of the loop (to the right) is called retrograde motion. *(R. H. Baker, and L. Fredrick,* An Introduction to Astronomy, *7th ed., Van Nostrand Reinhold, 1968.)*

The Origin of Astronomy

Astronomy is one of the oldest of the sciences because answers were once needed for such simple questions as: What time is it? When will you be back? What direction should we travel? The motions of the sun provided some answers before watches, calendars, and the compass were invented. As seen from middle northern latitudes, the sun rises farthest to the northeast about June 22 and also sets farthest to the northwest (Figs. 17–8 and 17–9). On successive days, the sun then rises and sets progressively farther toward the south until about December 22 when it rises farthest to the southeast and sets farthest to the southwest. During this 6-month interval, the sun has also been rising later each morning and setting earlier each evening; the amount of daylight has been dwindling. At the equinoxes, about the 22nd of March and September, the sun rises due east and sets due west. The altitude of the noontime sun changes progressively also; it is 47 degrees higher on June 22 than on December 22 (Fig. 17–10). From December to June these changes are reversed.

However, two factors cause some of the changes pictured in the preceding paragraph to be less uniform than described. The noon

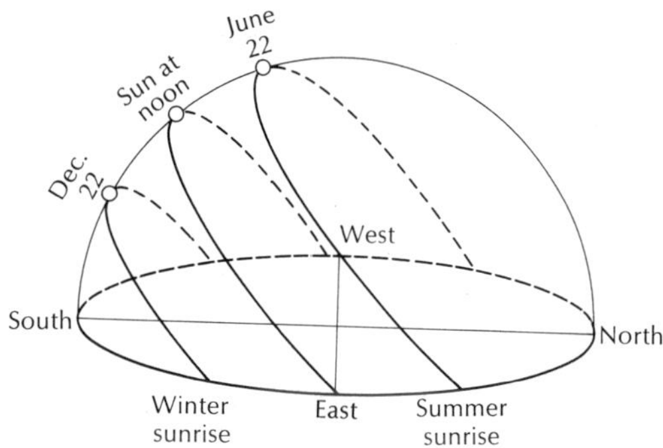

17-8. The sun's daily path through the sky during winter and summer as seen from our northern latitudes. The sun rises due east and sets due west only twice each year—spring and fall. Such changes in the sun's path are caused by the Earth's tilt, rotation, and revolution. *(R. H. Baker and L. Fredrick,* An Introduction to Astronomy, *Van Nostrand Reinhold, 1968.)*

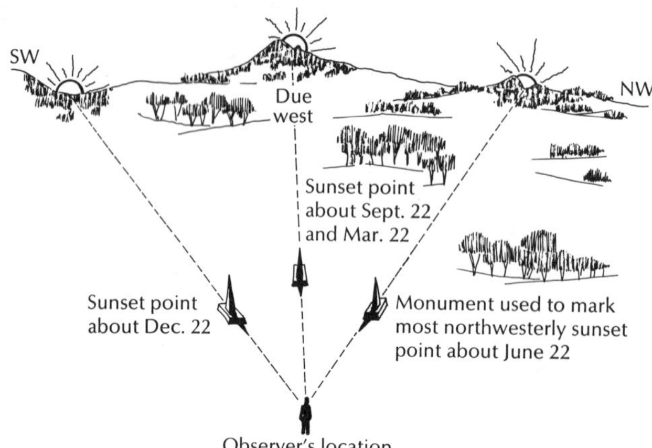

17-9. The shift of the sunset point was first plotted thousands of years ago to subdivide time into years. Impressive monuments were erected in some countries to mark various sunset or sunrise points.

sun changes altitude from one day to the next at a faster rate at an equinox than it does at a solstice (ecliptic and celestial equator are nearly parallel at a solstice). In addition, the noon sun is fast or slow in certain parts of a year (Fig. 17–3), which influences the times of sunrise and sunset. To illustrate, at a latitude of 40 degrees north, the earliest sunset occurs about 2 weeks before December 22 (when the noon sun is fast), and the latest sunrise occurs about 2 weeks after December 22 (when the noon sun is slow). However, the least amount of daylight does occur about December 22.

A vertical object such as a stick (gnomon) can thus be used to subdivide time into years and lesser units. The shadow points north at noon (as observed from the middle northern latitudes). This noontime shadow is shortest when the sun is highest about June 22 and longest about December 22. The sun to this day functions as a clock or calendar in some regions.

The splendor, beauty, and mystery of the heavens (undimmed by electric lights and smog) doubtless appealed to prehistoric man. Showers of meteors, the sudden appearance of a bright comet, a spectacular eclipse, stirred, frightened, and excited him. When men began to live in groups, knowl-

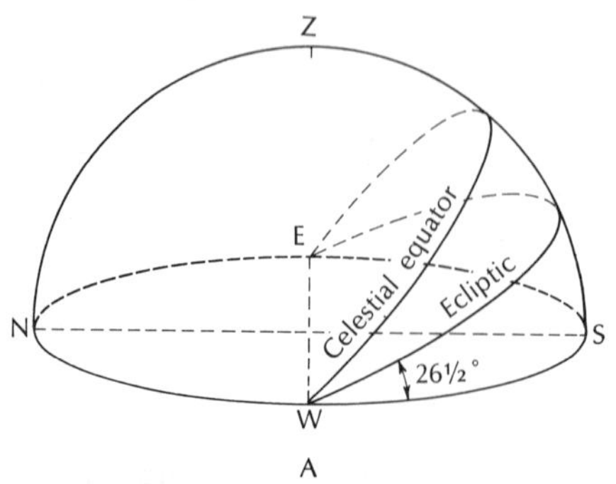

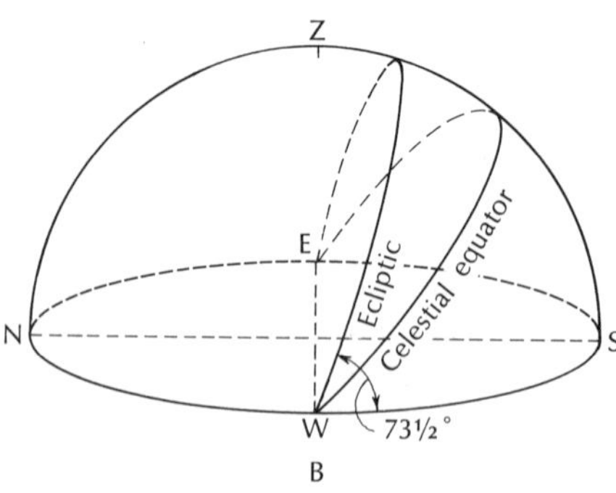

17-10. Relationship between ecliptic and horizon at 40 degrees north latitude (except for the precessional effect, the celestial equator remains fixed in space). The ecliptic has the least inclination to the horizon at sunset about September 22 (A) and the steepest inclination at sunset about March 22 (B). Since the sun, moon, and planets are always on or near the ecliptic, their paths through an observer's sky vary from low to high during a year. *(R. Baker and L. Fredrick,* An Introduction to Astronomy, *Van Nostrand Reinhold, 1968.)*

edge of the heavens became necessary. Dates had to be arranged for future hunts and campaigns, and the phases of the moon provided a ready calendar; observers learned that there were nearly 30 days between two successive full moons, and that approximately 12 full moons elapsed before a certain star rose again in the same place at the same time. On this basis the year was subdivided into twelve 30-day months.

Shepherds may have subdivided their night watches by the movements of familiar stars or constellations. The light of the moon was once much more important in the conduct of life than it is today: e.g., evening journeys were best undertaken at times of full moon. The pole star was an important aid in determining directions.

Specific seasonal events came to be associated with the positions of certain stars in the sky: the dates to pick elderberries or to plant corn, the mating seasons for cattle and deer, and the seasons of floods and frosts were all associated with the locations of certain bright stars in the sky that marked the time of the year. From this association of a star position with a seasonal event, it was only a step to ascribe the actual cause of the event to the star: e.g., the flooding of the Nile may have been attributed to the rising of Sirius in the east at dawn.

Geocentric and Heliocentric Explanations for Observed Celestial Motions

The age-old view that the Earth is at the center of the universe can still teach a lesson. Here is an erroneous concept that yet was believed almost without challenge for more than 2000 years. It is one of many examples in the history of science of abandoned theories that once went unchallenged. If the long argument between the *geocentric* (Earth-centered) and *heliocentric* (sun-centered) theories can teach us this, it is worth following.

The classic Greeks held a geocentric theory: the Earth was round, motionless, at the center of the universe. The sun, the moon, and the five bright planets known to them—the "seven stars" for which the days of the week are named—all revolved about the Earth. The universe itself was a huge hollow sphere which rotated completely from east to west once each day. Some scholars believed all stars to be the same distance from the Earth and fastened firmly to the inside of the sphere—perhaps like the heads of great golden nails.

Celestial objects within the sphere—the sun, moon, and planets—accompanied it in this daily movement and were themselves attached to hollow, transparent spheres. However, they lagged behind the stars in the daily westward rotation. Thus the sun (Fig. 17–11), moon and planets shifted eastward relative to the stars on the celestial sphere. As a matter of observation, the moon lags behind the stars the greatest distance in one day, the sun a lesser distance, and the planets shift eastward or westward at rates which vary.

A few of the Greeks realized that celestial motions could be explained on the basis of a moving Earth, but their concepts of sizes and distances in the universe had far

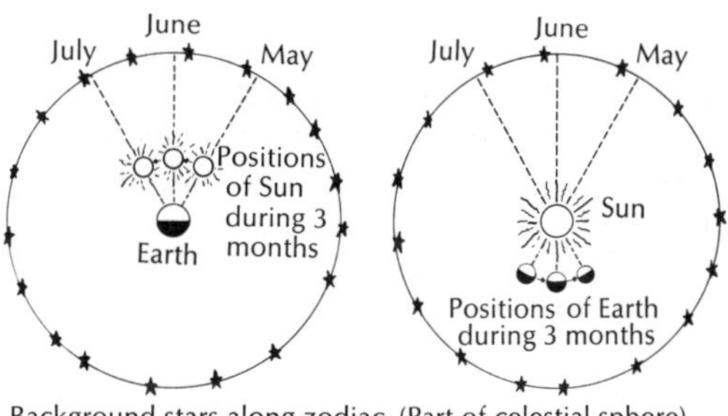

17-11. Eastward shift of the sun around the celestial sphere. The moon's eastward shift may be explained in a similar manner. Orientation: looking down on the north pole. A. Geocentric explanation: the sun and stars daily revolve westward about a round motionless Earth. B. Heliocentric explanation: the Earth rotates and revolves.

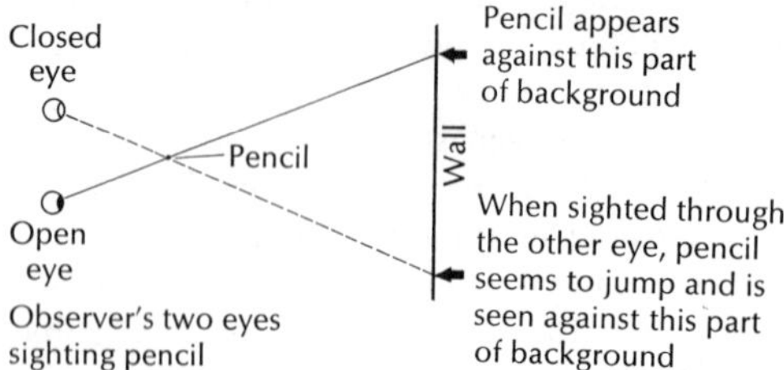

17-12. Parallax illustrated. Comparable changes in astronomy result from a shift in the observer's position on the Earth, from two observers making simultaneous observations from two different locations, and from changes in the position of the Earth in its orbit.

too small a scope. When the idea was brought forward in ancient times, as indeed it was, many difficulties kept it from being plausible. If the Earth is moving so rapidly, some argued, where are the powerful winds which should exist? Why are men not flung into space? And why do objects thrown into the air always return to the ground at the point where they were projected upward?

An apparently valid scientific test of the theory of the Earth's revolution around the sun was made (probably by Aristotle in the fourth century B.C.). If the Earth actually revolved about the sun, according to Aristotle, then the Earth's distance from the stars on the celestial sphere would vary, and the positions of some of these stars should appear to shift. This displacement is called *parallax*—the apparent shift of a near object against a distant background because the observer's position (or line of sight) has shifted (Figs. 17–12 and 17–13).

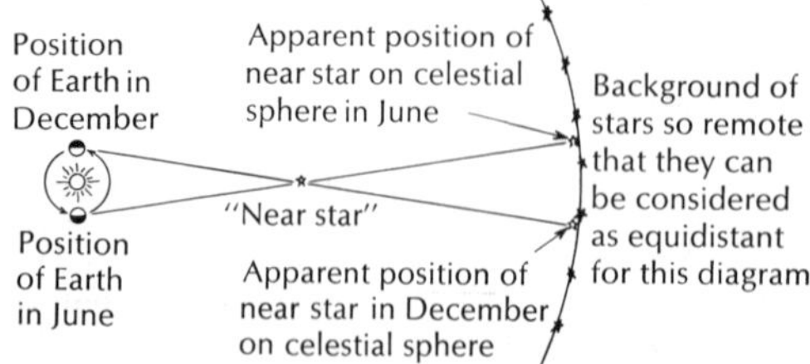

17-13. Parallactic displacement of a near star against a background of distant stars.

Aristotle searched for evidence of parallactic displacement, found none, and concluded that the Earth was motionless. However, his reasoning was correct, and parallax does occur; but the stars are so remote that parallactic displacements caused by the Earth's revolution about the sun were too small for him to measure. The first parallactic displacements were measured about 1835-1840. Measuring the largest one is approximately equivalent to measuring the diameter of a quarter at a distance of several miles. Parallax likewise provides a trigonometric method for obtaining the distances to the nearer stars p. 529).

In the second century after Christ, Ptolemy, the last of the great ancient astronomers, summarized, refined, and added to the geocentric theory, which afterward bore his name. He published a book, known by

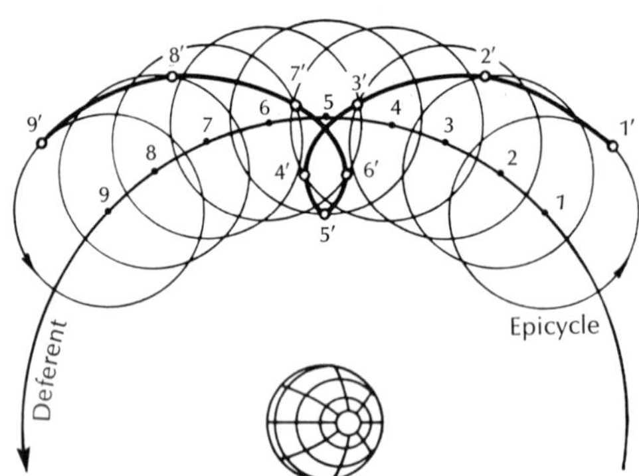

17-14. Geocentric explanation for planetary motions. Each 24 hours, it was supposed, the planet revolved almost completely around the Earth from east to west, but lagged slightly behind the stars and so shifted slowly eastward in front of them. To account for this and the occasional retrograde loop, the planet was conceived as revolving about an imaginary point which in turn revolved about the Earth (along the deferent in the sketch), like a point on the rim of a spinning wheel revolving about the Earth. Thus the planet moved through a looped path in space. When the center of the epicycle is at 1, the planet on the epicycle is at 1′. Other numbers show the locations of the center of the epicycle and the planet at regular intervals. Eastward is in the counterclockwise direction. *(R. Baker,* Astronomy, *Van Nostrand Reinhold, 1964.)*

its Arabic name as *Almagest,* which remained the outstanding astronomical work for nearly 1500 years.

Not all the Greek astronomers had been devoted to the geocentric theory. Aristarchus determined about 270 B.C. that the sun was much larger than the Earth and advocated a heliocentric theory. He noted that the first-quarter moon was due south at sunset, evidence that sunlight strikes both the Earth and the moon at the same angle (Fig. 22–5). The sun must thus be very far away. Furthermore, to appear about the size of the nearby moon, the sun has to be a huge object. Therefore, to Aristarchus it seemed more logical that observed celestial motions resulted from the Earth's rotation and revolution than from the movement of the sun and celestial sphere.

Eratosthenes measured the size of the Earth during this period (Fig. 20–3), and Hipparchus measured the distance to the moon (Fig. 21-2), discovered the precession of the equinoxes, used epicycles in explaining the motions of planets (Fig. 17–14), and classified the stars according to brightness. His contributions to astronomy may have been the greatest of all those by ancient investigators. Hipparchus also made and recorded a number of observations especially for the use of future astronomers; he realized that certain celestial motions were so slow that several lifetimes would be necessary to map them.

The geocentric explanation persisted almost without challenge from the sixth century B.C. until the time of Copernicus. The theory lasted so long because it was based on straightforward observation of celestial phenomena, and because with its aid accurate predictions concerning future planetary positions and the times of eclipses could be made.

Founders of Modern Astronomy

Five men had a very powerful impact on astronomy and science in general in the sixteenth and seventeenth centuries: Copernicus, Tycho Brahe, Kepler, Galileo, and Newton. During these centuries, new instruments were produced: e.g., the telescope, microscope, barometer, pendulum clock, vacuum pump, and thermometer. More precise quantitative measurements became possible. Mathematics and science were gradually accepted as subjects worthy of study by top scholars. Conclusions that had been accepted for centuries because they had been reached by some ancient authority such as Aristotle were now questioned and discarded if necessary. Experimentation and observation to test hypotheses became more common. Cause and effect relationships were sought to explain natural phenomena; inquiring men searched for the fundamental natural laws that would explain the universe they observed, and the regularities of many celestial motions indicated that such laws existed. Thus the origin of astronomy is in a sense the story of the development of science itself.

Copernicus

In the sixteenth century Nicolaus Copernicus (1473–1543), a Polish churchman-astronomer, stated the heliocentric (Copernican) theory: the Earth rotates and revolves. This new and startling hypothesis—in direct opposition to an explanation that had not been seriously questioned for 2000 years—needed strong supporting evidence to be accepted. Copernicus lacked the convincing data, although a moving Earth accounted better for the varying brightnesses of the planets. He could merely state that his theory was simpler (e.g., see explanation for retrograde motion, Fig. 17–15) and that it seemed more logical for the tiny Earth to rotate and revolve than for the entire universe to do so.

As Copernicus presumed circular orbital motion, predictions of future astronomical events were not significantly more accurate than those based on the Ptolemaic theory, and some epicycles were still necessary to explain planetary motions. Copernicus did predict that Mercury and Venus would show

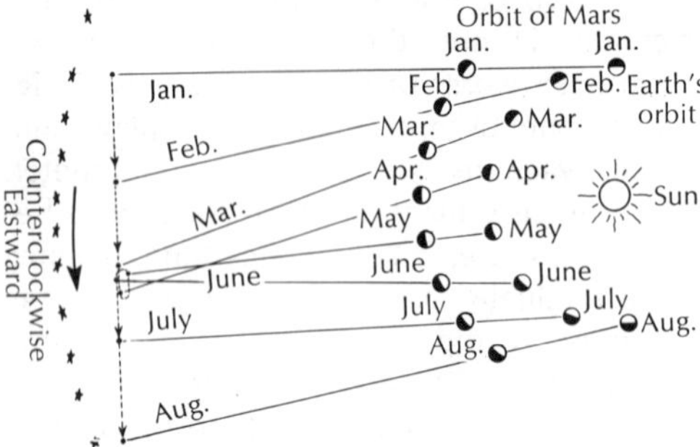

17-15. Heliocentric explanation for planetary motions (Mars). The apparent positions of Mars among the stars, as viewed from the Earth, are plotted for each month. The Earth revolves at a faster rate than Mars and covers a greater angular distance in any one month. Thus the Earth occasionally overtakes Mars and passes between it and the sun. At such times Mars appears to shift westward in front of the stars and to follow a looped path (Fig. 17-7). At all other times Mars shifts eastward against the stellar background at rates which vary. For Mars the retrograde motion may last about two months and occurs about every other year.

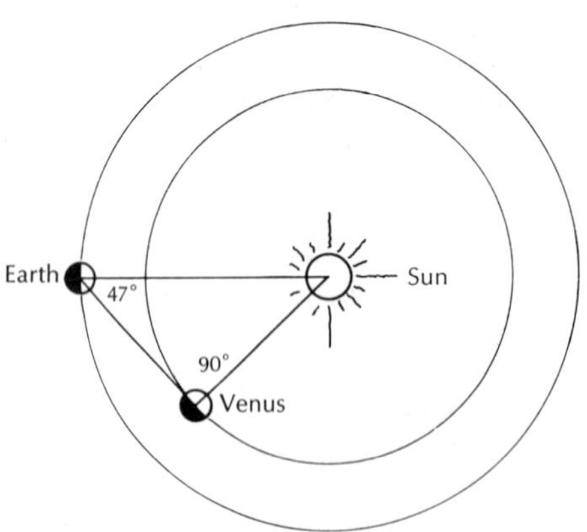

17-16. Method used by Copernicus to ascertain the distance to Venus. A right-angled triangle is formed by the sun, Earth, and Venus when Venus attains its maximum angular displacement from the sun in its first- or third-quarter phase.

$$\text{Sine } 47° = \frac{\text{distance of Venus from sun}}{\text{distance of Earth from sun}} = 0.73$$

Therefore distance of Venus from sun = 0.73 × distance of Earth from sun

(R. H. Baker, Astronomy, *Van Nostrand Reinhold, 1964.)*

phases and calculated the distances of Mercury and Venus from the sun in terms of the astronomical unit (Fig. 17–16).

Tycho Brahe

At the age of sixteen, Brahe (a Danish nobleman, 1546–1601) realized that the Ptolemaic-Copernican controversy could not be resolved without accurate information on the positions of the planets, sun, moon, and stars. Accumulating precise quantitative data as a basis for later theorizing, now commonplace in scientific research, was an entirely new technique in the sixteenth century. Brahe was determined to make the necessary observations and did so throughout his lifetime. Like Aristotle, Brahe searched for parallactic displacement and found none. Therefore, he thought that the sun and moon revolved about the Earth, although he had the other planets revolving about the sun.

Kepler

Johannes Kepler (1571–1630), a German mathematician-astronomer, was a student of Tycho Brahe at Prague and had access to his observations after Brahe's death. These two scholars complemented each other's talents nicely—Brahe, the observer, and Kepler, the theorizer. No hypothesis then known was completely satisfactory in accounting for planetary motions, and Kepler spent many years trying to devise a better explanation. He attempted to reconcile the motions of planets with circular orbits, epicycles, and other geometric figures. By using Brahe's precise observations of Mars over a span of two decades, Kepler finally hit upon elliptical paths as the explanation. He eventually developed three laws that precisely described the motions of the planets, for which he was dubbed, "Legislator of the Heavens."

(1) *Every planet follows an elliptical orbit about the sun; the sun is located at one focus of the ellipse* (Fig. 17–17).

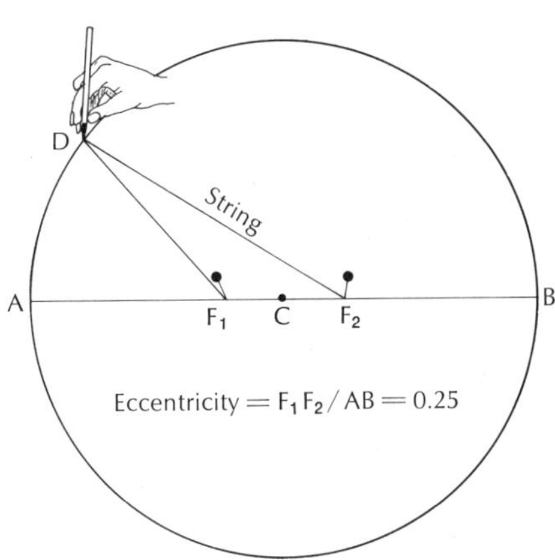

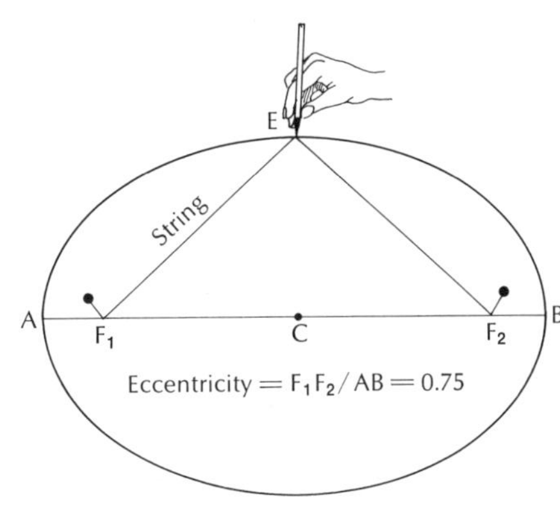

17-17. An ellipse can be drawn by means of a string tied to two pins. The length of the string equals the length of the major axis (AB) and is the same in each of these ellipses. Note that eccentricity describes the shape of an ellipse, whereas the major axis describes its size. Since the length of the string is a constant, F_1D plus F_2D equals F_1E plus F_2E, and this holds true for all other points on the two ellipses.

An *ellipse* is a closed curve which is the path of a point moving in such a way that the sum of its distances from two fixed points, the foci, is a constant. An ellipse is also produced by the closed intersection of a plane with a right circular cone (Fig. 23–15). A circle may be regarded as a special type of ellipse; both foci are superposed at the center. AB in the figure represents the major axis of the ellipse, and either AC or CB is the semimajor axis or mean distance. As the distance between the foci increases, the ellipse gets flatter and is said to increase in *eccentricity*—numerically equal to the distance between the foci divided by the major axis ($F_1F_2 \div AB$) or to the distance from the center to a focus divided by the semimajor axis ($F_1C \div AC$). The eccentricity of an ellipse may range from zero (a circle) to 1 (a parabola).

In the solar system, the sun is at a focal point for each of the elliptical orbits followed by the celestial objects that revolve about it: planets, asteroids, comets, and meteoroids. *Aphelion* refers to the point in any one orbit that is most distant from the sun, whereas *perihelion* refers to the point that is closest to the sun. *Apogee* and *perigee* are similar terms applied to objects orbiting the Earth (helio- means sun; geo- means earth). Such points occur at opposite ends of the major axis.

Kepler learned that Mars moves in an elliptical path by laboriously plotting points in its orbit. But first he had to calculate its sidereal period as 687 days (Fig. 17–18). This was based upon Brahe's observations which showed that 780 days elapsed between successive oppositions of Mars (Fig. 17–7). Kepler plotted about forty points in the orbit of Mars by using forty pairs of observations, each pair made 687 days apart. Kepler could then readily see that the orbit of Mars was not a circle, that the sun was not centered, and that the orbital speed of Mars varied.

(2) *A line joining the center of each planet with the center of the sun sweeps over equal areas in equal periods of time* (Fig. 17–19).

The length of an imaginary line from the center of the Earth to the center of the sun is longest at aphelion (when the Earth's orbital speed is slowest) and shortest at perihelion (when the Earth's orbital speed is fastest). Thus the length of the line varies inversely as the Earth's orbital velocity, and

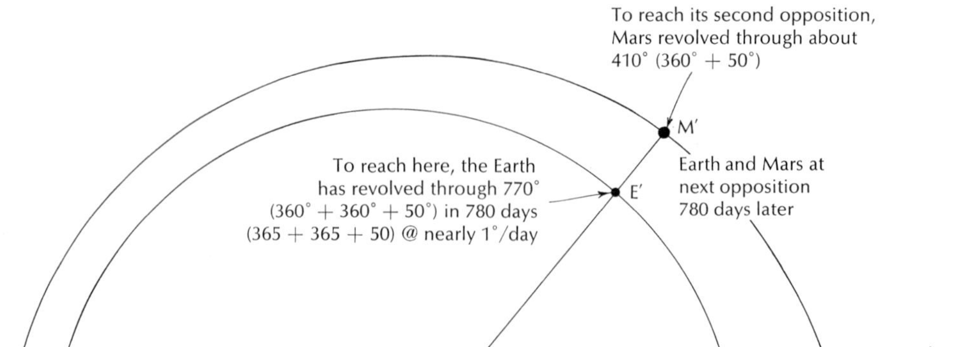

To reach its second opposition, Mars revolved through about 410° (360° + 50°)

M′

To reach here, the Earth has revolved through 770° (360° + 360° + 50°) in 780 days (365 + 365 + 50) @ nearly 1°/day

E′

Earth and Mars at next opposition 780 days later

Earth and Mars at first opposition

Sun

50°

43°

Line of sight from Earth to Mars at first opposition

E

M

Line of sight from Earth to Mars 687 days later when Mars had returned to its original position

Position of Earth 687 days later (it is about 43 days — thus 43° — shy of completing a second orbit)

E″

17-18. Technique used by Kepler to calculate: (1) the sidereal period of Mars and (2) a point in its orbit. Since Mars revolves through about 410 degrees in 780 days, it must revolve through 360 degrees in 687 days. Thus Mars returns to the same position in its orbit at 687-day intervals. Therefore, lines of sight from the Earth to Mars, made 687 days apart, intersect at a spot in the orbit of Mars.

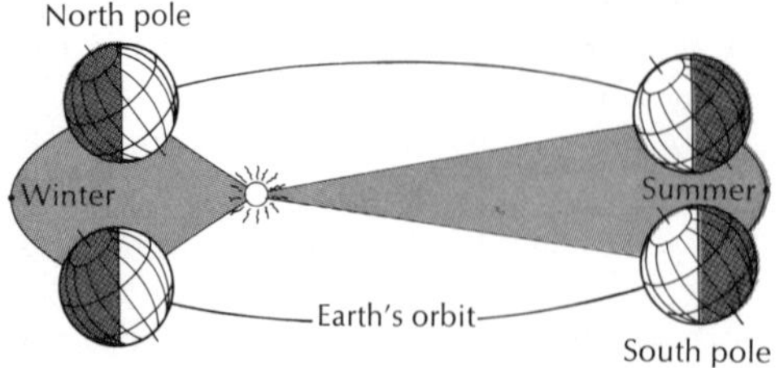

17-19. Following Kepler's equal areas law, the Earth revolves fastest in winter when it is closest to the sun and slowest during summer. *(R.H. Baker,* Astronomy, *8th ed., Van Nostrand Reinhold, 1964.)*

equal areas of space are swept over by this line in equal periods of time. The equal areas law holds for all orbiting bodies: for a satellite around a planet, for a comet or meteoroid about the sun, and for an artificial satellite about the Earth.

(3) *The squares of the periods of revolution of any two planets are in the same ratio as the cubes of their mean distances from the sun:*

$$\frac{(\text{Period of planet A})^2}{(\text{Period of planet B})^2} = \frac{(\text{dist. of planet A})^3}{(\text{dist. of planet B})^3}$$

A and B in law (3) may represent any two planets. If B represents the Earth, then both denominators will be equal to 1: the period of revolution of the Earth is 1 year, and its distance from the sun is 1 astronomical unit (nearly 93 million miles). Thus if a planet occurs at a mean distance of 4 astronomical units from the sun, its period of revolution must be 8 years.

The law indicates that with increasing distances from the sun, planets not only require longer periods of time to make complete revolutions about the sun, but also move more slowly in their orbits; e.g., Mercury has an average orbital velocity of about 30 mi/sec, the Earth about 18½ mi/sec, and Pluto approximately 3 mi/sec. As a similar relationship holds for any one planet in its own orbit—it moves fastest when closest to the sun—a fundamental relationship is evident between distance to the sun and orbital velocity. In fact, this relationship applies to all orbiting bodies: e.g., to the moons of Jupiter and to the particles forming Saturn's rings.

Galileo

Galileo Galilei (1564–1642), the Italian astronomer-physicist, was converted to the Copernican system early in his career and is perhaps best known for his use of the telescope in astronomy.

Near the beginning of the seventeenth century, Galileo learned that a Dutch spectacle-maker had constructed an instrument which magnified distant objects. The value of such an instrument in astronomy was immediately realized by Galileo, and he soon built a number of telescopes of his own. His first telescope magnified only three times, but a later one magnified 30 times. A modern pair of binoculars is a better instrument than Galileo's first telescope.

Imagine the surge of intense interest and excitement which Galileo felt as he pointed this new instrument at various celestial objects. He observed celestial phenomena which strongly supported the heliocentric theory and discovered that the Milky Way consisted of the combined light of thousands of stars, each invisible without optical assistance. He discovered four of Jupiter's moons (Fig. 17–20), and eight other moons have been found since. Clearly the Earth was not the center of all celestial motion, and Jupiter and its moons resembled a miniature solar system. Perhaps the Earth and the other planets moved about the sun in a similar manner.

Other startling discoveries followed at once. Through the telescope, the moon was seen to have an irregular surface marked by craters, mountains, and other topographic features; it definitely was not smooth. Galileo developed a method for measuring the heights of mountains on the moon. On the sun's surface he saw small black spots. During a series of observations, these spots moved as a group from one side of the sun to the other (Fig. 24–3), and Galileo concluded that the sun's rotation causes this. Most devastating to the geocentric theory was Galileo's discovery that Venus and Mercury showed phases during which they appeared to change in size (Figs. 23–4 and 23-5) — incompatible with circular orbits about the Earth.

17-20. Jupiter and its satellites emerging from behind the moon on January 16, 1947. The moon is illuminated by earthshine (Fig. 22-6). Three of the satellites are plainly visible. A fourth appears as a speck near the left edge of Jupiter's disk; it is just emerging from behind Jupiter. Jupiter was behind the moon for about 50 minutes because the moon shifts eastward about the width of a full moon in an hour. Note that the moon shows no sign of an atmosphere. *(Photographed by Paul E. Rogues, Griffith Observatory.)*

Galileo's support of the Copernican theory eventually led him into conflict with the church, and he is reported to have remarked that the Bible was written to help men get to heaven, not to tell men how the heavens move. Apparently Galileo was unaware of Kepler's three laws because he had read only the preface to Kepler's book.

Newton

It remained for Sir Isaac Newton (1642–1727), whose name is as central in astronomy as in physics and mathematics, to explain why planets follow paths about the sun. The force that keeps the planets moving continuously in their orbits had puzzled inquisitive men for hundreds of years. Kepler had thought that some force must be pushing the planets around. Galileo anticipated Newton by proposing that no such force was necessary, but realized that a force was needed to keep the planets from moving out into space along a straight line.

Experience on Earth suggests that objects move only when forces are applied to them and that the motions soon stop when the forces are no longer applied. However, we live on an Earth where air resistance, surface irregularities, and internal moving parts all produce friction. What would happen to a moving object if friction could be eliminated, as it is in interplanetary or interstellar space? Newton supplied an answer in his first law of motion (Galileo had already grasped the essence of this law): *Every object remains at rest or in uniform motion in a straight line unless it is compelled to change that state by a force impressed upon it.* Thus, in the absence of friction, uniform motion in a straight line is as natural a state as rest, and no force is needed to account for its continuation. Yet something is left unexplained. A planet does not move in a straight line at a uniform speed through space. Instead it follows an elliptical orbit about the sun. Some force must be acting upon it.

Several men at this time were pondering the mysterious force that kept a planet in its orbit, and they wondered if this force might be caused by the sun and be inversely proportional to the square of the distance of a planet from the sun. However, they could think of no proof (calculus had to be invented first); the genius of Newton was necessary to state and prove the great unifying law of universal gravitation: *Every particle of matter in the universe attracts every other particle with a force which is proportional to the product of their masses, and inversely proportional to the square of the distance between them.*

In equation form, the attraction between any two spherical objects is shown by

$$F \propto \frac{m_1 \times m_2}{d_2}$$

where F is the gravitational attraction, the symbol $\propto$ means "is proportional to," m_1 is the mass of one object, m_2 is the mass of the second object, and d is the distance between their centers (a sphere acts as if all of its gravitational attraction were concentrated at its center). As an equality, we have

$$F = \frac{G \times m_1 \times m_2}{d_2}$$

where G is the gravitational constant. In the metric system (dynes, grams, and centimeters) G is equal to 6.67×10^{-8} a very small quantity. Two 1-gram masses 1 centimeter apart have an attraction of 6.67×10^{-8} dynes. It follows, therefore, that the Earth is attracted gravitationally (perturbed) by each member of the solar system, but the sun is so very massive that its pull far exceeds the combined attractions of all of the other members.

In order to force a moving object such as a planet to curve out of a straight path, a force must be exerted at an angle to this path. A planet's momentum or inertia tends to make it move in a straight line at a uniform rate. On the other hand, the sun exercises a gravitational pull directly toward itself on a planet. Thus a planet is deflected from a straight line into an elliptical orbit (Fig. 17–21). In other words, to keep a planet from moving off along a tangent, an unbalanced force toward the sun is required (a centripetal or "center-seeking" force) and this is furnished by the gravitational attraction of the sun.

In moving forward along its orbital path, a planet "falls" continuously toward the sun. For the Earth, this distance is less than ⅛ inch in 1 second, and during this second the Earth moves forward about 18½ miles in its orbit. The moon likewise falls continuously toward the Earth as it revolves, and Newton used this motion to prove the inverse square part of his law of universal gravitation. The moon's mean distance from the Earth (about 240,000 miles) is about 60 times longer than the Earth's radius. Thus the Earth's gravitational attraction at the moon's distance should be only $(1/60)^2$ of what it is for an object at the Earth's surface—it should be about 3600 times smaller. Therefore, the moon should fall toward the Earth 3600 times more slowly than does a falling object at the Earth's surface. Newton was able to show that this is the case. However, Newton could not explain why gravitational attraction occurs, and indeed it is mysterious today.

Circular and elliptical orbits are the only closed curves along which objects can revolve repeatedly. At aphelion (position A in Fig. 17–22) a planet's speed is too small to carry it in a circular orbit about the sun, and so it is pulled inward by the sun's gravitational attraction and moves to position B. The planet is now moving more rapidly because the acceleration toward the sun has a component along the orbit in the direction the planet is moving. The planet's speed increases as it falls closer to the sun, and its orbit is curved sharply—the sun's gravitational attraction increases rapidly as the distance decreases. At perihelion (position C) the planet's speed exceeds that for a circular path about the sun at this distance, and the planet swings outward from the sun to position D. At D the sun's gravitational pull is slowing down the planet. Therefore, it is deflected toward the sun in its orbit and reaches aphelion again with precisely the same speed that it had

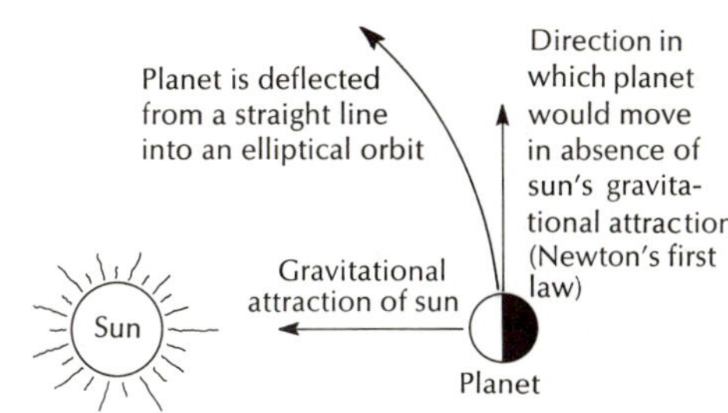

17-21. Forces governing planetary orbits.

at the start. Thus it follows the same orbital path time after time (see also Fig. 18–4).

If the sun's gravitational attraction were somehow increased, the average distance of the Earth from the sun would decrease, and its average speed would increase. If the Earth were somehow stopped motionless and then released, it should move directly toward the sun.

According to Newton's second law of motion: *The product of the mass of an object times its acceleration varies directly as the resultant force, and the change in motion takes place in the direction of that force.* This is familiar to many as the equation $F = ma$, where F is the force (in dynes),

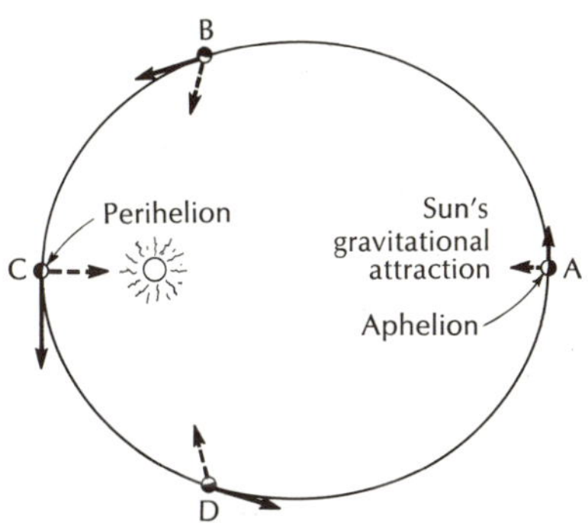

17-22. Motion in an elliptical orbit. The lengths of the solid arrows tangential to the orbital positions A, B, C, and D are proportional to the speeds of the planet at these positions, and the lengths of the dashed arrows are proportional to the gravitational pull of the sun, but neither is to scale. *(After McLaughlin,* Introduction to Astronomy, *1961, p. 160.)*

m is the mass (in grams), and *a* is the acceleration (in centimeters per second squared) or change in the velocity per unit of time. Note that acceleration may consist of a change in direction without a change in speed—e.g., revolution in a circular orbit. This law indicates that a certain force will accelerate a large mass more slowly than a small one.

According to Newton's third law of motion: *To every action there is always an equal and contrary reaction.* Familiar examples are the recoil of a rifle and the movement of a small boat as one dives off. This has led to a common error—the assumption that there is an outward force on a planet (a centrifugal or "center-fleeing" force) that is equal in magnitude but opposite in direction to the inward gravitational pull of the sun. Two equal and opposite forces do exist, but the two forces act on two different objects. The gravitational attraction of the sun produces an inward, unbalanced (centripetal force) on a planet. This is a very large force because of the enormous mass of the sun.

Following Newton's third law of motion, the gravitational attraction of a planet produces an equal and opposite force on the sun, but the sun's motion is changed only a small amount because it is so very massive. Thus the sun and Earth attract each other with equal and opposite forces, but the sun is accelerated very little because it is 333,000 times more massive than the Earth.

In the familiar and comparable example of a ball being whirled around a circular path at the end of a string, equal and opposite forces also exist, but there is no outward force on the ball. There is, however, an unbalanced force on the ball that is directed inward toward the hand. In a sense, one end of the string pulls the ball inward (the centripetal force*), the other end of the string pulls the hand outward (this is the actual centrifugal force). Thus if the string is cut, the ball flies off along a tangent to its former path; it does not fly directly outward.

* For circular motion, it can be shown that $F = \frac{mv^2}{r}$ where F is the centripetal force, m is the mass of the moving object, v is its speed along the circumference of the circle, and r is its radius.

Newton's laws apply to all freely falling bodies. If air resistance is disregarded, bodies of different masses located at the same distance from the Earth will fall toward it at the same rate; e.g., a 100-ton object would fall at the same rate as 1-ton and 1-gram objects. This holds because differences in the masses of most such falling objects near the Earth's surface are negligible as compared to the total mass of the Earth itself. However, 1-gram or 100-ton objects in orbit about the Earth at the moon's distance would fall at a slightly slower rate than the moon (p. 538).

Newton suggested that stars must be situated at tremendous distances from the sun not to be influenced by its gravitational attraction. The sun, which he considered a star, would have to be thousands of times as far from the Earth as it is, he said, in order to appear as dim as the stars. Newton was thus one of the first scientists to realize the enormous distances which lie between the Earth and the stars. Lagrange said of him: "Newton was the greatest genius that ever existed and the most fortunate, for we cannot find more than once a system of the world to establish."

Evidence Supporting the Laws of Kepler and Newton

A striking proof of the law of universal gravitation occurred about the middle of the nineteenth century. The planet Uranus had been discovered more or less by accident in 1781 (the discovery was a by-product of astronomical research and not the result of a planned search). In fact, Uranus had been observed and its position recorded on a number of occasions during the preceding ninety years—but as a star, not as a planet. Subsequently an orbit was calculated for it, and for a few decades the location of

Uranus in the sky checked accurately with the positions which had been calculated for it.

In 1781 Uranus was too far from Neptune to be influenced noticeably by its gravitational attraction (Fig. 17–23). But Uranus moves more rapidly than Neptune in its orbit, and as it approached Neptune, began to deviate noticeably from the orbit calculated—the gravitational pull of Neptune was much greater at the shorter distance. (Neptune's orbit is also affected by the gravitational attraction of Uranus). To be sure, the deviations were slight, but they seemed too large to be mathematical errors.

Two young men, Adams of England and Leverrier of France, were convinced that the gravitational pull of an unknown planet located somewhere beyond Uranus was causing the orbital deviation of Uranus. Independently, each undertook the difficult task of calculating the position of a planet of unknown size and distance (but see p. 569) whose gravitational pull might be diverting Uranus slightly from its normal path. Each succeeded.

Perhaps the most convincing evidence supporting the validity and accuracy of Kepler's and Newton's laws is their successful use in the prediction of the precise positions of celestial bodies for years into the future—e.g., the split-second timing of a total solar eclipse a decade from now—and their application in determining the orbits of artificial satellites and space travel.

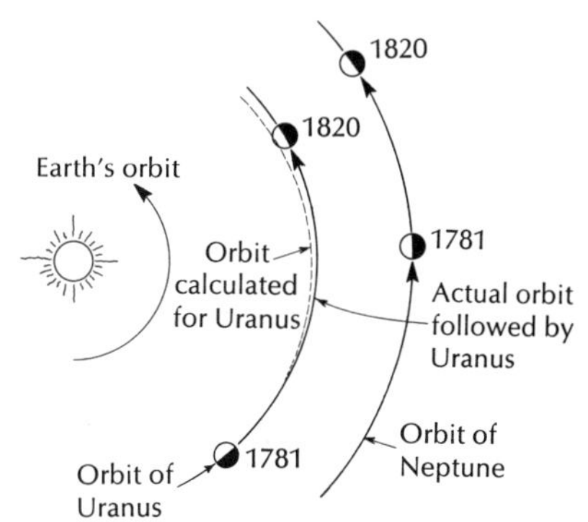

17-23. The discovery of Neptune.

Value of Astronomy to Man

Today there is widespread, intense interest in space travel, manned orbital vehicles, communication satellites, missiles, the possibility of the existence of life elsewhere in the universe, and similar items. Yet the basic contributions of astronomy to man and to science are still not generally appreciated.

Astronomy was one of the first of the sciences to develop, and practical applications were made long ago. The motions of celestial objects provided clocks, calendars, and aids in navigation. Recently there have been many more applications—e.g., the study of the disrupting effects of solar activity on long-range radio communication (p. 600) and on radar.

But even more fundamental values accrue from the study of astronomy, and one of the most important to scientists and scholars is the satisfaction of man's curiosity about the nature of the universe. Some men at great risk have climbed the highest mountains for a somewhat similar reason—"because," as Hillary said of Mount Everest, "it is there."

Again, various sciences today are intimately interrelated and overlap. That most practical of disciplines, mechanical engineering, could not develop until Newton's laws were formulated; yet Newton worked out the laws to explain the motions of the celestial objects he was studying. Observation of the motions of celestial objects led to the elimination of much superstition and to the idea that generalized natural laws existed and could be discovered by experimentation, calculation, and observation. Astronomy is an excellent subject for the application of mathematics and has resulted in many advances in this field.

The study of the nuclear reactions which occur under tremendous temperatures and pressures in the interiors of stars has been of great importance in man's study of nuclear reactions on the Earth. Moreover, instruments designed for the study of celestial objects have become very useful in other studies on the Earth—e.g., in optics and spectroscopy.

Perhaps most important to grasp is the vital necessity of basic research in all of the sciences, not alone in astronomy—research undertaken to discover basic principles, not to make money or to solve some specific engineering problem. It is impossible to say what benefits will accrue from any particular fundamental research project, but in the past such studies have led to discoveries that have proved to be of great practical and economic benefit. This has happened so frequently that we seem justified in assuming that it will continue to happen in the future. And the study of astronomy—impractical when viewed superficially—has resulted in some of the most important and basic discoveries concerning matter, energy, and motion.

Exercises and Questions for Chapter 17

1. What changes, if any, would occur in the lengths of the solar and sidereal days if the:
   (A) Earth's mean orbital speed were increased?
   (B) Earth revolved westward (clockwise) instead of eastward?
   (C) Earth rotated westward (clockwise) instead of eastward?
2. To show the diurnal motions of stars, make four time-exposure photographs. Point the camera toward the north for one, and then east, south, and west. Exposures of 5 to 10 minutes are long enough.
   (A) What measurement can you make on Fig. 17–1 to verify that the time exposure was about 1⅔ hours?
3. Observe the diurnal and annual motions of the sun.
   (A) Erect a gnomon (vertical stick on a flat surface) and mark the end of its noon shadow at intervals.
   (B) Note changes in the places and times of sunrise and sunset.
   (C) Note changes in the altitude of the noon sun.
4. With the aid of a Star Finder, plot the Big Dipper's position on a semicircular star-map sketch at 8 PM on November 1. Next, calculate—without using the Star Finder—the Big Dipper's position on:
   (A) February 1 at 8 PM.
   (B) November 1 at 2 AM.
   (C) May 1 at 2 AM.
   (D) August 1 at 8 PM.
5. Same as number 4 above, but substitute Orion on November 15 at 9 PM.
   (A) February 15 at 9 PM.
   (B) November 16 at 3 AM.
   (C) May 15 at 9 PM.
   (D) November 16 at 9 AM.
6. Assume that a star rises tonight at 7 PM. When:
   (A) did it rise 1 month ago (assume 30 days)?
   (B) did it rise 2 months ago?
   (C) will it rise 3 months from now?

7. Assume that a star sets at 8 PM tonight. When:
   (A) did it set 20 nights ago?
   (B) will it set 15 nights from now?
8. State the position taken by each of the following concerning the geocentric-heliocentric controversy and give at least one type of evidence or argument cited by each: Aristarchus, Aristotle, Copernicus, Brahe, Kepler, and Galileo.
9. One of Jupiter's moons is at a distance of about 260,000 miles and has a 2-day period of revolution. Why is its orbital speed so much greater than that of the Earth's moon?
10. Assume that a comet orbits the sun in an elliptical path that has a major axis of 100 astronomical units and an eccentricity of 0.2.
    (A) How far is the sun from the center of this ellipse?
    (B) What is the aphelion distance of the comet? Its perihelion distance?
11. Calculate the eccentricity of the Earth's elliptical orbit about the sun. Use 91.5 and 94.5 million miles as the perihelion and aphelion distances respectively.
12. Assume that a meteoroid completes an elliptical path about the sun in 27 years.
    (A) What is its mean distance from the sun (in astronomical units)?
    (B) If the eccentricity of its orbit is 0.2, what is its aphelion distance? Its perihelion distance?
13. Assume that a comet follows an elliptical path about the sun with perihelion and aphelion distances of 40 and 160 astronomical units respectively. Calculate its:
    (A) mean distance.
    (B) eccentricity.
    (C) period of revolution.
14. Assume that an astronaut has the same weight on a planet with a radius of about 24,000 miles as he does on the Earth. How does the mass of this planet compare with that of the Earth?
15. Assume that an object weighs 7200 lbs. at the Earth's surface. If this object were shifted to a distance of about 240,000 miles from the center of the Earth:
    (A) what would it weigh?
    (B) at what rate would it fall toward the Earth's surface (near the surface it would fall at 32 ft/sec/sec)?
16. Assume that a fictitious planet orbits the sun. How does the gravitational attraction between the sun and this planet compare with that between the sun and Earth if the planet's distance is:
    (A) 20 astronomical units and its mass is the same as the Earth's?
    (B) 10 astronomical units and its mass is 200 times the Earth's?
17. Assume that an object weighs 100 lbs. at the Earth's surface. What would it weigh on another planet if this planet had a radius of about:
    (A) 8000 miles and a mass four times that of the Earth?
    (B) 20,000 miles and a mass 100 times that of the Earth?
    (C) 400 miles and a mass 1/100 that of the Earth?

18

Man-Made Satellites and Space Travel

Launching an Artificial Satellite Into an Orbit About the Earth

The successful launching of a spacecraft into an orbit about the Earth, to the moon (Fig. 18–1), or to a more remote destination in space involves many factors: Newton's laws of universal gravitation and motion, Kepler's laws of planetary motion, air resistance, jet propulsion, guidance, and other factors. Two results must be achieved. A spacecraft has to be lifted above the dense lower part of the Earth's atmosphere, which extends outward for hundreds of miles, but which is exceedingly thin at altitudes greater than 100 miles or so. When located high enough above the surface, the spacecraft must then be accelerated to a very great speed in a direction nearly parallel to the Earth's surface.

As a spacecraft speeds through space, it falls continuously toward the Earth, but its momentum prevents it from being pulled directly into the Earth (Fig. 17–21). If the spacecraft's lateral motion is fast enough, it orbits the Earth as an artificial satellite. A slower speed at launchtime, or reduced speeds produced eventually by air resistance, may slow the spacecraft's forward speed sufficiently to bring it down through the denser air in a meteorlike plunge, unless a braking action has been arranged.

Speeds of 18,000 mi/hr (5 mi/sec; 8 km/sec) or more are necessary to keep a satellite orbiting the Earth for weeks, months, and years at an altitude of a few hundred miles. The multistage rocket is used to attain such speeds. The first stage lifts a spacecraft vertically and relatively slowly through the dense lower atmosphere. The other stages are then fired in turn to accelerate the payload to greater and greater speeds as the rocket rises higher and higher into thinner and thinner air. If the rocket were given a 5 mi/sec velocity at the Earth's surface, friction would immediately reduce its speed and might be great enough to vaporize it. The rocket would burn up as a meteor. Furthermore, the waste weight of each burned-out stage can be successively dropped.

A rocket is fired vertically, but its path is so controlled that it turns gradually away from the vertical. By the time the target altitude is reached, the last stage has been slanted into a nearly horizontal path about parallel to the Earth's surface directly beneath it. The last stage is then fired and this accelerates the payload to the necessary speed to stay in orbit for some time.

The effect of the Earth's gravitational attraction on freely falling objects is independent of other motions that they may have. In a familiar demonstration, several balls are made to leave a table top and

18-1. Apollo 12 LM above the lunar surface, November 1969. *(NASA: AS12-51-7507)*.

fall to the floor. One ball merely falls vertically, whereas the others are moving horizontally at different speeds as they leave the table top. Yet the balls all reach the floor at the same time. Although air resistance slows the forward progress of the forward-moving balls, it does not make them fall; the gravitational pull of the Earth does this. Therefore, all of the balls fall at the same rate ($d = \frac{1}{2}gt^2$, where d is the distance in feet, g is the acceleration of 32 ft/sec/sec due to the Earth's gravitational pull, and t is the time in seconds). Thus a freely falling object has dropped 16 feet at the end of the first second, 64 feet at the end of two seconds, and about 1 mile at the end of 18 seconds. However, the fastest moving ball strikes the floor farthest from the edge of the table, and the other balls strike in between. For this reason, a bomb dropped from a plane does not strike the Earth directly below, because it falls with the motion of the plane superposed on the falling motion caused by the Earth's gravitational attraction.

Therefore, if a man-made satellite is to be put into an orbit about the Earth, it must have enough lateral speed to offset the effect of falling. Imagine a launch platform located 1 mile above the ocean and assume that the Earth is a smooth, airless sphere (Fig. 18–2). Because the Earth is about 8000 miles in diameter, its surface curves about 1 mile in 90 miles; i.e., a horizontal line that is tangent to the ocean's curved surface at one point is 1 mile above this surface about 90 miles away. Next, assume that a bullet is dropped vertically from this platform at the same time that four other bullets are fired horizontally at different speeds. The bullets arrive at points *a, b, c, d,* and *e* simultaneously because they all fall 1 mile in about 18 seconds. Moreover, bullet e is in orbit about the Earth because its lateral speed of about 5 mi/sec is enough to give it a curved path that exactly parallels the curvature of the Earth's surface 1 mile beneath it. The bullet is being constantly accelerated by the Earth's gravitational pull, but in this case the acceleration involves only a change in direction. Thus the bullet revolves in a circular orbit at a uniform speed (we ignore air resistance for this).

Since the Earth does have an atmosphere and is not a smooth sphere, let us now imagine another launch platform located at an altitude of 100 to 200 miles above the Earth's surface where the air is very thin and air resistance is slight (but it is not negligible). Next, assume that bullets are fired laterally from this platform (i.e., parallel to the Earth's surface below) but at speeds of less than 5 mi/sec (Fig. 18–3; upper diagram). Each bullet starts along an elliptical path that has the Earth's center at its more distant focus. However, each path is soon interrupted by the Earth's surface. The starting point is the apogee position and the perigee position would be inside the Earth and 180 degrees from the launch point.

In inserting a spacecraft into an orbit about the Earth, the speed and direction of move-

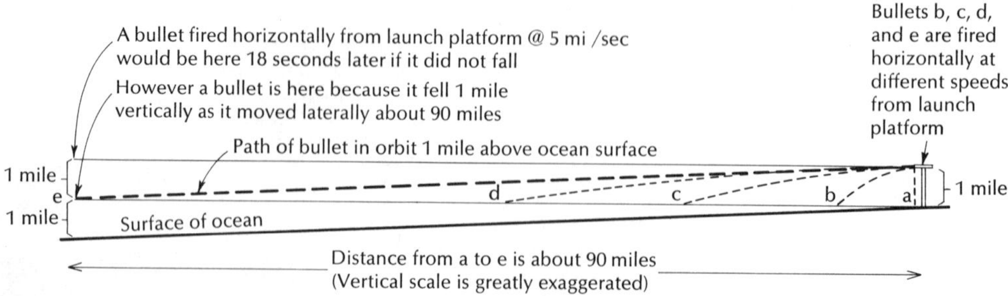

18-2. The Earth's gravitational attraction on freely falling objects is independent of other motions that they may have.

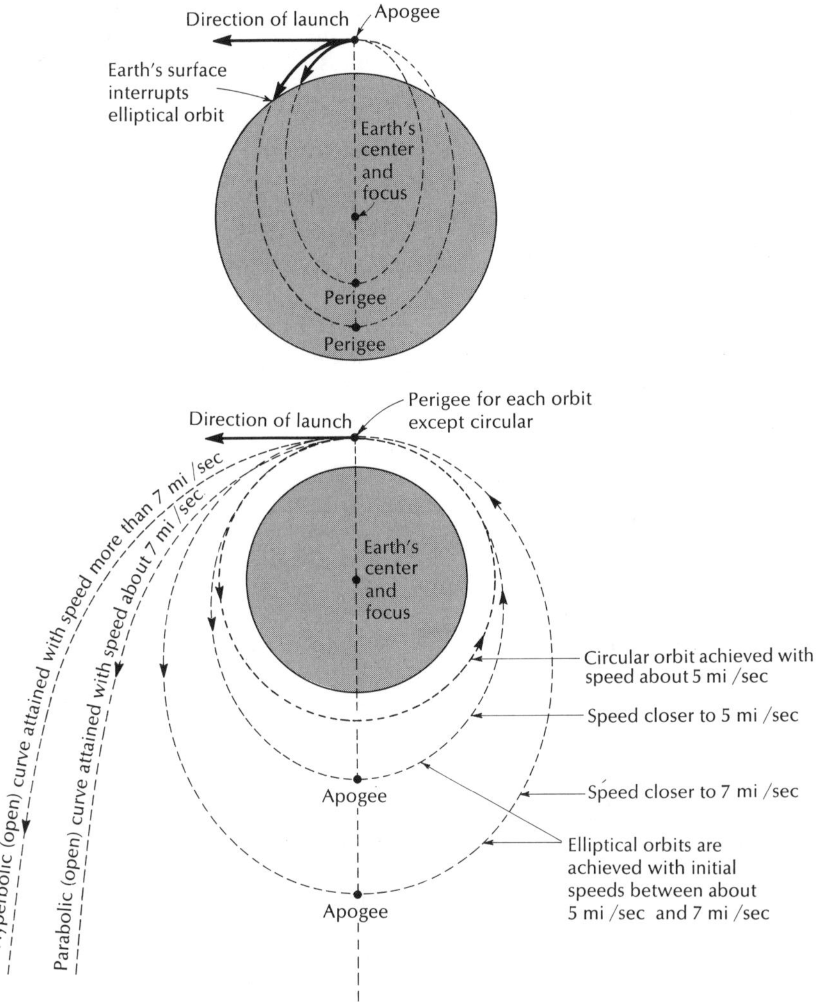

18-3. Bullets are fired horizontally at different speeds at an altitude of a few hundred miles and achieve different flight paths.

ment at burnout (when the rocket engine shuts off and the spacecraft begins to coast in unpowered flight) are both critical factors. In the lower diagram of Fig. 18–3, we see the paths of spacecraft that have been launched from the same place (say at an altitude of 300 miles), in the same direction, but at different speeds. About 5 mi/sec produced a circular orbit. At speeds between 5 mi/sec and 7 mi/sec, the spacecraft achieve elliptical orbits, and the Earth's center is located at the near focus of each ellipse. The launch point becomes the perigee position, and the apogee point is 180 degrees from the burnout position—the greater the speed, the farther away is the apogee point.

At speeds of 7 mi/sec or more, open curves are achieved, and the spacecraft escape from the Earth's gravitational control into space along parabolic or hyperbolic curves.

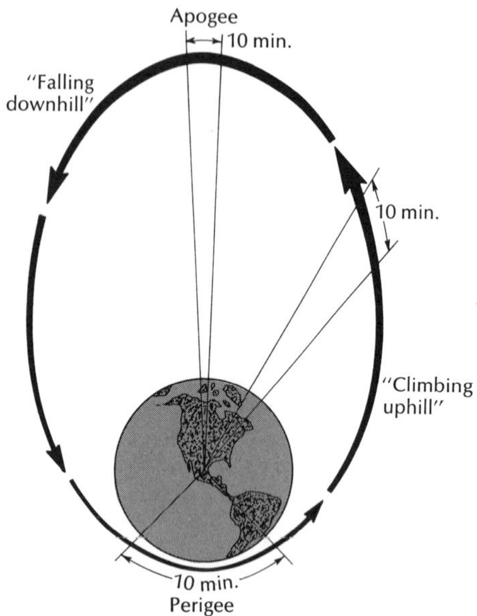

18-4. Elliptical orbit for a satellite. The satellite "falls downhill" when the Earth's gravitational attraction acts in the same general direction as the satellite is moving; therefore, it speeds up. The closer a satellite orbits the Earth, the faster it must move in order to stay in orbit. See Table 18-1. The satellite also moves according to Kepler's equal areas law (Fig. 17-19). (*Courtesy H. L. Goodwin,* Space: Frontier Unlimited, *Van Nostrand Reinhold, 1962.*)

Objects at high altitudes may thus be accelerated laterally to orbital speeds by the firing of a rocket for some seconds (Figs. 18–4 and 18–5). They may then coast all the way around the Earth, again and again, without additional rocket firings. The point at which the rocket was fired becomes a point in the orbit of the object. In the absence of friction, the object must return to this point—i.e., if the only factors are the gravitational pull of the Earth and the object's lateral momentum. For elliptical orbits, this point is either the apogee or perigee position, and a point 180 degrees away (i.e., half way around) is the other position—perigee or apogee.

The *primary* is the term used for the massive object (in this case the Earth) about which the orbital motion occurs. The orbital speed necessary to achieve a circle or an ellipse of a certain shape depends upon the mass of the primary and the distance from it. For more massive primaries, the speeds must be greater. For greater distances, the speeds are less. The gravitational pull of the primary must be offset. Thus where it is great, the lateral speed must be great as an offset—for very massive primaries or for short distances.

To achieve a circular orbit, the direction of firing is also critical; it must be at 90 degrees to a line joining the satellite with the center of the Earth (i.e., it must be parallel to the Earth's surface directly beneath it).

Once a satellite is in orbit about the Earth, air resistance gradually reduces its speed, unless all parts of the orbital path are so remote from the Earth that atmospheric friction is negligible. When there is any friction, no matter how slight, the

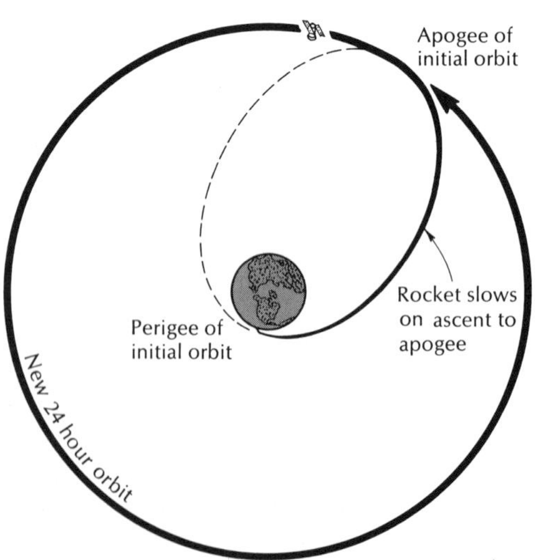

18-5. "Kick in the apogee" to achieve a circular orbit. At the moment the final rocket stage and its attached satellite reach the apogee position of the initial orbit, they are moving parallel to the Earth's surface far below (but too slowly to stay in orbit at this altitude). The final stage is then fired and increases the speed of the satellite enough to place it in a new circular orbit. (*H. L. Goodwin,* Space: Frontier Unlimited, *Van Nostrand Reinhold, 1962.*)

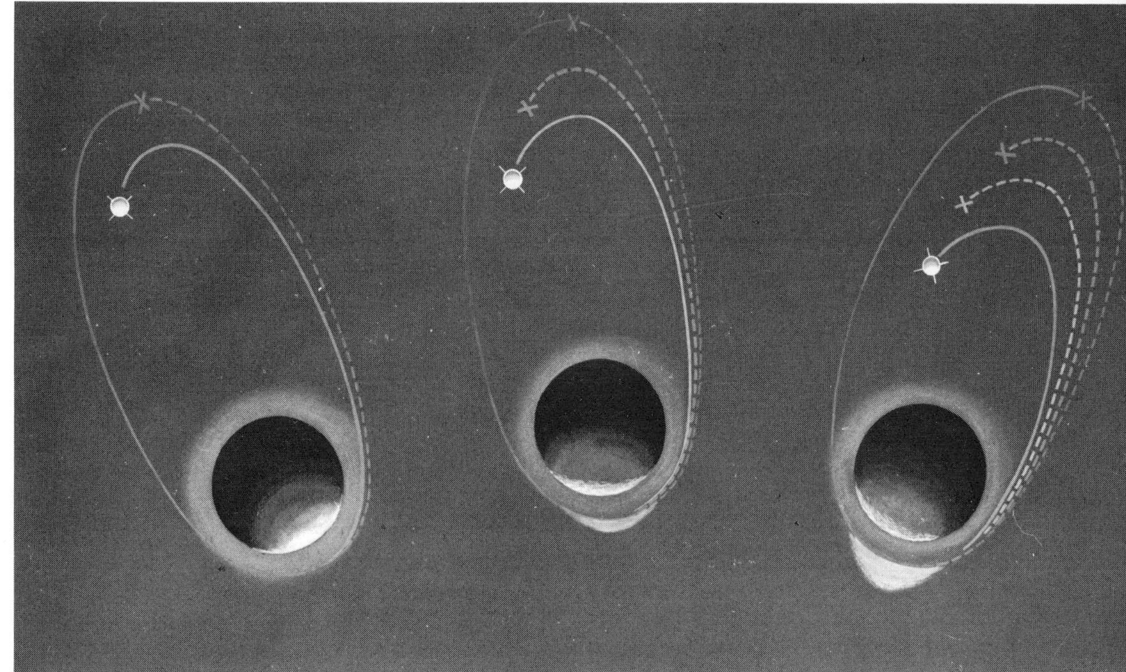

18-6. Atmospheric drag changes the apogee position more than the perigee position. *(Courtesy NASA.)*

orbit changes at each revolution. It changes slowly at first, and then more and more rapidly as the shrinking orbit brings the satellite into denser air. At first the perigee position remains about the same, and the altitude of the apogee point is reduced on each revolution. A nearly circular orbit eventually results, and its radius is then gradually reduced (Fig. 18–6).

A satellite moves according to Kepler's equal areas law; therefore, its greatest speed is attained when it is closest to the Earth. However, air resistance is also greatest at perigee, and this reduces the speed at each passage through the perigee position (i.e., the satellite speeds up as it approaches perigee but not so much as it would in the absence of air resistance). Thus the satellite does not gain the momentum necessary to swing out as far on the next revolution, and its apogee point shifts closer to the Earth.

The period of an artificial satellite is directly related to the size of its orbit (Table 18–1). In other words, more remote satellites travel more slowly and take longer to complete one revolution. This follows from Kepler's three laws. At a distance of about 22,300 miles above the Earth's surface, the period is 24 hours; therefore, a satellite in a circular orbit over the Earth's equator at this altitude remains "fixed" in space directly above a particular spot on the Earth beneath it (Fig. 18–7).

The speed of a satellite moving in an elliptical orbit about the Earth is given by the following equation: $V = \sqrt{GM} \sqrt{\frac{2}{R} - \frac{1.}{A}}$ The speed of the satellite is shown by V; G is the gravitational constant; M is the sum of the masses of the primary and satellite (generally negligible); R is the distance of the satellite from the primary at the moment the speed is calculated; and A is the mean distance (semimajor axis) of the satellite. This equation also holds for any other primary-

TABLE 18-1
ALTITUDE, VELOCITY, AND PERIOD OF ORBITING SATELLITES

APPROXIMATE ALTITUDE OF ORBITING SATELLITE ABOVE THE EARTH'S SURFACE	APPROXIMATE VELOCITY NECESSARY TO KEEP FROM FALLING INTO THE EARTH	APPROXIMATE PERIOD OF REVOLUTION
Several feet	19,000 mi/hr	84 min
100 miles	18,000 mi/hr	90 min
1000 miles	15,700 mi/hr	120 min
22,300 miles	6,900 mi/hr	24 hours
239,000 miles	2,300 mi/hr	About 1 month (the moon illustrates this orbit)
ASSUMPTIONS: The Earth has no atmosphere, the Earth is a true sphere, and the orbits are circular.		

satellite pair such as a planet about the sun. Appropriate units must be used.

For a circular orbit, R equals A, and $V = \sqrt{GM/R}$. In words, the speed of a satellite in a circular orbit is inversely proportional to the square root of its distance and directly proportional to the square root of the mass of its primary. Also, the square of the velocity varies directly as the mass of the primary and inversely as its distance.

According to Ahrendt* this circular velocity in miles per second equals $\sqrt{\frac{9.56 \times 10^4}{r}}$ (i.e., GM equals 9.56×10^4 when miles and

* Ahrendt, Myrl H., *The Mathematics of Space Exploration,* Holt, Rinehart, and Winston, 1965.

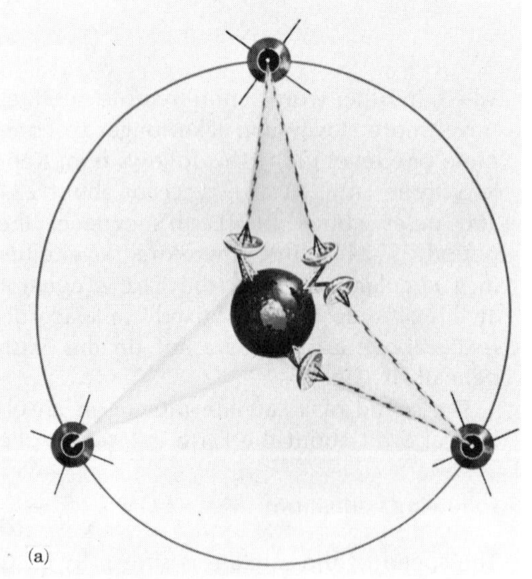

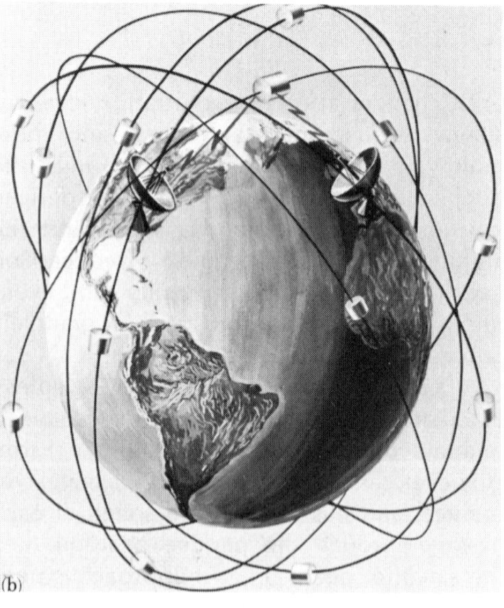

18-7. Communication satellites at an altitude of 22,300 miles in synchronous orbits (A) and at an altitude of about 5000 miles (B). To an observer on the Earth, the satellites in (A) would not appear to move. Either system can be used to provide worldwide television broadcasts and other communication services. The satellites are equipped with radio receivers and transmitters and other instruments so that they can receive, store, amplify and retransmit messages. They function as microwave towers in space. Enough "active-repeater" satellites are placed in orbit so that at least one is always present in space above any one area. *(Courtesy NASA.)*

seconds are used as the units). The radius (r) is measured in miles and the velocity (v) in mi/sec. Thus a satellite at an altitude of 1000 miles (use 3960 miles as the Earth's radius) would have a speed of $\sqrt{\frac{9.56 \times 10^4}{4960}} = \sqrt{19.3} =$ nearly 4.4 mi/sec, or about 15,-800 mi/hr. To make one orbit at this altitude would require about 2 hours (the circumference equals $2\pi r$, which equals $2 \times 3.14 \times 4960$, or about 31,150 miles).

To escape from its primary, an object must move away on an open path that is a parabola or hyperbola. The mean distance A is thus equal to infinity and $1/A$ becomes zero. Therefore, the escape velocity equals $\sqrt{\frac{2GM}{R}}$; in other words, it is equal to the $\sqrt{2}$ or about 1.4 times the speed needed for a circular orbit at this distance.

The direction in which a spacecraft is launched is an important factor; less energy is needed if the direction is the same as that of the Earth's rotation and revolution. Points on the Earth's surface move in an eastward direction at velocities that vary with latitude; they are greatest at the equator (about 1000 mi/hr or 0.3 mi/sec) and decrease toward the poles. All objects at the equator are moving eastward at this rate, and this motion is independent of other motions the objects may have. Thus a ball thrown vertically upward falls back to the starting point. The Earth does not leave such objects behind by rotating eastward from under them. Thus a spacecraft launched in a general eastward direction at the equator is already moving at a rapid speed. If 5 mi/sec is the velocity the spacecraft must eventually attain to stay in orbit, it already has 0.3 mi/sec of this; therefore, it must be given an additional speed of 4.7 mi/sec. If it were launched toward the west, the necessary speed would be 5.3 mi/sec. The direction of launch becomes even more important in interplanetary travel.

If a satellite were launched in a due eastward direction at the equator, it would move in space directly above the equator. It could not be used to photograph or measure phenomena at higher latitudes. On the other hand, if a satellite were launched along a meridian, it would eventually pass over all parts of the Earth, but much of its time would be spent above the unpopulated polar regions. Thus it is more common from the middle northern latitudes to launch satellites toward the southeast so that they pass back and forth over the heavily populated regions between the middle latitudes and the equator. However, the direction of launch will depend upon the purpose for which the satellite was designed.

Observing Man-Made Satellites from the Earth

Most satellites are seen by reflected sunlight and thus cannot be viewed during the day with the unaided eye against the bright background of diffused sunlight (Fig. 18–8). The moon itself is inconspicuous in the daytime sky. Most satellites are also relatively close to the Earth and nearly half of their orbits lie within the Earth's shadow; thus we can observe satellites near the times of sunrise and sunset. An observer on the darkened Earth near the end of evening twilight may see a satellite shining in the sunlight high overhead and moving eastward in a clear sky. This satellite may then disappear abruptly while still high above the horizon because it entered the Earth's shadow. The reflected light of a satellite may fluctuate if it has an asymmetrical shape and spins or tumbles as it revolves. In addition, a satellite has a lighted half and a darkened half and thus it exhibits some phase changes.

For watchers in the middle northern latitudes, satellites may move across the sky from southwest to northeast or from northwest to southeast. However, if a satellite is in a polar orbit, one will see it move from north to south or from south to north. A satellite may be visible at different times on any one night and for a number of nights in succession, but it never follows precisely

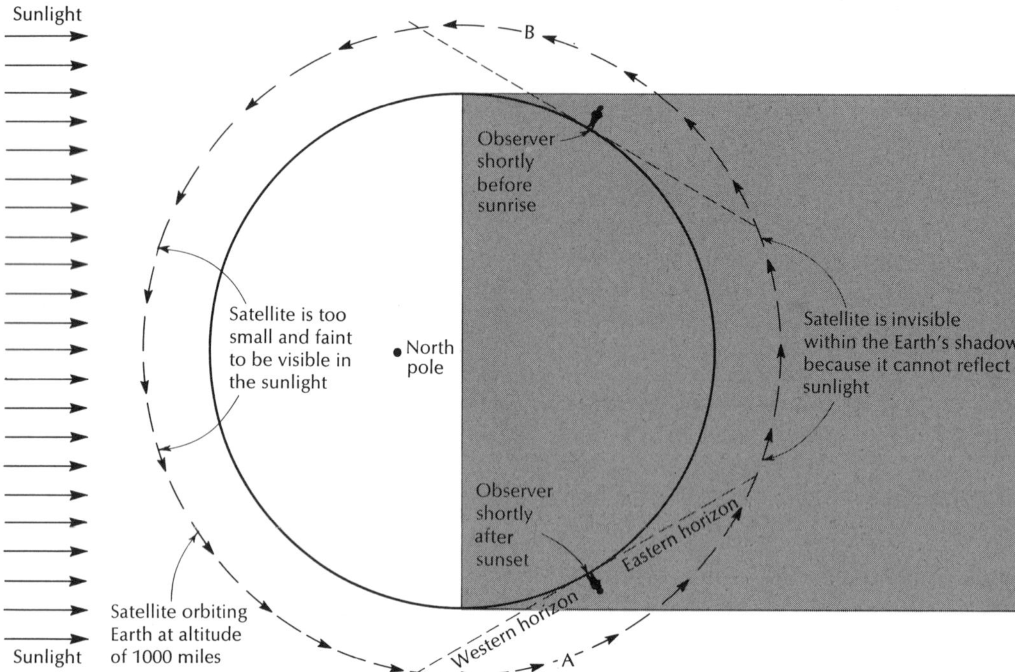

18-8. Visibility of a satellite from the Earth. A satellite is shown at a 1000-mile altitude moving eastward about the Earth. It is visible when in the general vicinity of points A and B. Slant of Earth's axis (indicated by position of north pole) shows that summer is occurring in the Northern Hemisphere.

the same path from one trip to the next. A sinuous line must be drawn on a map to show a satellite's orbital path projected downward to the Earth's surface directly beneath it.

This can be readily understood with the aid of models: a globe for the Earth and a hoop for the orbit of the satellite. Once the satellite is in orbit, its orbital plane remains nearly fixed in space relative to the stars (the plane actually shifts slowly, but assume now that it is fixed). Place the globe inside the hoop, tilt the hoop at a moderate angle to the equator, keep the hoop motionless, and spin the globe.

Explanations for some of the motions of satellites should now be apparent. A relatively low satellite completes a revolution in about 1½ hours, whereas the Earth rotates once in about 24 hours. Therefore, on successive revolutions at 1½-hour intervals, the satellite moves in about the same path, but the Earth has turned under it. As successive passes are viewed from any one location during a single night, the satellite is seen to follow parallel paths, but each will be higher or lower in the sky than the preceding one. If the direction is from southwest to northeast for an observer on one side of the Earth, on the opposite side of the Earth for another observer less than 1 hour later, the direction will be from northwest to southeast.

If the Earth were a perfect sphere of uniform density or composed of uniform concentric shells, the orbital plane of a satellite would remain fixed in space. However, the Earth's crust is not uniform in density and the Earth has an equatorial bulge. Therefore, the orbital plane of a satellite slowly shifts in space, changing in a manner that is similar to the precession of the equinoxes

(p. 523). The orbital plane maintains the same angle to the equator, but slowly turns in space; the rate may be several degrees in one day.

The amount of such a shift can be calculated approximately before the satellite is launched and precisely after it has been tracked during several revolutions. If slight deviations from the calculated orbit are observed, they can be interpreted to yield information about the shape of the Earth and the distribution of matter within it.

Space Travel

Even a very slight acquaintance with astronomical distances indicates that space travel in the foreseeable future will be limited to the vicinity of the solar system. To journey to Pluto at an assumed uniform rate of 5 mi/sec would take more than 20 years. (A higher velocity would be needed to escape from the Earth.) At this same rate, a trip to Alpha Centauri, the nearest star, would last for 150,000 years. Thus trips to even the nearest stars are beyond our reach at present, and trips to distant stars or other galaxies are quite impossible. Let us consider some of the problems and principles involved in trips to the moon, Venus, and Mars.*

A Trip to the Moon

A speed of nearly 7 mi/sec (approximately 25,000 mi/hr) must be attained by an object located a few hundred miles above the Earth's surface if it is to move outward from the Earth for a distance of about 240,000 miles and reach the moon. However, the object does not travel to the moon in about 10 hours, because the Earth's gravitational pull causes a gradual decrease in speed throughout the journey. Burnout occurs near the Earth, and a spacecraft then coasts for the remaining distance in unpowered flight.

A spacecraft which left the Earth at a speed of nearly 7 mi/sec would be barely moving at the point about 25,000 miles from the moon where the gravitational pull of the moon equals that of the Earth. From this distance onward, the moon's gravitational pull will gradually increase the spacecraft's speed. If no action is taken to slow the spacecraft (such as firing retrorockets) its speed will increase and it will strike the moon moving at about 1½ mi/sec. This equals the escape velocity from the airless moon.

Two widely separated objects (e.g., two stars or a spacecraft and the moon), approaching each other in space on a noncollision course, cannot link up into a mutually orbiting pair unless one of them is slowed by some outside force. Their mutual gravitational attraction gives them speeds in excess of escape velocity. In firing retrorockets to slow a spacecraft so that it can land on the moon, as much fuel is used in landing as in the subsequent blastoff.

At launching, a spaceship is not aimed directly at the moon because the moon will not be at this position when the spaceship eventually arrives—duck hunters may have an advantage in understanding this aspect of space travel. The spaceship is aimed at the point in space where the moon will be located at the time of intersection. Spacecraft can be made to approach or land on the moon along different types of orbital paths (Fig. 18–9).

Manned spacecraft are generally much heavier than unmanned ones, and each added pound of payload means a large increase in the size of the rocket needed to propel it through space. (The rocket functions as an expensive truck to move the payload from one place to another.) Thus unmanned spacecraft are much easier to deal with in many respects, and observations of widely diverse kinds can be made automatically by instruments in the spacecraft and the results radioed to the Earth. Man would be a nuisance in many spacecraft.

* Stong, C. L., Simple ways to calculate the orbits of space vehicles, *Scientific American,* January 1969.

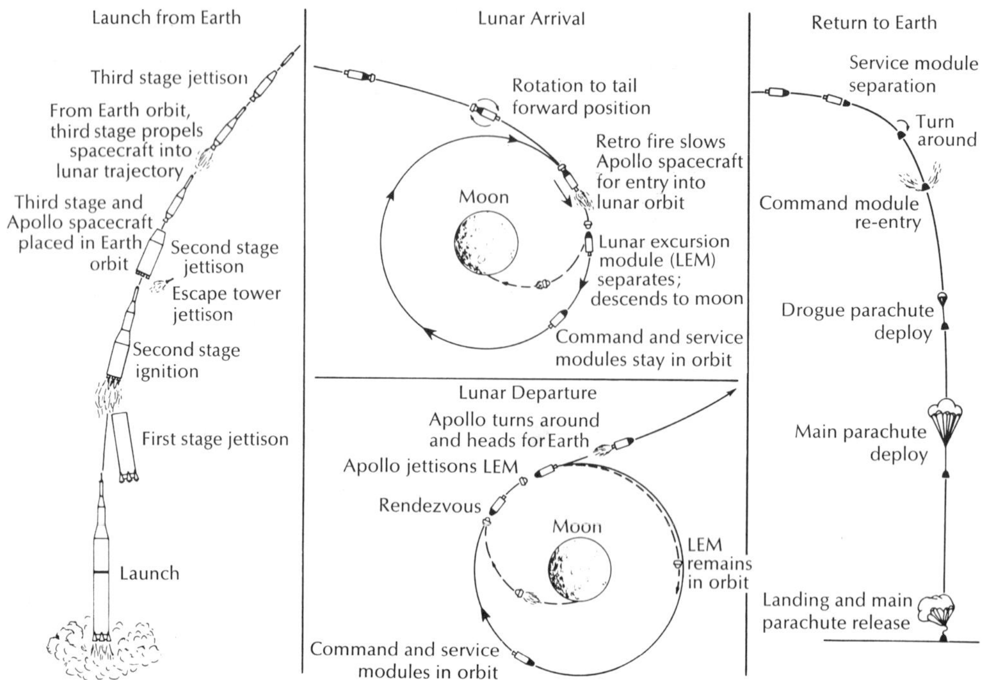

18-9. Sequence of major events in Apollo lunar exploration mission. *(Courtesy NASA.)*

Trips to Venus and Mars

At their closest approaches to the Earth, Venus and Mars are still well over a hundred times farther from the Earth than the moon is, but the energy necessary to propel spacecraft to these planets along minimum-energy flight paths is not much greater than that required for reaching the moon. However, the navigational precision must be of an extremely high order and involves four dimensions: three space coordinates and time. A spacecraft must arrive at precisely the right point in a vast region of space at precisely the right time (in 1 minute, Venus and Mars move about 1300 miles and 900 miles respectively).

Basic data include: (1) The Earth's average speed is about 18½ mi/sec, and all objects on the Earth are moving at this rate independently of other motions they may have. (2) Planets closer to the sun revolve more rapidly, whereas those farther away revolve less rapidly. (3) A spacecraft may be shifted out of the Earth's orbit into a new orbit with a mean distance that is either smaller or larger than 1 astronomical unit. This is done by decreasing or increasing the 18½ mi/sec speed the spacecraft had prior to launch because it was a part of the Earth. (4) The spacecraft must escape from the Earth's gravitational control.

For example, if the Earth's speed were somehow increased a few mi/sec by the application of a force for a brief time, the Earth would coast outward from the sun into a larger orbit. It would be slowed in its outward journey by the sun's gravitational attraction, and the Earth would be moving slowest in this new and enlarged orbit at its new aphelion point—located half way around (180 degrees) from the point where its speed was increased. The Earth would now fall back toward the sun, and its speed would increase until it reached a maximum back at the starting point (its new perihelion point). The Earth would thus be in a new orbit at a greater mean distance, and its average speed in the larger orbit would be less than its present average speed.

Thus the general scheme is rather simple. A spacecraft must have a speed at the start that is faster than the Earth's to spiral outward to the orbit of Mars (Figs. 18–10 and 23–6), or smaller than the Earth's to spiral inward toward the orbit of Venus. One must also consider the position of the Earth in space at the time a spacecraft reaches Mars or Venus; if the Earth is too far away, data radioed from the spacecraft may not reach us.

Since a spacecraft must first escape from the Earth, an initial velocity of about 7 mi/sec must be attained at an altitude of a few hundred miles (to eliminate the effects of air resistance). This takes a spacecraft along an open curve—parabola or hyperbola—outward to a distance of a million miles or so where the Earth's gravitational pull is so weak that it has no appreciable effect. Such a spacecraft slows continuously as it coasts outward through frictionless space, although the Earth's gravitational pull becomes less as the distance becomes greater. The spacecraft uses up the 7 mi/sec initial velocity to escape from the Earth (i.e., its speed relative to the Earth has become zero). However, it still has the 18½ mi/sec velocity that it had when launched from the Earth. Thus the spacecraft and the Earth, about a million miles apart, would revolve about the sun in similar orbits.

The spacecraft has now escaped from the Earth, but it is still about 1 astronomical unit from the sun. To change this distance, the speed of the spacecraft must be increased or decreased (Fig. 18–11). An increase of about 2 mi/sec would send the

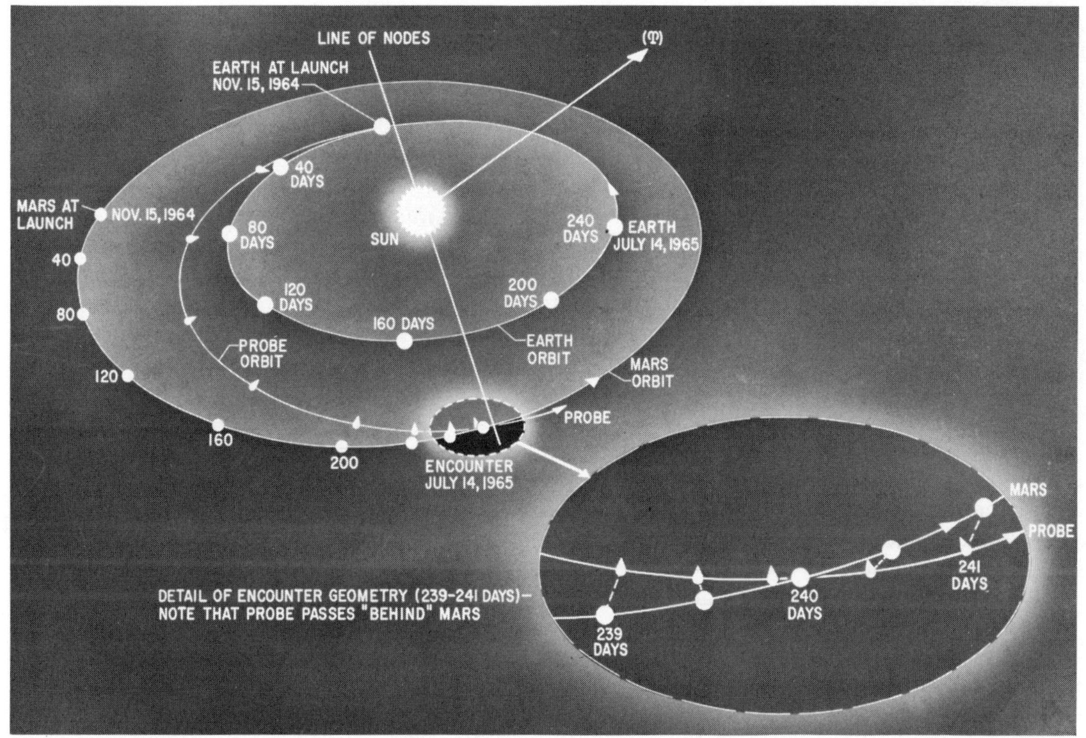

18-10. Typical 1964 Mars trajectory. A probe launched in November of 1964 would be expected to encounter Mars about eight months later. *(Courtesy NASA.)*

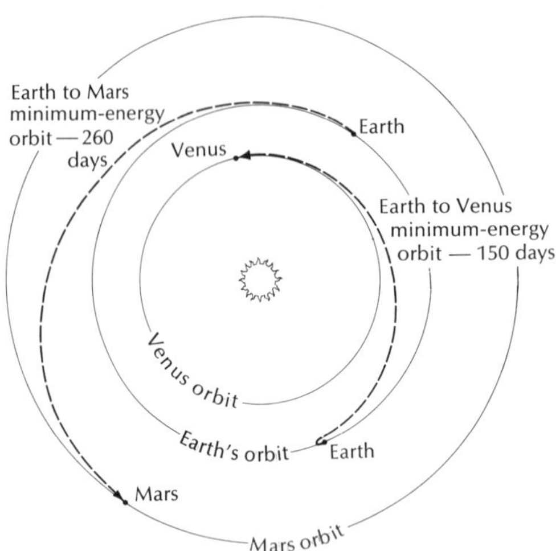

18-11. Minimum-energy flight paths to Venus and Mars (also called Hohmann orbits). *(H. L. Goodwin,* Space: Frontier Unlimited, *Van Nostrand Reinhold, 1962.)*

spacecraft out to the orbit of Mars, whereas a decrease of about 2 mi/sec would send it spiraling in toward Venus. This is accomplished by launching the spacecraft with an initial speed of about 9 mi/sec—in the same direction as the Earth is revolving to reach Mars, but "backwards" to reach Venus.

However, backwards does not mean a launch toward the west. Rather a spacecraft destined for Venus would be launched eastward—as would one programmed for Mars—and it might be placed in a parking orbit a few hundred miles above the Earth's surface. The timing of the subsequent rocket firing to accelerate the spacecraft to about 9 mi/sec would then determine whether it goes to Mars or Venus. If burnout occurs when the spacecraft is behind the Earth (on the sunset side) the spacecraft will go to Mars because it will then curve around the Earth in a hyperbolic path and will be moving in about the same direction as the Earth at the time it escapes from the Earth. The situation is reversed, and a Venus orbit is achieved, if the burnout occurs when the spacecraft is ahead of the Earth (on the sunrise side). However, as viewed from the sun, the spacecraft does not actually go backwards as it escapes from the Earth, rather it moves counterclockwise around the sun more slowly than the Earth.

To escape from the solar system, a spacecraft would have to be launched eastward with an initial speed of about 14½ mi/sec (to attain the escape velocity at 1 astronomical unit of 26 mi/sec). Subtract 7 mi/sec from the 14½ mi/sec for use in escaping from the Earth. This leaves 7½ mi/sec to be added to the 18½ mi/sec the spacecraft had at launch and totals 26 mi/sec. In other words, if the Earth's speed were increased to about 26 mi/sec, it too would leave the solar system. Escape velocity increases with the mass of the primary and decreases with distance from it.

The minimum-energy orbital path of a spacecraft from the Earth to another planet involves an unpowered coasting flight after burnout at the starting point. The path is a tangent to the Earth's orbit at the starting point and is a tangent to the orbit of the other planet half way round from the starting point. The starting point in the Earth's orbit will be the new perihelion point if the spacecraft's orbital speed exceeds that of the Earth. However, it is the aphelion point if the spacecraft's speed is somewhat less than 18½ mi/sec.

If less than this minimum quantity of energy is used at burnout, say for a trip to Mars, then the spacecraft will be moving too slowly to shift outward far enough from the sun to intersect the orbit of Mars at its aphelion point (180 degrees away from its starting point). On the other hand, if more than the minimum amount of energy is used, the larger burnout speed thus attained would cause the spacecraft to cross the orbit of Mars before it reaches the aphelion point. The elliptical orbit followed by such a spacecraft would be larger and have a more distant aphelion point than that for a minimum-energy orbit—hence the need for more energy.

As an illustration, assume that we want to send a spacecraft along a minimum-energy orbit to a fictitious planet that circles the sun in the plane of the ecliptic at a distance of 17 astronomical units (A.U.'s). How long will the trip take? What will be the eccentricity of the spacecraft's orbit? We know that the perihelion point in the spacecraft's orbit is 1 A.U. because it is launched from a point in the Earth's orbit (Fig. 18–12). We know also that its aphelion point will be 17 A.U.'s from the sun and located 180 degrees from the perihelion point. We know further that the spacecraft will steadily slow down as it moves outward from the sun.

Let us first assume that the planet will be nowhere near the aphelion point when the spacecraft reaches there. In this case, the spacecraft returns to its perihelion point. It is in an orbit about the sun with a mean distance of 9 A.U. Thus its period must be 27 years ($P^2 = D^3$) and its eccentricity 0.89 (16/18). Therefore, the journey from Earth to planet will take about 13½ years. At launch, the planet must be so located that it will need to revolve for 13½ years to reach the spacecraft's aphelion position.

Weightlessness has been illustrated in the following manner. Imagine that two satellites are following identical orbits about the Earth or some other celestial body; one would fall toward the Earth at the same rate as the other. If one satellite is then placed inside the other, they would still fall at the same rate. Moreover, if the inner satellite is replaced by a man, the rate of free fall remains the same. The man is weightless inside the satellite unless it is made to spin or change its acceleration.

A spacecraft might be used to detect the existence of life on another planet (Fig. 18–13).

As shown by the successful Mariner flights to Mars (and of other spacecraft to Venus), by the late 1960's man had achieved the capability of sending instruments to the nearby planets, operating them by remote control, and then returning useful precise data and photographs to the Earth. Solar system exploration seems destined for a marked increase in the 1970's, partly because unique opportunities exist during these years for the exploration of the more remote planets by utilizing a so-called gravity-assist technique. To illustrate, if a spacecraft is to be sent to Uranus along a minimum-energy orbit, a 16-year journey is required. However, if enroute to Uranus the spacecraft follows an orbit that brings it close to Jupiter, then Jupiter's gravitational attraction will change both the direction and speed of the spacecraft—i.e., the spacecraft moves

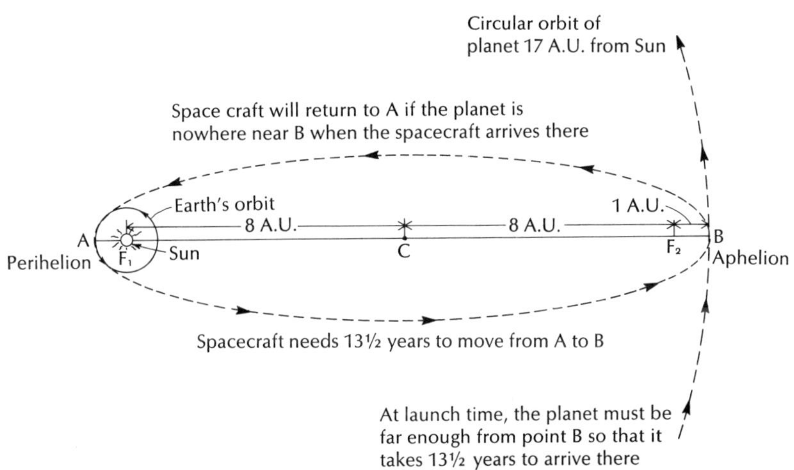

18-12. Minimum-energy orbit for a spacecraft to a planet 17 astronomical units from the sun. The spacecraft's path is a tangent to the Earth's orbit at A (its perihelion position) and a tangent to the planet's orbit at B (its aphelion position).

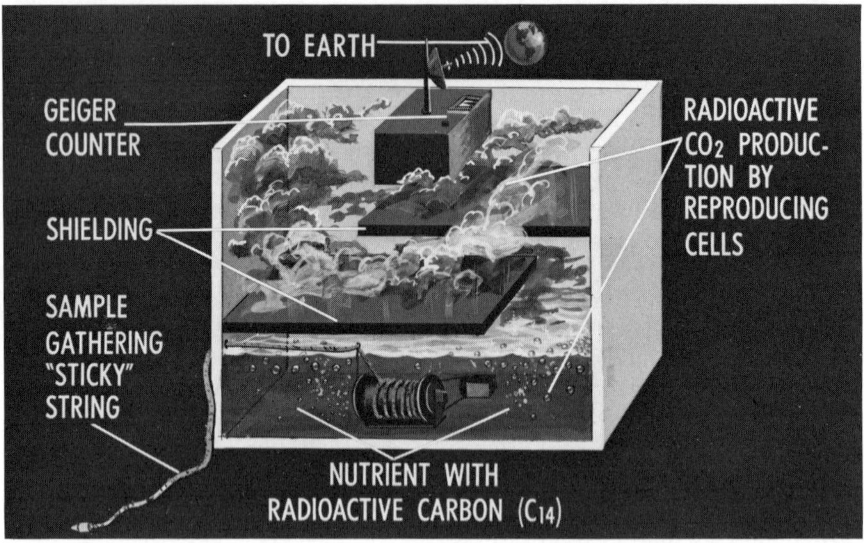

18-13. Device for detection of possible life on other planets. Planetary material would be drawn by an adhesive into a nutrient solution containing radioactive carbon. Forms of life may ingest the solution and give off radioactive carbon dioxide which would register on a Geiger counter. These data would then be radioed to the Earth. *(Courtesy NASA.)*

faster as it approaches Jupiter, it then swings around the planet, and next it slows as it leaves Jupiter and heads for Uranus. The time to reach Uranus by the Jupiter-assist route is reduced from 16 years to 6 years and the burnout speed is slightly less. It follows that Uranus must be located in an appropriate part of its orbit relative to Jupiter, and this occurs about once every 14 years.

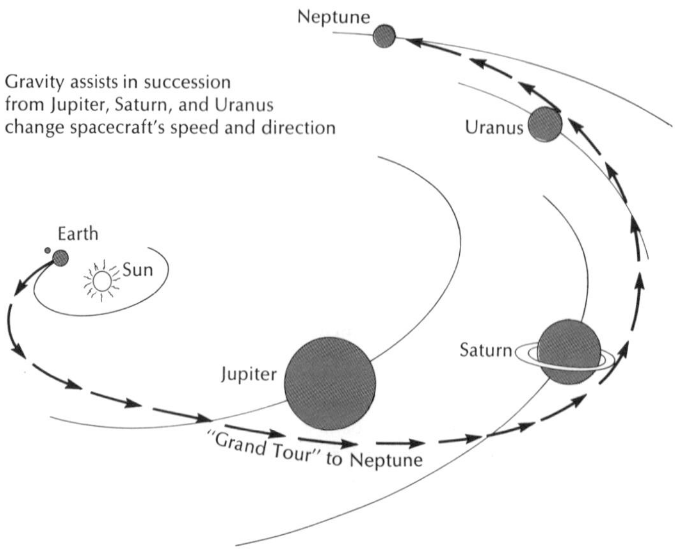

18-14. Highly schematic illustration of planet alignment that makes a "Grand Tour" possible. *(After artist's drawing,* American Scientist, *March-April 1970.)*

It so happens that during the late 1970's the major planets and Pluto are favorably located for other two-planet missions, as well as for some three-planet missions, plus one four-planet mission which has been appropriately called the "grand tour." A spacecraft launched in 1977 can be directed in such a manner that it will visit in turn each of the four major planets (Fig. 18–14) using gravity assists in succession from Jupiter, Saturn, and Uranus to finally reach Neptune. The opportunity to do this occurs only once every 175 years.

Exercises and Questions for Chapter 18

1. Assume that a spacecraft is to be placed into an orbit about the Earth.
   (A) Why is a multistage rocket used in its launching?
   (B) Why is an eastward launching direction favorable?
   (C) Why is a precisely circular orbit difficult to achieve?
   (D) Discuss advantages and disadvantages of equatorial vs. polar orbits; of an orbit that is inclined about 45 degrees to the Earth's equatorial plane.
   (E) Discuss how such a spacecraft might best be observed with the unaided eye.
2. Assume that a comet follows an elliptical path about the sun with a major axis of 20 astronomical units and an eccentricity of 0.5. What is its:
   (A) perihelion distance?
   (B) aphelion distance?
   (C) period of revolution?
3. A spacecraft is fired from the Earth into an elliptical orbit about the sun. It coasts outward from the Earth to an aphelion distance of 7 astronomical units and then falls back toward the sun.
   (A) What is the eccentricity of its elliptical orbit?
   (B) How long does it need to move from aphelion to perihelion?
4. Assume that an asteroid follows an elliptical path about the sun. Its aphelion and perihelion distances are 6 and 2 astronomical units respectively. What is its:
   (A) eccentricity?
   (B) period of revolution?
5. Assume that a spacecraft is launched to go from Earth to Mars along a minimum-energy orbit. Assume also that Mars will be 1.52 astronomical units from the sun when the spacecraft reaches it (at the aphelion point in the spacecraft's elliptical orbit around the sun).
   (A) What is the spacecraft's perihelion distance? Its aphelion distance? Its mean distance?
   (B) Assume that the launching time was inaccurate and that Mars was nowhere near the aphelion point when the spacecraft arrived. How long would the spacecraft need for one complete trip about the sun?
   (C) How long is needed for the trip from Earth to Mars?

(D) Why is the spacecraft not shot directly toward Mars when Mars is nearest the Earth (i.e., when Mars is at opposition)?

6. Assume that a spacecraft is to be sent along a minimum-energy path to Jupiter. Describe the path that it would follow and explain why all other paths would require an expenditure of more energy.
7. Why does it take about the same amount of energy to send a spacecraft to Venus as it does to Mars? Describe necessary speeds and launch directions in each case.
8. Does it take more energy to launch a spacecraft into an orbit so that it crashes into the sun or to launch it into an orbit that will eventually take it about 100 light-years from the Earth? Approximately what launch speeds and directions would be needed in each case?
9. Assume that a spacecraft is to be launched from the Earth into a clockwise, circular orbit about the sun. Discuss necessary speed and direction to do this.

   (A) What is the minimum launch speed from the Earth to make a spacecraft escape from the solar system in the clockwise direction?

19

Tools of the Astronomer

Until the 1940's, nearly all of our knowledge of the universe had been won from a study of the light from celestial objects and chiefly by means of three instruments: the telescope, the camera, and the spectroscope. Other portions of the electromagnetic spectrum are now being studied, and radio astronomy is a most important new development (Fig. 19–1).

Light and the Electromagnetic Spectrum

Although light is the main source of our information about stars and galaxies, the nature of light itself is still something of a mystery. Light has certain properties, such as interference and diffraction, that are best explained by considering it as a type of wave. However, light also has other properties, such as the photoelectric effect, that are best explained by considering it as a stream or aggregate of tiny, discrete units of energy called *photons* or light quanta. Different kinds of photons carry different amounts of energy and correspond to waves of different lengths. The amount of energy varies inversely as the wavelength.

According to the wave theory, light consists of certain electromagnetic waves which travel through empty space or homogeneous material in a straight line and at the rate of about 186,000 mi/sec (300,000 km/sec) in a vacuum. Other electromagnetic waves are either too long or too short for our eyes to detect, but in a vacuum they all travel at the same speed and form a continuous series (Fig. 19–2). Wavelengths range from as long as 15 miles to as short as one billionth of an inch. The light that we see forms only a very small fraction of this series, with the visible waves ranging between 1/30,000 and 1/70,000 of an inch in length.

Light apparently is produced by the quick jumps of electrons from higher to lower energy levels (p. 771). According to the inadequate but vivid Bohr model, the jumps would be toward the nuclei of their atoms. Such electrons had absorbed energy which moved

19-1. The world's largest radio telescope in Arecibo, Puerto Rico, was dedicated in November 1964 but is not yet completed in this photograph. A $2\frac{1}{2}$-ton dump truck illustrates the huge size of the reflector which covers an area of 18.5 acres. The transmitter hangs 435 feet above the center of the reflector. The spherical steel-mesh surface of this stationary instrument is located in a reshaped natural depression. The telescope may be "pointed" as much as 20 degrees from the zenith by shifting instruments at the focus. *(Courtesy Commonwealth of Puerto Rico.)*

them to higher energy levels (Bohr model: away from the nucleus) a fraction of a second before because they had been bombarded by some particle, such as another electron, or by other electromagnetic radiations.

Ordinary white light is a mixture of various electromagnetic waves which the human

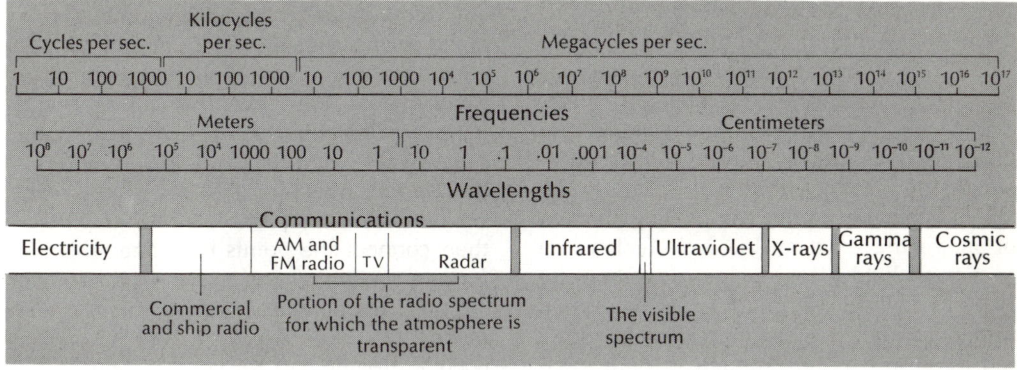

19-2. Range of wavelengths in electromagnetic radiation. There are 2.54 cm in 1 inch. Kilocycle and megacycle refer to 1000 and 1,000,000 cycles per second respectively. *(H. L. Goodwin,* Space: Frontier Unlimited, *Van Nostrand Reinhold, 1962.)*

eye can see individually as the different colors of the rainbow: red, orange, yellow, green, blue, and violet. Sunlight also includes waves that are too long or too short for our eyes to detect (p. 646). With a simple prism, white light (sunlight) can be separated into its component colors. The visible spectrum ranges from the short-wave, high-energy radiations that produce a violet color to the lower energy, longer wavelengths that have a red color (Fig. 19–3).

If an object is not itself a source of light, it is visible because it reflects or reradiates light from another source. The color of such an object depends upon which wavelengths it reflects, and hence does not absorb. In its simplest aspect, an object is some color, say yellow, because its atomic structure causes the reflection of light of the wavelength identified with yellow and the absorption of light of all other wavelengths. Radiation of various wavelengths within the visual range can produce spectral colors ranging all the way from the reddest red to the bluest blue. But our eyes also have the ability to blend two or more spectral colors and produce a sensation of a completely different color. Thus the blending of pure blue and pure yellow produces the sensation of green, even though the eye may not be receiving any radiation of the wavelength corresponding to a pure spectral green at all.

White involves the reflection of all waves and black the absorption of all waves. Thus black corresponds to the absence of all radiation in the visible wavelengths and is not a true color. This explains why black clothing tends to be uncomfortably warm when worn on a hot sunny day. However, texture is about as important as color in determining the amount of radiation that is absorbed or reflected—rough textures cause more absorption, whereas smooth, polished surfaces cause more reflection.

The color of an object that is its own source of light depends chiefly upon its temperature. As the temperature of an object is raised, changes occur in the type

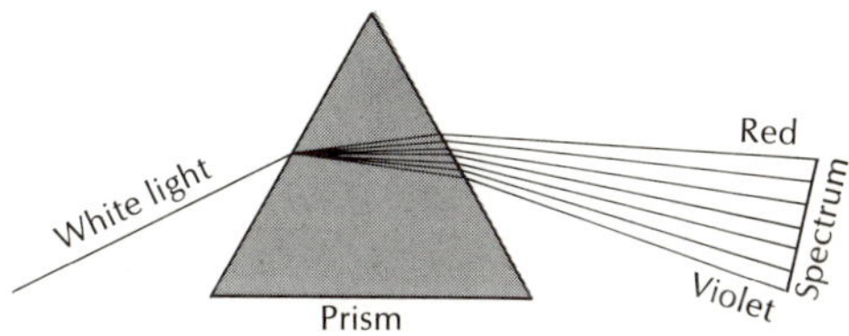

19-3. Origin of a spectrum. White light passes through a glass prism and is refracted or bent twice: once as it enters the prism and once as it leaves. However, the amount of refraction depends on the wave length: short waves are refracted most and long waves are refracted least. Thus the mixture of waves is spread out into a spectrum, the familiar rainbow of colors.

and amount of electromagnetic energy that it radiates. More radiation of all wavelengths is emitted at higher temperatures, but the proportion of shorter wavelengths (high-energy photons) increases much more rapidly than does the quantity of longer wavelengths (low-energy photons).

In a familiar example, the heating element of an electric stove on low heat feels warm to a hand held a few inches away because it emits invisible infrared radiation. When turned to high, the heating element glows red because it is hot enough to emit high-energy photons corresponding to red light in addition to the low-energy photons corresponding to infrared radiation. At still higher temperatures, the radiation would become stronger, and the color would change in turn to orange, to yellow, and then to green. Temperatures higher yet produce radiations which are strongest in the blue and violet wavelengths.

This relationship between temperature and color enables astronomers to determine the surface temperatures of the stars. Yellowish stars like the sun have surface temperatures of approximately 6000°K (Fig. 28–26). The surface temperatures of reddish stars are lower, and those of bluish stars are higher.

The Radiation Laws

To understand the radiation laws and spectral energy curves, imagine the following laboratory experiment. An object is at a certain temperature, say 4000°K, and we arrange to measure the total amount of radiation that it emits at each of a number of wavelengths. We find that more energy is radiated at about 0.7 microns than at any other wavelength (Fig. 19–4; 1 micron = 0.001 mm). We plot the intensity at each wavelength as a point on the graph and then connect the points by a line. This line forms a *spectral energy curve* for that object at that temperature. Measurements then made with the object at 5000°K and at 6000°K provide data for two additional spectral energy curves.

Two important radiation laws have resulted from quantitative studies of the spectral energy curves for so-called *black bodies* —objects which emit the maximum amount of radiation in each wavelength that is possible at the existing temperature (not necessarily black in appearance). Thus the spectral energy curves for all black bodies at a given temperature will be the same. Objects that are not black bodies radiate less than this maximum quantity at certain or all wavelengths.

According to Wien's law, *the wavelength in which the maximum amount of energy is radiated from a black body varies inversely as the absolute (Kelvin) temperature.* For example, assume that an object at 6000°K radiates the maximum amount of energy at 0.5 microns. By use of Wien's law, we can calculate that at 3000°K it will radiate at a maximum at 1 micron (6000 : 3000 = X : 0.5). At 12,000°K, the maximum radiation will occur at 0.25 microns.

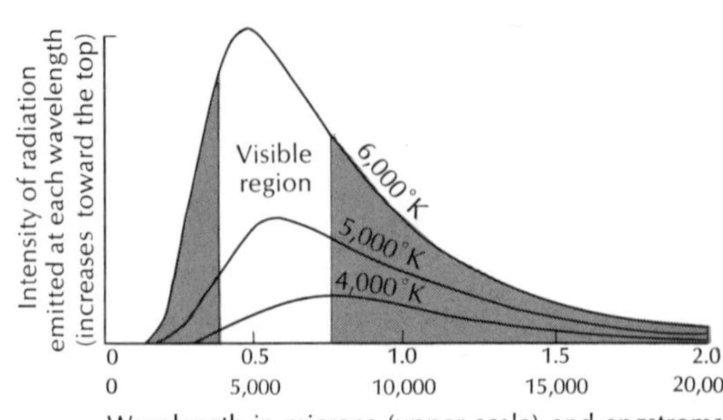

19-4. Spectral energy curves for a perfect radiator. Radiation intensity is generalized (100,000,000 angstroms equal 1 cm; 10,000 angstroms equal 1 micron; and 1000 microns equal 1 mm). The area under a curve represents the total quantity of radiation emitted at that temperature. Note that for high temperatures this is much larger than for lower temperatures and that the peak of the curve is shifted toward the shorter wavelengths. (R. H. Baker, *Astronomy,* 8th ed., Van Nostrand Reinhold, 1964.)

According to the Stefan-Boltzmann law, *the rate at which energy of all wavelengths is radiated by a black body is directly proportional to the fourth power of the absolute temperature*. Thus a black body at 400°K will radiate 16 times (2^4) more energy per unit area than at 200°K. At 600°K it would radiate 81 times (3^4) more energy than at 200°K, etc.

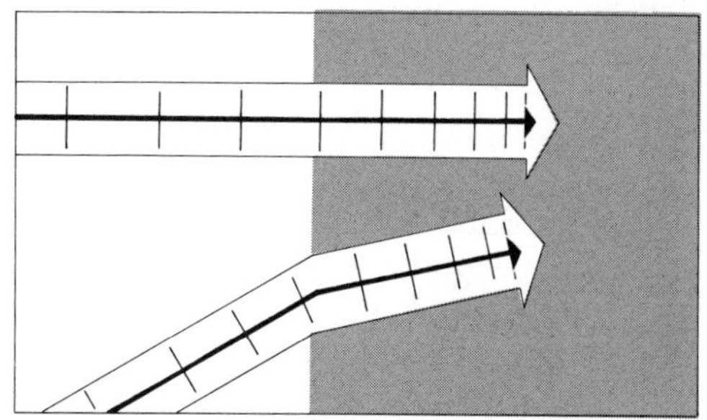

19-5. Refraction occurs when light passes at a slant into a different medium in which its speed is changed. No refraction occurs if the angle is 90 degrees (upper part of diagram). In this illustration, wave fronts move from left to right from a faster medium (wave fronts are farther apart) to a slower medium. The bending is toward a perpendicular to the boundary at the point of penetration. In the reverse direction, the bending is away from the perpendicular. (R. H. Baker and L. W. Fredrick, *An Introduction to Astronomy*, Van Nostrand Reinhold, 1968.)

The Telescope and Camera In Astronomy

Telescopes involve lenses and mirrors and the refraction and reflection of light waves. Waves of light from a source such as a star travel outward in all directions from this source. If we consider a minute portion of this radiation, we may think of it as a ray of light that travels in a straight line as it passes through a homogeneous medium or through a different medium that it enters at a 90-degree angle. However, if the ray passes at a slant into another type of medium, its speed and direction are changed, and we say that it is *refracted* or bent (Fig. 19–5). The lower part of a wave front is slowed as it enters a denser medium, whereas the upper part is still moving at the faster rate. This shifts the direction of movement of the wave front.

If the ray encounters a smooth, polished surface such as a mirror, it is bounced back or *reflected*. The angle at which a ray is reflected is the same as that at which it strikes a surface (angle of incidence).

A convex lens or a concave mirror—the *objective* of a telescope—can be so shaped that it will concentrate at a point all of the parallel light waves that fall on it from a source (Fig. 19–6). This point, where all of the once-parallel rays are converged, is called the *focal point*, and the distance to it from the center of a convex lens or mirror is called the *focal length*. The size of a telescope is generally given by the diameter of its objective (e.g., a 3-inch or a 200-inch telescope).

In contrast to a point source such as a star, light from a distant area source such as the moon may also enter a telescope. Light originating from any one point on the moon approaches the telescope along parallel lines and is focused at a single point. However, light waves coming from other points on the moon's surface approach the telescope from slightly different directions and are thus focused at different points, al-

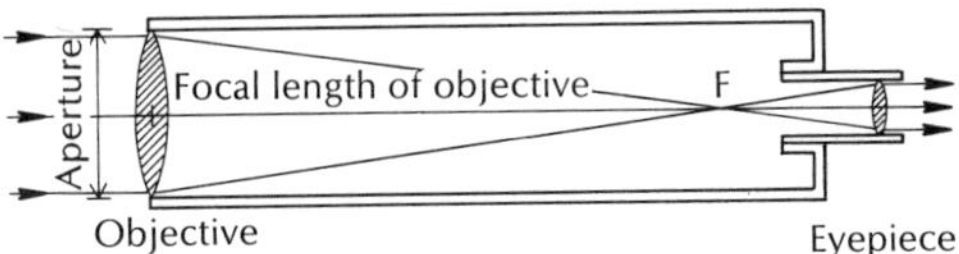

19-6. A simple refracting telescope. Note that all of the parallel rays falling on the large lens also pass through the eyepiece; thus they are concentrated. The large convex lens is called the objective and the small one is the eyepiece; their distance apart is equal to the sum of their focal lengths when the object being viewed is very far away. In most modern telescopes the objective and eyepiece are both compound lenses involving different shapes and materials. Various eyepieces can be used to achieve different magnifications. (R. H. Baker, *Astronomy*, 8th ed., Van Nostrand Reinhold, 1964.)

though all in the same plane—the *focal plane* of the telescope (i.e., at the focal point F in Fig. 19–6 and oriented perpendicular to the long axis). If a screen or a photographic plate is placed in the focal plane, an image of the object is formed on it.

The eyepiece is a special kind of magnifying glass used by the observer to study the image in the focal plane. Light rays diverge as they pass through the focal plane and must be refracted again to be parallel when they reach the eye. The eyepiece causes this bending and is similar to the objective of a refractor. However, the eyepiece is small enough so that the entire beam of parallel rays emerging from it can enter the pupil of the eye.

The closer an eyepiece is located to the image (i.e., the shorter its focal length) the greater is the magnification. However, increased magnification has disadvantages because distortions are also magnified (e.g., diffraction causes each star to be a small disk instead of a point) and because it means that any two points on the image have a wider separation. This shrinks the field of view and decreases the amount of light from any one part of the image. Further, an object will be in its field of view for a shorter time unless the telescope has a drive mechanism to compensate for the rotation of the Earth.

Magnification is the ratio of the focal length of the objective to the focal length of the eyepiece. Thus various eyepieces can be used to achieve different magnifications. Maximum magnification is about 50 times the diameter of the objective; e.g., a 3-inch refractor has an upper magnification limit of about 150 times, and generally the limit is quite a bit less than this. If the objective's focal length is 45 inches, then an eyepiece with a focal length of 1 inch will magnify 45 times, and a ½-inch eyepiece will magnify 90 times, etc.

The size of an image is directly proportional to the focal length of a telescope. Thus a small refractor is more or less comparable to a somewhat larger reflector both in performance and in price. A refractor's longer focal length (commonly 15 times the diameter of its objective whereas that of a reflector tends to be 5 to 8 times longer) means greater magnification per inch of objective, but it gathers less light (somewhat offset by the blocking of light by supports for the secondary mirror within a reflector). As compared to a reflector, a small refractor tends to be more convenient to point and look through, it tends to stay in adjustment longer, and it does not have a mirror that needs an occasional recoating (coatings last longer now because of recent improvements).* A sturdy tripod is important. A solid tube is unnecessary and is eliminated in large telescopes to save weight. However, one is useful to keep out stray light.

Since the parallel rays from any one point are focused from all parts of a lens or mirror, a complete image is obtained even if some of the light is blocked, but the image is then less bright. Thus the small reflecting mirror and supports located near the opening of a reflecting telescope do not produce "holes" in the image. The light that falls on the 200-inch mirror is reduced about 15% by such obstructions. Mirror supports also produce the four spikes commonly seen extending from bright overexposed stars on photographs.

The light-gathering power of a telescope varies as the square of the diameter of the objective (i.e., directly as its area). This can be compared to that of the unaided eye if we assume an average diameter of the pupil of about 1/5 inch (also changes in size with the intensity of the light). Thus a 1-inch diameter lens can gather 25 times as much light as the unaided eye, and the 200-inch (508 cm) Hale telescope at Mount Palomar, the world's largest optical telescope through the 1960's, can concentrate

* See *Consumer Reports* for November 1967 for an evaluation of small refractors and reflectors. See also a series of articles under the heading of Backyard Astronomer in *Natural History* by James S. Pickering: Choice of a Telescope, November 1966; Setting up a Telescope, January 1967; Finding Celestial Objects, March 1967; and The Sun and Double Stars, August-September 1967.

about 1 million times as much light as the unaided eye (Russian astronomers may have a 6-meter or 236-inch reflector available in the early 1970's). Objects much too faint to be seen without optical assistance are brilliantly visible when their light rays are funneled into this telescope. For many astronomical problems, the light-gathering capacity of a telescope is much more important than magnification. The *resolving power* of a telescope is also important—i.e., its capacity to separate two closely spaced objects—and it is greater with larger diameters.

A refracting telescope has a lens at one end through which light passes and bends on its way to the eyepiece at the other end. But the different lengths of waves in the light are bent at different angles and are not focused at the same point by an objective consisting of a single lens; red is focused farthest from the lens and violet closest to it (the lens has somewhat the same effect as two prisms placed base to base; see Fig. 19–3). Therefore, the image as seen through the eyepiece is in focus for one color but not for the other colors, and this blurs the image. The effect is called *chromatic aberration* and can be largely corrected in an achromatic telescope by using compound lenses of different shapes and materials. If the telescope is to be used visually, it is more important to have the yellow region of the spectrum in focus because the human eye is particularly sensitive to this light (the out-of-focus, blue-violet light may produce a halo about a star). On the other hand, the shorter blue and violet waves are more important when a telescope is to be used chiefly as a camera. In reflecting telescopes, there is no color distortion.

The parabolic mirror of a reflector forms good quality images if light rays enter nearly parallel to the telescope axis; thus its field of view is limited. Another type of telescope (Schmidt) was developed for use in photographing large regions of the sky—say a square 7 degrees on a side—with relatively short time exposures. It uses a mirror, a thin lens or correcting plate, and a curved plate holder.

All large optical telescopes are reflectors for a number of reasons—diameter of the largest refractor is 40 inches (102 cm). No chromatic aberration occurs in reflecting telescopes, and their shorter focal lengths reduce costs. Light is reflected from the surface of a mirror (actually from a very thin coating of aluminum or other material covering this surface). Therefore, the quality of the glass or Pyrex constituting the mirror does not have to be as high as it must in a refractor. Large mirrors can now be made of quartz (Fig. 19–7) and other materials which have extremely low volume changes with temperature changes (lower than Pyrex, which in turn is better than glass). Thus their precisely polished surfaces are not distorted by day-night temperature changes. The mirror of a reflector can be supported across the back as well as around its edge and has to be polished on only one side. A mirror does not have to be made of glass, because the light never actually touches it; but glass can be made homogenous and can be ground and polished to a very smooth surface. Furthermore, it does not tarnish like a metal.

The larger telescopes are used for photography much more than for visual observation and might well be regarded as special kinds of cameras. The negatives themselves are commonly used and bright stars appear as dark spots on them. The camera has the advantage of providing a permanent record of what is seen, unbiased by any personal prejudice, and it can also photograph objects in both infrared and ultraviolet light that cannot be seen by man.

Furthermore, many remote celestial objects are too faint for the eye to see even through the 200-inch telescope, and time exposures must then be made to extend outward the boundaries of the observable universe. Light from a very distant star, funneled through a telescope onto a film, may be too faint to create an image in a minute or an hour. However, if the time exposure is continued for a few hours or

19-7. New 61-inch astrometric reflector at the U.S. Naval Observatory near Flagstaff, Arizona. This is the first big reflector planned for the utmost accuracy in positional measurements. The large mirror is made of quartz, which has a very low coefficient of expansion. Thus rapid temperature changes near sunrise or sunset do not cause distortions. The observer is seated on a converted fork lift that he can raise or lower and move about the observatory floor. *(Official U.S. Navy photograph.)*

more, the star's photograph may eventually be obtained. Fogging of the entire negative by general skylight limits the length of time that a film can be exposed, but a successful 80-hour exposure was once made of the spectrum of a remote galaxy.

To observe objects at increasingly greater distances, astronomers have utilized different techniques. The amount of light that falls on a photographic plate or eyepiece has been increased by building larger and larger telescopes. Long time exposures have been made, and faster, more sensitive film has been developed. To make better use of the light available, an electronic device, called an image converter, has been constructed, and is being improved. It is much more sensitive to light than any known film (principle is similar to photoelectric effect, p. 510). Theoretically a 20-inch telescope

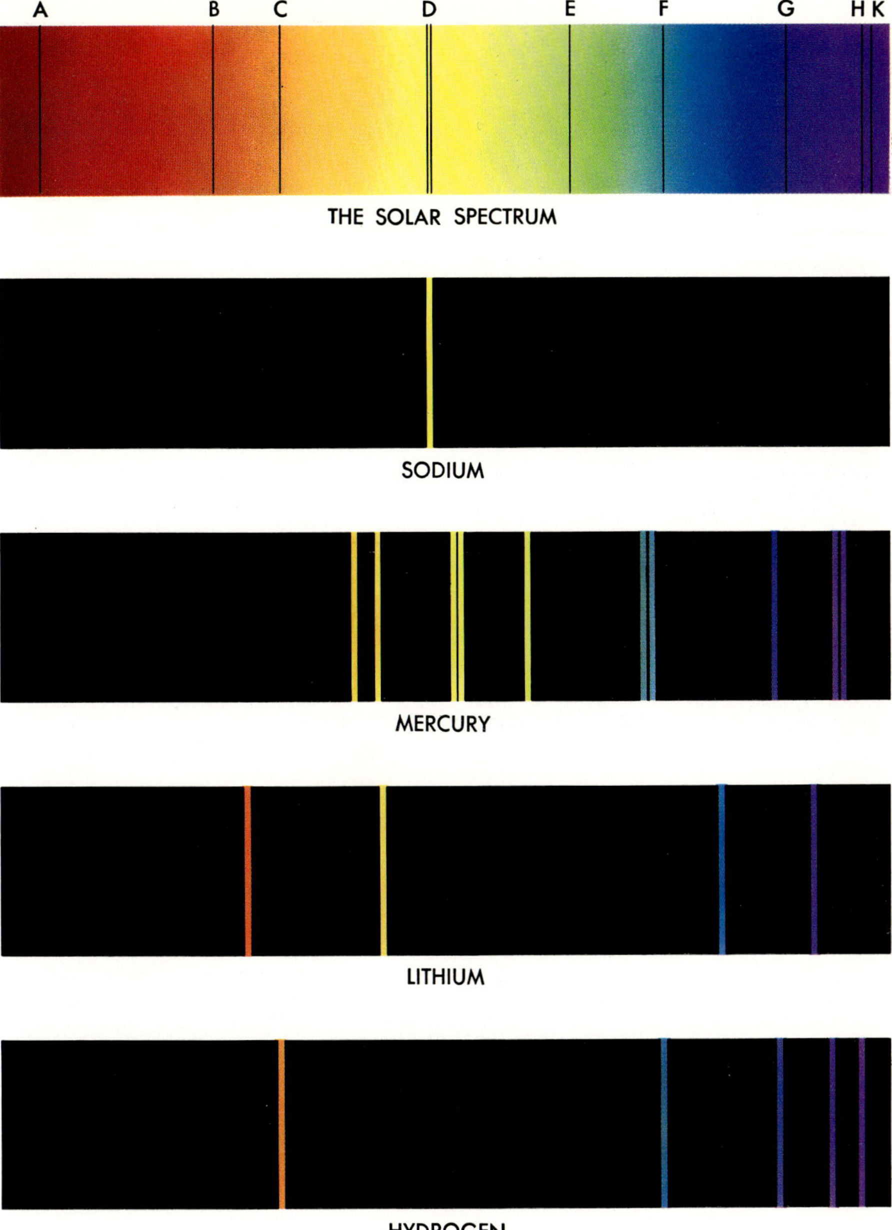

19-13. Four bright-line spectra and one dark-line spectrum. The solar spectrum at the top shows a few of the dark absorption lines which make it possible for us to identify elements present in the sun's atmosphere. Note that each of the four elements—sodium, mercury, lithium, and hydrogen—produces a distinctly different, bright-line spectrum or "fingerprint." The single yellow sodium line actually consists of two yellow lines very close together and somewhat difficult to separate. Their positions are shown by the two sodium absorption lines at D in the solar spectrum.

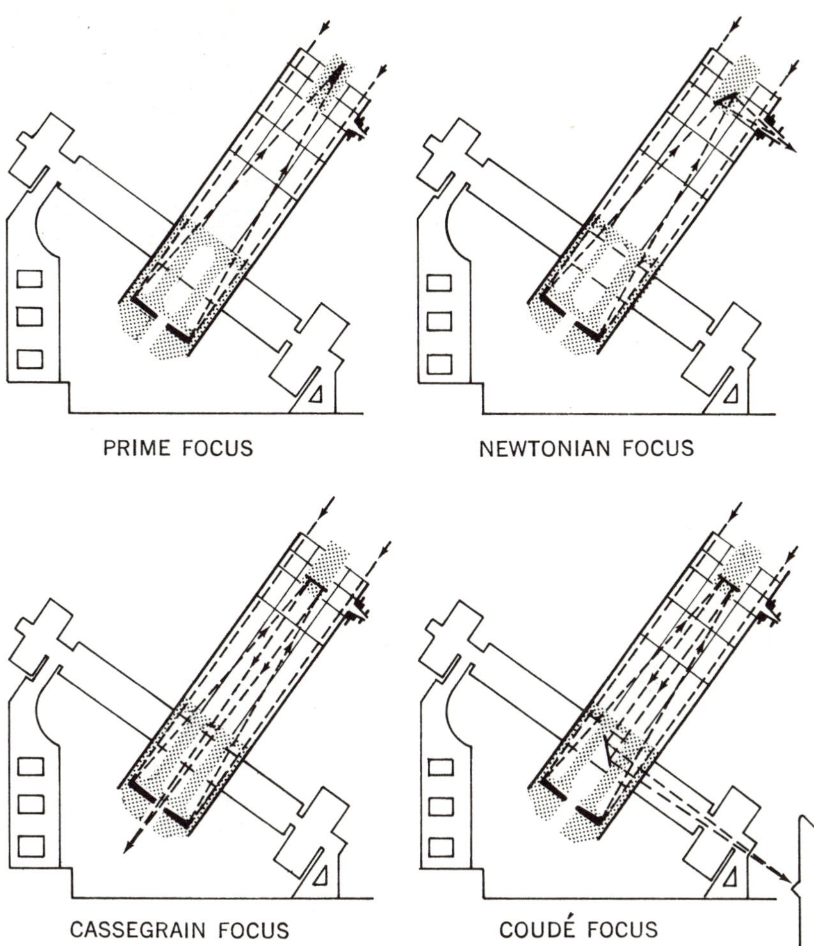

19-8. Four ways to focus a reflecting telescope. In all four arrangements the large objective mirror is at the bottom and a smaller reflecting mirror or an observer's cage is at the top. The polar axis of the mounting (dark bar from upper left to lower right) is oriented precisely parallel to the Earth's axis. (1) Prime focus: An astronomer can work at an observer's cage located at the focal point of the mirror. This is possible only in large instruments such as the 200-inch Hale and 120-inch Lick telescopes. (2) Newtonian focus: This is the common arrangement in small reflecting telescopes. An eyepiece is located in a tube at the side near the upper end. (3 and 4) Cassegrain focus and Coudé focus: Light is reflected from a smaller convex mirror in the upper end back through a hollow polar axis. This locates the focal point below the base of the telescope where it is more convenient to attach a spectroscope, camera, or other equipment. In the Coudé focus, the focal point is located in a laboratory at the lower right which can be darkened for spectroscopic study. (From *Frontiers in Space,* Mount Wilson and Palomar Observatories.)

with this attachment might detect objects as far away as a 200-inch telescope could without it.

Large telescopes are oriented on axes (plural of *axis*) and electrically operated so that they can be centered exactly upon a celestial object and follow it, without deviation, on its daily westward journey across the skies (Fig. 19–8). One axis (polar) is oriented parallel to the Earth's axis, and the telescope tube is rotated about it to follow the diurnal motions of celestial objects, thus compensating for the Earth's rotation. A telescope can also be pointed higher or lower in the sky on a second axis. Astronomers use a system of coordinates similar to terrestrial latitude and longitude to locate celestial bodies.

Three very large optical telescopes, including the 200-inch giant Hale reflector (Fig. 19–9), are located in California. Such large observatories are commonly placed in areas far from city lights and smog (Los Angeles has grown too large to please the astronomers at Mount Wilson), in climates that permit good seeing on many nights, and in high altitudes above the dense atmosphere of sea level. However, a large-telescope building boom is in progress, and several giant new instruments having diameters of about 150 inches are scheduled for completion in the early 1970's. Some of these are being located in the Southern Hemisphere (observatories 2 miles above sea level in the Chilean Andes have unusually calm, clear, transparent air far from city lights).

19-9. Dome of 200-inch telescope on Mount Palomar in California seen by moonlight. The entire top part of the dome can rotate on tracks so that the telescope may be pointed at any part of the sky that is suitably high above the horizon. Note the observer's cage near the upper end of the cylindrical steel network that constitutes the tube of this telescope. *(Courtesy Mount Wilson and Palomar Observatories photograph.)*

Experiments are underway that apply the techniques of automation to large telescopes. Such telescopes can be aimed and regulated by remote control with an accuracy of 1 second of arc and are particularly useful for making a series of brightness measurements of stars at different wavelengths. An Orbiting Astronomical Observatory satellite was successfully launched in December of 1968 into an orbit nearly 500 miles above the Earth's surface and functioned even better than expected (however, OAO I and OAO III, carrying a 36-inch telescope, failed). It carried eleven telescopes, the largest 16 inches in diameter; some were coupled with cameras that photograph the sky, and others were linked to instruments that measure the light and spectra of stars. Study of ultraviolet radiation from celestial objects was emphasized since such light cannot penetrate to surface telescopes (p. 651). Astronomers have likened this to Galileo's "first look." Plans call for placing other telescopes in Earth orbit in the early 1970's (Fig. 19–10) and for studying other regions of the electromagnetic spectrum that cannot penetrate the Earth's atmosphere.

Radio Astronomy

In addition to visible light, celestial objects such as stars, planets, gas clouds, and galaxies also produce long, invisible, electromagnetic waves that penetrate the Earth's atmosphere through the so-called radio window. These are intermediate in length between light waves and standard radio waves (from a fraction of a centimeter to 30 meters or so). Shorter wavelengths tend to be absorbed by the atmosphere, and longer ones are reflected.

19-10. Artist's concept of the components of Skylab in Earth orbit in the early 1970's (workshop, airlock, multiple docking adapter, Apollo Telescope Mount, and command service module). According to present plans, a crew will be launched by rocket from the Earth, dock with the space station, spend a month or two making observations, and then return to Earth in the command module. The telescope, controlled from panels inside Skylab, will make possible the undistorted study of celestial objects at many different wavelengths—not limited to wavelengths that can penetrate to bottom of Earth's atmosphere. *(Courtesy NASA.)*

Radio astronomy is a relatively new and rapidly developing branch of astronomy. In one type of radio telescope (Fig. 19–1), a large antenna (reflector or "dish") reflects the radio waves that strike it and focuses them on a small antenna whose position corresponds to that of the photographic plate located in the focal plane of an optical telescope. The current thus induced in the antenna is conducted to a receiver and recorded. As in an ordinary radio set, only one wavelength at a time can be utilized, not a range of wavelengths as in visible light. However, the receiver may be tuned quickly from one radio wavelength to another.

Unwanted radio signals or noise must be screened out, and these come from many sources: electrical motors, car ignitions, lightning flashes, man-made radio signals, moving electrons, and the radiation that comes from the ground, the antenna itself, and in fact from all warm objects. Thus location in a mountain-encircled depression far from industrial activity is favorable.

A reflector is commonly of metal and may be an open fencelike mesh if the openings are kept much smaller than the wavelengths being studied. The strength of the signal on a certain wavelength from a certain part of the sky is recorded on a registering device, and a telescope may be used in different ways: to sweep across a certain part of the sky, to track one portion continuously, or to remain motionless and record the changes as the sky "rotates by."

Radio telescopes function as well during the day as at night and can receive electromagnetic waves that penetrate readily through clouds in the Earth's atmosphere and also through the gas and dust clouds that occur here and there in space, particularly along the equatorial planes of spiral galaxies. These are advantages over the optical telescopes—full-time, all weather astronomy and greater penetration of space. However, the target objects are never actually seen and are difficult to locate precisely. Thus the radio telescope does not supplant the optical telescope; each has a useful function, and together they help enlarge our knowledge of the universe.

Radio signals from space have two general types of origin: thermal (radiation from warm objects, p. 496) and non-thermal (radiation produced by moving electrons and ionized particles). The nature of the thermal radiation is a measure of the temperature of the emitting body. The 21-centimeter signal (a wavelength of about 8 inches) from neutral hydrogen atoms in space has proved to be of great importance. This is a sharply limited spectral "line" and not a wide band of different wavelengths. It is capable of showing a Doppler effect (p. 508).

Among the advances already credited to radio telescopes are: (1) measurements of the surface temperatures of planets; (2) the discovery that some galaxies are very powerful sources of radio signals (Fig. 19–11); (3)

19-11. A very powerful source of radio waves whose nature is uncertain (NGC 5128). Among the suggested interpretations: a spiral galaxy seen edgewise may be colliding with an elliptical galaxy, or an explosion of some kind may be taking place. *(Courtesy Mount Wilson and Palomar Observatories.)*

data about the size, shape, and number of spiral arms of our own and other galaxies; (4) data about the amount of matter scattered in clouds of gas and dust in space; (5) quasi-stellar objects (p. 634); (6) data concerning the distribution of galaxies in space—including evidence against the Steady State hypothesis (p. 633); and (7) the possible existence of cosmic, black body, background radiation (p. 633).

Pinpoint resolving power is lacking, even in a very large radio telescope. However, this can be overcome by a technique known as interferometry (also used in optical telescopes). Radio signals from a discrete source tend to reach one of two telescopes—perhaps miles apart but connected by electrical cable—a tiny fraction of a second before the other; i.e., the signals are not in phase. The amount of wave interference thus produced varies quantitatively with the angle at which the radio signals approach the antennae. In a recent large-scale adaptation, data from independent radio telescopes located on different continents have been synchronized by means of very precise atomic oscillator clocks, and a resolution of 1/1000 second of arc has apparently been achieved.

The Spectroscope in Astronomy

The amount of information about an object that can be obtained by studying its light borders on the incredible, and the spectroscope is a key instrument (called a spectrograph if a photographic plate is substituted for the eyepiece, as is the common practice in astronomy). A spectroscope spreads apart the different wavelengths present in the entering light to form a spectrum which can then be studied to determine the presence or absence of certain key wavelengths. This may be accomplished by means of a glass prism (Fig. 19–12) or by using a diffraction grating (produces a spectrum in a different way).

Laboratory study has shown that there are three main types of spectra (Fig. 19–13): continuous, bright-line or emission, and dark-line or absorption. A *continuous spectrum* is produced when light enters a spectroscope from any solid, liquid, or com-

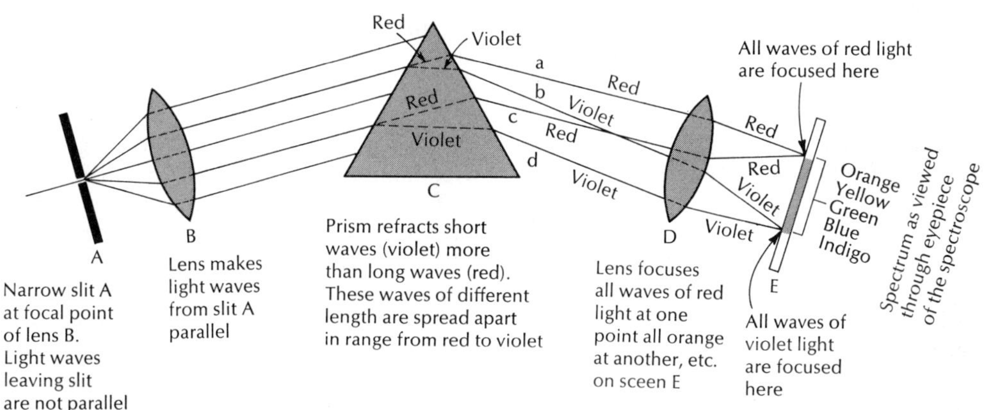

19-12. Paths of light waves inside a spectroscope. Highly schematic. Begin at the left at A and follow the paths of the light waves to the focal plane at E. An eyepiece is located still farther to the right for viewing this focal plane (the distance to the focal plane is equal to the focal length of the eyepiece). A photographic plate can be placed at E to make the instrument into a spectrograph. Many shades of each color occur, and the statement that all waves of red light are focused at one point should be interpreted broadly. It should be apparent that dark lines will occur in the spectrum if certain wavelengths or colors are missing.

pressed gas that has been heated enough to become luminous. In other words, a continuous band of colors like the rainbow is visible at the focal plane of a spectroscope if light from such a source is passed through the slit and viewed through the eyepiece. This indicates that such light consists of a mixture of components of all wavelengths from the longest red to the shortest violet (i.e., none of the wavelengths of visible light are missing, otherwise there would be dark gaps in the spectrum).

On the other hand, a *bright-line or emission spectrum* is seen through the eyepiece if light from a luminous but noncompressed gas is passed through the slit of a spectroscope—e.g., from a neon sign. Such a spectrum consists of a number of bright, colored lines arranged against a dark background—i.e., the colored lines are separated by black, vacant areas (Fig. 19–13). Evidently the light from such a source consists of a mixture of a relatively few colors (wavelengths). Most of the wavelengths of visible light are not produced by this particular gas, hence the black background. Furthermore, each element in this state always produces the same characteristic pattern of bright lines whether the experiment is performed in a lab in China or Peru, in the winter or summer, in 1930 or 1970. And of greatest importance, the luminous, noncompressed gaseous phase of each one of the known elements always produces an emission spectrum that is distinctively different from the spectrum of all other elements. There are characteristic, diagnostic features in the bright-line spectrum of each element—the numbers, locations, and intensities of the lines for a particular element form a pattern that is different from the pattern of any other element. Thus elements can be identified by their bright-line spectra because each has a distinguishing set of "fingerprints."

The bright spectral lines are images of the slit, which is made very narrow to prevent overlapping. However, in a continuous spectrum, so many images are produced that they do overlap and merge together. It should be noted that the lines are just images of the slit itself. If the shape of the slit were changed into a tiny triangle or circle, then the spectra would consist of tiny bright colored triangles or circles—not lines. Spectral lines are formed by atoms (p. 771), whereas molecules form bands.

In the third type of spectrum—the *dark-line or absorption spectrum*—the characteristic pattern of an element appears as dark lines against an otherwise continuous colored background (solar spectrum in Fig. 19–13, opposite, p. 500). The pattern, number, and locations of the dark lines of a particular element are identical with the lines in its bright-line spectrum, but a dark line stands in the place of each colored line in the spectrum (e.g., for sodium vapor, two dark lines appear in the yellow part of the spectrum).

Laboratory experiments have shown that two conditions are needed to produce a dark-line spectrum. The light must come from a continuous source, and it must pass through a relatively cooler gas before reaching a spectroscope. This cooler gas absorbs from the light the very same waves it would emit if it were hot enough to produce a bright-line spectrum (the energy that is absorbed is emitted at once, but in all directions, and sometimes in wavelengths different from those absorbed—hence the dark lines).

A dark-line spectrum thus shows the chemical composition of the cooler, light-absorbing gas located somewhere between a continuous source and a spectroscope. It tells us nothing, however, about the chemical composition of the light source itself, which could be any luminous solid, liquid, or compressed gas.

Sunlight originates from very hot gases that are under great pressure. Therefore, sunlight is a continuous source, and all possible wavelengths in the visible part of the spectrum are present at the sun's surface (photosphere). However, this light travels from the photosphere outward through relatively cooler gases in the lower part of the sun's atmosphere. These cooler gases absorb certain wavelengths, and thus dark lines or gaps are produced in the solar spectrum. Well over 20,000 dark lines originating from more than 60 elements have been identified in the atmosphere of the sun.

Cool gases in the Earth's own atmosphere also absorb certain wavelengths, but these are chiefly bands produced by molecules and not lines; thus they can be distinguished from lines produced on the sun (for another method, see Fig. 19–18). The chemical composition of the atmosphere of a planet can also be determined spectroscopically to some extent because light from the sun passes part way through this atmosphere before it is reflected to the Earth. In such cases, astronomers must distinguish between spectra produced by the sun's atmosphere, by the atmosphere of the planet, and by the atmosphere of the Earth. However, if the temperature of the atmosphere of a planet is very low, the spectrum of its gases in the visible region may become very weak or even disappear.

The bands produced by certain molecules in the Earth's atmosphere, such as water vapor and oxygen, may obscure identical bands produced by the same molecules in the atmosphere of a planet and thus prevent their detection. This has inspired spectrographic study from high-altitude balloons that can rise above the dense lower part of the Earth's atmosphere (Fig. 19–14), and measurements are desirable from instru-

19-14. The Stratoscope II balloon system a few moments after launch in Texas. Later, the 36-inch-aperture telescope came down in Tennessee after being lowered gently by balloons. The object of the initial flight was to make an infrared study of Mars. Total weight to be lifted was about 13,000 pounds; the payload was 6300 pounds of this total. The guidance system of the telescope is designed to track an object in the sky and hold it fixed in the image plane to within 0.02 second of arc for as long as one hour. During a time-exposure photograph, the balloon may float at an altitude of about 80,000 feet. *(N.S.F. photograph.)*

ments located at an even higher altitude. Therefore, satellites containing telescopes, spectographs, and other instruments are being launched into orbits about the Earth.

Thus elements can be identified in the sun, stars, nebulae, and galaxies by the characteristic spectra they produce—most commonly these are dark-line spectra. Spectra are also of great value in industry and research in identifying the elements present in certain specimens. A small quantity of the specimen is vaporized (e.g., in a carbon arc), the light is examined in a spectroscope, and the pattern of lines is noted and checked against a well-organized file catalogue. Very minute quantities of an element can be detected readily and quickly in this manner. The relative proportions of the various elements can also be estimated by studying the intensities of the lines they produce.

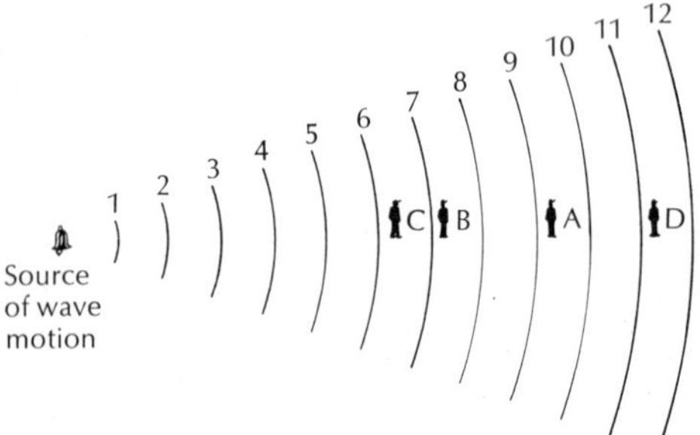

19-15. Doppler effect explained schematically. Assume that light or sound waves are produced periodically and move outward in all directions from the source area. The waves are numbered for convenience and are assumed to move by the observer at A at the rate of 6/sec. During the next second, therefore, waves 9, 8, 7, 6, 5, and 4 pass by this observer. However, if during a second the observer moves from A to B, waves 3 and 2 will also pass by him. Thus eight waves move by the observer during this second, and to him each wave seems shorter than it did when the distance remained unchanged. If the observer remains at B, the rate returns to 6/sec. On the other hand, if the observer moves to D during a second, waves 4 and 5 will not reach him, and the rate is reduced to 4/sec. To the observer, each wave seems longer. The magnitude of the apparent change in wavelength is directly proportional to the rate at which the distance is changing; e.g., if the observer moves more rapidly in a second and travels from A to C instead of from A to B, nine waves reach him during this second, and the waves seem shorter than they did at the rate of 8/sec.

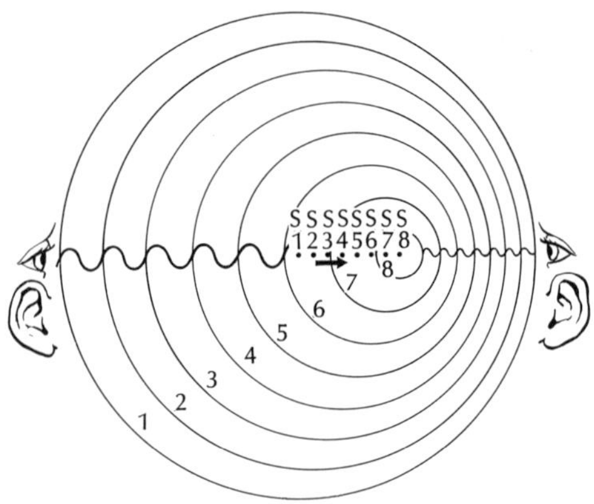

19-16. A Doppler effect may also be produced by movement of the source. Circle 1 represents a wave crest that originated at source S_1 and moved radially outward. Circle 2 is a wave crest that originated a moment later at S_2. Circles 3, 4, 5, etc. originated in turn at positions S_3, S_4, S_5, etc. The waves represent schematically either light or sound. Note that the wavelengths seem greater to the observer at the left (distance is increasing) and shorter to the observer at the right (distance is decreasing).

The type of spectrum produced by a particular element tends to vary with the temperature, and the spectrum at a very high temperature may be entirely different from that at a low temperature. Temperature differences seem to account for the main differences observed in stellar spectra (p. 607).

The Doppler Effect

The Doppler effect may be used to determine whether the distance between the Earth and a star is increasing or decreasing as well as the rates at which such changes may occur. So-called red and violet shifts are involved (Figs. 19–15 and 19–16). We can draw the following conclusions concerning the Doppler effect. If the distance between a source of waves and an observer is decreasing, the waves appear shorter to the observer, even though no actual change in wavelength occurs. If the distance is increasing, the waves appear longer. The observer, the source, or both may move; the Doppler effect cannot be used to distinguish among these three possibilities. Finally, the faster the distance changes, the shorter (if decreasing) or longer (if increasing) the waves appear to the observer.

A familiar form of Doppler effect is the change in pitch that a car horn apparently undergoes as it moves first toward and then away from us. The sound has a higher pitch as the distance decreases (waves appear shorter) and a lower pitch as the distance increases (waves appear longer). We ignore the change in loudness that also occurs. For the occupants of the car, no change occurs in the pitch.

An apparent change in the wavelength of visible light produces a corresponding change in its color. Motion toward a spectroscope seems to shorten all of the visible waves and shifts the color of each a bit toward the violet end of the spectrum. On the other hand, motion away from the spectroscope seems to increase the length of each wave and thus shift it toward the red end of the spectrum.

Astronomers compare the dark-line spectrum of a star with a laboratory spectrum (Fig. 19–17). To do this, the top and bottom portions of a sheet of film are exposed to a light source in the observatory (e.g., an iron arc which produces a bright-line spectrum). Next, the middle part of the film is exposed to light from a star. The spectra can thus be compared readily even though one is formed by emission and the other by absorption. When such comparisons were made for particular stars, the "fingerprints" of a number of elements were recognized, but the lines were offset a bit relative to the laboratory spectrum. For example, each spectral line of a certain star might be shifted precisely the same distance toward the right. For some stars the shift was toward the red end of the spectrum, but for other stars the shift was toward the violet end, and the amounts differed from star to star.

These shifts can be explained readily as Doppler effects: some stars are moving away from us, but others are moving toward us, and the distances are changing at different rates for different stars. The Earth's orbital motion about the sun also produces a Doppler shift. Note that in a red shift, the spectral lines do not all become red even in an emission spectrum. For stars, no color changes tend to occur at all because they commonly produce absorption spectra; each dark line is shifted toward the red end of the spectrum.

As evidence supporting the Doppler interpretation of red and violet shifts, we consider two motions of the sun: rotation and movement through space. Rotation is shown by the movement of sunspots (Fig. 24–3) and also by the spectroscope (Fig. 19–18). Since one side of the sun is spinning toward the Earth, its distance is thereby de-

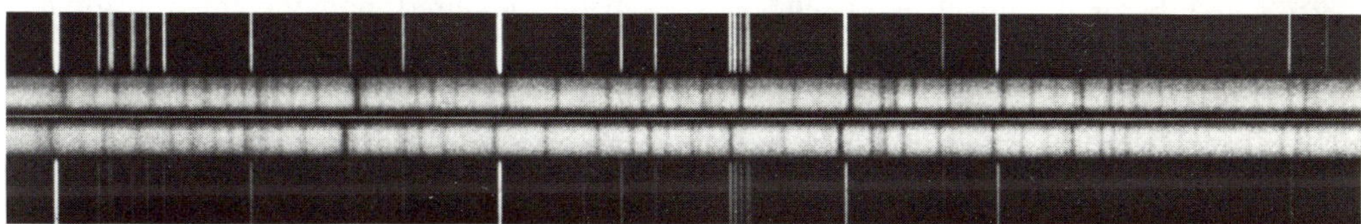

19-17. Doppler displacements in the dark-line (absorption) spectra of the brighter member of the double star Castor in the constellation Gemini. The top and bottom bands are part of a laboratory-produced, bright-line, comparison spectrum. The other two bands are absorption spectra of the star made on two different occasions. In the upper of these, the matching dark lines are displaced to the right—the red end of the spectrum—because the orbital motion of the star is away from the Earth. In the lower spectrum, the matching dark lines are displaced to the left—the violet end of the spectrum—because the orbital motion of the star is now toward the earth. *(Courtesy Lick Observatory.)*

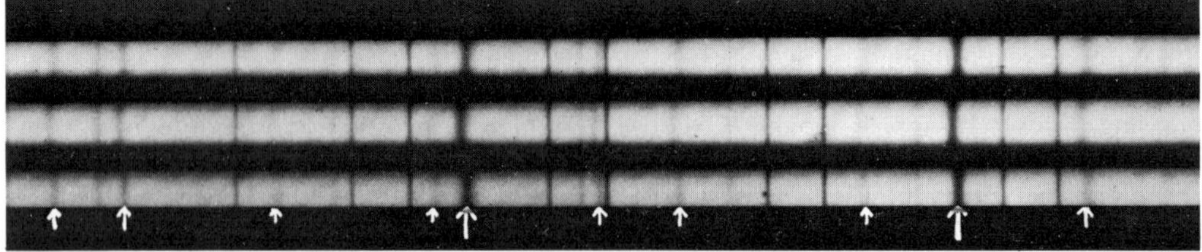

19-18. The sun's rotation affects its spectrum. Solar lines are marked by arrows and show Doppler shifts. Absorption lines produced by the Earth's atmosphere are unmarked and do not show a Doppler effect. The top and bottom spectra are duplicates and show a red shift (toward the right); the middle spectrum is of the opposite edge of the sun and shows a violet shift. *(Courtesy Mount Wilson Observatory.)*

creasing, and a violet shift is observed when a telescope with an attached spectroscope is pointed toward it. At the same time, since the opposite side is spinning away from us, its distance is increasing, and a red shift is observed. The central portion of the sun shows no shift in the positions of the dark lines because the distance between us and this part of the sun remains the same. Such Doppler effects can also be detected for Jupiter and other planets and can be used to determine whether or not a star is rotating (however, the lines are broadened, not shifted, p. 605).

The solar system appears to be moving toward the constellation Hercules at about 12 mi/sec relative to the stars that are nearest us in space. Doppler effects are noted: most of the stars in the vicinity of Hercules show violet shifts of about 12 mi/sec, whereas red shifts of about 12 mi/sec are common for stars in the opposite direction. Star motions corroborate this spectral evidence. We know that as we drive toward a group of houses, they seem to spread apart, and as we drive away the houses appear to come together. Stars in the vicinity of Hercules appear to be diverging, whereas stars on the opposite side of the sky seem to be converging. Therefore, the sun is moving toward Hercules. These movements have been detected by comparing the positions of stars on two or more occasions, preferably a few decades apart, because the movements are very small.

The law of the galactic red shifts applies only to galaxies and is discussed on p. 630.

Measuring the Intensity of Light

The first comparisons of the apparent brightnesses of stars and other objects in the sky were made visually. Later, the images of these objects were compared on photographic plates (negatives), and this method is still employed, particularly to measure the brightnesses of large numbers of stars. The image of a bright star on a photographic negative plate is both larger and darker than the image of a fainter star (Fig. 21–6).

A more accurate method involves the photoelectric cell, but only one star at a time can be measured in this manner. The cell is placed at the focal point of a telescope where the light from a star is concentrated upon it. The light falls on a photosensitive surface within the cell and causes electrons to be ejected from the surface (here we think of light as a stream of photons); the number of electrons ejected is directly proportional to the intensity of the light. These electrons constitute an electric current which can be measured, and the intensity of this current is a very accurate measure of the brightness of the star. The current is multiplied many times by using a succession of sensitive surfaces. Light from the sky also has an effect and must be estimated and subtracted. The employment of filters with this method, and with the photographic technique, makes it possible to measure the intensity of light in various wavelengths, including the infrared and ultraviolet. Thus the temperature of a star can be calculated (p. 496).

Exercises and Questions for Chapter 19

1. Three persons wearing white, yellow, and blue shirts respectively enter a room illuminated entirely by yellow light. What colors are their shirts in this room?
2. Assume that the Earth has a mean surface temperature of 300°K and that its peak radiation occurs at a wavelength of 10 microns. What is the temperature of another body that radiates its maximum amount of energy at a wavelength of:
   (A) 0.5 microns?
   (B) 5 microns?
   (C) 30 microns?
3. At what wavelength would a star radiate its maximum amount of energy if its surface temperature is:
   (A) 12,000°K?
   (B) 30,000°K?
4. Two objects are identical in size and nature (assume that each acts as a black body). How much more energy does the hotter body radiate if their temperatures in degrees Kelvin are:
   (A) 400 vs. 200?
   (B) 6000 vs. 2000?
   (C) 50,000 vs. 10,000?
5. List some favorable, some unfavorable, and some unique features of:
   (A) reflecting telescopes.
   (B) refracting telescopes.
   (C) radio telescopes.
6. Describe favorable conditions for the location of a large optical telescope; of a large radio telescope.
7. A friend has a few hundred dollars to spend on a telescope and asks for your advice in making the selection. What useful suggestions can you make? Reasons?
8. How does the light-gathering power of a 10-inch reflector compare with that of the unaided eye (use ¼ inch as the diameter of the pupil)?
   (A) Same as above but substitute the 200-inch telescope and use ⅕ inch for the pupil of the eye.
9. Assume that you have a telescope with an objective with a focal length of 80 inches. What magnification will you get if you use an eyepiece with a focal length of:
   (A) 2 inches?
   (B) ½ inch?
10. If a moth lands on the lens of your refractor when you are observing the moon, will you see it as a gigantic moth standing on the moon's surface? Explain.
11. Why do stars have different colors?
12. Describe the necessary conditions to produce a:
    (A) continuous spectrum.
    (B) bright-line spectrum.
    (C) dark-line spectrum.

13. Relative to the Doppler effect:
    (A) why do wavelengths appear shorter if the distance is decreasing?
    (B) why does a larger Doppler effect result if the distance is changing rapidly than if it is changing slowly?
14. Why have astronomers made spectrographic measurements of the atmospheres of planets from instruments suspended beneath high-altitude balloons?
    (A) Why is it advantageous to study the radiation from a star by means of telescopes, spectrographs, and other equipment in a spacecraft that is orbiting the Earth at an altitude of a few hundred miles?
15. Six spectra are shown schematically in the accompanying diagram. The rectangles at the top and bottom represent comparison spectra produced by a light source in a laboratory. Spectra A, B, C, and D represent the

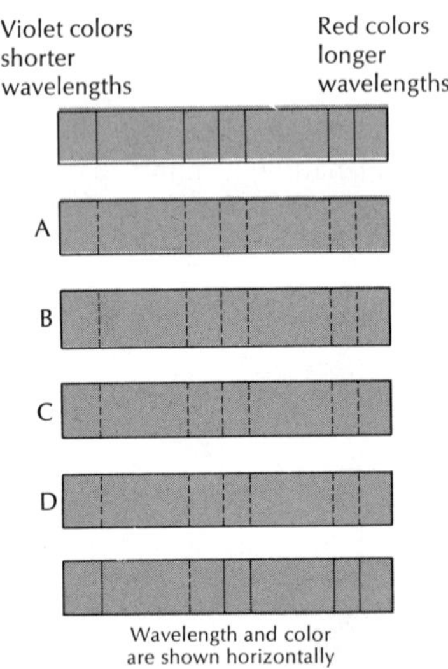

spectra of four stars. Six fictitious spectral lines are shown in the laboratory comparison spectra. These are dashed in the spectra of the stars to show their locations in the absence of a Doppler effect. In the spectrum of each star, draw in the six spectral lines in the proper positions to show that the distance between this star and the Earth is:
    (A) Decreasing slowly for star A.
    (B) Decreasing rapidly for star B.
    (C) Increasing slowly for star C.
    (D) Increasing rapidly for star D.
16. Is a star showing a red shift necessarily farther from the Earth than one showing a violet shift?
    (A) Are so-called red and violet shifts in the spectra of stars actually color changes?

20

The Planet Earth

The Spherical Earth

Striking proof of the Earth's spherical shape is now available (Fig. 20–1), but ancient scholars considered the Earth to be round and cited supporting evidence. The moon passes through a portion of the Earth's shadow at each lunar eclipse, and repeated observations show that this shadow is always curved in the same arc (Fig. 20–2). The deck of a departing ship disappears in the distance before the superstructure, and changes occur in the positions of stars as an observer travels north or south; e.g., the North Star moves higher and higher in the sky the farther one goes north (Fig. 20–5).

A Greek scholar, Eratosthenes, in the 3d century B.C. measured the size of the Earth. He knew that the circumference of a circle equals $2\pi r$, and he calculated the length of an arc of the Earth's circumference (Fig. 20–3). From this he obtained the radius and diameter. Although the exact length of the unit he used is uncertain, Eratosthenes may have been within 5 percent of the true value. We admire the mind that was capable of conceiving this measurement.

Locations on the Earth's surface can be given by their latitude and longitude (Fig. 20–4). The Earth's equator is an example of a *great circle* (i.e., a circle formed by the intersection of a plane with the surface of a sphere, providing that the plane passes through its center). The plane of the equator is perpendicular to the Earth's axis of rotation and passes through the center of the Earth midway between the north and south poles. Lines of latitude (parallels) trend east-west parallel to the equator. Thus the latitude of a certain point is the arc between this point and the equator (i.e., the angle at the center of the Earth between the equatorial plane and a radius of the Earth from this point). Latitude is measured either north or south of the equator and ranges from 0 degrees at the equator to 90 degrees at the poles. Lines of longitude (meridians) trend north-south and are perpendicular to the

20-1. Apollo 8 photograph of Earth, December 1968. *(NASA: AS8-16-2593.)*

lines of latitude. They converge at the North and South Poles and by convention are measured in angular degrees from the meridian that passes through Greenwich, England—up to 180 degrees east or west of this meridian.

Astronomical measurements of changes in the altitude of the celestial poles with changes in latitude have shown that the Earth is not a true sphere; it is flattened at the poles and bulged at the equator. An observer's latitude exactly equals the altitude of the celestial pole (Fig. 20–5), a fact that is of basic importance in navigation and geodesy. Since Polaris is about 1 degree from the north celestial pole, this statement is nearly true for Polaris also. Therefore, if the Earth were a true sphere, one could

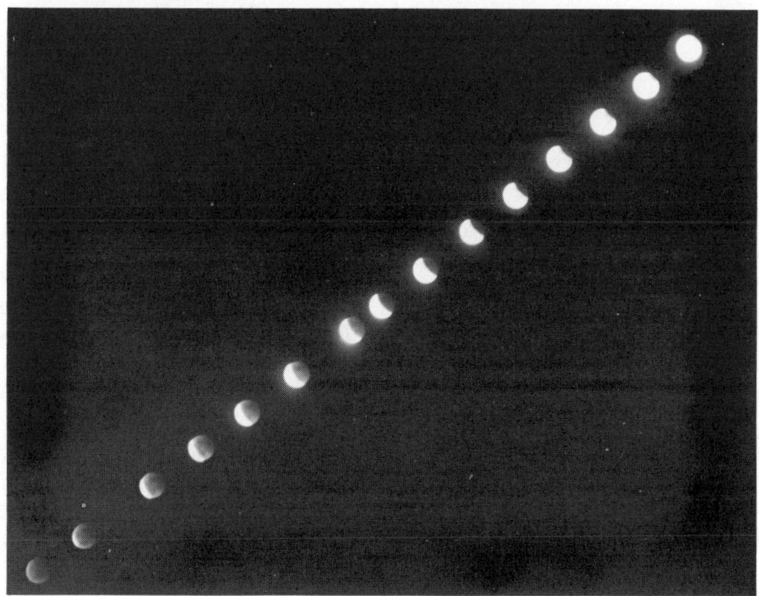

20-2. An eclipse of the moon shows that the Earth's surface is curved. The camera was pointed eastward at the rising moon and exposed at about five-minute intervals (29 Jan. 1953 from New York). The moon is totally eclipsed at the lower left and out of the dark part of the Earth's shadow in the upper right about $1\frac{1}{4}$ hours later. Note that the moon shifts eastward continuously at a rate of about $\frac{1}{2}$ degree (1 diameter) per hour. This eastward shift is superposed on the diurnal westward movement. *(Photograph by Neil Croom.)*

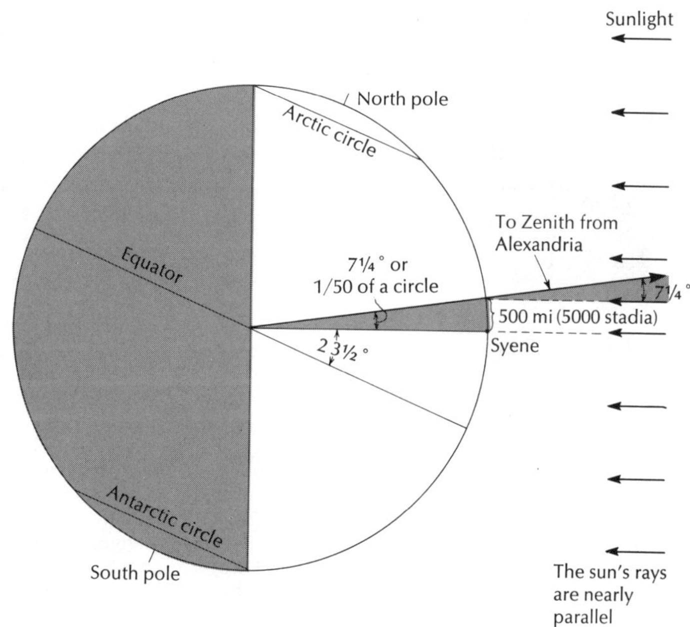

20-3. Method used by Eratosthenes about 235 BC to measure the size of the Earth. At noon at the summer solstice (about June 22) the sun is at the zenith at Syene but makes an angle of $7\frac{1}{4}$ degrees ($\frac{1}{50}$ of a circle) with the zenith at Alexandria (located about north of Syene). Zenith lines from Syene and Alexandria, projected inward along radii to the Earth's center, meet at an angle of $7\frac{1}{4}$ degrees (if two parallel lines are cut by a third straight line, their corresponding angles are equal). Thus 5000 stadia (about 500 miles)—the measured distance between Syene and Alexandria—represents approximately $\frac{1}{50}$ of the Earth's circumference (25,000 miles or 250,000 stadia). However, Syene is a bit too far north for the sun to be precisely at the zenith, and Alexandria is not due north of Syene.

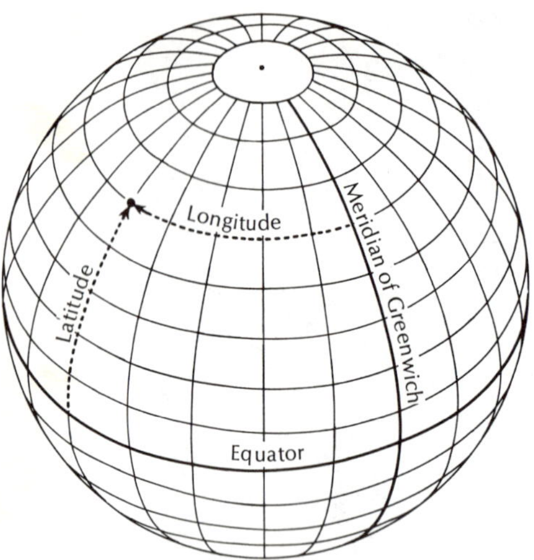

20-4. Latitude and longitude show locations on the Earth. The meridians in the diagram are 15 degrees apart (the angle through which the Earth rotates in one hour), and the parallels of latitude are 10 degrees apart. Thus the point shown on the Earth has a latitude of 50 degrees north and a longitude of 75 degrees west. (Courtesy R. H. Baker and L. W. Fredrick, *An Introduction to Astronomy,* Van Nostrand Reinhold, 1968.)

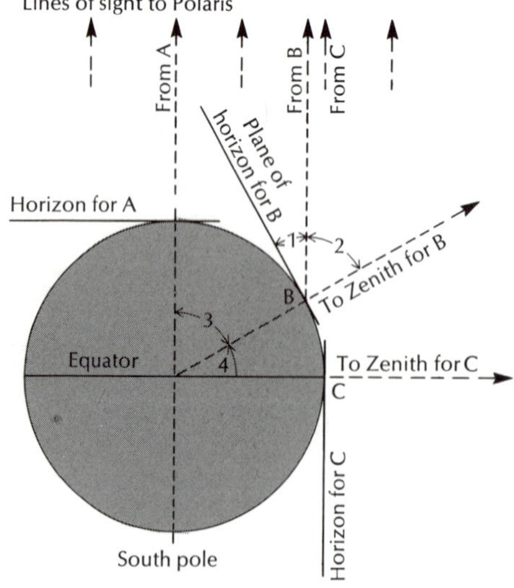

20-5. Stars appear at different altitudes in the sky at different latitudes because the Earth is spherical. Polaris (the North Star) has been chosen as an illustration, and rays of light from it are essentially parallel throughout the entire solar system. Polaris makes an angle of about 90 degrees with the horizon at the north pole (Polaris is 1 degree from the north celestial pole), an angle of about 30 degrees at B, and is on the northern horizon as viewed from C. If the Earth were flat and in the plane of the equator, Polaris would be directly overhead from all parts of the Earth. To prove that the latitude of B (angle number 4) is approximately equal to the altitude of Polaris above the horizon (angle number 1), we know that angle 2 equals angle 3 (why?), and that the sum of angles 3 and 4 is 90 degrees. Also, the sum of angles 1 and 2 is 90 degrees. Thus angle 4 equals angle 1.

start at the equator where the altitude of Polaris (actually the north celestial pole) is 0 degrees and travel due northward along a meridian for 1/90 of the distance from the equator to the north pole. At this new location, Polaris would be exactly 1 degree above the northern horizon. Traveling northward another 1/90 would raise the altitude of Polaris to 2 degrees, another 1/90 to 3 degrees, etc. This change in the altitude of Polaris is caused by the Earth's curvature. Therefore, wherever the Earth's surface is curved more than average, as it is at the equator, there this north-south distance must be less than average.

According to precise measurements, an observer has to move an average distance of about 69 miles (111 km) in a north-south direction to change the altitude of Polaris by 1 degree. However, this distance is somewhat greater near the poles and less near the equator (about 69.4 vs. 68.7 miles). Thus the Earth's surface is curved less at the poles and more at the equator.

Self-Gravity Makes Planets and Stars Spherical

Self-gravity causes the Earth and other large celestial objects to be spherical—e.g., the moon, planets, sun, and stars. For an isolated body in space, the gravitational attraction of all of its matter—its self-gravity—is everywhere directed inward and acts to concentrate the greatest amount of matter into the smallest possible volume. The resulting figure is a sphere, the most economical space saver of all three-dimensional figures. If some gigantic force could distort the Earth's shape, gravitational attraction would slowly level irregularities and again make the Earth spherical.

On the other hand, few celestial objects are perfect spheres; they rotate, and spinning on an axis causes polar regions to flatten out and equatorial zones to bulge. The size of the equatorial bulge is approximately proportional to the rate of rotation, although other factors such as density and distribution of mass also enter in (e.g., Jupiter spins very rapidly and has noticeably flattened polar regions, whereas Venus spins more slowly and is more nearly a sphere). Thus self-gravity acts to make a sphere of a planet or star, whereas rotation tends to keep it from being truly spherical.

The Earth's equatorial diameter is about 27 miles (42 km) longer than its polar diameter, but the difference is relatively very small, about 7927 miles vs. 7900 miles (12,756 vs. 12714 km). On a model Earth about 25 feet in diameter, it would amount to less than an inch, and the highest mountains would not extend one-quarter of an inch above sea level.

The polar flattening caused by spinning makes sea level about 13½ miles closer to the center of the Earth at the poles than it is at the equator. Because of this, a body weighs more at the poles than at the equator (the gravitational attraction of a sphere acts as if it were all concentrated at the center). However, rotation also has an additional effect that has been illustrated (Earth Science Curriculum Project) by imagining a man standing on a spring-type scale in an elevator on a motionless Earth. The man on the spring scale weighs less when the elevator is accelerating downward than he does when the elevator is motionless or moving downward at a uniform rate—some of the Earth's gravitational attraction (the cause of his weight) is used up in keeping the man on the scales when the acceleration is occurring. In a similar manner, an object at the surface of the rotating Earth must continuously be pulled inward; i.e., it experiences a continuous acceleration (a change of direction in this case) otherwise it would move off into space tangentially to the surface. This inward acceleration is greatest at the equator where the Earth's linear rate of rotation is greatest and is zero at the poles. Thus the greatest reduction in gravity occurs at the equator, and it bulges outward.

In other words, for a moving object to deviate from a straight-line path, a force must be exerted on it. The inertia of Earth materials would make them move tangentially outward if no inward-directed centripetal force were exerted on them continuously (toward the axis, not the Earth's center). We can think of this centripetal force as using up a part of the Earth's gravitational attraction. Therefore, where this required centripetal force is greatest (at the equator), the Earth's gravitational pull is least. In combination, these two effects of rotation would reduce the weight of a 1000-pound polar bear (at the north pole) by about 5 pounds at the equator (ignore other factors such as loss of mass by perspiration).

We need to distinguish carefully between the terms mass and weight. *Mass* can be defined as the amount of matter in an object, which depends upon the density and volume of this object. In other words, mass refers to the total number of subatomic particles (protons, neutrons, and electrons) contained in an object. The inertia of a large mass—its resistance to any change in its motion or state of rest—is greater than that of a small mass; this property also defines mass. A greater force must be applied to accelerate a larger mass to the same velocity as a smaller mass, and this holds true whether these masses are at the Earth's surface or in interstellar space. The mass of an object such as an iron ball, therefore, is the same anywhere in the universe (it must stay intact, however).

Weight, on the other hand, results from gravitational pull and depends upon mass (directly) and distance (inversely as the square of the distance). To illustrate, we compare the weight of a man on the Earth with his weight on some other planet (assume that his mass remains the same). The man would weigh the same as he does on Earth on a planet with an 8000-mile radius and a 4-times greater mass (he would also weigh the same on a planet with a 24,000-mile radius and a mass 36 times greater than that of the Earth. In each case, the larger radius tends to make him weigh less (by 2^2 and 6^2), but this is offset by the larger masses (4 and 36).

On the moon a man weighs about one-sixth of his weight on the Earth, and Apollo astronauts learned that they could move fastest by using "a two-footed hop, 6 to 8 feet a hop." If one could exist without a cumbersome space suit, he might be able to jump over a small house (depending upon his skill and muscular development), put the shot the length of a football field, hit a golf ball a mile or so, and send a rifle bullet 100 to 200 miles above the surface. Even enclosed in a space suit, an astronaut easily carried equipment weighing about 190 pounds on Earth (30 pounds on the moon).

The irregular manner in which many of the smaller asteroids reflect sunlight suggests that they are not spherical in shape. Apparently the self-gravity of a small asteroid, some miles in diameter, is too weak to cause its strong rocky or metallic material to flow into a spherical shape.

The Earth's Mass

The mass of the Earth has been measured a number of times (Fig. 20–6). In the Jolly measurement, the lead and the mercury are spheres because the gravitational attraction of a sphere acts as if it were all concentrated at the center. The gravitational constant cancels out since it occurs on each side; it is not shown. The arm on the right is longer to keep the lead ball as far as possible from the weights on the pan at the left. The total mass of the Earth equals about 6×10^{21} metric tons (1 metric ton = 1000 kilograms = 2204.6 pounds). Since we know the Earth's volume, we can calculate that an average sample of the Earth weighs 5.52 times as much as an equal volume of water (its specific gravity); in other words, the Earth's average density is 5.52 g/cc.

Evidence of the Earth's Rotation

Until the true magnitude of celestial distances was known, diurnal motion could not be used as proof of the Earth's rotation. However, to circle the Earth once in 24 hours, any object more than 27 astronomical units from the Earth would have to travel with a speed greater than the velocity of light. Neptune, Pluto, and all of the stars (except the sun) are beyond the critical range, and the remote galaxies are very far beyond it.

Several lines of evidence prove that the Earth rotates. Its equatorial bulge has been measured, and a Coriolis effect occurs (p. 658). An object dropped from a skyscraper always lands a short distance east of the spot that was located vertically below it (in the absence of air currents or other disturbances), because falling is independent of other motions an object may have (p. 477). The top of the skyscraper and its base are rotated by the Earth in precisely the same time, but the top travels a greater distance and moves more rapidly eastward. As the dropped object falls, it retains this faster eastward movement and thus lands a bit to the east of a point that was located directly below it.

An experiment by Foucault in 1851 provided dramatic evidence that the Earth turns on an axis. He suspended a heavy iron ball at the end of a 200-foot wire fastened to the roof of the Pantheon, a large building in Paris. A pointer was attached to the base of the weight, and a ring of sand was placed beneath the pendulum so that lines were drawn in the sand as the pendulum swung back and forth. The pendulum's inertia keeps it swinging in the original direction, and there is no tendency to change direction because the ground beneath is turning. However, little by little, the direction of the pendulum's swing appears to turn in a clockwise direction as indicated by successive lines in the sand—the Earth's counterclockwise rotation is actually turning the ground beneath the pendulum (Fig. 26–16).

This phenomenon is most easily understood if we imagine a pendulum suspended from a point in a roof over the north pole. In this case, the plane of the pendulum's swing appears to shift clockwise with respect to the Earth's surface through 360 degrees in 23 hours and 56 minutes. A similar shift happens at the south pole, but the direction is counterclockwise. At latitudes between the poles and the equator, more than 24 hours are needed for a pendulum's plane to rotate through 360 degrees; it takes about 31 hours in Paris and nearly 37 hours in New York (Fig. 20–7). The apparent rotation becomes increasingly slower toward the equator where no rotation occurs at all because the surface does not twist beneath a pendulum as it does at other latitudes.

All motion is relative. Perhaps you have sat in a railroad car at a station beside a second train on a nearby track. When one

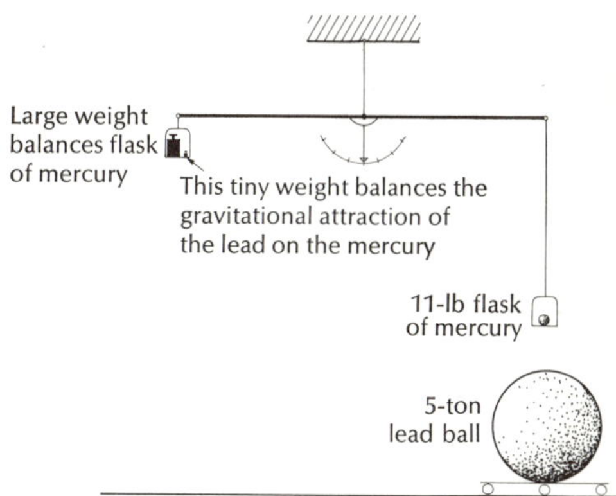

20-6. Jolly's method of measuring the mass of the Earth (1881). A large weight on the left pan exactly balances an 11-lb spherical flask of mercury on the right pan. The Earth's gravitational attraction pulls down equally on the two objects. Then a 5-ton lead ball is moved under the mercury. Its gravitational attraction is sufficient to pull the mercury down a short distance, and a small weight must be added on the left to restore the balance. Very accurate measurements are required. The following equations can then be set up. As the only unknown, x, is the mass of the Earth, it can be determined:

$$\frac{(\text{Small weight}) \cdot x}{(\text{Earth's radius})^2} = \frac{(\text{mass of mercury} \cdot (\text{mass of lead})}{(\text{distance between their centers})^2}$$

20-7. The Foucault pendulum in the U.N. General Assembly Building, New York. The 12-in. sphere is kept swinging by an electromagnet at the center of a 6-ft metal ring. The plane of the swing shifts clockwise, and at this latitude a cycle is completed in about 36 hrs and 45 mins. (see Fig. 26-16). *(Courtesy United Nations.)*

train starts, it is often impossible to tell, without using nearby buildings as reference points, which train is actually moving. The sphere of the Earth and the celestial sphere are like the two trains, except that there are no fixed reference points anywhere. Everything on the Earth moves together—air, water, buildings, people. Even though the rate exceeds 1000 mi/hr at the equator, we have no way of detecting the movement, and therefore it seems to be the celestial sphere which is moving. No wonder the belief in a motionless Earth persisted for more than 2000 years!

Evidence of the Earth's Revolution

As powers of observation grew, revolution of the Earth about the sun was proved in at least three ways involving parallax, a Doppler shift, and aberration. Near stars show a parallactic displacement against the background of more remote stars (p. 464 and Fig. 17–13). As the Earth revolves toward a certain star, its distance decreases; but about 6 months later, as the Earth revolves away from this star, the distance increases (the actual motion of the star has to be considered also). This produces a Doppler effect in the spectrum of the star (p. 508).

The aberration of starlight was discovered by Bradley about 1727 (Fig. 20–8). Many of us have noticed that the direction in which raindrops fall appears to be changed by our own motion. The drops seem to slant in the direction in which we move, and they slant more if we move faster. Light rays from a star also seem to slant, but in opposite directions at 6-month intervals. A star at the ecliptic pole makes a tiny annual circle with a radius of 20.5 seconds of arc. As small as it is, this angle is still more than 20 times larger than the largest parallax. Stars in the plane of the ecliptic move back and forth in a straight line each year, whereas stars located at other angles follow tiny elliptical orbits (each semimajor axis is 20.5 seconds in length). To determine the effects of aberration, precise measurements had to be made of the angles between stars at different times of a year and then compared.

Seasons of the Year

The seasons of the year—fall, winter, spring, and summer—are caused by the Earth's revolution about the sun and by the tilt of its axis to this orbital plane. The varying distance of the Earth from the sun, about 94.5 to 91.5 million miles, is not an important

20-8. Aberration of raindrops and starlight. (R. H. Baker and L. W. Fredrick, *An Introduction to Astronomy*, 7th ed., Van Nostrand Reinhold, 1968.)

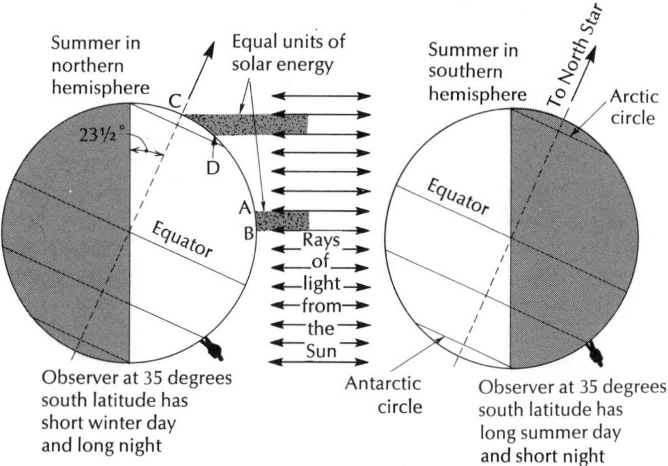

20-9. Winter and summer in a side view of the Earth. Since an observer on the Earth is rotated parallel to the equator; one can readily see that summer days are longer than summer nights, and that the sun cannot set at a pole during its summer nor rise at a pole during its winter. The sun's rays are most concentrated when they strike the Earth's surface vertically (AB) and less concentrated when they approach at an angle (CD) because a larger area must be covered with the same amount of energy. The beam of a flashlight can be concentrated or spread out in much the same way.

factor in seasonal changes (e.g., the Northern Hemisphere has winter when the Earth is closest to the sun).

The intensity of solar energy depends in large part upon the angle at which it strikes the surface—the steeper the angle, the more concentrated it is (Fig. 20–9). Solar energy is concentrated most on the Northern Hemisphere during its summer season when the north end of the axis is tilted toward the sun. More hours of daylight also occur at this time, and thus summers are warmer than winter. Simultaneously the Southern Hemisphere has winter. This angle of tilt is always approximately the same, and the north end of the axis points about at Polaris throughout a year (i.e., the axis remains parallel to itself all year long). Therefore, 6 months later when the Earth is on the opposite side of its orbit, the north end of the axis is tilted away from the sun, and the Northern Hemisphere has its winter. At the midpoints between these two extremes, the axis is tilted neither toward nor away from the sun, and fall or spring occur.

When the sun and Earth are viewed from a point above the north pole (Fig. 16–2), we note that half of the Earth is always illuminated by sunlight. In addition, a dot representing the north pole occurs in the same relative position within a circle representing the Earth in all orbital locations. This dot is in the daytime side during summer and in the nighttime side during winter (note that the dot is never in the center of the circle). At about the 22d of March and September, this dot is located exactly on the line separating day from night and the equinoxes (equal night) occur. Daylight and nighttime (includes twilight) are then equal in length everywhere on Earth (except at poles; refraction effects ignored). As compared with spring and summer, fall and winter are shorter seasons in the Northern Hemisphere because the Earth revolves fastest at these times, and the distances are shorter.

As shown in Fig. 20–10, the planes of the ecliptic and the celestial equator intersect at an angle of 23½ degrees along a line that points in very nearly the same direction throughout a year. In other words, the Earth's equatorial plane stays parallel to itself as the Earth revolves about the sun (it must because the axis points at Polaris all year long). The *equinoxes* are the two points on the celestial sphere marked by the ends of this line (i.e., where the ecliptic and celestial equator intersect). As the sun shifts eastward along the ecliptic, it gradually approaches one of these points. When the sun, as viewed from the Earth, is exactly in front of an equinox, then the fall

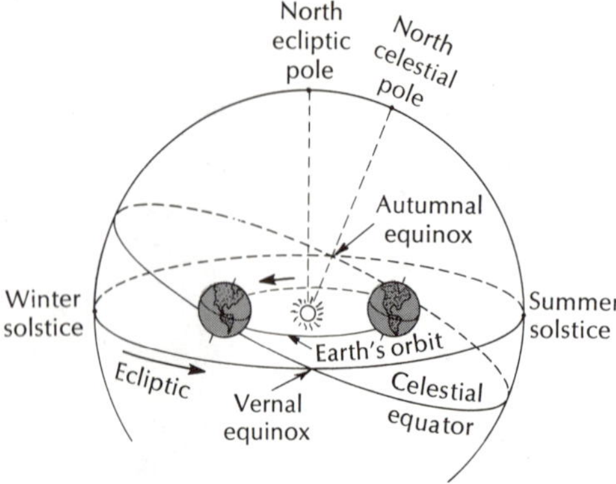

20-10. The equinoxes are the two points on the celestial sphere where the ecliptic intersects the celestial equator. Spring begins when the sun —shifting eastward along the ecliptic—reaches the vernal equinox. (R. H. Baker and L. W. Fredrick, *An Introduction to Astronomy,* Van Nostrand Reinhold, 1968.)

or spring season begins (depends upon the hemisphere and the equinox). Thus this occurs on a certain day at a particular hour, minute, and second.

The *solstices* are the two points on the ecliptic located midway between the equinoxes. At these positions, the sun is located farthest above or below the celestial equator —hence the term solstice which means "the sun stands still." For a few days, the sun seems to remain at about the same altitude.

The motions of the sun are quite different at different latitudes (pp. 449 and 523). At the north pole about June 22, the sun remains at an altitude of 23½ degrees as it shifts all the way around the horizon, and there is no night (Fig. 20–11). However, about December 22, the sun remains 23½ degrees below the horizon, and there is no daylight. On about the 22d of March and September, the sun stays on the horizon all day, and very long sunrises and sunsets occur.

The latitudes of 23½ degrees (the Tropic of Cancer in the north and of Capricorn in the south) delimit the zone in which the noon sun is at the zenith once each year (twice at the equator—once at each equinox). At 23½ degrees north latitude, this occurs about June 22, and at 23½ degrees south latitude, this occurs about December 22.

If the Earth's axis were perpendicular to its orbital plane, there would be no seasons. Each degree of latitude would experience the same sort of weather all year long— an extended combination of fall and spring. Locations near the equator would be warmest, and those near the poles would be coldest.

20-11. The "midnight sun." Photograph was exposed at about 15-minute intervals starting at 11:15 PM on 19 June 1967 at Eagle Summit, Alaska. (Courtesy P. Sheridan; from R. H. Baker and L. W. Fredrick, *Astronomy,* 9th ed., Van Nostrand Reinhold, 1971.)

The Precession of the Equinoxes

The gravitational pull of the moon and sun on the Earth's equatorial bulge causes the Earth's axis to shift like that of a spinning top (Fig. 20–12). If the Earth were not rotating, this gravitational pull on its equatorial bulge would re-orient the Earth, and its axis would become about perpendicular to the plane of the ecliptic. In other words, the equatorial bulge would then be parallel to the plane of the ecliptic (not quite true because the moon's orbital plane is inclined about 5 degrees to the ecliptic and the moon's effect is stronger). However, because the Earth is spinning, its axis remains always tilted at an angle of about 23½ degrees to a perpendicular to the plane of the ecliptic. Meanwhile the axis shifts slowly westward —it is said to precess—and describes a cone-shaped path in space. Therefore, the positions of the celestial poles and celestial equator shift gradually on the celestial sphere. The equinoxes are thereby shifted westward (backward) along the ecliptic and make the year of the seasons—from equinox to equinox—about 20 minutes shorter than the period of the Earth's revolution relative to the stars.

This circular path of the celestial poles is completed in nearly 26,000 years, and no one star such as Polaris is permanently located near a celestial pole (a few decades from now, Polaris will be within ½ degree of the pole). As another effect, a winter constellation now will become a summer constellation half a precessional cycle from now.

Other slow periodic movements of the Earth also occur. Tilt of the axis—now nearly 23½ degrees—changes from less than 22 to more than 24 degrees and back during a 40,000-year period. Moreover, the aphelion and perihelion positions—as well as the eccentricity of the Earth's elliptical orbit —change very slowly, and a complete cycle approximates 100,000 years in each case. These combine with the axial "wobble" to make the precession of the equinoxes a 21,000-year cycle.

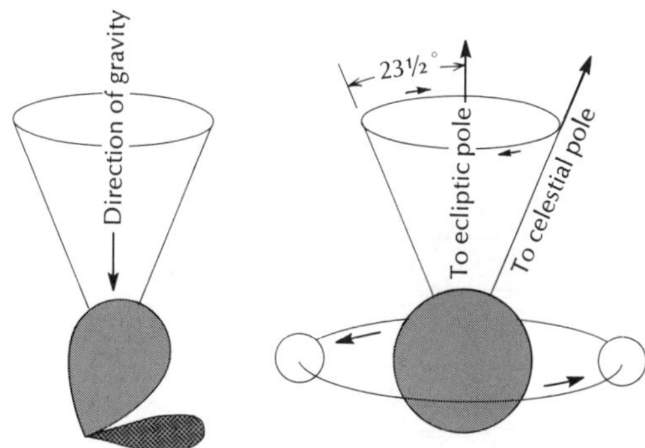

20-12. The Earth's axis shifts like that of a spinning top. (*Courtesy* R. H. Baker and L. W. Fredrick, *An Introduction to Astronomy,* Van Nostrand Reinhold, 1968.

Altitude of the Ecliptic

From any one location on the Earth, the celestial equator maintains the same angle with the horizon throughout a year, and this angle is the complement of the latitude (p. 452). For example, at 40 degrees north latitude, the celestial equator makes an angle of 50 degrees with the horizon in the south; at 75 degrees north latitude, this angle is 15 degrees. However, the ecliptic may be as much as 23½ degrees either above or below the celestial equator. Thus the altitudes of the sun, moon and planets vary considerably because they are always on or near the ecliptic.

To illustrate, at 40 degrees north latitude, the sun is 73½ degrees above the southern horizon at noon about June 22; but it is 47 degrees lower, or 26½ degrees above the southern horizon, at noon about December 22. A side-view sketch of the Earth, sun, and moon will show that the full moon, always opposite the sun, is high above the horizon at midnight in the winter when the noon sun is low and vice versa (Fig. 20–13).

Some Atmospheric Phenomena

Among other phenomena, the atmosphere is responsible for the aurorae, for halos and

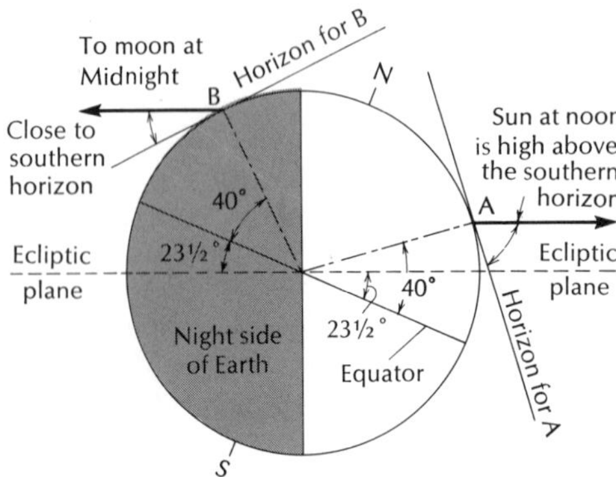

20-13. Altitude of the moon and sun above the southern horizon. Side view. The sun is always on the ecliptic, and the moon is always within 5 degrees of it. An observer at 40 degrees north latitude is shown in the summer at noon (A) and midnight (B). The ecliptic plane is above (north) the celestial equator at A and below it at B. Thus the sun is high in the south at noon for this observer, but the full moon at midnight is rather low in the sky.

coronas (p. 703), for meteors (p. 587), for the twinkling of stars, for atmospheric refraction, for the twlilight, and for the blue sky and colorful sunsets and sunrises.

Blue Sky and Reddish Sunsets

The longer red wavelengths of visible light (p. 495) penetrate the atmosphere more readily than the shorter blue wavelengths which tend to be scattered. As each tiny particle scatters light (these are chiefly gas molecules in the atmosphere), it seems to act as a kind of radiation source and diffuses the light in all directions. Therefore, the sky is a bluish color because we see it by this indirect, scattered, diffused sunlight among which the shorter waves predominate. In fact, daylight consists of this diffused sunlight. No energy is transferred during the scattering process; it involves only a redistribution in space.

On the other hand, we see the sun by light that comes directly from it to our eyes. However, this direct sunlight is minus some of the shorter waves that were diffused out of it during passage through the atmosphere. Thus if we were above the atmosphere or on the airless moon, the sun would look less yellow because its light would contain more of the shorter wavelegths. For an observer on the Earth at sunrise or sunset, the sun's rays must penetrate the greatest amount of atmosphere, and more scattering and diffusion can take place. In contrast to noontime when sunlight tends to approach the surface at a steep angle, at sunset the sunlight approaches nearly parallel to the surface and passes through many more miles of atmosphere to reach the surface. Thus the long red wavelengths predominate as this light enters one's eyes, and the sun looks reddish. A great deal of dust in the atmosphere near the horizon causes the greatest diffusion and produces spectacular, colorful sunsets and sunrises (generally not so colorful because less dust tends to be present at sunrise—nights are generally less windy than days). Occasionally at midday, when considerable intervening dust and haze are present, the sun takes on an orange hue.

As seen from an orbiting satellite above the bulk of the atmosphere, the sky approaches the black of interstellar space, even though the sun is shining and stars are visible. To understand this, imagine that you are looking out of a window away from the sun. Its rays are streaming by the satellite, but too few molecules are present in the atmosphere to bend or reflect these rays to you. Therefore, none of this sunlight reaches your eyes, and black shows an absence of visible radiation. However, light from stars located in this direction can enter your eyes, and thus you see stars in a black sky in the daytime if you look away from the sun.

Twilight

Twilight is the diffused sunlight that we see after sunset and before sunrise. Just after the sun has set, its light will still be streaming through the upper atmosphere and will be scattered so that some of it still reaches the surface. Obviously, this scattered light will not be visible when the sun is far below the horizon. The duration of twilight at sunset (sunrise is similar) depends upon the length of time taken by the sun to sink through a certain angle. The time is shortest when the sun's path is vertical to the horizon (throughout the year at the equator) and longest when the path is slanted at a small angle to the horizon. Three types of twilight (civil, nautical, and astronomical) have been defined; they last until the sun is 6, 12, and 18 degrees respectively below the horizon.

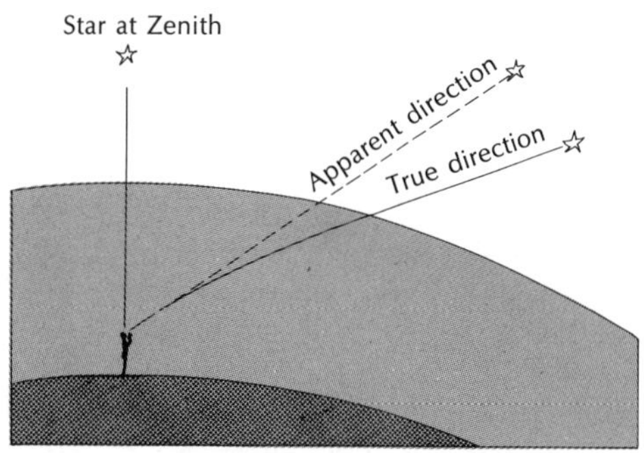

20-14. Atmospheric refraction makes celestial objects appear higher above the horizon than their actual positions, and this effect increases greatly near the horizon. (R. H. Baker, *Astronomy*, Van Nostrand Reinhold, 1964.)

Atmospheric Refraction

We are all familiar with the apparent bend in a straight stick when part of it is seen under water. The bending of light waves (refraction) as they penetrate the Earth's atmosphere is similar and causes an apparent increase in the altitude of celestial objects (Fig. 20–14). We see such objects in the direction their light rays are moving as they approach the eye—not in the unrefracted direction they followed above the atmosphere. The light from an object on the horizon has a larger amount of atmosphere to penetrate than does the light from objects at higher altitudes; thus the amount of refraction increases as the altitude of the object decreases.

This can be checked by measuring the angular separation of two stars when they are high in the sky and again when one of them is near the horizon (the angle is smaller here). For this reason the altitude of a star above the horizon is best obtained by measuring the angle between it and the zenith and subtracting this quantity from 90 degrees. Sextant observations in navigation are always corrected for refraction.

The sun appears flattened when seen near the horizon because its bottom part is refracted upward more than its top part; its size parallel to the horizon remains unchanged. The sun likewise appears larger when near the horizon, but this is an optical illusion: the moon and constellations also appear larger when near the horizon. Refraction of light at the horizon amounts to approximately half a degree, about the angular diameter of the sun. Because of this, the sun is visible a few minutes before it actually rises above the eastern horizon and also a few minutes after it actually sets below the western horizon (effect is smallest when sun's path is steepest).

Twinkling

Stars twinkle because their light must pass through turbulent air of different densities and temperatures before it reaches us at the Earth's surface. Thus it is refracted in slightly different directions, and some interference occurs among the waves. Some wavelengths are strengthened momentarily, whereas others are weakened. This causes the amount of light that enters the eyes to vary slightly from one moment to the next both in intensity and in the apparent direction in which the star is seen.

On the other hand, the bright planets tend not to twinkle because they are area sources of light, not point sources like the stars; variations from different portions of the disk tend to cancel out. Near the horizon, however, small planets may twinkle vigorously like stars.

Exercises and Questions for Chapter 20

1. Assume that astronaut A lands on a planet 800 miles due north of astronaut B. At night they learn that a certain star is at B's zenith but that it makes an angle of 6 degrees with A's zenith. The astronauts weigh the same as they did on Earth. Relative to the Earth, what is this planet's:
   (A) diameter?
   (B) mass?
2. Assume that two observers are on the same north-south meridian on a planet and observe a star that is at the zenith of one observer and 30 degrees from the zenith for the second observer. The observers are 500 miles apart. Calculate the planet's:
   (A) circumference.
   (B) radius.
   (C) mass if the observers weigh only half as much as they did on Earth.
3. Saturn's total mass is about 95 times larger than that of the Earth, yet an object at its surface would weigh about the same as at the Earth's surface. Why?
4. What would be the expected change in the Earth's shape in each of the following? Assume that the Earth:
   (A) stops rotating.
   (B) rotates more rapidly.
   (C) rotates at the same rate but is located 10 astronomical units from the sun.
   (D) has much more of its mass concentrated within its core.
5. Make a top-view sketch of Earth and sun at the beginning of each of the four seasons. Do this by drawing four circles about equidistant from a point representing the sun.
   (A) Within one of the circles, mark a dot so that it shows the approximate location of the north pole on June 22.
   (B) Mark dots for the north pole in each of the other circles. Dates are about September 22, December 22, and March 22. Are all four dots in the same relative position? Does the Earth's axis point in about the same direction all year long?
   (C) How can you distinguish fall from spring? Hint: since the Earth revolves in the counterclockwise direction, you know which season occurred 3 months earlier and can thus determine the direction in which the axis is slanted.
6. Describe the shadow cast by a vertical stick from sunrise to sunset if the stick is located at the equator about the 22d of March or September.
7. Describe the daily motion of the sun and the relative amount of daytime and/or nighttime if an observer is located at a latitude of:
   (A) 90°N about June 22.
   (B) 90°S about June 22.
   (C) 90°N about March 22.
   (D) 40°N about June 22.
   (E) 40°S about June 22.

8. Draw a side view sketch of the Earth and sun to show the season that now occurs at perihelion in the Northern Hemisphere.
   (A) Make a second sketch to show the season in the Northern Hemisphere that will occur at perihelion some 13,000 years from now.
   (B) What effect, if any, may this have on winters and summers in the Northern Hemisphere?
9. (After Wyatt) On March 22 you are temporarily blinded by the image of the sun reflected into your eyes by the windshield of a car approaching you on a level road from the north (is the direction necessary?). Assume that the windshield is a plain piece of glass pitched at an angle of 45 degrees from the vertical.
   (A) What time of day is it?
   (B) What is the altitude of the sun?
   (C) At what latitude are you?
   (D) If the date were December 22, at what latitude would you be?
10. Assume that a pilot flies a jet plane high above the equator in a certain direction at a certain rate. He notes that his solar time does not change (e.g., it is always noontime or sunset). What is his direction and speed?
11. Explain why twilight is shorter at the equator than at higher latitudes.
12. Why does the color of the sky become darker at increased altitudes?
    (A) Why is the sky black to a space traveler even though he is in direct sunlight?
    (B) Why is the sky blue as seen from the Earth's surface?
    (C) Why does the sun appear reddish at sunset?
    (D) Why does the sun appear to change shape as it nears the horizon?

21

Some Measurements of Celestial Objects

Trigonometric methods allow lengths and distances (Fig. 21–1) to be measured by a sort of remote control that involves the solution of triangles. For example, in determining the distance across a large, steep-walled canyon, a surveyor can carefully measure a certain distance (a base line) along one rim. The two ends of this base line and a conveniently chosen point on the opposite rim form a triangle. The surveyor then measures the angles at each end of the base line and can calculate the distance across the canyon either by constructing a scaled diagram or by using trigonometry—e.g., by applying the law of sines: in a plane triangle the lengths of any two sides have the same ratio to each other as do the sines of the angles opposite those sides. A base line must be adequate in length for reliable results. If the canyon is nearly 20 miles across, a satisfactory base line would be about 1 mile in length. However, when such a technique is applied to the nearer stars, we find that the comparable base line is only a few inches in length—(the angles are measured at 6-month intervals from opposite sides of the Earth's orbit).

Following the surveyor's technique, man has devised measuring rods mighty enough to span the inaccessible distances of celestial space, and the first nearer measurements were made centuries ago. It was necessary to work in steps that were gradually extended outward: the size of the Earth was measured first (Fig. 20–3), next the distance to the moon was measured, and then in turn the distances to the nearer stars, to remote stars, to the nearer galaxies, and finally to the very remote galaxies.

Distance to the Moon and Its Size

In the 2d century B.C. the talented Alexandrian astronomer, Hipparchus, measured the distance to the moon (Fig. 21–2) and also the size of the moon (Fig. 21–3). This latter method has wide application: the linear diameter of any celestial object can be

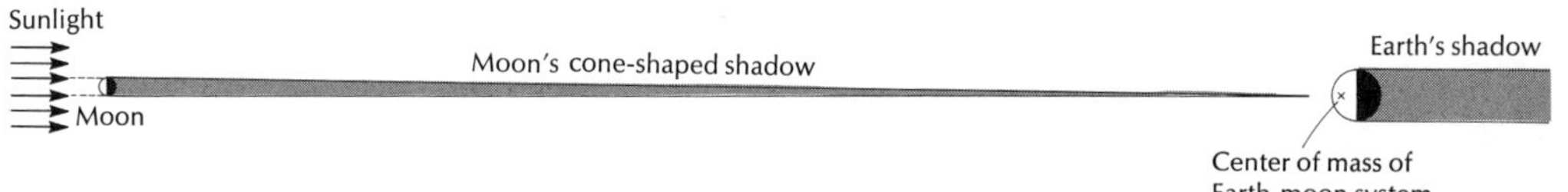

21-1. Moon's long, slender, cone-shaped shadow drawn to scale. It averages 232,000 miles in length, whereas the mean distance from the center of the Earth to the center of the moon is about 239,000 miles; this is the semimajor axis of the moon's elliptical orbit. The center of mass of the Earth-moon system is located within the Earth about 2900 miles from its center; this is the balance point about which each revolves, once each month. The motion of this center of mass around the sun forms the smooth elliptical curve generally shown as the Earth's orbit.

obtained by multiplying its angular diameter (expressed in radians) by its distance. The unit of length chosen for the distance (e.g., light-year or kilometer) will also be the unit for the diameter. One radian equals about 57.33 degrees or 206,265 seconds of arc. For example, an object that is 206,265 light-years away has a linear diameter of 1 light-year if its angular diameter is 1 second of arc.

When the distance to the moon became known, the distance from the Earth to the sun was calculated to be about 5 million miles because Aristarchus had previously estimated (erroneously) it as about 20 times the distance from the Earth to the moon. This figure of 5 million miles was current for many centuries.

Distance to the Sun

If the sun is observed simultaneously from two widely separated positions on the Earth, it shows a parallactic displacement, but the maximum difference in direction, obtained by two observers at opposite ends of a diameter of the Earth, would be only 1/200 of a degree of arc, a very small angle to measure with precision. Thus it is difficult to measure directly the distance to the sun: the sun is not a point source of light, its center is difficult to locate precisely, and stars are not visible during the day as a background for reference points (exception—total solar eclipses).

However, relative distances in the solar system can all be calculated in terms of the astronomical unit by applying the laws of Kepler and Newton. Therefore, once the actual distance of one of the members has been obtained, the distances of the other members can be determined. A few asteroids approach close enough to the Earth at times for accurate measurements of their geocentric parallaxes (Fig. 21–4). The parallaxes of Mars at opposition, and of Venus as it passes directly between Earth and sun, have also been used to determine the astronomical unit.

To illustrate the general method involved, we know the sidereal period of Mars (Fig. 17–18), and we can use Kepler's third law ($P^2 = D^3$) to determine its distance from the sun in terms of the astronomical unit (this is about 1.53 A.U.'s). Next, we determine the distance at opposition between Mars and the Earth by measuring its geocentric parallactic displacement. Assume that this is about 49 million miles (range is 35 to 63 million miles, Fig. 23–6). Thus the distance of Mars from the sun is 1 A.U. plus about 49 million miles. We can now set up and solve the following equation to obtain the approximate number of miles in 1 astronomical unit: 1 A.U. + 49,000,000 = 1.53 A.U. More accurate results are obtained by using a small, relatively nearby asteroid in place of Mars.

Distances to the Nearer Stars by Parallax

Once the distance from the Earth to the sun had been determined and powerful

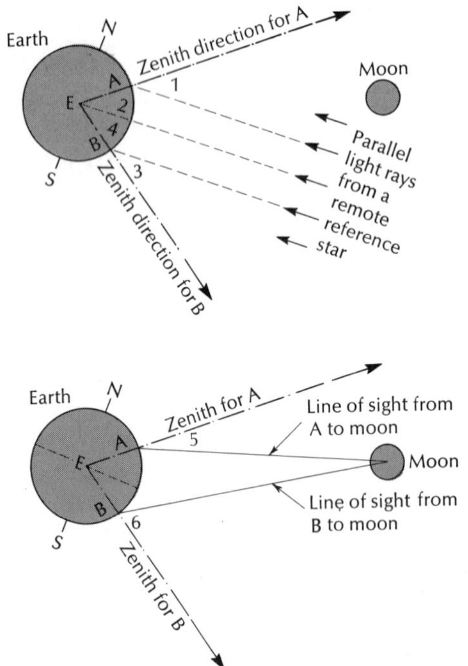

21-2. Measurement of the distance from the Earth to the moon (Hipparchus used a similar method). Diagram is a side view and not to scale. In each diagram, two observers are located on the same north-south meridian at points A and B (they could be in the same hemisphere). Thus the observers have the same celestial meridian. The upper diagram indicates how the observers determine the angle AEB at the Earth's center between their respective radii. Angle AEB is the sum of the observer's latitudes, but these were not then known. At the same moment, each observer measures the angle from his zenith to a certain reference star located on his (their) celestial meridian (angles 1 and 3). The sum of these two angles equals the sum of angles 2 and 4. Why? Next, in the lower diagram, each observer simultaneously measures the angle between his zenith and the moon when it is on his celestial meridian (angles 5 and 6). Since the radius of the Earth was known at this time, it was possible to draw a diagram carefully to scale: lines of sight from A and B intersect at the center of the moon. *(After Rogers.)*

telescopes had been constructed, it became possible to measure the distances to the nearer stars by the parallax method (Fig. 17–13). Astronomers measure stellar parallax by taking lines of sight to a near star at

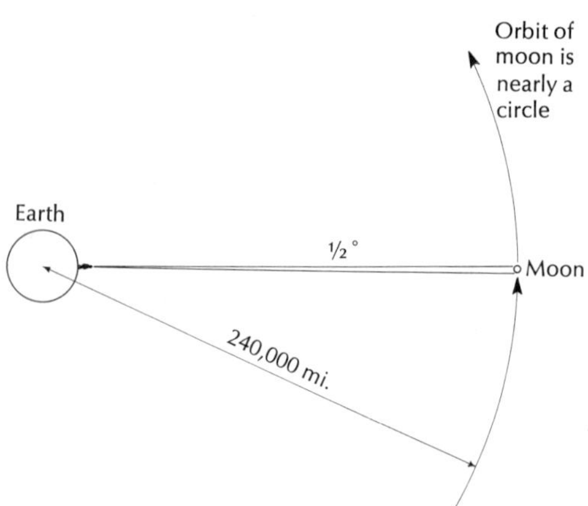

21-3. A method of ascertaining the size of the moon. The entire distance around the moon's orbit can be calculated because the orbit is nearly circular; it is approximately equivalent to the circumference of a circle ($2\pi r$) whose radius is 240,000 miles. The angular diameter of the moon in the sky is measured and is about $\frac{1}{2}$ degree (measuring it from the surface rather than from the Earth's center does not make a significant difference). As the moon's orbit is very large, an arc of $\frac{1}{2}$ degree is nearly a straight line and thus about equal to the moon's diameter. Thus:

$$\text{Moon's diameter} = \tfrac{1}{2} \cdot \frac{2\pi r}{360}$$

$$= \tfrac{1}{2} \cdot \frac{2 \times 3.14 \times 240{,}000}{360} \text{ miles}$$

which is about 2000 miles. An accurate determination would have to consider the moon's orbit as elliptical, its angular diameter as slightly more than $\frac{1}{2}$ degree, and certain other refinements.

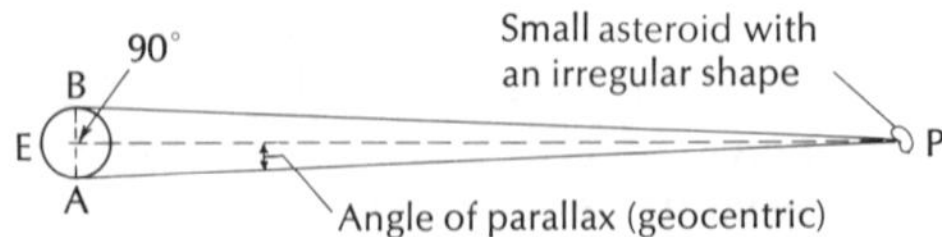

21-4. Finding the distance of an asteroid from the Earth. The angle of geocentric parallax (the angle made by the Earth's equatorial radius as viewed from the asteroid) can be determined by observations from points A and B or by two observations about twelve hours apart from the same position. In the right triangle EAP, the distance EA is known, and the distance EP can be calculated by trigonometry.

time intervals of six months when the Earth is on opposite sides of its orbit. The angle at the star formed by the intersection of these two lines of sight can be determined by photographing a certain area of the sky at six-month intervals. Astronomers refer to half of this angle as the star's parallax. *Heliocentric parallax* may also be defined as the maximum angular separation of the sun and Earth as seen from a star. This angle of parallax varies inversely as the distance of a celestial object from the Earth, and thus it is largest for the nearest objects.

Nearer stars have slightly different positions on photographs taken six months apart. However, the vast majority of stars are so distant that they can be used as fixed reference points—their parallactic displacements are too tiny to be measurable. The parallax angle can be measured directly on a set of photographic plates, because the focal length of a telescope determines the scale of a photographic plate (i.e., the distance on a plate corresponding to 1 degree on the celestial sphere equals 0.01744 times the focal length of the telescope used).

Many of the nearer stars also show proper motion—the slow drift of a star across the celestial sphere due to its own motion through space (p. 605). However, if photographs are taken at intervals of six months over a period of two or three years, the effect of proper motion can be separated from that of parallax.

In measuring the distance from the Earth to a near star, we again resort to the construction and solution of triangles. Geometrically the two points on the Earth's orbit and the star form an isosceles triangle in which the length of the base is twice the distance from the Earth to the sun. This triangle can be subdivided into two congruent right triangles in which a side (base) and an angle (the measured parallax) are known. The length of the hypotenuse, or the distance from the star to the Earth, can then be calculated.

The largest angle of parallax is still less than 1 second of arc—that of the nearest star at a distance of about 4.3 light-years. This is comparable to measuring the diameter of a silver dollar about a dozen miles away and is the reason why Brahe and Aristotle could not detect parallactic displacements among the stars. Therefore, stars must be within some 100 to 300 light-years from the Earth for their distances to be determined by trigonometric parallax.

If an object were located at 3.26 light-years from the Earth, its angle of parallax would be exactly 1 second of arc. Astronomers use this as a unit of distance and call it the *parsec* (from *par*allax *sec*ond).

Distances to Remote Stars

Most of the methods used in modern times for ascertaining distances to remote stars and galaxies depend directly or indirectly upon the inverse square law of light intensity: *the apparent brightness of a source of light varies inversely as the square of its distance from the observer.* To illustrate the basic principle, imagine that in one direction on a dark night you see the faint headlights of a car. You conclude that it is far away. In another direction, the headlights of a car appear moderately bright, and you conclude that it is closer to you. To reach these conclusions, you assumed that the two cars had headlights of equal luminosity, and that a source of light appears dimmer as its distance from you increases.

The inverse square relationship (Fig. 21–5) is a fundamental one that applies to a number of phenomena: electrical forces, gravitational forces, the intensity of light, and magnetism. Why the intensity varies inversely as the square of the distance is readily seen if we compare the surface areas ($4\pi r^2$) of a number of spheres; e.g., spheres with radii of 1, 2, 4, 8, and 16 meters have surface areas equal to 4π (1, 4, 16, 64, and 256) square meters respectively. Thus doubling the radius increases the surface area by four times. Now let us superpose the spheres so that they share a common center, and imagine a source of light radiating in all directions at this center. It is apparent

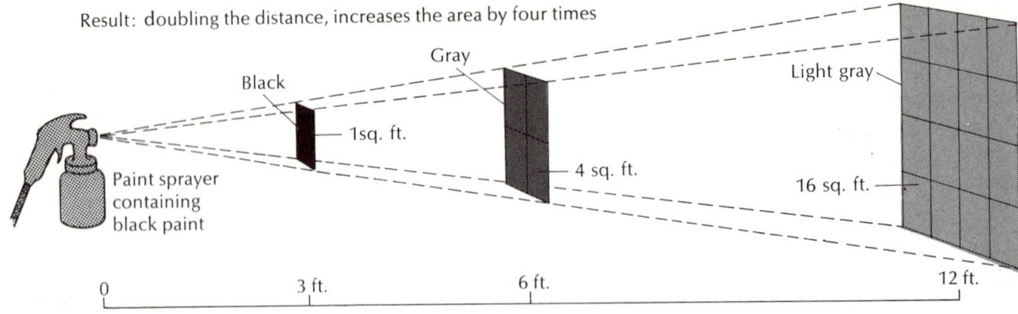

21-5. The apparent brightness of a source of light varies inversely as the square of its distance from an observer. Suppose we place a 1-ft square a distance of 3 ft from a paint sprayer and spray it with black paint. If straight lines are drawn from the sprayer opening through the corners of this square and on into space, we find that the lines are 1 ft apart at the 3-ft distance, 2 ft apart at the 6-ft distance, 4 ft apart at a distance of 12 ft, etc. This follows from the principle of similar triangles. Doubling the distance doubles the length of a side of the square and thus increases its area by four times. Now we take another square that is 2 ft on a side (4 sq ft of area), place it 6 ft from the paint sprayer and spray it with black paint (same quantity as before). It follows that all of the paint that would fall on a 1-ft square at a distance of 3 ft must now be spread across an area of 4 sq ft, and only $\frac{1}{4}$ as much paint falls on each square foot of the larger square.

that this light must illuminate four times more area each time the radius is doubled; therefore, its intensity is reduced four times per unit area.

In the absence of intervening gas and dust, the observed brightness of a star depends upon two factors: its actual or intrinsic brightness (this can be called its luminosity) and its distance from the observer. Remote stars appear faint even if they are actually very luminous.

The inverse-square relationship makes it possible to determine the distance of a remote object by comparing the apparent brightness of the object, which astronomers can measure readily, with its actual luminosity which is known or assumed. Conversely, if the apparent brightness and the distance are known, the actual luminosity can be calculated. Thus in determining the distance to a star by this method, three factors are involved: the actual luminosity of the star (i.e., its absolute magnitude, p. 533), its apparent brightness as observed from the Earth, and its distance. If any two of these are known, the third can be calculated.*

* The equations used by astronomers follow: $M = m + 5 + 5 \log p$ and $M = m + 5 - 5 \log D$. M is the absolute magnitude; m is the apparent magnitude; p is the parallax in seconds of arc; and D is the distance in parsecs.

Consider as an illustration an observer and two assistants carrying flashlights of identical luminosity. On a dark night the assistants move away from the observer in opposite directions. If they stop at equidistant points, the apparent brightnesses of the flashlights will be the same. Now assistant A moves twice as far away as assistant B, then 3, 4, and 5 times this distance. At each stop the brightness of his flashlight is compared with that of assistant B, with the following results. At twice the distance, it appeared 4 times fainter (2^2); and at 3, 4, and 5 times the distance, it appeared 9, 16, and 25 times fainter (3^2, 4^2, and 5^2). Thus we can determine the relative and actual distances of the two assistants by applying the inverse square relationship. For example, if we compare two equally luminous sources at distances of say 12 feet and 4 feet, we have the following relationship:

$$\frac{\text{brightness at 12 feet}}{\text{brightness at 4 feet}} = \frac{(4)^2}{(12)^2} = \frac{16}{144} = \frac{1}{6}$$

Stellar Magnitudes

In comparing the apparent brightnesses of stars as observed from the Earth, astron-

Earth photographed from Applications Technology Satellite in synchronous orbit at an altitude of 22,300 miles on 10 November 1967. South America and the African bulge (right) are conspicuous, but parts of Europe, North America, and the Greenland Ice Sheet are also visible. Antarctica is blanketed by clouds. A tropical storm (bottom, center—note clockwise circulation) has a cold front extending into Argentina. An eastward-moving cold front stretches from the Great Lakes to Mexico. *(Courtesy NASA.)*

omers refer to their *apparent magnitudes* and use a scale that was originated by Hipparchus about 150 B.C. He divided stars into six groups in order of their apparent brightnesses as seen from the Earth. Stars in Group 1 are the brightest and said to have an apparent magnitude of +1; stars in Group 6 are the faintest stars visible to the unaided eye and are said to have an apparent magnitude of +6.

Subsequently, this scale has been extended in both directions (0 magnitude is brighter than +1, −1 magnitude is brighter than 0, etc.) and the differences between magnitudes set at precisely $\sqrt[5]{100}$ or 2.512 times. This makes a difference of 2 magnitudes equal to 2.512 times 2.512 (about 6¼) and a difference of 5 magnitudes equal to precisely 100 (2.512^5). A difference of 3 or 4 magnitudes amounts to a difference in brightness of about 16 and 40 times. A large difference in magnitude can be handled readily by using multiples of five; e.g., the sun is about 14 magnitudes brighter than the full moon, and this amounts to 100 × 100 × 40 or about 400,000 times.

The apparent brightnesses of stars may be measured by various instruments and also by comparing their images on a photographic negative (Fig. 21–6). To compare the relative luminosities of stars, astronomers calculate their *absolute magnitudes*—i.e., the apparent brightness a star would have if it were located at the standard distance of 32.6 light-years (10 parsecs) from the Earth. The absolute magnitude of the sun is about +5.

A few examples of apparent magnitudes follow (at their greatest brilliance if a variation occurs): the sun (−26.8), the moon (−12.6), Mercury (−1.9), Venus (−4.4), Mars (−2.8), Jupiter (−2.5), Saturn (−0.4), and Sirius, the brightest star in the sky (−1.4).

Estimating the Actual Luminosities of Stars

One difficulty in using the inverse-square law of light intensity in calculating the distances to remote stars lies in determining the actual luminosities of such stars (i.e., their absolute magnitudes). Here astronomers have shown a great deal of ingenuity, and several methods have been discovered. One involves a star's spectrum and H-R diagrams (p. 609); another involves the mass-luminosity

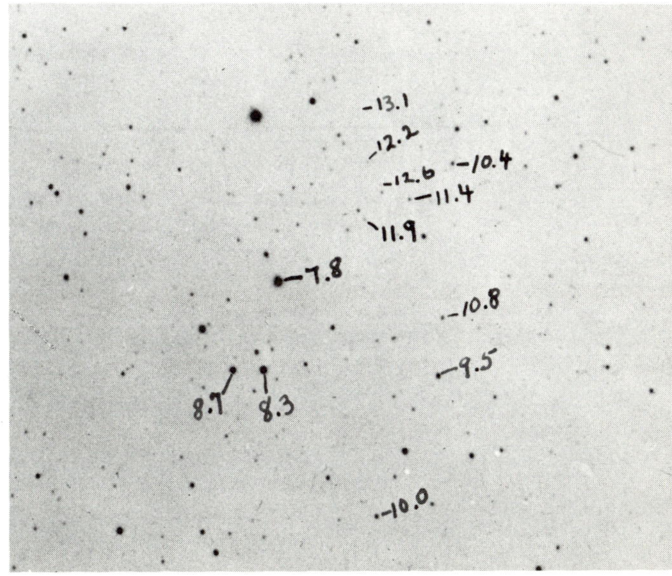

21-6. A method of comparing the apparent brightnesses of stars. Brighter stars leave larger and darker images on a photographic plate. Astronomers commonly work with negatives in their studies rather than with prints made from them. *(The magnitude sequence was worked out by Leon Campbell.)*

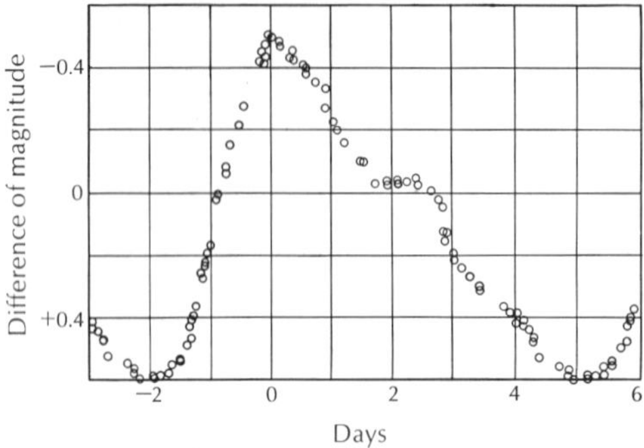

21-7. Light curve of a classical Cepheid having a period of 7 days (Eta Aquilae). Note that the brightness increases more rapidly than it decreases. This variation is repeated week after week. The light curves of Cepheids with shorter or longer periods are shaped somewhat differently but are likewise repeated time after time. *(Photoelectric light curve by C. C. Wylie.)*

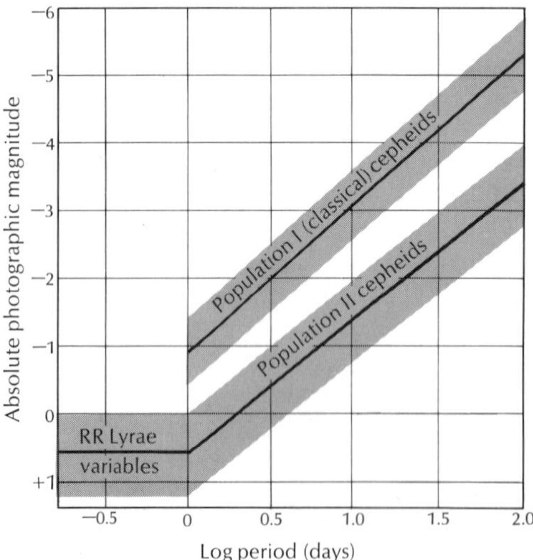

21-8. Period-luminosity relationship for Cepheids. Magnitude increases toward the top, and period increases toward the right. More luminous Cepheids have longer periods than less luminous ones, and Population I Cepheids are about four times more luminous ($1\frac{1}{2}$ magnitudes) than corresponding Population II Cepheids. Enough scatter occurs in the period-luminosity relationship to form bands about one magnitude wide rather than straight lines. *(From preliminary data supplied by H. C. Arp.)*

relationship (p. 607); and one of the most useful and first to be applied successfully involves stars that vary in brightness in a regular manner (Cepheid variables).

Apparently Cepheids vary in brightness because they expand and contract periodically; i.e., the diameter and surface temperature of an average Cepheid may change about 10%, and its color and spectrum also change. Such periodic changes may also take place during an unstable stage in the life cycles of some stars (p. 612).

The two Magellanic Clouds contain Cepheid stars and are small neighboring galaxies (between 150,000 to 200,000 light-years away) that can be seen as faint patches of light with the unaided eye from the Southern Hemisphere. They are so far away relative to their sizes that all of the stars in one of the Clouds are at approximately equal distances from us—in a similar vein, all of the bears in Yellowstone National Park are roughly the same distance from Portland, Maine. Thus any differences in brightness in the stars in a Cloud, as viewed from the Earth, must be due primarily to differences in their absolute magnitudes.

In the early 1900's it was noticed (by Leavitt and Shapely) that certain stars in the Magellanic Clouds changed in brightness systematically during periods ranging from several days to a few weeks (Fig. 21–7). Careful study of many photographs taken at different times showed that each of these was a Cepheid variable star (Cepheids had been discovered earlier in our own galaxy). Each Cepheid required a certain amount of time to change from minimum brightness to maximum, and then back to minimum again —its *period*. Furthermore, a *period-luminosity* relationship was also discovered: i.e., the length of the period is related directly to the luminosity. In other words, more luminous Cepheids have longer periods than less luminous Cepheids (Fig. 21–8).

At this stage, however, the Cepheids could not be used for distance determinations because their absolute magnitudes were unknown (distances to the Magellanic Galaxies were not then known). This problem was attacked by attempts to determine the absolute magnitudes of Cepheid variables within our own galaxy (assumed to be similar). If their distances could be calculated, their apparent magnitudes could then be measured, and their absolute magnitudes worked out. However, no Cepheid is close enough to the Earth for an accurate measurement of its parallactic displacement. Nevertheless, several Cepheids are close enough so that their distances could be estimated approximately by an indirect method involving their proper motions (p. 605, inaccuracies showed up later). Thus the absolute magnitudes of a number of Cepheids within our galaxy were eventually calculated and the data plotted on a period-luminosity graph.

Following this, it was assumed that all Cepheids with the same period would have the same absolute magnitude, whether located in our galaxy or in other galaxies. On the basis of this assumption (later proved incorrect) the distance to a remote Cepheid could readily be calculated: the period of the remote Cepheid and its apparent magnitude were measured. Next, its absolute magnitude was read off the period-luminosity graph. Now we have two of the three variables in the distance equation (footnote, p. 532)—the absolute and apparent magnitudes—and can solve the equation for the distance.

Since Cepheids are yellowish supergiant stars about 1000 to 10,000 times more luminous than the sun, they can be seen at great distances in our telescopes. Thus they provide a very long arm for measuring interstellar distances (up to about 20 million light-years). However, Cepheids are relatively rare stars, and only a few dozen galaxies containing visible Cepheids occur within this range—galaxies are far apart.

Distances to Remote Galaxies

The fainter a galaxy, the more distant it may be, although galaxies vary so widely in absolute magnitudes that such estimates are quite uncertain. Therefore, distances to relatively nearby galaxies were calculated by using the Cepheid variable method (the distance to a Cepheid within a galaxy is taken as approximately the distance of the galaxy itself). For galaxies too remote for Cepheids to be visible, one may substitute such very luminous objects as blue supergiants, globular clusters, and supernovae (very rare). We must first determine the absolute magnitudes of such objects within our own galaxy and then assume that these hold true for other galaxies. Such assumptions may, of course, not be valid and put errors in our distance measurements.

Many galaxies are too remote for even these bright objects to be visible, and to estimate their distances, the absolute magnitudes of a number of nearby galaxies were calculated—each one as an entity. It was hoped that such absolute magnitudes might be combined to produce a representative average. Remote galaxies would then be assumed to have this average absolute magnitude, and measurements of their apparent magnitudes would provide enough data to calculate their distances.

However, entire galaxies show too great a variation in absolute magnitudes for this technique to be used, and astronomers have attempted to overcome this problem by working with clusters of galaxies. They assume that the ten or so brightest galaxies in any one cluster will have about the same average absolute magnitude as the same number of brightest galaxies from any other large cluster. Their apparent magnitudes are then measured and compared. The faintest ones are assumed to be the most remote, and the equation is used to determine their actual distances.

Cautions Concerning Stellar Distances

Calculations of distances in astronomy involve a number of assumptions which have a tendency to pyramid as distances become greater; e.g., light has been assumed to come through space without any diminution of its brightness by intervening dust or gas. This assumption is in error and more so in some directions than in others. If light from a star is dimmed by obscuring matter, our measurement of its apparent brightness will be too small, and its calculated distance will be too great—the magnitude of the error depends upon the quantity of obscuring matter. This is most concentrated within the spiral arms and along the equatorial plane of our galaxy—in our neighborhood. Similarly, we overestimate the distance to a traffic signal on a foggy night. Therefore, opportunity exists for the presence of large errors in these successive extrapolations.

Calibration of the period-luminosity relationship for Cepheid stars—i.e., determining their absolute magnitudes—has posed problems, and may still be uncertain. Prior to 1952, conclusions were reached concerning these absolute magnitudes, and the data were used to calculate distances to galaxies here and there in the universe. By 1952, it was realized that the absolute magnitudes of the Cepheids are actually four times higher (i.e., 1.5 magnitudes) than previously estimated. Therefore, the distances to these galaxies all had to be doubled, and additional upward revisions have been made since 1952. This revision also changed estimates of the diameters of galaxies (reduced by half) and of the age of the universe (doubled, and subsequently revised upward).

One important verification has been made of the 1952 revision of the absolute magnitudes of Population I Cepheids. Some of them have been discovered as members of relatively nearby clusters whose distances could be obtained by reference to H-R diagrams (p. 609) and by a study of the motions of the stars within a cluster. Since the apparent magnitudes of these cluster Cepheids could be obtained, their absolute magnitudes could be calculated by means of the distance equation, because the distances to the clusters were known (all stars within a particular cluster are at approximately the same distance from us). These results checked rather well with the 1952 figures.

Since 1952, another group of variable stars has been discovered—the Population II Cepheids. Each Population II Cepheid has an absolute magnitude that is about four times less luminous (nearly 1.5 magnitudes) than its counterpart in the Population I (classical) group. Thus the Population II Cepheids at present occupy approximately the same position on the period-luminosity graph as did the Population I group prior to 1952.

Sizes and distances within our own galaxy were not affected by the 1952 correction because a different type of star had been used (R. R. Lyrae, p. 615). Changes in the estimated distance to the Andromeda galaxy are instructive. In 1929 this distance was determined by the Cepheid method to be about 900,000 light-years. In 1944 this calculated distance was reduced to approximately 750,000 light-years by a correction for intervening gas and dust. In 1952 this revised figure was doubled to about 1½ million light-years. Since 1952 the distance has been revised again to approximately 2.3 million light-years. Thus distance determinations to remote objects are still uncertain, and the answer to "How big is the universe?" may not be known until some time well in the future, if ever.

Measuring the Masses of Some Celestial Objects

We determine the mass of a celestial object by measuring its gravitational pull on another object, and again we must start with the Earth and work outward to stars and galaxies. After the mass of the Earth was determined (Fig. 20–6), it became possible to measure the mass of the moon. If two celestial objects mutually revolve about a point—the center of mass or gravity for this two-body system—their masses and distances from the center are related as follows: *The mass of one body multiplied by its distance from the center equals the mass of the other body multiplied by its distance from the center.* Two boys balanced on a seesaw illustrate the principle.

In other words, if the Earth and moon were attached to the ends of an imaginary weightless rod, they would balance each other at this center of mass, which is located approximately 2900 miles (4670 km) from the Earth's center, or about 1000 miles below the surface on the side toward the moon (Fig. 21–1). The Earth and moon mutually revolve about this point, and it is the path of this center of mass that is shown when a smooth elliptical curve is drawn for the Earth's orbit about the sun. The center of the Earth is sometimes closer to the sun than this smoothly moving center of mass and sometimes more distant; the Earth's path about the sun is slightly wavy.

To detect these motions, precise measurements were necessary of the positions and velocities of the Earth and moon relative to the sun and other objects; e.g., during each month the sun goes through a cycle relative to the Earth's position in which it appears to be slightly closer to the Earth at one time and then to be further away two weeks later. It also shifts slightly forward and backward at two-week intervals as the center of the Earth is ahead or behind the center of mass. In other words, the sun seems to follow a circular path in the sky with a radius of about 2900 miles.

Since the moon's mean distance from the Earth is nearly 239,000 miles (from center to center), the moon is about 236,100 miles from the center of mass. This means that the center of the moon is about 81½ times farther from the center of mass of the Earth-moon system than is the center of the Earth (236,100 vs. 2900). Therefore, the mass of the moon is about 81.5 times less than the mass of the Earth. Since we know the size of the moon, we calculate its average specific gravity as 3.3, much less than that of the Earth.

This same relationship can be used to determine the relative masses of another planet and one of its satellites and for a pair of mutually revolving stars if their distances from the center of mass can be determined. Furthermore, if the actual mass of one member of a pair is known, that of the other member can be calculated as was done for the Earth-moon system.

Masses of the Sun, Planets, and Stars

Newton learned that Kepler's third law should be restated as follows: *The squares of the periods of any two planets, each multiplied by the combined mass of the sun and planet, are proportional to the cubes of their mean distances from the sun.* In fact, any two pairs of mutually revolving bodies may be compared by means of this law; it is not restricted to the sun and planets.

Newton's modification makes only a slight difference when two planets are compared. The most massive of the planets, Jupiter, is still 1000 times less than the mass of the sun. Thus the mass of any planet is so small compared to the mass of the sun that it can be disregarded. This leaves the mass of the sun in both the numerator and the denominator, and it cancels out. For such purposes, Kepler's third law as originally stated (without considering the masses of the sun and planets) is satisfactory.

To obtain the mass of the sun, we apply Newton's version of Kepler's third law. The Earth-moon system can be used as one pair and the sun and any one of the planets—say Mercury—as the other pair. Thus we have an equation with a single unknown, the mass of the sun:

$$\frac{P^2\,(\text{mass}_{\text{sun}} + \text{mass}_{\text{Mercury}})}{P^2\,(\text{mass}_{\text{Earth}} + \text{mass}_{\text{moon}})} = \frac{D^3}{D^3}$$

This unknown that we obtain by solving the equation is actually the sum of the masses of the sun and Mercury (two unknowns) but Mercury's mass is relatively so very small that the sum is essentially the mass of the sun itself. To illustrate, if we let a pile of nearly 7 million marbles represent the mass of the sun, only a single marble is needed for Mercury. We must use the same units for period, mass, and distance in the numerator as we do in the denominator. In the numerator, the period and distance are those of Mercury about the sun; in the denominator, the period and distance are those of the moon about the Earth.

Note that this revision of Kepler's third law accounts for a very slight difference in the orbits of objects of different masses about a common primary. To illustrate, assume that a marble orbits the Earth at exactly the moon's distance. Since the distances are identical, this means that: P^2 (mass $_{\text{Earth}}$ + mass $_{\text{moon}}$) = P^2 (mass $_{\text{Earth}}$ + mass $_{\text{marble}}$). Thus the moon must have a slightly smaller period and must move slightly faster in its orbit. In other words, the gravitational pull between the Earth and moon is greater than that between the Earth and marble; thus the moon must move faster if it is to stay in the same orbit about the Earth. Conversely, if the periods of the moon and marble were identical, the marble's distance would have to be slightly greater.

In a similar manner, astronomers can calculate the mass of any planet with a satellite. Let us use Mars and one of its satellites to illustrate. Again we must solve for a double unknown, but the mass of Mars' satellite is relatively so very small that it too can be disregarded:

$$\frac{P^2\,(\text{mass}_{\text{Mars}} + \text{mass}_{\text{satellite}})}{P^2\,(\text{mass}_{\text{Earth}} + \text{mass}_{\text{moon}})} = \frac{D^3}{D^3}$$

In like manner, one may determine the combined masses of a pair of mutually revolving stars (binaries). We shall do this in terms of the sun's mass as the unit and by using the sun-Earth as the comparison pair in the denominator. In this case the arithmetic simplifies because the denominator becomes equal to one: the period is 1 year; the distance is 1 A.U.; and the mass of the sun-Earth pair is taken as 1. Therefore, the sum of the masses of the two stars equals the cube of their mean distance divided by the square of their period. Thus one must be able to measure the period and mean distance of a pair of mutually orbiting stars to obtain the sum of their masses. Furthermore, if the center of mass of the system can be located, then the mass of each star can be determined individually.

A study of the mass and luminosity of a number of stars has shown that the luminosity of a star varies directly (but exponentially) as its mass—the most massive stars are the most luminous (p. 607); however, this applies only to main sequence stars.

Mass of Our Galaxy

We can use the sun's orbit about the center of our galaxy to make a rough approximation of the mass of the whole galaxy. We shall assume that the galaxy's entire mass is concentrated at its center (this ignores the gravitational influence of stars farther out than the sun, but the effect is relatively small) and that the sun follows a circular orbit.

The relationship between the mass of a primary and the orbital speed of a satellite is basic in this calculation (p. 480). Apparently the sun takes about 200 million years to complete one orbit. However, if the mass of the galaxy were larger, the sun would have to move faster to stay in its present orbit, and its period would be shorter. In like manner, stars closer to the center revolve faster and those farther out revolve more slowly (p. 624). We shall use the sun-galaxy pair in the numerator of Newton's modification of Kepler's third law and the sun-Earth pair in the denominator (this becomes one and is eliminated). We shall use 2×10^9 A.U.'s as the distance of the sun from the center (about 33,000 light-years):

$$\text{mass}_{\text{galaxy}} + \text{mass}_{\text{sun}} = \frac{D^3}{P^2} = \frac{(2 \times 10^9)^3}{(2 \times 10^8)^2}$$

$$= \frac{8 \times 10^{27}}{4 \times 10^{16}} = 2 \times 10^{11}$$

Thus the mass of our entire galaxy is approximately equal to the combined masses of 2×10^{11} suns.

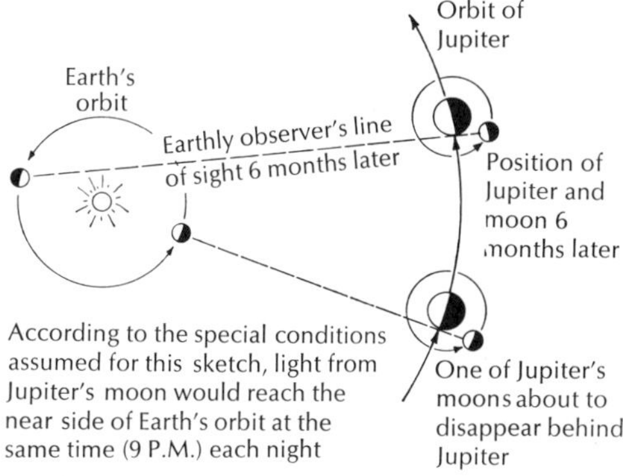

21-9. Roemer's method of determining the speed of light. An observer on the Earth notes that a moon disappears behind Jupiter at a certain time on one evening, e.g. 9:00 PM. By measuring the length of time until it disappears again, the observer determines its period of revolution (assume it is 24 hours). The moon's period of revolution should be constant, and it should continue to disappear at 9:00 PM on succeeding nights. However, it does not do so. Roemer recorded the moon's times of disappearance at intervals of l month as follows: 9:00, 9:03, 9:06, 9:09, 9:12, 9:15, 9:17 (6 mos later) 9:15, 9:12, 9:09, 9:06, 9:03, and 9:00 (1 yr later). The 17-min delay 6 mos later (his figure actually exceeded this by several minutes) was caused by the greater distance that light from the moon then had to travel to reach the Earth. The series of times should run over 13 mos, the synodic period of Jupiter, rather than 12 mos.

Two Other Measurements

In the late 1600's as he was studying the satellites of Jupiter, Roemer observed that light needed about 17 minutes to span the Earth's orbit (Fig. 21–9). Since this distance was known approximately, he could then calculate the speed of light (distance ÷ time = speed).

At noon at the equinoxes when the sun's rays parallel the plane of the equator, one's latitude can be determined by measuring the angle made by a stick and its shadow (Fig. 21–10).

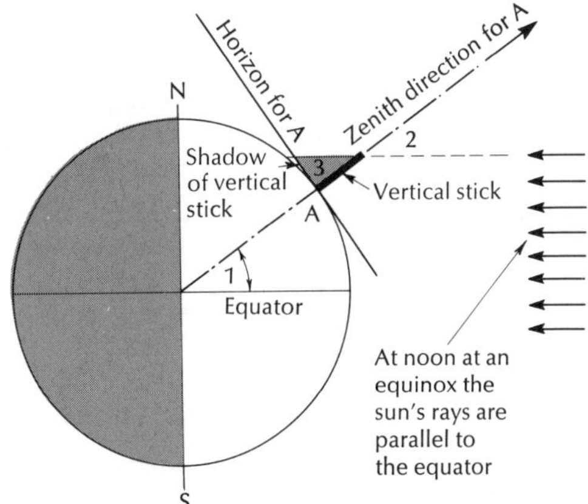

21-10. Measurement of latitude at noon at an equinox. Side view; diagram is obviously not to scale. We plan to measure the latitude of A which is equal to angle 1 at the center of the Earth. By geometry, angles 1, 2, and 3 are equal. One method of measuring angle 3 follows. Be certain the stick is vertical (use a makeshift plumb bob) and measure the lengths of the stick and shadow (the surface should be level; average several measurements). Then the tangent of angle 3 is equal to the length of the shadow divided by the length of the stick. This can be solved readily: substitute your measured lengths, divide, and look up the answer in a table of tangents.

Exercises and Questions for Chapter 21

1. What is the linear diameter of a galaxy that is 2 million light-years away and has an angular diameter of 4 degrees?
2. A moon, as viewed from a certain planet, has an angular diameter of 1 degree at a distance of 100,000 miles. Calculate its diameter.

3. At a horizontal distance of 1000 feet, a certain flagpole subtends an angle of 3°. How tall is the flagpole?
4. The top of a volcanic island is 1 mile above sea level and is just visible to an observer on a ship about 90 miles away. From this information calculate the size of the Earth (assume a spherical Earth and no light refraction).
5. At noon, standard time, at an equinox a 50-foot flagpole casts a shadow that points toward the north and is 50 feet long. What is the geographic latitude of the flagpole?
   (A) What is the approximate latitude if a flagpole is 100 feet tall and its noon shadow is about 57.7 feet in length at an equinox?
6. Star A is 2000 miles in diameter and 100 light-years away. Star B is 20,000 miles in diameter and 1000 light-years away. The stars have identical surface temperatures. What are their relative brightnesses as viewed from the Earth (assume no intervening matter)?
7. The apparent magnitudes of two stars are given in each of the following. In each case, which star is the brighter? How many times brighter?
   (A) +1 vs. +2
   (B) −2 vs. −4
   (C) +6 vs. +9
   (D) −10 vs. −6
   (E) +5 vs. +10
   (F) −3.5 vs. +3.5
8. In each of the following, calculate how many times the brightness of one object exceeds that of the other.
   (A) The sun at −26.5 vs. the full moon at −12.5.
   (B) Star A has an apparent magnitude of zero; that of star B is +22.
9. In each of the following, calculate the number of magnitudes involved.
   (A) One star appears 160,000 times brighter than another.
   (B) One star appears 4 billion times brighter than another.
10. Assume that a small moon orbits a planet at a mean distance of about 240,000 miles with a period of about 273 days. How does the mass of this planet compare with that of the Earth-moon pair?
11. The mass of Jupiter is about 1/1000 that of the sun, and its mean distance is about 5 astronomical units. Determine the approximate location of the center of mass of the sun-Jupiter pair.
12. Cepheids A and B each have 10-day periods. Calculate their relative distances in each of the following (assume no intervening matter).
   (A) A and B belong to the same population and have the same apparent magnitude.
   (B) A is Population I and B is Population II. Their apparent magnitudes are the same.
   (C) A and B belong to the same population, but A appears 16 times fainter than B.
   (D) A and B belong to the same population, but A appears 100 times brighter than B.
13. Assume that two cepheids belong to the same population and have identical 9-day periods. However, cepheid A has an apparent magnitude of +1, whereas that of cepheid B is +6. What are their relative distances?

14. Assume that a small planet revolves about a star at a mean distance of about 24 million miles in a period of about 27⅓ days. What is the approximate mass of this star relative to the mass of the Earth-moon pair as one?
15. Assume that two stars form a binary system (i.e., they mutually revolve about a center of mass). Their period of revolution is 4 years, and their mean distance apart is 8 astronomical units.
    (A) What is the sum of the masses of the two stars relative to the sun-Earth system as one?
    (B) Assume that the smaller star is seven times farther from the center of mass than the larger. What is the relative mass of each star if the sun-Earth pair is taken as one?
16. Professor Peter van de Kamp, Swarthmore, Pennsylvania, has suggested the following exercise. Delta (δ) Cephei is a pulsating Cepheid variable star located at the point of a small isosceles triangle in the constellation Cepheus (see accompanying sketch from *Sky and Telescope* June 8, 1953). The other two stars in the small triangle do not vary in brightness and may be used as comparison stars in estimating the apparent brightness of Delta Cephei on successive nights (if weather permits).

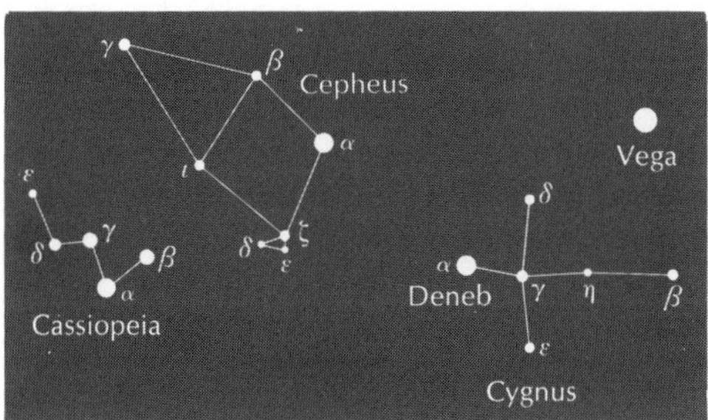

Delta Cephei never appears brighter than Zeta (ζ) Cephei nor fainter than Epsilon (ε) Cephei. The value of the brighter comparison star can be taken as A and the fainter as E. The relative brightness of Delta Cephei can then be estimated on a 9-point scale such as A, A-B, B, B-C, C, C-D, D, D-E, and E. Results should be plotted on a graph (show brightness on the vertical axis and time in days along the horizontal axis). Cloudy weather probably will interfere with observation. However, reasonably good weather and diligent observation over a period of several weeks can result in enough data to approximate the light-curve and period of Delta Cephei. Only one observation is made per night; its time is probably not a significant factor.

(A) If location, time, and weather combine to make observations difficult, assume that the following estimates have been made. Plot the data and estimate the period.

Oct.	1	A	Oct.	11	E	Oct.	21	D-E	Nov.	1	D
	2	B-C		12	A-B		22	E		2	E
	3	C-D		13	B		23	C		3	D
	4	D-E		14	C		24	A-B		4	A
	5	E		15	D		25	B-C		5	B-C
	6	C		16	E		26	D		6	C-D
	7	A-B		17	D		27	D-E		7	D-E
	8	B-C		18	A		28	E		8	E
	9	D		19	B-C		29	A-B		9	C
	10	D-E		20	C-D		30	B		10	A-B
							31	C		11	B-C

(B) Calculate the distance to Delta Cephei. Assume that its mean apparent magnitude is +4.65 and that its mean absolute magnitude is −3.5.

17. Do the laboratory exercise on variable stars in the M15 globular cluster that was published in the October 1967 *Sky and Telescope*. Copies may be obtained from Sky Publishing Corporation, 49–50–51 Bay State Road, Cambridge, Massachusetts 02138.

22

The Moon

"That's one small step for a man, one giant leap for mankind," said Neil Armstrong at 10:56 P.M. on July 20, 1969 as he became the first man to step upon the moon. At this most dramatic moment in space exploration, perhaps one out of every four persons on Earth excitedly either watched or listened. "We came in peace for all mankind," was written on the Apollo 11 landing module from which Edwin Aldrin, his companion astronaut, would soon emerge while the third member of the team, Michael Collins, piloted the command module around the moon. This first landing of men on the moon climaxed the American Apollo Program, and a successful second landing was soon to follow (Apollo 12 in November 1969). Thus the moon—the Earth's only major satellite (Figs. 22–1 and 22–2) became an object of exciting interest in the 1960's as the target of space exploration programs which increased steadily in capability and scope.

Prior to these landings, many unmanned spacecraft, both American and Russian, had circled the moon and had crashed and soft-landed upon it. Although instruments in these spacecraft had relayed much new data to the Earth, many questions concerning the moon and its long history remained unanswered, and it was said that the many new photographs resembled mirrors: each viewer saw his own theories reflected back to him. Now that more than 100 pounds of lunar soil and rock (Fig. 22–3) have been carried back to the Earth and examined with all the varied instruments available to hundreds of the world's top scientists, many questions still remain unanswered. However, some of the questions are now new ones, and many new facts and data, some unexpected, have placed constraints on the theories that are yet to develop. But before reviewing this new information, let us look at the moon as it has been viewed over the centuries through Earth-based eyes and telescopes.

The Earth and moon have some aspects of a double planet. The moon is larger in comparison with the Earth than any other satellite compared with its primary, although Jupiter, Saturn, and Neptune all

22-1. LM in ascent near Mare Smythii Region with planet Earth in the distance. This July 1969 Apollo 11 photograph contrasts light-colored highland terrain with a dark-colored mare that has fewer craters and stands at a lower altitude. *(NASA: AS11-44-6642.)*

have larger satellites. The moon has a diameter of 2160 miles (3476 km) and revolves in an elliptical orbit about the Earth at a mean distance of nearly 239,000 miles (384,404 km); the range is about 221,000 to 253,000 miles (356,400 to 466,700 km).

An average sample of the moon (if we could take one) would prove to be 3.34 times as massive as an equal volume of water, whereas the Earth's average specific gravity is 5.52. The moon's low specific gravity (similar to the Earth's upper mantle) seems to rule out the presence of a dense central core like that of the Earth and sug-

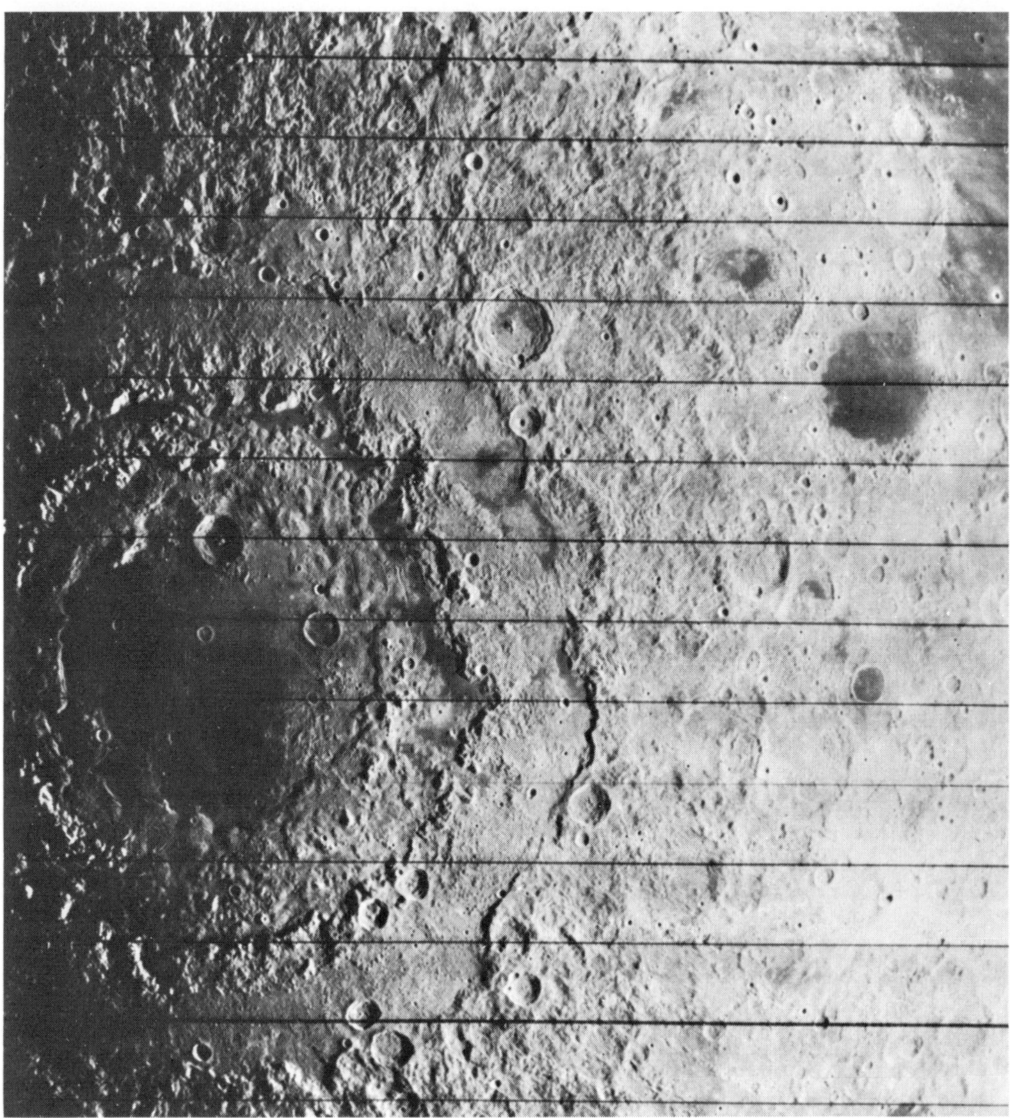

22-2. Orientale Basin on the far side of the moon. (NASA, Langley Research Center. Courtesy R. H. Baker and L. W. Fredrick, 7th ed., *An Introduction to Astronomy,* Van Nostrand Reinhold, 1968.)

gests an overall difference in chemical composition. The moon also lacks a magnetic field like that of the Earth's (p. 298).

The apparent angular diameter of the moon varies as much as 12% indicating that the moon's distance from the Earth varies and that its path about the Earth is an ellipse and not a circle. This change in the moon's diameter should not be confused with the apparent, but illusory, increase in size that celestial bodies undergo when viewed near the horizon.

The distance to the moon was first measured by Hipparchus long ago (Fig. 21–2). In another method, a parallactic displacement on the celestial sphere is noted (Fig. 21–4)

22-3. Stereomicroscope photograph of Apollo 11 contingency sample. Spherules apparently represent lunar rock debris that was melted by meteoroid impact and then solidified during flight. The largest spherule is $\frac{1}{2}$ mm in diameter, and colors include black, red, yellow, brown, and clear. Small rock fragments are also visible. (Striped background is surface of aluminum container.) *(NASA, S-69-45182.)*

if the moon's direction in space is determined simultaneously from two widely separated points, or from the same point at intervals of say 12 hours. The moon's *geocentric parallax* can be defined as the maximum angle made by the Earth's equatorial radius as viewed from the moon at its mean distance from the Earth. It is almost 1 degree or twice the angular diameter of the moon. This angle and the Earth's radius are known elements of a long, slender right triangle, one of whose sides extends from the center of the Earth to the center of the moon. Trigonometry can then be used to calculate the distance, as can a method involving geometry (Fig. 21–2).

In 1946 the U. S. Army Signal Corps directed radar pulses at periodic intervals toward the moon and were able to detect faint signals 2.56 seconds later. This provides another means of measuring the mean distance to the moon very accurately (238,857 miles ± about 1 mile), because the radar pulses travel at the speed of light and can be measured in millionths of a second. Such precision is needed in calculating orbits for lunar-bound spacecraft. However, similar measurements made by reflecting a laser beam from a special mirror set upon the moon's surface by Apollo astronauts are even more precise (384,404.4 km), and perhaps the mean distance can be determined within inches. Comparison of such laser measurements over a decade or more may then measure the exact rate of the moon's recession from the Earth.

Phases of the Moon

As the moon is a nearly spherical body in space, half of it must always be illuminated by sunlight, whereas the other half is darkened and turned away from the sun (but no part has daylight continuously because the moon rotates, Fig. 22–9). The moon's revolution about the Earth brings varying proportions of the lighted half into view and the phases result (Fig. 22–4). A new moon occurs when the moon is located between the Earth and the sun, and the illuminated half is turned completely away from the Earth. About two weeks later, when the Earth is located between the sun and moon, the entire illuminated half faces toward the Earth, and the full moon phase occurs.

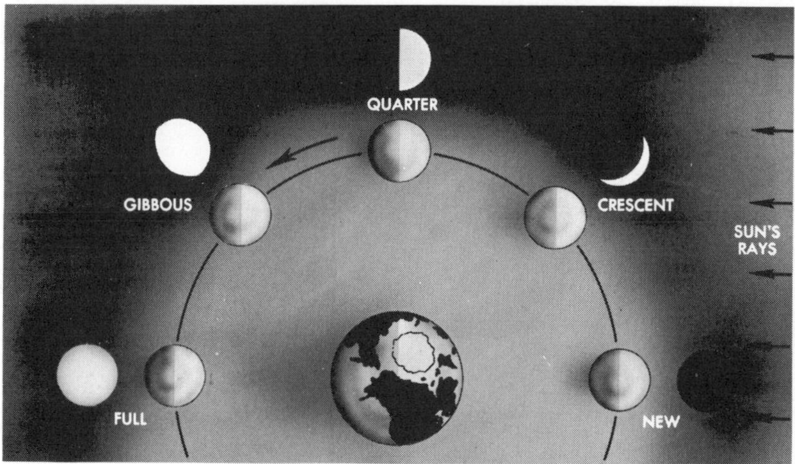

22-4. Phases of the moon. The outer figures represent its appearance from the Earth. (R. H. Baker and L. W. Fredrick, *An Introduction to Astronomy*, 7th ed., Van Nostrand Reinhold, 1968.)

The new moon is nearly in line with the sun and invisible in its glare. In this phase, the moon rises and sets with the sun and thus is above the horizon only during the daylight hours. Therefore, for an earthly observer, the cycle of the phases does not actually begin until a thin crescent, about two days old, becomes visible low in the western sky at sunset. Almost a week later, a half moon (first quarter phase) rises in the east at about noontime, is high in the southern sky at sunset, and sets in the west at midnight (Fig. 22–5). At the end of another week, the full moon rises in the east as the sun sinks in the west. It is above the southern horizon at midnight and sets at sunrise. At the third quarter phase, the half moon rises at midnight and sets at noontime. The cycle is completed in about 29½ days.

Earthshine (Fig. 22–6) may be observed on the moon during the crescent phase that occurs either before or after the new moon phase. Light rays that cause earthshine travel first from the sun to the Earth, are next reflected to the darkened part of the moon, and then are reflected back to the Earth again.

A knowledge of the positions of the moon in the sky during its different phases

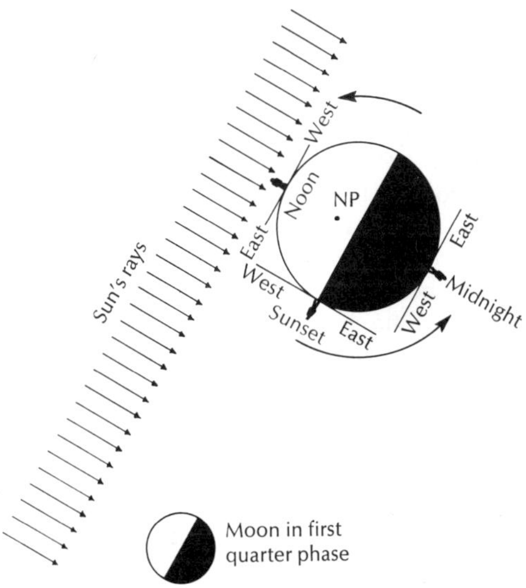

22-5. Location of the moon in the observer's sky at first quarter. Orientation: in space looking down on the Northern Hemisphere in summer (shown by the position of the north pole). Three observers are pictured on the Earth: at noon, sunset, and midnight. The moon is on the eastern horizon (rising) for the noon observer, high in the south for the sunset observer, and on the western horizon (setting) for the midnight observer. In six hours the Earth's counterclockwise rotation will spin the noon observer into the sunset position, and six hours more will bring him to midnight.

22-6. Earthshine faintly illuminates most of the moon, whereas direct sunlight illuminates the bright crescent. As viewed through a telescope in astronomy, objects are commonly inverted and reversed. The horns of the crescent actually point upward from the horizon and away from the sun. *(Courtesy Yerkes Observatory.)*

is sometimes useful in dating events. A college girl attending the annual spring-vacation frolic in Florida once wrote home: "The moon was a breathtakingly beautiful sight as it rose above the Atlantic last night." Since the moon at this time was midway between the third quarter and new moon phases, the words would have told understanding parents that their daughter was still up at about 3:00 A.M.!

The Moon's Rotation and Revolution

As it revolves about the Earth, the moon shifts eastward about one diameter per hour against the celestial background. Since this is more rapid than the sun, the *lunar day* (Fig. 22–7)—the time for one complete rotation of the Earth relative to the moon—is longer than the solar and sidereal days and averages 24 hours and 51 minutes. Stated another way, when the moon rises on a certain night, it occurs in front of a particular location on the celestial sphere. About 24 hours later, this same location on the celestial sphere rises again, but the moon is no longer in front of it. On the average, the moon is about 13 degrees to the east, and thus it rises about 51 minutes later each day.

However, the length of the lunar day varies during a year (from about 24.3 to 25.3 hours at 40 degrees north latitude); e.g., when the moon is full or nearly full in September and October (respectively, the Harvest Moon and Hunter's Moon), it may rise as little as 20 minutes later on successive nights. Thus moonlight may brighten the countryside during the early evening hours for several nights in a row. This happens because the ecliptic then makes a small angle with the western horizon at sunset, and thus the moon's eastward shift places it only a short distance below the horizon from one night to the next (Fig. 17–10). The opposite effect occurs near the spring equinox when the ecliptic slants more steeply toward the western horizon.

At the full moon phase, the moon is

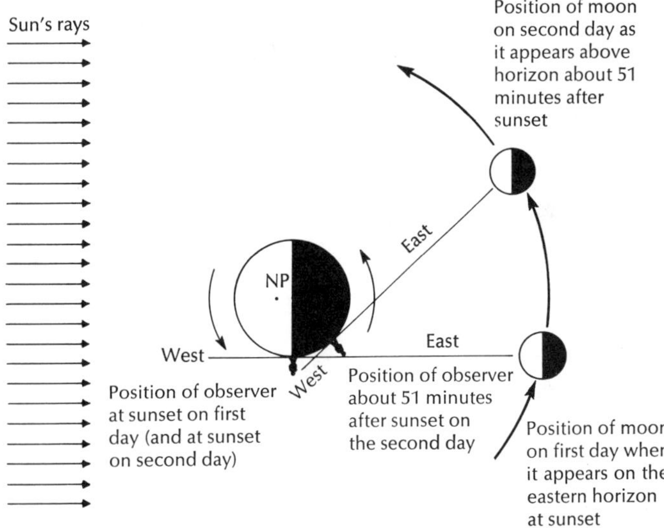

22-7. The moon rises later each day. The full moon appears on the eastern horizon at sunset on one day for an observer. The Earth rotates the observer back to this sunset position in about 24 hr. Meanwhile, the moon has moved forward in its orbit, and it is no longer visible at sunset. The Earth must spin 51 mins longer on the average to move the observer into position to see the moon rise in the east again. What is the season of the year in the Southern Hemisphere at this time?

above the horizon for a length of time approximately in inverse proportion to the sun. During long winter nights, the full moon rises earlier and farthest to the northeast, moves higher in the sky, and sets later than it does during the summer months; the sun's path is the reverse of this (Fig. 20–13).

The time between two full moons or two new moons approximates 29½ days, and this completes a month relative to the sun (the *synodic month;* "synod" means meeting). On the other hand, the moon completes an eastward swing of 360 degrees relative to the stars in about 27⅓ days (the *sidereal month*). The orbital movement of the Earth causes this difference (Fig. 22–8).

Approximately the same hemisphere of the moon faces the Earth at all times; in other words, from the Earth one can see the face of the "man in the moon" but never the back of his head. This occurs because the moon rotates through 360 degrees in the 27⅓ days needed for a sidereal revolution of 360 degrees (Fig. 22–9). However, each month any one point on the moon has about 2 weeks of daytime in which the sun is visible (as time is measured on the Earth) and about 2 weeks of night-

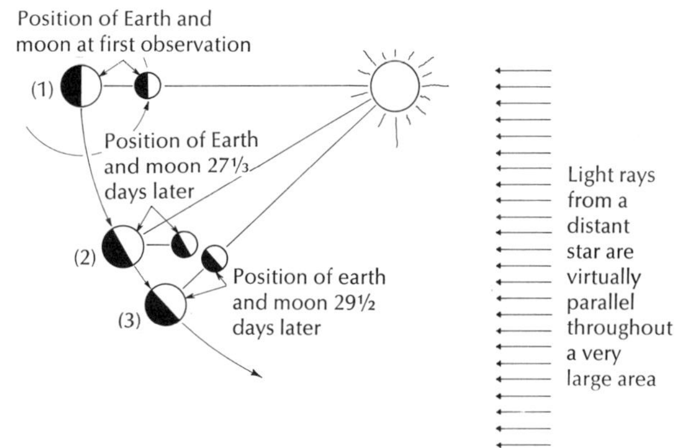

22-8. Sidereal and synodic months. (1) At the first observation (upper left), the moon is in its new phase, and the Earth, moon, sun, and reference star are in line in space. (2) About 27⅓ days later, after the moon has revolved through 360 degrees around the Earth, the sidereal month is completed. Earth, moon, and reference star are once more aligned, but not so the sun. The moon is a crescent. (3) Approximately 2 days later—about 29½ days after (1)—the synodic month ends. The moon is again in a line between the Earth and the sun in the new moon phase.

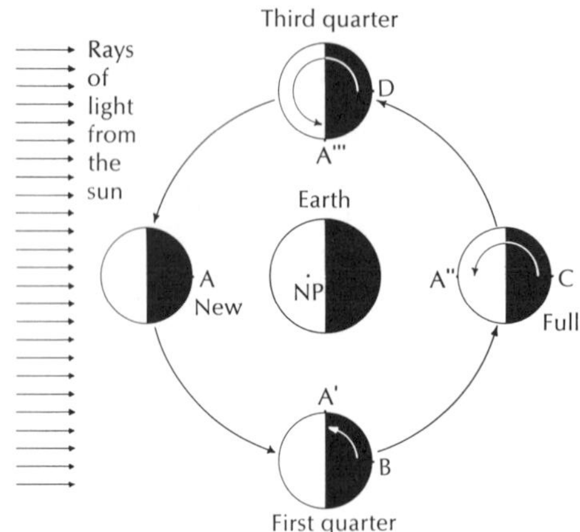

22-9. The moon always presents about the same side to the Earth. Begin with the new moon phase. Point A (nighttime) is in the middle of the hemisphere that is turned toward the Earth. In other phases this same point is indicated at A′, A″ (daytime), and A‴. Points B, C, and D show the positions that would be occupied by point A if the moon did not rotate. The small curved arrows inside the moon show the amount of rotation since the new moon phase. In any period of time, this amount of rotation equals the amount of revolution.

time. To illustrate this phenomenon, try following a circular path about a friend, but keep the same side of your face turned toward the friend in the center at all times. In order to do this, you will have to rotate once during each complete circular trip.

One can speculate that the moon's rate of rotation was once more rapid and that its spinning has since been slowed by the Earth's tidal effect. Since the moon may at

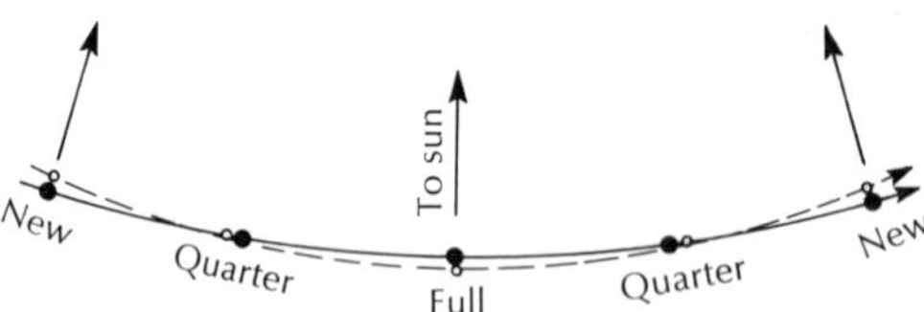

22-10. The orbits of the Earth and moon around the sun. (R. H. Baker, *Astronomy,* Van Nostrand Reinhold, 1964.)

one time have been closer to the Earth, this effect would then have been particularly powerful. Moreover, in a reciprocal effect, the tidal attraction of the moon is slowing the Earth's rate of rotation (p. 566). Like our moon, other satellites in the solar system also have periods of rotation that equal their periods of revolution, and thus each directs the same hemisphere towards its primary.

Although the moon rotates and revolves in a manner to keep the same side facing the Earth, observations spanning a period of time encompass about 59% of its surface. Since the moon's axis is tilted about 6½ degrees to its orbital plane, the moon's north pole is tilted toward the Earth at one time, whereas 2 weeks later its south pole slants toward the Earth. Moreover, the moon's orbital velocity changes while its rotation remains about uniform, thus we see around the edges.

The moon and the Earth are so close to each other and so far from the sun that the moon's path, like that of the Earth, is always concave toward the sun. The moon and Earth revolve around their mutual center of gravity (p. 537), and as seen from a point in space, their orbits would nearly coincide. The moon would periodically overtake the Earth and would later in its turn be passed by the Earth (Fig. 22–10). The Earth's mean orbital speed about the sun approximates 18½ mi/sec, and the moon revolves about the Earth at about ½ mi/sec, or about 18 to 19 mi/sec relative to the sun.

When the Earth's gravitational pull coincides generally with the direction in which the moon is moving—as it does from new moon to full moon—the moon speeds up. When the Earth's pull tends to oppose the direction in which the moon is moving—as it does from full moon to new moon—the moon slows down. However, a lag occurs, and although the moon is being gradually slowed from the full phase onward, nevertheless its onward velocity does not fall below that of the Earth until after it reaches the third-quarter position.

The Moon Lacks an Atmosphere

Before men walked on the moon, there was considerable evidence indicating that it has no appreciable atmosphere: clouds are not seen when the moon is observed through powerful telescopes; shadows are distinct and black; no signs of twilight have been observed; and the light of a star is blotted out abruptly as it disappears behind the moon.

The lack of a noticeable, permanent atmosphere is explained by the moon's small gravitational attraction. If the moon at one time did have gas molecules surrounding it, as seems likely, they have long since escaped its gravitational pull by virtue of their rapid movement. Certain craters may at times have a very faint, hazy appearance as if veiled by gases, and the central peak in one such crater may have become brighter as it was observed.

In 1969 as Apollo astronauts moved about observing the moon's surface and setting up equipment, they were enclosed in pressurized, air-conditioned suits to protect them in the unearthlike, hostile lunar environment: the effect of zero lunar atmospheric pressure, high daytime temperatures, and the dangerous solar ultraviolet radiations that the Earth's atmosphere screens out. Normal conversation, of course, was impossible because sound waves need a medium for their transmission. The astronauts weighed only one-sixth of their weights on the Earth (p. 518).

Except in the immediate neighborhood of the sun, the sky was black and stars were visible. If the astronauts had stayed longer, they would have seen the Earth, as a moon in the lunar sky, go through phases without rising and setting. The Earth appears larger and very much brighter than the moon does to an earthly observer, and it is visible only from the side of the moon that always faces the Earth. As indicated by vibrations detected by lunar-based seismometers, sizable meteoroids occasionally strike the lunar surface, and smaller meteoroids are presumably much more numerous. However, in the absence of an atmosphere, no meteors can flash as "shooting stars" across the lunar sky. Large temperature variations (range is about 500°F—about 215°F to −280°F) occur between day and night, but weather predictions are simple: 2 weeks (as measured on the Earth) of clear and hot weather are always followed by 2 weeks that are clear and cold.

Lunar Topography

The most conspicuous topographic features on the moon are its dark areas, craters, mountains, and at times rays (Figs. 22-15 through 22–18). Lesser land forms include large cracks, rilles, grooves, fault-produced cliffs, and small wrinkles and domes. Strong shadows are present when the moon is in or near a quarter phase, and topographic features such as craters and mountains are then conspicuous. Although a full moon shows less detail, its rays do stand out strikingly (Fig. 22–11). A full moon is about nine times as bright as a half moon, partly because the shadowed portions of the half moon reflect no light to the Earth.

Large dark areas form the so-called lunar seas or *maria* (singular mare) and extend for hundreds of miles. Maria are plainly visible to the unaided eye and make the facial features of the "man in the moon." Some of these relatively smooth areas are circular, but others are irregular in outline. Maria are generally depressed relative to the rest of the surface, but their surfaces occur at different levels. Partially submerged craters, smaller craters, rays, cliffs, rilles, and small wrinkles and domes occur within them. Maria apparently are former depressions that have been flooded by flow after flow of very fluid larva. However, they were once thought to be dust-filled depressions by some scientists. Maria contrast quite distinctly with the lunar *highlands* which are higher in altitude, lighter colored, mountainous, cratered, and less smooth. The

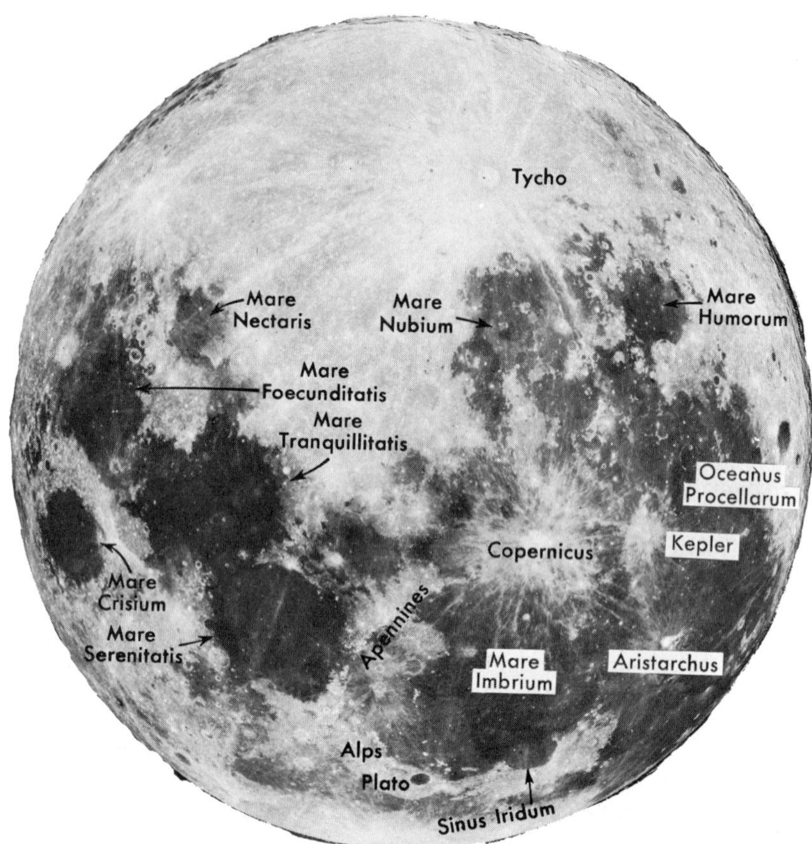

22-11. The full moon. The seas and bright rays are conspicuous, but the mountains and craters are not. Tycho appears near the top and Copernicus slightly below the center. *(Courtesy Mount Wilson and Palomar Observatories.)*

smoother, less-cratered maria are somewhat younger features since they overlap onto the slopes of the more heavily cratered highlands.

Lunar mountains generally occur in curved elongated ranges that margin the maria, but isolated peaks also exist. Some mountains project 4 to 5 miles or more above their surroundings, comparable to the higher mountains on the Earth. The mountains are jumbled masses that show little resemblance to the fold and fault block mountains known on the Earth.

Thousands of *craters* make pockmarks on the moon's surface. A typical lunar crater is circular in outline and surrounded by a ridge which has a steep, sometimes terraced, inner slope and a gentle outer slope. In most craters, the floor is lower than the adjoining terrain, but in a few it is higher. In a typical crater, the volume of material in its encircling ridge about equals the volume of the crater itself. A number of craters contain central peaks and look somewhat like Mexican sombreros. The largest approximate 150 miles in diameter (250 km—the so-called "walled plains"), and two or more craters may overlap on the lunar surface. The origin of the craters has been controversial, and formation by meteoroid (asteroid) bombardment has competed as a theory with one of origin by igneous activity.

However, an impact origin for most of the craters now seems established. Presumably many craters, especially the larger ones, formed long ago when meteoroid-asteroid debris was more abundant in the Earth-moon region of the solar system. Moreover, such features were probably once common on the Earth but have subsequently been destroyed by erosion—in contrast, on the moon they have been preserved with relatively little change for billions of years.

Lunar *rays* are light-colored streaks that range up to several miles in width and hundreds of miles in length. They radiate outward from certain craters such as Tycho and Copernicus, show no shadows, and seem to cross all types of topography without deviation. Secondary craters occur only within the rays and were first discovered in photographs taken by unmanned Ranger satellites. These have a different shape from the primary craters (shallower, smoother, elongated, with gently sloping walls). Rays and secondary craters evidently have been produced by the impacts of debris splashed out of the primary craters.

Rilles are interesting and controversial features (Fig. 22–18). They are relatively long, narrow trenches (2 miles wide, ½ mile deep, and 100 to 150 miles long might be representative) that may look somewhat like a sinuous, stream-eroded Earth valley (less common) or trend in a straight or irregular manner. They cut indiscriminately across the lunar surface and tend to be located near the margins of the maria. The straight rilles are probably grabenlike features formed by faulting, but the winding ones are more difficult to account for. Perhaps they are collapsed lava tubes (p. 122)—i.e., they formed because lava flowing in a channel crusted over, and this roof subsequently collapsed when the underlying lava drained out of the tunnel.

Faulting has taken place on the moon and dislocated the surface to form cliffs such as the Straight Wall, some 70 miles or more in length, and perhaps up to 1000 feet in height. A number of small domes and cones can be observed and probably represent small volcanoes. The surfaces of the maria are somewhat wrinkled in places as if molten material had flowed beneath them.

Lunar History

Prior to the Apollo landings, hypotheses concerning the origin of the moon fell into three main groups, although at least one worker had decided that if the moon were not already in existence he would conclude that such a body could not exist: (1) the moon may once have been part of the Earth; (2) the moon may have formed as a companion of the Earth from the nebula that probably produced the solar system (p. 309); or (3) the moon may have formed elsewhere within the solar system and subsequently was captured by the Earth. Woven into these hypotheses was the "hot" vs. "cold" moon debate. In the one view, the moon formed by accretion of cold particles or cooled off soon after its formation and ever since has remained a cold, inert mass of rock from its center to its surface. In the other view, the moon's interior is molten, at least in part, and volcanic activity and lava flows have occurred again and again.

The nature and ages of the Apollo samples of lunar rocks and rock debris hold important clues concerning the origin of the moon and its subsequent history. Rock specimens from the Sea of Tranquillity (Apollo 11) are either basaltic igneous rocks or breccias made primarily of fragments of these basaltic rocks, and the lunar soil consists chiefly of tiny particles of such rocks and glassy material with a similar chemical composition —thus basaltic rocks predominate by far. The most common minerals in the rocks are pyroxene, plagioclase feldspar, ilmenite, and olivine (Figs. 22–12 and 22–13). Textures are typically igneous, grain sizes range from fine to medium, and vesicular and equigranular specimens are included. Thus many specimens resemble those that might be picked up from basaltic lava flows on the Earth. However, there are key differences in

22-12. Photomicrograph of Apollo 11 sample showing major minerals: plagioclase feldspar (white), ilmenite (black), and pyroxene. *(NASA, S-69-47903.)*

content such as the glassy spherules (Fig. 22–3) and meteorite fragments (Fig. 22–14), in the overall chemical composition, and in the widespread presence of impact metamorphism. Moreover, the Tranquillity rocks formed in the almost complete absence of molecular oxygen and water and have not subsequently been acted upon by these materials. Relative to comparable terrestrial rocks and to meteorites, the lunar rocks have a high proportion of elements such as titanium and zirconium but a low proportion of sodium and potassium.

Radioactive age determinations have shown that the oldest material from the Tranquillity site is about 4.6 billion years old which corresponds strikingly with the 4.6 billion year age of the oldest meteorites found on Earth and with the commonly accepted age of the Earth itself. This is an exciting discovery since the oldest rocks yet found on Earth are about 3.5 billion years

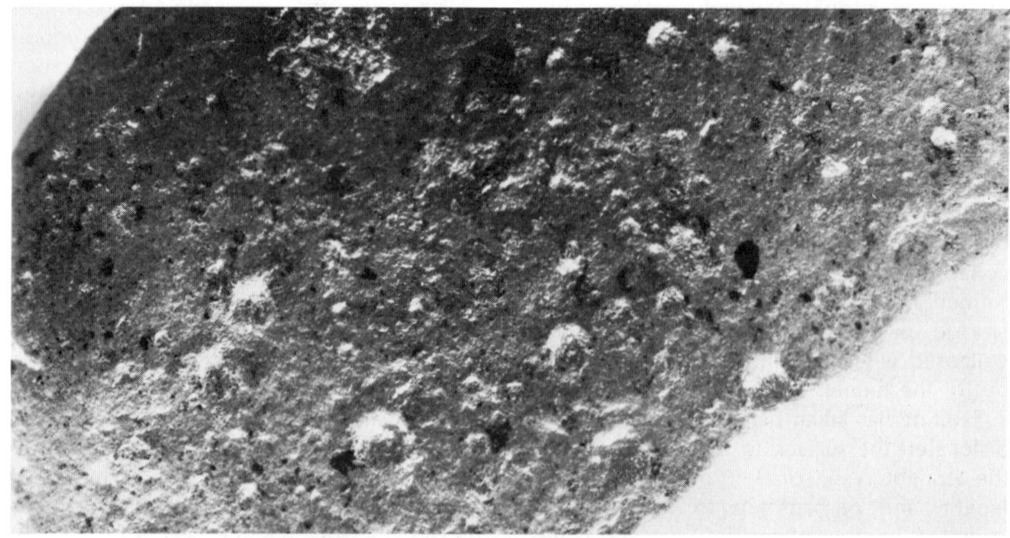

22-13. Breccia (Apollo 11 sample). Note glass-lined pits and glass encrustations. *(NASA: S-69-47905.)*

of age, which leaves a 1-billion year gap in the history of the Earth—perhaps lunar history can fill in some of the details of this gap. Other Apollo 11 specimens have an age of about 3.7 billion years, and some Apollo 12 samples from the Ocean of Storms appear to be about 3.3 to 3.4 billion years old. They are also lower in titanium and richer in iron and nickel than the Apollo 11 samples.

Wood* has interpreted a surprising amount of lunar history from a spoonful of the Apollo 11 rocks, and we shall follow his "sermon in stones." According to Wood, the lunar surface (Figs. 22–15 and 22–16) closely resembles a shell-pocked World War I battlefield and is a rolling, hummocky, impact-riddled terrain lacking the jagged rocky ridges once thought to be common. Loose debris produced by meteoritic bombardment, the *lunar regolith,* covers the surface nearly everywhere and averages about 10 to 20 feet (3 to 6 meters) or more in thickness in the maria but is perhaps two to three times thicker in the lunar highlands where the cratering has continued for a longer time. Craters 100 to 180 feet in diameter (30 to 55 meters) seem to have bedrock floors that coincide more or less with the bedrock-regolith boundary because they are shaped differently from smaller craters that do not penetrate downward into the bedrock—i.e., they have shallow, flat or humped floors and steep rims.

When a relatively small meteroid hits the lunar surface (Figs. 22–17 and 22–18) three changes occur. Material at the point of impact is melted, and droplets and irregular blobs are tossed outward to solidify in flight as tiny glass spheres or to spatter and harden into glassy coatings on fragments at the lunar surface. Such glassy material is abundant in the soil samples brought back from the moon. Beyond this relatively small melted zone, the pressure of the impact fuses the loose soil into a consolidated mass—the lunar breccia. Still farther from the impact point, the regolith debris is smashed, shattered, and scattered outward to form a new crater. As meteoroid after meteoroid collides with the moon, first here and then there, the average size of the regolith fragments diminishes, and the regolith itself becomes thicker, although the rate slows with time as fewer craters penetrate into the bedrock. Regolith debris is

* Wood, John A., "The Lunar Soil," *Scientific American,* August 1970.

22-14. Mound on lunar surface; Apollo 12, November 1969 photograph. *(NASA: AS12-46-6827.)*

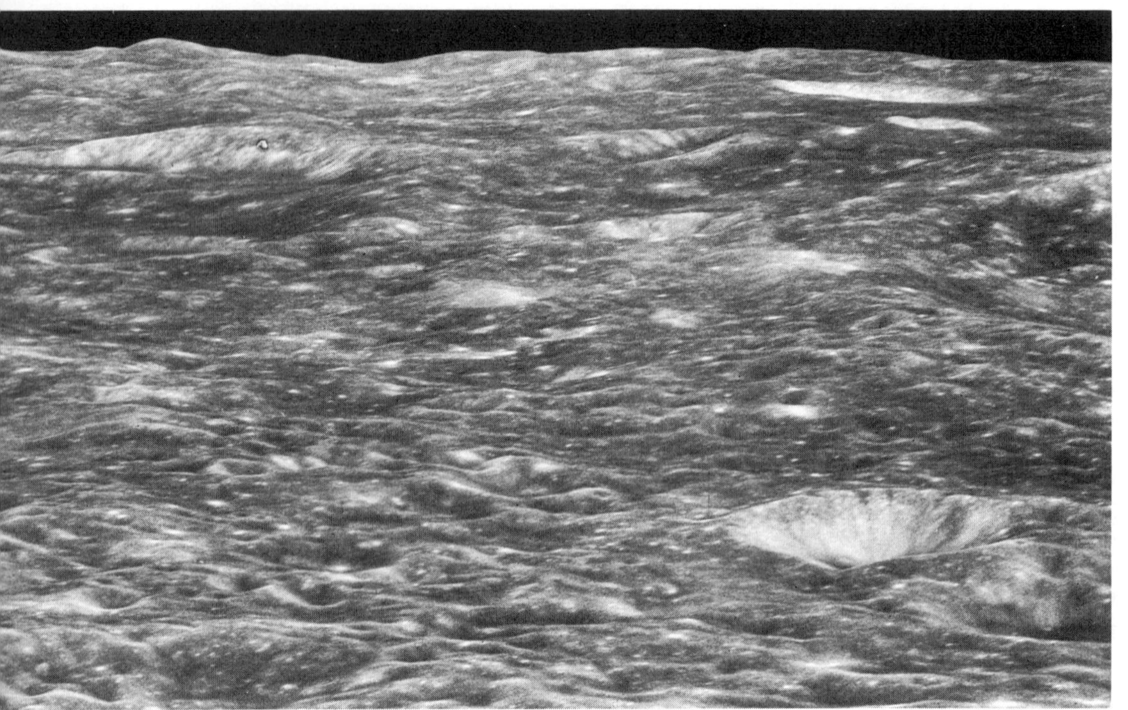

22-15. Far side of the moon as photographed from Apollo 8 in December 1968. Note rolling, hummocky, impact-riddled, light-colored highlands stark against the black sky. *(NASA: AS8-14-2453.)*

thus mixed rather well by the cratering process, although most of the debris at any one spot probably is similar in composition to the bedrock beneath it. However, some debris ejected explosively at a crater travels for many miles, and thus a small proportion of highland material can be expected in maria regolith and vice versa. Meteoritic fragments also occur within the regolith.

Although most particles in the Apollo 11 regolith samples are quite small, about 25% exceed ¼ mm in diameter and thus are actually tiny rocks and record aspects of lunar history. Most particles are similar to terrestrial basalts, but a few resemble terrestrial anorthosites (formed chiefly of the plagioclase variety of feldspar). The abundant basaltic fragments are dark colored because of their high ilmenite content, and Wood suggests that they come from the maria. Therefore, the less common, light-colored anorthosites probably are fragments of the lunar highlands. Glasses and breccias of anorthositic composition occur along with the fragments of anorthositic igneous rocks, although glasses and breccias of basaltic composition are more common, which is to be expected in maria samples. Measurements from an unmanned Surveyor spacecraft that had landed earlier in the lunar highlands also suggest an anorthositic composition. Therefore, if the maria are chiefly basaltic, whereas the highlands are chiefly anorthositic material, this may account for their relative elevations. The lunar basalts have a specific gravity of about 3.3 whereas that of anorthosite is about 2.9.

So-called *"mascons"* (mass concentrations) have been detected within the maria but do not occur within the highlands. The mascons were discovered by noting slight deviations in the orbits of spacecraft in orbit about the moon—such spacecraft are pulled closer to its surface wherever the mass of underlying rock is greater than average. This happens over the central portions of some of the maria, although the cause of the excess mass is uncertain (perhaps the buried remains of an asteroid whose collision with the moon produced the mare depression). Therefore, if the highlands consisted of the same material as the maria, their higher elevations should cause an extra gravitational tug on the spacecraft, but such deviations have not been observed. This suggests that the highlands are made of less dense rock which offsets their larger volume. Since the average highland surface occurs nearly a mile (1.4 km) above the mean level of the maria, this shell of less dense rock needs to be about 6 miles (10 km) thick in order to "float" at its present level above a zone of heavier rocks (see isostasy, p. 165).

Magmatic differentiation (p. 132) is a process which occurs if a magma cools slowly so that certain types of minerals crystallize before others. If the first-formed crystals are heavier than the magma, they sink; if less dense, they rise to the top of the

22-16. December 1968, Apollo 8 photograph showing the far side of the moon, highland terrain, and the Crater Tsiolkovsky. *(NASA: AS8-12-2196.)*

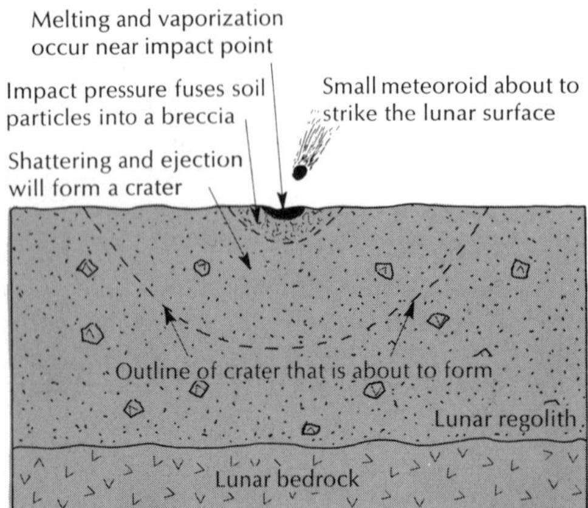

22-17. Formation of small lunar impact crater with effects confined to the regolith. Breccia forms by fusion of lunar soil but is immediately shattered and excavated by the impact. *(After John A. Wood,* Scientific American, *Aug. 1970.)*

magma body. One can speculate, therefore, that all or part of the outer moon was once molten material some 4½ billion years ago. As this magma subsequently cooled, plagioclase crystals formed here and there within it and then floated to the top to form a lunar crust. The widespread highlands—interspersed by maria on the side facing the Earth—may represent this crust. If the chemical composition of the moon is similar to that of chondritic meteorites, then plagioclase-forming ingredients may not be abundant in lunar material. Therefore, a source layer more than 100 miles thick (200 km) might be required to produce enough plagioclase crystals to float upward and form a crust 6 miles thick. Heat energy released during accretion (p. 00) or during radioactive disintegration might have caused the molten condition.

According to Dr. Wood, preliminary study of Apollo 12 and Apollo 14 specimens has revealed a new type of lunar igneous rock (norite or KREEP) and glass that is probably both ancient and abundant. It consists of approximately equal amounts of plagioclase and pyroxene (similar to pyroxenes in highland anorthosites) and may have crystallized from the magma that remained after the anorthosite had formed. This new rock may be part of an ancient lunar crust and form a layer that everywhere underlies the anorthosite. Specimens may have reached the Apollo 14 Fra Mauro site in the highlands because they were thrown there during the huge cratering event that produced the Mare Imbrium basin and thus excavated rocks that ordinarily would be buried some miles beneath the surface—in a similar fashion, the Apollo 12 site is crossed by a ray from the crater Copernicus.

Although the outer part of the moon may once have been molten, its whole interior now seems to be solid. Evidence: the moon is bulged more than its present slow rate of rotation can account for, especially on the side facing the Earth. Presumably the bulge formed when the moon rotated faster and has been preserved because the moon's interior is rigid material not hot enough to flow. Mascons furnish similar evidence of this rigidity.

The significant differences in chemical composition between the Earth and moon suggest that the moon is no "Adam's rib," and that it may not have formed in the vicinity of the Earth. However, if the moon did originate elsewhere within the solar system, then mechanical problems are posed by its subsequent capture by the Earth into a nearly circular orbit near the plane of the ecliptic. Therefore, it has been suggested that a number of small satellites once orbited the Earth and that the moon collided with one or more of these when it approached the Earth. Such collisions might have slowed the moon for capture by the Earth, increased its total mass, made its orbit gradually more circular, and might account for chemical differences from one part of the moon to another. Possibly the maria originated over an extended period of time by the collision of asteroid-sized objects with the moon to produce large depressions and huge quantities of molten material. Thus igneous rocks at the moon's surface have formed by the cooling of molten material, but we still do not know whether this

22-18. Craters and rills are prominent in this Apollo 8, December 1968 photograph of the lunar surface. *(NASA: AS8-13-2225.)*

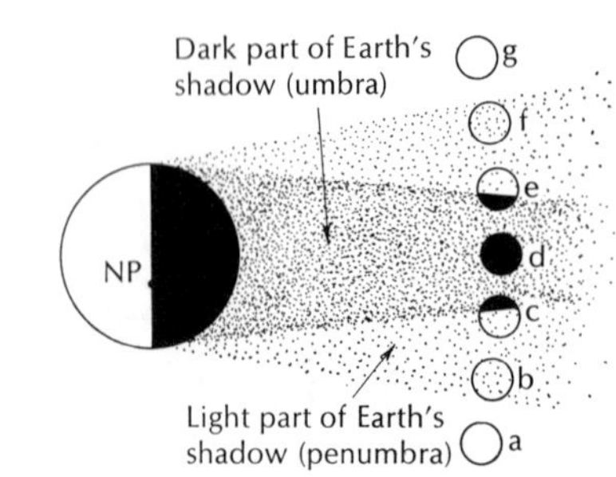

22-19. Top view of a lunar eclipse. In *a* and *g* the moon is shown full, before and after moving through the Earth's shadow. In b and f the moon is in the zone from which some of the sun's light is excluded, but the change in the moon's appearance is barely noticeable. In c and e the moon is partially eclipsed, and in d the eclipse is total. What is the season in the Northern Hemisphere?

molten material was formed by volcanic activity similar to that which occurs on the Earth, by meteoroid-asteroid bombardment, by heat energy released during accretion, or by some combination of these or other processes. Many questions remain unanswered.

Individual rock specimens picked up from the moon tend to have rounded upper surfaces and flat or irregular lower surfaces.

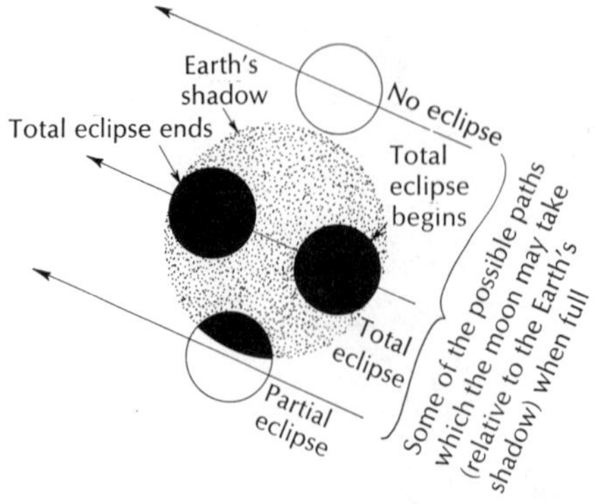

22-20. Side view of a lunar eclipse showing the Earth's shadow and moon in cross section. The stippled circle represents the Earth's shadow (umbra) at the moon's distance from the Earth.

The upper surfaces also have a sandblasted appearance suggesting that particle impacts cause some erosion on the moon. Moreover, inert gases found within particles of lunar soil and breccia show that the moon has long been bombarded by the solar wind (p. 600).

Lunar Eclipses

Rays of light from the sun are not quite parallel. Therefore, spherical bodies in space, such as the Earth and moon, cast long, tapering, conelike shadows which always point directly away from the sun. The shadows are invisible unless some object moves through them. The moon is eclipsed whenever the Earth comes directly between it and the sun (Figs. 20–2 and 22–19). The moon is always full at the time of a lunar eclipse, and the eclipse is visible to everyone on the nighttime side of the Earth, unless clouds hide it. For this reason, lunar eclipses are more familiar to the average person than are solar eclipses, even though the latter are actually more frequent (about four out of seven).

The diameter of the Earth's shadow at a distance of 239,000 miles is about 5700 miles (9177 km), approximately 2.6 times the moon's diameter; thus total lunar eclipses can occur whenever the moon passes centrally through this shadow (Fig. 22–20). However, the moon's orbital plane is inclined approximately 5 degrees to the Earth's orbital plane, and thus an eclipse does not occur each time the moon is full (Fig. 22–21).

The longest lunar eclipses extend for nearly 4 hours (this does not consider the penumbra): a partial eclipse phase of about 1 hour precedes and follows the totality portion, which may last for 1⅔ hours. As the moon moves eastward more rapidly than the Earth's shadow, the moon's eastern side is always darkened first. Even when totally eclipsed, the moon is usually visible as a reddish disk, because the long red wavelengths in sunlight can penetrate an atmosphere more readily. Some red light passes completely through the Earth's atmosphere and is refracted inward so that it reaches the moon, even within the umbra. Reflection of some of this red light back to the Earth gives the moon its ruddy appearance. However, the moon may be blackish and nearly or completely disappear during an occasional eclipse when dust (chiefly volcanic?) and other materials in the atmosphere interfere more than usual with the passage of sunlight.

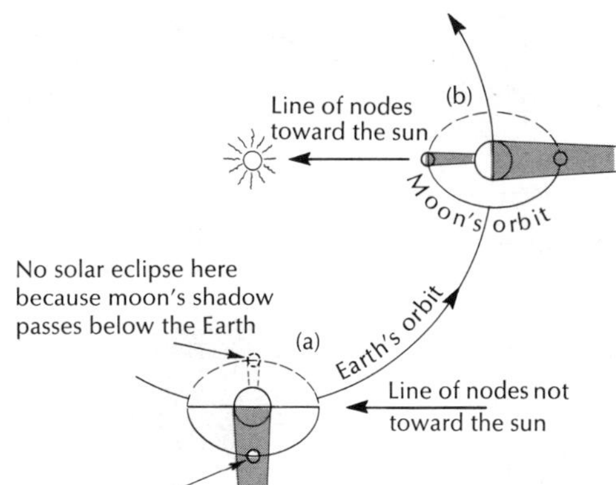

22-21. Why eclipses are infrequent. The moon's orbital plane is inclined about 5 degrees to the ecliptic and intersects it at two opposite points called the nodes (similar to the equinoxes). The dashed half of the moon's orbit is below the ecliptic. Eclipses can only occur when the moon is new or full and when the line of nodes points toward the sun as it does at (b). *(R. H. Baker,* Astronomy, *Van Nostrand Reinhold, 1964.)*

Solar Eclipses

A total solar eclipse is a magnificent celestial phenomenon and a rare occurrence in the lives of most of us. Solar eclipses occur when the new moon comes directly between the Earth and the sun (Figs. 22–22 to 22–24). However, a solar eclipse does not take place at each new moon for the same reason that a lunar eclipse does not occur at each full moon (Fig. 22–21)—the sun, moon, and Earth are not in line in space. During the new moon phase, the tip of the moon's shadow may sweep above or below the Earth.

The moon's shadow has an average length of about 232,000 miles (373,500 km) which is about 7000 miles (11,270 km) shorter than the mean distance between the moon and the Earth. However, along a portion of its elliptical orbit the moon is closer than 232,000 miles to the Earth, and thus the tip of its shadow touches the Earth. Its largest shadow is about 167 miles (269 km) in diameter, but usually the shadow is much smaller.

The shadow moves eastward more rapidly than the Earth rotates eastward. Therefore, during a total solar eclipse, the shadow overtakes the Earth and passes across it along a path extending from west to east. The rate at which the shadow crosses the Earth's surface varies from a minimum of about 1000 miles (1600 km) per hour at the equator to a maximum of about 5000 miles (8000 km) per hour at the poles. The longest total solar eclipse lasts about 7 minutes. This is the time needed by the rapidly moving circular shadow, 167 miles or less in diameter, to pass a particular spot on the Earth.

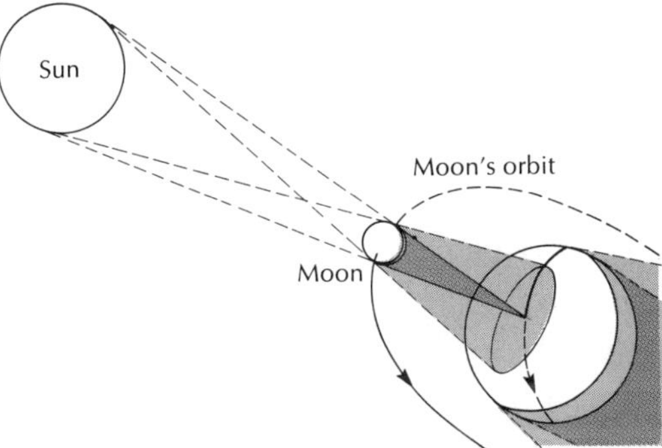

22-22. Conditions for a total solar eclipse. The moon's shadow moves in an eastward direction across the Earth's surface and the eclipse is total for observers located within the path of the small dark shadow (umbra) and partial for observers within the lightly shaded area (penumbra). *(R. H. Baker and L. W. Fredrick,* An Introduction to Astronomy, *Van Nostrand, 1968.)*

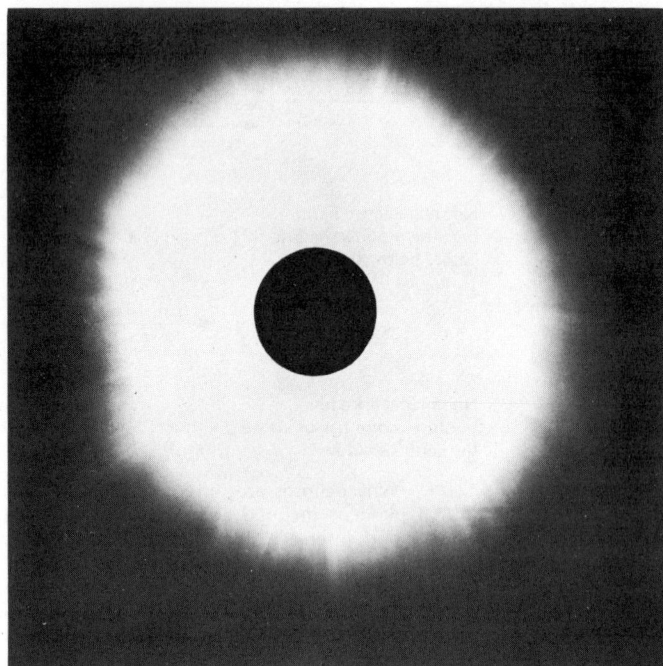

22-23. The total solar eclipse of 20 May 1947. The corona tends to have a circular outline during a sunspot maximum. See Fig. 24-2 for a view of the corona during a sunspot minimum. *(Courtesy Yerkes Observatory.)*

A total solar eclipse (when the observer is entirely within the umbra) is preceded and followed by partial phases (when the observer is within the penumbra) that each last for about 1 hour. Partial eclipses also occur without an accompanying total phase; i.e., only the penumbra touches the Earth. Partial eclipses occur more frequently than total eclipses and can be seen from a larger area.

The line of nodes (Fig. 22–21) points at the sun at intervals of about 6 months, and thus it can point toward the sun three times in one calendar year if the first time is early in January. Since the time interval during which a solar eclipse may occur tends to exceed 1 month—i.e., alignment of Earth, moon, and sun is close enough for an eclipse throughout this interval— two partial eclipses are possible each time the line of nodes points toward the sun, and at least one must occur (either partial or total). Therefore, at least two solar eclipses must occur each year, about 6 months apart, and five may occur. If two partial solar eclipses occur about 1 month apart, a total eclipse of the moon will take place midway between the two. However, any one calendar year may also go by without a lunar eclipse of any kind (Table 22–1). This happens because the time interval during which a lunar eclipse can occur is less than a month. Thus the full moon phase may not take place during this interval when the line of nodes is pointing nearly toward the sun.

The moon may come directly between the sun and Earth but at too great a distance for the tip of its shadow to graze the Earth's surface. At such times a total solar eclipse cannot take place. However, a person situated on the Earth in the area formed by the projection of this shadow sees an *annular eclipse* (Fig. 22–22): the dark body of the moon does not completely cover the sun, and a thin outer ring remains visible. To illustrate this, close one eye and hold a coin between the other eye and something large and round, like a clock. If the coin is close enough to your eye it will completely cover the clock (total eclipse);

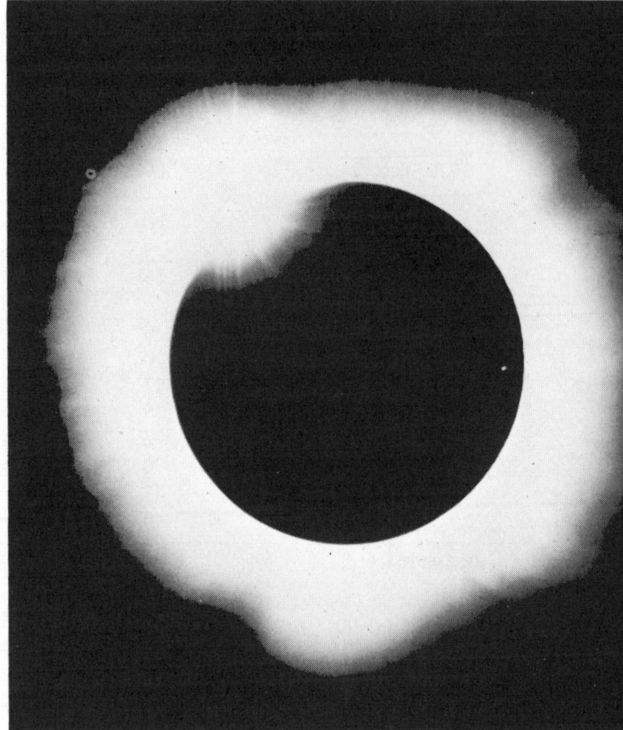

B

A

22-24. Solar eclipse of 7 March 1970. (A) The eclipse was only partial as viewed from Rifton, New York. (B) It was total and a magnificent spectacle as seen in a clear sky from Virginia Beach as totality ended, and the diamond ring effect occurred. *(A, courtesy David Moody; B, courtesy Crow's Nest Observatory and Raphael Warshaw.)*

however, if the coin is then moved farther from your eye, an outer ring of the clock will be visible (annular eclipse).

The changes which accompany a total solar eclipse begin with a partial eclipse that lasts for about 1 hour; the moon's dark body causes a circular indentation on the west side of the sun. This indentation grows

TABLE 22-1*

UMBRAL LUNAR ECLIPSES VISIBLE IN THE UNITED STATES AND CANADA			TOTAL SOLAR ECLIPSES	
DATE	DURATION OF TOTALITY		DATE	REGION
1970, Feb. 21	0 hr	0 min	1970, Mar. 7	Mexico, Florida, Carolinas, Newfoundland
1970, Aug. 17	0	0		
1971, Feb. 10	1	23		
1972, Jan. 30	0	36	1972, July 10	N.E. Asia, Alaska, Canada, Nova Scotia
1972, July 26	0	0		
1973, Dec. 10-11	0	0	1973, June 30	N.E. So. Am., Africa
1974, Nov. 29	1	17	1974, June 20	Australia
1975, May 25	1	29		
1975, Nov. 18-19	0	42	1976, Oct. 23	Central So. Am., Hawaii
1977, Apr. 4	0	0	1977, Oct. 12	Africa, Australia
1978, Mar. 24	1	31		
1979, Sept. 6	0	46	1979, Feb.26	Canada, Greenland

* Data courtesy of R. Duncombe, U.S. Naval Observatory.

larger and larger until only a thin crescent of the sun remains visible. The light reaching the Earth is now of a different quality because it comes entirely from the outer parts of the sun. Just before the sun is completely hidden, the continuous crescentic sliver that remains is subdivided because the sun's light passes only through depressions along the moon's silhouette (Bailey's beads). With increasing darkness, temperatures drop, dew may form, bright stars and planets become visible, and animals may act as if night were approaching. After totality, these phenomena are repeated in the reverse order. The appearance of the totally eclipsed sun—its chromosphere, prominences, and corona—is discussed in Chapter 24.

Tides

Tides in the Earth's oceans are caused by the gravitational attractions of the moon and the sun, but the effect of the moon is more than double that of the sun (Fig. 22–25). It was realized long ago that the motions of the moon and the magnitude of the tides are directly related. (1) The largest tides occur at the new and full phases, whereas the smallest tides occur during the quarter phases. (2) The time between two high tides or between two low tides in many coastal areas averages 12 hours and 25 minutes. (see Fig. 22–26). In other words, a point on the Earth directly under the moon at one time is directly opposite it on the far side of the Earth 12 hours and 25 minutes later. Thus we experience successive high tides as the Earth rotates us through the nearly stationary tidal bulges (they shift eastward 13 degrees per day).

Though two high tides and two low tides are common every 24 hours and 50 minutes in many parts of the Earth, the intervals from high to low and from low to high may be quite unequal, and the heights of successive high tides and low tides may show wide variation.

Tides several inches in magnitude are also produced in the Earth's solid body, but they can be detected only with very sensitive instruments. The surface of the ocean may rise and fall a few feet or less in mid-ocean, and here a typical tide is a wave with a period of 12 hours and 25 minutes and a length that extends halfway around the Earth. However, the funneling effect of bays and other factors cause a much greater change in sea level along certain coasts, and changes as much as 70 feet have been reported.

The following explanation of the tides is incomplete and simplified. The Earth's rigid body acts as a unit in responding to the gravitational pull of the moon—as if all of the Earth's mass were concentrated at its center. However, water at the Earth's surface is free to move, and its response to the pull of the moon varies with the square of its distance. Thus the moon pulls water on the near side more strongly than it pulls the solid Earth. Similarly, the moon's pull on the water on the opposite side of the Earth is less strong than the pull on the rigid Earth. Therefore, with respect to the rigid Earth, the water on the side facing the moon tends to flow toward the moon, and the water on the other side flows away from the moon. Thus the moon simultaneously causes high tides on both the near and far sides of the Earth (Figs. 22–25 and 22–26). As the Earth rotates, any point on the Earth's surface is carried from high to low tide.

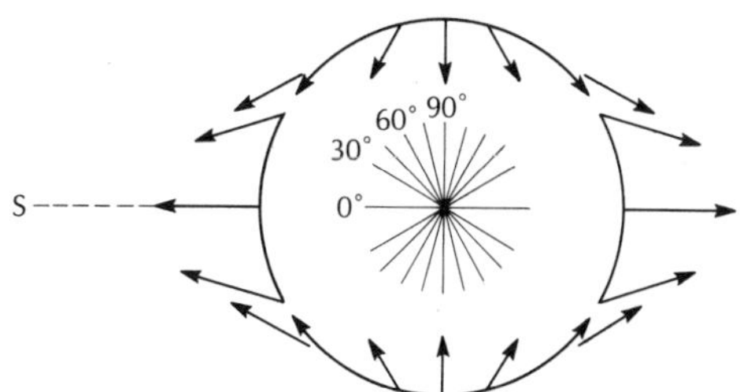

22-25. Schematic view of tide-producing force exerted by one celestial body (S) on another (e.g., by the moon or sun on the Earth). Note that this force is about parallel with the surface in places and has a horizontal component that parallels the surface in other places. Therefore, although the tide-producing force itself is very small compared with the Earth's total gravitational pull, it can cause water to move laterally. *(R. W. Fairbridge, ed.,* The Encyclopedia of Oceanography, *Van Nostrand Reinhold, 1966.)*

The tidal bulges do not occur exactly along the line extending from the center of the moon through the center of the Earth but somewhat ahead of this line in the counterclockwise direction. If friction did

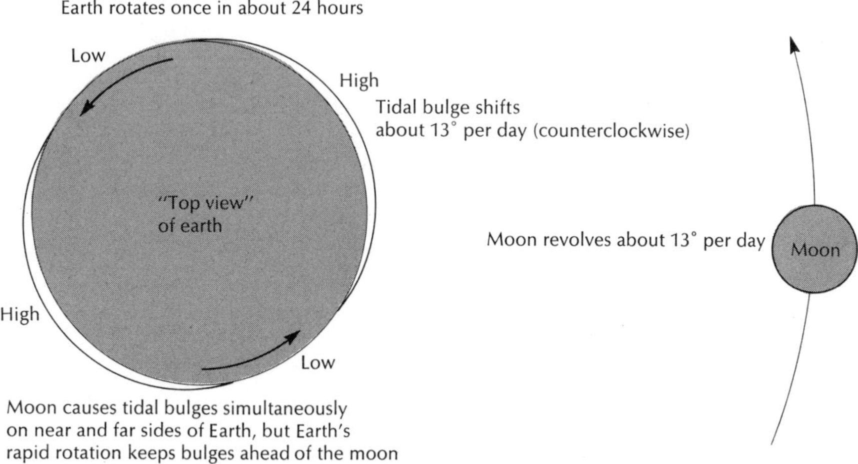

22-26. Schematic, simplified view of tides (height greatly exaggerated). The moon's eastward (counterclockwise) revolution about the Earth shifts the tidal bulges eastward about 13 degrees per day. Thus the time for an observer to be rotated by the Earth from the high tide on the near side to the high tide on the far side is about 12 hr and 25 min. If the moon did not revolve about the Earth, this time would be reduced to about 12 hr.

not occur between the Earth's surface and the water in the oceans, the tidal bulges would remain aligned with the moon. On the other hand, if the moon is ignored and friction is assumed to be infinite, the tidal bulges would shift eastward at the same rate as the Earth rotates. In actuality, the tidal bulges reach equilibrium between these opposing forces, and an observer on the Earth passes "beneath" the moon before he passes through the high tide zone.

Although high tides caused by the moon on the near and far sides of the Earth tend to be equal, successive high tides at a particular locality may not be of the same magnitude. One factor is the variation in the position of the moon north (above) and south of the plane of the Earth's equator (produced by the Earth's axial tilt and the inclination of the moon's orbital plane). Thus the equal tidal bulges on opposite sides of the Earth tend to occur at different latitudes because they are not aligned parallel to the equator. However, when the moon is located in the plane of the equator, then successive high tides tend to be more nearly equal.

The total gravitational attraction between the sun and the Earth exceeds that between the moon and the Earth by about 175 times. However, tides are caused by a difference in gravitational attraction on the near and far sides of a body, not by the total pull on it. Quantitatively, the tide-producing force varies inversely as the cube of the distance. Thus the relatively short distance separating the Earth and the moon makes its tide-raising force more than double that of the distant sun.

During times of new and full moon (actually 1 to 2 days later because the tidal bulges are not aligned with the moon), the tidal effects of the moon and the sun reinforce each other and cause exceptionally great tides (the *spring tides*). During the quarter phases of the moon, the tidal effect of the sun opposes that of the moon, and so tides are weaker (the *neap tides*).

The size of the tides is also affected by the moon's varying distance from the Earth —tides are greatest when the distance is least. Tides are also affected by the size, shape, and extent of a body of water and the manner in which the water in it oscillates—the soup in a bowl carried by an unsteady hand shifts back and forth in a somewhat similar manner. The greatest tides occur when the moon is closest to the Earth during its new or full moon phases *(perigee tides)*.

Tides and the Earth's Rotation

Tidal friction between water and ocean floors apparently is causing the Earth to rotate more slowly with the passage of time, and the day may now be approximately 1.6 (or 1.8) seconds longer than it was 100,000 years ago. At first glance this retardation seems to be too small to be measured. However, a year now would be 365 × 1.6 seconds longer than a year 100,000 years ago, and each of the 100,000 years would have been slightly longer than the preceding year. If a very precise clock had been started 2000 years ago (based upon the Earth's rate of rotation then), it would be ahead of present-day clocks (based upon the Earth's present rate of rotation) by about 3 hours.

This change is sufficiently great to be checked by the records of ancient eclipses. Using the present length of the day, modern astronomers calculated the times and places at which certain ancient total solar eclipses should have occurred. They found that the times and places did not check; e.g., a total solar eclipse which took place 2000 years ago would have occurred about 3 hours earlier and 45 degrees west of the predicted time and place (in 3 hours the Earth rotates ⅛ of the angular distance around its axis). Such differences are apparent from ancient records. The times and places of ancient eclipses do check, however, if the gradual lengthening of the day is considered.

Fossil corals* seem to provide additional evidence for a gradual increase in the length of the day. Some present-day corals, living

* Runcorn, S. K., "Corals as Paleontological Clocks," *Scientific American*, October 1966.

in oceans with seasonal temperature changes, show annual growth rings, and some of these show daily growth ridges superimposed on the annual rings. These can be counted with the aid of a good magnifying glass and average 365 per year, although some uncertainty is involved in the counting, and the range in numbers is about from 360 to 370.

Similar annual and daily growths apparently have been recognized in Devonian corals that lived nearly 400 million years ago, but the number of daily growths is about 400 (or 410) per year. Since the Earth's period of revolution probably has not changed, this may indicate that the days were shorter during the Devonian (about 400 per year) and have since increased in length. Still older corals show more than 400 daily growths per year, whereas younger corals show fewer, and the data from corals of different ages seem to be compatible with a gradual increase in the length of the day (not uniform because of presumed changes in moon's distance and in area occupied by shallow seas).

The gradual lengthening of the day has important implications concerning geologic history (p. 313) and the origin of the moon (p. 553).

Exercises and Questions for Chapter 22

1. A typical student at a New York State College, diligently studying his favorite science course at 2:00 A.M., looked out of the dormitory window and noted the beauty of a campus illuminated by light from the moon in the eastern sky. What approximate phase was the moon in at this time?
2. One evening at midnight you notice a bright starlike object in the south. Which of the following might the object be: Mercury, Venus, Mars, Jupiter, Uranus, Neptune, Polaris, Rigel (bluish-white star in Orion), and Sirius (brightest star in Canis Major)?
3. What changes, if any, would take place in the lengths of the sidereal and synodic months if the moon revolved from east to west about the Earth (assume no other changes from existing conditions)?
4. Assume that a moon has a mean distance of 239,000 miles from a planet. How would the mass of this moon compare to that of the planet if their center of mass were located:
   (A) 2900 miles from the center of the planet?
   (B) 23,900 miles from the center of the planet?
   (C) 119,500 miles from the center of the planet?
5. Assume that an observer is located in the central portion of that part of the moon that is turned toward the Earth. At this observation site:
   (A) What temperature changes would occur during a month?
   (B) How frequently would the sun rise and set? the Earth?
   (C) Would constellations have the same shape as when viewed from the Earth?
   (D) Would stars appear to move eastward, westward, or remain motionless?
   (E) Would the observer have a sidereal day and a solar day as on the Earth? Explain.
6. If the sun were about twice as far away from the Earth and moon as it is at present, how would this generally affect the number and length

of total solar eclipses? Assume the same masses for the sun, Earth, and moon as at present and the same distance between the Earth and moon.

7. Calculate the length of the Earth's shadow. Draw circles for the sun and Earth. Next, draw a line from the center of the sun through the center of the Earth and extend the line beyond the Earth. Draw a second line that is tangential to both sun and Earth. This intersects the first line at the tip of the Earth's shadow. You now have two similar right triangles—radii of the Earth and sun form the bases of the triangles—and can solve for the length of the shadow because the radii of sun and Earth and the astronomical unit are known.
8. Why do solar and lunar eclipses tend to occur at intervals of either a few weeks or about six months?
   (A) How could five solar eclipses occur in the same calendar year?
9. Compare total lunar and solar eclipses relative to each of the following:
   (A) Phase of moon at time eclipse occurs.
   (B) Maximum length of totality.
   (C) Approximate size of area from which eclipse is visible.
   (D) Appearance during totality.
10. Make one "top-view" sketch that shows all of the following:
    (A) Sun, moon, and Earth during the first quarter phase.
    (B) Mercury as a morning star in the full moon phase.
    (C) Mars located due south of an observer on the Earth at midnight.
    (D) Venus as an evening star in such a position that it sets earlier on successive evenings.
11. Plot on a star-map sketch the eastward shift of the moon against the background of stars as its phase changes from a crescent to a full moon. Measure the moon's sidereal period by noting its position relative to some reference stars and again nearly 1 month later as it completes its 360° eastward shift.
12. Explain each of the following concerning tides:
    (A) Why do 12 hours and 25 minutes commonly elapse between two high tides?
    (B) Why may successive high tides be of different magnitude at a certain point?
    (C) Why can predictions be made years in advance that particularly high tides will occur on certain dates whereas other high tides cannot be predicted even a few days in advance?
    (D) Why does a day on the Earth appear to be increasing in length?
    (E) Why is the moon more effective than the sun in causing tides on the Earth?
13. In what ways would a hike on the moon's surface probably differ from one on the Earth's surface?
14. An interesting, informative astronomy laboratory exercise was described in *Sky and Telescope* in April 1964 (reprints may be on sale). This involves the determination of the shape of the moon's elliptical orbit by measurements of the angular diameter of the moon on a series of lunar photographs that come with the exercise. The eccentricity of the moon's orbit can be calculated from the data.
15. Describe the "true" path of the moon in space relative to the Earth and sun. Explain why the moon's orbital speed varies.

23

The Solar Family

The solar family includes the nine major planets, their thirty-two known satellites, the asteroids, comets, meteoroids, and countless dustlike particles (see Fig. 23–1 for some members of the solar family). The sun's gravitational pull shapes and controls the elliptical orbits of each of these diverse objects and holds them together so that they move as a unit through space. As viewed from above the Earth's north pole, eastward (counterclockwise) motion is by far the most common within the system and is called *direct* motion. The opposite, or westward movement, is termed *retrograde*. All planets revolve eastward about the sun as do most of the satellites about their primaries. Counterclockwise rotation is also common: by the sun, Earth, moon, all planets except Uranus and Venus, and most of the satellites. However, the entire solar system itself—isolated by great distances from all other celestial objects—moves clockwise about the galactic center. Table 23–1 lists physical data concerning the planets.

The orbital planes of the planets all pass through the sun (at a focal point of each ellipse) and lie close to the plane of the ecliptic. Seven orbital planes are inclined 3 degrees or less to the ecliptic, but those of Mercury and Pluto are inclined at greater angles. The orbits are nearly circular, but again Mercury and Pluto deviate most. Mean orbital speeds decrease with distance from the sun; they range from nearly 30 mi/sec for Mercury to nearly 3 mi/sec for Pluto.

The distances of the planets from the sun show an interesting numerical relationship known as Bode's Law,* and when Uranus was discovered in 1781, it fitted nicely into the scheme. This success then led to a search for a planet that the law indicated should be located between Mars and Jupiter, and eventually the asteroids were discovered. Neptune and Pluto,

* (1) List the planets in order of increasing distance from the sun. (2) Write 0 beneath Mercury, 3 beneath Venus, and then double the number for each succeeding planet. (3) Write the number 4 beneath each planet. (4) Add these two numbers for each planet. (5) Divide by 10 to obtain the distance in astronomical units.

23-1. A display of the Leonid meteors photographed from Kitt Peak in Arizona in November 1966. "Pouring out of the Big Dipper were 43 Leonids in 43 seconds," wrote the observer. *(Courtesy* Sky and Telescope *and David McLean.)*

however, are far from the locations predicted by Bode's scheme (Pluto is about where Neptune is supposed to be).

Based upon distance from the sun, we have two groups: the two *inferior* planets (closer to the sun than the Earth) and the six *superior* planets. Their motions as observed from the Earth are quite different (Fig. 23–2), although planets are always located near the ecliptic.

The orbital position of an inferior or superior planet may be related to the angle between two lines from the Earth—one to the sun and one to the planet (these lines, of course, move but we consider the motion of a planet relative to them). When the angle is 0 degrees, a planet is said to be in *conjunction* (superior or inferior). When the angle is 90 degrees, a planet (superior only) is at *quadrature*; when this angle is 180 degrees, the planet is at *opposition*. When the angle is acute (between conjunction and quadrature) a planet is said to have a certain *elongation*—say 30 degrees—

TABLE 23-1

PLANETS	MEAN DISTANCE FROM SUN: MILLIONS OF MILES	MILLIONS OF KM	A.U.	PERIOD OF REVOLUTION: SIDEREAL	SYNODIC (days)	MEAN ORBITAL SPEED: MI/SEC	KM/SEC
Mercury	36	58	0.39	88 days	116	29.5	47.8
Venus	67	108	0.72	225 days	584	21.8	35.1
Earth	93	149.5	1.00	365¼ days	—	18.5	29.8
Mars	142	229	1.52	687 days	780	14.9	24.2
Jupiter	483	778	5.20	12 years	399	8.1	13.1
Saturn	886	1426	9.55	29.5 years	378	6.0	9.7
Uranus	1780	2870	19.2	84 years	369.7	4.2	6.8
Neptune	2790	4494	30.0	165 years	367.5	3.3	5.4
Pluto	3670	5900	39.5	248 years	366.7	2.9	4.7

TABLE 23-1 (Continued)

PLANETS	ECCENTRICITY OF ELLIPTICAL ORBIT	INCLINATION OF ORBITAL PLANE TO ECLIPTIC	MEAN DIAMETER: MILES	KM	RELATIVE MASS (Earth is 1)	DENSITY (g/cm³)	ESCAPE VELOCITY: MI/SEC	KM/SEC	PERIOD OF ROTATION	NUMBER OF KNOWN SATELLITES
Mercury	0.206	7° 0′	3000	4830	0.06	5.3	2.6	4.3	59 days	0
Venus	0.007	3° 24′	7600	12,200	0.8	5.1	6.5	10.3	243 days (retrograde)	0
Earth	0.017	0° 0′	7913	12,740	1.0	5.5	7.0	11.2	23 hrs 56 min	1
Mars	0.093	1° 51′	4200	6762	0.1	4.0	3.2	5.1	24 hrs 37 min	2
Jupiter	0.048	1° 18′	87,000	140,070	318	1.3	37.1	57.5	9 hrs 50 min	12
Saturn	0.056	2° 29′	75,000	120,750	95	0.7	22.3	35.4	10 hrs 14 min	10
Uranus	0.047	0° 46′	30,000	48,300	15	1.6	13.9	21.9	10 hrs 49 min	5
Neptune	0.009	1° 46′	29,000	46,700	17	2.3	15.4	24.4	15 hrs 48 min	2
Pluto	0.250	17° 9′	about 4000	about 6000	0.18 (?)	(?)	(?)	(?)	6.4 days	(?)

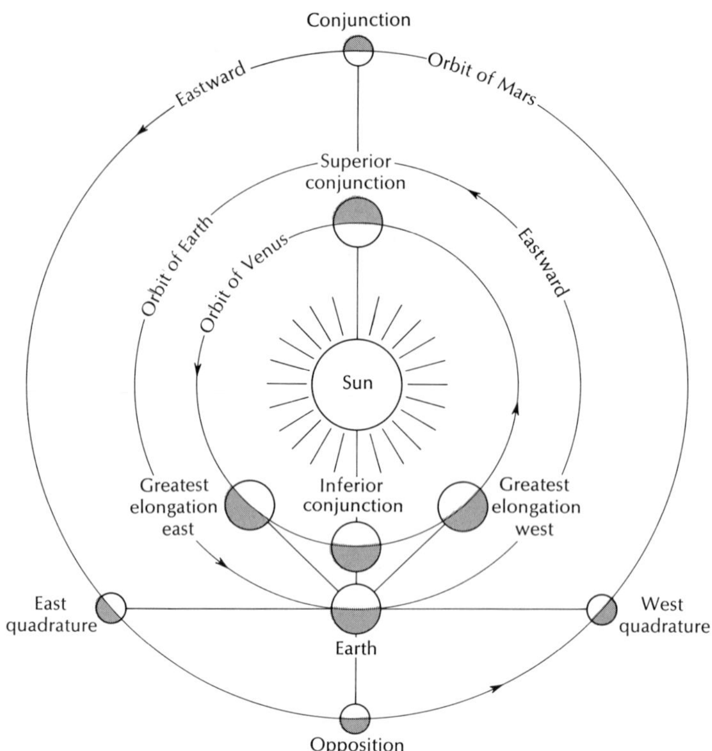

23-2. Orbital positions of inferior and superior planets. As observed from the Earth, east is to the left of the sun and west is toward the right. *(Courtesy U. S. Navy Hydrographic Office.)*

and this may be either east or west of the sun. The changing distances between the Earth and the other planets affect their apparent sizes (Fig. 23–3) and brightnesses.

Mercury and Venus, the two inferior planets, go through phases like those of the moon (Fig. 22–4), but superior planets show mainly the full phase. We shall refer to Venus only, but the explanation is generally the same for Mercury. In Figs. 23–2 and 23–4, the Earth is kept stationary for simplification, and Venus is shown in different positions. As Venus speeds more rapidly in its orbit, it must eventually pass between the Earth and the sun (inferior conjunction) and then in turn reach greatest western elongation, opposition, greatest eastern elongation, and return to conjunction.

Like the moon and other spherical bodies in space, half of Venus is always illuminated by sunlight, whereas half is always in darkness. Therefore, the phases of Venus—like those of the moon—depend upon the proportion of the lighted half that can be seen from the Earth. This is zero at new moon and nearly 100 percent at approximately the full phase (Venus is not visible at opposition).

The diameter of Venus appears smallest when it is farthest from the Earth as a full moon (Fig. 23–5). On the other hand, Venus seems to be about six times larger when it is relatively near the Earth as a thin crescent. This is evidence—first noted by Galileo—that Venus revolves about the sun and not about the Earth (in contrast, the moon's diameter changes by no more than 12 percent during its cycle of phases).

Venus occurs as a morning star for about 9 months and then becomes an evening star for the next 9 months to complete a cycle relative to the sun (synodic). Note (Fig. 23–2) that the first and third quarter positions—at the greatest western and eastern elongations respectively—are not midway between conjunction and opposition. At a quarter phase, a 90-degree angle occurs at Venus between lines to it from the Earth and sun. At such times, the angular separation of Venus and the sun is at a maximum (up to 47 degrees for Venus and 28 degrees for Mercury).

Venus begins its 9-month interval as an evening star as a full moon. Its angular separation from the sun is then small, and Venus sets soon after sunset. However, this angle between the sun and Venus gradually increases during the next several months because Venus moves in its orbit farther east of the sun. When Venus reaches greatest eastern elongation, it is then relatively high above the western horizon at sunset and sets longest after the sun. During the shorter interval between its third quarter and new moon phases, Venus sets earlier on successive evenings. Venus then makes a conspicuous change in an observer's sky: it shifts from an evening star above the western horizon (prior to its new moon phase) to a morning star above the eastern horizon (after new moon).

Venus and Mercury were each given two names by ancient scholars who believed that a morning star was one planet, whereas an evening star was an entirely different planet. The distance between Venus and the Earth varies from about 26 to 160 million miles (42 to 258 million km), and thus its brightness varies considerably. The thick, opaque atmosphere of Venus reflects about 70 to 80% of the sunlight that falls upon it, the highest for any planet; in contrast, the Earth, moon, and Mercury reflect about 30, 7, and 6% respectively. When brightest as a crescent, Venus is about 13 times brighter than Sirius, the brightest of the stars. Venus then causes objects on Earth to show slight shadows on a moonless night,

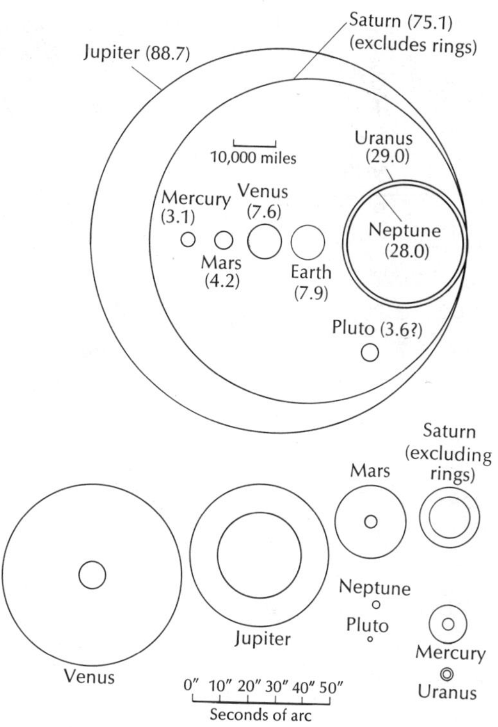

23-3. Actual and apparent diameters of the planets. Above, numbers show diameters in thousands of miles (some discrepancies occur with figures used in Table 23-1). Below, maximum and minimum apparent angular diameters of the planets as observed from the Earth. *(A. G. Smith and T. D. Carr,* Radio Exploration of the Planetary System, *Van Nostrand Reinhold.)*

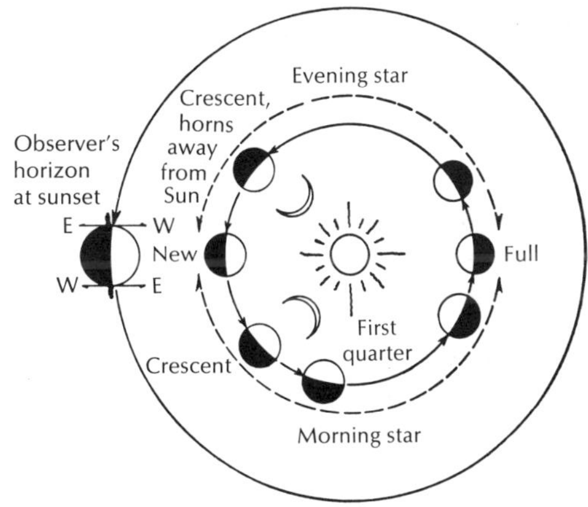

23-4. Phases of Venus.

23-5. Venus in different phases as seen through a telescope. *(Courtesy Lowell Observatory.)*

and it can be seen during the day by observers who know just where to look for it. At one time during World War II, Venus was the target for some unsuccessful anti-aircraft fire!

Based upon size and density, the planets also subdivide into two groups. The four *terrestrial* planets are relatively small and dense (Mercury, Venus, Earth, and Mars). In contrast, the four *major* planets (Jupiter, Saturn, Uranus, and Neptune) are large, rotate rapidly, have conspicuously flattened poles, and relatively low densities. Pluto would seem to belong with the terrestrial group, but some of its physical properties are not yet known. Density differences, such as those between the Earth and Jupiter, indicate differences in chemical composition. However, presumably all of the planets formed at about the same time, in the same way, and out of similar material—like that making up the sun (p. 311).

Mercury

Of all the planets, Mercury (Table 23–1) is the smallest, although Pluto's size is not known exactly; in fact, Jupiter and Saturn each has a moon bigger than Mercury. Mercury is also closest to the sun, has the greatest orbital speed, and the shortest period of revolution.

With the unaided eye, Mercury is always too near the sun for favorable viewing. In other words, we tend to see it only near the horizon during twilight. Thus we must look obliquely through a great thickness of atmosphere, and clouds and dust frequently interfere with clear observation. In a period of about 1 year, Mercury is an evening star three times and a morning star three times. Mercury is best seen when the ecliptic makes its steepest angle with the horizon; this occurs when Mercury is an evening star in the fall and a morning star in the spring.

Astronomers obtain their best telescopic views of Mercury during the day when it is highest above the horizon. Mercury shows phases similar to those of Venus (p. 00).

Mercury's surface reflects light in about the same manner as does the moon, and thus it may have similar topographic features. Mercury also seems to lack an appreciable atmosphere and for the same reasons: low self-gravity and high surface temperatures. At perihelion, temperatures in the central part of the lighted side of Mercury may approach 800°F—hot enough to melt lead and tin. Mercury is then about 15 million miles closer than at aphelion and receives about twice as much solar radiation.

Mercury was long thought to rotate and revolve in the same amount of time (88 days) and to have one side perpetually facing the sun. In the middle 1960's, however, radar observations showed Mercury's period of rotation to be about 59 days. Thus Mercury rotates three times during two orbits about the sun.

Venus

In both size and mass (Table 23–1), Venus is slightly less than the Earth. The surface of this nearly spherical planet is perpetually hidden by a dense covering of clouds of unknown composition and thickness, and its atmosphere is characterized by high temperatures, a great abundance of carbon dioxide, and high pressures. Much data have been gathered about Venus in recent years —from the Earth as well as from spacecraft sent to Venus—but many interpretations are still controversial.

In October 1967 two spacecraft, Venera 4 (Russian) and Mariner 5 (U.S.A.), arrived at Venus after 4-month journeys. An insulated capsule from Venera 4 was parachuted into Venus's atmosphere and descended toward its surface; Mariner 5 passed within 2500 miles (4000 km) of Venus's surface. Signals from the descending capsule of Venera 4 indicated that temperatures increased from about 105°F to 535°F (40°C to 280°C), that carbon dioxide makes up 90 to 95% of the atmosphere, that oxygen and water vapor together make up less than 1.5% whereas nitrogen is less than 7%, and that atmospheric pressure at the surface of Venus is 15 to 22 times greater than it is at the Earth's surface (but 100 times greater according to one interpretation of Mariner 5 data). No evidence for the existence of a radiation belt or magnetic field was detected, but Venus does have an ionosphere of electrons and ions. Presumably this is produced, like the Earth's, by the interaction of high-energy solar radiation with its upper atmosphere (p. 654). The peak intensity seems to occur at an altitude of about 60 miles (100 km), whereas the Earth's peak is nearly three times higher.

Mariner 5's results generally tended to confirm those of Venera 4, but questions were raised. Did the last signals from Venera 4 actually come from a point at or near its surface? Was Venera 4 crushed during descent by atmospheric pressure, or did it possibly land on a mountain high above the mean surface level? Such questions may have been answered on 15 December 1970 when the Russian Venera 7 apparently became the first spacecraft to "soft land" via parachute onto the surface of Venus—it stopped abruptly at a speed of 37 mi/hr (60 km/hr). Three preceding Venera spacecraft (numbers 4, 5, and 6) had apparently each stopped signalling at an altitude of 10 to 13 miles (17 to 22 km). Surface temperature was reported to be 474°C ± 20°C (887°F or 747°K) and surface pressure about 1300 lb/in.^{2} Parachute and spacecraft were designed to function effectively under such extreme conditions.

The equatorial surface temperature of Venus, prior to the Venera-Mariner flights, had been calculated as about 800°F (427°C) based upon a study of very short radio waves (microwaves) that could originate at the surface and pass upward through the clouds more or less unchanged. However, all of the detected radiation was assumed to come from the surface. If the atmosphere also emitted such radiation, then surface temperature estimates would have to be reduced by a corresponding amount. We can at least partially explain Venus's high temperatures: on the average, it receives about twice as much solar radiation per unit area as does the Earth, and the large quantities of carbon dioxide should produce a strong greenhouse effect (p. 647).

If carbon dioxide does make up about 90% of the atmosphere of Venus (previous estimates had been only a few percent), then comparison with the Earth suggests that water should also be abundant. However, we would not expect abundant free oxygen, because it would be very active chemically at such high temperatures and would occur chiefly in chemical combination with other elements. Reciprocally, we would expect the Earth to contain similar large amounts of carbon dioxide, which it does, but in chemical union with carbonate rocks and not free in the atmosphere. Presumably the carbon dioxide and other gases have been released gradually from the interior of Venus as a result of igneous activity, as apparently has happened on the Earth (p. 312). Such carbon dioxide, water, and other materials are assumed to have been present in the gas and dust from which the planets were made (p. 309) and to have been trapped within the planets at the time of their origin. However, water has been released on the Earth in large quantities, and presumably this should have happened on Venus also. If spread uniformly over a smooth surface, water would form a continuous layer on the Earth more than 1½ miles thick. Corresponding amounts of water might be expected on Venus, but at its high temperatures, the water should occur as atmospheric vapor rather than as a liquid at its surface. However, such quantities of water vapor seem to be missing and no entirely acceptable explanation is now available.

The period of rotation of Venus has been difficult to determine because its surface is completely obscured by clouds and no permanent markings can be detected visually. Its nearly spherical shape had suggested slow rotation and a period of a few weeks or more was generally assumed. However, recent radar studies indicate a very slow rotation in the retrograde direction of about 243 days. Venus may present very nearly the same side toward the Earth at successive inferior conjunctions.

Venus transits the sun an average of only two times each century. At such times it can be seen through a smoked glass without a telescope as a round dark spot which moves slowly across the sun's surface. The next transit is predicted for June 8, 2004. Mercury transits the sun about thirteen times each century, but transits of Mercury cannot be seen by the unaided eye.

Mars

Mars is the small reddish planet (Table 23–1) named after the bloody god of war. Its rotation and axial tilt are similar to the Earth's, but the Martian seasons last nearly twice as long because Mars' period of revolution is nearly 2 Earth years. Distances to Mars from the Earth at opposition range from about 35 to 63 million miles (Fig. 23–6; 56 to 97 million kms). The closest oppositions recur at intervals of about 15 years (August 1971 is the most recent), but less favorable oppositions take place every 780 days.

Mars has two small satellites. Recent data indicate that the smaller is about 5 miles across, whereas the larger is 14 miles along one diameter but 11 miles along a diameter that is perpendicular to this. According to one interpretation, such an elongated irregular shape may mean that the two moons are captured asteroids rather than satellites that formed by accretion within a nebula surrounding the protoplanet Mars (p. 309). One satellite makes a complete west-to-east revolution in about 7½ hours (however, its synodic period is nearly 11 hours). An observer on the more slowly rotating Mars would see this moon rise in the west and set in the east more than twice each Martian day (e.g., from midnight to midnight). No other known satellite does this.

The surface features of Mars can be seen

from the Earth more clearly than those of any other planet because of its relative nearness and rarefied atmosphere (Figs. 23–7, 23–8, and 23–9). Photographs taken on film sensitive to atmosphere-piercing red light show permanent markings, whereas blue-sensitive films show a larger, nearly featureless disk—the shorter wavelengths are scattered by its atmosphere and do not generally penetrate to the surface. The chief constituent of the atmosphere of Mars appears to be carbon dioxide, whereas water vapor and oxygen seem to be very scarce, and the proportion of nitrogen is uncertain (unlikely to appear in spectrum).

Mars has white polar caps which at times may cover an area of a few million square miles. However, they are thought to be very thin, with some drifts perhaps a few meters thick, because they melt or sublimate rapidly and because source material seems very limited. According to an older view, the ice caps are made of ice crystals, but 1970 opinion favors solid carbon dioxide as the material. Seasons are more extreme in the southern hemisphere (closest to the sun during its summer), and its polar cap may disappear completely during its summer. The polar caps show the same patterns as they advance and retreat in each Martian year; thus the patterns may be related to topography.

About two-thirds of the surface of Mars is reddish-orange—probably dry, dusty desert areas—and the rest is a grayish color that varies seasonally in darkness. No water bodies are visible or thought to exist. Huge yellowish clouds, probably very large dust storms, are visible above the deserts at times, obscuring the surface as they move across it. These dust storms apparently change the appearance of the darker areas on occasions by depositing reddish dust upon parts of them. Wind may be an important geological agent of erosion and deposition on Mars.

As a polar cap disappears, a darkened zone develops near it and advances (at 25 to 30 mi/day) toward and even beyond the equator. Some astronomers also describe this as a color change—bluish green in the Martian spring and brownish in the fall—but others observe only a change in darkness. The change is attributed by some to sea-

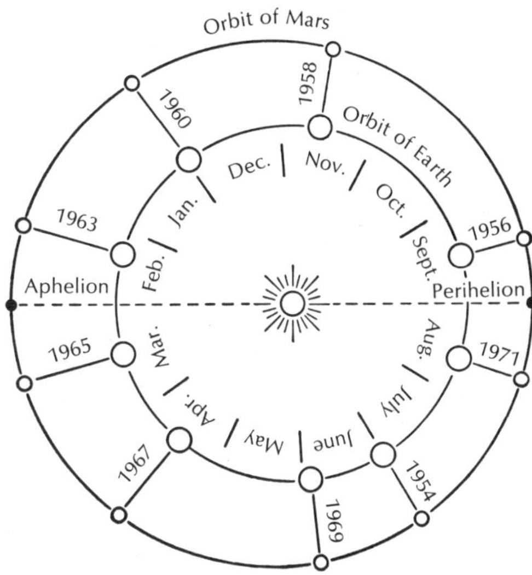

23-6. Mars at various oppositions from 1954 to 1971. Oppositions occur about 780 days apart, the time taken by the Earth to gain one lap on Mars. *(R. H. Baker,* Astronomy, *Van Nostrand Reinhold, 1964.)*

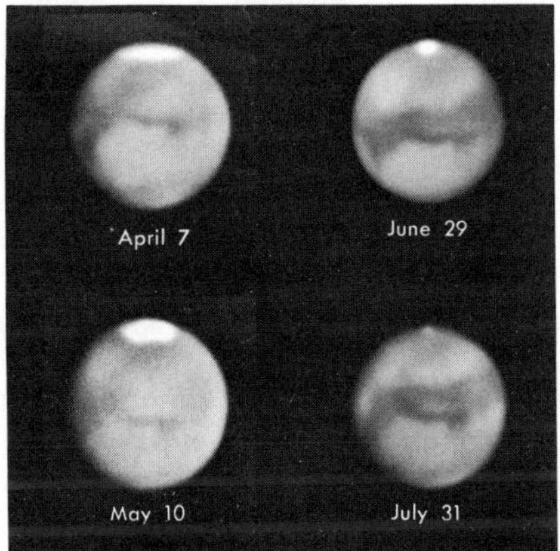

23-7. Seasonal changes on Mars. South is at the top. The four photographs are assigned dates in the Martian years and show the shrinking of the south polar cap that accompanies the arrival of its summer season, as well as the gradual darkening of the dark markings. *(Courtesy Lowell Observatory.)*

23-8. Mars. Camera in Mariner 6 was 2150 miles away, and area is about 550 by 420 miles. The large ring plain in the lower right is about 160 miles across. *(Frame 21, NASA.)*

sonal growth of vegetation, by others to chemical changes in certain salts which might take up water in the summer and give it off in the winter, and by still others to wind activity.

In 1877 an Italian astronomer discovered new surface features on Mars which he described as fine dark lines crossing the orange areas. He called them *canali,* meaning "channels," and the word was transliterated as the English "canals." An American astronomer, Percival Lowell, believed that these lines were straight, arranged in a geometrical manner, and hence were artificially produced by some type of intelligent being. According to his hypothesis, such beings would live near the equator for warmth, and the scarcity of water would force them to use great irrigation ditches to lead the melting ice and snow in the spring from the polar caps toward the equator. However, liquid water is far too scarce and presumably could not flow along the surface because it would freeze or evaporate—the overall, average surface temperature of Mars is about −60°F. Recent photographs indicate that such fanciful canals do not exist.

Conditions would certainly be most rigorous and difficult for life as we know it on Mars. Temperatures are always low—even at the equator the noon temperature probably averages only 40°F (4°C), and this drops far below freezing at night. Surface atmospheric pressure is probably 100 times less than at the Earth's surface—perhaps the equivalent on the Earth of an altitude more than twice the height of Mt. Everest. As a spacecraft crosses behind Mars, its radio signals have to pass through different levels of the Martian atmosphere in succession, and changes in the signal measure the atmospheric pressure.

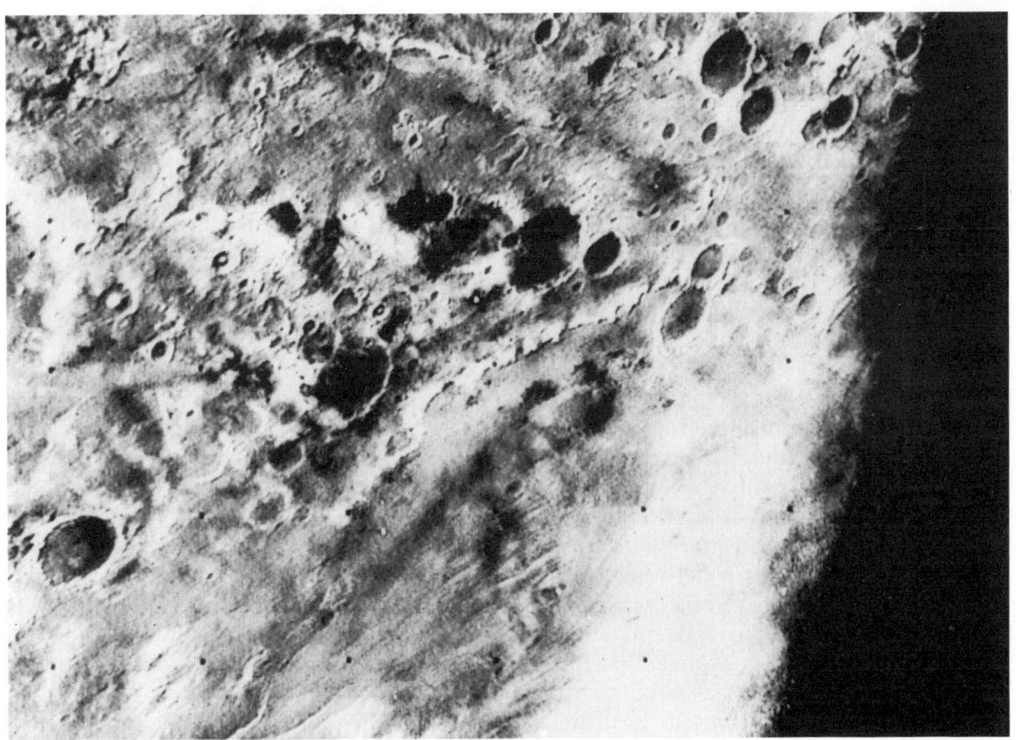

23-9. Mars viewed by Mariner 7 camera in August 1969. The field of view is about 1200 miles across; it is centered at a latitude of 79 degrees south and it includes the south pole. The terminator is at the right. *(Frame 19, NASA.)*

On 14 July 1965, the Mariner 4 spacecraft passed within 6118 miles of the surface of Mars, and 22 pictures were taken and relayed to the Earth. One of these (Fig. 23–7) was described at the time as among the most important scientific photographs ever made. The technologically superior Mariners 6 and 7 made highly successful flybys in 1969, and additional intensive study occurred during the close approach of Mars in 1971. The Mariner photographs* show a cratered surface that bears a surprising resemblance to that of the moon. The smaller Martian craters resemble bowl-shaped, primary impact craters on the moon, but the larger ones have lower rims and flatter bottoms—perhaps some kind of erosion or deposition has modified their shapes. The largest craters are much larger than those on the moon—one has a diameter of about 300 miles (500 km).

* Leighton, Robert B., "The Surface of Mars," *Scientific American*, May 1970.

Three types of Martian terrain have been described: in one, craters are abundant and conspicuous; another is featureless; and the third is described as chaotic. Presumably the entire surface was once abundantly cratered, which means that some form of erosion or deposition has almost entirely erased the craters from the featureless terrain and largely done so within the jumbled ridges and valleys of the chaotic terrain. Mars seems to lack such lunarlike features as rays and strings of secondary craters (easily eroded or buried).

The nearness of Mars to the asteroid belt had suggested that it would have craters, but their excellent preservation is remarkable and unexpected. Collisions with asteroids large enough to produce the larger craters have probably been quite rare since early in the history of the solar system. which suggests that the cratered surface of Mars is very old. Preservation of such an ancient surface indicates that Mars may never have had an atmosphere much denser than the present very thin one, nor has it had enough water to form streams, lakes, or oceans. Otherwise, erosional effects should be more conspicuous.

The Mariner 4 flyby indicated that Mars lacks a magnetic field and a radiation belt. By inference, therefore, it probably lacks a molten metallic core like the Earth's (p. 162). Since mountain chains, ocean basins, and continents are not visible in the pictures, Mars's internal structure may be less dynamic than that of the Earth.

Nearly all hypotheses concerning the origin of life involve processes occurring in a water solution. Thus the apparent absence of water on Mars indicates that it may be a dead planet and that it has been so for a very long time. Mars seems to be a much less hospitable place for life than previously thought. It resembles the moon, and possibly Mercury, much more than it does the Earth. Results from the Mariner spacecraft enhance the uniqueness of the Earth as an inhabited planet.

The Four Major Planets and Pluto

Four giant planets occur in orbits between Mars and Pluto: Jupiter, Saturn, Uranus, and Neptune (Table 23–1). We see only the outer portions of their atmospheres, which seem similar, and their temperatures are all very low—from about −200°F to −300°F (−130° C to −185°C). Spectroscopic studies show that methane (CH_4) and ammonia (NH_3) are present in the atmospheres of Jupiter and Saturn, but on Jupiter they probably occur only as minor constituents in an atmosphere dominated by molecular hydrogen and helium. However, no ammonia can be detected in the very cold atmospheres of Uranus and Neptune (perhaps frozen so that it cannot be detected spectroscopically) but methane absorption bands are prominent. Gaseous bands are distinct in the atmospheres of Jupiter and Saturn but lacking or very faint in Uranus and Neptune.

The specific gravities of Uranus and Neptune are slightly higher than those of Jupiter and Saturn, even though they are compressed less by their smaller masses. Thus they probably have different and denser materials at their centers (perhaps rock cores). Calculations based upon their shapes and rates of rotation suggest that the masses of all four planets are even more strongly concentrated toward their centers than occurs in the Earth.

Hydrogen and helium probably compose the great bulk of the major planets. Only these two elements, apparently the most abundant in the universe, seem light enough to account for their very low specific gravities. However, most of the hydrogen and helium in their interiors probably occurs as solids rather than as gases (a liquid phase may occur between them). The low temperatures make spectroscopic identification of molecular hydrogen in the atmospheres difficult, but it has been done for Jupiter and Uranus. Although methane has been identified in the atmospheres of each of the major planets, it probably occurs as a minor constituent of them.

Life as we know it would seem impossible on the major planets because of their low temperatures and poisonous atmospheres.

Jupiter

Jupiter (Fig. 23–10) is the giant among the planets, and its volume and mass exceed the combined volumes and masses of the other planets. Next to Venus, and occasionally Mars, Jupiter is the brightest of the planets; rarely is it fainter than Sirius. Its size and satellites make observation exciting with even a small telescope.

23-10. Jupiter, 24 October 1952. The Great Red Spot and a satellite and its shadow appear near the top of the photograph, which was made with blue-sensitive film. *(Courtesy Mount Wilson and Palomar Observatories.)*

Of its twelve known moons, the four largest can be seen readily through a small telescope (Fig. 17–20) and form a sort of miniature solar system, first observed by Galileo. These four largest satellites have diameters ranging from 2000 to 3200 miles. They revolve eastward about Jupiter and have orbits that are nearly circular and nearly in Jupiter's equatorial plane (Fig. 23–11). Their mean distances range from about ¼ to 1⅕ million miles and their periods of revolution from about 2 to 16 days. Jupiter's outermost four small satellites revolve in the retrograde direction. Their periods of revolution are about 2 years, and their distances approximate 13 to 15 million miles.

Jupiter's bands have various colors (including reds, browns, and bluish white), occur parallel to the equator, and can be seen through a small telescope; they are somewhat analogous to wind belts on the Earth. Minor changes in these belts occur constantly, particularly near the equator, but the major zones are rather stable as viewed from the Earth. Evidently we see only the outer part of Jupiter's dense thick atmosphere, but certain markings in this remain visible for many years; e.g., the Great Red Spot has been visible for a century or more, although its size, color, intensity, and position have all changed somewhat. Its nature is uncertain. Jupiter rotates very rapidly, but at different rates at different latitudes; thus it is flattened noticeably at the poles.

Jupiter is a strong source of polarized radio signals, and the radiation may be produced by the movement of electrons trapped within Jupiter's magnetic field. Jupiter may radiate away more energy than it receives from the sun; if so, the generating source within it is unknown.

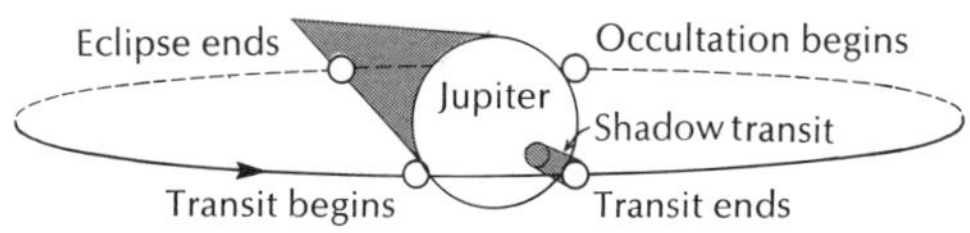

23-11. A satellite of Jupiter in different parts of its orbit shortly after Jupiter has passed opposition. *(R. H. Baker,* Astronomy, *Van Nostrand Reinhold, 1964.)*

23-12. Saturn as photographed through the 100-inch reflector. Note the flattening of the polar region. *(Courtesy Mount Wilson and Palomar Observatories.)*

Saturn

Saturn (Table 23–1 and Fig. 23–12), unique in its possession of a system of rings, is one of the most beautiful of celestial objects viewable through a telescope. To the unaided eye it appears as a bright, yellowish, starlike object in the night sky, and it was the most distant planet known to ancient astronomers. Saturn's low specific gravity is also unique because it is less than one for the planet as a whole, and thus an average sample, if it could be obtained, would float on water.

The planes of Saturn's rings coincide with its equatorial plane which is always tilted about 27 degrees to Saturn's orbital plane. The rings are paper-thin because their thickness is less than 10 miles, and may even be measured in inches, whereas the outside diameter of the outer ring is about 171,000 miles (275,000 km). Thus they become invisible, or almost so, when viewed precisely edge-on (Fig. 23–13), and Saturn appears fainter at such times. In fact, Saturn is about half as bright with its rings edge-on as when they are fully open (other viewing conditions being equal). Saturn's moons are thus most conspicuous when the rings are edge-on, and Saturn's tenth known satellite was discovered on such an occasion late in 1966. This faint object may be a few hundred miles in diameter and may orbit Saturn in 18 hours at a distance of about 100,000 miles, just outside the outermost ring.

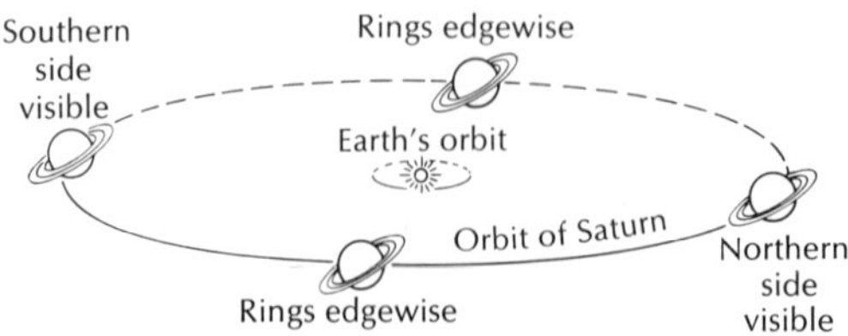

23-13. Saturn's rings at different angles. *(R. H. Baker,* Astronomy, *Van Nostrand Reinhold, 1964.)*

The rings probably consist of myriads of small, ice-coated rock fragments and dust particles revolving as miniature satellites in four orbital groupings about Saturn (the fourth, innermost, very faint ring was discovered in 1969). However, only two rings show clearly. The bright middle ring is about 16,000 miles wide and is separated from the outer ring (about 10,000 miles wide) by a 3000-mile gap. Although the third, inner ring is quite faint, it may be visible on a photograph where it obscures part of Saturn. The particles comprising Saturn's rings have the same velocities that large satellites would have at these distances. The rings are not solid because the light of a star, as seen through them, is dimmed but not blotted out, and the inner parts of the rings revolve more rapidly than the outer parts. The rings may constitute planetary material which was too close to its parent body to consolidate into a single large satellite, or possibly one or more satellites may have approached too close to Saturn and been torn to pieces by its tidal attraction (less likely).

Uranus, Neptune, and Pluto

Uranus and Neptune are twin planets (Table 23–1; see discoveries p. 472), but Uranus is unusual in having its equatorial plane tilted about 82 degrees to the plane of its orbit. Its rotation is retrograde (can also be considered as a 98-degree tilt and a counterclockwise rotation). Uranus's five satellites revolve in the same direction as the planet rotates and in the plane of its equator. Uranus thus has a unique seasonal change; its axis is nearly parallel to its orbital plane and points continuously in a fixed direction in space. At one time one end of the axis points directly at the sun, but half an orbital revolution later it points almost directly away from the sun.

Pluto is least known of the planets. Pluto's eccentricity is large enough so that at times it is closer to the sun than is Neptune. However, the large inclination of its orbital plane keeps Pluto from ever approaching close to Neptune. Data concerning Pluto's size, mass, and density have been contradictory but may recently have been resolved. Supposed discrepancies in the orbit of Uranus had led to a search for an unknown planet (nearly seven times more massive than the Earth) beyond Neptune and its subsequent discovery. But Pluto turned out to be much smaller and fainter than expected. Apparently the Uranus data were unreliable and Pluto's discovery was more the result of an intensive search than of an accurate prediction. A recent study of the perturbation effect of Pluto on Neptune indicates that its mass is about 0.18 that of the Earth.

An attempt was made in 1965 to estimate the diameter of Pluto by observing its occultation (Fig. 23–11) of a star. Since the orbital velocity of Pluto is known, the duration of the occultation would be a measure of its diameter. Unfortunately, the occultation could not be seen at any of the participating observatories. However, this does indicate that Pluto's diameter is probably less than 4200 miles, and this small size can account for its very low brightness. If one or more planets still occur beyond Pluto, discovery will be most difficult.

Asteroids

Asteroids are the minor planets (planetoids). They revolve eastward about the sun, and the majority have orbits located between Mars and Jupiter and periods of 3½ to 6 years. The largest of these bodies probably exceeds 400 miles (700 km) in diameter, and the smallest visible ones are 1 to 2 miles across. The number in each size increases as the size decreases; therefore, asteroids too small to be visible are probably very abundant. Many of the smaller asteroids seem to have irregular shapes (p. 518), but the largest ones may be nearly spherical.

The orbits of more than 1500 asteroids have been calculated, but thousands more are thought to exist. The orbits of most of the asteroids are not more irregular than those of Mercury and Pluto, but some are much more so.

Asteroids may be detected photographically by using a time exposure of an hour or so. The camera is rotated so that the background stars appear as dots; asteroids then show up as short streaks because they have a relatively rapid motion of their own with respect to the stellar background.

The total mass of all the asteroids is probably much less than that of the Earth. Perhaps they originated because planetary material between Mars and Jupiter collected into a number of large-size, spherically shaped asteroids, two or more of which subsequently collided. More frequent collisions would then be expected among the fragments, especially the smaller ones, and irregular shapes and orbits could result. The Earth probably encounters such objects each day and all or most meteorites may come from this source—one meteorite was observed to fall in Oklahoma on 3 January 1970 and had an orbit traceable to the asteroid belt.

Some asteroids large enough to be observed approach to within a million miles of the Earth (within 600,000 miles for at least one). At such close distances their geocentric parallaxes can be determined precisely, making them very valuable in the calculation of other distances in astronomy.

Comets

Comets probably constitute the most unusual members of the solar system. A well-developed comet (Fig. 23–14) has a long tail, which points approximately away from the sun, and a brighter head section (coma) that contains a small bright nucleus (presumably contains most of the matter present in a comet). However, not all comets have tails, and a nucleus is not always visible. The size, shape, and brightness of a comet vary as it approaches and leaves the vicinity of the sun. Thus the most permanent feature of a comet is its orbit, but this also changes if a comet approaches close enough to a planet to be deflected by its gravitational pull. Comet orbits are oriented at all angles to the plane of the ecliptic, and comets show no preference at all for direct motion.

In size, comets are enormous; the average head may be about as large as Jupiter (in volume), and tail sections may extend for millions of miles. In mass, however, they are very small. No comet has been known to affect the orbit of a planet or a satellite, although comets themselves have been influenced greatly by these bodies. When viewed through the tail of a comet, or even through its head, the light of a star is barely dimmed, and apparently the Earth has passed through a comet's tail with no noticeable effect (the particles were too small to act as meteors).

Comets are quite different from meteors in size, distance, and appearance in the sky. A bright comet covers a rather large part of the sky, moves very slowly with respect to the background stars, and may be visible on successive nights for weeks or even months. Most comets are many millions of miles away; they do not occur within the Earth's atmosphere as once believed. Tycho Brahe discovered this when he was unable to detect a parallactic displacement for a bright comet. In contrast, meteors are short-lived and occur within the Earth's atmosphere (p. 587).

According to the "dirty snowball" hypothesis, the nucleus of a comet may be a porous structure a few miles or less in diameter (possibly larger). It may consist of stony or metallic fragments embedded in ices of methane, ammonia, and water. When far removed from the sun, a comet may consist chiefly of a nucleus. However, as a comet approaches the sun, it encounters more intense solar radiation, which causes expulsion of gas and dust from the nucleus (explosively at times). This material then diffuses into space to enlarge the head and form a tail. A comet thus becomes larger and brighter as it approaches the sun. Its light is partly reflected sunlight and partly reradiated sunlight (more important closer to the sun).

Tails seem to be of two types: straight ones made almost entirely of gases and curved dust-particle tails. When dust particles are ejected from a nucleus so that their distances from the sun change, their orbital speeds and periods of revolution must also change; thus a tail made of such particles becomes curved.

Electromagnetic radiation and electrified particles from the sun (the solar wind) push material in a comet's tail away from the sun. Solar radiation pressure on extremely tiny particles is greater than its gravitational attraction on them. Material expelled from the head of a comet to form its tail cannot be recovered, because the comet's gravitational attraction is too weak. Each particle must continue to revolve in an orbit about the sun.

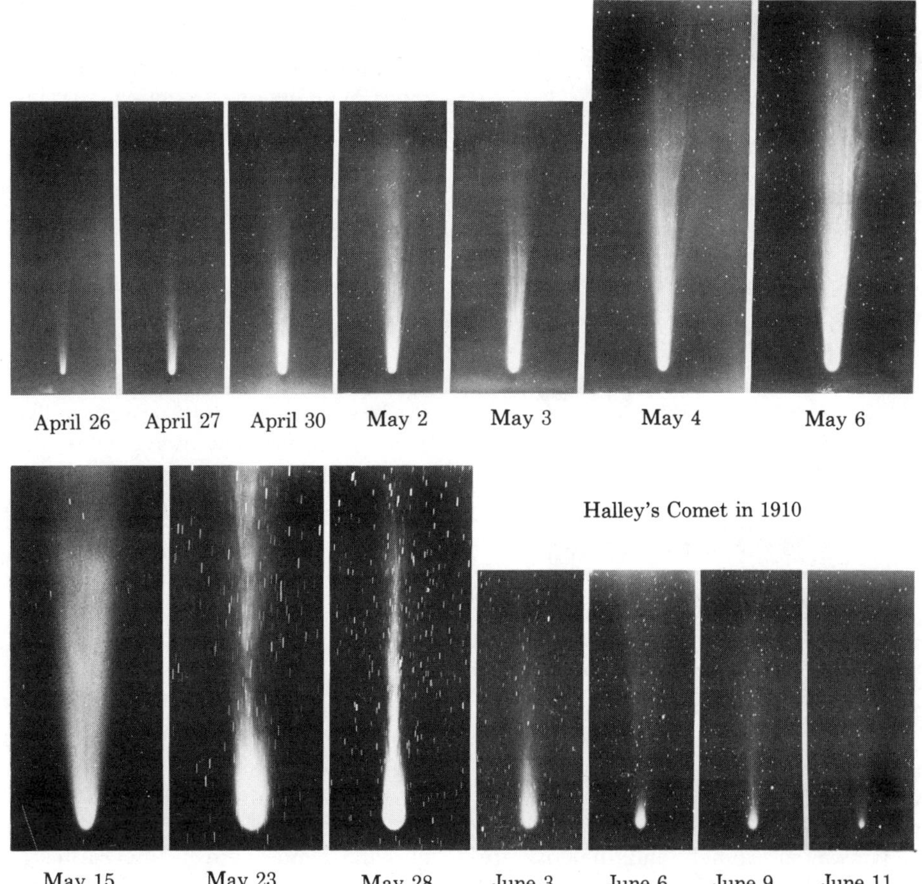

23-14. Halley's comet. Fourteen successive views from April 26 to June 11, 1910. On May 15 the comet subtended an angle of about 40 degrees on the celestial sphere. *(Courtesy Mount Wilson and Palomar Observatories.)*

Comets tend to have very elongated orbits and thus spend long periods of time far out from the sun. In following Kepler's equal areas law, a comet moves much more slowly at aphelion than at perihelion. Perhaps most comet nuclei represent debris left over from the formation of the solar system and occur here and there within a large spherical halo centered on the sun and planets. When disturbed enough by the gravitational pull of the planets or some passing star, they may start inward slowly and eventually swing around the sun. We would know that an object such as a comet is not a member of the solar system if its speed exceeds escape velocity at its particular distance or if its path is either a parabola or a hyperbola (open curves). In either case, the object would swing once around the sun and would then disappear into interstellar space (unless the gravitational pull of another body subsequently affected it).

Objects can follow only four types of curved paths about a celestial body such as the sun (Fig. 23–15). To revolve repeatedly about the sun, an object must follow a circular or elliptical orbit. To move only once about the sun, an object must follow a

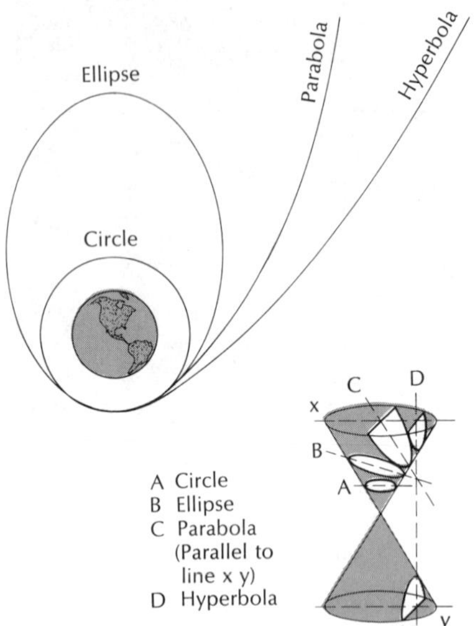

23-15. Satellite orbits and trajectories shown as conic sections. (H. L. Goodwin, *Space: Frontier Unlimited,* Van Nostrand Reinhold, 1962; adapted from a NASA diagram.)

parabola or hyperbola. Note that all four curves are alike near perihelion—about the only part observable from the Earth. So far as is known, all comets and meteors are members of the solar system.

Comets that can be seen readily with the unaided eye and excite some public interest average about one in every 5 to 10 years. Spectacular comets are much less common, and probably Halley's comet is the most famous. With but one exception it has been observed at intervals of about 75 years at every perihelion passage since 240 BC. It is scheduled to reappear about 1986 (Fig. 23–16).

Meteoroids, Meteors, and Meteorites

Meteors are the familiar objects known as shooting stars, but they are very unlike stars. Three terms are now used: meteoroids, meteors, and meteorites. *Meteoroids* are solid particles, commonly no larger than sand grains or small pebbles, that travel in orbits about the sun (size calculated from luminosity which is proportional to mass and square of velocity, but density of air and compactness of meteor are also factors). The laws of Kepler and Newton describe and explain their orbital motions. Some meteoroids weigh many tons, and no arbitrary limit separates large meteoroids from small asteroids.

Meteoroids are invisible until they approach close enough to the Earth to be pulled into the atmosphere by the Earth's gravitational attraction (unnecessary if they were on a collision course). Entering the atmosphere at great speed, they are heated to incan-

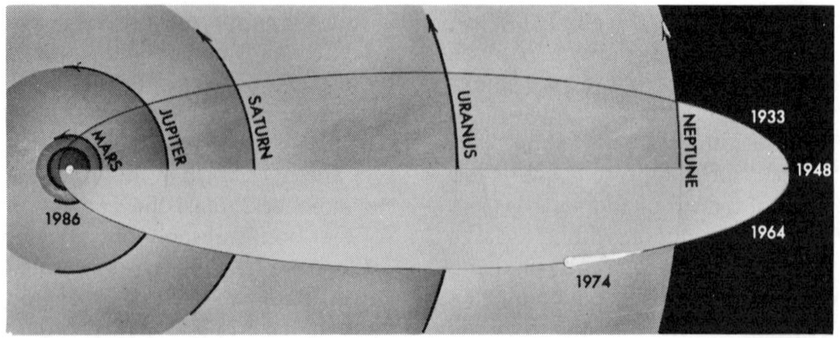

23-16. Orbit of Halley's comet. It revolves in the retrograde (clockwise) direction with an eccentricity of 0.97, a perihelion distance of 0.6 A.U., and an inclination of 18 degrees to the ecliptic. *(R. H. Baker and L. W. Fredrick,* An Introduction to Astronomy, *Van Nostrand Reinhold, 1968.)*

descence as they collide with molecules in the air. The term *meteor* is applied to this luminous phase, and a fireball is an especially bright meteor—a few are even visible in daytime. Temperatures rise high enough to vaporize most meteors completely, and the brighter ones leave trails of gases which may remain visible for a number of minutes. However, most meteors are visible only for seconds. The air through which the meteor travels is ionized and made luminous by radiation produced when the free electrons recombine with the ions. We see only the cylindrical trail of ionized gases produced by a meteor, not the meteoroid itself.

Most meteors become visible at altitudes of 60 to 70 miles (97 to 113 km) and disappear at altitudes of 40 to 50 miles (65 to 80 km). Their average speeds may approach 26 mi/sec (42 km/sec), the escape velocity at 1 A.U. However, the effects of the Earth's orbital speed, direction of movement, and gravitational attraction must be allowed for. Millions of meteors may enter the Earth's atmosphere daily, but most of them are completely volatilized before they reach the surface. We are fortunate indeed to have an air umbrella.

The term *meteorite* refers to the solid object that has reached the Earth's surface —probably it was the size of a basketball or larger before it entered the atmosphere. About three-fourths of the known elements have been identified in meteorites, which are metallic or stony or a mixture of the two. Metallic meteorites consist largely of iron (80 to 95% or so) with some nickel, whereas stony meteorites are somewhat similar to certain dark-colored, heavy igneous rocks. Spectroscopic study of meteors as they volatilize at high altitudes and other evidence indicates that stony meteoroids are much more numerous than the metallic ones, even though metallic meteorites are more abundant in museums (probably because they are more readily recognized). Meteorites may give a clue to the chemical composition of the Earth's interior.

Characteristically, meteorites have a thin, blackened crust pitted with irregular holes that look something like thumbprints and have been caused by differential fusion; some portions were vaporized more readily than adjoining portions. The freshly ground surface of a stony meteorite commonly shows a light grayish color and metallic specks. Surprisingly, most meteorites are not at high temperatures when they strike the Earth's surface. They come from the extreme cold of outer space, and only their surfaces are heated as they penetrate the Earth's atmosphere. If large enough, they reach the surface before volatilizing completely.

Meteoroids appear to consist of two quite different types. One type apparently has its source in the asteroid belt, and fragments of these form the stony and metallic meteorites. Comets apparently are the source of the meteoroids of the other type, which evidently do not form meteorites. Meteoroids of this second type may consist of particles of friable frozen gases which vaporize completely in the Earth's atmosphere. Comet-associated meteoroids travel around the sun in both the direct (eastward) and retrograde directions and approach the sun at all angles to the plane of the ecliptic. Asteroid-associated meteoroids seem to have orbital motions similar to those of the asteroids: the motion is direct, of rather small eccentricity, and occurs somewhat near the plane of the ecliptic.

More meteors are commonly visible in the early morning hours than are visible before midnight, and they tend to differ in color and length of trail (Fig. 23–17). The rate at which a meteor descends through the Earth's atmosphere determines the amount of friction and heat developed. Speeds vary from 7 to 45 mi/sec (11 to 72 km/sec); in contrast, a rifle bullet commonly moves at about 0.5 mi/sec. Even a motionless object in space, subsequently captured by the Earth, would atttain a speed of 7 mi/sec before crashing into its surface—this is the Earth's escape velocity. High speeds produce great friction, high temperatures, and bluish-white colors; such meteors tend to vaporize rapidly and have short trails. Slower speeds

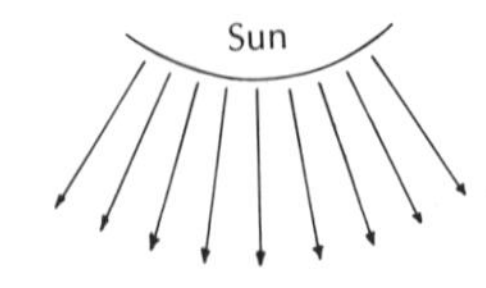

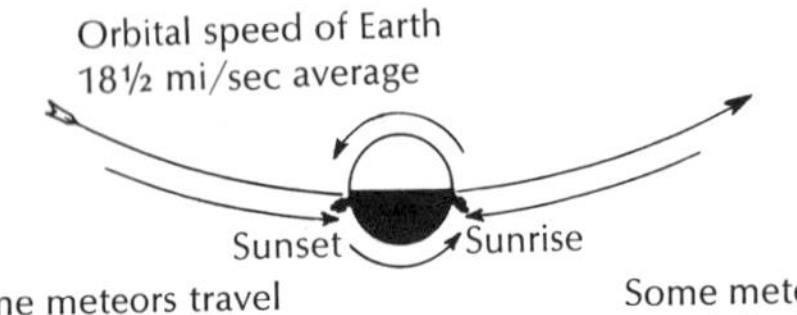

23-17. Meteor colors. Meteoroids travel both clockwise and counterclockwise in orbits about the sun at rates which may be faster or slower than the speed of the Earth.

cause less friction, lower temperatures, and reddish colors.

Most meteors noted by an observer in the early evening hours are traveling more rapidly than the Earth and in the same direction. Thus the "front end" of the meteor collides rather gently with the "rear bumper" of the Earth and a reddish meteor results. We subtract the Earth's orbital velocity of 18.5 mi/sec from that of the meteor and add the effect of the Earth's gravitational pull. On the other hand, in the hours from midnight to sunrise, an observer sees some meteors which are traveling in the opposite direction from the Earth. There is a head-on collision and bluish-white meteors result. We add the Earth's orbital speed and gravitational pull to the speed of the meteor. But meteors may also move eastward more slowly than the Earth; these are overtaken gradually and produce a reddish light.

23-18. The radiant of a meteor shower. *(R. H. Baker,* Astronomy, *Van Nostrand Reinhold, 1964.)*

We tend to see more meteors per hour after midnight for at least two reasons. (1) Since the luminosity of a meteor is proportional to the square of its speed (other factors being equal) tiny, slow-moving, before-midnight meteoroids may produce insufficient light to be visible, whereas fast-moving, after-midnight meteoroids of the same size do become visible. (2) We can see all the meteoroids that approach the Earth (no matter what their speeds) as well as the slow-moving ones overtaken by the Earth.

Occasional showers of meteors are visible (Fig. 23–18). The meteors causing a shower are actually following parallel paths but seem to come from one spot in the sky (radiant) for the same reason that railroad tracks seem to diverge from a distant point. The best views tend to occur when a radiant is highest above the horizon after midnight (other conditions being equal). Certain showers occur each year and are named for the constellations from which they seem to radiate: e.g., the Perseids in August, the Orionids in October, and the Geminids in December (Fig. 23–19).

The Leonid meteor shower (Fig. 23–1) of November 1966 was a most spectacular display of celestial fireworks at certain times and in certain places. Observers just before dawn in the western United States were startled and delighted by an intense rain of many thousands of meteors, perhaps as many as 40 per second during a short in-

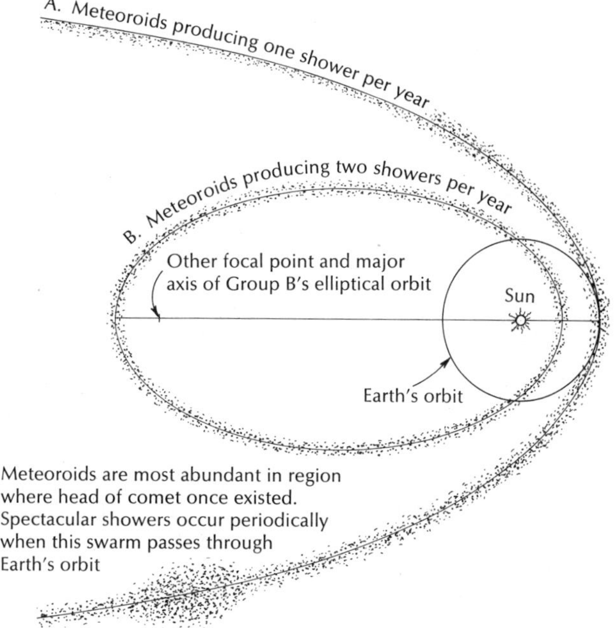

23-19. Origin of annual and semiannual meteor showers. The Earth's orbit intersects the orbit of the Group A meteoroids at one location (their perihelion position), and a meteor shower occurs each year at the time the Earth passes through this part of its orbit. Exceptional showers occur periodically. Group B meteoroids produce two showers per year (e.g., the Orionids in October and the Aquarids in May probably represent debris from Halley's Comet).

terval at the peak of the shower. Some were fireballs or left trails like rockets; others were small. Together they produced a display that will be long remembered. However, clouds prevented observations in parts of the eastern United States, where the most intense part of the shower apparently took place after sunrise.

The Leonids seem to radiate from the constellation Leo and showers take place each November, with spectacular displays tending to occur about every 33 years. Evidently the Leonid meteoroids are strewn all along the orbit of a former comet, and a swarm is located where the nucleus of the comet used to be—the Earth thus encounters this swarm at 33-year intervals. The 1966 shower rivaled that of 1833, but the Leonids were disappointing in 1932 and 1899.

The height of a meteor in the sky can be calculated by triangulation (p. 528) if its direction and altitude are measured simultaneously by two observers a known distance apart. This can be done by photography, and the speed of the meteor can also be calculated. One method involves a shutter that opens and closes many times a second—the longer the gap between successive images of the meteor, the faster it is moving (when distances and directions are equal). The ionized trails left by meteors can also be detected by radar and thus observed during the day.

No human is known to have been killed by a meteorite (but an Alabama woman was bruised by one in 1954), and the chances that an individual may be hit are extremely slight, although some animals have been killed and buildings have been hit (e.g., a garage in Illinois in 1938). More recently, near Hartford, Connecticut on 8 April 1971, a 12-ounce stony meteorite, 2½ inches in diameter, crashed through the roof of a house and was found in the living room ceiling.

Two spectacular meteoritic events have occurred in Siberia during the present century. On June 30, 1908 the Earth collided with a massive body (perhaps the nucleus of a comet) which apparently exploded before reaching the surface. No meteoritic debris has been found in the impact area (central Siberia, Tunguska fall), and depressions once interpreted as meteorite craters

23-20. Meteor Crater, Arizona, exceeds 4000 feet in diameter and is nearly 600 feet deep. Many fragments of metallic meteorites have been found in the vicinity of the crater, and additional material may be buried far beneath the surface. *(Photograph by John Forrell, Fort Worth, Texas.)*

are now ascribed to terrestrial processes. According to reports, the blast was heard 400 miles away, windows were broken at a distance of 50 miles, trees were blown over within a radius of 20 or more miles, and several hundred reindeer were killed. If the fall had occurred about 5 hours later, it might have struck Leningrad. A very bright daytime fireball occurred on February 12, 1947 and devastated the uninhabited side of a mountain in Siberia with iron-nickel meteorites, the largest weighing 1 to 2 tons. Craters dot the impact area.

Meteor Crater in Arizona (Fig. 23–20) undoubtedly was made by the impact of a large meteoroid or a group of smaller ones, and other impact craters are known on the Earth; some are very old and larger than Meteor Crater. The intense shock and high temperatures produced by the impact of a meteorite with the Earth's surface may form a special mineral (coesite) and structures called shatter cones, and these may still provide evidence of the event after erosion has destroyed much of the crater itself.

A multitude of depressions (Fig. 23–21) in the Carolinas and adjoining coastal areas to the north and south may also have been caused by meteoritic bombardment, but this is controversial. Temperatures caused by the impact of large swift meteors are probably high enough to volatilize completely the rocks, soil, and water at such spots, and the sudden great expansion necessitated by the change to a gaseous state causes the explosion. The column of air directly in the path of a meteor does not have time to move completely aside; it is compressed into an air-shock wave that affects a much larger area than that actually hit by the meteorite.

Interplanetary space along the plane of

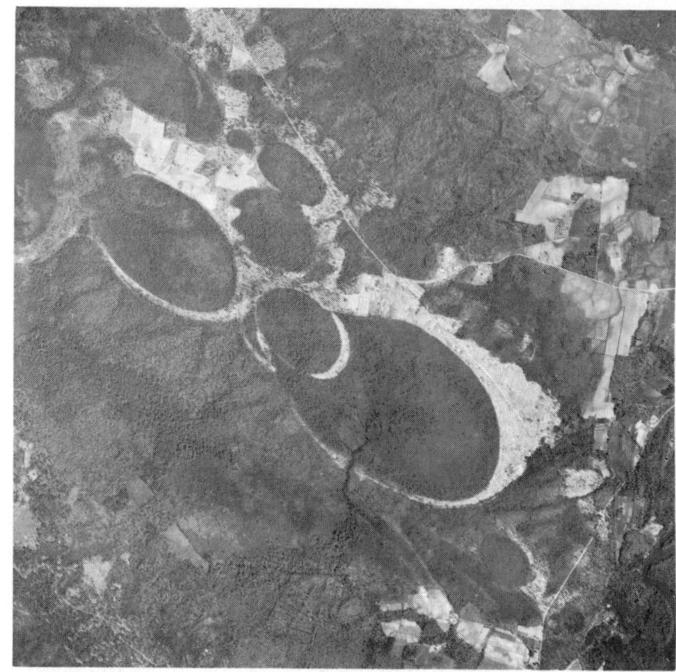

23-21. A few of the Carolina bays. Nineteen bays are recognized here, eighteen in a strip about a mile wide, extending northwest-southeast through the picture. Some overlap. Light-colored areas are sand rims which are best developed on southeastern sides. *(Courtesy Professor C. E. Prouty, Michigan State University.)*

the ecliptic apparently is occupied by multitudes of particles of dust or of larger size which orbit the sun and reflect its light. The quantity of debris seems to increase toward the sun, and little appears to exist beyond the orbit of Jupiter. Such debris apparently accounts for the zodiacal light (Fig. 23–22) which is visible only on a clear

23-22. The zodiacal light and Comet Ikeya-Seki, 31 October 1965. The comet's tail is brighter than the zodiacal light. Note the sickle part of Leo in the upper left (the bright star Regulus forms the end of the handle). *(Photograph by H. Gordon Solberg, Jr., Las Cruces, New Mexico.)*

dark night and is so faint that it is commonly difficult to observe. However, from a darkened ship at sea, it is reported to be a striking sight, somewhat comparable to the Milky Way in brightness. The zodiacal light forms a wedge-shaped area of illumination along the ecliptic that points away from the sun and is widest and brightest near the horizon. Particles nearer the sun reflect more light than do more distant particles—thus the wedgelike appearance. In northern latitudes, it is best viewed when the ecliptic is most steeply inclined to the horizon (in the west after sunset in the spring, and in the east before sunrise in the autumn). It extends for 30 degrees or so above the horizon, but some observers at the equator report that at times it extends as a faint band from horizon to horizon.

Exercises and Questions for Chapter 23

1. Make one top-view sketch that shows each of the following celestial objects in the position indicated. Try to show the planets about at their relative distances from the sun, although this means that some may be off your paper in some positions.
   (A) Sun, Earth, and moon in the third quarter phase.
   (B) Mark a dot within the circle representing the Earth so that it shows the beginning of spring in the Northern Hemisphere.
   (C) Mercury as a morning star in the full moon position.
   (D) Mercury at maximum western elongation.
   (E) Venus as an evening star that sets later on successive nights.
   (F) Venus at inferior conjunction.
   (G) Mars in such a position that it is retrograding.
   (H) Jupiter at opposition.
   (I) Mars due south at midnight.
   (J) Mars at eastern quadrature.
   (K) Saturn at conjunction.
2. A number of hypotheses have been suggested to account for the Christmas Star. Discuss (see *Sky and Telescope,* December 1968, as one reference.).
3. Observe each of the planets visible this semester—with the unaided eye and through a telescope.
   (A) How can a planet be distinguished from a star (without optical assistance)?
   (B) Mark the position of each planet on a star map by carefully noting its alignment and angular distances from nearby stars. Make additional observations once or twice a month and plot any changes observed.
   (C) Draw a top-view diagram to show approximately where these planets are located relative to the Earth and sun.
4. What causes a planet to retrograde?
   (A) Does Mars retrograde more or less frequently than Saturn? Discuss.
   (B) When does Venus retrograde?

5. What evidence indicates that smaller asteroids have irregular shapes, whereas the largest asteroids are spherical?
   (A) Why does shape seem to depend upon size?
   (B) What relationship seems to exist between meteorites and asteroids?
   (C) How do the orbits of asteroids comonly differ from those of comets?
6. List as many unique features for each of the planets as you can.
7. Discuss the possibility of life as we know it existing on one of the other planets in the solar system.
8. What causes the light emitted by a meteor?
   (A) How do before-midnight meteors tend to differ from after-midnight meteors? Why?
   (B) What causes a meteor shower?
9. What significance may meteorites have concerning the composition of the Earth's interior?
   (A) The age of the Earth?
10. Describe and explain the changes that occur typically as a comet approaches the sun.
    (A) What kind of evidence would indicate that a particular comet was not a member of the solar system?
11. List as many differences as you can between comets and meteors.

24

The Sun and Other Stars

By terrestrial standards, the sun is very far away, but by astronomical standards (Fig. 24–1) it is our near neighbor—the only star that we can study in detail and that is close enough to have a visible disk. Fortunately, the sun seems to be average in many of its characteristics, and thus what we learn about the sun can probably be extrapolated in varying degrees to other stars. If the Earth is scaled down to the size of a grapefruit, then the sun becomes a 50-foot sphere a mile away.

The sun's mass exceeds those of the Earth and Jupiter by about 333,000 and 1000 times respectively. Therefore, it contains nearly all of the matter in the solar system and exerts a strong gravitational control over it. The sun's 864,000-mile diameter (1.35 million km) exceeds those of the Earth and Jupiter by about 109 and 10 times respectively. Thus its volume is greater than theirs by about 1.3 million and 1000 times. Although the sun's average density about equals Jupiter's, it is much less than the Earth's (1.4 vs. 5.5 g/cm^3).

By weight, about 96 to 99% of the sun seems to be made of hydrogen and helium (hydrogen probably makes up about 60 to 80% of the total). Therefore, all of the other elements together may make up only 1 to 4% of its total mass. The sun's surface gravity is 28 times that of the Earth. Thus a 200-pound Texan would weigh 5600 pounds on the sun but would, of course, be vaporized instantly!

The sun, as a hot gaseous body, has no physical features corresponding exactly to the solid body and atmosphere of the Earth. Direct radiation (p. 646) from the sun comes from the *photosphere,* a layer about 100 to 200 miles thick, which has gradational boundaries, and which we see as the sun's disk. The photosphere is opaque to radiation from the interior, and we cannot look through it into the underlying gases.

Above the photosphere, a progressively thinner gaseous atmosphere—the chromosphere and corona—extends outward for more than a million miles, but it can be seen by the unaided eye only during a total

24-1. Panorama of the Milky Way, equatorial view. The great dark rift in Cygnus and Aquila is left of center. In the lower right we see the Magellanic Clouds. *(Prepared by Martin and Tatyana Kesküla. Courtesy Lund Observatory, Sweden.)*

solar eclipse. The *chromosphere* overlies the photosphere and is a few thousand miles in thickness. During a total solar eclipse, it appears as a relatively thin reddish ring surrounding the dark body of the moon. Reddish jetlike columns of hot gas (spicules) rise upward through the chromosphere and give it an irregular, jagged outline (best seen around the sun's edge). Individual spicules last only for minutes, but many are in existence at any one time. The pearly white corona (Fig. 24–2) forms the very diffused, faintly luminous, outer part of the sun's atmosphere.

The sun rotates on its axis from west to east but at different rates at different latitudes; thus it cannot be a solid body. The rate is fastest at the equator (about 25 days) and slowest near the poles (about 33 days at latitude 75 degrees). The cause of this variation is uncertain. The movement of sunspots (Fig. 24–3) and the Doppler effect (Fig. 19–18) show that rotation occurs.

Observing the Sun

Never look directly at the sun with the unaided eye or through telescopes or binoculars lacking special filters (the retina is likely to be burned causing a permanent blind spot). However, it can be observed safely through heavily smoked glass or two thicknesses of blackened photographic film. The most convenient way of observing the sun with a telescope is to project its image onto a piece of white cardboard held a foot or so behind the eyepiece (use an angle eyepiece or shade the cardboard—unfortunately, the image lacks sharpness, and the concentrated solar beam may damage the eyepiece). A group can observe simultane-

24-2. The sun's corona near the minimum point in the sunspot cycle. *(Courtesy Yerkes Observatory.)*

ously in this manner. Large fixed telescopes have been devised for observing the sun—light is reflected by movable mirrors into a stationary objective and tube. These have very long focal lengths and form large images of the sun (up to 34 inches in diameter).

The *coronagraph* is a telescope of a special type that artificially eclipses the sun. A black disk is set in the telescope's focal plane and blocks off the light from the photosphere, thus permitting observation of the chromosphere, prominences, and inner corona during the day. Such an instrument must be sited at a high altitude where the air is less dense and very clear and must have an objective of particularly high quality to prevent the scattering of light.

The sun has also been photographed by instruments attached to unmanned high-altitude balloons (Figs. 19–14 and 24–4), from rockets which rise above the dense air of the Earth's lower atmosphere, and from telescopes in satellites orbiting the Earth. Some of these show remarkable detail.

We shall use 6000°K (5727°C or 10,340°F) as the average temperature of the sun's surface, but the photosphere is actually a layer about 200 miles in thickness that varies in temperature from about 7000°K (12,140°F) at the bottom to 4500°K (7600°F) at the top. Each square inch of the photosphere radiates about 10,000 calories per second—an enormous amount of energy. The total radiation remains about constant, although fluctuations do occur in the high-energy ultraviolet and X-ray portion (p. 600).

Sunlight is so bright that a large spectrograph can be used to study it; the resulting spectrum can be spread across many feet and shows thousands of dark absorption (Fraunhofer) lines. About 67 of the elements known on the Earth also occur on the sun. Many of the remaining elements are either very rare on the Earth (thus their spectra would be very weak) or have spectra that are difficult to observe. Many of the sun's spectral lines are still not identified because they are produced under conditions quite different from those on the Earth.

For a few seconds at the time of a solar eclipse, the chromosphere produces a bright-line spectrum. (Fig. 24–5); thus the chromosphere is known as the reversing layer because it commonly produces absorption lines instead of emission lines. This occurs when the dark body of the moon covers the photosphere but not the chromosphere.

Photosphere and Atmosphere

Through the telescope the sun's yellowish disk is noticeably darker around its outer margin (limb-darkening). Light from the outer edge of the sun travels to us obliquely through its atmosphere and absorption occurs (sunsets are reddish for a somewhat similar reason). Thus we see deeper and hotter regions of the sun at the center of its disk, and we observe higher-level, cooler, less bright regions near the outer margin.

24-5. Flash spectrum of the chromosphere during the total solar eclipse of January 24, 1925. This was photographed with a slitless spectrograph near the end of totality. The chromosphere of the sun was then narrow enough to act as its own slit; thus the images are crescents and not straight lines. Note that the chromosphere, which normally causes dark lines in the solar spectrum, here produces a bright-line spectrum. Gases in the lower chromosphere are actually very hot, but they are "cool" relative to the still hotter gases of the photosphere. The longer lines were made by longer crescents which extend higher into the sun's atmosphere. In fact, the differences in the lengths of the crescentic lines make it possible to map the relative distribution of certain elements in the solar chromosphere. The breaks in the crescentic lines are caused by irregularities in the moon's surface. *(Courtesy Mount Wilson and Palomar Observatories.)*

cool, and sink. Larger and longer-lived convectional cells (supergranules) are superposed on the smaller ones.

Sunspots are perhaps the most familiar of solar features and appear through a telescope as relatively small dark areas, circular to irregular in shape, that are located within about 5 to 40 degrees of the sun's equator. The largest of the spots have diameters that exceed 50,000 miles. The temperature of a sunspot center is about 4500°K (hotter than many stars) or approximately 1500°K (2700°F) lower than the sun's surface elsewhere. Therefore, sunspots appear dark only because we see them against a brighter, hotter background. The inner part of a sunspot (umbra) is cooler and darker than its outer part which commonly shows a radial structure. Bright regions (faculae) tend to enclose sunspots but are not conspicuous and are best observed near the edge of the sun.

The number of sunspots visible on the sun's disk shows a strong periodic variation, with a maximum number of spots occurring about every 11 years on the average (Fig. 24–6). Spots tend to increase in number more rapidly than they decrease in a single cycle, and cycles themselves vary both in magnitude and in length. The International Geophysical Year in 1957-58 was timed to coincide with a maximum of sunspot activity, and additional observations were made in 1964, the year of the "quiet sun." Half of all sunspots exist for a few days or less, but some have lasted for several months.

A sunspot cycle begins with the appearance of a few spots at solar latitudes of about 30 degrees north and south, and as the cycle progresses, new spots originate successively closer to the equator. Each spot lives out its life at approximately the latitude in which it was formed. At the next minimum, the last spots of one cycle will be forming near the equator at the same time that the first spots of the new cycle are forming at about 30 degrees latitude. The cause of this cycle is unknown as is the cause of the sunspots themselves; however,

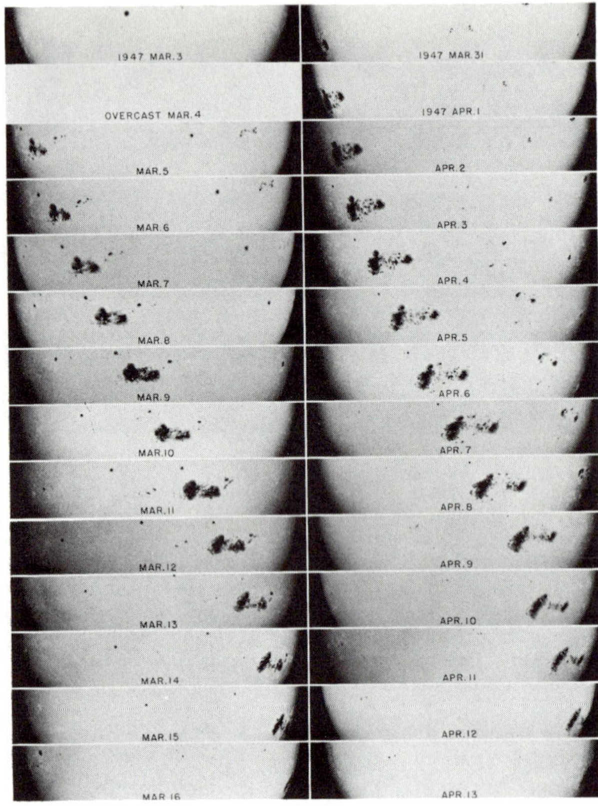

24-3. Sunspots as eviden sun's rotation. This group days and made four trans tained a very large size be seen without a telesc that spots appear, disap change shape. The sizes of may be estimated by imag large the Earth would app side a spot; more than 10 beads on a string would to stretch across the sun. *Mount Wilson and Paloma tories.)*

The photosphere is mottled by bright round *granules* (Fig. 24–4); each is several hundred miles across and exists for only a few minutes. Each granule is hotter (about 100°K) than the photosphere imn surrounding it and evidently is p convectional circulation—the top o column of hotter gases that then sp

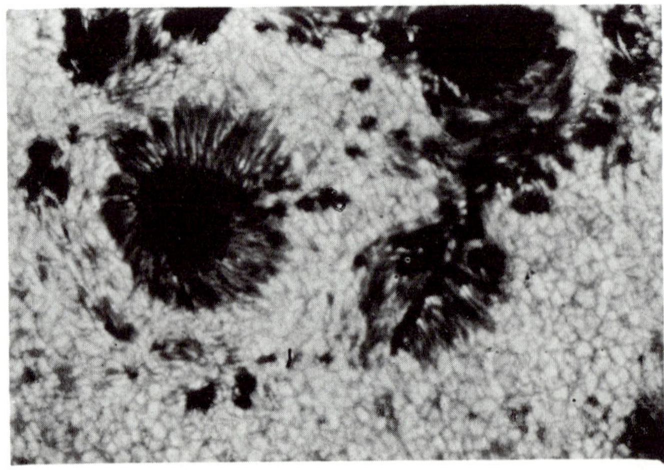

24-4. Sunspots and photographed from manned balloon at a of about fifteen mil August 1957. *(Courtes Elmer Corporation.)*

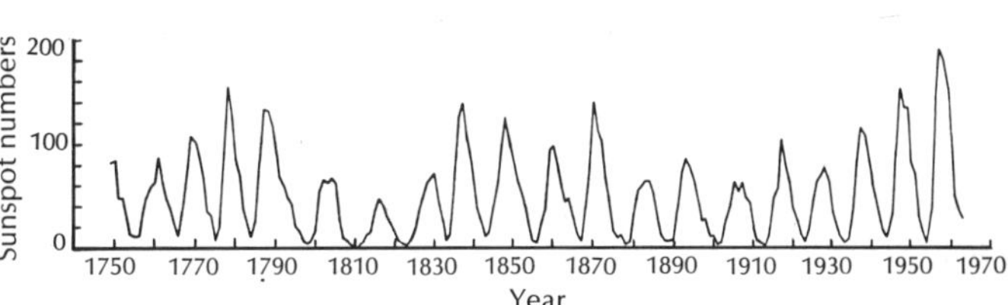

24-6. Annual mean sunspot numbers. The index plotted here (Wolf number) is based upon an equation that includes the daily number of individual sunspots, groups of sunspots, and a calibration constant for the equipment used. (*Courtesy R. W. Fairbridge, ed.,* The Encyclopedia of Atmospheric Sciences and Astrogeology, *Van Nostrand Reinhold, 1967.)*

a sunspot is commonly regarded as a vortex of ionized gases moving at about 1 to 2 mi/sec. The sun tends to be most active during sunspot maxima.

Sunspots also have magnetic fields whose polarities (whether the north or south pole faces the Earth) change systematically in several ways. The spectral lines in sunspots are each split into two or more lines in which the light is strongly polarized. Laboratory study shows that this effect (the Zeeman effect) is produced by light that originates from within a strong magnetic field.

Prominences (Fig. 24–7) are visible during total solar eclipses or in the coronagraph and consist of great reddish gas clouds that erupt from the chromosphere or stream downward toward it. They are striking features against the white background of the corona and some extend outward for several hundred thousand miles. Prominences are widely varied in shape and activity and are puzzling features. Occasionally they seem to form high above the sun's disk, and downward-moving material appears to be more abundant than that visibly erupted. They seem to be somewhat cooler and denser regions where light is emitted as free electrons are captured by atoms and ions. Prominences are most readily seen at the

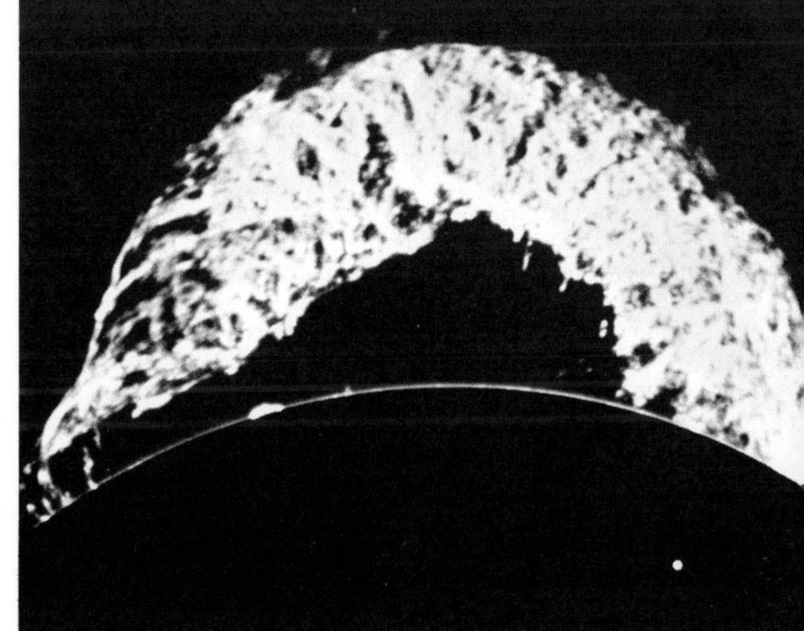

24-7. The very large, spectacular solar prominence of 4 June, 1946 rose more than a million miles above the sun's surface. The white dot in the lower right suggests the size of the Earth. *(Courtesy High Altitude Observatory, Boulder, Colorado.)*

margin of the sun's disk, and motion pictures of prominences have been made by time-lapse photography.

A *solar flare* (p. 656) originates within the chromosphere and involves an explosive brightening of a relatively small part of the sun's atmosphere. An enormous quantity of electromagnetic radiation and charged particles (corpuscular radiation) is emitted abruptly. Flares are commonly associated with sunspot groups and may extend outward for thousands of miles and last from several minutes to a few hours. Flares are rarely seen during observations of the sun in direct light but may be photographed in light from the red spectral line of hydrogen. They affect the Earth by sending outward intense ultraviolet radiations (may exceed normal output of entire sun), X-rays, charged particles, and radio waves.

The extremely diffused gases within the corona (Fig. 24–2) have temperatures approximating 1,000,000°K indicating that individual atoms are moving very rapidly. However, the total heat energy present is relatively small, because density is so low. The cause of the high temperature is unknown. The inner yellowish part can be viewed or photographed through the coronagraph, but the outer, pearly white portion can be seen only at a total solar eclipse (emits about as much light as the full moon). The corona may extend for millions of miles and merge with the zodiacal light. The shape of the corona varies with the sunspot cycle. At sunspot minima (Fig. 24–2), long streamers extend outward from the equatorial region and shorter, slightly curved streamers radiate outward from the polar regions like magnetic lines of force. In contrast, at sunspot maxima (Fig. 22–23) the corona is nearly uniform and circular in outline.

When the sun is studied by radio telescope, it appears much larger because some radio waves originate within the corona. Short radio waves can escape from deep within the corona, whereas longer radio waves can escape only from the upper part of the corona. Thus astronomers can obtain data from different depths within the corona by studying it in different radio wavelengths. Radio waves are emitted continuously at a fairly uniform rate, but much stronger emissions are superposed on them in an irregular manner, particularly at sunspot maxima. Some outbursts are associated with large sunspot groups, some with solar flares, and some have no evident relationship to other solar activity.

Some Solar-Terrestrial Relationships

Solar activity and its terrestrial effects increase in number and intensity at sunspot maxima. Although the sun's total radiation output tends to remain about constant (p. 646), fluctuations do occur in the quantity of ultraviolet, X-rays, and charged particles that it emits (all are most intense at sunspot maxima).

Radio fade-outs may take place throughout the lighted half of the Earth whenever an intense solar flare occurs near the sun's meridian (Fig. 26–11 and p. 655), because the radio-wave reflecting layers in the ionosphere are changed by the abrupt increase in ultraviolet radiation.

The greatly increased quantities of charged particles (chiefly protons and electrons) also associated with a strong flare can cause ionospheric storms which influence radio reception all over the Earth and may last for days. The particles generally take about one day to move from the sun to the Earth, but some very high-speed particles can make the trip in about an hour.

A so-called swift wind of hydrogen (stream of protons) blows continuously through the solar system, augmented by electrons and perhaps a few other charged particles (at about 250 mi/sec at a distance of 1 A.U.). Measurements from space probes have confirmed the existence of this *solar wind* (a picturesque but misleading term) and have shown that it is turbulent and moves faster when the sun is more active.

Evidence for its existence had included the auroras, terrestrial magnetic storms, fluctuations in cosmic ray activity, and the behavior of a comet's tail (always points away from the sun, p. 584). Since comets have been acting as "solar wind socks" for centuries, and in different parts of the solar system, they furnish evidence of the continuous existence of streams of particles that radiate outward in all directions from the sun.

Charged particles from the sun are involved in the origin of the northern and southern lights (auroras) and solar electromagnetic radiation produces a continuous, but very faint, airglow. Auroras and airglow are most pronounced at sunspot maxima. Protons and electrons from the sun and other sources are captured by the Earth's magnetic field and spiral along magnetic lines of force from one hemisphere to the other (Figs. 26–12 and 26–13). When such charged particles subsequently escape from the magnetosphere—as tends to happen following a strong influx from the sun (say 24 hours after a strong solar flare)—they pass downward through the Earth's atmosphere toward the magnetic poles. In so doing, the charged particles ionize some of the atoms in their paths (chiefly nitrogen and oxygen). When the free electrons thus produced recombine with the ions, the varicolored lights observed in the *auroras* are emitted.

Auroras are most common between magnetic latitudes of 65 to 70 degrees (where the outer magnetic lines of force come down to the surface at steep angles) and occur simultaneously in each hemisphere. At times of exceptionally strong solar activity, however, charged particles can penetrate deeper into the Earth's magnetic field. They spiral along lines of force that intersect the Earth's surface farther from the magnetic poles. Such particles may then be discharged at lower latitudes, and an occasional aurora can be seen as much as 40 degrees from the north magnetic pole (located on the north shore of Hudson Bay at 70 degrees north latitude).

Conversely, auroras tend not to occur at and near the magnetic poles because the lines of force which strike the surface in these regions at near 90-degree angles curve far outward into space where their intensity is too weak to capture the charged particles. In other words, if a charged particle approaches the Earth in the plane of the equator, the deeper it penetrates the magnetic field, the lower the latitude at which discharge tends to occur.

Auroras have been produced artificially by exploding a nuclear device at a high altitude (e.g., Project Starfish, July 9, 1962, at 250 miles). This explosion introduced high-energy electrons and other charged particles into the Earth's magnetic field because most by-product radioactive nuclei give off electrons (beta particles), and by-product free neutrons subsequently decay into protons and electrons. The device was exploded in the northern hemisphere and auroras were produced simultaneously in both hemispheres.

An *airglow* also occurs and resembles an aurora in some respects. However, the airglow is very faint, can be detected only with sensitive instruments, shows little variation with latitude, occurs continuously (but is stronger when sun is most active), and seems to result from solar electromagnetic radiation. The airglow exceeds the combined light of all the stars and is presumably stronger in daylight (not visible because of diffused sunlight). Airglow radiation in the infrared exceeds that in the visible light. The light comes chiefly from atomic and molecular oxygen, nitrogen, hydroxyl (OH), and sodium (unexplained). Peak intensity may occur at an altitude of 60 to 100 miles or so.

Magnetic storms or disturbances in the Earth's magnetic field are also most pronounced at sunspot maxima. Particles are ejected from the sun and arrive at the Earth a day later to produce electric currents at high altitudes which change the Earth's magnetic field. Such changes may occasionally disrupt long-distance communication via telephone and telegraph.

The sunspot cycle (or cycles?) may or may not have an effect upon the Earth's weather; this is controversial. There is some evidence that tree rings in certain areas are wider at times of sunspot maxima and that some lakes are slightly higher. Perhaps this results from somewhat increased precipitation, which in turn may be caused by "solar cloud seeding"—particles come from the sun and indirectly provide the nuclei necessary for snowflakes or raindrops to form. The variability of our weather makes such correlations difficult.

Origin of Solar (Stellar) Energy

The total quantity of energy radiated by the sun is enormous and seems to remain relatively constant; in fact, the geologic record appears to show that the sun's total energy output has not fluctuated very much during the past few billion years. The source of the sun's energy has long been mysterious, and even now it may be incompletely or imperfectly known. We do know that the sun is actually too hot to burn because burning requires the formation of molecules, and few molecules can remain intact at such high temperatures. Furthermore, the energy liberated by chemical reactions in combustion is far too small to account for the vast amounts of energy released by the sun. However, decades of astronomical research and observation have led to certain conclusions concerning the manner in which stars form, produce energy, and evolve, although many unanswered questions remain.

Stars have formed in the past, and are probably forming right now, from the matter that makes up the clouds of interstellar gas and dust (chiefly hydrogen, with some helium)—i.e., from the denser portions of the bright and dark nebulae. For a star to develop, gravitational attraction (self-gravity) must cause such a cloud, or a portion of it, to contract, but the mass of this contracting cloud must be within a certain critical range for a single star to form. More than this amount of matter results in the formation of multiple stars and of star clusters. Less than this amount probably results in the formation of planets and satellites. Thus planets much larger than Jupiter probably cannot exist, because small cool reddish stars would form instead.

Gravitational contraction was once considered the source of solar energy, but it proved inadequate. However, it is a source of energy at certain stages in the life cycle of a star; e.g., it provides energy when it shrinks a cloud of gas and dust into a sphere and raises central temperatures and pressures sufficiently to initiate the thermonuclear reactions that are now considered to be the main source of solar (and stellar) energy. The cloud is then transformed into a star.

Thermonuclear reactions, as the term indicates, involve interactions between atomic nuclei at high temperatures, and the processes are known as fusion and fission. For nuclear reactions to occur, the nuclei must collide; but this is prevented by a number of factors. (1) Nuclei are kept widely separated within their electron "shields." (2) Each nucleus has a positive electric charge, and like charges repel. To overcome this repulsion, the nuclei must be made to move at very great speeds, as occurs at extremely high temperatures or in particle accelerators in laboratories. Repulsion is not a problem if one of the bombarding particles is a neutron. (3) The tiny sizes of nuclei and the relatively vast distances that separate them make collisions infrequent.

When nuclei do collide, they react and change in some manner. A tendency exists to either build larger nuclei from smaller ones (*fusion*) or for larger nuclei to be subdivided into smaller ones (*fission*). Iron and neighboring elements on the periodic chart seem to have the most stable nuclei, and the by-products of fusion and fission tend to approach the iron nucleus in size.

Whenever a nuclear reaction occurs, some matter is always changed into energy. Therefore, the sum of the masses of the by-product nuclei is always less than the sum of the masses of the reacting nuclei. To illustrate, the total mass of four individual particles—two protons and two neutrons—is greater than their combined mass as a helium nucleus (also two protons and two neutrons). Thus matter was transformed into energy and released when this nuclear combination occurred. The relative proportions in all nuclear reactions are precisely shown by Einstein's equation: $E = mc^2$ (E is the resulting energy in ergs, m is the mass in grams that is changed into energy, and c stands for the speed of light in cm/sec). Thus the transformation of only a small amount of matter results in the production of an enormous quantity of energy.

Thermonuclear reactions occur only within the central cores of the sun and other stars, where protons, electrons, and atomic nuclei are crowded together, move very rapidly, and collide frequently—thus pressures and temperatures are very great. The interior of the sun has to be very hot to prevent its collapse under the continuous squeezing of its self-gravity. Temperatures of the order of 12 to 15 million degrees Kelvin probably occur at its center to keep its atomic particles moving fast enough to exert an outward counterbalancing pressure.

An average sample of the sun's core may be 100 times denser than water, but still it behaves like a gas, and densities in other stars are much greater. Densities decrease outward to the sun's surface, where they may be more than a thousand times less than that of sea level air on the Earth. According to Wyatt, perhaps 95% of the sun's mass is closer to its center than to its surface—despite the much larger volume of this outer part. At the extreme temperatures and pressures existing deep within stars, many of the nuclei may have lost the electrons that normally occur with them or lose and gain them rapidly. Such matter can have an extraordinarily high density because the nuclei are tiny, and many can be crowded into a small space if their electron shells are missing or incomplete. Yet a nucleus contains nearly all of the mass of an atom.

Since the sun consists almost entirely of hydrogen and helium, reactions have been worked out which involve chiefly these two elements and which could take place in the central cores of stars. The proton-proton reaction and the carbon cycle seem to be the most likely ones (Appendix). The end result of each series of reactions is the conversion of hydrogen to helium and the transformation of matter into energy. Thus one can think of the central core of a star as a gigantic nuclear reactor in which the equivalent of numerous so-called hydrogen bombs are exploding every second. The core does not disperse violently into space, however, because of the powerful restraining effect of the star's self-gravity.

Equilibrium exists inside the sun and within most stars between the inward-directed force of gravitational contraction and the outward-directed forces of radiation (most important in the very hot stars) and gas pressures. If the sun were to expand, its atomic particles would become separated more widely and fewer nuclear reactions would occur. Expansion also results in a drop in temperature. Thus the outward-directed pressures would lessen and gravitational contraction would become the dominating force and cause contraction.

On the other hand, if gravitational contraction squeezes the central core beyond the equilibrium position, the atomic particles are forced closer together and the rate of nuclear activity correspondingly speeds up. This increases the outward-directed pressures. Thus relatively slight fluctuations back and forth across an equilibrium position probably occur continuously in stable stars such as the sun.

Energy probably is transferred from the core to the surface in the following manner. High-energy photons (p. 493) are associated with the shorter electromagnetic wavelengths, whereas low-energy photons are associated with the longer electromagnetic wavelengths. Thus the high-energy photons of gamma radiation and X-rays probably predominate in the radiations that originate within the core. As these photons move outward, they must be absorbed and re-emitted innumerable times and in random directions, thus some are directed back toward the core. Furthermore, one high-energy photon may be absorbed and two or more low-energy photons emitted in its place (the sum of their energies equals that of the single photon absorbed). Thus at successively greater distances from the center, the numbers of photons tend to increase, whereas their average energies tend to decrease. In other words, photons of different energies predominate at different distances from the center.

If the sun is to remain in equilibrium, the amount of energy released in the core in one second must be exactly equal to the amount of energy radiated by the photosphere in one second. If we imagine the sun as consisting of concentric shells like an onion, then each of these shells must be transmitting energy at exactly the same rate as every other shell (the shells are fictitious). But the areas of the various shells are proportional to the squares of their radii ($A = 4\pi R^2$). Thus as the shells become larger, the amount of radiation to be transmitted per square foot diminishes, and the nature of the radiation changes—more of the lower energy photons are produced. At the photosphere, the photons of visible light are radiated outward into space together with those of higher and lower energies.

Energy is also transferred by convection in some sections between the core and photosphere, and this may be a cause of the turbulence that seems to exist below the photosphere.

It follows that the mass of the sun must be less now than it was yesterday and much less than it was a year ago, since about 8 billion pounds of matter are transformed into energy each second. Every 50 million years the sun loses a mass equivalent to that of the Earth, but enough hydrogen fuel is left to fire the solar furnace for some billions of years into the future.

Other Suns

"A sad spectacle! If they be inhabited, what a scope for misery and folly; if they be not inhabited, what a waste of space." (Carlyle contemplating the stars)

The number of stars (Fig. 24–8) visible with the aid of a large telescope is practically limitless, but we can see only 6000 or so with the unaided eye—and less than half at any one time. Families of planets may move about many of these distant suns, and conditions may be favorable for life on some of them. We think that planets are common objects, but no telescope is powerful enough to prove or disprove this conjecture directly.

Of great interest, therefore, was the announced discovery in 1963 by Peter van de Kamp of a large planet orbiting about a small red star (Barnard's star, about 6 light-years away, and invisible to the naked eye). According to this announcement, the planet's mass is about 1½ times that of Jupiter, its period of revolution approximates 24 years, and its mean distance is 4 astronomical units (Kepler's third law; mass of Barnard's star assumed to be 0.15 that of the sun). The planet itself has not been seen—it is much too faint. Instead, it was detected by the "gravitational wobble" that it produced in the motion of Barnard's star. This deviation from a straight line amounted to a maximum of 1/10,000 of an inch (hundreds of photographic plates taken decades apart were compared). This is near the limit of detection: a more massive primary, a less massive planet, and a greater distance between them, would all result in a smaller "wobble." Continued observation has led to a recent alternative interpretation that two planets, not one, circle Barnard's star at distances of about 12 and 26 astronomical units—the mass of each about equals that of Jupiter.

Other stars presumably have sunspots, prominences, coronas, and similar features, but we cannot actually observe such features since stars are only point sources of light. Although stars range widely in size, density, surface temperature, and luminosity, they differ less in mass and chemical composition.

Stellar Motions

The *proper motion* of a star is its very small, annual shift on the celestial sphere measured in seconds of arc. This is caused by the star's own motion in space and was detected in the early 1700's by comparing the positions of certain stars with those recorded some 1500 years earlier by Ptolemy. Thus stars are not "fixed" but move at rates measured in tens of miles per second. Yet they are so remote that constellations remain unchanged in shape to the unaided eye for many centuries. The largest recorded proper motion is that of Barnard's star: about 1 degree on the celestial sphere in 3½ centuries.

The size of a star's proper motion results from the combined effect of three independent variables. When the other variables are kept equal, proper motion varies inversely with distance, increases with speed, is largest when the direction of movement is 90 degrees to the line of sight. For any one star, the relationship between distance and proper motion is quite unreliable, but it does correlate approximately with distance when statistical averages are determined for groups of stars (see distances to Cepheid stars, p. 535).

Stars rotate on their axes, and this widens their spectral lines. Unlike the sun (p. 510), a spectroscope cannot be aimed first at one side of a star and then at the other; nevertheless, red and violet Doppler shifts do occur simultaneously, and this broadens the spectral lines. In general, the width of a spectral line is proportional to the rate of rotation, but the orientation of a star's axis in space is also a factor; e.g., if the axis points directly at the Earth, no rotational widening is observed.

24-8. Star trails in the southwestern sky produced by the constellation Orion and other stars. This photograph was made in the same way as Fig. 17-1. Notice that the stars seem to move in a clockwise direction when the observer faces south. Orion was imagined by ancient star-watchers to represent a hunter: Betelgeuse is a shoulder star, Rigel is one knee, and three stars in a row form a belt that supports a dagger containing the Great Nebula. Differences in stellar colors are illustrated by red Betelgeuse and blue Rigel. Each is a large, highly luminous star several hundred light years away; in fact, Betelgeuse may be one of the largest stars known. The three belt stars point westward toward the Hyades and eastward at Sirius in Canis Major. Orion is a strikingly beautiful sight on a clear dark night. *(Photograph, John Stofan, Teaneck, N.J.)*

The average rotational velocity for white stars may approximate 30 mi/sec (48 km/sec) but that of bluish stars tends to be higher, and that of yellow and red stars tends to be lower (perhaps unseen planets have absorbed much of their angular momentum).

Stellar Sizes and Densities

Stars differ widely in their overall average densities: at one extreme are stars several hundred thousand times denser than water, whereas at the other extreme are stars several hundred thousand times less dense than water. Supergiant stars have average densities in their outer parts that approach the best vacuums that can be produced on the Earth and have diameters that exceed the sun's by several hundred times or more. On the other hand, some stars are as small as Mercury and Mars (or even smaller, see p. 618) and these are amazingly dense objects called white dwarfs. An average cubic inch of the matter in a white dwarf, if it could be moved to the Earth's surface, might weigh a ton or more. Such stars presumably are made of so-called degenerate matter in which atoms have been stripped of their space-consuming electron shells. Thus the mass of a sun can be packed into a volume equal to that of a small planet. However, the great majority of the stars are neither supergiants nor white dwarfs. They are more like the sun in both size and density, although the average densities of small reddish stars (lower end of main sequence) may be 100 times that of the sun, and these are very abundant.

The diameters of stars can be calculated by at least three methods. A special type of instrument (stellar interferometer) has been used with the largest telescopes to measure the angular diameters of a small number of relatively near, very large stars. The actual diameters of these stars can then be calculated if their distances are known (p. 527).

In eclipsing binary systems, two stars revolve around their common center of gravity and alternately eclipse each other. In some instances, the diameters of the individual stars can be determined from the duration of the eclipses.

The diameters of most stars, however, are calculated indirectly from their surface temperatures and absolute magnitudes (can be determined if distance is known, p. 532). The amount of light radiated by a star depends upon its: (1) surface area which varies directly as the square of the radius or diameter (the area of a sphere equals $4\pi r^2$); and (2) surface temperature—the amount of light emitted per unit area of surface varies directly as the fourth power of the absolute temperature (p. 496).

We can compare other stars to the sun, and we find that their luminosities are proportional to the squares of their radii (relative to the sun as 1) multiplied by the fourth power of their absolute temperatures (relative to the sun as 6000°K)—in symbols ($L \propto R^2 \times T^4$).

To illustrate, the luminosity of a star that has a radius of 864,000 miles and a surface temperature of 18,000°K is 324 times that of the sun ($L \propto 2^2 \times 3^4$).

To calculate the radius of a star in this way, assume that the star has a surface temperature of 12,000°K and is 64 times more luminous than the sun. Therefore, its radius must be twice that of the sun ($64 \propto R^2 \times 2^4$).

Stellar Temperatures and Luminosities

The colors of stars (more accurately, their spectral energy curves, Fig. 19–4) indicate that their surface temperatures vary from a maximum of 50,000°K or more (blue stars) to a minimum of about 2000°K or less (red stars). More stars are closer to the minimum than to the maximum. The radiation from a star whose surface is much less than 2000°K would be too weak to detect visually unless the star were relatively very close. Such very cool stars would radiate chiefly in the infrared region, and a number have been discovered. They seem to be extremely large objects.

Stars have a very great range in luminosities. At one extreme are very luminous blue stars (also supergiant reddish stars) that are thousands of times more luminous than the sun. At the other extreme are small cool reddish stars that are thousands of times less luminous than the sun.

Small, low-luminosity, reddish stars are by far the most common type of star in our immediate neighborhood of space. According to Peter van de Kamp, 59 stars are now known to occur within 5 parsecs of the Earth (about 16 light-years, Fig. 24–9). This number includes the sun but not the seven unseen companions (like those of Barnard's star). Of the 59, 31 seem to be single stars, 11 are double stars, and 2 are triple-star systems. Sirius, with a diameter double that of the sun, is the largest and brightest star in the group. Its luminosity (23 times that of the sun) actually exceeds the combined luminosities of all the other stars. This suggests that most of the starlight in the universe may be emitted by a relatively small number of hot, massive stars.

If this is a representative stellar sample, then small reddish low-luminosity stars are by far the most numerous in the universe. However, we can never actually know this because such stars, when far away, are too faint to be visible. Thus star counts at great distances overemphasize the proportion of brighter stars. A recent star count by W. J. Luyten of a larger region does not change the 5-parsec results appreciably (556 stars occur within 10 parsecs of the sun; their total masses and luminosities exceed those of the sun by 268 and 235 times respectively).

The luminosities of stars may be compared with that of the sun, or their absolute magnitudes (p. 533) may be computed and then compared. The absolute magnitude of the sun is +4.86.

Stellar Masses and Compositions

Stars apparently vary less in mass and chemical composition than in most of their other physical properties. Stars are known that have masses more than 50 times greater than the sun, but these are exceptional, as are stars that are much less massive than the sun. Stars indeed are rare which are 5 times more massive or 10 times less massive than the sun. However, only the masses of double stars can be calculated directly (p. 538).

The mass of a star seems to be one of its most fundamental properties and the determining factor in the type of star it is, how much light it radiates, and even apparently how fast it "ages" and how long it "lives."

An important *mass-luminosity relationship* exists: the greater the mass of a star, the greater is its luminosity (varies approximately as the mass raised to the third or fourth power). Perhaps 90 percent of all the stars that are visible from the Earth conform to this relationship; these are the so-called main sequence stars (p. 609). The exceptions include the white dwarfs (more massive than expected) and the red giants (somewhat less massive than expected) that we shall encounter later.

Stars consist primarily of hydrogen and helium (perhaps 96 to 99% or more by weight) and hydrogen probably makes up about 50 to 80% of this total. All of the other elements together may make up only 1 to 4% or less of the matter present in stars, and of these, the lighter elements generally tend to be the more abundant. However, the chemical composition of a star does change with time because hydrogen is transformed into helium and other nuclear transformations occur.

Stellar Spectra

The spectrum of a star yields an amazing variety of information: e.g., its temperature, structure, chemical composition, rate of rotation, whether it is approaching the Earth or receding from it, and its magnetic characteristics. A star may be classified by its spectrum (Fig. 24–10), which depends upon

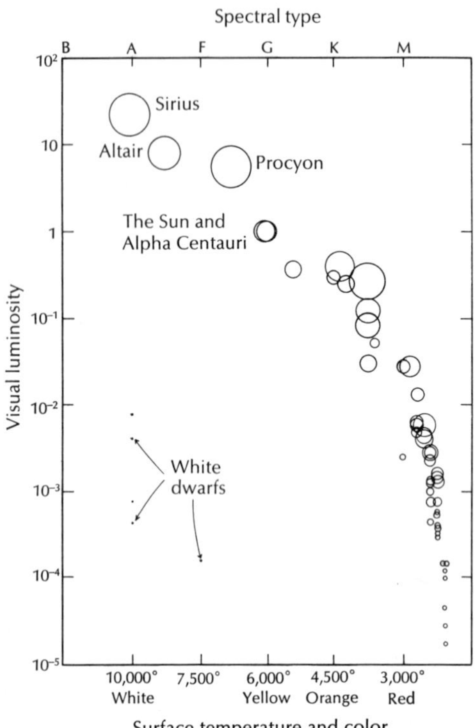

24-9. The fifty-five nearest stars (now updated to fifty-nine) are arranged on the graph according to visual luminosity, spectral class, and comparative diameters (the size of a circle is proportional to the size of the star it represents). The sun's luminosity is taken as 1. *(Courtesy Peter van de Kamp, Sproul Observatory, Swarthmore, Pa.)*

the star's temperature (most important factor by far), chemical composition, and physical structure. Seven types (O, B, A, F, G, K, M) include nearly all stars, and the following mnemonic sentence keeps these in order: *Oh Be A Fine Girl, Kiss Me!* Members of one type have temperatures, colors, and spectral lines which are all transitional to those of neighboring types. The sun is a yellowish, class G star.

Two stars may have identical chemical compositions but yet have distinctly different spectra because their surface temperatures are different; e.g., the spectra of molecules are observed only in relatively cool stars because molecules dissociate into their constituent atoms at higher temperatures. In addition, more energy is required to ionize some elements (e.g., helium) than others (e.g., the metals), and the spectrum of an ionized atom tends to be quite different from that of neutral atoms of the same element. Furthermore, when hydrogen becomes ionized by losing its one electron, it can no longer produce any spectral lines at all. Thus no spectral lines of hydrogen may be observed in a very hot star that is yet three-fourths hydrogen.

The Hertzsprung-Russell (H-R) Diagram

The H-R type of diagram (Fig. 24–11; named after the two astronomers who developed it) is constructed by plotting the absolute magnitudes of stars (vertical axis) against their spectral types on a graph. Temperature, color, and spectral class all change gradually along the horizontal axis from the hottest, blue, class 0 stars at the left to the coolest, red, class M stars at the right (color magnitude diagrams are similar and are used more commonly because the color of a faint star can generally be measured accurately with a photoelectric cell, whereas its spectrum is barely visible on film). Since luminosities increase toward the top of the graph, the vertical axis also shows the relative masses of the main sequence stars (they increase toward the top following the mass-luminosity relationship). However, Fig. 24–11 overemphasizes the brighter stars. In a more representative sample (Fig. 24–9), by far the greatest number of stars would probably occur in the lower right part of the main sequence, and white dwarfs would probably be much more numerous.

Note that most of the stars in Fig. 24–11 cluster along a zone that extends from the upper left downward to the lower right. Large, hot, bluish, massive stars occur at the top and small, cool, reddish, low-mass stars occur at the lower right. These are the end members of a completely gradational sequence that forms the *main sequence*—an appropriate term since 90 percent or so of all known stars probably form part of it.

Unusual stars called *white dwarfs* (very small hot stars of low luminosity) occur near the bottom of the diagram. Probably these should be more numerous, but they can be observed only if they are relatively close to the Earth. Stars that occur above the main sequence in the upper right part of the H-R diagram are appropriately called *giants* and supergiants (absolute magnitudes of −2 or above—an arbitrary criterion). They are very large in volume.

Reddish stars on the right side of the diagram are subdivided into two groups, one of much greater luminosity than the other. Since stars of the same color (say type M) emit the same quantity of radiation per unit area, the luminous M stars must be very large in size—giants or supergiants.

H-R diagrams can be plotted for samples of stars from different regions of space, and thus we have different types of H-R diagrams which reveal certain basic relationships among stars.

Once developed, H-R diagrams can be used to determine the distances to stars (known as the method of *spectroscopic parallaxes*). Assume that a certain star has a spectrum showing that it belongs to a particular class of main sequence star. We then read its absolute magnitude directly off an H-R graph and arrange to measure its apparent magnitude. We thus have enough data to calculate the distance (p. 532).

Star Clusters

The study of star clusters has been important in the development of current theories concerning stellar evolution and in discovering the shape of our galaxy and our location within it. The members of a cluster of stars are closer together than average and move as a unit through space, although some individual motion may be superposed upon the group motion. Presumably all of the stars in any one cluster formed at about the same time and from the same cloud of gas and dust. Thus their chemical compositions at the time of origin were probably very similar. This seems the only possible explanation for their association in space and the common motions they share. Although distances between stars in a particu-

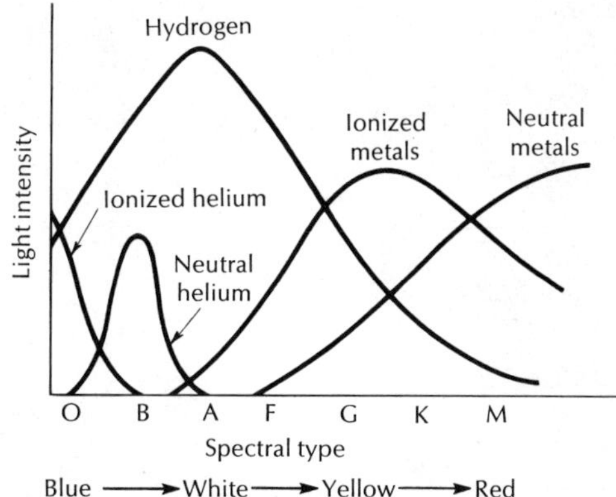

24-10. The spectra of stars. Spectral class, color, and temperature (highest at the left) are all plotted along the horizontal axis. *(After Inglis.)*

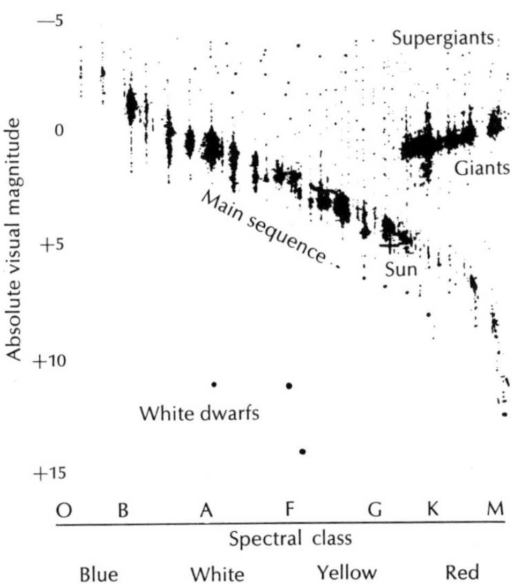

24-11. H-R diagram for stars in our section of the galaxy (part of a spiral arm with Population I stars). *(Modified from a diagram by W. Gyllenberg, Lund Observatory.)*

24-12. Globular cluster M 13 in Hercules. It has a diameter of about 150 light-years and is some 30,000 light-years away, but these figures are only approximations. *(Courtesy Mount Wilson and Palomar Observatories.)*

lar cluster are less than average, nevertheless the members are still too widely separated to have been pulled together by their gravitational attraction.

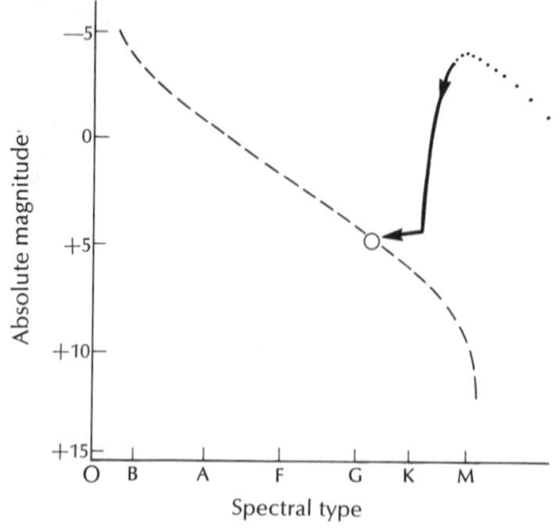

24-13. Likely, pre-main sequence, evolutionary track for star with mass similar to that of sun. *(R. H. Baker and L. W. Fredrick,* An Introduction to Astronomy, *Van Nostrand Reinhold, 1968.)*

Star clusters are of two main types. *Galactic clusters* are most abundant along the equatorial plane (Milky Way) of our galaxy; hence their name. They have irregular shapes, and stars tend not to be more concentrated toward their centers. The Pleiades (Fig. 25–3) and Hyades are familiar examples. The number of stars in a galactic cluster commonly range from a few tens to a few thousands.

In contrast, *globular clusters* (Fig. 24–12) have spherical shapes and stars are more concentrated toward their centers. Although they appear close together, stars within a globular cluster may still average a light-year or so apart, and collisions are probably very rare. Globular clusters are difficult to photograph—if exposure is correct for the brighter central portion, then the fainter outer part remains invisible, and vice versa. Globular clusters have diameters of the order of 100 to 200 light-years, and a single cluster may contain as many as a few hundred thousand stars. Globular clusters form a "halo" around our galaxy (Fig. 25–1).

Life Cycles of Stars

The stars in a single cluster presumably formed at about the same time and from the same type of material. Therefore, any differences that we now observe among its members are presumably due to differences in their original masses and to differences in their evolutionary developments. We may think of stars as having genetic ages (the present stages in their evolutionary cycles such as young, mature, and old) and chronological ages (the actual numbers of years since they originated). The arrangement of stars in H-R diagrams of different clusters can thus provide clues concerning the life cycles of stars.

The evolution of stars is such a slow process when compared to the lifetime of one man (or even of all recorded history) that we cannot hope to see changes occur in a particular star from one year or decade to the next. We must observe stars in their present stages and then try to deduce the different steps in their development. An analogy may be helpful. If one were to walk for an hour through a forest, he could observe trees in different stages of their life cycles: seeds, tiny saplings, young trees, big trees, dead trees, and partly decayed logs. The observer could not watch a tree evolve through its cycle during that hour, but he might be astute enough to deduce such a cycle from his observations of different trees in different stages of their life cycles. So it may be with astronomers and stars.

The mass of a star is basic to its origin and subsequent development (chemical composition is a lesser factor but can produce differences). According to current ideas, a star originates from cool diffused interstellar material that contracts into a large sphere under its own self-gravity. As contraction takes place, temperatures and pressures in the interior increase—part of the energy released by contraction goes to heating the interior and part is radiated away. Eventually thermonuclear reactions can begin in its central core, and a star has come into being. Equilibrium develops between the inward-directed and outward-directed forces, and contraction ceases.

The star is now on the main sequence, and it passed through the contracting stage (Fig. 24–13) relatively rapidly. Therefore, few stars are observed in this stage today, and few points representing them occur on an H-R diagram. Stars apparently contract gravitationally at different rates which are determined by their masses. Large massive stars, with very strong self-gravities, contract rapidly to the main sequence stage, whereas small low-mass stars contract slowly.

Imagine that we can follow the life cycle of a star (Fig. 24–14), that we collect data on it at say 1000-year intervals, and that we then plot these as points on an H-R diagram—one point for each 1000 years. The points will be relatively far apart on the graph when evolutionary changes are most rapid but close together when the rate of change is slow. In fact, point after point will be superposed during the long-lasting main sequence stage. Our first points, representing cool pre-star interstellar matter, will be off the graph in the lower right. With time, successive points will form a curved line that eventually reaches the main sequence (Fig. 24–13). The luminosity of a star during this relatively brief contracting stage may attain a peak much greater than it has subsequently as a main sequence star because its enormous surface area more than offsets its somewhat cooler surface temperature. The line of points may curve sharply downward just before the main sequence is reached

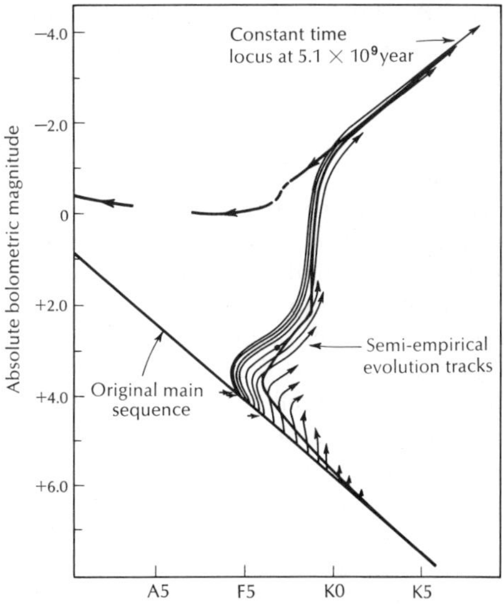

24-14. Empirically determined evolutionary tracks for the Globular Cluster M 3 (by A. Sandage). Arrows show probable changes since the stars left the original main sequence about 5 billion years ago. The sun's absolute visual magnitude is +4.8, (same as its absolute bolometric magnitude which includes all radiation emitted by a star, not just that in the visible region). Sun-sized stars in M 3 show relatively little movement on the H–R diagram during the 5 billion years, whereas somewhat more massive stars become red giants during this interval. Still more massive stars have advanced beyond the red giant stage (downward and to the left), and some may now be at or near the white dwarf stage (not shown). (R. H. Baker and L. W. Fredrick, *Astronomy,* Van Nostrand Reinhold, 1971.)

—i.e., the luminosity decreases abruptly—because the size of the star decreases rapidly without much change in its surface temperature. Until a star reaches the main sequence, its source of energy is primarily that released during contraction.

Location on the main sequence seems to be determined almost entirely by the mass of the parent cloud of gas and dust: a large mass contracts into a large hot blue star of type 0, a medium-sized mass contracts into a yellowish type-G star like the sun, and most abundant of all, a small mass produces a small cool red star of type M or K. In still smaller masses, thermonuclear reactions may never be initiated, and further contraction may produce a so-called black dwarf.

The mass of a star also determines the rate at which its thermonuclear fuel is consumed because a star's luminosity measures the nuclear reactions occurring within it. Thus hot massive blue stars consume their fuel at a very rapid rate—in a few million years or less—and stars such as Rigel may not have been in existence when the first men walked upon the Earth. Such massive stars have large, hot, dense cores; thus nuclear reactions occur at a very rapid rate throughout a large volume, and certain reactions take place that are impossible under less extreme conditions. On the other hand, a star like the sun may require some 10 billion years or more to use up the hydrogen in its core, and still smaller stars probably require much longer. Until a critical portion of the hydrogen in the core has been used up, a star remains in equilibrium and stays about on the main sequence (however, as helium accumulates in its core, the volume and temperature of a star apparently increase gradually). This lasts for a long time for most stars, and thus we find the great majority on the main sequence when we construct H-R diagrams.

The H-R diagram for a very young cluster might show some stars on the upper part of its main sequence (large masses and rapid contraction) whereas less massive stars would not yet have reached the main sequence and would be represented by points to its right. In some stellar groups in nebulous regions of space, certain stars have been observed to vary irregularly in brightness (because they are not yet in equilibrium?), and a few stars have been observed on more recent photographs which seem not to have been present on older photographs. Perhaps intervening gas and dust have dispersed rapidly enough to reveal them, or perhaps some parts of the contracting phase are actually quite rapid. These are interpreted as signs of youth.

Figure 24–14 illustrates likely subsequent stages in the life cycle of a star about as massive as the sun, and stages for stars of greater or lesser mass would perhaps be somewhat similar. When most of the hydrogen within its core has been transformed into helium, nuclear reactions slow down, and the star contracts. Central temperatures and pressures are thereby increased; gravitational potential energy is changed into kinetic energy; and part of this energy is then radiated outward. This initiates nuclear reactions in the hydrogen shell surrounding the "burned-out" core, and this new release of energy may be the reason the star expands to form a red giant. The star's outer surface cools as it expands—thus it reddens—but the total luminosity increases because its surface area becomes so large. Thus the line of points representing successive stages in the life cycle of this star slants upward and to the right on an H-R diagram.

Following the red giant stage, a star probably changes into a white dwarf, but just how this is accomplished is still quite uncertain. The star may go through an unstable period during which it pulsates as a variable star (does this also precede the red giant stage?) and an eruptive stage during which it ejects matter into space (becomes a nova, supernova, or planetary nebula?). High-mass stars may actually shift back and forth a few times between the Cepheid and red giant phases, thus producing the scatter observed in the period-luminosity relationship. Moreover, in contrast to stars of lower mass, massive stars may leave the main sequence by shifting almost horizontally to the right—thus they fit the mass-luminosity relationship better.

After a star becomes a white dwarf, it presumably must cool slowly, with corresponding color changes, into a dark body. In this view a white dwarf represents the "burned-out" core of a former main-sequence star. Nuclear reactions have ceased, and gravitational attraction has squeezed the atomic nuclei into a very dense ball.

When H-R diagrams for different clusters are compared, any differences should depend primarily upon age differences. Therefore, the larger the portion of the upper part of the main sequence that is missing, the older is a cluster. Stars that once were present in the upper main sequence of an old cluster are probably represented now by different kinds of stars elsewhere on the diagram, especially in the white dwarf region. Stars that evolve less rapidly are now red giants, whereas stars on the lower half of the main sequence evolve so slowly that they have yet to leave it.

Current theory indicates that the sun has been a main-sequence star for nearly 5 billion years and predicts that it may continue to remain one for another 5 billion years or so. Following this, the sun may expand into a star with a diameter a few tens of times larger than it has at present. The outer surface of the sun should then be cooler (it may become a red giant) and much closer to the Earth. Temperatures on the Earth may then be high enough to boil away the oceans and melt metals such as lead. Later, after possible stages as a pulsating variable star and an explosive nova, or a planetary nebula, the sun should become a fantastically dense white dwarf star about the size of the Earth and gradually cool off. Life as we know it is possible only during the long, main sequence, mature stage of the sun's development.

Origin of the Heavy Elements

The heavy elements (those beyond helium) may have been formed primarily by nuclear transformations that took place within the cores of stars from material that was once chiefly hydrogen, but may also have contained some helium and a smattering of other elements. At temperatures and pressures far exceeding those now inside the sun, three helium nuclei may fuse into a carbon nucleus (atomic number 6)—perhaps attained during or near the end of the red giant stage, or perhaps attained only in more massive stars at such a stage. Successive helium nuclei may then be added to a carbon nucleus, each increasing the atomic number by two and the mass number by four, and thus oxygen and neon (atomic numbers 8 and 10 respectively) may form. Carbon nuclei themselves may fuse to produce magnesium nuclei (atomic number 12). At different stages, a single neutron may also be added to a nucleus—this may change into a proton (thus raising the atomic number by one) and an electron (ejected). Thus may the nuclei, up to and including iron (atomic number 26) be built up.

We may picture different nuclear reactions taking place simultaneously within a massive star, but at different distances from its center. Perhaps iron forms at the center at the same time that magnesium is produced farther out. Meanwhile carbon and helium may be forming within shells of lesser pressure and temperature that are still farther outward.

Iron nuclei are the most stable and energy in great quantities must be added to either build up or break apart the iron nuclei. However, this may occur at some stage in the life cycle of a star, or of some stars (perhaps the supernovae) and thus all of the natural elements may be produced within stellar interiors (but not all within the life cycle of a single star).

As stars evolve, they seem eventually to pass through an eruptive stage in which matter from the interior is ejected into space (from deep within interior when eruptions are most violent). Perhaps on the average, a star ejects half of its mass before it ends as a white dwarf, and this ejected material increases the proportion of heavier elements in the interstellar matter. New stars (so-called second generation) may subsequently form from this "contaminated" gas and dust, and still heavier elements may be produced within them before their white dwarf stages. In this manner, some of the ejected debris of an older star forms the parent material for a younger star. Most of the heavy elements might then be expected in a third generation star, and perhaps the sun is such a star. If there is any validity to this speculation concerning the origin of the heavier elements, a startling conclusion follows: the atoms in you and me, as well as those in the sun, may once have been part of a star!

Stellar Populations

The distribution and abundance of stars of different types vary within a galaxy as shown by differences among the H-R diagrams for groups of stars located in different regions. The most important causes of the differences seem to be the presence or absence of interstellar gas and dust and the absolute and genetic ages of the groups.

Accordingly, two broad groupings have been recognized: Population I and Population II. Stars of Population I characteristically occur in the spiral arms of galaxies and in other regions where interstellar gas and dust are abundant. According to one definition, they are stars that have formed from so-called contaminated interstellar matter. The heavier elements seem to be most abundant in stars of Population I, and the sun belongs to this group (or to a somewhat similar intermediate group). The brightest stars of Population I are large, very luminous blue stars, and Cepheids of this group have luminosities that are four or more times as great as those in Population II. Some Population I stars may be old, whereas others are much younger, and new stars are probably forming today.

The stars of Population II are located in regions of space where interstellar gas and dust are scarce or absent: in the nuclei of spiral galaxies, in the areas between the spiral arms, in elliptical galaxies, and in globular clusters. Red giants and supergiants are the brightest stars of Population II, and the heavier elements are scarce. Population II stars are interpreted as old stars that probably formed from nearly pure hydrogen, or from a hydrogen-helium mixture. Star formation stopped in such regions after the supplies of interstellar matter had been used up, and such groups now lack massive stars such as the blue giants that evolve rapidly. In fact, in a very old grouping, the entire upper half of the main sequence may be missing.

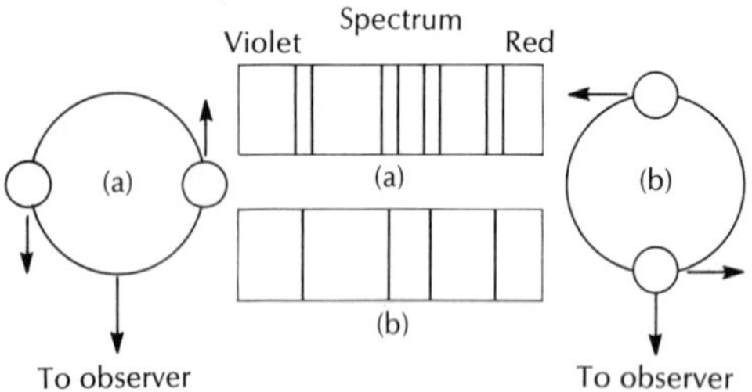

24-15. Spectrum of a spectroscopic binary (Mizar). Many double stars cannot be separated visually because they are either too close together or too far away. However, they can sometimes be detected spectroscopically by their Doppler effect if their orbital planes are so oriented that during one part of their revolution one star approaches the Earth (violet shift) at the same time that the other star recedes from the Earth (red shift). This was occurring when the upper spectrum was photographed. The lines are double. On the other hand, the lines of the two stars are superposed in the lower spectrum because the stars were then moving perpendicular to the line of sight and neither was approaching or receding *(R. H. Baker,* Introduction to Astronomy, *Van Nostrand Reinhold.)*

Double and Multiple Stars

Perhaps half of all stars are members of double-star systems (binaries) or less commonly of groups of three or more stars. The stars in a binary or multiple-star system revolve about their mutual center of mass and are relatively close together—in fact, some pairs may actually be in contact. A number of pairs can be separated visually or photographically, whereas others can be detected only by spectroscopic study (Fig. 24–15). If one star of a binary is much brighter than its companion, the only spectrum observed is that of the brighter star. At one time it shows a red shift, but half an orbital revolution later it shows a violet shift.

Still other binary systems are recognizable because their orbits are oriented in such a manner that one star periodically eclipses the other, either partially or entirely (Fig. 24–16). In this manner the period of revolution can be determined. Double stars are important in astronomy because they provide a means of measuring the mass of a star (p. 538).

True binaries should be distinguished from two stars which appear close only because they are almost on the same line of sight from the Earth (an optical double). The second star (Mizar) from the end of the handle of the Big Dipper illustrates both types. It forms an optical double with a very faint star (Alcor) that can be seen with the unaided eye near it on the celestial sphere. On the other hand, Mizar is a binary, but this can be observed only through a telescope.

Sirius is another double star, and its companion is a white dwarf. The dwarf's presence was detected "gravitationally" before it was first observed; the gravitational attraction of the dwarf causes Sirius to weave slightly in space.

If two stars are sufficiently close, gravitational attraction may cause gases to stream out of one or both stars. Such gases can be detected if they cause extra absorption lines when they are located between one of the stars and the Earth. This illustrates one of the general techniques used by astronomers in their search for an understanding of stars. The spectrum of a star is photographed on a number of occasions, and when these pictures of different dates are brought together for study, some unusual lines may be observed, or certain lines may be present

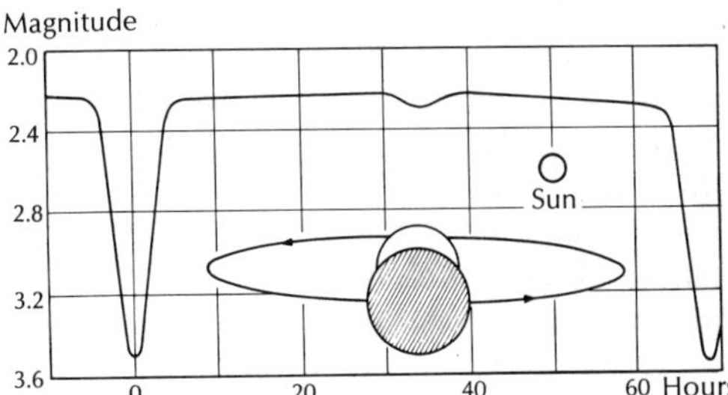

24-16. Light variation of an eclipsing binary (Algol in Perseus). The size of the sun is shown on the same scale. One star is somewhat smaller but much brighter than the other, and the two are about 13 million miles apart. Algol "winks" about once every seventy hours because the fainter star partially eclipses its brighter companion at this time. A smaller "wink" occurs midway when the brighter star partially eclipses the fainter one. In this part of its orbit, the fainter star acts somewhat as a "full moon" and is made more luminous by light from the brighter star. Thus the light curve rises slightly before the smaller "wink" and declines slightly after it. *(Determined by Joel Stebbins.)*

at one time and not at another. The possible conditions on the star that could produce the observed changes or lines in its spectrum must then be interpreted. In this manner, interpretations that seem appropriate one year are discarded in a following year because new data have been obtained or a new interpretation seems more probable. Changes and revisions in certain aspects of astronomy are very rapid.

Variable Stars

The light of a star may vary regularly or irregularly and for a number of reasons including the following: (1) it may pulsate regularly as a Cepheid (p. 534) or as similar stars with longer or shorter periods than Cepheids; (2) the variation may be caused by an eclipse of one member of a binary system by the other (not a true variable); and (3) certain stars brighten explosively (novae and supernovae).

R R Lyrae stars vary regularly like Cepheids but have periods of about 24 hours or less. They are of particular importance in estimating stellar distances because they all seem to have the same average luminosity regardless of period. Furthermore, they are about 100 times as luminous as the sun and can be observed at great distances. They were important yardsticks in measuring our galaxy.

Novae and supernovae are stars that brighten tens of thousands of times in an explosive outburst. The name means "new stars," but novae are small hot stars both before and after an eruption. After an outburst a nova may take a number of years to return approximately to its former luminosity, and fluctuations in luminosity may occur during this period. Some novalike stars erupt at intervals of a few decades or so, whereas the typical novae have been observed to erupt only once. However, this does not eliminate the possibility of repeated eruptions at long intervals. Spherical shells of expanding gases have been detected around some novae, but only a relatively small fraction of the mass of a nova is blown away. One can speculate that novae form one stage in the life cycle of some stars and precede the white dwarf stage. Astronomers do not know precisely why a nova erupts.

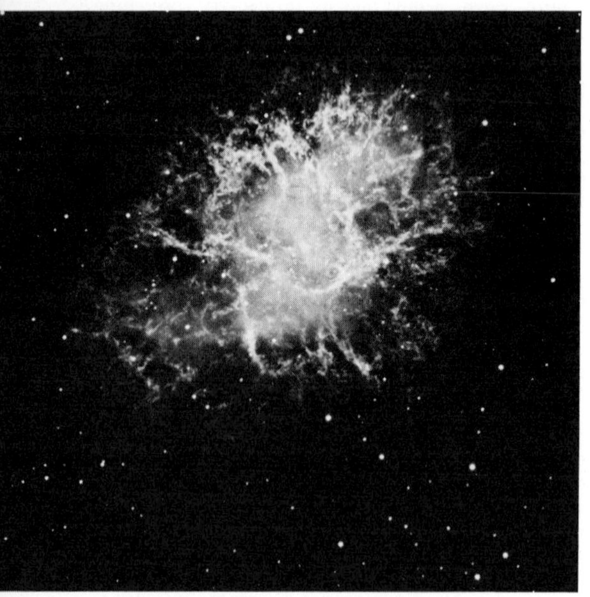

24-17. The gaseous expanding shell of a supernova (the Crab Nebula in Taurus), photographed on film sensitive to red light. A bright "new" star was recorded by Chinese astronomers on 4 July 1054. Apparently it could be seen during the day for a few weeks and at night for a year or so. The actual explosion, however, occurred about 3000 BC because the Crab Nebula is estimated to be some 4000 light-years away. Photographs taken in different years show that the Nebula is visibly expanding; the rate has been measured and when projected backward into time, it indicates that the eruption appeared some 900 years ago, at a date that corresponds to the Chinese record. A white dwarf star may be located at the center of the nebula, which is a strong source of X-rays and radio waves. *(Courtesy Mount Wilson and Palomar Observatories.)*

The explosion of a supernova is phenomenal: at greatest brilliance its luminosity may approach that of the entire galaxy in which it is located, and a sizable fraction of its mass is blown off. Three supernovae have been observed in our galaxy and evidence for others may have been discovered: one was seen and recorded by Chinese and Japanese astronomers in 1054 (Fig. 24–17); the others were observed by Tycho Brahe in 1572 and by Kepler in 1604.

Planetary Nebulae

These objects have misleading names and are not related to planets, although they do have a certain resemblance, when viewed through a small telescope, to the hazy greenish disk of Neptune. Actually, they are enormous, expanding, gaseous envelopes that surround small, very hot stars. Their diameters span thousands of light-years. Much ultraviolet light is radiated by a hot blue star at the center, and this causes fluorescence: gases in the nebula absorb the ultraviolet and reradiate visible light. The gaseous shell is continuous, although it may appear like a ring (Fig. 24–18; the ring nebula in Lyra is another example). When our line of sight is 90 degrees to the shell, we may not see its very diffused gases; however, when we look obliquely through the shell near the edge, it appears brightest and is then visible.

A planetary nebula differs from the gaseous envelope around a nova in having more mass (perhaps 10 to 20% of that of the sun), a slower rate of expansion, and a longer life. However, the lifetime of a nebula may only be 100,000 years or so—expansion will so dilute the gases within this interval that they then become invisible. Thus the planetary nebulae we observe today are not those that could have been seen 100,000 years ago. Presumably, therefore, they are relatively common and occur during one phase in the life cycles of many stars—perhaps just prior to the white dwarf stage. They are part of a recycling process in which stars form from interstellar matter, subsequently eject gases to this interstellar material, and from which other stars may eventually form.

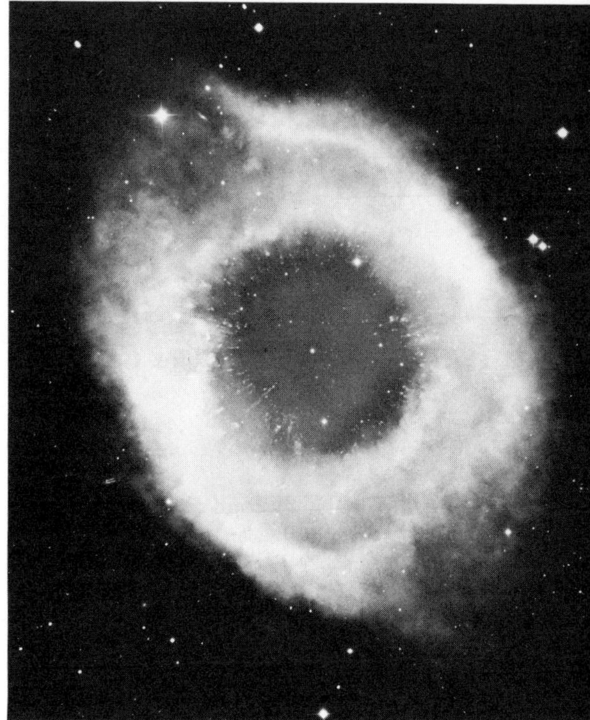

24-18. Planetary nebula in Aquarius. Photographed with the 200-inch Hale reflector on film sensitive to red light. Note the peculiar spoke-like details near the inner rim of the ring. *(Courtesy Mount Wilson and Palomar Observatories.)*

Pulsars

Pulsars (pulsating radio sources) are an exciting astronomical discovery of the late 1960's. The sharp, intense radio pulses from a particular pulsar are repeated with almost incredible regularity over very short periods of time (of the order of 1 to 2 seconds or less); in fact, the pulses are repeated so faithfully that they have been timed to a ten-millionth of a second. Some have been detected to slow down very slightly in a year. Moreover, the Doppler effect of the Earth's orbital motion has been noted.

The pulses are emitted at different wavelengths, each of which is polarized, and similar pulses in the visible region have since been discovered from a pulsar located within the Crab Nebula. Presumably the different wavelengths of a single pulse are emitted simultaneously but the shorter radio wavelengths arrive before the longer radio wavelengths, which apparently are slowed by free electrons in space. In fact, it may be possible to estimate the density of free electrons in different directions from the Earth by this slowing effect (i.e., in the space between the Earth and a pulsar).

The source of the pulses is uncertain and has generated much thought. They were first nicknamed the LGM's (Little Green Men) because their clock-ticking regularity suggested to some observers the possibility of artificial signals from some advanced civilization—perhaps space beacons set up to guide interstellar travelers. According to the explanation favored in 1971, the pulses are produced by the very rapid rotation of a so-called *neutron star,* and the length of a pulse is the time for one rotation. A neutron star, as the name suggests, may be made chiefly of neutrons, and may be one of the densest objects imaginable (under extreme pressures, protons may unite with electrons to form neutrons). These are packed so very tightly that an object perhaps 25 miles (40 km) in diameter may contain the mass of an entire star. Pulsars may emit some very high-energy cosmic rays and may be the remnants of supernovae whose inner parts imploded as their outer parts exploded. It has also been suggested that an implosion might proceed beyond the neutron star stage to produce a so-called *black hole*. Such fantastically dense objects would have so large a surface gravitational attraction that radiation could not leave them—we could not observe them directly.

Exercises and Questions for Chapter 24

1. How many motions does the sun have in space? What is the frame of reference for each?
2. How does the Doppler effect furnish evidence that the sun is:
   (A) rotating?
   (B) moving toward Vega-Hercules?
   (C) revolving about the center of the Milky Way Galaxy?
3. The sun with a surface temperature of 6000°K radiates the maximum amount of energy at a wavelength of 0.5 micron. What is the surface temperature of a star that radiates its maximum energy at a wavelength of:
   (A) 1 micron?
   (B) ¼ micron?
   (C) 0.1 micron?
4. How does the luminosity of a star compare with that of the sun if the star's surface temperature is:
   (A) 6000°K and its diameter is 864,000 miles?
   (B) 24,000°K and its diameter is 2,592,000 miles?
   (C) 3000°K and its diameter is 432,000 miles?
5. How does the sun produce energy?
   (A) How is this energy transferred from the sun's interior to its surface?
6. Assume that a nearby star "wobbles" as it moves through space because an unseen planet is revolving about it. How can astronomers estimate the mass of this planet? What data are needed?
7. How does an astronomer determine for a star:
   (A) its surface temperature?
   (B) its rate of rotation?
   (C) its chemical composition?
   (D) its mass?
   (E) its distance (distinguish between near and remote stars)?
   (F) whether or not any change is occurring in its distance from Earth?
8. Relative to the physical properties of stars:
   (A) Which show the greatest range of values?
   (B) Which show the least range of values?
9. What kinds of stars are most common in the universe? Discuss pertinent data and problems.

10. Why is the mass of a star considered to be one of its most fundamental properties?
11. Describe in general terms some of the differences that occur among stellar spectra. Why do stars have different spectra?
12. Why are studies of clusters of stars very important in a consideration of stellar evolution?
13. Describe the stages that are thought to occur in the life cycle of a star that has a mass about equal to that of the sun.
    (A) Which stages are least well known?
    (B) Which stages may last the longest?
14. Draw H-R diagrams for two clusters of stars and label important features. Assume that one cluster is relatively young and that the other is relatively old.
15. In what kinds and regions of galaxies do Population I stars tend to occur?
    (A) Population II?
    (B) What are some of the chief differences and similarities?
    (C) Why do they occur?
16. What is meant by a spectroscopic binary and how is one detected?

25

Nebulae, Galaxies, and the Universe

Interstellar Matter

This is the parent material of stars, and in certain regions of space it acts as a "cosmic smog" by obscuring or altering the light of remote stars and galaxies behind it (Fig. 25–1). However, the interstellar matter is so diffused that even the denser portions constitute better vacuums than man can produce on Earth. Interstellar matter consists of gas and very tiny dust particles of uncertain composition. The gas is chiefly hydrogen and helium and may have a total mass exceeding that of the dust by 100 or more times.

Two sources are suggested: gas and dust that have existed for a very long time but never formed into stars and matter that has been ejected into space from the atmospheres and interiors of stars.

Some polarization has been detected in the light from stars (perhaps caused by tiny elongated intervening dust particles which are aligned imperfectly along magnetic lines of force in space). A rough correlation exists between the amount of polarization and the amount of reddening (p. 622) in a particular direction.

Astronomers are uncertain about the total quantity of matter present in the universe as interstellar gas and dust or as intergalactic matter. In the spiral arm of a galaxy it may be a large fraction of the total amount of matter present in the stars within the arm. In the galaxy as a whole, however, it is probably much less.

Bright and Dark Nebulae

These clouds of gas and dust appear to be similar, but the bright nebulae (Figs. 25–2 and 25–3) are illuminated by stars within them, whereas the dark nebulae lack such bright stars and reveal their existence by obscuring the light of the stars behind them. However, at one time these dark regions were incorrectly interpreted as vacant areas of space. The clouds of dark cosmic smog must be relatively close to us;

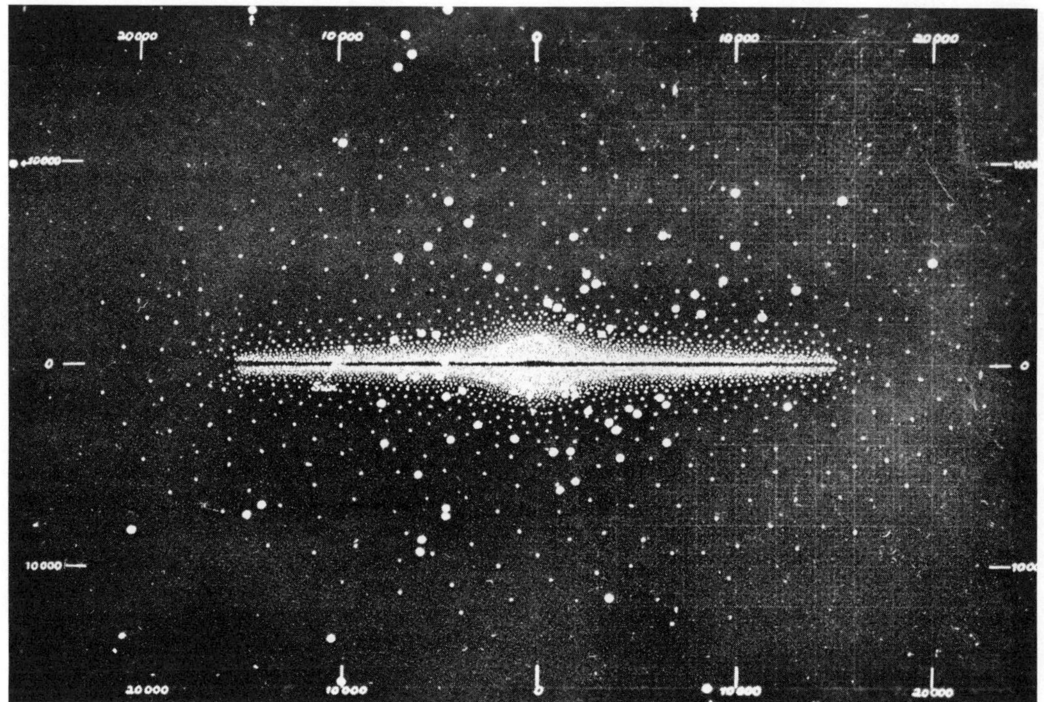

25-1. Schematic edgewise sketch of our galaxy. Most stars are probably grouped along its equatorial plane, especially within the spiral arms (not shown) and nucleus. However, some outlying stars are scattered around the main structure. Dark obscuring matter is shown concentrated along the equatorial plane (Milky Way) but is not a continuous uniform sheet. Globular clusters (large white spots) form a nearly spherical halo around the galaxy; they appear to increase in number toward the center. Estimated distances are shown along the margins; the unit is the parsec (p. 531, 1 parsec = 3.26 light-years). *(Courtesy Yerkes Observatory.)*

otherwise, stars would be abundant between them and us. Conspicuous dark zones occur within the Milky Way of our galaxy and along the equatorial planes of other spiral galaxies.

The bright nebulae are illuminated in two different ways, depending upon the temperature of the stars within them (Fig. 25–3). High-temperature stars, such as type 0, radiate much ultraviolet light which is absorbed by gases in the nebulae and then emitted as visible light (fluorescence) producing emission nebulae. Such nebulae have bright-line spectra. On the other hand, reflection nebulae are illuminated by somewhat cooler stars and have dark-line spectra similar to those of the stars whose light they reflect. Planetary nebulae are a type of bright emission nebula (p. 616).

Invisible Interstellar Gas and Dust

Bright and dark nebulae are most abundant in the spiral arms and along the equatorial planes of spiral galaxies. They occur within very widespread regions of even more diffused gas and dust—so rarefied, in fact, that it cannot be observed directly. However, this invisible interstellar material can be detected because: (1) the gas produces certain dark absorption lines in the spectra of distant stars (different from those produced by the star's hotter atmosphere and narrower be-

25-2. The Great Nebula in Orion. This is a striking example of a bright nebula. It is about 1600 light-years away and constitutes the middle "star" of Orion's sword. The nebula appears somewhat larger than the full moon when viewed through a large enough telescope and seems to have an actual diameter of nearly 30 light-years. Its total mass is much greater than that of the sun. *(Courtesy Lick Observatory.)*

cause the interstellar gases are very cold), (2) the dust reddens their light, and (3) neutral hydrogen atoms emit radio waves 21 cm in length (p. 504).

Certain narrow dark lines occur in the spectra of some stars and tend to be stronger for the more distant stars. In some instances, these lines have Doppler displacements different from those of the other dark lines in a star's spectrum. Thus the narrower dark lines are produced by interstellar gas and not in a star's atmosphere. Moreover, in some directions more than one such cloud occurs, and each may be moving in a different direction at a different rate as shown by the superposed Doppler effects.

The very tenuous interstellar dust reddens the light of remote stars and other galaxies, because the shorter blue waves are scattered more readily than the longer red ones. The reddening is greatest along the equatorial plane of our galaxy and decreases to a minimum 90 degrees from this plane. The reddening may be detected by comparing the spectral type of a star with its observed color; e.g., if a star belongs to spectral class 0 (blue) and yet appears yellowish, we know that its light has been reddened.

Our Galaxy

Perhaps we can actually learn more about our own galaxy by first examining the Andromeda galaxy (Fig. 25–4), which appears to be more or less a twin, and which we can see from the outside rather than from within. It is visible to the unaided eye on a clear dark night as a faint patch of light located in the direction of the constellation Andromeda but more than 2 million light-years beyond it in space. If we could see all of the galaxy, it would span about 4 degrees (the width of eight full moons); however, with the unaided eye, we see only the bright nucleus. This giant spiral is the most remote object visible without optical assistance. In it astronomers have recognized Population I and II stars, Cepheids, novae, bright and dark nebulae, globular clusters, and a Milky Way of its own. Radio waves have been detected which seem similar to those emanating from our galaxy.

The solar system is located in a similar large spiral galaxy (p. 624) whose equatorial plane is marked by the Milky Way (Figs. 24–1 and 25–5), that faint band of light which crosses the sky in the neighborhood of constellations such as Auriga, Perseus, Cassiopeia, Cygnus, Sagittarius, and Scorpio. The Milky Way is a beautiful sight when viewed under the proper conditions—a clear

25-3. Nebulae surrounding stars of the Pleiades. Contrast this telescopic view with that of the unaided eye. The spectrum of this bright nebula is the same as the spectrum of the stars that illuminate it, thus the nebula shines by reflected starlight. *(Photograph by E. E. Barnard.)*

moonless night far from artificial lights. Its light is produced by the combined radiations of millions of stars, each too faint to be visible to the naked eye. All individual stars seen without optical assistance are part of our galaxy, and most of them are relatively quite near us.

Determining the shape of our galaxy is difficult since we are located within it and is similar to obtaining an over-all view of a house from a single spot in one room. We may make comparisons with what appear to be similar galaxies located far away in space, and we may make star counts at progressively greater distances in different directions.

We may point a telescope in any direction within the plane of the Milky Way and take a series of photographs with successively longer time exposures (a way of increasing the distance which can be observed each time). We find that stars continue to be abundant at greater distances within this plane. However, if another series of photographs is taken with the telescope turned 90 degrees to the plane of the Milky Way, stars become progressively less numerous at approximately equal distances on either side, and eventually no additional stars belonging to our galaxy are encountered. Continued mapping has shown that our galaxy has a flattened disklike structure that bulges into a nucleus at the center (the Earth's equatorial plane makes an angle of 62 degrees with this disk). Spiral arms extend from the nucleus and wind around it as conspicuous elongated parts of the flattened disk. A large, roughly spherical halo or corona is

25-4. The great spiral galaxy in Andromeda (Messier 31). Two small elliptical galaxies are near neighbors: directly above the nucleus and to its left. The galaxy's equatorial plane is inclined about 13 degrees from an edgewise orientation. The foreground stars are part of our galaxy. *(Courtesy Mount Wilson and Palomar Observatories.)*

centered on the nucleus and contains globular clusters, individual stars, and widely dispersed hydrogen gas.

According to older views, the sun was near the center of our galaxy. Later it was learned that the galactic center is very far away and located in the direction of Sagittarius, whereas Auriga is in the opposite direction. The first clue to this and to our peripheral location within the galaxy was given by Shapley's study of the distribution of the globular clusters in space just prior to 1920. More than a hundred of these clusters are now known. They are scattered throughout a spherical region of space that is centered in the Sagittarius direction, and they become more numerous toward this center. Shapley decided that this must be the center of the galaxy and boldly predicted that other elements of the galaxy would also be symmetrical about this center; this has been verified.

Our galaxy is about 100,000 light-years in diameter, and its nucleus may reach a maximum thickness of 10,000 light-years; perhaps this is double the thickness of the Milky Way at our great distance of about 33,000 light-years (10,000 parsecs) from the center. The solar system seems to be near its equatorial plane.

We cannot actually see through the nucleus to the other side; in fact, it is very difficult just to see into the near edge of the nucleus—i.e., to find "peep holes" through the clouds of intervening gas and dust. This cosmic smog makes a large portion of our galaxy forever invisible optically. Yet comparison with similar large spirals, such as the Andromeda galaxy, indicates that the unseen part should be similar to the visible part. Radio astronomy is much less hampered by this cosmic smog and extends our knowledge into the farther reaches of the galaxy.

Population I stars occur in the spiral arms where interstellar matter is abundant. The large, luminous, short-lived, type-0, blue stars of this group are visible for long distances, and their distribution in space has aided in mapping the locations of the spiral arms in our galaxy and in other galaxies. Population II stars occur in the globular clusters and in the nucleus of our spiral.

As viewed from a position far beyond the North Star, our galaxy rotates in a clockwise direction. Stars in its equatorial plane seem to revolve in nearly circular paths about the nucleus and to stay approximately within this plane. Stars nearer the nucleus revolve more rapidly than those out near the edge. Relative to the center of the galaxy, the sun appears to move at about 150 mi/sec (240 km/sec) toward the constellation Cygnus (about 90 degrees from the center). The sun may revolve once about the center in about 200 million years.

25-5. Great Rift in the Milky Way from Sagittarius to Cassiopeia. *(Mosaic, from Mount Wilson and Palomar Observatories.)*

The globular clusters and other halo objects all apparently revolve about the center of the galaxy in greatly elongated elliptical orbits, most of which are steeply inclined to the plane of the Milky Way. Thus occasionally they pass near the nucleus or into it. The clusters and other halo objects presumably follow Kepler's equal areas law relative to the galactic center; therefore, we see the majority of them when they are moving most slowly at their greatest distances from the nucleus.

Radio waves map a galaxy similar in shape to the optical one. A radio telescope is made to sweep across the sky and the intensity of a selected radio wavelength is recorded continuously. The most intense signals clearly coincide with the Milky Way, but no Great Rift appears because the radio waves readily penetrate the cosmic smog that hinders visual observation. The center of the radio galaxy is likewise in Sagittarius, and spiral arms have been traced.

Other Galaxies

Galaxies (p. 444) appear to be distributed throughout the universe without diminishing in numbers to the limits of observation. In diameter, they range from several thousand light-years for the smallest to 150,000 light-years and more for the largest (the outer parts become so faint that it is difficult to decide where a galaxy actually ends). Most galaxies seem to occur in association with other galaxies, and these range from pairs to groups of several thousand or more. Our galaxy is part of a small cluster called the Local Group which until recently had 17 known members: 3 spirals, 4 irregular, and 10 elliptical (including 6 so-called dwarfs). Distance across the cluster is about 3 million light-years.

However, according to a 1971 announcement, the Local Group may actually have another member or two. Two faint red patches had been detected earlier near the equatorial plane of our galaxy in the general direction of Cassiopeia. Although the objects are reddened and nearly obscured by galactic dust, subsequent study with large optical and radio telescopes indicates that these are galaxies (named Maffei 1 and 2 after their discoverer). The brighter patch may be a giant elliptical galaxy about 3 million light-years away, whereas the fainter appears to be a large spiral about 9 million light-years away (first estimated as same distance as Maffei 1). If these distances are

25-6. Edgewise view of a spiral galaxy (Class Sa). It shows a relatively large nucleus and a dark equatorial band of obscuring matter. The arms are relatively short and tightly coiled around the central disk, but are not readily observed in the edgewise view. NGC 4594 in Virgo. *(Courtesy Mount Wilson and Palomar Observatories.)*

correct, Maffei 1 is a member of the Local Group—but way out in its suburbs—whereas Maffei 2 is too remote.

Although galaxies have widely different shapes and sizes, they can be classified into three main types: spiral, elliptical, and irregular. Except for the globular variety of elliptical galaxy, their appearances depend greatly upon their orientation in space: their equatorial planes may be oriented at any angle to our line of sight, from 0 degrees (an edgewise view) to 90 degrees.

Rather small elliptical galaxies seem to be most abundant in our neighborhood of space and probably are the most common type of galaxy in the universe, but they are too faint to be detected at greater distances. Thus studies of more remote galaxies emphasize the large luminous kinds, among which the spirals predominate.

Spiral galaxies are subdivided into two main types: normal and barred. Each type in turn shows gradations that can be organized into three subtypes (Figs. 25–6, 25–7, and 25–8). In the normal spirals these range from galaxies with relatively large nuclei and short, tightly coiled arms (class Sa) to galaxies with relatively small nuclei and large, loosely coiled arms (class Sc). Commonly two main arms emerge from opposite sides of the nucleus and coil around it in similarly shaped curves, but more than two arms and branches from the two main arms have been observed. A spiral galaxy rotates; i.e., all of its stars and interstellar matter revolve about its center of mass. Stars nearer the center revolve more rapidly than those farther out, and thus the arms gradually become more tightly wound around the nucleus. Population I stars occur

25-7. The middle type of normal spiral galaxy (Class Sb). The Andromeda galaxy and our galaxy seem to belong to this group. The central disks are relatively smaller and the spiral arms are longer and more open than in class Sa. M81 in Ursa Major. *(Courtesy Lick Observatory.)*

in the spiral arms, as does much interstellar matter, but Population II stars occur in the nucleus and in the globular clusters.

The barred spirals are less abundant than the normal spirals and have spiral arms that project from the ends of a bar that extends through the nucleus (Fig. 25–9). The bar and the nucleus appear to turn in space as a unit. They also show gradations from tightly coiled arms to widely open arms. The cause of the bar is not known.

Elliptical galaxies are reddish in color (red giants contribute much of their light) and range in shape from spherical to flattened disks (Fig. 25–10). They have no spiral arms, appear symmetrical, and stars become less numerous at increasing distances from their centers. Elliptical galaxies contain Population II stars, and interstellar matter tends to be very scarce within them, especially the dust. The orientation of an elliptical galaxy in space may be impossible to determine; e.g., if we observe a flattened elliptical galaxy at a 90-degree angle to its equatorial plane, it will appear circular. The smallest elliptical galaxies (several thousand light-years in diameter) are yet much larger than the biggest globular clusters (perhaps several hundred light-years in diameter) which they resemble somewhat. Elliptical galaxies have a very great range in size; the smallest are considerably smaller than the smallest spirals, yet giant ellipticals are much bigger than giant spirals.

Irregular galaxies appear to be the least numerous of the three types and are unsymmetrical objects that show no definite structures such as nuclei and spiral arms. Our two nearest galactic neighbors are located between 150,000 and 200,000 light-years away and have been classified by some astronomers as irregular galaxies—the Large

25-8. A spiral galaxy with a relatively small nucleus and long prominent arms (Class Sc). M51 in Canes Venatici. A smaller irregular galaxy appears below the larger spiral and stars occur in the zone between the two as a sort of bridge. *(Courtesy Mount Wilson and Palomar Observatories.)*

and Small Magellanic Clouds. However, certain characteristics of the Large Magellanic Cloud (to the unaided eye, 7 degrees in angular diameter) suggest that it may be related to the barred spirals. Population I stars predominate in irregular galaxies, and the young, relatively rare, but highly luminous blue giants and supergiants give them a bluish appearance. Interstellar matter is abundant (perhaps it ranges up to 50% of the total); however, some Population II stars and globular clusters also occur, and little dust is present in some irregular galaxies.

The Evolution of Galaxies

One expects an individual galaxy to evolve with time because decreases must occur in the amount of hydrogen fuel left within each star and also in the total quantity of interstellar matter (diminished by the formation of new stars, but partially replenished by ejections from stars). The gradational types observed within the elliptical and spiral classes had also suggested to astronomers a few decades ago that galaxies might evolve (Fig. 25–11). A spherical galaxy was thought to be the beginning structure. With time, it changed into a progressively more flattened, disklike shape (perhaps because it contracted and rotated faster). Further modification produced a spiral with a very large nucleus, and this in turn changed slowly as the nucleus became smaller and the arms larger. Continued modification might then change it into an irregular galaxy.

Subsequent discovery of Population I vs. II stars and their interpretation as young vs. old groupings seem to make this sequence quite unlikely—elliptical galaxies consist of older stars and have little interstellar matter. However, the sequence might still evolve in the opposite direction from irregular galaxies (Population I with interstellar matter) into spirals and then into ellipticals.

On the other hand, angular momentum (p. 310), mass, and density of the parent interstellar matter may all be factors in the type of galaxy that forms. The largest elliptical galaxies have far more mass than any of the spirals, whereas some of the spirals have far more angular momentum than do the elliptical galaxies (however, mass and angular momentum are difficult to measure accurately). This seems to rule out the evolution of one from the other because the total mass and angular momentum of an isolated galaxy should remain about the same (mass, of course, is used up during nuclear reactions).

Perhaps galaxies do not evolve through a sequence that includes all known varieties —in fact, most galaxies might even have approximately the same age. In this view, some regions of interstellar matter may form directly into elliptical galaxies, whereas others may form directly into the spirals,

25-9. A barred spiral galaxy (NGC 1300 in Eridanus) photographed with the 200-inch Hale telescope. This is an intermediate type: the arms are neither tightly coiled nor widely open. *(Courtesy Mount Wilson and Palomar Observatories.)*

and the rate of rotation of the parent mass might be a determining factor. More rapid rotation might keep the interstellar matter spread out sufficiently for a flattened structure with spiral arms to develop.

If total angular momentum is relatively low, different kinds of galaxies might still originate if they form from interstellar matter of different densities. Star formation in a galaxy developing out of denser material may be relatively rapid, the interstellar matter may be used up fairly soon, and an elliptical galaxy of Population II stars may result. On the other hand, star formation in less dense material may proceed more slowly to produce an irregular galaxy that contains Population I stars and interstellar matter (with high angular momentum, a spiral might have resulted instead). Much is still to be learned about the evolution of galaxies.

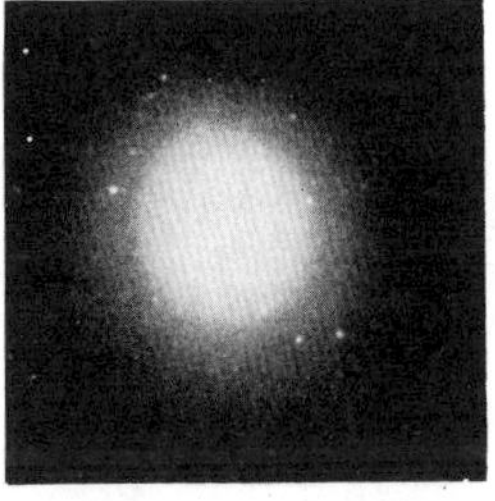

25-10. Two elliptical galaxies illustrate end members of a gradational series. The nearly spherical galaxy at the left (NGC 4278) is at one end of the gradational series of elliptical galaxies, and the elongated flattened galaxy at the right (NGC 3115 in Sextans) is near the other end. *(Courtesy Mount Wilson and Palomar Observatories.)*

25-11. Part of a cluster of galaxies in Corona Borealis at a distance of about 120 million light-years. Type SO galaxies are most abundant here; they are more flattened than the ellipticals but lack spiral arms. *(Photographed with the 200-inch Hale telescope, Palomar Observatory.)*

The Concept of an Expanding Universe

The dark lines in the spectra of galaxies are shifted toward the red end of the spectrum (Fig. 25–12), and the amount of the red shift is about proportional to the distance of a galaxy from the Earth: i.e., a near galaxy shows a small red shift, whereas a remote galaxy shows a large red shift. This relationship is called the *law of galactic red shifts* (Fig. 25–13), and it is tentatively interpreted as a Doppler effect. Therefore, each galaxy may actually be moving away from the Earth (i.e., from our galaxy) with a rate of recession that is directly proportional to the magnitude of its red shift. A general correlation also exists between the apparent magnitude of a galaxy and its red shift—the fainter the galaxy, the larger its red shift tends to be (absolute magnitudes of galaxies do vary greatly, but the fainter ones cannot be observed at great distances).

A galaxy's speed of recession, therefore, is equal to a constant times its distance from us. However, the exact value of this constant (if it is precisely a constant) is uncertain. The speed of recession may increase at a rate of about 100 km/sec per million parsecs (nearly 20 mi/sec per million light-years) but this is a provisional figure and perhaps should be reduced considerably. The largest rate of recession measured to date exceeds 40% of the speed of light (excluding quasars, p. 634). Such indicated speeds make astronomers quite uneasy about interpreting the red shift as a Doppler effect, but no more plausible alternative has been proposed.

Assume for the moment that this interpretation is correct and that distances separating galaxies have been increasing at these rapid rates for a long, long time. It follows that last year all of the galaxies were closer together and that they were still closer the year before. If we choose a particular galaxy, determine its distance, and then divide this distance by its speed of recession, we obtain a date in the past when it and our galaxy should have been close together—about 10 billion years ago. For all other galaxies—at different distances, in different directions, moving at different speeds—the result is also about 10 billion years ago. Thus all of the matter in the present galaxies may have been packed into one small part of the universe some 10 billion years ago. From this location, the matter then moved outward, and in time formed into stars and galaxies (e.g., the sun and its planets may have originated nearly 5 billion years ago). A unique

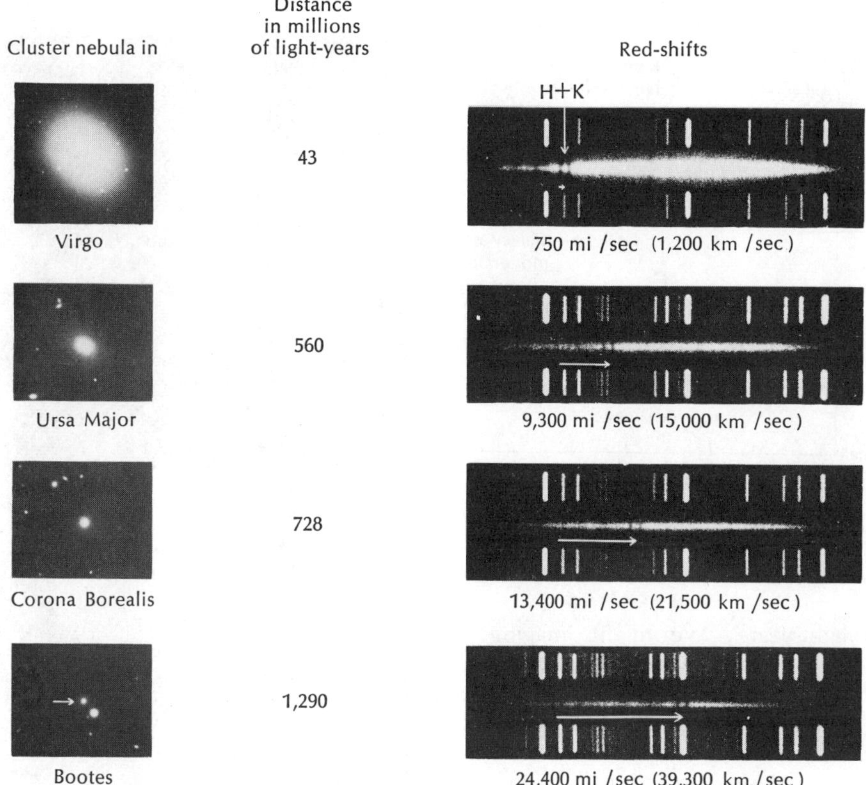

25-12. Photographs, distances (provisional), and spectra of four galaxies. At increased distances, galaxies appear smaller and fainter. A comparison spectrum is shown above and below the spectrum of each galaxy. The horizontal arrows show the shifts of the H and K lines of calcium. Although these H and K lines do not appear in the comparison spectra, their precise wavelengths are known, as are the wavelengths of the lines that do appear in the comparison spectra. Therefore, we know where the H and K lines would be located if no Doppler red shift occurred. The spectrum of a galaxy is a blend of the lines of its most luminous stars. *(Courtesy Mount Wilson and Palomar Observatories.)*

event or cataclysm may thus have occurred some 10 or so billion years ago, and this interpretation is known as the *Big Bang* hypothesis. According to this view, matter that subsequently developed into the galaxies began to move outward simultaneously but in different directions and at different rates. As objects in space should continue to move at uniform speeds in straight lines in the absence of other forces, galaxies with slow speeds have traveled shorter distances than those with rapid speeds. This accounts for the law of the galactic red shifts.

Galaxies within a cluster appear to be held together by the gravitational attraction of all of the matter within the cluster* (distances between neighboring galaxies within a cluster may average 1 to 2 million light-years). Therefore, the sizes of the clusters probably remain the same, whereas the distances separating clusters of galaxies are probably increasing (distances between neighboring clusters may average 100 to 200

* However, some calculations show that the total mass of all the member galaxies is far too small to hold a cluster together. Thus very large quantities of intergalactic matter, not readily visible from the Earth, may actually occur within a cluster—and perhaps outside of the clusters also.

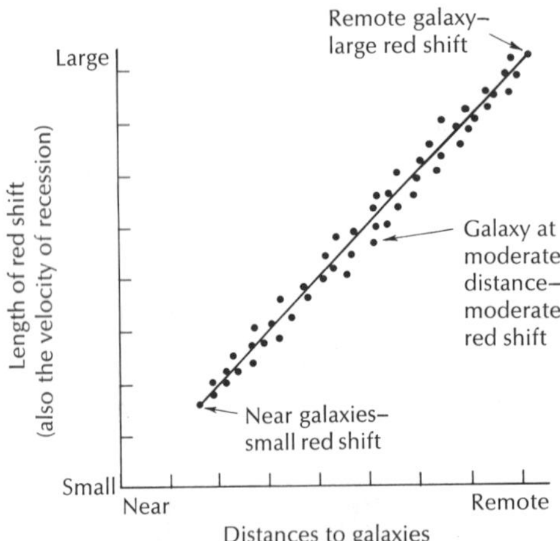

25-13. Schematic diagram of the law of galactic red shifts.

million light-years). We might imagine a cluster as a swarm of bees moving in a straight line at a uniform rate through space. Random motion occurs within the swarm as the bees fly to and fro, and under such conditions, galaxies occasionally interact (Fig. 24–14).

According to this interpretation, the clusters of galaxies are all moving away from us, but this does not necessarily put our galaxy in a unique position. If a balloon (universe) covered by uniformly spaced dots (galaxies) is inflated, each dot moves away from its neighbors, and from any one dot all other dots appear to be receding (the dots expand in this analogy, but the clusters of galaxies probably do not). Similarly all galaxies may be moving away from each other (Fig. 25–15).

The date of the Big Bang has been shifted a number of times since about 1950 and is still a very tentative figure. Each time we change the estimated distances to the galaxies (p. 536) we must make a corresponding change in the date of the Big Bang—i.e., the greater the distance, the longer the time necessary for a galaxy to cover it (the size of the red shift remains unchanged). Thus the date has jumped from the 2-billion-year figure generally accepted about 1950, to 4 billion years, to 6 billion years, to the current 10 billion years.

25-14. Interacting galaxies (NGC 4038 and 4039). *(Photographed with the 48-inch Schmidt telescope, Palomar Observatory.)*

Of course, we have no information at all concerning what, if anything, preceded the postulated Big Bang. However, according to one view, a Big Squeeze may have preceded the Big Bang; i.e., the mutual gravitational attraction of material widely distributed in space caused it to contract and collapse into a relatively small, tremendously dense region. From this center it subsequently exploded outward. In still another modifica-

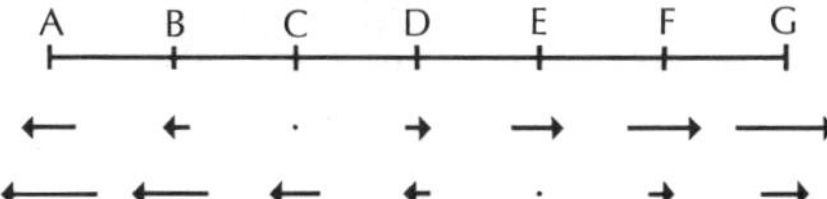

25-15. Schematic two-dimensional illustration showing that our galaxy may not be the center of the outward movement. The arrows are vectors representing the positions and motions of galaxies; longer arrows show higher speeds. In the middle line, point C shows the true center of outward motion. In the bottom line at E, observations are made from a galaxy that is actually moving toward the right; however, it appears to be at the center of the outward motion (make appropriate additions or subtractions of the vectors).

tion, a pulsating or *oscillating universe* is imagined—an endless alternation of big squeezes and big bangs. Some astronomers think that this is compatible with data presently available. In this version, it is postulated that the mutual gravitational attraction of all of the matter in the universe would gradually slow down and then stop the outward motion. If this were to happen, all of the matter should then move at accelerating rates back toward the center of mass.

However, are galaxies now too far apart and moving too rapidly for gravitational attraction ever to bring them together? In some calculations, the total quantity of matter observable in galaxies is far too small to do this—thus intergalactic matter, not yet detected, would have to be abundant. The sizes of the red shifts of very remote galaxies should provide key data. To illustrate, assume that a Big Bang occurred an even 10 billion years ago and that the outward movement of the galaxies has been slowing down gradually ever since. Let us now compare a nearby region of the universe with a remote area. The law of galactic red shifts was developed by observing relatively nearby galaxies—say those within about 1 billion light-years. Therefore, we observe such galaxies as they were 9 billion years or more after the Big Bang, and their speeds of recession would presumably have been much reduced during 9 billion years of "gravitational drag." Assume next that we can observe a very remote galaxy 8 billion light-years away. We would be seeing this galaxy as it was 2 billion years after the Big Bang and after it had been slowing down for only 2 billion years. Therefore, it would be moving faster than we would have predicted by a direct extension of the law of the galactic red shifts (Fig. 21–14); i.e., points for such galaxies would be higher on the graph than we had anticipated. In other words, a line on the graph trending centrally through the dots would curve upward. According to some astronomers, such has been observed, but the data cannot be conclusive, because determination of the distances to remote galaxies is still quite unreliable.

According to the *Steady State* or Continuous Creation hypothesis, the data should be interpreted in a quite different way, although the Doppler interpretation of the red shifts is accepted. In this view, the universe had no beginning, will have no end, and is uniform in time and space. As expansion occurs, new matter (hydrogen) is created in intergalactic space to maintain a uniform density. New stars and galaxies form and evolve from this new matter and replace those that have expanded outward. Thus individual stars and galaxies change, but the over-all nature (concentration of galaxies) of the universe remains uniform.

Observations of certain kinds should eventually make it possible to discard some of these hypotheses and may already have done so; e.g., observations with both optical and radio telescopes of quasars (p. 634) and other objects seem to show that the universe is not uniform in space and time. Furthermore, in an exciting unexpected development, cosmic background radiation surviving from the *"primeval fireball"* of the Big Bang was apparently discovered in 1965.* The discovery of such radiation had been predicted. According to theory, the initial very high-energy gamma radiation produced at the time of the Big Bang should gradually change into low-energy longer

* P. J. E. Peebles and David T. Wilkinson, "The Primeval Fireball," *Scientific American,* June 1967.

wavelengths during the ensuing expansion. The radiation, billions of years later, should now occur in the radio and microwave regions of the electromagnetic spectrum. Detection is difficult since the radiations are much weaker than other sources of microwave radiation and seem to come toward Earth uniformly from all directions (isotropic). But detection and measurements have now been made at several wavelengths which seem to fit the spectral energy curve produced by black body radiation at about 3°K. If this radiation is eventually verified as the residue of a primeval fireball, then a great step forward will have been made in our understanding of the universe.

Quasi-stellar radio sources or *quasars* became a major problem (and opportunity) for astronomers in the 1960's. They can be described as starlike objects of unknown nature, located at uncertain distances, with very large red shifts. Apparently they have a nonuniform distribution in space because quasars with small red shifts are unknown. In fact, the smallest red shift for a quasar about equals the biggest red shift so far observed for a so-called normal galaxy, whereas the largest quasar red shift may be 5 to 6 times larger (maximum observed distances for such galaxies is 1 to 2 billion light-years). Moreover, following one interpretation of available data, quasars seem to be most abundant at a distance of about 8 to 9 billion light-years (distances calculated from law of galactic red shifts). In addition, the number of quasars per unit volume of the universe appears to diminish at increasingly shorter distances, and may also decrease beyond 8 to 9 billion light-years. If true, quasars are objects associated closely with the early days of the universe.

Quasars are very hot, blue objects that emit much ultraviolet and infrared radiation. Some quasars also emit strong radio signals, but most do not. Relatively rapid light fluctuations have been observed from some quasars—by 10% or more during a period of several days to several weeks or longer. This suggests a very small size for an object perhaps 100 times more luminous than an entire galaxy. Following one interpretation, the average linear diameter of a quasar may be several hundred light-years, but some are much smaller, and relatively small radiation subsources probably occur within them that are only light-days or light-weeks in diameter.

If the red shifts are interpreted as resulting from the expansion of the universe, then the most remote of these objects may be about 9 billion light-years away and receding from us at more than 80% of the speed of light. If the object is relatively small and this remote, then we do not know how it produces the enormous amount of energy that it emits. The interpretation that quasar red shifts result from an expanding universe was strengthened recently when a quasar was discovered that apparently is a member of a remote cluster of galaxies—its red shift about equals those of galaxies within the cluster.

On the other hand, quasars may be relatively nearby objects. According to one view, they were ejected at very high speeds from the nucleus of our galaxy some millions of years ago when gravitational collapse occurred there. However, no quasars with blue shifts have been detected, and such should occur if quasars had also been ejected from neighboring galaxies. Thus the quasars' distances must be determined before their role can be elucidated, although evidence seems to be accumulating that they really are as remote as their red shifts indictate. Perhaps they represent some unique kind of celestial object that existed only during the earlier part of the expansion of the universe, or perhaps they represent an early stage in the development of some galaxies. At any rate, most of the quasars that we observe probably no longer exist—at least as quasars.

Although much uncertainty exists, it seems that the evidence against the Steady State view is now very strong and that a Big Bang did occur. What happened prior to this, we do not know, but some support exists for the concept of an oscillating universe. Whether space is finite or infinite in extent is another question that has no definite answer at present.

Does Life Exist Elsewhere?

Uncertainties abound because conclusions must be based upon assumptions concerning the origin of planets, of life itself, and of the conditions under which we think life can exist.* Intelligent life elsewhere in the solar system seems highly improbable, although primitive vegetation may exist on Mars. A suitable home for life as we know it, is probably a planet with a moderate gravitational attraction. The gravitational pull of a Jupiter-size planet is sufficient to retain hydrogen and other light elements in its atmosphere; thus poisonous gases such as methane and ammonia would probably be present. On the other hand, a small planet such as Mercury cannot hold an atmosphere.

This moderate-size planet should follow a stable orbit about a star at an appropriate distance so that it is neither too hot nor too cold; i.e., it must lie within the inhabitable zone that surrounds each star. Temperatures, of course, decrease outward from a star, and if a certain star is hot enough, its inhabitable zone will be quite wide; planets of the proper size might by chance exist within this zone. On the other hand, the inhabitable zone of a small red star, apparently the most abundant type of star, would be rather small, and relatively few planets of the proper size could be expected within such zones.

If the orbits of such planets are to be stable and remain within the inhabitable zones at all times, they probably should revolve about single stars; revolution around one member of a binary star or multiple star group would probably result in unstable orbits. Perhaps half of all stars are double stars or members of multiple-star systems and thus unlikely locations for inhabited planets.

How abundant are planets? How long a time is necessary for life to evolve? Would evolutionary development produce similar end results if it occurred elsewhere in the universe on a planet similar to the Earth? These are questions without definite answers.

Let us assume for the moment that planets are common by-products in the formation of stars (p. 310). How long does it take for intelligent life to develop? On the Earth, evolutionary development seems to have taken a few billion years. If this is assumed to be an average, then we eliminate as possible candidates for inhabited planets all hot massive stars that evolve rapidly. It may seem something of an anticlimax now to conclude that intelligent beings would most probably exist on planets like the Earth revolving about stars like the sun—if intelligent life does exist elsewhere in the universe. However, such stars are very numerous in our galaxy, and the universe contains many galaxies.

What are the chances of finding planets inhabited by intelligent beings around any of the sunlike stars within several dozen light-years of the Earth? If they exist, how would we know about it? If life exists elsewhere and has evolved in some such manner as it has on the Earth, it seems highly unlikely that any near stars would have planets on which life and civilization had reached approximately the same stage as on the Earth. For one reason, stars have formed at different times in the past; thus planets and life, if they exist elsewhere, must also have formed at different times. Consider also the technological developments of the past few centuries, of the past few decades, and even of the past few years. It is perhaps more likely that an age of dinosaurs would be occurring on one, whereas amphibians would be evolving from a type of lung fish on another. On the other hand, more advanced civilizations than our own may be present somewhere. In the geological history of the Earth, various organisms have evolved, culminated, and declined—

* Some of the ideas in this unit are based upon an article by Su-Shu Huang, "Life Outside the Solar System," *Scientific American*, April 1960, pp. 55-63.

some apparently abruptly and unaccountedly. Will man have a similar fate on the Earth?

The possibility of intelligent beings on a planet located within a few tens of light-years from the Earth and capable of sending radio messages into space seems very, very small. However, a "listening" program was undertaken by means of a radio telescope in a project aptly called Project Ozma. It was unsuccessful as expected, but the thinking it provoked probably made the project worthwhile. In fact, the question was raised: If we detect messages from outer space, should we answer? Would other intelligent beings be hostile?

Although radio waves travel at the speed of light, communication would be frustratingly slow even with intelligent beings on relatively near stars. One generation might ask a question and their great grandchildren receive the reply. But what question can be asked and in what language? One suggestion is to send the value of pi, a numerical constant, and hope for a continuation of numbers; e.g., QUESTION: 3.14159; ANSWER: 2653589; pi is equal to 3.14159265-3589. The question of intelligent life elsewhere in space is intriguing, exciting, and profound, but it seems unanswerable for the present at least.

Relativity

Gravitational attraction plays an important role in explaining certain astronomical phenomena. It can be measured precisely, its effects can be predicted and explained, but its ultimate cause is probably as mysterious to us now as it was to Newton. Gravitational attraction seems to function instantaneously over immense distances and is unaffected by the presence or absence of barriers between two objects. How can this be?

Einstein has developed a competing hypothesis (one aspect of relativity) that explains the phenomenon of gravitational attraction in another way, but the two give identical results except when extremely large masses and very great velocities are involved. In such cases Einstein's explanation appears to fit the facts better. Relativity deals with fundamental ideas concerning time, space, motion, energy, mass, and "gravitation." Some of the ideas of relativity are difficult to grasp because they go contrary to our common sense, but we shall discuss a few aspects.

According to the ideas of relativity, the velocity of light is the maximum velocity possible and is independent of the motion of the source or observer. For example, imagine that two objects are in line, approaching each other, and each moves at a velocity equal to 80 percent of the speed of light. An observer on one measures the rate at which light approaches him from the other object. Much to his surprise, he finds that its velocity is the same as it would be if he were motionless. Furthermore, the distance between the two objects does not decrease at 1.6 times the speed of light, although it does decrease at a rate that exceeds 80 percent of the speed of light. This does not make sense. If two cars are in line and on a collision course, and if each moves at 40 mi/hr, we know that the distance between them will decrease at the rate of 80 mi/hr. But this does not hold true for light; its velocity is independent of other motions objects may have and is a constant. A famous experiment was performed by Michelson and Morley which showed that the velocity of the Earth had no effect upon the velocity of light from stars—i.e., whether the Earth moved toward, away from, or sideways to the light from a star.

Another aspect of relativity involves changes in mass at very high velocities. According to this view, the mass of an object increases rapidly and greatly at very high velocities that approach the speed of light. Thus no object having mass can be accelerated to the speed of light because its mass would increase to infinity in so doing. In certain laboratory experiments this effect has actually been detected: a tiny particle accelerated to 86 percent of the speed of light doubles in mass.

Light traveling outward from a very massive or dense source, such as a white dwarf star, should lose energy according to the ideas of relativity, and this loss of energy should show up as a "gravitational redshift" (different from a Doppler effect). This effect has been observed as predicted. Furthermore, light traveling through a very strong gravitational field should be bent as if the light had mass; this effect has also been detected, although it is very slight.

The increase in mass and other effects caused by velocities that approach the speed of light are most unusual when related to our everyday experiences; e.g., clocks slow down, and measuring sticks become shortened in the direction of motion. To illustrate, it is estimated that a ruler 1 foot long would be shortened to 6 inches if it moved at a rate equal to 90 percent of the speed of light. Furthermore, it has been predicted that passengers on a very high-speed spaceship would age much more slowly on a long journey than the friends they left behind on the Earth. Thousands of years might elapse, as time is measured on the Earth, before the journey was completed, whereas this span of time might be measured in decades by the passengers of the spaceship.

However, all of these unusual effects occur only when velocities are very great—far beyond those attainable now or in the foreseeable future. Newton's law of universal gravitation continues to explain most of the everyday phenomena with which we are familiar. The ultimate cause of Einstein's theoretical explanation is likewise unknown: a gravitational field is replaced by a curvature of the space that surrounds objects. The space-curvature explanation shows its greater accuracy only when extremely high speeds and large masses are involved.

Another important aspect of relativity, and that for which it was named, involves the frame of reference to which a certain motion is related and also the relationship between time and space that we have discussed as "seeing the past." Three dimensions are necessary to locate a spot on the Earth; they correspond to height, width, and breadth. But we must also use a fourth dimension, time, to pinpoint a certain event; e.g., if we arrange to meet someone inside a deep mine we have to specify when as well as where, and we make use of four dimensions of space-time. The Earth appears to us to be stationary and other celestial objects seem to move about it, but this is not so—the apparent motion is relative. The motion of the sun depends upon the stars we select as reference points: relative to nearby stars the sun is moving in a certain direction at a certain rate; however, relative to the center of our galaxy, the direction and rate are quite different. To what frame of reference can we refer the motion of our galaxy, if it is moving? We conclude eventually that no fixed frame of reference applies throughout the entire universe. Space, motion, and time are relative and depend upon the frame of reference that we choose. Thus we live in a multidimensional, space-time universe that may be finite or infinite in extent and whose beginning and ultimate destiny we do not know.

Exercises and Questions for Chapter 25

1. Interstellar matter—even more diffused than in the bright and dark nebulae—exists in certain regions. What is the evidence for the existence of the:
   (A) dust?
   (B) gas?

2. Describe the shape, size, and general nature of the galaxy in which we are located.
   (A) How much of it can be studied with optical telescopes?
   (B) With radio telescopes?
3. Relative to galaxies:
   (A) Describe the main types.
   (B) Which kind is most abundant?
   (C) How may galaxies evolve?
4. Relative to the concept of an expanding universe:
   (A) On what evidence is the concept based?
   (B) What is expanding?
   (C) What rates are involved?
   (D) How does the big-bang hypothesis attempt to account for the expansion?
   (E) The steady-state hypothesis?
   (F) The oscillating universe hypothesis?
5. In the late 1960's the big-bang hypothesis became favored over the steady-state hypothesis. Why? What evidence would favor the oscillating hypothesis?
6. What major changes have been made during the last few decades concerning the age of the universe? Why?
7. Discuss the differences in magnitude that are involved in interplanetary, interstellar, and intergalactic space travel.
8. In considering the possible existence of life as we know it on some other planet:
   (A) How common are planets?
   (B) Why is the size of a planet a critical factor? Its location?
   (C) Which type of star would be most favorable as the primary for an inhabited planet?

Part III Meteorology

26

The Atmosphere

Some people are weatherwise,
But most are otherwise.
—Benjamin Franklin

No one would deny that weather plays an important part in our daily lives, but few people recognize the full extent of its influence. Just as a sudden shower has dampened many a family picnic, so rains and storms (Fig. 26–1) at crucial times have determined who would be victorious or vanquished, and thus have changed the course of world history.

We are affected individually. Weather influences our moods, what we buy, and how we act (e.g., impulsive crimes are most frequent on hot summer nights). Freezing rain may make driving dangerous. A heavy fog may ground the plane we were scheduled to take. Unless we locate our houses with care, prevailing winds may blow toward us the smells and smokes from dumps and factories or the leaves from a neighbor's lawn. A glorious fall day may make a drive through the country an experience to be long remembered.

Larger groups are affected also. An ice storm may bring down the power lines that service entire communities. A heavy snow can cripple the transportation of vital foodstuffs to a large city. Flooding rivers and hurricanes have destroyed whole communities. In the face of the tremendous power of great storms, man is helpless. He can only hope for adequate warning so that he can retreat to a safe place.

The influence of the weather will abide, since we can probably never hope to control the weather on a large scale. We can, however, hope to study it and predict it ever more accurately and over longer periods of time, with the aid of the science called meteorology.

Meteorology is the science of the atmosphere and of weather phenomena—rain, snow, tornadoes, dew, clouds, wind, and sunshine. Understanding them depends upon a knowledge of air, heat, and water as they function on a rotating, revolving, sun-warmed Earth. Let us consider first the atmosphere.

Composition and Density of the Atmosphere

The atmosphere is the sea of air which envelops the Earth. It has no distinct outer boundary and becomes exceedingly diffused in its outermost parts where it fades away into the even more diffused medium of interplanetary space. This gaseous shell is held to the Earth by gravitational attraction and has a layered structure (its origin is considered in Chapter 12). Within it, man lives and breathes.

Air consists of a mixture of gases whose nature, density, and physical state change with altitude (Fig. 26–2). Variations in chemical composition, pressure, temperature, and various electrical and magnetic phenomena have been used to subdivide the atmosphere into a number of concentric shells, which may have gradational boundaries and uncertain locations. Moreover, the shells based upon one criterion do not necessarily coincide with those based upon another criterion.

An average sample of pure dry air (i.e., without moisture, suspended particles, or pollutants) is colorless and odorless and is made up chiefly of molecules of nitrogen (about 78% by volume) and oxygen (about 21%) and atoms of argon (nearly 1%). These three gases appear to be rather uniformly mixed within the first 60 to 70 miles (100 to 113 km) or so (the *homosphere*).

However, water vapor, carbon dioxide (about 3/100 of 1%), dust particles, and ozone are present in minor and variable amounts in the lower part of the homosphere. Water vapor is invisible, colorless, and odorless and may compose as much as 4 or 5% of the volume of this lower air under very warm and humid conditions. Although these four substances occupy a relatively small volume of the total atmosphere, their roles in causing weather phenomena and in the heating of the atmosphere far transcend their quantity. Mi-

26-1. Nearly vertical view of cloud-outlined eddies (vortices) produced in the atmosphere downwind from mountainous Guadaloupe Island in the eastern Pacific Ocean. Weak convective currents in the lower atmosphere produce the cellular structures visible in the stratocumulus clouds. In a closed cell, the air ascends near its center and descends near its margins. In an open cell, the circulation is reversed; it has a clear center and clouds for walls. *(Gemini XII, 13 November 1966, S66-63494, NASA.)*

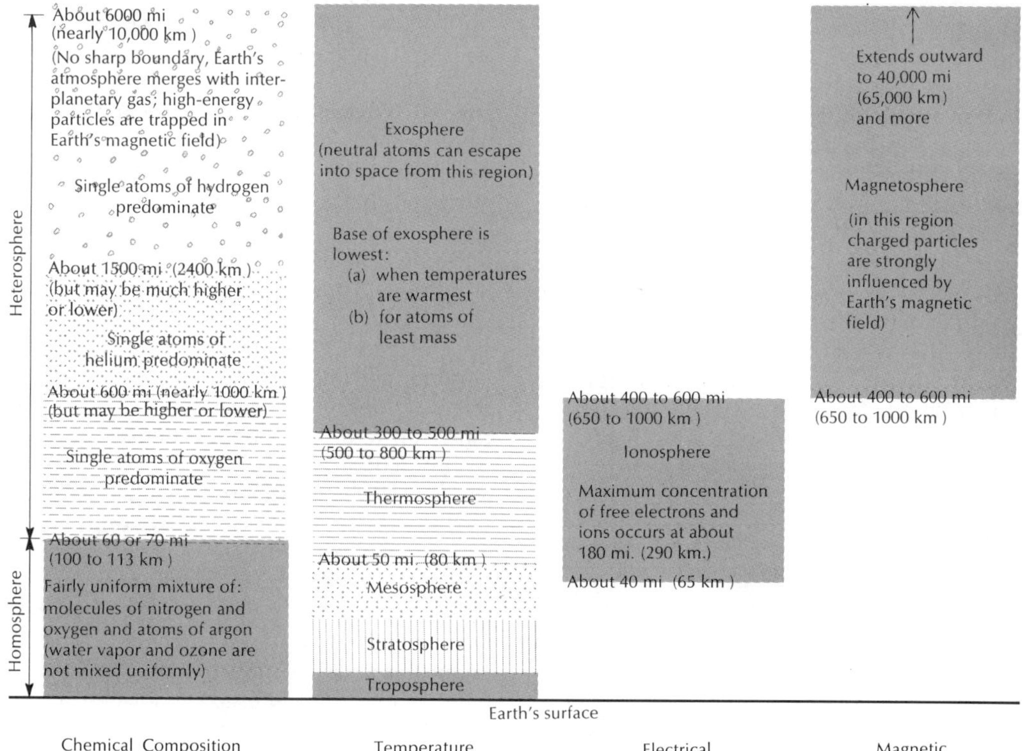

26-2. Subdivisions of the atmosphere are based upon different properties and compared schematically in this diagram (not to scale).

nute amounts of at least fifteen other gases make up less than 4/100 of 1% of the atmosphere.

To emphasize the well-mixed nature of the homosphere and the very widespread scope of atmospheric circulation, we follow Shapley's saga of the argon atoms as related in *Beyond the Observatory*. An estimated 3×10^{19} atoms of argon are inhaled or exhaled with each deep breath. The argon atoms in your next exhalation—Shapley's Breath X—will in succeeding weeks and months diffuse throughout the entire lower atmosphere. No matter where you are on the Earth's surface one year later Shapley estimates, 24 hours of breathing will probably include about 15 of the same argon atoms from Breath X. Put in another way, each and every inhabitant of the Earth in one day's breathing a year later will probably inhale about 15 of the argon atoms from your single Breath X.

According to Shapley, this rebreathing of argon atoms links us to the past and to the future—e.g., we inhale argon atoms once exhaled by Christ, by Shakespeare, and by Washington. It also emphasizes the importance of keeping the atmosphere from being unduly polluted or poisoned, not just in one's own country but everywhere on Earth; e.g., a nuclear explosion in the atmosphere above one country can scatter radioactive atoms into all the other countries of the Earth.

Overlying the homosphere is a zone extending to an altitude of about 600 miles in which single atoms of oxygen predominate, although some nitrogen and oxygen molecules occur in the denser air near its base, and single atoms of nitrogen are present at higher altitudes. Helium atoms seem to be the most abundant material in the region between approximately 600 miles and 1500 miles, and hydrogen atoms probably predominate in the zone overlying the helium shell. Perhaps the hydrogen zone can be considered as extending to an altitude of 6000 miles or so, but it merges almost imperceptibly with the medium of interplanetary space. However, electrified particles are controlled by the Earth's magnetic field at still greater distances and may be considered an extension of the Earth's atmosphere. These shells do not have fixed thicknesses or locations and at any one time the actual altitudes may differ considerably from the averages given above.

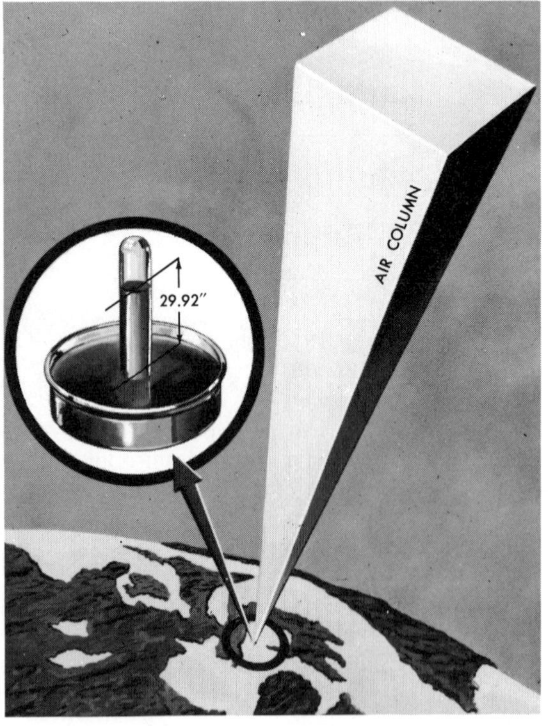

26-3. At sea level the pressure exerted by a column of mercury 29.92 in. (76 cm) in length equals the pressure exerted by a similar column of air that extends to the top of the atmosphere. At altitudes above sea level, the mercury column is shorter. (AF Manual 105-5, *Weather for Aircrews.)*

Thus the atmosphere as a whole has a density stratification; the heaviest atoms and molecules predominate near the surface, and the lightest atoms occur near the top. However, this does not mean that the lightest atoms occur only at the top of the atmosphere because they also occur at lower levels. Rather, the lighter the atom, the thicker is the vertical zone through which it diffuses. Thus only hydrogen atoms can spread upward into the topmost level. Density decreases geometrically with increasing altitude, and most of the total mass occurs relatively near the surface (Table 26–1). This rapid upward decrease in density is due to the compressibility of air; the air at any one level is squeezed by the air that overlies it. Therefore, it is squeezed most and has the greatest density at the bottom.

To illustrate this great variation in density with altitude, we might note that a cubic yard of air at sea-level pressure contains about the same amount of matter as 1 cubic mile of the very diffused air located at an altitude of 100 to 200 miles. Again, at sea level pressures any one molecule probably moves less than 1/100,000 inch before it collides with another molecule; however, at an altitude of 60 miles the average distance between collisions may be a number of inches, and at 200 miles it is hundreds of feet.

Atmospheric pressure is caused by the mass of all the atoms and molecules in the atmosphere being pressed down upon the Earth's surface by the gravitational pull of the Earth. Man is not crushed by the atmosphere because the pressure inside his body is equal to that outside.

A column of air 1 sq in. in cross section and extending to the top of the atmosphere

TABLE 26-1

Of the total mass of the atmosphere	
About 50%	occurs in the lowermost 3½ miles
About 75%	occurs in the lowermost 7 miles
About 90%	occurs in the lowermost 10 miles
About 99%	occurs in the lowermost 20 miles
About 99.9%	occurs in the lowermost 30 miles
About 99.999%	occurs in the lowermost 50 miles

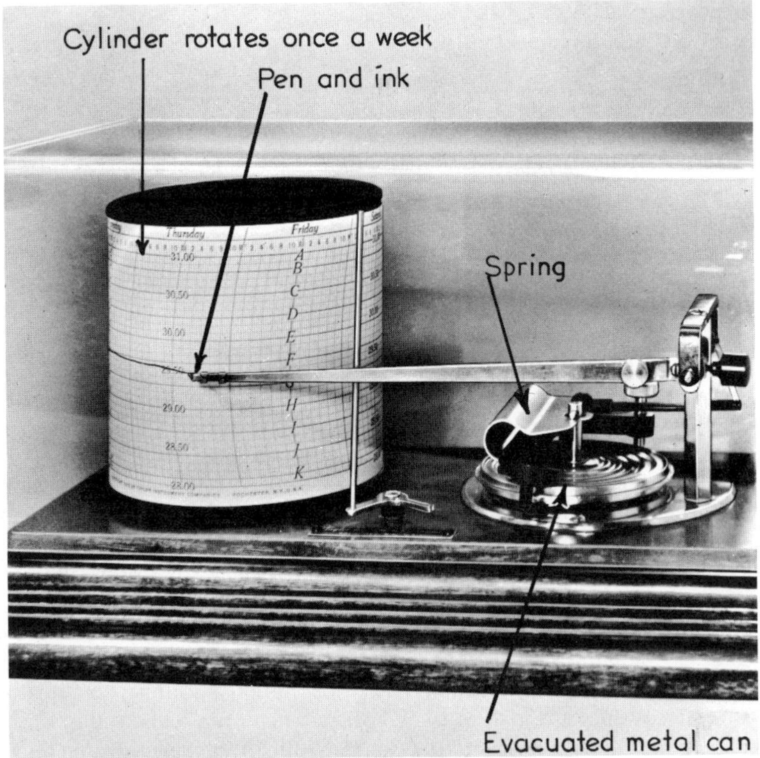

26-4. A barograph functions as a self-recording aneroid barometer. Air pressure is measured by means of a partially evacuated, airtight metal can which is kept from collapsing by a spring. When atmospheric pressure increases, the top of the can is depressed, and the motion is transmitted to a pointer. When atmospheric pressure decreases, the top of the can expands, and the pointer moves in the opposite direction. A clock mechanism is wound to make the drum rotate for a week. *(Courtesy Taylor Instrument Co.)*

weighs 14.7 lb at sea level. Therefore, atmospheric pressure at sea level is nearly 15 lb/sq in. or about 1 ton/sq ft (1013.2 millibars, p. 742). A column of mercury that is 1 sq in. in cross section and 29.92 in. long likewise weighs 14.7 lb at sea level (Fig. 26–3).

Atmospheric pressure may be measured by mercurial and aneroid barometers (Fig. 26–4). Aneroid barometers may be used in airplanes, but they are then calibrated to show altitude rather than pressure and are called altimeters. Altimeter readings are based upon assumed standard atmospheric conditions which tend not to coincide exactly with local conditions. Therefore, the altitude shown by an altimeter may be inaccurate unless frequent adjustments are made for local weather conditions.

If a mercurial barometer is lifted to an altitude of about 3½ miles, its column falls to 15 inches because only half of the total mass of the atmosphere occurs above it. At altitudes of 7 and 20 miles, the lengths of the mercury column will be about 7½ inches and 3/10 of an inch respectively.

Barometric pressure is shown on weather maps by contour-like lines called *isobars* (Gr.: *iso,* "equal"; *baros,* "weight"). All locations on the Earth's surface along an iso-bar have the same barometric pressure (the readings are corrected to sea level). Changes in barometric pressure are an important index to changes in the weather.

Heating of the Atmosphere

The temperature of an object is related directly to the rate of movement or vibration of the atoms and molecules which compose it. At higher temperatures the molecules of a body move more rapidly than they do at lower temperatures. In the language of molecules, one desires a drink of slowly moving water molecules on a hot day and longs for air made of rapidly moving molecules during cold winter months. Thus an account of the heating of the atmosphere is an account of how its atoms and molecules obtain the energy for more rapid movement. This is primarily by radiation.

The atmosphere is heated chiefly from the bottom by solar radiation that enters at the top. To understand how this heating occurs, we must first consider the nature of solar radiation and the radiation laws (p. 496). Second, we must keep in mind the atmosphere's chemical composition and its geometric decrease in density with altitude.

A third group of items involves the greenhouse effect and the interaction between solar and terrestrial radiation and atoms and molecules of different kinds. Particular photons or wavelengths can be absorbed by certain atoms and molecules but not by others. Where this absorption occurs, the atmosphere tends to be heated. The absorbing atoms and molecules later emit the same or different photons. Thus a high-energy photon may be absorbed, and two or more photons may be emitted. The energy of the absorbed photon equals the sum of the energies of the emitted photons. In other words, high-energy short wavelengths may be absorbed, and a larger number of low-energy longer wavelengths may be emitted (fluorescence is a good illustration here).

Before interaction with the Earth's atmosphere, about 41% of solar radiation is in the form of visible light, about 9% consists of ultraviolet and X-rays, and about 50% consists of infrared and longer wavelengths. Maximum intensity is in the blue-green part of the electromagnetic spectrum at a wavelength of about 0.5 microns. Since the bulk of solar radiation occurs at wavelengths shorter than about 3 or 4 microns, it can be considered as short-wave radiation. In contrast, terrestrial radiation occurs primarily within the 3- to 80-micron range with a peak at about 11 microns; it can be called long-wave radiation.

Each 24 hours the Earth receives from the sun an amount of energy equivalent to the burning of 560 billion tons of coal, yet the Earth intercepts only a tiny fraction of the total energy that the sun radiates in all directions. However, this quantity appears to remain nearly uniform, although fluctuations do occur in the high-energy ultraviolet and X-ray region. The solar energy that reaches the outermost part of the Earth's atmosphere amounts to nearly 2 calories/sq cm/min (when the sun is located at its mean distance from the Earth and when the energy falls on a surface oriented perpendicular to the radiation). This amount is called the *solar constant*. Solar radiation, called *insolation* (incoming solar radiation), includes radiation from the sky (indirect) as well as radiation which has come directly from the sun.

Of 100 units of solar radiation intercepted by the Earth, an estimated 30 units are reflected or backscattered to space without heating either the surface or the atmosphere (Fig. 26–5). This ratio of reflected radiation to total radiation is known as the *albedo* (pronounced: ăl·bē′·dō) of an object. Thus the Earth's albedo is now calculated at about 30 percent, although a figure of 35 to 40 percent was used until recently when measurements from weather satellites became available. The older calculations underestimated the amount of radiation absorbed in the tropics because the average cloud cover of the tropics was overestimated (clouds cause most of the backscattering).

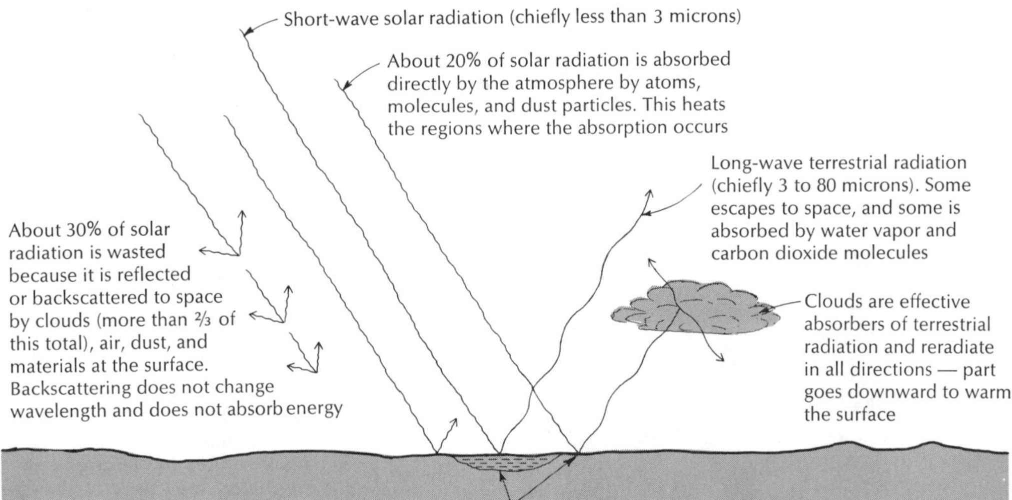

26-5. Heating of the Earth's atmosphere

An estimated 20 units out of the 100 are absorbed directly by the atmosphere by atoms, molecules, and dust particles. This heats the regions where the absorption occurs and includes nearly all of the ultraviolet and X-rays. These 20 units of short-wave radiation are subsequently emitted as 20 equivalent units of long-wave radiation.

The 50 remaining units of solar radiation are absorbed by materials at the surface such as rocks, vegetation, and water and emitted as long-wave terrestrial radiation. Some terrestrial radiation is absorbed on its journey, especially by water vapor and carbon dioxide molecules, and thus the temperature of the atmosphere is increased. Therefore, the atmosphere is heated chiefly from the bottom because terrestrial radiation is the main heat source and because the numbers of water vapor and carbon dioxide molecules decrease rapidly upward.

Secondary heat sources occur at higher altitudes where the very high-energy short-wave radiation is absorbed. Thus temperatures do not decrease continuously or uniformly upward from the Earth's surface and the overall result of this absorption and reradiation is the existence of three warm zones in the Earth's atmosphere (Fig. 26–6): in the lower troposphere, in the vicinity of the stratopause, and in the thermosphere. Two cold regions separate the three warm zones.

To summarize, of 100 units of solar radiation, 30 units are backscattered to space. The remaining 70 units are absorbed, 20 units by the atmosphere directly and 50 units at the surface. These 70 units are subsequently reradiated to space—after a succession of absorptions and reradiations—as 70 equivalent units of long-wave radiation. Thus incoming and outgoing radiation appear to balance and the Earth as a whole seems to remain at a uniform temperature.

The trapping of solar energy is akin to what is called the *greenhouse effect*. The glass of a greenhouse is readily penetrated by (is transparent to) short-wave solar radiation. This heats the air inside the greenhouse, which cannot escape through the walls and roof, and thus the heat energy tends to remain inside.* According

* Benford, Gregory, and Book, David, "Sky Color," *Natural History*, Feb. 1971.

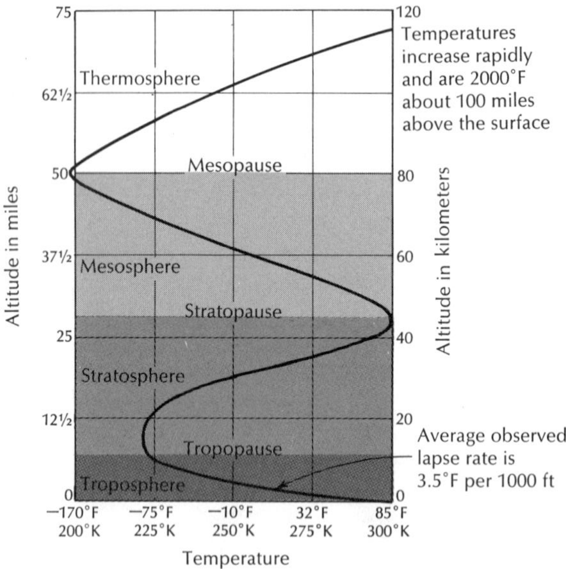

26-6. Main units of the atmosphere based upon variations in temperature with altitude (generalized). The atmosphere is heated chiefly from the bottom but higher temperatures occur at certain altitudes because ultraviolet and shorter radiations from the sun are absorbed at these altitudes. The altitude of the tropopause varies with the latitude and the season; it ranges from about four to twelve miles. Note the uniform temperatures in the lower stratosphere; they vary with latitude and range from about −50° to −100F°. *(Temperature-altitude graph after Butler.)*

to an older explanation, materials inside the greenhouse absorb the solar energy and reradiate long-wave radiation that cannot readily penetrate back through the glass. Thus this infrared radiation is trapped inside and raises the temperature. To show that the opaqueness of glass to infrared radiation is not the key factor, a greenhouse was once built with panes of rock salt (transparent to infrared), and it functioned well. This phenomenon is familiar to anyone who has entered a car which has been parked for some time in the sun with its windows closed. The Earth has been called "the greenhouse with a blue roof."

Fossil Fuels and Carbon Dioxide

Carbon dioxide is an important absorber of terrestrial long-wave radiation and is one of the by-products of the burning of coal, oil, and natural gas. These are the fossil fuels which originated long ago from organic matter. In a sense, they represent stored sunlight, since solar radiation was necessary for the existence and growth of the organisms from which they were formed. Since the fossil fuels are being consumed at rates that increase from one year to the next, very large quantities of carbon dioxide have been added to the atmosphere within the past 100 years, and the addition of still larger quantities appears to lie immediately ahead. Will these additions affect temperatures on the Earth?

In geologic terms, the consumption of the fossil fuels is a unique event. These fossil fuels accumulated during the 600 million years of the Paleozoic, Mesozoic, and Cenozoic Eras but are now being used up during a few centuries. Accordingly, we live in an *Age of Fossil Fuels.* Will this unique event in the Earth's long history—a sudden large increase in the quantity of carbon dioxide —cause temperatures to increase because absorption of out-going terrestrial radiation is thereby increased? If the atmosphere is warmed substantially, will sea level rise as glaciers melt, and will climatic belts shift and make certain regions inhabitable as others become undesirable as a home for man? The results to date are uncertain and predictions differ concerning future changes. The oceans contain about fifty times as much carbon dioxide as the atmosphere and may be able to absorb a large portion of the carbon dioxide released by combustion.

Mean annual temperatures on the Earth apparently have risen about 1°F during the past 100 years, and this is about the change one would expect from the increase in carbon dioxide during this same period. In a few thousand years according to one line of reasoning, temperatures may stabilize several degrees or more above their present levels. The oceans will absorb some of the new supplies of carbon dioxide, but enough will be added to the air to produce the increase in temperature.

However, the worldwide rise in temperatures of the past century may have been reversed about two decades ago. Moreover, since opposite reactions can result from the same change, the increased cloudiness which presumably would accompany increased temperatures, might increase the backscattering of solar radiation enough to prevent some of the rise in temperature. Furthermore, climatic changes such as those of the past century may actually result from several different factors rather than from a single influence. When these factors all act in the same direction, a significant climatic change occurs; when they tend to offset each other, no trend may occur. Thus the warming tendency produced by increased quantities of carbon dioxide may be temporarily overwhelmed at the moment by other factors whose nature is still obscure.

However, the drop in temperatures since the 1940's has also been attributed to an increase in atmospheric pollution. In this view, the pollutants may have increased the Earth's albedo and thus started a cooling trend. On the other hand, the pollutants might have an opposite effect by absorbing terrestrial radiation and reradiating some of this energy downward to warm the surface. Thus we lack a clearcut answer to the carbon dioxide question.

Layers and Structure of the Atmosphere

Temperature variations with altitude provide one criterion for subdividing the atmosphere into different layers (Figs. 26–6 and 26–7). There exist three warm zones with intervening cold regions.

The atmosphere is heated primarily from the bottom by terrestrial radiation (the greenhouse effect), and this accounts for the existence of the first warm layer. Temperatures tend to decrease from the surface upward at the average rate of about 3.5°F/1000 ft (2°C/1000 ft; 0.6°C/100 m; 6.5°C/km) for a distance, which is called the average *observed lapse rate*. In other words, if one were lifted by balloon and measured the air temperature at 1000-foot intervals, he would be ascertaining the observed lapse rate. This rate applies to air that is not rising or sinking appreciably and varies considerably at different times and places (e.g., temperatures actually increase upward within an inversion —a negative lapse rate—and may exceed 7°F/1000 ft at times). Vertical currents result in expansion or compression of the air, and the resulting temperature changes are known as adiabatic (p. 680).

This upward decrease or lapse in temperature tends to stop some miles above the surface; above this level, temperatures tend to remain fairly uniform with increasing altitude for a number of miles. The term *tropopause* is used for this transitional zone where temperatures stop decreasing upward. Above it is the stratosphere, and beneath it is the troposphere.

Nearly all weather phenomena—clouds, precipitation, storms, and the like—are confined to the lowest layer of the atmosphere called the *troposphere* (Gr.: "turning layer," pronounced trŏp·ō·sfēr). The troposphere is thickest at the equator (some 11 to 12 miles) and thinnest at the poles (some 4 to 5 miles). Its thickness correlates fairly well with temperature; the troposphere is thickest at the latitudes (more important) and during the seasons (less important) when temperatures are warmest. Perhaps increased turbulence produced by more intense solar radiation causes this. It is interesting that lower temperatures occur at the tropopause above the equator (lower than −100°F) than above the poles (perhaps an average of

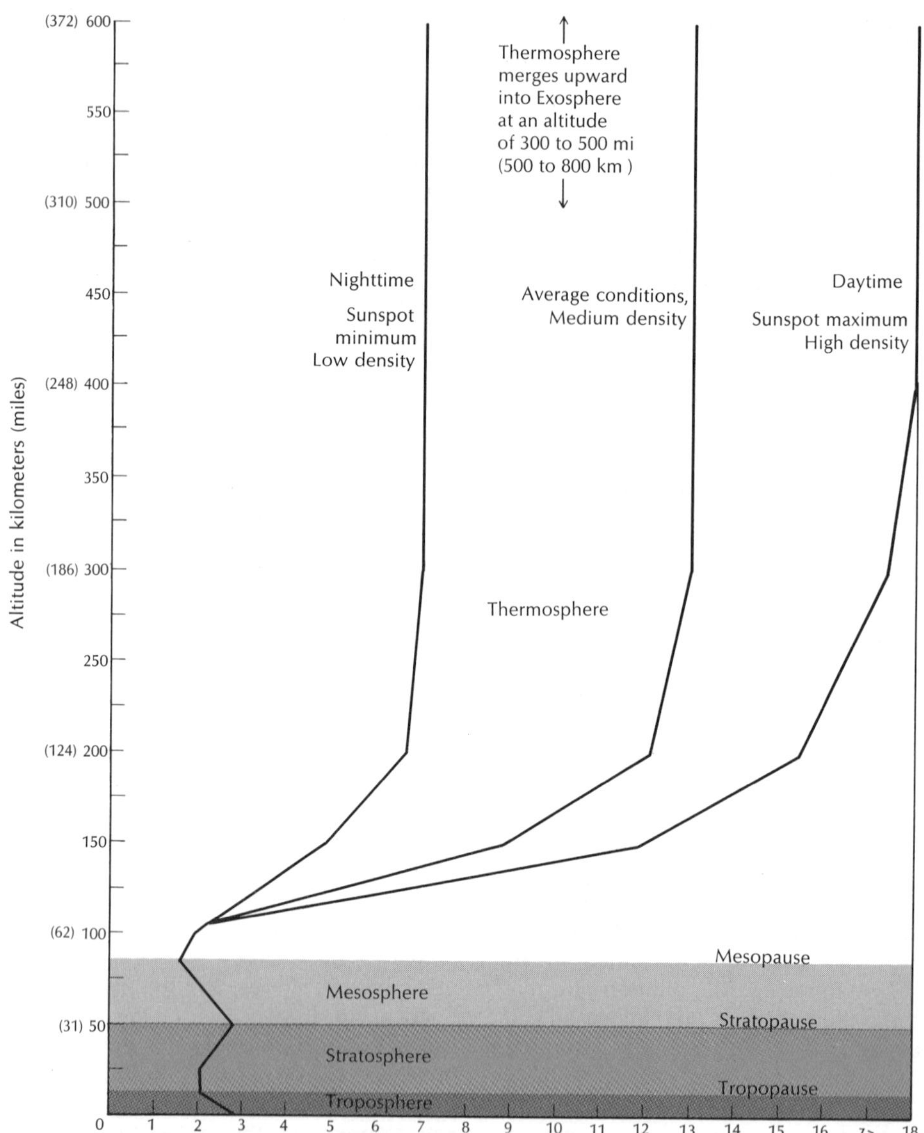

26-7. Temperature variations with altitude under three different sets of conditions (generalized)—two extremes and an average. Note that at any one time under a given set of conditions, the temperature of the thermosphere tends to remain about uniform above an altitude of a few hundred kilometers. However, at these high altitudes, temperatures may be about 1100°K (2000°F) higher in daytime at a sunspot maximum than they are in nighttime at a sunspot minimum. Reason: so few atoms occur that relatively small variations in solar energy (particularly in the X-ray and ultraviolet radiations) produce very large changes in temperature.

−70°F). The average upward decrease of 3.5°F/1000 ft extends to greater heights at the equator to cause this.

The tropopause is located at an average altitude of 7 to 8 miles (12 km) in the middle latitudes (Fig. 26–25). However, a conspicuous tropopause is missing at times, because the troposphere blends gradually into the stratosphere. At other times more than one tropopause may exist at different altitudes. These may overlap, and gaps may occur between them (Fig. 26–20).

Above the lower stratosphere, temperatures tend to increase upward to a maximum of about 80°F at an altitude of nearly 30 miles (45–50 km). The middle of this second warm layer is taken as the *stratopause*. The *stratosphere,* located between the tropopause and stratopause, thus consists of two parts: a lower zone which tends to have about uniform temperatures at any one time and latitude, and an upper zone in which temperatures increase upward.

The existence of this second warm layer is caused primarily by the absorption of high-energy ultraviolet radiation by ozone, the three-atom molecule of oxygen which has a faint bluish color and a distinctive odor. Ozone forms if a single atom of oxygen unites with a molecule of oxygen. Two single atoms of oxygen are produced when an oxygen molecule absorbs certain wavelengths in the ultraviolet region. Furthermore, if an ozone molecule unites with a single oxygen atom, it disintegrates into two molecules of oxygen. Thus in the vicinity of the stratopause, oxygen exists in three forms: as single atoms and as molecules of oxygen and ozone. These form and disintegrate as they absorb solar ultraviolet radiation and convert it into longer wave heat energy. From the stratopause, temperatures decrease upward for a distance because less oxygen occurs at higher altitudes. On the other hand, temperatures decrease downward because progressively less ultraviolet radiation is present; some was absorbed at higher altitudes.

Ozone molecules can absorb solar radiation in three regions of the electromagnetic spectrum: two in the ultraviolet and one in the infrared. Therefore, even though ozone constitutes only a very small percentage of the gases present in the stratosphere, yet its existence is very important for life on the Earth. If ozone were missing, our skins would be burned and our eyes blinded by the ultraviolet radiation that it absorbs (thus the present concern over possible major effects of high-flying supersonic transports on this ozone). If all of the ozone could be concentrated at sea level pressures, it would probably make a shell only a fraction of an inch in thickness.

Ozone tends to be most abundant (about 1 molecule in 100,000) at an altitude of about 16 miles (25 km) even though the warm zone that it causes occurs much higher at the stratopause. Apparently there are two reasons for this seeming discrepancy. First, the atmosphere is so diffused at the higher altitudes that much less heat energy is needed to produce the high temperatures. Second, ozone forms and disintegrates at a faster rate near the stratopause and thus absorbs more solar radiation. In fact, ozone apparently cannot form in quantity at an altitude of 16 miles because too little ultraviolet penetrates down to this level. Ozone, therefore, forms at the higher altitudes but is then moved by circulation of the atmosphere to lower altitudes where it is stored for a time.

Flying in the stratosphere is possible over the polar regions, where the troposphere is only 4 or 5 miles thick, but is more difficult nearer the equator. The absence of strong turbulence in the stratosphere makes such flying desirable. The relative stability of the stratosphere seems due to its structure: a colder, denser layer underlies a warmer, less dense layer—a stable arrangement.

Above the stratopause, temperatures decrease again to a minimum of about −175°F at an altitude of approximately 50 miles. Above 50 miles, temperatures increase again outward to regions where the temperature may exceed 1500 to 2000°F. The term *mesosphere* (Gr.: "middle layer") has been suggested for the zone located between 30 and

50 miles, the term *thermosphere* (Gr.: "warm layer") for the high-temperature zone above the mesosphere (to an altitude of 350 miles or so), and the term *exosphere* (Gr.: "outer layer") for the region above the thermosphere.

The third of the atmosphere's three warm layers occurs in the thermosphere and apparently is caused by the absorption of very high-energy ultraviolet radiation and X-rays, chiefly by single atoms of oxygen. Temperatures above the lower thermosphere tend to remain essentially the same with altitude at any one time and place—the upper thermosphere is said to be isothermal above an altitude of about 150 miles (240 km)—but great variation occurs from one time to another (Fig. 26–7). The variation is due to two factors. Fluctuations occur in the quantity of solar ultraviolet and X-ray radiation and in atmospheric densities. These fluctuations change from day to night and also during the sunspot cycle. Second, gases at these high altitudes are extremely diffused, which means that the total amount of heat energy present is small even though temperatures are high. Thus relatively small fluctuations in the high-energy part of solar radiation result in large changes in temperature in the upper thermosphere.

An unshielded man in the thermosphere and in the Earth's shadow would freeze despite the very high temperatures. At the surface, the temperature of an unheated object is about the same as the temperature of the air around it. Energy is exchanged continuously between the air and the object, and a balance is attained. However, atoms and molecules are relatively so far apart in the thermosphere that heat energy would be transferred in one direction only —away from the man. Temperatures under such conditions do not have the same significance as they do at the surface; they are called kinetic temperatures and indicate only the average rate of movement of the gases involved.

The exosphere is the region from which neutral atoms can escape from the Earth's gravitational pull into space (charged particles become part of the magnetosphere, p. 655). To escape from the Earth, an atom must attain a speed of about 7 mi/sec. On the average, atoms in the exosphere move at slower rates than 7 mi/sec, but an occasional one attains this speed. This is most likely to happen for the lightest atoms and when temperatures are highest. Thus hydrogen and helium tend to escape more or less continuously from the Earth and are replaced more or less continuously. New supplies of hydrogen atoms are made available by the action of high-energy solar radiation on molecules such as water vapor and methane in the lower atmosphere. Some hydrogen also comes from the sun as protons (hydrogen nuclei) in the solar wind (p. 600). Helium atoms apparently are produced by the release of alpha particles (helium nuclei) during radioactive disintegration of elements such as uranium at the surface.

Some Methods of Studying the Atmosphere at Higher Altitudes

Unmanned balloons, rockets, and artificial satellites provide the main means of studying the atmosphere at successively higher altitudes, and aircraft can be used at lower altitudes. *Radiosonde balloons* are important in obtaining weather data up to altitudes of about 15 to 20 miles or more (24 to 32 km). A ventilated package containing various weather instruments and a radio transmitter is lifted by a balloon and automatically records and transmits data back to the surface concerning barometric pressure, humidity, and temperature. Since the balloon rises at a known rate, its movement can be used to determine winds aloft. Special balloons have reached higher altitudes with heavier payloads—up to 29 miles (46 km) with a 360-pound payload in 1966.

A promising new development known as GHOST (Global Horizontal Sounding Technique) involves balloons which are designed to float at a certain pressure such as 100 mb for periods of several months or more and thus to travel around the Earth a number of times. In early 1969 one such balloon was still operating after drifting more than 25 times around the Southern Hemisphere at an altitude of about 10 miles (16 km) over a period of about 1½ years. Such balloons carry small, light-weight, weather-instrument packages which supposedly constitute a negligible hazard to aircraft. The balloons are being tested in the Southern Hemisphere to reduce political problems and also the chances of collisions with aircraft. However, icing has been a problem at lower altitudes, and Mother Ghost is a newer technique. "Mother" is a carrier balloon designed to float at about 15 miles (24 km). It carries 100 or so small dropsondes that can be released by remote control (like radiosondes but descend by parachute when dropped).

A cooperative World Weather Watch is now underway in which Ghost balloons play a part. However, fewer balloons may be needed because infrared sensors on orbiting weather satellites can now be used to obtain temperatures at different altitudes (Fig. 26–25)—also wind and pressure data at middle latitudes. Additional weather measurements may come from a network of automatic weather stations attached to buoys anchored on a grid pattern over the oceans. Weather data from these sources can then be relayed via a network of orbiting weather satellites to appropriate ground stations. Such global coverage, added to data obtained at ground weather stations and by ships and planes, may finally furnish the meteorologist with a better understanding of weather processes and provide the means for longer and more accurate forecasts.

Light-weight rockets can be used to lift instruments to obtain atmospheric data at otherwise inaccessible altitudes between 25 to 125 miles (40 to 200 km) approximately —too high for balloons and yet too low for orbiting weather satellites. For example, grenades may be ejected and fired at different altitudes during a rocket flight, and the behavior of the smoke or sound from such explosions can be interpreted to give information about temperatures and winds at different altitudes (Fig. 26–8). A radar target or a metallized parachute or balloon may be ejected at high altitudes and tracked as it descends.

In one ingenious type of experiment (the principle is also used in orbiting satellites) a large evacuated ball is ejected from a rocket, and a small ball is located at the center of the larger one. This inner ball is freed automatically during descent and the two balls would fall at the same rate in the absence of atmospheric drag. However, drag slows the larger ball, and the inner one—falling at a faster rate in a vacuum—eventually makes contact with the bottom of the larger ball. The time of contact is radioed to the surface, and the time and distance figures (tracked by radar) can be interpreted to obtain temperatures and densities at different levels. Higher temperatures and densities produce greater drag. The inner ball can be reset and released a number of times during a single fall.

Sodium vapor clouds provide additional

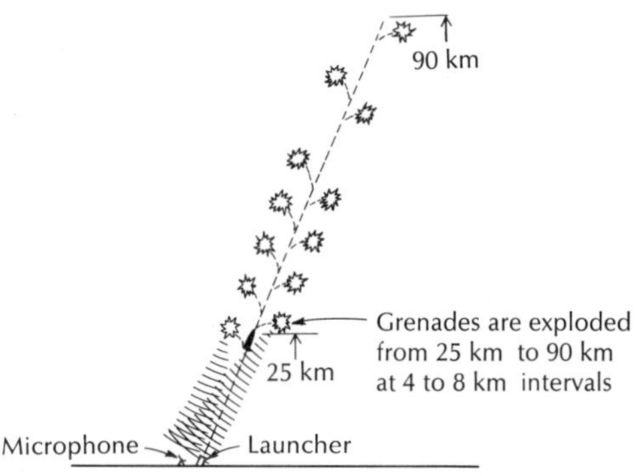

26-8. Grenade explosions aid in measuring winds and temperatures at high altitudes. A number of grenades are ejected and exploded during the upward flight. The times and directions of the sound arrivals are then recorded by microphones on the ground. These data permit calculations of the winds and temperatures in the sections of the atmosphere occurring between each pair of grenades. *(Courtesy NASA.)*

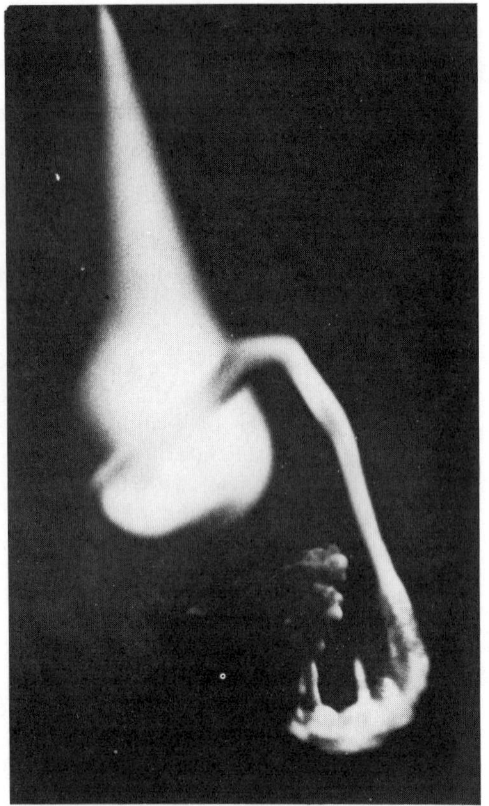

26-9. Sodium vapor clouds give information about high-altitude winds. The sodium vapor is ejected during the flight of a rocket, and the resulting clouds are photographed simultaneously from different ground stations. Thus their altitudes and locations can be determined by triangulation techniques. Data are obtained concerning winds and diffusion rates at altitudes from about forty to more than 100 miles. The firings are made at dawn or twilight: the sodium trail is illuminated by sunlight, but the amount of scattered sunlight is low. *(Courtesy NASA.)*

information concerning high-altitude winds (Fig. 26–9) as does the movement of faint clouds (notilucent) at altitudes of 50 miles or so. Such clouds may consist of dust particles rather than of ice crystals.

Before rockets became available, the behavior of sound waves and meteors provided data on temperatures at altitudes beyond the reach of balloons; e.g., the noise from a large explosion could be detected along the ground for a certain distance and then again at a much greater distance with a no-sound zone between (Fig. 26–10 illustrates a similar effect). Furthermore, the sound arrived later than expected at the greater distances. This phenomenon was explained as an effect of refraction—the existence of a warm zone at high altitudes which speeded up the sound waves and bent them back toward the surface. Since they traveled a greater distance than the ground wave, they arrived later than expected.

At higher altitudes, atmospheric drag on orbiting satellites furnishes data concerning temperatures and densities at these levels; e.g., distance changes in the orbits of seven satellites were once noted about a day after the occurrence of a solar flare. Particles ejected from the sun at the time of the flare finally reached the Earth and increased the drag on the satellites. This decreased their speeds, which changed their orbits (p. 481). Artificially produced auroras (p. 601) and the reflection of radio signals (p. 655) are other means of obtaining data about the upper atmosphere.

The Ionosphere

Solar radiation, particularly its high-energy ultraviolet and X-rays, produces electrified particles in the upper atmosphere by detaching electrons from some of the atoms. This region of electrically charged particles is called the *ionosphere,* and it extends from a base at an altitude of 30 to 40 miles (about 50 to 65 km) to a top at 400 to 600 miles (about 645 to 965 km). These free electrons and ions—atoms that have gained or lost electrons and thus have either a negative or positive electric charge—act as mirrors at certain altitudes and for certain frequencies, and they reflect radio waves back to the surface (Fig. 26–10). Thus long distance radio reception is made possible.

The amount of energy needed to detach electrons from atoms tends to vary from one kind of atom to another. Moreover, the amount of very high-energy solar radiation

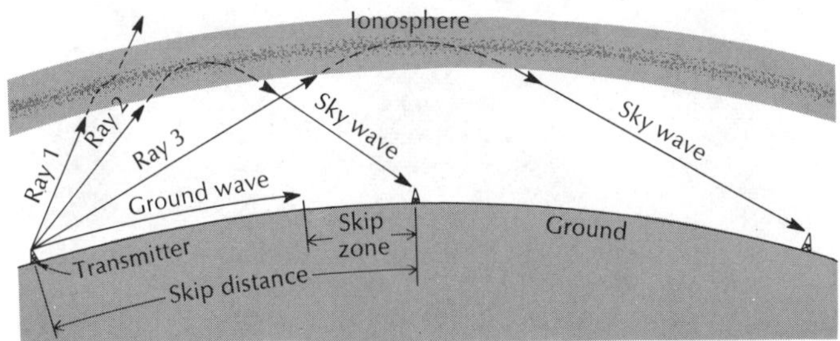

26-10. Effect of the ionosphere on radio waves. Ray 1 enters the ionosphere at a steep angle and is not reflected back to the surface. However, its direction is changed (refraction) as it moves outward. Rays 2 and 3 enter the ionosphere at smaller angles and are reflected back to the surface (they are called sky waves). The critical angle between reflection or no reflection depends upon the intensity of the ionization and wavelength (shorter wavelengths cause smaller critical angles). The skip distance for a particular wavelength is the minimum distance from the transmitter at which its sky waves can be received. The ground wave travels from the transmitter along the surface and weakens with distance. Therefore, if the distance traveled by a ground wave is less than its skip distance, a skip zone is produced in which no signal occurs. Wave interference and variations in ionospheric intensity cause fluctuations in the strength of the signal at a radio receiver. Extreme variations may cancel the signal and cause a radio blackout. *(Courtesy U.S. Navy Hydrographic Office.)*

decreases downward, and the chemical composition of the atmosphere varies. Therefore, more electrons and ions occur at some levels than at others, and the ionosphere consists of a number of fairly distinct layers which, however, change with changes in solar radiation. Fewer and weaker reflecting layers occur on the night side of the Earth because some of the free electrons recombine with positively charged ions to form electrically neutral atoms. The reflecting layers reappear during the next day when solar radiation again causes ionization.

Changes in the ionosphere cause changes in the reflection of radio signals, and fading or blackouts may occur; e.g., the increased quantity of ultraviolet and X-rays at the time of a solar flare may cause a radio blackout (Fig. 26–11). As another example, signals from distant radio stations tend to be stronger at night because the lowest layer (D) of the ionosphere disappears at night—during the day this D layer absorbed the signals from distant radio stations.

The ionosphere has a peak intensity at an altitude of about 180 miles (290 km); i.e., the concentration of free electrons and ions per unit volume is greater here than it is at either higher or lower altitudes. Two factors influence this concentration: (1) the amount of high-energy solar radiation available to cause ionization (this decreases downward) and (2) the density of the atmosphere. The concentration decreases upward from 180 miles because relatively few atoms and molecules occur at very high altitudes. Therefore, although a high percentage of atoms are ionized, the actual number present is comparatively low. Atoms and molecules are quite far apart throughout the ionosphere, and thus atoms stay ionized for a relatively long time. At lower altitudes, more frequent collisions occur among atoms and electrons, and the ions are transformed into electrically neutral atoms.

The Magnetosphere

Electrically charged particles from space that have been captured by the Earth's magnetic field (Fig. 11–19) form a huge doughnut-shaped halo about the Earth. This halo is called the *magnetosphere*. It begins at an altitude of about 600 miles (965 km); below

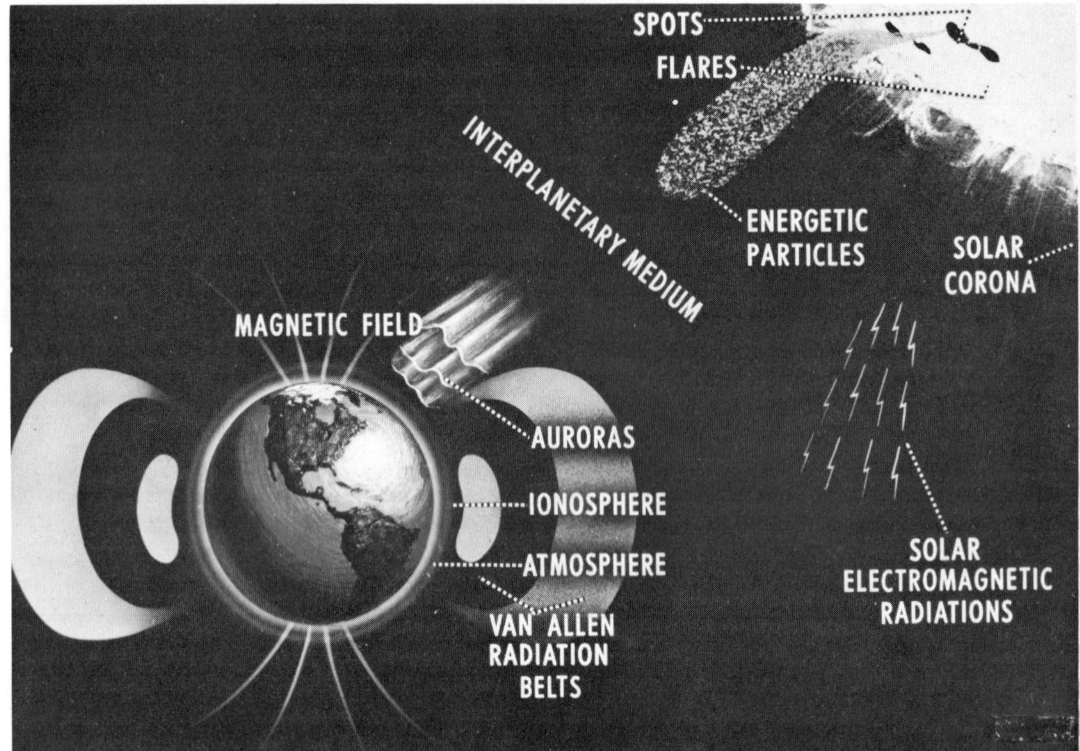

26-11. Sun-Earth phenomena. Continuous release of energy by nuclear fusion within the sun causes turbulence at its surface. At times the turbulence is very great and clouds of charged particles and electromagnetic radiation stream outward into space. The term solar flare is applied to such eruptions, and the charged particles are called solar cosmic rays. When the flare is properly situated, the clouds of particles reach the Earth and interact with its atmosphere. At such times, magnetic storms, radio blackouts, auroral displays, and other phenomena are produced. *(Courtesy NASA.)*

this level, the charged particles interact too frequently with other particles and atoms to remain trapped. The magnetosphere extends outward to an altitude of about 40,000 miles (65,000 km) on the side toward the sun, but it extends for much greater distances on the side away from the sun (Fig. 26–12). The charged particles are chiefly protons and electrons that come from the sun or originate as by-products from the interaction of cosmic rays with the atmosphere.

The ions and electrons that enter the Earth's atmosphere from the sun and cosmic radiation are trapped for a time by the Earth's magnetic field. They spiral around the magnetic lines of force and move quickly from the Northern to the Southern Hemisphere and back (Fig. 26–13). If an electron or ion comes close to the surface during its spiraling descent, it is likely to collide with an atmospheric atom or molecule and lose so much energy that it escapes (is discharged) from the magnetosphere. The magnetosphere is constantly discharging particles into the lower atmosphere, particularly near the magnetic poles. Such discharges are especially common after a solar flare and may produce auroras (p. 601).

The magnetosphere was first called the Van Allen radiation belts (Fig. 26–11) after the physicist who discovered them; they were an unexpected result of the 1957–58 International Geophysical Year. High-energy protons have a peak concentration in the magnetic field at an altitude of about 2000 miles (3200 km) and form the inner, relatively stable Van Allen belt. Cosmic radiation may be the source of the protons. A concentration of high-energy electrons occurs in the outer belt and fluctuates in intensity with solar activity. However, vast numbers of low-energy protons and electrons occur within these belts and also throughout the magnetosphere. Thus it is probably best to view the magnetosphere as one large halo which varies at different altitudes. Suitable protection must be provided during manned flights through these zones of high-energy particles.

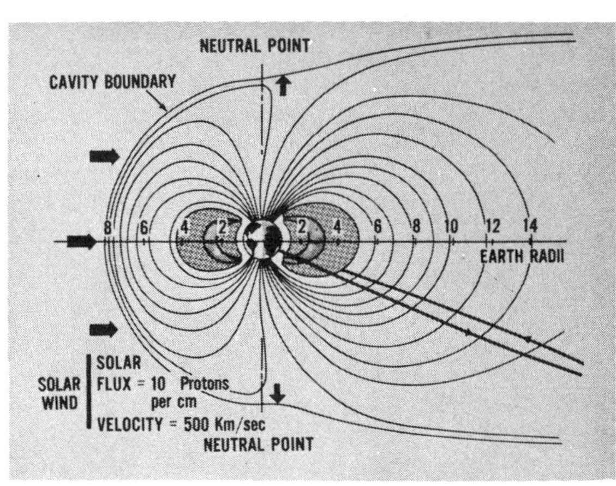

26-12. Interaction between the Earth's magnetic field and the solar wind. The solar wind consists of a stream of low-energy charged particles that are ejected continuously by the sun. The particles vary in number with the rate of solar activity and may travel at a few hundred miles per second. The solar wind distorts the Earth's magnetic field, which is compressed on the sunward side and lengthened on the side away from the sun. The magnetosphere presumably would be symmetrical in the absence of a solar wind. As the solar wind moves past the Earth, a wake is produced, somewhat similar to that formed by a boulder in a stream. A shock wave also forms in front of the magnetosphere boundary on the sunward side of the Earth, and a zone of turbulence occurs between this shock wave and the outer boundary of the magnetosphere. A cloud of charged particles ejected from the sun has a magnetic field and can deflect cosmic radiation. The heavy lines in the lower right show the outward and inward paths of an orbiting satellite. *(Courtesy NASA.)*

The Circulation of the Atmosphere

The scale of movement within the atmosphere is extreme—from tiny eddies to global wind systems. However, the basic cause of all wind and movement is temperature differences. These produce pressure differences, which in turn cause air movements.

We shall consider the general circulation of the atmosphere from a global view and with emphasis upon mean conditions. Some basic ideas are involved: heated air rises; air moves horizontally from high to low pressure areas; and moving air is deflected.

The intensity of solar radiation varies: daily, seasonally, and with latitude (p. 521), and it is absorbed and reflected by different types of surfaces and topographic features. This results in temperature differences from one location or region to another. Since heated air expands, it becomes less dense and more buoyant, and it is pushed upward by nearby heavier, cooler air. On a global scale then, we expect air to rise at the equator where solar radiation is most intense and to sink at the poles.

Air has a tendency to move horizontally from an area of higher pressure to one of lower pressure (i.e., down the pressure gradient). The greater the difference in pressure, the more rapidly does the air tend to move. Except initially, however, air does not move directly from high- to low-pressure areas (Fig. 26–14).

Moving objects—bullets, planes, air masses—are deflected to the right in the direction of movement in the Northern Hemisphere but to the left in the Southern Hemisphere (the Coriolis effect, Fig. 26–15). The Coriolis effect is not caused by a force that pushes

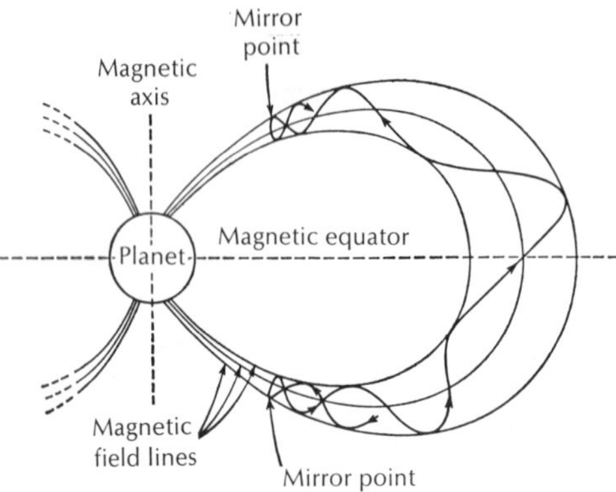

26-13. Trajectory of an ion trapped in the magnetic field of a planet. A radiation belt is formed by many such ions oscillating in similar trajectories. Auroras occur in the regions closer to the Earth where the ions may be disturbed out of their spiral orbits. To understand the trajectory, assume that a charged particle approaches the Earth's magnetic equator at exactly a 90 degree angle to the magnetic lines of force—it is moving parallel to the plane of the magnetic equator. The path of the charged particle would become curved as it crossed the magnetic lines of force until it moved in a circular path. The plane of this circle would be parallel to the plane of the magnetic equator. The diameter of the circle, and its location within the magnetic field, would depend upon the energy of the charged particle and thus the depth it could penetrate into the magnetic field. However, most charged particles approach the magnetic lines of force at an angle and thus develop a spiralling trajectory toward one pole or the other. (A. G. Smith and T. D. Carr, *Radio Exploration of the Planetary System*, Van Nostrand Reinhold, 1967.)

moving objects to the right in the Northern Hemisphere. Rather, the moving object tends to proceed in a straight line (Newton's first law of motion, p. 470) and the Earth rotates underneath it. The true motion would be apparent from a position in space not involved in the rotation of the Earth.

The phenomenon can be illustrated with an old phonograph record, a ruler, and a piece of chalk. While the record spins, try to draw a straight line along the ruler and across the record, either inward or outward. In each case a deflection to the right occurs (if the record is rotated counterclockwise) and the chalk mark on the record is curved. Rotate the record clockwise, and a deflection to the left results.

The Coriolis effect is perhaps most easily understood if we imagine an observer with a rocket located exactly at the north pole. The Earth's linear rate of rotation is zero here, but it increases to a maximum of about 1000 mi/hr at the equator. The rocket is fired at a target some tens of miles to the south. As viewed from space, the rocket moves in a straight line; it leaves the pole with no eastward rotation. The target, however, is being rotated eastward. By the time the rocket reaches its latitude, the target will have been shifted some distance to the east, and the rocket misses it. To observers on the Earth, however, the rocket misses the target by appearing to curve toward the west—i.e., it appears to be deflected toward the right. That the target actually moved out of the path of the rocket is difficult to grasp.

If the positions of the rocket and target are reversed, a deflection to the right will again occur. However, this time the target has no eastward motion, whereas the rocket does. The rocket tends to retain the eastward rate of rotation it had at the latitude of firing. Consequently it shifts continuously eastward on its northward journey; it again misses the target and appears to have curved to the right. The effect at different latitudes is illustrated in Fig. 26–16.

The magnitude of the Coriolis effect increases with both latitude (it is zero at the equator) and velocity; e.g., the deflection is twice as large at twice the speed. However, its effect is negligible or very small upon objects moving short distances and on local winds. Therefore, wind directions at the surface tend to be different from those at higher altitudes where greater deflection occurs because of higher velocities.

Air moving about parallel with the surface beneath it is called wind, and winds are designated according to the directions from which they blow—i.e., a north wind blows from the north. Weather vanes point into the wind, and their use may have initiated such terminology.

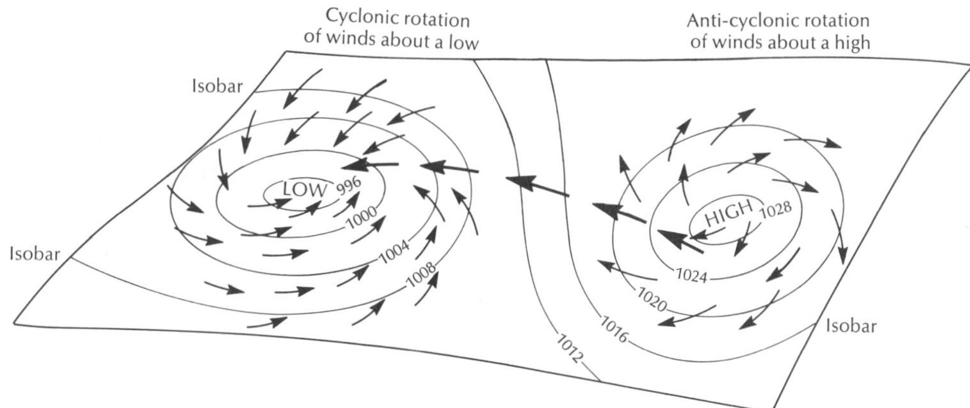

26-14. The relationship of surface winds to pressure patterns. Surface winds tend to make angles of 10 to 50 degrees with the isobars (at higher altitudes, they tend to move parallel to the isobars). Over the oceans this angle might average 10 to 15 degrees or less, whereas over the lands it ranges between 20 and 50 degrees approximately. Cyclones and anticyclones are synonymous with lows and highs. A cyclone is not a hurricane. Note the counterclockwise circulation around the low and the clockwise circulation around the high. These directions are reversed in the Southern Hemisphere and result from the Coriolis deflection. (AF Manual 105-5, *Weather for Aircrews*.)

The global pattern of wind systems may not be apparent where it is obscured by movements that occur locally, nor at any given time (synoptic), but it will show up in mean records accumulated over a period of time. At the equator, where insolation is most intense, the lower troposphere is heated, expands, becomes less dense, and rises. This warm moist air cools as it rises (Fig. 26–17) and produces a belt of low pressure at the surface along the equator. However, pressures decrease upward less rapidly in warm air than in cool air (p. 664). Thus pressures at an altitude of several miles above the equator are actually greater than they are at these same altitudes at higher latitudes (where surface pressures exceed those at the equator). Therefore, the upward movement ceases, and the air spills out of the equatorial region toward the north and toward the south.

Close to the Earth's surface, air moves toward the equator to replace the rising air. If the Earth were not rotating, and if there were no complications resulting from the presence of both continents and oceans, a huge simple convectional circulation would probably result. Air would rise at the equator, sink over the poles, and move horizontally between these latitudes.

But the Coriolis effect resulting from the spinning of the Earth deflects eastward much of the air that starts northward or southward from the equatorial region in the upper troposphere. By the time this air has reached the vicinity of the 30-degree latitudes, much of it may be moving almost due eastward (but high above the Earth's surface). As air piles up here, pressures increase; thus it sinks to form the high pressure zones whose centers occur at a latitude of about 30 degrees. These areas of descending air currents and greater surface pressures (the *horse latitudes*) are not continuous; they are subdivided into large cells (centers of high pressure) which tend to be located over the oceans and are best developed near their eastern margins.,

In the Northern Hemisphere air moves along the Earth's surface from the higher pressures of the horse latitudes toward the lower pressures at the equator. This air

In the Northern Hemisphere, the Coriolis deflection is always to the right; the effect increases toward the north pole

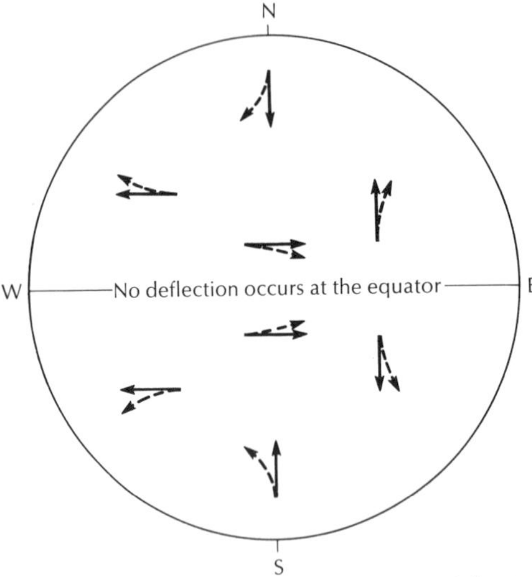

In the Southern Hemisphere, the Coriolis deflection is always to the left; the effect increases toward the south pole

26-15. A deviation of projectiles (or other moving objects) is produced by the Earth's rotation. The heavy lines show the direction of firing; the broken lines, the direction of flight as modified by the Earth's rotation. Whether the projectile is fired north or south, east or west, it always deviates to the right in the Northern Hemisphere and to the left in the Southern Hemisphere. However, the deflection is negligible over short distances.

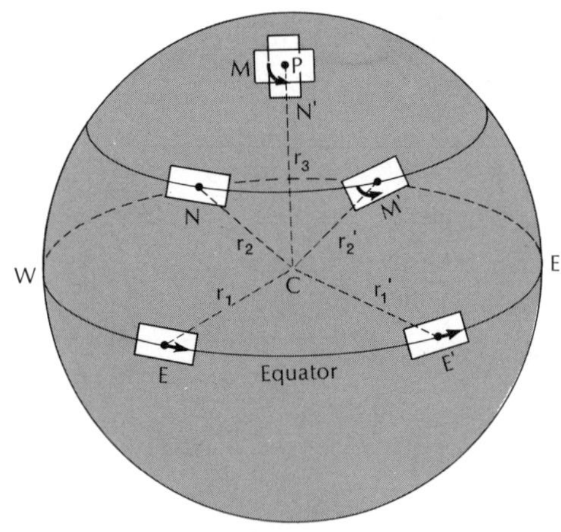

26-16. The Coriolis effect at different latitudes. Since the turning of the Earth causes the Coriolis effect, this effect will be largest where the surface "twists" most rapidly underneath moving objects (at the poles); on the other hand, no effect will occur where the surface does not "twist" underneath a moving object (at the equator). Imagine that the three rectangles represent the floors of large rooms each perpendicular to a radial axis from the center to its location (Cr_1, Cr_2, and Cr_3). At the pole the floor rotates through 360 degrees in nearly twenty-four hours around the axis r_3. At the equator, on the other hand, the floor has no horizontal rotation around the axis r_1. At M, the floor twists more slowly around axis r_2 than it does at the pole. The amount of twisting, therefore, increases with latitude (with the sine of the latitude). The behavior of a Foucault pendulum shows that such twisting occurs (p. 518). (After William L. Donn, *Meteorology*, 3rd ed., McGraw-Hill Book Co., 1965.)

is likewise deflected to the right (again the Coriolis effect) and forms the *northeast trade wind belt.* Air also moves along the Earth's surface northward from the horse latitudes and is deflected to the right to form the *belt of westerlies* (southwesterlies in the Northern Hemisphere).

The circulation of the Earth's atmosphere at high latitudes is less well known, but air apparently is cooled and compressed above the polar regions (the center of high pressure in the winter is over northern Siberia, not over the Arctic Ocean). Therefore it sinks and moves southward from the polar region along the surface. It is deflected to the right in the Northern Hemisphere to produce winds known as the *polar easterlies.*

Cold air moving southward along the surface from the high-latitude regions meets the northern margin of the belt of westerlies at a latitude of 50 to 60 degrees, with a range of 30 to 70 degrees. The line of intersection or convergence is known as the *polar front* (Fig. 26–18), and along it warmer air in the belt of westerlies tends to be lifted above the colder heavier air from the north. The polar front is quite irregular and frequently develops huge bulges that extend southward into the middle latitudes. The air that rises as a result of convergence

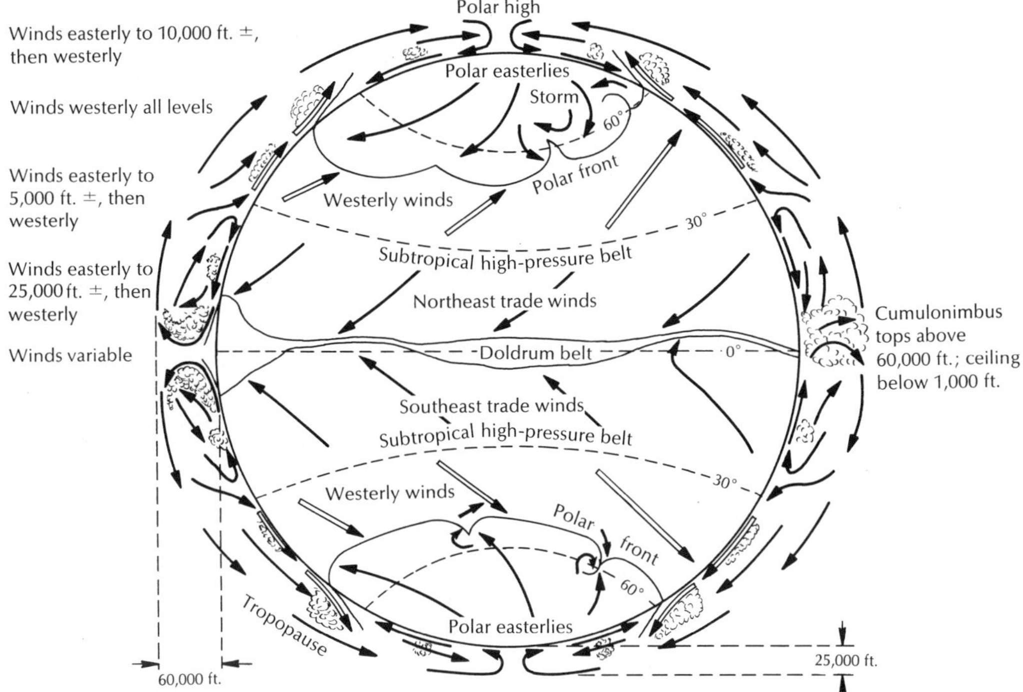

26-17. Idealized pattern of atmospheric circulation. The Earth is shown without continents for simplicity. The subtropical high-pressure belts are not continuous. (TM 1-300, *Meteorology for Army Aviation.*)

along the polar front produces lower barometric pressures at latitudes of 55 to 65 degrees, but this is far from a continuous belt. The polar front is important for understanding weather phenomena in the middle latitudes (p. 729).

Similar circulation in the Southern Hemisphere, combined with a deflection to the left, produce comparable belts of trade winds, westerlies, and perhaps polar easterlies.

The northeast and southeast trade winds

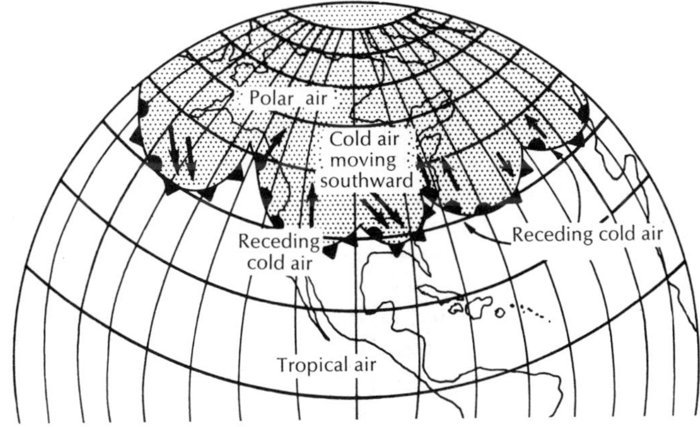

26-18. The polar front. (TM 1-300, *Meteorology for Army Aviation.*)

converge along the equator (doldrum belt) to form the *intertropical convergence zone* (p. 661), the so-called meteorological equator. Convergence generally causes upward movement, cloudiness and precipitation. Over the oceans this zone tends to be located along the equator in January but at a latitude of about 10 degrees north in July. The seasonal shift is greater over the continents—e.g., over Africa the zone may move from 15 to 20 degrees south latitude in January to 15 to 20 degrees north latitude in July. Thus a location at low latitudes in Africa may experience two rainy seasons each year as the intertropical convergence zone advances beyond this location and later retreats across it.

Surprisingly, maximum surface temperatures develop closer to the tropics (during their summer seasons) than to the equator: more hours of daylight occur on a particular day (always about 12 hours at the equator), and the sun's rays are more concentrated for a longer sustained period (sun is near zenith for a more extended period).

The other wind and pressure belts likewise shift seasonally 5 to 10 degrees or more—the shift is toward the polar regions in summer and toward the equator in winter. Temperatures in the troposphere above the Antarctic seem to average about 20°F (11°C) colder than those above the Arctic (heat energy is transmitted from Arctic ocean water through the thin ice to the atmosphere). Therefore, a stronger temperature gradient occurs between the south pole and the equator than between the north pole and the equator. This leads to stronger winds in the Southern Hemisphere than in the Northern Hemisphere (the great expanse of ocean is another factor) and a greater shift of its belts toward the equator; e.g., the average location of the center of the horse latitudes is about 30 degrees south latitude but 35 or more degrees north latitude.

Above the surface wind belts, winds from the west predominate throughout most of the troposphere, especially poleward from latitudes of 30 degrees or so, and reach their peak development in the jet streams. These west winds apparently result from a general poleward drift in the middle and upper troposphere coupled with a Coriolis deflection to the east.

Evidence of a convectional cell on either side of the equator seems reliable, although it is probably more complex than is shown in Fig. 26–17. Such cells are needed to transfer heat energy away from the equator. Similar cells were once thought to be associated with the belts of westerlies and polar easterlies, but apparently they do not exist. They are not needed to account for the transfer of heat energy to higher latitudes; the movement of air masses and lows and highs can do this.

The pressure areas and wind systems are more continuous and better developed in the Southern Hemisphere than in the Northern Hemisphere because of the greater proportion of water to land. Variations occur within a certain belt because of the presence of land or sea, of high and low topography, and of low- and high-pressure systems (cyclones and anticyclones). The irregularities are especially prominent in the belt of westerlies where a succession of lows and highs move within the belt and obscure the overall pattern (similar to the eddies in a river). Mark Twain had the middle latitude westerlies in mind when he said, "If you do not like our weather, wait five minutes."

Jet Streams

According to the World Meteorological Organization: "A *jet stream* is a strong narrow current, concentrated along a quasi-horizontal axis in the upper troposphere or in the stratosphere, characterized by strong vertical and lateral wind shears and featuring one or more velocity maxima." A jet stream is further described by the WMO as thousands of kilometers in length, hundreds of kilometers in width, and some kilometers in depth. On the average within a jet stream, wind speeds may change in a vertical direction by some 20 to 35 mi/hr per mile and by some 15 to 20 mi/hr per 100 miles along the axis. To be classified as a jet stream, winds along the axis must attain a minimum speed of about 70 mi/hr (113 km/hr).

A *polar front jet stream* and a *subtropical jet stream* occur in each hemisphere and blow eastward in great sweeping curves near the tropopause (Fig. 26–19). They may extend all the way around the Earth at certain times and shift considerably from north to south from season to season and even in the same season from one year to the next. Jet streams seem to be the concentrated portions of the westerly winds that are common in the upper troposphere. The winds blow in gusts, and vary in speed along the trend of a jet stream as well as outward from the central core. These zones of higher velocity may shift from one region to another. Thus irregularity in velocity and location is characteristic of a jet stream.

Planes which fly eastward in a jet stream may have their speeds increased by 100 mi/hr or more, and planes flying westward must overcome strong headwinds. However, flights are now so numerous that planes must stay within designated corridors regardless of the effect of a jet stream.

26-19. Schematic cross section of a jet stream. A jet stream may encircle the Earth. Note the core of very fast winds (1 knot means 1 nautical mile per hour; 100 knots approximates 115 miles/hr). (AF Manual 105-5, *Weather for Aircrews.*)

The exact relationship between jet streams and weather changes is still somewhat uncertain. However, the locations of the jets influence the paths followed by the lows and highs so common in the belt of westerlies. Furthermore, movements within a jet stream may initiate the development of a low or high at the surface (p. 731).

To account for the origin of a jet stream we first note that constant pressure surfaces (say those at 500 and 300 mb) are closer together in cold dense air than they are in warm, less dense air (Fig. 26–20). In other words, pressure decreases upward more rapidly in cold air than in warm air. Let us assume that temperatures decrease uniformly from the equator to the poles and apply this generalization. At the surface, pressures are greater at the poles than at the equator. However, pressures decrease upward least rapidly in the warm air above the equator. Therefore, at an altitude of several miles, pressures are greater above the equator than they are at an equal altitude above the poles. Therefore, at these high altitudes in the troposphere, air tends to move poleward, and the Coriolis effect produces winds blowing generally toward the east. Speeds are greatest near the tropopause in the vicinity of the subtropical highs

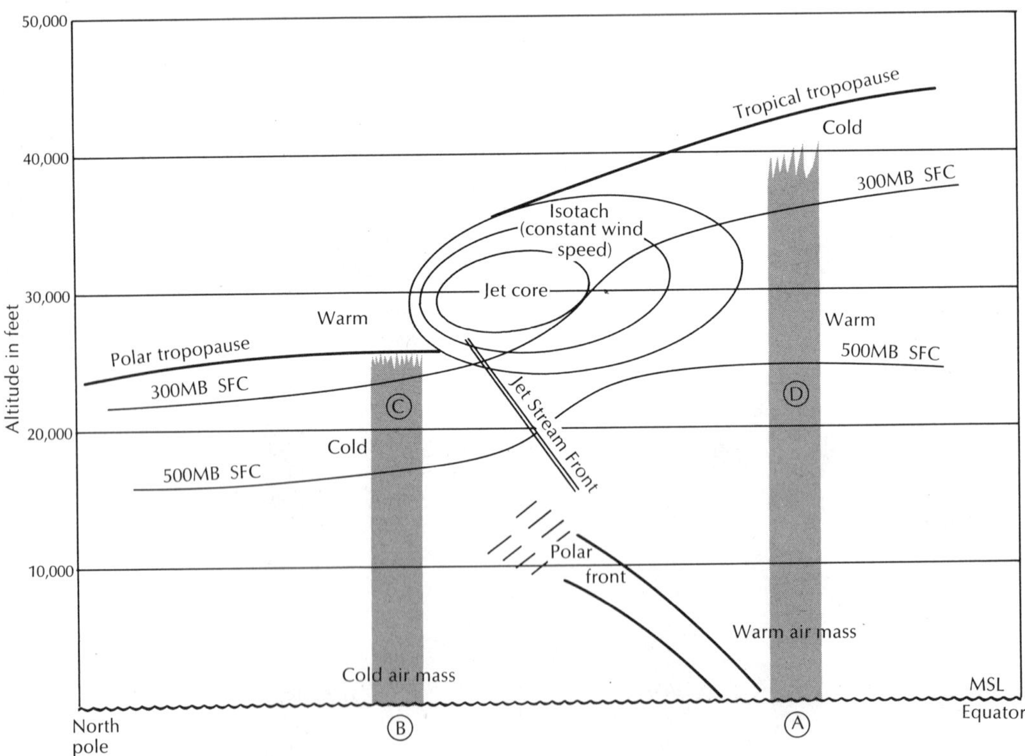

26-20. Location of the polar front jet stream at a gap in the tropopause above the polar front. Vertical scale is exaggerated. Isotachs are lines of equal wind speed. MB refers to millibars and SFC to surface. Pressures decrease upward more rapidly in cold air than in warm. Thus the vertical distance separating the 500 and 300 mb pressure surfaces in the cold air column at C is shorter than the vertical distance between them in the warm air column at D. Therefore, the pressure at D is greater than at C. This causes air to move horizontally from D toward C, and a Coriolis deflection produces a west wind. These west winds attain their peak speeds above the polar front, and form the polar front jet stream because the temperature differences—and thus the pressure differences—are greatest here. (AF Manual 105-5, *Weather for Aircrews.)*

(horse latitudes) and polar front to produce these two jet streams.

A *stratospheric polar night jet stream* occurs near the stratopause above a polar region during winter, and it apparently forms for a similar reason. During winter, temperatures in the stratosphere become colder above the poles (sun is below the horizon for six months) than above the equator. Therefore, pressures are greater near the stratopause above the equator (they decrease upward at a slower rate to cause this) than above the poles. This causes a poleward flow high in the stratosphere which the Coriolis effect changes into a west wind. This is most concentrated near the stratopause above the polar regions in winter.

In summer, winds near the stratopause above a polar region blow generally toward the west but usually at less than jet-stream speeds. The explanation is similar to that for the winter winds but more or less reversed. In summer in the stratosphere, temperatures are warmer above the poles (sun is above the horizon for six months) than above the equator. Since pressures decrease upward least rapidly in this warm air, pressures are greater near the stratopause above a polar region in summer than they are at this same level above the equator. A high-altitude drift toward the equator is thus initiated which is deflected toward the west.

Weather Conditions at Different Latitudes

Circulation of the atmosphere (Fig. 26–21) aids in causing certain types of weather at certain latitudes. At the equator, where warm moist air rises, it expands and cools, and heavy precipitation is common, with high temperatures and gentle variable winds. Warm air has a much greater capacity for containing water vapor than does cold air (p. 677) and thus a greater amount of precipitation tends to occur at the equator than at higher, colder latitudes. The northeast and southeast trade winds converge upon this zone and push upward the warm moist air along the equator. The converging air masses are similar and thus no conspicuous front occurs. Sailing vessels may dally within this intertropical convergence zone while the wind "blows up the mast" and not against the sails. Seasonal shifting of this zone away from the equator allows the trade winds from one hemisphere to continue across the equator and be deflected in the opposite direction in the other hemisphere.

In the horse latitudes relatively dry air descends, is compressed and warmed, and then diverges toward the north and south along the surface. As the air is warmed, its capacity to hold water vapor increases; its relative humidity decreases; it is "dried out"; and very little precipitation results. Thus many of the world's desert areas are located at latitudes of about 25 to 30 degrees. In the trade wind belt, winds blow more uniformly and fair weather tends to predominate except where air is lifted over mountain barriers. Trade winds are widespread and cover nearly half of the Earth's surface.

Weather in the belts of westerlies is famous for shifts from fair to stormy and back again as storm systems (lows or cyclones) form and move along the region of the polar front. During the cooler months when the polar front and its accompanying storm systems are best developed, these shift toward the equator, and the lower middle latitudes undergo more frequent periods of precipitation and changeable weather. Cyclonic storms apparently occur at times in polar regions.

Thus global circulation of the Earth's atmosphere tends to cause convergence of different air masses at certain latitudes which results in upward movement, expansion, cooling, and precipitation along the line of convergence (unless temperatures are so low that the air can hold relatively little water vapor). At other latitudes, subsidence and divergence occur, and the resulting compression and warming allow very little precipitation.

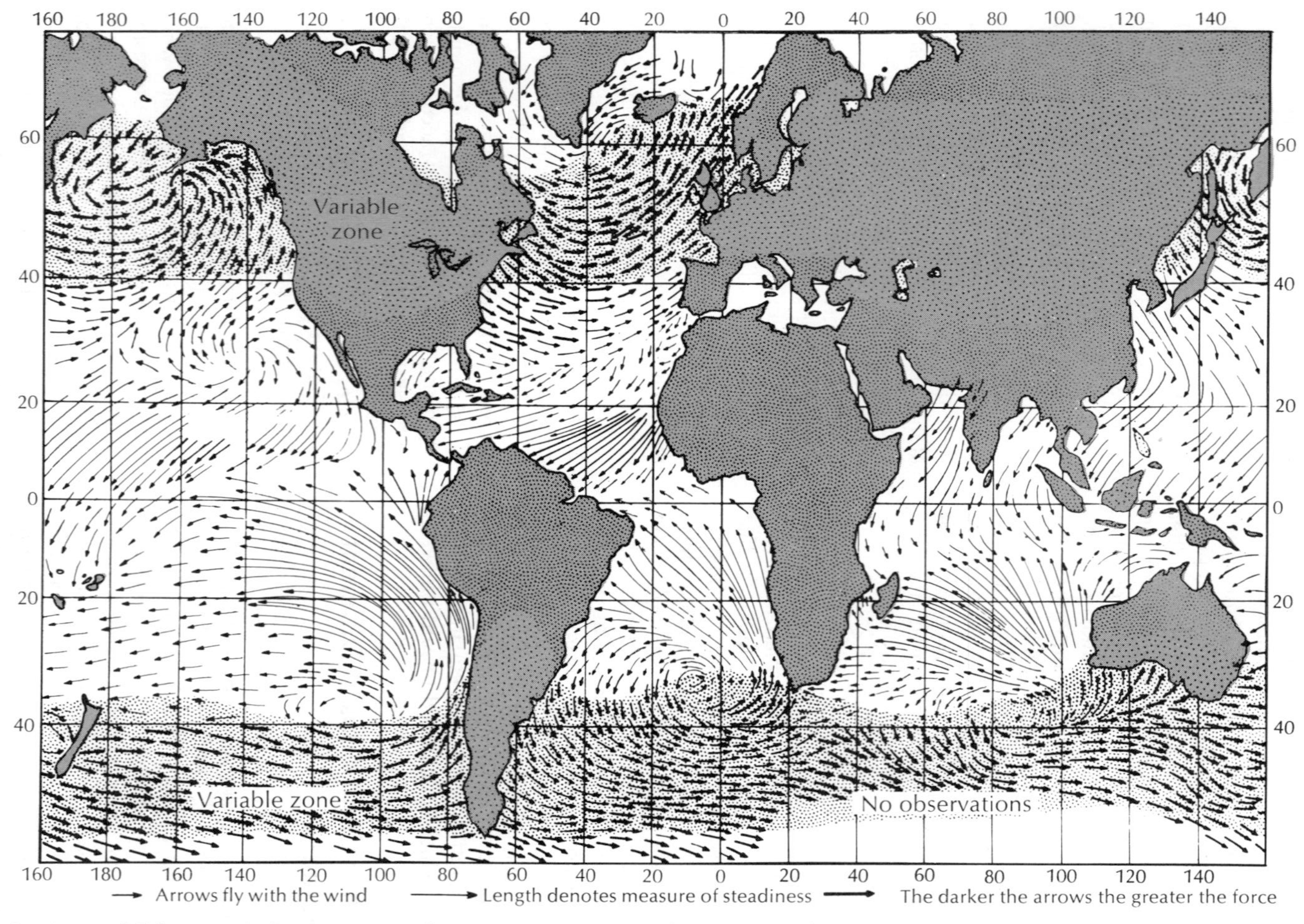

26-21. January and February wind systems over the oceans. Compare with Fig. 15-11 showing surface currents in the oceans. (TM 1-300, *Meteorology for Army Aviation.*)

Will a cool summer in one area be balanced by a warm winter in the same area? By a warm summer a year later? If drought strikes one area during a summer, must it be followed by a wet spell? Immediately or long after? Temperatures and precipitation vary greatly at any one place—daily, seasonally, and annually—but they do tend to fluctuate around a statistical mean. However, the balancing out of temperature and precipitation variations seems to take place simultaneously and on a worldwide basis. Solar radiation seems to remain quite uniform. Therefore, if local conditions function to make temperatures hotter than usual at one place, temperatures must be colder than usual elsewhere at this same time.

It is estimated that the water vapor molecules in the atmosphere are replaced via evaporation, condensation, and precipitation about once every 10 days or so and that the total amount of water vapor in the atmosphere probably remains about the same at all times; this quantity depends largely upon solar energy. Thus if more precipitation is occurring in one area at one time, less water vapor is available for precipitation elsewhere at this same time.

Heating of Land and Water

A number of weather phenomena can be explained by the fact that land heats up and cools off more quickly than water. Several factors are involved. The heating or cooling of soil is limited to a thin surface layer, whereas circulation causes the temperature of the water to be nearly uniform throughout, and evaporation cools the surface. More heat energy is also required to raise the temperature of a given weight of water a given number of degrees than to raise the temperature of an equal weight of soil an equal number of degrees (the specific heat of water is very high, about three to five times more than that of dry soil). As a housewife noted, it takes longer to bring water to a boil than to bring an equal volume of thick soup to a boil.

In coastal regions during the warmer months, a *sea breeze* commonly blows from the sea onto the land during the day (Fig. 26–22). It develops 3 to 4 hours after sunrise and grows strongest in the early afternoon. At night the wind may reverse its direction and blow from land to sea *(land breeze)*. During the day the air over coastal land areas commonly becomes warmer than the air above the adjoining water. Cooler air above the ocean thus moves inland to push upward the warmer lighter air over the land.

At night the more rapid cooling of the land chills the air above it, whereas the temperature of the water varies little. The cooler heavier land air then moves seaward to push up lighter warmer nocturnal air. The land breeze tends to peak at sunrise and to be weaker than the sea breeze.

Sea breezes are most pronounced (depth, speed, and inland extent) where temperatures are high—during the hottest time of the day and nearer the equator. In middle latitudes, sea breezes tend to be shallow (up to 1000 feet or so) and gentle (up to about 12 mi/hr) and affect a narrow belt along the coast (as much as 10 miles or so). In the tropics, comparable maxima might be 4000 feet, 25 mi/hr, and 100 miles.

Edinger has described land and sea breezes observed from a ship approaching Hawaii. Upon awakening at sunrise he was chilled by air that had cooled at higher altitudes on the island at night and was then moving down the slopes and out to sea. By noon a line of clouds had formed part way up the slopes. This was the leading edge of the sea breeze made visible by cloud droplets produced by expansion and cooling during the upslope movement.

The monsoon seasons of India and of other areas may be explained in a somewhat similar manner but are on an annual cycle rather than a daily one. During hot summer weather, air above India and the great land mass of central Asia is heated strongly and is pushed upward by moisture-laden air which moves northward from the Indian Ocean across the land. Altitudes increase

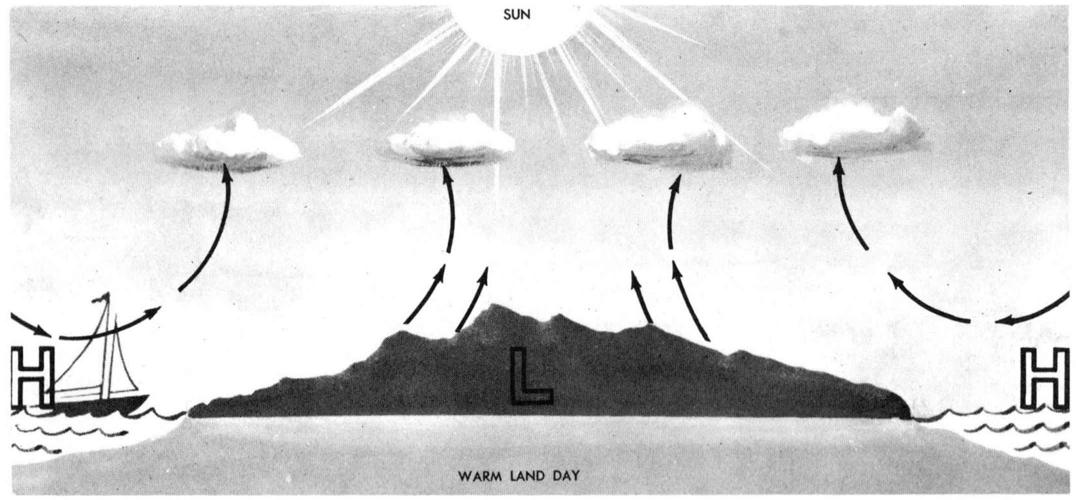

26-22. The sea breeze (onshore). It is best developed when the ocean water is cold and the land is warm. At times it may attain velocities of 20 to 25 miles/hr. Its arrival may cause the relative humidity to increase and temperatures to drop. A lake breeze is a similar phenomenon on a smaller scale. (AF Manual 105-5, *Weather for Aircrews.)*

gradually in the direction of the massive Himalaya Mountains. The moist northward moving air rises higher and higher and cools as it rises. Chilling of the air at higher altitudes results in enormous precipitation, and as much as 1000 inches of water have fallen on Cherrapunji, India in 1 year. In contrast, the average rainfall for the United States is approximately 30 inches per year and that for the Earth as a whole about 30 to 40 inches per year.

The winter part of the monsoon circulation (outward movement of cold heavy air from central Asia) is deflected from India by the mountainous areas between India and the continental interior. Thus Cherrapunji averages only 1 inch of rainfall per month during December and January. Asia is the only continent in which the effects of the monsoon are especially prominent.

Other Winds

Mountain and *valley breezes* also occur. Cold heavy air may move down a mountain slope at night, and a narrow valley between high mountain slopes will experience a pronounced mountain breeze. On the other hand, during the day air may blow up a mountain slope (Fig. 26–23). Mountain breezes, particularly during the colder months, tend to be stronger than valley breezes because they flow in the direction of gravitational attraction. Cold heavy air may also flow off the edge of a plateau and be funneled along valleys to lower altitudes (gravity winds). A relatively warm dry wind moving down a mountain slope is called a chinook or foehn (Fig. 27–7).

Turbulence is produced as air moves across an irregular surface, whether this consists of hills and depressions or buildings and trees. The turbulence increases with wind speed and with the sizes of obstructions. The height of the affected zone may exceed the height of the obstruction by five or six times.

Winds tend to be slower at low altitudes (the effect of surface friction) and to increase in velocity at higher altitudes. However, changes in wind speed with increased

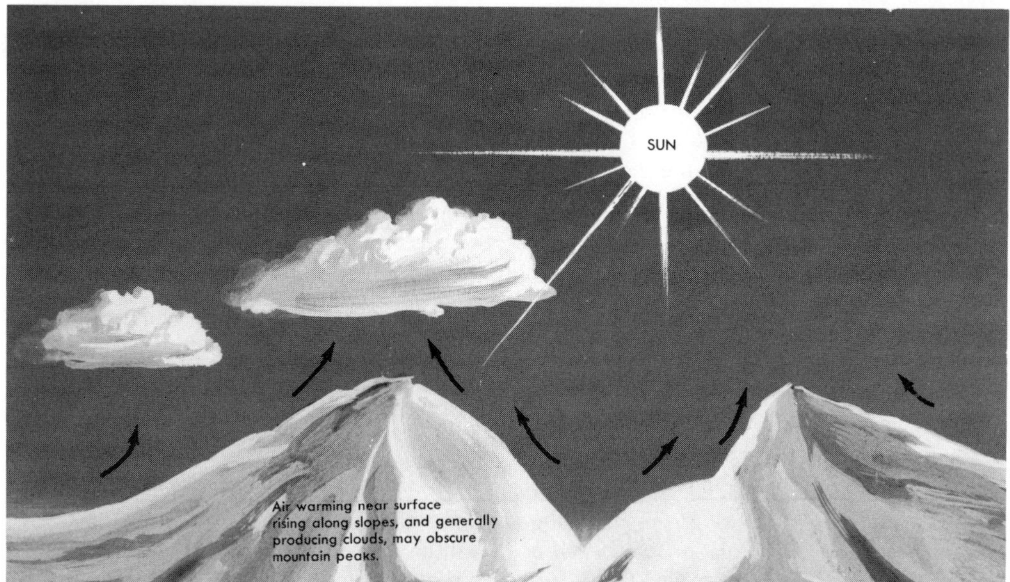

26-23. Valley breeze. The slopes of a mountain are heated on a sunny day to higher temperatures than occur at comparable altitudes above adjacent valleys and lowlands. The sun's rays in early morning may strike some mountain slopes at nearly 90 degrees and thus be more concentrated than on the more level ground where they strike at a small angle. Air in contact with the mountain slopes is thus warmed and rises up the slope; it appears to come from adjacent lowlands as a valley breeze. These temperature differences tend to be reversed at night because radiational cooling occurs along the mountain slopes. (AF Manual 105-5, *Weather for Aircrews.*)

altitude are greatest very near the surface; e.g., speeds at a height of 30 to 35 feet may be double those 1 to 2 feet above ground level. Speeds at 100 feet may be only 20% greater than those at 30 to 35 feet, and increases in speed above 100 feet are generally even slower. Another factor in the development of higher velocities at higher altitudes is the decreased density of the air—a smaller force is needed to attain a certain speed.

Wind direction also tends to change at higher altitudes. Nearer the surface where friction is appreciable, winds are slower and, over land, tend to blow across the isobars at angles of 30 to 40 degrees; they slant in toward a low-pressure center (Fig. 26–14). At altitudes of a few thousand feet, where friction with the surface is negligible, the winds tend to be stronger and to blow nearly parallel to the isobars—the Coriolis deflection is greater.

Less friction results as air moves across a water surface; thus wind velocities tend to be greater at sea than on land, and wind directions make angles of 10 to 15 degrees or less with the isobars. Onshore winds along a coast thus tend to have a greater velocity than after they have moved some miles inland over an irregular surface.

Under certain conditions very high wind velocities may occur at a mountain top because air is funneled between horizontally moving air above and the mountain mass below. In a similar manner, a river or a glacier flows more rapidly through a constricted portion of its channel.

Within an air mass there is likely to be a daily cycle of wind velocities; strongest in the afternoon when the surface is heated most and convectional circulation is most pronounced, and weakest at about sunrise. March is a windy month in many parts of the United States because temperature contrasts between air masses are most pronounced about this time. Masses of cold polar air still exist toward the north, but the sun has had a chance to warm the air toward the south.

Daily and Seasonal Lags in Temperature

Let us look at a daily cycle of temperature changes for a period during which the overlying air is not replaced by a warmer or cooler air mass. During the day the ground is heated by insolation, and it also radiates heat energy into space. But it receives more heat energy than it radiates away until about 3:00 P.M. Thus the warmest part of the day comes at about 3:00 P.M., much as a room is warmest shortly after the fire in the stove has gone out.

The ground radiates heat energy all night and continues to radiate away more heat energy than it receives until shortly after sunrise, the coolest part of the day. Likewise, in the Northern Hemisphere the warmest month of the year tends to be July and the coldest January, out of step with June and December, the months of maximum and minimum insolation. The time lag is even longer over the oceans: the warmest part of any given day tends to occur at about 5:00 P.M., and the warmest month tends to be August.

The oceans have a great moderating effect upon temperatures. They are warmer in the winter than the air above them, and cooler in the summer. Since there is a tendency toward equilibrium between air and water temperatures, and since the amount of heat energy present in ocean water is so very great, a small change in the temperature of the water results in a very large change in the temperature of the air. Thus the water acts to cool the air during the day and to heat it during the night, and fluctuations are quite limited. Daily temperature changes of more than a degree or two Fahrenheit are uncommon over the oceans. In some parts in the middle latitudes the mean temperature difference between winter and summer may be as little as 10 Fahrenheit degrees. On the other hand, temperature variations tend to be at a maximum in the middle latitudes at a location far from large bodies of water in the interior of a large continent.

The amount of heat energy received by the Earth at the equator exceeds that radiated away, whereas the heat energy received in the polar regions is less than that radiated away. The dividing line between these regions may occur at a latitude of about 35 degrees. Since the Earth as a whole appears to be neither cooling off nor warming up appreciably at present (temperature fluctuations over longer periods of time have oc-

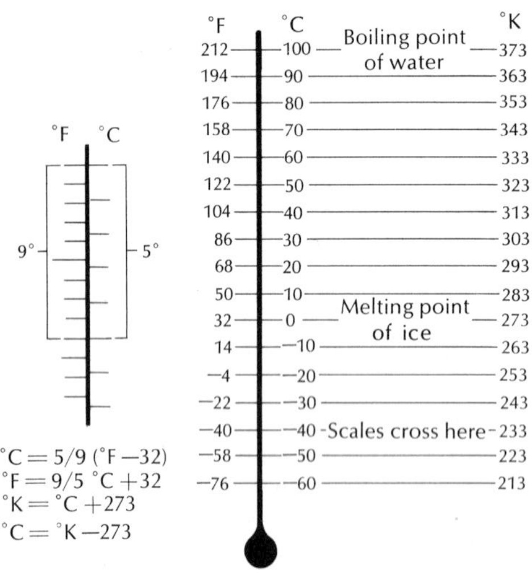

26-24. Temperature conversion diagram. One degree on the Kelvin scale has the same magnitude as one degree on the Celsius (Centigrade) scale. However, zero on the Kelvin scale is the lowest temperature possible (atoms and molecules are motionless; this is about −273°C and −460°F). Ice melts at 273°K and water boils at 373°K. Thus to convert a Celsius reading into the Kelvin scale, one adds 273 to the Celsius number. (AF Manual 105-5, *Weather for Aircrews.*)

curred in the past), there must be a transfer of heat energy from lower to higher latitudes. This is accomplished in several ways: (1) by circulating air, (2) by north-south ocean currents, e.g., the warm Gulf Stream and cold Labrador Current, and (3) by changes in the physical states of water; for example, ice forms in one region and melts in another, or evaporation takes place at a low latitude and condensation at a high latitude. Each such change absorbs or releases heat energy.

Various types of instruments are used to measure temperature: the familiar liquid-in-glass thermometer, the bimetallic temperature-sensing element of a thermostat, the maximum-minimum thermometer, and the automatically recording thermograph. Three scales are commonly used (Figs. 26–24 and 28–26): Celsius (Centigrade is the older name), Fahrenheit, and the Absolute or Kelvin scale.

Wind Observations

At the surface, wind directions (always stated as direction from which the wind blows) and speeds are measured by weather vanes and anemometers. Winds at higher altitudes may be measured by tracking balloons that are precisely inflated so that they rise at a known rate. The tracking may be done visually or by radar. The *radiosonde* (Fig. 26–25) is such a balloon, and it also carries equipment that measures the temperature, pressure, and humidity of the air that it moves through.

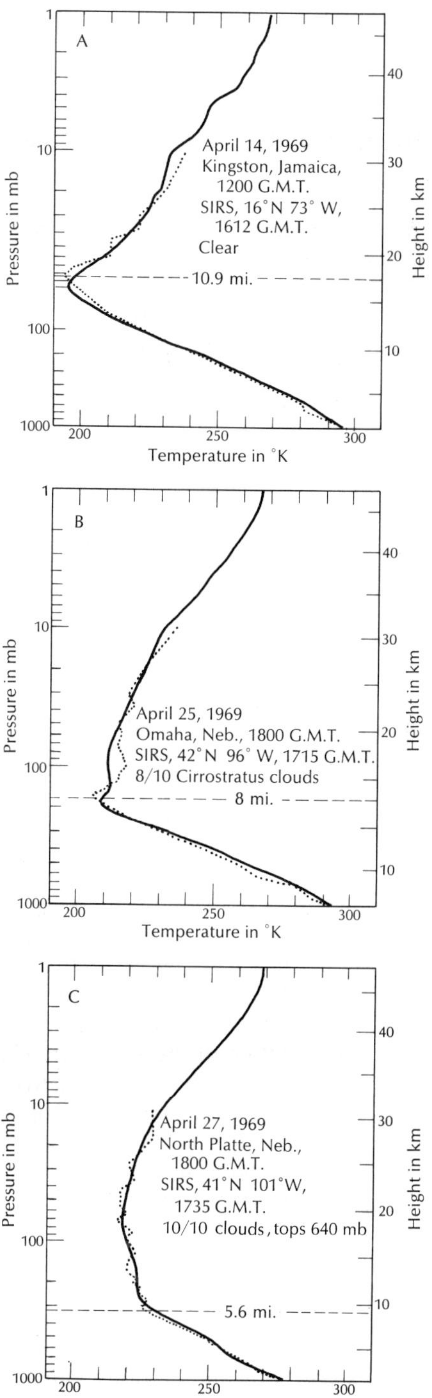

26-25. Vertical temperature profiles (solid lines) derived from data recorded by an infrared spectrometer in a Nimbus III satellite in April 1969. Dotted lines show radiosonde measurements made simultaneously as checks at the test locations. Note location of tropopause (higher at lower latitudes) and altitude at which the radiosonde balloons burst. *(Courtesy D. Q. Wark and D. T. Hilleary, ESSA.)*

Exercises and Questions for Chapter 26

1. Based upon chemical composition, what are the main units of the atmosphere?
2. Why does atmospheric pressure decrease upward? Why is the rate of decrease not uniform?
3. Assume that the following temperatures represent average conditions and plot them on a temperature-altitude graph. Draw dashed lines to represent the tropopause, stratopause, and mesopause. Label the troposphere, stratosphere, mesosphere and thermosphere.

Altitude	*Temp.*	*Altitude*	*Temp.*	*Altitude*	*Temp.*
0 km	288°K	12 km	217°K	48 km	283°K
1	282	14	217	52	283
2	275	18	217	60	254
3	269	20	217	70	210
4	262	25	217	80	166
5	256	30	231	90	166
8	238	40	261	100	199
10	223			105	224

4. The first few hundred miles of the Earth's atmosphere is subdivided into three warm zones separated by two cold zones.
   (A) Approximately where are these zones located?
   (B) Account for each of the five zones.
5. Why may the altimeter in a plane be functioning properly and yet show an inaccurate altitude?
6. Why does ozone form most rapidly at an altitude of 25 to 30 miles and yet be most abundant at an altitude of about 16 miles?
7. At any one level within the upper thermosphere, temperatures may be hundreds of degrees different from one time to another. Why is the temperature fluctuation so much greater than it is at the surface?
8. Is the bottom of the exosphere closer to the surface when temperatures within it are high or low (relatively)? Why?
   (A) Is the bottom of the exosphere at any one time the same level for hydrogen atoms as for helium atoms? Explain.
9. Describe some of the methods used to obtain data about the atmosphere at levels ranging from the surface to an altitude of a few hundred miles.
10. Why may the strength of a radio signal tend to change during a night?
11. The ionosphere has a peak intensity at an altitude of about 180 miles.
    (A) Why does its intensity tend to decrease upward from this level?
    (B) Why does its intensity tend to decrease downward from this level?
12. Explain why the Coriolis effect is zero at the equator and increases toward the poles.
    (A) Why is the deflection to the right in the northern hemisphere?
13. Why does a jet stream tend to occur above the polar front?

14. Why does the stratopause jet stream change direction from winter to summer?
15. A low-pressure system occurs at 40 degrees latitude. Describe the associated winds (relative to the isobars) near the surface if:
    (A) the low is in the northern hemisphere over the land.
    (B) the low is in the southern hemisphere over the ocean.
16. Describe in a generalized way the Earth's global rainfall pattern and explain why more precipitation occurs at some latitudes than at others.

27

Water and Air

Some Physical Properties of Water

Water is far the most abundant substance in and on the Earth in the outermost three-mile zone below sea level. There water is estimated to be some three times as abundant as all other substances together. The physical properties of this water, together with that present in the atmosphere (Fig. 27–1), have great significance in meteorology.

• Water reaches its greatest density at 4°C (nearly 40°F) and expands upon freezing. Therefore, lakes freeze from the top rather than from the bottom as they would if ice did not float.

• The amount of heat energy necessary to raise the temperature of 1 gram of water 1 Celsius degree is called its *specific heat* and is very high relative to that of other substances—e.g., it is several times greater than that of dry soil or aluminum and about ten times greater than the specific heat of iron, zinc, or copper. Therefore, large bodies of water moderate temperature changes in the air over nearby land areas, especially if prevailing winds are from the water onto the land. The thermal capacity (mass times specific heat) of the oceans is enormous; they absorb huge quantities of heat energy in the summer and release it during the winters. For this reason, mean temperatures on an oceanic island such as Bermuda undergo relatively little change during a year.

• Heat energy must be added to produce phase changes going from solid to liquid to gas; on the other hand, heat energy is liberated in phase changes going from gas to liquid to solid. Such phase changes for water involve relatively very large amounts of heat energy (Fig. 27–2). To change 1 gram of ice at 0°C into 1 gram of water (also at 0°C) requires the addition of 80 calories of heat energy. This is called the *heat of fusion* (melting). One *calorie* is the amount of heat energy needed to raise the temperature of 1 gram of water 1 Celsius degree at 15°C. This same large quantity of energy, known as the *heat of crystallization* (freezing), must be released whenever water

27-1. Early morning fog at Norwich, England. Tree tops and other tall objects project above this shallow fog that was produced by radiational cooling of the ground during the night. Air in contact with the ground was chilled, became denser, and drained into depressions. Air along the ground is thus stable (shallow surface inversion) and smoke emitted at the surface is prevented from rising. However, warm gases emitted from the tall power station chimney are above the stable layer and rise further before spreading out horizontally. (Photograph by Charles E. Brown, from F. H. Ludlam and R. S. Scorer, *Cloud Study, a Pictorial Guide,* John Murray, London, 1966.)

changes into ice. Thus the formation of ice on water bodies in the winter and the melting of ice and icebergs during the warmer months limit the temperature changes of large water bodies during a year.

The *heat of vaporization* is nearly seven times as great, because about 540 calories of heat energy must be added to 1 gram of water at 100°C to change it into 1 gram of water vapor also at 100°C. On the other

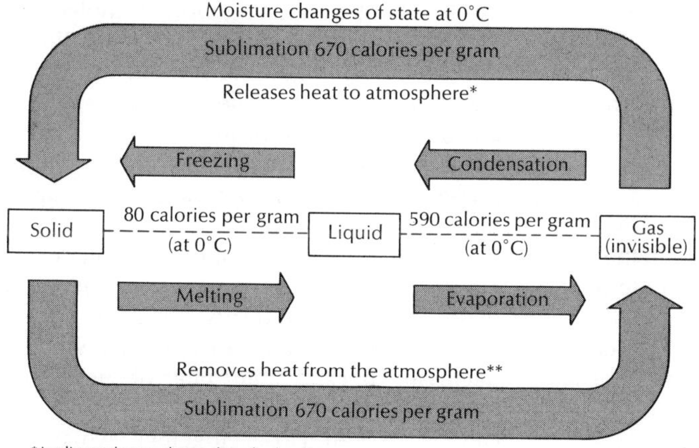

27-2. Heat energy exchanges involved in the phase changes of water. Note that these exchanges are all based on a temperature of 0°C. (TM 1-300, *Meteorology for Army Aviation.)*

hand, when the change is from the gas to the liquid, this same quantity of energy is liberated (the *heat of condensation*). In comparison, only 100 calories of heat energy are absorbed in raising the temperature of 1 gram of water from 0°C to 100°C. Whenever condensation produces dew, fog, or clouds, this heat of condensation is liberated. The use of water in heating and cooling systems depends upon these properties much more than upon its abundance and availability.

Although the energy exchanges shown in Fig. 27–2 are all based on a temperature of 0°C, phase changes can occur over a wide range of temperatures; e.g., if evaporation takes place at temperatures of 0°C, 40°C, and 100°C, then the energy that must be absorbed per gram equals about 590, 575, and 540 calories respectively. Thus less total energy is absorbed when 1 gram of water is evaporated directly at 0°C than when it is evaporated by first raising its temperature to 100°C and then vaporizing it at this temperature.

Still greater quantities of energy are involved when water molecules go directly from solid to gas without passing through the liquid phase (about 670 calories of heat energy are absorbed per gram if the ice is at 0°C). The process may also be reversed, although the 670 calories are then released, and each is called *sublimation*. The energy exchanges associated with phase changes involve the bonds between water molecules (see below).

• The solvent power of water is large both in the numbers of substances it can dissolve and also in their quantities.

• The compressibility of water is less than that of steel.

• The surface tension of water is very great, and this is of importance in the formation of raindrops and in various life processes involving capillary action.

• The transparency of water to electromagnetic radiation is likewise great. Thus sunlight diffuses downward through the top 600 feet or so of ocean water.

• The extreme mobility of water in all three phases should be emphasized because this makes possible the transportation of very large quantities of heat energy from one area to another—in general this is from lower to higher latitudes: water may evaporate at one latitude and condense at another; ice may form at one place and melt at another; surface and deep-water currents may shift individual water molecules from one part of an ocean to another, or to a different ocean.

The physical properties of water are largely determined by the manner in which two hydrogen atoms join with one oxygen atom to form a water molecule. The oxygen atom needs two electrons to fill out its second electron shell (p. 772), and each hydrogen atom shares an electron with it. We can picture two small hydrogen atoms, located 105 degrees apart on one side of an oxygen atom, giving it two positive electric charges. The opposite side of the oxygen atom appears to have two negative charges, which makes a water molecule polar—i.e., it has a positive electric pole and a negative electric pole.

Water molecules are joined one to another by strong so-called hydrogen bonds—i.e., an electrical attraction occurs between the positive pole of one molecule and the negative pole of another molecule. Each molecule acts as if it had four charges located at four "corners." When all four charges of each molecule are attracting opposite charges on neighboring molecules, they are approximately fixed in a three-dimensional pattern, and we call this phase ice. Individual molecules can vibrate, but are not free to twist or move. This is a more open structure than occurs in the liquid phase when bonds form and break apart many times each second. On the average, therefore, molecules in the liquid phase are closer together than they are in the solid phase, and liquid water is denser than ice.

Water density increases as temperature decreases because the rate of molecular movement slows down, and the molecules can be packed closer together, which increases the density. However, below a peak density at 4°C, some of the molecules unite into icelike clusters which decreases the average density, and these clusters become larger and more numerous closer to 0°C.

As the temperature of a mass of ice is raised closer to 0°C, its molecules vibrate faster. By the time they reach 0°C, the molecules vibrate rapidly enough to break their bonds. The molecules are no longer being held in fixed positions, and we say that the ice is melting. If ice and water occur together, heat energy added to them goes into breaking the hydrogen bonds rather than into causing the molecules to move more rapidly, thus raising the temperature. Therefore, the temperature does not change until all of the bonds are broken and all of the ice melted. For water to evaporate into the gaseous phase, all of the intermolecular bonds must be broken, which requires even more energy than is needed to cause melting.

Water: Its Condensation and Evaporation

Water vapor enters the air because evaporation and sublimation (passage directly from the solid to the gaseous state) occur. The sun shines on dew-covered ground, on a wet bathing suit, or on a freshly watered lawn and soon these objects are dry because the liquid water has been changed into water vapor. The moisture in freshly laundered clothes may freeze on a line on a cold day, and the ice subsequently may sublimate. On a hot dry summer day, ½ to ¾ of an inch of water may be evaporated from the surface of a lake; and in one month the level of the lake may be lowered more than 20 inches. The transpiration of plants is also a very large source of water vapor. Almost as much water vapor is added to the air over a field of lush vegetation on a hot sunny day as is evaporated from the surface of a nearby lake.

If water is placed in the bottom of a container and the top is sealed, certain changes occur in the water and air in the container. Water molecules leave the surface of the water and evaporate into the air as a gas; this absorbs energy. However, only the fastest moving molecules escape, which means that the average rate of motion of the molecules remaining in the liquid is thus decreased. Therefore, evaporation is a cooling process.

Occasionally, a water vapor molecule returns to the liquid, but for a time more water vapor molecules are added to the air than are subtracted from it. This raises the vapor pressure of the water vapor, which depends upon the number of water molecules in the air and upon their average rate of motion (their temperature); it is independent of the pressure exerted by other gases in the air. The vapor pressure of the water increases as evaporation continues, and this increase causes more and more of the gas molecules to return to the liquid state. Eventually an equilibrium is reached, and as many molecules return to the water surface as leave it; the air is then said to be saturated at this temperature and pressure.

If the temperature is now raised, the average rate of molecular motion is increased, and less tendency exists for the faster moving molecules to return to the liquid phase. Thus, more water vapor must be added to the air to make it saturated. Here we have one of the most important relationships in meteorology—that between the temperature of the air and its capacity to hold vapor: *warm air can hold more moisture than an equal volume of cool air.* The proportion is not direct because the capacity of a certain volume of air to hold water vapor increases very rapidly at higher temperatures (Fig. 27–3). On a warm summer afternoon, the air may contain about five times as much water vapor as does air at freezing temperatures in the winter.

Thus temperatures in the lower atmosphere can be a limiting factor in the amount of precipitation that occurs at a certain time and place—especially at high latitudes and within continental interiors during winter. Precipitation is very low at

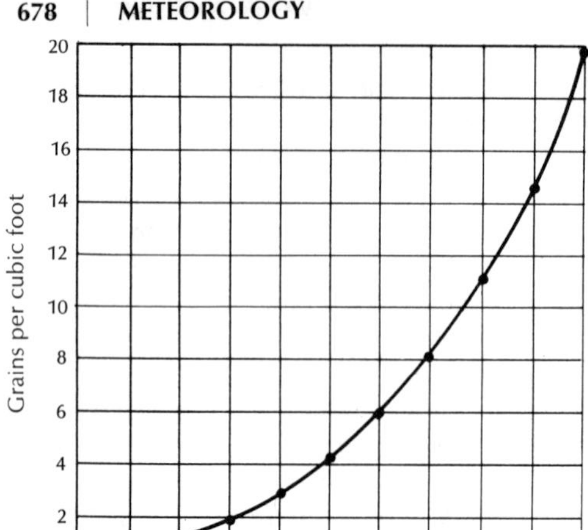

27-3. Different amounts of water vapor are needed to saturate 1 ft^3 of air at different temperatures (there are 7000 grains in 1 lb, Avoirdupois weight). Conclusion: a certain volume of warm air can contain much more water vapor without saturation than can an equal volume of cool air.

high latitudes where the air is generally quite cold, and precipitation is high near the equator where temperatures are high (the maximum zone, like that of the so-called heat equator, tends to occur in the lower latitudes of the Northern Hemisphere). However, other factors are also important, and the lifting and chilling of air masses that lead to precipitation may result largely from convectional circulation, from movement across mountain barriers (orographic), and in association with low-pressure systems (cyclonic circulation) in the belts of westerlies. Thus precipitation processes are too diverse to correlate entirely with temperatures and latitude—especially over land areas—but secondary precipitation peaks do occur in the middle latitudes where cyclonic circulation is best developed and zones of low precipitation do occur at subtropical latitudes (p. 665).

In *condensation,* many invisible and widely separated water vapor molecules join together to form a tiny visible droplet. However, these molecules need some kind of surface to condense upon: onto the windshield of a car, a blade of grass, or a tiny invisible particle floating in the atmosphere (a condensation nucleus). Condensation commonly does not begin until the air has become saturated with water vapor and may not begin even then, for supersaturation can occur.

A parcel of air may become saturated because it is cooled or because water vapor is added to it. When a given parcel of air is cooled at a constant pressure, without the addition or subtraction of water vapor, the temperature will reach a point at which the air is saturated and below which condensation will occur. This critical temperature is called the *dew point.* At the same temperature, the dew point is higher in moist air than in dry, and a smaller drop in temperature causes saturation.

In the atmosphere, the random motion of water vapor molecules results in frequent collisions, but the molecules do not aggregate into droplets unless a *condensation nucleus* is present to form a group and hold it together (unnecessary at extremely low temperatures). The most effective nuclei have an affinity for water and can cause condensation even in unsaturated air. These are called hygroscopic substances; they are readily wet by water and dissolved in water. Nuclei are always present in the air, and smaller nuclei are much more abundant than larger ones. However, the very smallest and most numerous of the nuclei are not effective as condensation nuclei, and many of the larger ones are not effective until conditions of supersaturation have been attained.

The nuclei are of many kinds, and the minute particles enter the air as by-products of combustion, as sea salt from the oceans (witness what may happen to a salt shaker on a humid day), because volcanoes have erupted, as dust picked up by the wind, as fragments of meteors (probably minor), and in other ways. The nuclei are several times more abundant over continents than over oceans and are most abundant over large industrialized areas or immediately downwind from them.

Cooling decreases molecular motion, and molecules cohere more readily or adhere to some nucleus upon collision. If many water vapor molecules are present in a certain amount of air, chances are opportune for numerous collisions and for the development of tiny liquid droplets upon the ever-present condensation nuclei. On the other hand, a rise in temperature increases the rate of molecular movement and tends to disrupt the liquid droplets. Factors favorable for condensation, therefore, are: (1) the presence of abundant water molecules, (2) the presence of suitable condensation nuclei, and (3) a decrease in temperature.

The bathroom mirror fogs when one takes a hot shower because more water vapor molecules have been added to the air. You "see your breath" on a cold day because you puff out warm moist air that immediately cools below its dew point and condensation results. For a similar reason, tiny water droplets accumulate on the outside of a cold glass of water on a hot humid summer day, and pipes "sweat." A jet plane leaves a vapor trail at high altitudes where the air is very cold, chiefly because water vapor is added to the air as fuel is consumed.

Dew, frost, fog, and clouds result from condensation, commonly because air has been chilled until saturation occurred. If air is cooled below the dew point by contact with a cool surface, water vapor will condense as dew directly upon that surface—rock, grass, tractor, or building; the dew does not fall. If the temperature is below 0°C (32°F), delicate feathery white frost crystals tend to develop instead. If the temperature falls below freezing after dew has formed, the water droplets freeze into tiny pellets of ice.

Precipitation in the form of drizzle, hail, rain, sleet, and snow involves the active falling of moisture condensed from the air (cloud droplets are too small to fall rapidly) and is discussed below.

Humidity

Absolute humidity refers to the mass of water vapor actually present in a unit volume of air and may be measured in grams per cubic meter or other units (Fig. 27–3). On the other hand, *relative humidity* is the ratio between the amount of water vapor actually present in a certain volume and the amount that would have to be present in this space to cause saturation under the existing conditions of temperature and pressure. For example, air at a relative humidity of 75% contains three-fourths of the total amount of water vapor required for saturation. A relative humidity of 100% indicates that the air is saturated, and the temperature of the air is at the dew point. Either an increase in temperature or a decrease in atmospheric pressure increases the amount of moisture that can be present within a certain volume (the relative humidity is decreased; Fig. 27–4). On the other hand, a drop in temperature or an increase in at-

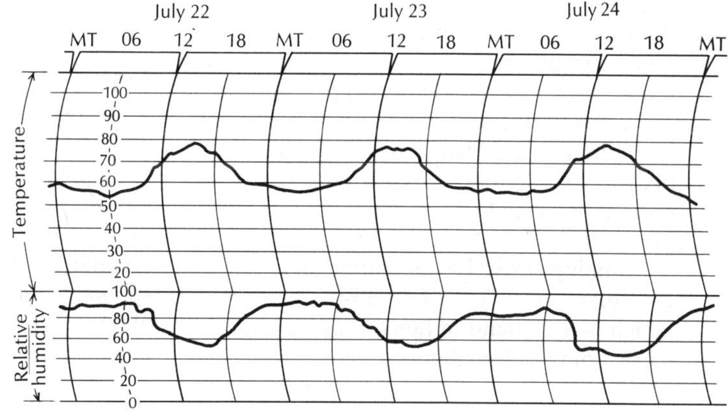

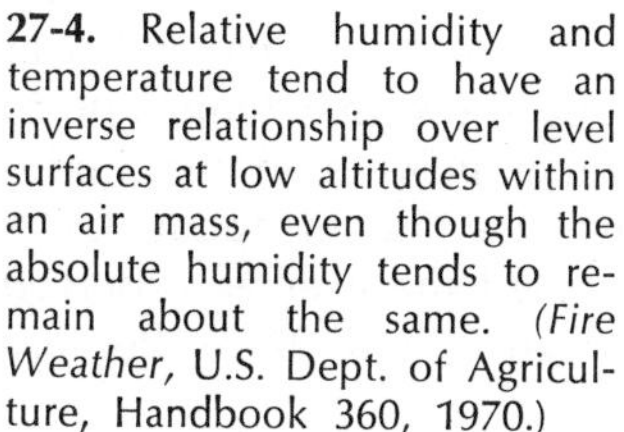
27-4. Relative humidity and temperature tend to have an inverse relationship over level surfaces at low altitudes within an air mass, even though the absolute humidity tends to remain about the same. (*Fire Weather,* U.S. Dept. of Agriculture, Handbook 360, 1970.)

mospheric pressure raises the relative humidity.

Specific humidity is the ratio between the mass of the water vapor in a parcel of air to the total mass of the air in the parcel (including the water vapor). Specific humidity does not change as air rises or falls with consequent volume changes (assuming that no water vapor is added to or subtracted from the air). Absolute humidity, on the other hand, varies with changes in altitude, even though no change occurs in moisture content. Thus specific humidity is a more useful measurement for meteorologists than absolute humidity.

Relative humidity may be less than 100% during a rainstorm because the rain originated within a saturated cloud situated some thousands of feet above the surface and then fell through unsaturated air near the ground. High relative humidity is uncomfortable on warm days because the rate of evaporation, which is a cooling process, becomes slow when the air is nearly saturated with water vapor.

A breeze aids evaporation by preventing saturated air from stagnating, and a windy cold day feels particularly chilly because the rate of evaporation of moisture from one's skin is increased. A swimmer feels chilly as he steps out of the water and rapid evaporation occurs; yet he feels quite comfortable a moment later after a brisk rub with a towel has removed excess water. In the heating of buildings, lower temperatures and higher humidity are less expensive and healthier than higher temperatures coupled with low humidity. The air in some homes during the winter is so dry that rapid evaporation occurs and one feels cool, despite rather high temperatures. Cold winter air can contain little moisture even if its relative humidity is high. Therefore, when this air is heated to summer temperatures inside a house, its relative humidity becomes very low.

Humidity may be measured by a hygrometer (Greek: "moisture measure") which uses a strand of human hair (or horsehair) from which the oils have been removed. The instrument is simple and inexpensive but is not particularly precise and must be calibrated occasionally. The hair becomes shorter in dry air and longer in moist air.

Humidity may also be measured by a pair of thermometers, one of which has a wet cloth around its bulb. If the humidity of the air is 100%, evaporation does not occur, and the temperature readings of the two thermometers will be the same. However, if humidity is not 100%, evaporation and cooling take place at the wet bulb, and dry air causes a considerable drop in the temperature of the wet-bulb thermometer. Thus the difference in the readings of the wet- and dry-bulb thermometers can be calibrated to indicate the humidity of the air (Table 27–1; the greater the spread, the drier the air). To prevent air from stagnating around the thermometers, they may be whirled in the air or fanned.

A temperature-humidity index has been developed which relates human discomfort to various combinations of temperature and relative humidity. Along similar lines, a windchill scale has been worked out which relates cooling, or the rate of heat removal from the human body, to various combinations of wind speed and low temperatures. An increase in wind speed of 10 mi/hr seems to approximate the effect produced by a 20°F drop in temperature, although magnitudes vary under different conditions; e.g., 0°F with a 20 mi/hr wind may produce about the same discomfort as does −45°F with very little wind.

Lapse Rates and Adiabatic Heating and Cooling

Temperature measurements made at different altitudes, at different seasons, and over both land and sea show that temperatures commonly decrease upward in the troposphere at an average rate of about 3.5°F/1000 ft. This is called the average *observed lapse rate* (Figs. 26–6 and 27–5) because

TABLE 27-1

DATA FOR DETERMINING RELATIVE HUMIDITY (in percent)

Difference in degrees between wet-bulb and dry-bulb thermometers.

Air Temperature (reading of dry-bulb thermometer) in degrees Fahrenheit	1	2	3	4	5	6	7	8	9	10	11	12	13	14	15	16	17	18	19	20	21	22	23	24	25	26	27	28	29	30
30°	89	78	68	57	47	37	27	17	8																					
32°	90	79	69	60	50	41	31	22	13	4																				
34°	90	81	72	62	53	44	35	27	18	9	1																			
36°	91	82	73	65	56	48	39	31	23	14	6																			
38°	91	83	75	67	59	51	43	35	27	19	12	4																		
40°	92	84	76	68	61	53	46	38	31	23	16	9	2																	
42°	92	85	77	70	62	55	48	41	34	28	21	14	7																	
44°	93	85	78	71	64	57	51	44	37	31	24	18	12	5																
46°	93	86	79	72	65	59	53	46	40	34	28	22	16	10	4															
48°	93	87	80	73	67	60	54	48	42	36	31	25	19	14	8	3														
50°	93	87	81	74	68	62	56	50	44	39	33	28	22	17	12	7	2													
52°	94	88	81	75	69	63	58	52	46	41	36	30	25	20	15	10	6													
54°	94	88	82	76	70	65	59	54	48	43	38	33	28	23	18	14	9	5												
56°	94	88	82	77	71	66	61	55	50	45	40	35	31	26	21	17	12	8	4											
58°	94	89	83	77	72	67	62	57	52	47	42	38	33	28	24	20	15	11	7	3										
60°	94	89	84	78	73	68	63	58	53	49	44	40	35	31	27	22	18	14	10	6	2									
62°	94	89	84	79	74	69	64	60	55	50	46	41	37	33	29	25	21	17	13	9	6	2								
64°	95	90	85	79	75	70	66	61	56	52	48	43	39	35	31	27	23	20	16	12	9	5	2							
66°	95	90	85	80	76	71	66	62	58	53	49	45	41	37	33	29	26	22	18	15	11	8	5	1						
68°	95	90	85	81	76	72	67	63	59	55	51	47	43	39	35	31	28	24	21	17	14	11	8	4	1					
70°	95	90	86	81	77	72	68	64	60	56	52	48	44	40	37	33	30	26	23	20	17	13	10	7	4	1				
72°	95	91	86	82	78	73	69	65	61	57	53	49	46	42	39	35	32	28	25	22	19	16	13	10	7	4	1			
74°	95	91	86	82	78	74	70	66	62	58	54	51	47	44	40	37	34	30	27	24	21	18	15	12	9	7	4	1		
76°	96	91	87	83	78	74	70	67	63	59	55	52	48	45	42	38	35	32	29	26	23	20	17	14	12	9	6	4	1	
78°	96	91	87	83	79	75	71	67	64	60	57	53	50	46	43	40	37	34	31	28	25	22	19	16	14	11	9	6	4	1
80°	96	91	87	83	79	76	72	68	64	61	57	54	51	47	44	41	38	35	32	29	27	24	21	18	16	13	11	8	6	4
82°	96	91	87	83	79	76	72	69	65	62	58	55	52	49	46	43	40	37	34	31	28	25	23	20	18	15	13	10	8	6
84°	96	92	88	84	80	77	73	70	66	63	59	56	53	50	47	44	41	38	35	32	30	27	25	22	20	17	15	12	10	8
86°	96	92	88	84	80	77	73	70	66	63	60	57	54	51	48	45	42	39	37	34	31	29	26	24	21	19	17	14	12	10
88°	96	92	88	85	81	78	74	71	67	64	61	58	55	52	49	46	43	41	38	35	33	30	28	25	23	21	18	16	14	12
90°	96	92	88	85	81	78	74	71	68	64	61	58	56	53	50	47	44	42	39	37	34	32	29	27	24	22	20	18	16	14

From S. N. Namowitz and D. B. Stone, *Earth Science*, American Book Company, 1969.

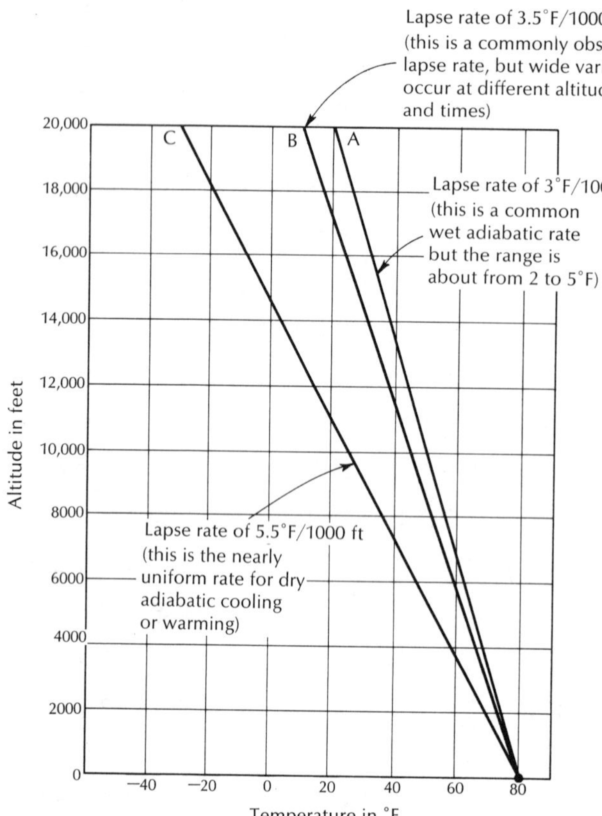

27-5. Three typical lapse rates plotted on a temperature-altitude graph. Relative to the horizontal axis, the so-called steepest lapse rates (i.e., those that show the largest temperature changes per 1000 feet) actually have the least steep slopes, a paradox which may lead to some confusion. On the graph, (C) shows the steepest lapse rate, and (A) has the least steep lapse rate.

temperatures tend to lapse or decrease upward, and it relates to air that is not actively rising or falling. In other words, the observed lapse rate refers to the actual temperature differences that occur at successive altitudes above a certain location.

In contrast, *adiabatic* cooling or heating occurs in a parcel of air (a conveniently small portion of the atmosphere) that moves upward or downward like a balloon that rises or sinks through still air. In the adiabatic process, it is assumed that a parcel's size and rate of vertical movement are such that no exchange of heat energy occurs between it and the enclosing air that it moves through—either upward or downward. The change occurs within the parcel and results entirely from the change in volume that accompanies the change in altitude. As air rises, it must push upward or aside the air which it replaces. This requires energy and results in a drop in the temperature of the rising air. Rising air cools at the so-called *dry adiabatic lapse rate* (Fig. 27–5), nearly uniform at 5.5°F/1000 ft of upward movement (about 1°C/100m). As air descends, it is compressed and warmed adiabatically at the same rate.

As a parcel of air rises and cools, its capacity to hold water vapor decreases and its relative humidity increases. Eventually, at saturation (i.e., at the dew point and 100 percent relative humidity) condensation commonly occurs, and the latent heat energy of condensation is released. This is a very large quantity (590 calories per gram of water condensed at 0°C). The heat energy thus released partly offsets the cooling effect due to expansion and the rising parcel cools at a slower, less steep rate—at the

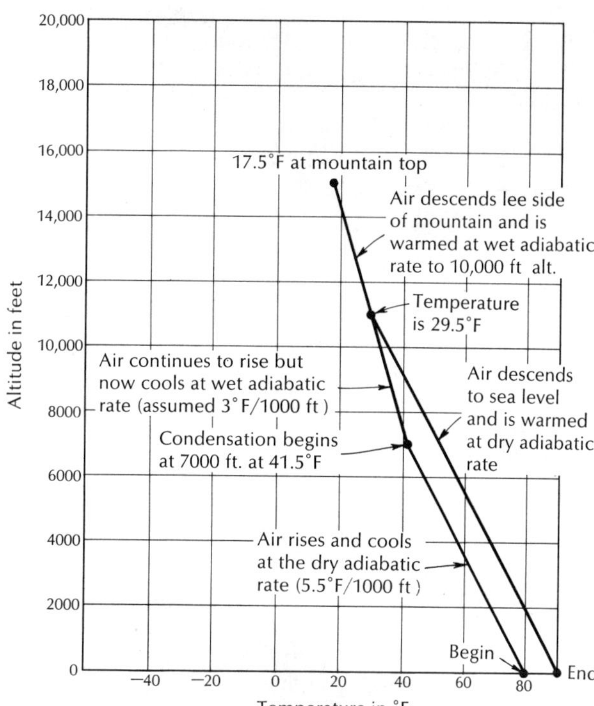

27-6. Temperature changes occur as air moves across a mountain range, and these are plotted on a temperature-altitude graph (example is fictitious). Air temperature is 80°F at sea level at the start. The air cools at the dry adiabatic rate of 5.5°F/1000 ft to an altitude of 7000 ft where condensation begins. The air then cools at an average assumed wet adiabatic rate of 3°F/1000 ft. Precipitation occurs. At the mountain top, the temperature is 17.5°F. As the air descends on the lee side, it warms at an assumed wet adiabatic rate of 3°F/1000 ft because evaporation occurs. However, evaporation stops at 11,000 ft, and the air warms at the dry adiabatic rate of 5.5°F/1000 ft from 11,000 ft downward. Therefore, it has a temperature of 90°F at sea level. This is the way in which some warm dry winds develop (*chinooks,* Fig. 27-7.)

wet adiabatic lapse rate, which varies with the temperature and moisture content. The wet adiabatic lapse rate may average about 3°F/1000 ft in warmer air, but it ranges from about 2 to 5°F/1000 ft; it is steepest in very cold and therefore dry air, in which it approximates the dry adiabatic lapse rate.

When a parcel of air subsides, it is warmed at the dry adiabatic lapse rate if no evaporation occurs. When evaporation occurs as air descends, large quantities of heat energy are absorbed, which partly offsets the warming caused by compression. Thus the temperature of the descending air increases less rapidly. Because of these adiabatic processes, air on the lee side of a mountain may have a higher temperature than that on the windward side at the same altitude (Figs. 27–6 and 27–7).

Rising masses or parcels of air are called *thermals,* and they attain their peak velocities in thunderstorms, where speeds may at times approach 150 to 200 mi/hr. They form as a part of a convectional circulation because of localized heating at the bottom or localized cooling at the top (less common). A parcel of air may move across a warmer part of the Earth's surface or across warm water and thus be heated more than the surrounding air. Air above a forest fire does this. The heated air expands, becomes lighter, and is pushed upward by surrounding heavier air; cumulus clouds may form in this manner (Figs. 27–8 and 27–9). In fact, some fires result in sufficient uplift to cause precipitation that aids in extinguishing them.

A *dew point lapse rate* also occurs in a rising parcel of air; i.e., the dew point tends to decrease by about ½ to 1°F/1000 ft of upward movement. A parcel of air at the surface, under a given set of conditions, has a certain dew point—i.e., it must be cooled to a certain critical temperature for saturation and condensation to occur. If this same parcel is now lifted (say 1000

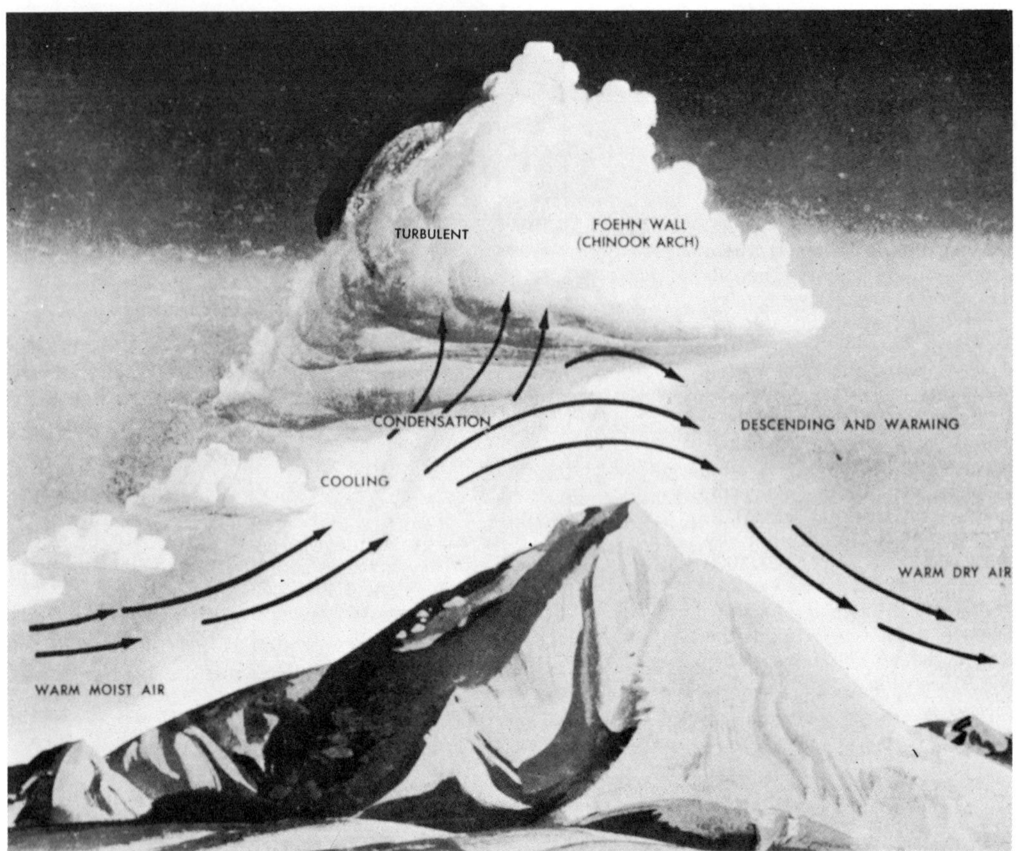

27-7. Warm dry downslope (foehn or chinook) winds may be produced on the lee side of a mountain. (AF Manual 105-5, *Weather for Aircrews.*)

feet), it expands and fewer water vapor molecules will occur per unit volume. With less water vapor present per unit volume, saturation will take place only at a colder temperature.

Stability and Instability

The buoyancy of a rising parcel of air depends upon the difference in density between it and the air that it rises through (Fig. 27–10 and *Archimedes' principle:* the buoyant force on a floating or submerged object is equal to the mass of the fluid that is displaced). The buoyancy tends to be greatest when a parcel of rising warm moist air cooling at the slow wet adiabatic lapse rate, moves upward through air that has a steep observed lapse rate. It should be emphasized that in this and succeeding examples, a small parcel of air, like a balloon, is assumed to rise or sink through a portion of the atmosphere that is motionless (at least vertically). A second assumption is that the air within the rising or sinking parcel does not exchange mass or heat energy across its boundaries.

The differences in density are caused chiefly by differences in temperature, but a difference in moisture content is also a factor. At comparable temperatures and pressures, moist air is lighter than dry air because each water vapor molecule in the air takes the place of a more massive oxygen or nitrogen molecule (*Avogadro's law:* equal volumes of gases have equal numbers of molecules if they have the same temperatures and pressures).

The stability or instability of the atmosphere is an important factor in determining the type and magnitude of the weather changes that occur in air that rises or falls. The processes involved can be illustrated by imagining what would happen under different conditions to a small parcel of air that is made to move upward or downward through the atmosphere. Such vertical movements do not start spontaneously; they must be initiated by a push of some kind (e.g., by temperature differences, as part of a convectional circulation, because of turbulence caused by winds moving across a rough surface, or because of convergence or divergence). Under different conditions, the air is said to show absolute stability, neutral stability, absolute instability, conditional instability, and convective instability.

(1) If the air is in a state of *absolute stability,* a parcel, upon being displaced, tends to return its original level (Fig. 27–11A). Thus stable air moved downward becomes more buoyant and therefore returns to its former level.

To illustrate, assume: (a) a surface temperature of 80°F, (b) that the air within the moving parcel is dry enough to cool or warm at the dry adiabatic lapse rate of 5.5°F/1000 ft, and (c) that the lower troposphere at this location has an observed lapse rate of 3.5°F/1000 ft. A balloon, representing this parcel of air, is filled with the 80°F air at ground level and pushed upward for 2000 feet (if temperatures within a balloon are the same as air temperatures outside, the balloon will not rise—unless it is filled with a lighter gas). Temperature within the balloon at 2000 feet is 69°F (80° minus 11°); temperature in the enclosing air at 2000 feet is 73°F (80° minus 7°). Therefore, the air in the balloon is colder and denser. The balloon sinks back to the surface when the upward push stops.

27-8. Cumulus cap cloud on top of smoke column rising from forest fire, Bitterroot National Forest, Montana. 5 August 1961. *(Courtesy U.S. Forest Service.)*

To illustrate movement in the downward direction, assume that a balloon is filled with air at the 2000 foot level (73°F) and pushed downward to the surface. Here the temperature in the balloon is 84°F (73° plus 11°). Since it is now 4°F warmer than the surface air around it, the balloon rises back to the 2000 foot level when released.

(2) If upon being displaced, the parcel remains at the level to which it was moved by a push of some sort, then it is in a state of

27-9. Underwater nuclear explosion at Bikini produces a huge mushroom cloud. Note the flat bases of the scattered cumulus clouds—one in the foreground hides part of the top of the mushroom. They show clearly the level at which condensation occurs as individual columns of air rise, expand, and cool to the dew point. *(Official Navy photograph.)*

neutral stability (Fig. 27–11B)—after the pushing effect stops, the parcel neither returns to its original level nor continues in the direction in which it was being displaced.

(3) The air is in a state of *absolute instability* if upon being displaced, the parcel continues (after the force is removed) to move at an accelerating rate in the direction it was being displaced, either up or down (Fig. 27–11C). Thus unstable air moved downward, becomes less buoyant and continues to sink. A steep lapse rate favors instability.

To illustrate, assume: (a) the air is very dry, (b) the surface temperature is 80°F, and (c) the observed lapse rate is 7.5°F/1000 ft. A balloon filled with air at ground level (80°F) needs a push to start upward, but then rises at an accelerating rate with no additional pushing. At 2000 feet, the temperature within the balloon is 4°F warmer than the enclosing air (69°F vs. 65°F). At 4000 feet, the difference has increased to 8°F and the balloon rises faster (the temperature difference increases @ 2°F/1000 ft in this instance). This upward movement will continue until conditions change.

(4) *Conditional instability* (Fig. 27–11D) exists when the observed lapse rate in the air is less than the dry adiabatic lapse rate and more than the wet adiabatic lapse rate. A common observed lapse rate is 3.5°F/1000 ft. Since rising unsaturated air cools at 5.5°F/1000 ft, it becomes less buoyant as it rises. However, if it has been given a strong upward push, the air may rise high enough that condensation begins and heat energy is released. If the wet adiabatic rate of cooling upon ascent is less than the observed lapse rate, the temperature difference between the rising parcel and enclosing air will steadily diminish until the two become equal. At higher altitudes the parcel becomes warmer than the enclosing air and rises at an accelerating rate. This is known as conditional instability because it depends upon (is conditioned by) the cooling of the rising parcel at a wet adiabatic lapse rate that is less than the observed lapse rate.

Under favorable circumstances (steep lapse rate in the surrounding air and much moisture in the rising parcel) such an upward movement, as part of a giant cumulonimbus cloud, may penetrate a few miles into the stratosphere. However, the buoyancy decreases rapidly as the air moves upward through the stable stratosphere (cooler and heavier at the bottom).

Locally, certain weather phenomena can be correlated with stability and instability (Fig. 27–12). Stable conditions are indicated when clouds are layered and show little evidence of vertical motion, when smoke rises to a certain level and then spreads laterally, when haze and smoke reduce visibility near the surface, when winds are

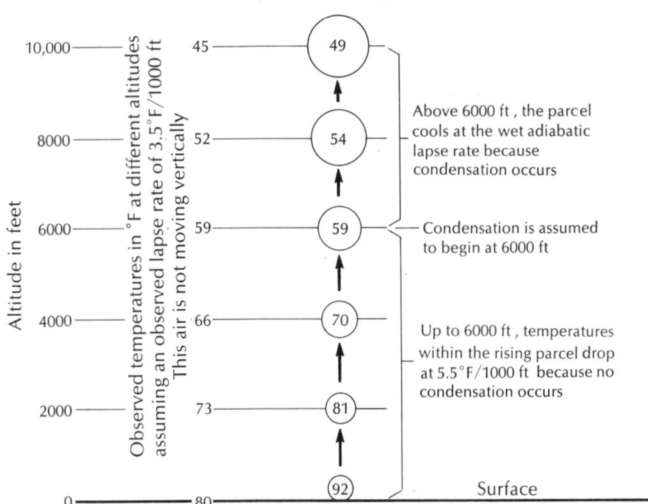

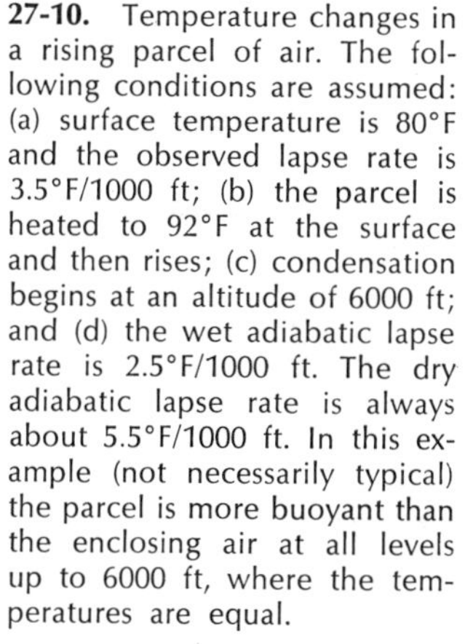

27-10. Temperature changes in a rising parcel of air. The following conditions are assumed: (a) surface temperature is 80°F and the observed lapse rate is 3.5°F/1000 ft; (b) the parcel is heated to 92°F at the surface and then rises; (c) condensation begins at an altitude of 6000 ft; and (d) the wet adiabatic lapse rate is 2.5°F/1000 ft. The dry adiabatic lapse rate is always about 5.5°F/1000 ft. In this example (not necessarily typical) the parcel is more buoyant than the enclosing air at all levels up to 6000 ft, where the temperatures are equal.

steady, and when fog develops. On the other hand, unstable conditions are suggested when cumuliform clouds show strong vertical development, when smoke rises high above a surface, when winds are gusty (unless caused by surface irregularities) and dust devils develop, and when visibility is good.

Certain types of problems in meteorology that involve vertically moving parcels of air can be answered by using the following generalized formula:

$$\text{The altitude (in thousands of feet) at which the temperature difference between a parcel of air and the enclosing air becomes zero} = \frac{\text{the difference in temperature at the start between the parcel and the enclosing air}}{\text{the amount this difference is decreased per 1000 feet of vertical movement}}$$

For example, assume: (a) a surface temperature of 60°F, (b) the air is very dry, (c) an observed lapse rate of 3.5°F/1000 ft, and (d) that a parcel of air is heated to 70°F. How high will the parcel rise? It rises until its temperature equals that of the enclosing air, which is at an altitude of 5000 feet. The temperature difference of 10°F is reduced by 2°F/1000 ft (5.5°F minus 3.5°F).

For a different example, assume: (a) a surface temperature of 70°F, (b) a dew point of 52°F, and (c) a dew point lapse rate of 1°F/1000 ft. At what level will a cloud form if part of this air is pushed upward? For condensation to occur, the temperature difference of 18°F between air temperature and dew point must be decreased to zero. A rising parcel of this air cools at 5.5°F/1000 ft, whereas the dew point decreases by 1°F/1000 ft. Therefore, the temperature difference between the two is decreasd @ 4.5°F/1000 ft. Thus a cloud forms at 4000 feet (18 ÷ 4.5).

(5) A stable layer of air may develop a state of absolute instability if the entire layer is lifted mechanically under conditions which steepen its lapse rate (Fig. 27–13). This process is called *convective instability*. The observed lapse rate of the layer will be steeper after it has been lifted mechanically if the lower part of the layer cools less than the upper part. A favorable condition is the presence of more moisture in the lower part of the uplifted layer than in its upper part. Thus as uplift proceeds, condensation begins at a low altitude in the lower part of the layer, which thus cools slowly during

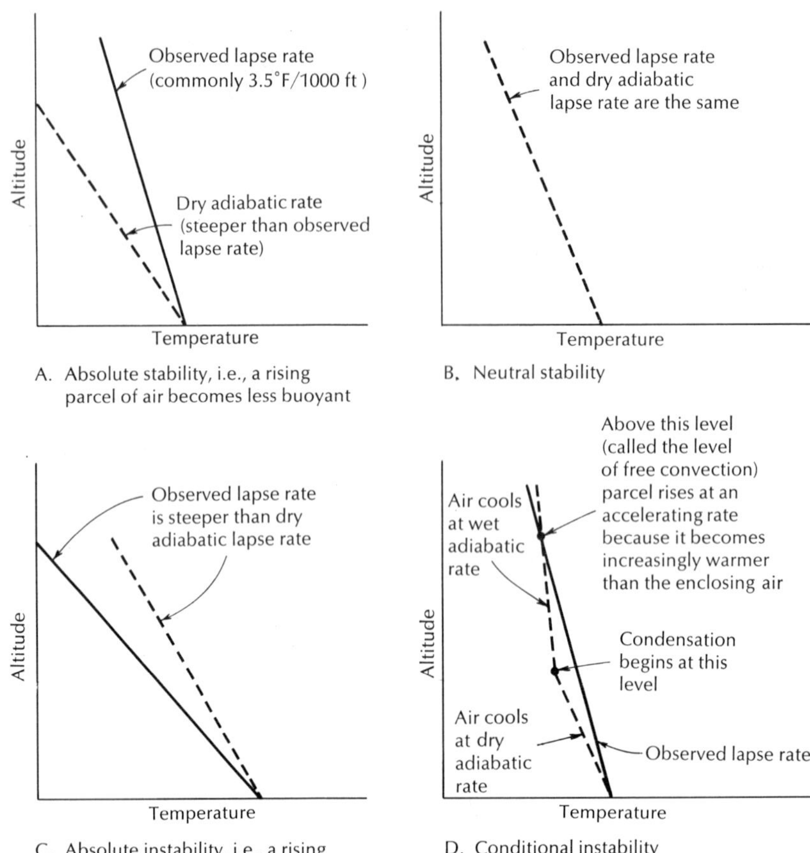

27-11. Conditions illustrating stability and instability are plotted on four temperature-altitude graphs. A rising parcel of air cools at the adiabatic rates shown by the dashed lines. The temperature of the air the parcel rises through is shown by the solid lines (except in B where the two lines coincide). The steep observed lapse rate shown in (C) is common during the warmer months in desert areas but uncommon elsewhere.

most of the uplift. On the other hand, the drier upper part of the layer cools for a greater distance at the steeper dry adiabatic lapse rate, and its temperature drops by a larger amount. The top of the layer also rises more than does the bottom, which increases the amount of adiabatic cooling.

Temperature Inversions

Temperatures tend to decrease upward from the Earth's surface in the troposphere at the average rate of approximately 3.5°F/1000 ft (the average observed lapse rate). However, exceptions to this general upward decrease in temperature frequently occur—locally, temporarily, and at different altitudes—and these are called *inversions:* temperatures increase upward for a certain distance before they resume the normal lapse in temperature at still higher altitudes (Fig. 27–14). To bring about a temperature inversion, either the lower part of a mass of air must be chilled or an upper portion must be warmed.

Shallow temperature inversions are associated with the formation of some types of

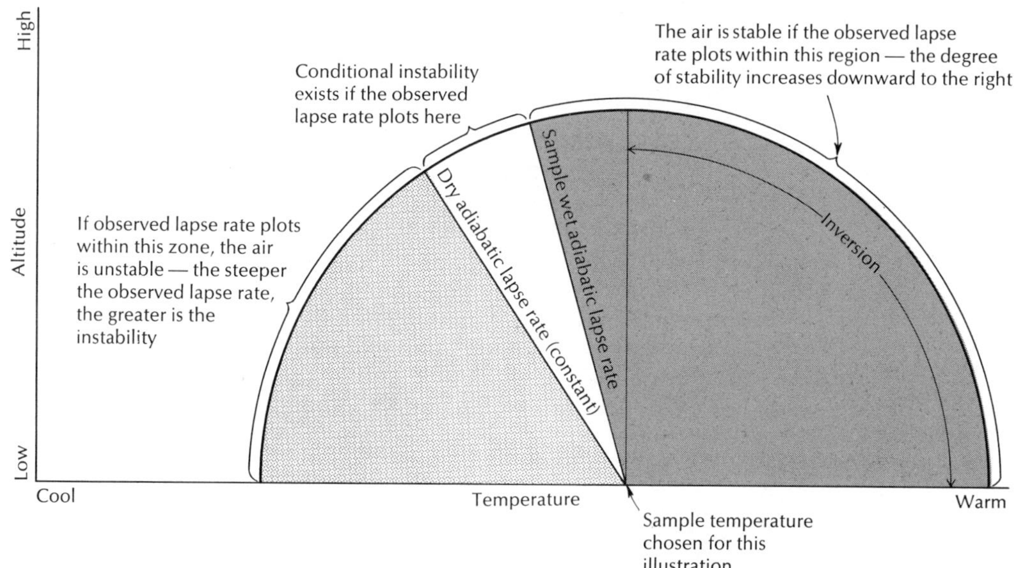

27-12. The degree of stability or instability of the air at a particular time and place can be determined by plotting the observed lapse rate and wet adiabatic lapse rate at this time and place on the graph, together with the dry adiabatic lapse rate.

fog (Fig. 27–15) and form because air is chilled near the surface.

Inversions may originate mechanically by adiabatic temperature changes in vertical air currents caused by turbulence, convectional circulation, and subsidence. If air moves upward in a certain region, it may do so because winds move along an irregular surface and produce turbulence in the lower air. Air may also move upward as part of a convectional circulation because it has been warmed at the bottom, or because air in an adjacent area was cooled at a higher altitude and sank. In either case, the upward movement of air tends to stop at a certain altitude because the forces causing it are limited—for example, the velocity of the wind producing the turbulence or the factors causing convectional circulation.

The adiabatic cooling and heating (if at the dry rate) associated with these vertical movements change the lapse rate that existed before the circulation occurred (Fig. 27–16). If air is lifted from the surface to some higher altitude, it cools at the dry adiabatic rate of 5.5°F/1000 ft as it rises (assume that condensation does not occur because the upward movement is too small or the air is too dry). Therefore, by the time this air arrives at the higher altitude, its temperature is lower than that of the air already present at this altitude—assume the observed lapse rate is about 3.5°F/1000 ft. Thus air at the surface is *potentially cooler* than air at some altitude above it—if it is lifted to this altitude, it cools adiabatically on the way up, and its temperature decreases more rapidly than the average observed lapse rate.

On the other hand, air at a certain altitude may be potentially warmer than the air beneath it. If this high-altitude air sinks, its temperature increases at the dry adiabatic rate (assume no evaporation occurs), and this exceeds the average observed lapse rate. Vertical motions of this sort within a layer of air of a certain depth cause the lapse rate to steepen—the air in the mixed layer becomes cooler at the top and warmer at the bottom. Thus an inversion occurs be-

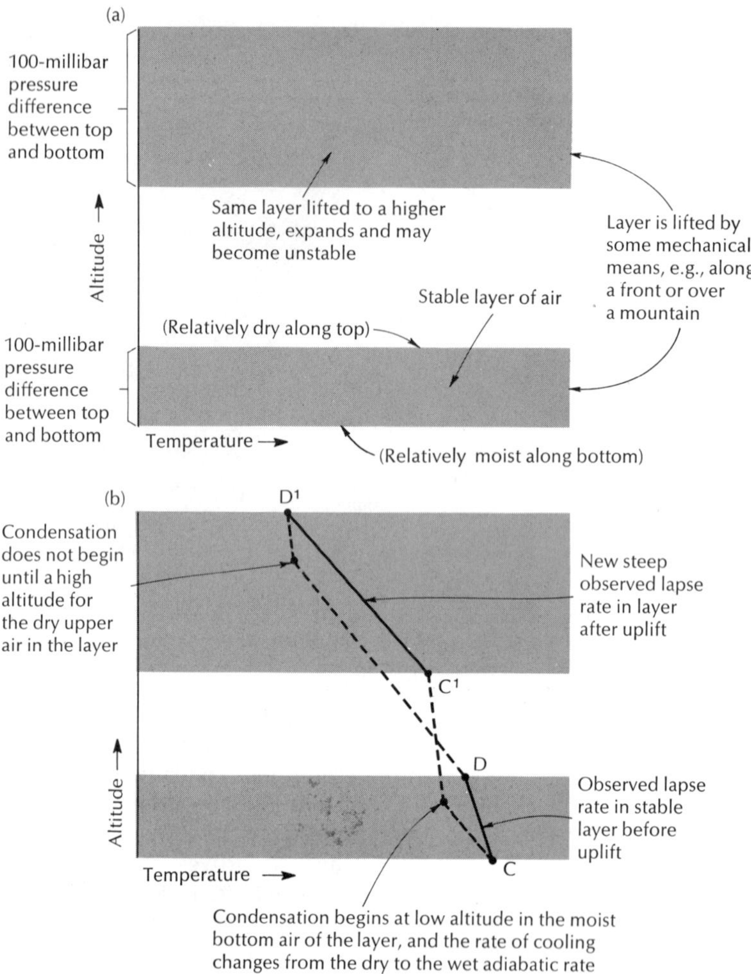

27-13. Temperature-altitude graphs illustrate convective instability. Note that the geometrical decrease of pressure with altitude causes the top of the layer to rise more than the bottom. In B we can follow the upward movement and adiabatic cooling of two parcels: C moves to C′ and D moves to D′; thus parcel D cools much more than parcel C.

tween the top of this mixed layer and the unaffected air above it.

Subsidence may also produce an inversion, and this downward movement of air may occur as part of a stagnating high pressure system. The air commonly is prevented from sinking all of the way to the surface because of upward movements in air nearer the surface caused by turbulence and convection. Thus the air sinks a certain distance and is heated adiabatically. The bottom of this subsiding layer is compressed proportionally more than the top. Temperatures at the base of the subsiding zone exceed those at the top of the nonsinking layer beneath it (the descending air is *potentially warmer)* and an inversion occurs in the transition zone between them.

The magnitude of such an inversion is increased if the subsidence has been large, if its downward motion has been stopped by turbulence in the air beneath it (the top

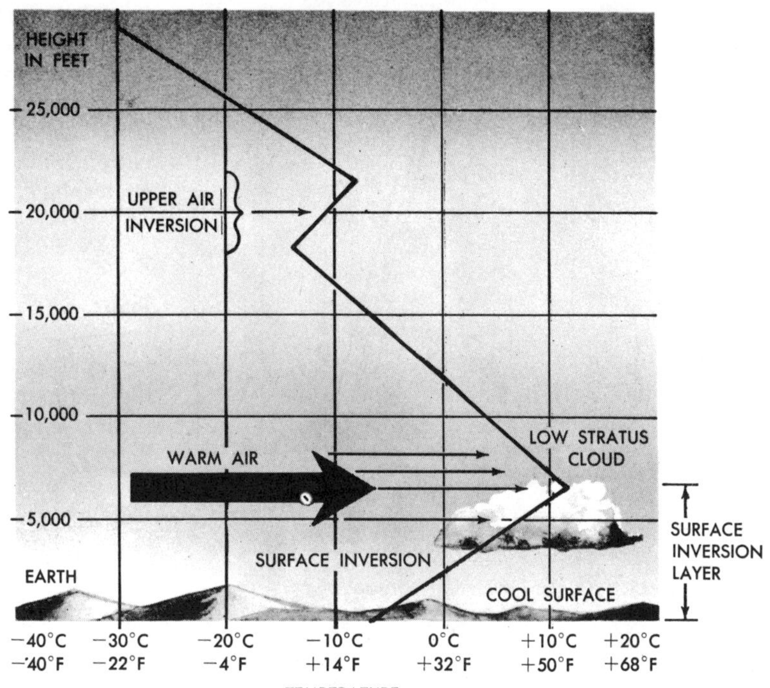

27-14. Surface and upper air inversions. (AF Manual 105-5, *Weather for Aircrews.*)

of this turbulent zone determines the altitude of the inversion), and if considerable radiational cooling occurs in the air immediately beneath the inversion.

An inversion is an extreme case of absolute stability (Fig. 27–11A) and acts as a lid to prevent parcels of air from moving upward through the inversion zone. Such rising parcels of air must expand and cool as they rise. If such a parcel begins to move upward through the inversion zone, it will continue to cool, whereas the air within the inversion is warmer at successively higher levels. Thus rising parcels quickly lose whatever buoyancy they have and stop rising. Cloud layers may develop beneath an inversion, and dust and smoke may spread laterally along it.

Inversions are most important factors in air pollution, especially an inversion that develops at low altitude at a time when winds are light. Pollutants discharged into the atmosphere by man's activities at the Earth's surface are then confined to the relatively thin zone between the inversion and the surface. If little horizontal movement is occurring at the time, the pollutants tend to remain in the area and to become ever more concentrated. At the moment there is little, if anything, that man can do to prevent such an inversion from developing. However, meteorologists can generally forecast the conditions under which inversions do develop, and thus man can be warned that it is especially vital for him to reduce the quantity of the pollutants discharged into the air at this time and place.

Inversions also occur along the boundaries of converging air masses (fronts, p. 722) wherever warm light air has moved upward above cool heavy air.

Surface inversions tend to occur in the polar regions, especially in winter (p. 719), and upper air inversions are common beneath subsiding air associated with the horse latitudes and outer parts of the trade wind

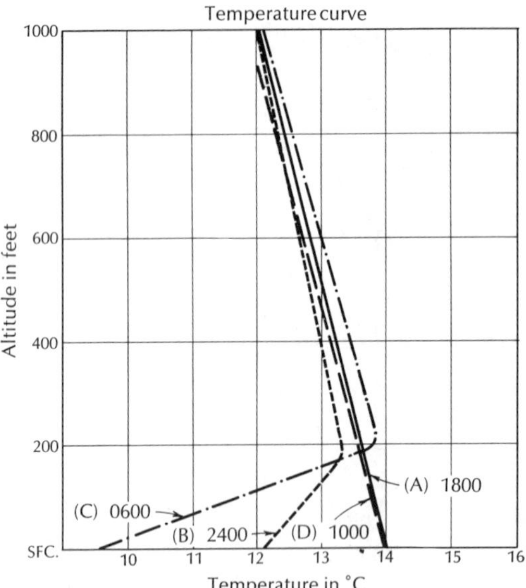

27-15. Surface inversion produced by radiational cooling during the night. SFC stands for surface, 0600 for 6 AM, 2400 for midnight, etc. Note that surface temperatures dropped about 4.5°C (8°F) from 6 PM to 6 AM. AF Manual 105-5, *Weather for Aircrews.)*

belts (p. 665). This trade wind inversion is one of the most widespread and tends to be located at a higher altitude above areas of convergence and at a lower altitude above regions of divergence. In middle and lower latitudes, surface inversions may develop over the oceans during summer because near-surface air is kept relatively cool by the water while the overlying air is heated. This is particularly true in regions of cool surface currents. One might also think of the entire stratosphere as a gigantic upper-air inversion.

Fog*

A fog (Fig. 27–1) is a cloud, commonly of the stratus type, that occurs at or near the Earth's surface and reduces visibility to less than 1 km (about 0.62 mile). No physical difference occurs between the stratus cloud seen by an observer from a valley (sometimes called a high fog) and the fog encountered by a nearby mountain climber, although clouds of other types may also form fog at high altitudes. Thus one man's fog may be another man's cloud.

Commonly, fogs consist of very small droplets of water (about 20 microns or .001 inch in diameter, Fig. 27–26) which condensed from and remain suspended in the air or rise and fall very slowly. Some fogs

* Myers, Joel N., "Fog," *Scientific American,* Dec. 1968.

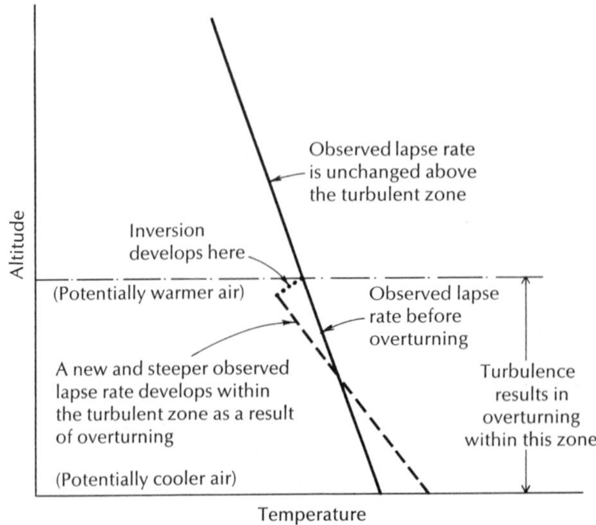

27-16. Development of an upper-air inversion.

consist of tiny ice crystals, though liquid supercooled droplets may exist well below the freezing temperature.

Two processes account for the formation of most fogs: (1) air along the ground may be cooled below its dew point, and (2) enough water vapor may be added to the air to saturate it. In either case the temperature and the dew point are made equal. In advection, radiation, and upslope fogs the cooling process predominates. In evaporation (steam) and frontal (rain) fogs, the addition of water vapor predominates. Often the two processes go on simultaneously.

Factors favorable for the formation of fog are: (1) high relative humidity, i.e., the difference between the air temperature and the dew point must be small, commonly less than 4°F; (2) clear skies; (3) the presence of condensation nuclei; and (4) the absence of strong winds (upslope fogs are an exception here).

Radiation Fogs

These tend to form on "C" nights (clear, calm, and cool) when much terrestrial radiation can escape upward through the atmosphere. Ground temperatures decrease steadily throughout a night because terrestrial radiation is a continuous process. Sometime before sunrise, therefore, the air along the ground may be cooled below its dew point, and condensation may occur to form a radiation fog, which is thus associated with a local, shallow, and temporary temperature inversion (Fig. 27–15).

Radiation fogs commonly do not develop on cloudy nights because clouds can absorb terrestrial radiation and reradiate some of it back to the surface. A clear night following a cloudy day is favorable for the development of a radiation fog because temperatures will be relatively low at sunset, and the humidity may be relatively high.

If air is nearly calm during a night, only its very bottom portion will be chilled, and condensation will produce a very shallow fog. However, a slight wind across an uneven surface results in mixing and turbulence, and this may extend the cooling effect upward from the surface for a few tens to a few hundreds of feet; if condensation occurs, a thicker fog is formed. With stronger winds and greater turbulence, no one parcel of air may stay in contact with the ground long enough for cooling to produce fog. However, this air may rise, cool, and produce a cloud.

Low areas are favorable locations for radiation fogs to develop because cold heavy air tends to drain downward into them from higher ground where radiation cooling is more rapid. Also, more water may be evaporated into the air over depressions than over high ground.

Radiation fog is commonly rather shallow and tall buildings or higher elevations often project above it. Radiation fogs may be thickest a short time after sunrise after solar radiation has produced some wind and turbulence.

Fogs begins to disperse generally some time after sunrise, and they tend to disperse from the bottom upward by absorbing terrestrial radiation. Although a fog may reflect well over half of the solar radiation striking it—and thus it tends to resist dispersion—nevertheless some radiation penetrates the fog and warms the surface, and this increases the rate of terrestrial radiation. Water vapor evaporated from the fog may subsequently be carried upward to recondense and form a stratus cloud.

Fogs also disappear if they are warmed adiabatically by moving to a lower altitude. Thus fogs tend not to form on the downwind side of a mountain.

Fogs formed by radiational cooling are limited to land areas because the daily fluctuation in temperature at the surface of a large body of water is very slight. But such fogs may form over land and be shifted via a gentle wind out to sea. Radiation fogs are most common in the late summer and fall in regions such as the northeastern United States. Winds are commonly too strong in winter and overall temperatures are too high in the summer. In the fall, the nights are long enough for considerable radiational cooling to occur, yet the air is still warm enough to have a high moisture content. Radiation fogs are associated with stable air and fair weather.

High-inversion fogs constitute a longer lasting type of radiation fog most common during the winter months. A layer of stable cool moist air (maritime polar) accumulates below a layer of warmer drier air. The lower heavier air is thus trapped beneath the inversion, and radiational cooling over a period of days and weeks may lower its temperature sufficiently for a stratus cloud to develop. This cloud radiates heat energy, undergoes additional cooling at night, and thus may cause condensation in the air beneath it. It may thicken downward in this manner to form a surface fog at night. This is a common type of fog in certain valleys, such as the San Joaquin Valley in California.

Upslope Fogs

These form in areas where stable moist air moves along the Earth's surface as part of the general circulation and is forced by the topography in its path to rise to higher altitudes. The upward movement results in cooling at the dry adiabatic rate before condensation begins. In a region of hills and valleys, upslope fog may form wreaths about each of the hills at a certain altitude, whereas the sides and bottoms of valleys below this condensation level are free of fog. The Great Plains region of North America has a surface that slopes gradually upward toward the west and the Rocky Mountains. Whenever stable moist air blows westward across this upward-sloping surface, conditions are favorable for the development of upslope fog. Radiation cooling may also be a factor at times.

Upslope fogs are not restricted to nighttime formation and may develop in the presence of winds of 20 to 30 mi/hr or more. Up to a certain velocity, fog may form because the faster the air moves upslope, the faster it cools. However, with very high winds the turbulence is too great for fog to form.

Advection Fogs

Advection means "being carried in from elsewhere," and advection fogs may form over land or sea wherever relatively warm moist air moves across a cooler surface. The frequent and persistent fogs off Newfoundland where the cold Labrador Current and the warm Gulf Stream meet are of this type (sea fog). Another example is the fog that is common along the California coast, where onshore winds move across a cool coastal current. Warm moist air moving northward in the winter across a snow-covered land favors advection fog, furthered also by the considerable radiational cooling during the winter night.

Advection fogs may form off the New England coast during the warmer months of the year because the prevailing winds are out to sea and warm moist air moves eastward over the cooler ocean water. Such fogs may be blown a short distance inland in the afternoon if a sea breeze develops (p. 667).

The seasonal temperature lag of large water bodies such as the Great Lakes favors the formation of advection fogs. In the spring and early summer, the air warms more rapidly than the water. Fogs may form over the cooler water if warm moist air moves across it. On the other hand, in the fall the water remains relatively warm after the land areas around the lakes have cooled. Therefore, if air moves across the water, it will be warmed and moistened. Later, as this air moves across the cooler land, condensation may occur and fog may develop.

Evaporation Fogs

An evaporation fog may develop if water vapor is added to a mass of relatively cool air in sufficient quantity to saturate it without much change in temperature. To see such a fog, put a pan of hot water in cold air or in another pan of cold water, or run the hot shower in a cold bathroom (according to reports, a familiar sight in unheated English bathrooms). An evaporation fog will form if cool air moves across a warmer water body, and rapid evaporation from the warmer water saturates the colder air (the temperature difference probably should be some 20 to 25°F). Such fog is also called steam fog (or arctic sea smoke at high latitudes) because it looks like steam or smoke rising from the water surface (Fig. 15–1).

Another type of evaporation fog (the frontal variety) involves the forward margins of air masses, especially warm fronts. Under these conditions rain may fall from a high-altitude warm air mass downward through a cooler stable air mass along the surface. Sufficient evaporation may occur to produce stratus clouds or fog in the cooler air.

Smog and Polluted Air in Our Effluent Society

In smog (smoke and fog, Fig. 1–3) many of the tiny particles added to the air by the diverse activities of a large city act as condensation nuclei for the development of tiny fog droplets. Over an industrial area, such nuclei tend to be so abundant that the fog droplets are smaller than elsewhere (competition for the available water vapor molecules tends to prevent growth of larger droplets). Thus they fall more slowly, and the fogs persist longer. The mixture of smoke and fog also persists by masking the surface from insolation that could warm the ground and the air above it. Thus relatively cool, heavy, stable, smog-filled air may blanket an area and accumulate industrialized effluents until moved along by strong winds or washed out by heavy rains.

According to Kneese, the substances found in polluted air may be placed into two groups. Stable primary substances are not changed in the air and consist of dusts, smokes, droplets, and fumes. These cause corrosion and dirty buildings, reduce sunlight, and affect life processes. Secondary pollutants are produced by photochemical interactions (sunlight) with primary pollutants within the atmosphere. These are less well known, less predictable, and probably more dangerous. For example, sulfur dioxide may be oxidized to sulfur trioxide and then react with water vapor to form sulfuric acid.

According to a recent estimate, air pollution in the United States is produced approximately by: the motor vehicle, 60%; industry, 17%; power plants, 14%; and space heating and waste disposal, 9%.

People in sizable numbers have died when exposed to particularly concentrated smogs for several days. More than 4000 smog-caused deaths apparently occurred in London during two weeks in December of 1953. In Donora, Pennsylvania in 1948 more than a thousand people were made ill and twenty died. A three-day smog occurred in November of 1966 in New York City and may have caused 168 deaths (based upon a statistical study of 1966 deaths vs. long-time averages).

In the heavily industrialized Ruhr Basin in Germany eight million people are daily subjected to a rain of 1½ million tons of dust, ashes, and carbon and some 4 million tons of sulfur dioxide. Thus a mild Pompeian burial occurs day and night and skies are almost always dark. Waiters in restaurants in Duisburg are reported to change their collars three times each day. Statistics seem to show that the health and physical development of the children in the Ruhr have been impaired.

Urbanization, industrialization, the rush of modern living, increased consumption of motor vehicle fuel (tripled during the past 30 years), and population growth all combine to increase the problems posed by dirty air—"man's insult to man, his sin of emission." In an apt analogy, it has been pointed out that a man who is running raises more dust than one who is walking. Many men running, therefore, stir up whole clouds of dust.

Some of the serious problems involved are illustrated by a reddish cloud that appeared over southern Ohio on 24 January 1965 (as reported in the *New York Times* of 28

February 1965). A fallout of reddish dust from the cloud soon coated roofs, the ground, cars, and people—as much as 9 tons/sq mi. The source of the dust apparently was hundreds of miles away in the southwestern part of the United States where windstorms had put the dust into the air. The prevailing westerlies had then carried it eastward. Analysis of the dust disclosed alarming quantities of pesticides and various agricultural chemicals that had been used on farms in states such as Texas and Oklahoma weeks, months, or even years earlier. Such chemicals (more than 45,000 pesticide formulas are registered with the Department of Agriculture) retain some of their potency for years after application. Although the danger from any single episode such as that of January 24 is limited, the gradual accumulation of poisonous chemicals in water supplies over a number of years may very well be harmful. It is alarming to think that poisons used on farms in Texas may gradually be contaminating water supplies in Ohio.

New York City has been described as located at the terminus of a 3000-mile sewer of atmospheric filth that starts as far away as California and grows like a dirty snowball all the way. It is said that the city and neighboring areas trade smoke and dirty air as winds shift. An average of 60 tons of heavy dust falls on each square mile of New York City each month. A 1965 report indicated that just by breathing a New Yorker inhales daily as much benzopyrene (described as a cancer-inducing hydrocarbon) as he would by smoking two packs of cigarettes each day.

It has been said that Los Angeles would be uninhabitable if its air were as dirty as New York air—also, that Los Angeles would be only half as smoggy if it were twice as windy. Thus meteorological conditions vary from city to city. Edinger (1967) has discussed the special problems of Los Angeles in a most interesting manner. A very large high pressure system is located to the west over the ocean and produces two important effects. Prevailing winds swirl in a clockwise direction around this high and tend to blow cool moist air from the northwest onto the land. An upper-air inversion is produced beneath this high (p. 690), and its base is located some 1000 feet or so above the water surface. Thus about 1000 feet of cool moist air—his so-called marine layer—rests upon the water. Immediately above the marine layer (i.e., within the inversion), temperatures rise abruptly (as much as 20°F or more within a hundred yards or so) and relative humidity drops abruptly; this is hot, dry desert air.

The shallow marine layer and the overlying inversion and desert air are present above Los Angeles about 275 days each year. Most inhabitants live, work, and play entirely immersed within this shallow, cool, moist layer of air. Into this marine layer—and confined to it by the inversion above—go vast quantities of pollutants; the gasoline motor is the chief offender. On days when the winds are relatively strong and/or the inversion is relatively high, diffusion keeps concentrations bearable. Bad attacks occur on days with little wind and a low inversion.

Statistics seem to show a correlation between deaths caused by cancer of the respiratory tract and population concentrations—twice as many occur in large cities as in rural areas; rates for smaller cities and urban areas fall in between. The U.S. Public Health Service has reported that a six-year study in New Orleans showed that asthma attacks resulting in hospital admissions rose and fell almost consistently with the number of times the New Orleans Fire Department was called to put out dump fires.

The greatest damage to health and property probably arises from frequent exposure to polluted air at many locations, rather than from the occasional extreme concentration that causes death and newspaper headlines. Smog and polluted air are already a serious problem, indeed a critical one in many areas, and it is growing worse. Estimates of total annual damage in the United States to livestock, vegetation, materials, and property range up to $11 billion and more (possible effects on human health are excluded in such estimates). The technology and understanding to eliminate most of this polluted air is available today. The total estimated cost of cleaning our air—as expensive as it would be—would yet be only a fraction of the total damage (perhaps like getting $100 each time one spends a few $10 bills). Corrective measures have been and are being taken but not nearly on the scale necessary. Suggested remedies such as elimination of gasoline-powered auto traffic within city limits indicate the size and nature of the problems involved.

Weather data from LaPorte, Indiana, which is located 30 miles downwind (east) of Chicago, indicate that man's industrial activity may influence the weather. Precipitation has increased 30 to 40% since 1925, and precipitation has tended to correlate with the rate of industrial activity in the Chicago area where heat, moisture, and pollutants are added to the atmosphere in large quantities. Relative to nearby weather stations in Indiana, Illinois, and Michigan, LaPorte tends to have more thunderstorms, and more days with hail.

Clouds

About half of the Earth's surface at any one time may be covered by clouds, but only 6% or so of the surface may be receiving precipitation; thus non-precipitating clouds are by far the most common. Clouds consist of moisture which has condensed from the air onto tiny condensation nuclei to form minute water droplets (Fig. 27–26) or sublimated as small ice crystals. Very slight upward currents are sufficient to keep them from falling, or the droplets rise and fall gently. If the air were completely calm, the droplets would fall slowly and evaporate.

Like fog, clouds commonly form because air containing water vapor and suitable condensation nuclei has been cooled below the dew point. This cooling frequently results from upward motion which produces expansion and adiabatic cooling. Clouds may also form because enough water vapor is added to a parcel of air to cause saturation—e.g., by raindrops that fall from a higher altitude and evaporate on the way down (Fig. 28–7). Droplets may fall out of the lower portion of a cloud into unsaturated air below where they evaporate. In the meantime new cloud droplets are forming within the cloud, and so the cloud is maintained for a time.

Clouds are of three general types: *stratus* (layered, blanketlike) and *cumulus* (flat-bottomed, globular or heaped masses), and *cirrus* (curl or streak). Other terms used in naming clouds are: *alto* (high), *nimbus* (rain cloud), and *fracto* (broken). In combination with wind and pressure observations, clouds are useful local indicators of future weather changes; they constitute a form of writing in the sky that man can interpret as signs of coming weather phenomena. Since the amount of water vapor decreases upward within the troposphere, the highest clouds are relatively thin and diffused, whereas enough moisture may be present at lower levels for clouds to be both thicker and denser.

Upward movements of air have several principal causes. In convection, a part of the Earth's surface—say a patch of bare soil or a paved runway—may be heated more than the area surrounding or adjoining it. The air above this part becomes warmer and lighter and is pushed upward by neighboring cooler heavier air (Fig. 27–17). Convectional circulation occurs also on a global scale, as in the doldrums. On a smaller scale, a heated chimney or radiator in a room causes upward movement. Prevailing winds may move air along the surface and upward over mountain barriers (Fig. 27–7). Upward movements are also associated with low pressure systems, fronts, and the convergence of air masses (Chapter 28).

Clouds are classified into four families based upon their altitudes and into ten main types (Fig. 27–18). The highest clouds consist chiefly of ice crystals, whereas the others consist of liquid droplets. The standardized classification of clouds with abbreviated definitions which appears below has been reproduced from a weather map pub-

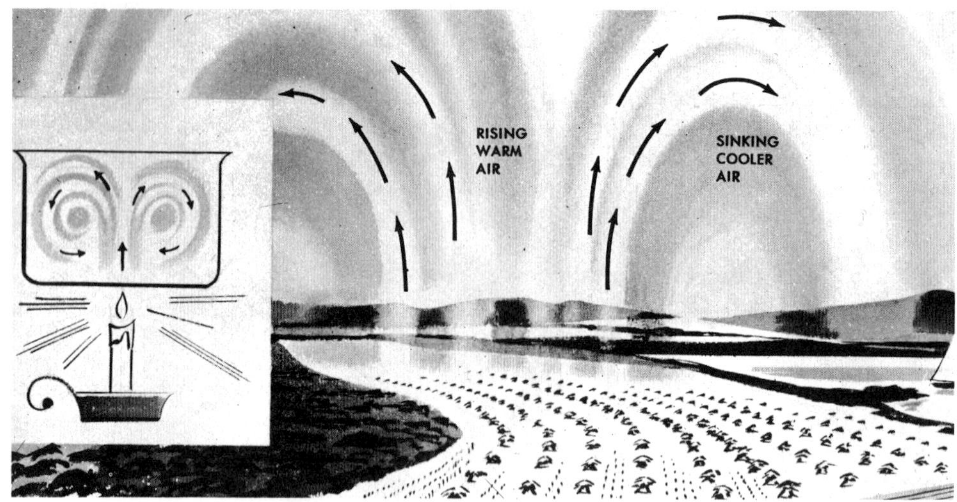

27-17. Vertical and horizontal currents caused by unequal surface heating. The term convection is usually used for the vertical movements in such a circulation and the term advection for the horizontal movements. (AF Manual 105-5, *Weather for Aircrews.*)

lished by the U.S. Weather Bureau on January 24, 1946:

In the International System there are ten principal kinds of clouds. Their names, classification and mean heights are shown in the following table. The mean heights are for temperate latitudes and refer not to sea level but to the general level of land in the region. There is nearly always some variation from the mean height, and in certain cases there may be large departures. Thus, cirrus clouds may sometimes be observed as low as 10,000 feet in temperate regions and at lower levels in higher latitudes.

Family A: High Clouds (mean lower level, 20,000 feet)
1. Cirrus (Fig. 27–19)
2. Cirrocumulus
3. Cirrostratus

Family B: Middle Clouds (mean upper level, 20,000 feet; mean lower level, 6500 feet)
4. Altocumulus
5. Altostratus

Family C: Low Clouds (mean upper level, 6500 feet; mean lower level, close to surface)
6. Stratocumulus (Figs. 27–20 and 27–21)
7. Stratus
8. Nimbostratus

Family D: Clouds with Vertical Development (mean upper level, that of cirrus; mean lower level, 1600 feet)
9. Cumulus (Figs. 27–7, 27–9, and 27–19)
10. Cumulonimbus (Figs. 27–22 and 27–23)

Cloud Definitions*

1. Cirrus—Detached clouds of delicate and fibrous appearance, usually without shading, generally white in color, often of silky appearance. They are always composed of ice crystals. [Cirrus clouds have a greater variety of forms than other types: isolated tufts, lines resembling strands of hair and forming a curl at one end, branching featherlike plumes, and bands that stretch across the sky and seem to converge at a distant point on the horizon. This is an effect of perspective, because the bands may actually be parallel. Because of their high altitude, cirrus

* Abridged; material in brackets is supplied by the author.

clouds may reflect yellow and red colors at sunset and sunrise.]

2. Cirrocumulus—A layer or patch composed of small white flakes or of very small globular masses, which are arranged in groups or lines, or more often in ripples resembling those of the sand on the seashore. [Cirrocumulus clouds are rare. They should be associated with cirrus and cirrostratus clouds and are smaller and higher than altocumulus clouds with which they are commonly confused. Most of the individual puffs do not exceed 1 degree in angular diameter; i.e., they are not more than double the apparent size of the sun or full moon.]

3. Cirrostratus—A thin whitish veil which does not blur the outlines of the sun or moon, but usually gives rise to halos. [Cirrostratus clouds are almost always so relatively thin and diffused that objects at the surface cast shadows unless the sun is near the horizon. They may give the sky a milky look. They are thicker and lower and form a more uniform sheet than cirrus clouds. If cirrostratus clouds are observed a few hours after cirrus clouds, precipitation will probably occur within the next 24 hours.]

4. Altocumulus—A layer (or patches) composed of rather flattened globular masses, the smallest elements of the regularly arranged layer being small and thin. These elements are arranged in groups, in lines, or waves, following one or two directions and are sometimes so close together that their edges join. [Altocumulus clouds may or may not show shadows and coronas (p. 703) and are commonly associated with fair weather. Most of the regularly arranged heaps range from 1 to 5 degrees in angular diameter. Altocumulus grade toward altostratus clouds when they are packed closely and toward cumulus clouds when they

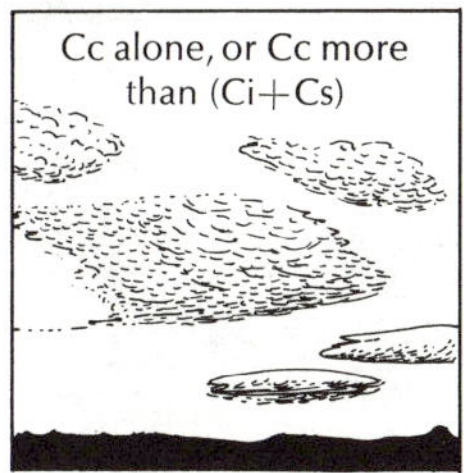

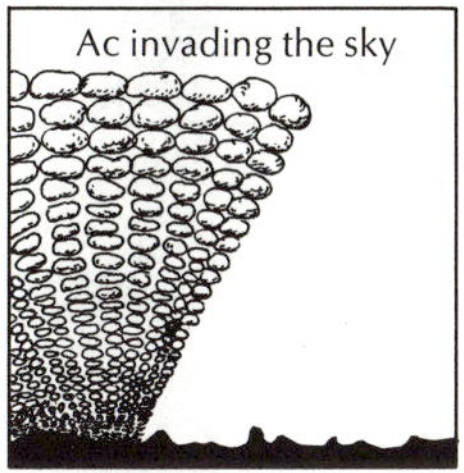

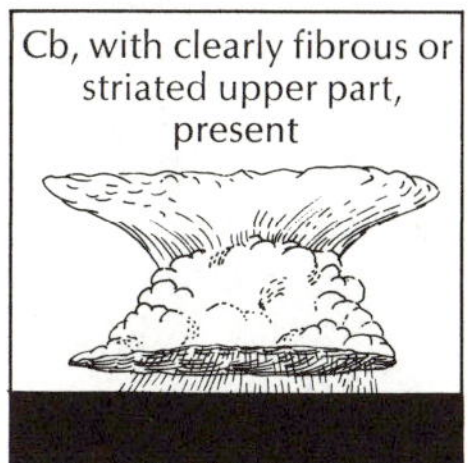

27-18. Cloud Sketches. Ci, Cc, and Cs are abbreviations for cirrus, cirrocumulus, and cirrostratus, respectively. Ac and Sc refer to altocumulus and stratocumulus, whereas Cu and Cb stand for cumulus and cumulonimbus. *(Cloud Atlas,* World Meteorological Organization, 1956.)

27-19. Cirrus clouds extend across the Red Sea and Nile Valley. They occur near the core of a jet stream and show its direction. Note rock structures and low-level cumulus clouds. *(Gemini XII, 15 November 1966, S66-63530, NASA.)*

show vertical development. Altocumulus clouds may form a dense layer showing definite relief on its lower surface. Altocumulus and altostratus clouds may be transformed one into the other during a day as turbulence increases or decreases.]

5. Altostratus—Striated or fibrous veil, more or less gray or bluish in color. This cloud is like thick cirrostratus, but without halo phenomena; the sun or moon shows vaguely, with a faint gleam as though through ground glass. [If altostratus clouds are observed soon after cirrus and cirrostratus clouds, precipitation is probable within the next 12 hours or so. Altostratus clouds commonly form by a lowering of cirrostratus clouds; they hide the sun and the moon only in their darker portions. Altostratus clouds are thick enough to keep objects at the surface from casting shadows, a feature which distinguishes them from cirrostratus clouds. As a warm front approaches, one may observe the gradual disappearance of the shadows as cirrostratus is replaced by altostratus. Thus altostratus may merge into cirrostratus at one extreme and into nimbostratus at the other extreme. Nimbostratus clouds are darker than altostratus clouds and hide the sun and moon in all parts. Precipitation may develop within and fall from altostratus clouds.]

6. Stratocumulus—A layer (or patches) composed of globular masses or rolls; the smallest of the elements are fairly large;

27-20. Summer afternoon photograph of clouds taken from a plane at an altitude about 3000 feet above a lake in Sweden. Cumulus clouds occur beneath an extensive layer of shallow stratocumulus. The base of the stratocumulus clouds is at the bottom of a very stable layer of air (upper air inversion). The cumulus clouds rise to the stable layer and spread out. No cumulus are forming over the cooler waters of the lake, except for those that rise intermittently above the islands. The lower portions of two of these (in center and to the right) have disappeared, leaving cut-off, mushrooming heads. (Photograph by L. Larsson, from F. H. Ludlam and R. S. Scorer, *Cloud Study, A Pictorial Guide*, John Murray, London, 1966.)

they are soft and gray with darker parts. The elements are arranged in groups, in lines, or in waves, aligned in one or two directions. Often the rolls are so close that their edges join. [Stratocumulus clouds may cover the entire sky and are most common in winter. They may form from stratus clouds by increased convectional circulation or vice versa.]

7. Stratus—A low uniform layer of cloud resembling fog but not resting on the ground. [The sky may have a hazy appearance and drizzle, a very fine mistlike rain, may fall. When broken into fragments, the cloud is called fractostratus.]

8. Nimbostratus—A low, formless, and rainy layer, of a dark color, usually nearly uniform; when it gives precipitation it is in the form of continuous rain or snow. [At times it may be difficult to distinguish nimbostratus from stratus. Nimbostratus clouds appear formless and nearly uniform, whereas stratus shows some contrasts and lighter sections. Nimbostratus clouds may also have a very irregular lower surface and commonly show streaks of precipitation that trail beneath and do not reach the ground. Nimbostratus is a thick cloud; its base commonly occurs at relatively low altitudes and its top at middle or high altitudes.]

27-21. Cells and billows occur in this deck of stratocumulus clouds over eastern Mexico. Note folded rock structures and bend in the trend of the folds. *(Gemini VII, 9 Dec., 1965, S65-63889, NASA.)*

9. Cumulus—Dense clouds with vertical development; the upper surface is dome shaped and exhibits rounded protuberances, while the base is nearly horizontal. [These white cottony puffs are the visible evidence of columns of air rising above the condensation level. They may seem to merge into stratocumulus when viewed near the horizon. If fragmented by turbulence, they are called fractocumulus.]

10. Cumulonimbus—Heavy masses of cloud with great vertical development, whose cumuliform summits rise in the form of mountains or towers, the upper parts having a fibrous texture and often spreading out in the shape of an anvil (Fig. 29–6). [The following are associated with cumulonimbus clouds: thunderstorms, lightning, hail, strong updrafts and downdrafts, sudden heavy showers, and tornadoes.]

Remember that all gradations occur, that more than one kind of cloud may be in the sky at one time, and that clouds form and disappear rapidly. Identification may be difficult unless one has been observing the development of the clouds and is aware of the physical conditions involved in their formation. Decide which clouds may be hardest to tell apart, e.g., cirrostratus vs. altostratus, and then concentrate upon their distinguishing characteristics.

27-22. Cumulonimbus clouds showing anvil tops. *(Courtesy U.S. Weather Bureau.)*

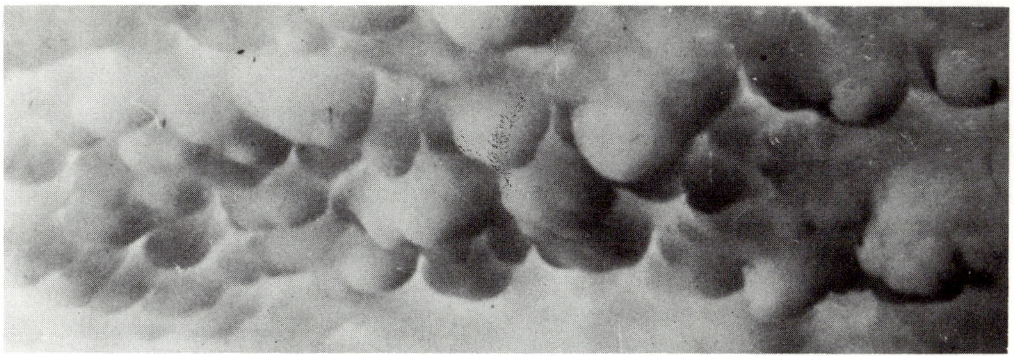

27-23. Mammatus. Such rounded dark masses occur along the lower surfaces of cumulonimbus clouds (less frequently, stratocumulus) and tend to be associated with strong turbulence—sometimes they are called a "tornado sky." *(Courtesy NOAA.)*

Clouds may disperse for several reasons: precipitation may remove surplus water; the clouds may be warmed by solar and terrestrial radiation which decreases their relative humidity; surrounding drier air may mix with the clouds, especially near the margins; and if downward air currents develop, the air will be warmed by compression.

Clouds are useful indicators of wind direction and speed at different altitudes. On a breezy crisp day with scattered cumulus puffs overhead, turn your back to the wind and note the direction of cloud movement. Commonly this will be to your right, evidence of a larger Coriolis effect and wind speed at the higher altitude.

Layered clouds tend not to remain stable. The top part radiates heat energy into space and cools. The bottom part absorbs terrestrial radiation and warms up. Thus convectional circulation begins.

The angle at which one views clouds changes their appearances. For example, assume that a cirrostratus cloud covers most of the sky and is of uniform thickness. This cloud will look thin when observed directly overhead but thicker (thus denser and darker) when viewed in the distance along a line of sight that passes diagonally upward through the cloud. For a similar reason, two cumulus puffs will be distinctly separated if located overhead, but will appear to merge when viewed at a slant in the distance. Thus cirrus clouds seem to merge into cirrostratus when seen near the horizon and cumulus into stratocumulus.

Halos and Coronas

Halos (Fig. 27–18) occur when cirrus type clouds occur between an observer and the sun or moon. A halo is a ring of light surrounding the sun or moon with a radius of 22 degrees; however, only cirrostratus clouds are widespread enough to produce a complete ring. Less commonly, a second halo with a radius of 46 degrees may also form. Halos are caused by the bending or refraction of sunlight passing through ice crystals. The orientation of the ice crystals determines the radius of the halo, which may be whitish or colored reddish on the inside.

On the other hand, an altocumulus or altostratus cloud located between the sun or moon and an observer may produce a corona, a small whitish or colored ring, usually with a radius less than about 5 degrees, that immediately surrounds the sun or moon. Although commonly whitish, red may occur on the outside and blue on the inside. Coronas are produced by the diffraction of light around tiny liquid droplets.

27-24. Mountain or lee wave clouds (altocumulus lenticularis). *(Official photograph, Environmental Science Service Administration.)*

Mountain Wave Clouds

When wind flows at an angle of about 90 degrees across an obstruction such as a mountain, a number of standing waves may be produced downwind (also called lee and mountain waves). Such a series of waves have parallel crests and troughs, tend to be located in the middle and upper troposphere, and occur within 100 to 300 miles downwind from the obstruction. At levels with sufficient moisture, condensation will occur in the rising air on the upwind side of a crest, whereas evaporation will occur in the descending air on the downwind side of the crest. A lenticular-shaped cloud (its base may be nearly flat) may thus develop —generally a type of altocumulus (Figs. 27–24 and 27–25). Such a cloud will appear to remain almost stationary along a crest. In a deep wave, lenticular clouds may occur at different levels stacked one above the other. However, if the air is very dry, no clouds will form along the standing wave crests.

The cloud droplets do not remain stationary; it is only the outline of the cloud that does this. Air streams continuously through each standing wave, and the cloud droplets that form on the upwind side of a crest are soon evaporated on the downwind side.

Downdrafts and turbulence associated with mountain waves pose problems for aircraft. Particularly dangerous are: the cap cloud (a mountain occurs within it), the roll cloud, and the strong downdraft on the lee side of the mountain. The roll cloud tends to develop at about the same altitude as the obstruction, whereas the lenticular clouds tend to occur at altitudes of several miles.

Origin of Precipitation

Size relationships are important in considering how cloud droplets develop into raindrops. The diameter of a typical cloud droplet exceeds that of the average condensation nucleus by about 100 times. On the

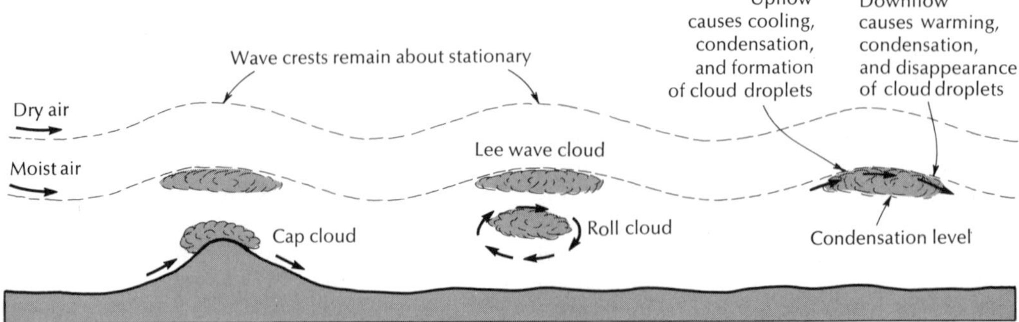

27-25. Mountain waves and clouds. Dashed lines suggest direction of air flow. Between the surface and a wave crest, there may be circular motion (rotor) in a vertical plane and a so-called roll cloud may occur at the top. Locally the wind direction may be reversed along the surface.

other hand, the diameter of a typical raindrop exceeds that of a typical cloud droplet by about 100 times (Fig. 27–26). The differences are even greater when volumes are considered (the volume of a sphere varies as the cube of its radius or diameter): about a million cloud droplets are needed to make one raindrop. Think of a raindrop as a basketball and a cloud droplet as a BB shot, and then ask a question. How does nature take one million BB shot and in a short time fashion them into a basketball?

Condensation of water vapor molecules on suitable condensation nuclei produces cloud droplets. This is a necessary first step. However, some clouds have been seen to form and produce rain in an hour or two. This is far too short an interval for cloud droplets to grow into raindrops by a continuation of the condensation process—the raindrops are much too large, and the process is far too slow. Two types of processes are thought to be involved in the origin of raindrops, and we can speak of *warm rain* involving the coalescence of cloud droplets and *cold rain* involving the melting of snowflakes.

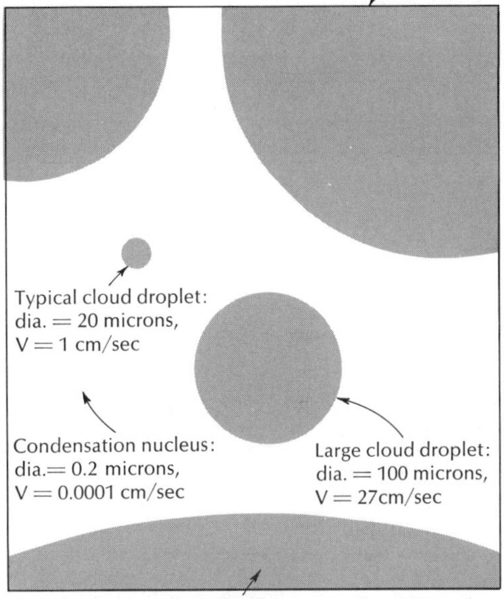

27-26. Comparative sizes and terminal falling velocities of some particles involved in condensation and precipitation. Diameters (d) are given in microns (1 micron equals 1/1000 mm) and the terminal velocities (V) in cm/sec. A human hair is about 100 microns in diameter. Note the diameters of condensation nuclei, typical cloud droplets, and typical raindrops (0.2 microns vs. 20 microns vs. 2000 microns). *(After McDonald.)*

Warm Rain

Some raindrops form within a cloud by the coalescence of cloud droplets, and cloud droplet size is important in this process. If the droplets in a cloud are uniform in size, air movements cause them to rise and fall at approximately equal rates and relatively few collisions result. Such a cloud can undergo a considerable amount of turbulence without an appreciable number of collisions among the cloud droplets. On the other hand, if a mixture of large and small droplets is present, then the droplets move at different rates, and the larger droplets grow larger by amassing the smaller droplets that collide with them.

Collision and collection may occur at the bottom of a descending large droplet which falls faster than smaller droplets and also at the top where air resistance is reduced in its wake and adjacent droplets tend to fall faster and overtake it. However, droplets occasionally bounce apart from a collision without coalescing.

If strong updrafts are present as in cumulus clouds, coalescence may occur in a different manner. After a few raindrops form in the cloud, they remain suspended or fall slowly, and the tiny cloud droplets stream upward past them. Collisions and coalescence follow. When a raindrop reaches about ⅕ or ¼ of an inch in diameter, surface tension can no longer keep it as a single drop, and it subdivides into two or more smaller drops. These in turn grow larger and subdivide again. Such a chain reaction may produce many raindrops in a short time. Observations indicate that clouds

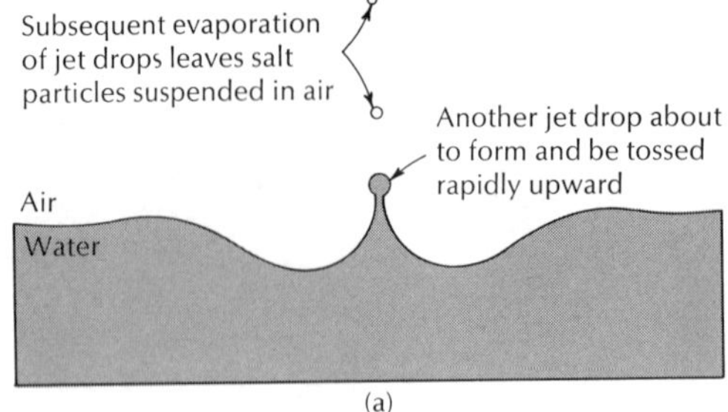

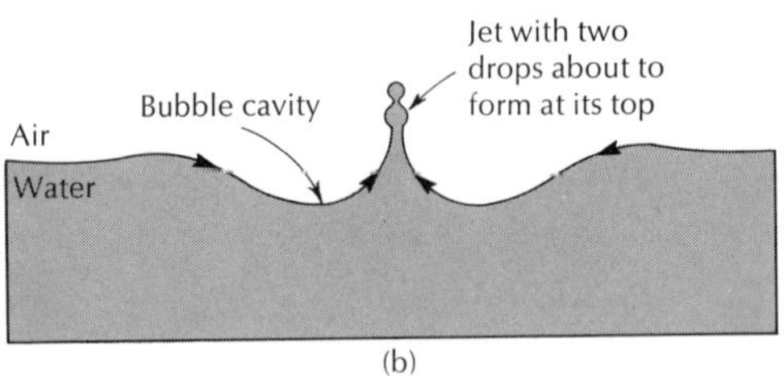

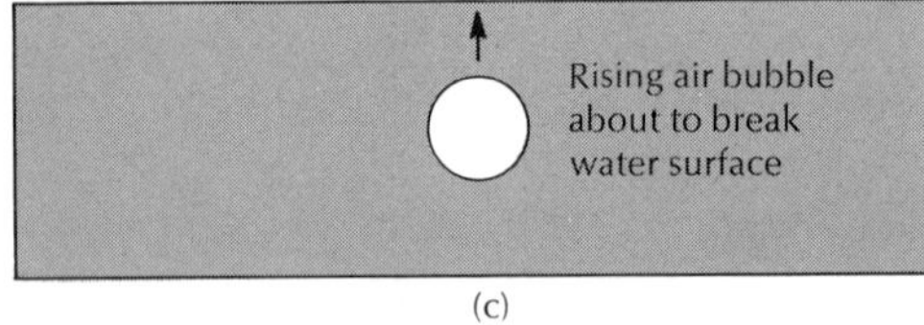

27-27. Bursting bubble produces a bubble jet and jet drops. Schematic sketches of high-speed, close-up photos of bursting bubbles. Diameters of jet drops (commonly five per bubble) tend to be about one-tenth the diameter of the bubble that produced them. Smaller, more numerous so-called bubble-film droplets also form. *(Sketched from photos by Charles Kientzler.)*

at warmer temperatures must have a minimum thickness of 1 mile or so to produce precipitation. Evidently the coalescence process needs to span a considerable vertical range to be effective. Evaporation may take place between the base of a cloud and the ground, and thus some precipitation never falls all of the way to the Earth's surface.

But how do large cloud droplets form—ones large enough and that develop rapidly enough to initiate the coalescence process? One answer involves condensation on giant salt nuclei (about 1 to 10 microns in diameter); this process produces large cloud droplets 100 microns in diameter (five times larger than average). Electrical charges may also play a role in the coalescence of cloud droplets.

Blanchard (1969) discusses some experiments that are pertinent. Fog droplets were collected on spider webs and evaporated in a drying chamber to study the condensation nuclei at their centers. Next, moist air was added, and new droplets were grown on the nuclei. Larger droplets developed on larger nuclei, and the rate of growth slowed with increasing size.

The effect of electric fields was studied using close-up, high-speed photography. In one ingenious experiment, two vibrating hypodermic needles were used to squirt two streams of water droplets along paths that intersected. In the presence of an electric field, the droplets tended to coalesce; without the field, they tended to bounce apart.

Extensive experimentation showed that giant salt nuclei get into the air from ocean water when bubbles break at the surface (Fig. 27–27). Breaking waves and rain produce many of the bubbles. As a bubble of air breaks at the surface, a tiny cavity develops, water seems to move downward toward the center of this tiny basin and then upward at the center in the form of a slender cylindrical jet of water which tosses several tiny droplets into the air. When these subsequently evaporate, they leave minute salt particles (the giant salt nuclei) floating in the atmosphere.

Cold Rain

Cloud droplets, unlike large water bodies, remain liquid at temperatures well below freezing (from 0°C to −40°C; from 32°F to −40°F), and supercooled droplets below the freezing temperature are actually quite common at high altitudes. Extremely tiny droplets (1/1000 mm in diameter) of pure water have been cooled in laboratory experiments to −40°C before they froze spontaneously. However, this cannot be done if suitable, so-called *freezing nuclei* such as ice crystals are present.

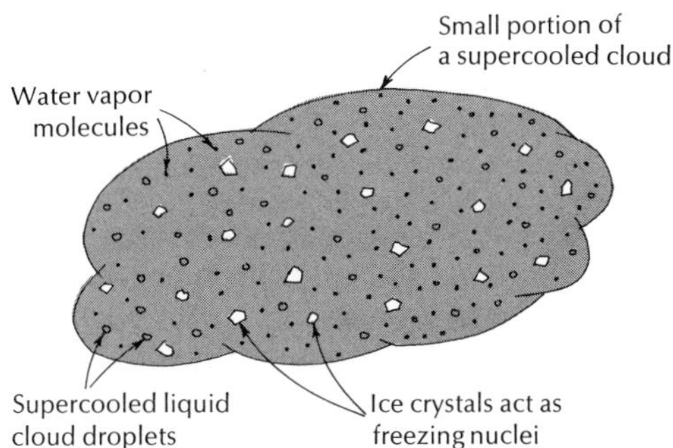

27-28. Origin of cold rain. Diagram is schematic. The cloud is saturated with water vapor relative to the supercooled liquid cloud droplets but supersaturated relative to the ice crystals. Therefore, evaporation occurs from the liquid droplets to maintain saturation and replace the molecules sublimating onto the ice crystals. Thus moisture is transferred to the ice crystals which grow rapidly at the expense of the liquid droplets. They unite as they fall to form snowflakes which then melt to produce raindrops.

Supercooled clouds occur in middle and high latitudes and in such clouds, liquid water droplets are common at temperatures down to about −20°C; occasionally they occur as low as −35°C or so. Therefore, although ice melts at 0°C, liquid cloud droplets freeze over a range of about 40 Celsius degrees and are common in the warmer portion of this range. However, if ice crystals or other suitable freezing nuclei such as silver iodide particles (geometrically similar to ice crystals) form or are introduced into such supercooled clouds, the ice crystals grow larger at the expense of the liquid droplets. Eventually they fall at different rates in a fluttering motion and collisions at lower altitudes and warmer temperatures cause numerous snow crystals to fuse together into snowflakes. At still lower levels, these melt to form raindrops. For this reason, snow may accumulate on a hilltop while rain falls in a nearby valley.

In this second method of raindrop formation, scattered ice crystals develop in a saturated, supercooled cloud and then grow by sublimation as the liquid droplets in the cloud evaporate (Fig. 27–28). Water molecules can be detached more readily from liquid droplets than from ice crystals; therefore, the water vapor pressure over water at below-freezing temperatures is greater than that over ice. Thus when air is saturated relative to water, it is supersaturated relative to ice. Under such conditions of supersaturation, water molecules in the gaseous state attach themselves directly to the ice crystals. To replace these molecules and maintain the saturated equilibrium condition relative to liquid water, other water molecules evaporate from the liquid droplets. In a rather short time, the ice crystals grow large enough to fall through the cloud and unite to form snowflakes. Thus coalescence becomes the most important growth process in the later stages of the ice crystal–snowflake origin of raindrops.

Radar and Precipitation

Radar has aided greatly in the study of precipitation as well as in the location and tracking of hailstorms, hurricanes, and tornadoes. Energy impulses (electromagnetic waves longer than visible light) are sent out as a beam in a certain direction by the radar transmitter, and objects struck by the beam reflect back a tiny amount of the energy (Fig. 27–29). The length of the round trip is timed precisely and the direction and distance to the reflecting object are determined.

The wavelengths chosen for the impulse must represent a compromise: longer wavelengths penetrate clouds better but are reflected only by the larger raindrops, snowflakes, or hail. Most radar sets can detect only the larger precipitation particles, but they permit a meteorologist to look inside or X-ray a cloud. Thus the shapes and locations of the precipitation regions within clouds can be located and their development and movement traced, even though the clouds themselves do not generally show up on the radarscope. In this manner, it has been shown that some rain is formed by the

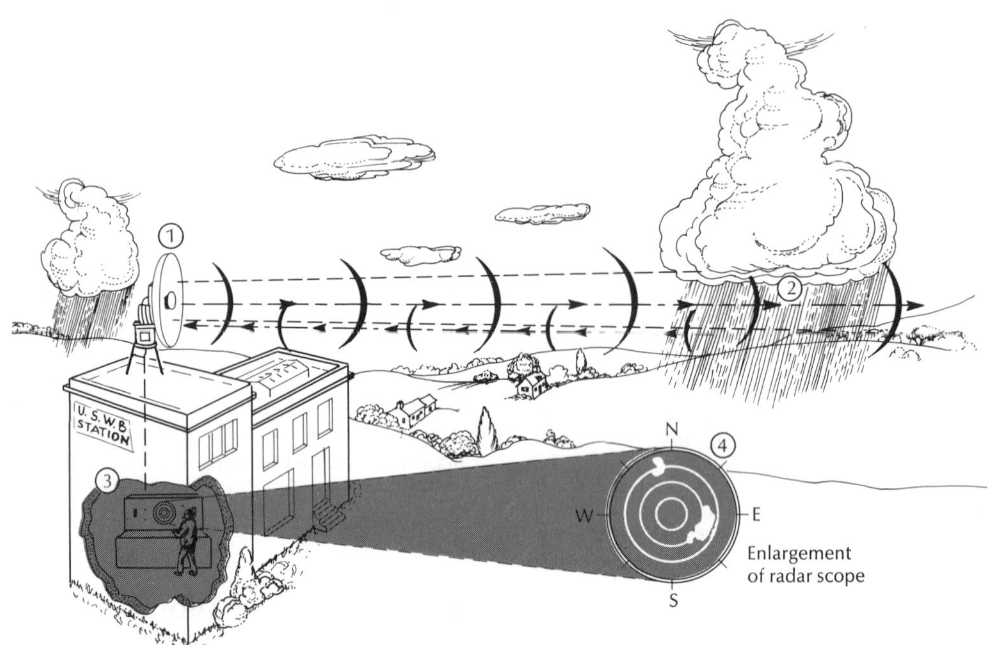

27-29. Powerful bursts of radiofrequency energy (large arrows) are emitted from the rotating radar antenna (1), and a portion (small arrows) is reflected back by precipitation particles in the storm (2). The pulses are transposed onto a radar console (3) by electronic circuitry. The direction and distance of the storm can be read from the radarscope (4). A second storm is also shown to the northnorthwest. *(From Daily Weather Map of 23 August 1962, courtesy U.S. Weather Bureau.)*

melting of snowflakes—note the *bright-echo band* in Fig. 27–30.

Although nimbostratus clouds have a uniform appearance as seen from below, radar study has shown that some consist of three main layers: rain clouds in the lower part, clouds of snowflakes in the middle portion, and ice crystal clouds in the upper and coldest zone. In the parts of the cloud where precipitation is heaviest, these zones merge into one continuous mass. These findings support the cold rain process.

Radar has also shown that other raindrops did not form this way but by the coalescence of liquid droplets because the precipitation developed within clouds at levels where temperatures were warmer than 32°F.

Attempts are being made to use radar to estimate the quantity of precipitation, but difficulty is being encountered in correlating the strength of the signal with numbers and sizes of raindrops present and with their rate of fall.

Weather Modification*

Modern attempts to modify clouds and increase rainfall are based upon the processes that seem to produce cold rain and warm rain. Thus certain techniques are tried if supercooled clouds are present, but different techniques are used where only warm

* 1. Tribus, Myron, "Physical View of Cloud Seeding," *Science*, 10 April 1970. 2. Simpson, Joanne, and Woodley, W. L., "Seeding Cumulus in Florida: New 1970 Results," *Science*, 9 April 1971.

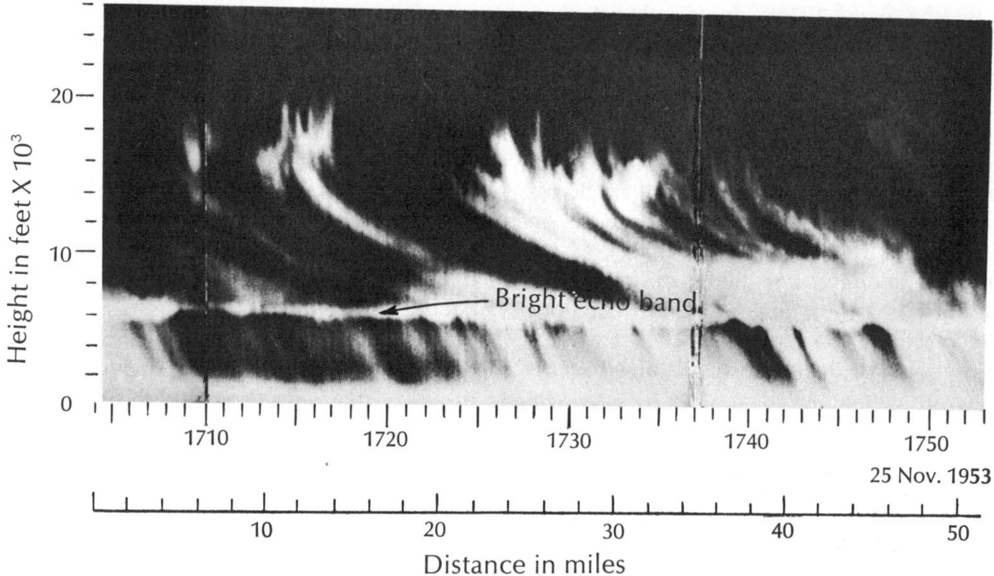

27-30. Height-versus-time record of radar echoes from a cloud system that passed above Montreal, Canada. A vertically pointing 3-cm radar set was used. In calculating the distance scale, the weather pattern was assumed to move at a velocity of about 60 mph, which was the wind velocity at 18,000 ft. Note the bright-echo band located at an altitude of about 6000 ft; this occurs just beneath the melting level of 32°F. Echo streaks slant more steeply below the bright-echo band than they do above it. Experimentation has shown that small reflecting particles (i.e., smaller than the wavelength being used) are about five times more reflective as liquids than as solids. Thus the radar record of this storm seems to show that snowflakes formed in cold air at high altitudes and fluttered slowly downward. They melted at an altitude of about 6000 ft as they fell into warmer air. The raindrops thus formed gave a much stronger radar reflecting signal, fell more rapidly, and produced steeper echo streaks. *(Courtesy Stormy Weather Group, McGill University.)*

clouds exist (temperatures are above freezing).

Precipitation does not fall from many supercooled clouds, which suggests that suitable freezing nuclei are absent in effective numbers. Thus man has attempted to add such nuclei to the clouds directly (e.g., by adding particles of silver iodide from the ground or air) or indirectly by adding pellets of Dry Ice (solid carbon dioxide at a temperature of about −79°C). As a pellet falls through a cloud, it chills the air in its path enough to produce ice crystals, and these spread through the cloud as freezing nuclei. Each pellet of Dry Ice may cause the development of millions of tiny ice crystals. Silver iodide and Dry Ice are effective only in supercooled clouds.

In convective clouds at above-freezing temperatures, different techniques have been tried in attempts to initiate the coalescence process by producing a number of large cloud droplets that will fall at different rates from those of the bulk of the cloud droplets; e.g., giant salt nuclei have been added to the clouds, and water droplets have been sprayed into the clouds from a plane.

The extreme variability of precipitation—from one area to another, from one day to the next, and from year to year—makes it difficult to assess results. Furthermore, cloud seeding is commonly done under conditions in which precipitation is possible. Therefore, if precipitation does occur, it was not necessarily a result of the seeding, and some experiments have emphasized randomization of choice.

By the early 1970's, meteorologists had increased their knowledge of the processes by which precipitation is produced in clouds sufficiently to consider some aspects of weather modification quite promising. Random seeding of random clouds has tended to be ineffective—rather, particular clouds have to be chosen, and they have to be seeded at particular times and places. Mathematical models may be used to predict the results of seeding. If the predictions check with the actual results, then an understanding of the processes involved seems to have been achieved.

Natural freezing nuclei occur in clouds, and those effective at low temperatures (say colder than −20°C to −25°C) tend to be abundant. Therefore, seeding such clouds in an attempt to increase precipitation may actually have the opposite effect. So many freezing nuclei would then compete for the available moisture that most individual crystals would remain small and produce only small raindrops or snowflakes, which would tend to evaporate before they reached the surface. Such seeding can apparently decrease precipitation or redistribute it farther downwind.

Favorable circumstances for increasing precipitation occur where moisture-laden prevailing winds must climb over a mountain barrier, and ground-based silver iodide generators may actually be located within such clouds. Such conditions exist in the western third of the United States (Coast Ranges and Rocky Mountains) and perhaps precipitation can be increased by 10 to 15 percent or more under these conditions. In the growing of many crops, a small increase in precipitation at a critical time may make a great deal of difference in the success or failure of the crop, and such modification seems possible in certain areas.

An experiment to increase snowfall in Colorado (to increase subsequent runoff) has produced promising results to date. By selective seeding on days when cloud temperatures were favorable (between −12°C and −25°C), when few natural nuclei were present, and when moisture was abundant, the snowfall was apparently increased by 2 or 3 times. Seeding on cold days, on the other hand, tended to reduce snowfall.

Another recent experiment to increase precipitation has involved the so-called dynamic seeding of single cumulus clouds in Florida. Large quantities of silver iodide crystals have been dropped (via pyrotechnic flares, Fig. 27–31) into the supercooled tops of such clouds. Extra amounts of latent heat energy apparently were released as a result of this massive seeding and tended to increase cloud buoyancy—the clouds tended to grow larger and more rapidly, and lasted longer, than if they had not been seeded. Increased precipitation has resulted during the 1968 to 1970 experiments—probably some three times more rain came from the seeded clouds than from nearby unseeded control clouds. The increase in rainfall apparently is caused by the increased sizes of the clouds rather than from any significant modification of the processes taking place within them. Fair days were more favorable than rainy days.

The very largest effects of seeding cumulus clouds have come from so-called mergers, although results are still tentative. In one experiment, two clouds were a short distance apart, and one was seeded. As the clouds grew, they merged resulting in a dramatic increase in their size and precipitation (relative to nearby, unseeded, control clouds)—on that particular day, the merged system produced as much rainfall as 36 unseeded single clouds.

Man may unknowingly be seeding the atmosphere by driving his car because even a partial, monomolecular layer of iodine on the surface of a submicroscopic particle of lead oxide from an automobile exhaust can turn the particle into an active freezing nucleus. When introduced into a laboratory-produced supercooled cloud, such nuclei grow into large ice crystals quickly by sublimation of water vapor molecules onto them. The process has been described as akin to a golf ball swelling into an Empire State Building in a matter of seconds. One gram of iodine may activate as many as 10^{18}

27-31. Development of a seeded cloud on 20 August 1963 over the Caribbean Sea. Top left, time of seeding. Top right, 9 minutes later just before maximum vertical growth. Bottom left, 19 minutes after seeding. The horizontal "explosion" is under way. Bottom right, 38 minutes after seeding, the cloud has attained giant proportions. The pyrotechnic silver iodide generator shown in the center was developed by the Naval Ordance Test Station. A number of these generators are dropped at intervals from a plane and in seconds fill a cloud with a vast number of large silver iodide particles. (J. S. Malkus and R. H. Simpson, "Modification Experiments on Tropical Cumulus Clouds," *Science,* **145,** 541-548, 7 Aug. 1964.)

lead oxide particles. Thus the addition of iodine to supercooled clouds containing concentrations of lead oxide particles may be an effective new technique in cloud seeding.

Fog dispersal (Fig. 27–32) is another facet of weather modification. We recognize two kinds of fog: supercooled liquid droplets at below-freezing temperatures make up a *cold fog,* whereas liquid droplets at above-freezing temperatures make up a *warm fog.* Attempts to disperse fog involve attempts to redistribute the moisture by "making big ones out of little ones." This improves visibility, and big enough particles may fall out of the fog. Success has been achieved with cold fogs that are "snowed out" by adding silver iodide or Dry Ice to them.

Warm fogs may make up about 95% of the fog reported at airports in the conterminous United States and have proved more difficult to disperse. Some success has been achieved with chemicals that promote coalescence and also with so-called fog

27-32. Fog dispersion over airport runway near Elmira, New York, 16 October 1968. Top photograph was taken 1 min prior to aerial seeding of the fog with sodium chloride. The white line in the middle photograph is a salt plume from seeding aircraft. Planes and hangars show clearly in the bottom photograph taken 15 min after seeding. The test was conducted by Cornell Aeronautical Laboratory for NASA. *(Courtesy NASA.)*

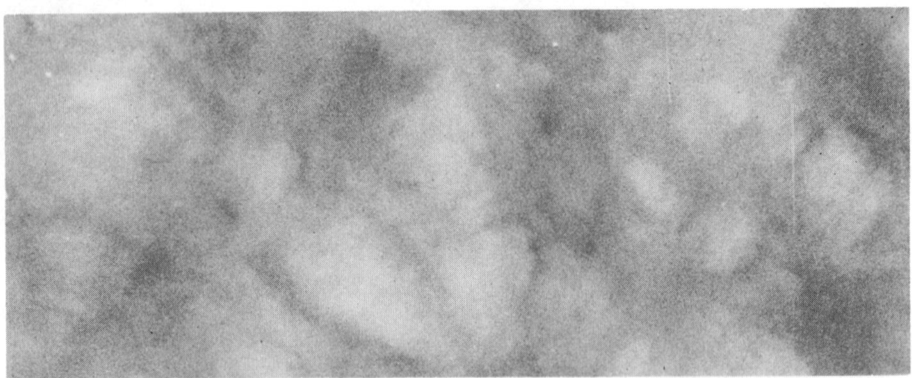

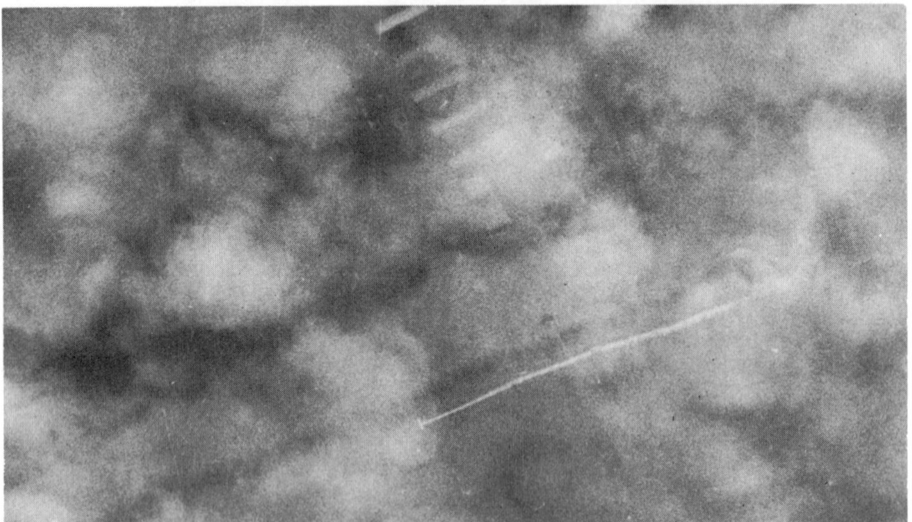

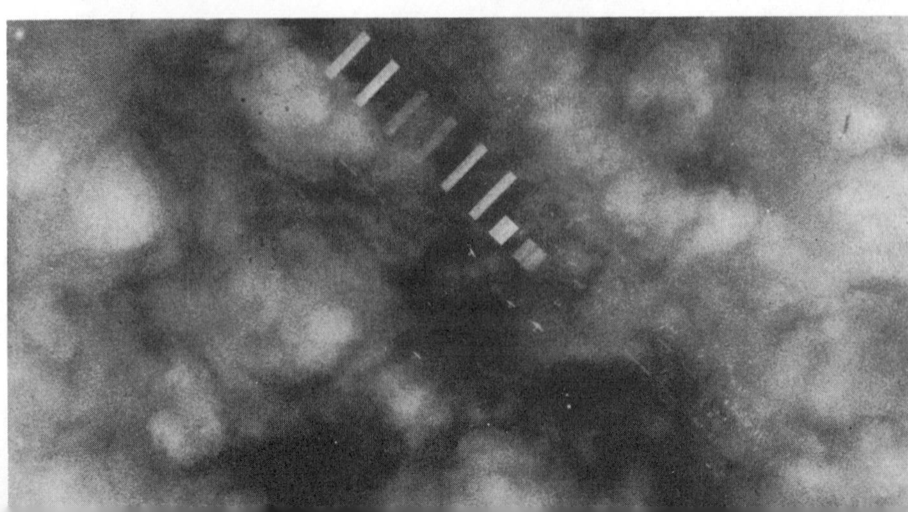

brooms. These are rotating frames strung vertically with many fine nylon lines. The brooms have dispersed laboratory fogs, apparently by jostling droplets into collisions and coalescence. In fact, similar frames located in an arid part of coastal Chile in the path of fogs which frequently move inland in the evening may provide a source of drinking water (water droplets drip down the nylon lines into receptacles).

Helicopters can be used to dispel shallow warm fogs if dry air overlies the fog; their rotors force the dry air downward to mix with the fog layers and cause the droplets to evaporate (used in rescue of downed pilots in Vietnam). Localized heating at an airport has also been used to clear fogs, but the method is expensive and creates turbulent landing conditions.

Cloud seeding downwind from Lake Erie may have aided ski areas located still farther east by increasing the number of ice crystals in clouds that are in the process of producing snow. The extra numbers of freezing nuclei provide competition for the available supply of water vapor molecules and resulting ice crystals tend to be smaller. Thus they float longer and are carried farther by the wind; some apparently reached the ski areas.

Various attempts at suppressing hail have been made. The basic plan involves seeding a potentially dangerous cumulonimbus cloud with silver iodide in an attempt to increase the number of freezing nuclei that compete for the available supply of moisture. Thus many small hailstones (or snowflakes) may form rather than a relatively few large ones, and these may melt between cloud and ground. Also, the impact energies of hailstones correlate approximately with the fourth powers of their diameters. To be effective, the silver iodide must be added to a cloud at a certain critical time, and artillery shells and rockets have been used to do this.

A program to reduce cloud-to-ground lightning discharges to prevent some forest fires is also underway by the U. S. Forest Service. Laboratory experiments have shown that irregular ice crystals decrease the spark potential across a gap by up to 40%. Very large additions of silver iodide particles may produce such ice crystals in effective quantities. Another approach attempts to prevent electrical charges in a storm from building to the lightning stage by adding chaff (strands of metallized nylon thread) to the clouds; the chaff may weaken the electrical field by causing many local discharges.

Experiments to modify hurricanes are also being made (p. 765).

Types of Precipitation

Sleet—frozen raindrops—can form if raindrops fall through a subfreezing layer of air near the surface and are frozen into ice beads (see warm and occluded fronts, p. 725). These bounce when they hit the ground and may form a white layer.

Glaze is an ice coating and not a type of precipitation. It forms when cold drizzle or raindrops (perhaps supercooled) strike near-freezing or subfreezing objects at the surface and freeze into clear ice. Such freezing rain occurs commonly at a particular location as a transitional stage between rain and snow. Glaze makes walking and driving treacherous, and occasional ice storms cause a great deal of damage.

The ice may coat wires, twigs, trees, cars, and other objects, and the mass of the ice may at times be some twenty times greater than that of the wire or twig it surrounds. Such cylindrical deposits around wires have been measured that were 8 inches in diameter and 11 lbs/ft. As much as 5 tons of ice may coat a single evergreen tree about 50 feet tall. Wires and power poles break, trees bend, branches snap off, and houses may be without electricity for hours or even for days. A single ice storm may cause millions of dollars worth of damage.

Rain. Surface tension (molecular attraction) causes the familiar spherical shapes assumed by droplets of water or of mercury. It shrinks the size of a water droplet so as to leave the fewest number of molecules exposed in the surface layer; i.e., the droplet behaves as if squeezed by an elastic film that pushes inward toward the center from all sides (a stronger pull is exerted on any one water molecule at the surface of a drop by its neighboring water molecules than by relatively distant gas molecules in the surrounding air). In other words, a sphere packs the greatest amount of matter into the least possible space. However, surface-tension forces act effectively only across very short distances, and thus droplets cannot exceed a certain critical size. Although the surface tension of a drop increases with its surface area, and thus with the square of its radius, the volume of a drop increases as the cube of its radius. Thus the relative amount of surface tension available to hold a drop together decreases rapidly with size. Furthermore, the larger drops fall faster, and this tends to disrupt them. Although the smallest raindrops are spherical, the largest—like hamburger buns—are flattened at the bottom and rounded at the top (Fig. 7–35). Raindrops range in diameter from about 1/100 to ¼ of an inch, and their rate of fall in still air varies directly with the drop size. The largest drops fall from cumulonimbus clouds because these have strong upward currents that support the raindrops until they attain a maximum size. Heavy downpours occur from such clouds when the upward currents stop or shift locations, and maximum rates of precipitation tend to result more from increases in the sizes of the drops than from increases in their numbers. The speed at which a raindrop approaches the surface in a downdraft is large—the speed at which the raindrop falls in still air (up to 25 mi/hr or so) must be added to the speed of the downdraft. Rain washes impurities from the air—dust, pollen, soot—and may occasionally be colored by these materials.

Drizzle consists of tiny raindrops (Fig. 27–26) produced in stratus clouds which are relatively thin and located near the Earth's surface. Drizzle droplets fall very slowly or seem to float in the air. Apparently they are small and numerous because the cloud depth is shallow and the droplets fall at approximately similar rates—too slowly for coalescence to produce larger drops. Scattered drizzle-size droplets suggest an origin as raindrops at high altitude and strong modification by evaporation between cloud and surface: some drops shrank a good deal, whereas others completely disappeared.

Most methods of measuring the amount of precipitation are based upon the simple technique of leaving a pail in the open to collect rain or snow. However, this method shows only the total amount. The rate of accumulation may be learned from a weighing gauge which records continuously the increase in the total quantity of the precipitation collected.

Extreme variations may occur in the distribution of rainfall within even small areas, especially in warmer weather. Showers which can wet one field on a farm may miss the adjoining one completely; the shower boundaries can be quite abrupt and the entire shower area is relatively small. Thus a very large number of rain gauges distributed at close intervals are apparently necessary to give accurate total quantities.

One method of measuring the sizes of raindrops involves stretching an old nylon stocking on a frame, coating it with powdered sugar, and exposing it very briefly during a rain storm. High-speed, close-up photography has shown that the diameters of the spots on the stocking are about equal to the diameters of the raindrops that produced them.

Some of the heaviest observed amounts of rainfall reported by the U.S. Weather Bureau are 12 inches in 42 minutes at Holt, Missouri, on June 22, 1947; nearly 31 inches in about 4½ hours at Smethport, Pennsylvania, on July 18, 1942; about 102 inches in 4 days in July 1876 at Cherrapunji, India; and 884 inches in 6 months at Cherrapunji, India, from April to September in 1861. In one location in northern Chile, on the other hand, rainfall averaged only .02 inch per year over a 43-year period.

Hail consists of rounded particles of ice which develop within and fall from cumulonimbus clouds. A hailstone* grows primarily by additions of supercooled liquid droplets to a nucleus—perhaps a frozen raindrop. Larger hailstones commonly exhibit an onionlike structure of alternating shells of clear and opaque ice. The clear ice tends to consist of larger crystals in thicker layers, whereas the thinner opaque layers are made of smaller ice crystals and contain more air bubbles. Larger hailstones may have been shifted about between ascending and descending currents, but smaller stones may form during a single descent.

A key factor in determining whether a hailstone layer is transparent or opaque seems to be the rate at which the supercooled droplets freeze after impact. If freezing is rapid, only small crystals have time to form, air bubbles tend to be trapped, and opaque ice results. If freezing is relatively slow, then larger crystals have time to form, and air bubbles tend to escape—this produces a transparent layer. The latent heat energy of crystallization is released as the droplets freeze, and several factors may influence the freezing rate; e.g., the cloud environment may be basic. If temperatures are warmer than −20°C to −25°C, larger crystals may result, whereas colder temperatures tend to produce smaller crystals. On the other hand, the size of the supercooled droplets may be important—smaller droplets tend to freeze more quickly upon impact. Furthermore, the concentration of water droplets may be important—a hailstone may gather so much moisture when it moves through a region with abundant droplets that freezing is slow. At any rate, hailstone growth seems to be complex, and no simple general pattern has apparently yet been discovered.

Successive concentric shells may be added to a hailstone until it becomes too large to be supported by upward-moving air currents and falls to the ground. Since very large hailstones occasionally fall, they show that extremely powerful updrafts occur in some thunderstorms; e.g., a hailstone about 5½ inches (14 cm) in diameter fell in Nebraska in July 1928. However, only one hailstone perhaps out of 100 to 1000 hailstones exceeds 1 inch in diameter.

* Knight, Charles, and Nancy, "Hailstones," *Scientific American*, April 1971.

On the average, hailstones may cause about $300 million of damage to agricultural crops, buildings, cars, and other property each year in the United States (in a typical year, this exceeds the destruction produced by tornadoes). Hailstones are most common in the middle latitudes, apparently because temperatures are too high nearer the equator and too little moisture occurs in the cooler air nearer the poles. For the United States, May seems to be the month with the largest quantity of hail, although thunderstorms are far more frequent in July. The air is warm enough to contain adequate supplies of moisture in May, and yet freezing temperatures occur at moderate altitudes. However, the very largest stones tend to fall during the warmer months when the moisture-carrying capacity of the air is at a maximum. Hailstorms do not last long, cannot be predicted precisely, and are most capricious in their extent; one man's crops may be destroyed and his neighbor's remain unharmed.

Snow crystals are crystals of ice that form in the air by sublimation. Most snow crystals are variations on a six-sided pattern, but three of the sides are suppressed in some crystals and a three-pointed structure results. Some snow crystals are needlelike. It has been reported that no two snow crystals are precisely identical, but this overemphasizes very minor differences. The regularity shown in snow crystal photographs may be somewhat misleading because a photographer tends to choose the well-shaped crystals.

Snowflakes consist of aggregates of snow crystals and are largest in near-freezing air because individual snow crystals can more readily fuse together at such temperatures. According to Tannehill, huge clots of snow, 15 inches across and 8 inches thick, fell in Montana in 1887! If true, this must have been a fantastic sight. Temperatures are never too low for snow to form, but very cold air has a very low moisture-carrying capacity. On the average, about 10 inches of snow will melt down to 1 inch of water, but the range is very wide (from about 6 to 30 inches).

Snow is a very efficient insulator, especially a new fluffy snow with many tiny air pockets; e.g., temperature measurements at one station showed −19°F at 3 feet above the snow surface, −27°F at the snow surface, and +24°F 7 inches below the snow surface. The upper surface of the snow reflects a very high percentage of the solar radiation falling on it, and the snow itself radiates away heat energy. On the other hand, the heat energy radiated by the ground beneath the snow is trapped. For this reason, water may drip from a snow-covered roof at temperatures below freezing (accentuated by a poorly insulated house). Sled dogs are comfortable if they bury themselves in the snow at night, and temperatures are surprisingly warm inside an igloo.

Although the term blizzard is popularly applied to any heavy, somewhat windy snowfall, a blizzard has been defined as a violent (35 mi/hr or more), intensely cold (20°F or colder) wind, laden with snow mostly or entirely picked up from the ground. It may or may not snow during a blizzard.

According to the U.S. Weather Bureau, the greatest recorded seasonal snowfalls include the 884 inches at Tamarack, California, that accumulated during the winter of 1906–1907 and the 1000 inches of snow that piled up at the 5500-foot altitude on Mt. Rainier in Washington from July 1955 to June 1956. During the New York snowstorm of December 26–27, 1947, 26 inches of snow fell in 24 hours, a record for the city. More than $8,000,000 was spent in removing the snow from the metropolitan area. Additional heavy snowfalls include the 87 inches that fell at Silver Lake, Colorado, in 27½ hours on April 14–15, 1921, and the 108 inches that accumulated at Tahoe, California, in four days in January 1952.

Exercises and Questions for Chapter 27

1. You blow up, seal, and release a balloon on a calm day. Will it rise more readily on a cold day or a warm day? Explain.
2. How many calories of heat energy must be added to 10 grams of ice at 0°C to change it to 10 grams of steam at 100°C? Assume that the ice does not sublimate and that the water (after the ice has melted) does not evaporate until it reaches 100°C.
3. If the surface temperature on a farm in Iowa is 73°F, what is the temperature likely to be at an altitude of 30,000 feet (on the average)?
4. Assume that: (1) surface air temperature is 60°F; (2) relative humidity is 33⅓%; (3) the dew point lapse rate is 1°F/1000 ft; and (4) a parcel of this air is pushed upward.
   (A) At what altitude should clouds begin to form?
   (B) Why does the dew point tend to lapse upward?
   (C) At what altitude would clouds have formed if the surface temperature had been 70°F and the relative humidity 75%?
5. Assume that: (1) surface air temperature is 60°F; (2) the observed lapse rate is 3.5°F/1000 ft; (3) a small parcel of this air is heated to 72°F and rises; and (4) the air is too dry for condensation to occur.

(A) How high will the parcel rise?
(B) What is the temperature at the altitude at which the upward movement stops?
(C) How high would a parcel rise if the air temperature were 80°F and the parcel were heated to a temperature of 96°F?

6. Assume that: (1) the surface temperature at Las Vegas is 110°F; (2) the temperature at an altitude of 5000 feet is 75°F; (3) the observed lapse rate is uniform; and (4) that a parcel of the surface air is given an upward push.
   (A) What is the difference in temperature between the parcel and the air around it at the 2000-foot level? At the 4000-foot level? What is happening to the buoyancy of the parcel as altitude increases?
   (B) What is the term for this condition?
   (C) What would the temperature have to be at the 5000-foot level for the condition of neutral stability to occur?
7. Assume that: (1) air temperature is 80°F at sea level on the windward side of a cone-shaped mountain that forms an island; (2) this air rises along the side of the cone to the top where the temperature is 40°F; (3) condensation began at an altitude of 5000 feet and continued to the top; and (4) the wet adiabatic lapse rate is 2.5°F/1000 ft.
   (A) What is the altitude of the top of the mountain?
8. Assume that: (1) surface air temperature is 70°F; (2) a small parcel of this air is forced to rise; (3) condensation begins at the 5000-foot level; (4) the wet adiabatic lapse rate is 2°F/1000 ft; and (5) the observed lapse rate is 3.5°F/1000 ft.
   (A) What is the minimum level to which the parcel must be lifted before it can rise at an accelerating rate without any additional pushing?
   (B) What is the term used for this condition within the atmosphere?
9. How does a halo differ from a corona, and what may be the significance of a halo as a sign of forthcoming weather changes?
10. If a pail of water were emptied from the top of a skyscraper, what would be experienced by persons on the street directly below?
11. Why does an inversion act as a barrier to any parcel of air that starts to rise upward through the inversion?
12. Describe some of the factors which seem to be involved in the origin of "cold rain" and "warm rain." What role may electric charges play in the origin of raindrops?
13. How does the role played by a condensation nucleus differ from that played by a freezing nucleus in the origin of clouds and raindrops?
14. What is the significance of the bright echo band that has been observed on the radarscope of a vertically scanning radar during a storm?
15. What strategy might a meteorologist adopt in attempting to:
   (A) disperse a warm fog?
   (B) disperse a cold fog?
   (C) increase precipitation from a supercooled cloud?
   (D) decrease the amount of damage caused by hailstones?

28

Air Masses, Lows and Highs, Weather Maps, and Weather Forecasting

Air Masses

An *air mass* comprises a huge section of the lower troposphere (Fig. 28–1) and may extend for hundreds of miles. At any one altitude above the surface, an air mass has nearly uniform temperature and moisture conditions, or these change gradually from one portion to another. Although an air mass tends to be homogeneous horizontally, its temperature and moisture commonly are different at different altitudes.

Four types of air masses are of particular importance in causing weather changes in the middle latitudes: *continental polar, maritime polar, continental tropical,* and *maritime tropical.* These terms indicate the chief physical properties of the air masses: hot if tropical and cold if polar; moist if the air accumulated over water (maritime) and dry if the source region was a land area. Arctic and equatorial air masses have also been named, but these are similar to polar and tropical air masses respectively and have less effect upon the United States. Another type of air mass (superior) is discussed briefly below.

Large sections of the troposphere are more commonly "air-conditioned" in certain regions (air mass source regions: Fig. 28–2) where air rests upon, or moves slowly across, a large uniform surface, either land or sea. Such favorable areas are those of high pressure where air sinks and diverges slowly along the surface. This commonly occurs along the horse latitudes and poleward from the polar front, and for this reason air masses are of two main types: polar and tropical.

As air moves across a large area with rather uniform surface conditions (particularly temperature and moisture), it may reach equilibrium with the surface beneath it in a few days or a week or two. Further changes in the air mass are quite slow. Equilibrium develops between the air and the surface by exchanges involving turbulence, convection, evaporation, condensation, and radiation (from the air as well as from the sun and Earth's surface). Equilibrium is attained more rapidly over a warm surface than over a cold surface because of mixing produced by turbulence and convection.

If the surface is cold—e.g., the snow and ice of the polar regions in winter—more heat and energy is radiated into space than is received at the surface. Clouds may reradiate some of this energy back to the surface, but little sunshine occurs during winter (long winter nights at high latitudes). Thus the surface becomes colder. The air overlying the surface likewise becomes chilled and for a similar reason—it radiates away more heat energy that it receives. Although radiational cooling of this sort is a relatively slow process, a stable air mass eventually forms above the snow and ice (cold and bottom-heavy), and an inversion commonly develops (p. 691). Temperatures increase from the surface upward to a maximum at an altitude of about ½ to 1 mile; above this, temperatures decrease again.

28-1. Hurricane Camille over Virginia on 19 August 1969. This storm caused damage of $1.4 billion in the United States. Note another hurricane over the Atlantic and cloud systems elsewhere, especially the counterclockwise circulation around the lows in the upper left and upper right. *(Courtesy NOAA.)*

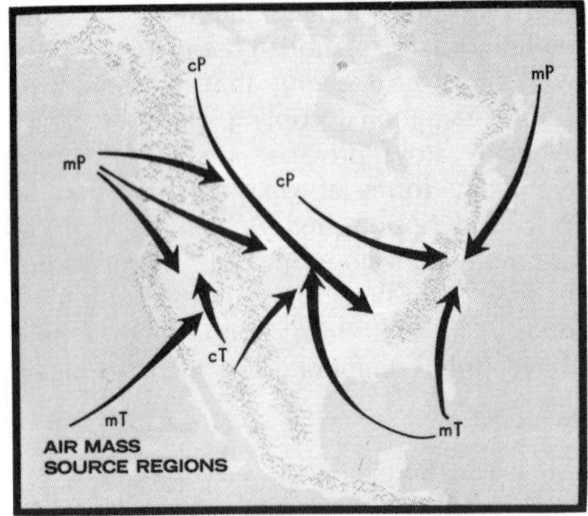

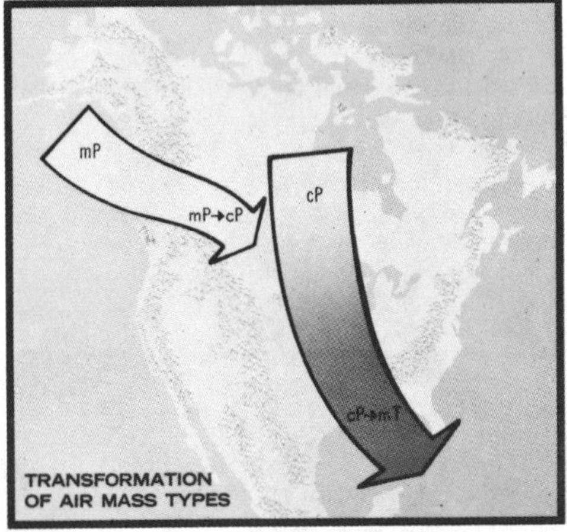

28-2. (A) Air mass source regions and common directions of movement. (B) An air mass may move from one source region, stagnate over a second, and be changed into a different type of air mass. (*Fire Weather,* U.S. Dept. of Agriculture, Handbook 360, 1970.)

Although little moisture occurs in continental polar air, its relative humidity may be high because of the low temperatures.

If the surface is warm and moist, then other processes assume prominence in creating an equilibrium. Air is mixed because certain areas are heated more strongly than others. This causes a circulation which distributes the heat energy upward through a greater thickness of air. If evaporation occurs, water vapor is added to the air, and heat energy is absorbed in the process; on the other hand, if condensation occurs, then heat energy is released. Precipitation may take place, and this reduces the moisture content of the air.

Such air masses may move hundreds and even thousands of miles from their source regions and still retain enough of their original characteristics to be recognizable.

When continental polar air leaves its source area in Canada to travel southward into the United States, its arrival can be recognized readily even by the nonmeteorologist. It produces a cold wave in the winter and sends fuel trucks scurrying quickly from one house to another. In the summer its vigorous gusty arrival from the northwest may herald a welcome relief from a preceding heat wave. Occasionally tongues of continental polar air reach southward into Florida. The Earth's surface in the United States commonly is warmer than such air, and some parts are warmer than others. As a result, the bottom air is heated and rises, not uniformly, but in great ascending bubbles or columns (thermals) something like the boiling of water on a much enlarged scale.

A typical spring day in New York inside a mass of polar Canadian air would dawn with cloudless blue skies. During the morning, solar radiation becomes increasingly intense and initiates the upward movement of large blobs of air. These upward-moving air currents become visible when they reach altitudes at which condensation occurs; white cottony puffs of cumulus clouds develop. By late afternoon the sky may be newly covered by such clouds. However, after sunset, the upward movement decreases, the clouds disappear by evaporation, and a brilliant, cool, starlit evening results.

Pilots and their passengers know when they fly through such currents by the upward bump as they enter and the sudden drop as they leave. Glider pilots can attain great altitudes by circling within these rising masses of air. Dust is carried upward and a clear sparkling day results. Precipitation is not common from such clouds, for little moisture is present. However, cumulonimbus clouds and thunderstorms may develop.

A different reaction is involved when maritime tropical air moves northward and northeastward from the Gulf of Mexico into the United States. The ground in winter is colder than such an air mass, the bottom air becomes chilled, stability develops, and moisture readily condenses into clouds of the stratus type. Precipitation is common along the forward edge of such an air mass (see description of a warm front below).

The shape of the North American continent and its topography—wide in the north, narrow in the south, and mountainous along its eastern and western margins—explains why continental polar and maritime tropical air masses are the two most important to affect the United States, especially its central and eastern portions. The central portion of North America has been described as a great atmospheric highway, stretching from the Arctic to the Gulf, along which air masses of a strikingly different nature migrate from their source regions into the middle latitudes and then move generally eastward in alternation. Thus cold dry weather at a midwest location may be replaced in a matter of hours by warm humid weather or vice versa.

However, maritime polar air is also a frequent visitor. It sometimes comes off the Atlantic Ocean onto the northeastern United States as part of a storm called a *northeaster* (because moisture-laden, precipitating winds, swirling counterclockwise around a coastal low, blow off the ocean from the northeast onto the land). However, maritime polar air more commonly originates as a continental polar air mass over Siberia. In moving eastward across the Pacific in the belt of westerlies, its lower portion is warmed and becomes more moist. Under these conditions, cumulus clouds and showers are common, and heavy precipitation occurs on the western sides of the Coastal Ranges and Rocky Mountains.

Continental tropical air is less common in the United States, but it may develop in the summer over the arid and semiarid southwestern states when the horse latitudes have shifted northward. It tends to cause droughts and to remain in the general area in which it formed.

Superior air masses are relatively warm and very dry and are common at altitudes of a mile or more at different times of the year above large areas of the United States. Although superior air masses may occasionally rest on the surface (their physical conditions are then similar to continental tropical air), they are most common as upper-air masses and have upper-air sources. They form at times from air that has subsided along the high-pressure horse latitudes and tend to produce an upper-air inversion (p. 691). A maritime tropical air mass commonly occurs between a superior air mass and the surface. Superior air masses are warm and dry, especially in their lower portions, because the air in them has subsided and thus been compressed, heated, and "dried out."

The type of weather an air mass brings to a certain location depends upon several factors: the conditions in the source area that produced the air mass, the conditions in the areas over which it has traveled since leaving the source region, and finally the rate at which it has moved. As an air mass moves away from a source area, it may be modified by heating or cooling from the surface, by the addition or subtraction of moisture, and by mixing resulting from turbulence, subsidence, and uplift. It may rise because its lower layers are heated, because it crosses a mountain, or because of the convergence of unlike air masses. Cloudiness and precipitation are likely if the uplifted air mass contains abundant moisture.

Air-Mass vs. Frontal Weather

Most of the United States and the southern part of Canada occurs within the belt of westerlies and has prevailing winds that blow from the southwest and west toward the east. Air from the westerlies, blowing up over the polar front (Fig. 26–18) accumulates at high latitudes. This builds up atmospheric pressures enough to cause frequent outbreaks of huge blobs of continental polar air toward the south, thus shifting the polar front far to the south at this time and place. At other times, pressures in the vicinity of the horse latitudes build up sufficiently to push maritime tropical air northward across the United States and into Canada. Thus the polar front advances and retreats. Once these contrasting air masses move into the United States, they are shifted across it in a generally eastward direction, and a few days of continental polar air mass weather tend to alternate with a few days of maritime tropical weather. Maritime polar air masses from the Pacific are also frequent visitors, but these have been modified somewhat by the time they arrive in the eastern United States.

Because air masses are very large—in some instances they span more than 1000 miles—a few days may be needed for any one air mass to pass by a given locality. During this passage, one experiences *air-mass weather;* the same general weather conditions are repeated each day, although gradual modifications occur from one day to the next as the physical conditions of the air mass change slowly.

Major weather changes occur as the rear section of one air mass moves by a certain location and is replaced by the leading edge of a different air mass. In other words, changeable weather is associated with the relatively narrow, elongated zone along which two different masses of air converge. Thus the more sudden weather changes tend to occur along the margins of air masses, and we may speak of air-mass weather vs. frontal weather.

Fronts

Fronts are the transitional zones or boundaries that occur between two different air masses. Although shown on weather maps as lines, fronts are mixed zones that range in width from 3 to 50 miles, more or less. Temperature and moisture conditions tend to be noticeably different on opposite sides of a front, because the two air masses do not mix readily, and the transition zone (frontal surface) between them is relatively narrow. Lines with appropriate symbols (Figs. 28–22 to 28–27) are drawn on weather maps to show where the base of a front touches the surface. Commonly these lines are convex in the direction of movement and may extend for hundreds of miles. A front may slope upward from the surface for 1 to 2 miles or less, or it may extend to the upper troposphere. One must learn to think of fronts in three dimensions and to visualize the orientation of a frontal surface sloping upward from its location on a map.

Different types of fronts have been recognized: cold, warm, stationary, occluded

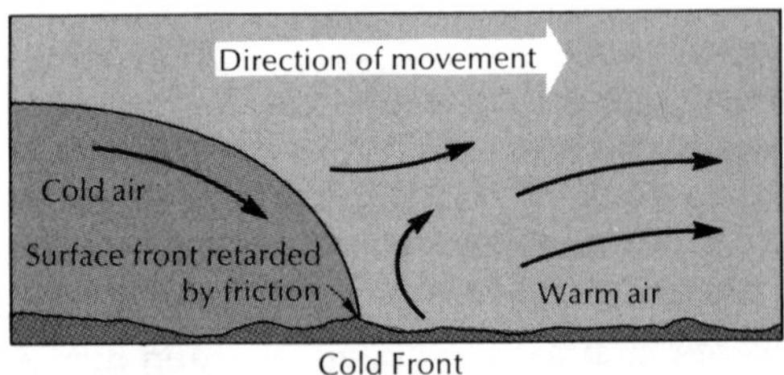

Cold Front

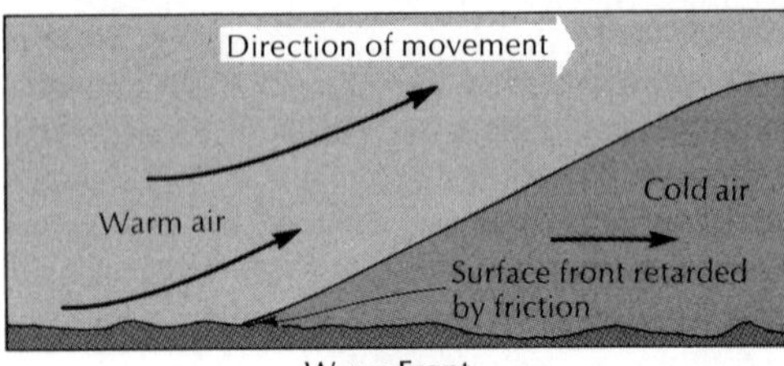

Warm Front

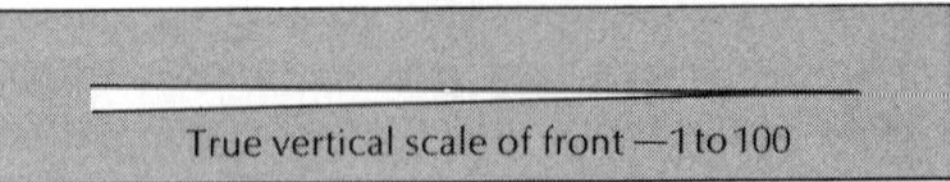

28-3. Vertical cross sections showing frontal slopes. Note that surface friction acts to steepen a cold front but makes a warm front less steep. Note also that a cold front slants backward, whereas a warm front slants forward. Cold fronts and warm fronts have slopes that range from about 1 to 40 through 1 to 400. To emphasize the vertical exaggeration, a 1 to 100 slope is drawn to true scale at the bottom. (TM 1-300, *Meteorology for Army Aviation.*)

28-4. Cold front. *(Official photograph, Environmental Science Services Administration.)*

(two types), and upper air fronts, and each of these may be prominent and persistent or weak and evanescent. An important factor here is the stability or instability (p. 684) of the uplifted air mass. In all fronts the colder denser air occurs at the surface beneath the frontal zone, and warmer lighter air is located above it; therefore, an inversion occurs at a front.

The forward margin of a moving mass of polar air is aptly termed a *cold front,* and certain weather changes are characteristically associated with it (Figs. 28–3 and 28–4). Because cold air is heavier than warm air, it stays near the surface and wedges under the warmer air that occurs ahead of it. This warmer air is shoved upward along the frontal zone and it rises, expands and cools. Towering cumulus and cumulonimbus clouds tend to develop along the forward margin of the cold air, and sudden heavy precipitation is characteristic of the passage of a cold front. From its line of contact with the Earth's surface, the cold front slants backward and upward over the cold air mass at a slope that may average about 1 mile in a vertical direction for each 40 to 80 or more miles along the Earth's surface. Cold fronts tend to move southeastward into the United States.

The amount of precipitation and turbulence associated with the arrival of a cold front depends upon the nature of the two air masses involved and the velocity of the front. Extreme conditions result if very warm, moist, unstable air is underrun by fast-moving, very cold air. If the front moves rapidly (up to 60 mi/hr on occasions but commonly less than half of this rate), the slope at its forward margin is steepened be-

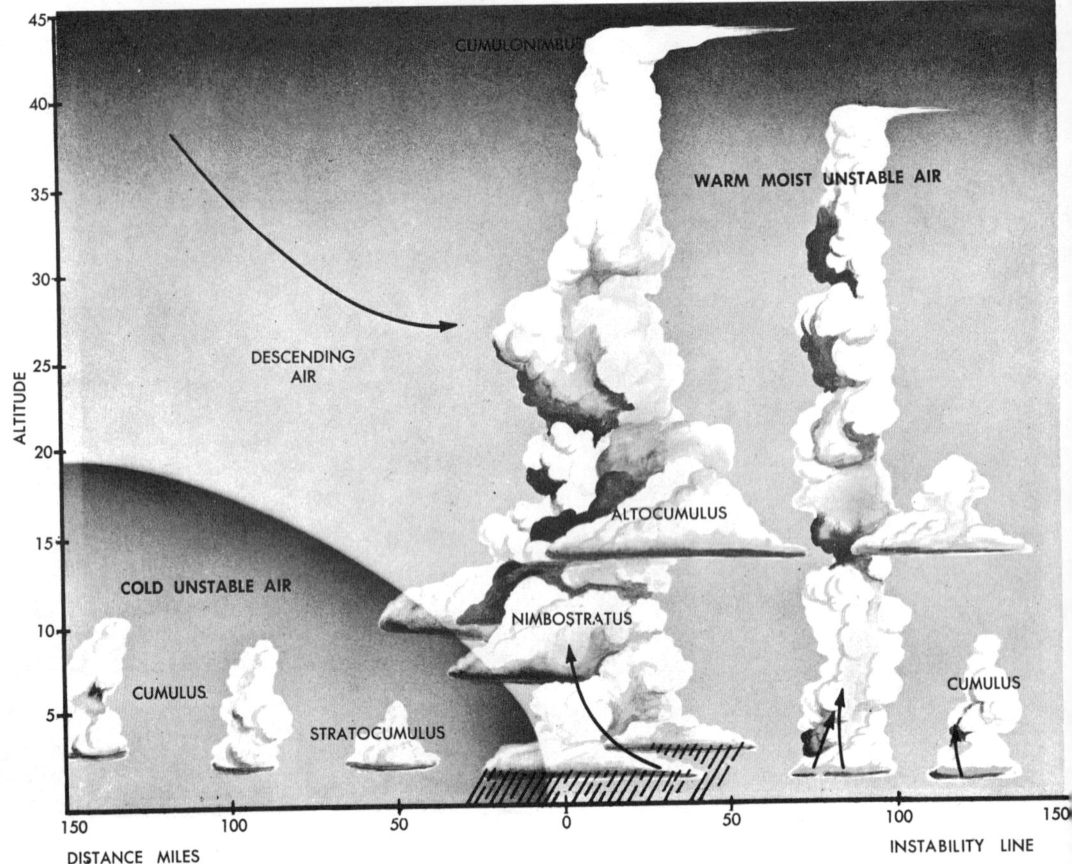

28-5. Fast-moving cold front. An instability (squall) line is shown ahead of the cold front. (AF Manual 105-5, *Weather for Aircrews.*)

cause of friction with the ground, and the warm air is shoved upward vigorously in a relatively narrow zone located chiefly in advance of the cold front (Fig. 25–5).

On the other hand, if a cold front moves slowly, its slope will be less steep, and the warm air ahead of it can slide upward gradually along the frontal surface. Thus layered clouds can form and extend for many miles behind the base of the front (Fig. 28–6). If relatively little difference occurs in the temperature and moisture content of the two air masses, or if the warmer air is dry, then the cold front may pass almost unnoticed.

Clouds commonly develop all along a cold front because of the upward-moving air currents it causes, although they may be absent in places if the air is very dry. Precipitation tends to occur in association with thunderstorms scattered here and there along the front. Newspapers may report erroneously that such showers have "cooled the air" instead of stating correctly that the arrival of cooler air has caused the precipitation.

In a *warm front* (Fig. 28–7) the warm air mass advances at a more rapid rate than that of the colder air in front of it. Thus the warmer lighter air glides gradually upward above the colder air along a surface that tends to slope at 1:100 or 1:300 or more.

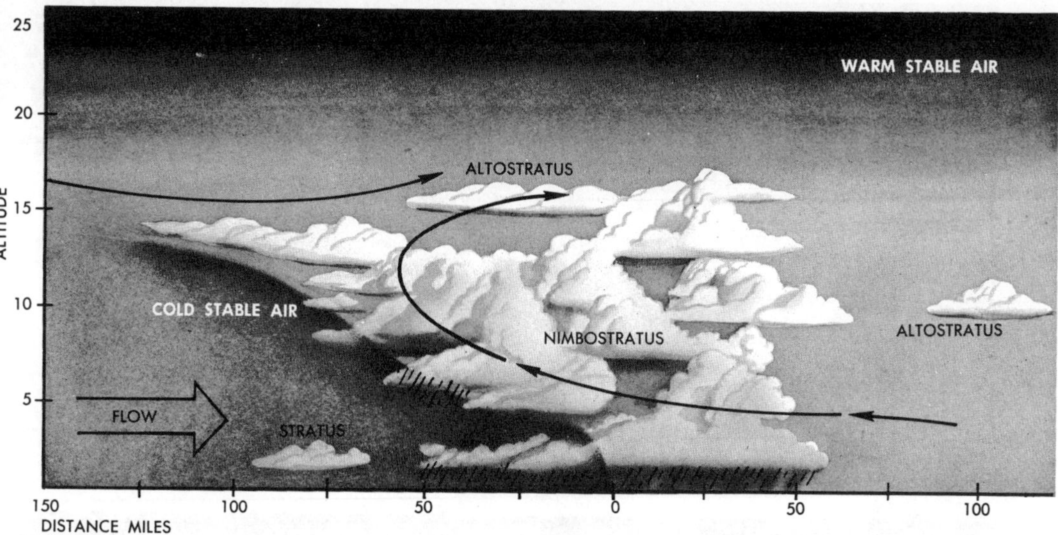

28-6. Slow-moving cold front. (AF Manual 105-5, *Weather for Aircrews.*)

Surface friction tends to slow the motion of the base of a warm front and this decreases its slope. Thus the cirrus clouds at the top of a warm front may be at an altitude of several miles and 500 to 1000 miles ahead of its base.

Layered clouds tend to be most common along a warm front. As a warm front approaches, one observes that thinner, higher types of clouds are gradually replaced by lower, denser varieties. Cirrus is followed by cirrostratus and perhaps by cirrocumulus —the mackerel sky that may be a sign of approaching precipitation. Alto-type clouds come next, and precipitation may begin from the altostratus. Precipitation continues from the nimbostratus clouds that are next in the procession and that may extend as much as a few hundred miles ahead of the base of the warm front.

Rain falling through the cooler air beneath a sloping frontal zone may add enough moisture by evaporation to saturate the cooler air in places and produce fog and stratus or stratocumulus clouds (stratocumulus forms instead of stratus if sufficient convection is occurring in the cool air). Warm front precipitation generally covers a wider area, lasts longer, and is less intense than that associated with a cold front. Many of the same phenomena occur with the cold front, but they are concentrated in a relatively narrow zone.

If warm moist unstable air moves upward along a warm front, cumulus and cumulonimbus clouds may also form; thus thunderstorms occasionally occur in association with warm fronts. As with cold fronts, no two warm fronts are identical, and the weather conditions produced depend upon the rate at which the air masses move and the extent of the temperature and moisture differences between the two air masses (Fig. 28–7). Warm fronts in the United States commonly move eastward and northeastward.

At times a warm air mass may be in contact with a cold air mass and neither air mass actively advances or retreats. In such instances the frontal surface is called a *stationary front*. On a weather map a maritime tropical air mass may be separated from a continental polar air mass by a long sinuous line. This line is marked by warm front symbols in an area where the warm air is advancing; it shows cold front symbols in an

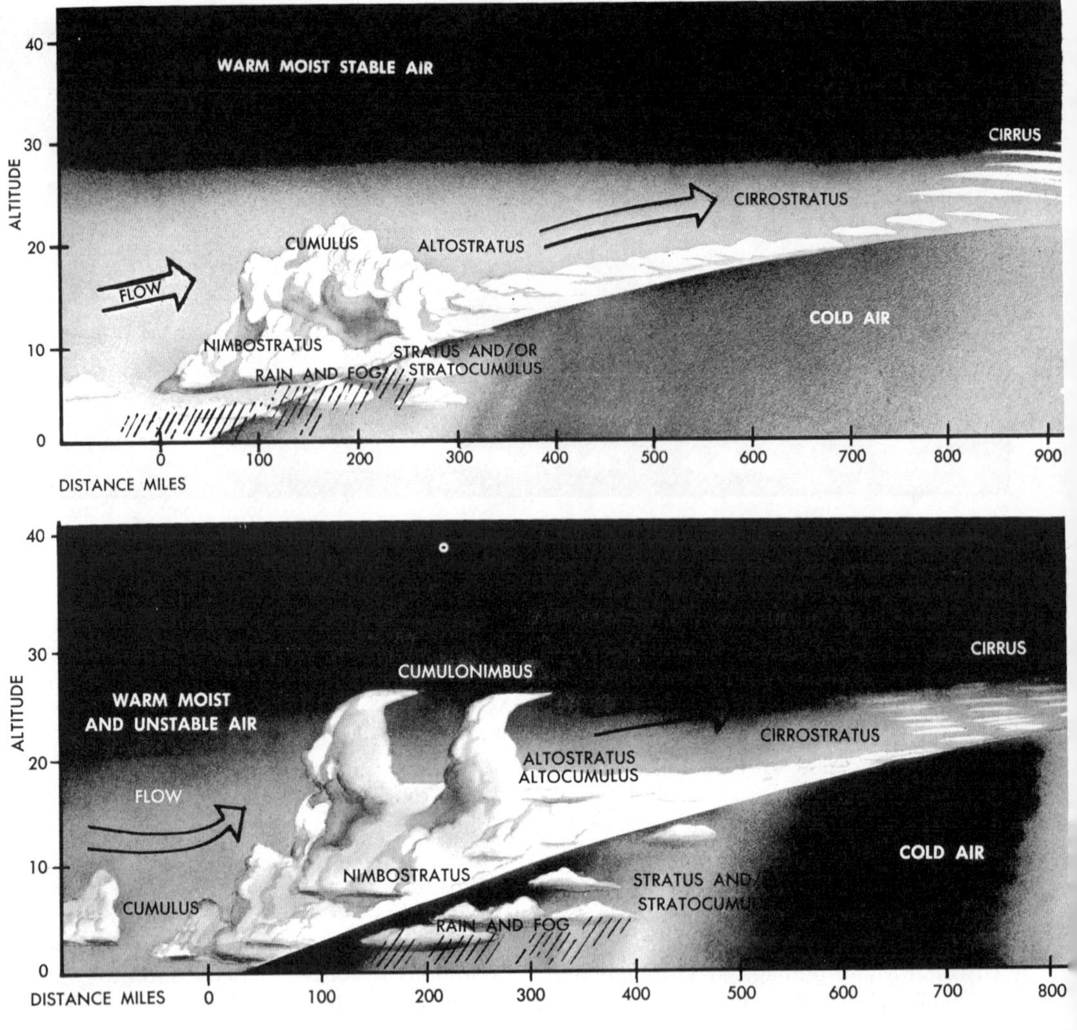

28-7. Two typical warm fronts. (AF Manual 105-5, *Weather for Aircrews.*)

adjoining area where the cold air is moving forward; and it has stationary front symbols in still another area where neither air mass is advancing.

Occluded fronts form because cold fronts tend to move eastward faster than warm fronts (partly because cold air can displace warm air along the surface more readily). Occluded fronts are thus associated with a cold front, a warm front, and three air masses: one warm and two polar (one colder and denser than the other). The warm air mass is located between the two polar air masses, and in the United States all three are moving generally eastward. The faster, eastward-moving cold front may overtake the base of the warm front (its northern portion, Fig. 28–12). The warm air mass is thereby lifted from the surface, the so-called occlusion process, and two types occur. The type is determined by the difference in density (due primarily to temperature differences) between the two polar air masses. If the air behind the cold front is colder

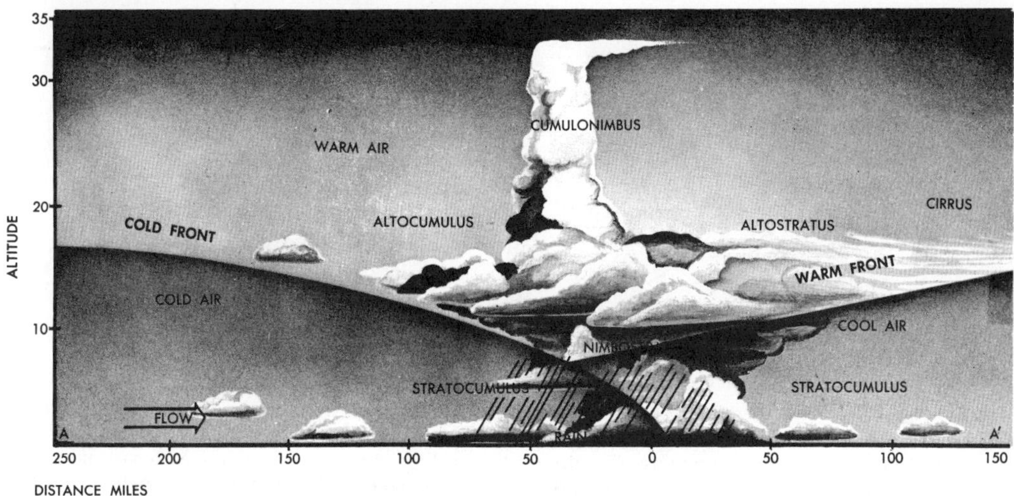

28-8. A cold-front type of occlusion. (AF Manual 105-5, *Weather for Aircrews.*)

and denser than the polar air located east of it and beneath the warm front, then a *cold front occlusion* (Fig. 28–8) occurs—the dense air from the west wedges beneath the other two air masses. On the other hand, if the polar air in the west has the lesser density, then it will move eastward and upward along the warm front and thus over the second polar air mass; this is a *warm front occlusion* (Fig. 28–9).

An *upper air front* is shown in Fig. 28–10 and a *prefrontal squall line* in Figs. 28–5 and 28–11. The cause of squall lines is uncertain, but they are accompanied by violent thunderstorms and tornadoes. A squall line may be a few hundred miles or more in

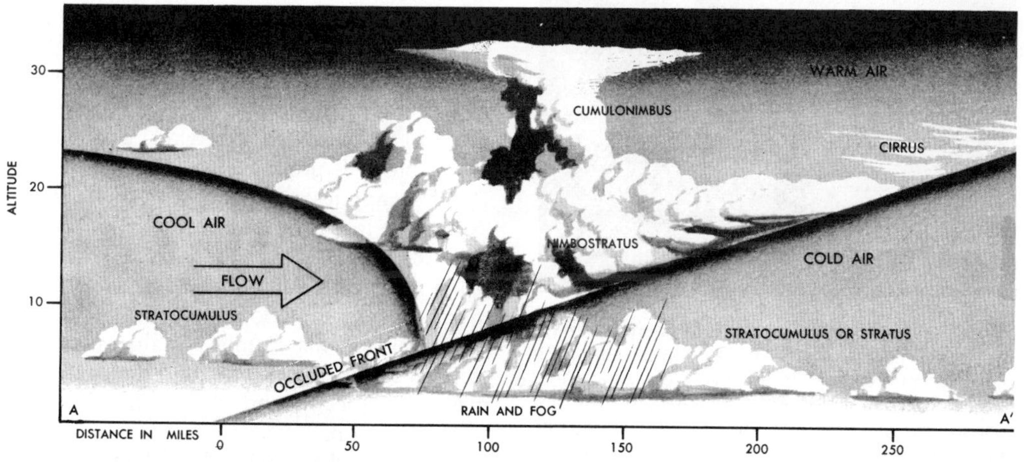

28-9. A warm-front type of occlusion. (AF Manual 105-5, *Weather for Aircrews.*)

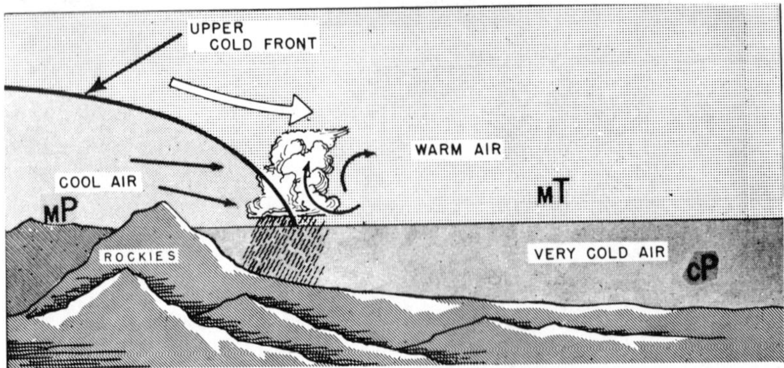

28-10. Upper air cold front. A mass of maritime polar air moves eastward across the Rocky Mountains but cannot descend through very cold and dense continental polar air that blankets the surface eastward from the mountains. Therefore, the maritime polar air continues eastward along the upper boundary of the layer of continental polar air and interacts with warm air in its path (overlying the continental polar air) in a typical cold-front fashion. *(Courtesy U.S. Weather Bureau.)*

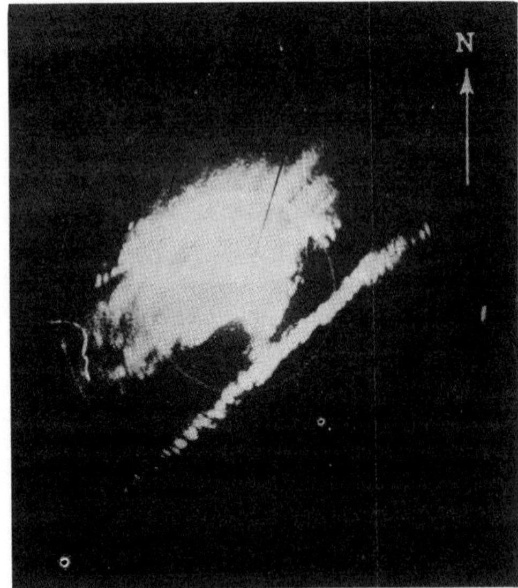

28-11. A well-defined squall line is shown as it was observed on the Hatteras, N.C., radar PPI scope (Plan Position Indicator). The concentric circles are 50 miles apart. As shown, the northeast-southwest trending squall line is nearly separated from a large rain area to the west and north. Both were moving eastward. *(Courtesy U.S. Weather Bureau.)*

length and tends to parallel a cold front that occurs behind it. The squall line may occur some 50 to 300 miles ahead (in the United States, to the east of the cold front). Squall lines often last less than 24 hours and move at 20-25 mi/hr.

To locate a front on a weather map from data collected simultaneously at many weather stations, one attempts to locate a line with marked weather differences on either side: temperature, wind, humidity, clouds, and barometric pressure. Higher altitude temperature measurements may be more representative than surface measurements, which are affected by local heating, cooling, and topography. Changes in wind direction are rather reliable indicators. If one's back is toward the wind, any associated front will probably be toward the left. This also applies approximately to the location of a low-pressure center (Figs. 28–13 and 28–21). Humidity differences may be diagnostic, especially the dew point. This tends to remain about constant within a particular air mass unless water vapor is gained or lost, even though temperatures may vary locally and during a day at a single location.

Isobars tend to make sharp bends where they cross fronts (more so for cold fronts than for warm fronts); thus in constructing a weather map, one should attempt to locate fronts before drawing in isobars. Within the warm sector (p. 732) the isobars tend to be fairly straight and to be oriented about parallel to the direction in which the low-pressure center is moving.

Lows and Highs

Barometric readings from widely scattered stations show that atmospheric pressure varies considerably at any one time at different stations and at any one station over a period of time. However the pressure differences between two cities such as New Haven and Baltimore are commonly less than the difference in pressure between the top and bottom floors of a tall building. Horizontal pressure changes near sea level seldom are greater than about 5 mb/100 miles (3 mb/100 km).

When lines (*isobars*) are drawn on a map through the locations of stations which have the same barometric pressure (after corrections which reduce all of the readings to sea level), definite centers of low or high pressure commonly appear. These pressure systems are large; diameters range from a few hundred miles to more than 1000 miles. The *lows (cyclones)* and *highs (anticyclones)* have characteristic conditions of wind, temperature, clouds, and precipitation associated with them.

Lows probably form in more than one way, and a common manner of origin seems to be in association with the polar front (Fig. 28–12). According to an older view, a stationary front may separate continental polar air in the north from maritime tropical air in the south. The air masses move parallel to the front and do not mix readily. However, at some irregularity along the boundary, the warm air may begin to flow toward the cold air and upward. This northward bulge develops into a warm front. The cold air on the north side of the irregularity pushes southward, forming a cold front. The reduction in pressure is associated with: the substitution of warm light air for cold heavy air ahead of the warm front, the upward movement of the warm moist air, and condensation in this rising air which releases heat energy that aids the upward movement.

From the region surrounding the developing low-pressure center, air moves inward along the Earth's surface and is deflected to the right of the direction of movement in the Northern Hemisphere until it is blowing at an angle of 10 to 50 degrees to the isobars. Thus a counterclockwise circulation originates about the low-pressure area and winds spiral inward and upward around the center.

Lows tend to be elliptical in shape and to be elongated in a northeast-southwest direction. Eastward and northeastward movement of lows in the United States averages about 20 mi/hr in the summer and 30 mi/hr in the winter (about 500 to 700 mi/day). Clouds and precipitation are common because warm moist air moves upward near the low-pressure center.

It was once thought that lows developed because of convectional circulation in a warm air mass. This hypothesis was discarded when it was learned that lows are more common in winter than in summer and that they form over the oceans as well as over the lands—neither condition is favorable for the heating of the bottom of an air mass to produce convection. Also, two or more air masses and two or more frontal systems (warm, cold, and occluded) are associated with mature lows.

High-pressure areas are commonly larger than low-pressure areas, but the differences in pressure from their margins to their centers is usually less. In a high-pressure system, air subsides at the center and diverges along the Earth's surface outward from the center; it is deflected to the right and spirals outward and around the center in a clockwise circulation in the Northern Hemisphere. A high-pressure center commonly forms as part of a continental polar air mass that moves southeastward into the United States from Canada and thence east-

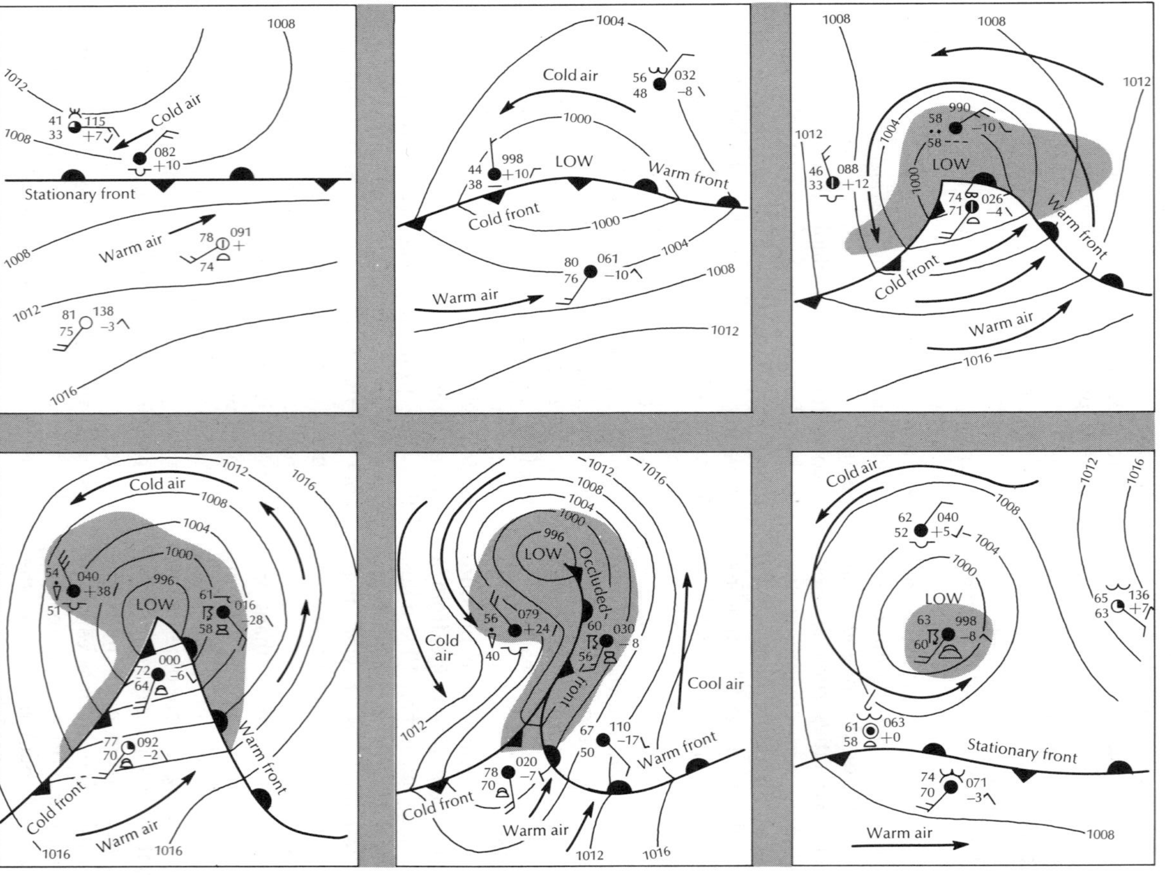

28-12. Six stages in the development of an idealized, middle latitude low (cyclone). Occlusion occurs because cold fronts generally move eastward faster than warm fronts. The warm air sector is thus narrowed and eventually eliminated at the surface. The whirling counterclockwise circulation then weakens in the colder air. Note the location of the low-pressure center at the intersection of cold and warm fronts and later at the northern end of the occluded front. Although many paths are possible, stage 7 might be in Texas, stage 10 near the Great Lakes 24 to 48 hr later, and stage 11 over the Gulf of St. Lawrence. *(Courtesy U.S. Weather Bureau.)*

ward over the Atlantic. Other high-pressure centers develop farther south in warmer air, but less is known about them. They may produce long hot dry spells during the summer months. As subsidence and divergence occur in highs, they tend to be associated with fair weather.

According to some rather recent ideas, the polar front jet stream (p. 663), some miles above the surface, may actually be the primary cause for the development of some (perhaps most) lows and highs. Air movements along fronts may actually be secondary effects. Following this view, moving air currents within a jet stream cause centers of divergence to develop at certain times and locations, whereas centers of convergence form at other times. If a center of divergence originates within the polar front jet stream,, air is transferred outward from this center, and pressures will drop all of the way to the surface. Less air now occurs in the column above the surface at this location. Since this diverging air must be replaced by air from elsewhere, new supplies tend to come from the area beneath the center of divergence, and this produces an upward current that leads to the formation of a low-pressure system.

In a similar manner, if convergence occurs in some part of the jet stream, the accumulation of air at this center will increase pressures all of the way down to the surface and cause subsidence. A high-pressure system may form at the surface in this manner.

As air moves horizontally from high- to low-pressure areas because of the pressure differential, winds are strong when the difference in pressure is great. If the pressure lines (isobars) are close together on a map, winds are strong, and the pressure system is well developed. On the other hand, if the pressure lines are far apart, winds tend to be gentle, and the pressure system is not intensively developed.

Along the surface, winds tend to blow at acute angles to the pressure lines (about 30 to 40 degrees on the average). At a higher altitude, winds tend to parallel isobars—velocities are greater and the Coriolis effect is more pronounced. Winds also tend to blow more nearly parallel to the isobars over water than over land.

Highs and lows are less common and less intense during the summer months than in the winter and move at a slower rate and along different paths. In like manner, the polar front in winter is located far to the south of its average summer position. In the Southern Hemisphere, movement of air in the low- and high-pressure systems is reversed—clockwise in a low and counterclockwise in a high.

After occlusion, the low-pressure system tends to weaken (i.e., pressures increase at the center). Cold dense air has replaced warm light air along the surface, which increases pressure, and the supply of warm moist air has been cut off from the center.

A Simplified Weather Map

Pressure systems, air masses, fronts, and characteristic weather conditions as they tend to develop in the United States are shown schematically in Fig. 28–13. An observer is located in the eastern part of the United States. A low-pressure center has developed along the boundary separating continental polar from maritime tropical air. Assume that on a particular day such as Monday, this mature low-pressure system is situated about 600 miles west of the observer. A warm front extends southeastward from the center, a cold front extends southwestward, and a center of high pressure is located to the northwest in the continental polar air mass. The air masses, their accompanying warm and cold fronts, and the pressure centers are all moving toward the east and northeast. Weather data collected during preceding years suggest that the center of the low will pass north of the observer's location and that a movement of about 600 miles per day (nearly 1000 km) is typical. The high will probably move southeastward and then eastward; its center may pass south of the observer.

A person at the observation point com-

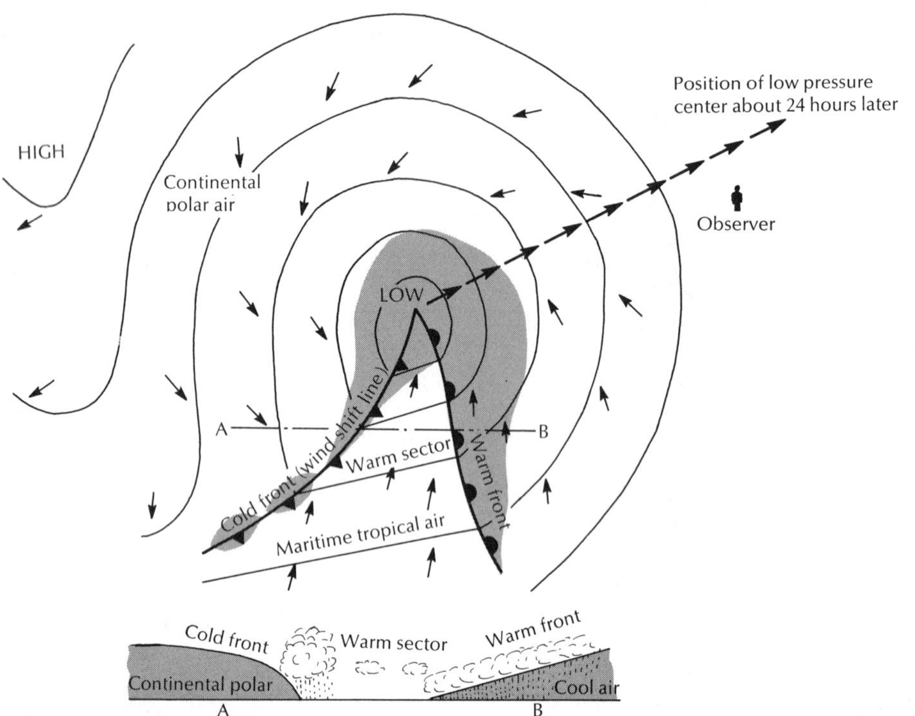

28-13. Diagrammatic weather map shows continental polar and maritime tropical air masses associated with well-developed high- and low-pressure areas with accompanying cold and warm fronts. The future path of the low-pressure center is plotted, with its probable location the next day marked. The high-pressure center will move southeastward and then follow the low to the east. Note the counterclockwise circulation which has developed around the low-pressure system and the clockwise circulation around the high. AB is a cross section through the cold and warm fronts. The cold front commonly overtakes the northern part of the warm front, thereby forming an occlusion. Precipitation is likely within shaded areas. Scattered showers may occur along the cold front.

monly experiences the following weather changes. First, certain phenomena take place which are associated with the approach of a warm front and low-pressure system: the barometer falls slowly; temperature and humidity remain about the same or rise gradually, winds are from the east or southeast, and a few high harmless-looking cirrus clouds appear in the west. Several hours later the clouds are thicker and lower (altostratus follows cirrostratus) and the wind blows from the southeast or south. Precipitation may begin from the altostratus, and it continues and increases when the nimbostratus clouds arrive. The sky then becomes dark and overcast, and rain or snow falls steadily for a number of hours.

A warm front commonly extends for 500–1000 miles (800 to 1600 km) from its foot at the Earth's surface to its top where wisps of cirrus clouds are visible. Therefore, the cirrus and cirrostratus clouds may appear some 12 hours or more ahead of the rain. On Tuesday the lower part of the warm front passes the observer and is followed by the *warm sector*. This is the wedge-shaped area overlain by warm air and located between a warm front to the east and a cold front to the west. It becomes warm and humid, and the sky may clear. The barometer is low and steady, and the wind blows from the south or southwest.

At a later hour on Tuesday the cold front approaches the observer. Its presence is heralded by a high cloud bank capped here and there by huge towering masses (cumulonimbus) which appear in the west and move eastward. Thunderstorms and brief heavy downpours are common as the cold front passes over an area, although the storms occur here and there along the front.

By Wednesday the sky has cleared, and by afternoon it contains white puffs of fair-weather cumulus clouds which sparkle against the blue background. The barometer is rising, and humidity is low; it is cool, and the wind blows vigorously from the northwest. The observer can now predict with some confidence that he will experience these same general weather conditions (air-mass weather) for another day or so as the continental polar mass moves by, although each day should be a little warmer than the previous one (especially after the center of the high has passed eastward, and southerly winds replace the northerly ones). However, a watchful eye should be kept on the barometer, the wind, and on the southwestern sky for the first signs of the approach of the next low. Fig. 28–14 shows a likely later stage after occlusion has occurred.

Many variations centering about these general weather changes are possible, and

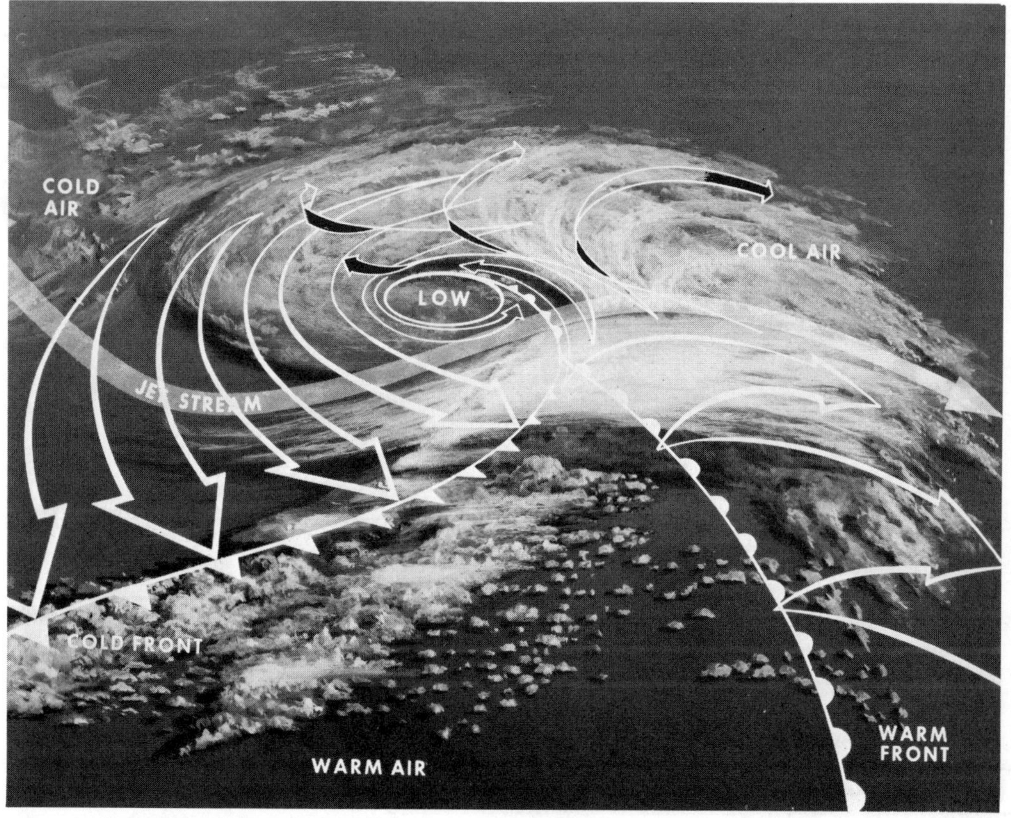

28-14. Artist's view of clouds and fronts associated with a mature low-pressure system. *(Courtesy NOAA.)*

the changes may be quite different locally from those described. The fronts may move at different rates of speed in different directions; the low- and high-pressure systems may be weakly or strongly developed; and the air masses themselves may vary in physical properties. For example, in the Kansas flood of 1951, the front part of a cold air mass remained nearly motionless over Kansas for about four days instead of moving eastward as normally happens. During this time, warm moist air from the Gulf of Mexico moved northward and upward over the cold air, shedding torrents of rain on the surface below.

The rate at which a storm moves is an important factor in the quantity of precipitation that falls on a given location, and a slow rate usually means a greater accumulation. Moreover, a relatively small change in the direction of a storm during the colder months can determine whether a certain city receives snow, sleet, or rain (Fig. 28–15). Thus not all weather forecasts are accurate.

The belt of westerlies has been likened to a huge river which flows in an easterly direction across the United States. In this river, lows and highs are like gigantic moving eddies with their counterclockwise and clockwise circulations. In fact, the eddies are so numerous that they tend to obscure the general eastward movement of the river

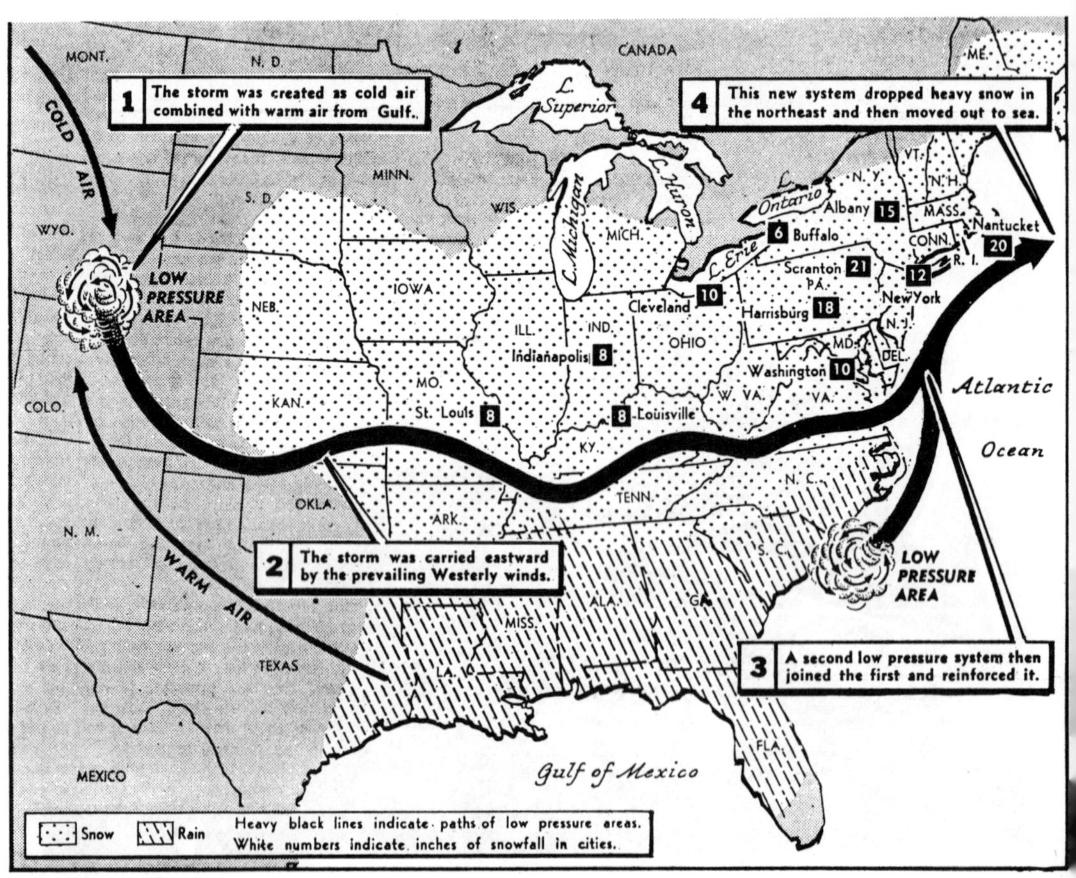

28-15. The development and path of a snowstorm, January 10 to 14, 1964—a fairly typical winter storm system. (Courtesy *The New York Times*, Jan. 19, 1964.)

itself. The eddies develop somewhat periodically and follow each other eastward, some persisting much longer than others. This eastward procession of lows and highs and the passage of large masses of different kinds of air cause the alternation of weather conditions common to the middle latitudes.

Although the general trend of movement is eastward, from the Mississippi basin the most frequented tracks of the weather disturbances are northeastward across the New England states or northeastward to the Great Lakes and then along the St. Lawrence Valley (Fig. 28–16). Some storms cross the Atlantic and travel eastward as far as Siberia.

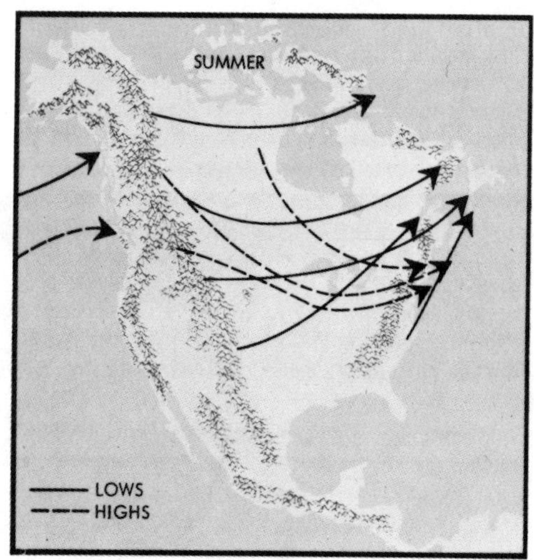

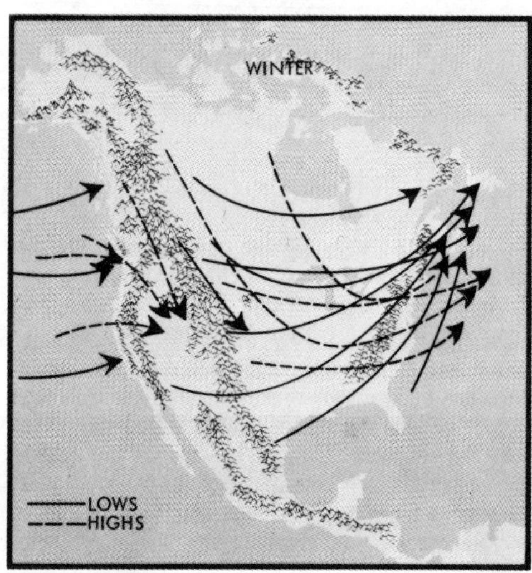

28-16. Common paths of lows and highs are farther north in summer than in winter when the systems tend to be better developed and to move faster. (*Fire Weather,* U.S. Dept. of Agriculture, Handbook 360, 1970.)

Weather Forecasting

Weather forecasting is based upon information concerning the whereabouts of different air masses, pressure systems, fronts, and their movements at any one time. The most important elements of the weather are (1) temperature of the air, (2) atmospheric pressure, (3) direction and speed of the wind, (4) humidity, (5) type and amount of cloudiness, and (6) amount of precipitation. This information is then intepreted in the light of past experience which indicates the paths followed most frequently by low- and high-pressure systems and the rates at which they commonly travel at different times of the year. Here electronic computers are of great value.

In addition, the characteristic reactions of different air masses to various ground conditions must be known as well as a knowledge of winds and atmospheric conditions aloft. In general, accurate predictions depend upon weather data recorded simultaneously by widely distributed weather observers, then collected rapidly, and immediately analyzed by experienced meteorologists at a central office (the synoptic weather map). Such data can now be fed into computers which can then print automatically many of the maps needed in weather forecasting.

Weather forecasting as a public service began about 100 years ago. The invention of the telegraph (about 1840) and radio (about 1900—needed for data from the oceans) were necessary prerequisites to modern forecasting methods. The first radiosonde weather records were made in the 1930's. Despite all that has been done, weather observing stations are still relatively far apart and records are very sparse from a number of very large areas.

Local detailed forecasts in the belt of westerlies for a period of 12 to 18 hours are considered to be about 85 percent accurate, and forecasts up to 36 hours in advance may attain an accuracy of about 75 percent. According to one view, there may

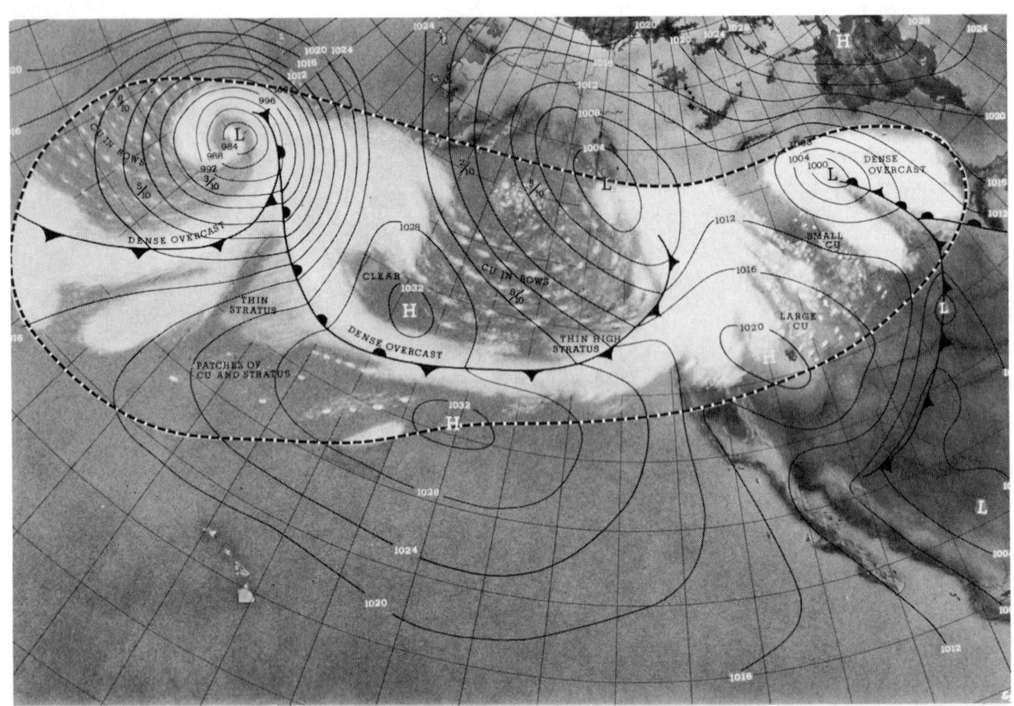

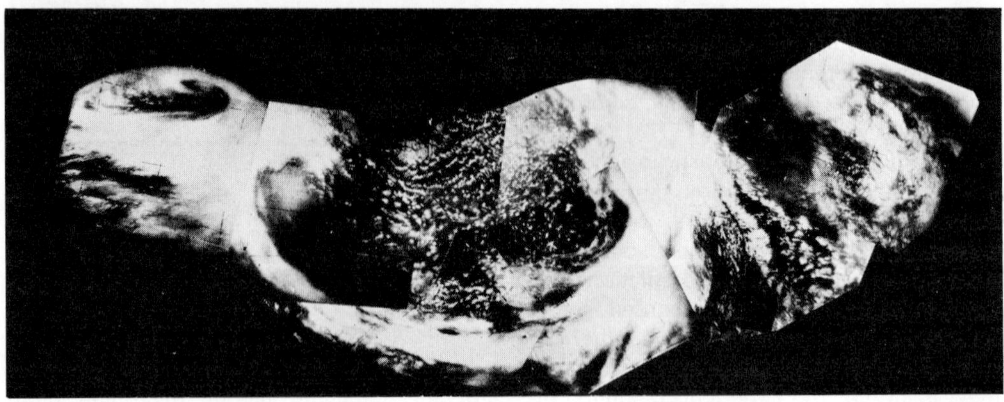

28-17. On May 20, 1960, photographs taken by the Tiros I weather satellite showed a series of storms extending from north of Japan to the eastern portion of the United States. The picture at the top shows the surface weather map for the eastern Pacific Ocean and western North America on May 20. Superimposed on the weather map is an artist's drawing of the clouds shown in the actual Tiros photographs below. Few routine weather observations are available from the vast areas of the Pacific Ocean. Without the Tiros photographs, weathermen might have been unaware of the position, or even the existence, of these storms. *(Courtesy U.S. Department of Commerce, Weather Bureau.)*

be a built-in limitation of 2 or 3 days or so in forecasting specific local weather changes because such changes depend upon the origin and behavior of lows and highs, which sometimes develop in a number of hours and may complete their cycles in a few days. It seems difficult to predict in detail the behavior of a pressure system that has yet to be born. However, other meteorologists disagree that such a limitation exists.

Generalized forecasts, from 5 days to a month in advance, are now being made by the U. S. Weather Bureau, and accuracy is somewhat better than 50 percent. To illustrate the uncertainties, by the late 1960's no country reportedly had yet used the long-range weather forecasting techniques of another country.

Weather Satellites

The revolutionary and dramatic nature of weather observations from orbiting satellites is obvious, and we no longer marvel at a TV weather program that includes a photograph of the clouds that covered an entire country a few hours earlier. Nearly continuous and global coverage of major weather systems is the exciting achievement of the newer weather satellites (28–17). Thus a tropical storm over a remote section of the ocean may now be tracked and studied, as well as the existence and movement of lows, highs, fronts, and jet streams. Such observation is a necessary prerequisite to a better understanding of weather phenomena. Tiros I (Television Infrared Observation Satellite—Fig. 28–18) was launched in April 1960, proved to be highly successful, and has been followed by other Tiros satellites and by more advanced satellites.

According to Johnson,* round-the-clock global coverage has been nearly achieved by the two-satellite ESSA system that began in 1966. These satellites are in nearly polar orbit and are approximately synchronous with the sun's apparent motions; i.e., observations are made over a particular country or region at the same solar time day after day. The two satellites have different functions, provide some duplication in case of equipment malfunction, and are replaced whenever necessary.

Data for the whole Earth are daily fed into computers and processed into three projections showing all major weather systems for that day—polar stereographic projections of the northern and southern hemispheres and a Mercator projection from 35 degrees north latitude to 35 degrees south latitude. Precomputed latitude and longitude lines, as well as other geographic data, are then superposed onto the projections (Fig. 28–19).

Among other achievements, correlation

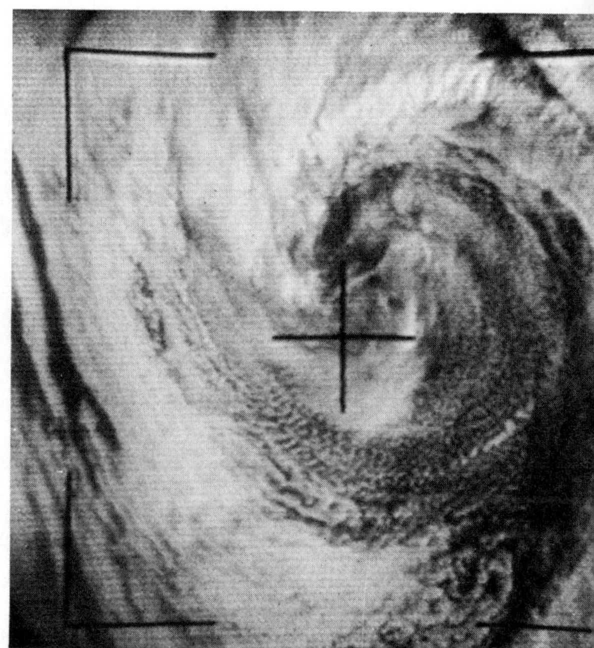

28-18. Tiros photograph of cyclonic system in the Central Pacific. *(Tiros VII, 6 April 1964, U.S. Weather Bureau.)*

* Johnson, Arthur W., "Weather Satellites: II," *Scientific American,* January 1969. ESSA stands for Environmental Science Services Administration, which in 1970 became part of NOAA, the National Oceanic and Atmospheric Administration.

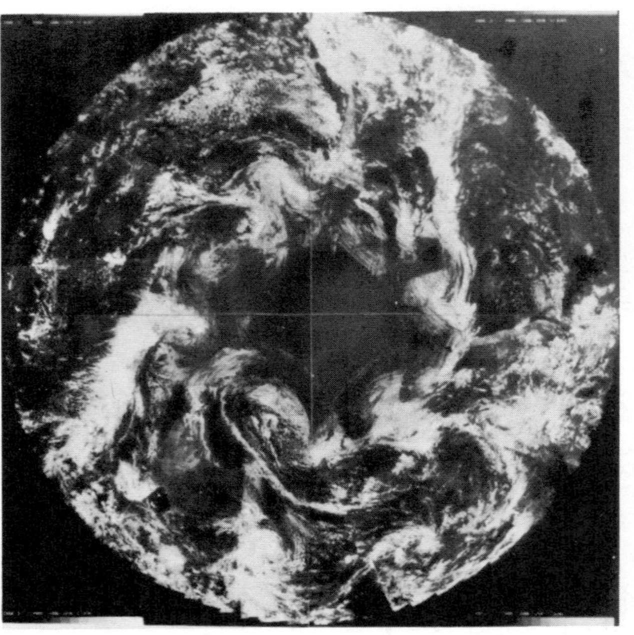

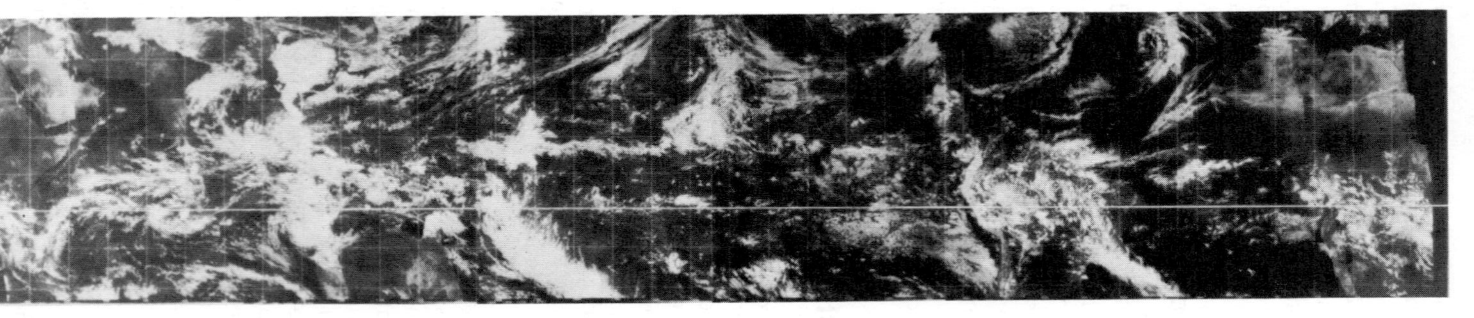

28-19. ESSA satellite computerized mosaic of the world's weather for one day (1 June 1967). Note counterclockwise circulation around lows in the Northern Hemisphere and clockwise circulation in the Southern Hemisphere. *(Courtesy NOAA.)*

of weather satellite photographs with near-surface measurements has made it possible to estimate approximately the maximum wind speeds in spiral weather systems (like tropical storms) from their cloud patterns. Jet streams have been detected visually because high cirrus clouds stop abruptly at the margin of a jet stream, and areas covered by snow and ice are being mapped.

The experimental Nimbus satellites have demonstrated that infrared sensing techniques (Fig. 28–20) can provide much useful information and make nighttime observations possible. Here the relationship between temperature and radiation (p. 496) is involved: more energy is radiated at higher temperatures than at lower, and the nature of the radiation is different. Moreover, radiation observed from cloud tops records their temperatures and thus their approximate altitudes. Ocean currents differing in temperature from surrounding water can also be mapped, and daily observation yields velocity data. Satellites have been placed in orbit (Applications Technology Satellite) at altitudes of about 22,000 miles where their periods of revolution equal the Earth's period of rotation—thus one seems to remain motionless in the sky above a certain point.

Weather forecasting is now being established on a global basis. Although weather data are freely exchanged internationally, weather analysis has tended to be on a national basis. Since the weather that affects any one country often approaches it from some other country or from the oceans, adequate forecasting demands the availability of nearly global weather data and its study on the same worldwide basis. Thus a World Weather Watch has been initiated (p. 653) and the weather satellites play a vital role in this program.

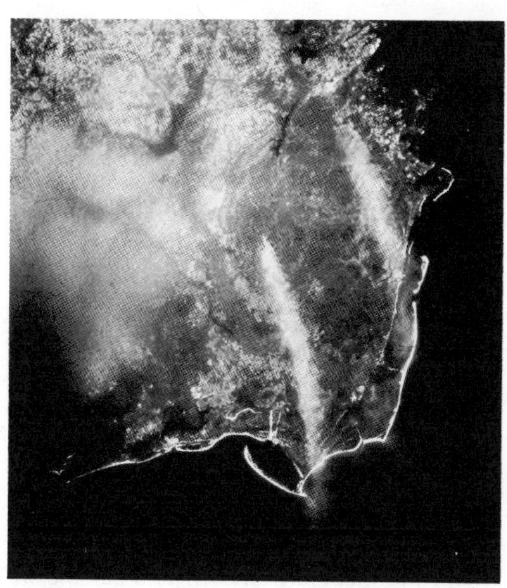

28-20. Infrared photo shows smoke drifting over the Gulf of Mexico from forest fires southwest of Tallahassee, Florida. *(Gemini VII, 7 Dec. 1965, S65-64053, NASA.)*

Weather Maps

A copy of the Daily Weather Map produced by the U. S. Weather Bureau and other data are shown in Figs. 28–21 through 28–27. The material that follows has been copied from legends (1968) prepared for the Daily Weather Map which is now mailed as a weekly series* from Monday through Sunday. Weather data for any one day are shown on a single page which contains a surface weather map, a 500-millibar chart, a highest and lowest temperatures chart, and a daily precipitation chart.

Weather maps showing the development and movement of weather systems are among the principal tools used by the weather forecaster. Of the several types of maps used, some portray conditions near the surface of the earth, while others depict conditions at various heights in the atmosphere. Some cover the entire Northern Hemisphere, while others cover only local areas as required for special purposes. The

* In 1969 the subscription price was $4.50 per year. A single explanatory sheet was available free, and bulk copies were priced at 50 copies for $2.30. Send requests to: Environmental Science Services Administration, Publications Section, AD 143, Rockville, Maryland 20852. Make checks payable to Superintendent of Documents.

maps used for daily forecasting by the Weather Bureau are similar in many respects to the printed Daily Weather Map. At Weather Bureau offices, maps showing conditions at the earth's surface are drawn four times daily. Maps of upper level temperature, pressure, and humidity are prepared twice each day. . . .

Principal Surface Weather Map

To prepare the surface map and present the information quickly and pictorially, two actions are necessary: (1) Weather observers at many places must go to their posts at regular times each day to observe the weather [at 6-hour intervals] and send the information by wire or radio to the offices where the maps are drawn; and (2) the information must be quickly transcribed to the maps. In order that the necessary speed and economy of space and transmission time may be realized, codes have been devised for sending the information and for plotting it on the maps.

Codes and Map Plotting

A great deal of information is contained in a brief coded weather message. If each item were named and described in plain language, a very lengthy message would be required; and it would be confusing to read and difficult to transfer to a map. Use of a code permits the message to be condensed to a few five-figure numeral groups, each figure of which has a meaning depending on its position in the message. Persons trained in

28-21. Daily Weather Map for November 29, 1963. *(Courtesy U.S. Weather Bureau.)*

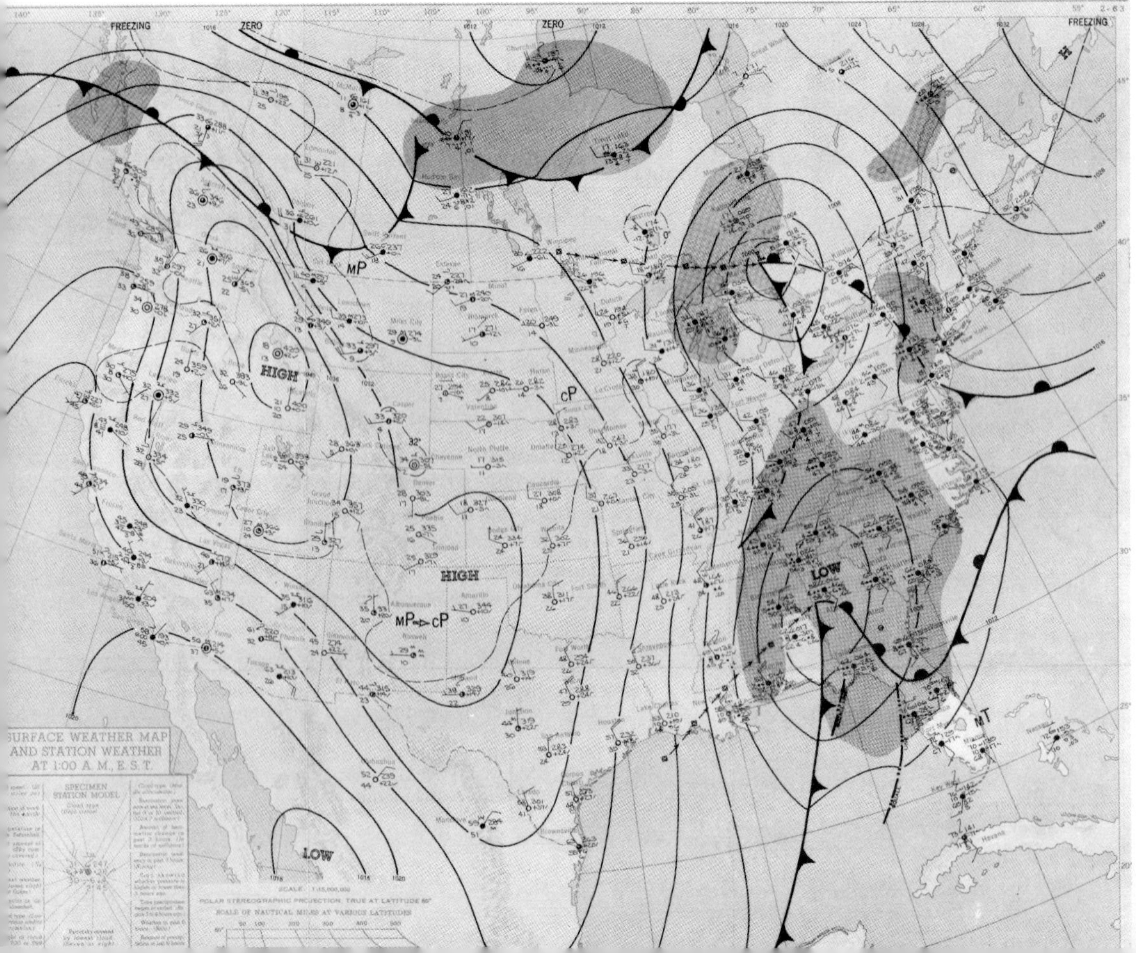

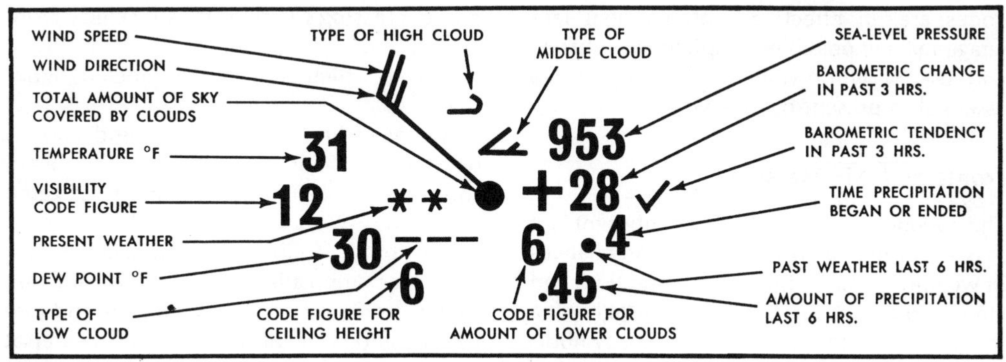

28-22. Weather data plotted around a specimen station circle. (AF Manual 105-5, *Weather for Aircrews.*)

the use of the code can read the message as easily as plain language.

The location of the reporting station is printed on the map as a small circle (the station circle). A definite arrangement of the data around the station circle, called the station model, is used [Fig. 28–22]. . . .

Both the code and the station model are based on international agreements. Through such standardized use of numerals and symbols, a meteorologist of one country can use the weather reports and weather maps of another country even though he does not understand the language. Weather

	Sky coverage
	No clouds
	Less than one-tenth or one-tenth
	Two-tenths or three-tenths
	Four-tenths
	Five-tenths
	Six-tenths
	Seven-tenths or eight-tenths
	Nine-tenths or overcast with openings
	Completely overcast
	Sky obscured

28-23. Symbols used on station circles to show sky coverage in tenths. *(Courtesy U.S. Weather Bureau.)*

	Miles (statute) per hour	Knots		Miles (statute) per hour	Knots
	Calm	Calm		44-49	38-42
	1-2	1-2		50-54	43-47
	3-8	3-7		55-60	48-52
	9-14	8-12		61-66	53-57
	15-20	13-17		67-71	58-62
	21-25	18-22		72-77	63-67
	26-31	23-27		78-83	68-72
	32-37	28-32		84-89	73-77
	38-43	33-37		119-123	103-107

28-24. Wind direction is shown on a weather map by the orientation of the shaft; wind blows along a shaft toward a station circle. Wind speed is shown in knots (nautical miles per hour; 1 nautical mile equals about 1 1/6 statute miles). A half feather, full feather, and flag represent 5, 10, and 50 knots ±2 knots, respectively. Feathers are placed on the left side of a shaft when facing downwind. *(Courtesy U.S. Weather Bureau.)*

codes are, in effect, an international language making possible complete interchange and use of worldwide weather reports so essential in present-day activities. . . .

Fronts and Air Masses

The boundary between two different air masses is called a "front." Important changes in weather, temperature, wind direction, and clouds, often occur with the passage of a front. Half circles and/or triangular symbols are placed on the lines representing fronts to indicate the kind of front. The side on which the symbols are placed indicates the direction of frontal movement. The boundary of relatively cold air of polar origin advancing into an area occupied by warmer air, often of tropical origin, is called a "cold front." The boundary of relatively warm air advancing into an area occupied by colder air is called a "warm front." The line along which a cold front has overtaken a warm front at the ground is called an "occluded front." A boundary between two air masses, which shows at the time of observation little tendency to advance into either the warm or cold areas, is called a "stationary front." Air mass boundaries are known as "surface fronts" when they intersect the ground, and as "upper air fronts" when they do not. Surface fronts are drawn in solid black, fronts aloft are drawn in outline only.

Front symbols are given below:

Cold front (surface) ▲▲▲▲
Warm front (surface) ◆◆◆◆
Occluded front (surface) ◆▲◆▲
Stationary front (surface) ▼◆▼◆
Warm front (aloft) ◠◠◠◠
Cold front (aloft) △△△△

A front which is disappearing or is weak and decreasing in intensity is labeled "Frontolysis."

A front which is forming is labeled "Frontogenesis."

A "squall line" is a line of thunderstorms or squalls usually accompanied by heavy showers and shifting winds, and is indicated as—..—..—..—.

The paths followed by individual disturbances are called storm tracks and are shown as → →. The symbols ⊠ indicate past positions of the low pressure center at 6-hour intervals.

"High" (H) and "Low" (L) indicate the centers of high and low barometric pressure.

Solid lines are isobars and connect points of equal sea level barometric pressure. The spacing and orientation of these lines on weather maps are an indication of speed and direction of wind flow. In general, wind direction is parallel to these lines with low pressure to the left of an observer looking downwind. (Over land, wind directions tend to make average angles of 30 to 40 degrees with the isobars, but these angles range from 0 to 90 degrees.) Speed is directly proportional to the closeness of the lines (termed pressure gradient). Isobars are labeled in the metric unit, millibars, and may be converted to inches of mercury by use of the scale [Fig. 28–25].

Isotherms are lines connecting points of equal temperature. Two isotherms are drawn on the large surface weather map when applicable. The freezing or 32°F. isotherm is drawn as a dashed line (— — —), the 0° F. isotherm is drawn as a dash-dot (—•—•—•) line.

Masses of air are classified to indicate their origin and basic characteristics. For example, the letter P (Polar) denotes relatively cold air from northern regions, and the letter T (Tropical) denotes relatively warm air from

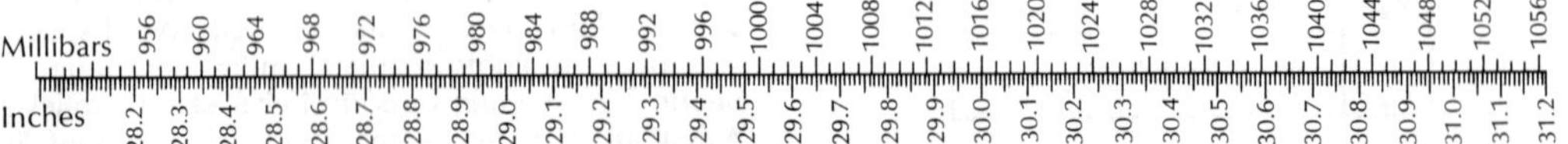

28-25. Conversion scale relating atmospheric pressure in millibars to its equivalent in inches of mercury. *(Courtesy U.S. Weather Bureau.)*

southerly regions. Letters placed before P and T indicate air of maritime characteristics (m) or continental characteristics (c). Letters placed after P and T show that the air mass is colder (k) or warmer (w) than the surface over which it is moving. A plus sign (+) between two air mass symbols indicates mixed air masses, and an arrow → between two symbols indicates a transitional air mass changing from one type to the other. Two air mass symbols, one above the other and separated by a line, indicate one air mass aloft and another at lower levels. Air mass symbols are formed by combinations of the following letters:

m = Maritime; c = Continental; A = Arctic; P = Polar, T = Tropical; E = Equatorial; S = Superior (a warm, dry air mass having its origin aloft);
k = colder, and w = warmer than the surface over which the air mass is moving.

Areas where precipitation is occurring at the time of observation are shaded.

Auxiliary Maps

Temperature Map. Temperature data are entered from selected weather stations in the United States. The figures entered above the station dot denote maximum temperatures reported from these stations during the 24 hours ending 1:00 a.m., E.S.T.; the figures entered below the station dot denote minimum temperature during the 24 hours ending at 1:00 p.m., E.S.T., of the previous day. The letter "M" denotes missing data.

Shaded areas labeled "HIGHER" or "LOWER" indicate the areas where temperatures recorded at 1:00 a.m., E.S.T., are at least 10° warmer or colder than 24 hours ago.

Precipitation Map. Precipitation data are entered from selected weather stations in the United States. When precipitation has occurred at any of these stations in the 24-hour period ending at 1:00 a.m., E.S.T., the total amount, in inches and hundredths, is entered above the station dot. When the figures for total precipitation have been compiled from incomplete data and entered on the map, the amount is underlined. "T" indicates a trace of precipitation, and the letter "M" denotes missing data.

The geographical areas where precipitation has fallen during the 24 hours ending at 1:00 a.m., E.S.T., are shaded.

500-Millibar Map. Contour lines, isotherms, and wind arrows are shown on the insert map for the 500-millibar contour level (for 7:00 a.m. E.S.T.). Solid lines are drawn to show height above sea level and are labeled in feet. Dashed lines are drawn at 5° intervals of temperature, and labeled in degrees Centigrade. A temperature conversion table is shown (Fig. 28–26). True wind direction is shown by "arrows" which are plotted as flying with the wind. The wind speed is shown by flags and feathers, each flag representing 50 knots, each full feather 10 knots, and each half-feather 5 knots. [10 knots is equal to approximately 11½ mi/hr].

Weather Lore

Keen observers in different countries over the centuries have noted certain phenom-

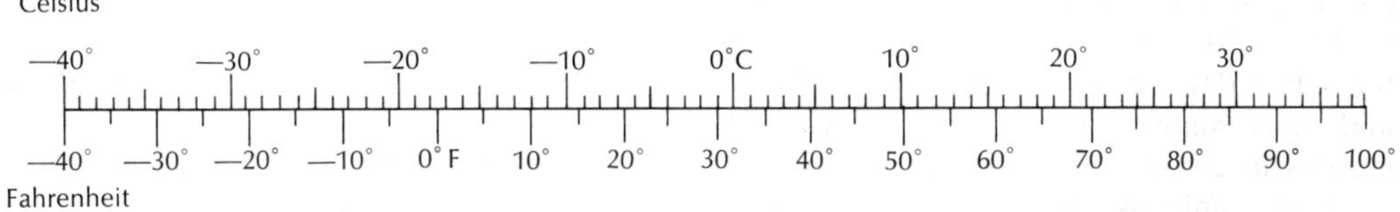

28-26. Temperature conversion scale. *(Courtesy U.S. Weather Bureau.)*

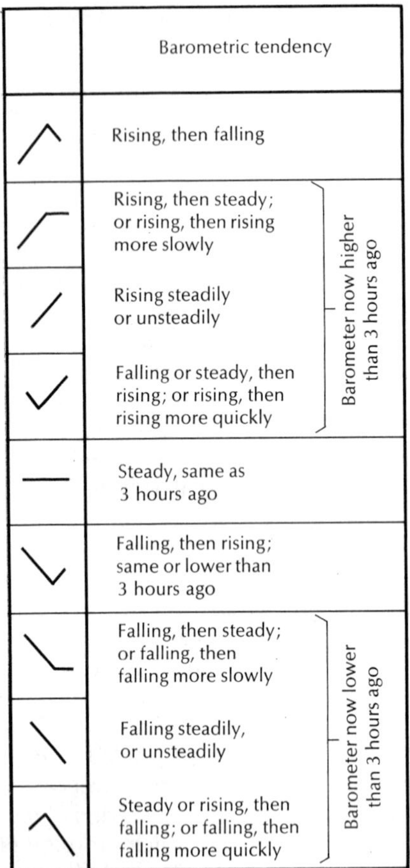

28-27. Barometric tendency. *(Courtesy U.S. Weather Bureau.)*

ena that have been helpful in predicting the local weather. But superstition has also been so intermixed with observation that much weather lore should be ignored. In this category are proverbs relating weather changes to the positions of planets, stars, and moon. Other proverbs relate coming weather changes with the reactions of people, plants, and animals, and here we need to distinguish between short-range forecasts for a few days or less (some have validity) and long-range forecasts for weeks and months in advance (none have validity).

Certain animals supposedly have forewarning of a coming severe winter and act accordingly; e.g., they may grow a heavier coat of fur or store more food. Relatively little research has been done concerning this subject, but organisms apparently have no special abilities to foretell the weather months in advance. Certainly the meteorologist cannot do this successfully. An animal may grow a heavier coat of fur in a particular year, but this undoubtedly depends upon an ample food supply or some other biological factor.

On the other hand, certain animals and plants, as well as people with certain ailments, do detect slight changes in temperature, humidity, atmospheric pressure and the like that commonly precede the arrival of a low-pressure system.

A number of fairly reliable weather proverbs are based upon cloud observations and apply particularly to the belt of westerlies in the Northern Hemisphere. They are related to the succession of clouds that occurs as a warm front approaches, to the drop in barometric pressure that precedes the arrival of a low-pressure system, and to the general movement of weather changes from west to east. The following illustrate this type of proverb:

1. Mackerel scales and mares' tails
   Make lofty ships carry low sails.
2. Trace in the sky the painter's brush;
   The winds around you soon will rush.
3. The moon with a circle brings water in her beak.
4. Rain long foretold, long last;
   Short notice, soon past.
5. A rainbow in the morning is the shepherd's warning;
   A rainbow at night is the shepherd's delight.
6. When the grass is dry at morning light
   Look for rain before the night.
   When the dew is on the grass,
   Rain will never come to pass.
7. When the ditch and pond offend the nose,
   Then look for rain and stormy blows.
8. But I know ladies by the score
   Whose hair, like seaweed, scents the storm;

Long, long before it starts to pour
Their locks assume a baneful form.

9. Rain before seven, shine before eleven.

The first three proverbs listed above involve the cirrus-type clouds that herald the approach of a warm front and emphasize the importance of clouds as indicators of coming weather changes. Cirrostratus clouds produce the halo or circle mentioned in proverb 3. In proverb 4, the precipitation that accompanies a warm front lasts longer than the convectional shower that may develop rather suddenly on a warm afternoon.

Proverb 5 is based upon two facts: one sees a rainbow by looking away from the sun, and weather changes tend to approach from the west. Thus a shepherd observing a rainbow in the morning is looking toward the west at air that contains raindrops—the storm is approaching him. In the evening the storm is in the east and already past him.

Proverb 6 relates conditions favorable for the formation of dew (stable, cloudless, dry air so that radiational cooling occurs at night) to the different conditions that favor precipitation on a large scale (unstable moist air which tends to prevent the ground from cooling off enough at night for dew to form).

Proverb 7 relates to the drop in atmospheric pressure that commonly precedes precipitation—certain gases with unpleasant odors may be dissolved in water and some of these escape if atmospheric pressure is reduced enough. In proverb 8, human hair stretches in moist air and shortens in dry air (the basis of the hair hygrometer). If precipitation over a particular location lasts only for half a day or so, then proverb 9 has some validity. If rain begins at some time during a night, it will probably stop before noon of the next day.

Exercises and Questions for Chapter 28

1. Describe the types of exchanges that may occur between the Earth's surface and the lower atmosphere to effect a condition of near-equilibrium and produce an air mass.
   (A) In what geographic locations do air masses develop most commonly?
   (B) What symbols are used on the U. S. Weather Bureau maps for air masses?
2. Why is an upper air inversion common at a latitude of about 30 degrees?
   (A) Why is a surface inversion common in the winter in polar regions?
3. Why does atmospheric pressure decrease at the center of a developing low (cyclone)?
   (A) Older view?
   (B) More recent view?
4. Do surface atmospheric pressures at the centers of high-pressure systems (anticyclones) tend to be greater in winter or summer? Why?
   (A) For low-pressure systems?
5. Why is weather in the middle latitudes so changeable?
6. Arrange a number of daily weather maps in chronological order with the most recent map on the bottom (a set of seven such maps is published weekly by the U. S. Weather Bureau, and several sets covering different seasons could be used here). Study the top map and predict the changes

which you think are likely to occur on the weather map for the next day. Check and repeat. Presumably you will make mistakes, and unexpected weather changes may occur (perhaps because insufficient data were available for you to forecast such changes). However, you should learn a good deal about weather changes.

7. What observations can you make locally that are most useful in forecasting coming weather changes?
   (A) What special influence, if any, does your local topography have on weather phenomena?
8. In each of the following examples, certain weather phenomena are assumed to occur in a certain region at a certain time. In each case, weather data are plotted around four station circles. You are to choose the station which shows the most likely set of weather elements under the weather conditions assumed. Next, explain why at least one of the weather elements shown at each of the other three stations is improbable.
   (A) A radiation fog developed during the nighttime in September at this weather station in Massachusetts. Assume that the data were recorded at 6 A.M.

(1) (2) (3) (4)

   (B) The center of a large high-pressure system is over this weather station in New York in October at noontime.

(1) (2) (3) (4)

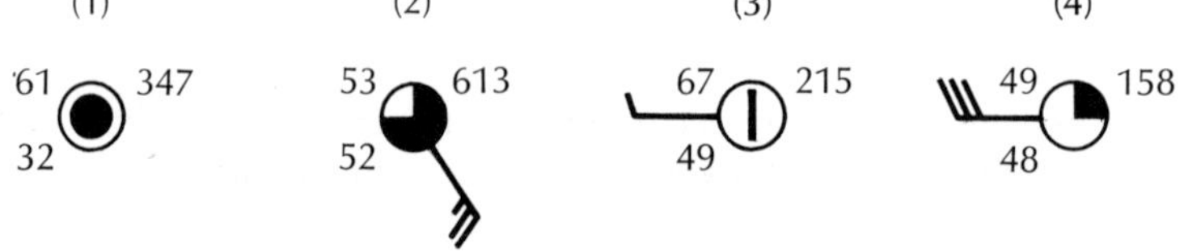

   (C) The center of a strong low-pressure system is 200 miles due north of this weather station in Pennsylvania in November at 10 A.M.

(1) (2) (3) (4)

   (D) This weather station is on a ship on the ocean at 45 degrees south latitude at 8 A.M. in August. The center of a large intense low-pressure system is 100 miles due north of the ship at this moment.

(1) (2) (3) (4)

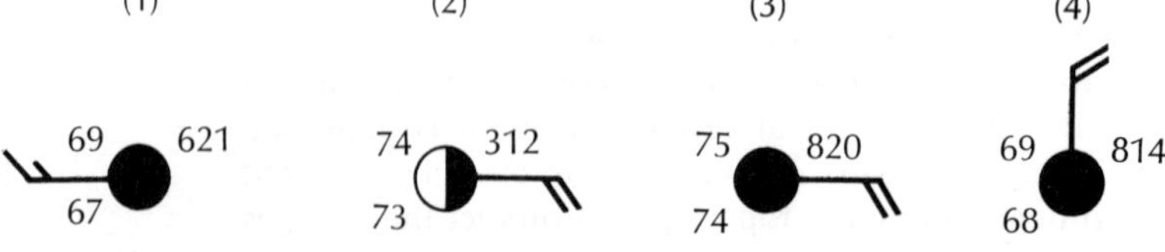

(E) The center of a large well developed high-pressure system is about 200 miles due north of this weather station in Vermont in February at midnight.

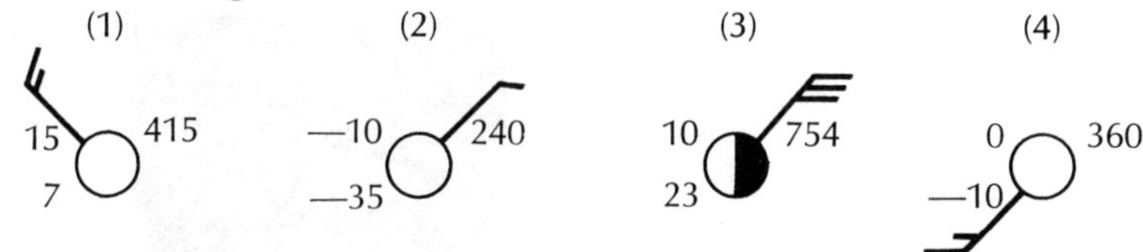

(F) In September at 9 A.M. the temperature at this New York weather station was 70°F; wind blew from the south. An occluded front of the warm-front type passed by the station about noontime. At about 4 P.M. the most likely conditions at this station would be those shown at station:

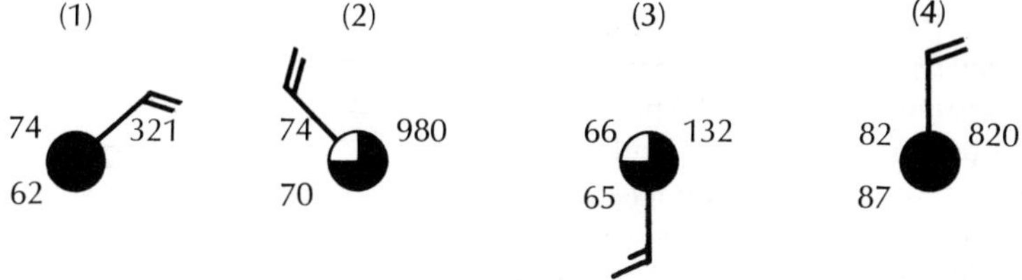

9. Assume that the center of a large high-pressure system is about 200 miles northwest of a New York weather station on September 30 at 3 P.M. Choose a set of weather conditions that would be appropriate at this station at this time and plot them around a weather station circle as they would be shown on a U. S. Weather Bureau map (wind direction, wind speed, temperature, dew point, barometric pressure, barometric tendency, and total amount of cloud cover). Next, repeat this for three other station circles, but in each case make at least one of the weather elements an unlikely one (as if you were making out a multiple choice question).

(A) Do the same as requested above but assume that the center of an intense low-pressure system is located about 75 miles south of a weather station in Pennsylvania in January at noontime.

29

Storms: Thunderstorms, Tornadoes, and Hurricanes

A thunderstorm is an intense, local, rain storm of short duration which is accompanied by thunder and lightning and sometimes by hail (Figs. 29–1 and 29–2). Within a thunderstorm are fast-moving, vertically-directed air currents, which are mainly upward in the early stage of development and downward in the dissipating stage. However, both upward and downward currents exist simultaneously in adjoining sections of a storm during its middle and most violent stage—marked by heavy rains, high winds, hail, lightning, and occasionally by tornadoes. Gusts and turbulence are particularly strong in the boundary zones between adjacent upward and downward currents.

Lt. Col. Rankin,* a jet pilot of the U. S. Marine Corps, has been able to describe vividly the inside of a thunderstorm because he parachuted directly into a towering one at about 6:00 P.M. on July 26, 1959, in the vicinity of Norfolk, Virginia. Because of sudden engine trouble, he was explosively ejected from his fast-moving plane at an altitude of about 9 miles into very cold and diffused air (about −70°F; atmospheric pressure approximately one-seventh of that at sea level). Unprotected in a summer-weight flying suit, he was partially frozen and suffered extreme pain from the sudden decompression; his eyes felt as if they were being pulled out of their sockets, his stomach was distended, his ears seemed about to burst open, and cramps affected his entire body.

Rankin's parachute was set to open automatically at an altitude of 10,000 feet, thus he fell for more than 7 miles (at a terminal velocity of about 2 mi/min) downward into the blackish, ominous-looking, seething clouds of the thunderstorm. He was pelted by hailstones before his chute opened in the center of the storm. Subsequently Rankin was pulled explosively upward by ascending currents, dropped downward in others, and slammed, pounded, and stretched in the turbulent air. Lightning was all around him, and thunder actually shook his teeth; furthermore, he was

* Lt. Col. William H. Rankin, *The Man Who Rode the Thunder*, Prentice-Hall, 1960.

almost drowned by the torrential rains. Finally he fell out of the bottom of the storm and landed. After hospitalization and a remarkable recovery, Lt. Col. Rankin flew again within a month.

More than 40,000 thunderstorms may occur during an average day on the Earth, and some 1500 to 2000 may be taking place at any one time. Thunderstorms commonly result from the rapid upward movement of a parcel of warm moist air; thus they tend to be associated with high temperatures. They are most frequent in the afternoon over land areas in the lower latitudes and under similar conditions in the middle latitudes during the warmer months of the year. They are uncommon at high latitudes. The latent heat energy released during condensation is a main factor in their formation, and only warm air can contain large quantities of water vapor. Steep observed lapse rates favor thunderstorm development and may be produced by heating from below by terrestrial radiation or by cooling at the top by radiation to space.

A thunderstorm cell 3 miles in diameter may contain some 500,000 tons or so of ice, snow, and water that have condensed or sublimated. The latent heat thus released would be sufficient to raise the temperature of nearly 6 billion pounds of water from the freezing to the boiling point.

Thunderstorms occur in association with cumulonimbus clouds and are characterized by local extent, short lifetime, gusty winds, thunder and lightning, and heavy rains. Occasionally, tornadoes originate in association with very large and violent thunderstorms. Hail falls only from thunderstorms, but many thunderstorms do not produce hail that falls to the surface, especially storms in the lower latitudes. However, hail may exist aloft at times in such storms. For the United States as a whole, July is the month with most thunderstorms, perhaps more than during all of the other months combined, and Florida is the state with the greatest number (Fig. 29–3). The contrasts in temperature between land and sea play an important role here.

Over the oceans, thunderstorms are less common and tend to occur when the water is warm relative to the air above it—in the nighttime or during the winter months.

29-1. Lightning. *(Courtesy NOAA.)*

29-2. The large cloud has just developed into cumulonimbus. Cumulus clouds occur at a lower altitude. *(Photograph by F. Ellerman.)*

Lapse rates may steepen at night in air over warm ocean water because bottom temperatures remain about the same, whereas temperatures at higher altitudes become colder.

Thunderstorms occur under conditions of two different types. Air-mass thunderstorms are more common and occur within a warm moist air mass by localized convectional circulation. These develop most frequently on warm humid afternoons. Frontal thunder-

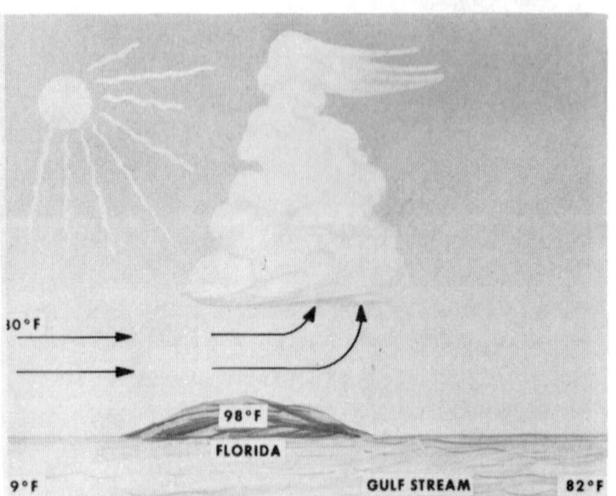

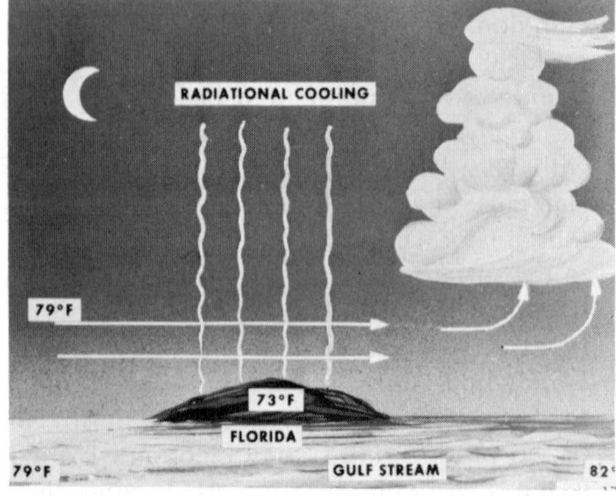

29-3. Convective thunderstorms of a type common in Florida. Sea breezes from the Atlantic and Gulf of Mexico may converge in central Florida and contribute to daytime thunderstorm formation. (TM 1-300, *Meteorology for Army Aviation.*)

storms may occur at any hour during any season as a result of the uplift of warm moist air along an occluded or cold front. Such storms may be scattered here and there along a front because locally the air is lifted more vigorously, contains more water vapor, or is less stable. Similarly, warm moist air may be lifted enough to produce thunderstorms as it crosses a mountain barrier, although convectional circulation may also be involved here. During the daytime, air may be heated strongly on the slopes and crest of a mountain by contact with and radiation from the mountain surface. In contrast, air at the same altitude above adjacent valleys and lowlands tends to be cooler.

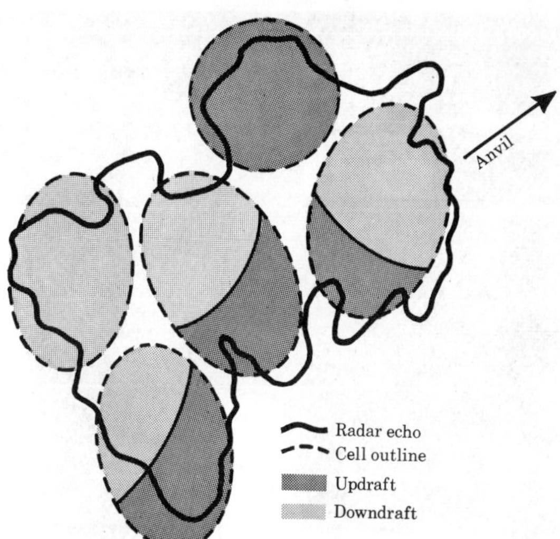

29-4. A thunderstorm commonly contains a number of cells in different stages of development. Radar echoes are obtained from raindrops, ice crystals, and hailstones within a cloud, but not from cloud droplets (with radar wavelengths commonly used). Precipitation and strong vertical air movements tend to be absent from portions of a thunderstorm cloud outside of the cells, but enough raindrops are commonly present to produce a weak radar signal. *(After Byers.)*

Life Cycle of a Thunderstorm Cell

A thunderstorm probably consists of one or more units called cells which range from about 1 to 5 miles in diameter (Fig. 29–4). According to this model, each cell passes through three main phases: cumulus, mature, and dissipating (Figs. 29–5 and 29–6). Thus at any one time a thunderstorm may contain several different cells, and these may be in different stages of their development. The entire life cycle of any one cell is commonly played out within an hour or so. However, the storm itself may last much longer than this because new cells form and replace those that have dissipated.

In the *cumulus stage* of a thunderstorm cell, the chief movement of air is upward and the air within a rising parcel at any one level tends to be warmer than the air outside of it (note the upward bulge in the isotherms). Probably this stage lasts for 10 to 15 minutes after the top of the cloud has reached the freezing level. Velocities are greater in the upper parts of such cumulus clouds, because the upward-moving air cools rather slowly at the wet adiabatic lapse rate. Thus it becomes more buoyant with altitude (the observed lapse rate in the air around the cell must be relatively steep). However, velocities tend to vary from one part of a cloud to another and also to change with time.

Large quantities of air are drawn into an upward moving current from the sides (entrainment), and the mass of a rising parcel of air may be doubled by this influx from the sides during an upward movement of 3 to 4 miles. If the new air is relatively dry, some evaporation will occur as it mixes with the moist cloud air, and the buoyancy will be decreased.

Precipitation develops in the upward-rushing air. Ice crystals and snowflakes form at the colder temperatures that occur in the upper parts of a cumulus cloud where supercooled cloud droplets tend to be common, whereas raindrops form in the warmer air at lower altitudes. Cumulus clouds of this type are quite common, but only a small percentage continue to grow into full-fledged thunderstorms. Some cells fail to

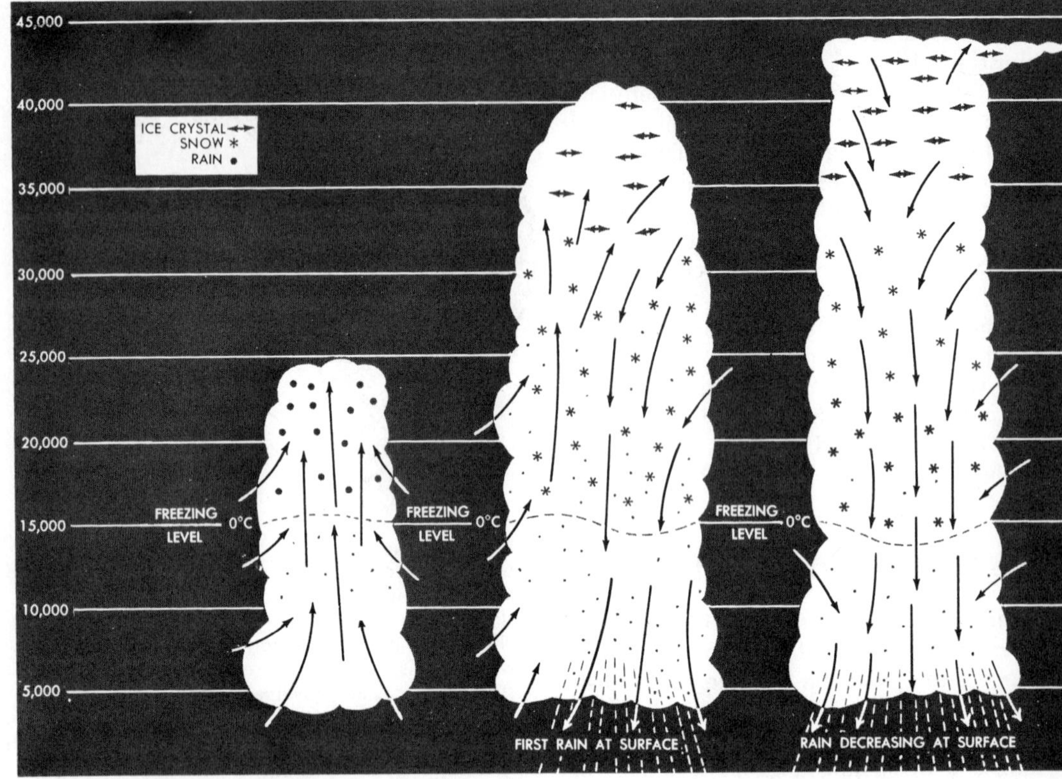

29-5. Three stages in the life cycle of a thunderstorm: cumulus (left), mature, and dissipating (right). (AF Manual 105-5, *Weather for Aircrews.*)

develop further because they contain too little water vapor, are "diluted" by additions of dry air (entrainment), or because they rise upward into air that is more stable and drier, especially into a strong inversion.

When precipitation particles become too large to be kept suspended in the updrafts, they begin to fall, and when they reach the ground the *mature stage* of a cell is said to begin. Friction between the air and the falling raindrops causes some of the air in a cloud to be dragged downward. Downdrafts occur first in the lower sections of a cloud but extend upward and increase somewhat in velocity as time passes. However, they do not extend to the very top of a cell because the slow fall of ice crystals and snowflakes exerts little downward drag on this upper air.

This moist air in the downdraft is compressed and warmed during its subsidence but at the wet adiabatic rate (evaporation occurs and has a cooling effect). Therefore, its temperature increases only at a slow rate during descent, and at any one level the air in a downdraft tends to be colder and less buoyant than surrounding air at the same level (note the downward bulge of the isotherm in Fig. 29–5). These speeds may increase downward, and the most powerful currents of air develop during the mature stage; these are upward in one part of a cell and downward in another part. At this stage, the top part of the cloud (now a cumulonimbus) may extend all of the way to the stratosphere, or even penetrate for some distance into it (Fig. 29–7).

Hail may form within the cloud at this

time and is evidence of the great speed of the upward moving air currents: e.g., a 3-inch hailstone falls at the rate of about 6000 ft/min; therefore, to become this size, it must have been supported for a time in air that was moving upward at more than a mile a minute.

Cumulonimbus clouds are the sources of some very intense rainfalls and of maximum size raindrops, and these are a direct result of the strong updrafts that occur within them. Large quantities of precipitation can be supported within an upward-rushing air current, and the particles can grow to maximum sizes. However, when a downward movement begins in one part of a thunderstorm cell, the enclosed raindrops fall quickly upon the surface below—the rate at which drops would fall in still air must be added to the downward velocity of the air itself.

The downward moving air current in the precipitation zone spreads out as it reaches the surface, particularly in the direction in which the entire storm is moving. This cool air (temperature drops of 10°F or more have been recorded) functions as a miniature cold air mass, and its forward portion resembles a small cold front (Fig. 29–8). It produces one of the characteristic features of a thunderstorm—the strong, gusty, cool wind that precedes the arrival of the precipitation and may attain velocities greater than 50 mi/hr (80 km/hr). The passage of the cool air may cause an abrupt increase in barometric pressure of ½ to 1 millibar and may result in an abrupt change in wind direction. A new thunderstorm cell may be formed downwind as this cool air plows ahead and shoves upward the air in its path. However, other factors are also involved in the development of a new cell.

The mature stage may last for 15 to 30 minutes or so. During this stage, the downdrafts become more extensive as air is dragged into a cell from all sides. When the downdrafts have spread throughout the entire lower region of the cell, the third or *dissipating stage* in its life cycle has been reached. The supply of upward moving air

29-6. Single-cell type thunderstorms in cumulus stage (above) and dissipating stage (below). *(Courtesy U.S. Weather Bureau.)*

has been cut off, and condensation and precipitation stop. Relatively little vertical movement occurs in the uppermost portion of a cell during its dissipating stage, and strong winds may spread the cirrus clouds into a characteristic anvil top, especially in a downwind direction (Fig. 29–6). Sometimes alto-type clouds form an anvil at lower altitudes.

During the cumulus and mature stages, stronger winds at higher altitudes may tilt the entire top of a cloud column downwind. Precipitation can then fall out of the

29-7. Tops of cumulonimbus clouds viewed from an altitude of 8 miles near the Shetland Islands (distant clouds on the right are over mountains in Norway). The anvil tops of the two nearer cumulonimbus clouds are unsually symmetrical. A third anvil is located just beyond the second one. (Photograph by RAF, from F. H. Ludlam and R. S. Scorer, *Cloud Study*.)

upper portion of the cloud and not pass downward through its lower portion. The upward movement of air within the lower section of the cloud is thus unimpeded, the updraft part of its life cycle is prolonged, and many raindrops may evaporate before they reach the ground. A violent tilted updraft may fling hailstones outward from the top of a mature cell, and they fall a few miles ahead of a thunderstorm.

The electrical and sound phenomena associated with thunderstorms may be quite violent. *Lightning* (Fig. 29–1) is the flash in the sky caused by a very rapid flow of electric current, and *thunder* is the sound produced by the intensely heated, rapidly expanding air in the vicinity of a lightning flash. Lightning discharges may occur from one part of a cloud to another, from one cloud to another, and from a cloud to the Earth's surface.

Centers of electric charge are generated or induced within clouds and on the Earth (Fig. 29–9). When they become powerful enough, the air between two nearby centers of unlike charge can no longer act as an insulator. This air is ionized, and the ions move as an electric current toward the centers of charge. Positively charged ions move in one direction, and negatively charged ions move in the opposite direction. All of this happens in a fraction of a second, but very high-speed photography shows that a lightning flash consists of a series of steps rather than of one continuous stroke. It is more like an electric arc than a single spark.

The actual lightning channel itself may be only a few inches wide (indicated by the damage it does on occasions). However, it is extremely hot (perhaps double the temperature of the sun's surface) and violently heats the air in its path—thus metal cores in certain rubber-coated wires have been completely vaporized with little harm to the rubber insulation. Damage to trees may result from the effect of the very high temperatures on moisture in the wood—water may be disassociated into hydrogen and oxygen and these may be ignited explosively.

The sound waves produced by a lightning flash move at about 1100 ft/sec—approximately 1 mile in 5 seconds. An observer can estimate his distance from a lightning flash by measuring the time interval between seeing the flash and hearing the thunder (the light travels at about 186,000 mi/sec). If the lightning flash is more than 5 or 6 miles away, however, its thunder may not be heard. So-called heat lightning (seen but not heard) is caused by distant lightning flashes.

Thunder should not be feared; if you hear it, the lightning flash which caused it is over, and you are safe so far as it is concerned. Thunder may occur as a single loud clap if the sound waves from a single lightning flash arrive at a certain place at about the same time, or it may rumble for different reasons. Sound waves from a long horizontal stroke arrive at different time

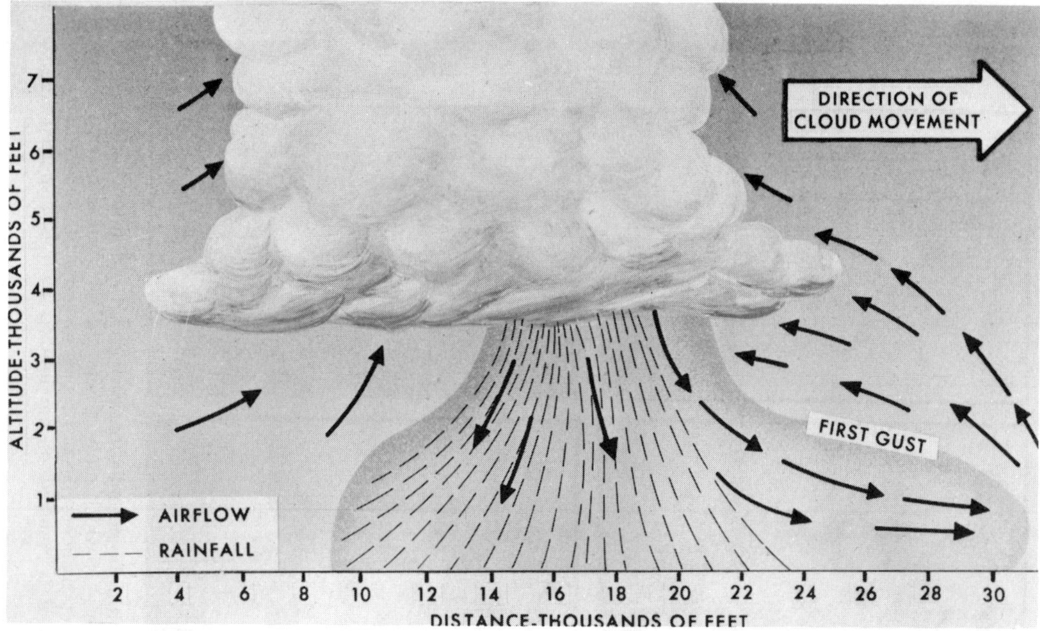

29-8. Movement of air beneath a thunderstorm cell in the mature stage. (TM 1-300, *Meteorology for Army Aviation.*)

intervals, and some may be reflected among clouds or from mountains.

One is relatively safe within most closed cars, planes, or metal-framed buildings because lightning discharges tend to follow the metal and are conducted around the occupants inside. Isolated tall objects are particularly inviting targets for lightning discharges. For this reason one should not take shelter under a single large tree, play golf, or sail a boat during a thunderstorm. Well-grounded lightning rods located on the tops of isolated houses provide some protection against lightning discharges, and they function in two ways. Electric charges escape into the air at the sharp points of the metal rods, which keeps large induced charges from building up. If lightning does strike the house, the electric current tends to follow the metal rods (good conductors) downward into the ground. Television antennas, of course, should be well grounded.

Church bells were formerly rung during thunderstorms in certain parts of Europe, and the job of bell-ringer was a dangerous one. Churches are often located in prominent spots, and the bell at the top of a church had a rope that was pulled by the bell-ringer. If struck by lightning, the electrical current tended to go down the rope, through the bell-ringer, and into the ground.

Precisely how particles with like electric charges are concentrated in certain parts of a cloud is uncertain, but the upper portion of a cumulonimbus cloud is likely to be charged positively, whereas the middle portion commonly has a negative charge. A smaller area of positive charge may also form in the base of a cloud, and this induces a negative charge on the surface beneath it. The charges may well form in more than one way.

Tornadoes

Tornadoes (Spanish: *tronada,* "thunderstorm") or twisters are smaller and more intense than either thunderstorms or hurricanes. In fact, they are the most violent

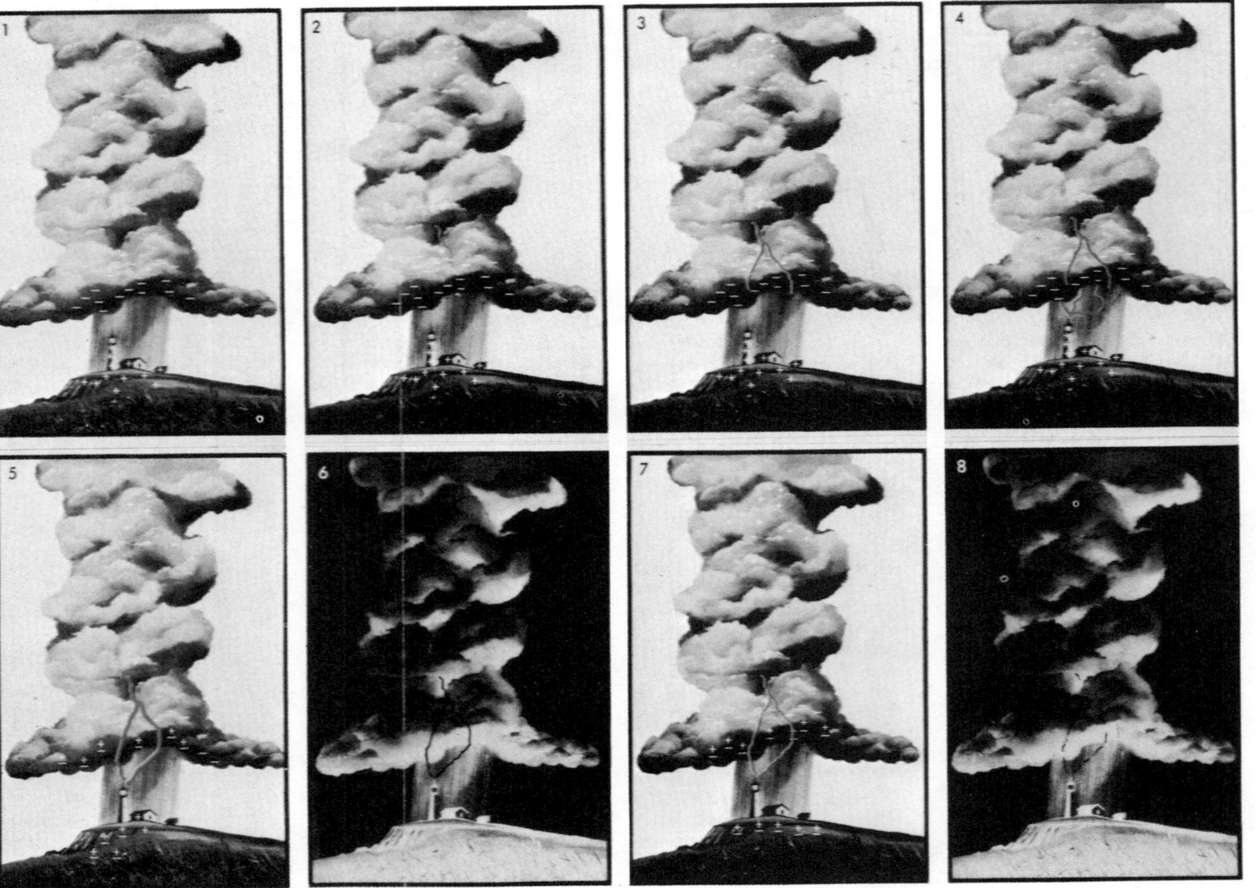

29-9. Lightning. As the thunderstorm induces a growing positive charge in the Earth, the potential between cloud and ground increases (1) until pilot leader starts a conductive channel toward ground (2) followed by step leaders (3) which move downward for short intervals (4) until they are met by streamers from ground. The return stroke from the ground illuminates branches (5) and seems to come from the cloud. The main stroke is followed by sequence of dart leaders and returns (6,7) until the potential is reduced or the ionized path is dispersed (8). Elapsed time: about one second. *(Courtesy ESSA.)*

storms known on Earth. They are more abundant in the United States than in most regions and are most common in the afternoon in the spring and early summer in the Mississippi Valley region (Kansas, Iowa, Texas, and Oklahoma have had the greatest number). However, every state in the United States has had a tornado.

Tornadoes form in association with especially violent thunderstorms and consist of a center of low pressure (a vortex) around which air swirls violently (commonly counterclockwise in the Northern Hemisphere). From a distance a tornado looks like a long, dark, funnel-shaped or ropelike cloud which hangs like an elephant's trunk from the base of a giant cumulonimbus cloud. A funnel appears to form within a cloud and then descends toward the ground, but not all funnels reach the surface (Fig. 29–10). The outer boundary of a funnel may be sharp or indistinct. Commonly the diameter of a funnel cloud is less than ¼ mile, but its havoc-wreaking trail along the surface is generally wider than this because winds are also very strong around the cloud.

The droplets of a funnel cloud, like those of other clouds, formed because moist air was cooled below the dew point by expansion. Since barometric pressure inside a funnel is extremely low, air is drawn into it from all sides and expands suddenly. Violently uprushing air within a funnel cloud is likewise cooled adiabatically as it ascends. Sometimes the middle portion of a tornado is not visible—dust and debris have darkened its lower portion, whereas cloud droplets have formed by condensation only within its upper portion.

The top part of a tornado may move faster than the portion near the surface, and thus after a time it does not hang vertically downward from the cumulonimbus cloud that spawned it. The average tornado lasts for less than 1 hour, and the length of its path approximates 10 to 40 miles or less, but some have traveled for more than 100 miles. In the United States the common direction is toward the northeast, and the average rate seems to be some 25 to 40 mi/hr, but faster and slower rates have been recorded. The average time that a tornado exists above any one spot is less than 1 minute, and a twisting funnel may rise here and there above the surface and spare the buildings and people in its path. Some 600 to 700 tornadoes may occur in the United States in an average year, but not all are reported, especially in sparsely settled areas.

The destruction produced by a tornado

29-10. Four photographs of a destructive tornado at Gothenburg, Kansas. *(Photographs by Mrs. Ray Homer. Courtesy U.S. Weather Bureau.)*

is caused by its violent winds, by the tremendous lifting effect of its updrafts, and by the explosive effect of a sudden drop in atmospheric pressure outside of buildings. Winds may whirl around the low pressure center of a tornado at rates up to 200 to 400 mi/hr or even more. Anemometers are destroyed, but certain phenomena indicate the extreme velocities involved; e.g., straws have been blown into tree trunks. Updrafts associated with a tornado may attain speeds of 100 to 200 mi/hr, and thus it acts as a giant vacuum cleaner.

Barometric pressure may drop greatly as the center of a tornado passes—equivalent to 1 to 2 inches on a mercurial barometer. Thus within 10 to 20 seconds pressures may change by as much as 400 lb/sq ft, and closed buildings explode outward, and roofs are blown off.

Debris whirled along by the wind adds to the danger and may include tractors, animals, bricks, glass, rocks, and trees. Observers have reported loud roaring sounds like the combined rumblings of many trains and the thunder of a jet squadron, as well as hissing and buzzing noises. A few observers have looked upward into the center of a tornado and lived to report that a clear area extended upward for half a mile or so.

Just how tornadoes originate is uncertain, and their precise locations cannot be forecast. Tornadoes develop out of thunderstorms, especially from those in middle latitudes where temperature differences tend to be at a maximum. However, perhaps 1000 thunderstorms do not produce a tornado for each one that does. According to Vonnegut,* tornadoes tend to originate from particularly large and violent thunderstorms. These may tower 8 to 12 miles above the surface, and thus their tops have penetrated for a few miles into the stratosphere. Their lightning discharges produce almost continuous bursts of radio static (perhaps at a rate 100 times greater than in non-tornado thunderstorms). On radar, a small hooklike echo has sometimes been observed in such storms and may represent the tornado itself.

Three conditions are known to be favorable for the development of tornadoes in the central United States, and area forecasts can be made. (1) A layer of warm moist air occurs along the surface, has a thickness of about 1 to 2 miles, and is moving northward from the Gulf of Mexico. (2) Above this is a thick layer of drier air that has crossed the Rockies. The lower part of this air was warmed adiabatically as it descended eastward from the mountains, but its upper part is still cold, and thus it has a steep lapse rate. (3) A jet stream blowing eastward occurs a few miles above the surface.

If some of the maritime tropical air is shoved upward, its ascent will be accelerated, because it will cool at the wet adiabatic rate (slow) and rise upward through air that has a steep lapse rate. The maritime tropical air thus may become increasingly buoyant at higher altitudes and rise explosively. Air drawn inward from the sides and subjected to the Coriolis effect causes the whirling motion. The conservation of angular momentum is also involved: as the air spirals inward, it whirls faster.

Many weird events happen in tornadoes. At times, horses, cars, and people have been picked up by these violent storms, carried some distance, and then lowered to the ground more or less undamaged. In such cases, the objects are whirled up by violent updrafts and lowered slowly through lesser currents surrounding a funnel cloud. In other cases, fences may be rolled into giant balls and cars tumbled into shapeless masses. Chickens and turkeys may have their feathers stripped away, and some have lived through this experience. The reduced pressures may cause corks to pop from bottles.

Two kinds of waterspouts (Fig. 29–11) occur. One kind develops out of a giant cumulonimbus cloud as does a tornado over land, although it tends to be somewhat less violent. The water level beneath such a tornado funnel may rise by a foot or two, and large quantities of spray are lifted from

* Vonnegut, Bernard, "Inside the Tornado," *Natural History*, April 1968.

29-11. Waterspouts. *(Courtesy NOAA.)*

the surface. The other kind of waterspout is relatively harmless and not a tornado; it is similar to the dust devils that form occasionally over land. These whirls begin at the surface and extend upward and may form beneath a cloudless sky. A small cloud sometimes forms above this type of waterspout at the altitude where condensation occurs in the whirling, upward-moving air. The whirling motion may be clockwise or counterclockwise, and humidity differences may be a factor in starting the local upward current (moist air is less dense than dry air under similar conditions).

The "storm cellar" of the midwesterner is the safest place in a tornado. In the United States the storms usually move northeastward; thus a place on the floor along the southwest wall and under a bed or table provides some protection. Windows and doors should be opened on the northeast side. If a tornado occurs at night, the first warning may be a loud roar. This means the tornado is about to strike, and one has very little time to do anything about it. A prone position in a hole offers the best chance of survival.

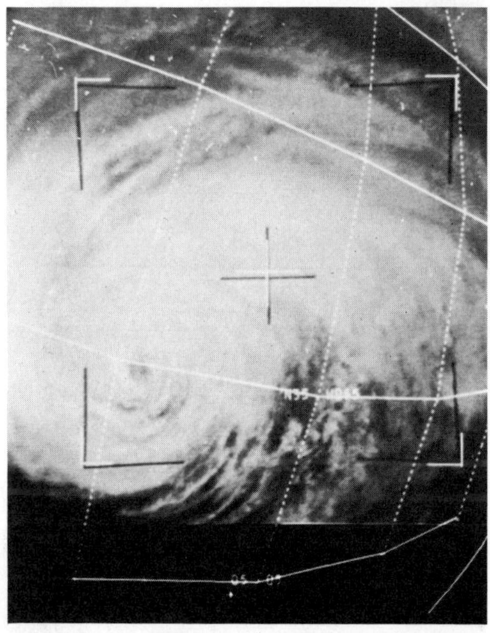

29-12. Hurricane Faith, 1 September 1966, about 300 miles off Cape Hatteras, North Carolina. *(Courtesy ESSA.)*

Hurricanes

Hurricane, typhoon, and tropical cyclone (Figs. 29–12 and 29–13) are synonymous names used in different countries for a particularly violent type of circular storm characterized by torrential rains, thunder and lightning, and powerful winds that swirl about a central area of low barometric pressure (the eye). Giant cumulonimbus clouds form a towering wall of thunderstorms around the eye and are arranged in spiral bands elsewhere within the storm (Fig. 29–14). A hurricane somewhat resembles a whirling phonograph record because it is actually quite thin (several miles from bottom to top vs. several hundred miles in diameter). Hurricanes have characteristic features and follow certain paths, but many variations occur. Wind speeds increase from

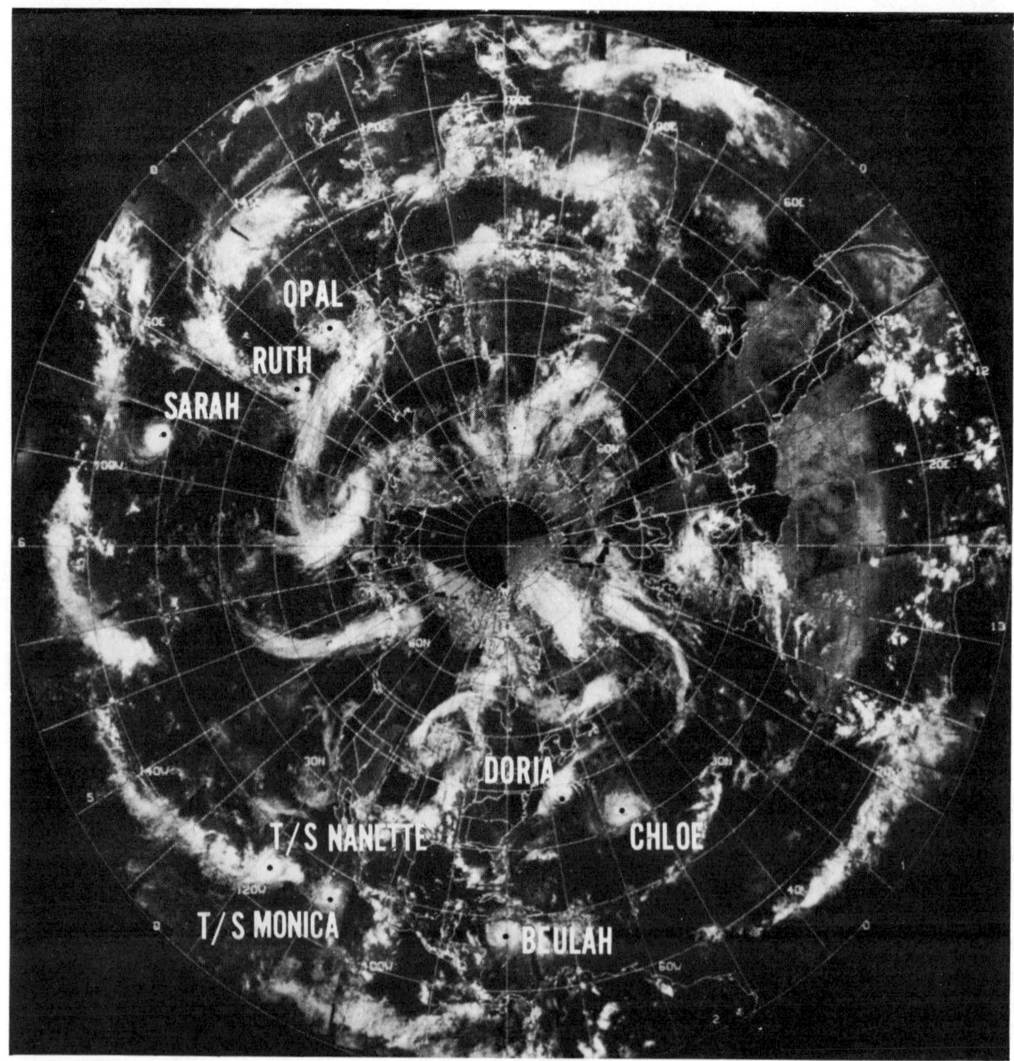

29-13. Six hurricanes and two tropical storms can be seen in this satellite photo taken on 14 September 1967. The small black circles locate the centers of the storms. *(Courtesy NOAA.)*

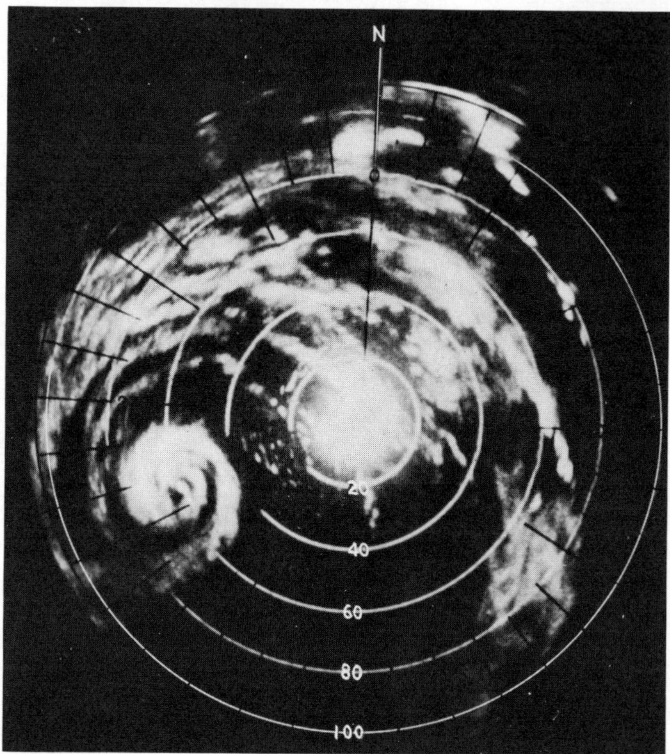

29-14. Photograph of radarscope showing a hurricane. The concentric circles are spaced 20 miles (32 km) apart. Note the eye located 60 miles westsouthwest of the center. The white areas show regions of precipitation and the locations of cumulonimbus clouds within the hurricane (clouds without precipitation do not show). Note their spiral arrangement and maximum development in a ring enclosing the eye. *(Courtesy U.S. Weather Bureau.)*

the outer margins inward to a maximum within the eye-wall clouds, and speeds here may exceed 150 to 200 mi/hr, although this is quite exceptional.

A *hurricane* has closed concentric isobars that are closer together nearer the eye (i.e., the pressure gradient steepens toward the eye), a conspicuous rotary motion, and wind speeds of 74 mi/hr (64 knots or 120 km/hr) or more. The central region in which hurricane-force winds occur is relatively small (say 100 miles in outer diameter), whereas winds exceeding 40 mi/hr may occur throughout a much larger area (say 400 miles in diameter). A *tropical storm* is a preceding, less intense stage which is characterized by closed isobars, rotary motion, and wind speeds of 39–73 mi/hr (34–63 knots or 63–118 km/hr). A tropical storm may liberate as much energy via condensation as does a hurricane, but it lacks a ring of very intense eye-wall winds; its energy is less concentrated. Tropical depressions and tropical disturbances (no closed isobars and little or no rotary circulation at the surface) seem to be earlier and weaker stages.

About half of all tropical storms seem to develop into full-fledged hurricanes, but only a fraction of the disturbances and depressions continue growing. The numbers of tropical storms reported have increased greatly in the past decade or so, especially since the advent of the weather satellites, but this seems due to better reconnaissance rather than to greater actual frequency.

In recent years, the U. S. Weather Bureau has been applying feminine names to hurricanes; the first hurricane of the season in the North Atlantic receives a name beginning with the letter A, the second one with B, etc. A new set of names is made up for each year.

Hurricanes originate near the equator over warm water—about 80°F (27°C) or warmer.

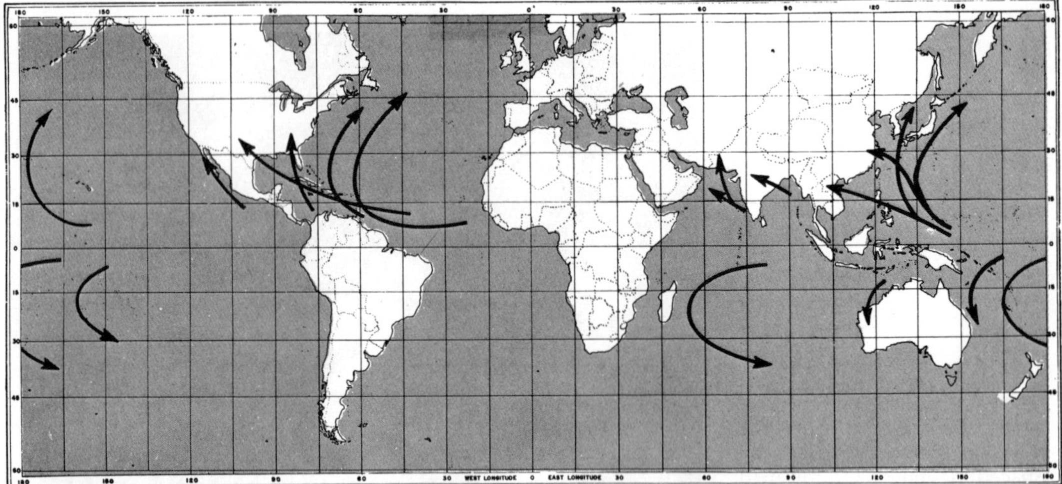

29-15. Arrows show regions where hurricanes tend to form (at latitudes of 5 to 15 degrees) and the paths they tend to follow. The water may be too cold for hurricanes to form in the South Atlantic. (AF Manual 105-5, *Weather for Aircrews*.)

However, they cannot originate at the equator where the Coriolis effect is zero. Hurricanes are most common and last longest (a 12-day average for August storms) during late summer and early fall when surface ocean water is warmest. At such times, the trade-wind belts may converge far enough from the equator to be influenced by the Coriolis deflection.

As tropical storms develop and grow into hurricanes, they tend to move slowly westward and slightly away from the equator across the trade-wind belts (Fig. 29–15). Next they tend to cross the horse latitudes and move in a clockwise direction (Northern Hemisphere) at increasing speeds around the western ends of the semipermanent high pressure cells located over the oceans at latitudes of 30 to 35 degrees. The southwestern portion of the North Pacific spawns the greatest number of hurricanes (typhoons), and these affect places such as the Philippines, Taiwan, and Japan.

Hurricanes that affect the eastern United States form over the eastern Atlantic near the Cape Verde Islands and move westward toward the West Indies. Some hurricanes then continue westward across the Gulf of Mexico, whereas others curve northward toward Florida and thence northeastward along the coast (Fig. 29–16). Most of these hurricanes continue curving northeastward across the Atlantic and do not damage the northeastern United States, but some do, and these are long remembered. Irregular looping movements also develop here and there and make precise prediction difficult. An entire hurricane may move at 10 to 30 mi/hr (16 to 48 km/hr), but rates of 60 mi/hr have been observed, and 12 mi/hr may be an average, particularly at lower latitudes. A full-fledged hurricane commonly lasts from several days to more than a week.

Relatively little wind occurs within the eye of a hurricane, and clouds are scattered or absent. Eyes may average 10 to 20 miles (16 to 32 km) in diameter but some have spanned 40 miles, and the diameter of an eye tends to increase with altitude. Temperatures tend to be about 5° to 30°F (3° to 16°C) warmer at all levels within an eye as compared to similar levels near the margins. A solid wall of towering thunderstorm clouds rings an eye. Similar cumulonimbus clouds with their intense rainfall are arranged in spiral bands about an eye. Any one thunderstorm cell exists for only a short time, but others tend to form here and there along the line, and thus the spiral structure is preserved.

Barometric pressure within an eye commonly does not drop much below 28 inches (as measured by a mercurial barometer), but pressures between 26 to 27 inches have been recorded. The difference in pressure between the center and margin of a hurricane, therefore, does not tend to exceed 1 to 2 inches, but barometers have been observed to fall more than 1 inch in an hour. Pressure changes produced by the passage of a hurricane are relatively slow in contrast to those caused by tornadoes; closed buildings do not tend to explode outward.

Rainfall at any one location along the path of a hurricane may average 6 to 12 inches, but much heavier rains can occur if a hurricane moves very slowly. A rainfall of 6 feet 4 inches was once recorded in 24 hours at Baguio in the Philippines, and on another occasion about 8 feet of rain deluged Jamaica in 4 days. Such rains can produce disastrous flash floods. Perhaps one-fourth of all precipitation in areas such as the southeastern United States, Japan, India, and southeastern Asia may be produced by hurricanes and tropical storms.

The arrival of the eye at a land location means a brief respite from the fury of a storm (for 1 to 2 hours or more) but then the winds abruptly resume their violent movement (more destructive than a similar wind that has built up gradually), although now they blow in the opposite direction. For a ship at sea, however, there is little relief within the eye, for monstrous waves occur there. Apparently waves cannot grow very high in the region surrounding an eye because the tops of the waves are blown off by the strong winds.

The central core of a hurricane (Fig. 29–17) is warmer at all levels than is its peripheral zone at comparable levels. This makes air near the center less dense and more buoyant than the cooler air near the margin; therefore, it rises. Air moves from

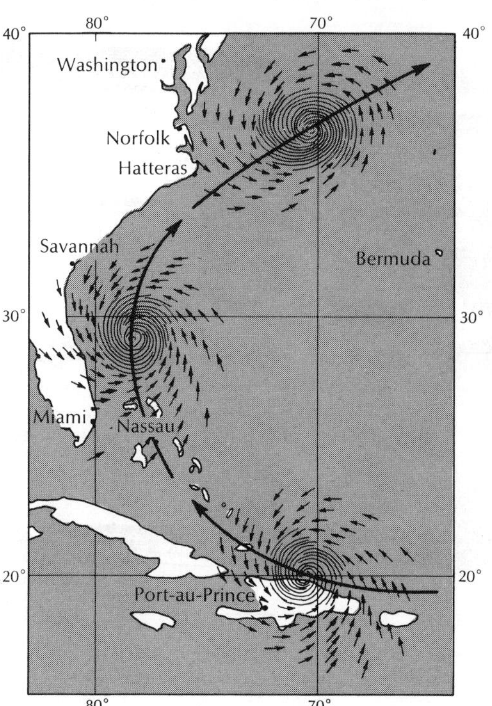

29-16. Typical track and wind system of a hurricane in the North Atlantic. *(Courtesy U.S. Weather Bureau.)*

the perimeter inward toward the low-pressure center to replace this rising air and undergoes a Coriolis deflection. Moreover wind speeds increase toward the center because the pressure gradient steepens in that direction. Since greater speeds mean greater Coriolis deflection, these inward-swirling winds move more nearly parallel to the isobars as they approach the center. Eventually the winds are moving about parallel to the isobars, and thus do not quite get to the center. This leaves a relatively quiet central zone—the eye—in which air tends to descend, and around which air swirls violently upward.

The main source of energy of a hurricane is the latent heat energy released during condensation, and the bulk of this occurs from the warm moist air that rises within the ring of towering thunderstorms enclosing an eye. However, some energy is also released by condensation within similar

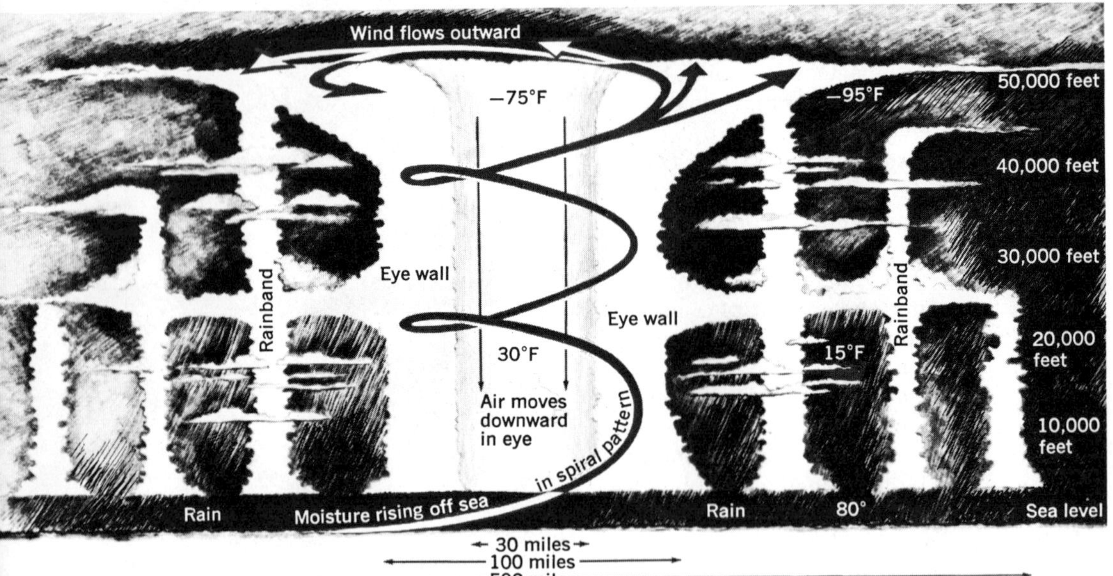

29-17. The hurricane model. *(Courtesy Robert H. and Joanne Simpson.)*

cumulonimbus clouds scattered in the spiral rainbands that nearly parallel the swirling winds. Upward motion in the eye-wall clouds is aided by the heat energy released during condensation, by convergence along the surface toward the eye, and by a divergent movment at the top. This divergence moves air outward, and the Coriolis deflection shapes it into a clockwise swirl. This air is transported horizontally outward at high altitudes until it sinks beyond the margin of the storm. Without this divergence at the top, air would accumulate at the center, barometric pressures would thus increase, and the entire circulation would be slowed and perhaps halted.

Of about 100 initial disturbances, perhaps only one eventually becomes a full-fledged hurricane, but the exact sequence of events leading to this development is still unknown. Involved may be the more or less chance vertical alignment of two atmospheric disturbances: one produces convergence and an inward counterclockwise swirl along the surface; the other produces divergence and an outward clockwise swirl at the top. The divergent disturbance at the top may be fully as important as the convergent disturbance along the surface.

A hurricane weakens as it moves across land or colder ocean water because its supply of moisture is greatly reduced. Friction as winds move across irregular land surfaces is a smaller factor. In addition, rain cools the surface, which in turn cools the lower air and decreases its upward movement. Raindrops falling on the ocean's surface mix with the upper water and have little effect upon its temperature.

The heat energy released by condensation in one hurricane in one day may be equivalent to that released by the explosion of 8000 nuclear bombs—each the equal of 1 million tons of TNT explosive. Thus man cannot hope to overpower a hurricane. However, he can hope to understand it better and perhaps to control its development by causing a relatively small modification at a critical time and place. Project Stormfury (Essa and the U. S. Navy) is such a program.

One approach involves a massive seeding of the outer eye-wall clouds with silver iodide. This may sharply increase condensation and release enough heat energy to produce more upward motion farther from the center. In effect, this might shift the eye-wall zone of intense winds farther from the center and thereby reduce the violence of its winds (p. 310, the conservation of angular momentum is involved—an increase in distance from the center is offset by a decrease in speed). The eye-wall zone of hurricane Debbie was seeded twice in August 1969 (also part of its spiral rainband system) and on each occasion maximum wind speeds weakened several hours after seeding (by about 30 and 15% respectively). The results were considered quite promising.

Another approach involves seeding thunderstorm clouds in the spiral rainbands in an attempt to increase the upward motion farther from the eye-wall zone (presumably this would reduce upward motion within the eye-wall zone). Another possibility: decrease the amount of evaporation that occurs from the warm ocean water beneath the storm, thus cutting off the storm's fuel supply. This might be done by spreading an inhibiting chemical over the sea surface (difficult where waves are vigorous). It has also been suggested that an incipient disturbance might be hampered in its development by seeding nearby areas to produce competing disturbances.

Men and animals have been killed by hurricanes and, on certain tragic occasions, in very large numbers. Most of the deaths have occurred in low-lying coastal areas and have been caused by drowning. Onshore winds may pile up water along a coast, and irregularities in the coastline may concentrate this upon certain areas. Furthermore, a large drop in barometric pressure at a storm center may cause the water level to rise several feet higher. Huge wind-driven waves are superimposed upon these high waters and smash against a shore to add to the destruction.

The East Pakistan flood of December 1970 —one of the Earth's all-time natural disasters with its 250,000 to 300,000 or more deaths—involved most of these conditions. The cyclone itself was not exceptional (120 mi/hr winds), but it helped to produce a wall of water about 18 feet above mean sea level—nearly twice as high as the highest ground on many of the very heavily populated islands and coastal areas that it swept across. With limited communication and transportation facilities, evacuation was impossible for most of the population.

For ships in the Northern Hemisphere, the right side of a hurricane (facing in the direction of its movement) is more severe than the left side because the forward motion of the storm is added to the counterclockwise movement of the winds.

Hurricanes can now be tracked by means of planes, radar, and weather satellites, and their paths can be predicted with some precision. Specially reinforced and equipped planes fly into and out of a storm, locate the eye, and report upon its intensity and direction of motion. Special balloons carrying self-contained, automatic instruments may be dropped into an eye and drift with it, making vital observations and reporting them by radio. Nearly continuous observation of critical areas by weather satellites in recent years is probably of greatest value. Thus hurricanes tend not to come without warning. People have time to protect themselves and their possessions: they can board up windows, move boats and planes, and evacuate low-lying, exposed locations. However, property damage is still very great, and damage caused by a single hurricane in the United States has exceeded 1 billion dollars.

Carla, the third tropical storm of 1961, grew into a particularly large, violent hurricane that caused an estimated damage of ½ billion dollars along the Texas-Louisiana coast. At the height of the storm, hurricane-force winds extended outward for 300 miles from the 30-mile-wide eye at its center. The hurricane approached the Louisiana coast and seemed ready to strike land on Saturday night (September 9). However, a high-pressure system to the north and east blocked its forward progress. Clockwise circulation around this high then shifted the hurricane gradually westward along the coast until it went inland on Monday near the southern tip of Texas. The delay had given time for successful evacuation of many thousands of persons living on very low ground near the coast.

At Port Lavaca, Texas, a wind gauge blew away at 153 mi/hr and the wind-propelled rain formed sheets of water parallel to the surface. According to one observer: "This isn't rain, it's water from the next county." Treetops touched the ground and remained horizontal for a time. As a minor result of the storm, the higher ground and debris crawled with snakes. The Houston zoo reportedly collected nearly half a ton of snakes. Others were discovered in buildings, beds, machinery, and cupboards, and one even crawled out of a bathtub drain to share the tub momentarily with the departing occupant.

About forty deaths were caused by the storm, but about one-third of these were due to tornadoes associated with the hurricane. However, many thousands would undoubtedly have drowned if they had not been warned in time and moved inland ahead of the flooding waters. The hurricane subsequently weakened and changed into a low-pressure system that moved northeastward.

Hurricanes resemble the low-pressure systems (extratropical cyclones) of the middle latitudes in some ways but differ from them in many features. Similarities: winds converge and spiral inward and upward about a center of low barometric pressure; widespread areas of cloudiness and precipitation develop; and the direction of motion is counterclockwise in the Northern Hemisphere.

Some differences: hurricanes are considerably smaller but much more violent and have an eye at their centers. Hurricanes occur within one type of air mass, whereas a low is usually associated with two or more air masses and with two or more frontal systems. Hurricanes lack these phenomena. Hurricanes are more symmetrical and have steeper pressure gradients. Thus their isobars form near-circles and are closer together. Precipitation is much heavier from a hurricane. Lows are best developed during the colder months of the year, whereas hurricanes are late summer and early fall developments. However, hurricanes commonly change into extratropical cyclones as they weaken and encounter cold fronts in the middle latitudes.

Exercises and Questions for Chapter 29

1. Where and when are thunderstorms most frequent?
   (A) Why?
   (B) What exceptions occur?
2. Why are sudden, heavy, relatively brief, local downpours commonly associated with thunderstorms?
3. Describe some reasons why cumulus clouds commonly do not continue to develop and thus evolve into the mature and dissipating stages of a thunderstorm.
4. What causes the downdraft in the mature stage of a thunderstorm?
   (A) Why does it tend to spread across the lower part of a cell to form the dissipating stage?
   (B) Why do isotherms tend to bulge downward where they cross a downdraft?
   (C) Why does an anvil top tend to develop on a cumulonimbus cloud during the dissipating stage?

5. A thunderstorm cell 3 miles in diameter may contain 500,000 tons of water that have condensed (assume that it is all water, although some would probably be snow and ice).
   (A) How much latent heat energy does this quantity of condensation liberate?
   (B) How many tons of water could be raised from the freezing point to the boiling point by this quantity of heat energy (assume no evaporation)? There are 453.6 grams per pound, but use 450.
6. Assume that 1 inch of rain falls on an area. How many tons of water per square mile does this amount to? One cubic foot of water weighs 62.4 pounds, and there are 63,360 inches in a mile—however, use 60 and 60,000 in your calculations.
7. What causes the funnel cloud of a tornado?
8. What conditions seem to favor the origin of tornadoes? Where and when in the United States do these conditions occur most frequently?
9. Compare and contrast a low-pressure system with a hurricane.
10. Describe the place of origin, paths, and rates of movement followed by hurricanes that affect the United States.
11. Compare and contrast the destructiveness of a tornado with that of a hurricane.

Appendix: Atoms and Related Phenomena

Matter

You and I, rocks, air, water, the Earth, moon, sun, stars, and other objects in the universe are made of matter, which we may describe as something that occupies space. An understanding of the earth sciences, therefore, is intimately interwoven with an understanding of the nature of matter and with certain other topics from the sciences of chemistry and physics: e.g., atoms, elements, compounds, mass, energy, heat, radioactivity, chemical reactions, nuclear reactions, and electromagnetic radiations. A simplified discussion of some of these topics is the subject matter of this appendix.

Matter may exist in three states—gas, liquid, and solid—and transformations from one state to another may occur under proper conditions of temperature and pressure. One type of matter may differ from another in color, hardness, mass, density, and temperature, and some types of matter react chemically when they come in contact with other types. However, matter itself is difficult to define precisely. Matter exerts a gravitational pull on other objects and is in turn attracted by them. Matter has inertia; i.e., if at rest, it tends to remain in a state of rest; and if in motion, it tends to resist any change in this motion.

Atoms are the basic units that make up matter, but atoms themselves are divisible into a large number of still smaller units. Atoms are in constant motion at all temperatures above absolute zero (—273°C, about −460°F), the rate of motion is more rapid at higher temperatures; in fact, the rate of motion determines the temperature. At a given temperature, the molecules in a gas have an average rate of motion, but some molecules move more rapidly than this average rate and others less rapidly. The rate decreases as temperatures drop, and all motion stops at —273°C, which is thus the lowest temperature attainable.

A solid tends to maintain its shape and volume because the atoms in it are in contact with one another and vibrate back and forth in fixed positions within a rigid, three-dimensional pattern. On the other hand, a liquid has a certain volume but is shaped by its container; its atoms are in contact, but they are not part of a rigid pattern. In contrast, a gas diffuses throughout the container that shapes it. The atoms in a gas may be combined into molecules, but the molecules are in motion and are spaced relatively far apart. Thus gases are readily compressed, whereas solids and liquids are not. Gas pressure results from the impacts of moving molecules against objects. The pressure is greater at high temperatures and pressures where molecular motion is more rapid, impacts are more powerful, and more molecules occur within a unit volume.

A solid may be changed into a liquid (it melts) if temperatures are increased enough. At such high temperatures its atomic units move too rapidly to be kept in a rigid pattern. On the other hand, a gas condenses into a liquid if temperatures are decreased enough and pressures are increased; the rate of movement and the spaces separating the molecules are both decreased. About 98–99% of all of the matter in the universe probably consists of two gases; hydrogen (by far the more abundant) and helium.

Atoms

Although atoms cannot be observed directly, they have been studied indirectly in many ingenious ways: e.g., by radioactivity, spectral emissions and absorptions, behavior in magnetic and electric fields, and tracks produced by particles moving through cloud chambers (Fig. 1). According to the precise quantitative data thus obtained, an atom consists of two main parts. A *nucleus* contains almost all of the mass of the atom, and a system of electrons moves about the nucleus. Many elementary particles make up an atom, but only three of these need to be considered here: the proton, neutron, and electron. They may be described in terms of mass and electric charge.

An electron has about the least mass of any elementary particle (except some that have none). Thus its mass (at rest, 9.1083×10^{-28} gram) can be used as a unit of mass in describing the masses of other particles. An electron has a unit negative electric charge, whereas a proton has the same quantity of electric charge, but it is positive; thus the charge of one neutralizes that of the other. The rest mass of a *proton* is 1836.12 times larger than that of an electron, whereas that of a *neutron* is 1838.65 times larger. Protons and neutrons (electrically neutral) occur within the nuclei of atoms.

The number of protons within a nucleus determines the type of element to which an atom belongs. If only one proton occurs in the nucleus of an atom (the simplest kind of atom), the element is hydrogen; if the number of protons is 2, the element is helium; 3, lithium; 6, carbon; 8 oxygen; 82, lead; and 92, uranium. Thus all the atoms of any one element have the same number of protons, which is called the *atomic number* of that element. In an electrically neutral atom, the number of protons and electrons are equal.

More than 100 elements are known (105 in 1970), but only 92 of these occur naturally on the Earth; the others have been made artificially by nuclear transformations (some may also form in the interiors of certain stars). An element cannot be subdivided into something smaller than an atom and still retain its characteristic properties.

The sum of the protons and neutrons within the nucleus of an atom is the *mass number* of that atom. Commonly the atomic number is shown as a subscript preceding the chemical symbol, whereas the mass number is a superscript following the chemical symbol. Thus we have $_2He^4$ because helium has an atomic number of 2 and a mass number of 4; i.e., it has two protons and two neutrons in its nucleus and two electrons normally move about the nucleus.

Although all atoms of a particular element must have an identical number of protons, there may be some variation in the number of neutrons present. These are the different isotopes of an element; e.g., hydrogen has

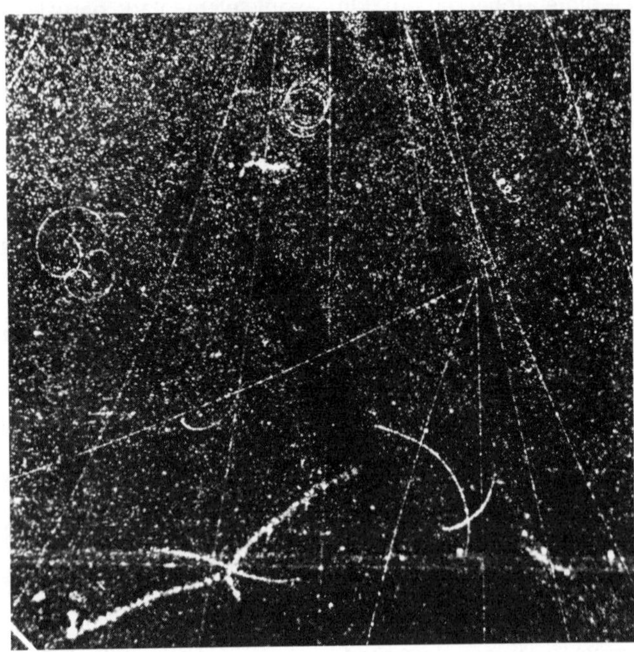

1. Cloud-chamber photograph of a neutron-proton collision (Brookhaven Cosmotron). Air within a cloud chamber is made supersaturated with water vapor so that condensation will take place if suitable nuclei are present. Therefore, if an atomic particle such as a proton speeds through the chamber, it produces ions along its path, and the ions act as condensation nuclei. A "fog" results along the track of a speeding particle and such trails can be seen and photographed. Occasionally particles collide, and the resulting tracks provide clues concerning the nature of the colliding particles and their by-products. In this illustration, most of the tracks are produced by protons and mesons that enter from the top. Note, however, that three tracks begin at one point in right center. This records the collision of a neutron that entered at the top (it did not leave a streak) with a proton. The proton then moved toward the bottom (center) and two mesons left streaks on either side of it. *(Courtesy Brookhaven National Laboratory.)*

three isotopes. Each hydrogen isotope has one proton and one electron (if it is not ionized), but the number of neutrons varies from none to two. Therefore all hydrogen isotopes have atomic number 1, but their mass numbers may be 1, 2, or 3.

The *atomic weight* of an element refers to its total mass. It is generally not a whole number because it consists of a weighted average of the several isotopes of that element that occur in nature.

An atom has been likened to the solar system, although the analogy is strictly limited. Most of the mass of the solar system is in the sun (nucleus); planets revolve in orbits about the sun (electrons move about a nucleus but not in fixed orbits); the gravitational attraction of the sun keeps the planets in their orbits (the positive electric charge on the nucleus attracts the negative electric charges on the electrons); and the solar system is chiefly space as is an atom.

Atomic Structure and Spectra

In 1913 the Danish physicist Niels Bohr advanced the theory that electrons in an atom move in certain definite orbits about a central nucleus. Although electrons do not move in definite orbits, we can think of them as moving at certain average distances from a nucleus. We can speak of these average distances as *energy levels,* and we imagine the higher energy levels to be located farther from a nucleus. Bohr incorporated part of the quantum theory into this explanation: an atom can radiate away or absorb energy only in discrete units called *photons* or quanta. As long as an electron moves at the same energy level, it does not gain or lose energy. However, if it shifts to a higher energy level because it receives energy from some outside source, then it tends within a tiny fraction of a second to return to its former energy level and release this energy.

The energy added to an orbiting electron may come from a photon or an electron from the outside. If the source is an electron that enters the atom from the outside, then some of its energy of motion will be transferred if it collides with an orbiting electron. After the collision, the bombarding electron departs minus the energy it gave up to the orbiting electron; thus it moves more slowly. If the source is a photon carrying the proper quantity of energy, the entire photon is absorbed and disappears. However, only certain photons can be absorbed by any one atom. This absorbed energy tends to be given off at once, perhaps in the form of a similar photon, or perhaps in the form of a number of photons of lower energy.

When an orbiting electron absorbs energy in this manner, it is said to become *"excited."* We imagine the electron as immediately jumping to a higher energy level farther from the nucleus (Fig. 2). However, this electron tends to remain at the higher energy level for only a small fraction of a second. It then jumps back to a lower energy level closer to the nucleus and gives off the energy it had just absorbed (as a photon of visible light if the proper quantity of energy is involved). According to this view, only certain discrete quantities of energy can be absorbed or emitted because an electron can make only certain prescribed jumps from one energy level to another. Different jumps are possible; e.g., from a fourth energy level to the third, or from fourth to second, or from fourth to first; a combination of jumps may also result, such as 4 to 3, 3 to 2, and finally 2 to 1. A photon of energy is emitted at each jump, and the quantity depends upon the difference in energy levels.

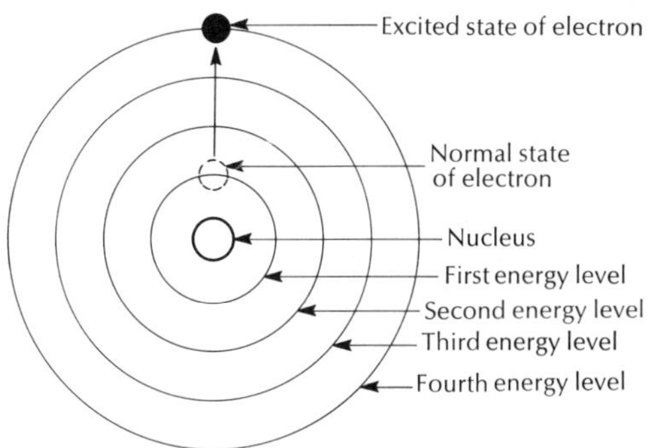

2. Schematic view of an electron in an excited state. Energy from the outside was absorbed by this electron, and it shifted to a higher energy level. Whether the excited electron jumps from its normal state to the second, third, or fourth energy level depends upon the amount of energy it absorbs. However, it can absorb only certain discrete quantities below the critical quantity that causes it to leave the atom entirely—the process of ionization. An excited electron tends to return immediately to its normal state, and it may do so in one jump (thereby emitting the same amount of energy that it had absorbed), or it may return by jumping in succession from one energy level to the next. Each jump then results in the emission of a photon, but of less energy than that emitted in the longer single jump; e.g., the energy absorbed in one jump from 1 to 4, would equal the total energy emitted in the three jumps from 4 to 3, 3 to 2, and 2 back to 1.

This theory was partly developed to explain the existence of a series of spectral lines (p. 506) that had been observed in 1885 in the visible spectrum of hydrogen (Fig. 3). Other series of lines were also discovered in the hydrogen spectrum and these were related and explained in a similar manner.

The amount of energy carried by a certain photon (equivalent to a particular wavelength) varies inversely as the wavelength—red light has less energy than blue, which in turn has less energy than violet. A single

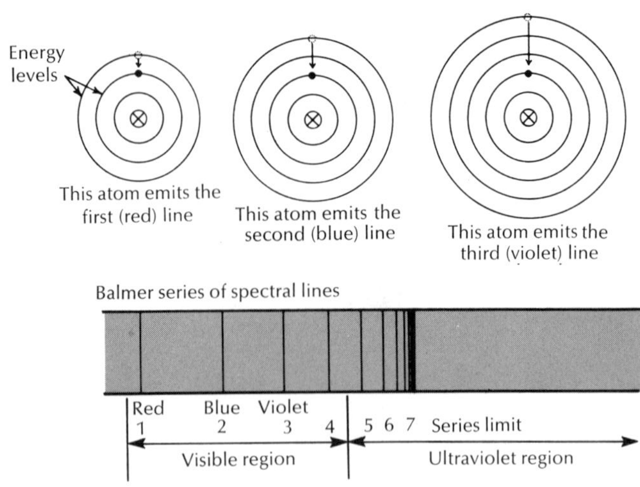

3. Schematic view of the relationship between electronic transitions and spectral lines (Balmer series). This group of lines is located in the visible and near-ultraviolet region and is imagined to result from electrons dropping back to the second energy level. Balmer worked out an equation by means of which the wavelengths of the spectral lines of hydrogen can be calculated. In modified form, the equation is

$$\frac{1}{L} = K\left(\frac{1}{n_1^2} - \frac{1}{n_2^2}\right)$$

L is the wavelength of a spectral line; K is a constant; n_1 is an integer whose value is 2 for the Balmer series; n_2 is another integer whose lowest value is 3, but which may have successively larger values of 4, 5, 6, etc. The smallest value (3) gives the longest wavelength (the red line, which is associated with the least energy). The value of 4 gives the blue line, etc.

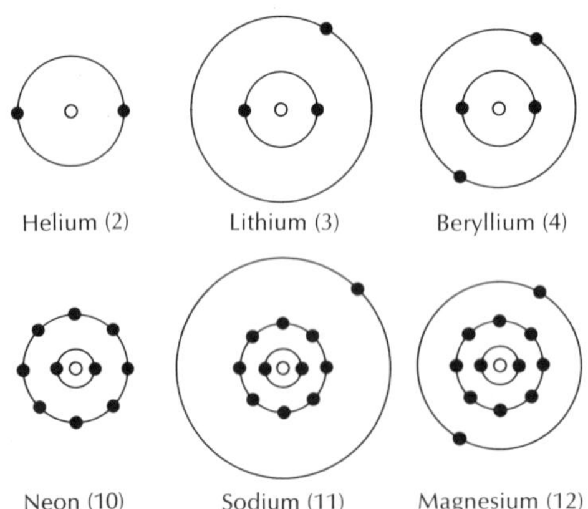

4. Schematic view of the electron structures of some lighter atoms. Atoms with similar outermost electron configurations tend to have similar chemical properties. Thus helium and neon are inert gases, because their outermost electron shells are filled. Lithium and sodium, on the other hand are active metals, because each tends to lose one electron. (R. H. Baker, *Astronomy*, 8th ed., Van Nostrand Reinhold, 1964.)

spectral line results from each jump from a higher energy level to a lower one. Since each element has a unique configuration of electrons, each element as a glowing gas produces a unique spectrum.

Chemical Reactions

Chemical reactions are considered very briefly here, chiefly to distinguish them from nuclear reactions. Electrons move about atomic nuclei and are arranged in a systematic manner (Fig. 4). Atoms are most stable when they have a certain number of electrons in their outermost shells; this is two for the first shell and eight for most of the others. The stable atoms with these configurations are the inert gases.

If an atom loses or gains one or more electrons, it becomes an *ion* and is said to be ionized; it no longer is electrically neutral. A number of elements have only a few electrons in their outermost shells (chiefly the metals), and a tendency exists for these electrons to be lost. On the other hand, other atoms have more than four electrons in their outermost shells and they tend to gain electrons (chiefly the nonmetals). Thus several elements (a family) have only one electron in their outermost shells, and these elements tend to have similar chemical properties; another family consists of elements with two electrons in their outermost shells, etc. In fact, this is the basis for the arrangement of elements (Table 1) into a periodic chart. The atoms are arranged in order of increasing atomic number from 1 to 103. However, they are divided into a number of horizontal rows (periods) in such a manner that elements with similar properties (i.e., similar electronic configurations in their outermost shells) fall into vertical columns.

The size of an atom does not increase steadily with increasing atomic number; e.g., an atom of gold (atomic number 79) is smaller in volume than an atom of sodium (atomic number 11). The 79 protons in each gold nucleus produce a large electrical attraction for the electrons moving about them, which tends to shrink the sizes of their electron shells. The smallest atoms have diameters only about one-fifth those of the largest.

Chemical reactions involve changes in the outermost electron shells. In one type of reaction, or chemical bonding, the atoms of an element with only one electron in the outermost shell combine readily with the atoms of another element that has seven electrons in the outermost shell (p. 65). Sodium and chlorine exemplify this type of chemical union, which involves ionization. The atoms of sodium each lose an electron and become positively charged ions, whereas the atoms of chlorine each gain an electron and become negatively charged ions. The opposite electric charges on the ions thus bind them into the three-dimensional structure of sodium chloride. In another type of chemical bonding, electrons are shared rather than transferred.

Nuclear Reactions

Nuclear reactions produce the energy that is radiated into space by the sun and other stars and involve a great deal more energy than is liberated during chemical reactions. Nuclear reactions are discussed briefly on p. 602, but the proton-proton reaction and carbon cycle are described below. Three main steps occur in the proton-proton reaction:

$$
\begin{aligned}
{}_1H^1 + {}_1H^1 &\rightarrow {}_1H^2 + \beta^+ \\
{}_1H_2 + {}_1H^1 &\rightarrow {}_2He^3 + \gamma \\
{}_2He^3 + {}_2He^3 &\rightarrow {}_2He^4 + 2\ {}_1H^1
\end{aligned}
$$

In the first step, two hydrogen nuclei (each is a single proton) unite to form a hydrogen isotope with a mass number of 2 (this is the deuteron). A positron (a positively charged electron, symbol β^+) is liberated and represents a release of energy. In the second step, the deuteron combines with a proton to produce a helium isotope with a mass number of 3. Again energy is liberated, this time in the form of gamma radiation (γ). In the third step, a helium nucleus with a mass number of 4 is formed and two protons are liberated. The proton-proton reaction is probably most important in the sun and other stars of relatively low temperature.

The carbon cycle is probably important in the hotter stars and involves the six steps listed below:

$$
\begin{aligned}
{}_6C^{12} + {}_1H^1 &\rightarrow {}_7N^{13} + \gamma \\
{}_7N^{13} &\rightarrow {}_6C^{13} + \beta^+ \\
{}_6C^{13} + {}_1H^1 &\rightarrow {}_7N^{14} + \gamma \\
{}_7N^{14} + {}_1H^1 &\rightarrow {}_8O^{15} + \gamma \\
{}_8O^{15} &\rightarrow {}_7N^{15} + \beta^+ \\
{}_7N^{15} + {}_1H^1 &\rightarrow {}_6C^{12} + {}_2He^4
\end{aligned}
$$

Table 1. THE CHEMICAL ELEMENTS

Element	Sym-bol	Atomic Number	Atomic Weight	Element	Sym-bol	Atomic Number	Atomic Weight
Hydrogen	H	1	1.0080	Iodine	I	53	126.90
Helium	He	2	4.003	Xenon	Xe	54	131.30
Lithium	Li	3	6.939	Cesium	Cs	55	132.91
Beryllium	Be	4	9.012	Barium	Ba	56	137.34
Boron	B	5	10.811	Lanthanum	La	57	138.91
Carbon	C	6	12.01	Cerium	Ce	58	140.12
Nitrogen	N	7	14.007	Praseodymium	Pr	59	140.91
Oxygen	O	8	15.999	Neodymium	Nd	60	144.24
Fluorine	F	9	18.999	Promethium	Pm	61	(147)
Neon	Ne	10	20.183	Samarium	Sm	62	150.35
Sodium	Na	11	22.99	Europium	Eu	63	151.96
Magnesium	Mg	12	24.32	Gadolinium	Gd	64	157.25
Aluminum	Al	13	26.98	Terbium	Tb	65	158.92
Silicon	Si	14	28.01	Dysprosium	Dy	66	162.50
Phosphorus	P	15	30.97	Holmium	Ho	67	164.93
Sulfur	S	16	32.064	Erbium	Er	68	167.26
Chlorine	Cl	17	35.453	Thulium	Tm	69	168.93
Argon	Ar	18	39.948	Ytterbium	Yb	70	173.04
Potassium	K	19	39.102	Lutetium	Lu	71	174.97
Calcium	Ca	20	40.08	Hafnium	Hf	72	178.49
Scandium	Sc	21	44.96	Tantalum	Ta	73	180.95
Titanium	Ti	22	47.90	Tungsten	W	74	183.85
Vanadium	V	23	50.94	Rhenium	Re	75	186.2
Chromium	Cr	24	52.00	Osmium	Os	76	190.2
Manganese	Mn	25	54.94	Iridium	Ir	77	192.2
Iron	Fe	26	55.85	Platinum	Pt	78	195.09
Cobalt	Co	27	58.93	Gold	Au	79	196.97
Nickel	Ni	28	58.71	Mercury	Hg	80	200.59
Copper	Cu	29	63.54	Thallium	Tl	81	204.37
Zinc	Zn	30	65.37	Lead	Pb	82	207.19
Gallium	Ga	31	69.72	Bismuth	Bi	83	208.98
Germanium	Ge	32	72.59	Polonium	Po	84	(210)
Arsenic	As	33	74.92	Astatine	At	85	(210)
Selenium	Se	34	78.96	Radon	Rn	86	(222)
Bromine	Br	35	79.91	Francium	Fa	87	(223)
Krypton	Kr	36	83.80	Radium	Ra	88	(226)
Rubidium	Rb	37	85.47	Actinium	Ac	89	(227)
Strontium	Sr	38	87.62	Thorium	Th	90	232.04
Yttrium	Y	39	88.91	Protactinium	Pa	91	(231)
Zirconium	Zr	40	91.22	Uranium	U	92	238.03
Niobium	Nb	41	92.91	Neptunium	Np	93	(237)
Molybdenum	Mo	42	95.94	Plutonium	Pu	94	(242)
Technetium	Tc	43	(99)	Americium	Am	95	(243)
Ruthenium	Ru	44	101.1	Curium	Cm	96	(247)
Rhodium	Rh	45	102.90	Berkelium	Bk	97	(247)
Palladium	Pd	46	106.4	Californium	Cf	98	(249)
Silver	Ag	47	107.87	Einsteinium	Es	99	(254)
Cadmium	Cd	48	112.40	Fermium	Fm	100	(263)
Indium	In	49	114.82	Mendelevium	Md	101	(256)
Tin	Sn	50	118.69	Nobelium	No	102	(256)
Antimony	Sb	51	121.75	Lawrencium	Lw	103	(257)
Tellurium	Te	52	127.60				

ASTRONOMY

Abell, G., *Exploration of the Universe,* 2nd ed., Holt, Rinehart and Winston, New York, 1969.

——— *Exploration of the Universe: Brief Edition,* Holt, Rinehart and Winston, New York, 1969.

Abetti, Giorgio, *The History of Astronomy,* Henry Schuman, 1952.

Ahrendt, M. H., *The Mathematics of Space Exploration,* Holt, Rinehart and Winston, New York, 1965.

Aller, L. H., *Atoms, Stars, and Nebulae,* rev. ed., Harvard University Press, Cambridge, Mass., 1971.

Annual Review of Astronomy and Astrophysics, Annual Reviews, Inc., Palo Alto, Calif., 0000.

Baker, R. H., and Fredrick, L. W. *Astronomy,* 9th ed., Van Nostrand Reinhold, New York, 1971.

———*An Introduction to Astronomy,* 7th ed., Van Nostrand, New York, 1968.

Baldwin, R. B., *A Fundamental Survey of the Moon,* McGraw-Hill Book Co., New York, 1966.

Bergamini and the editors of *Life, The Universe,* Time, Inc., New York, 1962.

Brandt, J. C. *The Sun and Stars,* McGraw-Hill Book Co., New York, 1966.

Bray, R. J., and Loughhead, R. E., *Sunspots,* John Wiley & Sons, New York, 1964.

Dixon, Robert T., *Dynamic Astronomy,* Prentice-Hall, Englewood Cliffs, N.J., 1971.

Glasby, John S., *Variable Stars,* Harvard University Press, Cambridge, Mass., 1969.

Glasstone, S., *Sourcebook on the Space Sciences,* Van Nostrand, New York, 1965.

Harsanyi, Zsolt, *The Star Gazer,* translated by Paul Tabor, Putnam, New York, 1939.

Hawkins, G. S., *Meteors, Comets, and Meteorites,* McGraw-Hill Book Co., New York, 1965.

Hodge, Paul W., *The Revolution in Astronomy,* Holiday House, New York, 1970.

——— *Concepts of the Universe,* McGraw-Hill Book Co., New York, 1969.

Howard, N. E., *Standard Handbook for Telescope Making,* Crowell, New York, 1959.

Huffer, C. M., Trinklein, F. E., and Bunge, M., *An Introduction to Astronomy,* Holt, Rinehart and Winston, New York, 1967.

Inglis, S. J., *Planets, Stars, and Galaxies,* 2nd ed., John Wiley & Sons, 1967.

Jackson, J. H., *Pictorial Guide to the Planets,* Thomas Y. Crowell, New York, 1965.

Jastrow, R., *Red Giants and White Dwarfs,* revised ed., Harper and Row, New York, 1971.

Jennison, R. C., *An Introduction to Radio Astronomy,* Philosophical Library, New York, 1966.

King, Henry C., *The History of the Telescope,* Sky Publishing Corp., Cambridge, Mass., 1955.

Koestler, A., *The Sleepwalkers,* Macmillan, New York, 1959.

Kopal, Zdenok, *Widening Horizons,* Taplinger Publishing Co., New York, 1970.

Kosofsky, L. J., and El-Baz, Farouk, *The Moon as Viewed by Lunar Orbiter,* NASA Special Publication 200, 1970.

Kraus, J. D., *Radio Astronomy,* McGraw-Hill Book Co., New York, 1966.

Krinov, E. L., *Giant Meteorites,* Pergamon Press, Elmsford, N.Y., 1966.

Land, Barbara, *The Telescope Makers,* Thomas Y. Crowell, New York, 1968.

Lovell, B., and Lovell, J., *Discovering the Universe,* Harper and Row, New York, 1963.

Lowman, P. D., *Lunar Panorama,* Wletflugbild, Feldmeilen/Zurich, 1969.

Lundquist, C. A., *Space Science,* McGraw-Hill Book Co., New York, 1966.

Mayall, Wyckoff, and Polgreen, *The Sky Observer's Guide,* Golden, 1965.

Meadows, A. J., *Early Solar Physics,* Pergamon Press, Elmsford, N.Y., 1970.

Menzel, D. H., Whipple, F. L., and deVaucouleurs, G., *Survey of the Universe,* Prentice-Hall, Englewood Cliffs, N.J., 1970.

Menzel, D. H., *Astronomy,* Random House, New York, 1970.

Menzel, Donald H., *A Field Guide to the Stars and Planets,* Houghton Mifflin, Boston, Mass., 1964.

Miczaika, G., and Stinton, W., *Tools of the Astronomer,* Harvard University Press, Cambridge, Mass., 1961.

Minnaert, M., *The Nature of Light and Color in the Open Air,* Dover Publications, New York, 1954.

——— *Practical Work in Elementary Astronomy,* D. Reidl Publishing Co., Dordrecht, Holland, 1969.

Moore, Patrick, *The Atlas of the Universe,* Rand McNally & Co., Skokie, Ill., 1970.

Mutch, T. A., *Geology of the Moon,* Princeton University Press, Princeton, N.J., 1970.

Page, T. (ed.), *Stars and Galaxies,* Prentice-Hall, Englewood Cliffs, N.J., 1962.

Page, T., and Page, L. W., (eds.), *Neighbors of the Earth,* Macmillan, New York, 1965.

——— *Wanderers in the Sky,* Macmillan, New York, 1965.

Pannekoek, A., *A History of Astronomy,* Interscience Publishers, New York, 1961.

Pecker, Jean-Claude, *Space Observatories,* Springer-Verlag, 1970.

——— *Experimental Astronomy,* Springer-Verlag, 1970.

Rey, Hans A., *The Stars, a New Way to See Them,* Houghton Mifflin, Boston, Mass., 1967.

Richardson, R. S., *Mars,* Harcourt, Brace and World, New York, 1964.

——— *Getting Acquainted with Comets,* McGraw-Hill Book Co., New York, 1967.

Ronan, Colin, *The Astronomers,* Hill and Wang, New York, 1964.

Ronan, Colin A. *Discovering the Universe,* Basic Books, New York, 1971.

Roth, Günter D, *Handbook for Planet Observers,* Van Nostrand Reinhold, New York, 1970.

Sandage, A., *The Hubble Atlas of Galaxies,* Carnegie Institution of Washington, Washington, D.C., 1961.

Sarton, G., *A History of Science,* 2 vols., John Wiley & Sons, New York, 1964, 1965.

Schatzman, E. L., *The Structure of the Universe,* World University Library, 1968.

Science **167**, Jan. 30, 1970 (entire issue devoted to Apollo 11 reports).

Scientific American Readings, *Frontiers in Astronomy,* W. H. Freeman, San Francisco, Calif., 1956-1970.

Shapley, H., *Galaxies,* rev. ed., Harvard University Press, Cambridge, Mass., 1963.

——— and Howarth, H. E. (eds.), *Source Book in Astronomy,* McGraw-Hill Book Co., New York, 1929.

——— (ed.), *Source Book in Astronomy, 1900-1950,* Harvard University Press, Cambridge, Mass., 1960.

Steinberg, J. L., and Lequex, J., *Radio Astronomy,* McGraw-Hill Book Co., New York, 1963.

Struve, O., and Zebergs, V., *Astronomy of the Twentieth Century,* Macmillan, New York, 1962.

Sutton, R. M., *The Physics of Space,* Holt, Rinehart and Winston, New York, 1965.

Taylor, R. J., *The Stars: Their Structure and Evolution,* Springer-Verlag, 1970.

Vehrenberg, Hans, *Atlas of Deep-Sky Splendors,* Sky Publishing Corp., Cambridge, Mass., 1967.

von Braun, W., and Ordway, F. I., III, *History of Rocketry and Space Travel,* Thomas Y. Crowell, New York, 1966.

Whipple, F. L., *Earth, Moon, and Planets,* 3rd ed., Harvard University Press, Cambridge, Mass., 1968.

Wood, John A., *Meteorites and the Origin of Planets,* McGraw-Hill Book Co., New York, 1968.

Woodbury, D. O., *The Glass Giant of Palomar,* Dodd Mead, New York, 1948.

Wyatt, S. P., *Principles of Astronomy,* 2nd ed., Allyn and Bacon, Boston, Mass., 1971.

Young, L. B., (ed.), *Exploring the Universe,* 2nd ed., American Foundation for Continuing Education, McGraw-Hill Book Co., New York, 1971.

GEOLOGY

Adams, A. B., *Eternal Quest,* G. P. Putnam's Sons, New York, 1969.
Adams, F. D., *The Birth and Development of the Geological Sciences,* Dover, New York, 1938.
Ahrens, L. H., *Distribution of the Elements in Our Planet,* McGraw-Hill Book Co., New York, 1965.
Albritton, C. C., Jr., *The Fabric of Geology,* Addison-Wesley, Reading, Mass., 1963.
Beerbower, J. R., *Search For the Past,* Prentice-Hall, Englewood Cliffs, N.J., 1960.
Berry, W. B. N., *Growth of a Prehistoric Time Scale,* W. H. Freeman, San Francisco, Calif., 1968.
Black, Rhona M., *The Elements of Paleontology,* Cambridge University Press, 1971.
Bloom, A. L., *The Surface of the Earth,* Prentice-Hall, Englewood Cliffs, N.J., 1969.
Blyth, F. G. H., *Geological Maps and Their Interpretation,* Edward Arnold Ltd., London, 1965.
Bott, M. H. P., *The Interior of the Earth,* St. Martin's Press, New York, 1971.
Bullard, F. M., *Volcanoes: In History, In Theory, In Eruption,* University of Texas Press, Austin, Texas, 1962.
Chorley, R. J. (ed.), *Water, Earth, and Man,* Methuen, 1969.
Clark, D. L., *Fossils, Paleontology and Evolution,* Wm. C. Brown Co., Dubuque, Iowa, 1968.
Clark, S. P., *Structures of the Earth,* Prentice-Hall, Englewood Cliffs, N.J., 1970.
Cloud, Preston (ed.), *Adventures in Earth History,* W. H. Freeman, San Francisco, Calif., 1970.
Colbert, E. H., *Dinosaurs: Their Discovery and Their World,* Dutton, New York, 1961.
Davis, S. N., and DeWiest, R. J. M., *Hydrogeology,* John Wiley & Sons, New York, 1966.
Davis, W. M., *Geographical Essays,* Dover Publications, New York, 1954.
Dott, R. H., Jr., and Batten, R. L., *Evolution of the Earth,* McGraw-Hill Book Co., New York, 1971.
Dunbar, C. O., and Waage, K. M., *Historical Geology,* 3rd ed., John Wiley & Sons, New York, 1969.
Dyson, J. L., *The World of Ice,* Alfred A. Knopf, New York, 1962.
Eardley, A. J., *General College Geology,* Harper and Row, New York, 1965.
Easterbrook, D. J., *Principles of Geomorphology,* McGraw-Hill Book Co., New York, 1969.
Eicher, D. L., *Geologic Time,* Prentice-Hall, Englewood Cliffs, N.J., 1968.
Fairbridge, R. W. (ed.), *The Encyclopedia of Geomorphology,* Reinhold, New York, 1968.
Faul, H., *Ages of Rocks, Planets, and Stars,* McGraw-Hill Book Co., New York, 1966.
Fenton, C. L., and Fenton, M. A., *Giants of Geology,* Doubleday, New York, 1952.
——— *The Fossil Book,* Doubleday, New York, 1958.
——— *The Rock Book,* Doubleday, New York, 1940.
Flint, R. F., *Glacial and Quaternary Geology,* John Wiley & Sons, New York, 1971.
Foster, R. J., *Physical Geology,* Charles E. Merrill, Columbus, Ohio, 1971.
Gass, I. G. et al., *Understanding the Earth: A Reader in the Earth Sciences,* Artemis Press, 1971.
Glasson, K. R., and McDonnell, K. S., *Graded Exercises in Geological Mapping,* F. W. Cheshire, Melbourne, 1968.
Green, J., and Short, N. M., *Volcanic Landforms and Surface Features,* Springer-Verlag, 1971.
Harbaugh, J. W., *Stratigraphy and Geologic Time,* Wm. C. Brown Co., Dubuque, Iowa, 1968.
Heller, R. L., *Geology and Earth Sciences Source Book,* 2nd ed., Holt, Rinehart and Winston, New York, 1970.
Herbert, D., and Bardossi, F., *Kilauea: Case History of a Volcano,* Harper and Row, New York, 1968.
Holmes, A., *Principles of Physical Geology,* 2nd ed., Ronald Press, New York, 1965.
Hurlbut, C. S., Jr., *Minerals and Man,* Random House, New York, 1969.
Iacopi, R., *Earthquake Country,* Lane Book Co., Menlo Park, Calif., 1964.
Kay, M., and Colbert, E., *Stratigraphy and Earth History,* John Wiley & Sons New York, 1965.
King, P. B., *The Evolution of North America,* Princeton University Press, Princeton, N.J., 1959.
Kummel, B., *History of the Earth,* 2nd ed., W. H. Freeman, San Francisco, Calif., 1970.
Lahee, F. H., *Field Geology,* 6th ed., McGraw-Hill Book Co., New York, 1961.
Laporte, L. F., *Ancient Environments,* Prentice-Hall, Englewood Cliffs, N.J., 1968.
Leet, L. D., and Judson, S., *Physical Geology,* 4th ed., Prentice-Hall, Englewood Cliffs, N.J., 1971.
Lobeck, A. K., *Things Maps Don't Tell Us,* Macmillan, New York, 1956.
Longwell, C. R., Flint, R. F., and Sanders, J. E., *Physical Geology,* John Wiley & Sons, New York, 1969.
Lung, R., and Proctor, R. (eds.), *Engineering Geology in Southern California,* Association of Engineering Geologists, 1966.

Maps and Aerial Photographs—Sources of Information and Materials, Association of American Geographers, 1970.

Mason, B., *Principles of Geochemistry,* 3rd ed., John Wiley & Sons, New York, 1966.

Mather, K. F., *The Earth Beneath Us,* Random House, New York, 1964.

——— and Mason, S. L., *A Source Book in* Geology, McGraw-Hill Book Co., New York, 1939.

Mather, K. F. (ed.), *Source Book in Geology 1900-1950,* Harvard University Press, Cambridge, Mass., 1967.

Matthews, W. H., III, *Invitation to Geology: The Earth Through Time and Space,* Natural History Press, New York, 1971.

McDonald, G. A., and Abbott, A. T., *Volcanoes in the Sea: The Geology of Hawaii,* University of Hawaii Press, Honolulu, Hawaii, 1970.

McDonnell, K. S., and Massey, D. G., *Enquiring into the Earth,* Longman's Croydon, Victoria, Australia, 1968.

Mears, Brainerd, Jr., *The Nature of Geology: Contemporary Readings,* Van Nostrand Reinhold, New York, 1970.

——— *The Changing Earth,* Van Nostrand Reinhold, New York, 1970.

Merrill, G. P., *The First One Hundred Years of American Geology,* Hafner Publishing Co., Darien, Conn., 1969.

Mineral Facts and Problems, Bureau of Mines Bulletin 650, Supt. of Documents, 1970.

Morisawa, Marie, *Streams: Their Dynamics and Morphology,* McGraw-Hill Book Co., New York, 1968.

Oakeshott, G. B., *California's Changing Landscapes—A Guide to the Geology of the State,* McGraw-Hill Book Co., New York, 1971.

Palmer, E. L., *Fossils,* D. C., Heath Co., Lexington, Mass., 1965.

Park, C. F., Jr., and MacDiarmid, R. A., *Ore Deposits,* 2nd ed., W. H. Freeman, San Francisco, Calif., 1970.

Putnam, W. C., and Bassett, Ann, *Geology,* 2nd ed., Oxford, 1971.

Raup, D. M., and Stanley, S. M., *Principles of Paleontology,* W. H. Freeman, San Francisco, Calif., 1971.

Rittman, A., *Volcanoes and Their Activity,* John Wiley & Sons, New York, 1962.

Rodgers, John, *The Tectonics of the Appalachians,* Wiley-Interscience, New York, 1970.

Schwarzbach, M., *Climates of the Past,* Van Nostrand, New York, 1963.

Shelton, J. S., *Geology Illustrated,* W. H. Freeman, San Francisco, Calif., 1966.

Shimer, J., *This Sculptured Earth: The Landscape of America,* Columbia University Press, New York, 1959.

Sinkankas, J., *Mineralogy: A First Course,* Van Nostrand, New York, 1966.

——— *Mineralogy for Amateurs,* Van Nostrand, New York, 1964.

Skinner, B. J., *Earth Resources, Prentice-Hall,* Englewood Cliffs, N.J., 1969.

Spencer, E. W., *Introduction to the Structure of the Earth,* McGraw-Hill Book Co., New York, 1969.

Stacey, F. D., *Physics of the Earth,* John Wiley & Sons, New York, 1969.

Stokes, W. L., and Judson, S., *Introduction to Geology: Physical and Historical,* Prentice-Hall, Englewood Cliffs, N.J., 1968.

Swinton, W. E., *The Dinosaurs,* 2nd ed., John Wiley & Sons, New York, 1970.

Thornbury, W. D., *Regional Geomorphology of the United States,* John Wiley & Sons, New York, 1965.

Thornbury, W. D., *Principles of Geomorphology,* 2nd ed., John Wiley & Sons, New York, 1969.

Tucker, R. H. et al, *Global Geophysics,* American Elsevier, New York, 1970.

Tuttle, S. D., *Landforms and Landscapes,* Wm. C. Brown, Dubuque, Iowa, 1970.

Twidale, C. R., *Structural Landforms,* M.I.T. Press, Cambridge, Mass., 1971.

Wegner, A. L., *The Origin of Continents and Oceans,* translated from 4th German ed. by John Biram, Dover Publications, New York, 1966.

Weitz, J. L., *Your Future in Geology,* Richards Rosen Press, New York, 1966.

Woodford, A. O., *Historical Geology,* W. H. Freeman, San Francisco, Calif., 1965.

Wyllie, P. J., *The Dynamic Earth,* Wiley-Interscience, New York, 1971.

METEOROLOGY

Atkinson, B. W., *The Weather Business,* Doubleday Science Series, New York, 1969.

Aviation Weather, U.S. Government Printing Office, Washington, D.C., 1965.

Barrett, E. C., *Viewing Weather from Space,* Praeger, New York, 1967.

Barry, R. G., and Chorley, R. J., *Atmosphere, Weather, and Climate,* Holt, Rinehart, and Winston, New York, 1970.

Battan L. J., *Cloud Physics and Cloud Seeding,* Doubleday, New York, 1962.

——— *The Nature of Violent Storms, Doubleday,* New York, 1962.

——— *Radar Observes the Weather,* Doubleday, New York, 1962.

——— *The Unclean Sky,* Doubleday, New York, 1966.

——— *The Dirty Air,* Doubleday, New York, 1968.

——— *Harvesting the Clouds,* Doubleday, New York, 1969.

Bentley, W., and Humphreys, W., *Snow Crystals,* Dover, New York, 1962.

Blair, T. A., and Fite, R. C., *Weather Elements,* 5th ed., Prentice-Hall, Englewood Cliffs, N.J., 1965.

Blanchard, D. C., *From Raindrops to Volcanoes,* Doubleday, New York, 1967.

Byers, H. R., *Elements of Cloud Physics,* University of Chicago Press, Chicago, Ill., 1965.

——— *General Meteorology,* McGraw-Hill Book Co., New York, 1959.

——— *The Thunderstorm,* U.S. Government Printing Office, Washington, D.C., 1949.

Craig, R. A., *The Edge of Space,* Doubleday, New York, 1968.

Day, J. A., and Sternes, G. L., *Climate and Weather,* Addison-Wesley, Reading, Mass., 1970.

Dobson, G., *Exploring the Atmosphere,* Clarendon Press, Oxford, 1963.

Donn, W. L., *Meteorology,* 3rd ed., McGraw-Hill Book Co., New York, 1965.

Dunn, G. E., and Miller, B. I., *Atlantic Hurricanes,* Louisiana State University Press, Baton Rouge, La., 1964.

Edinger, J. G., *Watching for the Wind,* Doubleday, New York, 1967.

Fairbridge, R. W. (ed.), *The Encyclopedia of Atmospheric Sciences and Astrogeology,* Reinhold, New York, 1967.

Flohn, H., *Climate and Weather,* World University Library, 1969.

Flora, S. D., *Tornadoes of the United States,* University of Oklahoma Press, Norman, Okla., 1954.

——— *Hailstorms of the United States,* University of Oklahoma, Press, Norman Okla., 1956.

Grant, H. D., *Cloud and Weather Atlas,* Coward-McCann, New York, 1944.

Kimble, G., *Our American Weather,* McGraw-Hill Book Co., New York, 1955.

Ludlam, F. H., and Scorer, R. S., *Cloud Study, A Pictorial Guide,* John Murray, London, 1966.

Mason, B. J., *Clouds, Rain, and Rainmaking,* Cambridge University Press, 1962.

Middleton, W. E. K., *Invention of the Meteorological Instruments,* Johns Hopkins Press, Baltimore, Md., 1969.

Miller, A., and Thompson, J. C., *Elements of Meteorology,* Charles E. Merrill, Columbus, Ohio, 1970.

Neuberger, H., and Nicholas, G., *Manual of Demonstrations and Experiments for Teaching Meteorology,* Pennsylvania State University Press, University Park, Pa., 1962.

Neuberger, Hans, *Introduction to Physical Meteorology,* The Pennsylvania State University Press, 1965.

Neuberger, H., and Cahir, J., *Principles of Climatology,* Holt, Rinehart and Winston, New York, 1969.

Overman, M., *Water,* Doubleday, New York, 1969.

Petterssen, S., *Introduction to Meteorology,* 3rd ed., McGraw-Hill Book Co., New York, 1969.

Reiter, E. R., *Jet Streams,* Doubleday, New York, 1967.

Riehl, H., *Introduction to the Atmosphere,* McGraw-Hill Book Co., New York, 1965.

Scorer, R., and Wexler, H., *Cloud Studies in Colour,* Pergamon Press, Elmsford, N.Y., 1967.

Sloane, Eric, *Folklore of American Weather,* Meredith Press, Des Moines, Iowa, 1963.

Stewart, G., *Storm,* Modern Library, Random House, New York, 1947.

Sutcliffe, R. C., *Weather and Climate,* W. W. Norton, New York, 1966.

Thompson, P., O'Brien, R., and the editors of *Life, Weather,* Time, Inc., New York, 1965.

Trewartha, G. T., *Introduction to Climates,* McGraw-Hill Book Co., New York, 1968.

Tricker, R. A. R., *The Science of the Clouds,* American Elsevier, New York, 1970.

——— *An Introduction to Meteorological Optics,* American Elsevier, New York, 1970.

Viemeister, P., *The Lightning Book,* Doubleday, New York, 1961.

Widger, W. K., Jr., *Meteorological Satellites,* Holt, Rinehart and Winston, 1966.

World Meteorological Organization, Geneva, Switzerland, *International Cloud Atlas,* Vols. 1 and 2, 1956.

OCEANOGRAPHY

Barber, N. F., *Water Waves,* Wykeham Publications, 1969.

Bardach, J. E., *Harvest of the Sea,* Harper and Row, New York, 1968.

Bascom, W., *Waves and Beaches,* Doubleday, New York, 1964.

Behrman, D., *The New World of the Oceans; Men and Oceanography,* Little Brown, Boston, Mass., 1969.

Carson, Rachel, *The Sea Around Us,* Oxford University Press, 1951.

Coker, R., *This Great and Wide Sea,* Harper Torchbooks, New York, 1962.

Cowen, R., *Frontiers of the Sea,* Doubleday, New York, 1960.

Cromie, W. *Exploring the Secrets of the Sea,* Prentice-Hall, Englewood Cliffs, N.J., 1962.

Duxbury, Alyn C., *The Earth and Its Oceans,* Addison-Wesley, Reading, Mass, 1971.

Engel, L., and the editors of *Life, The Sea,* Time, Inc., New York, 1961.

Exploiting the Ocean, Marine Technology Society, 1966.

Gordon, B. L., Man and the Sea: Classic Accounts *of Marine Explorations,* Natural History Press, New York, 1970.

Groen, P., *The Waters of the Sea,* Van Nostrand, New York, 1967.

Gross, M. G., *Oceanography,* Charles E. Merrill, Columbus, Ohio, 1967.

Fairbridge, R. W. (ed.), *Encyclopedia of Oceanography,* Reinhold, New York, 1966.

Heezen, B., Tharp, M., and Ewing, M., *The Floors of the Oceans,* Geological Society of America Special Paper 65, 1959.

Heezen, B. C., and Hollister, C. D., *The Face of the Deep,* Oxford University Press, 1971.

Hill, M. N. (ed.), *The Sea,* Vols. 1, 2, 3, and 4, John Wiley & Sons, New York, 1970.

Hull, S., *The Bountiful Sea,* Prentice-Hall, Englewood Cliffs, N. J., 1964.

Keen, M. J., *An Introduction to Marine Geology,* Pergamon Press, Elmsford, N.Y., 1968.

King, C., *An Introduction to Oceanography,* McGraw-Hill Book Co., New York, 1963.

Long, E. J. (ed.), *Ocean Sciences,* U.S. Naval Institute, 1964.

McLellan, H. J., *Elements of Physical Oceanography,* Pergamon Press, Elmsford, N.Y., 1965.

Menard, H., *Marine Geology of the Pacific,* McGraw-Hill Book Co., New York, 1964.

Menard, H. W., *Anatomy of an Expedition,* McGraw-Hill Book Co., New York, 1969.

Mero, J., *Mineral Resources of the Sea,* American Elsevier, New York, 1965.

Miller, R. C., *The Sea,* Random House, New York, 1966.

Pickard, G. L., *Descriptive Physical Oceanography,* Pergamon Press, Elmsford, N.Y., 1963.

Ross, D. A., *Introduction to Oceanography,* Appleton-Century Crofts, New York, 1970.

Science and the Sea, U.S. Naval Oceanographic Office, Washington, D.C., 1967.

Scientific American, "The Ocean," W. H. Freeman, San Francisco, Calif., 1969.

Shepard, F. P., *The Earth Beneath the Sea,* 2nd ed., Johns Hopkins Press, Baltimore Md., 1967.

——— *Submarine Geology,* 2nd ed., Harper and Row, New York, 1963.

Shepard, F. P., and Wanless, H., *Our Changing Coastlines,* McGraw-Hill Book Co., New York 1971.

Stewart, H. B., Jr., *Deep Challenge,* Van Nostrand, New York, 1966.

Turekian, K. K., *Oceans,* Prentice-Hall, Englewood Cliffs, N.J., 1968.

Weyl, P. K., *Oceanography: An Introduction to the Marine Environment,* John Wiley & Sons, New York, 1970.

Yasso, W. E., *Oceanography: A Study of Inner Space,* Holt, Rinehart and Winston, New York, 1965.

GENERAL EARTH SCIENCE

Beiser, A. and eds. of *Life, The Earth,* Time, Inc., 1962.

Cortright, E. M., (ed.), *Exploring Space with a Camera,* NASA SP-168, 1968.

Earth Photographs from Gemini VI through XII, NASA SP-171, 1968.

Earth Science Curriculum Project, *Investigating the Earth,* Houghton Mifflin, Boston, Mass., 1967.

Ericson, D., and Wollin, G., *The Deep and the Past,* Alfred A. Knopf, New York, 1964.

Greenhood, D., *Mapping,* University of Chicago Press, Chicago, Ill., 1964.

Lunar Photographs from Apollos 8, 10, and 11, Superintendent of Documents, Washington, D.C., 1971.

Mason, B., and Melson, W. G., *The Lunar Rocks,* Wiley-Interscience, New York, 1970.

Nicks, Oran W., *This Island Earth,* NASA, 1970.

Rogers, E. M., *Physics for the Inquiring Mind,* Princeton University Press, Princeton, N.J., 1960.

Scientific American Reprints, see list by Freeman.

Shapley, H., *Beyond the Observatory,* Charles Scribner's Sons, New York, 1967.

Spar, J., *Earth, Sea, and Air,* 2nd ed., Addison-Wesley, Reading, Mass., 1965.

Spencer, E. W., *Geology: A Survey of Earth Science,* Crowell, New York, 1965.

Strahler, A. N., *The Earth Sciences,* 2nd ed., Harper and Row, New York, 1971.

Young, L. (ed.), *The Mystery of Matter,* American Foundation for Continuing Education, Oxford University Press, 1965.

PERIODICALS

Astronomy and Space, David and Charles, Devon, England.

Bulletin of American Meteorological Society, 45 Beacon St., Boston, Mass. 02108.

Bulletin of Geological Society of America, P.O. Box 1719, Boulder, Colo. 80302.

Daily Weather Maps, Weekly Series, Supt. of Documents, Washington, D.C. 20402.

Earth Science, Earth Science Publ. Corp., Box 550, Downers Grove, Ill. 60515.

Geotimes, American Geological Institute, 2201 M St. N.W., Washington, D.C. 20037

Griffith Observer, Griffith Observatory, Los Angeles, Calif.

Journal of Atmospheric Sciences, Am. Meteorological Society, 45 Beacon St., Boston, Mass. 02108.

Journal of Geological Education, National Association of Geology Teachers, 2201 M St. N.W., Washington, D.C. 20037.

Leaflets and Publications of Astronomical Society of the Pacific, California Academy of Sciences, Golden Gate Park, San Francisco, Calif. 94118.

Monthly Weather Review, National Oceanic and Atmospheric Administration, Rockville, Md. 20852.

National Geographic Society, 17th and M Sts. N.W., Washington, D.C. 20036.

Natural History, Am. Museum of Nat. History, Central Park West at 79th St., New York, N.Y. 10024.

Nature, Macmillan Journals Ltd., 4 Little Essex Street, London, WC2R 3LF.

Observer's Handbook, Royal Astronomical Society of Canada, Toronto.

Review of Popular Astronomy, Sky Map Publications, St. Louis, Mo.

Science, 1515 Massachusetts Ave., N.W. Washington, D.C. 20005.

The Sciences, New York Academy of Sciences, 2 East 63rd St., New York, N.Y. 10021.

Scientific American, 415 Madison Ave., New York, N.Y. 10017.

Sky and Telescope, Sky Publishing Corp. 49-51 Bay State Road, Cambridge, Mass. 02138.

Weather, Royal Meteorological Society, London, England.

Weatherwise, American Meteorological Society, 45 Beacon St., Boston, Mass. 02108.